System Dynamics

Fourth Edition

William J. Palm III
University of Rhode Island

SYSTEM DYNAMICS

Published by McGraw-Hill Education, 2 Penn Plaza, New York, NY 10121.

This book is printed on acid-free paper.

1 2 3 4 5 6 7 8 9 LWI 24 23 22 21 20

ISBN 978-1-260-57076-2
MHID 1-260-57076-2

Cover Image: ©*Chesky/Shutterstock*

mheducation.com/highered

To my wife, Mary Louise; and to my children, Aileene, Bill, and Andrew.

CONTENTS

PREFACE

System dynamics deals with mathematical modeling and analysis of devices and processes for the purpose of understanding their time-dependent behavior. While other subjects, such as Newtonian dynamics and electrical circuit theory, also deal with time-dependent behavior, system dynamics emphasizes methods for handling applications containing multiple types of components and processes such as electromechanical devices, electrohydraulic devices, and fluid-thermal processes. Because the goal of system dynamics is to understand the time-dependent behavior of a system of interconnected devices and processes as a whole, the modeling and analysis methods used in system dynamics must be properly selected to reveal how the connections between the system elements affect its overall behavior. Because systems of interconnected elements often require a control system to work properly, control system design is a major application area in system dynamics.

TEXT PHILOSOPHY

This text is an introduction to system dynamics and is suitable for such courses commonly found in engineering curricula. It is assumed that the student has a background in elementary differential and integral calculus and college physics (dynamics, mechanics of materials, thermodynamics, and electrical circuits). Previous exposure to differential equations is desirable but not necessary, as the required material on differential equations, as well as Laplace transforms and matrices, is developed in the text.

The decision to write a textbook often comes from the author's desire to improve on available texts. The decisions as to what topics to include and what approach to take emerge from the author's teaching experiences that give insight as to what is needed for students to master the subject. This text is based on the author's forty-four years of experience in teaching system dynamics.

This experience shows that typical students in a system dynamics course are not yet comfortable with applying the relevant concepts from earlier courses in dynamics and differential equations. Therefore, this text reviews and reinforces these important topics early on. Students often lack sufficient physical insight to relate the mathematical results to applications. The text therefore uses everyday illustrations of system dynamics to help students to understand the material and its relevance.

If laboratory sessions accompany the system dynamics course, many of the text's examples can be used as the basis for experiments. The text is also a suitable reference on hardware and on parameter estimation methods.

MATLAB® AND SIMULINK®[1]

MATLAB and Simulink are used to illustrate how modern computer tools can be applied in system dynamics.[2] MATLAB was chosen because it is the most widely

[1]MATLAB and Simulink are registered trademarks of The MathWorks, Inc.

[2]The programs in this text are based on the following software versions, or higher versions: Version 9.6 of MATLAB, Version 9.3 of Simulink, and Version 10.6 of the Control Systems Toolbox.

used program in system dynamics courses and by practitioners in the field. Simulink, which is based on MATLAB and uses a diagram-based interface, is increasing in popularity because of its power and ease of use. In fact, students convinced the author to use Simulink after they discovered it on their own and learned how easy it is to use! It provides a useful and motivational tool.

It is, however, not necessary to cover MATLAB or Simulink in order to use the text, and it is shown how to do this later in the Preface.

TEXT OVERVIEW

Chapter 1 introduces the basic terminology of system dynamics, covers commonly used functions, and reviews the two systems of units used in the text: British Engineering (FPS) units and SI units. These are the unit systems most commonly used in system dynamics applications. The examples and homework problems employ both sets of units so that the student will become comfortable with both. Chapter 1 also covers some basic methods for parameter estimation. These methods are particularly useful for obtaining numerical values of spring constants, damping coefficients, and other parameters commonly found in system dynamics models. The chapter also contains introductions to differential equations and to MATLAB, and it presents the first of the text's several case studies: design of motion-control systems. The material on function identification and least-squares fitting, formerly in Chapter 1 in the third edition, is now in Appendix C.

Chapter 2 covers differential equations in more depth, and develops the Laplace transform method for solving differential equations with applications to equations having step, ramp, sine, impulse, and other types of forcing functions. It also introduces transfer function models.

Chapter 3 covers rigid-body dynamics, including planar motion. This chapter continues the motion-control case study by showing how to select a suitable motor and gear system.

Chapter 4 covers modeling of mechanical systems having stiffness and damping, and it applies the analytical methods developed in Chapter 2 to solve the models. This chapter then introduces the second case study: design of vehicle suspensions.

Chapter 5 develops block diagrams and the state-variable model, which is useful for certain analytical techniques as well as for numerical solutions. The optional sections of this chapter introduce Simulink, which is based on block-diagram descriptions, and apply the chapter's concepts using MATLAB. This chapter concludes with further coverage of the vehicle suspension case study.

Chapter 6 treats modeling of electric circuits, operational amplifiers, electromechanical devices, sensors, and electroacoustic devices. It also discusses how motor parameters can be obtained, and it returns to the motion-control case study and shows how to analyze motor and amplifier performance.

Part I of Chapter 7 covers fluid systems. Part II covers thermal systems. These two parts are independent of each other. A background in fluid mechanics or heat transfer is not required to understand this chapter, but students should have had elementary thermodynamics before covering the material on pneumatic systems in Section 7.5.

Chapters 8 and 9 cover analysis methods in the time domain and the frequency domain, respectively.

Chapter 8 integrates the modeling and analysis techniques of earlier chapters with an emphasis on understanding system behavior in the time domain, using step, ramp, and impulse functions primarily. The chapter covers step-response specifications such as maximum overshoot, peak time, delay time, rise time, and settling time.

Chapter 9 demonstrates the usefulness of the transfer function for understanding and analyzing a system's frequency response. It introduces Bode plots and shows how they are sketched and interpreted to obtain information about time constants, resonant frequencies, and bandwidth. The chapter returns to the vehicle-suspension case study, and shows how to use frequency response methods to evaluate suspension performance.

Chapters 10, 11, and 12 deal with a major application of system dynamics, namely, control systems. Chapter 10 is an introduction to feedback control systems, including the PID control algorithm applied to first- and second-order plants. The chapter concludes with thorough coverage of feedback control applied to the motion-control case study.

Chapter 11 deals with control systems in more depth and includes design methods based on the root locus plot and practical topics such as compensation, controller tuning, actuator saturation, reset windup, and state-variable feedback, with emphasis on motion-control systems. Chapter 12 covers series compensation methods and design with the root locus plot and the Bode plot.

Chapter 13 covers another major application area, vibrations. Important practical applications covered are vibration isolators, vibration absorbers, modes, and suspension system design. This chapter is now on the text website to allow room for the new case studies in earlier chapters.

ALTERNATIVE COURSES IN SYSTEM DYNAMICS

The choice of topics depends partly on the desired course emphasis, partly on the students' background in differential equations and dynamics, and partly on whether the course is a quarter or semester course.

Fluid and thermal systems are covered in Chapter 7, which has been shortened in this edition. Some students may have had courses in fluid mechanics and heat transfer, but probably have not been exposed to the system dynamics viewpoint, which focuses on the analogies between fluid and thermal resistance and capacitance and the corresponding electrical concepts. The theory and methods of the remaining chapters do not depend on Chapter 7, but some examples do.

In the author's opinion, a basic semester course in system dynamics should include most of the material in Chapters 1 through 7, and Chapters 9 and 10. At the author's institution, the system dynamics course is a junior course required for mechanical engineering majors, who have already had courses in dynamics and differential equations. It covers Chapters 1 through 10, with brief coverage of Chapter 7 and Chapter 8, and with some MATLAB and Simulink sections omitted. This optional material is then covered in a senior elective course in control systems, which also covers Simulink, and Chapters 11 and 12.

The text is flexible enough to support a variety of courses. The sections dealing with MATLAB and Simulink are at the end of the chapters and may be omitted. If students are familiar with Laplace transform methods and linear differential equations, Chapter 2 may be covered quickly. If students are comfortable with rigid-body planar motion, Chapter 3 may be used for a quick review.

GLOSSARY AND APPENDICES

There is a glossary containing the definitions of important terms, five appendices, and an index. Appendices D and E are on the text website.

Appendix A is a collection of tables of MATLAB commands and functions, organized by category. The purpose of each command and function is briefly described in the tables.

Appendix B is a brief summary of the Fourier series, which is used to represent a periodic function as a series consisting of a constant plus a sum of sine terms and cosine terms. It provides the background for some applications of the material in Chapter 9.

Appendix C covers function identification, and shows how to use MATLAB to fit models to scattered data using the least-squares method.

Appendix D is a self-contained introduction to MATLAB, and it should be read first by anyone unfamiliar with MATLAB if they intend to cover the MATLAB and Simulink sections. It also provides a useful review for those students having prior experience with MATLAB.

Appendix E covers numerical methods, such as the Runge-Kutta algorithms, that form the basis for the differential equation solvers of MATLAB. It is not necessary to master this material to use the MATLAB solvers, but the appendix provides a background for the interested reader.

Answers to selected homework problems are given following Appendix C.

CHAPTER FORMAT

The format of each chapter follows the same pattern, which is

1. Chapter outline
2. Chapter objectives
3. Chapter sections
4. MATLAB sections (in most chapters)
5. Simulink section (in most chapters)
6. Chapter review
7. References
8. Problems

This structure has been designed partly to accommodate those courses that do not cover MATLAB and/or Simulink, by placing the optional MATLAB and Simulink material at the end of the chapter. Chapter problems are arranged according to the chapter section whose concepts they illustrate. All problems requiring MATLAB and/or Simulink have thus been placed in separate, identifiable groups.

OPTIONAL TOPICS

In addition to the optional chapters (11, 12, and 13), some chapters have sections dealing with material other than MATLAB and Simulink that can be omitted without affecting understanding of the core material in subsequent chapters. All such optional material has been placed in sections near the end of the chapter. This optional material includes:

1. Function discovery, parameter estimation, and system identification techniques (Sections 8.4 and 9.6)
2. General theory of partial-fraction expansion (Section 2.7)
3. Impulse response (Sections 2.6 and 4.6)
4. Motor performance (Section 6.6)
5. Sensors and electroacoustic devices (Section 6.8)

DISTINGUISHING FEATURES

The following are considered to be the major distinguishing features of the text.

1. **MATLAB.** Standalone sections in most chapters provide concise summaries and illustrations of MATLAB features relevant to the chapter's topics.
2. **Simulink.** Standalone sections in Chapters 5 through 12 provide extensive Simulink coverage not found in most system dynamics texts.
3. **Parameter estimation.** Coverage of function discovery, parameter estimation, and system identification techniques is given in Sections 1.3, 8.4, 9.6, and Appendix C. Students are uneasy when they are given parameter values such as spring stiffness and damping coefficients in examples and homework problems, because they want to know how they will obtain such values in practice. These sections show how this is done.
4. **Motor performance evaluation.** Section 6.6 discusses the effect of motor dynamics on practical considerations for motor and amplifier applications, such as motion profiles and the required peak and rated continuous current and torque, and maximum required voltage and motor speed. These considerations offer excellent examples of practical applications of system dynamics but are not discussed in most system dynamics texts.
5. **System dynamics in everyday life.** Commonly found illustrations of system dynamics are important for helping students to understand the material and its relevance. This text provides examples drawn from objects encountered in everyday life. These examples include a storm door closer, fluid flow from a bottle, shock absorbers and suspension springs, motors, systems with gearing, chain drives, belt drives, a backhoe, a water tower, and cooling of liquid in a cup.
6. **Case studies and theme applications.** Two common applications provide themes for case studies, examples, and problems throughout the text. These are motion-control systems, such as a conveyor system and a robot arm, and vehicle suspension systems.

ACKNOWLEDGMENTS

I want to acknowledge and thank the many individuals who contributed to this effort. At McGraw-Hill, my thanks go to the editors who helped me through four editions: Tom Casson, who initiated the project, Jonathan Plant, Lora Neyens, Bill Stenquist, and Thomas Scaiffe. I am grateful to Tina Bower and Laura Bies for their patience and help with the fourth edition.

The University of Rhode Island provided an atmosphere that encourages teaching excellence, course development, and writing, and for that I am appreciative.

I am grateful to my wife, Mary Louise; and my children, Aileene, Bill, and Andrew, for their support, patience, and understanding through forty years of textbook creation. Finally, thanks to my grandchildren, Elizabeth, Emma, James, and Henry, for many enjoyable diversions!

William J. Palm III
Kingston, Rhode Island
March 2019

ABOUT THE AUTHOR

William J. Palm III is Professor Emeritus of Mechanical, Industrial, and Systems Engineering at the University of Rhode Island. In 1966 he received a B.S. from Loyola College in Baltimore, and in 1971 a Ph.D. in Mechanical Engineering and Astronautical Sciences from Northwestern University in Evanston, Illinois.

During his forty-four years as a faculty member, he has taught nineteen courses. One of these is a junior system dynamics course, which he developed. He has authored nine textbooks dealing with modeling and simulation, system dynamics, control systems, vibrations, and MATLAB. These include *MATLAB for Engineering Applications*, fourth edition (McGraw-Hill, 2019), *A Concise Introduction to MATLAB* (McGraw-Hill, 2008), and *Differential Equations for Engineers and Scientists* (McGraw-Hill, 2013) with Yunus Çengel. He wrote a chapter on control systems in the *Mechanical Engineers' Handbook*, fourth edition (M. Kutz, ed., Wiley, 2014), and was a special contributor to the fifth editions of *Statics* and *Dynamics,* both by J. L. Meriam and L. G. Kraige (Wiley, 2002).

Professor Palm's research and industrial experience are in control systems, robotics, vibrations, and system modeling. He was the Director of the Robotics Research Center at the University of Rhode Island from 1985 to 1993, and is the co-holder of a patent for a robot hand. He served as Acting Department Chair from 2002 to 2003. His industrial experience is in automated manufacturing; modeling and simulation of naval systems, including underwater vehicles and tracking systems; and design of control systems for underwater vehicle engine test facilities.

Affordability & Outcomes = Academic Freedom!

You deserve choice, flexibility, and control. You know what's best for your students and selecting the course materials that will help them succeed should be in your hands.

That's why providing you with a wide range of options that lower costs and drive better outcomes is our highest priority.

Students—study more efficiently, retain more, and achieve better outcomes. Instructors—focus on what you love—teaching.

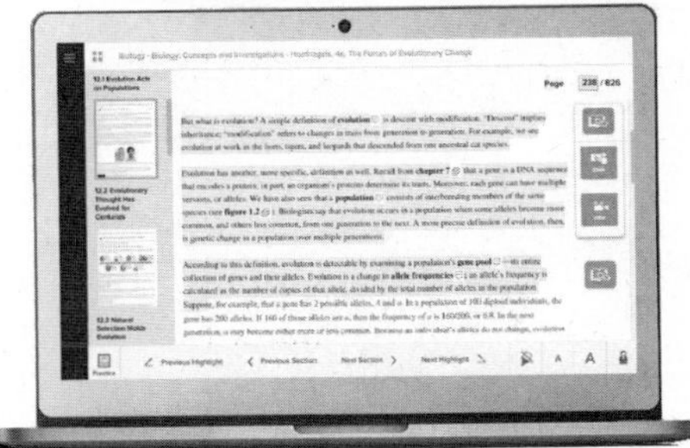

Laptop: McGraw-Hill Education

They'll thank you for it.

Study resources in Connect help your students be better prepared in less time. You can transform your class time from dull definitions to dynamic discussion. Hear from your peers about the benefits of Connect at **www.mheducation.com/highered/connect/smartbook**

Make it simple, make it affordable.

Connect makes it easy with seamless integration using any of the major Learning Management Systems—Blackboard®, Canvas, and D2L, among others—to let you organize your course in one convenient location. Give your students access to digital materials at a discount with our inclusive access program. Ask your McGraw-Hill representative for more information.

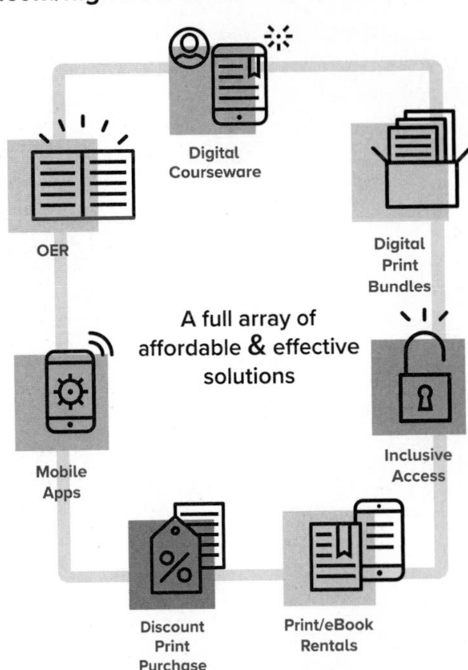

Learning for everyone.

McGraw-Hill works directly with Accessibility Services Departments and faculty to meet the learning needs of all students. Please contact your Accessibility Services office and ask them to email accessibility@mheducation.com, or visit **www.mheducation.com/about/accessibility.html** for more information.

Learn more at: www.mheducation.com/realvalue

Rent It

Affordable print and digital rental options through our partnerships with leading textbook distributors including Amazon, Barnes & Noble, Chegg, Follett, and more.

Go Digital

A full and flexible range of affordable digital solutions ranging from Connect, ALEKS, inclusive access, mobile apps, OER and more.

Get Print

Students who purchase digital materials can get a loose-leaf print version at a significantly reduced rate to meet their individual preferences and budget.

CHAPTER 1

Introduction

CHAPTER OUTLINE

CHAPTER OBJECTIVES

When you have finished this chapter, you should be able to

1. Define the basic terminology of system dynamics.
2. Apply the basic steps used for engineering problem solving.
3. Apply the necessary steps for developing a computer solution.
4. Use units in both the FPS and the SI systems.
5. Develop linear models from given algebraic expressions.
6. Use direct integration to solve dynamics problems involving a differential equation in which the derivative can be isolated.
7. Model and design a simple motion-control system for a single rotational load.
8. Use MATLAB to perform simple calculations and plotting, and use the MATLAB help system.

This chapter introduces the basic terminology of system dynamics, which includes the notions of *system*, *static* and *dynamic elements*, *input*, and *output*. Because we will use both the foot-pound-second (FPS) and the metric (SI) systems of units, the chapter introduces these two systems. Developing mathematical models of input-output relations is essential to the applications of system dynamics. Therefore, we begin our study by introducing some basic methods for developing algebraic models of static elements. We show how to use the methods of function identification and parameter estimation to develop models from data, and how to fit models to data that have little scatter. ■

1.1 INTRODUCTION TO SYSTEM DYNAMICS

This text is an introduction to system dynamics. We presume that the reader has some background in calculus (specifically, differentiation and integration of functions of a single variable) and in physics (specifically, free body diagrams, Newton's laws of motion for a particle, and elementary dc electricity). In this section we establish some basic terminology and discuss the meaning of the topic "system dynamics," its methodology, and its applications.

1.1.1 SYSTEMS

The meaning of the term *system* has become somewhat vague because of overuse. The original meaning of the term is a *combination of elements intended to act together to accomplish an objective*. For example, a link in a bicycle chain is usually not considered to be a system. However, when it is used with other links to form a chain, it becomes part of a system. The objective for the chain is to transmit force. When the chain is combined with gears, wheels, crank, handlebars, and other elements, it becomes part of a larger system whose purpose is to transport a person.

The system designer must focus on how all the elements act together to achieve the system's intended purpose, keeping in mind other important factors such as safety, cost, and so forth. Thus, the system designer often cannot afford to spend time on the details of designing the system elements. For example, our bicycle designer might not have time to study the metallurgy involved with link design; that is the role of the chain designer. All the systems designer needs to know about the chain is its strength, its weight, and its cost, because these are the factors that influence its role in the system.

With this "systems point of view," we focus on how *connections* between the elements influence the *overall* behavior of the system. This means that sometimes we must accept a less-detailed description of the operation of the individual elements to achieve an overall understanding of the system's performance.

Figure 1.1.1 illustrates a liquid-filled tank with a volume inflow f (say in cubic feet per second). The liquid height is h (say in feet). We see in Example 1.4.2 that the functional relationship between f and h has the form $f = bh^m$, where b and m are constants. We would not call this a "system." However, if two tanks are connected as shown in Figure 1.1.2, this connection forms a "system." Each tank is a "subsystem"

Figure 1.1.1 The effect of liquid height h on the out flow rate f.

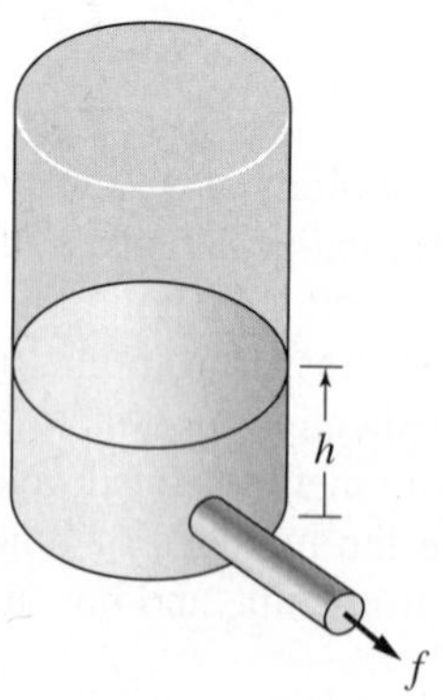

Figure 1.1.2 Two connected tanks.

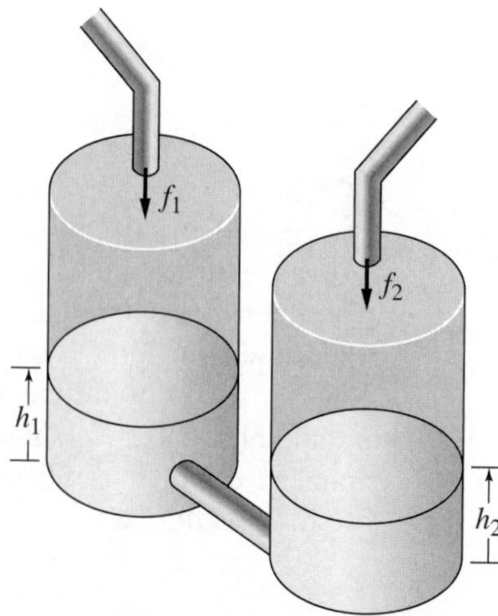

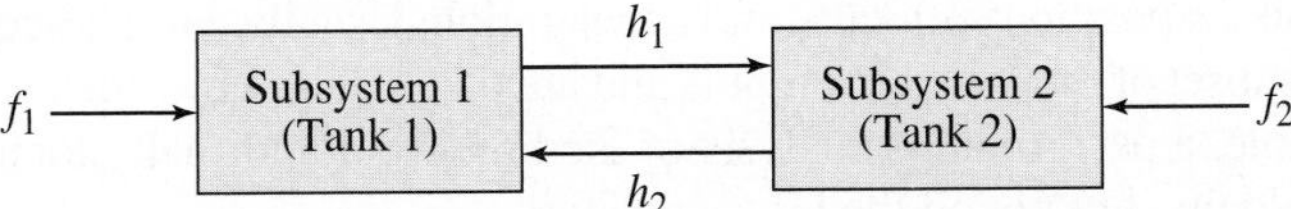

Figure 1.1.3 A system diagram illustrating how the two liquid heights affect each other.

whose liquid height is influenced by the other tank. We can obtain a differential equation model for each height by using the single-tank relationship $f = bh^m$ and applying the basic physical principle called conservation of mass to express the connection between the two tanks. This results in a model of the entire system.

We often use diagrams to illustrate the connections between the subsystems. Figure 1.1.2 illustrates the physical connection, but Figure 1.1.3 is an example of a diagram showing that the height h_1 affects the height h_2, and vice versa. (The flow goes from the higher height to the lower one.) Such a diagram may be useful for a nontechnical audience, but it does not show *how* the heights affect each other. To do that, we will use two other types of diagrams—called *simulation* diagrams and *block diagrams*—to represent the connections between the subsystems and the *variables* that describe the system behavior. These diagrams represent the differential equation model.

1.1.2 INPUT AND OUTPUT

Like the term "system," the meanings of *input* and *output* have become less precise. For example, a factory manager will call a meeting to seek "input," meaning opinions or data, from the employees, and the manager may refer to the products manufactured in the factory as its "output." However, in the system dynamics meaning of the terms, an *input* is a *cause*; an *output* is an *effect* due to the input. Thus, one input to the bicycle is the force applied to the pedal. One resulting output is the acceleration of the bike. Another input is the angle of the front wheel; the output is the direction of the bike's path of travel.

The behavior of a system element is specified by its *input-output relation*, which is a description of how the output is affected by the input. The input-output relation expresses the cause-and-effect behavior of the element. Such a description, which is represented graphically by the diagram in Figure 1.1.4, can be in the form of a table of numbers, a graph, or a mathematical relation. For example, a force f applied to a particle of mass m causes an acceleration a of the particle. The input-output or causal relation is, from Newton's second law, $a = f/m$. The input is f and the output is a.

The input-output relations for the elements in the system provide a means of specifying the connections between the elements. When connected together to form a system, the inputs to some elements will be the outputs from other elements.

The inputs and outputs of a system are determined by the selection of the system's boundary (see Figure 1.1.4). Any causes acting on the system from the world external to this boundary are considered to be system inputs. Similarly, a system's outputs are the outputs from any one or more of the system elements that act on the world outside the system boundary. If we take the bike to be the system, one system input would be the pedal force; another input is the force of gravity acting on the bike. The outputs

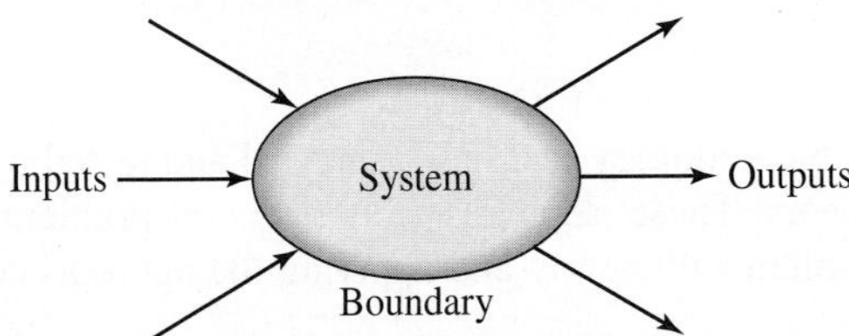

Figure 1.1.4 A system input-output diagram, showing the system boundary.

may be taken to be the bike's position, velocity, and acceleration. Usually, our choices for system outputs are a subset of the possible outputs and are the variables in which we are interested. For example, a performance analysis of the bike would normally focus on the acceleration or velocity, but not on the bike's position.

Sometimes input-output relations are reversible, sometimes not. For example, we can apply a current as input to a resistor and consider the resulting voltage drop to be the output ($v = iR$). Or we can apply a voltage to produce a current through the resistor ($i = v/R$). However, acceleration is the cause of a change in velocity, but not vice versa. If we integrate acceleration a over time, we obtain velocity v; that is, $v = \int a\,dt$. Whenever an output of an element is the time integral of the input and the direction of the cause-effect relation is not reversible, we say that the element exhibits *integral causality*. We will see that integral causality constitutes a basic form of causality for all physical systems.

Similar statements can be made about the relation between velocity and displacement. Integration of velocity produces displacement x: $x = \int v\,dt$. Velocity is the cause of displacement, but not vice versa.

Note that the mathematical relations describing integral causality can be reversed; for example, we may write $a = dv/dt$, but this does not mean that the cause-and-effect relation can be reversed.

1.1.3 STATIC AND DYNAMIC ELEMENTS

When the present value of an element's output depends only on the present value of its input, we say the element is a *static* element. For example, the current flowing through a resistor depends only on the present value of the applied voltage. The resistor is thus a static element. However, because no physical element can respond instantaneously, the concept of a static element is an approximation. It is widely used, however, because it results in a simpler mathematical representation; that is, an algebraic representation rather than one involving differential equations.

If an element's present output depends on past inputs, we say it is a *dynamic element*. For example, the present position of a bike depends on what its velocity has been from the start.

In popular usage, the terms static and dynamic distinguish situations in which no change occurs from those that are subject to changes over time. This usage conforms to the preceding definitions of these terms if the proper interpretation is made. A static element's output can change with time only if the input changes and will not change if the input is constant or absent. However, if the input is constant or removed from a dynamic element, its output can still change. For example, if we stop pedaling, the bike's displacement will continue to change because of its momentum, which is due to past inputs.

A *dynamic system* is one whose present output depends on past inputs. A *static system* is one whose output at any given time depends only on the input at that time. A static system contains all static elements. Any system that contains at least one dynamic element must be a dynamic system. *System dynamics*, then, is the study of systems that contain dynamic elements.

1.1.4 MODELING OF SYSTEMS

Table 1.1.1 contains a summary of the methodology that has been tried and tested by the engineering profession for many years. These steps describe a general problem-solving procedure. Simplifying the problem sufficiently and applying the appropriate

Table 1.1.1 Steps in engineering problem solving.

1. Understand the purpose of the problem.
2. Collect the known information. Realize that some of it might turn out to be not needed.
3. Determine what information you must find.
4. Simplify the problem only enough to obtain the required information. State any assumptions you make.
5. Draw a sketch and label any necessary variables.
6. Determine what fundamental principles are applicable.
7. Think generally about your proposed solution approach and consider other approaches before proceeding with the details.
8. Label each step in the solution process.
9. If you use a program to solve the problem, hand check the results using a simple version of the problem. Checking the dimensions and units, and printing the results of intermediate steps in the calculation sequence can uncover mistakes.
10. Perform a "reality check" on your answer. Does it make sense? Estimate the range of the expected result and compare it with your answer. Do not state the answer with greater precision than is justified by any of the following:
    a. The precision of the given information.
    b. The simplifying assumptions.
    c. The requirements of the problem.

    Interpret the mathematics. If the mathematics produces multiple answers, do not discard some of them without considering what they mean. The mathematics might be trying to tell you something, and you might miss an opportunity to discover more about the problem.

fundamental principles is called *modeling*, and the resulting mathematical description is called a *mathematical model*, or just a *model*. When the modeling has been finished, we need to solve the mathematical model to obtain the required answer. If the model is highly detailed, we may need to solve it with a computer program.

Modeling is the art of obtaining a quantitative description of a system or one of its elements that is simple enough to be useful for making predictions and realistic enough to trust those predictions. For example, consider a potato being heated in an oven. The oven designer wants to design an oven that is powerful enough to bake a potato within a prescribed time (Figure 1.1.5). Note that because the oven has yet to be designed, we cannot do an experiment to obtain the answer. Potatoes vary in size and shape, but a good estimate of the required oven power can be obtained by modeling the potato as a sphere having the thermal properties of water. Then, using the thermal systems methods given in Chapter 7, we can predict how long it will take to bake the potato.

It often is necessary to choose between a very accurate but complicated model and a simple but not so accurate model. Complicated models may be difficult to solve, or they may require experimental data that are unavailable or hard to find. There usually is no "right" model choice because it depends on the particular situation. We aim to

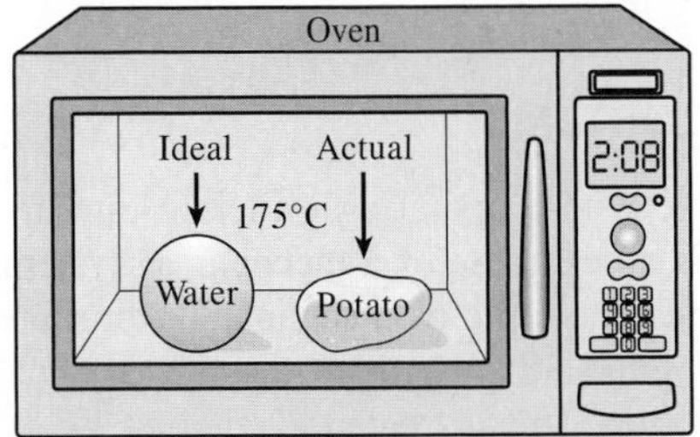

Figure 1.1.5 A potato modeled as a sphere of water.

choose the simplest model that yields adequate results. Just remember that the predictions obtained from a model are no more accurate than the simplifying assumptions made to develop the model. That is why we call modeling an art; it depends partly on judgment obtained by experience.

The form of a mathematical model depends on its purpose. For example, design of electrical equipment requires more than a knowledge of electrical principles. An electric circuit can be damaged if its mounting board experiences vibration. In this case, its force-deflection properties must be modeled. In addition, resistors generate heat, and a thermal model is required to describe this process. Thus, we see that devices can have many facets: thermal, mechanical, electrical, and so forth. No mathematical model can deal with all these facets. Even if it could, it would be too complex, and thus too cumbersome, to be useful.

For example, a map is a model of a geographic region. But if a single map contains all information pertaining to the roads, terrain elevation, geology, population density, and so on, it would be too cluttered to be useful. Instead, we select the particular type of map required for the purpose at hand. In the same way, we select or construct a mathematical model to suit the requirements of a particular study.

The examples in this text follow the steps in Table 1.1.1, although for compactness the steps are usually not numbered. In each example, following the example's title, there is a *problem statement* that summarizes the results of steps 1 through 5. Steps 6 through 10 are described in the *solution* part of the example. To save space, some steps, such as checking dimensions and units, are not always explicitly displayed. However, you are encouraged to perform these steps on your own.

1.1.5 MATHEMATICAL METHODS

Because system dynamics deals with changes in time, mathematical models of dynamic systems naturally involve differential equations. Therefore, we introduce differential equation solution methods starting in this chapter. Additional methods, such as those that make use of computers, are introduced in subsequent chapters.

1.1.6 CONTROL SYSTEMS

Often dynamic systems require a *control system* to perform properly. Thus, proper control system design is one of the most important objectives of system dynamics. Microprocessors have greatly expanded the applications for control systems. These new applications include robotics, mechatronics, micromachines, precision engineering, active vibration control, active noise cancellation, and adaptive optics. Recent technological advancements mean that many machines now operate at high speeds and high accelerations. It is therefore now more often necessary for engineers to pay more attention to the principles of system dynamics. Starting in Chapter 10, we apply these principles to control system design.

1.1.7 APPLICATIONS IN MECHANICAL SYSTEMS

Mechanical systems are loosely defined as those whose operating principles are primarily Newton's laws of motion. The bicycle is an example of a mechanical system. Chapters 3 and 4 deal with mechanical systems. The topic of mechanical vibrations covers the oscillations of machines and structures due either to their own inherent flexibility or to the action of an external force or motion. This is treated in Chapter 13.

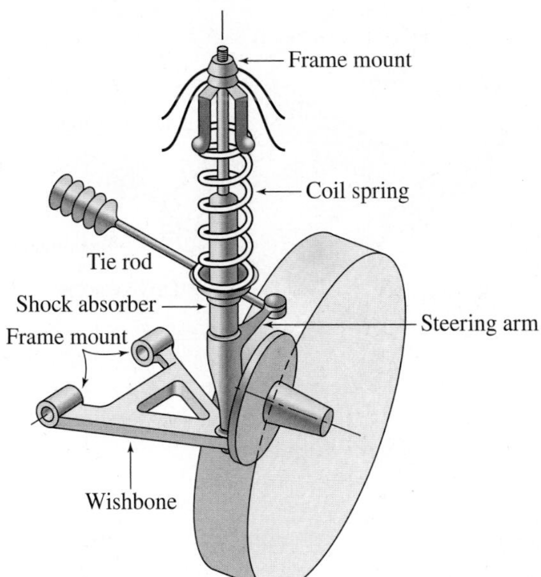

Figure 1.1.6 A vehicle suspension system.

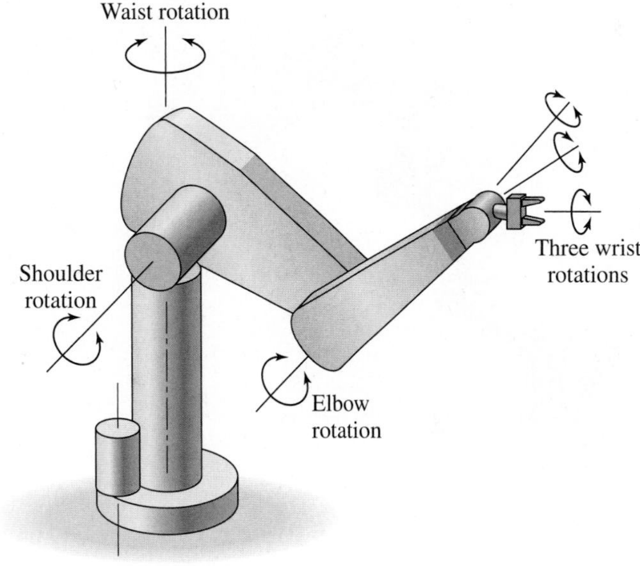

Figure 1.1.7 A robot arm.

One of our major theme applications in mechanical systems is vehicle dynamics. This topic has received renewed importance for reasons related to safety, energy efficiency, and passenger comfort. Of major interest under this topic is the design of vehicle suspension systems, whose elements include various types of springs and shock absorbers (Figure 1.1.6). *Active* suspension systems, whose characteristics can be changed under computer control, and vehicle-dynamics control systems are undergoing rapid development, and their design requires an understanding of system dynamics.

1.1.8 APPLICATIONS IN ELECTRICAL AND ELECTROMECHANICAL SYSTEMS

Electromechanical systems contain both mechanical elements and electrical elements such as electric motors. Two common applications of system dynamics in electromechanical systems are in (1) motion-control systems and (2) vehicle dynamics. Therefore, we will use these applications as major themes in many of our examples and problems. Chapter 6 introduces electrical and electromechanical systems.

Figure 1.1.7 shows a robot arm, whose motion must be properly controlled to move an object to a desired position and orientation. To do this, each of the several motors and drive trains in the arm must be adequately designed to handle the load, and the motor speeds and angular positions must be properly controlled. Figure 1.1.8 shows a typical motor and drive train for one arm joint. Knowledge of system dynamics is essential to design these subsystems and to control them properly.

Mobile robots are another motion-control application, but motion-control applications are not limited to robots. Figure 1.1.9 shows the mechanical drive for a conveyor system. The motor, the gears in the speed reducer, the chain, the sprockets, and the drive wheels all must be properly selected, and the motor must be properly controlled for the system to work well. In subsequent chapters we will develop models of these components and use them to design the system and analyze its performance.

Figure 1.1.8 Mechanical drive for a robot arm joint.

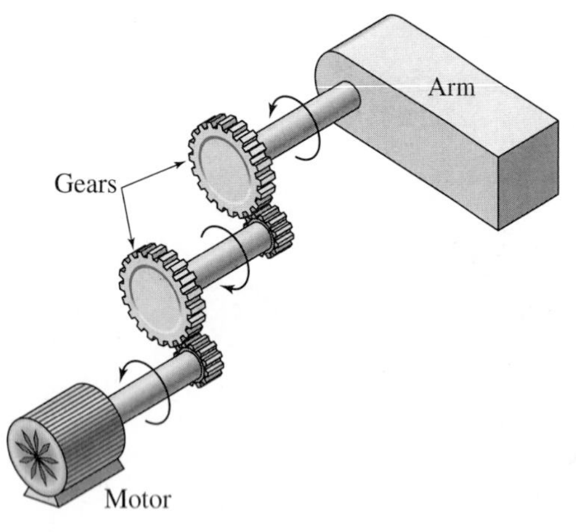

Figure 1.1.9 Mechanical drive for a conveyor system.

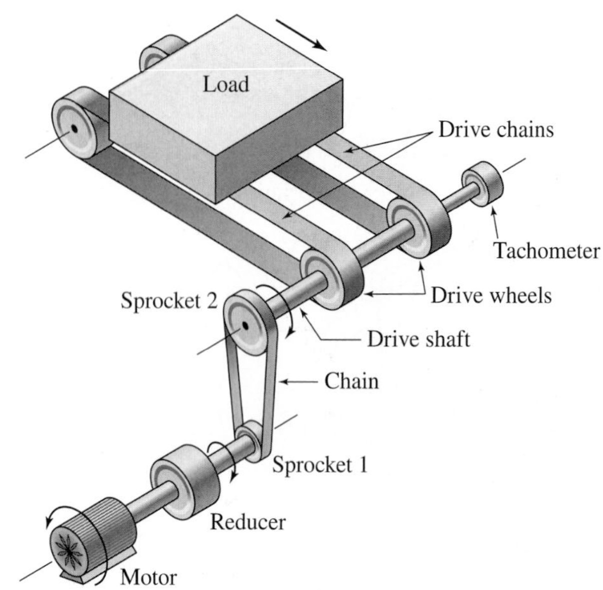

1.1.9 APPLICATIONS IN FLUID SYSTEMS

A fluid system is one whose operation depends on the flow of a fluid. If the fluid is *incompressible*, that is, if its density does not change appreciably with pressure changes, we call it a liquid, or a *hydraulic* fluid. On the other hand, if the fluid is *compressible*, that is, if its density does change appreciably with pressure changes, we call it a gas, or a *pneumatic* fluid.

Figure 1.1.10 shows a commonly seen backhoe. The bucket, forearm, and upper arm are each driven by a *hydraulic servomotor*. A cutaway view of such a motor is shown in Figure 1.1.11. We will analyze its behavior in Chapter 7. Compressed air

Figure 1.1.10 A backhoe.

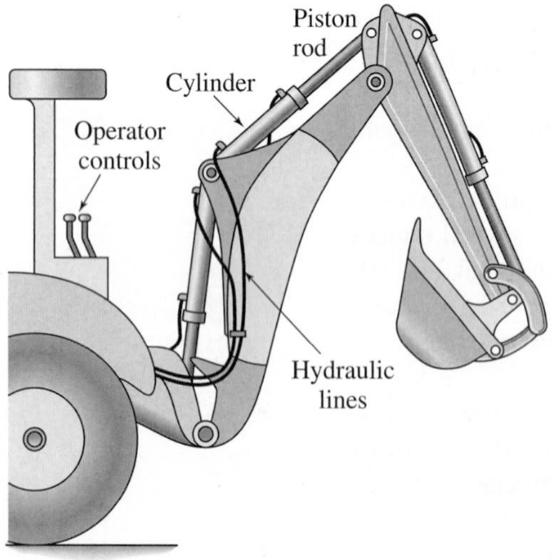

Figure 1.1.11 A hydraulic servomotor.

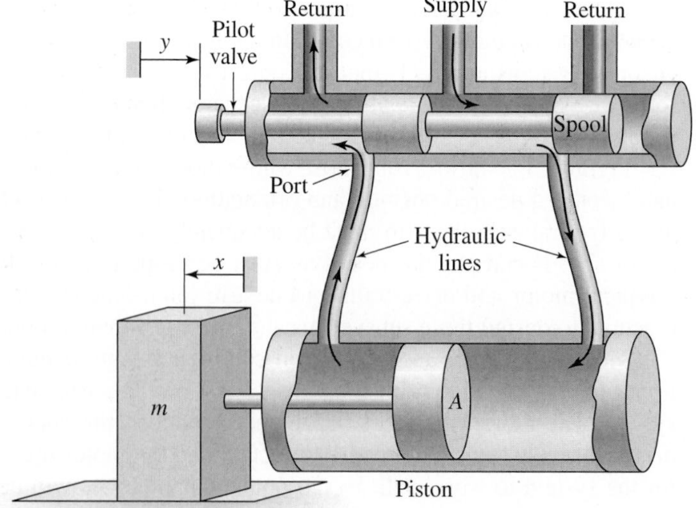

Table 1.1.2 Steps for developing a computer solution.

1. State the problem concisely.
2. Specify the data to be used by the program. This is the "input."
3. Specify the information to be generated by the program. This is the "output."
4. Work through the solution steps by hand or with a calculator; use a simpler set of data if necessary.
5. Write and run the program.
6. Check the output of the program with your hand solution.
7. Run the program with your input data and perform a reality check on the output.
8. If you will use the program as a general tool in the future, test it by running it for a range of reasonable data values, and perform a reality check on the results. Document the program with comment statements, flow charts, pseudo-code, or whatever else is appropriate.

cylinders and the common storm door closer are examples of pneumatic systems, and we encounter them in Chapter 7.

1.1.10 APPLICATIONS IN THERMAL SYSTEMS

A thermal system is one whose behavior depends primarily on the exchange of heat. The oven-potato application we saw earlier is an example of a thermal system. Many thermal systems involve fluid flow, such as with a steam engine or an air conditioner, and so we often speak of *thermo-fluid systems*. These examples also have mechanical components such as pistons, and so we could refer to them as thermo-fluid-mechanical systems, although we rarely use such cumbersome terminology. The designation as thermal, fluid, or mechanical depends on what aspect of the system we are analyzing. Thermal systems are first treated in Chapter 7, with more applications covered in later chapters.

1.1.11 COMPUTER METHODS

The computer methods used in this text are based on MATLAB and Simulink.®[1] If you are unfamiliar with MATLAB, Appendix D on the textbook website contains a thorough introduction to the program. No prior experience with Simulink is required; we will introduce the necessary methods as we need them. For the convenience of those who prefer to use a software package other than MATLAB or Simulink, we have placed all the MATLAB and Simulink material in optional sections at the end of each chapter. They can be skipped without affecting your understanding of the following chapters. If you use a program, such as MATLAB, to solve a problem, follow the steps shown in Table 1.1.2.

1.2 UNITS

In this book we use two systems of units, the *FPS* system and the metric SI. The common system of units in business and industry in English-speaking countries has been the foot-pound-second (FPS) system. This system is also known as the U.S. customary system or the British Engineering system. Much engineering work in the United States has been based on the FPS system, and some industries continue to use it. The metric Système International d'Unités (SI) nevertheless is becoming the worldwide standard.

[1] Simulink is a registered trademark of The MathWorks, Inc.

Until the changeover is complete, engineers in the United States will have to be familiar with both systems.

In our examples, we will use SI and FPS units in the hope that the student will become comfortable with both. Other systems are in use, such as the meter-kilogram-second (mks) and centimeter-gram-second (cgs) metric systems and the British system, in which the mass unit is a pound. We will not use these, in order to simplify our coverage and because FPS and SI units are the most common in engineering applications. We now briefly summarize these two systems.

1.2.1 FPS UNITS

The FPS system is a *gravitational* system. This means that the primary variable is force, and the unit of mass is derived from Newton's second law. The *pound* is selected as the unit of force and the *foot* and *second* as units of length and time, respectively. From Newton's second law of motion, force equals mass times acceleration, or

$$f = ma \tag{1.2.1}$$

where f is the net force acting on the mass m and producing an acceleration a. Thus, the unit of mass must be

$$\text{mass} = \frac{\text{force}}{\text{acceleration}} = \frac{\text{pound}}{\text{foot}/(\text{second})^2}$$

This mass unit is named the *slug*.

Through Newton's second law, the weight W of an object is related to the object mass m and the acceleration due to gravity, denoted by g, as follows: $W = mg$. At the surface of the earth, the standard value of g in FPS units is $g = 32.2$ ft/sec^2.

Energy has the dimensions of mechanical work; namely, force times displacement. Therefore, the unit of energy in this system is the *foot-pound* (ft-lb). Another energy unit in common use for historical reasons is the *British thermal unit* (Btu). The relationship between the two is given in Table 1.2.1. Power is the rate of change of energy with time, and a common unit is *horsepower*. Finally, temperature in the FPS system can be expressed in degrees *Fahrenheit* or in absolute units, degrees *Rankine*.

1.2.2 SI UNITS

The SI metric system is an *absolute* system, which means that the mass is chosen as the primary variable, and the force unit is derived from Newton's law. The *meter* and the

Table 1.2.1 SI and FPS units.

	Unit name and abbreviation	
Quantity	**SI Unit**	**FPS Unit**
Time	second (s)	second (sec)
Length	meter (m)	foot (ft)
Force	newton (N)	pound (lb)
Mass	kilogram (kg)	slug
Energy	joule (J)	foot-pound (ft-lb), Btu (= 778 ft-lb)
Power	watt (W)	ft-lb/sec, horsepower (hp)
Temperature	degrees Celsius (°C), degrees Kelvin (K)	degrees Fahrenheit (°F), degrees Rankine (°R)

Table 1.2.2 Unit conversion factors.

Length	1 m = 3.281 ft	1 ft = 0.3048 m
	1 mile = 5280 ft	1 km = 1000 m
Speed	1 ft/sec = 0.6818 mi/hr	1 mi/hr = 1.467 ft/sec
	1 m/s = 3.6 km/h	1 km/h = 0.2778 m/s
	1 km/hr = 0.6214 mi/hr	1 mi/hr = 1.609 km/h
Force	1 N = 0.2248 lb	1 lb = 4.4484 N
Mass	1 kg = 0.06852 slug	1 slug = 14.594 kg
Energy	1 J = 0.7376 ft-lb	1 ft-lb = 1.3557 J
Power	1 hp = 550 ft-lb/sec	1 hp = 745.7 W
	1 W = 1.341×10^{-3} hp	
Temperature	T°C = 5 (T°F − 32)/9	T°F = 9T°C/5 + 32

second are selected as the length and time units, and the *kilogram* is chosen as the mass unit. The derived force unit is called the *newton*. In SI units the common energy unit is the newton-meter, also called the *joule*, while the power unit is the joule/second, or *watt*. Temperatures are measured in degrees Celsius, °C, and in absolute units, which are degrees *Kelvin*, K. The difference between the boiling and freezing temperatures of water is 100°C, with 0°C being the freezing point.

At the surface of the earth, the standard value of g in SI units is $g = 9.81$ m/s^2.

Table 1.2.2 gives the most commonly needed factors for converting between the FPS and the SI systems.

1.2.3 OSCILLATION UNITS

There are three commonly used units for frequency of oscillation. If time is measured in seconds, frequency can be specified as *radians/second* or as *hertz*, abbreviated Hz. One hertz is one cycle per second (cps). The relation between cycles per second f and radians per second ω is $2\pi f = \omega$. For sinusoidal oscillation, the *period* P, which is the time between peaks, is related to frequency by $P = 1/f = 2\pi/\omega$. The third way of specifying frequency is revolutions per minute (rpm). Because there are 2π radians per revolution, one rpm $=$ $(2\pi/60)$ radians per second.

1.3 DEVELOPING LINEAR MODELS

A *linear* model of a static element has the form $y = mx + b$, where x is the input and y is the output of the element. As we will see in Chapter 2, solution of dynamic models to predict system performance requires solution of differential equations. Differential equations based on linear models of the system elements are easier to solve than ones based on nonlinear models. Therefore, when developing models we try to obtain a linear model whenever possible. Sometimes the use of a linear model results in a loss of accuracy, and the engineer must weigh this disadvantage with advantages gained by using a linear model. In this section, we illustrate some ways to obtain linear models.

1.3.1 DEVELOPING LINEAR MODELS FROM DATA

If we are given data on the input-output characteristics of a system element, we can first plot the data to see whether a linear model is appropriate, and if so, we can extract a suitable model. Example 1.3.1 illustrates a common engineering problem—the estimation of the force-deflection characteristics of a cantilever support beam.

EXAMPLE 1.3.1

A Cantilever Beam Deflection Model

■ Problem

The deflection of a cantilever beam is the distance its end moves in response to a force applied at the end (Figure 1.3.1). This distance is called the *deflection* and it is the output variable. The applied force is the input. The following table gives the measured deflection x that was produced in a particular beam by the given applied force f. Plot the data to see whether a linear relation exists between f and x.

Force f (lb)	0	100	200	300	400	500	600	700	800
Deflection x (in.)	0	0.15	0.23	0.35	0.37	0.5	0.57	0.68	0.77

■ Solution

The plot is shown in Figure 1.3.2. Common sense tells us that there must be zero beam deflection if there is no applied force, so the curve describing the data must pass through the origin. The straight line shown was drawn by aligning a straightedge so that it passes through the origin

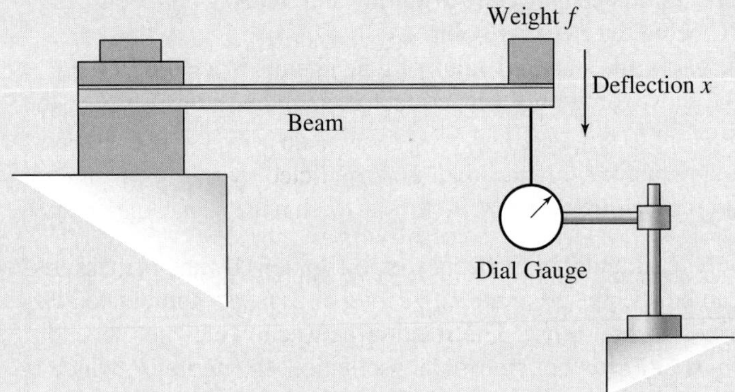

Figure 1.3.1 Measurement of beam deflection.

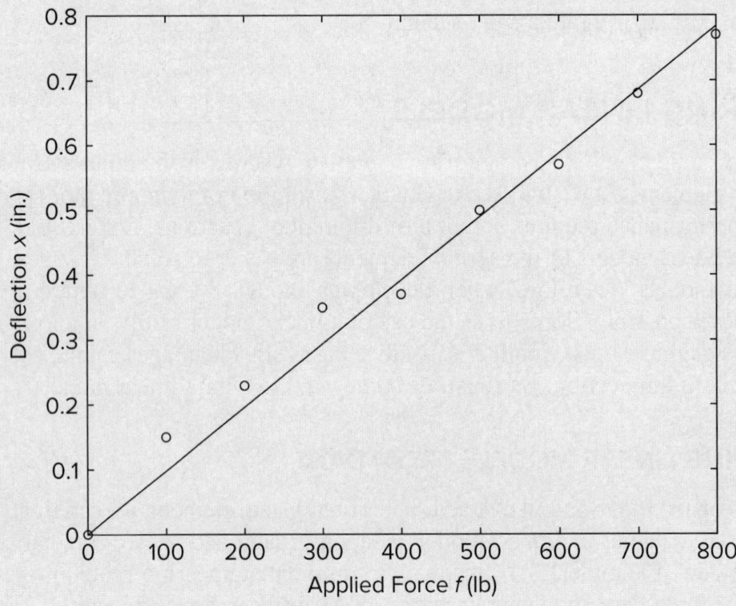

Figure 1.3.2 Plot of beam deflection versus applied force.

and near most of the data points (note that this line is subjective; another person might draw a different line). The data lie close to a straight line, so we can use the linear function $x = af$ to describe the relation. The value of the constant a can be determined from the slope of the line. Choosing the origin and the last data point to find the slope, we obtain

$$a = \frac{0.78 - 0}{800 - 0} = 9.75 \times 10^{-4} \text{ in./lb}$$

As we will see in Chapter 4, this relation is usually written as $f = kx$, where k is called the beam *stiffness*. Thus, $k = 1/a = 1025$ lb/in.

Once we have discovered a functional relation that describes the data, we can use it to make predictions for conditions that lie *within* the range of the original data. This process is called *interpolation*. For example, we can use the beam model to estimate the deflection when the applied force is 550 lb. We can be fairly confident of this prediction because we have data below and above 550 lb and we have seen that our model describes these data very well.

Extrapolation is the process of using the model to make predictions for conditions that lie *outside* the original data range. Extrapolation might be used in the beam application to predict how much force would be required to bend the beam 1.2 in. We must be careful when using extrapolation, because we usually have no reason to believe that the mathematical model is valid beyond the range of the original data. For example, if we continue to bend the beam, eventually the force is no longer proportional to the deflection, and it becomes much greater than that predicted by the linear model. Extrapolation has a use in making tentative predictions, which must be backed up later on by testing.

In some applications, the data contain so much scatter that it is difficult to identify an appropriate straight line. In such cases, we must resort to a more systematic and objective way of obtaining a model. This topic is treated in Appendix C.

1.3.2 LINEARIZATION

Not all element descriptions are in the form of data. Often we know the analytical form of the model, and if the model is nonlinear, we can obtain a linear model that is an accurate approximation over a limited range of the independent variable. Examples 1.3.2 and 1.3.3 illustrate this technique, which is called *linearization*.

Linearization of the Sine Function

EXAMPLE 1.3.2

■ Problem

We will see in Chapter 3 that the models of many mechanical systems involve the sine function $\sin\theta$, which is nonlinear. Obtain three linear approximations of $f(\theta) = \sin\theta$, one valid near $\theta = 0$, one near $\theta = \pi/3$ rad (60°), and one near $\theta = 2\pi/3$ rad (120°).

■ Solution

The essence of the linearization technique is to replace the plot of the nonlinear function with a straight line that passes through the reference point and has the same slope as the nonlinear function at that point. Figure 1.3.3 shows the sine function and the three straight lines obtained with this technique. Note that the slope of the sine function is its derivative, $d\sin\theta/d\theta = \cos\theta$, and thus the slope is not constant but varies with θ.

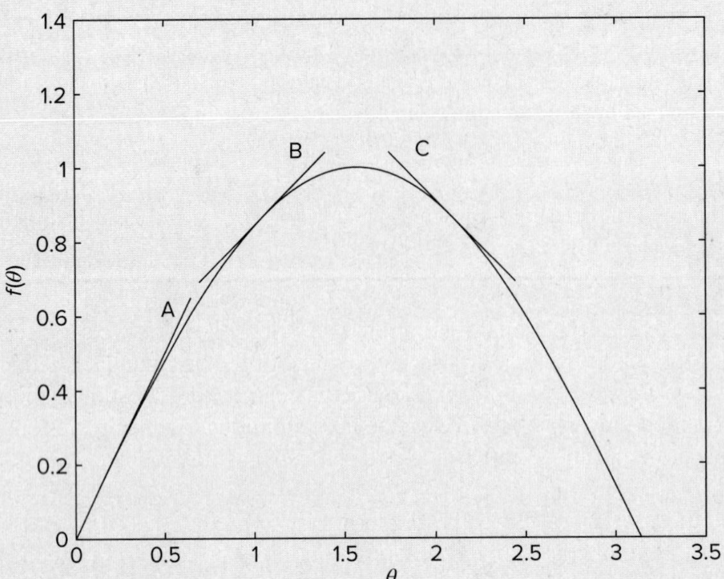

Figure 1.3.3 Three linearized models of the sine function.

Consider the first reference point, $\theta = 0$. At this point the sine function has the value $\sin 0 = 0$, the slope is $\cos 0 = 1$, and thus the straight line passing through this point with a slope of 1 is $f(\theta) = \theta$. This is the linear approximation of $f(\theta) = \sin\theta$ valid near $\theta = 0$, line A in Figure 1.3.3. Thus, we have derived the commonly seen small-angle approximation $\sin\theta \approx \theta$.

Next consider the second reference point, $\theta = \pi/3$ rad. At this point the sine function has the value $\sin\pi/3 = 0.866$, the slope is $\cos\pi/3 = 0.5$, and thus the straight line passing through this point with a slope of 0.5 is $f(\theta) = 0.5(\theta - \pi/3) + 0.866$, line B in Figure 1.3.3. This is the linear approximation of $f(\theta) = \sin\theta$ valid near $\theta = \pi/3$.

Now consider the third reference point, $\theta = 2\pi/3$ rad. At this point the sine function has the value $\sin 2\pi/3 = 0.866$, the slope is $\cos 2\pi/3 = -0.5$, and thus the straight line passing through this point with a slope of -0.5 is $f(\theta) = -0.5(\theta - 2\pi/3) + 0.866$, line C in Figure 1.3.3. This is the linear approximation of $f(\theta) = \sin\theta$ valid near $\theta = 2\pi/3$.

In Example 1.3.2 we used a graphical approach to develop the linear approximation. The linear approximation can also be developed with an analytical approach based on the Taylor series. The Taylor series represents a function $f(\theta)$ in the vicinity of $\theta = \theta_r$ by the expansion

$$f(\theta) = f(\theta_r) + \left(\frac{df}{d\theta}\right)_{\theta=\theta_r}(\theta - \theta_r) + \frac{1}{2}\left(\frac{d^2 f}{d\theta^2}\right)_{\theta=\theta_r}(\theta - \theta_r)^2 + \cdots$$
$$+ \frac{1}{k!}\left(\frac{d^k f}{d\theta^k}\right)_{\theta=\theta_r}(\theta - \theta_r)^k + \cdots \tag{1.3.1}$$

Consider the nonlinear function $f(\theta)$, which is sketched in Figure 1.3.4. Let $[\theta_r, f(\theta_r)]$ denote the reference operating condition of the system. A model that is linear can be obtained by expanding $f(\theta)$ in a Taylor series near this point and truncating

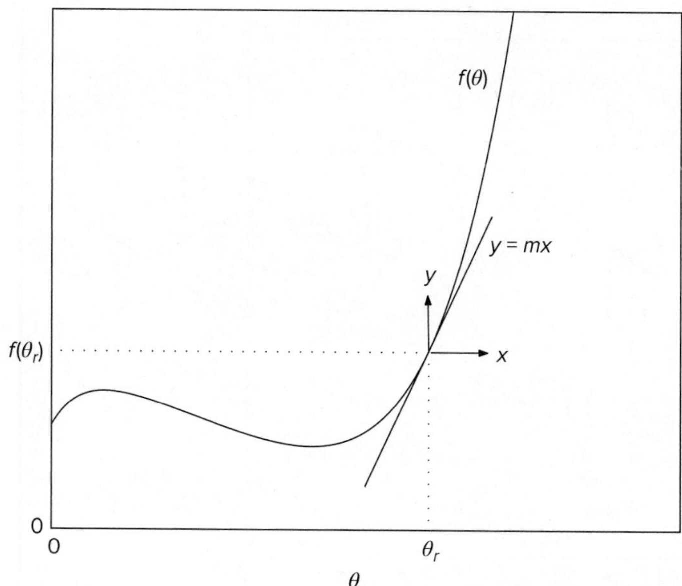

Figure 1.3.4 Graphical interpretation of function linearization.

the series beyond the first-order term. If θ is "close enough" to θ_r, the terms involving $(\theta - \theta_r)^i$ for $i \geq 2$ are small compared to the first two terms in the series. The result is

$$f(\theta) = f(\theta_r) + \left(\frac{df}{d\theta}\right)_r (\theta - \theta_r) \tag{1.3.2}$$

where the subscript r on the derivative means that it is evaluated at the reference point. This is a linear relation. To put it into a simpler form, let m denote the slope at the reference point.

$$m = \left(\frac{df}{d\theta}\right)_r \tag{1.3.3}$$

Let y denote the difference between $f(\theta)$ and the reference value $f(\theta_r)$.

$$y = f(\theta) - f(\theta_r) \tag{1.3.4}$$

Let x denote the difference between θ and the reference value θ_r.

$$x = \theta - \theta_r \tag{1.3.5}$$

Then (1.3.2) becomes

$$y = mx \tag{1.3.6}$$

The geometric interpretation of this result is shown in Figure 1.3.4. We have replaced the original function $f(\theta)$ with a straight line passing through the point $[\theta_r, f(\theta_r)]$ and having a slope equal to the slope of $f(\theta)$ at the reference point. Using the (y, x) coordinates gives a zero intercept, and simplifies the relation.

Linearization of a Square-Root Model — EXAMPLE 1.3.3

■ Problem

We will see in Chapter 7 that the models of many fluid systems involve the square-root function $\sqrt{h}$, which is nonlinear. Obtain a linear approximation of $f(h) = \sqrt{h}$ valid near $h = 9$.

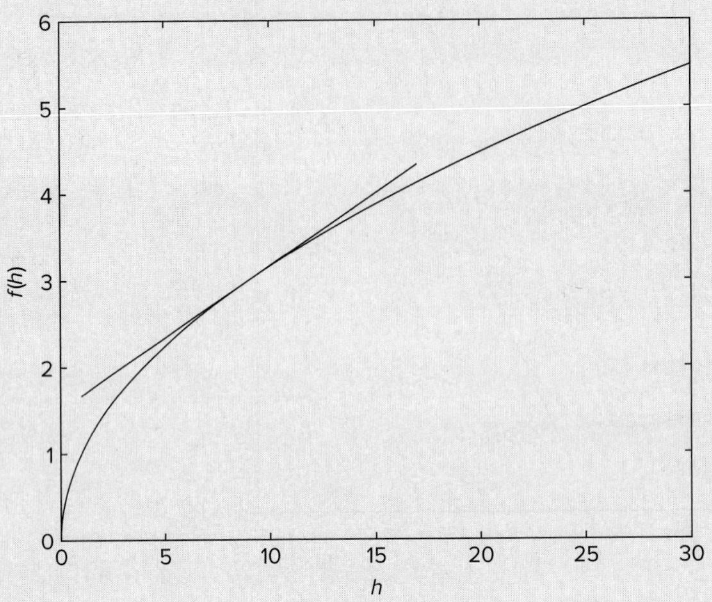

Figure 1.3.5 Linearization of the square-root function.

■ **Solution**

The truncated Taylor series for this function is

$$f(h) = f(h_r) + \left.\frac{d(\sqrt{h})}{dh}\right|_r (h - h_r)$$

where $h_r = 9$. This gives the linear approximation

$$f(h) = \sqrt{9} + \left.\frac{1}{2}h^{-1/2}\right|_r (h - 9) = 3 + \frac{1}{6}(h - 9)$$

This equation gives the straight line shown in Figure 1.3.5.

Sometimes we need a linear model that is valid over so wide a range of the independent variable that a model obtained from the Taylor series is inaccurate or grossly incorrect. In such cases, we must settle for a linear function that gives a conservative estimate.

EXAMPLE 1.3.4

Modeling Fluid Drag

■ **Problem**

The drag force on an object moving through a liquid or a gas is a function of the velocity. A commonly used model of the drag force D on an object is

$$D = \frac{1}{2}\rho A C_D v^2 \quad (1)$$

where ρ is the mass density of the fluid, A is the object's cross-sectional area normal to the relative flow, v is the object's velocity relative to the fluid, and C_D is the *drag coefficient*, which is usually determined from wind-tunnel or water-channel tests on models. Curve A in Figure 1.3.6 is a plot of this equation for an Aerobee rocket 1.25 ft in diameter, with $C_D = 0.4$, moving through the lower atmosphere where $\rho = 0.0023$ slug/ft^3, for which equation (1) becomes

$$D = 0.00056v^2 \quad (2)$$

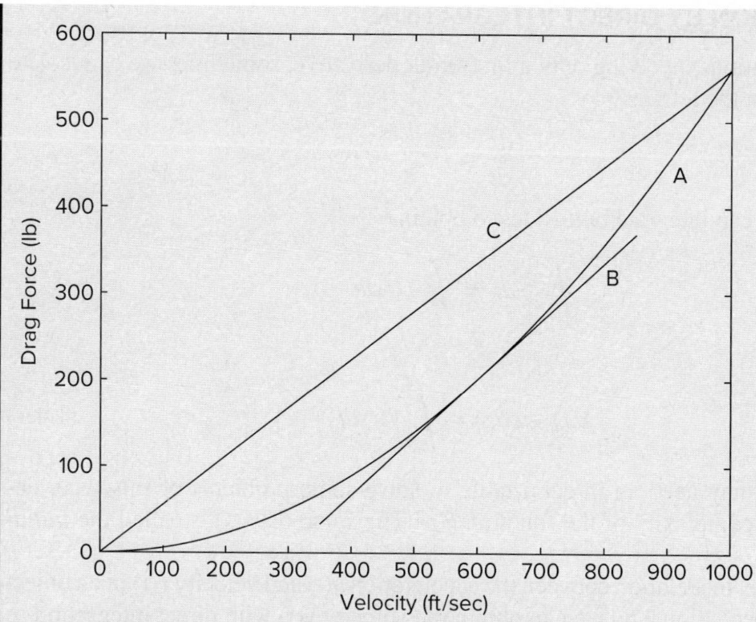

Figure 1.3.6 Models of fluid drag.

a. Obtain a linear approximation to this drag function valid near $v = 600$ ft/sec.
b. Obtain a linear approximation that gives a conservative (high) estimate of the drag force as a function of the velocity over the range $0 \leq v \leq 1000$ ft/sec.

■ Solution

a. The Taylor series approximation of equation (2) near $v = 600$ is

$$D = D|_{v=600} + \left.\frac{dD}{dv}\right|_{v=600} (v - 600) = 201.6 + 0.672(v - 600)$$

This straight line is labeled B in Figure 1.3.6. Note that it predicts that the drag force will be negative when the velocity is less than 300 ft/sec, a result that is obviously incorrect. This illustrates how we must be careful when using linear approximations.

b. The linear model that gives a conservative estimate of the drag force (that is, an estimate that is never less than the actual drag force) is the straight-line model that passes through the origin and the point at $v = 1000$. This is the equation $D = 0.56v$, shown by the straight line C in Figure 1.3.6.

1.4 INTRODUCTION TO DIFFERENTIAL EQUATIONS

A *differential equation* is simply an equation involving derivatives with respect to the independent variable. In system dynamics this variable is often the time variable, usually denoted by t. This section considers some simple problems in which the derivative can be isolated and the basic methods of integral calculus can be applied to obtain a solution. We also discuss the role of the *initial conditions* in obtaining a solution. The approach is illustrated with a simple dynamics application: the challenge of bringing a planetary lander to a soft landing. Time derivatives are often abbreviated with the dot notation, as

$$\dot{x} = \frac{dx}{dt} \qquad \ddot{x} = \frac{d^2x}{dt^2}$$

1.4.1 SOLUTION BY DIRECT INTEGRATION

With some equations involving only a first-order derivative, sometimes we can isolate the derivative on the left, as

$$\frac{dx}{dt} = f(t)$$

In this case, we can integrate both sides to obtain

$$\int_{x(t_0)}^{x(t)} dx = \int_{t_0}^{t} f(t)\,dt$$

or

$$x(t) - x(t_0) = \int_{t_0}^{t} f(t)\,dt \tag{1.4.1}$$

Obviously you may need an integral table to solve some problems of this type, depending on the complexity of the function $f(t)$. The value of $x(t_0)$ is called the *initial condition*.

For example, the relation between the acceleration $a(t)$ and velocity $v(t)$ of an object is a differential equation, and we can obtain the velocity $v(t)$ with direct integration as follows:

$$\frac{dv}{dt} = a(t), \qquad v(t) - v(t_0) = \int_{t_0}^{t} a(t)\,dt$$

The relation between the velocity $v(t)$ and the displacement $x(t)$ is also a differential equation, and we can obtain the displacement $x(t)$ as follows:

$$\frac{dx}{dt} = v(t), \qquad x(t) - x(t_0) = \int_{t_0}^{t} v(t)\,dt$$

Suppose that the acceleration is $a(t) = 6t^2$, $t_0 = 0$, and the initial velocity is $v(0) = 5$. Then

$$v(t) - 5 = \int_0^t 6t^2\,dt = 2t^3$$

Thus, $v(t) = 5 + 2t^3$. If the initial displacement is $x(0) = 7$,

$$x(t) - 7 = \int_0^t \left(5 + 2t^3\right) dt = 5t + \frac{t^4}{2}$$

1.4.2 APPLICATION TO A PLANETARY LANDER

From basic physics recall that Newton's law of motion for a mass m translating in one dimension x is

$$m\frac{d^2m}{dt^2} = m\ddot{x} = \sum f_i$$

where $\ddot{x}$ is the acceleration and $\sum f_i$ is the algebraic sum of all forces acting in the x direction.

For example, consider the problem of a rocket or planetary lander attempting to make a "soft" landing, which is defined as the lander having zero velocity when

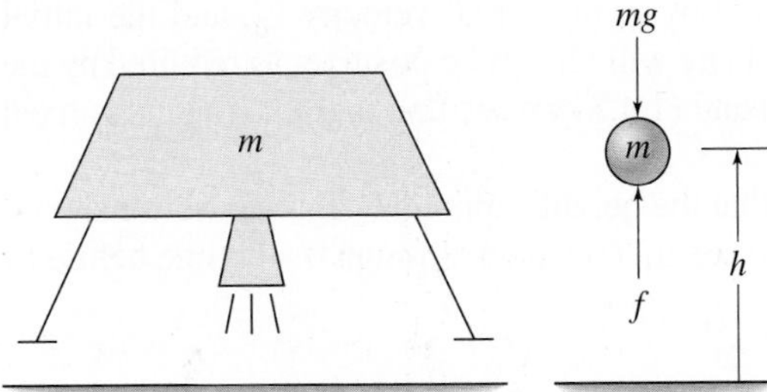

Figure 1.4.1 Simple model of the vertical motion of a lander.

touchdown occurs at $h = 0$ (Figure 1.4.1). A simple representation of this situation is given by Newton's law:

$$m\ddot{h} = f - mg$$

where mg is the lander weight.

Soft Landing with Constant Thrust We will use this model to answer the following question: Is it possible to achieve a soft landing using *constant* thrust? To answer this, we first simplify the problem, if possible, to analyze only the most important aspects of the problem. This is often hard to do, and experience definitely helps (which you hopefully will obtain from all of our examples!). So we will assume that the acceleration g due to gravity is constant; this means that the lander does not have a great change in altitude. We neglect any atmospheric drag. Can you think of other simplifications being used in the following model?

Express the thrust as a function of the weight as follows: $f = \beta mg$, where β is a constant that we need to compute. This lets us temporarily eliminate the mass from the model as follows:

$$m\ddot{h} = \beta mg - mg = (\beta - 1)mg = \alpha mg$$

or, after cancelling m, we obtain $\ddot{h} = \alpha g$, where α is a new constant: $\alpha = \beta - 1$. Integrating both sides once gives the velocity.

$$\dot{h}(t) = \alpha gt + \dot{h}(0) = \alpha gt - v_0$$

where $v_0 = -\dot{h}(0)$ is the initial vertical velocity *downward.* To achieve a soft landing, $\dot{h} = 0$, and the previous equation shows that the time to land t_f is

$$t_f = \frac{v_0}{\alpha g} \tag{1.4.2}$$

Integrating the $\dot{h}$ expression gives

$$h(t) = \alpha g \frac{t^2}{2} - v_0 t + h_0$$

where h_0 is the initial height. Substituting $t = t_f$ into this equation gives

$$h(t_f) = \frac{\alpha g}{2}\left(\frac{v_0}{\alpha g}\right)^2 - \frac{v_0^2}{\alpha g} + h_0 = -\frac{v_0^2}{2\alpha g} + h_0 = 0$$

This gives

$$\alpha = \frac{v_0^2}{2gh_0} \tag{1.4.3}$$

Thus, we see that the value of α is fixed by g, the initial velocity v_0, and the initial height h_0. The required thrust $f = (\alpha + 1)mg$ will always be positive, as required by the physics of rocket engines. From (1.4.2) and (1.4.3) we see that the time to land is fixed by v_0 and h_0.

The mathematics does not realize that the height cannot be negative, so we should check for that possibility by checking to see if $h(t)$ passes through 0 any time before t_f. Setting $h(t)$ equal to zero gives

$$0 = \alpha g\frac{t^2}{2} - v_0 t + h_0$$

From the quadratic formula and using (1.4.3), we have

$$t = \frac{v_0 \pm \sqrt{v_0^2 - 2\alpha g h_0}}{\alpha g} = \frac{v_0}{\alpha g}$$

which is the time to land, t_f. Thus, $h(t) \geq 0$.

EXAMPLE 1.4.1

Constant Thrust Example

■ Problem

Suppose that $m = 2500$ kg and $g = 2$ m/s^2 (which indicates a planetary body smaller than the Earth). Suppose also that the initial downward velocity is $v_0 = 5$ m/s and the initial height is $h_0 = 500$ m. Find the thrust required for a soft landing and the time to land.

■ Solution

From (1.4.3) we have

$$\alpha = \frac{25}{2(2)(500)} = 0.0125$$

Thus, the thrust required is $f = \beta mg = (\alpha + 1)mg = 1.0125(2500)(2) = 5062.5$ N. The time to land is found from (1.4.2) to be $t_f = 5/[(0.0125)(2)] = 200$ s.

Note some of the other simplications made in obtaining this result. We assumed the lander mass is constant, but in fact it decreases with time due to fuel consumption. We also assumed purely vertical motion, whereas the lander could be descending from orbit. We also assumed that g is constant, whereas in fact g varies with altitude (but this is a good assumption when the altitude is only a few hundred meters).

Linear Velocity Profile We approached the lander problem by assuming that the thrust is constant. We could have approached the problem by assuming that the velocity $\dot{h}(t)$ is a certain function, called a *velocity profile*. For example, we could have assumed that the velocity is a linear function of time; that is,

$$\dot{h}(t) = at + b$$

where a and b are constants to be found. Differentiating this gives the acceleration $\ddot{h} = a$, and substituting this into the equation of motion gives

$$m\ddot{h} = ma = f - mg$$

This shows that $f = m(a + g)$, which is constant. Thus, the assumed linear velocity profile gives the same results as found with the constant thrust assumption.

Quadratic Velocity Profile Using constant thrust enables a soft landing to be achieved but we cannot specify the time of flight t_f. Noting that constant thrust results in a linear velocity profile, we can try another velocity profile that may let us also set a desired flight time. So we try a quadratic profile of the form

$$\dot{h}(t) = at^2 + bt + c$$

where we can easily show that $c = -v_0$ and $\ddot{h} = 2at + b$. Substituting this into the equation motion shows that the thrust must now be a linear time function: $f = m(2at + b + mg)$. One of the chapter problems asks you to compute the required values of a and b.

1.5 A CASE STUDY IN MOTION CONTROL

Motion-control engineering is a specialty within the field of automation and involves the design of systems for controlling position or speed of machines. The main components of such systems are a motion controller (such as electronics or a computer), a power amplifier, and an actuator or prime mover to do the work. Typical actuators include hydraulic or pneumatic pistons and electric motors. Motion control is widely used in robotics and computer numerically controlled (CNC) machine tools, and in assembly lines for textiles, packaging, and printing. The term "motion controller" is also used to describe a type of video game controller used to provide input, but this application is not discussed here.

Various kinds of mechanical components are used with the actuator to achieve the desired motion. These include gears, power screws, and belts or chains. Figure 1.5.1 shows a typical motion-control system designed to translate an object (called the "load") over a certain distance with a desired speed. In Chapter 3 we will return to this system to analyze how well it functions. We will show how to model it simply, as if only a single rotational object (called the "equivalent load" or "reflected load") is connected to the motor shaft.

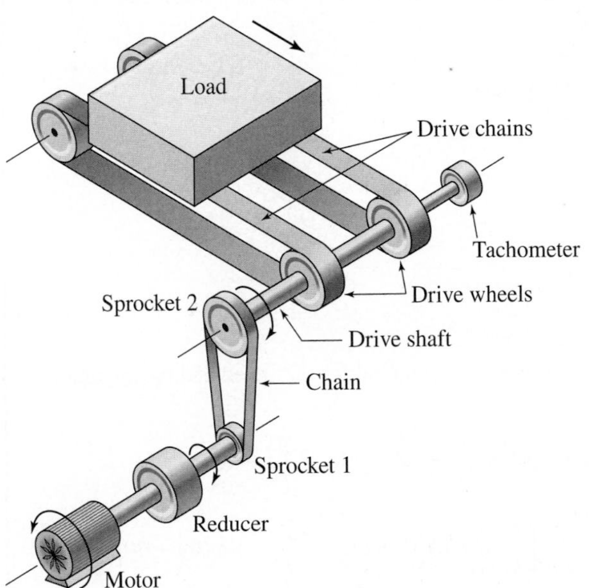

Figure 1.5.1 Mechanical drive for a conveyor system.

If we consider the dynamics of a rigid body whose motion is constrained to allow only rotation about an axis through a nonaccelerating point, then from basic physics we know that the equation of motion is

$$I\frac{d\omega}{dt} = I\dot{\omega} = M$$

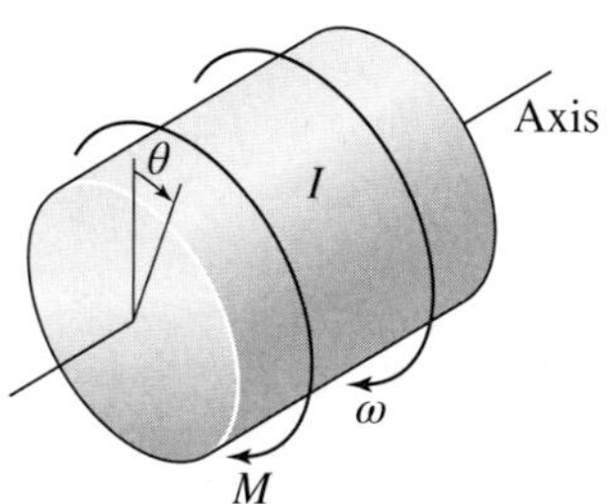

Figure 1.5.2 An object rotating about a fixed axis.

where ω is the angular velocity of the mass about an axis through a point fixed in an inertial reference frame and attached to the body (or the body extended), I is the mass moment of inertia of the body about that point, and M is the sum of the moments applied to the body about that point. This situation is illustrated in Figure 1.5.2. The angular displacement is θ, always measured in radians, and $\omega = d\theta/dt = \dot{\theta}$, whose units are radians per unit time. The term *torque* and thus the symbol T are often used instead of *moment* and M. When the context is unambiguous, we use the term *inertia* as an abbreviation for *mass moment of inertia*. In Chapter 3 we consider the case of more general rotation and show how I is calculated for specific body shapes.

EXAMPLE 1.5.1

Constant Torque Solution

■ Problem

Figure 1.5.3 illustrates a basic example of motion control where a motor supplies a torque T to an object, called the "load," whose inertia is I_L. The rotating part of the motor is called the "armature" and has an inertia I_m (in Chapter 6 we will study motor design and performance in more detail). If the shaft connecting the motor to the load does not twist, then the motor and load rotate at the same speed ω and the combined inertia of the system will be the sum: $I = I_m + I_S + I_L$ where I_m, I_S, and I_L are the inertias of the motor, shaft, and load, respectively. The equation of motion is given by

$$(I_m + I_S + I_L)\dot{\omega} = T \quad \text{or} \quad I\dot{\omega} = T \tag{1}$$

Obtain the speed and displacement as functions of time.

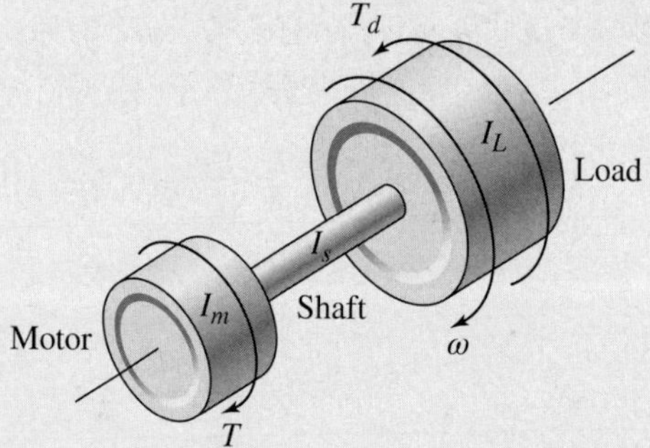

Figure 1.5.3 Simple model of a mechanical drive with a rotational load.

■ Solution

If the motor torque is constant, we can solve the equation for the speed and the angular displacement as functions of time, as follows:

$$\omega(t) = \int_0^t \dot{\omega}(t)\,dt + \omega(0) = \int_0^t \frac{T}{I}\,dt + \omega(0) = \frac{T}{I}t + \omega(0)$$

$$\theta(t) = \int_0^t \omega(t)\,dt + \theta(0) = \int_0^t \left[\frac{T}{I}t + \omega(0)\right]dt + \theta(0) = \frac{T}{I}\frac{t^2}{2} + \omega(0)t + \theta(0)$$

If the system starts from rest at $\theta(0) = 0$, then $\omega(0) = 0$ and

$$\omega(t) = \frac{T}{I}t, \qquad \theta(t) = \frac{T}{I}\frac{t^2}{2}$$

So a constant applied torque produces a linear velocity profile and a parabolic displacement profile.

Trapezoidal Velocity Profile In many motion-control applications we want to accelerate the system to a desired speed and run it at that speed for some time before decelerating to a stop, as illustrated by the trapezoidal speed profile shown in Figure 1.5.4. The constant-speed phase of the profile is called the *slew* phase. An example requiring a slew phase of specified duration and speed is a conveyor system in a manufacturing line, in which the item being worked on must move at a constant speed for a specified time while some process is applied, such as painting or welding. The speed profile can be specified in terms of rotational motor speed ω or in terms of the translational speed v of a load.

In other applications, the duration and speed of the slew phase is not specified, but we want the system to move (or rotate) a specified distance (or angle) by the end of one cycle. An example of this is a robot arm link that must move through a specified angle. The acceleration and deceleration times, and the duration and speed of the slew phase must be computed to achieve the desired distance or angle. Often the profile is followed by a zero-speed phase to enable the motor to cool before beginning the next cycle. The profile in Figure 1.5.4 assumes that the system begins and ends at rest. Note also that $t_f - t_2$ must equal t_1 if the profile is symmetric.

Available motor torque is aways limited, and the manufacturer will specify the maximum available torque T_{max}, which is the torque the motor can provide for a short time. The manufacturer will also state the *rated continuous torque*, also called *duty cycle torque*. This is the torque the motor can provide continuously without being damaged, for example, by overheating. The average torque required for a specific application is computed by the *root mean square* method and is denoted T_{rms}. The profile must not require an average torque greater than the rated continuous torque. The root mean square average is defined as

$$T_{rms} = \sqrt{\frac{1}{t_f}\int_0^{t_f} T^2(t)\,dt} \qquad (1.5.1)$$

Thus, because of the squared torque, the rms average gives equal weight to positive and negative values of torque.

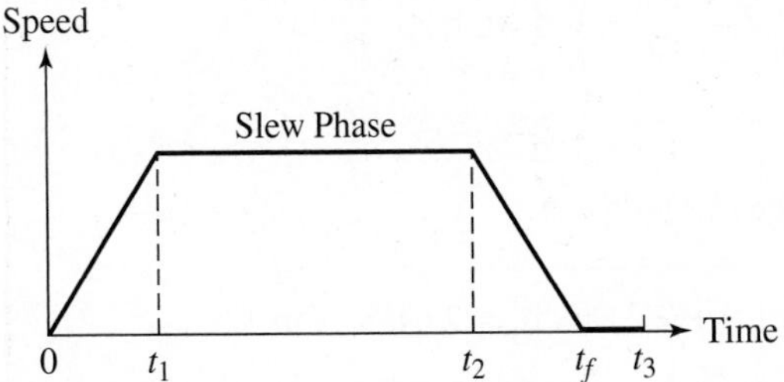

Figure 1.5.4 A trapezoidal speed profile.

EXAMPLE 1.5.2

Maximum Versus Average Required Torque

■ Problem

Suppose that $I_m = 0.003$, $I_S = 0.001$, and $I_L = 0.006$ so that $I = I_m + I_S + I_L = 0.01$ kg·m² for the system shown in Figure 1.5.3. Compute how much motor torque is required to achieve the velocity profile shown in Figure 1.5.4, where $t_1 = 0.5$ s, $t_2 = 2.5$ s, and $t_f = 3$ s, and the slew speed is 150 rad/s.

■ Solution

From Figure 1.5.4, we obtain the required angular acceleration from the slope of the velocity curve. This gives

$$\dot{\omega} = \begin{cases} \frac{150}{0.5} = 300 & 0 \le t \le 0.5 \\ 0 & 0.5 < t < 2.5 \\ -\frac{150}{0.5} = -300 & 2.5 \le t \le 3 \end{cases}$$

Thus, the required torque is

$$T = I\dot{\omega} = \begin{cases} 3 & 0 \le t \le 0.5 \\ 0 & 0.5 < t < 2.5 \\ -3 & 2.5 \le t \le 3 \end{cases}$$

So the maximum required torque is 3 N·m.

The rms average torque is calculated as follows:

$$T_{rms} = \sqrt{\frac{1}{3}\left[\int_0^{0.5} 3^2 dt + \int_{0.5}^{2.5} 0^2 dt + \int_{2.5}^{3} (-3)^2 dt\right]} = \sqrt{\frac{1}{3}\left[3^2(0.5) + 0 + (-3)^2(0.5)\right]}$$

$$= \sqrt{3} = 1.7321 \text{ N·m}$$

Note that T_{rms} is obviously never greater than the maximum torque.

EXAMPLE 1.5.3

Friction Effects

■ Problem

Suppose the load in Example 1.5.2 is acted on by a torque $T_d = 2$ N·m caused by Coulomb (dry) friction. Recalculate the maximum and rms torques.

■ Solution

The equation of motion is now

$$I\dot{\omega} = T - T_d \qquad \omega > 0$$

Thus,

$$T = I\dot{\omega} + T_d = \begin{cases} 3 + 2 = 5 & 0 \le t \le 0.5 \\ 2 & 0.5 < t < 2.5 \\ -3 + 2 = -1 & 2.5 \le t \le 3 \end{cases}$$

So the maximum required torque is 5 and the rms torque is

$$T_{rms} = \sqrt{\frac{1}{3}\left[(5)^2(0.5) + (2)^2(2) + (-1)^2(0.5)\right]} = 2.6458 \text{ N·m}$$

GENERAL RESULTS FOR A TRAPEZOIDAL VELOCITY PROFILE

The results of the two previous examples can be generalized as follows. The equation of motion for the system in Figure 1.5.5 is

$$I\dot{\omega} = T - T_d \tag{1.5.2}$$

where I is the system inertia at the motor shaft, T is the motor torque, and T_d is a disturbance torque such as that caused by friction.

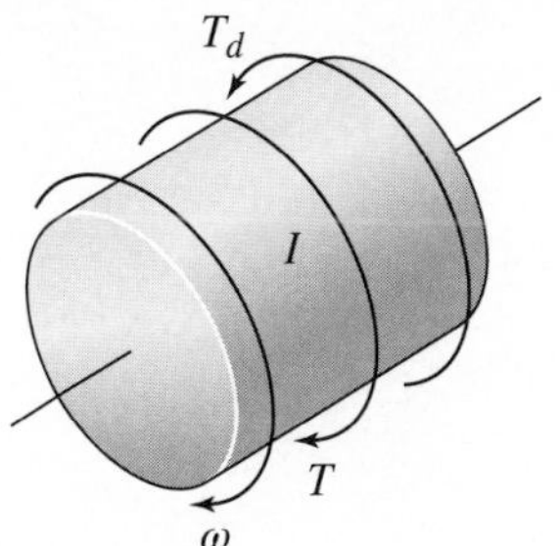

Figure 1.5.5 A rotational system with a single inertia.

Maximum Motor Speed for a Specified Displacement Suppose that the application requires that the load rotate through a specified displacement θ_f in the specified time t_f. For this application the trapezoidal profile will specify the *angular* velocity rather than a translational velocity. Thus, the total angular displacement θ_f is the area under the trapezoidal profile. Assuming that $\theta(0) = 0$ and because $t_f - t_2 = t_1$, we have

$$\theta(t_f) = \theta_f = \int_0^{t_f} \omega(t)dt = 2\left(\frac{1}{2}\omega_{max}t_1\right) + \omega_{max}(t_2 - t_1) = \omega_{max}t_2$$

Thus, the maximum required motor speed is given by

$$\omega_{max} = \frac{\theta_f}{t_2} \tag{1.5.3}$$

So if we specify the displacement θ_f and the time t_2, the maximum required speed is determined from this equation.

Maximum Motor Torque for a Specified Displacement Let α denote the angular acceleration $\dot{\omega}$. The maximum required angular acceleration for the trapezoidal profile is

$$\alpha_{max} = \frac{\omega_{max}}{t_1} = \frac{\theta_f}{t_1 t_2} \tag{1.5.4}$$

Using (1.5.2) through (1.5.4), the maximum required motor torque is given by

$$T_{max} = I\alpha_{max} + T_d = I\frac{\omega_{max}}{t_1} + T_d = I\frac{\theta_f}{t_1 t_2} + T_d \tag{1.5.5}$$

RMS Motor Torque The rms torque is calculated from (1.5.1) and (1.5.2):

$$T_{rms}^2 = \frac{1}{t_f}\int_0^{t_f} T^2(t)\,dt = \frac{1}{t_f}\int_0^{t_f} (I\alpha + T_d)^2\,dt$$

or

$$T_{rms}^2 = \frac{1}{t_f}\int_0^{t_1} (I\alpha_{max} + T_d)^2\,dt + \frac{1}{t_f}\int_{t_1}^{t_2} (0 + T_d)^2\,dt + \frac{1}{t_f}\int_{t_2}^{t_f} (-I\alpha_{max} + T_d)^2\,dt$$

This reduces to

$$T_{rms}^2 = \frac{1}{t_f}\left[2I^2\alpha_{max}^2 t_1 + T_d^2\,(t_1 + t_2)\right]$$

or, since $\alpha_{max} = \theta_f/t_1 t_2$ and $t_1 + t_2 = t_f$,

$$T_{rms} = \sqrt{\frac{2I^2\theta_f^2}{t_f t_1 t_2^2} + T_d^2} \tag{1.5.6}$$

Table 1.5.1 Motor Requirements for Trapezoidal Speed Profile.

Profile times t_1, t_2, t_f	See Figure 1.5.4
Motor displacement	θ_f = area under speed profile
Load torque felt at motor	T_d
Maximum Speed	$\omega_{max} = \frac{\theta_f}{t_2}$
Maximum Acceleration	$\alpha_{max} = \frac{\theta_f}{t_1 t_2}$
Maximum Torque	$T_{max} = I\frac{\theta_f}{t_1 t_2} + T_d$
RMS Average Torque	$T_{rms} = \sqrt{\frac{2I^2\theta_f^2}{t_f t_1 t_2^2} + T_d^2}$

These equations are summarized in Table 1.5.1 and are used to compute the motor requirements for the trapezoidal speed profile for an application where the total angular displacement is specified. To determine whether a given motor will be satisfactory, the calculated values of ω_{max}, T_{max}, and T_{rms} are compared to the manufacturer's data on maximum speed, peak torque, and rated continuous torque.

EXAMPLE 1.5.4

Motor Calculations for a Trapezoidal Profile

■ Problem

A certain system like that shown in Figure 1.5.3 has a load inertia $I_L = 6 \times 10^{-3}$ kg·m^2, a shaft inertia of $I_S = 5 \times 10^{-4}$ kg·m^2, and a motor inertia of $I_m = 10^{-3}$ kg·m^2. A friction torque of 2 N·m acts on the load. It is desired to rotate the load through 180° $= \pi$ rad in one-half second from start to stop. The load should be rotated at a constant speed for 0.4 s during the slew phase. Compute the motor's maximum speed, the maximum required torque, and the rms average torque.

■ Solution

The total load displacement is given as $\theta_f = \pi$ rad. We also have $t_f = 0.5$ s and a slew time of $t_2 - t_1 = 0.4$ s. Using a symmetrical velocity profile requires that $t_1 = 0.05$ and $t_2 = 0.45$. The total inertia is $I = 6 \times 10^{-3} + 5 \times 10^{-4} + 10^{-3} = 7.5 \times 10^{-3}$. From Table 1.5.1 we have $\omega_{max} = \pi/0.4 = 6.9813$ rad/s or $60\omega_{max}/(2\pi) = 66.6667$ rpm. So the load will rotate at that speed for 0.4 s.

We are given that $T_d = 2$ N·m. So the maximum torque required is $T_{max} = 3.0472$ N·m and the rms average torque is $T_{rms} = 2.0541$ N·m.

Lead Screws A lead screw or power screw (Figure 1.5.6) is used to convert the rotational motion of a motor into linear motion of a load mounted on the nut. They are easy

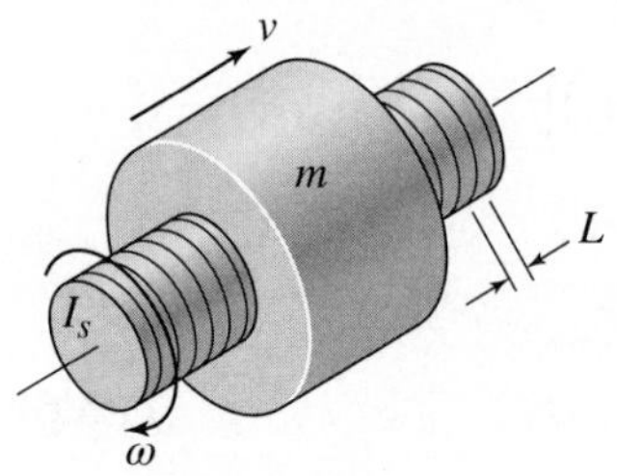

Figure 1.5.6 A lead screw.

to design and manufacture and can carry large loads with high precision. However, because of the large contact area between the sliding surfaces of the nut and screw, they can have large frictional energy losses.

As will be discussed in more detail in Chapter 3, we can model the effect of the translating load by computing an equivalent inertia. We do this by first obtaining an expression for the kinetic energy of the load. If the distance between two adjacent threads is L (called the *screw lead*), then a screw rotation through an angle θ in radians will produce a load translation x, where $x = L\theta/(2\pi)$, and a translational velocity of $\dot{x} = v = L\omega/(2\pi)$, where $\omega = \dot{\theta}$. Thus, the kinetic energy KE of the load is

$$KE = \frac{1}{2}mv^2 = \frac{1}{2}m\left(\frac{L\omega}{2\pi}\right)^2$$

whereas the kinetic energy of a rotating inertia I_L is

$$KE = \frac{1}{2}I_L\omega^2$$

Equating the two expressions for kinetic energy shows that

$$KE = \frac{1}{2}m\left(\frac{L\omega}{2\pi}\right)^2 = \frac{1}{2}I_L\omega^2$$

Solving for I_L gives the expression for the equivalent inertia of the load:

$$I_L = \frac{mL^2}{4\pi^2} \tag{1.5.7}$$

This is the inertia that the motor "feels" due to the load.

The total inertia felt by the motor is the sum of the motor's own inertia I_m, the equivalent load inertia I_L, and the inertia of the screw, which we obtain by modeling the screw as a cylindrical rod. The formula for the screw inertia is

$$I_S = \frac{1}{2}m_S R^2 \tag{1.5.8}$$

where m_S and R are the mass and radius of the cylinder.

Sizing a Motor for a Lead Screw — EXAMPLE 1.5.5

■ Problem

Neglect friction and estimate the torque required to accelerate a 100-kg load mass from rest to 0.2 m/s in 0.1 s, using a steel lead screw of length 1.1 m and radius 0.013 m. The screw lead is $L = 0.005$ m/rev. The motor inertia is 2×10^{-4} kg·m^2. Take the density of steel to be 7700 kg/m^3.

■ Solution

The equivalent inertia of the load mass is, from (1.5.7),

$$I_L = \frac{100(0.005)^2}{4\pi^2} = 6.3326 \times 10^{-5}$$

The mass of the screw is

$$m_S = \rho V = 7700\pi(0.013)^2(1.1) = 4.4970 \text{ kg}$$

and the inertia of the screw is

$$I_S = \frac{1}{2}4.4970(0.013)^2 = 3.8 \times 10^{-4}$$

So the total system inertia is

$$I = I_m + I_L + I_S = 2 \times 10^{-4} + 6.3326 \times 10^{-5} + 3.8 \times 10^{-4} = 6.4333 \times 10^{-4} \text{ kg·m}^2$$

From the equation of motion, the required torque is $T = I\dot{\omega}$. The linear displacement x of the mass is related to the rotation angle θ of the screw by $x = L\theta/2\pi$. Thus, the linear acceleration $\ddot{x}$ is related to the angular acceleration $\ddot{\theta} = \dot{\omega}$ by $\dot{\omega} = 2\pi\ddot{x}/L$.

The given linear acceleration is $\ddot{x} = (0.2 - 0)/0.1 = 2$ m/s^2. Thus,

$$\dot{\omega} = \frac{2\pi(2)}{0.005} = 2513 \text{ rad/s}^2$$

Thus, the required torque is $T = 6.4333 \times 10^{-4}(2513) = 1.6168$ N·m.

The calculations given in Table 1.5.1 can also be applied to a lead screw system using a trapezoidal translational velocity profile, as long as the specified translational displacement, speed, and acceleration are first converted to their equivalent rotational quantities using the screw lead L.

Further Studies in Motion Control In Chapter 3 we will continue our study of motion control by showing how various mechanical elements such as gears and belts can be modeled with an equivalent inertia. Then in Chapter 6 we will show how an electric motor can be selected to provide the required speed and torque. Sometimes external effects such as friction are difficult to predict, and Chapter 10 shows how to design a computer control system to counteract such effects.

1.6 MATLAB REVIEW

This section reviews enough of MATLAB to enable the reader to perform simple calculations and plotting. Further information can be obtained with the MATLAB help system (click on **Help** on the menu bar).

Basic Calculations MATLAB displays the prompt (≫) to indicate that it is ready to receive instructions. To divide 8 by 10, type `8/10` and press Enter. You will see

```
≫8/10
ans = 0.8000
```

MATLAB uses high precision for its computations, but by default it usually displays its results using four decimal places except when the result is an integer. MATLAB assigns the most recent answer to a variable called `ans`, which is an abbreviation for *answer*. You can use the variable `ans` for further calculations; for example, using the MATLAB symbol for multiplication (*), we obtain

```
≫5*ans
ans = 4
```

You can assign the result to a variable of your own choosing, say, `r`, as follows:

```
≫r = 8/10
r = 0.8000
```

This is called an *assignment statement*. The variable or answer, and only the variable or answer, is always on the left of the = symbol, which is called the *assignment* or *replacement* operator. It cannot be used the same way as the equals sign is used in mathematics. For example, the mathematics equation $x = x + 2$ is invalid because it implies that $0 = 2$, but such a statement is valid in MATLAB as long as x has been given a value in a previous statement.

MATLAB uses the symbols `+ - * / ^` for addition, subtraction, multiplication, right division, and exponentiation (power) of scalars. For example, typing `x = 8/4 + 3*5 - 2^3` returns the answer `x = 9`. The *forward slash* (`/`) represents *right division*. *Left division* is denoted by the *backslash* (`\`). The left division operator is useful for solving sets of linear algebraic equations, as we will see. For example, `7/2 = 2\7 = 3.5`. The slash always leans toward the divisor.

Following a set of rules called *precedence*, mathematical expressions are evaluated starting from the left, with the exponentiation operation having the highest order of precedence, followed by multiplication and division with equal precedence, followed by addition and subtraction with equal precedence. Parentheses can be used to alter this order. Evaluation begins with the innermost pair of parentheses and proceeds outward. For example, consider the session:

```
>>8/4*2+8/(4*2)+3*4^2 +(3*4)^2 +27^1/3 +27^(1/3)
```

This calculation produces the following sum:

```
4 + 1 + 48 + 144 + 9 + 3 = 209
```

You can put several commands on the same line. Separate them with a comma if you want to see the results of the previous command or a semicolon if you want to suppress the display.

Built-in Constants and Functions MATLAB has several predefined special constants, such as the built-in constant `pi`. The symbols `i` and `j` denote the imaginary unit, $i = j = \sqrt{-1}$. Use them to create and represent complex numbers, such as `x = 5 + 8i`. Note that an asterisk is not needed between `i` or `j` and a number, although it is required with a variable, such as `x = 5 - i*y`. This convention can cause errors if you are not careful. For example, the expressions `y = 7/2*i` and `x = 7/2i` give two different results: $y = (7/2)i = 3.5i$ and $x = 7/(2i) = -3.5i$.

MATLAB has hundreds of built-in functions. One of these is the *square root* function, `sqrt`. A pair of parentheses is used after the function's name to enclose the value—called the function's *argument*—that is operated on by the function. For example, to compute the square root of 9 and assign its value to the variable r, you type `r = sqrt(9)`. For example, to compute $\sin x$, where x has a value in radians, you type `sin(x)`. To compute $\cos x$, type `cos(x)`. The exponential function e^x is computed from `exp(x)`. The natural logarithm, $\ln x$, is computed by typing `log(x)`. Note the spelling difference between mathematics text, ln, and MATLAB syntax, `log`. You compute the base-10 logarithm by typing `log10(x)`.

The inverse sine, or arcsine, is obtained by typing `asin(x)`. It returns an answer in radians, not degrees. The function `asind(x)` returns degrees. The inverse tangent, or arctangent, is obtained by typing `atan(x)`. It returns an answer in radians, not degrees. The function `atand(x)` returns degrees. You must be careful when using either inverse tangent function. For example, typing `atand(1)` returns 45 degrees but the tangent of -135 degrees is also 1. So you must know the correct quadrant in order to interpret the answer correctly. MATLAB has the four quadrant inverse tangent function, `atan2(y,x)`, that automatically computes the radian angle in the correct quadrant of a line from the origin (0, 0) to a point whose coordinates are (x, y). The function `atan2d(y,x)` returns the answer in degrees. So typing `atan2d(-1,-1)` returns -135 degrees.

Arrays A numerical *array* is an ordered collection of numbers (a set of numbers arranged in a specific order). An example of an array is one that contains the numbers 0, 4, 3, and 6, in that order. With one exception noted later, you must use *square brackets* to define the array **x** to contain this collection. You must separate the elements by either commas or spaces or both. For example, type `x = [0, 4, 3, 6]`. Commas are preferred to improve readability and avoid mistakes. Note that the array **y** defined as `y = [6, 3, 4, 0]` is *not* the same as **x** because the order is different.

You need not type all the numbers in the array if they are regularly spaced. Instead, you type the first number and the last number, with the spacing in the middle, separated by colons. For example, the numbers 0, 0.1, 0.2, …, 10 can be assigned to the array `u` by typing `u = 0:0.1:10`. Note that parentheses are not needed. To compute $w = 5 \sin u$, type `u = 0:0.1:10; w = 5*sin(u)`, where the semicolons suppress the display. You can see the seventh value by typing `u(7)`.

The `linspace` function also creates a linearly spaced row vector, but instead you specify the number of values rather than the increment. The syntax is `linspace(x1, x2,n)`, where `x1` and `x2` are the lower and upper limits and `n` is the number of points. If `n` is omitted, 100 linearly spaced points are generated. The `logspace` function creates an array of logarithmically spaced elements. Its syntax is `logspace(a, b, n)`, where `n` is the number of points between 10^a and 10^b. If `n` is omitted, the number of points defaults to 50.

Arrays with more than one column are called *row arrays*. You can create *column arrays*, which have more than one row, by using a semicolon to separate the rows. For example, typing `r = [0; 4; 3; 6]` creates a column array with four rows and one column.

Polynomial roots can be found with the `roots(a)` function, where `a` is the polynomial's coefficient array, starting with the coefficient of the highest power of the variable. The result is a *column* array that contains the polynomial roots. For example, to find the roots of $x^3 - 7x^2 + 40x - 34 = 0$, the session is

```
>>a = [1,-7,40,-34];
>>roots(a)
ans = 3.0000 + 5.000i
      3.0000 - 5.000i
      1.0000
```

The roots are $x = 1$ and $x = 3 \pm 5i$. The two commands could have been combined into the single command `roots([1,-7,40,-34])`.

The `poly(r)` function computes the coefficients of the polynomial whose roots are specified by the array `r`. The result is a row array that contains the coefficients. For example, to find the polynomial whose roots are 1 and $3 \pm 5j$, the session is

```
>>p = poly([1,3+5j,3-5j])
p = 1  -7  40  -34
```

Thus, the polynomial is $x^3 - 7x^2 + 40x - 34$.

An array having rows and columns is a two-dimensional array that is sometimes called a *matrix*. The most direct way to create a matrix is to type the matrix row by row, separating the elements in a given row with spaces or commas and separating the rows with semicolons. Brackets are required. For example, typing `A = [2,4,10;16,3,7]` creates the following matrix:

$$\mathbf{A} = \begin{bmatrix} 2 & 4 & 10 \\ 16 & 3 & 7 \end{bmatrix}$$

Array Operations Array addition and subtraction for arrays of the same size can be easily done by adding or subtracting the corresponding components. However, multiplication of two *arrays* is not so straightforward. MATLAB defines *element-by-element multiplication* or *array multiplication* only for arrays that are the same size. The definition of the product `x.*y`, where `x` and `y` each have n elements, is

$$\mathbf{x}.^{*}\mathbf{y} = [x(1)y(1), x(2)y(2) \ldots, x(n)y(n)]$$

if **x** and **y** are row vectors. For example, if `x = [2, 4, -5]` and `y = [-7, 3, -8]`, then `z = x.*y` is computed as follows:

$$z = [2(-7), 4(3), -5(-8)] = [-14, 12, 40]$$

If **u** and **v** are column vectors, the result of `u.*v` is a column vector. Note that `.*` is treated as *one* symbol.

The generalization of array multiplication to arrays with more than one row or column is straightforward. Both arrays must be the same size. The array operations are performed between the elements in corresponding locations in the arrays. For example, if

$$\mathbf{A} = \begin{bmatrix} 11 & 5 \\ -9 & 4 \end{bmatrix} \qquad \mathbf{B} = \begin{bmatrix} -7 & 8 \\ 6 & 2 \end{bmatrix}$$

then `C = A.*B` gives this result:

$$\mathbf{C} = \begin{bmatrix} -7(11) & 5(8) \\ -9(6) & 4(2) \end{bmatrix} = \begin{bmatrix} -77 & 40 \\ -54 & 8 \end{bmatrix}$$

Here are some more examples of array operations. In division, the slash slants toward the divisor. Array right division: `[2,24]./[4,8] = [2/4,24/8] = [0.5, 3]`. Array left division: `[2, 5].\[6, 10] = [2\6, 5\10] = [3, 2]`. Array exponentiation: `[3, 5].^2 = [3^2, 5^2] = [9, 25]`. Also: `[3,5].^[2,3] = [3^2,5^3] = [9, 125]`.

Vectorized Functions The built-in MATLAB functions such as `sqrt(x)` and `exp(x)` automatically operate on array arguments to produce an array result the same size as the array argument `x`. Thus, these functions are said to be *vectorized* functions. Thus, when multiplying or dividing these functions, or when raising them to a power, you must use element-by-element operations if the arguments are arrays. For example, to compute $z = (e^y \sin x)\cos^2(x)$, you must type `z = exp(y).*sin(x).*(cos(x)).^2`. Obviously, you will get an error message if the size of `x` is not the same as the size of `y`. The result `z` will have the same size as `x` and `y`.

Matrix Operations In the matrix product **AB**, the number of columns in **A** must equal the number of rows in **B**. The product **AB** has the same number of rows as **A** and the same number of columns as **B**. For example,

$$\begin{bmatrix} 6 & -2 \\ 10 & 3 \\ 4 & 7 \end{bmatrix} \begin{bmatrix} 9 & 8 \\ -5 & 12 \end{bmatrix} = \begin{bmatrix} 6(9) - 2(-5) & 6(8) - 2(12) \\ 10(9) - 5(3) & 10(8) + 3(12) \\ 4(9) - 5(7) & 4(8) + 7(2) \end{bmatrix} = \begin{bmatrix} 64 & 24 \\ 75 & 116 \\ 1 & 116 \end{bmatrix}$$

Use the operator `*` to perform matrix multiplication. The following MATLAB session shows how to perform the matrix multiplication shown previously.

```
>>A = [6, -2; 10, 3; 4, 7];
>>B = [9, 8; -5, 12];
>>A*B
```

Solving Linear Equations Left division is used to solve sets of linear equations. For example, the set

$$4x + 3y = 23, \qquad 8x - 2y = 6$$

can be written in matrix form as

$$\begin{bmatrix} 4 & 3 \\ 8 & -2 \end{bmatrix} \begin{bmatrix} x \\ y \end{bmatrix} = \begin{bmatrix} 23 \\ 6 \end{bmatrix}$$

These can be solved as follows:

```
>>A = [4, 3; 8, -2];
>>B = [23; 6];
>>Solution = A\B
Solution = 2
           5
```

Plots Let us plot the function $y = 5\cos(2x)$ for $0 \leq x \leq 7$. We choose to use an increment of 0.01 to generate a large enough number of x values in order to produce a smooth curve. The function `plot(x,y)` generates a plot with the x values on the horizontal axis (the abscissa) and the y values on the vertical axis (the ordinate). The session is

```
>>x = 0:0.01:7;
>>y = 3*cos(2*x);
>>plot(x,y),xlabel('x'),ylabel('y')
```

The plot appears on the screen in a *graphics window*. The `xlabel` function places the text in single quotes as a label on the horizontal axis. The `ylabel` function performs a similar function for the vertical axis. If a hard copy of the plot is desired, the plot can be printed by selecting **Print** from the **File** menu in the graphics window.

Sometimes it is useful or necessary to obtain the coordinates of a point on a plotted curve. The function `ginput` can be used for this purpose. Place it at the end of all the plot and plot formatting statements, so that the plot will be in its final form. The command `[x,y] = ginput(n)` gets `n` points and returns the x and y coordinates in the vectors `x` and `y`, which have a length `n`. Position the cursor using a mouse, and press the mouse button.

M-Files MATLAB uses two types of M-files: *script files* and *function files*. You can use the Editor built into MATLAB to create M-files. The symbol % designates a *comment*, which is not executed by MATLAB. Comments are used mainly in script files for the purpose of documenting the file. To create this new M-file when in the Command window, select **New Script** from the **HOME** tab. You will then see a new edit window and the EDITOR tab will appear.

The `input` function displays text on the screen, waits for the user to enter something from the keyboard, and then stores the input in the specified variable. For example, the command `x = input('Please enter the value of x:')` causes the message to appear on the screen. If you type 5 and press **Enter**, the variable `x` will have the value 5.

EXAMPLE OF A SCRIPT FILE

The following is a simple example of a script file that shows the preferred program style. The speed of a falling object dropped with no initial velocity is given as a function of time t by $v = gt$, where g is the acceleration due to gravity. In SI units, $g = 9.81$ m/s^2. We want to compute and plot v as a function of t for $0 \le t \le t_{\text{final}}$, where t_{final} is the final time entered by the user. The script file is the following.

```
% Program Falling_Speed.m: plots speed of a falling object.
% Created on July 8, 2019
% Input Variable:
% tfinal = final time (in seconds)
% Output Variables:
% t = array of times at which speed is computed (seconds)
% v = array of speeds (meters/second)
% Acceleration in SI units
g = 9.81;
% Input section:
tfinal = input('Enter the final time in seconds:');
% Calculation section:
% Creates an array of 500 time values.
t = linspace(0,tfinal,500);
v = g*t;
% Output section:
plot(t,v),xlabel('Time (s)'),ylabel('Speed (m/s)')
```

After creating this file, you save it with the name Falling_Speed.m. To run it, you type Falling_Speed (without the .m) in the Command window at the prompt. You will then be asked to enter a value for t_{final}. After you enter a value and press **Enter**, you will see the plot on the screen.

Special Matrices The function `size(A)` returns a row vector `[m n]` containing the size of the $m \times n$ array **A**. Sometimes we want to initialize a matrix to have all zero elements. The `zeros` function creates a matrix of all zeros. Typing `zeros(n)` creates an $n \times n$ matrix of zeros, whereas typing `zeros(m,n)` creates an $m \times n$ matrix of zeros, as will typing `A(m,n) = 0`. Typing `zeros(size(A))` creates a matrix of all zeros having the same dimension as the matrix **A**. This type of matrix can be useful for applications in which we do not know the required dimension ahead of time. The syntax of the `ones` function is the same, except that it creates arrays filled with 1s.

For example, to create and plot the function

$$f(x) = \begin{cases} 10 & 0 \le x \le 2 \\ 0 & 2 < x < 5 \\ -3 & 5 \le x \le 7 \end{cases}$$

the script file is

```
x1 = 0:0.01:2;
f1 = 10*ones(size(x1));
x2 = 2.01:0.01:4.99;
f2 = zeros(size(x2));
x3 = 5:0.01:7;
f3 = -3*ones(size(x3));
f = [f1, f2, f3];
x = [x1, x2, x3];
plot(x, f),xlabel('x'),ylabel('y')
```

(Consider what the plot would look like if the command `plot(x,f)` were replaced with the command `plot(x1,f1,x2,f2,x3,f3)`.)

Polynomial Operations We can describe a polynomial in MATLAB with a row vector whose elements are the polynomial's coefficients, *starting with the coefficient of the highest power of x*. For example, the vector `[4,-8,7,-5]` represents the polynomial $4x^3 - 8x^2 + 7x - 5$. To add two polynomials, add the arrays that describe their coefficients. If the polynomials are of different degrees, add zeros to the coefficient array of the lower-degree polynomial. Use the `conv` function to multiply polynomials and use the `deconv` function to perform synthetic division. Consider the product of the polynomials $f(x)$ and $g(x)$:

$$f(x)g(x) = (9x^3 - 5x^2 + 3x + 7)(6x^2 - x + 2) = 54x^5 - 39x^4 + 41x^3 + 29x^2 - x + 14$$

The MATLAB session is

```
>>f = [9,-5,3,7];
>>g = [6,-1,2];
>>product = conv(f,g)
product = 54 -39 41 29 -1 14
```

The `polyval(a,x)` function evaluates a polynomial at specified values of its independent variable `x`, which can be a matrix or a vector. The polynomial's coefficient array is `a`. The result is the same size as `x`. For example, to plot the polynomial $f(x) = 9x^3 - 5x^2 + 3x + 7$ for $-2 \leq x \leq 5$, you type

```
>>x = -2:0.01:5;
>>f = polyval([9,-5,3,7], x);
>>plot(x, f),xlabel('x'),ylabel('f(x)')
```

MATLAB Help The `help` function is the most basic way to determine the syntax and behavior of a particular function. For example, typing `help log10` in the Command window produces documentation for the function. The `lookfor` function allows you to search for functions on the basis of a keyword. For example, typing `lookfor sine` produces over a dozen matches, depending on which toolboxes you have installed. From this list you can find the correct name for the sine function.

1.7 CHAPTER REVIEW

This chapter introduced the basic terminology of system dynamics, which includes the notions of system, static and dynamic elements, input, and output. The chapter also introduced the foot-pound-second (FPS) and the metric (SI) systems of units, which will be used throughout this text. We developed methods for obtaining linear algebraic models of input-output relations, and we reviewed simple differential equations and their application in dynamics. We then introduced a case study of a motion-control system, and concluded the chapter with a review of the basics of MATLAB.

Review of Objectives

Now that you have finished this chapter, you should be able to

1. Define the terms: static and dynamic elements, static and dynamic systems, input, and output.
2. State the principle of integral causality.

3. State the basic steps used for engineering problem solving.
4. List the steps in developing a computer solution.
5. Use both FPS and SI units.
6. Develop linearized models from given algebraic expressions.
7. Solve simple differential equations by direct integration.
8. Interpret solutions in terms of simple dynamics problems.
9. Apply a trapezoidal velocity profile to design a simple motion-control system.
10. Use MATLAB to do simple calculations and plotting, and use the MATLAB help system.

PROBLEMS

Section 1.2 Units

1.1 Calculate the weight in pounds of an object whose mass is 3 slugs. Then convert the weight to newtons and the mass to kilograms.

1.2 Folklore has it that Sir Isaac Newton formulated the law of gravitation supposedly after being hit on the head by a falling apple. The weight of an apple depends strongly on its variety, but a typical weight is 1 newton! Calculate the total mass of 100 apples in kilograms. Then convert the total weight to pounds and the total mass to slugs.

1.3 A ball is thrown a distance of 50 feet, 5 inches. Calculate the distance in meters.

1.4 An American football field is 100 yards long. How long is it in meters?

1.5 How long is a 100-meter dash in feet?

1.6 Convert 50 feet per second to miles per hour.

1.7 The speed limit in some countries is 100 kilometers per hour. How fast is that in miles per hour?

1.8 How many 60 watt lightbulbs are equivalent to one horsepower?

1.9 Convert the temperature of 70°F to °C.

1.10 Convert 30°C to °F.

1.11 A particular motor rotates at 3000 revolutions per minute (rpm). What is its speed in rad/sec, and how many seconds does it take to make one revolution?

1.12 The displacement of a certain object is described by $y(t) = 23 \sin 5t$, where t is measured in seconds. Compute its period and its oscillation frequency in rad/sec and in hertz.

1.13 A car is driven at 40 miles per hour over a road whose surface is sinusoidal with a period of 30 feet. How many times per second will the car go over the peak of a bump?

1.14 The displacement in meters of a certain vibrating mass is described by $x(t) = 0.005 \sin 6t$. What is the amplitude and frequency of its velocity $\dot{x}(t)$? What is the amplitude and frequency of its acceleration $\ddot{x}(t)$?

Section 1.3 Developing Linear Models

1.15 The distance a spring stretches from its "free length" is a function of how much tension is applied to it. The following table gives the spring length y that was produced in a particular spring by the given applied force f. The spring's free length is 4.7 in. Find a functional relation between f and x, the extension from the free length ($x = y - 4.7$).

Force f (pounds)	Spring length y (inches)
0	4.7
0.47	7.2
1.15	10.6
1.64	12.9

1.16 The following "small-angle" approximation for the sine is used in many engineering applications to obtain a simpler model that is easier to understand and analyze. This approximation states that $\sin x \approx x$, where x must be in radians. Investigate the accuracy of this approximation by creating three plots. For the first plot, plot $\sin x$ and x versus x for $0 \leq x \leq 1$. For the second plot, plot the approximation error $\sin(x) - x$ versus x for $0 \leq x \leq 1$. For the third plot, plot the percent error $[\sin(x) - x]/\sin(x)$ versus x for $0 \leq x \leq 1$. How small must x be for the approximation to be accurate within 5%?

1.17 Obtain two linear approximations of the function $f(\theta) = \sin\theta$, one valid near $\theta = \pi/4$ rad and the other valid near $\theta = 3\pi/4$ rad.

1.18 Obtain two linear approximations of the function $f(\theta) = \cos\theta$, one valid near $\theta = \pi/3$ rad and the other valid near $\theta = 2\pi/3$ rad.

1.19 Obtain a linear approximation of the function $f(h) = \sqrt{h}$, valid near $h = 25$.

1.20 Obtain two linear approximations of the function $f(r) = r^2$, one valid near $r = 5$ and the other valid near $r = 10$.

1.21 Obtain a linear approximation of the function $f(h) = \sqrt{h}$, valid near $h = 16$. Noting that $f(h) \geq 0$, what is the value of h below which the linearized model loses its meaning?

1.22 The flow rate f in m^3/s of water through a particular pipe, as a function of the pressure drop p across the ends of the pipe (in N/m^2) is given by $f = 0.002\sqrt{p}$. Obtain a linear model of f as a function of p that always *underestimates* the flow rate over the range $0 \leq p \leq 900$.

Section 1.4 Introduction to Differential Equations

1.23 Solve each of the following problems by direct integration.

a. $$4\dot{x} = 3t \qquad x(0) = 2$$

b. $$5\dot{x} = 2e^{-4t} \qquad x(0) = 3$$

c. $$3\ddot{x} = 5t \qquad x(0) = 2 \qquad \dot{x}(0) = 7$$

d. $$4\ddot{x} = 7e^{-2t} \qquad x(0) = 4 \qquad \dot{x}(0) = 2$$

e. $$\ddot{x} = 0 \qquad x(0) = 2 \qquad \dot{x}(0) = 5$$

1.24 Suppose that for a certain lander, $m = 5000$ kg and $g = 2$ m/s^2. Suppose also that the initial downward velocity is $v_0 = 10$ m/s and the initial height is $h_0 = 1000$ m. Find the thrust required for a soft landing and the time to land.

1.25 Using the linear velocity profile $\dot{h}(t) = at + b$ for a lander, obtain the equations for a and b, and the required thrust to achieve a soft landing.

1.26 Suppose that we back off from the requirement that a lander land with zero velocity and instead require that lander's velocity at touchdown be no greater than a specified value v_f. Derive the corresponding equations for the time to land t_f and the required thrust.

1.27 a. Use the quadratic velocity profile of the form

$$\dot{h}(t) = at^2 + bt + c$$

where $c = -v_0$ and obtain the equations to solve for a and b. Show that the thrust must be a linear time function: $f = m(2at + b + mg)$. Show that the flight time t_f may be specified.

b. Find the thrust $f = m(2at + b + g)$ using the values $m = 2500$ kg, $g = 2$ m/s^2, $h_0 = 500$ m, and $v_0 = 5$ m/s. Consider two flight times: (1) $t_f = 200$ s and (2) $t_f = 100$ s.

Section 1.5 A Case Study in Motion Control

1.28 Suppose that $I_m = 0.005$, $I_S = 0.002$, and $I_L = 0.008$ kg·m^2 for the system shown in Figure 1.5.3. Compute how much motor torque is required to achieve the velocity profile shown in Figure 1.5.4, where $t_1 = 0.1$ s, $t_2 = 4.5$ s, and $t_f = 4.6$ s, and the slew speed is 200 rad/s.

1.29 Suppose the load in Problem 1.28 is acted on by a friction torque $T_F = 3$ N·m. Recalulate the maximum and rms torques.

1.30 Redo the calculations in Example 1.5.4 except that it is desired to rotate the load through five revolutions (10π rad) in one-half second from start to stop. Compute the motor's maximum speed, the maximum required torque, and the rms average torque.

1.31 A certain system like that shown in Figure 1.5.3 has a load inertia $I_L = 10^{-2}$ kg·m^2, a shaft inertia of $I_S = 6 \times 10^{-4}$ kg·m^2, and a motor inertia of $I_m = 2 \times 10^{-3}$ kg·m^2. A friction torque of 2 N·m acts on the load. It is desired to rotate the load through two revolutions (4π rad) in one-half second from start to stop. The load should be rotated at a constant speed for 0.6 s during the slew phase. Compute the motor's maximum speed, the maximum required torque, and the rms average torque.

1.32 Neglect friction and estimate the torque required to accelerate a 50-kg load mass from rest to 0.3 m/s in 0.2 s, using a steel lead screw of length 0.7 m and radius 0.01 m. The screw lead is $L = 0.008$ m/rev. The motor inertia is 2×10^{-4} kg·m^2. Take the density of steel to be 7700 kg/m^3.

Section 1.6 MATLAB Review

1.33 Suppose that $x = 2$ and $y = 5$. Use MATLAB to compute the following.

a. $\dfrac{yx^3}{x - y}$ b. $\dfrac{3x}{2y}$ c. $\dfrac{3}{2}xy$ d. $\dfrac{x^5}{x^5 - 1}$

1.34 Suppose that $x = -7 - 5j$ and $y = 4 + 3j$. Use MATLAB to compute the following.

a. $x + y$ b. xy c. x/y

1.35 Use MATLAB to compute the following.

a. $(3 + 6j)(-7 - 9j)$ b. $\frac{5 + 4j}{5 - 4j}$ c. $\frac{3}{2}j$ d. $\frac{3}{2j}$

1.36 Evaluate the following expressions in MATLAB, for the values $x = 5 + 8j$, $y = -6 + 7j$.

a. $u = x + y$

b. $v = xy$

c. $w = x/y$

d. $z = e^x$

e. $r = \sqrt{y}$

f. $s = xy^2$

1.37 Use MATLAB to calculate:

a. e^2 b. $\log 2$ c. $\ln 2$

1.38 Use MATLAB to calculate:

a. $\cos(\pi/3)$

b. $\cos 80°$

c. $\cos^{-1} 0.7$ in radians

d. $\cos^{-1} 0.6$ in degrees

1.39 Use MATLAB to calculate:

a. $\tan^{-1} 2$ b. $\tan^{-1} 100$

b. The angle corresponding to $x = 2$, $y = 3$

c. The angle corresponding to $x = -2$, $y = 3$

d. The angle corresponding to $x = 2$, $y = -3$

1.40 Suppose x takes on the values $x = 1, 1.2, 1.4, \ldots, 5$. Use MATLAB to compute the array y that results from the function $y = 7 \sin 4x$. Use MATLAB to determine how many elements are in the array y and the value of the third element in the array y.

1.41 Use MATLAB to calculate the roots of $13x^3 + 182x^2 - 184x + 2503 = 0$.

1.42 Use MATLAB to calculate the roots of $70x^3 + 24x^2 - 10x + 20 = 0$.

1.43 Use MATLAB to find the polynomial that has the roots $x = -2 \pm 5j, -7$.

1.44 Use MATLAB to plot the functions $u = 2 \log_{10}(60x + 1)$ and $v = 3 \cos 6x$ over the interval $0 \le x \le 2$. Properly label the plot and each curve. The variables u and v represent speed in miles per hour; the variable x represents distance in miles.

1.45 The planetary lander treated in Example 1.4.1 has a height versus time profile of $h(t) = 0.0125t^2 - 5t + 500$. Use the `polyval` function to plot $h(t)$ for $0 \le t \le 200$ s to verify that $h(t) \ge 0$.

1.46 Create a program to compute the thrust required for a soft landing for a planetary lander and the time to land. The given inputs are the lander mass m, the gravitational constant g, the downward velocity v_0, the initial height h_0.

1.47 Example 1.5.2 results in the following velocity profile. Obtain a plot of $\omega(t)$.

$$\omega = \begin{cases} 300t & 0 \le t \le 0.5 \\ 150 & 0.5 < t < 2.5 \\ -300(t-2.5) + 150 & 2.5 \le t \le 3 \end{cases}$$

1.48 Code Table 1.5.1 in MATLAB, use the results of Example 1.5.2 to check your program, and use the program to solve Problem 1.31.

1.49 Create a program to calculate the torque required for a lead screw to accelerate a load mass m from rest to speed v_s in time t_1, using a steel lead screw of length L and radius r. The screw lead is L. The given motor inertia is I_m. Neglect friction and take the density of steel to be 7700 kg/m^3.

CHAPTER 2

Dynamic Response Methods

CHAPTER OUTLINE

CHAPTER OBJECTIVES

When you have finished this chapter, you should be able to do the following:

1. Apply separation of variables or the Laplace transform method, whichever is appropriate, to obtain the solution of a linear differential equation model.
2. Identify the free, forced, transient, and steady-state components of the complete response.
3. Be able to identify and interpret the time constant, oscillation frequency, and damping ratio.
4. When applying the Laplace transform method, be able to perform the appropriate expansion and apply the appropriate transform properties to obtain the inverse transform.
5. Obtain transfer functions from models expressed as single equations or as sets of equations.
6. Evaluate the effects of pulse and impulse inputs, and input derivatives on the response.
7. Use MATLAB to apply the chapter's methods.

Dynamic models are differential equations that describe a dynamic system. In this chapter, we develop methods for obtaining analytical solutions to differential equations commonly found in engineering applications. In Section 2.1, we introduce some important concepts and terminology associated with differential equations, and present methods for quickly solving simple differential equations. This section introduces the concepts of free, forced, transient, and steady-state response. Section 2.2 explains the important concepts of stability, time constant, and natural frequency.

An important tool for obtaining solutions in general is the Laplace transform, introduced in Section 2.3 and applied in Section 2.4. The transform also forms the basis for the important concept of the transfer function, covered in Section 2.5. The effects of pulse and impulse inputs and input derivatives on a system's dynamic response are treated in Section 2.6.

Section 2.8 shows how to use MATLAB to obtain partial-fraction expansions and to obtain responses to step, impulse, and other input function types. A review of the chapter's main concepts is given in Section 2.10. ■

2.1 SOLVING DIFFERENTIAL EQUATIONS

An *ordinary differential equation* (*ODE*) is an equation containing ordinary, but not partial, derivatives of the dependent variable. Because the subject of system dynamics is time-dependent behavior, the independent variable in our ODEs will be time t. We will often denote the time derivative with an overdot, as

$$\dot{x} = \frac{dx}{dt} \qquad \ddot{x} = \frac{d^2x}{dt^2}$$

In a standard form for expressing an ODE, all functions of the dependent variable are placed on the left-hand side of the equal (=) sign, and all isolated constants and isolated functions of time are placed on the right-hand side. The quantities in the right-hand side are called the *input*, or *forcing function*. The dependent variable is called the solution or the *response*. For example, consider the equation $3\ddot{x} + 7\dot{x} + 2t^2x = 5 + \sin t$, where x is the dependent variable. The input is $5 + \sin t$ and the response is $x(t)$. If the right-hand side is zero, the equation is said to be *homogeneous*; otherwise, it is *nonhomogeneous*.

2.1.1 INITIAL CONDITIONS

An example of an ODE is $2\dot{x} + 6x = 3$, where x is the dependent variable. "Solving the equation" means to obtain the function $x(t)$ that satisfies the equation. For this example, the function is $x(t) = Ce^{-3t} + 0.5$, where C is a constant. We cannot determine a numerical value for C unless we are given a specified value for x at some time t. Most commonly x is specified at some starting time, usually denoted t_0. The specified value of x at t_0 is denoted x_0 and is called the *initial condition*. Often the starting time t_0 is taken to be at $t = 0$.

When solving differential equations, you need never wonder if your answer is correct, because you can always check your answer by substituting it into the differential equation and by evaluating the solution at $t = t_0$.

2.1.2 CLASSIFICATION OF DIFFERENTIAL EQUATIONS

We can categorize differential equations as *linear* or *nonlinear*. Linear differential equations are recognized by the fact that they contain only linear functions of the dependent variable and its derivatives. Nonlinear functions of the *independent* variable do *not* make a differential equation nonlinear. For example, the following equations are linear:

$$\dot{x} + 3x = 5 + t^2 \qquad \dot{x} + 3t^2x = 5 \qquad 3\ddot{x} + 7\dot{x} + 2t^2x = \sin t$$

whereas the following equations are nonlinear:

$$2\ddot{x} + 7\dot{x} + 6x^2 = 5 + t^2, \text{ because of } x^2$$

$$3\ddot{x} + 5\dot{x}^2 + 8x = 4, \text{ because of } \dot{x}^2$$

$$\ddot{x} + 4x\dot{x} + 3x = 10, \text{ because of } x\dot{x}$$

The equation $\dot{x} + 3t^2x = 5$ is a *variable-coefficient* differential equation, so named because one of its coefficients is a function of the independent variable t. By contrast, the equation $\dot{x} + 2x = 5$ is a *constant-coefficient* differential equation. When solving constant-coefficient equations, the initial time t_0 can always be chosen to be 0. This simplifies the solution form.

The *order* of the equation is the order of the highest derivative of the dependent variable in the equation. The equation $3\ddot{x} + 7\dot{x} + 2x = 5$ is thus called a *second-order* differential equation.

A model can consist of more than one equation. For example, the model

$$3\dot{x}_1 + 5x_1 - 7x_2 = 5$$

$$\dot{x}_2 + 4x_1 + 6x_2 = 0$$

consists of two equations that must be solved *simultaneously* to obtain the solution for the two dependent variables $x_1(t)$ and $x_2(t)$. The equations are said to be *coupled*. Although each equation is first order, the set can be converted into a single differential equation of second order. Thus, solving a set of two coupled first-order equations is equivalent to solving a single second-order equation. In general, a coupled set of differential equations can be reduced to a single differential equation whose order is the sum of the orders of the individual equations in the set.

2.1.3 SEPARATION OF VARIABLES FOR FIRST-ORDER MODELS

You can solve the equation

$$\frac{dx}{dt} = g(t)f(x) \tag{2.1.1}$$

for $x(t)$ with the initial condition $x(0)$ by separating the variables x and t as follows. First write the equation as

$$\frac{dx}{f(x)} = g(t)\,dt$$

Then integrate both sides to obtain

$$\int_{x(0)}^{x(t)} \frac{dx}{f(x)}\,dx = \int_0^t g(t)\,dt \tag{2.1.2}$$

The success of the method obviously depends on being able to evaluate the integrals.

Note that the equation of the form

$$\frac{dx}{dt} + g(x) = f(t)$$

does not separate.

We now illustrate the method with several examples drawn from various applications in system dynamics.

Simple Growth Equation Many processes can be described by the simple growth law

$$\frac{dN}{dt} = kN \tag{2.1.3}$$

where k is a constant. Solving with separation of variables, we obtain

$$\int_{N(t_0)}^{N(t)} \frac{dN}{N} = \int_{t_0}^{t} k\,dt$$

or

$$\ln N|_{N(t_0)}^{N(t)} = k(t - t_0)$$

$$\ln N(t) - \ln N(t_0) = \ln \frac{N(t)}{N(t_0)} = k(t - t_0)$$

Finally,

$$\frac{N(t)}{N(t_0)} = e^{k(t-t_0)} \tag{2.1.4}$$

Usually, but not always, we use $t_0 = 0$ as the start of the process. In this case the solution is usually written as

$$N(t) = N(0)e^{kt} \tag{2.1.5}$$

If $k > 0$, this solution represents *exponential growth* and is often used to model population growth of animals and bacteria in an unrestricted environment. It also is used to calculate the growth of investments when interest is compounded continuously.

If $k < 0$, this solution represents *exponential decay* and is often used to model the decay of a radioactive substance and the speed of an object subject to a fluid drag force. Figure 2.1.1 illustrates the solution behavior for several cases.

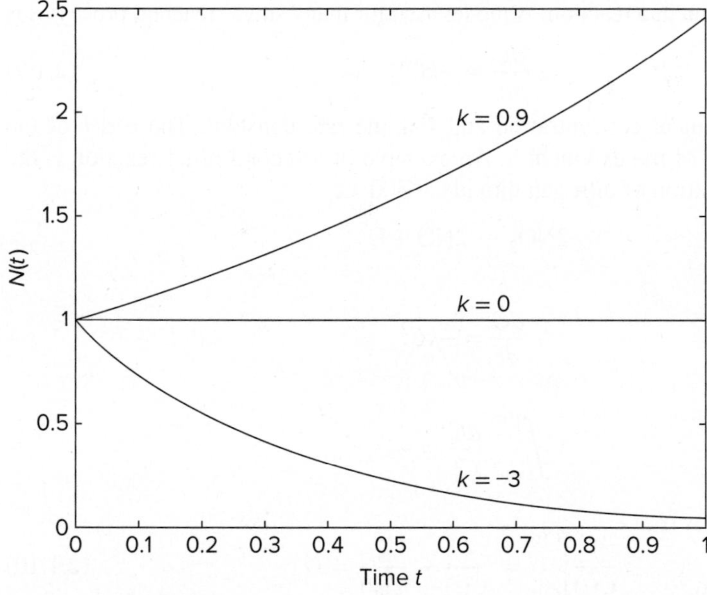

Figure 2.1.1 Behavior of (2.1.5) for several values of k.

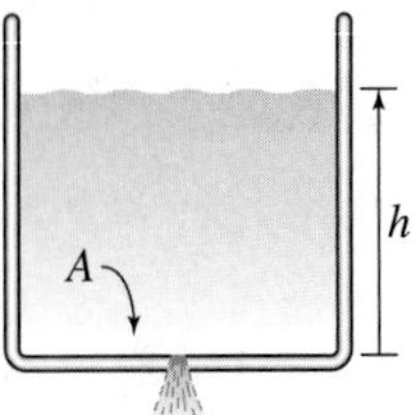

Figure 2.1.2 A leaking liquid container.

Hydraulic Application In Chapter 7 we will study hydraulic systems, and one important principle of hydraulics is Torricelli's law, which states that the flow rate of a liquid through an opening in a container is proportional to the square root of the height of the liquid surface above the opening. From conservation of mass we see that the liquid volume V in the tank will change as follows:

$$\frac{dV}{dt} = -k\sqrt{h} \tag{2.1.6}$$

For a tank having vertical sides and bottom area A, the liquid volume is linearly related to the liquid height h as $V = Ah$ (see Figure 2.1.2). For such a tank, the model becomes

$$A\frac{dh}{dt} = -k\sqrt{h}$$

or

$$\frac{dh}{dt} = -\frac{k}{A}\sqrt{h} = -b\sqrt{h} \tag{2.1.7}$$

Separating the variables gives

$$\int_{h(0)}^{h(t)} \frac{dh}{\sqrt{h}} = -bt$$

$$2\sqrt{h}\Big|_{h(0)}^{h(t)} = -bt$$

$$h(t) = \left[\sqrt{h(0)} - \frac{b}{2}t\right]^2 \tag{2.1.8}$$

Note that this model predicts that the time to empty is $t = 2\sqrt{h(0)}/b$. In Chapter 7 we will see how the parameter b is related to the density of the liquid and the size of the opening.

Chemical Reactions Chemists and engineers must be able to predict the changes in chemical concentration in a reaction. A model used for many single-reactant processes is

$$\frac{dC}{dt} = -kC^n \tag{2.1.9}$$

where C is the chemical concentration and k is the rate constant. The *order* of the reaction is the value of the exponent n. An example of a second-order reaction is the gas-phase decomposition of nitrogen dioxide at 300°C.

$$2NO_2 \rightarrow 2NO + O_2$$

The model is

$$\frac{dC}{dt} = -kC^2$$

or

$$\int_{C(0)}^{C(t)} \frac{dC}{C^2} = -kt$$

or

$$\frac{1}{C(t)}\Bigg|_{C(0)}^{C(t)} = \frac{1}{C(t)} - \frac{1}{C(0)} = kt \tag{2.1.10}$$

General Linear First-Order Form Throughout our study we will see many models of the following form:

$$\frac{dx}{dt} + ax = b \tag{2.1.11}$$

where a and b are constants, and $a \neq 0$. Separation of variables can be used to solve such an equation.

Solution of a Linear Equation — EXAMPLE 2.1.1

■ Problem

Use separation of variables to obtain the solution of (2.1.11).

■ Solution

Separating the variables and integrating gives

$$\int_{x(0)}^{x(t)} \frac{dx}{b - ax} = t$$

or

$$\ln[b - ax(t)] - \ln[b - ax(0)] = -at$$

$$b - ax(t) = e^{-at}[b - ax(0)]$$

Solve for $x(t)$.

$$x(t) = \frac{b}{a} + \left[x(0) - \frac{b}{a}\right]e^{-at} \tag{2.1.12}$$

This can be rearranged as

$$x(t) = x(0)e^{-at} + \frac{b}{a}(1 - e^{-at}) \tag{2.1.13}$$

Figure 2.1.3 shows the reponse for the specific case: $x(0) = 3$, $a = 2$, and $b = 20$.

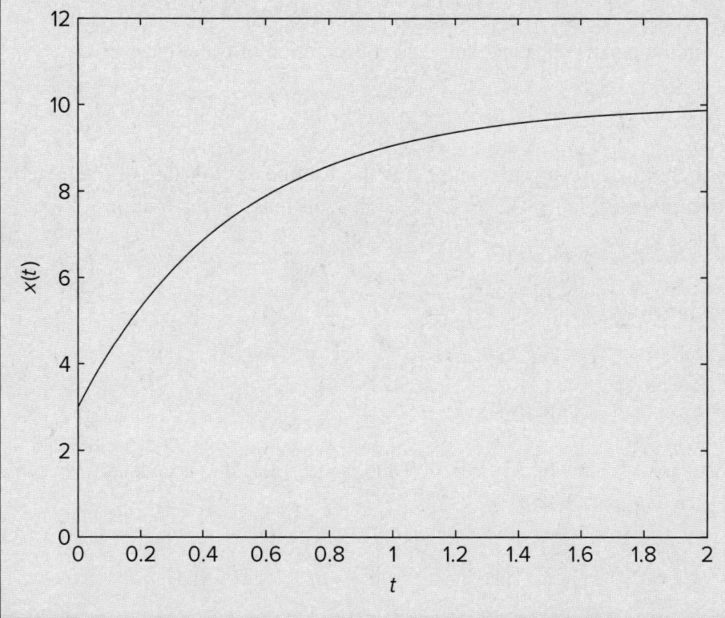

Figure 2.1.3 Solution of (2.1.13) for $x(0) = 3$, $a = 2$, and $b = 20$.

EXAMPLE 2.1.2 A Mixing Process

■ Problem

A mixing tank is shown in Figure 2.1.4. Pure water flows into the tank of volume $V = 600$ m^3 at the constant volume rate of 5 m^3/s. A solution with a salt concentration of s_i kg/m^3 flows into the tank at a constant volume rate of 2 m^3/s. Assume that the solution in the tank is well mixed so that the salt concentration in the tank is uniform. Assume also that the salt dissolves completely so that the volume of the mixture remains the same. The salt concentration s_o kg/m^3 in the outflow is the same as the concentration in the tank. The input is the concentration $s_i(t)$, whose value may change during the process, thus changing the value of s_o. Obtain a dynamic model of the concentration s_o.

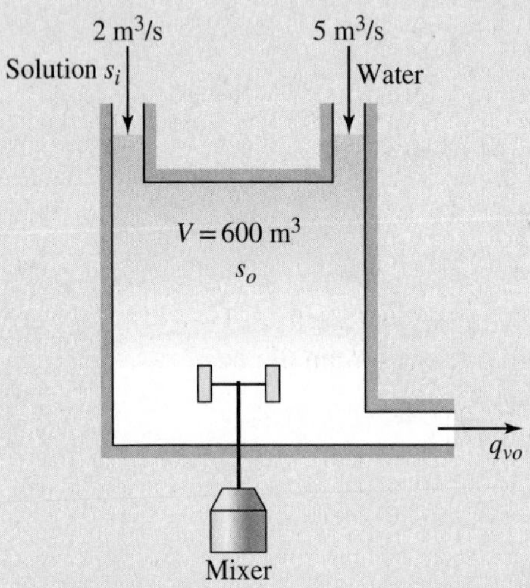

Figure 2.1.4 A mixing process.

■ Solution

Two mass species are conserved here: water mass and salt mass. The tank is always full, so the mass of water m_w in the tank is constant, and thus conservation of water mass gives

$$\frac{dm_w}{dt} = 5\rho_w + 2\rho_w - \rho_w q_{vo} = 0$$

where ρ_w is the mass density of fresh water, and q_{vo} is the volume outflow rate of the mixed solution. This equation gives $q_{vo} = 5 + 2 = 7$ m^3/s. The salt mass in the tank is $s_o V$, and conservation of salt mass gives

$$\frac{d}{dt}\left(s_o V\right) = 9(5) + 2s_i - s_o q_{vo} = 2s_i - 7s_o$$

or, with $V = 600$,

$$600\frac{s_o}{dt} = 2s_i - 7s_o$$

This has the same form as (2.1.11) where $a = 7/600$ and $b = s_i/300$. If s_i is constant, we can solve for $s_o(t)$ using (2.1.12).

$$s_o(t) = \frac{2s_i}{7} + \left[s_o(0) - \frac{2s_i}{7}\right] e^{-7t/600}$$

2.1.4 TYPES OF RESPONSE

The linear first-order model has the general form

$$\frac{dx}{dt} + ax = f(t) \tag{2.1.14}$$

where $f(t)$ is called the *input* or *forcing function*. If $f(t)$ is a constant, which we have denoted by b, and if a is a nonzero constant, then the solution is given by (2.1.13) and can be displayed as

$$x(t) = \underbrace{x(0)e^{-at}}_{\text{free response}} + \underbrace{\frac{b}{a}(1 - e^{-at})}_{\text{forced response}} \tag{2.1.15}$$

It consists of the sum of two terms. The first term depends on the initial condition $x(0)$ but not on the input b. This part of the solution is called the *free response*. The second term depends on the forcing function but not on the initial condition $x(0)$. This part of the solution is called the *forced response*. It is a general property of linear differential equations that the response is always the sum of the free and the forced responses. This property enables us to simplify the analysis by examining each response separately, and then adding the two to obtain the complete solution.

It is extremely useful to distinguish between the free and the forced responses because this separation enables us to focus on the effects of the input by temporarily setting the initial conditions to zero and concentrating on the forced response. When we have finished analyzing the forced response, we can obtain the complete response by adding the free response to the forced response.

The solution can also be rearranged as follows:

$$x(t) = \underbrace{\frac{b}{a}}_{\text{steady state}} + \underbrace{\left[x(0) - \frac{b}{a}\right] e^{-at}}_{\text{transient}} \tag{2.1.16}$$

Note that the solution consists of two parts, one that disappears with time due to the e^{-at} term, and one that remains. These terms are called the *transient* and the *steady-state* responses, respectively. Both responses need not occur; it is possible to have one without the other. For example, there is no transient response in the preceding solution if $x(0) = b/a$, and there is no steady-state response if $b = 0$. The complete or total response is the sum of the transient and steady-state responses. This is a property of linear differential equations called *superposition*.

To summarize, we can separate the total response as follows:

- Transient Response The part of the response that disappears with time.
- Steady-State Response The part of the response that remains with time.
- Free Response The part of the response that depends on the initial conditions.
- Forced Response The part of the response due to the forcing function.

The Step Function and Step Response The *step function* is so named because its shape resembles a stair step (see Figure 2.1.5). If height of the step is 1, the function is called the *unit-step function*, denoted by $u_s(t)$. It is defined as

$$u_s(t) = \begin{cases} 0 & t < 0 \\ 1 & t > 0 \end{cases} \tag{2.1.17}$$

and is undefined and discontinuous at $t = 0$.

Figure 2.1.5 The unit-step function.

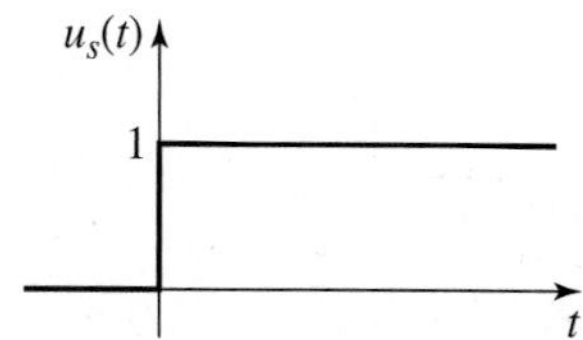

Suppose a function $f(t)$ is zero for $t < 0$ and a nonzero constant, say M, for $t > 0$. Then we can express it as $f(t) = Mu_s(t)$. The value of M is the magnitude of the step function; if $M = 1$, the function is the unit-step function. The forced response due to a step function is called the *step response*.

2.1.5 TRIAL-SOLUTION METHOD

We can use the results of Example 2.1.1 to gain insight into the solution of the equation $\dot{x} + ax = b$, where $a \neq 0$. Its solution has the form

$$x(t) = C + De^{st} \tag{2.1.18}$$

where C, D, and s are constants to be determined. We can verify that this form is the solution by substituting $x(t)$ into the differential equation, as follows:

$$\dot{x} + ax = sDe^{st} + a\left(C + De^{st}\right) = (s + a)De^{st} + aC = b$$

The only way this equation can be true is if $s + a = 0$ and $aC = b$. Thus, $s = -a$ and $C = b/a$. The remaining constant, D, can be determined from the initial value $x(0)$ as follows. Substituting $t = 0$ into the solution form gives $x(0) = C + De^0 = C + D$. Thus, $D = x(0) - C = x(0) - b/a$, and the solution can be written as

$$x(t) = \frac{b}{a} + \left[x(0) - \frac{b}{a}\right] e^{-at}$$

The exponential coefficient s is called the *characteristic root*, and its equation $s + a = 0$ is called the *characteristic equation*. Characteristic roots are of great use in determining the form of the trial solution. This insight leads to the trial-solution method for solving equations. It can be used to solve higher-order equations as well. The method is useful for quickly obtaining solutions of common ODEs whose solution forms are already known from experience.

2.1.6 SECOND-ORDER LINEAR MODELS

Throughout this text we will encounter many models that are second-order differential equations. Most of them are linear with constant coefficients. An example is the mass shown in Figure 2.1.6(a). It is attached to a linear spring and slides on a lubricated surface. We will study systems like this in more detail in Chapter 4, but for now we can obtain the following equation of motion from the free body diagram shown in part (b).

$$m\ddot{x} + c\dot{x} + kx = f(t) \tag{2.1.19}$$

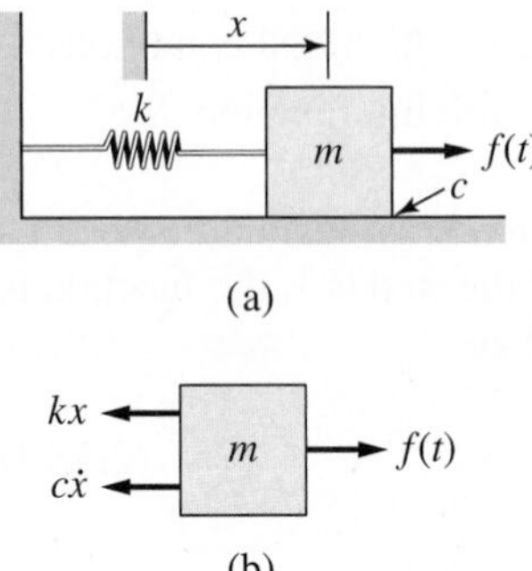

Figure 2.1.6 A mass-spring system with damping.

Let us use this form to represent any second-order linear equation. For now, let us consider the case where the input f is constant. Following the trial-solution method, we try a solution of the form

$$x(t) = C_1 e^{s_1 t} + C_2 e^{s_2 t} + C_3 \tag{2.1.20}$$

This gives

$$\dot{x} = s_1 C_1 e^{s_1 t} + s_2 C_2 e^{s_2 t}$$

and

$$\ddot{x} = s_1^2 C_1 e^{s_1 t} + s_2^2 C_2 e^{s_2 t}$$

Substitute these into the differential equation and collect the exponentials.

$$(ms_1^2 + cs_1 + k)C_1 e^{s_1 t} + (ms_2^2 + cs_2 + k)C_2 e^{s_2 t} + kC_3 = f$$

This shows that the trial solution is correct as long as

$$ms_1^2 + cs_1 + k = 0$$

$$ms_2^2 + cs_2 + k = 0$$

$$C_3 = \frac{f}{k} \tag{2.1.21}$$

Thus, both values of s must satisfy the characteristic equation

$$s^2 + cs + k = 0 \tag{2.1.22}$$

This is a quadratic equation and its two roots can be found with the quadratic formula. These roots can be either (1) real and distinct, (2) real and equal (repeated), or (3) complex conjugates. The trial solution gives the correct solution unless the two roots are identical. The correct form in this case is given in the following list.

One of the following three forms can be used to find the solutions:

Distinct, real roots: $s = s_1, s_2$

$$x(t) = C_1 e^{s_1 t} + C_2 e^{s_2 t} + C_3 \tag{2.1.23}$$

Repeated roots: $s = s_1, s_1$

$$x(t) = C_1 e^{s_1 t} + C_2 t e^{s_1 t} + C_3 \tag{2.1.24}$$

Complex-conjugate roots: $s = \sigma \pm j\omega$

$$x(t) = e^{\sigma t}\left(C_1 \sin \omega t + C_2 \cos \omega t\right) + C_3 \tag{2.1.25}$$

Let us now see how the constants C_1 and C_2 are found.

Step Response with Two Distinct, Real Roots

EXAMPLE 2.1.3

■ Problem

Find the step response of the following equation.

$$10\ddot{x} + 70\dot{x} + 100x = 200u_s(t) \qquad x(0) = \dot{x}(0) = 0$$

■ Solution

First let us simplify the numbers by dividing by the coefficient of $\ddot{x}$.

$$\ddot{x} + 7\dot{x} + 10x = 20u_s(t) \qquad x(0) = \dot{x}(0) = 0$$

Next find the roots from

$$s^2 + 7s + 10 = 0$$

They are $s = -2$ and $s = -5$. Using the trial solution (2.1.23) with $C_3 = 20/10 = 2$, we have

$$x(t) = C_1 e^{-2t} + C_2 e^{-5t} + 2$$

Now find $x(0)$ and $\dot{x}(0)$.

$$x(0) = C_1 + C_2 + 2 = 0$$
$$\dot{x}(0) = -2C_1 - 5C_2 = 0$$

The solution is $C_1 = -10/3$ and $C_2 = 4/3$, and

$$x(t) = -\frac{10}{3}e^{-2t} + \frac{4}{3}e^{-5t} + 2$$

The solution is plotted in Figure 2.1.7. The transient response consists of the exponential terms. The steady-state response is 2. Note that we cannot identify the free response or the forced response once we insert numerical values for the initial conditions and the input.

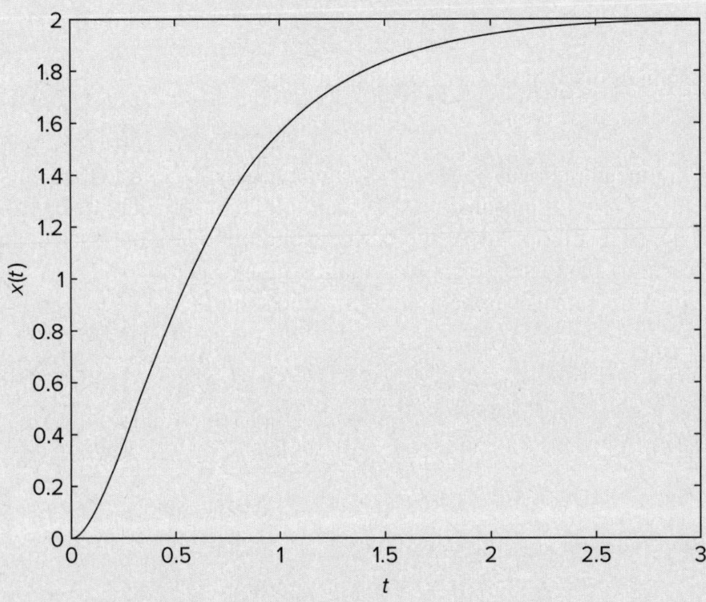

Figure 2.1.7 Response plot for Example 2.1.3.

2.1.7 SOLUTIONS FROM TABLES

The trial-solution method can be applied in general to obtain the solutions for all three root types. Many of the solutions needed in this text can be found from Tables 2.1.1 and 2.1.2. Table 2.1.1 summarizes the step response for first- and second-order linear equations. Note that the initial conditions are *zero* for this table (thus the table gives only the forced response). No algebra is required to use this table other than that used to find the values of a and b.

Table 2.1.2 summarizes the complete solution form for first- and second-order linear equations having a constant input. This table can be used to find the *free* response

Table 2.1.1 Step response for zero initial conditions.

	ODE	Roots	$x(t)$
1.	$\dot{x} + ax = Mu_s(t)$	$s = -a$	$x(t) = \dfrac{M}{a}(1 - e^{-at})$
2.	$\ddot{x} + (a+b)\dot{x} + abx = Mu_s(t)$	$s = -a,\ -b$ $a \neq b$	$x(t) = M\left[\dfrac{e^{-at}}{a(a-b)} + \dfrac{e^{-bt}}{b(b-a)} + \dfrac{1}{ab}\right]$
3.	$\ddot{x} + 2a\dot{x} + a^2x = Mu_s(t)$	$s = -a,\ -a$	$x(t) = \dfrac{M}{a^2}\left[1 - (at+1)e^{-at}\right]$
4.	$\ddot{x} + b^2x = Mu_s(t)$	$s = \pm bj \quad b > 0$	$x(t) = \dfrac{M}{b^2}(1 - \cos bt)$
5.	$\ddot{x} + 2a\dot{x} + (a^2+b^2)x = Mu_s(t)$	$s = -a \pm bj \quad b > 0$	$x(t) = \dfrac{M}{a^2+b^2}\left[1 - \left(\dfrac{a}{b}\sin bt + \cos bt\right)e^{-at}\right]$

Table 2.1.2 Solution forms for a constant input.

Equation	Solution form
First order: $\dot{x} + ax = b \quad a \neq 0$	$x(t) = \dfrac{b}{a} + Ce^{-at}$
Second order: $\ddot{x} + a\dot{x} + bx = c \quad b \neq 0$	
1. $(a^2 > 4b)$ distinct, real roots: s_1, s_2	$x(t) = C_1e^{s_1t} + C_2e^{s_2t} + \dfrac{c}{b}$
2. $(a^2 = 4b)$ repeated, real roots: s_1, s_1	$x(t) = (C_1 + tC_2)e^{s_1t} + \dfrac{c}{b}$
3. $(a = 0, b > 0)$ imaginary roots: $s = \pm j\omega$, $\omega = \sqrt{b}$	$x(t) = C_1\sin\omega t + C_2\cos\omega t + \dfrac{c}{b}$
4. $(a \neq 0, a^2 < 4b)$ complex roots: $s = \sigma \pm j\omega$, $\sigma = -a/2$, $\omega = \sqrt{4b - a^2}/2$	$x(t) = e^{\sigma t}(C_1\sin\omega t + C_2\cos\omega t) + \dfrac{c}{b}$

by setting $c = 0$ and evaluating the constants (C or C_1 and C_2). It can also be used to obtain the *step* response for nonzero initial conditions. In both cases, some algebra is required. We illustrate this process with several examples.

These tables can be used to solve quickly many of the problems in this text, but we emphasize that they are limited to linear first- and second-order equations with constant coefficients and constant input. To obtain solutions of other types of equations, especially of higher order or with time-varying inputs, we depend on the Laplace transform method covered in Section 2.3.

Step Response with Complex Roots

EXAMPLE 2.1.4

■ Problem

A certain mass-spring system has a mass $m = 5$, a damping constant $c = 30$, a spring constant $k = 170$, and an external force $f = 340$, in consistent units. The equation of motion is

$$5\ddot{x} + 30\dot{x} + 170x = 340u_s(t) \qquad x(0) = \dot{x}(0) = 0$$

Find the step response.

■ Solution

We need to find the step response with zero initial conditions, so Table 2.1.1 is the easiest to use. In the table the coefficient of $\ddot{x}$ is 1, so we divide our equation by 5 to obtain

$$\ddot{x} + 6\dot{x} + 34x = 68u_s(t)$$

The characteristic equation is $s^2 + 6s + 34 = 0$ and its roots are $s = -3 \pm 5j$. Thus, $a = 3$, $b = 5$, and $M = 68$ in entry 5 in Table 2.1.1. Substituting these into the solution gives

$$x(t) = \frac{68}{3^2 + 5^2}\left[1 - \left(\frac{3}{5}\sin 5t + \cos 5t\right)e^{-3t}\right]$$

or

$$x(t) = 2\left[1 - \left(\frac{3}{5}\sin 5t + \cos 5t\right)e^{-3t}\right]$$

The solution is plotted in Figure 2.1.8. The transient response consists of the two exponentials and the steady-state response is 2. Note that the transient response dies out before oscillations due to the sine and cosine functions can appear.

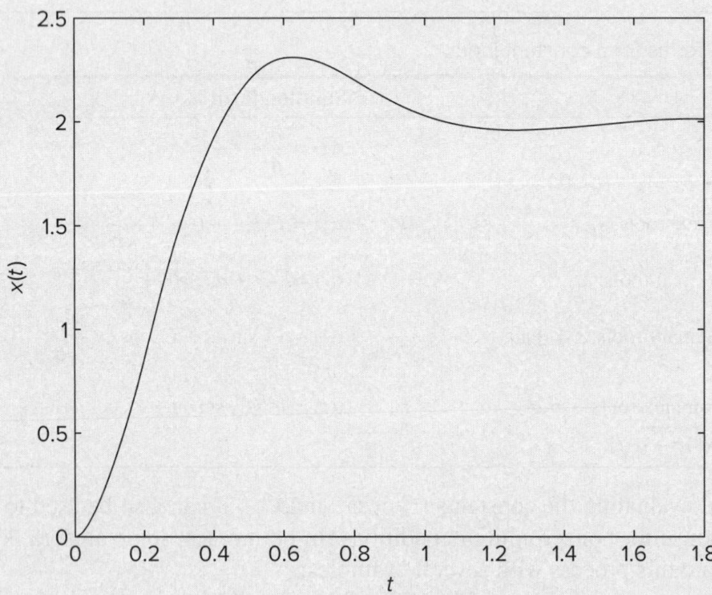

Figure 2.1.8 Response plot for Example 2.1.4.

EXAMPLE 2.1.5

Complete Response with Imaginary Roots

■ Problem

A certain mass-spring system with no damping has a mass $m = 1$, a spring constant $k = 16$, and an external force $f = 144$, in consistent units. The equation of motion is

$$\ddot{x} + 16x = 144u_s(t)$$

(a) Obtain the solution if both initial conditions are zero. (b) Obtain the solution for the initial conditions $x(0) = 12$, $\dot{x}(0) = 4$.

■ Solution

a. Characteristic polynomials are easily obtained from the differential equation by replacing $\ddot{x}$ with s^2, $\dot{x}$ with s, and x with 1. The characteristic equation here is

$$s^2 + 16 = 0$$

and the roots are $s = \pm 4j$. Since the initial conditions are zero, Table 2.1.1 is the easiest to use. Entry 4 in Table 2.1.1 gives $b = 4$ and $M = 144$. The forced response is

$$x(t) = \frac{144}{4^2}(1 - \cos 4t) = 9(1 - \cos 4t)$$

This response is plotted in Figure 2.1.9. Since the initial conditions are zero, the free response is zero and thus the forced response is the same as the complete response, which is also the steady-state response.

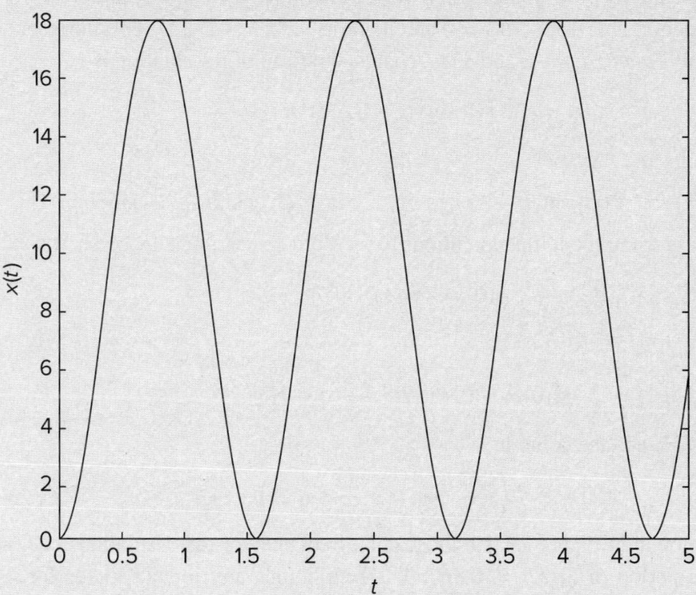

Figure 2.1.9 Response plot for Example 2.1.5(a).

b. We cannot use Table 2.1.1 for this part because the intial conditions are not zero. From entry 3 in Table 2.1.2, with $b = 16$, $\omega = 4$, and $c = 144$, the solution form is

$$x(t) = C_1 \sin 4t + C_2 \cos 4t + \frac{144}{16}$$

This gives

$$x(0) = C_2 + 9 = 12$$

Thus $C_2 = 3$. Also,

$$\dot{x}(t) = 4C_1 \cos 4t - 4C_2 \sin 4t$$

so that

$$\dot{x}(0) = 4C_1 = 4$$

This gives $C_1 = 1$. The solution is

$$x(t) = \sin 4t + 3 \cos 4t + 9$$

This is the complete response and also the steady-state response.

EXAMPLE 2.1.6

Complete Response with Complex Roots

■ Problem

Find the complete response of the following equation.

$$\ddot{x} + 4\dot{x} + 53x = 15u_s(t) \qquad x(0) = 8 \qquad \dot{x}(0) = -19$$

■ Solution

We cannot use Table 2.1.1 for this part because the intial conditions are not zero. Comparing this equation with the form in Table 2.1.2, we see that $a = 4$, $b = 53$, and $c = 15$. The characteristic equation is $s^2 + 4s + 53 = 0$ and its roots are $s = -2 \pm 7j$. This corresponds to entry 4 in the table, where $\sigma = -2$ and $\omega = 7$. Thus, the form of the solution is

$$x(t) = e^{-2t}\left(C_1 \sin 7t + C_2 \cos 7t\right) + \frac{15}{53}$$

and

$$\dot{x}(t) = -2e^{-2t}\left(C_1 \sin 7t + C_2 \cos 7t\right) + e^{-2t}\left(7C_1 \cos 7t - 7C_2 \sin 7t\right)$$

Now evaluate these using the initial conditions

$$x(0) = 8 = C_2 + \frac{15}{53}$$

which gives $C_2 = 409/53$. Also,

$$\dot{x}(0) = -19 = -2\left(\frac{409}{53}\right) + 7C_1$$

Thus, $C_1 = -27/53$, and the solution is

$$x(t) = e^{-2t}\left(\frac{409}{53} \cos 7t - \frac{27}{53} \sin 7t\right) + \frac{15}{53}$$

The solution is plotted in Figure 2.1.10. The radian frequency of the oscillations is 7, which corresponds to a period of $2\pi/7 = 0.867$. The oscillations are difficult to see for $t > 2$ because $e^{-2t} < 0.02$ for $t > 2$. So, for most practical purposes, we may say that the response is essentially constant at the steady-state response $15/53$ for $t > 2$.

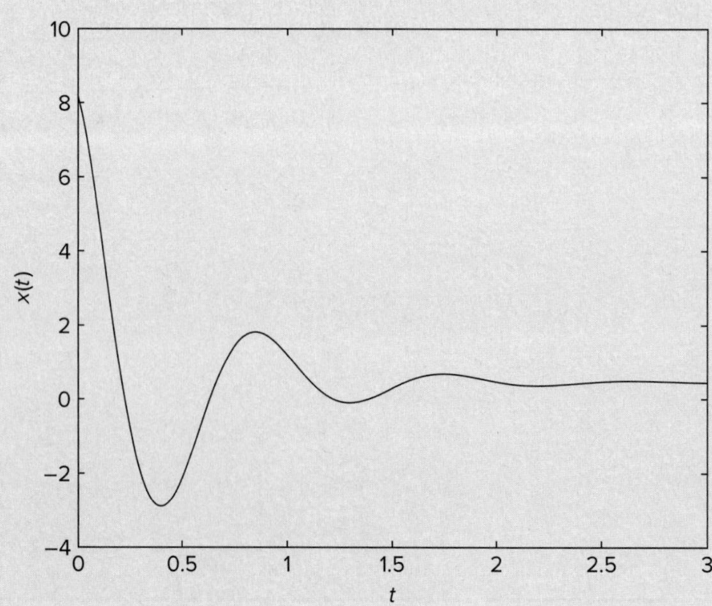

Figure 2.1.10 Response plot for Example 2.1.6.

Sine Form of the Response

EXAMPLE 2.1.7

■ Problem

Thus far, our solutions having complex roots have been expressed as the sum of a sine and a cosine, each multiplied by an exponential function. It is sometimes more convenient to express the result entirely as either a sine or a cosine. For example, the step response of a second-order system with complex roots $s = -a \pm jb$ has the form

$$x(t) = e^{-at}B\sin bt + e^{-at}C\cos bt + D = e^{-at}(B\sin bt + C\cos bt) + D \tag{2.1.26}$$

a. Show how this form may be expressed as

$$x(t) = Ae^{-at}\sin(bt + \phi) + D \tag{2.1.27}$$

b. The equation

$$\ddot{x} + 4\dot{x} + 53x = 15 \qquad x(0) = 8 \qquad \dot{x}(0) = -19$$

has the complete response

$$x(t) = -\frac{27}{53}e^{-2t}\sin 7t + \frac{409}{53}e^{-2t}\cos 7t + \frac{15}{53}$$

Express the response in the form given by (2.1.27).

■ Solution

a. Use the identity

$$\sin(bt + \phi) = \cos\phi\sin bt + \sin\phi\cos bt$$

Comparing this with (2.1.26), we obtain

$$B = A\cos\phi \tag{2.1.28}$$

$$C = A\sin\phi \tag{2.1.29}$$

Square both sides of each equation, add the results, and use the identity $\cos^2\phi + \sin^2\phi = 1$.

$$B^2 + C^2 = A^2\left(\cos^2\phi + \sin^2\phi\right) = A^2$$

We arbitrarily choose A to be the positive square root (this fixes the quadrant of ϕ).

$$A = +\sqrt{B^2 + C^2} \tag{2.1.30}$$

Note that

$$\cos\phi = \frac{B}{A} \tag{2.1.31}$$

and

$$\sin\phi = \frac{C}{A} \tag{2.1.32}$$

These two relations determine the quadrant of ϕ, since $A > 0$.

The angle ϕ can be determined from (2.1.31), (2.1.32), or from

$$\tan\phi = \frac{\sin\phi}{\cos\phi} = \frac{C/A}{B/C} = \frac{C}{B} \tag{2.1.33}$$

b. For this problem,

$$B = -\frac{27}{53} \qquad C = \frac{409}{53}$$

and

$$A = +\sqrt{B^2 + C^2} = 7.7338$$

From (2.1.31) and (2.1.32), we see that $\cos \phi < 0$ and $\sin \phi > 0$. Thus, ϕ is in the second quadrant and from (2.1.33)

$$\phi = \tan^{-1}\left(\frac{409/53}{-27/53}\right) = 1.6367 \text{ rad}$$

Thus, the response can be expressed as

$$x(t) = 7.7338e^{-2t}\sin(7t + 1.6367) + \frac{15}{53}$$

2.1.8 ROOTS AND COMPLEX NUMBERS

Many of the equations of system dynamics are second-order or higher, and thus we will need to work with equations, such as roots of characteristic polynomials, that involve complex numbers. Table 2.1.3 gives some properties of complex numbers. You should review these properties if you have not used complex numbers in a while. For example, the complex number $z_1 = 3 + 4j$ can be represented in polar form by computing its magnitude $|z_1|$ using the Pythagorean theorem (see Figure 2.1.12).

$$|z_1| = \sqrt{3^2 + 4^2} = 5$$

The angle θ_1 is found from trigonometry.

$$\theta_1 = \tan^{-1}\left(\frac{4}{3}\right) = 0.927 \text{ rad} = 53.1^\circ$$

So the polar representation of z_1 is

$$z_1 = 5\angle 0.927 \text{ rad} = 5\angle 53.1^\circ$$

and is illustrated in Figure 2.1.12.

Using trigonometry with Figure 2.1.11, we see that

$$x = |z|\cos\theta \qquad y = |z|\sin\theta$$

and thus

$$z = |z|\cos\theta + |z|\sin\theta j = |z|(\cos\theta + j\sin\theta)$$

Using the Euler identity (Table 2.1.4),

$$e^{j\theta} = \cos\theta + j\sin\theta$$

we have

$$z = |z|(\cos\theta + j\sin\theta) = |z|e^{j\theta}$$

This is the complex exponential representation of the complex number z.

Suppose that $z_2 = -3 + 4j$. Then from Figure 2.1.13,

$$z_2 = -3 + 4j = 5\angle 2.214 \text{ rad} = 5\angle 126.9^\circ$$

Table 2.1.3 Roots and complex numbers.

The quadratic formula

The roots of $as^2 + bs + c = 0$ are given by

$$s = \frac{-b \pm \sqrt{b^2 - 4ac}}{2a}$$

For complex roots, $s = -\sigma \pm j\omega$, the quadratic can be expressed as

$$as^2 + bs + c = a\left[(s + \sigma)^2 + \omega^2\right] = 0$$

Complex numbers

Rectangular representation:

$$z = x + jy, j = \sqrt{-1}$$

Complex conjugate:

$$\bar{z} = x - jy$$

Magnitude and angle:

$$|z| = \sqrt{x^2 + y^2} \qquad \theta = \angle z = \tan^{-1}\frac{y}{x} \text{ (See Figure 2.1.11)}$$

Polar and exponential representation:

$$z = |z|\angle\theta = |z|e^{j\theta}$$

Equality: If $z_1 = x_1 + jy_1$ and $z_2 = x_2 + jy_2$, then

$$z_1 = z_2 \text{ if } x_1 = x_2 \text{ and } y_1 = y_2$$

Addition:

$$z_1 + z_2 = (x_1 + x_2) + j(y_1 + y_2)$$

Multiplication:

$$z_1 z_2 = |z_1||z_2|\angle(\theta_1 + \theta_2)$$

$$z_1 z_2 = (x_1 x_2 - y_1 y_2) + j(x_1 y_2 + x_2 y_1)$$

Complex-conjugate multiplication:

$$(x + jy)(x - jy) = x^2 + y^2$$

Division:

$$\frac{1}{z} = \frac{1}{x + yj} = \frac{x - jy}{x^2 + y^2}$$

$$\frac{z_1}{z_2} = \frac{|z_1|}{|z_2|}\angle(\theta_1 - \theta_2)$$

$$\frac{z_1}{z_2} = \frac{x_1 + jy_1}{x_2 + jy_2} = \frac{x_1 + jy_1}{x_2 + jy_2}\frac{x_2 - jy_2}{x_2 - jy_2} = \frac{(x_1 + jy_1)(x_2 - jy_2)}{x_2^2 + y_2^2}$$

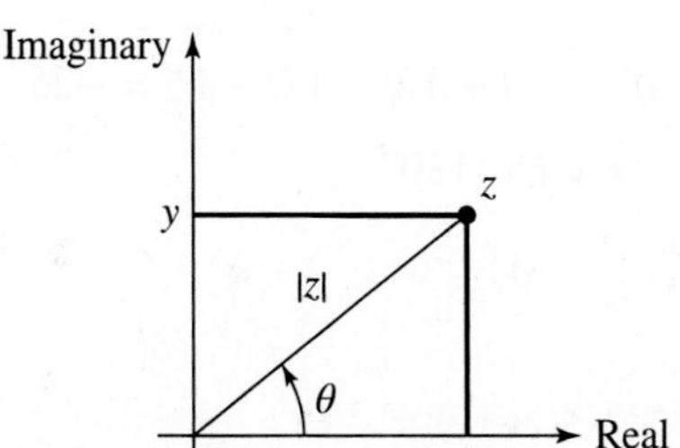

Figure 2.1.11 Polar representation of a complex number $z = x + jy$.

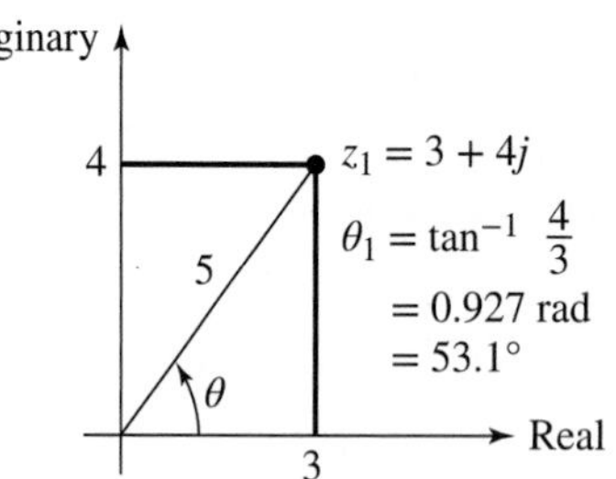

Figure 2.1.12 Polar representation of the complex number $z_1 = 3 + 4j$.

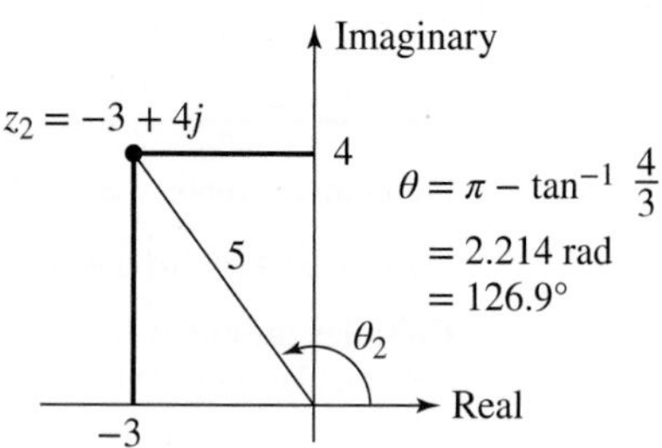

Figure 2.1.13 Polar representation of the complex number $z_2 = -3 + 4j$.

Table 2.1.4 The exponential function.

Taylor series

$$e^x = 1 + x + \frac{x^2}{2} + \frac{x^3}{6} + \cdots + \frac{x^n}{n!} + \cdots$$

Euler's identities

$$e^{j\theta} = \cos\theta + j\sin\theta$$
$$e^{-j\theta} = \cos\theta - j\sin\theta$$

Limits

$\lim_{x\to\infty} xe^{-x} = 0$ if x is real.

$\lim_{t\to\infty} e^{-st} = 0$ if the real part of s is positive.

If a is real and positive,

$e^{-at} < 0.02$ if $t > 4/a$.

$e^{-at} < 0.01$ if $t > 5/a$.

The *time constant* is $\tau = 1/a$.

Addition and subtraction of two complex numbers is most easily done with the rectangular representation, as

$$z_1 + z_2 = 3 + 4j + (-3 + 4j) = 0 + 8j = 8j$$
$$z_1 - z_2 = 3 + 4j - (-3 + 4j) = 6 + 0j = 6$$

However, multiplication and division are best done using the polar or complex exponential forms, if they are already available.

$$z_1z_2 = (3 + 4j)(-3 + 4j) = 3(-3 + 4j) + 4j(-3 + 4j) = -9 + 12j - 12j - 16 = -25$$
$$z_1z_2 = (5\angle 53.1°)(5\angle 126.9°) = 25\angle(53.1° + 126.9°) = 25\angle 180°$$

or

$$z_1z_2 = 25(\cos 180° + j\sin 180°) = 25(-1 + 0j) = -25$$

When dividing with the rectangular form, first multiply the numerator and denominator by the complex conjugate of the denominator (complex conjugate numbers have the same real part but opposite-signed imaginary parts).

$$\frac{z_1}{z_2} = \frac{3+4j}{-3+4j} = \frac{3+4j}{-3+4j}\frac{-3-4j}{-3-4j} = \frac{7-24j}{25} = 0.28-0.96j$$

$$\frac{z_1}{z_2} = \frac{5\angle 53.1°}{5\angle 126.9°} = 1\angle(53.1° - 126.9°) = 1\angle(-73.8°)$$

or

$$\frac{z_1}{z_2} = 1\left[\cos(-73.8°) + j\sin(-73.8°)\right] = 0.28 - 0.96j$$

Note that polar angles add with multiplication and subtract with division.

2.2 RESPONSE PARAMETERS AND STABILITY

Common differential equations encountered in system dynamics are linear, constant-coefficient equations, each with a *constant* on the right-hand side. Two common categories are:

1. First-order equation: $\dot{x} + ax = b \quad a \neq 0$
2. Second-order equation: $\ddot{x} + a\dot{x} + bx = c \quad b \neq 0$

The solution form for each case is given in Table 2.1.2. Note that the solution for the second-order case can have one of four possible forms, depending on the nature of the two roots. The case with imaginary roots is actually a special case of complex roots where the real part is zero. Table 2.1.2 does not give formulas for the undetermined constants in the solution, because often all we require is the general form of the solution. To determine the values of these constants, you must be given the initial conditions.

2.2.1 ASSESSMENT OF SOLUTION BEHAVIOR

Note that the characteristic equation can be quickly identified from the ODE by replacing $\dot{x}$ with s, $\ddot{x}$ with s^2, and so forth. For example, $3\ddot{x} + 30\dot{x} + 222x = 148$ has the characteristic equation $3s^2 + 30s + 222 = 0$ and the roots $s = -5 \pm 7j$. This is case 4 in Table 2.1.2, and the solution form is (since $148/222 = 2/3$)

$$x(t) = e^{-5t}(C_1 \sin 7t + C_2 \cos 7t) + \frac{2}{3}$$

From this form we can tell that the solution will oscillate with a radian frequency of 7. The oscillations will eventually disappear because of the exponential term e^{-5t}, which is less than 0.02 for $t > 4/5$. Thus, as $t \to \infty$, $x(t) \to 2/3$. Sometimes this is all the information we need about the solution, and if so, we need not evaluate the constants C_1 and C_2.

We have seen that responses can decay or increase exponentially and that they can oscillate with increasing, decreasing, or constant amplitudes. In this section, we introduce four parameters that help us describe the different types of responses. We also treat the related concept of *stability*, which is a desirable property for many dynamic systems.

2.2.2 THE TIME CONSTANT

The free response of the first-order model

$$\dot{x} + ax = b \tag{2.2.1}$$

may be written in the form

$$x(t) = x(0)e^{-at} = x(0)e^{-t/\tau} \tag{2.2.2}$$

where we have introduced a new parameter τ with the definition

$$\tau = \frac{1}{a} \quad \text{if} \quad a > 0 \tag{2.2.3}$$

The equation $\dot{x} + ax = b$ may be expressed in terms of τ by replacing a with $1/\tau$ as follows:

$$\tau\dot{x} + x = b\tau \tag{2.2.4}$$

The new parameter τ has units of time and is the model's *time constant*. It gives a convenient measure of the exponential decay curve and an estimate of how long it will take for the transient response to disappear, leaving only the steady-state response. The free response and the meaning of the time constant are illustrated in Figure 2.2.1. After a time equal to one time constant has elapsed, x has decayed to 37% of its initial value. We can also say that x has decayed *by* 63%. At $t = 4\tau$, $x(t)$ has decayed to 2% of its initial value. At $t = 5\tau$, $x(t)$ has decayed to 1% of its initial value. The time constant is defined only when $a > 0$. If $a \leq 0$, the free response does not decay to 0 and thus the time constant has no meaning.

The time constant is useful also for analyzing the response when the forcing function is a constant. We can express the total response given by (2.1.18) in terms of τ by substituting $a = 1/\tau$ as follows:

$$x(t) = \underbrace{b\tau}_{\text{steady state}} + \underbrace{[x(0) - b\tau]\,e^{-t/\tau}}_{\text{transient}} = x_{ss} + \left[x(0) - x_{ss}\right] e^{-t/\tau} \tag{2.2.5}$$

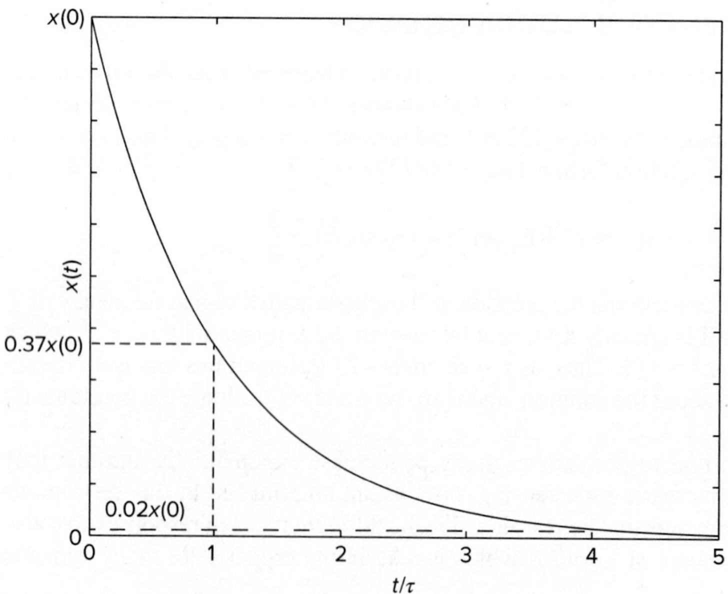

Figure 2.2.1 The free response $x(t) = x(0)e^{-t/\tau}$.

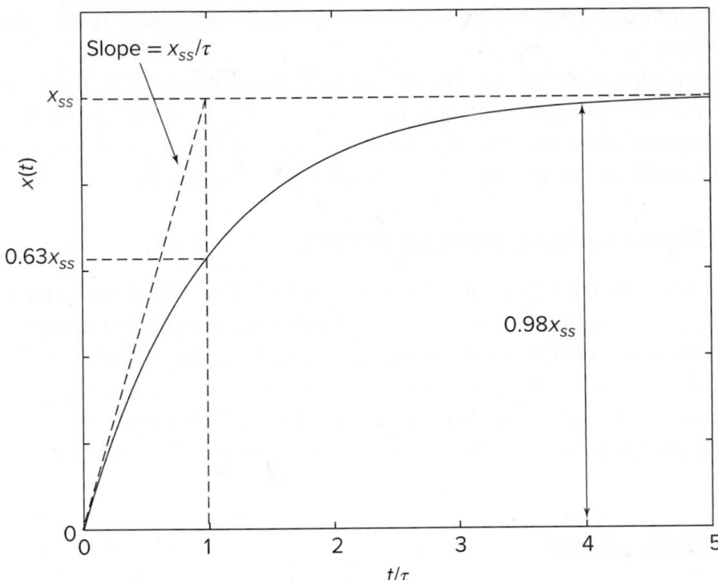

Figure 2.2.2 The step response $x(t) = x_{ss}(1 - e^{-t/\tau})$.

where $x_{ss} = b/a = b\tau$, which is the steady-state response. The response approaches the constant value $b\tau$ as $t \to \infty$. The forced response (for which $x(0) = 0$) is plotted in Figure 2.2.2. At $t = \tau$, the response is 63% of the steady-state value. At $t = 4\tau$, the response is 98% of the steady-state value, and at $t = 5\tau$, it is 99% of steady-state. Thus, because the difference between 98% and 99% is so small, for most engineering purposes we may say that $x(t)$ reaches steady-state at $t = 4\tau$, although mathematically, steady-state is not reached until $t = \infty$.

If $x(0) \neq 0$, the response is shifted by $x(0)e^{-t/\tau}$. At $t = \tau$, 37% of the difference between the initial value $x(0)$ and the steady-state value remains. At $t = 4\tau$, only 2% of this difference remains.

2.2.3 THE DOMINANT-ROOT APPROXIMATION

The time-constant concept is not limited to first-order models. It can also be used to estimate the response time of higher-order models. For example, the complete response of the equation $\ddot{x} + 10\dot{x} + 21x = 42$ has the form

$$x(t) = C_1 e^{-3t} + C_2 e^{-7t} + 2$$

Because e^{-7t} decays to zero faster than e^{-3t}, the term $C_2 e^{-7t}$ will disappear faster than the term $C_1 e^{-3t}$, provided that $|C_2|$ is not much greater than $|C_1|$. The time constant of the root $s = -7$ is $\tau_1 = 1/7$, and thus the term $C_2 e^{-7t}$ will be essentially 0 for $t > 4\tau_1 = 4/7$. Thus, for $t > 4/7$, the response is essentially given by $C_1 e^{-3t} + 2$.

The root $s = -3$ is said to be the *dominant root* because it dominates the response relative to the term corresponding to the root $s = -7$, and the time constant $\tau_2 = 1/3$ is said to be the *dominant-time constant*. We can estimate how long it will take for the transient response to disappear by multiplying the dominant-time constant by 4. Here the answer is $t = 4/3$.

We cannot make exact predictions based on the dominant root because the initial conditions that determine the values of C_1 and C_2 can be such that $|C_2| \gg |C_1|$, so that the second exponential influences the response for longer than expected. The dominant-root approximation, however, is often used to obtain a quick estimate of response time.

The farther away the dominant root is from the other roots (the "secondary" roots), the better the approximation.

The dominant-root approximation can be applied to higher-order models. If all the roots have negative real parts (some roots may be complex), the dominant root is the one having the largest time constant, and the time it will take the transient response to disappear can be estimated by multiplying the dominant-time constant by 4.

2.2.4 TIME CONSTANTS AND COMPLEX ROOTS

The equation $\ddot{x} + 6\dot{x} + 25x = 0$ has the complex roots $-3 \pm 4j$. These lead to the term e^{-3t} in the response. Thus, we may apply the concept of a time constant to complex roots by computing the time constant from the negative inverse of the real part of the roots. Here the model's time constant is $\tau = 1/3$, and thus the response is essentially at steady state for $t > 4\tau = 4/3$, as shown in Figure 2.2.3 for the case where $x(0) = 10$ and $\dot{x}(0) = 0$. Since complex roots occur only in conjugate pairs, each pair has the same time constant.

2.2.5 NATURAL FREQUENCY

The free response of the equation

$$m\ddot{x} + kx = 0 \tag{2.2.6}$$

is given by

$$x(t) = x(0)\cos\omega_n t + \frac{\dot{x}(0)}{\omega_n}\sin\omega_n t \tag{2.2.7}$$

where

$$\omega_n = \sqrt{\frac{k}{m}} \tag{2.2.8}$$

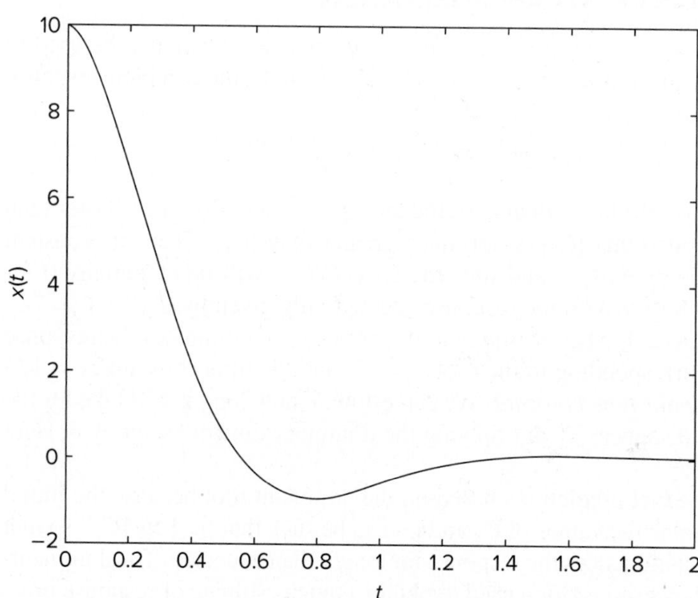

Figure 2.2.3 Response for $\ddot{x} + 6\dot{x} + 25x = 0$ where $x(0) = 10$ and $\dot{x}(0) = 0$.

The radian frequency of oscillation of the free response will be $\omega_n = \sqrt{k/m}$, and is called the *natural frequency*. Similarly, the natural period of oscillation is given by $P_n = 2\pi/\omega_n$.

The response is a constant-amplitude oscillation about 0. The amplitude depends on the initial conditions $x(0)$ and $\dot{x}(0)$, but the oscillation frequency and the period are independent of the initial conditions.

2.2.6 DAMPED NATURAL FREQUENCY

Suppose the equation contains a first derivative term, as $\ddot{x} + 6\dot{x} + 25x = 0$. The solution is

$$x(t) = x(0)e^{-3t}\cos 4t + \frac{\dot{x}(0) + 3x(0)}{4}e^{-3t}\sin 4t$$

The free response oscillates with a radian frequency of 4. The period of the oscillation is $2\pi/4 = \pi/2$, but the amplitude decays with time. Note that the addition of the derivative term $6\dot{x}$ changes the frequency from 5 to 4, and causes the oscillations to decay. The first-order derivative term is called the *damping term*.

Using our solution tables, we can generalize this result to the equation

$$m\ddot{x} + c\dot{x} + kx = 0 \tag{2.2.9}$$

The characteristic equation is

$$ms^2 + cs + k = 0 \tag{2.2.10}$$

and the roots are

$$s = \frac{-c \pm \sqrt{c^2 - 4mk}}{2m} = -a \pm bj \tag{2.2.11}$$

where we have assumed that the roots are complex (the only case to give oscillatory response) and

$$a = -\frac{c}{2m} \qquad b = \sqrt{\frac{k}{m} - \left(\frac{c}{2m}\right)^2}$$

The solution is

$$x(t) = e^{-at}\left[x(0)\cos bt + \frac{\dot{x}(0) + ax(0)}{b}\sin bt\right] \tag{2.2.12}$$

The frequency of oscillation of the free response is b, and is called the *damped natural frequency*. It is frequently denoted by the symbol ω_d.

$$\omega_d = \sqrt{\frac{k}{m} - \left(\frac{c}{2m}\right)^2} \tag{2.2.13}$$

From this we can see that the damping term $c\dot{x}$ decreases the oscillation frequency. The largest possible oscillation frequency occurs when $c = 0$, and in this case ω_d equals ω_n.

If c, the coefficient of the damping term, is large enough, then ω_d will be zero or imaginary. In this case, both roots are real and no oscillation occurs. The value of c for which $\omega_d = 0$ and both roots are real and equal is the value

$$c = 2\sqrt{mk} \tag{2.2.14}$$

This value is called the *critical damping value*, because it represents the dividing line between oscillatory and non-oscillatory free response. If $c < 2\sqrt{mk}$, oscillation occurs. If $c > 2\sqrt{mk}$, the response is exponential.

2.2.7 THE DAMPING RATIO

If both roots are negative or have negative real parts, the free response of a second-order equation can be conveniently characterized by the *damping ratio*, denoted by ζ (and sometimes called the *damping factor*). For the characteristic equation $ms^2+cs+k=0$, the damping ratio is defined as the ratio of the actual value of c to its critical value, as

$$\zeta = \frac{c}{2\sqrt{mk}} \tag{2.2.15}$$

This definition is not arbitrary but is based on the way the roots change from real to complex as the value of c is changed. Three cases can occur:

1. **Critically Damped Case:** Repeated roots occur if $c^2 - 4mk = 0$; that is, if $c = 2\sqrt{mk}$. This value of the damping constant is the critical damping constant and when c has this value the equation is said to be *critically damped.*
2. **The Overdamped Case:** If $c > 2\sqrt{mk}$, two distinct, real roots exist, and the equation is *overdamped.*
3. **The Underdamped Case:** If $c < 2\sqrt{mk}$, complex roots occur, and the equation is *underdamped.*

Note that

1. For the critically damped case, $\zeta = 1$.
2. Exponential behavior occurs if $\zeta > 1$ (the overdamped case).
3. Oscillations exist when and only when $\zeta < 1$ (the underdamped case).

If any root is positive or has a positive real part, the damping ratio is meaningless and therefore not defined. For example, because the equation $s^2 - 4s + 25 = 0$ has the roots $s = 2 \pm 5j$, its free response will increase without bound, and the damping ratio is negative and therefore meaningless.

The damping ratio can be used as a quick check for oscillatory behavior. For example, with the characteristic equation $s^2 + 5ds + 4d^2 = 0$, where $d > 0$, the damping ratio is $\zeta = 5d/(2\sqrt{4d^2}) = 5/4$. Because $\zeta > 1$, no oscillations can occur in the free response regardless of the value of $d > 0$ and regardless of the initial conditions.

Equation (2.2.10) has three parameters, m, c, and k. We can reduce the number of parameters to two by writing the characteristic equation in terms of the parameters ζ and ω_n. First divide (2.2.10) by m and use the facts that $\omega_n^2 = k/m$ and

$$2\zeta\omega_n = 2\left(\frac{c}{2\sqrt{mk}}\right)\left(\sqrt{\frac{k}{m}}\right) = \frac{c}{m}$$

The characteristic equation becomes

$$s^2 + 2\zeta\omega_n s + \omega_n^2 = 0$$

and the roots are

$$s = -\zeta\omega_n \pm j\omega_n\sqrt{1-\zeta^2} \tag{2.2.16}$$

where we have assumed that the roots are imaginary or complex and thus $\zeta \le 1$. We thus see that the damped frequency of oscillation can be expressed in terms of ζ and ω_n as

$$\omega_d = \omega_n\sqrt{1-\zeta^2} \tag{2.2.17}$$

When $\zeta \le 1$, this equation shows that the damped frequency is always less than the undamped frequency.

The real part of the root is the negative reciprocal of the time constant, and so we have

$$\tau = \frac{1}{\zeta\omega_n} \tag{2.2.18}$$

Remember that this formula applies only if $\zeta \le 1$. Table 2.2.1 summarizes the formulas for these parameters.

Table 2.2.1 Response parameters for second-order models.

1. Model:	$m\ddot{x} + c\dot{x} + kx = f(t)$
2. Characteristic Equation:	$ms^2 + cs + k = 0$
3. Natural Frequency:	$\omega_n = \sqrt{\dfrac{k}{m}}$
4. Damping Ratio:	$\zeta = \dfrac{c}{2\sqrt{mk}}$
5. Damped Natural Frequency:	$\omega_d = \omega_n\sqrt{1-\zeta^2}$
6. Time Constant (if $\zeta \le 1$):	$\tau = \dfrac{2m}{c} = \dfrac{1}{\zeta\omega_n}$

The solution of $\ddot{x} + 6\dot{x} + 25x = 0$ for $\dot{x}(0) = 0$ and $x(0) = 10$ is

$$x = 10e^{-3t}\cos 4t + \frac{15}{2}\sin 4t$$

This is plotted in Figure 2.2.3. For this differential equation the damping ratio is $\zeta = 6/(2\sqrt{25}) = 0.6 < 1$, so we would expect to see oscillations with a radian frequency of 4, but we do not. Why not? The answer lies in the relation of the time constant to the oscillation period. The roots are $s = -3 \pm 4j$, so the time constant is $\tau = 1/3$. The response will be essentially zero after $t = 4\tau = 4/3 = 1.33$, but the period is $2\pi/4 = 1.57$. Because the period is greater than 4τ, the oscillations do not have time to repeat before their amplitude becomes very small.

Oscillations will not be very visible in the response if the damped period $2\pi/\omega_d$ is greater than four time constants. This occurs when the following ratio is greater than 1.

$$R = \frac{2\pi/\omega_d}{4\tau} = \frac{2\pi}{\omega_d}\frac{1}{4/\zeta\omega_n} = \frac{2\pi}{\omega_n\sqrt{1-\zeta^2}}\frac{\zeta\omega_n}{4} = \frac{\pi}{2}\frac{\zeta}{\sqrt{1-\zeta^2}}$$

This ratio R will be greater than 1 if $\zeta > 0.537$.

2.2.8 STABILITY

We have seen free responses that approach 0 as $t \to \infty$. We have also seen free responses that approach a constant-amplitude oscillation as $t \to \infty$. The free response may also approach ∞. For example, the model $\dot{x} = 2x$ has the free response $x(t) = x(0)e^{2t}$ and thus $x \to \infty$ as $t \to \infty$.

Let us define some terms.[1]

Unstable A model whose free response approaches ∞ as $t \to \infty$ is said to be *unstable*.

Stable If the free response approaches 0, the model is *stable*.

Neutral Stability The borderline case between stable and unstable. Neutral stability describes a situation where the free response does not approach ∞ but does not eventually reach 0.

The stability properties of a linear model are determined from its characteristic roots. To understand the relationship between stability and the characteristic roots, consider some simple examples.

The first-order model $\dot{x} + ax = f(t)$ has the free response $x(t) = x(0)e^{-at}$, which approaches 0 as $t \to \infty$ if the characteristic root $s = -a$ is *negative*. The model is unstable if the root is *positive* because $x(t) \to \infty$ as $t \to \infty$.

A borderline case, called neutral stability, occurs if the root $s = -a$ is 0. In this case, $x(t)$ remains at $x(0)$. Neutral stability describes a situation where the free response does not approach ∞ but does not approach and eventually remain at 0.

Now consider the following second-order models, which have the same initial conditions: $x(0) = 1$ and $\dot{x}(0) = 0$. Their responses are shown in Figure 2.2.4.

1. The model $\ddot{x} - 4x = f(t)$ has the roots $s = \pm 2$, and the free response:

$$x(t) = \frac{1}{2}\left(e^{2t} + e^{-2t}\right)$$

2. The model $\ddot{x} - 4\dot{x} + 229x = f(t)$ has the roots $s = 2 \pm 15j$, and the free response:

$$x(t) = e^{2t}\left(\cos 15t - \frac{2}{15}\sin 15t\right)$$

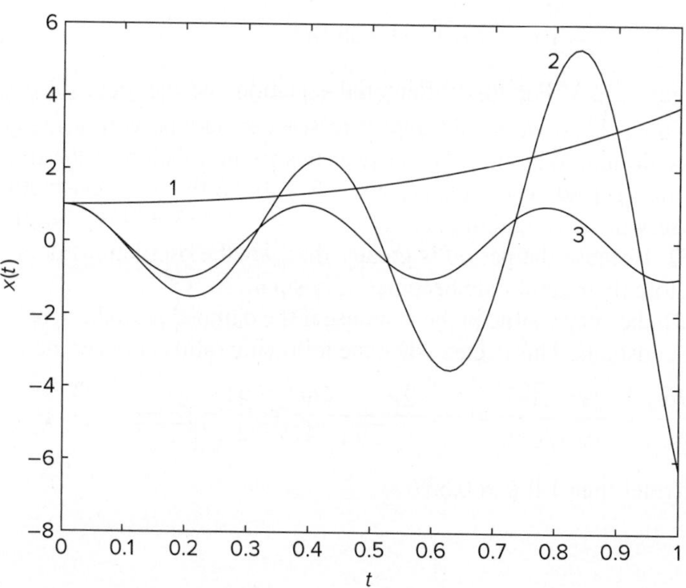

Figure 2.2.4 Examples of unstable and neutrally stable models.

[1]There is some variation in terminology related to stability. A stable model is sometimes said to be *asymptotically stable*. A neutrally stable model is said by some to be *critically stable*.

3. The model $\ddot{x} + 256x = f(t)$ has the roots $s = \pm 16j$, and the free response:

$$x(t) = \cos 16t$$

From the plots we can see that none of the three models displays stable behavior. The first and second models are unstable, while the third is neutrally stable. Thus, the free response of a neutrally stable model can either approach a nonzero constant or settle down to a constant-amplitude oscillation.

The effect of the real part of the characteristic roots can be seen from the free response form for the complex roots $\sigma \pm \omega j$ (entry 4 in Table 2.1.2, with $c = 0$):

$$x(t) = e^{\sigma t}(C_1 \sin \omega t + C_2 \cos \omega t)$$

Clearly, if the real part is positive (that is, if $\sigma > 0$), then the exponential $e^{\sigma t}$ will grow with time and so will the amplitude of the oscillations. This is an unstable case.

If the real part is 0 (that is, $\sigma = 0$), then the exponential becomes $e^{0t} = 1$, and the amplitude of the oscillations remains constant. This is the neutrally stable case. The imaginary part of the root is the frequency of oscillation; it has no effect on stability.

Model 1 is unstable because of the exponential e^{2t}, which is due to the positive root $s = +2$. The negative root $s = -2$ does not cause instability because its exponential disappears in time.

If we realize that a real number is simply a special case of a complex number whose imaginary part is 0, then these examples show that a linear model is unstable if at least one root has a positive real part. We will see that the free response of any linear, constant-coefficient model, of any order, consists of a sum of terms, each multiplied by an exponential. Each exponential will approach ∞ as $t \to \infty$, if its corresponding root has a positive real part.

A root is said to have a *multiplicity* k if it is repeated k times. For example, the roots $s = 0, 0$ have multiplicity 2, as do the roots $s = 4j, 4j, -4j, -4j$, which are the roots of $s^4 + 32s^2 + 256 = 0$. Consider the equation $\ddot{x} = 0$. Its roots are $s = 0, 0$ and its solution is $x(t) = x(0) + v_0 t$, where $v_0 = \dot{x}(0)$. Now, $|x(t)| \to \infty$ as $t \to \infty$ if $v_0 \neq 0$. Thus, the model $\ddot{x} = 0$ exhibits unstable behavior if $v_0 \neq 0$ and neutrally stable behavior if $v_0 = 0$. Since stability is a property of the model only, and not of the initial conditions, we cannot say that the model $\ddot{x} = 0$ is stable, or even neutrally stable.

When a model has roots on the imaginary axis that are of multiplicity k, these roots generate a term in the solution of the form t^{k-1}. So, if $k \geq 2$, the term t^{k-1} will grow with time and thus the model will be unstable. For example, a model with the characteristic equation

$$s^5 + 3s^4 + 32s^3 + 96s^2 + 256s + 768 = 0$$

has the roots $s = -3, 4j, 4j, -4j, -4j$. So the model will have solutions $x(t)$ containing the terms

$$C_1 e^{-3t}, \qquad C_2 t \sin 4t, \qquad C_3 t \cos 4t$$

If the initial conditions are such that either C_2 or C_3 is nonzero, then $|x(t)| \to \infty$.

Neutral stability is associated with roots having zero real parts, but these examples show that a model is not neutrally stable if it has roots of multiplicity 2 or greater on the imaginary axis. Such models are said to be *marginally stable*.

Thus, we can make the following statement about linear models.

Stability Test for Linear Constant-Coefficient Models

A constant-coefficient linear model is *stable* if and only if all of its characteristic roots have negative real parts.

The model is *neutrally stable* if one or more roots have a zero real part with no roots on the imaginary axis of multiplicity 2 or greater, and the remaining roots have negative real parts.

The model is *unstable* if *any* root has a positive real part.

If a linear model is stable, it is not possible to find initial conditions for which the free response approaches ∞ as $t \to \infty$. However, if the model is unstable, there might be initial conditions that result in a response that disappears in time. For example, the model $\ddot{x} - 4x = 0$ has the roots $s = \pm 2$, and thus is unstable. However, if the initial conditions are $x(0) = 1$ and $\dot{x}(0) = -2$, then the free response is $x(t) = e^{-2t}$, which approaches 0 as $t \to \infty$. The exponential e^{2t} corresponding to the root at $s = +2$ does not appear in the response because of the special nature of these initial conditions. We cannot rely on initial conditions to stabilize a system.

Because the time constant is a measure of how long it takes for the exponential terms in the response to disappear, we see that the time constant is not defined for neutrally stable and unstable cases.

2.2.9 A PHYSICAL EXAMPLE

A physical example illustrating the meaning of stability is shown in Figure 2.2.5. Suppose that there is some slight friction in the pivot point of the pendulum. In part (a) the pendulum is hanging and is at rest at $\theta = 0$. If something disturbs it slightly, it will oscillate about $\theta = 0$ and eventually return to rest at $\theta = 0$. We thus see that the system is *stable*.

If the friction is large enough, the pendulum will not oscillate before coming to rest. In both cases, however, the system is stable because the pendulum eventually returns to its rest position at $\theta = 0$.

Now suppose that there is no friction. The pendulum will oscillate with a constant amplitude forever about $\theta = 0$ if disturbed. This is an example of neutral stability.

If the pendulum is perfectly balanced at $\theta = \pi$, as in Figure 2.2.5b, it will never return to $\theta = \pi$ if it is disturbed. So, this equilibrium is an unstable one.

Figure 2.2.5 Pendulum illustration of stability properties.

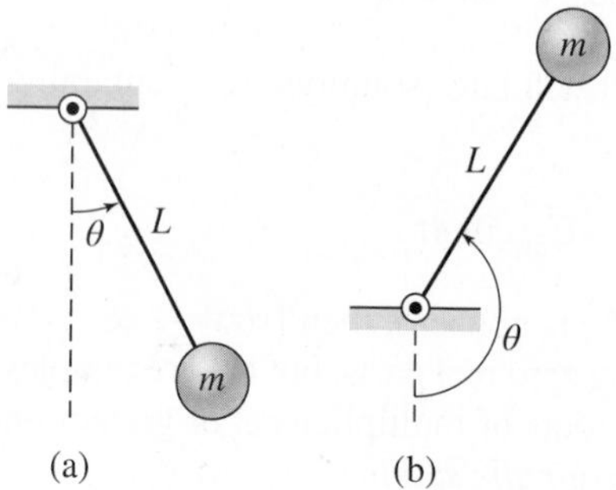

2.2.10 THE ROUTH-HURWITZ CONDITION

The characteristic equation of many systems has the form $ms^2 + cs + k = 0$. A simple criterion exists for quickly determining the stability of such a system. This is proved in a chapter problem. The condition states that the second-order system whose characteristic polynomial is $ms^2 + cs + k$ is stable if and only if m, c, and k have the same sign. This requirement is called the Routh-Hurwitz condition.

2.2.11 STABILITY AND EQUILIBRIUM

An *equilibrium* is a state of no change. The pendulum in Figure 2.2.5 is in equilibrium at $\theta = 0$ and when perfectly balanced at $\theta = \pi$. The equilibrium at $\theta = 0$ is stable, while the equilibrium at $\theta = \pi$ is unstable. From this we see that the same physical system can have different stability characteristics at different equilibria. So we see that stability is not a property of the system alone, but is a property of a specific equilibrium of the system. When we speak of the stability properties of a model, we are actually speaking of the stability properties of the specific equilibrium on which the model is based.

Figure 2.2.6 shows a ball on a surface that has a valley and a hill. The bottom of the valley is an equilibrium, and if the ball is displaced slightly from this position, it will oscillate forever about the bottom if there is no friction. In this case, the equilibrium is neutrally stable. The ball, however, will return to the bottom if friction is present, and the equilibrium is stable in this case.

If we displace the ball so much to the left that it lies outside the valley, it will never return. Thus, if friction is present, we say that the valley equilibrium is *locally stable* but *globally unstable*. An equilibrium is globally stable only if the system returns to it for *any* initial displacement.

The equilibrium on the hilltop is globally unstable because, if displaced, the ball will continue to roll down the hill.

For linear models, stability analysis using the characteristic roots gives global stability information. However, for nonlinear models, linearization about an equilibrium gives us only local stability information.

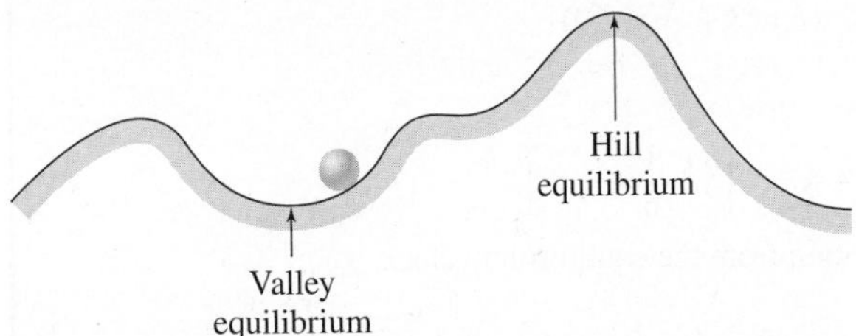

Figure 2.2.6 Surface illustration of stability properties.

2.2.12 LINEARIZATION

Linear dynamic models have several useful properties not found with nonlinear models, and thus we should try to model a system with a linear model whenever possible. Obtaining a closed-form solution for a nonlinear differential equation is usually difficult, and often impossible. In the next two examples we discuss the process of obtaining a model that is "approximately" linear; that is, a linear equation that approximately describes the system's behavior under certain conditions. This process is called *linearization*.

Engineering systems are designed to operate under certain specified conditions. For example, an aircraft is designed to cruise at a specified altitude and specified speed.

The specified altitude and speed are referred to as the aircraft's *reference operating condition*. Specifying this condition allows the designers to know what is required of their designs. This condition is sometimes called the *nominal* operating condition. Often the reference operating condition is an equilibrium condition, but not necessarily. Knowing the system's reference operating condition, we can replace the model's nonlinear functions with linear ones that are approximately correct near the reference operating condition. A systematic procedure for doing this is based on the Taylor series expansion. This was illustrated in Chapter 1 with Examples 1.3.2, 1.3.3, and 1.3.4.

We do not always need to solve the linearized model for the output as a function of time. Often we simply need to analyze the stability properties near equilibrium, or to estimate the time constant or oscillation frequency near equilibrium. The next two examples show how to do this.

EXAMPLE 2.2.1

Linearizing a First-Order Model

■ Problem

In Section 2.1 we saw that a model of liquid height h in a tank is obtained from Torricelli's law. Suppose that the model for a specific tank is

$$\frac{dh}{dt} = q - 2\sqrt{h} \tag{1}$$

where q is a volume flow rate into the tank. Assume that height is measured in meters and time in seconds.

If the inflow is held steady at $q = 6$, the height reaches and stays at the constant equilibrium value $h_e = 9$. Obtain a linearized model of the height to be used to estimate the time constant when the inflow q changes slightly from $q = 6$.

■ Solution

In Example 1.3.3 we found a linearized approximation to the square-root function for h near $h = 9$:

$$\sqrt{h} = 3 + \frac{1}{6}(h - 9)$$

Substitute this into (1):

$$\dot{h} = q - 2\left[3 + \frac{1}{6}(h - 9)\right] \tag{2}$$

We define x and u to represent deviation from the equilibrium values.

$$x = h - 9 \tag{3}$$

$$u = q - 6 \tag{4}$$

Note that $\dot{h} = \dot{x}$. Substituting this and (3) and (4) into (2) gives

$$\dot{x} = u - \frac{1}{6}x$$

The characteristic root is $s = -1/6$ and thus the time constant is $\tau = 6$ s. This shows that the equilibrium at $h = 9$ is stable, and that if the inflow q changes slightly from $q = 6$, the height will change slightly from $h = 9$ and it will take an amount of time equal to 4(6) = 24 s for the height to reach its new value.

Suppose that $h = 9$ and the inflow changes from $q = 6$ to $q = 6.5$. We can find the new steady-state height from (1) to be $h_e = 10.56$. It will take about 24 s to reach this new height.

EXAMPLE 2.2.2

Linearizing a Second-Order Model

■ Problem

As we will see in Chapter 3, the equation of motion of a pendulum is

$$I\ddot{\theta} = T - mgL \sin \theta$$

Let us simplify the notation by dividing by I and letting $M = T/I$ and $a = mgL/I$. Then the equation becomes

$$\ddot{\theta} = M - a \sin \theta \tag{1}$$

Obtain a linear model valid for small changes in θ near (a) $\theta = 0$, (b) $\theta = \pi/3$, and (c) $\theta = 2\pi/3$.

■ Solution

The value of M required to keep θ at an equilibrium θ_e is found by setting $\ddot{\theta} = 0$ to obtain

$$M_e = a \sin \theta_e \tag{2}$$

We define x and u to represent deviation from the equilibrium values.

$$x = \theta - \theta_e \tag{3}$$

$$u = M - M_e \tag{4}$$

In Example 1.3.2 we found that the linearized approximation to the sine function is

$$\sin \theta \approx \sin \theta_e + \cos \theta_e(\theta - \theta_e) \tag{5}$$

If we substitute (3), (4), and (5) into (1), we obtain

$$\ddot{x} = u + M_e - a\left[\sin \theta_e + \cos \theta_e(\theta - \theta_e)\right]$$

Two terms drop out by using (2):

$$\ddot{x} = u - a\cos \theta_e(\theta - \theta_e) = u - (a\cos \theta_e)x \tag{6}$$

Let $b = a\cos \theta_e$. Then (6) simplifies to

$$\ddot{x} + bx = u \tag{7}$$

which is a very simple equation!

a. For $\theta_e = 0$, $b = a\cos 0 = a$, and (7) becomes

$$\ddot{x} + ax = u$$

where it is important to note that $a > 0$. So the characteristic roots are purely imaginary: $s = \pm j\sqrt{a}$. If $M = M_e$, then $u = 0$ and the response is the free response:

$$x(t) = C_1 \sin \sqrt{a}t + C_2 \cos \sqrt{a}t$$

So $\theta(t)$ will oscillate about the equilibrium $\theta = 0$ with a frequency $\omega_n = \sqrt{a}$. The equilibrium is neutrally stable.

b. For $\theta_e = \pi/3$, $b = a\cos \pi/3 = 0.5a$, and (7) becomes

$$\ddot{x} + 0.5ax = u$$

The characteristic roots are purely imaginary: $s = \pm j\sqrt{0.5a}$. So the equilibrium $\theta = \pi/3$ is also neutrally stable, but $\theta(t)$ will oscillate about $\pi/3$ but with a frequency $\omega_n = \sqrt{0.5a}$.

c. For $\theta_e = 2\pi/3$, $b = a\cos 2\pi/3 = -0.5a$, and (7) becomes

$$\ddot{x} - 0.5ax = u$$

Now the characteristic roots are distinct and real: $s = \sqrt{0.5a}$ and $s = -\sqrt{0.5a}$. One root is negative and the other is positive. The positive root means that this equilibrium is unstable. If $M = M_e$, then $u = 0$ and the response is the free response:

$$x(t) = C_1 e^{\sqrt{0.5a}t} + C_2 e^{-\sqrt{0.5a}t}$$

So $x(t) \to \infty$, which in practical terms means that θ does not return to the value $\theta = 2\pi/3$.

2.3 THE LAPLACE TRANSFORM METHOD

The Laplace transform provides a systematic and general method for solving linear ODEs, and is especially useful either for nonhomogeneous equations whose right-hand side is a function of time or for sets of equations. Another advantage is that the transform converts linear differential equations into algebraic relations that can be handled easily.

Although named after Pierre-Simon Laplace, the transform actually is based on the work of Léonard Euler, from 1763, for solving second-order linear ODEs. The Laplace transform $\mathcal{L}[x(t)]$ of a function $x(t)$ is defined as follows:

$$\mathcal{L}[x(t)] = \lim_{T\to\infty}\left[\int_{0-}^{T} x(t)e^{-st}\,dt\right] \tag{2.3.1}$$

but is usually expressed more compactly as

$$\mathcal{L}[x(t)] = \int_{0-}^{\infty} x(t)e^{-st}\,dt \tag{2.3.2}$$

The variable of integration, t, is arbitrary, and the transform is a function of the parameter s, which is a complex number.

The lower limit of 0− is used so that we may use the transform to solve differential equations where

1. the forcing function is an impulse, or
2. the differential equation contains derivatives of a forcing function that is discontinuous at $t = 0$.

We will see examples of these cases in Sections 2.4 and 2.6. See also [Lundberg, 2007].

Unless it is explicitly stated otherwise, we will assume that $x(t) = 0$ for $t < 0$. In this case,

$$\int_{0-}^{0} x(t)e^{-st}dt = 0$$

and the Laplace integral becomes

$$\mathcal{L}[x(t)] = \int_{0}^{\infty} x(t)e^{-st}dt$$

A simpler notation for the transform uses the uppercase symbol to represent the transform of the corresponding lowercase symbol; that is,

$$X(s) = \mathcal{L}[x(t)] \tag{2.3.3}$$

The process of determining the time function $x(t)$ whose transform is $X(s)$ is denoted by

$$x(t) = \mathcal{L}^{-1}[X(s)]$$

where the symbol $\mathcal{L}^{-1}$ denotes the *inverse* transform.

When the limit in (2.3.1) is not finite, there is no Laplace transform defined for $x(t)$. For many common functions, we can choose s to obtain a finite limit. In practice, however, when solving ODEs, we need not be concerned with the choice for s, because the transforms of common functions that have transforms have been derived and tabulated. Table 2.3.1 is a table of commonly needed transforms.

Table 2.3.1 Table of Laplace transform pairs.

	$X(s)$	$x(t),\ t \geq 0$
1.	1	$\delta(t)$, unit impulse
2.	$\dfrac{1}{s}$	$u_s(t)$, unit step
3.	$\dfrac{c}{s}$	constant, c
4.	$\dfrac{e^{-sD}}{s}$	$u_s(t-D)$, shifted unit step
5.	$\dfrac{n!}{s^{n+1}}$	t^n
6.	$\dfrac{1}{s+a}$	e^{-at}
7.	$\dfrac{1}{(s+a)^n}$	$\dfrac{1}{(n-1)!}t^{n-1}e^{-at}$
8.	$\dfrac{b}{s^2+b^2}$	$\sin bt$
9.	$\dfrac{s}{s^2+b^2}$	$\cos bt$
10.	$\dfrac{b}{(s+a)^2+b^2}$	$e^{-at}\sin bt$
11.	$\dfrac{s+a}{(s+a)^2+b^2}$	$e^{-at}\cos bt$
12.	$\dfrac{a}{s(s+a)}$	$1-e^{-at}$
13.	$\dfrac{1}{(s+a)(s+b)}$	$\dfrac{1}{b-a}(e^{-at}-e^{-bt})$
14.	$\dfrac{s+p}{(s+a)(s+b)}$	$\dfrac{1}{b-a}[(p-a)e^{-at}-(p-b)e^{-bt}]$
15.	$\dfrac{1}{(s+a)(s+b)(s+c)}$	$\dfrac{e^{-at}}{(b-a)(c-a)}+\dfrac{e^{-bt}}{(c-b)(a-b)}+\dfrac{e^{-ct}}{(a-c)(b-c)}$
16.	$\dfrac{s+p}{(s+a)(s+b)(s+c)}$	$\dfrac{(p-a)e^{-at}}{(b-a)(c-a)}+\dfrac{(p-b)e^{-bt}}{(c-b)(a-b)}+\dfrac{(p-c)e^{-ct}}{(a-c)(b-c)}$
17.	$\dfrac{b}{s^2-b^2}$	$\sinh bt$

Table 2.3.1 (*Continued*)

$X(s)$	$x(t), t \geq 0$
18. $\dfrac{s}{s^2 - b^2}$	$\cosh bt$
19. $\dfrac{a^2}{s^2(s+a)}$	$at - 1 + e^{-at}$
20. $\dfrac{a^2}{s(s+a)^2}$	$1 - (at+1)e^{-at}$
21. $\dfrac{\omega_n^2}{s^2 + 2\zeta\omega_n s + \omega_n^2}$	$\dfrac{\omega_n}{\sqrt{1-\zeta^2}} e^{-\zeta\omega_n t} \sin \omega_n \sqrt{1-\zeta^2}\, t$
22. $\dfrac{s}{s^2 + 2\zeta\omega_n s + \omega_n^2}$	$-\dfrac{1}{\sqrt{1-\zeta^2}} e^{-\zeta\omega_n t} \sin \left(\omega_n \sqrt{1-\zeta^2}\, t - \phi\right), \ \phi = \tan^{-1} \dfrac{\sqrt{1-\zeta^2}}{\zeta}$
23. $\dfrac{\omega_n^2}{s\left(s^2 + 2\zeta\omega_n s + \omega_n^2\right)}$	$1 - \dfrac{1}{\sqrt{1-\zeta^2}} e^{-\zeta\omega_n t} \sin \left(\omega_n \sqrt{1-\zeta^2}\, t + \phi\right), \ \phi = \tan^{-1} \dfrac{\sqrt{1-\zeta^2}}{\zeta}$
24. $\dfrac{1}{s[(s+a)^2 + b^2]}$	$\dfrac{1}{a^2+b^2}\left[1 - \left(\dfrac{a}{b} \sin bt + \cos bt\right) e^{-at}\right]$
25. $\dfrac{b^2}{s(s^2+b^2)}$	$1 - \cos bt$
26. $\dfrac{b^3}{s^2(s^2+b^2)}$	$bt - \sin bt$
27. $\dfrac{2b^3}{(s^2+b^2)^2}$	$\sin bt - bt \cos bt$
28. $\dfrac{2bs}{(s^2+b^2)^2}$	$t \sin bt$
29. $\dfrac{s^2-b^2}{(s^2+b^2)^2}$	$t \cos bt$
30. $\dfrac{s}{\left(s^2+b_1^2\right)\left(s^2+b_2^2\right)}$	$\dfrac{1}{b_2^2 - b_1^2}\left(\cos b_1 t - \cos b_2 t\right), \quad \left(b_1^2 \neq b_2^2\right)$
31. $\dfrac{s^2}{(s^2+b^2)^2}$	$\dfrac{1}{2b}(\sin bt + bt \cos bt)$

For some relatively simple functions either the Laplace transform does not exist (such as for e^{t^2} and $1/t$), or it cannot be represented as an algebraic expression (such as for $1/(t+a)$). In the latter case, the integral must be represented as an infinite series, which is not very useful for our purposes.

The Laplace transform of a function $x(t)$ exists for all $s > \gamma$ if $x(t)$ is piecewise continuous on every finite interval in the range $t \geq 0-$ and satisfies the relation $|x(t)| \leq Me^{\gamma t}$ for all $t \geq 0-$ and for some constants γ and M [Kreyzig, 2011]. These conditions are sufficient but not necessary. For example, the function $1/\sqrt{t}$ is infinite at $t = 0$ but it has a transform, which is $\sqrt{\pi/s}$.

The *inverse* Laplace transform $\mathcal{L}^{-1}[X(s)]$ is that time function $x(t)$ whose transform is $X(s)$; that is, $x(t) = \mathcal{L}^{-1}[X(s)]$. It can be shown that if two continuous functions have the same transform, then the functions are identical. This is of practical significance because we will need to find the function $x(t)$, given $X(s)$.

2.3.1 TRANSFORMS OF COMMON FUNCTIONS

We now derive a few transforms to develop an understanding of the method.

EXAMPLE 2.3.1

Transform of a Constant

■ Problem

Suppose $x(t) = c$, a constant, for $t \geq 0$. Determine its Laplace transform.

■ Solution

It is implied that $c = 0$ for $t < 0$. From the transform definition, we have

$$\mathcal{L}[x(t)] = \lim_{T\to\infty}\left(\int_{0-}^{0} 0\,dt + \int_{0}^{T} ce^{-st}\,dt\right) = c\lim_{T\to\infty}\left(\int_{0}^{T} e^{-st}\,dt\right)$$

or

$$\mathcal{L}(c) = c\lim_{T\to\infty}\left(\frac{1}{-s}e^{-st}\Big|_{0}^{T}\right) = c\lim_{T\to\infty}\left(\frac{1}{-s}e^{-sT} + \frac{1}{s}e^{0}\right) = \frac{c}{s}$$

where we have assumed that the real part of s is greater than zero, so that the limit of e^{-sT} exists as $T \to \infty$. Thus, the transform of a constant c is c/s.

Suppose the function $x(t)$ is zero for $t < 0$ and a nonzero constant, say M, for $t > 0$. Then we can express it as $x(t) = Mu_s(t)$. The value of M is the *magnitude* of the step function; if $M = 1$, the function is the unit step function. From the results of Example 2.3.1, we can easily see that the Laplace transform of $x(t) = Mu_s(t)$ is M/s.

EXAMPLE 2.3.2

The Exponential Function

■ Problem

Derive the Laplace transform of the exponential function $x(t) = e^{-at}$, $t \geq 0$, where a is a constant.

■ Solution

It is implied that $x(t) = 0$ for $t < 0$. From the transform definition, we have

$$\mathcal{L}(e^{-at}) = \lim_{T\to\infty}\left(\int_{0-}^{0} 0\,dt + \int_{0}^{T} e^{-at}e^{-st}\,dt\right) = \lim_{T\to\infty}\left(\int_{0}^{T} e^{-(s+a)t}\,dt\right)$$

or

$$\mathcal{L}(e^{-at}) = \lim_{T\to\infty}\left[\frac{1}{-(s+a)}e^{-(s+a)t}\Big|_{0}^{T}\right] = \frac{1}{s+a}$$

2.3.2 THE LINEARITY PROPERTY

The Laplace transform is a definite integral, and thus it has the properties of such integrals (Table 2.3.2). For example, a multiplicative constant can be factored out of the integral, and the integral of a sum equals the sum of the integrals. These facts lead to the *linearity property* of the transform; namely, for the functions $f(t)$ and $g(t)$, if a and b are constants, then

$$\mathcal{L}[af(t) + bg(t)] = a\mathcal{L}[f(t)] + b\mathcal{L}[g(t)] = aF(s) + bG(s)$$

One use of the linearity property is to determine transforms of functions that are linear combinations of functions whose transforms are already known. For example,

Table 2.3.2 Properties of the Laplace transform.

$x(t)$	$X(s) = \int_0^\infty f(t)e^{-st}\,dt$
1. $af(t) + bg(t)$	$aF(s) + bG(s)$
2. $\dfrac{dx}{dt}$	$sX(s) - x(0)$
3. $\dfrac{d^2x}{dt^2}$	$s^2X(s) - sx(0) - \dot{x}(0)$
4. $\dfrac{d^nx}{dt^n}$	$s^nX(s) - \sum_{k=1}^{n} s^{n-k}g_{k-1}$ $g_{k-1} = \left.\dfrac{d^{k-1}x}{dt^{k-1}}\right\|_{t=0}$
5. $\int_0^t x(t)\,dt$	$\dfrac{X(s)}{s} + \dfrac{g(0)}{s}$ $g(0) = \left.\int x(t)\,dt\right\|_{t=0}$
6. $x(t) = \begin{cases} 0 & t < D \\ g(t-D) & t \geq D \end{cases}$ $= u_s(t-D)g(t-D)$	$X(s) = e^{-sD}G(s)$
7. $e^{-at}x(t)$	$X(s+a)$
8. $tx(t)$	$-\dfrac{dX(s)}{ds}$
9. $x(\infty) = \lim_{s\to 0} sX(s)$	
10. $x(0+) = \lim_{s\to\infty} sX(s)$	

For Entries 2, 3, 4, and 5, if $x \neq 0$ for $t < 0$, then replace the initial conditions at $t = 0$ with the pre-initial condtions at $0-$.

if $x(t) = 6 + 4e^{-3t}$, its transform is

$$X(s) = \frac{6}{s} + \frac{4}{s+3} = \frac{10s + 18}{s(s+3)}$$

The inverse transform also has the linearity property, so that

$$\mathcal{L}^{-1}[aF(s) + bG(s)] = a\mathcal{L}^{-1}[F(s)] + b\mathcal{L}^{-1}[G(s)] = af(t) + bg(t)$$

We can often avoid the integration operations by using the linearity property and suitable identities, as shown in Example 2.3.3.

EXAMPLE 2.3.3

The Sine and Cosine Functions

■ Problem

Derive the Laplace transforms of the exponentially decaying sine and cosine functions, $e^{-at}\sin\omega t$ and $e^{-at}\cos\omega t$, for $t \geq 0$, where a and ω are constants.

■ Solution

Note that from the Euler identity, $e^{j\theta} = \cos\theta + j\sin\theta$, with $\theta = \omega t$, we have

$$e^{-at}(\cos\omega t + j\sin\omega t) = e^{-at}e^{j\omega t} = e^{-(a-j\omega)t} \tag{1}$$

Thus, the real part of $e^{-(a-j\omega)t}$ is $e^{-at}\cos\omega t$ and the imaginary part is $e^{-at}\sin\omega t$. However, from the result of Example 2.3.2, with a replaced by $a - j\omega$, we have

$$\mathcal{L}\left[e^{-(a-j\omega)t}\right] = \frac{1}{s+a-j\omega} \tag{2}$$

In this form, we cannot identify the real and imaginary parts. To do so, we multiply the numerator and denominator by the complex conjugate of the denominator and use the fact that $(x - jy)(x + jy) = x^2 + y^2$ (see Table 2.1.3); that is,

$$\frac{1}{x-jy} = \frac{x+jy}{(x-jy)(x+jy)} = \frac{x+jy}{x^2+y^2}$$

Thus, with $x = s + a$ and $y = \omega$, equation (2) becomes

$$\mathcal{L}\left[e^{-(a-j\omega)t}\right] = \frac{1}{s+a-j\omega} = \frac{s+a+j\omega}{(s+a-j\omega)(s+a+j\omega)}$$

$$= \frac{s+a+j\omega}{(s+a)^2+\omega^2} = \frac{s+a}{(s+a)^2+\omega^2} + j\frac{\omega}{(s+a)^2+\omega^2}$$

From equation (1), we see that the real part of this expression is the transform of $e^{-at}\cos\omega t$ and the imaginary part is the transform of $e^{-at}\sin\omega t$. Therefore,

$$\mathcal{L}\left(e^{-at}\cos\omega t\right) = \frac{s+a}{(s+a)^2+\omega^2}$$

and

$$\mathcal{L}\left(e^{-at}\sin\omega t\right) = \frac{\omega}{(s+a)^2+\omega^2}$$

Note that the transforms of the sine and cosine can be obtained by letting $a = 0$. Thus,

$$\mathcal{L}(\cos\omega t) = \frac{s}{s^2+\omega^2} \quad \text{and} \quad \mathcal{L}(\sin\omega t) = \frac{\omega}{s^2+\omega^2}$$

Another property of the Laplace transform is called *shifting along the s-axis* or *multiplication by an exponential*. This property states that

$$\mathcal{L}\left[e^{-at}x(t)\right] = X(s+a) \tag{2.3.4}$$

To derive this property, note that

$$\mathcal{L}\left[e^{-at}x(t)\right] = \int_{0-}^{\infty} e^{-at}x(t)e^{-st}\,dt = \int_{0-}^{\infty} x(t)e^{-(s+a)t}\,dt$$

$$= X(s+a)$$

The Function te^{-at} — EXAMPLE 2.3.4

■ Problem

Derive the Laplace transform of the function te^{-at}, $t \geq 0$.

■ Solution

Here the function $x(t)$ is t, $X(s) = 1/s^2$, and thus from (2.3.4),

$$\mathcal{L}\left[e^{-at}x(t)\right] = \mathcal{L}\left(te^{-at}\right) = \left.\frac{1}{s^2}\right|_{s\to s+a} = \frac{1}{(s+a)^2}$$

Another property is *multiplication by t*. It states that

$$\mathcal{L}[tx(t)] = -\frac{dX(s)}{ds} \tag{2.3.5}$$

To derive this property, note that

$$\frac{d}{ds}X(s) = \frac{d}{ds}\left[\int_{0-}^{\infty} x(t)e^{-st}\,dt\right] = -\int_{0-}^{\infty} tx(t)e^{-st}\,dt = -\mathcal{L}[tx(t)]$$

EXAMPLE 2.3.5

The Function $t \cos \omega t$

■ Problem

Derive the Laplace transform of the function $t \cos \omega t$, $t \geq 0$.

■ Solution

Here the function $x(t)$ is $\cos \omega t$, $X(s) = s/(s^2 + \omega^2)$, and thus from (2.3.5),

$$\mathcal{L}[tx(t)] = \mathcal{L}(t \cos \omega t) = -\frac{d}{ds}\left(\frac{s}{s^2 + \omega^2}\right) = \frac{s^2 - \omega^2}{(s^2 + \omega^2)^2}$$

2.3.3 THE DERIVATIVE PROPERTY

To use the Laplace transform to solve differential equations, we will need to obtain the transforms of derivatives. Applying integration by parts to the definition of the transform, we obtain (assuming that $x(t) = 0$ for $t < 0$)

$$\begin{aligned}\mathcal{L}\left(\frac{dx}{dt}\right) &= \int_0^{\infty} \frac{dx}{dt}e^{-st}\,dt = x(t)e^{-st}\Big|_0^{\infty} + s\int_0^{\infty} x(t)e^{-st}\,dt \\ &= s\mathcal{L}[x(t)] - x(0) = sX(s) - x(0)\end{aligned} \tag{2.3.6}$$

This procedure can be extended to higher derivatives. For example, the result for the second derivative is

$$\mathcal{L}\left(\frac{d^2x}{dt^2}\right) = s^2X(s) - sx(0) - \dot{x}(0) \tag{2.3.7}$$

The general result for any order derivative is given in Table 2.3.2. If $x \neq 0$ for $t < 0$, then replace the initial conditions at $t = 0$ with the pre-initial conditions at $0-$.

2.3.4 THE INITIAL-VALUE THEOREM

Sometimes we will need to find the value of the function $x(t)$ at $t = 0+$ (a time infinitesimally greater than 0), given the transform $X(s)$. The answer can be obtained with the *initial-value theorem*, which states that

$$x(0+) = \lim_{t \to 0+} x(t) = \lim_{s \to \infty} [sX(s)] \tag{2.3.8}$$

where the limit $s \to \infty$ is taken along the real axis. The conditions for which the theorem is valid are that the latter limit exists and that the transforms of $x(t)$ and dx/dt exist. If $X(s)$ is a rational function and if the degree of the numerator of $X(s)$ is less than the degree of the denominator, then the theorem will give a finite value for $x(0+)$. If the degrees are equal, then the initial value is undefined and the initial-value theorem is invalid (see Section 2.6 for a discussion of this case). The proof of the theorem is obtained in Problem 2.31.

For the transform

$$X(s) = \frac{7s+2}{s(s+6)}$$

the theorem gives

$$x(0+) = \lim_{s\to\infty} \frac{7s+2}{s+6} = 7$$

This is confirmed by evaluating the inverse transform, $x(t) = 1/3 + (20/3)e^{-6t}$. (We state the inverse transform here for illustrative purposes only; normally we apply the theorem in situations where the inverse transform is not convenient to obtain.)

We will see important applications of this theorem in Section 2.6.

2.3.5 THE FINAL-VALUE THEOREM

To find the limit of the function $x(t)$ as $t \to \infty$, we can use the *final-value theorem*. The theorem states that

$$x(\infty) = \lim_{t\to\infty} x(t) = \lim_{s\to 0} sX(s) \tag{2.3.9}$$

The theorem is true if the following conditions are satisfied. The functions $x(t)$ and dx/dt must possess Laplace transforms, and $x(t)$ must approach a constant value as $t \to \infty$. The latter condition will be satisfied if all the roots of the denominator of $sX(s)$ have *negative* real parts. The proof of the theorem is obtained in Problem 2.32.

For example, if $X(s) = 1/s$, which corresponds to $x(t) = 1$, then

$$\lim_{s\to 0} sX(s) = \lim_{s\to 0} s\frac{1}{s} = 1$$

which is correct. (We state the inverse transform here for illustrative purposes only; normally we apply the theorem in situations where the inverse transform is not convenient to obtain.)

As another example, if

$$X(s) = \frac{7}{(s+4)^2 + 49}$$

then

$$\lim_{s\to 0} sX(s) = \lim_{s\to 0} \frac{7s}{(s+4)^2 + 49} = 0$$

which is correct [$X(s)$ has the inverse transform $x(t) = e^{-4t} \sin 7t$].

The function $x(t)$ will approach a constant value if all the roots of the denominator of $sX(s)$ have negative real parts. Thus, a common situation in which the theorem does not apply is a periodic function. For example, if $x(t) = \sin 5t$, then $X(s) = 5/(s^2+25)$, and

$$\lim_{s\to 0} sX(s) = \lim_{s\to 0} \frac{5s}{s^2+25} = 0$$

Thus, the limit exists but the result is incorrect, because $x(t)$ continually oscillates and therefore does not approach a constant value.

The theorem is not applicable to the transform

$$X(s) = \frac{9s+2}{s(s-8)}$$

because, after the s terms in $sX(s)$ are canceled, the denominator of $sX(s)$ is $s-8$, which has the positive root $s = 8$. Therefore, $x(t)$ does not approach a constant value as $t \to \infty$ [this can be observed from the inverse transform, which is $x(t) = -1/4 + (37/4)e^{8t}$].

2.4 SOLVING EQUATIONS WITH THE LAPLACE TRANSFORM

We now show how to solve differential equations by using the Laplace transform. Consider the linear first-order equation

$$\dot{x} + ax = f(t) \tag{2.4.1}$$

where $f(t)$ is the input and a is a constant. If we multiply both sides of the equation by e^{-st} and then integrate over time from $t = 0-$ to $t = \infty$, we obtain

$$\int_{0-}^{\infty} (\dot{x} + ax)\, e^{-st}\, dt = \int_{0-}^{\infty} f(t)e^{-st}\, dt$$

or

$$\mathcal{L}(\dot{x} + ax) = \mathcal{L}[f(t)]$$

Using the linearity property, this becomes

$$\mathcal{L}(\dot{x}) + a\mathcal{L}(x) = \mathcal{L}[f(t)]$$

If we are given that $x = 0$ for $t < 0$, then, using the derivative property and the alternative transform notation, the above equation can be written as

$$sX(s) - x(0) + aX(s) = F(s) \tag{2.4.2}$$

This equation is an algebraic equation that can be solved for $X(s)$ in terms of $F(s)$ and $x(0)$. Its solution is

$$X(s) = \frac{x(0)}{s+a} + \frac{1}{s+a}F(s) \tag{2.4.3}$$

The denominator of the first term on the right side is $s + a$, which is the characteristic polynomial of the equation.

The inverse operation gives

$$x(t) = \mathcal{L}^{-1}\left[\frac{x(0)}{s+a}\right] + \mathcal{L}^{-1}\left[\frac{1}{s+a}F(s)\right] \tag{2.4.4}$$

This equation shows that the solution consists of the sum of two terms. The first term depends on the initial condition $x(0)$ but not on the forcing function $f(t)$. This part of the solution is called the **free response.** From Table 2.3.1, entry 6, it is seen that the free response is $x(0)e^{-at}$.

The second term on the right side of (2.3.13) depends on the forcing function $f(t)$ but not on the initial condition $x(0)$. This part of the solution is called the **forced response.** It cannot be evaluated until $F(s)$ is available.

EXAMPLE 2.4.1 Step Response of a First-Order Equation

■ **Problem**

When the input is a step function, the response is sometimes called the *step response*. Suppose that the input $f(t)$ of the equation $\dot{x}+ax=f(t)$ is a step function of magnitude b whose transform is $F(s) = b/s$. Obtain the expression for the complete response.

■ **Solution**

From (2.4.4) the forced response is obtained from

$$\mathcal{L}^{-1}\left[\frac{1}{s+a}F(s)\right] = \mathcal{L}^{-1}\left(\frac{1}{s+a}\frac{b}{s}\right)$$

This transform can be converted into a sum of simple transforms as follows:

$$\frac{1}{s+a}\frac{b}{s} = \frac{C_1}{s} + \frac{C_2}{s+a} \tag{1}$$

To determine C_1 and C_2, we can use the *least common denominator* (*LCD*) $s(s+a)$ and write the expression as

$$\frac{1}{s+a}\frac{b}{s} = \frac{C_1(s+a)+C_2 s}{s(s+a)}$$

Comparing the numerators on the left and right sides, we see that this equation is true only if

$$b = C_1(s+a) + C_2 s = (C_1 + C_2)s + aC_1$$

for arbitrary values of s. This requires that $C_1 + C_2 = 0$ and $aC_1 = b$. Thus, $C_1 = b/a$ and $C_2 = -b/a$, and equation (1) becomes

$$\frac{1}{s+a}\frac{b}{s} = \frac{b}{a}\left(\frac{1}{s} - \frac{1}{s+a}\right)$$

Thus, using entries 2 and 6 in Table 2.3.1, we see that the forced response is

$$\frac{b}{a}\left(1 - e^{-at}\right)$$

The addition of the free and the forced responses gives the complete response:

$$x(t) = x(0)e^{-at} + \frac{b}{a}\left(1 - e^{-at}\right)$$

To solve a differential equation by using the Laplace transform, we must be able to obtain a function $x(t)$ from its transform $X(s)$. This process is called *inverting* the transform. In Example 2.4.1 we inverted the transform by expressing it as a sum of simple transforms that appear in our transform table. The algebra required to find the coefficients C_1 and C_2 is rather straightforward. This sum is called a *partial-fraction expansion*. The form of a partial-fraction expansion depends on the roots of the transform's denominator. These roots consist of the characteristic roots plus any roots introduced by the transform of the forcing function. When there are only a few roots, using the LCD method quickly produces a solution for the expansion's coefficients. The coefficients can also be determined by the general-purpose formulas, which are advantageous for some problems. These are discussed in Section 2.7.

A case requiring special attention occurs when there are repeated factors in the denominator of the transform. Example 2.4.2 shows how this case is handled.

Ramp Response of a First-Order Equation

EXAMPLE 2.4.2

■ **Problem**

Determine the complete response of the following model, which has a ramp input:

$$\dot{x} + 3x = 5t \qquad x(0) = 10$$

■ **Solution**

Applying the transform to the equation we obtain

$$sX(s) - x(0) + 3X(s) = \frac{5}{s^2}$$

Solve for $X(s)$.

$$X(s) = \frac{x(0)}{s+3} + \frac{5}{s^2(s+3)} = \frac{10}{s+3} + \frac{5}{s^2(s+3)}$$

[We could also have obtained this result directly by using (2.4.3) with $a = 3$ and $F(s) = 5/s^2$.] The free response is given by the first term on the right-hand side and is $10e^{-3t}$. To find the forced response, we express the second term on the right as

$$\frac{5}{s^2(s+3)} = \frac{C_1}{s^2} + \frac{C_2}{s} + \frac{C_3}{s+3}$$

As a general rule, the partial-fraction expansion must include one term for each distinct factor of the denominator; here, s and $s+3$. However, when there are repeated factors (here, the extra factor s), we must include additional terms, one for each extra factor that is repeated. We may now use the LCD method to obtain the coefficients C_1, C_2, and C_3.

$$\frac{5}{s^2(s+3)} = \frac{C_1(s+3) + C_2 s(s+3) + C_3 s^2}{s^2(s+3)} = \frac{(C_2 + C_3)s^2 + (C_1 + 3C_2)s + 3C_1}{s^2(s+3)}$$

Comparing the numerators we see that $C_2 + C_3 = 0$, $C_1 + 3C_2 = 0$, and $3C_1 = 5$. Thus, $C_1 = 5/3$, $C_2 = -C_1/3 = -5/9$, and $C_3 = -C_2 = 5/9$. The forced response is

$$C_1 t + C_2 + C_3 e^{-3t} = \frac{5}{3}t - \frac{5}{9} + \frac{5}{9}e^{-3t}$$

The complete response is the sum of the free and the forced response, and is

$$x(t) = 10e^{-3t} + \frac{5}{3}t - \frac{5}{9} + \frac{5}{9}e^{-3t}$$

The plot of the response is shown in Figure 2.4.1. Because $e^{-3t} < 0.02$ for $t > 4/3$, the response for $t > 4/3$ is approximately given by $x(t) = 5t/3 - 5/9$, which is the equation of a straight line with slope $5/3$ and intercept $-5/9$.

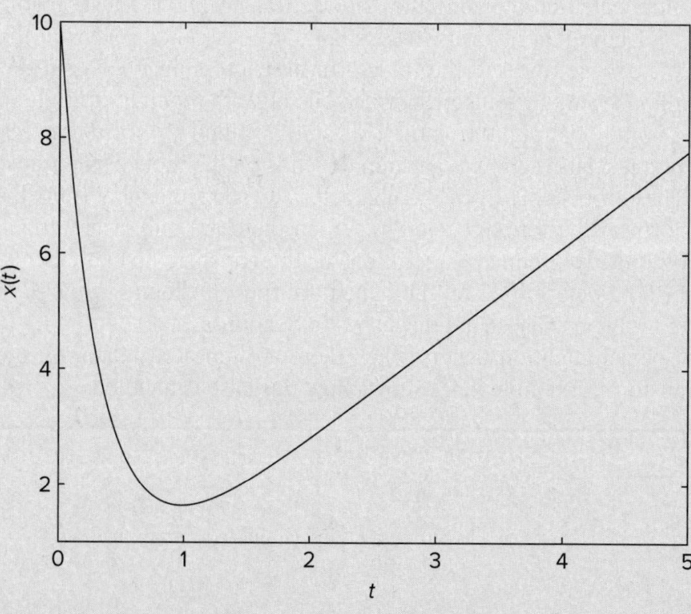

Figure 2.4.1 Response for Example 2.4.2.

Complex factors in the denominator of the transform can be often handled easily by using the fact that the complex conjugates $s = -a \pm bj$ correspond to the quadratic factor $(s + a)^2 + b^2$.

EXAMPLE 2.4.3

Transform Inversion for Complex Factors

■ Problem

Invert the following transform by representing it as the sum of terms that appear in Table 2.3.1:

$$X(s) = \frac{8s + 13}{s^2 + 4s + 53}$$

■ Solution

The roots of the denominator are $s = -2 \pm 7j$ and so the transform can be expressed as

$$X(s) = \frac{8s + 13}{(s + 2)^2 + 49}$$

We can express $X(s)$ as a sum of terms similar to entries 10 and 11 in Table 2.3.1, as follows (note that $a = 2$ and $b = 7$):

$$X(s) = \frac{8s + 13}{(s + 2)^2 + 49} = C_1 \frac{s + 2}{(s + 2)^2 + 49} + C_2 \frac{7}{(s + 2)^2 + 49} = \frac{C_1(s + 2) + 7C_2}{(s + 2)^2 + 49}$$

Comparing numerators, we see that

$$8s + 13 = C_1(s + 2) + 7C_2 = C_1 s + 2C_1 + 7C_2$$

This is true only if $C_1 = 8$ and $2C_1 + 7C_2 = 13$, or $C_2 = -3/7$. Thus,

$$x(t) = C_1 e^{-2t} \cos 7t + C_2 e^{-2t} \sin 7t = 8e^{-2t} \cos 7t - \frac{3}{7} e^{-2t} \sin 7t$$

EXAMPLE 2.4.4

Step Response of a Second-Order Equation

■ Problem

Obtain the complete response of the following model:

$$\ddot{x} + 4\dot{x} + 53x = 15u_s(t) \qquad x(0) = 8 \qquad \dot{x}(0) = -19$$

■ Solution

Transforming the equation gives

$$s^2X(s) - sx(0) - \dot{x}(0) + 4[sX(s) - x(0)] + 53X(s) = \frac{15}{s}$$

Solve for $X(s)$ using the given initial conditions.

$$X(s) = \frac{x(0)s + \dot{x}(0) + 4x(0)}{s^2 + 4s + 53} + \frac{15}{s(s^2 + 4s + 53)}$$

$$= \frac{8s + 13}{s^2 + 4s + 53} + \frac{15}{s(s^2 + 4s + 53)} \qquad (1)$$

The first term on the right of equation (1) corresponds to the free response and is the same transform inverted in Example 2.4.2.

The second term on the right of equation (1) corresponds to the forced response. We can combine the terms on the right side of equation (1) into a single term as follows:

$$X(s) = \frac{x(0)s^2 + [\dot{x}(0) + 4x(0)]s + 15}{s(s^2 + 4s + 53)} = \frac{8s^2 + 13s + 15}{s[(s + 2)^2 + 7^2]}$$

Inversion of this transform gives the complete response directly.

$$X(s) = \frac{8s^2 + 13s + 15}{s(s^2 + 4s + 53)} = \frac{C_1}{s} + C_2 \frac{s+2}{(s+2)^2 + 7^2} + C_3 \frac{7}{(s+2)^2 + 7^2}$$

$$= \frac{(C_1 + C_2)s^2 + (4C_1 + 2C_2 + 7C_3)s + 53C_1}{s[(s+2)^2 + 7^2]}$$

Comparing numerators on the left and right sides, we see that $C_1 + C_2 = 8, 4C_1 + 2C_2 + 7C_3 = 13$, and $53C_1 = 15$. Thus, $C_1 = 15/53$, $C_2 = 409/53$, and $C_3 = -27/53$, and the corresponding response is

$$x(t) = \frac{15}{53} + \frac{409}{53} e^{-2t} \cos 7t - \frac{27}{53} e^{-2t} \sin 7t$$

The response was plotted in Figure 2.1.10 in Section 2.1.

As Example 2.4.3 shows, the step response of a second-order equation with complex roots results in a transform of the following form:

$$\frac{As^2 + Bs + C}{s[(s+a)^2 + b^2]} = \frac{C_1}{s} + C_2 \frac{s+a}{(s+a)^2 + b^2} + C_3 \frac{b}{(s+a)^2 + b^2} \tag{2.4.5}$$

Using the same procedure employed in Example 2.4.4, we can show that the resulting coefficients are as follows:

$$C_1 = \frac{C}{a^2 + b^2} \qquad C_2 = A - C_1 \qquad C_3 = \frac{B - aA - aC_1}{b} \tag{2.4.6}$$

The response is

$$x(t) = C_1 + C_2 e^{-at} \cos bt + C_3 e^{-at} \sin bt \tag{2.4.7}$$

As we have seen in the previous three examples, sometimes the solution of a differential equation can be obtained directly from the table of transforms (Table 2.3.1) without the need for a partial fraction expansion. So you can save time by learning to recognize these forms.

EXAMPLE 2.4.5

Step Response with Two Distinct, Real Roots: General Case

■ Problem

Solve the following problem, which is the general case of the forced response to a unit-step input with two distinct, real roots: $s = -a, -b$.

$$\ddot{x} + (a+b)\dot{x} + abx = Mu_s(t) \qquad x(0) = \dot{x}(0) = 0$$

■ Solution

Transforming the equation gives

$$\left[s^2 + (a+b)s + ab\right]X(s) = \frac{M}{s}$$

Note that the characteristic polynomial can be factored as $s^2 + (a+b)s + ab = (s+a)(s+b)$. Thus,

$$X(s) = \frac{M}{s\left[s^2 + (a+b)s + ab\right]} = \frac{M}{s(s+a)(s+b)}$$

This has the form of entry 15 in Table 2.3.1 if we select $c = 0$. The table gives

$$x(t) = M\left[\frac{e^{-at}}{a(a-b)} + \frac{e^{-bt}}{b(b-a)} + \frac{1}{ab}\right]$$

Thus, we have derived the solution given in Table 2.1.1, entry 2.

Step Response with Two Repeated Roots: General Case

EXAMPLE 2.4.6

■ **Problem**

Solve the following problem, which is the general case of the forced response to a unit-step input with two repeated roots: $s = -a, -a$.

$$\ddot{x} + 2a\dot{x} + a^2x = Mu_s(t) \qquad x(0) = \dot{x}(0) = 0$$

■ **Solution**

Transforming the equation gives

$$(s^2 + 2as + a^2)X(s) = \frac{M}{s}$$

Note that the characteristic polynomial can be factored as $s^2 + 2as + a^2 = (s + a)^2$. Thus,

$$X(s) = \frac{M}{s(s^2 + 2as + a^2)} = \frac{M}{s(s + a)^2}$$

Except for the numerator, this has the form of entry 20 in Table 2.3.1. Because the numerator in the table is a^2, we write

$$X(s) = \frac{M}{s(s + a)^2} = \frac{M}{a^2}\frac{a^2}{s(s + a)^2}$$

The table gives

$$x(t) = \frac{M}{a^2}\left[1 - (at + 1)e^{-at}\right]$$

This solution is plotted in Figure 2.4.2 for the case where $a = 2$ and $M = 28/5$.

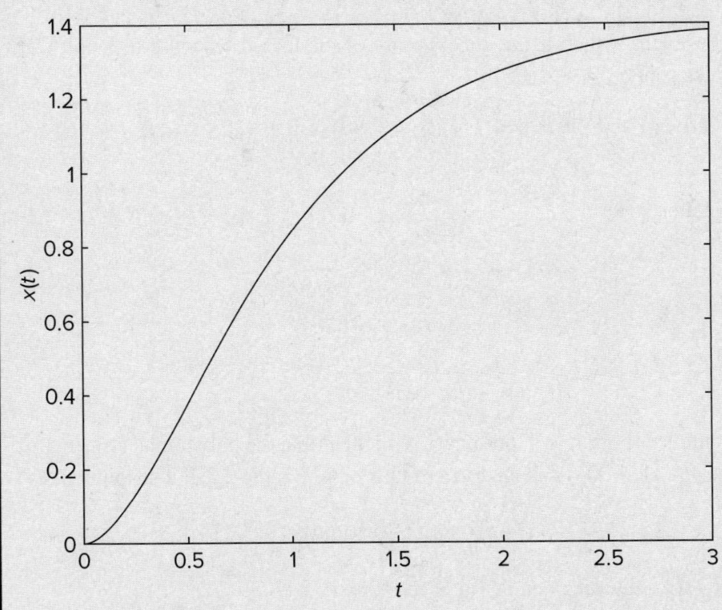

Figure 2.4.2 Response for Example 2.4.6.

EXAMPLE 2.4.7

Step Response with Two Imaginary Roots: General Case

■ Problem

Solve the following problem, which is the general case of the forced response to a unit-step input with two imaginary roots: $s = \pm bj$.

$$\ddot{x} + b^2 x = Mu_s(t) \qquad x(0) = \dot{x}(0) = 0$$

■ Solution

Transforming the equation gives

$$(s^2 + b^2)X(s) = \frac{M}{s}$$

or

$$X(s) = \frac{M}{s(s^2 + b^2)}$$

Except for the numerator, this has the form of entry 25 in Table 2.3.1. Because the numerator in the table is b^2, we write

$$X(s) = \frac{M}{s(s^2 + b^2)} = \frac{M}{b^2}\frac{b^2}{s(s^2 + b^2)}$$

The table gives

$$x(t) = \frac{M}{b^2}(1 - \cos bt)$$

Thus, we have derived the solution given in Table 2.1.1, entry 4.

EXAMPLE 2.4.8

Step Response with Two Complex Roots: General Case

■ Problem

Solve the following problem, which is the general case of the forced response to a unit-step input with two complex roots: $s = -a \pm bj$.

$$\ddot{x} + 2a\dot{x} + (a^2 + b^2)x = Mu_s(t) \qquad x(0) = \dot{x}(0) = 0$$

■ Solution

Transforming the equation gives

$$\left(s^2 + 2as + a^2 + b^2\right) X(s) = \frac{M}{s}$$

or

$$X(s) = \frac{M}{s\left(s^2 + 2as + a^2 + b^2\right)}$$

When dealing with complex roots, it is important to remember that the polynomial $(s+a)^2+b^2$ has the roots $s = -a \pm bj$. Here $X(s)$ has the form of entry 24 in Table 2.3.1. The table gives

$$x(t) = \frac{M}{a^2 + b^2}\left[1 - \left(\frac{a}{b}\sin bt + \cos bt\right) e^{-at}\right]$$

Thus, we have derived the solution given in Table 2.1.1, entry 5.

Initial Condition Effects

EXAMPLE 2.4.9

■ Problem

Solve the following equation with the given initial conditions.

$$\ddot{x} + a\dot{x} + bx = 0 \qquad x(0) = x_0 \qquad \dot{x}(0) = \dot{x}_0$$

■ Solution

Transforming the equation gives

$$[s^2X(s) - x_0 s - \dot{x}_0] + a[sX(s) - x_0] + bX(s) = 0$$

$$\left(s^2 + as + b\right) X(s) = x_0 s + \dot{x}_0 + a x_0$$

$$X(s) = \frac{x_0 s + \dot{x}_0 + a x_0}{s^2 + as + b}$$

We cannot go any further until we have numerical values for the constants a and b, because without them we do not know the type of roots (distinct real, complex, etc.). In addition, nonzero values for the initial conditions x_0 and $\dot{x}_0$ introduce a term $cs + d$ in the numerator, where c and d are nonzero constants.

Thus far in our examples, the forcing function was zero for $t < 0$. Sometimes, however, we encounter applications where this is not so. We will see two such applications in Chapters 4 and 6.

Forcing Function with a Nonzero Pre-Initial Condition

EXAMPLE 2.4.10

■ Problem

Solve the following equation with the given initial conditions.

$$\frac{dx}{dt} + x = \frac{df}{dt} \qquad x(0-) = 0$$

where

$$f(t) = \begin{cases} 1 & t < 0 \\ 0 & t > 0 \end{cases}$$

■ Solution

Transforming both sides of the equation gives

$$sX(s) - x(0-) + X(s) = sF(s) - f(0-)$$

The term *pre-initial condition* refers to conditions given at $t = 0-$. We are given that $x(0-) = 0$, and from the definition of $f(t)$ we see that $f(0-) = 1$ and $F(s) = 0$. Thus, we have

$$X(s) = -\frac{1}{s+1}$$

and

$$x(t) = -e^{-t} \qquad t > 0$$

From the initial-value theorem,

$$x(0+) = \lim_{s\to\infty} s\,X(s) = \lim_{s\to\infty} s\frac{-1}{s+1} = -1$$

which agrees with the solution for $x(t)$.

2.5 TRANSFER FUNCTIONS

The complete response of a linear ODE is the sum of the free and the forced responses. For zero initial conditions, the free response is zero, and the complete response is the same as the forced response. Thus, we can focus our analysis on the effects of the input only by taking the initial conditions to be zero temporarily. When we have finished analyzing the effects of the input, we can add to the result the free response due to any nonzero initial conditions.

The concept of the *transfer function* is useful for analyzing the effects of the input. Consider the model

$$\dot{x} + ax = f(t) \tag{2.5.1}$$

and assume that $x(0) = 0$. Transforming both sides of the equation gives

$$sX(s) + aX(s) = F(s)$$

Then solve for the ratio $X(s)/F(s)$ and denote it by $T(s)$:

$$T(s) = \frac{X(s)}{F(s)} = \frac{1}{s+a}$$

The function $T(s)$ is called the *transfer function* of (2.5.1).

The transfer function is the transform of the forced response divided by the transform of the input. It can be used as a multiplier to obtain the forced response transform from the input transform; that is, $X(s) = T(s)F(s)$. The transfer function is a property of the system model only. The transfer function is independent of the input function and the initial conditions.

The transfer-function concept is extremely useful for several reasons.

1. *Transfer Functions and Software.* Software packages such as MATLAB do not accept system descriptions expressed as single, higher-order differential equations. Such software, however, does accept a description based on the transfer function. In Section 2.9 we will see how MATLAB does this. In Chapter 5 we will see that the transfer function is the basis for a graphical system description called the *block diagram*, and block diagrams are used to program the Simulink dynamic simulation software. So the transfer function is an important means of describing dynamic systems.
2. *ODE Equivalence.* It is important to realize that the transfer function is equivalent to the ODE. If we are given the transfer function, we can reconstruct the corresponding ODE. For example, the transfer function

$$\frac{X(s)}{F(s)} = \frac{5}{s^2 + 7s + 10}$$

corresponds to the equation $\ddot{x} + 7\dot{x} + 10x = 5f(t)$. You should develop the ability to obtain transfer functions from ODEs and ODEs from transfer functions. This process is easily done because the initial conditions are assumed to be zero when working with transfer functions. From the derivative property, this means that to work with a transfer function you can use the relations $\mathcal{L}(\dot{x}) = sX(s)$, $\mathcal{L}(\ddot{x}) = s^2X(s)$, and so forth. Examples 2.5.1 and 2.5.2 will show how straightforward this process is.

3. *The Transfer Function and Characteristic Roots.* Note that the denominator of the transfer function is the characteristic polynomial, and thus the transfer function tells us something about the intrinsic behavior of the model, apart from the effects of the input and specific values of the initial conditions. In the previous equation, the characteristic polynomial is $s^2 + 7s + 10$ and the roots are -2 and -5. The roots are real, and this tells us that the free response does not oscillate and that the forced response does not oscillate unless the input is oscillatory. Because the roots are negative, the model is stable and its free response disappears with time.

2.5.1 MULTIPLE INPUTS AND OUTPUTS

Obtaining a transfer function from a single ODE is straightforward, as we have seen. Sometimes, however, models have more than one input or occur as sets of equations with more than one dependent variable. It is important to realize that there is one transfer function for each input-output pair. If a model has more than one input, a particular transfer function is the ratio of the output transform over the input transform, with all the remaining inputs ignored (set to zero temporarily).

EXAMPLE 2.5.1

Two Inputs and One Output

■ Problem

Obtain the transfer functions $X(s)/F(s)$ and $X(s)/G(s)$ for the following equation.

$$5\ddot{x} + 30\dot{x} + 40x = 6f(t) - 20g(t)$$

■ Solution

Using the derivative property with zero initial conditions, we can immediately write the equation as

$$5s^2X(s) + 30sX(s) + 40X(s) = 6F(s) - 20G(s)$$

Solve for $X(s)$.

$$X(s) = \frac{6}{5s^2 + 30s + 40}F(s) - \frac{20}{5s^2 + 30s + 40}G(s)$$

When there is more than one input, the transfer function for a specific input can be obtained by temporarily setting the other inputs equal to zero (this is another aspect of the superposition property of linear equations). Thus, we obtain

$$\frac{X(s)}{F(s)} = \frac{6}{5s^2 + 30s + 40} \qquad \frac{X(s)}{G(s)} = -\frac{20}{5s^2 + 30s + 40}$$

Note that the denominators of both transfer functions have the same roots: $s = -2$ and $s = -4$.

We can obtain transfer functions from systems of equations by first transforming the equations and then algebraically eliminating all variables except for the specified input and output. This technique is especially useful when we want to obtain the response of one or more of the dependent variables in the system of equations.

EXAMPLE 2.5.2

A System of Equations

■ Problem

a. Obtain the transfer functions $X(s)/V(s)$ and $Y(s)/V(s)$ of the following system of equations:

$$\dot{x} = -3x + 2y$$

$$\dot{y} = -9y - 4x + 3v(t)$$

b. Obtain the forced response for $x(t)$ and $y(t)$ if the input is $v(t) = 5u_s(t)$.

■ Solution

a. Here two outputs are specified, x and y, with one input, v. Thus, there are two transfer functions. To obtain them, transform both sides of each equation, assuming zero initial conditions.

$$sX(s) = -3X(s) + 2Y(s)$$

$$sY(s) = -9Y(s) - 4X(s) + 3V(s)$$

These are two algebraic equations in the two unknowns, $X(s)$ and $Y(s)$. Solve the first equation for $Y(s)$:

$$Y(s) = \frac{s+3}{2}X(s) \tag{1}$$

Substitute this into the second equation.

$$s\frac{s+3}{2}X(s) = -9\frac{s+3}{2}X(s) - 4X(s) + 3V(s)$$

Then solve for $X(s)/V(s)$ to obtain

$$\frac{X(s)}{V(s)} = \frac{6}{s^2 + 12s + 35} \tag{2}$$

Now substitute this into equation (1) to obtain

$$\frac{Y(s)}{V(s)} = \frac{s+3}{2}\frac{X(s)}{V(s)} = \frac{s+3}{2}\frac{6}{s^2 + 12s + 35} = \frac{3(s+3)}{s^2 + 12s + 35} \tag{3}$$

The desired transfer functions are given by equations (2) and (3). Note that denominators of both transfer functions have the same factors, $s = -5$ and $s = -7$, which are the roots of the characteristic equation: $s^2 + 12s + 35$.

b. From equation (2),

$$X(s) = \frac{6}{s^2 + 12s + 35}V(s) = \frac{6}{s^2 + 12s + 35}\frac{5}{s} = \frac{30}{s(s^2 + 12s + 35)}$$

The denominator factors are $s = 0$, $s = -5$, and $s = -7$, and thus the partial-fraction expansion is

$$X(s) = \frac{C_1}{s} + \frac{C_2}{s+5} + \frac{C_3}{s+7}$$

where $C_1 = 6/7$, $C_2 = -3$, and $C_3 = 15/7$. The forced response is

$$x(t) = \frac{6}{7} - 3e^{-5t} + \frac{15}{7}e^{-7t} \tag{4}$$

From (1) we have $y = (\dot{x} + 3x)/2$. From (4) we obtain

$$y(t) = \frac{9}{7} + 3e^{-5t} - \frac{30}{7}e^{-7t}$$

EXAMPLE 2.5.3

An Example of Subsystems: Two Coupled Tanks

■ **Problem**

Consider two brine tanks each containing 500 L (liters) of brine connected as shown in Figure 2.5.1. At any time t, the first and the second tank contain $x_1(t)$ and $x_2(t)$ kg of salt, respectively. The brine concentration in each tank is kept uniform by continuous stirring. Brine containing r kg of salt per liter is entering the first tank at a rate of 15 L/min, and fresh water is entering the second tank at a rate of 5 L/min. The incoming brine density $r(t)$ can be changed to regulate the process, so $r(t)$ is an input variable.

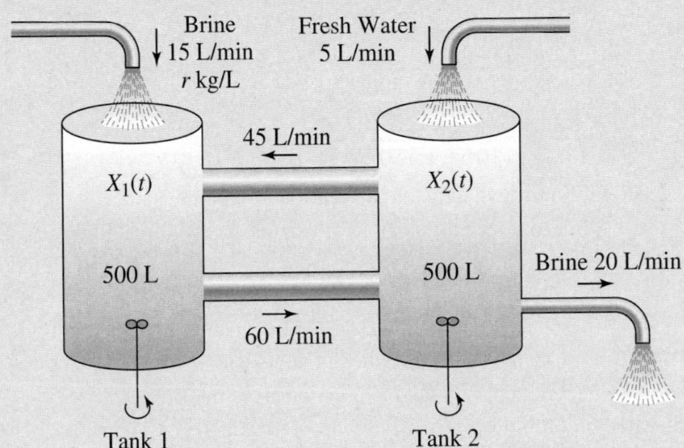

Figure 2.5.1 Mixing in two brine tanks.

Brine is pumped from the first tank to the second one at a rate of 60 L/min and from the second tank to the first one at a rate of 45 L/min. Brine is discharged from the second tank at a rate of 20 L/min.

a. Obtain the differential equations, in terms of x_1 and x_2, that describe the salt content in each tank as a function of time.
b. Obtain the transfer functions $X_1(s)/R(s)$ and $X_2(s)/R(s)$.
c. Suppose that $r(t) = 0.2$ kg/L. Determine the steady-state values of x_1 and x_2, and estimate how long it will take to reach steady state.

■ **Solution**

a. We assume that the liquid volume does not change when the salt is dissolved in it. We also observe that the total volume of the brine in each tank remains constant at 500 L because the incoming and the outgoing volume flow rates for each tank are equal. Therefore, each liter of brine in the first tank contains $x_1/500$ kg of salt, and salt leaves the first tank at a rate of $60(x_1/500)$ kg/min.

Considering that each liter of brine in the second tank contains $x_2/500$ kg of salt, the rate at which salt leaves the second tank and enters the first one is $45(x_2/500)$ kg/min. In addition, new brine enters the first tank at a rate of $15r$ kg/min because each liter of the new brine contains r kg of salt. A similar argument can be given for the second tank. Then the rates of change of the salt content of each tank, in kg/min, can be expressed as

$$\frac{dx_1}{dt} = 15r(t) - 60\frac{x_1}{500} + 45\frac{x_2}{500} \tag{1}$$

$$\frac{dx_2}{dt} = 0 + 60\frac{x_1}{500} - 65\frac{x_2}{500} \tag{2}$$

b. Multiplying each equation by 100 gives

$$100\frac{dx_1}{dt} = 1500r(t) - 12x_1 + 9x_2$$

$$100\frac{dx_2}{dt} = 12x_1 - 13x_2$$

Transforming each equation, using zero initial conditions, and collecting terms gives

$$(100s + 12)X_1(s) - 9X_2(s) = 1500R(s)$$

$$-12X_1(s) + (100s + 13)X_2(s) = 0$$

Solving these equations gives the transfer functions.

$$\frac{X_1(s)}{R(s)} = \frac{1500(100s + 13)}{10^4 s^2 + 2500s + 48} \quad (3)$$

$$\frac{X_2(s)}{R(s)} = \frac{1.8 \times 10^4}{10^4 s^2 + 2500s + 48} \quad (4)$$

Note that the numerators are different, whereas the denominators are the same.

c. The characteristic roots are the roots of the denominator, which are $s = -0.021$ and $s = -0.029$. Thus, the system is stable and the time constants are $\tau = 1/0.021 = 47.62$ and $\tau = 1/0.229 = 4.37$. The dominant-time constant is 47.62 minutes, so the time to reach steady state is approximately $4(47.62) = 190.5$ minutes. The salt content x_1 may actually take longer than this estimate because of the s term in the numerator of its transfer function. Determination of the actual value requires solving (1) and (2) or (3) and (4).

Applying the final-value theorem to (3) and (4) with $R(s) = 0.2/s$ gives the steady-state values.

$$x_{1ss} = \lim_{s\to 0} s\left\{\left[\frac{1500(100s + 13)}{10^4 s^2 + 2500s + 48}\right]\frac{0.2}{s}\right\} = \frac{1500(13)(0.2)}{48} = 81.25 \text{ kg}$$

$$x_{2ss} = \lim_{s\to 0} s\left\{\left[\frac{1.8 \times 10^4}{10^4 s^2 + 2500s + 48}\right]\frac{0.2}{s}\right\} = \frac{1.8 \times 10^4}{48}0.2 = 75 \text{ kg}$$

These are the steady-state values regardless of the initial values of x_1 and x_2.

This example illustrates a typical problem in system dynamics. The system consists of two subsystems, the two tanks. To model the entire system, we must first understand the dynamics of each subsystem. This was illustrated in Example 2.1.2, where we derive the model of a single tank. Only then can we build the model of the system having two tanks. Building a model of a more complex system, say one containing many tanks, would then proceed in a similar way.

2.6 PULSE AND IMPULSE INPUTS

In our development of the Laplace transform and its associated methods, we have assumed that the process under study starts at time $t = 0-$. Thus, the given initial conditions, say $x(0-)$, $\dot{x}(0-)$, … , represent the situation at the start of the process and are the result of any inputs applied prior to $t = 0-$. That is, we need not know what the inputs were before $t = 0-$ because their effects are contained in the initial conditions.

The effects of any inputs starting at $t = 0-$ are not felt by the system until an infinitesimal time later, at $t = 0+$. If the dependent variable $x(t)$ and its derivatives do not change between $t = 0-$ and $t = 0+$, the solution $x(t)$ obtained from the differential equation will match the given initial conditions when $x(t)$ and its derivatives are evaluated at $t = 0$. The results obtained from the initial-value theorem will also match the given initial conditions.

However, we will now investigate the behavior of some models for which $x(0-) \neq x(0+)$, or $\dot{x}(0-) \neq \dot{x}(0+)$, and so forth for higher derivatives. The initial-value theorem gives the value at $t = 0+$, which for some models is not necessarily equal to the value at $t = 0-$. In these cases, the solution of the differential equation is correct only for $t > 0$. This phenomenon occurs in models having impulse inputs and in models containing derivatives of a discontinuous input, such as a step function.

2.6.1 PULSE INPUTS

The Shifted Step Function — EXAMPLE 2.6.1

■ Problem

If the discontinuity in the unit-step function occurs at $t = D$, Figure 2.6.1, then the function $x(t) = Mu_s(t-D)$ is 0 for $t < D$ and M for $t > D$. The function $u_s(t-D)$ is called the *shifted* step function. Determine $X(s)$.

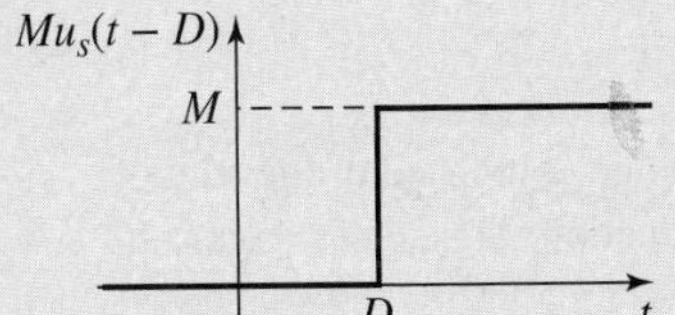

Figure 2.6.1 Shifted step function.

■ Solution

From the transform definition, we have

$$\mathcal{L}[x(t)] = \lim_{T\to\infty}\left[\int_{0-}^{T} Mu_s(t-D)e^{-st}\,dt\right] = \lim_{T\to\infty}\left(\int_{0-}^{D} 0e^{-st}\,dt + \int_{D}^{T} Me^{-st}\,dt\right)$$

or

$$\mathcal{L}[x(t)] = 0 + M\lim_{T\to\infty}\left(\frac{1}{-s}e^{-st}\Big|_{D}^{T}\right) = M\lim_{T\to\infty}\left(\frac{1}{-s}e^{-sT} + \frac{1}{s}e^{-sD}\right) = \frac{M}{s}e^{-sD}$$

Thus, the transform of the shifted unit-step function $u_s(t-D)$ is e^{-sD}/s.

Example 2.6.1 introduces a property of the transform called *shifting along the t-axis*. From this example, we see that the effect of the time shift D is to multiply the transform of the unshifted function by e^{-sD}. This illustrates the time-shifting property, which states that if

$$x(t) = \begin{cases} 0 & t < D \\ g(t-D) & t > D \end{cases}$$

then $X(s) = e^{-sD}G(s)$.

The Rectangular Pulse Function — EXAMPLE 2.6.2

■ Problem

The rectangular pulse function $P(t)$ is shown in Figure 2.6.2a. Derive the Laplace transform of this function (a) from the basic definition of the transform and (b) from the time-shifting property.

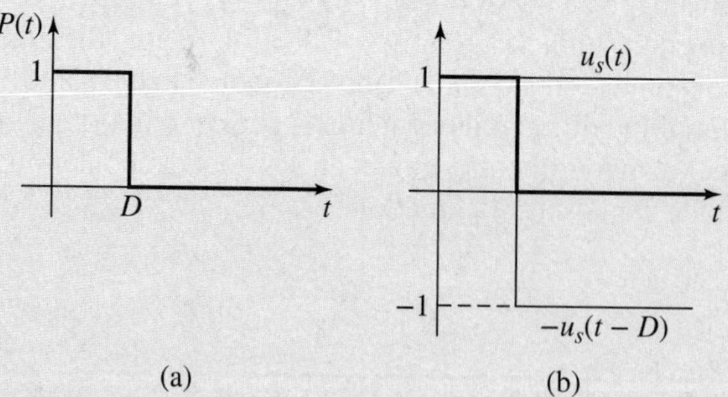

Figure 2.6.2 (a) Rectangular pulse function. (b) Rectangular pulse composed of two step functions.

■ **Solution**

a. From the definition of the transform,

$$\mathcal{L}[P(t)] = \int_{0-}^{\infty} P(t)e^{-st}\,dt = \int_{0}^{D} 1e^{-st}\,dt + \int_{D}^{\infty} 0e^{-st}\,dt = \int_{0}^{D} 1e^{-st}\,dt$$

$$= \left.\frac{e^{-st}}{-s}\right|_0^D = \frac{1}{s}\left(1 - e^{-sD}\right)$$

b. Figure 2.6.2b shows that the pulse can be considered to be composed of the sum of a unit-step function and a shifted, negative unit-step function. Thus, $P(t) = u_s(t) - u_s(t - D)$ and from the time-shifting property,

$$P(s) = \mathcal{L}[u_s(t)] - \mathcal{L}[u_s(t-D)] = \frac{1}{s} - e^{-sD}\frac{1}{s} = \frac{1}{s}\left(1 - e^{-sD}\right)$$

which is the same result obtained in part (a).

2.6.2 THE IMPULSE

An input that changes at a constant rate is modeled by the ramp function. The step function models an input that rapidly reaches a constant value, while the rectangular pulse function models a constant input that is suddenly removed. The *impulse* is similar to the pulse function, but it models an input that is suddenly applied and removed after a *very short* time. The impulse, which is a mathematical function only and has no physical counterpart, has an infinite magnitude for an infinitesimal time.

Consider the rectangular pulse function shown in Figure 2.6.3a. Its transform is $M(1 - e^{-sD})/s$. The area A under the pulse is $A = MD$ and is called the *strength* of the pulse. If we let this area remain constant at the value A and let the pulse duration D

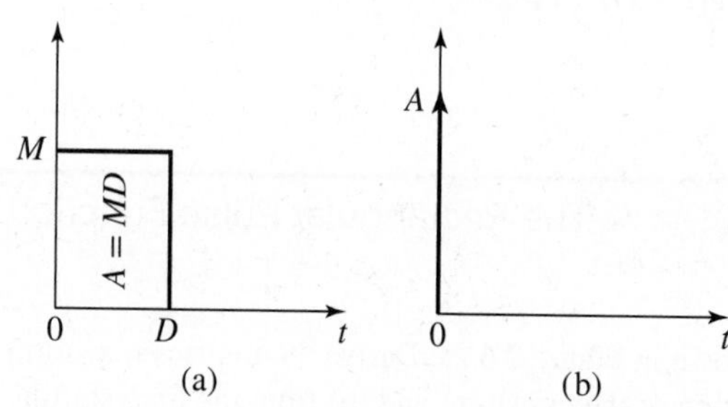

Figure 2.6.3 (a) The rectangular pulse. (b) The impulse.

approach zero, we obtain the impulse, represented in Figure 2.6.3b. Because $M = A/D$, the transform $F(s)$ is

$$F(s) = \lim_{D\to 0} \frac{A}{D}\frac{1 - e^{-sD}}{s} = \lim_{D\to 0} \frac{Ase^{-sD}}{s} = A$$

after using L'Hopital's limit rule. If the strength $A = 1$, the function is called a *unit* impulse.

The unit impulse, called the *Dirac delta function* $\delta(t)$ in mathematics literature, often appears in the analysis of dynamic systems. It is an analytically convenient approximation of an input applied for only a very short time, such as when a high-speed object strikes another object. The impulse is also useful for estimating the system's parameters experimentally and for analyzing the effect of differentiating a step or any other discontinuous input function.

In keeping with our interpretation of the initial conditions, we consider the impulse $\delta(t)$ to start at time $t = 0-$ and finish at $t = 0+$, with its effects first felt at $t = 0+$.

EXAMPLE 2.6.3

Impulse Response of a Simple First-Order Model

■ Problem

Obtain the unit-impulse response of the following model in two ways: (a) by separation of variables and (b) with the Laplace transform. The initial condition is $x(0-) = 3$. What is the value of $x(0+)$?

$$\dot{x} = \delta(t)$$

■ Solution

a. Integrate both sides of the equation to obtain

$$\int_{x(0-)}^{x(t)} dx = \int_{0-}^{t} \delta(t)\, dt = \int_{0-}^{0+} \delta(t) dt + \int_{0+}^{t} \delta(t) dt = 1 + 0$$

because the area under a unit impulse is 1. This gives

$$x(t) - x(0-) = 1 \qquad \text{or} \qquad x(t) = x(0-) + 1 = 3 + 1 = 4$$

This is the solution for $t > 0$ but not for $t = 0-$. Thus, $x(0+) = 4$ but $x(0-) = 3$, so the impulse has changed $x(t)$ instantaneously from 3 to 4.

b. The transformed equation is

$$sX(s) - x(0-) = 1 \qquad \text{or} \qquad X(s) = \frac{1 + x(0-)}{s}$$

which gives the solution $x(t) = 1 + x(0-) = 4$. Note that the initial value used with the derivative property is the value of x at $t = 0-$.

The initial-value theorem gives

$$x(0+) = \lim_{s\to\infty} sX(s) = \lim_{s\to\infty} s\frac{1 + x(0-)}{s} = 1 + 3 = 4$$

which is correct.

EXAMPLE 2.6.4

Impulse Response of a First-Order Model

■ Problem

Obtain the unit-impulse response of the following model. The initial condition is $x(0-)=0$. What is the value of $x(0+)$?

$$\frac{X(s)}{F(s)} = \frac{1}{s+5}$$

■ Solution

Because $f(t) = \delta(t)$, $F(s) = 1$, and the response is obtained from

$$X(s) = \frac{1}{s+5}F(s) = \frac{1}{s+5}$$

The response is $x(t) = e^{-5t}$ for $t > 0$. This gives

$$x(0+) = \lim_{t\to 0+} x(t) = \lim_{t\to 0+} e^{-5t} = 1$$

So the impulse input has changed x from 0 at $t = 0-$ to 1 at $t = 0+$. This same result could have been obtained from the initial-value theorem:

$$x(0+) = \lim_{s\to\infty} sX(s) = \lim_{s\to\infty} s\frac{1}{s+5} = 1$$

EXAMPLE 2.6.5

Impulse Response of a Simple Second-Order Model

■ Problem

Obtain the unit-impulse response of the following model in two ways: (a) by separation of variables and (b) with the Laplace transform. The initial conditions are $x(0-) = 5$ and $\dot{x}(0-) = 10$. What are the values of $x(0+)$ and $\dot{x}(0+)$?

$$\ddot{x} = \delta(t) \tag{1}$$

■ Solution

a. Let $v(t) = \dot{x}(t)$. Then equation (1) becomes $\dot{v} = \delta(t)$, which can be integrated to obtain $v(t) = v(0-) + 1 = 10 + 1 = 11$. Thus, $\dot{x}(0+) = 11$ and is not equal to $\dot{x}(0-)$.

   Now integrate $\dot{x} = v = 11$ to obtain $x(t) = x(0-) + 11t = 5 + 11t$. Thus, $x(0+) = 5$, which is the same as $x(0-)$.

   So for this model the unit-impulse input changes $\dot{x}$ from $t = 0-$ to $t = 0+$ but does not change x.

b. The transformed equation is

$$s^2X(s) - sx(0-) - \dot{x}(0-) = 1$$

or

$$X(s) = \frac{sx(0-) + \dot{x}(0-) + 1}{s^2} = \frac{5s + 11}{s^2} = \frac{5}{s} + \frac{11}{s^2}$$

which gives the solution $x(t) = 5 + 11t$ and $\dot{x}(t) = 11$. Note that the initial values used with the derivative property are the values at $t = 0-$.

The initial-value theorem gives

$$x(0+) = \lim_{s\to\infty} sX(s) = \lim_{s\to\infty} s\frac{5s + 11}{s^2} = 5$$

and because $\mathcal{L}(\dot{x}) = sX(s) - x(0-)$,

$$\dot{x}(0+) = \lim_{s\to\infty} s[sX(s) - x(0-)] = \lim_{s\to\infty} s\left(\frac{5s+11}{s} - 5\right) = 11$$

as we found in part (a).

EXAMPLE 2.6.6

Impulse Response of a Second-Order Model

Problem

Obtain the unit-impulse response of the following model. The initial conditions are $x(0-) = 0$, $\dot{x}(0-) = 0$. What are the values of $x(0+)$ and $\dot{x}(0+)$?

$$\frac{X(s)}{F(s)} = \frac{1}{2s^2 + 14s + 20}$$

Solution

Because $f(t) = \delta(t)$, $F(s) = 1$, and

$$X(s) = \frac{1}{2s^2 + 14s + 20}F(s) = \frac{1}{2s^2 + 14s + 20} = \frac{1}{6}\frac{1}{s+2} - \frac{1}{6}\frac{1}{s+5}$$

The response is $x(t) = (e^{-2t} - e^{-5t})/6$. This gives

$$x(0+) = \lim_{t\to 0+} x(t) = \lim_{t\to 0+}\left(\frac{e^{-2t} - e^{-5t}}{6}\right) = 0$$

and

$$\dot{x}(0+) = \lim_{t\to 0+} \dot{x}(t) = \lim_{t\to 0+}\left(\frac{-2e^{-2t} + 5e^{-5t}}{6}\right) = \frac{1}{2}$$

So the impulse input has not changed x between $t = 0-$ and $t = 0+$ but has changed $\dot{x}$ from 0 to 1/2. These results could have been obtained from the initial-value theorem:

$$x(0+) = \lim_{s\to\infty} sX(s) = \lim_{s\to\infty} s\frac{1}{2s^2 + 14s + 20} = 0$$

and, noting that $x(0-) = 0$,

$$\dot{x}(0+) = \lim_{s\to\infty} s[sX(s) - x(0-)] = \lim_{s\to\infty} s\frac{s}{2s^2 + 14s + 20} = \frac{1}{2}$$

In summary, be aware that the solution $x(t)$ and its derivatives $\dot{x}(t)$, $\ddot{x}(t)$, ... will match the given initial conditions at $t = 0-$ only if there are no impulse inputs and no derivatives of inputs that are discontinuous at $t = 0$.

If $X(s)$ is a rational function and if the degree of the numerator of $X(s)$ is less than the degree of the denominator, then the initial-value theorem will give a finite value for $x(0+)$. If the degrees are equal, then the initial value is undefined and the initial value theorem is invalid. The latter situation corresponds to an impulse in $x(t)$ at $t = 0$ and therefore $x(0+)$ is undefined. When the degrees are equal, the transform can be expressed as a constant plus a partial-fraction expansion. For example, consider the transform

$$X(s) = \frac{9s+4}{s+3} = 9 - \frac{23}{s+3}$$

The inverse transform is $x(t) = 9\delta(t) - 23e^{-3t}$ and therefore $x(0+)$ is undefined.

2.6.3 NUMERATOR DYNAMICS

The following model contains a derivative of the input $g(t)$:

$$5\dot{x} + 10x = 2\dot{g}(t) + 10g(t)$$

Its transfer function is

$$\frac{X(s)}{G(s)} = \frac{2s + 10}{5s + 10}$$

Note that the input derivative $\dot{g}(t)$ results in an s term in the numerator of the transfer function, and such a model is said to have *numerator dynamics*. So a model with input derivatives has numerator dynamics, and vice versa, and thus the two terms describe the same condition.

With such models we must proceed carefully if the input is discontinuous, as is the case with the step function, because the input derivative produces an impulse when acting on a discontinuous input. To help you understand this, we state without rigorous proof that the unit impulse $\delta(t)$ is the time derivative of the unit-step function $u_s(t)$; that is,

$$\delta(t) = \frac{d}{dt}u_s(t) \tag{2.6.1}$$

This result does not contradict common sense, because the step function changes from 0 at $t = 0-$ to 1 at $t = 0+$ in an infinitesimal amount of time. Therefore, its derivative should be infinite during this time. To further indicate the correctness of this relation, we integrate both sides and note that the area under the unit impulse is unity. Thus,

$$\int_{0-}^{0+} \delta(t)\,dt = \int_{0-}^{0+} \frac{d}{dt}u_s(t)\,dt = u_s(0+) - u_s(0-) = 1 - 0 = 1$$

Thus, an input derivative will create an impulse in response to a step input. For example, consider the model

$$5\dot{x} + 10x = 2\dot{g}(t) + 10g(t)$$

If the input $g(t) = u_s(t)$, the model is equivalent to

$$5\dot{x} + 10x = 2\delta(t) + 10u_s(t)$$

which has an impulse input.

Numerator dynamics can significantly alter the response, and the Laplace transform is a convenient and powerful tool for analyzing models having numerator dynamics.

EXAMPLE 2.6.7 A First-Order Model with Numerator Dynamics

■ Problem

Obtain the transfer function and investigate the response of the following model in terms of the parameter a. The input $g(t)$ is a unit-step function.

$$5\dot{x} + 10x = a\dot{g}(t) + 10g(t) \qquad x(0-) = 0$$

■ **Solution**

Transforming the equation with $x(0-) = 0$ and solving for the ratio $X(s)/G(s)$ gives the transfer function:

$$\frac{X(s)}{G(s)} = \frac{as + 10}{5s + 10}$$

Note that the model has numerator dynamics if $a \neq 0$.

For a unit-step input, $G(s) = 1/s$ and

$$X(s) = \frac{as + 10}{s(5s + 10)} = \frac{1}{s} + \frac{a - 5}{5}\frac{1}{s + 2}$$

Thus, the response is

$$x(t) = 1 + \frac{a - 5}{5}e^{-2t}$$

From this solution or the initial-value theorem, we find that $x(0+) = a/5$, which is not equal to $x(0-)$ unless $a = 0$ (which corresponds to the absence of numerator dynamics). The plot of the response is given in Figure 2.6.4 for several values of a. The initial condition is different for each case, but for all cases the response is essentially constant for $t > 2$ because the term e^{-2t} becomes small.

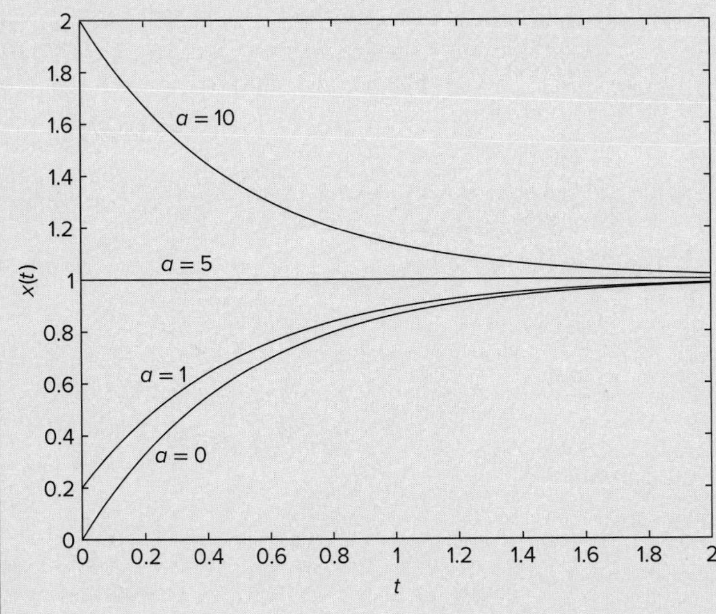

Figure 2.6.4 Plot of the response for Example 2.6.7.

A Second-Order Model with Numerator Dynamics

EXAMPLE 2.6.8

■ **Problem**

Obtain the transfer function and investigate the response of the following model in terms of the parameter a. The input $g(t)$ is a unit-step function.

$$3\ddot{x} + 18\dot{x} + 24x = a\dot{g}(t) + 6g(t) \qquad x(0-) = 0 \qquad \dot{x}(0-) = 0$$

■ **Solution**

Transforming the equation with zero initial conditions and solving for the ratio $X(s)/G(s)$ gives the transfer function:

$$\frac{X(s)}{G(s)} = \frac{as + 6}{3s^2 + 18s + 24}$$

Note that the model has numerator dynamics if $a \neq 0$.

For a unit-step input, $G(s) = 1/s$ and

$$X(s) = \frac{as + 6}{s(3s^2 + 18s + 24)} = \frac{1}{4}\frac{1}{s} + \frac{a-3}{6}\frac{1}{s+2} + \frac{3-2a}{12}\frac{1}{s+4}$$

Thus, the response is

$$x(t) = \frac{1}{4} + \frac{a-3}{6}e^{-2t} + \frac{3-2a}{12}e^{-4t}$$

From this solution or the initial-value theorem, we find that $x(0+) = 0$, which is equal to $x(0-)$, and that $\dot{x}(0+) = a/3$, which is not equal to $\dot{x}(0-)$ unless $a = 0$ (which corresponds to the absence of numerator dynamics). The plot of the response is given in Figure 2.6.5 for several values of a. Notice that a "hump" in the response (called an "overshoot") does not occur for smaller values of a and the height of the hump increases as a increases. However, the value of a does not affect the steady-state response.

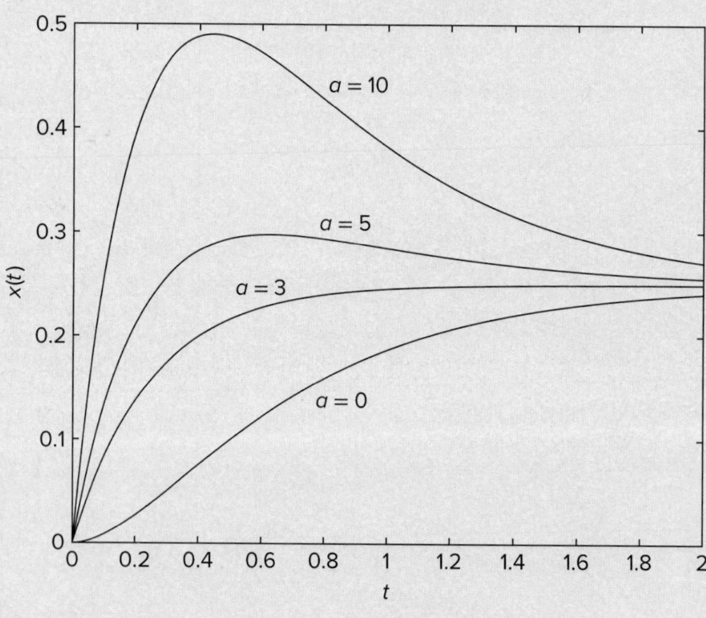

Figure 2.6.5 Plot of the response for Example 2.6.8.

2.7 PARTIAL-FRACTION EXPANSION

To solve a differential equation by using the Laplace transform, we must be able to obtain a function $x(t)$ from its transform $X(s)$. This process is called *inverting* the transform. Unless the transform is a simple one appearing in the transform table, it will have to be represented as a combination of simple transforms.

The expansions in Section 2.4 are simple examples of partial-fraction expansions. In practice, however, we might encounter higher-order system models or complicated inputs. Both situations require a general approach to obtaining the expansion, and this section develops such an approach.

Most transforms occur in the form of a ratio of two polynomials, such as

$$X(s) = \frac{N(s)}{D(s)} = \frac{b_m s^m + b_{m-1}s^{m-1} + \cdots + b_1 s + b_0}{s^n + a_{n-1}s^{n-1} + \cdots + a_1 s + a_0} \tag{2.7.1}$$

In all of our examples, $m \leq n$. If $X(s)$ is of the form (2.7.1), the method of partial-fraction expansion can be used. Note that we assume that the coefficient a_n is unity. If not, divide the numerator and denominator by a_n. The first step is to solve for the n roots of the denominator. If the a_i coefficients are real (as they will be for all our applications), any complex roots will occur in conjugate pairs.

There are two cases to be considered. The first is where all the roots are distinct; the second is where two or more roots are identical (repeated).

2.7.1 DISTINCT ROOTS CASE

If all the roots are distinct, we can express $X(s)$ in (2.7.1) in factored form as follows:

$$X(s) = \frac{N(s)}{(s+r_1)(s+r_2)\cdots(s+r_n)} \tag{2.7.2}$$

where the roots are $s = -r_1, -r_2, \ldots, -r_n$. This form can be expanded as

$$X(s) = \frac{C_1}{s+r_1} + \frac{C_2}{s+r_2} + \cdots + \frac{C_n}{s+r_n} \tag{2.7.3}$$

where

$$C_i = \lim_{s\to -r_i} \left[X(s)(s+r_i)\right] \tag{2.7.4}$$

Multiplying by the factor $(s + r_i)$ cancels that term in the denominator before the limit is taken. This is a good way of remembering (2.7.4). Each factor corresponds to an exponential function of time, and the inverse transform is

$$x(t) = C_1 e^{-r_1 t} + C_2 e^{-r_2 t} + \cdots + C_n e^{-r_n t} \tag{2.7.5}$$

EXAMPLE 2.7.1

A Third-Order Equation

■ **Problem**

Obtain the solution of the following problem:

$$10\frac{d^3x}{dt^3} + 100\frac{d^2x}{dt^2} + 310\frac{dx}{dt} + 300x = 750u_s(t)$$

$$x(0) = 2 \qquad \dot{x}(0) = 4 \qquad \ddot{x}(0) = 3$$

■ **Solution**

Using the Laplace transform method, we have

$$10\left[s^3X(s) - \ddot{x}(0) - s\dot{x}(0) - s^2x(0)\right] + 100\left[s^2X(s) - \dot{x}(0) - sx(0)\right]$$

$$+ 310[sX(s) - x(0)] + 300X(s) = \frac{750}{s}$$

Solving for $X(s)$ using the given initial values, we obtain

$$X(s) = \frac{2s^3 + 24s^2 + 105s + 75}{s(s^3 + 10s^2 + 31s + 30)} = \frac{2s^3 + 24s^2 + 105s + 75}{s(s+2)(s+3)(s+5)}$$

Since the roots of the cubic are $s = -2, -3$, and -5, the partial-fraction expansion is

$$X(s) = \frac{C_1}{s} + \frac{C_2}{s+2} + \frac{C_3}{s+3} + \frac{C_4}{s+5}$$

For this problem the Least Common Denominator (LCD) method requires a lot of algebra, and the coefficients can be more easily obtained from the formula (2.7.4). They are

$$C_1 = \lim_{s\to 0} sX(s) = \frac{5}{2}$$

$$C_2 = \lim_{s\to -2} (s+2)X(s) = \frac{55}{6}$$

$$C_3 = \lim_{s\to -3} (s+3)X(s) = -13$$

$$C_4 = \lim_{s\to -5} (s+5)X(s) = \frac{10}{3}$$

Thus, the answer is

$$x(t) = \frac{5}{2} + \frac{55}{6}e^{-2t} - 13e^{-3t} + \frac{10}{3}e^{-5t}$$

The plot of the response is shown in Figure 2.7.1. The response contains three exponentials. The terms e^{-3t} and e^{-5t} die out faster than e^{-2t}, so for $t > 4/3$, the response is approximately given by $x(t) = 5/2 + (55/6)e^{-2t}$. For $t > 2$, the response is approximately constant at $x = 5/2$. The "hump" in the response is produced by the positive values of $\dot{x}(0)$ and $\ddot{x}(0)$.

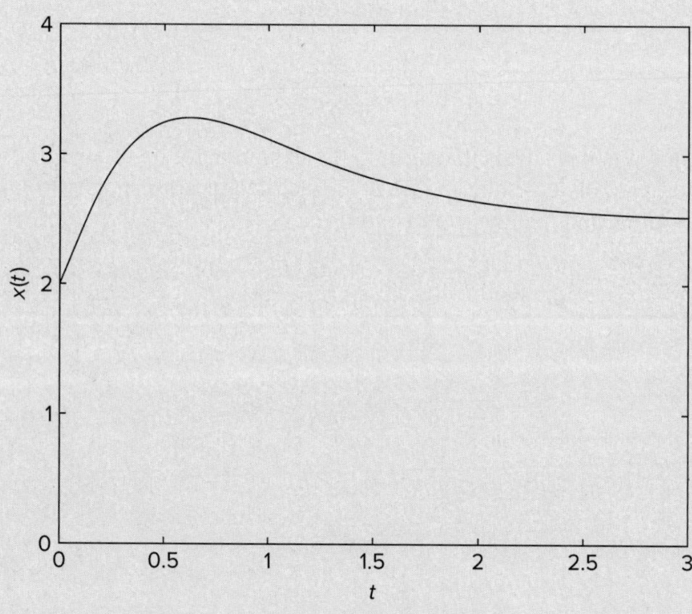

Figure 2.7.1 Response for Example 2.7.1.

2.7.2 REPEATED-ROOTS CASE

Suppose that p of the roots have the same value $s = -r_1$, and the remaining $(n-p)$ roots are distinct and real. Then $X(s)$ is of the form

$$X(s) = \frac{N(s)}{(s+r_1)^p(s+r_{p+1})(s+r_{p+2})\cdots(s+r_n)} \tag{2.7.6}$$

The expansion is

$$X(s) = \frac{C_1}{(s+r_1)^p} + \frac{C_2}{(s+r_1)^{p-1}} + \cdots + \frac{C_p}{s+r_1} + \cdots$$
$$+ \frac{C_{p+1}}{s+r_{p+1}} + \cdots + \frac{C_n}{s+r_n} \tag{2.7.7}$$

The coefficients for the repeated roots are found from

$$C_1 = \lim_{s\to -r_1} \left[X(s)(s+r_1)^p\right] \tag{2.7.8}$$

$$C_2 = \lim_{s\to -r_1} \left\{\frac{d}{ds}\left[X(s)(s+r_1)^p\right]\right\} \tag{2.7.9}$$

$$\vdots$$

$$C_i = \lim_{s\to -r_1} \left\{\frac{1}{(i-1)!}\frac{d^{i-1}}{ds^{i-1}}\left[X(s)(s+r_1)^p\right]\right\} \qquad i = 1, 2, \ldots, p \tag{2.7.10}$$

The coefficients for the distinct roots are found from (2.7.4). The solution for the time function is

$$f(t) = C_1\frac{t^{p-1}}{(p-1)!}e^{-r_1 t} + C_2\frac{t^{p-2}}{(p-2)!}e^{-r_1 t} + \cdots + C_p e^{-r_1 t} + \cdots$$
$$+ C_{p+1}e^{-r_{p+1}t} + \cdots + C_n e^{-r_n t} \tag{2.7.11}$$

One Negative Root and Two Zero Roots — EXAMPLE 2.7.2

■ Problem

The inverse Laplace transform of

$$X(s) = \frac{5}{s^2(3s+12)}$$

■ Solution

The denominator roots are $s = -12/3 = -4$, $s = 0$, and $s = 0$. Thus, the partial-fraction expansion has the form

$$X(s) = \frac{5}{s^2(3s+12)} = \frac{1}{3}\frac{5}{s^2(s+4)} = \frac{C_1}{s^2} + \frac{C_2}{s} + \frac{C_3}{s+4}$$

Using the coefficient formulas (2.7.4), (2.7.8), and (2.7.9) with $p = 2$ and $r_1 = 0$, we obtain

$$C_1 = \lim_{s\to 0}\left[s^2\frac{5}{3s^2(s+4)}\right] = \lim_{s\to 0}\left[\frac{5}{3(s+4)}\right] = \frac{5}{12}$$

$$C_2 = \lim_{s\to 0}\frac{d}{ds}\left[s^2\frac{5}{3s^2(s+4)}\right] = \lim_{s\to 0}\frac{d}{ds}\left[\frac{5}{3(s+4)}\right] = \lim_{s\to 0}\left[-\frac{5}{3}\frac{1}{(s+4)^2}\right] = -\frac{5}{48}$$

$$C_3 = \lim_{s\to -4}\left[(s+4)\frac{5}{3s^2(s+4)}\right] = \lim_{s\to -4}\left(\frac{5}{3s^2}\right) = \frac{5}{48}$$

The inverse transform is

$$x(t) = C_1 t + C_2 + C_3 e^{-4t} = \frac{5}{12}t - \frac{5}{48} + \frac{5}{48}e^{-4t}$$

EXAMPLE 2.7.3

Two Repeated Roots and One Distinct Root

■ Problem

Obtain the inverse Laplace transform of

$$X(s) = \frac{7}{(s+3)^2(s+5)}$$

■ Solution

The denominator roots are $s = -5$, $s = -3$, and $s = -3$. Thus, the partial-fraction expansion has the form

$$X(s) = \frac{7}{(s+3)^2(s+5)} = \frac{C_1}{(s+3)^2} + \frac{C_2}{s+3} + \frac{C_3}{s+5}$$

where

$$C_1 = \lim_{s\to-3}\left[(s+3)^2\frac{7}{(s+3)^2(s+5)}\right] = \lim_{s\to-3}\left(\frac{7}{s+5}\right) = \frac{7}{2}$$

$$C_2 = \lim_{s\to-3}\frac{d}{ds}\left[(s+3)^2\frac{7}{(s+3)^2(s+5)}\right] = \lim_{s\to-3}\frac{d}{ds}\left(\frac{7}{s+5}\right)$$

$$= \lim_{s\to-3}\left[\frac{-7}{(s+5)^2}\right] = -\frac{7}{4}$$

$$C_3 = \lim_{s\to-5}\left[(s+5)\frac{7}{(s+3)^2(s+5)}\right] = \lim_{s\to-5}\left[\frac{7}{(s+3)^2}\right] = \frac{7}{4}$$

The inverse transform is

$$x(t) = C_1te^{-3t} + C_2e^{-3t} + C_3e^{-5t} = \frac{7}{2}te^{-3t} - \frac{7}{4}e^{-3t} + \frac{7}{4}e^{-5t}$$

EXAMPLE 2.7.4

Exponential Response of a First-Order Model

■ Problem

Use the Laplace transform to solve the following problem.

$$\dot{x} + 5x = 7te^{-3t} \qquad x(0) = 0$$

■ Solution

Taking the transform of both sides of the equation, we obtain

$$sX(s) - x(0) + 5X(s) = \frac{7}{(s+3)^2}$$

Solve for $X(s)$ using the given value of $x(0)$.

$$X(s) = \frac{7}{(s+3)^2(s+5)}$$

The partial-fraction expansion was obtained in Example 2.7.3. It is

$$X(s) = \frac{7}{2(s+3)^2} - \frac{7}{4(s+3)} + \frac{7}{4(s+5)}$$

and the inverse transform is

$$x(t) = \frac{7}{2}te^{-3t} - \frac{7}{4}e^{-3t} + \frac{7}{4}e^{-5t}$$

The plot of the response is shown in Figure 2.7.2. The "hump" in the response is caused by the multiplicative factor of t in the input $7te^{-3t}$.

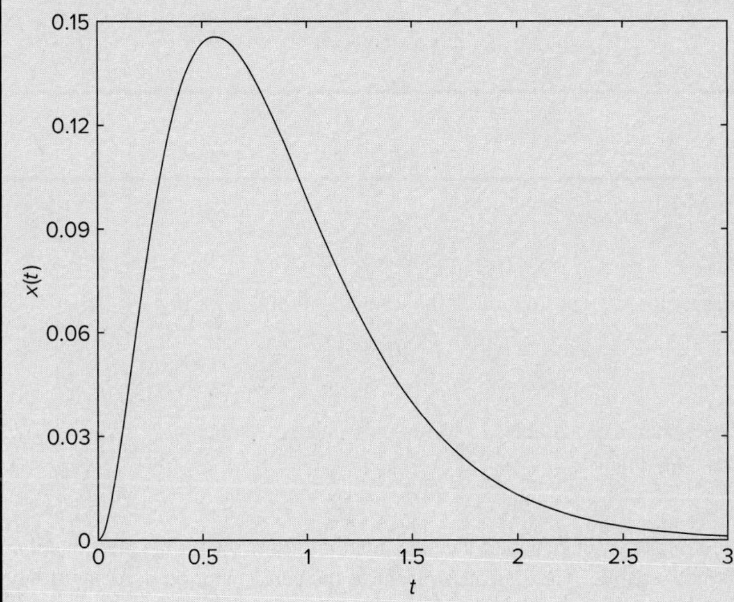

Figure 2.7.2 Response for Example 2.7.4.

Four Repeated Roots

EXAMPLE 2.7.5

■ Problem

Choose the most convenient method for obtaining the inverse transform of

$$X(s) = \frac{s^2 + 2}{s^4(s+1)}$$

■ Solution

There are four repeated roots ($s = 0$) and one distinct root, so the expansion is

$$X(s) = \frac{C_1}{s^4} + \frac{C_2}{s^3} + \frac{C_3}{s^2} + \frac{C_4}{s} + \frac{C_5}{s+1}$$

Because there are four repeated roots, use of (2.7.10) to find the coefficients would require taking the first, second, and third derivatives of the ratio $(s^2 + 2)/(s + 1)$. Therefore, the LCD method is easier to use for this problem. Using the LCD method we obtain

$$X(s) = \frac{C_1(s+1) + C_2 s(s+1) + C_3 s^2(s+1) + C_4 s^3(s+1) + C_5 s^4}{s^4(s+1)}$$

$$= \frac{(C_5 + C_4)s^4 + (C_4 + C_3)s^3 + (C_3 + C_2)s^2 + (C_2 + C_1)s + C_1}{s^4(s+1)}$$

Comparing numerators we see that

$$s^2 + 2 = (C_5 + C_4)s^4 + (C_4 + C_3)s^3 + (C_3 + C_2)s^2 + (C_2 + C_1)s + C_1$$

and thus $C_1 = 2$, $C_2 + C_1 = 0$, $C_3 + C_2 = 1$, $C_4 + C_3 = 0$, and $C_5 + C_4 = 0$. These give $C_1 = 2$, $C_2 = -2$, $C_3 = 3$, $C_4 = -3$, and $C_5 = 3$. So the expansion is

$$X(s) = \frac{2}{s^4} - \frac{2}{s^3} + \frac{3}{s^2} - \frac{3}{s} + \frac{3}{s+1}$$

The inverse transform is

$$x(t) = \frac{1}{3}t^3 - t^2 + 3t - 3 + 3e^{-t}$$

EXAMPLE 2.7.6

Series Solution Method

■ Problem

Obtain an approximate, closed-form solution of the following problem for $0 \le t \le 0.5$:

$$\dot{x} + x = \tan t \qquad x(0) = 0 \tag{1}$$

■ Solution

If we attempt to use separation of variables to solve this problem, we obtain

$$\frac{dx}{\tan t - x} = dt$$

so the variables do not separate. In general, when the input is a function of time, the equation $\dot{x} + g(x) = f(t)$ does not separate. The Laplace transform method cannot be used when the Laplace transform or inverse transform either does not exist or cannot be found easily. In this example, the equation cannot be solved by the Laplace transform method, because the transform of $\tan t$ does not exist.

An approximate solution of the equation $\dot{x} + x = \tan t$ can be obtained by replacing $\tan t$ with a series approximation. The number of terms used in the series determines the accuracy of the resulting solution for $x(t)$. The Taylor series expansion for $\tan t$ is

$$\tan t = t + \frac{t^3}{3} + \frac{2t^5}{15} + \frac{17t^7}{315} + \cdots \qquad |t| < \frac{\pi}{2}$$

The more terms we retain, the more accurate is the series. Also, the series becomes less accurate as the absolute value of t increases. To demonstrate the series solution method, let us use a series with two terms: $\tan t = t + t^3/3$. At the largest value of t, $t = 0.5$, the two-term series gives 0.5417 versus 0.5463 for the true value of $\tan 0.5$. So the two-term series is accurate to at least two decimal places over the range of t we are interested in ($0 \le t \le 0.5$).

Using the two-term series we need to solve the following problem:

$$\dot{x} + x = t + \frac{t^3}{3} \qquad x(0) = 0$$

Using the Laplace transform, we obtain

$$sX(s) + X(s) = \frac{1}{s^2} + \frac{1}{3}\frac{3!}{s^4}$$

or

$$X(s) = \frac{s^2 + 2}{s^4(s+1)}$$

The inverse transform was obtained in Example 2.7.5 and is

$$x(t) = \frac{1}{3}t^3 - t^2 + 3t - 3 + 3e^{-t}$$

We can expect this approximate solution of equation (1) to be accurate to at least two decimal places for $0 \le t \le 0.5$. Of course, greater accuracy can be achieved by retaining more terms in the Taylor series for $\tan t$.

2.7.3 COMPLEX ROOTS

When some of the roots of the transform denominator are complex, the expansion has the same form as (2.7.3), because the roots are in fact distinct. Thus, the coefficients C_i can be found from (2.7.4). However, these coefficients will be complex numbers, and the form of the inverse transform given by (2.7.5) will not be convenient to use. We now demonstrate two methods that can be used; the choice depends on whether or not you want to avoid the use of complex-valued coefficients.

Two Complex Roots **EXAMPLE 2.7.7**

■ **Problem**

Use two methods to obtain the inverse Laplace transform of

$$X(s) = \frac{3s+7}{4s^2+24s+136} = \frac{3s+7}{4(s^2+6s+34)}$$

■ **Solution**

a. The denominator roots are $s = -3 \pm 5j$. To avoid complex-valued coefficients, we note that the denominator of $X(s)$ can be written as $(s+3)^2 + 5^2$, and we can express $X(s)$ as follows:

$$X(s) = \frac{1}{4}\left[\frac{3s+7}{(s+3)^2+5^2}\right] \tag{1}$$

which can be expressed as the sum of two terms that are proportional to entries 10 and 11 in Table 2.3.1.

$$X(s) = \frac{1}{4}\left[C_1\frac{5}{(s+3)^2+5^2} + C_2\frac{s+3}{(s+3)^2+5^2}\right]$$

We can obtain the coefficients by noting that

$$C_1\frac{5}{(s+3)^2+5^2} + C_2\frac{s+3}{(s+3)^2+5^2} = \frac{5C_1 + C_2(s+3)}{(s+3)^2+5^2} \tag{2}$$

Comparing the numerators of equations (1) and (2), we see that

$$5C_1 + C_2(s+3) = C_2 s + 5C_1 + 3C_2 = 3s+7$$

which gives $C_2 = 3$ and $5C_1 + 3C_2 = 7$. Thus, $C_1 = -2/5$. The inverse transform is

$$x(t) = \frac{1}{4}C_1 e^{-3t}\sin 5t + \frac{1}{4}C_2 e^{-3t}\cos 5t = -\frac{1}{10}e^{-3t}\sin 5t + \frac{3}{4}e^{-3t}\cos 5t$$

b. The denominator roots are distinct and the expansion (2.8.3) gives

$$X(s) = \frac{3s+7}{4s^2+24s+136} = \frac{3s+7}{4(s+3-5j)(s+3+5j)}$$

$$= \frac{C_1}{s+3-5j} + \frac{C_2}{s+3+5j}$$

where, from (2.8.4),

$$C_1 = \lim_{s\to-3+5j}(s+3-5j)X(s) = \lim_{s\to-3+5j}\frac{3s+7}{4(s+3+5j)}$$

$$= \frac{-2+15j}{40j} = \frac{15+2j}{40}$$

This can be expressed in complex exponential form as follows (see Table 2.1.3):

$$C_1 = |C_1|e^{j\phi} = \left|\frac{15+2j}{40}\right|e^{j\phi} = \frac{\sqrt{229}}{40}e^{j\phi}$$

where $\phi = \tan^{-1}(2/15) = 0.1326$ rad.

The second coefficient is

$$C_2 = \lim_{s\to-3-5j}(s+3+5j)X(s) = \lim_{s\to-3-5j}\frac{3s+7}{4(s+3-5j)}$$

$$= \frac{2+15j}{40j} = \frac{15-2j}{40}$$

Note that C_1 and C_2 are complex conjugates. This will always be the case for coefficients of complex-conjugate roots in a partial-fraction expansion. Thus, $C_2 = |C_1|e^{-j\phi} = \sqrt{229}e^{-0.1326j}/40$.

The inverse transform gives

$$x(t) = C_1e^{(-3+5j)t} + C_2e^{(-3-5j)t} = C_1e^{-3t}e^{5jt} + C_2e^{-3t}e^{-5jt}$$

$$= |C_1|e^{-3t}\left[e^{(5t+\phi)j} + e^{-(5t+\phi)j}\right] = 2|C_1|e^{-3t}\cos(5t+\phi)$$

where we have used the relation $e^{j\theta} + e^{-j\theta} = 2\cos\theta$, which can be derived from the Euler identity (Table 2.1.4). Thus,

$$x(t) = \frac{\sqrt{229}}{20}e^{-3t}\cos(5t+0.1326)$$

This answer is equivalent to that found in part (a), as can be seen by applying the trigonometric identity $\cos(5t+\phi) = \cos 5t\cos\phi - \sin 5t\sin\phi$.

EXAMPLE 2.7.8

Free Response of a Second-Order Model with Complex Roots

■ Problem

Use the Laplace transform to solve the following problem:

$$4\ddot{x} + 24\dot{x} + 136x = 0 \qquad x(0) = \frac{7}{4} \qquad \dot{x}(0) = -\frac{11}{4}$$

■ Solution

Taking the transform of both sides of the equation, we obtain

$$4[s^2X(s) - x(0)s - \dot{x}(0)] + 24[sX(s) - x(0)] + 136X(s) = 0$$

Solve for $X(s)$ using the given values of $x(0)$ and $\dot{x}(0)$.

$$X(s) = \frac{4[x(0)s + \dot{x}(0)] + 24x(0)}{4s^2 + 24s + 136} = \frac{3s + 7}{4(s^2 + 6s + 34)}$$

The expansion was obtained in part (a) of Example 2.7.7. It is

$$X(s) = -\frac{1}{10}\left[\frac{5}{(s+3)^2 + 5^2}\right] + \frac{3}{4}\left[\frac{s+3}{(s+3)^2 + 5^2}\right]$$

and the inverse transform is

$$x(t) = -\frac{1}{10}e^{-3t}\sin 5t + \frac{3}{4}e^{-3t}\cos 5t$$

2.8 LAPLACE TRANSFORMS AND MATLAB

You can use MATLAB to easily compute the coefficients in the partial-fraction expansion. The appropriate MATLAB function is `residue`.

2.8.1 SYNTAX OF THE `residue` FUNCTION

If you have the MATLAB Symbolic Math Toolbox, you can use the function `laplace` and `ilaplace` to obtain transforms and inverse transforms symbolically.

Let $X(s)$ denote the transform. In the terminology of the `residue` function, the expansion coefficients are called the *residues* and the factors of the denominator of $X(s)$ are called the *poles*. The poles include the characteristic roots of the model and any denominator roots introduced by the input function. If the order m of the numerator of $X(s)$ is greater than the order n of the denominator, the transform can be represented by a polynomial $K(s)$, called the *direct term*, plus a ratio of two polynomials where the denominator degree is greater than the numerator degree. For example,

$$X(s) = \frac{6s^3 + 57s^2 + 120s + 80}{s^2 + 9s + 14} = 6s + 3 + \frac{9s + 38}{s^2 + 9s + 14} \tag{2.8.1}$$

or

$$X(s) = 6s + 3 + 5\frac{1}{s+7} + 4\frac{1}{s+2} \tag{2.8.2}$$

Here the direct term is the polynomial $K(s) = 6s + 3$.

The syntax of the `residue` function is as follows:

```
[r,p,K] = residue(num,den)
```

where `num` and `den` are arrays containing the coefficients of the numerator and denominator of $X(s)$. The output of the function consists of the array `r`, which contains the residues, the array `p`, which contains the poles, and the array `K`, which contains the coefficients of the direct term $K(s)$ in polynomial form.

Using (2.8.1) as an example, you would type

```
[r,p,K] = residue([6,57,120,80],[1,9,14])
```

The answer given by MATLAB is `r = [5, 4]`, `p = [-7, -2]`, and `K = [6, 3]`, which corresponds to (2.8.2). Note that the order in which the residues are displayed corresponds to the order in which the poles are displayed.

2.8.2 REPEATED POLES

Repeated poles are handled as follows. Consider the equation $\ddot{x} + 9\dot{x} + 14x = 3\dot{g} + 2g$, where $g(t) = 4e^{-7t}$. If the initial conditions are zero, the transform of the response is

$$X(s) = \frac{3s+2}{s^2+9s+14}\frac{4}{s+7} = \frac{12s+8}{(s+2)(s+7)^2}$$
$$= \frac{12s+8}{s^3+16s^2+77s+98}$$

The repeated poles are $s = -7, -7$; one of them is a characteristic root and the other is due to the input. To obtain the expansion, type

```
[r,p,K] = residue([12, 8],[1, 16, 77, 98])
```

The answer given by MATLAB is `r = [0.64, 15.2, -0.64]`, `p = [-7, -7, -2]`, and `K = []`. This corresponds to the expansion

$$X(s) = 0.64\frac{1}{s+7} + 15.2\frac{1}{(s+7)^2} - 0.64\frac{1}{s+2}$$

Note that for the residues due to repeated poles, the residue corresponding to the *highest* power is displayed as the *last* of those residues. The response here is

$$x(t) = 0.64e^{-7t} + 15.2te^{-7t} - 0.64e^{-2t}$$

2.8.3 COMPLEX POLES

Complex poles are handled as follows. Consider the equation $\ddot{x} + 6\dot{x} + 34x = 4\dot{g} + g$, where $g(t)$ is a unit-step function and the initial conditions are zero. The transform of the response is

$$X(s) = \frac{4s+1}{(s^2+6s+34)s} = \frac{4s+1}{s^3+6s^2+34s}$$

To obtain the expansion, type

```
[r,p,K] = residue([4, 1],[1, 6, 34, 0])
```

Observe that the last coefficient in the denominator is 0. The answer given by MATLAB is `r = [-0.0147-0.3912i,-0.0147+0.3912i,0.0294]`, `p = [-3+5i, -3-5i,0]`, and `K = []`. (Note that MATLAB uses the symbol i to represent the imaginary number $\sqrt{-1}$, whereas we have been using the symbol j.) The MATLAB results correspond to the expression

$$X(s) = \frac{-0.0147 - 0.3912j}{s+3-5j} + \frac{-0.0147 + 0.3912j}{s+3+5j} + \frac{0.0294}{s}$$

The response is

$$x(t) = (-0.0147 - 0.3912j)e^{(-3+5j)t} + (-0.0147 + 0.3912j)e^{(-3-5j)t} + 0.0294 \quad (2.8.3)$$

This form is not very useful because of its complex coefficients, but we can convert it to a more useful form by noting that the first two terms in the expansion have the form

$$\frac{C+jD}{s+a-jb} + \frac{C-jD}{s+a+jb} \quad (2.8.4)$$

which corresponds to the time function

$$(C + jD)e^{(-a+jb)t} + (C - jD)e^{(-a-jb)t}$$

Using Euler's identities: $e^{\pm jbt} = \cos bt \pm j \sin bt$, the previous form can be written as

$$2e^{-at}(C \cos bt - D \sin bt) \tag{2.8.5}$$

Using this identity with $C = -0.0147$ and $D = -0.3912$, we can write the response (2.8.3) as

$$x(t) = 2e^{-3t}(-0.0147 \cos 5t + 0.3912 \sin 5t) + 0.0294$$

2.8.4 SYMBOLIC TRANSFORMS IN MATLAB

The MATLAB Symbolic Math toolbox contains the `laplace` and `ilaplace` functions for obtaining the Laplace and inverse Laplace transforms. The choice of variable names is arbitrary; here we will use the standard notation for t as the time variable and s as the Laplace variable. You must first declare the variable with the `syms` command. Here are some examples.

Given

$$x(t) = 5e^{-2t} - 7e^{-3t}$$

find $X(s)$. The session is:

```
>>syms t
>>X = laplace(5*exp(-2*t)-7*exp(-3*t))
X =
5/(s + 2) - 7/(s + 3)
```

which gives

$$X(s) = \frac{5}{s+2} - \frac{7}{s+3}$$

Given

$$X(s) = \frac{4s+1}{s^3 + 6s^2 + 34s}$$

find $x(t)$. The program is:

```
>>syms s
>>x = ilaplace((4*s+1)/(s^3+6*s^2+34*s))
x =
1/34 - (exp(-3*t)*(cos(5*t) - (133*sin(5*t))/5))/34
```

which gives

$$x(t) = \frac{1}{34} - \frac{1}{34}e^{-3t}\left(\cos 5t - \frac{133}{5} \sin 5t\right)$$

2.9 TRANSFER-FUNCTION ANALYSIS IN MATLAB

Some of the functions from the MATLAB Control System Toolbox can be used to solve linear, time-invariant (constant-coefficient) differential equations. They are called ODE solvers and are powerful tools for studying dynamic systems.

2.9.1 THE tf AND tfdata FUNCTIONS

The ODE solvers in the Control System Toolbox can accept various descriptions of the equations to be solved. Here we will focus on the solvers that accept a transfer-function model of the system. An *LTI object* describes a linear, time-invariant model, or sets of equations, here referred to as the *system*. An LTI object can be created from different descriptions of the system, it can be analyzed with several functions, and it can be accessed to provide alternative descriptions of the system. For example, the equation

$$5\ddot{x} + 9\dot{x} + 4x = f(t)$$

is the reduced-form description of a particular system. From the reduced form we can immediately obtain the transfer function description of the model. It is

$$\frac{X(s)}{F(s)} = \frac{1}{5s^2 + 9s + 4} \tag{2.9.1}$$

To create an LTI object from a transfer function, you use the `tf(num,den)` function, where the array `num` is the array of coefficients of the numerator of the transfer function, arranged in order of descending powers of s, and `den` is the array of coefficients of the denominator of the transfer function, also arranged in descending order. The result is the LTI object that describes the system in the transfer-function form. For equation (2.9.1), the session is

```
>>sys1 = tf(1, [5, 9, 4]);
```

Here is another example. The LTI object `sys2` in transfer-function form for the equation

$$5\frac{d^3x}{dt^3} + 4\frac{d^2x}{dt^2} + 7\frac{dx}{dt} + 3x = 6\frac{d^2f}{dt^2} + 9\frac{df}{dt} + 2f \tag{2.9.2}$$

is created with the session:

```
>>sys2 = tf([6, 9, 2],[5, 4, 7, 3]);
```

As we will see in Chapter 5, we can also create an LTI object from descriptions of the system other than its transfer function. If the LTI object already exists, we can extract the coefficients of the numerator and denominator of the transfer-function model by using the `tfdata` function. Its syntax is `[num, den] = tfdata(sys).`

2.9.2 LINEAR ODE SOLVERS

The Control System Toolbox provides several solvers for linear models. These solvers are categorized by the type of input function they can accept: Some of these are a step input, an impulse input, and a general input function.

In their basic form, each of the following functions automatically puts a title and axis labels on the plot. You can change these by activating the Plot Editor or by right-clicking on the plot. This brings up a menu that includes the Properties as a choice. Selecting Properties enables you to change the labels as well as other features such as limits, units, and style.

The menu obtained by right-clicking on the plot also contains Characteristics as a choice. The contents of the subsequent menu depend on the particular function. When the `step` function is used, the Characteristics menu includes Peak Response, Rise Time, Settling Time, and Steady State. When you select Peak Response, for example, MATLAB identifies the peak value of the response curve and marks its location with a dot and dashed lines. Moving the cursor over the dot displays the numerical values of the peak response and the time at which it occurs. The Rise Time is the time required for the response to go from 10% to 90% of its steady-state value. The Settling Time is the time required for the response to settle within 2% of its steady-state value. You can change these percents by selecting the Characteristics tab under the Properties menu. The rise time and settling time are frequently used as measures of system performance and are discussed in Chapter 8.

The `step` Function The `step` function plots the unit-step response, assuming that the initial conditions are zero. The basic syntax is `step(sys)`, where `sys` is the LTI object. The time span and number of solution points are chosen automatically. To specify the final time `tfinal`, use the syntax `step(sys,tfinal)`. To specify a vector of times of the form `t = (0:dt:tfinal)`, at which to obtain the solution, use the syntax `step(sys,t)`. When called with left-hand arguments, as `[y, t] = step(sys,...)`, the function returns the output response `y`, and the time array `t` used for the simulation. No plot is drawn. The array `y` is $p \times q \times m$, where p is `length(t)`, q is the number of outputs, and m is the number of inputs.

The syntax `step(sys1,sys2,...,t)` plots the step response of multiple LTI systems on a single plot. The time vector `t` is optional. You can specify line color, line style, and marker for each system; for example, `step(sys1,'r', sys2, 'y--', sys3,'gx')`. The steady-state response and the time to reach that state are automatically determined. The steady-state response is indicated by a horizontal dotted line.

The plots generated by MATLAB for the following examples might be slightly different, depending on what version is used.

Step Response of a Second-Order Model

EXAMPLE 2.9.1

■ Problem

Consider the following model:

$$\frac{X(s)}{F(s)} = \frac{cs + 5}{10s^2 + cs + 5}$$

a. Plot the unit-step response for $c = 3$ using the time span selected by MATLAB.
b. Plot the unit-step response for $c = 3$ over the range $0 \le t \le 15$.
c. Plot the unit-step responses for $c = 3$ and $c = 8$ over the range $0 \le t \le 15$. Put the plots on the same graph.
d. Plot the step response for $c = 3$, where the magnitude of the step input is 20. Use the time span selected by MATLAB.

■ Solution

The MATLAB programs are shown below for each case.

a. This illustrates the use of the `step` function in its most basic form.

```
sys1 = tf([3,5],[10,3,5]);
step(sys1)
```

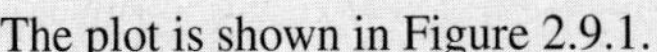

The plot is shown in Figure 2.9.1.

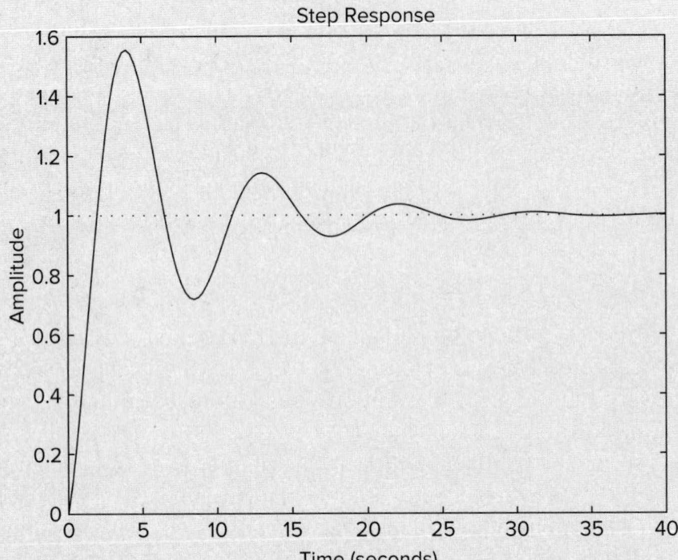

Figure 2.9.1 Response for part (a) of Example 2.9.1.

b. This illustrates how to use a user-selected time span and spacing.

```
sys1 = tf([3,5],[10,3,5]);
t = 0:0.01:15;
step(sys1,t)
```

The plot is shown in Figure 2.9.2.

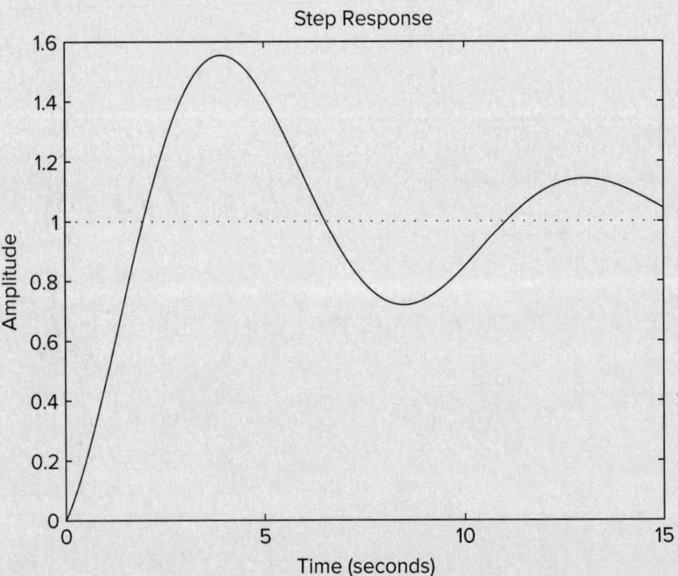

Figure 2.9.2 Response for part (b) of Example 2.9.1.

c. This illustrates how to plot two responses on the same graph.

```
sys1 = tf([3,5],[10,3,5]);
sys2 = tf([8,5],[10,8,5]);
t = 0:0.01:15;
step(sys1,sys2,'-',t),gtext('sys1'),gtext('sys2')
```

The plot is shown in Figure 2.9.3. The `gtext` function lets you place labels on the plot.

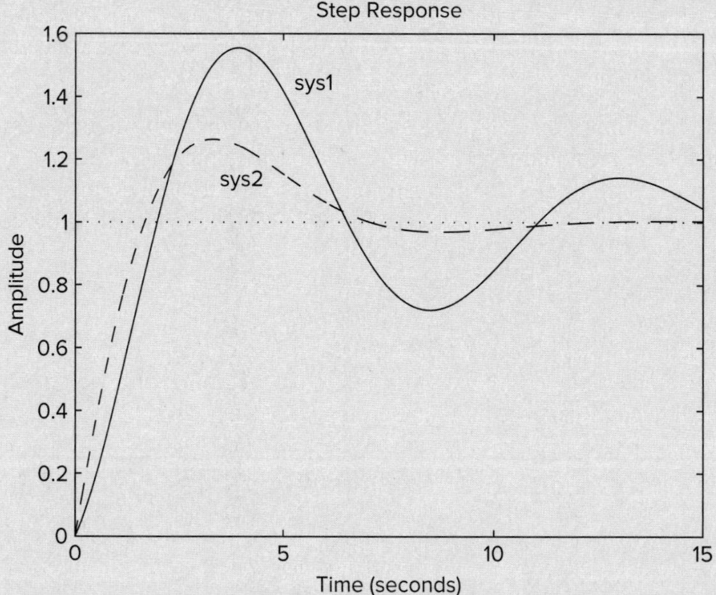

Figure 2.9.3 Response for part (c) of Example 2.9.1.

d. This illustrates how to obtain the response when the magnitude of the step input is not unity. The output of the `step(sys1)` function is for a unit-step input, and so it must be multiplied by 20. This multiplication can be performed within the `plot` function.

```
sys1 = tf([3,5],[10,3,5]);
[y, t] = step(sys1);
plot(t,20*y),xlabel('t'),ylabel('x(t)')
```

The plot is shown in Figure 2.9.4.

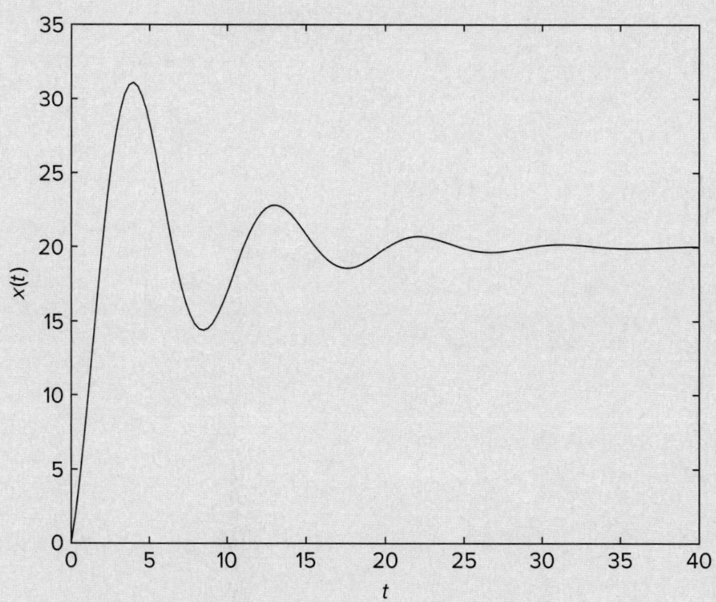

Figure 2.9.4 Response for part (d) of Example 2.9.1.

Note that when the `step` function is used without an assignment on the left-hand side of the equal sign, it automatically computes and plots the steady-state response, and puts a title and axis labels on the plot, with the assumption that the unit of time is seconds. When the form `[y, t] = step(sys)` is used, however, the steady-state response is not computed, and you must put the labels on the plot.

The `impulse` Function The `impulse` function plots the unit-impulse response, assuming that the initial conditions are zero. The basic syntax is `impulse(sys)`, where `sys` is the LTI object. The time span and number of solution points are chosen automatically. For example, the impulse response of (2.9.1) is found as follows:

```
>>sys1 = tf(1,[5, 9, 4]);
>>impulse(sys1)
```

The extended syntax of the `impulse` function is identical to that of the `step` function. The characteristics available with the `impulse` function by right-clicking on the plot are the Peak Response and the Settling Time.

EXAMPLE 2.9.2

Impulse Response of Second-Order Models

■ Problem

In Example 2.6.4 we obtained the response of the following model to a unit impulse:

$$\frac{X(s)}{F(s)} = \frac{1}{2s^2 + 14s + 20}$$

Our analysis showed that if $x(0-) = \dot{x}(0-) = 0$, then $x(0+) = 0$ and $\dot{x}(0+) = 1/2$. Use the `impulse` function to verify these results.

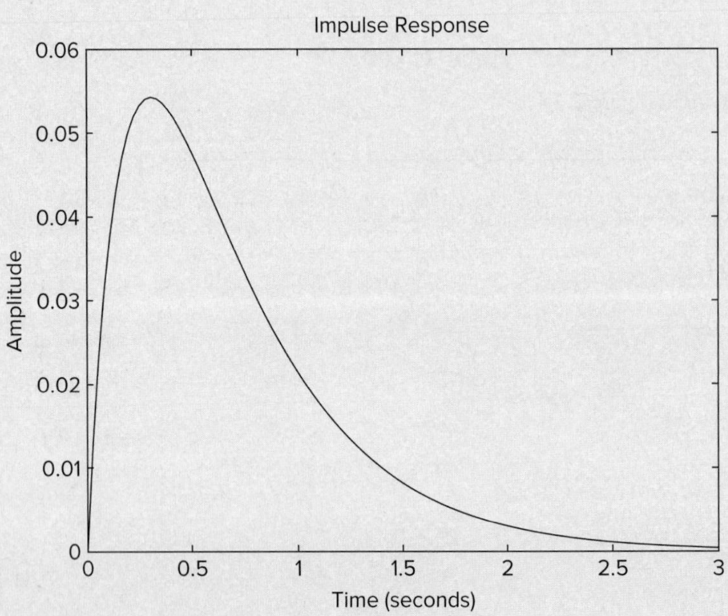

Figure 2.9.5 Impulse response of $x(t)$ for Example 2.9.2.

■ Solution

The session is shown here.

```
>>sys1 = tf(1,[2,14,20]);
>>impulse(sys1)
```

The plot is shown in Figure 2.9.5. From it we see that $x(0+) = 0$ and that $\dot{x}(0+)$ is positive as predicted. We are unable to determine the exact value of $\dot{x}(0+)$ from this plot, so we multiply the transfer function by s to obtain the transfer function for $v = \dot{x}$.

$$\frac{V(s)}{F(s)} = \frac{s}{2s^2 + 14s + 20}$$

We now use the `impulse` function on this transfer function.

```
>>sys2 = tf([1, 0],[2, 14, 20]);
>>impulse(sys2)
```

The plot is shown in Figure 2.9.6. From it we see that $\dot{x}(0+) = 0.5$ as predicted.

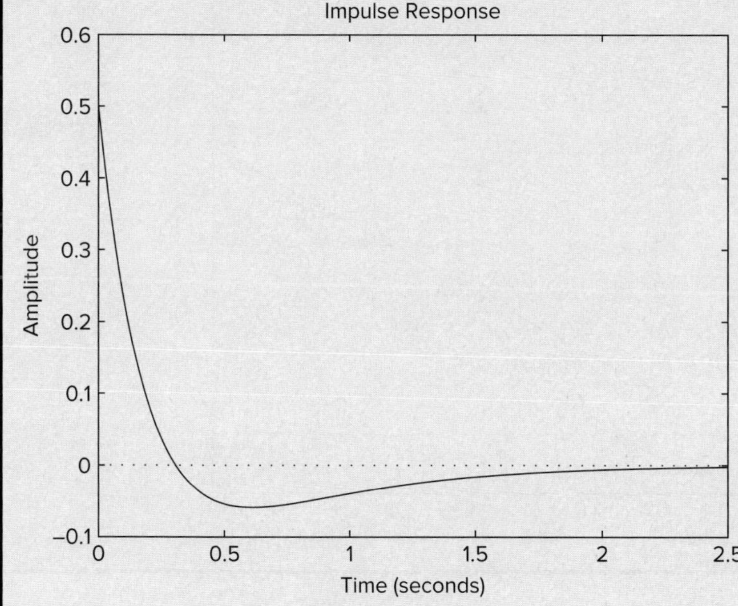

Figure 2.9.6 Impulse response of $v = \dot{x}(t)$ for Example 2.9.2.

The `lsim` Function The `lsim` function plots the response of the system to an arbitrary input. The basic syntax for zero-initial conditions is `lsim(sys,u,t)`, where `sys` is the LTI object, `t` is a time array having regular spacing, as `t = 0:dt:tf`, and `u` is a matrix with as many columns as inputs, and whose ith row specifies the value of the input at time `t(i)`.

The extended syntax of the `lsim` function is the same as that of the `step` function. When `lsim` is used without the left-hand arguments, the Peak Response is available from the Characteristics menu.

Ramp Response with the `lsim` Function — EXAMPLE 2.9.3

■ **Problem**

Plot the forced response of

$$\ddot{x} + 3\dot{x} + 5x = 10f(t)$$

to a *ramp* input, $u(t) = 1.5t$, over the time interval $0 \le t \le 2$.

■ **Solution**

We choose to generate the plot with 300 points. The MATLAB session is the following.

```
>>t = linspace(0,2,300);
>>f = 1.5*t;
>>sys = tf(10, [1, 3, 5]);
>>[y, t] = lsim(sys,f,t);
>>plot(t,y,t,f),xlabel('t'),ylabel('x(t) and f(t)'),...
    gtext('x(t)'),gtext('y(t)')
```

The plot is shown in Figure 2.9.7.

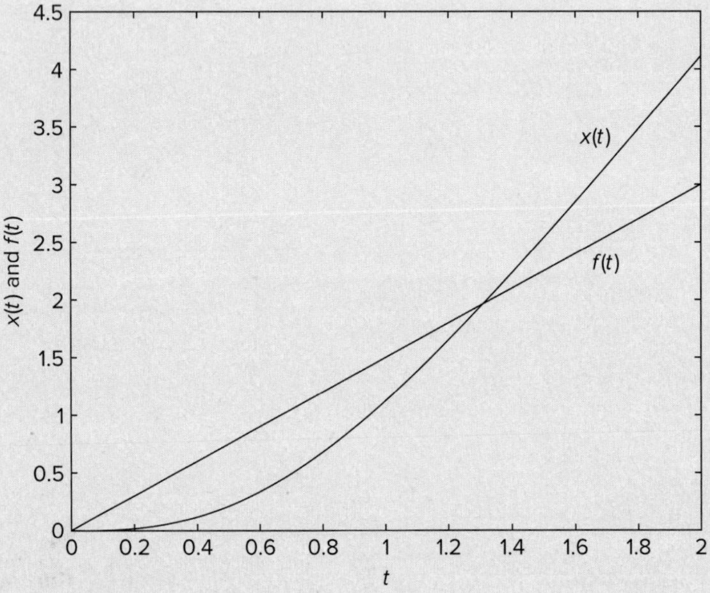

Figure 2.9.7 Ramp response for Example 2.9.3.

Sine Response with the `lsim` Command As another example, to find the response to $f(t) = 15\sin(3t)$, replace the second line in the previous program with `f = 15*sin(3*t);`.

Pulse Response with the `lsim` Command The pulse response of the system treated in Example 2.9.2 can be found with the following program.

```
t = linspace(0,3,300);
P = [ones(1,150),zeros(1,150)];
sys = tf(1,[2, 14, 20]);
[y,t] = lsim(sys,P,t);
plot(t,y),xlabel('t'),ylabel('x(t)')
```

2.10 CHAPTER REVIEW

Chapter 2 covers methods for solving ordinary differential equations. Now that you have finished this chapter, you should be able to do the following:

1. Choose and apply the separation-of-variables method, the trial-solution method, or the Laplace transform method to obtain the solution of a differential equation model.

2. When applying the Laplace transform method, be able to perform the appropriate expansion and apply the appropriate transform properties to obtain the inverse transform.
3. Identify the free, forced, transient, and steady-state components of the complete response.
4. Evaluate the effects of impulse inputs and input derivatives on the response.
5. Obtain transfer functions from models expressed as single equations or as sets of equations.
6. Use MATLAB to obtain inverse Laplace transforms and the forced response.

Perspective on Solving Differential Equations

In this chapter we have seen several methods for solving differential equations, and it is useful to understand the advantages of each method.

- The trial-solution method using exponential or harmonic functions is sometimes the easiest method, especially for first- and second-order linear equations with constant inputs.
- Separation of variables can be used to solve some linear and nonlinear ODEs, but the technique is limited to rather simple equations. For example, the method does not work for the equation $\dot{x} + g(x) = f(t)$, which does not separate.
- For some equations, the separation of variables method can be used to obtain a formal solution, but the resulting integral cannot be evaluated in terms of known elementary functions. An example is the equation $\dot{x} = \sin t^2$, for which the integral must be evaluated numerically.
- The Laplace transform method is especially advantageous when the mathematical model is linear, has constant coefficients, and either

    1. Is of order higher than three,
    2. Has a nonconstant input,
    3. Has input derivatives, or
    4. Consists of coupled differential equations.

- However, the Laplace transform method cannot be used when the Laplace transform or inverse transform either does not exist or cannot be found easily. For example, the equation $\dot{x} = \tan t$ can be solved by separation of variables but not by the Laplace transform method, because the transform of $\tan t$ does not exist.
- As shown in Example 2.7.3, an approximate closed-form solution can sometimes be obtained by using a series approximation. The number of terms used in the series determines the accuracy of the resulting solution.

When we cannot obtain a closed-form solution, we must solve the differential equation model using numerical methods, which are the subject of Chapter 5.

REFERENCES

[Kreyzig, 2011] Kreyzig, E., *Advanced Engineering Mathematics*, 10th ed., John Wiley and Sons, New York, 2011.

[Lundberg, 2007] Lundberg, K., Miller, H., and Trumper, D., Initial Conditions, Generalized Functions, and the Laplace Transform, *IEEE Control Systems Magazine*, 27(1), February 2007.

PROBLEMS

Section 2.1 Solving Differential Equations

2.1 Determine whether or not the following equations are linear or nonlinear, and state the reason for your answer.

a. $y\ddot{y} + 5\dot{y} + y = 0$ b. $\dot{y} + \sin y = 0$

c. $\dot{y} + \sqrt{y} = 0$ d. $\ddot{y} + 5t^2\dot{y} + 3y = 0$

e. $\ddot{y} + 3t^2 \sin y = 0$ f. $\dot{y} + e^t y = 0$

2.2 Solve each of the following problems by separation of variables.

a. $\dot{x} + 5x^2 = 25 \quad x(0) = 3$

b. $\dot{x} - 4x^2 = 36 \quad x(0) = 10$

c. $x\dot{x} - 5x = 25 \quad x(0) = 4$

d. $\dot{x} + 2e^{-4t}x = 0 \quad x(0) = 5$

2.3 Solve the following problems.

a. $5\dot{x} + 7x = 0 \quad x(0) = 4$ b. $5\dot{x} + 7x = 15 \quad x(0) = 0$

c. $5\dot{x} + 7x = 15 \quad x(0) = 4$

2.4 Solve the following problems.

a. $\ddot{x} + 10\dot{x} + 21x = 0 \quad x(0) = 4 \quad \dot{x}(0) = -3$

b. $\ddot{x} + 14\dot{x} + 49x = 0 \quad x(0) = 1 \quad \dot{x}(0) = 3$

c. $\ddot{x} + 14\dot{x} + 58x = 0 \quad x(0) = 4 \quad \dot{x}(0) = -8$

2.5 Solve the following problems.

a. $\ddot{x} + 7\dot{x} + 10x = 20 \quad x(0) = 5 \quad \dot{x}(0) = 3$

b. $5\ddot{x} + 20\dot{x} + 20x = 28 \quad x(0) = 5 \quad \dot{x}(0) = 8$

c. $\ddot{x} + 16x = 144 \quad x(0) = 5 \quad \dot{x}(0) = 12$

d. $\ddot{x} + 6\dot{x} + 34x = 68 \quad x(0) = 5 \quad \dot{x}(0) = 7$

2.6 Solve the following problems where $x(0) = \dot{x}(0) = 0$.

a. $3\ddot{x} + 30\dot{x} + 63x = 5$

b. $\ddot{x} + 14\dot{x} + 49x = 98$

c. $\ddot{x} + 14\dot{x} + 58x = 174$

2.7 Solve the following problems where $x(0) = \dot{x}(0) = 0$.

a. $\ddot{x} + 8\dot{x} + 12x = 60$

b. $\ddot{x} + 12\dot{x} + 144x = 288$

c. $\ddot{x} + 49x = 147$

d. $\ddot{x} + 14\dot{x} + 85x = 170$

2.8 Express the oscillatory part of the solution of the following problem in the form of a sine function with a phase angle.

$$\ddot{x} + 12\dot{x} + 40x = 10 \qquad x(0) = \dot{x}(0) = 0$$

2.9 If we invest \$10,000 at 5% compounded annually, at the end of one year it will be worth \$10,500. If the interest is compounded quarterly, at the end of one year (4 periods), it will be worth

$$10{,}000\left(1 + \frac{0.05}{4}\right)^4 = 10{,}509.45$$

Interest compounded continually is equivalent to an infinite number of compounding periods and the growth of the principal $P(t)$ obeys the equation

$$\frac{dP}{dt} = kP$$

where k is the yearly interest rate expressed as a decimal and time t is measured in years.

a. Compute the value of the investment after one year, using monthly compounding.

b. Compute the value of the investment after one year, using continuous compounding. How much more did you earn compared to annual compounding?

c. What rate must a bank pay, using continuous compounding, in order to achieve a savings equivalent to 5% annual compounding?

2.10 a. A microbiologist counts 100 bacteria in a certain culture. After 6 hours there are 450 bacteria. Assume that the population follows the basic growth law

$$\frac{dP}{dt} = kP$$

what is the value of k for this population?

b. In another experiment the bacteria population is observed to double in 6.5 hours. Estimate the size of the population after 36 hours.

2.11 In a certain chemical process, chemical A is converted to chemical B at a rate proportional to the square of the amount of A. It is observed that 0.06 kg (or 60 grams) of A have been converted to 0.01 kg (or 10 grams) of B after 1 hour. What amount of A is left after 2 hours?

2.12 A 60-gallon tank initially contains a mixture of 80% water and 20% alcohol. A mixture of 50% water and 50% alcohol flows in at 5 gallons per minute. The tank is well mixed and the outflow rate is also 5 gallons per minute. How many gallons of alcohol are in the tank after 20 minutes?

2.13 The ancients used water clocks to track time by designing a container shaped so that the water level falls at a constant rate c. Such a device is known as a Clepsydra clock. Use Torricelli's law to show that the container must be shaped so that the water volume is proportional to the fourth power of the radius.

2.14 We are not always able to solve for an unknown variable in terms of a formula. In such cases, the solution must be obtained by graphing, or with a numerical solution method, using a calculator or computer. The following problem illustrates this point. A certain rocket sled with a linear drag force has the equation of motion $2\dot{v} = 900 - 8v$, where $v = \dot{x}$. The sled starts from rest at $x(0) = 0$.

a. Solve for $x(t)$.

b. How long must the rocket fire before the sled travels 2500 meters?

2.15 The equation of motion for an ascending rocket having a linear drag force is

$$m\dot{v} = T - mg - cv$$

where T is the rocket's thrust and g is the acceleration due to gravity. Find the vertical velocity $v(t)$ and the rocket's height $h(t)$ given that $h(0) = v(0) = 0$.

Section 2.2 Response Parameters and Stability

2.16 Solve each of the following problems, and identify the transient and steady-state responses.

a. $\ddot{x} + 8\dot{x} + 15x = 30 \quad x(0) = 10 \quad \dot{x}(0) = 4$

b. $\ddot{x} + 10\dot{x} + 25x = 75 \quad x(0) = 10 \quad \dot{x}(0) = 4$

c. $\ddot{x} + 25x = 100 \quad x(0) = 10 \quad \dot{x}(0) = 4$

d. $\ddot{x} + 8\dot{x} + 65x = 130 \quad x(0) = 10 \quad \dot{x}(0) = 4$

2.17 Determine whether the following models are stable, unstable, or neutrally stable.

a. $4\dot{x} - 6x = 7$

b. $\ddot{x} - 8\dot{x} + 15x = 50$

c. $\ddot{x} - 2\dot{x} - 15x = 68$

d. $\dot{x} = 3$

e. $\ddot{x} + 9x = 5$

f. $\ddot{x} + 16\dot{x} = 7$

2.18 (a) Prove that the characteristic polynomial $ms^2 + cs + k$ represents a stable system if and only if m, c, and k have the same sign. (b) Derive the conditions for neutral stability.

2.19 For each of the following models, compute the time constant, if any. If no time constant exists, state why this is so.

a. $30\dot{x} + 5x = 0 \quad x(0) = 6$

b. $25\dot{x} + 6x = 15 \quad x(0) = 3$

c. $15\dot{x} + 3x = 0 \quad x(0) = -3$

d. $9\dot{x} - 3x = 0 \quad x(0) = 7$

2.20 Obtain the steady-state response of each of the following models, and estimate how long it will take the response to reach steady-state.

a. $15\dot{x} + 5x = 12 \quad x(0) = 0$

b. $15\dot{x} + 5x = 12 \quad x(0) = 3$

c. $15\dot{x} - 7x = 10 \quad x(0) = -5$

2.21 If applicable, compute ζ, τ, ω_n, and ω_d for the following characteristic polynomials. If not applicable, state the reason why.

a. $s^2 + 4s + 40$

b. $s^2 - 2s + 24$

c. $s^2 + 20s + 100$

d. $s + 10$

2.22 A model of liquid height h in a tank is obtained from Torricelli's law. Suppose that the model for a specific tank is

$$\frac{dh}{dt} = q - 5\sqrt{h}$$

where q is a volume flow rate into the tank. Assume that height is measured in meters and time in seconds. Suppose the inflow is held steady at $q = 20$. Find the equilibrium height and obtain a linearized model of the height to be used to estimate the time constant when the inflow q changes slightly from the equilibrium value.

2.23 The equation of motion of a certain pendulum is

$$\ddot{\theta} = M - a \sin \theta$$

Obtain a linear model valid for small changes in θ near (a) $\theta = \pi/4$ and (b) $\theta = 3\pi/4$. Investigate the stability of each equilibrium point.

2.24 The *logistic* equation is a model for some growth processes. It is

$$\dot{y} = ry\left(1 - \frac{y}{K}\right)$$

Find all the nonnegative equilibrium solutions, and obtain a linearized model using each equilibrium as a reference operating condition. Analyze the stability of each equilibrium, and compute the time constant if stable.

2.25 Suppose the drag force for a particular rocket sled is described by $5000v/(25 + v)$ N. The sled's mass is 1000 kg. The sled's equation of motion is

$$1000\dot{v} = f - \frac{5000v}{25 + v}$$

where f is the rocket's force. Estimate how long it will take for the sled's speed to reach a new steady-state value if the force f changes by 10%. Do this for two cases: (a) $v_r = 10$ m/s and (b) $v_r = 20$ m/s.

2.26 The characteristic polynomial of a certain system model is $s^2 + 10ds + 29d^2$, where d is a constant. (a) For what values of d is the system stable? (b) Is there a value of d for which the free response will consist of decaying oscillations?

Section 2.3 The Laplace Transform Method

2.27 Derive the Laplace transform of the ramp function $x(t) = mt$, whose slope is the constant m.

2.28 Extend the results of Problem 2.27 to obtain the Laplace transform of t^2.

2.29 Obtain the Laplace transform of the following functions.

a. $x(t) = 15 + 3t^2$

b. $x(t) = 8te^{-4t} + 2e^{-5t}$

c. $x(t) = te^{-2t} \sin 4t$

d. $x(t) = \begin{cases} 0 & t < 5 \\ t - 5 & t > 5 \end{cases}$

2.30 Use the initial- and final-value theorems to determine $x(0+)$ and $x(\infty)$ for the following transforms.

a. $X(s) = \dfrac{8}{2s + 3}$ b. $X(s) = \dfrac{7}{2s^2 + 6s + 3}$

2.31 Derive the initial-value theorem:

$$\lim_{s\to\infty} sX(s) = x(0+)$$

2.32 Derive the final-value theorem:

$$\lim_{s\to 0} sX(s) = \lim_{t\to\infty} x(t)$$

2.33 Derive the integral property of the Laplace transform:

$$\mathcal{L}\left[\int_0^t x(t)\, dt\right] = \frac{X(s)}{s} + \frac{1}{s}\int x(t)\, dt\bigg|_{t=0}$$

Section 2.4 Solving Equations with the Laplace Transform

2.34 Obtain the inverse Laplace transform $x(t)$ for each of the following transforms.

a. $X(s) = \dfrac{3}{s(s+2)}$ b. $X(s) = \dfrac{10s+7}{s(s+3)}$

c. $X(s) = \dfrac{4s+7}{(s+2)(s+5)}$ d. $X(s) = \dfrac{5}{s^2(s+4)}$

e. $X(s) = \dfrac{3s+2}{s^2(s+4)}$ f. $X(s) = \dfrac{12s+5}{(s+3)^2(s+7)}$

2.35 Obtain the inverse Laplace transform $f(t)$ for the following:

a. $\dfrac{2}{s^2+16}$ b. $\dfrac{5}{s^2+4} + \dfrac{4s}{s^2+4}$

c. $\dfrac{7}{s^2+6s+13}$ d. $\dfrac{4}{s(s+5)}$

e. $\dfrac{10}{(s+3)(s+6)}$ f. $\dfrac{5s+8}{(s+3)(s+7)}$

2.36 Obtain the inverse Laplace transform $f(t)$ for the following:

a. $\dfrac{5s}{s^2+9}$ b. $\dfrac{6}{s^2-9}$

c. $\dfrac{45}{s(s+3)^2}$ d. $\dfrac{2}{s(s^2+4s+13)}$

e. $\dfrac{20}{s(s+4)}$ f. $\dfrac{20s}{(s^2+4)^2}$

2.37 Obtain the inverse Laplace transform $x(t)$ for each of the following transforms.

a. $X(s) = \dfrac{7s+2}{s^2+6s+34}$ b. $X(s) = \dfrac{4s+3}{s(s^2+6s+34)}$

c. $X(s) = \dfrac{4s+9}{(s^2+6s+34)(s^2+4s+20)}$ d. $X(s) = \dfrac{5s^2+3s+7}{s^3+12s^2+44s+48}$

2.38 Use the Laplace transform to obtain the form of the solution of the following equation.

$$\ddot{x} + 4x = 6t$$

2.39 Use the Laplace transform to solve the following problem for $x(t)$.

$$\ddot{x} + 12\dot{x} + 40x = 3 \sin 5t \qquad x(0) = \dot{x}(0) = 0$$

2.40 Use the Laplace transform to solve the following problems.

a. $5\dot{x} + 7x = 0 \quad x(0) = 4$ b. $5\dot{x} + 7x = 15 \quad x(0) = 0$

c. $5\dot{x} + 7x = 15 \quad x(0) = 4$ d. $\dot{x} + 7x = 4t \quad x(0) = 5$

2.41 Use the Laplace transform to solve the following problems.

a. $\ddot{x} + 10\dot{x} + 21x = 0 \quad x(0) = 4 \quad \dot{x}(0) = -3$

b. $\ddot{x} + 14\dot{x} + 49x = 0 \quad x(0) = 1 \quad \dot{x}(0) = 3$

c. $\ddot{x} + 14\dot{x} + 58x = 0 \quad x(0) = 4 \quad \dot{x}(0) = -8$

2.42 Use the Laplace transform to solve the following problems.

a. $\ddot{x} + 7\dot{x} + 10x = 20 \quad x(0) = 5 \quad \dot{x}(0) = 3$

b. $5\ddot{x} + 20\dot{x} + 20x = 28 \quad x(0) = 5 \quad \dot{x}(0) = 8$

c. $\ddot{x} + 16x = 144 \quad x(0) = 5 \quad \dot{x}(0) = 12$

d. $\ddot{x} + 6\dot{x} + 34x = 68 \quad x(0) = 5 \quad \dot{x}(0) = 7$

2.43 Find the steady-state difference between the input $f(t)$ and the response $x(t)$, if $f(t) = 6t$.

$$\ddot{x} + 8\dot{x} + x = f(t) \qquad x(0) = \dot{x}(0) = 0$$

2.44 Determine the general form of the solution of the following equation, where the initial conditions $y(0)$ and $\dot{y}(0)$ have arbitrary values.

$$\ddot{y} + y = e^{-t}$$

Section 2.5 Transfer Functions

2.45 For each of the following equations, determine the transfer function $X(s)/F(s)$ and compute the characteristic roots.

a. $10\dot{x} + 14x = 15f(t)$ b. $6\ddot{x} + 60\dot{x} + 126x = 7f(t)$

c. $3\ddot{x} + 30\dot{x} + 63x = 17f(t)$ d. $2\ddot{x} + 28\dot{x} + 98x = 15f(t)$

e. $4\ddot{x} + 56\dot{x} + 232x = 8\dot{f}(t) + 3f(t)$ f. $10\dot{x} + 14x = 6\dot{f}(t) + 15f(t)$

2.46 Obtain the transfer functions $X(s)/F(s)$ and $Y(s)/F(s)$ for the following model.

$$4\dot{x} = y \qquad \dot{y} = f(t) - 5y - 17x$$

2.47 Obtain the transfer functions $X(s)/F(s)$ and $Y(s)/F(s)$ for the following model.

$$\dot{x} = -3x + 7y \qquad \dot{y} = f(t) - 8y - 3x$$

2.48 a. Obtain the transfer functions $X(s)/F(s)$ and $Y(s)/F(s)$ for the following model.

$$4\dot{x} = y \qquad \dot{y} = f(t) - 3y - 12x$$

b. Compute ζ, τ, ω_n, and ω_d for the model.

c. If $f(t) = u_s(t)$, will the responses $x(t)$ and $y(t)$ oscillate? If so, compute the radian oscillation frequency, and estimate how long it will take for the oscillations to disappear.

d. Suppose that $f(t) = u_s(t)$ and $x(0) = \dot{x}(0) = 0$. Obtain $x(t)$ and $y(t)$.

2.49 a. Obtain the transfer functions $X(s)/F(s)$ and $X(s)/G(s)$ for the following model.

$$\dot{x} = -4x + 2y + f(t) \qquad \dot{y} = -9y - 5x + g(t)$$

b. Compute ζ, τ, ω_n, and ω_d for the model.

c. If $f(t) = g(t) = 0$, estimate how long it will take for the response $x(t)$ and $y(t)$ to disappear for arbitrary initial conditions. Will the responses $x(t)$ and $y(t)$ oscillate? If so, compute the radian oscillation frequency, and estimate how long it will take for the oscillations to disappear.

Section 2.6 Pulse and Impulse Inputs

2.50 Obtain the Laplace transform of the function shown in Figure P2.50.

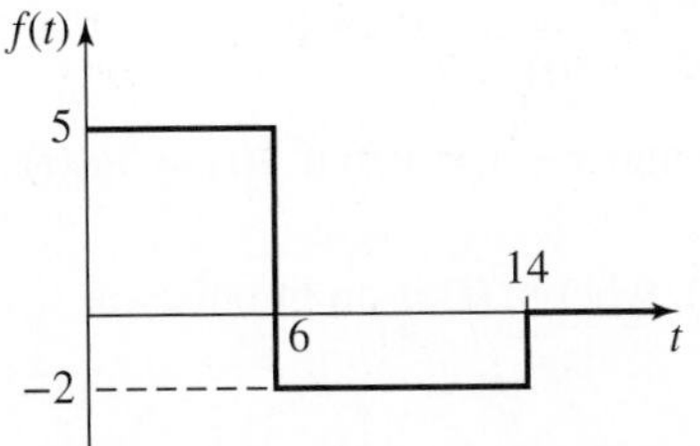

Figure P2.50

2.51 Invert the following transform.

$$X(s) = \frac{1 - e^{-3s}}{s^2 + 6s + 8}$$

2.52 Obtain the Laplace transform of the function plotted in Figure P2.52.

2.53 Obtain the Laplace transform of the function plotted in Figure P2.53.

Figure P2.52

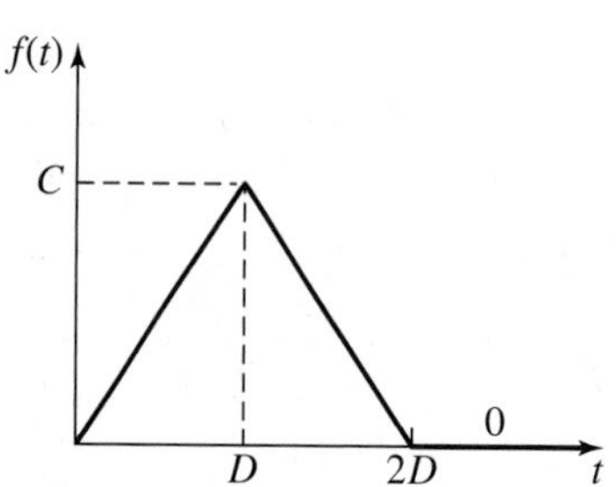

Figure P2.53

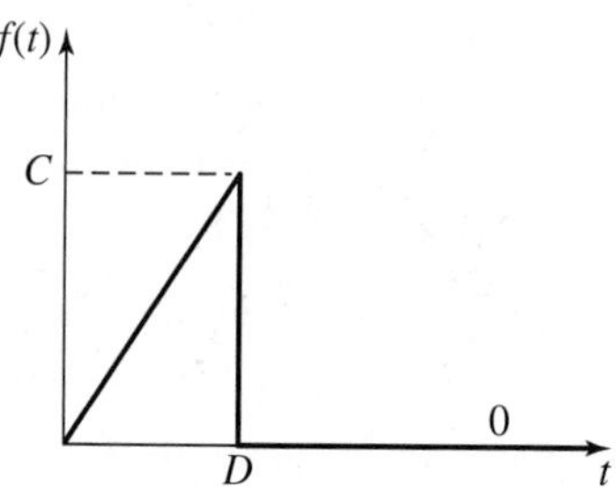

2.54 Obtain the Laplace transform of the function plotted in Figure P2.54.

Figure P2.54

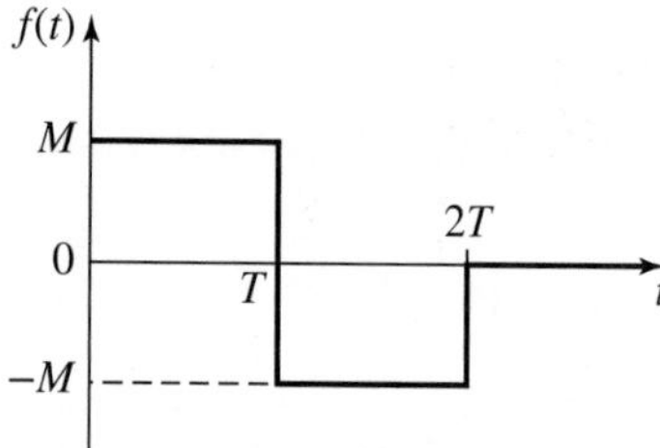

2.55 Obtain the response $x(t)$ of the following model, where the input $P(t)$ is a rectangular pulse of height 3 and duration 5.

$$4\dot{x} + x = P(t) \qquad x(0) = 0$$

2.56 Solve the following problems for $x(t)$. Compare the values of $x(0+)$ and $x(0)$. For parts (b) through (d), also compare the values of $\dot{x}(0+)$ and $\dot{x}(0)$.

a. $7\dot{x} + 5x = 4\delta(t) \quad x(0) = 3$

b. $3\ddot{x} + 30\dot{x} + 63x = 5\delta(t) \quad x(0) = \dot{x}(0) = 0$

c. $\ddot{x} + 14\dot{x} + 49x = 3\delta(t) \quad x(0) = 2 \quad \dot{x}(0) = 3$

d. $\ddot{x} + 14\dot{x} + 58x = 4\delta(t) \quad x(0) = 4 \quad \dot{x}(0) = 7$

2.57 Solve the following problems for $x(t)$. The input $g(t)$ is a unit-step function, $g(t) = u_s(t)$. Compare the values of $x(0+)$ and $x(0)$. For parts (c) and (d), also compare the values of $\dot{x}(0+)$ and $\dot{x}(0)$.

a. $7\dot{x} + 5x = 4\dot{g}(t) \quad x(0) = 3$

b. $7\dot{x} + 5x = 4\dot{g}(t) + 6g(t) \quad x(0) = 3$

c. $3\ddot{x} + 30\dot{x} + 63x = 4\dot{g}(t) \quad x(0) = 2 \quad \dot{x}(0) = 3$

d. $3\ddot{x} + 30\dot{x} + 63x = 4\dot{g}(t) + 6g(t) \quad x(0) = 4 \quad \dot{x}(0) = 7$

2.58 Compare the responses of $4\dot{x} + x = \dot{g}(t) + g(t)$ and $\dot{x} + x = g(t)$ if $g(t) = 5u_s(t)$ and $x(0-) = 0$.

2.59 Obtain the response of the model $3\dot{x} + x = f(t)$, where $f(t)$ is an impulse of strength 6 and $x(0-) = 4$.

2.60 Compare the pulse and impulse response of the model $\dot{y} = f(t) - 0.2y$ for two values of the pulse duration T. (a) $T = 1$ (b) $T = 10$

Section 2.7 Partial-Fraction Expansion

2.61 Invert the following transforms.

a. $\dfrac{16s^2 + 129s + 200}{s(s+5)(s+8)}$ b. $\dfrac{18s^2 + 172s + 394}{(s+3)(s+5)(s+7)}$

c. $\dfrac{25s^3 + 216s^2 + 500s + 288}{s(s+2)(s+4)(s+6)}$ d. $\dfrac{12s^2 + 125s + 1268}{(s+7)(s^2 + 8s + 116)}$

2.62 Invert the following transforms.

a. $\dfrac{3s+2}{s^2(s+10)}$ b. $\dfrac{5}{(s+4)^2(s+1)}$

c. $\dfrac{s^2 + 3s + 5}{s^3(s+2)}$ d. $\dfrac{s^3 + s + 6}{s^4(s+2)}$

2.63 Solve the following problems for $x(t)$.

a. $5\dot{x} + 3x = 10 + t^2 \quad x(0) = 2$ b. $4\dot{x} + 7x = 6te^{-5t} + e^{-3t} \quad x(0) = 5$

2.64 Obtain the inverse Laplace transform of

$$X(s) = \frac{30}{(s^2 + 6s + 34)(s^2 + 36)}$$

2.65 The Taylor series expansion for $\tan t$ is

$$\tan t = t + \frac{t^3}{3} + \frac{2t^5}{15} + \frac{17t^7}{315} + \cdots \qquad |t| < \frac{\pi}{2}$$

Use the first three terms in the series to obtain an approximate closed-form solution of the following problem over the interval $0 \le t \le 0.5$.

$$\dot{x} + x = \tan t \qquad x(0) = 0$$

Compare your answer at $t = 0.5$ with that obtained in Example 2.7.6, which was obtained by using two terms in the series.

2.66 Compare the LCD method with equation (2.7.4) for obtaining the inverse Laplace transform of

$$X(s) = \frac{7s + 4}{2s^2 + 16s + 30}$$

Section 2.8 Laplace Transforms and MATLAB

2.67 Use MATLAB to obtain the inverse transform of the following. If the denominator of the transform has complex roots, express $x(t)$ in terms of a sine and a cosine.

a. $X(s) = \dfrac{8s + 5}{2s^2 + 20s + 48}$ b. $X(s) = \dfrac{4s + 13}{s^2 + 8s + 116}$

c. $X(s) = \dfrac{3s + 2}{s^2(s+10)}$ d. $X(s) = \dfrac{s^3 + s + 6}{s^4(s+2)}$

e. $X(s) = \dfrac{4s + 3}{s(s^2 + 6s + 34)}$ f. $X(s) = \dfrac{5s^2 + 3s + 7}{s^3 + 12s^2 + 44s + 48}$

2.68 Use MATLAB to obtain the inverse transform of the following. If the denominator of the transform has complex roots, express $x(t)$ in terms of a sine and a cosine. Hint: You will find it convenient to use the `conv` function to multiply two polynomials.

a. $X(s) = \dfrac{5}{(s+4)^2(s+1)}$

b. $X(s) = \dfrac{4s+9}{(s^2+6s+34)(s^2+4s+20)}$

2.69 Use MATLAB to solve the following problem.

$$\dot{x} + 2x = te^{-3t}\sin 5t \quad x(0) = \dot{x}(0) = 0$$

Section 2.9 Transfer-Function Analysis with MATLAB

2.70 Use MATLAB to solve for and plot the unit-step response of the following models.

a. $3\ddot{x} + 21\dot{x} + 30x = f(t)$

b. $5\ddot{x} + 20\dot{x} + 65x = f(t)$

c. $4\ddot{x} + 32\dot{x} + 60x = 3\dot{f}(t) + 2f(t)$

2.71 Use MATLAB to solve for and plot the unit-impulse response of the following models.

a. $3\ddot{x} + 21\dot{x} + 30x = f(t)$

b. $5\ddot{x} + 20\dot{x} + 65x = f(t)$

2.72 Use MATLAB to solve for and plot the impulse response of the following model, where the strength of the impulse is 5.

$$3\ddot{x} + 21\dot{x} + 30x = f(t)$$

2.73 Use MATLAB to solve for and plot the step response of the following model, where the magnitude of the step input is 5.

$$3\ddot{x} + 21\dot{x} + 30x = f(t)$$

2.74 Use MATLAB to solve for and plot the response of the following models for $0 \le t \le 1.5$, where the input is $f(t) = 5t$ and the initial conditions are zero.

a. $3\ddot{x} + 21\dot{x} + 30x = f(t)$

b. $5\ddot{x} + 20\dot{x} + 65x = f(t)$

c. $4\ddot{x} + 32\dot{x} + 60x = 3\dot{f}(t) + 2f(t)$

2.75 Use MATLAB to solve for and plot the response of the following models for $0 \le t \le 6$, where the input is $f(t) = 6\cos 3t$ and the initial conditions are zero.

a. $3\ddot{x} + 21\dot{x} + 30x = f(t)$

b. $5\ddot{x} + 20\dot{x} + 65x = f(t)$

c. $4\ddot{x} + 32\dot{x} + 60x = 3\dot{f}(t) + 2f(t)$

2.76 Suppose a rectangular pulse $P(t)$ of height 5 and duration 4 is applied to the model $\dot{x} + 5x = P(t)$. The initial condition is $x(0) = 0$. Use MATLAB to plot the response.

2.77 Suppose a rectangular pulse $P(t)$ of height 3 and duration 2 is applied to the model $10\ddot{x} + 7\dot{x} + x = P(t)$. The initial conditions are $x(0) = \dot{x}(0) = 0$. Use MATLAB to plot the response.

C H A P T E R 3

Modeling of Rigid-Body Mechanical Systems

CHAPTER OUTLINE

CHAPTER OBJECTIVES

When you have finished this chapter, you should be able to

1. Obtain the equations of motion for an object consisting of a single mass undergoing simple translation or simple rotation.
2. Solve the equations of motion when the applied forces or moments are either constants or simple functions of time.
3. Apply the principle of conservation of mechanical energy to analyze systems acted on by conservative forces.
4. Apply the concepts of equivalent mass and equivalent inertia to obtain a simpler model of a multimass system whose motions are directly coupled.
5. Obtain the equation of motion of a body in planar motion involving simultaneous translation and rotation.
6. Select a suitable motor and gear system for a given motion-control system.

When modeling the motion of objects, the bending or twisting of the object is often negligible, and we can model the object as a rigid body. We begin this chapter by reviewing Newton's laws of motion and applying them to rigid bodies where the object's motion is relatively uncomplicated, namely, simple translations and simple rotation about a fixed axis. We then introduce energy-based methods and the concepts of equivalent mass and equivalent inertia, which simplify the modeling of systems having both translating and rotating components. Following that we treat the case of general motion in a plane, involving simultaneous translation and rotation.

In Chapter 4 we will consider systems that have nonrigid, or *elastic*, behavior. ■

3.1 TRANSLATIONAL MOTION

A *particle* is a mass of negligible dimensions. We can consider a body to be a particle if its dimensions are irrelevant for specifying its position and the forces acting on it. For example, we normally need not know the size of an earth satellite to describe its orbital path. *Newton's first law* states that a particle originally at rest, or moving in a straight line with a constant speed, will remain that way as long as it is not acted upon by an unbalanced external force. *Newton's second law* states that the acceleration of a mass particle is proportional to the vector resultant force acting on it and is in the direction of this force. *Newton's third law* states that the forces of action and reaction between interacting bodies are equal in magnitude, opposite in direction, and collinear. The third law is summarized by the commonly used statement that every action is opposed by an equal reaction.

For an object treated as a particle of mass m, the second law can be expressed as

$$m\mathbf{a} = m\frac{d\mathbf{v}}{dt} = \mathbf{f} \tag{3.1.1}$$

where **a** and **v** are the acceleration and velocity vectors of the mass and **f** is the force vector acting on the mass (Figure 3.1.1). Note that the acceleration vector and the force vector lie on the same line.

If the mass is constrained to move in only one direction, say along the direction of the coordinate x, then the equation of motion is the scalar equation

$$ma = m\frac{dv}{dt} = f \tag{3.1.2}$$

or

$$\frac{dv}{dt} = \frac{f}{m} = a \tag{3.1.3}$$

It will be convenient to use the following abbreviated "dot" notation for time derivatives:

$$\dot{x}(t) = \frac{dx}{dt} \qquad \ddot{x}(t) = \frac{d^2x}{dt^2}$$

Thus, we can express the scalar form of Newton's law as $m\dot{v} = f$.

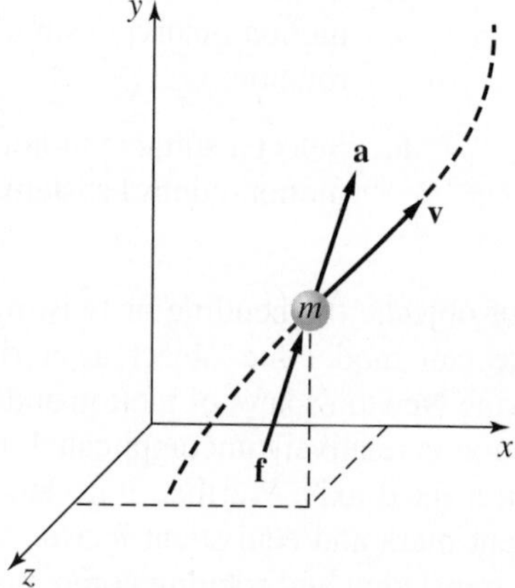

Figure 3.1.1 Particle motion showing the coordinate system, the applied force **f**, and the resulting acceleration **a**, velocity **v**, and path.

If we assume that the object is a rigid body and we neglect the force distribution within the object, we can treat the object as if its mass were concentrated at its mass center. This is the *point mass* assumption, which makes it easier to obtain the translational equations of motion, because the object's dimensions can be ignored and all external forces can be treated as if they acted through the mass center. If the object can rotate, then the translational equations must be supplemented with the rotational equations of motion, which are treated in Sections 3.2 and 3.4.

3.1.1 MECHANICAL ENERGY

Conservation of mechanical energy is a direct consequence of Newton's second law. Consider the scalar case (3.1.2), where the force f can be a function of displacement x.

$$m\dot{v} = f(x)$$

Multiply both sides by $v\,dt$ and use the fact that $v = dx/dt$.

$$mv\,dv = vf(x)\,dt = \frac{dx}{dt}f(x)\,dt = f(x)\,dx$$

Integrate both sides.

$$\int mv\,dv = \frac{mv^2}{2} = \int f(x)\,dx + C \tag{3.1.4}$$

where C is a constant of integration.

Work is force times displacement, so the integral on the right represents the total work done on the mass by the force $f(x)$. The term on the left-hand side of the equal sign is called the *kinetic energy* (KE).

If the work done by $f(x)$ is independent of the path and depends only on the end points, then the force $f(x)$ is derivable from a function $V(x)$ as follows:

$$f(x) = -\frac{dV}{dx} \tag{3.1.5}$$

Then, in this case, $f(x)$ is called a *conservative force*. If we integrate both sides of the last equation, we obtain

$$V(x) = \int dV = -\int f(x)\,dx$$

or from (3.1.4),

$$\frac{mv^2}{2} + V(x) = C \tag{3.1.6}$$

This equation shows that $V(x)$ has the same units as kinetic energy. $V(x)$ is called the *potential energy* (PE) *function*.

Equation (3.1.6) states that the sum of the kinetic and potential energies must be constant, if no force other than the conservative force is applied. If v and x have the values v_0 and x_0 at the time t_0, then

$$\frac{mv_0^2}{2} + V(x_0) = C$$

Comparing this with (3.1.6) gives

$$\frac{mv^2}{2} - \frac{mv_0^2}{2} + V(x) - V(x_0) = 0 \tag{3.1.7}$$

which can be expressed as

$$\Delta KE + \Delta PE = 0 \tag{3.1.8}$$

where the change in kinetic energy is $\Delta KE = m(v^2 - v_0^2)/2$ and the change in potential energy is $\Delta PE = V(x) - V(x_0)$. For some problems, the following form of the principle is more convenient to use:

$$\frac{mv_0^2}{2} + V(x_0) = \frac{mv^2}{2} + V(x) \tag{3.1.9}$$

In the form (3.1.8), we see that conservation of mechanical energy states that the change in kinetic energy plus the change in potential energy is zero. Note that the potential energy has a relative value only. The choice of reference point for measuring x determines only the value of C, which (3.1.7) shows to be irrelevant.

Gravity is an example of a conservative force, for which $f = -mg$. The gravity force is conservative because the work done lifting an object depends only on the change in height and not on the path taken. Thus, if x represents vertical displacement,

$$V(x) = mgx$$

and

$$\frac{mv^2}{2} + mgx = C \tag{3.1.10}$$

$$\frac{mv^2}{2} - \frac{mv_0^2}{2} + mg(x - x_0) = 0 \tag{3.1.11}$$

EXAMPLE 3.1.1

Speed of a Falling Object

■ Problem

An object with a mass of $m = 2$ slugs drops from a height of 30 ft above the ground (see Figure 3.1.2). Determine its speed after it drops 20 ft to a platform that is 10 ft above the ground.

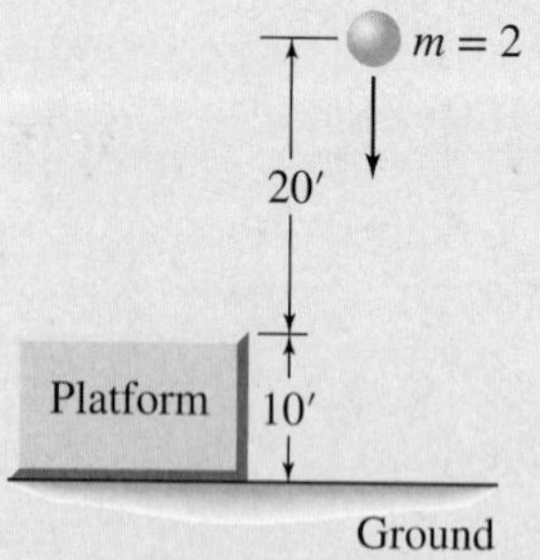

Figure 3.1.2 A falling object.

■ Solution

Measuring x from the ground gives $x_0 = 30$ ft and $x = 10$ ft at the platform. If the object is dropped from rest, then $v_0 = 0$. From (3.1.11)

$$\frac{m}{2}(v^2 - 0) + mg(10 - 30) = 0$$

or $v^2 = 40g$. Using $g = 32.2$ ft/sec^2, we obtain $v = \sqrt{392.4} = 19.81$ ft/sec. This is the speed of the object when it reaches the platform.

Note that if we had chosen to measure x from the platform instead of the ground, then $v_0 = 0$, $x_0 = 20$, and $x = 0$ at the platform. Equation (3.1.11) gives

$$\frac{m}{2}(v^2 - 0) + mg(0 - 20) = 0$$

or $v^2 = 40g$, which gives the same answer as before. When x is measured from the platform, (3.1.10) gives $C = 20mg$. When x is measured from the ground, $C = 30mg$, but the two values of C are irrelevant for solving the problem because it is the *change* in kinetic and potential energies that governs the object's dynamics.

3.1.2 CONSTANT FORCE CASE

For the point mass model, $ma = f$, (3.1.9) can be used to find the speed v as a function of displacement x. If f is a constant, then

$$\frac{mv^2}{2} = f(x - x_0) + \frac{mv_0^2}{2} \tag{3.1.12}$$

Noting that work equals force times displacement, the work done on the mass by the force f is $f(x - x_0)$. Thus (3.1.12) says that the final energy of the mass, $mv^2/2$, equals the initial energy, $mv_0^2/2$, plus the work done by the force f. This is a statement of conservation of mechanical energy.

3.1.3 DRY FRICTION FORCE

Not every constant force is conservative. A common example of a *non-conservative* force is the *dry* friction force. This force is non-conservative because the work done by the force depends on the path taken. The dry friction force F is directly proportional to the force N normal to the frictional surface. Thus, $F = \mu N$. The proportionality constant is μ, the *coefficient of friction.*

The dry friction force that exists before motion begins is called *static friction* (sometimes shortened to *stiction*). The static friction coefficient has the value μ_s to distinguish it from the *dynamic* friction coefficient μ_d, which describes the friction after motion begins. Dynamic friction is also called *sliding* friction, *kinetic* friction, or *Coulomb* friction. In general, $\mu_s > \mu_d$, which explains why it is more difficult to start an object sliding than to keep it moving. We will use the symbol μ rather than μ_d because most of our applications involve motion.

Because Coulomb friction cannot be derived from a potential energy function, the conservation of mechanical energy principle does not apply. This makes sense physically because the friction force dissipates the energy as heat, and thus mechanical energy, which consists of kinetic plus potential energy, is not conserved. Total energy, of course, is conserved.

EXAMPLE 3.1.2

Equation of Motion with Friction

■ Problem

Derive the equation of motion (a) for the block of mass m shown in Figure 3.1.3a and (b) for the mass m on an incline, shown in Figure 3.1.3b. In both cases, a force f_1, which is not the friction force, is applied to move the mass.

■ Solution

a. The free body diagrams are shown in Figure 3.1.3a for the two cases: $v > 0$ and $v < 0$. The normal force N is the weight mg. Thus, the friction force F is μN, or $F = \mu mg$. If $v > 0$, the equation of motion is

$$m\dot{v} = f_1 - \mu mg \qquad v > 0 \tag{1}$$

Dry friction always opposes the motion. So, for $v < 0$,

$$m\dot{v} = f_1 + \mu mg \qquad v < 0 \tag{2}$$

Equations (1) and (2) are the desired equations of motion.

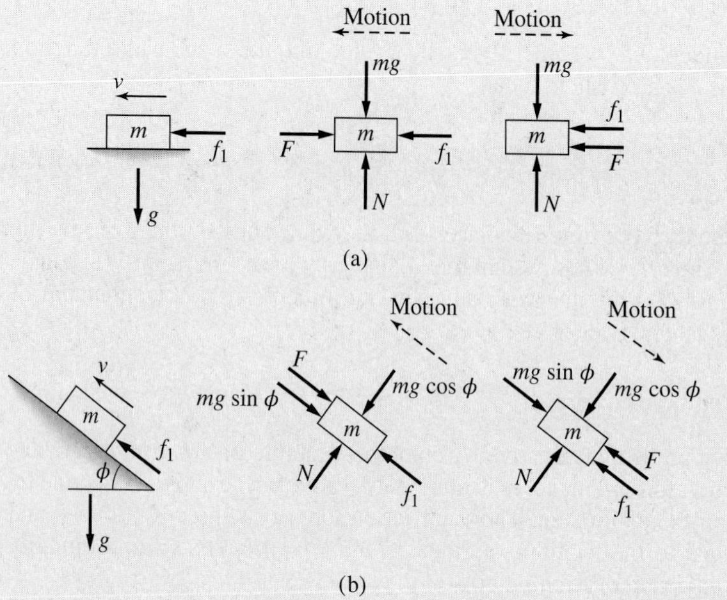

Figure 3.1.3 Motion with friction (a) on a horizontal surface and (b) on an inclined plane.

b. The friction force depends on the force normal to the surface. For the mass m shown in Figure 3.1.3b, the normal force N must equal $mg\cos\phi$ as long as the mass is in contact with the surface. The free body diagrams are shown in the figure for the two cases: $v > 0$ and $v < 0$. Newton's second law applied in the direction parallel to the surface gives, for $v > 0$,

$$m\dot{v} = f_1 - mg\sin\phi - \mu mg\cos\phi \qquad v > 0 \tag{3}$$

For $v < 0$,

$$m\dot{v} = f_1 - mg\sin\phi + \mu mg\cos\phi \qquad v < 0 \tag{4}$$

Equations (3) and (4) are the desired equations of motion.

EXAMPLE 3.1.3

Motion with Friction on an Inclined Plane

■ Problem

For the mass shown in Figure 3.1.3b, $m = 2$ kg, $\phi = 30°$, $v(0) = 3$ m/s, and $\mu = 0.5$. Determine whether the mass comes to rest if (a) $f_1 = 50$ N and (b) $f_1 = 5$ N.

■ Solution

Because the velocity is initially positive [$v(0) = 3$], we use equation (3) of Example 3.1.2:

$$2\dot{v} = f_1 - (\sin 30° + 0.5\cos 30°)(2)(9.81) = f_1 - 18.3$$

For part (a), $f_1 = 50$ and thus $\dot{v} = (50 - 18.3)/2 = 15.85$ and the acceleration is positive. Thus, because $v(0) > 0$, the speed is always positive for $t \geq 0$ and the mass never comes to rest.

For part (b), $f_1 = 5$, $\dot{v} = (5 - 18.3)/2 = -6.65$, and thus the mass is decelerating. Because $v(t) = -6.65t + 3$, the speed becomes zero at $t = 3/6.65 = 0.45$ s.

3.2 ROTATION ABOUT A FIXED AXIS

In this section, we consider the dynamics of rigid bodies whose motion is constrained to allow only rotation about an axis through a nonaccelerating point. In Section 3.4 we will treat the case of rotation about an axis through an accelerating point.

For planar motion, which means that the object can translate in two dimensions and can rotate only about an axis that is perpendicular to the plane, Newton's second law can be used to show that

$$I_O\dot{\omega} = M_O \tag{3.2.1}$$

where ω is the *angular velocity* of the mass about an axis through a point O fixed in an inertial reference frame and attached to the body (or the body extended), I_O is the *mass moment of inertia* of the body about the point O, and M_O is the sum of the moments applied to the body about the point O.

This situation is illustrated in Figure 3.2.1. The angular displacement is θ, and $\dot{\theta} = \omega$. The term *torque* and the symbol T are often used instead of *moment* and M. Also, when the context is unambiguous, we use the term "inertia" as an abbreviation for "mass moment of inertia."

3.2.1 CALCULATING INERTIA

The mass moment of inertia I about a specified reference axis is defined as

$$I = \int r^2\, dm \tag{3.2.2}$$

where r is the distance from the reference axis to the mass element dm. The expressions for I for some common shapes are given in Table 3.2.1.

If the rotation axis of a homogeneous rigid body does not coincide with the body's axis of symmetry, but is parallel to it at a distance d, then the mass moment of inertia about the rotation axis is given by the *parallel-axis theorem*,

$$I = I_s + md^2 \tag{3.2.3}$$

where I_s is the inertia about the symmetry axis (see Figure 3.2.2).

Figure 3.2.1 An object rotating about a fixed axis.

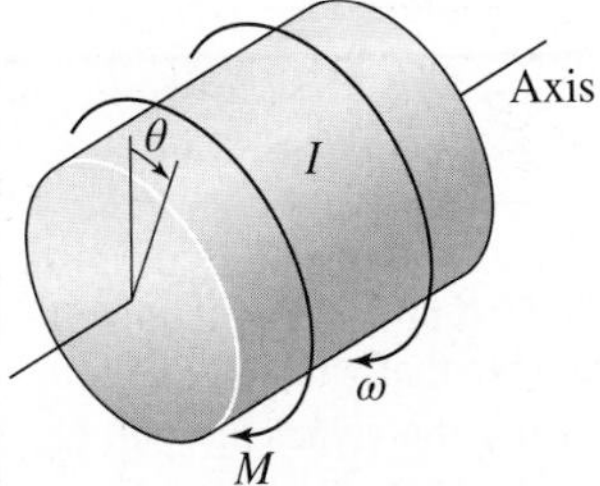

Figure 3.2.2 Illustration of the parallel-axis theorem.

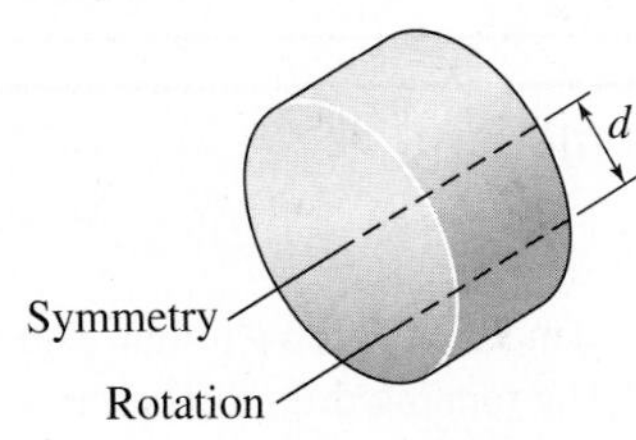

Table 3.2.1 Mass moments of inertia of common elements.

Sphere

$$I_G = \frac{2}{5}mR^2$$

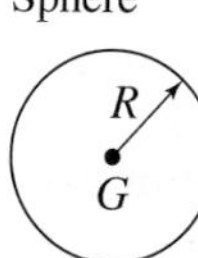

Mass rotating about point O

$$I_O = mR^2$$

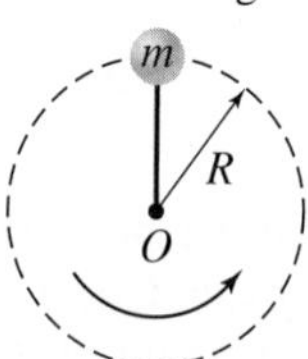

Hollow cylinder

$$I_x = \frac{1}{2}m(R^2 + r^2)$$

$$I_y = I_z = \frac{1}{12}m(3R^2 + 3r^2 + L^2)$$

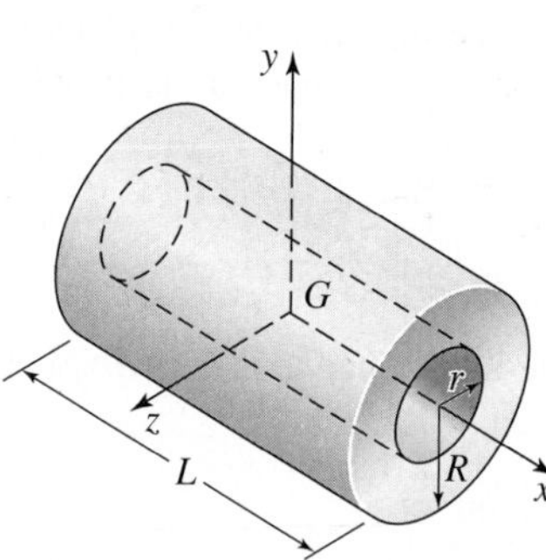

Rectangular prism

$$I_x = \frac{1}{12}m(b^2 + c^2)$$

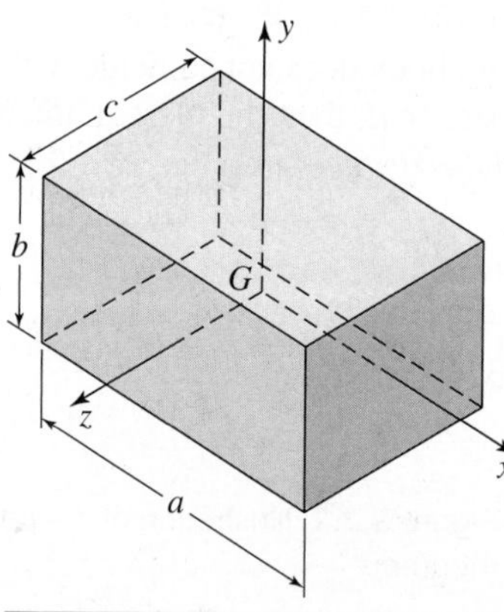

EXAMPLE 3.2.1

A Single-Drum Hoist

■ Problem

In Figure 3.2.3a, a motor supplies a torque T to turn a drum of radius R and inertia I about its axis of rotation. The rotating drum lifts a mass m by means of a cable that wraps around the drum. The drum's speed is ω. Discounting the mass of the cable, use the values $m = 40$ kg, $R = 0.2$ m, and $I = 0.8$ kg·m^2. Find the acceleration $\dot{v}$ if the torque $T = 300$ N·m.

Figure 3.2.3 (a) A single-drum hoist. (b) Free body diagram.

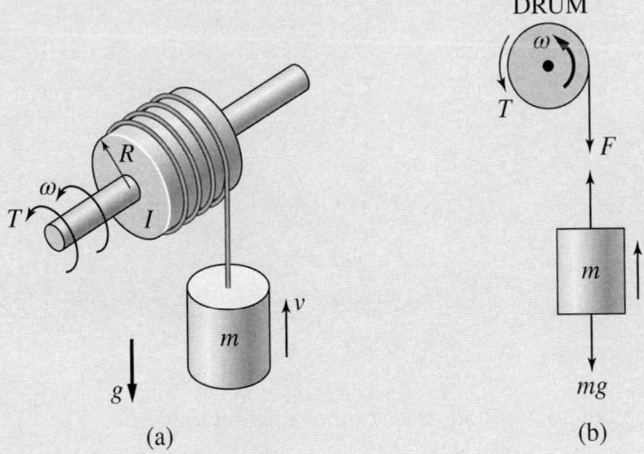

■ Solution

The free body diagrams are shown in Figure 3.2.3b. Let F be the tension in the cable. Because the drum rotation axis is assumed to be fixed, we can use (3.2.1). Summing moments about the drum center gives $M_O = 300 - 0.2F$. Because $I_O = I = 0.8$, we have

$$0.8\dot{\omega} = 300 - 0.2F \tag{1}$$

Summing vertical forces on the 40-kg mass m gives

$$40\dot{v} = F - 40(9.81) \tag{2}$$

Solve (2) for F and substitute for F in (1) to obtain

$$0.8\dot{\omega} = 300 - 8\dot{v} - 8(9.81) \tag{3}$$

Note that $v = R\omega = 0.2\omega$ to obtain $\dot{v} = 0.2\dot{\omega}$. Substitute this into (3) and collect terms to obtain $12\dot{v} = 300 - 8(9.81)$ or $\dot{v} = 18.46$ m/s^2.

Pendulum with a Concentrated Mass

EXAMPLE 3.2.2

■ Problem

The pendulum shown in Figure 3.2.4a consists of a concentrated mass m a distance L from point O, attached to a rod of length L. (a) Obtain its equation of motion. (b) Solve the equation assuming that θ is small.

■ Solution

a. Because the support at point O is assumed to be fixed, we can use (3.2.1). The free body diagram is shown in Figure 3.2.4b, where we have resolved the weight into a component parallel to the rod and one perpendicular to the rod. This makes it easier to compute the moment about point O caused by the weight. The parallel component has a line of action through point O and thus contributes nothing to M_O. The moment arm of the perpendicular component is L, and thus $M_O = -mgL \sin \theta$. The negative sign is required because the moment acts in the negative θ direction. From Table 3.2.1, $I_O = mL^2$, and thus the equation of motion is

$$mL^2\ddot{\theta} = -mgL \sin \theta$$

Figure 3.2.4 (a) Pendulum with a concentrated mass. (b) Free body diagram.

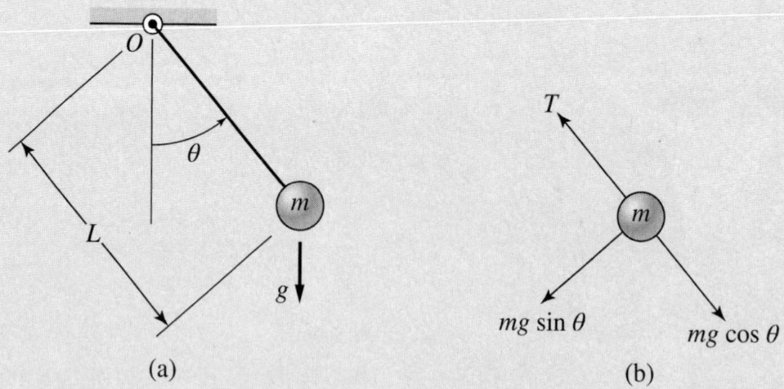

The product mL can be factored out of both sides of the equation to give

$$L\ddot{\theta} = -g \sin \theta \tag{1}$$

which is nonlinear and not solvable in terms of elementary functions.

b. If θ is small enough and measured in radians, then $\sin \theta \approx \theta$ (for example, if $\theta = 32° = 0.56$ rad, $\sin 0.56 = 0.5312$, an error of 5%). Then (1) can be replaced by the linear equation

$$L\ddot{\theta} = -g\theta \tag{2}$$

The characteristic roots are $s = \pm j\sqrt{g/L} = \pm j\omega_n$, where $\omega_n = \sqrt{g/L}$. The solution is

$$\theta(t) = \theta(0) \cos \omega_n t + \frac{\dot{\theta}(0)}{\omega_n} \sin \omega_n t$$

The pendulum will swing back and forth with a radian frequency of ω_n. By expressing the solution as a single sine function, we can obtain the following expression for the oscillation amplitude:

$$\theta_{\max} = \sqrt{\theta^2(0) + \frac{\dot{\theta}^2(0)}{\omega_n^2}}$$

So, before accepting any predictions based on this linear model, we should first check to see whether $\sin \theta_{\max} \approx \theta_{\max}$.

3.2.2 ENERGY AND ROTATIONAL MOTION

The work done by a moment M causing a rotation through an angle θ is

$$W = \int_0^\theta M \, d\theta \tag{3.2.4}$$

Multiply both sides of (3.2.1) by $\omega \, dt$, and note that $\omega = d\theta/dt$.

$$I\omega \, d\omega = M \, d\theta$$

Integrating both sides gives

$$\int_0^\omega I\omega \, d\omega = \frac{1}{2} I\omega^2 = \int_0^\theta M \, d\theta \tag{3.2.5}$$

We thus see that the work done by the moment M produces the kinetic energy of rotation: $\text{KE} = I\omega^2/2$.

3.2.3 PULLEY DYNAMICS

Pulleys can be used to change the direction of an applied force or to amplify forces. In our examples, we will assume that the cords, ropes, chains, and cables drive the pulleys without slipping and are inextensible; if not, then they must be modeled as springs. Figure 3.2.5 shows a pulley of inertia I whose center is fixed to a support. Assume that the tension forces in the cable are different on each side of the pulley. Then application of (3.2.1) gives

$$I\ddot{\theta} = RT_1 - RT_2 = R(T_1 - T_2)$$

An immediate result of practical significance is that the tension forces are approximately equal if $I\ddot{\theta}$ is negligible. This condition is satisfied if either the pulley rotates at a constant speed or if the pulley inertia is negligible compared to the other inertias in the system. The pulley inertia will be negligible if either its mass or its radius is small. Thus, when we neglect the mass, radius, or inertia of a pulley, the tension forces in the cable may be taken to be the same on both sides of the pulley.

The force on the support at the pulley center is $T_1 + T_2 + mg$. If the mass, radius, or inertia of the pulley are negligible, then the support force is $2T_1$.

Figure 3.2.5 Pulley forces.

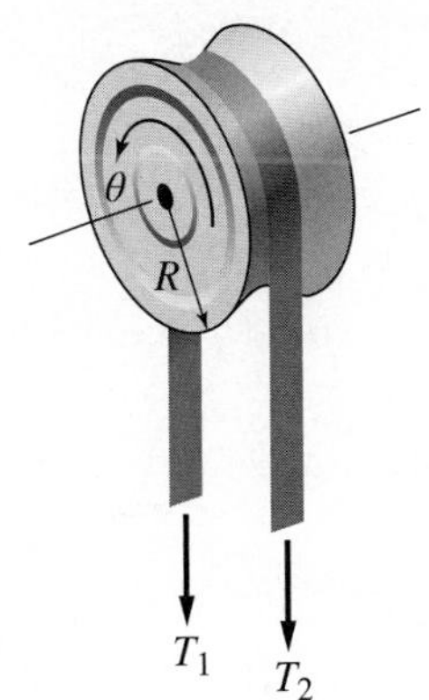

Energy Analysis of a Pulley System

EXAMPLE 3.2.3

■ Problem

Figure 3.2.6a shows a pulley used to raise the mass m_2 by hanging a mass m_1 on the other side of the pulley. If pulley inertia is negligible, then it is obvious that m_1 will lift m_2 if $m_1 > m_2$. How does a nonnegligible pulley inertia I change this result? Also, investigate the effect of the pulley inertia on the speed of the masses.

■ Solution

Define the coordinates x and y such that $x = y = 0$ at the start of the motion. If the pulley cable is inextensible, then $x = y$ and thus $\dot{x} = \dot{y}$. If the cable does not slip, then $\dot{\theta} = \dot{x}/R$. Because we were asked about the speed and because the only applied force is a conservative force (gravity), this suggests that we use an energy-based analysis. If the system starts at rest at $x = y = 0$, then the kinetic energy is initially zero. We take the potential energy to be zero at $x = y = 0$. Thus,

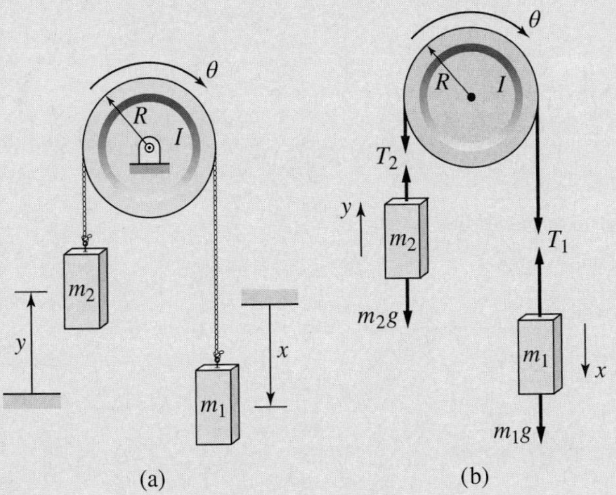

Figure 3.2.6 A pulley system for lifting a mass.

the total mechanical energy is initially zero, and from conservation of energy we obtain

$$\text{KE} + \text{PE} = \frac{1}{2}m_1\dot{x}^2 + \frac{1}{2}m_2\dot{y}^2 + \frac{1}{2}I\dot{\theta}^2 + m_2gy - m_1gx = 0$$

Note that the potential energy of m_1 has a negative sign because m_1 loses potential energy when x is positive.

Substituting $y = x$, $\dot{y} = \dot{x}$, and $\dot{\theta} = \dot{x}/R$ into this equation and collecting like terms gives

$$\frac{1}{2}\left(m_1 + m_2 + \frac{I}{R^2}\right)\dot{x}^2 + (m_2 - m_1)gx = 0$$

and thus

$$\dot{x} = \sqrt{\frac{2(m_1 - m_2)gx}{m_1 + m_2 + I/R^2}} \tag{1}$$

The mass m_2 will be lifted if $\dot{x} > 0$; that is, if $m_2 < m_1$. So the pulley inertia does not affect this result. However, because I appears in the denominator of the expression for $\dot{x}$, the pulley inertia does decrease the speed with which m_1 lifts m_2.

In Example 3.2.3, it is inconvenient to use an energy-based analysis to compute $x(t)$ or the tensions in the cable. To do this, it is easier to use Newton's law directly.

EXAMPLE 3.2.4

Equation of Motion of a Pulley System

■ Problem

Consider the pulley system shown in Figure 3.2.6a. Obtain the equation of motion in terms of x and obtain an expression for the tension forces in the cable.

■ Solution

The free body diagrams of the three bodies are shown in part (b) of the figure. Newton's law for mass m_1 gives

$$m_1\ddot{x} = m_1g - T_1 \tag{1}$$

Newton's law for mass m_2 gives

$$m_2\ddot{y} = T_2 - m_2g \tag{2}$$

Equation (3.2.1) applied to the inertia I gives

$$I\ddot{\theta} = RT_1 - RT_2 = R(T_1 - T_2) \tag{3}$$

Because the cable is assumed inextensible, $x = y$ and thus $\ddot{x} = \ddot{y}$. We can then solve (1) and (2) for the tension forces.

$$T_1 = m_1g - m_1\ddot{x} = m_1(g - \ddot{x}) \tag{4}$$

$$T_2 = m_2\ddot{y} + m_2g = m_2(\ddot{y} + g) = m_2(\ddot{x} + g) \tag{5}$$

Substitute these expressions into (3).

$$I\ddot{\theta} = (m_1 - m_2)gR - (m_1 + m_2)R\ddot{x} \tag{6}$$

Because $x = R\theta$, $\ddot{x} = R\ddot{\theta}$, and (6) becomes

$$I\frac{\ddot{x}}{R} = (m_1 - m_2)gR - (m_1 + m_2)R\ddot{x}$$

which can be rearranged as

$$\left(m_1 + m_2 + \frac{I}{R^2}\right)\ddot{x} = (m_1 - m_2)g \tag{7}$$

This is the desired equation of motion. We can solve it for $\ddot{x}$ and substitute the result into equations (4) and (5) to obtain T_1 and T_2 as functions of the parameters m_1, m_2, I, R, and g.

Equation (7) can be solved for $\dot{x}(t)$ and $x(t)$ by direct integration. Let

$$A = \frac{(m_1 - m_2)gR^2}{(m_1 + m_2)R^2 + I}$$

Then (7) becomes $\ddot{x} = A$, whose solutions are $\dot{x} = At + \dot{x}(0)$ and $x = At^2/2 + \dot{x}(0)t + x(0)$. Note that if we use the solutions to express $\dot{x}$ as a function of x, we will obtain the same expression as equation (1) in Example 3.2.3.

3.2.4 PULLEY-CABLE KINEMATICS

Consider Figure 3.2.7. Suppose we need to determine the relation between the velocities of mass m_A and mass m_B. Define x and y as shown from a common reference line attached to a fixed part of the system. Noting that the cable lengths wrapped around the pulleys are constant, we can write $x + 3y =$ constant. Thus $\dot{x} + 3\dot{y} = 0$. So the speed of point A is three times the speed of point B, and in the opposite direction.

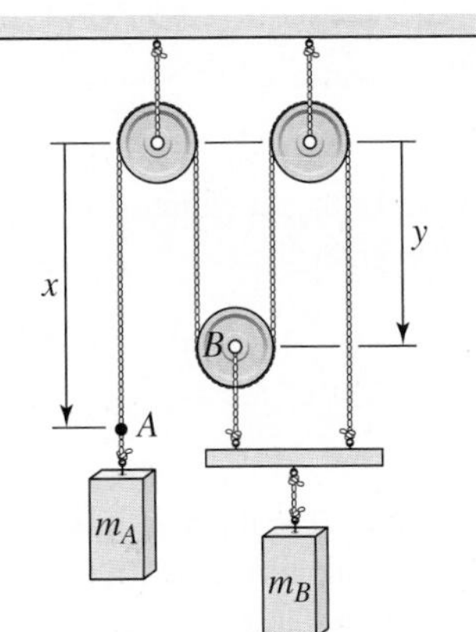

Figure 3.2.7 A multiple-pulley system.

In many problems, we neglect the inertia of the pulleys so as to keep the resulting model as simple as possible. As we will see, the models of many practical engineering applications are challenging enough without introducing pulley dynamics. Example 3.2.4 illustrates such an application.

EXAMPLE 3.2.5

A Pulley System

■ **Problem**

The two masses shown in Figure 3.2.8a are released from rest. The mass of block A is 60 kg; the mass of block B is 40 kg. Disregard the masses of the pulleys and rope. Block A is heavier than block B, but will block B rise or fall? Find out by determining the acceleration of block B by (a) using free body diagrams and (b) using conservation of energy.

Figure 3.2.8 (a) A pulley system. (b) Free body diagrams.

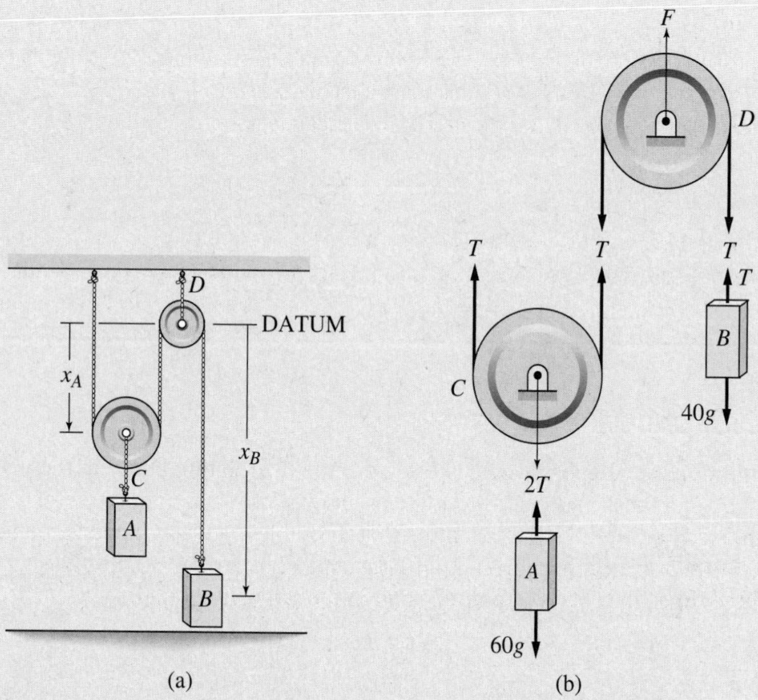

■ **Solution**

a. The free body diagrams are given in part (b) of the figure. We take the acceleration of each block to be positive downward. For block A, Newton's law gives

$$60\ddot{x}_A = 60g - 2T \tag{1}$$

where T is the tension in the rope. For block B,

$$40\ddot{x}_B = 40g - T \tag{2}$$

For the massless pulley C, $2T = T + T$, which gives no information. For the massless pulley D, $F = 2T$, which is useful only if we need to calculate the support force F.

If the unwrapped rope length below the datum line is L, then we have $2x_A + x_B = L$. Differentiating this twice gives $\ddot{x}_B = -2\ddot{x}_A$. Substituting this into (1) gives $T = 30g + 15\ddot{x}_B$, and from (2) we have

$$11\ddot{x}_B = 2g$$

or $\ddot{x}_B = 2g/11 = 1.784 \text{ m/s}^2$. The acceleration is positive, which means that block B is accelerating downward.

b. Choosing gravitational potential energy to be zero at the datum line, the total energy in the system is

$$\frac{1}{2}m_A\dot{x}_A^2 + \frac{1}{2}m_B\dot{x}_B^2 - m_A g x_A - m_B g x_B = \text{constant}$$

Differentiate both sides with respect to time to obtain

$$m_A\dot{x}_A\ddot{x}_A + m_B\dot{x}_B\ddot{x}_B - m_A g\dot{x}_A - m_B g\dot{x}_B = 0$$

However, because the unwrapped rope length below the datum line is L, then we have $2x_A + x_B = L$. This gives $\dot{x}_A = -\dot{x}_B/2$ and $\ddot{x}_A = -\ddot{x}_B/2$. Substitute these and cancel to obtain

$$\left(\frac{m_A}{4} + m_B\right)\ddot{x}_B - \left(\frac{m_A}{2} - m_g\right)g = 0$$

Substituting the given values for m_A and m_B gives the same answer as obtained in part (a): $11\ddot{x}_B = 2g$.

3.3 EQUIVALENT MASS AND INERTIA

Some systems composed of translating and rotating parts whose motions are directly coupled can be modeled as a purely translational system or as a purely rotational system, by using the concepts of equivalent mass and equivalent inertia. These models can be derived using kinetic energy equivalence.

Equivalent mass and equivalent inertia are complementary concepts. A system should be viewed as an equivalent mass if an external force is applied, and as an equivalent inertia if an external torque is applied. Examples 3.3.1–3.3.7 will illustrate this approach.

3.3.1 MECHANICAL DRIVES

Gears, belts, levers, and pulleys transform an input motion, force, or torque into another motion, force, or torque at the output. For example, a gear pair can be used to reduce speed and increase torque, and a lever can increase force.

Several types of gears are used in mechanical drives. These include helical, spur, rack-and-pinion, worm, bevel, and planetary gears. Other mechanical drives use belts or chains. We now use a spur-gear pair, a rack-and-pinion gear pair, and a belt drive to demonstrate the use of kinetic energy equivalence to obtain a model. This approach can be used to analyze other gear and drive types.

A pair of spur gears is shown in Figure 3.3.1. The input shaft (shaft 1) is connected to a motor that produces a torque T_1 at a speed ω_1, and drives the output shaft (shaft 2). One use of such a system is to increase the effective motor torque. The gear ratio N is defined as the ratio of the input rotation θ_1 to the output rotation θ_2. Thus, $N = \theta_1/\theta_2$. From geometry we can see that N is also the speed ratio $N = \omega_1/\omega_2$. Thus, the pair is a speed *reducer* if $N > 1$. The gear ratio is also the diameter ratio $N = D_2/D_1$, and the gear tooth ratio $N = n_2/n_1$, where n is the number of gear teeth.

If the gear inertias are negligible or if there is zero acceleration, and if we neglect energy loss due to friction, such as that between the gear teeth, then the input work $T_1\theta_1$

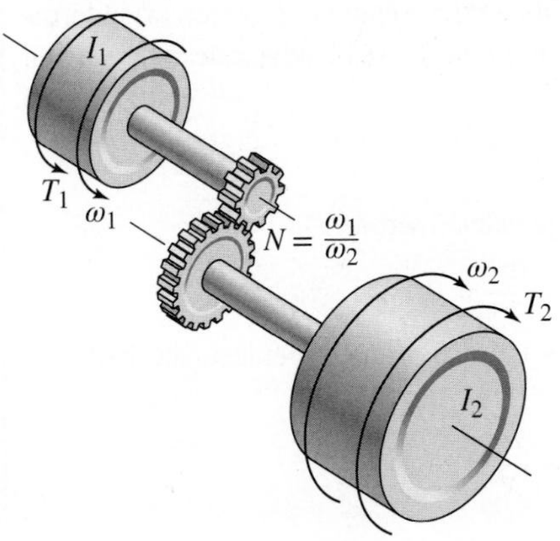

Figure 3.3.1 A spur-gear pair.

must equal the output work $T_2\theta_2$. Thus, under these conditions, $T_2 = T_1(\theta_1/\theta_2) = NT_1$, and the output torque is greater than the input torque for a speed reducer. For cases that involve acceleration and appreciable gear inertia, the output torque is less than NT_1.

EXAMPLE 3.3.1

Equivalent Inertia of Spur Gears

■ Problem

Consider the spur gears shown in Figure 3.3.1. Derive the expression for the equivalent inertia I_e felt on the input shaft, and obtain the equation of motion in terms of the speed ω_1.

■ Solution

Let I_1 and I_2 be the total moments of inertia on the shafts. Note that $\omega_2 = \omega_1/N$. The kinetic energy of the system is then

$$\text{KE} = \frac{1}{2}I_1\omega_1^2 + \frac{1}{2}I_2\omega_2^2 = \frac{1}{2}I_1\omega_1^2 + \frac{1}{2}I_2\left(\frac{\omega_1}{N}\right)^2$$

or

$$\text{KE} = \frac{1}{2}\left(I_1 + \frac{1}{N^2}I_2\right)\omega_1^2$$

Therefore, the equivalent inertia felt on the input shaft is

$$I_e = I_1 + \frac{I_2}{N^2} \tag{1}$$

This means that the dynamics of the system can be described by the model $I_e\dot{\omega}_1 = T_1 + T_2/N$. The torque T_2 is *not* the torque on the load shaft due to the torque T_1. Rather, T_2 is due to external causes. For example, if I_1 represents a motor, and I_2 represents a vehicle wheel, then T_2 would be due to road forces, or gravity. If the vehicle were going downhill, gravity would act to accelerate the vehicle ($\omega_2 > 0$), and the resulting torque T_2 would be positive. If the vehicle were going uphill, gravity would act to decelerate the vehicle ($\omega_2 < 0$), and the resulting torque T_2 would be negative.

EXAMPLE 3.3.2

A Speed Reducer

■ Problem

For the geared system shown in Figure 3.3.1, the inertias in kg·m^2 are $I_1 = 0.1$, for the motor shaft and $I_2 = 0.4$ for the load shaft. The motor speed ω_1 is five times faster than the load speed ω_2, so this device is called a *speed reducer*. Obtain the equation of motion (a) in terms of ω_1 and (b) in terms of ω_2, assuming that the motor torque T_1 and load torque T_2 are given.

■ Solution

For the given speed information, we have that $N = \omega_1/\omega_2 = 5$.

a. Referencing both inertias to shaft 1 gives the equivalent inertia

$$I_e = I_1 + \frac{I_2}{N^2} = 0.1 + \frac{0.4}{25} = 0.116$$

Note that the moment felt on shaft 1 due to T_2 is T_2/N. Summing moments on shaft 1 gives

$$I_e\dot{\omega}_1 = T_1 + \frac{T_2}{N}$$

or

$$0.116\dot{\omega}_1 = T_1 + \frac{T_2}{5} \tag{1}$$

b. Starting from basic principles, we express the kinetic energy of the system as

$$\text{KE} = \frac{1}{2}I_1\omega_1^2 + \frac{1}{2}I_2\omega_2^2 = \frac{1}{2}I_1\left(5\omega_2\right)^2 + \frac{1}{2}I_2\omega_2^2 = \frac{1}{2}(2.5 + 0.4)\,\omega_2^2 = \frac{1}{2}(2.9)\,\omega_2^2$$

Thus, the equivalent inertia referenced to shaft 2 is 2.9, and the equation of motion is

$$2.9\dot{\omega}_2 = NT_1 + T_2 = 5T_1 + T_2$$

Of course, we could have obtained this result directly from (1) with the substitution $\omega_1 = 5\omega_2$, if (1) were already available.

Note that with a speed reducer, the load speed is slower than the motor speed, but the effect of the motor torque on the load shaft is increased by a factor equal to the gear ratio N.

A Three-Gear System

EXAMPLE 3.3.3

■ Problem

For the system shown in Figure 3.3.2, assume that the shaft inertias are small. The remaining inertias in kg·m^2 are $I_1 = 0.005$, $I_2 = 0.01$, $I_3 = 0.02$, $I_4 = 0.04$, and $I_5 = 0.2$. The speed ratios are

$$\frac{\omega_1}{\omega_2} = \frac{3}{2} \qquad \frac{\omega_2}{\omega_3} = 2$$

Obtain the equation of motion in terms of ω_3. The torque T is given.

■ Solution

Note that

$$\omega_1 = \left(\frac{3}{2}\right)\omega_2 = \left(\frac{3}{2}\right)2\omega_3 = 3\omega_3$$

So this system is a speed reducer. Also note that $\omega_2 = 2\omega_3$. Either we may reference all inertias to shaft 1 and then use the relation $\omega_1 = 3\omega_3$, or we may reference everything to shaft 3 and note

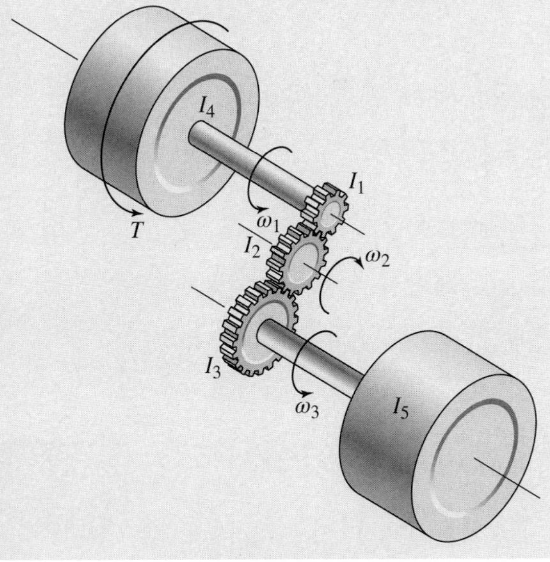

Figure 3.3.2 A three-gear system.

that the torque on shaft 3 due to T is $T/3$. Choosing the latter approach, we express the kinetic energy as

$$\text{KE} = \frac{1}{2}I_4\omega_1^2 + \frac{1}{2}I_1\omega_1^2 + \frac{1}{2}I_2\omega_2^2 + \frac{1}{2}I_3\omega_3^2 + \frac{1}{2}I_5\omega_3^2 = \frac{1}{2}\left(I_4 + I_1\right)\omega_1^2 + \frac{1}{2}I_2\omega_2^2 + \frac{1}{2}\left(I_3 + I_5\right)\omega_3^2$$

or

$$\text{KE} = \frac{1}{2}\left(I_4 + I_1\right)\left(3\omega_3\right)^2 + \frac{1}{2}I_2\left(2\omega_3\right)^2 + \frac{1}{2}\left(I_3 + I_5\right)\omega_3^2 = \frac{1}{2}(0.405)\,\omega_3^2 + \frac{1}{2}(0.04)\,\omega_2^2 + \frac{1}{2}(0.22)\,\omega_3^2$$

which reduces to

$$\text{KE} = \frac{1}{2}(0.665)\,\omega_3^2$$

Thus, the equivalent inertia referenced to shaft 3 is 0.665. Because the speed is reduced by a factor of 3 going from shaft 1 to shaft 3, the torque is increased by a factor of 3. Thus, the equation of motion is

$$0.665\dot{\omega}_3 = 3T$$

A spur-gear pair consists of only rotating elements. However, a rack-and-pinion consists of a rotating component (the pinion gear) and a translating component (the rack). The input to such a device is usually the torque applied to the shaft of the pinion. If so, then we should model the device as an equivalent inertia. The following example shows how to do this.

EXAMPLE 3.3.4

Equivalent Inertia of a Rack-and-Pinion

■ Problem

A rack-and-pinion, shown in Figure 3.3.3, is used to convert rotation into translation. The input shaft rotates through the angle θ as a result of the torque T produced by a motor. The pinion rotates and causes the rack to translate. Derive the expression for the equivalent inertia I_e felt on the input shaft. The mass of the rack is m, the inertia of the pinion is I, and its mean radius is R.

■ Solution

The kinetic energy of the system is (neglecting the inertia of the shaft)

$$\text{KE} = \frac{1}{2}m\dot{x}^2 + \frac{1}{2}I\dot{\theta}^2$$

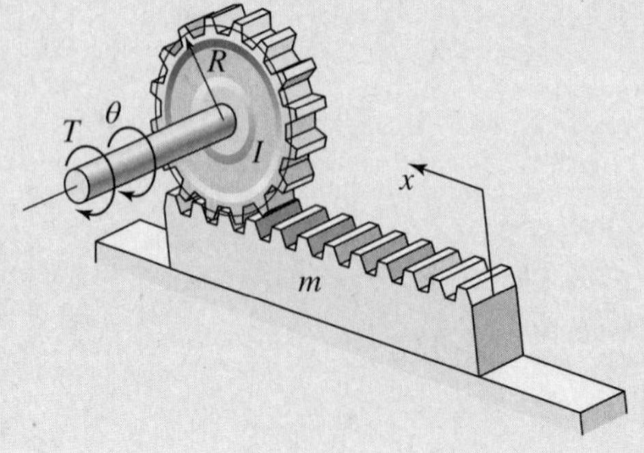

Figure 3.3.3 A rack-and-pinion gear.

where $\dot{x}$ is the velocity of the rack and $\dot{\theta}$ is the angular velocity of the pinion and shaft. From geometry, $x = R\theta$, and thus $\dot{x} = R\dot{\theta}$. Substituting for $\dot{x}$ in the expression for KE, we obtain

$$\text{KE} = \frac{1}{2}m(R\dot{\theta})^2 + \frac{1}{2}I\dot{\theta}^2 = \frac{1}{2}(mR^2 + I)\dot{\theta}^2$$

Thus, the equivalent inertia felt on the shaft is

$$I_e = mR^2 + I \tag{1}$$

and the model of the system's dynamics is $I_e\ddot{\theta} = T$, which can be expressed in terms of x as $I_e\ddot{x} = RT$.

Belt and chain drives are good examples of devices that can be difficult to analyze by direct application of Newton's laws but can be easily modeled using kinetic energy equivalence.

EXAMPLE 3.3.5

Equivalent Inertia of a Belt Drive

■ Problem

Belt drives and chain drives, like those used on bicycles, have similar characteristics and can be analyzed in a similar way. A belt drive is shown in Figure 3.3.4. The input shaft (shaft 1) is connected to a device (such as a bicycle crank) that produces a torque T_1 at a speed ω_1, and drives the output shaft (shaft 2). The mean sprocket radii are r_1 and r_2, and their inertias are I_1 and I_2. The belt mass is m.

Derive the expression for the equivalent inertia I_e felt on the input shaft.

■ Solution

The kinetic energy of the system is

$$\text{KE} = \frac{1}{2}I_1\omega_1^2 + \frac{1}{2}I_2\omega_2^2 + \frac{1}{2}mv^2$$

If the belt does not stretch, the translational speed of the belt is $v = r_1\omega_1 = r_2\omega_2$. Thus, we can express KE as

$$\text{KE} = \frac{1}{2}I_1\omega_1^2 + \frac{1}{2}I_2\left(\frac{r_1\omega_1}{r_2}\right)^2 + \frac{1}{2}m(r_1\omega_1)^2 = \frac{1}{2}\left[I_1 + I_2\left(\frac{r_1}{r_2}\right)^2 + mr_1^2\right]\omega_1^2$$

Therefore, the equivalent inertia felt on the input shaft is

$$I_e = I_1 + I_2\left(\frac{r_1}{r_2}\right)^2 + mr_1^2 \tag{1}$$

This means that the dynamics of the system can be described by the model $I_e\dot{\omega}_1 = T_1$.

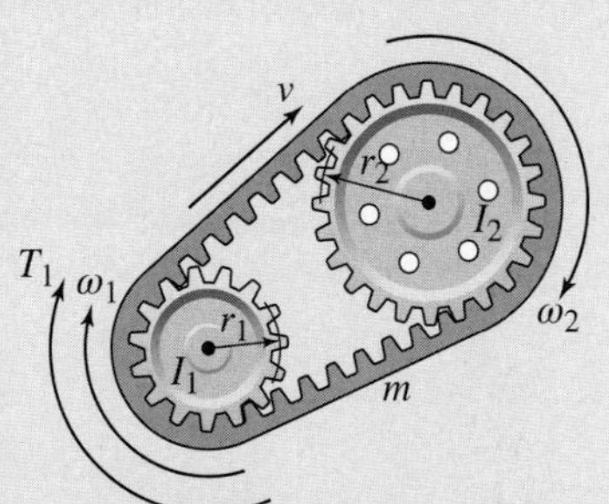

Figure 3.3.4 A belt drive.

Figure 3.3.5 Motion of a wheel.

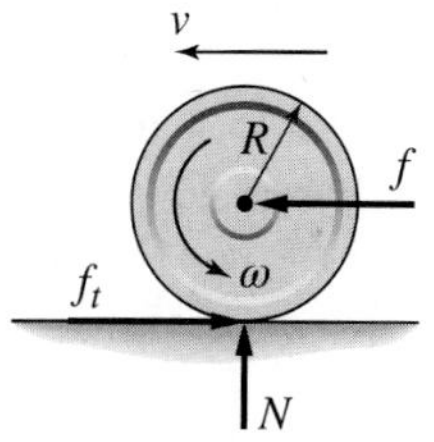

3.3.2 SLIDING VERSUS ROLLING MOTION

Wheels are common examples of systems undergoing general plane motion with both translation and rotation. The wheel shown in Figure 3.3.5 can have one of three possible motion types:

1. *Pure rolling motion.* This occurs when there is no slipping between the wheel and the surface. In this case, $v = R\omega$.
2. *Pure sliding motion.* This occurs when the wheel is prevented from rotating (such as when a brake is applied). In this case, $\omega = 0$ and $v \neq R\omega$.
3. *Sliding and rolling motion.* In this case $\omega \neq 0$. Because slipping occurs in this case, $v \neq R\omega$.

The wheel will roll without slipping (pure rolling) if the tangential force f_t is smaller than the static friction force $\mu_s N$, where N is the force of the wheel normal to the surface. In this case, the tangential force does no work because it does not act through a distance. If the static friction force is smaller than f_t, the wheel will slip.

EXAMPLE 3.3.6

A Wheeled Vehicle

■ Problem

An inextensible cable with a tension force $f = 400$ N is used to pull a two-wheeled cart on a horizontal surface (Figure 3.3.6a). The wheels roll without slipping. The cart has a mass $m_c = 100$ kg. Each wheel has a radius $R_w = 0.3$ m. Disregard the mass of the axle. The center of mass of the system is at point G. We wish to solve for the translational acceleration $\dot{v}$ of the cart.

a. Solve the problem assuming that the wheel masses are negligible.
b. Assuming that each wheel has a mass $m_w = 10$ kg and an inertia $I = 0.45$ kg·m^2 about its center, use the free body diagram method to solve the problem.

Figure 3.3.6 A two-wheeled cart.

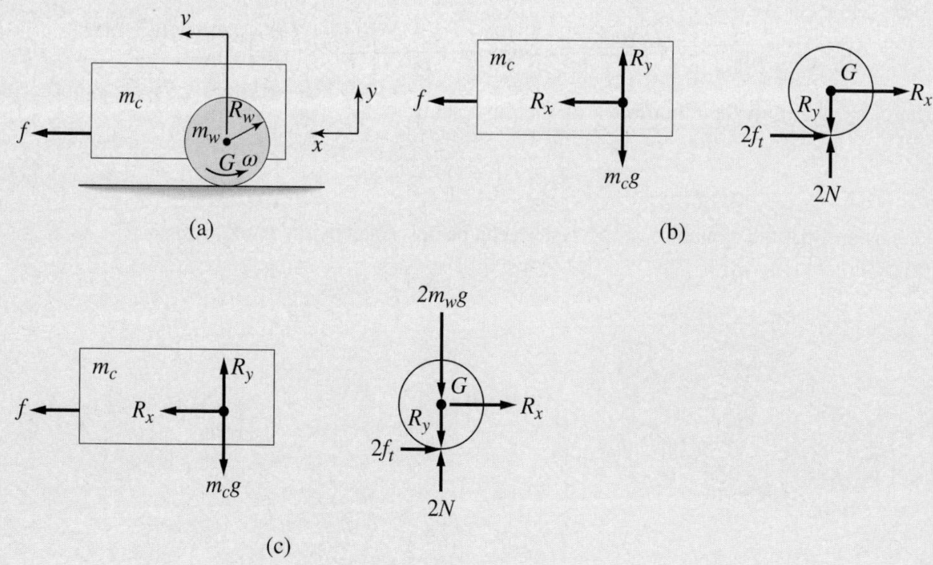

c. Solve the problem assuming that each wheel has a mass $m_w = 10$ kg and an inertia $I_w = 0.45$ kg·m^2 about its center. Use the energy equivalence method to solve the problem.

■ Solution

a. Discounting the wheel mass means that the wheel moment of inertia is also zero. The free body diagrams are shown in part (b) of the figure (treating the axle and the two wheels as a single rigid object). The forces R_x and R_y are the reaction forces between the axle and the cart body. For the cart mass, in the x direction

$$m_c \dot{v} = f + R_x \tag{1}$$

In the y direction, assuming that the wheels do not bounce so their vertical acceleration is zero, we have

$$0 = R_y - m_c g \tag{2}$$

which gives $R_y = m_c g$. For the massless wheel unit, the effect of the ground on *each* wheel can be resolved into a tangential component f_t and a normal component N. The equations of Statics give

$$\sum f_x = -R_x - 2f_t = 0 \qquad \text{(sum of forces in the } x \text{ direction is zero)} \tag{3}$$

$$\sum f_y = 2N - R_y = 0 \qquad \text{(sum of forces in the } y \text{ direction is zero)} \tag{4}$$

$$\sum M_G = 2f_t R_w = 0 \qquad \text{(sum of moments about point } G \text{ is zero)}$$

This implies that $f_t = 0$, because $R_w \neq 0$, and thus (3) implies that $R_x = 0$. From (2) and (4), we see that the normal force on each wheel is $N = m_c g/2$. Finally, from (1), the desired expression for the translational acceleration $\dot{v}$ of the cart is

$$m_c \dot{v} = f$$

or

$$\dot{v} = \frac{f}{m_c} = \frac{400}{100} = 4 \text{ m/s}^2.$$

Note that this equation is identical to the equation of motion of a wheel-less mass m_c being pulled by a force f on a *frictionless* surface.

b. The free body diagrams are shown in part (b) of the figure (again treating the axle and the two wheels as a single rigid object). For the cart mass, (1) and (2) still apply. However, for the axle-wheel system, we now have

$$\sum f_x = -R_x - 2f_t = 2m_w \dot{v} \tag{5}$$

$$\sum f_y = 2N - R_y - 2m_w g = 0 \qquad \text{(this assumes the wheels do not bounce)} \tag{6}$$

$$\sum M_G = 2f_t R_w = 2I_w \dot{\omega} \tag{7}$$

From (1),

$$R_x = m_c \dot{v} - f \tag{8}$$

Because $v = R_w \omega$, (7) gives

$$f_t = \frac{I_w \dot{\omega}}{R_w} = \frac{I_w \dot{v}}{R_w^2} \tag{9}$$

Using this result and (8) in (5), we obtain

$$\left(m_c + 2m_w + \frac{2I_w}{R_w^2}\right)\dot{v} = f$$

Substituting the given values gives the final answer: $130\dot{v} = 400$, or $\dot{v} = 3.077$ m/s^2. The effect of the wheel mass is to increase the vehicle mass by 20% from 100 to 120, but the wheel inertias increase the *equivalent mass* by 30% from 100 to 130. If needed, the expressions for the forces R_x, f_t, and N can be obtained by substituting this value of $\dot{v}$ into (6), (8), and (9), using $R_y = m_c g$.

c. Using the energy equivalence method, we write the kinetic energy expression for the entire vehicle.

$$\text{KE} = \frac{1}{2}m_c v^2 + \frac{1}{2}\left(2m_w v^2 + 2I_w \omega^2\right)$$

Because $R_w \omega = v$, this expression becomes

$$\text{KE} = \frac{1}{2}m_c v^2 + \frac{1}{2}\left(2m_w v^2 + 2I_w \frac{v^2}{R_w^2}\right) = \frac{1}{2}\left(m_c + 2m_w + \frac{2I_w}{R_w^2}\right)v^2$$

Thus, the equivalent mass is

$$m_e = m_c + 2m_w + \frac{2I_w}{R_w^2} = 130$$

and the equation of motion is $m_e \dot{v} = f$, which gives $\dot{v} = 3.077$ m/s^2, the same answer obtained in part (b).

Modeling Insight Example 3.3.6 shows that by assuming that the mass of the wheels is small compared with the total vehicle mass, we can ignore the rolling and translational motions of the wheels and thus treat the vehicle as a single rigid body translating in only one direction, with no rotating parts. In addition, it is important to realize that the assumption of negligible wheel mass implies that there is no tangential force on the wheels. We can see this from equation (9), which shows that f_t is zero if I_w is zero.

In Example 3.3.6, we modeled the wheeled vehicle as an equivalent translating mass. We can "lump" the rotating elements with the translating elements only when the translational and rotational motions are directly related to one another. If the vehicle's wheels slip, we cannot express the vehicle's translational speed in terms of the wheels' rotational speed. We must then treat the vehicle body and the wheels as separate masses, and apply Newton's laws to each separately. We must also take this approach if we need to compute any forces internal to the system.

3.4 GENERAL PLANAR MOTION

In Section 3.1 we limited our attention to systems undergoing pure translation, and in Section 3.2 we analyzed systems rotating about a single nonaccelerating axis. As demonstrated by the examples in Section 3.3, we can use energy equivalence to model a system as if it were in pure translation or pure rotation, but only if the motions of the rotating and translating components are directly coupled. We now consider the general case of an object undergoing translation and rotation about an accelerating axis.

We will restrict our attention to motion in a plane. This means that the object can translate in two dimensions and can rotate only about an axis that is perpendicular to the plane. Many practical engineering problems can be handled with the plane motion methods covered here. The completely general motion case involves translation in three dimensions, and rotation about three axes. This type of motion is considerably more complex to analyze, and is beyond the scope of our coverage.

3.4.1 FORCE EQUATIONS

Newton's laws for plane motion are derived in basic dynamics references, and we will review them here. We assume that the object in question is a rigid body that moves in a plane passing through its mass center, and that it is symmetrical with respect to that plane. Thus, it can be thought of as a slab with its motion confined to the plane of the slab. We assume that the mass center and all forces acting on the mass are in the plane of the slab.

We can describe the motion of such an object by its translational motion in the plane and by its rotational motion about an axis perpendicular to the plane. Two force equations describe the translational motion, and a moment equation is needed to describe the rotational motion. Consider the slab shown in Figure 3.4.1, where we arbitrarily assume that three external forces f_1, f_2, and f_3 are acting on the slab. Define an x-y coordinate system as shown with its origin located at a nonaccelerating point. Then the two force equations can be written as

$$f_x = ma_{G_x} \tag{3.4.1}$$

$$f_y = ma_{G_y} \tag{3.4.2}$$

where f_x and f_y are the net forces acting on the mass m in the x and y directions, respectively. The mass center is located at point G. The quantities a_{G_x} and a_{G_y} are the accelerations of the mass center in the x and y directions relative to the fixed x-y coordinate system.

3.4.2 MOMENT EQUATIONS

Recall that in Section 3.2 we treated the case where an object is constrained to rotate about an axis that passes through a fixed point O. For this case, we can apply the

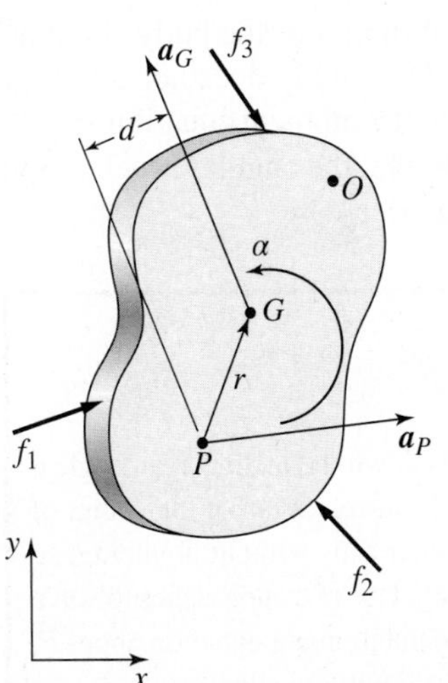

Figure 3.4.1 Planar motion of a slab.

following moment equation:

$$I_O\alpha = M_O \tag{3.4.3}$$

where α is the angular acceleration of the mass about the axis through point O, I_O is the mass moment of inertia of the body about the point O, and M_O is the sum of the moments applied to the body about the point O.

The following moment equation applies regardless of whether or not the axis of rotation is constrained:

$$M_G = I_G\alpha \tag{3.4.4}$$

where M_G is the net moment acting on the body about an axis that passes through the mass center G and is perpendicular to the plane of the slab. I_G and α are the mass moment of inertia and angular acceleration of the body about this axis. The net moment M_G is caused by the action of the external forces $f_1, f_2, f_3, \ldots$ and any couples applied to the body. The positive direction of M_G is determined by the right-hand rule (counterclockwise if the x-y axes are chosen as shown).

Note that point G in the preceding equation must be the mass center of the object; no other point may be used. However, in many problems the acceleration of some point P is known, and sometimes it is more convenient to use this point rather than the mass center or a fixed point. The following moment equation applies for an accelerating point P, which need not be fixed to the body:

$$M_P = I_G\alpha + ma_G d \tag{3.4.5}$$

where the moment M_P is the net moment acting on the body about an axis that passes through P and is perpendicular to the plane of the slab, a_G is the magnitude of the acceleration vector $\mathbf{a}_G$, and d is the distance between $\mathbf{a}_G$ and a parallel line through point P (see Figure 3.4.1).

An alternative form of this equation is

$$M_P = I_P\alpha + mr_x a_{P_y} - mr_y a_{P_x} \tag{3.4.6}$$

where a_{P_x} and a_{P_y} are the x and y components of the acceleration of point P relative to the x-y coordinate system. The terms r_x and r_y are the x and y components of the location of G relative to P, and I_P is the mass moment of inertia of the body about an axis through P. Note that in general, M_P does not equal M_G, and I_P does not equal I_G. If point P is fixed at some point O, then $a_{P_x} = a_{P_y} = 0$, and the moment equation (3.4.6) simplifies to (3.4.3), because $M_O = M_P$ and $I_O = I_P$. Note that the angular acceleration α is the same regardless of whether point O, G, or P is used.

EXAMPLE 3.4.1 A Rolling Cylinder

■ Problem

A solid cylinder of mass m and radius r starts from rest and rolls down the incline at an angle θ (Figure 3.4.2). The static friction coefficient is μ_s. Determine the acceleration of the center of mass a_{G_x} and the angular acceleration α. Assume that the cylinder rolls without bouncing, so that $a_{G_y} = 0$. Assume also that the cylinder rolls without slipping. Use two approaches to solve the problem: (a) Use the moment equation about G, and (b) use the moment equation about P. (c) Obtain the frictional condition required for the cylinder to roll without slipping.

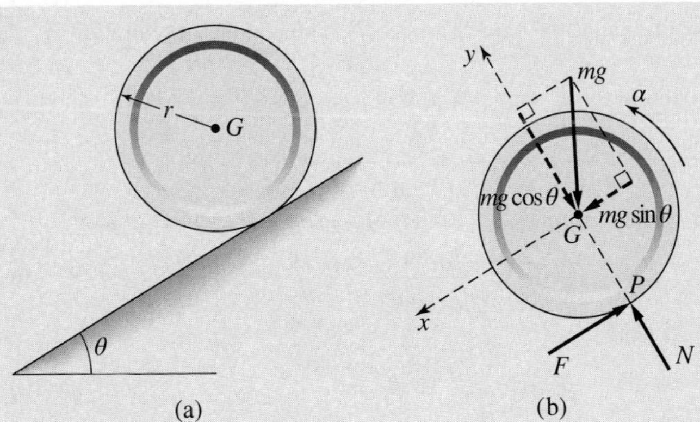

Figure 3.4.2 A cylinder rolling down an inclined plane.

■ Solution

The free body diagram is shown in the figure. The friction force is F and the normal force is N. The force equation in the x direction is

$$ma_{G_x} = f_x = mg\sin\theta - F \tag{1}$$

In the y direction,

$$ma_{G_y} = f_y = N - mg\cos\theta \tag{2}$$

If the cylinder does not bounce, then $a_{G_y} = 0$ and thus from (2),

$$N = mg\cos\theta \tag{3}$$

a. The moment equation about the center of mass gives

$$I_G\alpha = M_G = Fr \tag{4}$$

Solve for F and substitute it into (1):

$$ma_{G_x} = mg\sin\theta - \frac{I_G\alpha}{r} \tag{5}$$

If the cylinder does not slip, then

$$a_{G_x} = r\alpha \tag{6}$$

Solve this for α and substitute into (5):

$$ma_{G_x} = mg\sin\theta - \frac{I_G a_{G_x}}{r^2}$$

Solve this for a_{G_x}:

$$a_{G_x} = \frac{mgr^2\sin\theta}{mr^2 + I_G} \tag{7}$$

For a solid cylinder, $I_G = mr^2/2$, and the last expression reduces to

$$a_{G_x} = \frac{2}{3}g\sin\theta \tag{8}$$

b. We could have used instead the moment equation (2.4.5) about the point P, which is accelerating. This equation is

$$M_P = I_G\alpha + ma_{G_x}d$$

where $d = r$ and $M_P = (mg\sin\theta)r$. Thus,

$$mgr\sin\theta = I_G\frac{a_{G_x}}{r} + ma_{G_x}r$$

Solving this for a_{G_x} we obtain the same expression as (7). The angular acceleration is found from (6).

c. The maximum possible friction force is $F_{max} = \mu_s N = \mu_s mg \cos\theta$. From (4), (6), and (7),

$$F = \frac{I_G \alpha}{r} = \frac{I_G a_{G_x}}{r^2} = \frac{I_G mg \sin\theta}{I_G + mr^2}$$

If $F_{max} > F$, the cylinder will not slip. The condition of no slip is therefore given by

$$\mu_s \cos\theta > \frac{I_G \sin\theta}{I_G + mr^2} \tag{9}$$

For a solid cylinder, this reduces to

$$\mu_s \cos\theta > \frac{1}{3}\sin\theta \tag{10}$$

Simplifying Vehicle Models Example 3.3.6 treated a wheeled vehicle being pulled by an external force, such as from a cable tow. The example showed that by neglecting the mass of the wheels relative to the rest of the vehicle mass, we can ignore the rolling and translational motions of the wheels and thus treat the vehicle as a single rigid body translating in only one direction, with no rotating parts. The example also showed that the assumption of negligible wheel mass implies that there is no tangential force on the wheels. We will use this insight in the next example to simplify the analysis of the *undriven* wheels of a vehicle.

However, the situation is a little different for the *driven* wheels. In this case, a tangential traction force between the tire and the road is necessary to propel the vehicle, but if we neglect the driven wheel mass, we can ignore the rolling and translational motions of those wheels. Thus, the simplified vehicle model consists of a single translating mass.

Because it enables us to ignore the wheel motions, the assumption of negligible wheel mass also enables us to ignore the effects of the vehicle suspension. These concepts are explored in the following example.

EXAMPLE 3.4.2 Maximum Vehicle Acceleration

■ **Problem**

It is required to determine, as a function of the friction coefficient μ_s, the maximum acceleration of the rear-wheel drive vehicle shown in Figure 3.4.3. The vehicle mass is 1800 kg, and its dimensions are $L_A = 1.3$ m, $L_B = 1$ m, and $H = 0.5$ m.

Assume that the mass of the wheels is small compared with the total vehicle mass and neglect the effects of the vehicle suspension. These assumptions enable us to ignore the rolling and vertical motions of the wheels and thus to treat the vehicle as a single rigid body translating in only one direction, with no rotating parts. The assumption of negligible wheel mass implies that there is no tangential force on the front wheels.

Assume that each front wheel experiences the same reaction force $N_A/2$. Similarly, each rear wheel experiences the same reaction force $N_B/2$. Thus, the traction force $\mu_s N_B$ is the total traction force due to both driving wheels. The traction force is due to the torque T applied from the engine through the axles to the rear wheels. Thus, when the rear wheels are on the verge of slipping, $\mu_s N_B = T/R$, where R is the radius of the rear wheels.

■ **Solution**

The key to solving this problem is to recognize that the maximum traction force, and therefore the maximum acceleration, is obtained when the driving tires are just on the verge of slipping

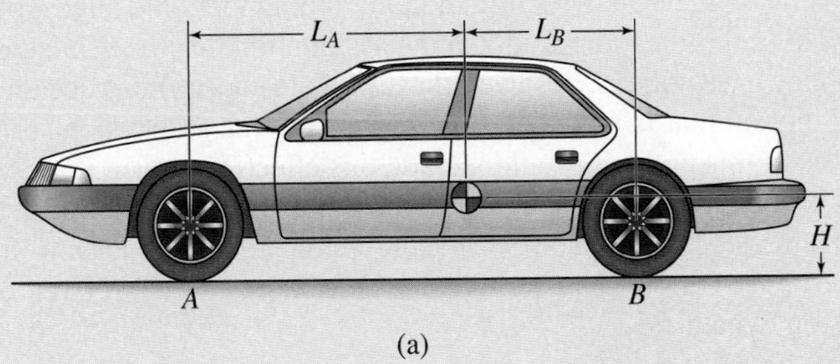

(a)

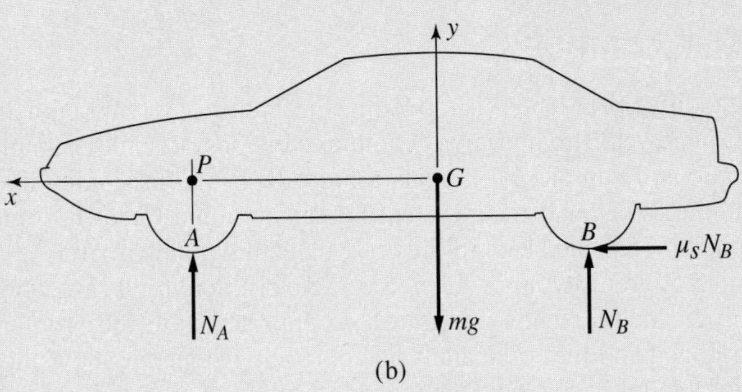

(b)

Figure 3.4.3 Vehicle acceleration.

relative to the road surface. In this condition, the friction force, which is the traction force, is given by $\mu_s N_B$. From the free body diagram shown in Figure 3.4.3b, Newton's law applied in the x direction gives

$$f_x = ma_{G_x} \quad \text{or} \quad \mu_s N_B = ma_{G_x} \tag{1}$$

In the y direction, if the vehicle does not leave the road, Newton's law gives

$$f_y = ma_{G_y} \quad \text{or} \quad N_A + N_B - mg = 0 \tag{2}$$

The moment equation about the mass center G gives

$$M_G = I_G \alpha \quad \text{or} \quad N_B L_B - \mu_s N_B H - N_A L_A = 0 \tag{3}$$

because $\alpha = 0$ if the vehicle body does not rotate.

Equation (1) shows that we need to find N_B to determine the acceleration a_{G_x}. Equations (2) and (3) can be solved for N_B. The solution is

$$N_B = \frac{mgL_A}{L_A + L_B - \mu_s H} = \frac{9.8(1.3)m}{1.3 + 1 - 0.5\mu_s} = \frac{25.5}{4.6 - \mu_s} m \tag{4}$$

and thus the maximum acceleration is

$$a_{G_x} = \frac{\mu_s N_B}{m} = \frac{25.5\mu_s}{4.6 - \mu_s}$$

An alternative approach to the problem is to use the moment equation (3.4.5) about the accelerating point P shown in the figure. This approach avoids the need to solve two equations to obtain N_B. This equation gives

$$M_P = I_G \alpha + ma_G d \quad \text{or} \quad \mu_s N_B H - N_B(L_A + L_B) + mgL_A = 0 \tag{5}$$

because $\alpha = 0$ and $d = 0$. This gives the same solution as equation (4).

Vehicle Suspension Effects The analysis in Example 3.4.2 ignored the effects of the vehicle suspension. This simplification results in the assumption that the vehicle body does not rotate. You may have noticed, however, that a vehicle undergoing acceleration will have a pitching motion that is made possible because of the suspension springs. So a complete analysis of this problem would include this effect. However, it is always advisable to begin with a simplified version of a problem, to make sure you understand the problem's basic features. If you cannot solve the problem with the suspension effects ignored, then you certainly will not be able to solve the more complex problem that includes the suspension dynamics!

3.5 ADDITIONAL EXAMPLES

This section provides additional practice in modeling mechanical systems by using six examples based on more realistic and more complex versions of some examples presented in earlier sections. Example 3.5.1 treats a more realistic, distributed-mass pendulum model. Example 3.5.2 analyzes a system for raising a mast and illustrates how a potentially complex model can be simplified by neglecting pulley inertia.

Examples 3.5.2 and 3.5.3 return to the wheeled vehicle problem for a case where the vehicle must climb an incline. They illustrate two approaches to the problem: the energy analysis approach and the force analysis approach. The latter shows how much more complex the model becomes if we must determine the internal reaction forces.

Example 3.5.4 shows how to use the equivalent inertia approach to obtain a simplified model of a realistic example of a robot arm link containing a gear drive. The final example analyzes a personal transporter whose model must describe the dynamics of a translational element connected to a rotational one.

EXAMPLE 3.5.1

A Rod-and-Bob Pendulum

■ Problem

The pendulum shown in Figure 3.5.1a consists of a concentrated mass m_C (the bob) a distance L_C from point O, attached to a rod of length L_R and inertia I_{RG} about its mass center. (a) Obtain its equation of motion. (b) Discuss the case where the rod's mass m_R is small compared to the concentrated mass.

Figure 3.5.1 A rod-and-bob pendulum.

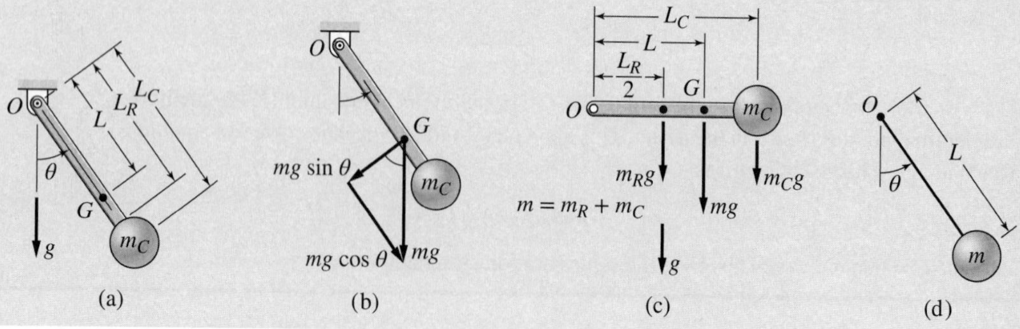

■ **Solution**

a. For the pendulum shown in Figure 3.5.1a, the inertia of the concentrated mass m_C about point O is $m_C L_C^2$ (see Table 3.2.1). From the parallel-axis theorem, the rod's inertia about point O is

$$I_{RO} = I_{RG} + m_R\left(\frac{L_R}{2}\right)^2$$

and thus the entire pendulum's inertia about point O is

$$I_O = I_{RO} + m_C L_C^2 = I_{RG} + m_R\left(\frac{L_R}{2}\right)^2 + m_C L_C^2$$

With the moment equation (3.2.1) about point O, the moment M_O is caused by the perpendicular component of the weight mg acting through the mass center at G (see Figure 3.5.1b). Thus, the desired equation of motion is

$$I_O\ddot{\theta} = -mgL\sin\theta \tag{1}$$

The distance L between point O and the mass center G of the entire pendulum is not given, but can be calculated as follows (Figure 3.5.1c). If the entire pendulum mass were concentrated at G, the weight force would produce the same moment about point O as the real pendulum. Thus, taking moments about point O, we have

$$mgL = m_C g L_C + m_R g\frac{L_R}{2}$$

where $m = m_C + m_R$. Solve for L to obtain

$$L = \frac{m_C L_C + m_R(L_R/2)}{m_C + m_R} \tag{2}$$

b. If we neglect the rod's mass m_R compared to the concentrated mass m_C, then we can take $m_R = I_{RG} = 0$, $m = m_C$, $L = L_C$, and $I_O = mL^2$. In this case, the equation of motion reduces to

$$L\ddot{\theta} + g\sin\theta = 0 \tag{3}$$

This is a model for a pendulum whose mass is concentrated at a distance L from the pivot point, like that shown in Figure 3.5.1d. Note that this equation of motion is independent of the value of m. This is the same result obtained in Example 3.2.2.

For small angles, $\sin\theta \approx \theta$ if θ is in radians. Substituting this approximation into (3) gives

$$L\ddot{\theta} + g\theta = 0 \tag{4}$$

This equation was solved in Example 3.2.2.

Lifting a Mast — EXAMPLE 3.5.2

■ **Problem**

A mast weighing 500 lb is hinged at its bottom to a fixed support at point O (see Figure 3.5.2). The mast is 70 ft long and has a uniform mass distribution, so its center of mass is 35 ft from O. The winch applies a force $f = 380$ lb to the cable. The mast is supported initially at the 30° angle, and the cable at A is initially horizontal. Derive the equation of motion of the mast. You may assume that the pulley inertias are negligible and that the pulley diameter d is very small compared to the other dimensions.

Figure 3.5.2 A system for lifting a mast.

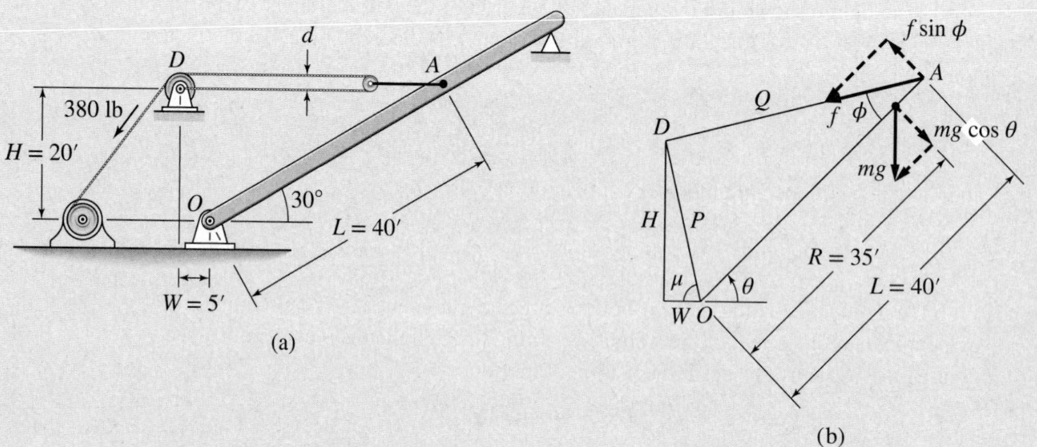

■ **Solution**

Part (b) of the figure shows the geometry of the mast at some angle $\theta > 30°$, with the diameter d neglected. From the law of sines,

$$\sin\phi = P\frac{\sin(180° - \mu - \theta)}{Q} = P\frac{\sin(\mu + \theta)}{Q}$$

From the law of cosines,

$$Q = \sqrt{P^2 + L^2 - 2PL\cos(180° - \mu - \theta)} = \sqrt{P^2 + L^2 + 2PL\cos(\mu + \theta)} \tag{1}$$

where $H = 20$ ft, $W = 5$ ft, and

$$\mu = \tan^{-1}\left(\frac{H}{W}\right) = \tan^{-1}\left(\frac{20}{5}\right) = 76° = 1.33 \text{ rad}$$

$$P = \sqrt{H^2 + W^2} = 20.6 \text{ m}$$

The moment equation about the fixed point O is

$$I_O\ddot{\theta} = -mgR\cos\theta + \frac{fLP}{Q}\sin(\mu + \theta) \tag{2}$$

The moment of inertia is

$$I_O = \frac{1}{3}m(70)^2 = \frac{1}{3}\frac{500}{32.2}70^2 = 25{,}400 \text{ slug-ft}^2$$

The force f at point A is twice the applied force of 380 lb, because of the action of the small pulley. Thus $f = 760$ lb. With the given values, the equation of motion becomes

$$25{,}400\,\ddot{\theta} = -17{,}500\cos\theta + \frac{626{,}000}{Q}\sin(1.33 + \theta) \tag{3}$$

where

$$Q = \sqrt{2020 + 1650\cos(1.33 + \theta)} \tag{4}$$

Equation (3) cannot be solved in closed-form to find $\theta(t)$, so it must be solved numerically using the methods to be introduced in Chapter 5. Consider how much more complex the model would have to be if we could not neglect the pulley inertia. We could no longer take the cable tension to be the same on each side of the pulley, and we would need to write two additional equations of motion, one for the pulley rotation and one for its translation.

A Vehicle on an Incline: Energy Analysis

EXAMPLE 3.5.3

■ Problem

A tractor pulls a cart up a slope, starting from rest and accelerating to 20 m/s in 15 s (Figure 3.5.3a). The force in the cable is f, and the body of the cart has a mass m. The cart has two identical wheels, each with radius R, mass m_w, and inertia I_w about the wheel center. The two wheels are coupled with an axle whose mass is negligible. Assume that the wheels do not slip or bounce. Derive an expression for the force f using kinetic energy equivalence.

■ Solution

The assumption of no slip and no bounce means that the wheel rotation is directly coupled to the cart translation. This means that if we know the cart translation x, we also know the wheel rotation θ, because $x = R\theta$ if the wheels do not slip or bounce. Because the input is the force f, we will derive an equivalent mass, and thus we will visualize the system as a block of mass m_e being pulled up the incline by the force f, as shown in Figure 3.5.3b. The wheels will roll without slipping (pure rolling) if the tangential force between the wheel and the surface is smaller than the static friction force. In this case the tangential force does no work because it does not act through a distance (see Section 3.4), and thus there is no energy loss due to friction. Therefore, in our equivalent model, Figure 3.5.3b, there is no friction between the block and the surface.

The kinetic energy of the system is

$$\text{KE} = \frac{1}{2}mv^2 + \frac{1}{2}\left(2m_w v^2\right) + \frac{1}{2}\left(2I_w \omega^2\right)$$

Because $v = R\omega$, we obtain

$$\text{KE} = \frac{1}{2}\left(m + 2m_w + 2\frac{I_w}{R^2}\right)v^2 \tag{1}$$

For the block in Figure 3.5.3b, $\text{KE} = 0.5m_e v^2$. Comparing this with equation (1) we see that the equivalent mass is given by

$$m_e = m + 2m_w + 2\frac{I_w}{R^2} \tag{2}$$

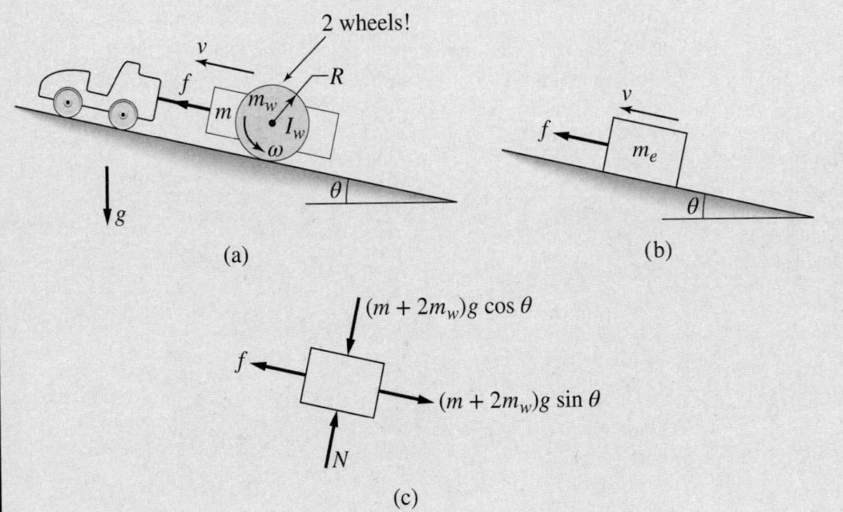

Figure 3.5.3 A vehicle on an incline.

The free body diagram is shown in part (c) of the figure. Note that the gravity force is determined, not from the equivalent mass m_e, but from the actual mass $m + 2m_w$. The reason for this can be seen from the potential energy expression. Assuming that the center of mass of the cart coincides with the axle location, we can express the potential energy of the system as $(m + 2m_w)gx \sin\theta$, where $\dot{x} = v$. Therefore, the actual mass $m + 2m_w$ must be used to compute the gravity force.

From the free body diagram we obtain the following equation of motion.

$$m_e\dot{v} = f - (m + 2m_w)g \sin\theta \tag{3}$$

where m_e is given by equation (2).

The acceleration is $\dot{v} = 20/15 = 4/3$ m/s^2. Substitute this into equation (3) and solve for f:

$$f = \frac{4}{3}m_e + (m + 2m_w)g \sin\theta \tag{4}$$

This is the force required to provide the specified acceleration.

EXAMPLE 3.5.4 A Vehicle on an Incline: Force Analysis

■ Problem

Consider again the system of Example 3.5.3, in which a tractor pulls a cart up a slope (Figure 3.5.3a). The force in the cable is f, and the body of the cart has a mass m. The cart has two identical wheels, each with radius R, mass m_w, and inertia I_w about the wheel center. The two wheels are coupled with an axle whose mass is negligible. Assume that the center of mass of the cart coincides with the axis of the axle. In Example 3.5.3, we assumed that the wheels do not slip. Now we want to develop a model that can be used to examine this assumption and also to compute the forces on the axle. Apply Newton's laws to develop a model of the system. (a) Show that the model gives the same result as equation (3) of Example 3.5.3, assuming no slip and no bounce. (b) Use the model to discuss the no-slip assumption.

■ Solution

a. The free body diagrams of the cart and the wheel-axle subsystem are shown in Figure 3.5.4. The forces R_x and R_y are the reaction forces between the cart and the axle. The forces f_t and N are the tangential and normal forces acting on both wheels as a result of their contact with the road's surface. Newton's second law applied to the cart gives

$$m\ddot{x} = R_x - mg \sin\theta + f \tag{1}$$

$$m\ddot{y} = R_y - mg \cos\theta \tag{2}$$

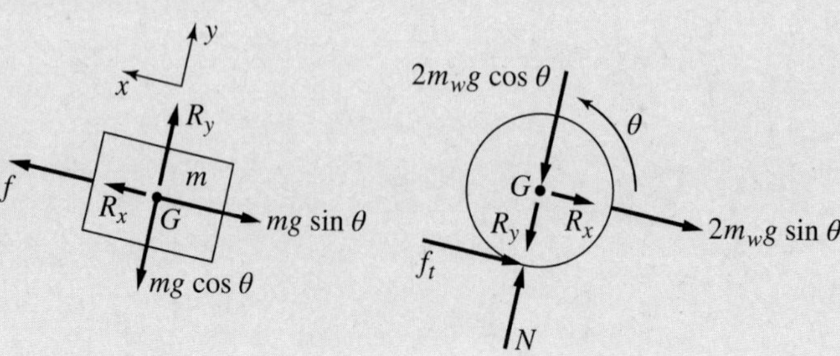

Figure 3.5.4 Free-body diagrams of the cart body and the wheel-axle subsystem.

Newton's second law applied to the wheel-axle subsystem gives

$$2m_w\ddot{x} = -R_x - 2m_w g \sin\theta - f_t \tag{3}$$

$$2m_w\ddot{y} = N - R_y - 2m_w g \cos\theta \tag{4}$$

Application of the moment equation about the center of mass of the wheel-axle subsystem gives

$$2I_w\dot{\omega} = Rf_t \tag{5}$$

The assumption of no slip means that

$$\ddot{x} = \dot{v} = R\dot{\omega} \tag{6}$$

Substitute this into equation (5) and solve for f_t:

$$f_t = \frac{2I_w}{R^2}\ddot{x} \tag{7}$$

Substitute this into equation (3) and solve for R_x:

$$R_x = -\left(2m_w + \frac{2I_w}{R^2}\right)\ddot{x} - 2m_w g \sin\theta \tag{8}$$

Substitute this into equation (1) and collect terms to obtain

$$\left(m + 2m_w + 2\frac{I_w}{R^2}\right)\ddot{x} = f - (m + 2m_w)g \sin\theta \tag{9}$$

which is the same as equation (3) of Example 3.3.1 since $\ddot{x} = \dot{v}$.

b. The assumption of no bounce means that $\ddot{y} = 0$. For this condition, equations (2) and (4) can be solved for N to obtain

$$N = (m + 2m_w)g \cos\theta \tag{10}$$

Slip occurs if the tangential force f_t is greater than the maximum static friction force, which is $\mu_s N$. Thus, if we solve equation (9) for $\ddot{x}$ and substitute it into equation (7), the resulting value of f_t must be no greater than $\mu_s N$ or else slip will occur. Thus, slip will not occur if

$$\mu_s(m + 2m_w)g \cos\theta \geq 2I_w \frac{f - (m + 2m_w)g \sin\theta}{(m + 2m_w)R^2 + 2I_w} \tag{11}$$

A Robot-Arm Link — EXAMPLE 3.5.5

■ **Problem**

A single link of a robot arm is shown in Figure 3.5.5. The arm mass is m and its center of mass is located a distance L from the joint, which is driven by a motor torque T_m through a pair of spur gears. We model the arm as a pendulum with a concentrated mass m. Thus, we take the arm's moment of inertia I_G to be zero. The values of m and L depend on the payload being carried in the hand and thus can be different for each application. The gear ratio is $N = 2$ (the motor shaft has the greater speed). The motor and gear rotation axes are fixed by bearings.

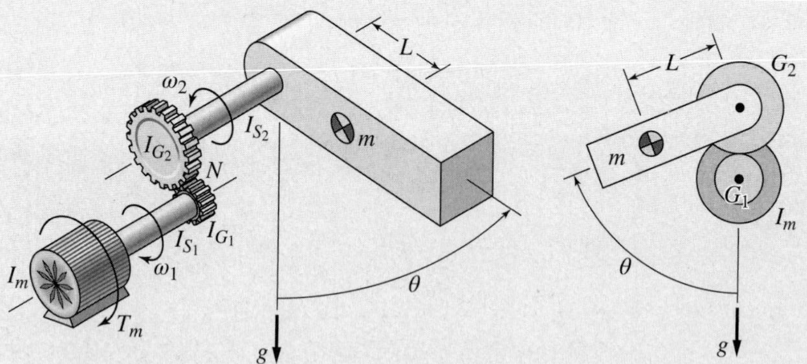

Figure 3.5.5 A robot-arm link.

To control the motion of the arm, we need to have its equation of motion. (a) Obtain this equation in terms of the angle θ. The given values for the motor, shaft, and gear inertias are

$$I_m = 0.05 \text{ kg·m}^2 \qquad I_{G_1} = 0.025 \text{ kg·m}^2 \qquad I_{S_1} = 0.01 \text{ kg·m}^2$$

$$I_{G_2} = 0.1 \text{ kg·m}^2 \qquad I_{S_2} = 0.02 \text{ kg·m}^2$$

(b) Solve this equation for the case where $T_m = 0.5$ N·m, $m = 10$ kg, and $L = 0.3$ m. Assume that the system starts from rest at $\theta = 0$ and that the angle θ remains small.

■ Solution

Our approach is to model the system as a single inertia rotating about the motor shaft with a speed ω_1. To find the equivalent inertia about this shaft, we first obtain the expression for the kinetic energy of the total system and express it in terms of the shaft speed ω_1. Note that the mass m is translating with a speed $L\omega_2$.

$$\text{KE} = \frac{1}{2}(I_m + I_{S_1} + I_{G_1})\omega_1^2 + \frac{1}{2}(I_{S_2} + I_{G_2})\omega_2^2 + \frac{1}{2}m(L\omega_2)^2$$

But $\omega_2 = \omega_1/N = \omega_1/2$. Thus,

$$\text{KE} = \frac{1}{2}\left[I_m + I_{S_1} + I_{G_1} + \frac{1}{2^2}(I_{S_2} + I_{G_2} + mL^2)\right]\omega_1^2$$

Therefore, the equivalent inertia referenced to the motor shaft is

$$I_e = I_m + I_{S_1} + I_{G_1} + \frac{1}{2^2}(I_{S_2} + I_{G_2} + mL^2) = 0.115 + 0.25mL^2$$

The equation of motion for this equivalent inertia can be obtained in the same way as that of a pendulum, by noting that the gravity moment $mgL\sin\theta$, which acts on shaft 2, is also felt on the motor shaft, but *reduced* by a factor of N due to the gear pair. Thus,

$$I_e\dot{\omega}_1 = T_m - \frac{1}{N}mgL\sin\theta$$

But $\omega_1 = N\omega_2 = N\dot{\theta}$. Thus,

$$I_e N\ddot{\theta} = T_m - \frac{1}{N}mgL\sin\theta$$

Substituting the given values, we have

$$2(0.115 + 0.25mL^2)\ddot{\theta} = T_m - \frac{9.8}{2}mL\sin\theta$$

or

$$(0.23 + 0.5mL^2)\ddot{\theta} = T_m - 4.9mL\sin\theta \qquad (1)$$

b. If $|\theta|$ is small, then $\sin\theta \approx \theta$ and the equation of motion becomes

$$(0.23 + 0.5mL^2)\ddot{\theta} = T_m - 4.9mL\theta \tag{2}$$

For the given values, the equation is

$$0.68\ddot{\theta} = 0.5 - 14.7\theta$$

The characteristic equation is $0.68s^2 + 14.7 = 0$ and the roots are $s = \pm 4.65j$. Following the method of Example 3.1.4, we obtain the following solution:

$$\theta(t) = \frac{0.5}{14.7}(1 - \cos 4.65t) = 0.034(1 - \cos 4.65t) \tag{3}$$

Thus, the model predicts that the arm oscillates with a frequency of 4.65 rad/s about $\theta = 0.034$ with an amplitude of 0.034 rad. To obtain the linear model (2), we used the approximation $\sin\theta \approx \theta$. Therefore, before we accept the solution (3), we should check the validity of this assumption. The maximum value that θ will reach is twice the amplitude, or $2(0.034) = 0.068$ rad. Therefore, because $\sin 0.068 = 0.068$ to three decimal places, our assumption that $\sin\theta \approx \theta$ is justified.

3.5.1 DYNAMICS OF A PERSONAL TRANSPORTER

Personal transporters are small vehicles designed to carry usually only one person. They have become more available because of the advent of less expensive sensors and more powerful microprocessor control systems to handle the complex calculations required to balance the vehicle. There are a variety of designs, including unicycles, but Figure 3.5.6 illustrates a two-wheel version. The transporter's motors drive the wheels to balance the vehicle with the help of a computer control system using tilt sensors and gyroscopes.

One type of gyroscope used is a *vibrating structure gyroscope* that functions much like the *halteres* of some winged insects. These are small knobbed cantilever–beam-like

Figure 3.5.6 A personal transporter.

structures that are flapped rapidly to maintain stability when flying. The vibrating beam tends to keep vibrating in the same plane even though its support rotates. When used as a sensor, the transducer attached to the beam detects the bending strain that results from the *Coriolis acceleration*. These sensors are simpler and less expensive than a rotating gyroscope.

We have restricted our coverage of dynamics to inertial (for example, non-rotating) coordinate systems. The Coriolis acceleration is a term that appears in the equations of motion of an object when expressed in a rotating frame of reference. The Coriolis acceleration depends on the velocity of the object. A reference frame fixed to the Earth is actually rotating and therefore non-inertial. Thus, a projectile, due to the Coriolis acceleration, appears to curve to the right in the northern hemisphere and to the left in the southern hemisphere. For many applications, the effect of the Coriolis term is negligible because of the relatively small velocities. However, in the vibrating gyroscope, the oscillation, and thus the velocity, of the beam are large, and thus the Coriolis effect is detectable.

The transporter will be balanced (kept nearly upright) as long as the wheels stay under the center of gravity. Thus, the transporter can be balanced by driving the wheels in the direction of leaning. This means that to accelerate forward, the person should lean forward. A representation of the transporter is shown in Figure 3.5.6. The drive motor applies a torque to the wheel-axle subsystem. The tangential force between the wheel and the road is f. This force acts in the opposite direction on the vehicle body and propels the transporter forward (to the left in the figure).

The dynamics of a personal transporter are similar to a classic control problem called the *inverted pendulum*.[1]

EXAMPLE 3.5.6

Transporter Equations of Motion

■ Problem

We may model the transporter as a cart of mass M (which includes the equivalent mass of the wheel-axle subsystem) and an inverted pendulum attached to the cart by a pivot at point P (see Figure 3.5.7a). The pendulum mass is m and its center of mass G is a distance L from P. The inertia of the pendulum about G is I_G. For generality we include an applied torque T about the pivot, which is due to a motor at the pivot in some applications. Derive the equations of motion with f and T as the inputs, and x and ϕ as the outputs.

■ Solution

The free body diagrams are shown in Figure 3.5.7b. First consider the pendulum. The vertical and horizontal components of the pendulum's mass center are $L \cos \phi$ and $x - L \sin \phi$, respectively. For the horizontal direction, Newton's law gives:

$$m\frac{d^2}{dt^2}(x - L \sin \phi) = H \tag{1}$$

where H is the horizontal component of the reaction force at the pivot. The moment equation (3) about the pendulum's pivot point P gives:

$$(I_G + mL^2)\ddot{\phi} - mL\ddot{x}\cos\phi = T + mgL\sin\phi \tag{2}$$

where $I_G + mL^2$ is the pendulum's moment of inertia about the pivot point.

[1]This problem is a good example of why you should be careful to check all technical information, especially any obtained from the Internet. The author found several websites, including an entry in a well-known online encyclopedia, that contained erroneous equations of motion for the inverted pendulum. The equations derived here are equivalent to those derived in [Cannon, 1967].

Figure 3.5.7 (a) Model of a personal transporter. (b) Free-body diagram of the base and the inverted pendulum.

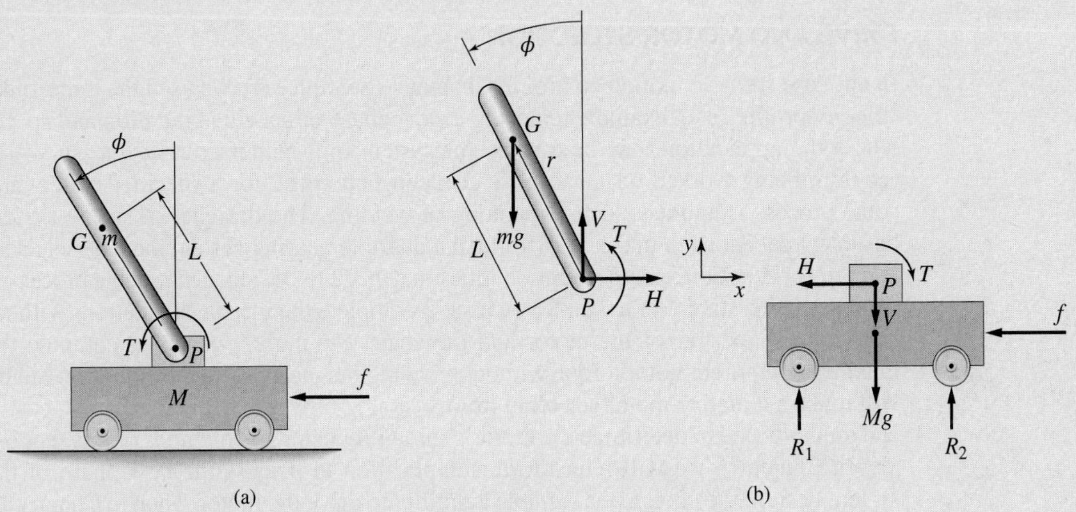

Now consider the base. Newton's law in the horizontal direction gives:

$$M\ddot{x} = -f - H \tag{3}$$

Because we are assuming that the base does not rotate or move vertically, we need not consider the moments and vertical forces on the base, unless we need to compute the reaction forces V, R_1, and R_2.

Evaluate the derivative in (1) to obtain:

$$m\ddot{x} - mL\frac{d}{dt}(\cos\phi\,\dot{\phi}) = m\ddot{x} - mL(-\sin\phi\,\dot{\phi}^2 + \cos\phi\,\ddot{\phi}) = H \tag{4}$$

Solve (3) for the reaction force H: $H = -f - M\ddot{x}$. Substitute this into (4) and collect terms to obtain:

$$(m + M)\ddot{x} - mL(\cos\phi\,\ddot{\phi} - \sin\phi\,\dot{\phi}^2) = -f \tag{5}$$

The equations of motion are (2) and (5).

3.6 A CASE STUDY IN MOTION CONTROL

So far we have considered how to model various types of mechanical drives: spur gears, rack-and-pinion, lead screws, belts, and chains, and there are many other types we have not considered: worm and planetary gears, differential gears, and so forth. Such a drive can be used to change the displacement, speed, torque, or force between the "input" and the "output." For example, a speed reducer increases the torque. A drive can be used to change the direction of rotation, or can be used to convert rotation into translation. Our interest here, however, is from the point of view of system dynamics. We do not cover the detailed design of these elements, such as the shape and stress analysis of the gear teeth, which can be quite involved, but rather in how diplacement, speed, and torque (or force) can be modified by using a mechanical drive. Thus, most of our interest is in the resulting speed ratio or transmission ratio, usually called the gear ratio even though the concept also applies to drive types other than gears.

DRIVE AND MOTOR SELECTION

In our case study in motion control in Chapter 1 (Section 1.5), we used the trapezoidal velocity profile as an example requiring a slew phase of specified duration and speed, where the application may be a conveyor system in a manufacturing line, in which the item being worked on must move at a constant speed for a specified time while some process is applied, such as painting or welding. The designer of such a system has many decisions to make: A motor and a possible gearing system must be selected, and the acceleration and deceleration times may need to be adjusted. Such a process is necessarily iterative. For example, we cannot complete the inertia calculations without knowing the inertias of the motor and the gears. Similarly, we cannot compute the maximum required motor torque without knowing the inertias. So the engineer usually will make a tentative motor selection from a catalog (always keeping an eye on cost!), and then proceed to determine the motor's suitability using the methods of this section.

In Chapter 6 we will return to this application to discuss the next phase of the system design: the choice of a suitable amplifier to drive the motor. Then in Chapter 10 we will design a controller for the system.

SELECTING A GEAR RATIO

Calculation of the desired transmission ratio sometimes can be quite simple. Some AC motors are most efficient when operating at a known fixed speed. If we know the required load speed, then the required transmission ratio is simply the ratio of these two speeds.

In some applications we want to maximize the load acceleration for a given motor and load, and proper selection of the transmission ratio can do that. Figure 3.6.1 represents a spur-gear transmission, but the optimization method can be applied to any type of transmission (lead screw, rack-and-pinion, belt drive, etc.). We assume that the inertias I_1, I_2, and the torques T_1, T_2 are given. With everything reflected to the load shaft, Newton's law gives

$$\dot{\omega}_2 = \frac{T_2 + NT_1}{I_2 + N^2 I_1}$$

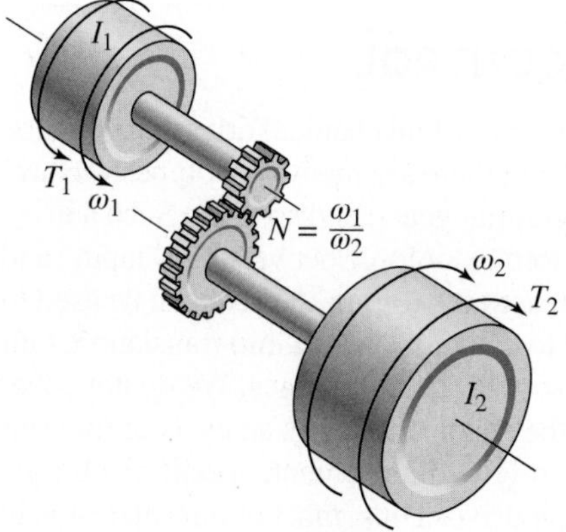

Figure 3.6.1 A spur-gear pair.

To maximize $\dot{\omega}_2$, differentiate the above expression with respect to N, set the derivative equal to 0. This gives

$$\frac{\partial \dot{\omega}_2}{\partial N} = \frac{(I_2 + N^2 I_1)T_1 - 2I_1 N(T_2 + NT_1)}{(I_2 + N^2 I_1)^2} = 0$$

This is true if the numerator is 0. Thus,

$$I_1 T_1 N^2 + 2I_1 T_2 N - I_2 T_1 = 0$$

The positive solution for N is

$$N = -\frac{T_2}{T_1} + \sqrt{\left(\frac{T_2}{T_1}\right)^2 + \frac{I_2}{I_1}} \tag{3.6.1}$$

This is the ratio that maximizes $\dot{\omega}_2$. (This can be confirmed to give a maximum rather than a minimum by showing that $\partial^2 \dot{\omega}_2 / \partial N^2 < 0$.)

If the load torque T_2 is 0, the optimal ratio for this case is denoted N_o and is: $N_o = \sqrt{I_2/I_1}$, or

$$I_1 = \frac{I_2}{N_o^2} \tag{3.6.2}$$

This says that the ratio that maximizes the load acceleration is the ratio that makes the reflected load inertia (felt on the motor shaft) equal to the inertia on the motor shaft. This is the principle of *inertia matching*. Note that in Figure 3.6.1 when T_2 resists the motor torque, $T_2 < 0$.

THE EFFECT OF LOAD TORQUE

If the actual ratio N differs from the value N_o such that

$$N = pN_o \tag{3.6.3}$$

then the efficiency E is

$$E = \frac{\text{actual } \dot{\omega}_2}{\max \dot{\omega}_2} = \frac{2p}{1 + p^2} \tag{3.6.4}$$

The plot of this efficiency is shown in Figure 3.6.2. Because $E(p) = E(1/p)$, the efficiency when "overgearing" is the same as when "undergearing." For example, if $p = 2$ or $p = 1/2$, $E = 0.8$, so the acceleration is 80% of the maximum possible.

Since $I_2/I_1 = N_0^2$, the effect of the load torque T_2 may be seen by expressing (3.6.1) in terms of N_o and the torque ratio $R = T_2/T_1$ as follows:

$$\frac{N}{N_o} = -\frac{R}{N_0} + \sqrt{\frac{R^2}{N_o^2} + 1} \tag{3.6.5}$$

A plot of the ratio N/N_o versus N_o for $R = \pm 0.1$ and $N_o \geq 2$ shows that N computed from (3.6.1) is within 5% of the value computed by inertia matching. For $R = \pm 0.5$, N computed from (3.6.1) is within 28% of the value computed by inertia matching. Keep in mind that if the torque T_2 represents only a friction torque, R should be small in a well-designed system. In other applications, such as with a robot arm or a crane, the torque T_2 may represent an inertial load such as the weight of an object being lifted by the device. In such applications, the torque ratio could be appreciable and (3.6.1) should be used.

Keep in mind that this method of selecting the gear ratio maximizes the load acceleration but may not give a suitable value for several reasons. An obvious reason

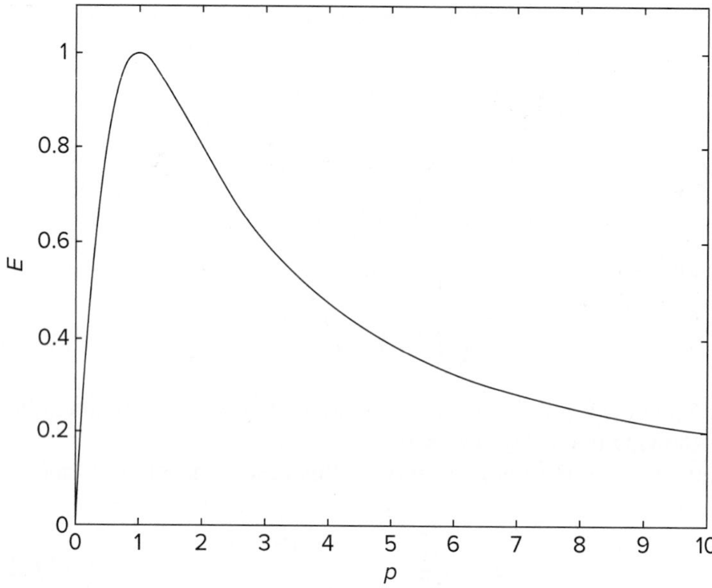

Figure 3.6.2 Gearing efficiency versus p.

is that the calculated value may not be feasible because of space limitations or cost. A more subtle reason is that the gear ratio can change system dynamics, such as vibration characteristics, that are not modeled here. This topic is better treated after we consider elastic behavior in Chapters 4 and 9.

EXAMPLE 3.6.1

A Conveyer System

■ Problem

Conveyor systems are used to produce translation of the load, as shown in Figure 3.6.3. The motor inertia is denoted by I_m. The reducer is a geared system that reduces the motor speed by a factor of N:1, where we must determine a suitable value for N. The inertia of the reducer, as felt on the motor shaft, is denoted by I_r. Disregard the inertias of the two sprockets, the chain, all shafts, and the tachometer, which is used to measure the speed for control purposes. (One of the chapter problems considers the effects of these masses.)

The only significant masses and inertias are the inertias of the four drive wheels (0.02 kg·m^2 each), the two drive chains (8 kg each), and the load mass (10 kg). The radius of sprocket 1 is 0.05 m, and that of sprocket 2 is 0.1 m. The drive wheel has a radius of 0.1 m, so the drive shaft speed ω_d is related to the load speed v as $v = 0.1\omega_d$, and is related to the speed of sprocket 1 by $\omega_1 = 2\omega_d$. The load friction torque measured at the drive shaft is 1.2 N·m.

The load must move at a speed of 1 m/s for 2 s, with acceleration and deceleration times chosen to be compatible with available motor torque. The desired speed profile of the load is shown in Figure 3.6.4 for assumed acceleration-deceleration times of 0.5 s. An available motor and reducer have the inertias $I_m = 0.003$ kg·m^2 and $I_r = 0.002$ kg·m^2. The reducer ratio is $N = 4$. The maximum motor torque is 3 N·m, its rated continous torque is 2 N·m, and its maximum speed is 100 rad/s.

Determine if this motor and reducer are suitable.

■ Solution

The speed ratio of the sprocket drive at the driving shaft is the ratio of the sprocket diameters, which is $0.1/0.05 = 2$. So the sprocket drive is a speed reducer when going from the motor to

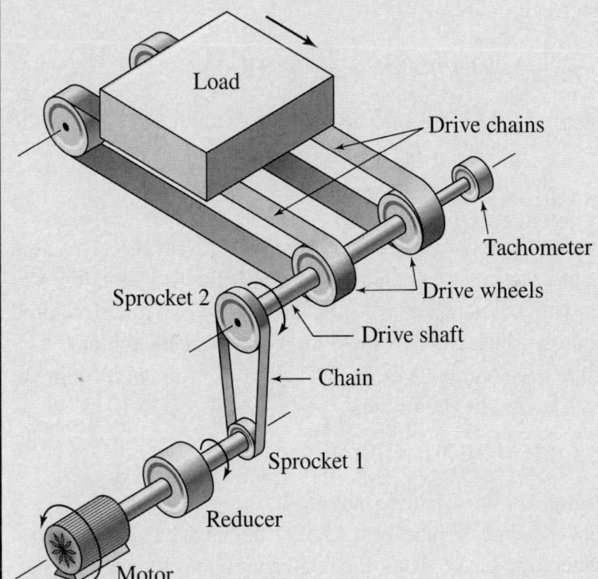

Figure 3.6.3 A conveyor drive system.

Figure 3.6.4 Load speed profile for Example 3.6.1.

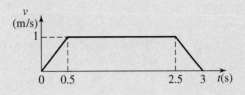

the drive shaft. Therefore, when referencing the significant inertias to the shaft of sprocket 1, we must divide the inertias by $2^2 = 4$.

First express the kinetic energy K of each component in terms of the speed ω_1 of sprocket 1.

Chains (2 of them):

$$K_{\text{chains}} = \frac{1}{2}[2(8)v^2] = \frac{1}{2}16(0.1\omega_d)^2 = \frac{1}{2}(0.16)\omega_d^2 = \frac{1}{2}(0.04)\omega_1^2$$

So the inertia due to both chains is $I_{\text{chains}} = 0.04$.

Drive Wheels (4 of them):

$$K_{\text{wheels}} = \frac{1}{2}[4I\omega_d^2] = \frac{1}{2}(0.08)\omega_d^2 = \frac{1}{2}(0.08)\frac{\omega_1^2}{4}$$

So the inertia due to the four drive wheels is $I_{\text{wheels}} = (0.08)/4 = 0.02$.

Load:

$$K_{\text{load}} = \frac{1}{2}10v^2 = \frac{1}{2}10(0.1\omega_d)^2 = \frac{1}{2}(0.025)\omega_1^2$$

So the inertia due to the load is $I_{\text{load}} = 0.025$.

The total inertia felt at the shaft of sprocket 1 due to the chains, drive wheels, and load is

$$I_1 = 0.04 + 0.02 + 0.025 = 0.085 \text{ kg}\cdot\text{m}^2$$

The friction torque felt at the shaft of sprocket 1 is $1.2/2 = 0.6$ N·m.

For inertia matching, from (3.6.2), using the given values for the motor and reducer, we have

$$N = \sqrt{\frac{I_1}{I_m + I_r}} = \sqrt{\frac{0.085}{0.003 + 0.002}} = 4.123$$

This ratio is close to the available value $N = 4$. From (3.6.3), $p = 4/4.123 = 0.9702$. Thus, using $N = 4$ instead of $N = 4.123$ still achieves 97% of the maximum possible acceleration.

Now that we have a value for N, we can compute the total inertia and load torque felt at the motor shaft. The equivalent inertia is

$$I = I_m + I_r + \frac{I_1}{N^2} = 0.003 + 0.002 + \frac{0.085}{16} = 0.0103$$

The load torque at the motor is $0.6/N = 0.6/4 = 0.15$. The equation of motion in terms of the motor speed ω_m is

$$0.0103\dot{\omega}_m = T_m - 0.15$$

The motor speed is related to the load speed v through the drive wheel radius (0.1 m), the sprocket speed ratio (2), and the reducer ratio $N = 4$ as follows: $\omega_m = 4(2)v/(0.1) = 80v$. From the given speed profile for v, the maximum load speed is 1 m/s, and thus the maximum required motor speed is $\omega_m = 80(1) = 80$ rad/s, which is below the limit of the available motor.

The maximum required angular acceleration is $\dot{\omega}_m = 80(1)/0.5 = 160$ rad/s^2. From the equation of motion, the maximum required motor torque is

$$T_{\max} = 0.0103\dot{\omega}_m + 0.15 = 0.0103(160) + 0.15 = 1.798 \text{ N·m}$$

which is less than the maximum torque of the available motor.

The torque during the constant speed phase must be 0.15 N·m and must be $0.0103(160) - 0.15 = 1.498$ N·m during the deceleration phase. Thus, the rated continuous torque must be

$$T_{\text{rms}} = \sqrt{\frac{1}{3}\left[(1.798)^2(0.5) + (0.15)^2(2) + (1.498)^2(0.5)\right]} = 0.9632 \text{ N·m}$$

which is less than the rated continuous torque of the available motor.

We could also have used Table 1.5.1 to compute the rated continuous torque. The area under the ω_m profile gives the total angular displacement of the motor shaft: $\theta_f = 0.5(8) + 2(80) = 200$ rad. We can use this value with $t_1 = 0.5$, $t_2 = 2.5$, and $t_3 = 3$ in the table as follows:

$$T_{\text{rms}} = \sqrt{\frac{2(0.0103)^2(200)^2}{3(0.5)(2.5)^2} + (0.15)^2} = 0.9632$$

So the available motor is capable of running at the maximum required speed and is able to deliver the values of $T_{\max}$ and T_{rms}, for the assumed acceleration-deceleration times of 0.5 s.

Using (3.6.1) to recompute the optimal reducer ratio with $T_1 = T_{\text{rms}}$ and $T_2 = 0.6$, we obtain

$$N = -\frac{0.6}{0.9632} + \sqrt{\left(\frac{0.6}{0.9632}\right)^2 + \frac{0.085}{0.005}} = 3.5469$$

which is close to our design value of 4.

3.7 CHAPTER REVIEW

In this chapter we reviewed Newton's laws of motion and applied them first to situations where the object's motion is relatively simple—simple translations or simple rotations—and then to the case of translation and rotation in a plane. We assumed that the masses are rigid and we restricted our analysis to forces that are constant or functions of time. We introduced the concepts of equivalent mass and equivalent inertia. These concepts simplify the modeling of systems having both translating and rotating components whose motions are directly coupled.

Now that you have finished this chapter, you should be able to

1. Obtain the equations of motion for an object consisting of a single mass undergoing simple translation or rotation, and solve them when the applied forces or moments are either constants or simple functions of time.
2. Apply the principle of conservation of mechanical energy to analyze systems acted on by conservative forces.
3. Apply the concepts of equivalent mass and equivalent inertia to obtain a simpler model of a multimass system whose motions are directly coupled.
4. Obtain the equation of motion of a body in planar motion, involving simultaneous translation and rotation.
5. Select a suitable motor and gear system for a given motion-control system.

REFERENCE

[Cannon, 1967] R. H. Cannon, Jr., *Dynamics of Physical Systems*, McGraw-Hill, New York, 1967.

PROBLEMS

Section 3.1 Translational Motion

3.1 Consider the falling mass in Example 3.1.1 and Figure 3.1.2. Find its speed and height as functions of time. How long will it take to reach (a) the platform and (b) the ground?

3.2 A baseball is thrown horizontally from the pitcher's mound with a speed of 90 mph. Neglect air resistance, and determine how far the ball will drop by the time it crosses home plate 60 ft away.

3.3 For the mass shown in Figure 3.1.3b, $m = 10$ kg, $\phi = 25°$, $v(0) = 2$ m/s, and $\mu = 0.3$. Determine whether the mass comes to rest if (a) $f_1 = 100$ N and (b) $f_1 = 50$ N. If the mass comes to rest, compute the time at which it stops.

3.4 A particle of mass $m = 25$ kg slides down a frictionless ramp starting from rest (see Figure P3.4). Suppose that $\theta = 30°$, $L = 8$ m, and $H = 3$ m. Compute the distance D of the impact point.

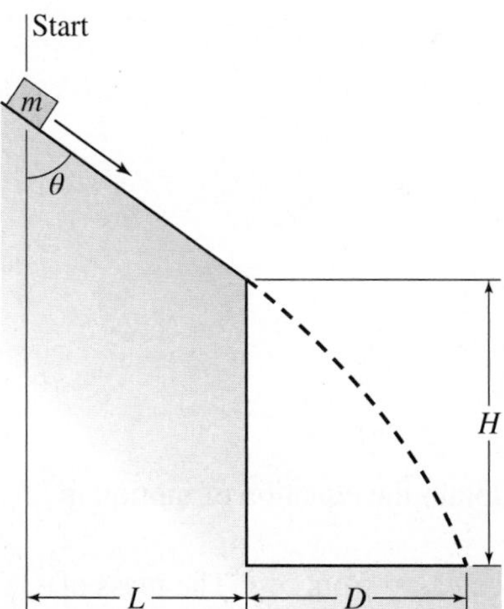

Figure P3.4

3.5 A particle of mass m slides down a frictionless ramp starting from rest (see Figure P3.4). The lengths L and H and the angle θ are given. Derive an expression for the distance D of the impact point.

3.6 A radar tracks the flight of a projectile (see Figure P3.6). At time t, the radar measures the horizontal component $v_x(t)$ and the vertical component $v_y(t)$ of the projectile's velocity and its range $R(t)$ and elevation $\phi(t)$. Are these measurements sufficient to compute the horizontal distance D from the radar to the launch point of the projectile? If so, derive the expression for D as a function of g and the measured values $v_x(t)$, $v_y(t)$, $R(t)$, and $\phi(t)$.

Figure P3.6

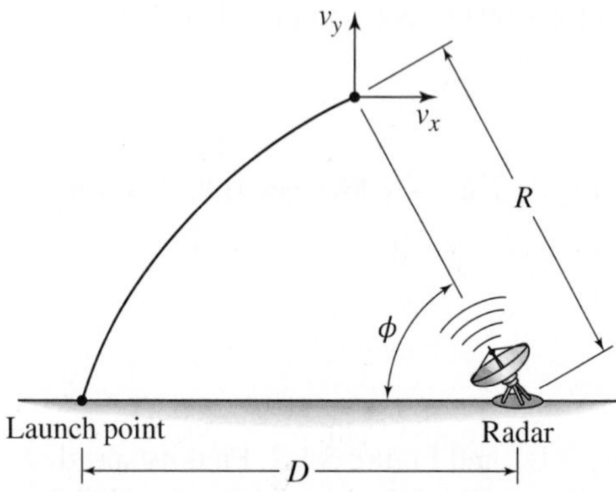

Section 3.2 Rotation About a Fixed Axis

3.7 Table 3.2.1 gives the inertia I_O for a point mass in rotation. Suppose instead that a sphere of radius r is in rotation about a fixed point O a distance R from its center of mass. Find the inertia I_O of the sphere about the point O.

3.8 A motor supplies a moment M to the pulley of radius R_1 (Figure P3.8). A belt connects this pulley to pulley A. Pulleys A and B form a rigid body with a common hub. Ignore the inertias of the three pulleys. Determine the acceleration of the mass m in terms of M, m, R_1, R_2, and R_3.

Figure P3.8

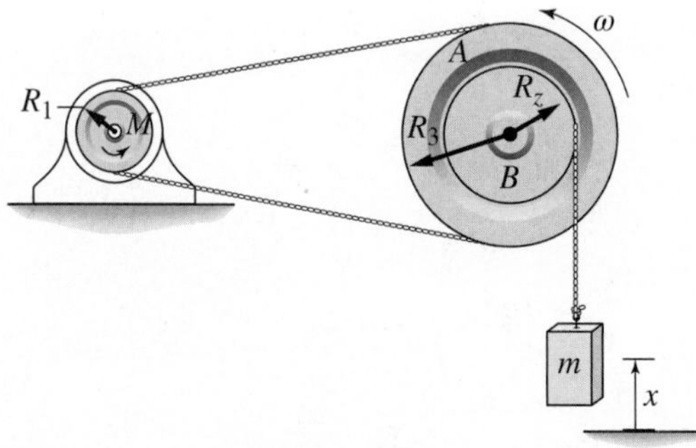

3.9 Figure P3.9 shows an inverted pendulum. Obtain the equation of motion in terms of the angle ϕ.

3.10 The two masses shown in Figure P3.10 are released from rest. The mass of block A is 100 kg; the mass of block B is 20 kg. Ignore the masses of the

Figure P3.9

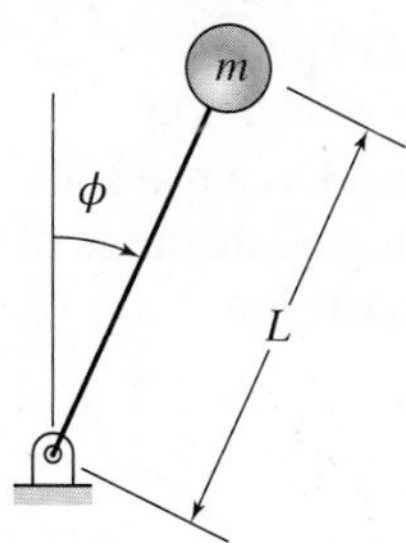

Figure P3.10

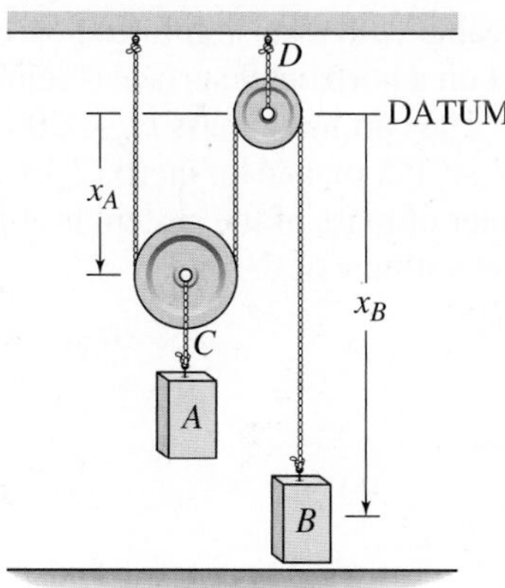

pulleys and rope. Determine the acceleration of block A and of block B. Do they rise or fall?

3.11 The motor in Figure P3.11 lifts the mass m_L by winding up the cable with a force F_A. The center of pulley B is fixed. The pulley inertias are I_B and I_C. Pulley C has a mass m_C. Derive the equation of motion in terms of the speed v_A of point A on the cable with the force F_A as the input.

Figure P3.11

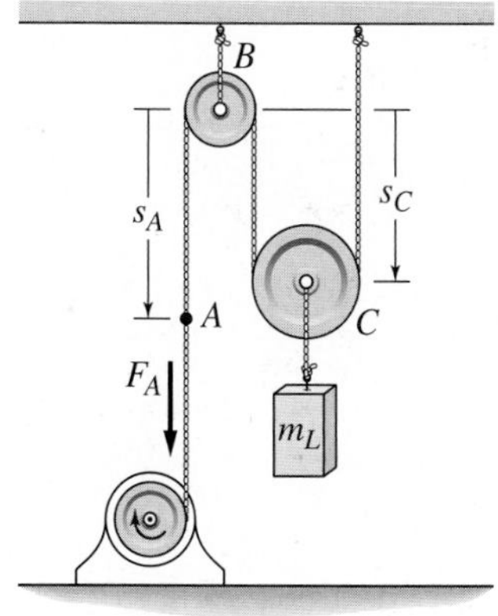

Figure P3.12

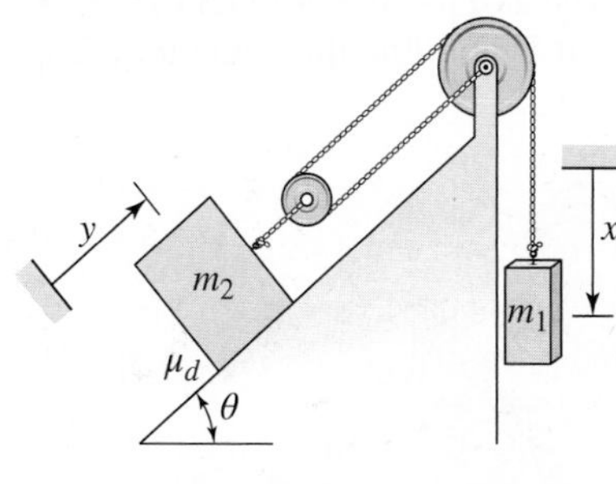

3.12 Instead of using the system shown in Figure 3.2.6a to raise the mass m_2, an engineer proposes to use two simple machines, the pulley and the inclined plane, to reduce the weight required to lift m_2. The proposed design is shown in Figure P3.12. The pulley inertias are negligible. The available horizontal space limits the angle of the inclined plane to no less than 30°.

a. Suppose that the friction between the plane and the mass m_2 is negligible. Determine the smallest value m_1 can have to lift m_2. Your answer should be a function of m_2 and θ.

b. In practice, the coefficient of dynamic friction μ_d between the plane and the mass m_2 is not known precisely. Assume that the system can be started to overcome static friction. For the value of $m_1 = m_2/2$, how large can μ_d be before m_1 cannot lift m_2?

Section 3.3 Equivalent Mass and Inertia

3.13 An inextensible cable with a tension force $f = 1000$ N is used to pull a two-wheeled cart on a horizontal surface (Figure P3.13). The wheels roll without slipping. The cart has a mass $m_c = 200$ kg. Each wheel has a mass of 16 kg, a radius $R_w = 0.5$ m, and an inertia $I_w = 2$ kg·m^2. Disregard the mass of the axle. The center of mass of the system is at point G. Compute the translational acceleration $\dot{v}$ of the cart.

Figure P3.13

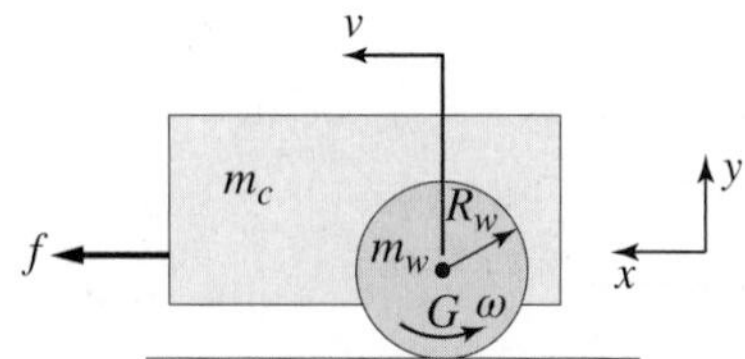

3.14 Consider the cart shown in Figure P3.13. Suppose we model the wheels as solid disks. Then the wheel inertia is given by $I_w = m_w R_w^2/2$. How small must the wheel mass m_w be relative to the cart body mass m_c so that the two wheels increase the equivalent mass m_e by no more than 10%?

3.15 Consider the spur gears shown in Figure P3.15, where $I_1 = 0.5$ kg·m^2 and $I_2 = 0.5$ kg·m^2. Shaft 1 rotates four times faster than shaft 2. The torques are given as $T_1 = 0.5$ N·m and $T_2 = -0.3$ N·m. Compute the equivalent inertia on shaft 2, and find the angular acceleration $\dot{\omega}_2$.

Figure P3.15

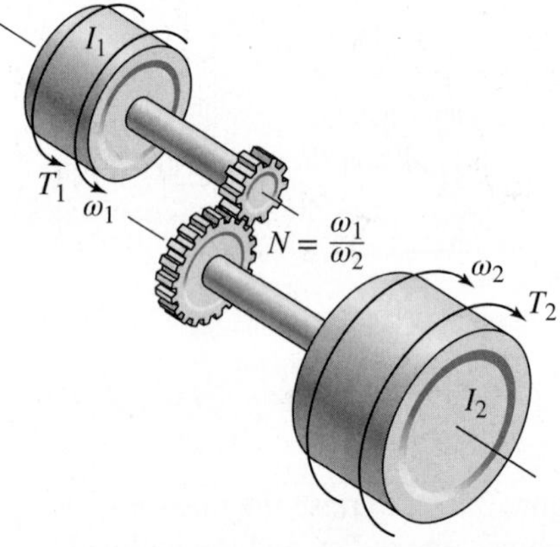

3.16 Consider the spur gears shown in Figure P3.15, where $I_1 = 0.2$ kg·m^2 and $I_2 = 3$ kg·m^2. Shaft 1 rotates twice as fast as shaft 2. The torques are given as $T_1 = 10$ N·m and $T_2 = 0$. (a) Compute the angular acceleration $\dot{\omega}_1$ assuming the shaft and gear inertias are negligible. (b) Suppose the gear on shaft 1 has an inertia 0.005 kg·m^2 and the gear on shaft 2 has an inertia 0.05 kg·m^2. Compute the angular acceleration $\dot{\omega}_1$.

3.17 Derive the expression for the equivalent inertia I_e felt on the input shaft, for the spur gears shown in Figure P3.15, where the combined gear-shaft inertias are I_{s_1} and I_{s_2}.

3.18 Draw the free body diagrams of the two spur gears shown in Figure P3.15. Use the resulting equations of motion to show that $T_2 = NT_1$ if the gear inertias are negligible or if there is zero acceleration. Here T_2 is taken to be the torque felt on shaft 2 due to the applied torque T_1.

3.19 The geared system shown in Figure P3.19 represents an elevator system. The motor has an inertia I_1 and supplies a torque T_1. Neglect the inertias of the gears, and assume that the cable does not slip on the pulley. Derive an expression for the equivalent inertia I_e felt on the input shaft (shaft 1). Then derive the dynamic model of the system in terms of the speed ω_1 and the applied torque T_1. The pulley radius is R.

3.20 Derive the expression for the equivalent inertia I_e felt on the input shaft, for the rack and pinion shown in Figure P3.20, where the shaft inertia is I_s.

Figure P3.19

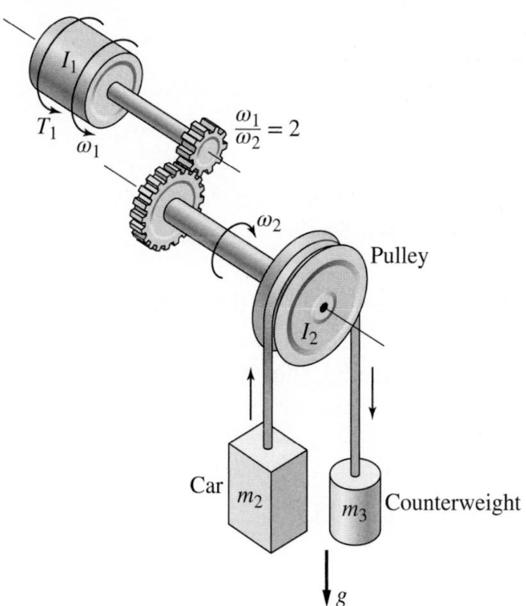

Figure P3.20

3.21 Derive the expression for the equivalent inertia I_e felt on the input shaft, for the belt drive treated in Example 3.3.5, where the shaft inertias connected to the sprockets are I_{s_1} and I_{s_2}.

3.22 For the geared system shown in Figure P3.15, proper selection of the gear ratio N can maximize the load acceleration $\dot{\omega}_2$ for a given motor and load. Note that the gear ratio is defined such that $\omega_1 = N\omega_2$. Assuming that the inertias I_1 and I_2 and the torques T_1 and T_2 are given,

a. Derive the expression for the load acceleration $\dot{\omega}_2$.

b. Use calculus to determine the value of N that maximizes $\dot{\omega}_2$.

3.23 For the geared system shown in Figure P3.23, assume that shaft inertias and the gear inertias I_1, I_2, and I_3 are negligible. The motor and load inertias in kg·m² are

$$I_4 = 0.03 \qquad I_5 = 0.15$$

Figure P3.23

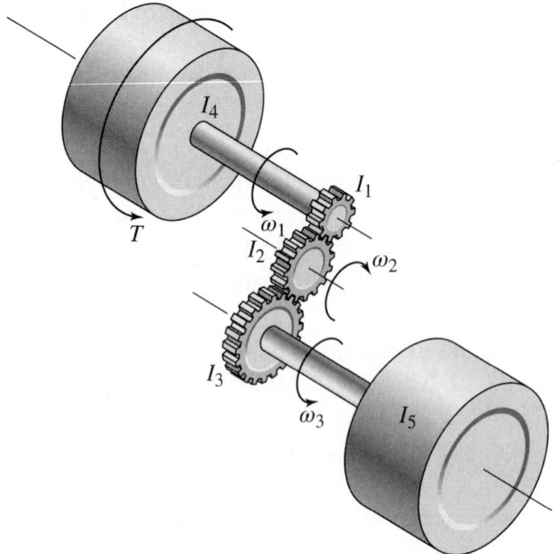

The speed ratios are

$$\frac{\omega_1}{\omega_2} = \frac{\omega_2}{\omega_3} = 1.8$$

Derive the system model in terms of the speed ω_3, with the applied torque T as the input.

3.24 For the geared system discussed in Problem 3.23, shown in Figure P3.23, the inertias are given in kg·m^2 as

$$I_1 = 10^{-3} \qquad I_2 = 3.84 \times 10^{-3} \qquad I_3 = 0.0148$$
$$I_4 = 0.03 \qquad I_5 = 0.15$$

The speed ratios are

$$\frac{\omega_1}{\omega_2} = \frac{\omega_2}{\omega_3} = 1.8$$

Derive the system model in terms of the speed ω_3, with the applied torque T as the input. The shaft inertias are negligible.

3.25 The geared system shown in Figure P3.25 is similar to that used in some vehicle transmissions. The speed ratios (which are the ratios of the gear radii) are

$$\frac{\omega_2}{\omega_1} = 2 \qquad \frac{\omega_3}{\omega_2} = \frac{3}{4} \qquad \frac{\omega_4}{\omega_3} = \frac{13}{15}$$

a. Determine the overall speed ratio ω_4/ω_1.

b. Derive the equation of motion in terms of the velocity ω_4, with the torque T_1 as the input. Neglect the inertias of the gears and the shafts.

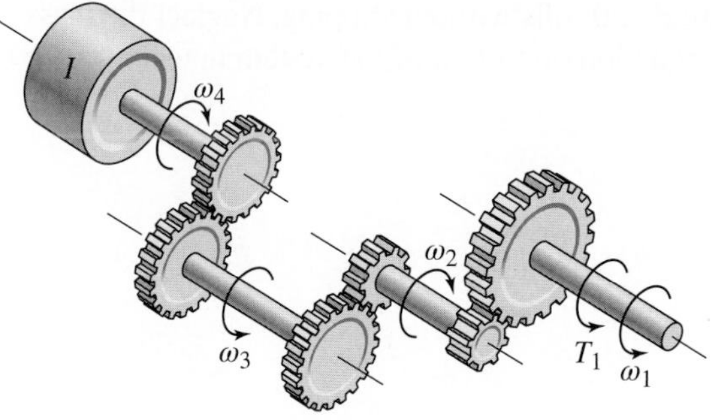

Figure P3.25

3.26 Consider the rack-and-pinion gear shown in Figure P3.20. Use the free body diagram method to obtain the expression for the acceleration $\ddot{x}$ in terms of the given quantities: R, T, m, and I.

3.27 The *lead screw* (also called a *power screw* or a *jack screw*) is used to convert the rotation of a motor shaft into a translational motion of the mass m (see Figure P3.27). For one revolution of the screw, the mass translates a distance L (called the *screw lead*). As felt on the motor shaft, the translating mass appears as an equivalent inertia. Use kinetic energy equivalence to derive an expression for the equivalent inertia. Let I_s be the inertia of the screw.

Figure P3.27

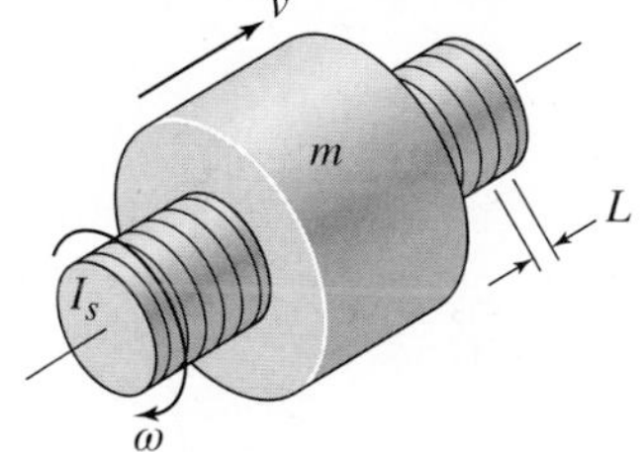

3.28 At time $t = 0$, the operator of the road roller disengages the transmission so that the vehicle rolls down the incline (see Figure P3.28). Determine an expression for the vehicle's speed as a function of time. The two rear wheels weigh 500 lb each and have a radius of 4 ft. The front wheel weighs 800 lb and has a radius of 2 ft. The vehicle body weighs 9000 lb. Assume the wheels roll without slipping.

3.29 Derive the equation of motion of the block of mass m_1 in terms of its displacement x (see Figure P3.29). The friction between the block and the surface is negligible. The pulley has negligible inertia and negligible friction. The cylinder has a mass m_2 and rolls without slipping.

Figure P3.28

Figure P3.29

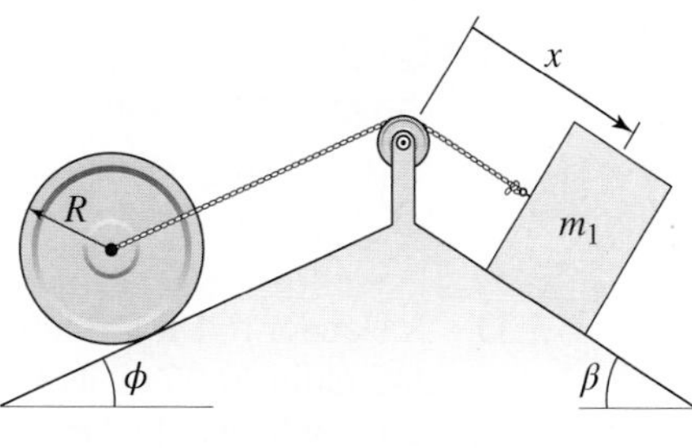

3.30 Assume the cylinder in Figure P3.30 rolls without slipping. Neglect the mass of the pulleys and derive the equation of motion of the system in terms of the displacement x.

Figure P3.30

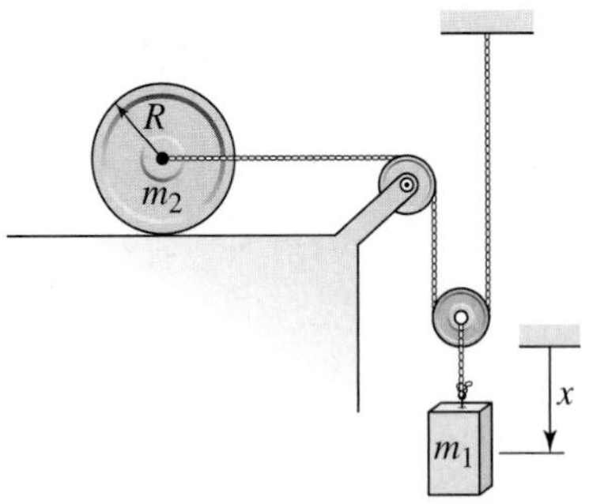

Section 3.4 General Planar Motion

3.31 A person pushes a roller of radius R and inertia $mR^2/2$, with a force f applied at an angle of ϕ to the horizontal (see Figure P3.31). The roller mass m is 80 kg and the roller radius is 0.4 m. Assume the roller does not slip. Derive the equation of motion in terms of (a) the rotational velocity ω of the roller and (b) the displacement x.

3.32 A slender rod 1.4 m long and of mass 20 kg is attached to a wheel of radius 0.05 m and negligible mass, as shown in Figure P3.32. A horizontal force f is applied to the wheel axle. Derive the equation of motion in terms of θ. Assume the wheel does not slip.

Figure P3.31

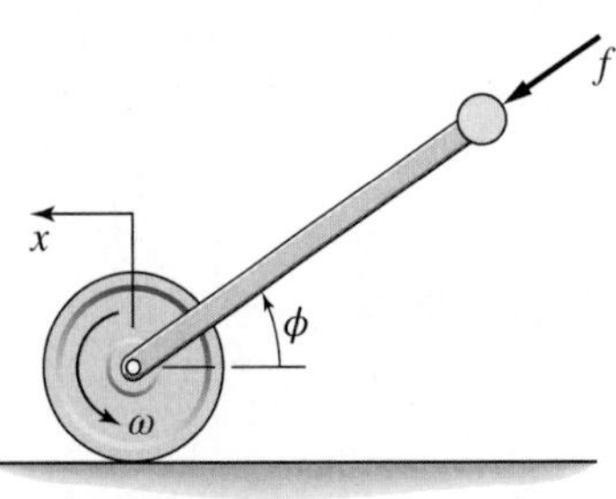

Figure P3.32

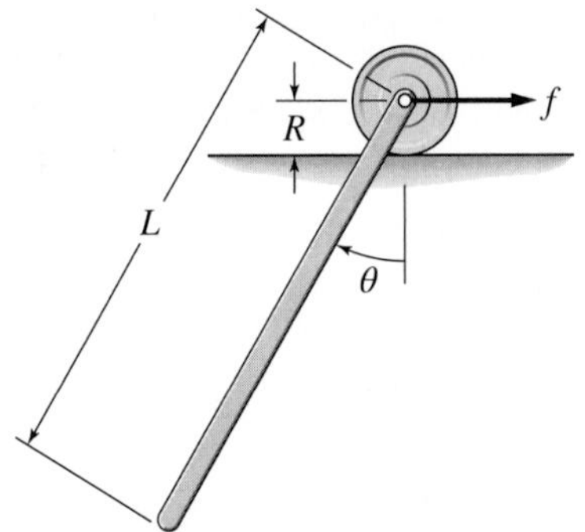

3.33 A slender rod 1.4 m long and of mass 20 kg is attached to a wheel of mass 3 kg and radius 0.05 m, as shown in Figure P3.32. A horizontal force f is applied to the wheel axle. Derive the equation of motion in terms of θ. Assume the wheel does not slip.

3.34 Consider the rolling cylinder treated in Example 3.4.1 and shown in Figure P3.34. Assume now that the no-slip condition is not satisfied, so that the cylinder slips while it rolls. Derive expression for the translational acceleration a_{G_x} and the angular acceleration α. The coefficient of dynamic friction is μ_d.

Figure P3.34

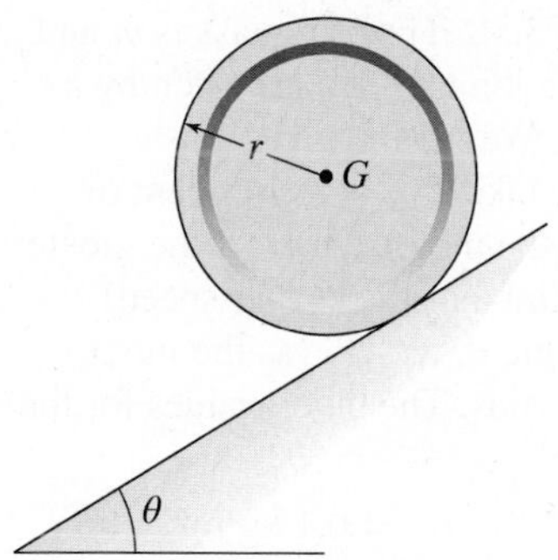

3.35 A hoop of mass m and radius r starts from rest and rolls down an incline at an angle θ. The hoop's inertia is given by $I_G = mr^2$. The static friction coefficient is μ_s. Determine the acceleration of the center of mass a_{G_x} and the angular acceleration α. Assume that the hoop rolls without bouncing or slipping. Use two approaches to solve the problem: (a) Use the moment equation about the mass center G and (b) use the moment equation about the contact point P. (c) Obtain the frictional condition required for the hoop to roll without slipping.

Section 3.5 Additional Examples

3.36 The pendulum shown in Figure P3.36 consists of a slender rod weighing 3 lb and a block weighing 10 lb.

a. Determine the location of the center of mass.
b. Derive the equation of motion in terms of θ.

Figure P3.36

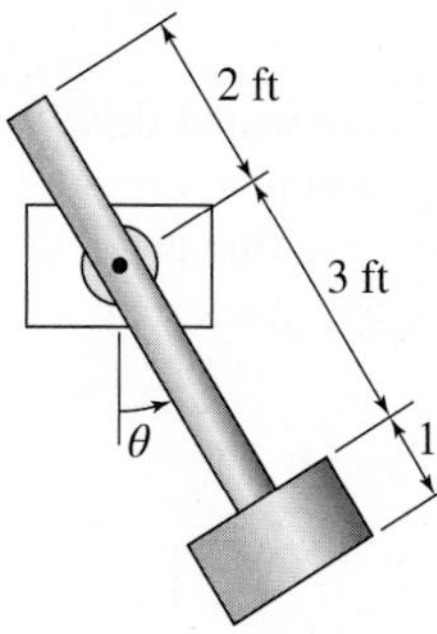

Figure P3.37

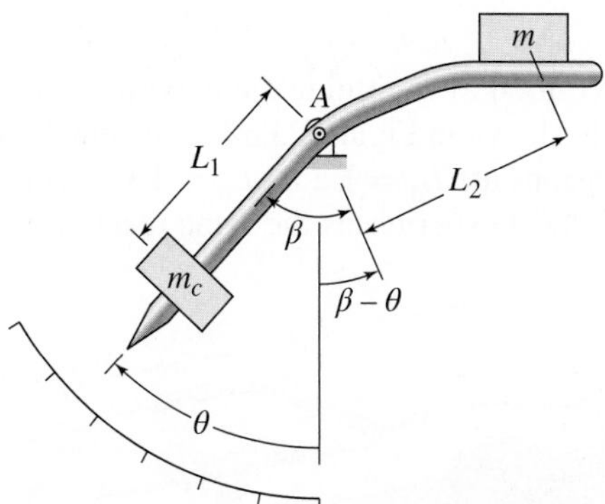

3.37 The scale shown in Figure P3.37 measures the weight mg of an object placed on the scale, by using a counterweight of mass m_c. Friction in the pivot point at A causes the pointer to eventually come to rest at an angle θ, which indicates the measured value of the weight mg. The angle β has a fixed value that depends on the shape of the scale arm. (a) Neglect the friction in the pivot and neglect the mass of the scale arm, and obtain the scale's equation of motion. (b) Find the equilibrium relation between the weight mg and the angle θ. (c) Find the weight mg if $m_c = 5$ kg, $L_1 = 0.2$ m, $L_2 = 0.15$ m, $\beta = 30°$, and $\theta = 20°$.

3.38 A single link of a robot arm is shown in Figure P3.38. The arm mass is m and its center of mass is located a distance L from the joint, which is driven by a motor torque T_m through two pairs of spur gears. We model the arm as a pendulum with a concentrated mass m. Thus, we take the arm's moment of inertia I_G to be zero. The gear ratios are $N_1 = 2$ (the motor shaft has the greater speed) and $N_2 = 1.5$ (the shaft connected to the link has the slower speed). Obtain the equation of motion in terms of the angle θ, with T_m as the input. Neglect the shaft inertias relative to the other inertias. The given values for the motor and gear inertias are

$$I_m = 0.05 \text{ kg·m}^2 \qquad I_{G_1} = 0.025 \text{ kg·m}^2 \qquad I_{G_2} = 0.1 \text{ kg·m}^2$$
$$I_{G_3} = 0.025 \text{ kg·m}^2 \qquad I_{G_4} = 0.08 \text{ kg·m}^2$$

The values for the link are

$$m = 10 \text{ kg} \qquad L = 0.3 \text{ m}$$

Figure P3.38

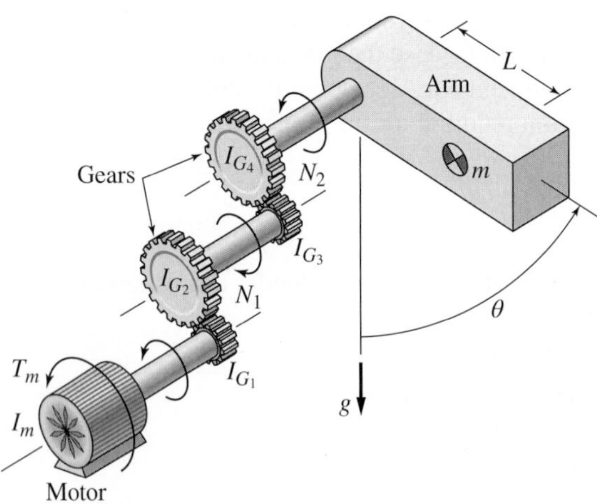

3.39 It is required to determine the maximum acceleration of the rear-wheel-drive vehicle shown in Figure P3.39. The vehicle mass is 1700 kg, and its dimensions are $L_A = 1.2$ m, $L_B = 1.1$ m, and $H = 0.5$ m. Assume that each front wheel experiences the same reaction force $N_A/2$. Similarly, each rear wheel experiences the same reaction force $N_B/2$. Thus, the traction force $\mu_s N_B$ is the total traction force due to both driving wheels. Assume that the mass of the wheels is small compared with the total vehicle mass; this assumption enables us to treat the vehicle as a translating rigid body with no rotating parts.

Figure P3.39

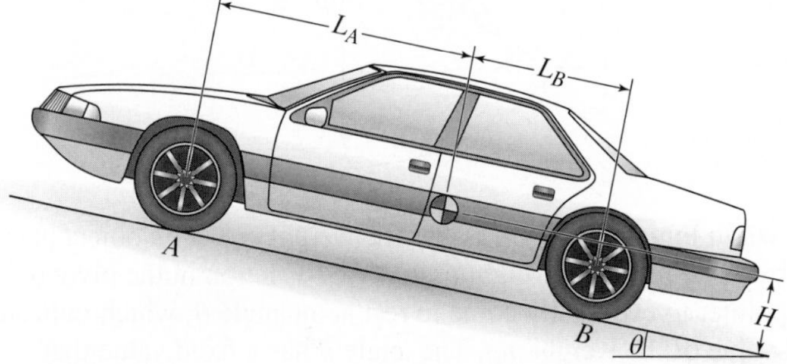

3.40 Figure P3.40 illustrates a pendulum with a base that moves horizontally. This is a simple model of an overhead crane carrying a suspended load with cables. The load mass is m, the cable length is L, and the base acceleration is $a(t)$. Assuming that the cable acts like a rigid rod, derive the equation of motion in terms of θ with $a(t)$ as the input.

3.41 Figure P3.41 illustrates a pendulum with a base that moves. The base acceleration is $a(t)$. Derive the equation of motion in terms of θ with $a(t)$ as the input. Neglect the mass of the rod.

3.42 The overhead trolley shown in Figure P3.42 is used to transport beams in a factory. The beam is rectangular, with a length of L. It is desired to limit the trolley horizontal acceleration a so that the beam does not swing too much. The beam starts from rest with $\theta(0) = 0$. (a) Use a small-angle approximation and determine the maximum value of θ if $L = 3$ m and $a = 2$ m/s^2. Once you obtain an answer, use it to check the validity of the small-angle approximation. (b) Solve the problem given in part (a), using symbolic values for L, a, and g. Does the answer depend on all three variables? If not, explain why.

Figure P3.40

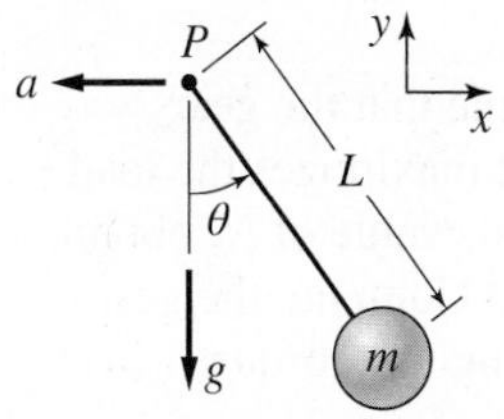

Figure P3.41

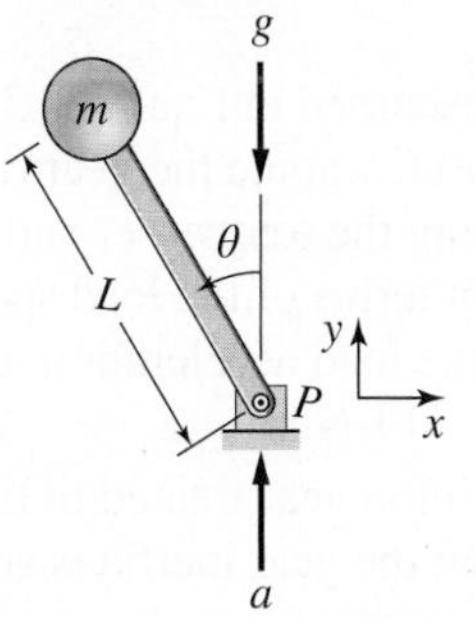

Figure P3.42

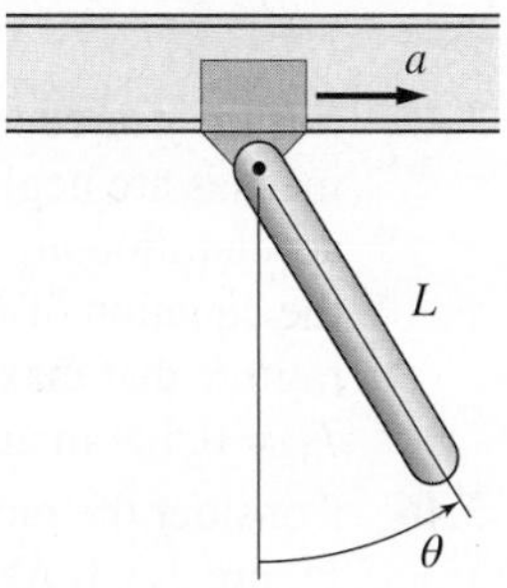

3.43 The analysis of the personal transporter in Example 3.5.6 assumed that the driving force f was the given input. Instead, model the system with the assumption that the input is the total torque T_w applied by the motor to the wheel-axle unit. The wheel radius is R. Obtain the equations of motion in terms of x and ϕ with T and T_w as the input.

3.44 The "sky crane" was a novel solution to the problem of landing the 2000 lb *Curiosity* rover on the surface of Mars. *Curiosity* hangs from the descent stage by 60-foot-long nylon tethers (Figure P3.44a). The descent stage uses its thrusters to hover as the rover is lowered to the surface. Thus the rover behaves like a pendulum whose base is moving, just like the pendulum analyzed in Problem 3.40 (see Figure P3.40). However, Problem 3.40 neglected the mass of the base, but the descent stage also has significant mass. So a better model is the one shown in Figure P3.44b. The rover mass is m_r, the descent stage mass is m_d, and the net horizontal component of the thruster forces is f. Among other simplifications, this model does not take into account vertical motion and any rotational motion.

Derive the equations of motion of the system in terms of the angle θ and the displacement x, with the force f as the given input.

Figure P3.44a

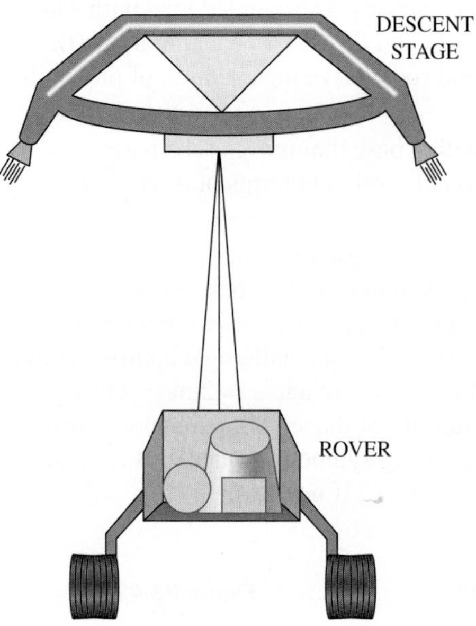

Figure P3.44b

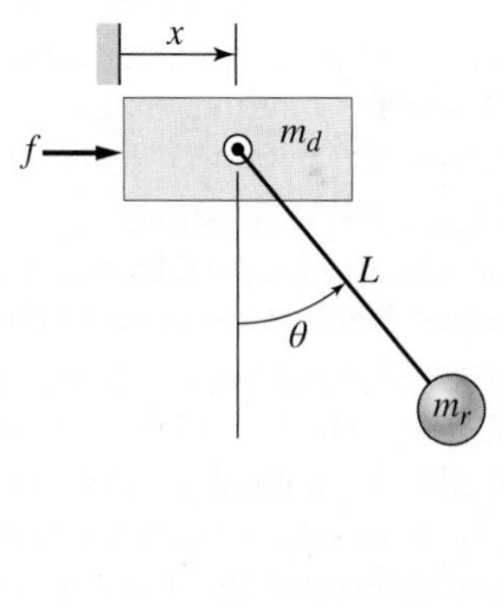

3.45 For the geared system discussed in Example 3.3.2, assume that the gear inertias are negligible. (a) Compute the gear ratio N that maximizes the load acceleration $\dot{\omega}_2$, neglecting the torques T_1 and T_2. For this value of N, obtain the equation of motion in terms of the load speed ω_2. (b) Compute the gear ratio N that maximizes the load acceleration $\dot{\omega}_2$, assuming that torques $T_1 = 0.5$ N·m and $T_2 = -0.1$ N·m.

3.46 Consider the rack-and-pinion gear treated in Example 3.3.4 and shown in Figure 3.3.3. Assume that the gear inertia is small compared to the inertia I, most of which is due to an object not shown on the shaft. Derive the formula for the gear radius R that maximizes the acceleration $\ddot{x}$.

3.47 Consider the elevator system treated in Problem 3.19 and shown in Figure P3.19. In that problem the gear ratio is $N = 2$. Suppose we want to use a gear ratio that maximizes the car acceleration. Obtain a formula for N in terms of R, I_1, I_2, m_2, and m_3.

3.48 For the conveyor system shown in Figure P3.48, the reducer reduces the motor speed by a factor of 2:1. The motor inertia is $I_1 = 0.003$ kg·m^2. Disregard the inertias of the reducer and the tachometer, which is used to measure the speed for control purposes. Ignore the inertias of the two sprockets, the chain, and all shafts. The only significant masses and inertias are the inertias of the four drive wheels (0.01 kg·m^2 each), the two drive chains (6 kg each), and the load mass (8 kg).

The radius of sprocket 1 is 0.04 m, that of sprocket 2 is 0.12 m. The drive wheel has a radius of 0.1 m. The load friction torque measured at the drive shaft is 0.9 N·m.

a. Derive the equation of motion of the conveyor in terms of the motor velocity, with the motor torque T_1 as the input.
b. Suppose the motor torque is constant at 1.2 N·m. Determine the resulting motor angular acceleration and load acceleration.

Figure P3.48a

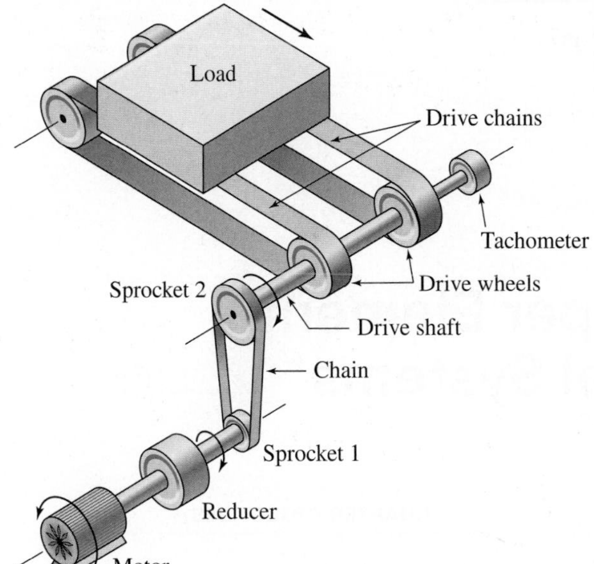

Figure P3.48b

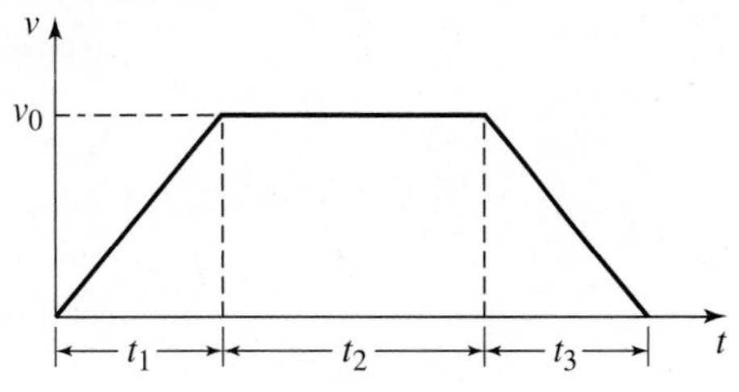

3.49 A conveyor drive system to produce translation of the load is shown in Figure P3.48a. The reducer is a gear pair that reduces the motor speed by a factor of 10:1. The motor inertia is $I_1 = 0.002$ kg·m^2. The reducer inertia as felt on the motor shaft is $I_2 = 0.003$ kg·m^2. Neglect the inertia of the tachometer, which is used to measure the speed for control purposes. The properties of the remaining elements are given here.

Sprockets:

Sprocket 1: radius = 0.05 m mass = 0.9 kg

Sprocket 2: radius = 0.15 m mass = 8.9 kg

Chain mass: 10.7 kg

Drive shaft: radius = 0.04 m mass = 2.2 kg

Drive wheels (four of them): radius = 0.2 m mass = 8.9 kg each

Drive chains (two of them): mass = 67 kg each

Load friction torque measured at the drive shaft: 54 N·m

Load mass: 45 kg

a. Derive the equation of motion of the conveyor in terms of the motor velocity ω_1, with the motor torque T_1 as the input.
b. Suppose the motor torque is constant at 10 N·m. Determine the resulting motor velocity ω_1 and load velocity v as functions of time, assuming the system starts from rest.
c. The profile of a desired velocity for the load is shown in Figure P3.48b, where $v_0 = 1$ m/s, $t_1 = t_3 = 0.5$ s, and $t_2 = 2$ s. Use the equation of motion found in part (a) to compute the required motor torque for each part of the profile.

CHAPTER 4

Spring and Damper Elements in Mechanical Systems

CHAPTER OUTLINE

CHAPTER OBJECTIVES

When you have finished this chapter, you should be able to

1. Model elements containing elasticity as ideal (massless) spring elements.
2. Model elements containing damping as ideal (massless) damper elements.
3. Obtain equations of motion for systems having spring and damper elements.
4. Apply energy methods to obtain equations of motion.
5. Obtain the free and forced response of mass-spring-damper systems.
6. Utilize MATLAB to assist in the response analysis.
7. Use simple models of vehicle suspensions to evaluate transient performance.

In Chapter 3 we applied Newton's laws of motion to situations in which the masses in question are assumed to be rigid and where the object's motion is relatively uncomplicated, namely, simple translations and simple rotation about a single axis. However, in many applications the mass either deforms somewhat under the action of forces or is connected to another object by an element that deforms. Such deformable elements exert a resisting force that is a function of displacement and are called *spring* elements or *elastic* elements. We treat the modeling of spring elements in Section 4.1 of this chapter, and in Section 4.2 we show how to obtain equations of motion for systems consisting of one or more masses and one or more spring elements.

In Chapter 3 we introduced energy-based methods and the concepts of equivalent mass and equivalent inertia, which simplify the modeling of systems having both translating and rotating components. In Section 4.3 we extend these methods and concepts to spring elements.

Velocity-dependent forces such as fluid drag are the subject of Section 4.4. Elements exerting a resisting force that is a function of velocity are called *damping* or *damper* elements. This section and Section 4.5 provide additional examples of how to model systems containing mass, spring, and damping elements.

MATLAB can be used to perform some of the algebra required to obtain transfer functions of multimass systems, and can be used to find the forced and free response. Section 4.7 shows how this is accomplished. Section 4.8 presents our second case study: design of a vehicle suspension system. ■

4.1 SPRING ELEMENTS

All physical objects deform somewhat under the action of externally applied forces. When the deformation is negligible for the purpose of the analysis, or when the corresponding forces are negligible, we can treat the object as a rigid body. Sometimes, however, an elastic element is intentionally included in the system, as with a spring in a vehicle suspension. Sometimes the element is not intended to be elastic, but deforms anyway because it is subjected to large forces or torques. This can be the case with the boom or cables of a large crane that lifts a heavy load. In such cases, we must include the deformation and corresponding forces in our analysis.

The most familiar spring is probably the helical coil spring, such as those used in vehicle suspensions and those found in retractable pens. The purpose of the spring in both applications is to provide a restoring force. However, considerably more engineering analysis is required for the vehicle spring application because the spring can cause undesirable motion of the wheel and chassis, such as vibration. Because the pen motion is constrained and cannot vibrate, we do not need as sophisticated an analysis to see if the spring will work.

Many engineering applications involving elastic elements, however, do not contain coil springs but rather involve the deformation of beams, cables, rods, and other mechanical members. In this section we develop the basic elastic properties of many of these common elements.

4.1.1 FORCE-DEFLECTION RELATIONS

A coil spring has a *free length*, denoted by L in Figure 4.1.1. The free length is the length of the spring when no tensile or compressive forces are applied to it. When a spring is compressed or stretched, it exerts a restoring force that opposes the compression or extension. The general term for the spring's compression or extension is *deflection*. The greater the deflection (compression or extension), the greater the restoring force. The simplest model of this behavior is the so-called *linear force-deflection model*,

Figure 4.1.1 A spring element.

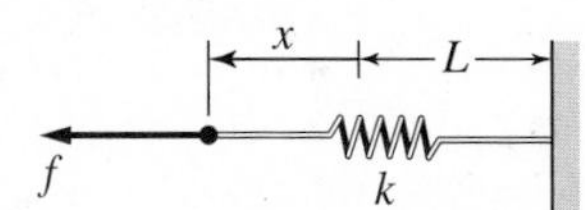

$$f = kx \tag{4.1.1}$$

where f is the restoring force, x is the compression or extension distance (the change in length from the free length), and k is the *spring constant*, or *stiffness*, which is defined to be always positive. Typical units for k are lb/ft and N/m. Some references, particularly in the automotive industry, use the term *spring rate* instead of spring constant. Equation (4.1.1) is commonly known as *Hooke's law*, named after Englishman Robert Hooke (1635–1703).

When $x = 0$, the spring assumes its free length. We must decide whether extension is represented by positive or negative values of x. This choice depends on the particular application. If $x > 0$ corresponds to extension, then a positive value of f represents the force of the spring pulling against whatever is causing the extension. Conversely, because of Newton's law of action-reaction, the force causing the extension has the same magnitude as f but is in the opposite direction.

The formula for a coil spring is derived in references on machine design. For convenience, we state it here without derivation, for a spring made from round wire.

$$k = \frac{Gd^4}{64nR^3}$$

where d is the wire diameter, R is the radius of the coil, and n is the number of coils. The *shear modulus of elasticity* G is a property of the wire material.

As we will see, other mechanical elements that have elasticity, such as beams, rods, and rubber mounts, can be modeled as springs, and are usually represented pictorially as a coil spring.

4.1.2 TENSILE TEST OF A ROD

A plot of the data for a tension test on a rod is given in Figure 4.1.2. The elongation is the change in the rod's length due to the tension force applied by the testing machine. As the tension force was increased, the elongation followed the curve labeled "Increasing." The behavior of the elongation under decreasing tension is shown by the curve labeled "Decreasing." The rod was stretched beyond its *elastic limit*, so that a permanent elongation remained after the tension force was removed.

For the smaller elongations the "Increasing" curve is close to a straight line with a slope of 3500 pounds per one thousandth of an inch, or 3.5×10^6 lb/in. This line is labeled "Linear" in the plot. If we let x represent the elongation in inches and f the tension force in pounds, then the model $f = 3.5 \times 10^6 x$ represents the elastic behavior of the rod. We thus see that the rod's spring constant is 3.5×10^6 lb/in.

This experiment could have been repeated using compressive instead of tensile force. For small compressive deformations x, we would find that the deformations are related to the compressive force f by $f = kx$, where k would have the same value as before. This example indicates that mechanical elements can be described by the linear

Figure 4.1.2 Plot of tension test data.

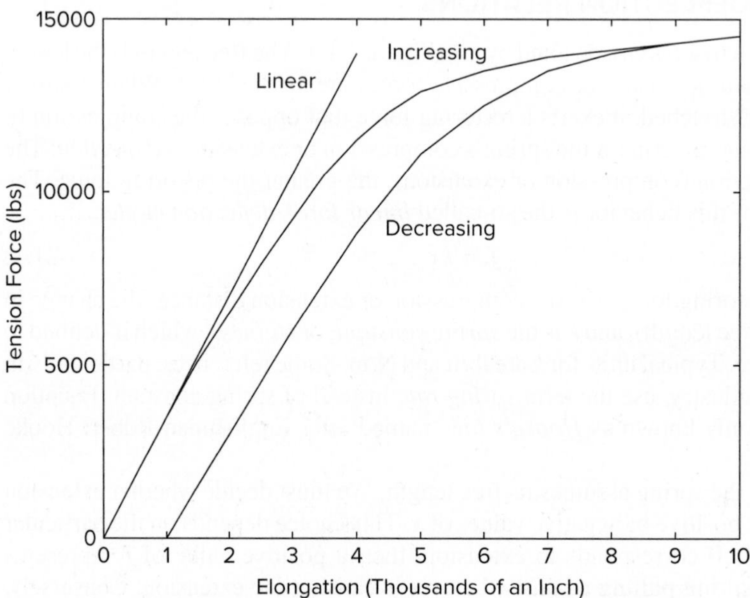

law $f = kx$, for both compression and extension, as long as the deformations are not too large, that is, deformations not beyond the elastic limit. Note that the larger the deformation, the greater the error that results from using the linear model.

4.1.3 ANALYTICAL DETERMINATION OF THE SPRING CONSTANT

In much engineering design work, we do not have the elements available for testing, and thus we must be able to calculate their spring constant from the geometry and material properties. To do this, we can use results from the study of mechanics of materials. Examples 4.1.1 and 4.1.2 show how this is accomplished.

EXAMPLE 4.1.1

Rod with Axial Loading

■ Problem

Derive the spring constant expression for a cylindrical rod subjected to an axial force (either tensile or compressive). The rod length is L and its area is A.

■ Solution

From mechanics of materials references, for example [Young, 2011], we obtain the force-deflection relation of a cylindrical rod:

$$x = \frac{L}{EA}f = \frac{4L}{\pi ED^2}f$$

where x is the axial deformation of the rod, f is the applied axial force, A is the cross-sectional area, D is the diameter, and E is the *modulus of elasticity* of the rod material. Rewrite this equation as

$$f = \frac{EA}{L}x = \frac{\pi ED^2}{4L}x$$

Thus, we see that the spring constant is $k = EA/L = \pi ED^2/4L$.

The modulus of elasticity of steel is approximately 3×10^7 lb/in.^{2}. Thus, a steel rod 20 in. long and 1.73 in. in diameter would have a spring constant of 3.5×10^6 lb/in., the same as the rod whose curve is plotted in Figure 4.1.2.

Beams used to support objects can act like springs when subjected to large forces. Beams can have a variety of shapes and can be supported in a number of ways. The beam geometry, beam material, and the method of support determine its spring constant.

EXAMPLE 4.1.2

Spring Constant of a Fixed-End Beam

■ Problem

Derive the spring constant expression of a fixed-end beam of length L, thickness h, and width w, assuming that the force f and deflection x are at the center of the beam (Figure 4.1.3).

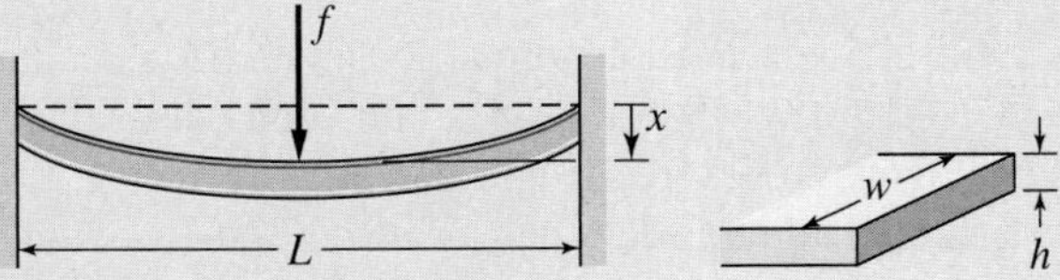

Figure 4.1.3 A fixed-end beam.

■ Solution

The force-deflection relation of a fixed-end beam is found in mechanics of materials references. It is

$$x = \frac{L^3}{192EI_A} f$$

where x is the deflection in the middle of the beam, f is the force applied at the middle of the beam, and I_A is the area moment of inertia about the beam's longitudinal axis [Young, 2011]. The area moment I_A is computed with an integral of an area element dA.

$$I_A = \int r^2 \, dA$$

Formulas for the area moments are available in most engineering mechanics texts. For a beam having a rectangular cross section with a width w and thickness h, the area moment is

$$I_A = \frac{wh^3}{12}$$

Thus the force-deflection relation reduces to

$$x = \frac{12L^3}{192Ewh^3} f = \frac{L^3}{16Ewh^3} f$$

The spring constant is the ratio of the applied force f to the resulting deflection x, or

$$k = \frac{f}{x} = \frac{16Ewh^3}{L^3}$$

Table 4.1.1 lists the expressions for the spring constants of several common elements. Note that two beams of identical shape and material, one a cantilever and one fixed-end, have spring constants that differ by a factor of 64. The fixed-end beam is thus 64 times "stiffer" than the cantilever beam! This illustrates the effect of the support arrangement on the spring constant.

A single leaf spring is shown in Table 4.1.1. Springs for vehicle suspensions are often constructed by strapping together several layers of such springs, as shown in Figure 4.1.4. The value of the total spring constant depends not only on the spring constants of the individual layers, but also on how they are strapped together, the method of attachment to the axle and chassis, and whether any material to reduce friction has been placed between the layers. There is no simple formula for k that accounts for all these variables.

Figure 4.1.4 A leaf spring.

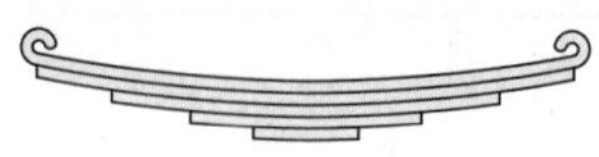

4.1.4 TORSIONAL SPRING ELEMENTS

Table 4.1.2 shows a hollow cylinder subjected to a twisting torque. The resulting twist in the cylinder is called *torsion*. This cylinder is an example of a *torsional* spring, which resists with an opposing torque when twisted. For a torsional spring element we will use the "curly" symbol shown in Figure 4.1.5. The spring relation for a torsional spring is usually written as

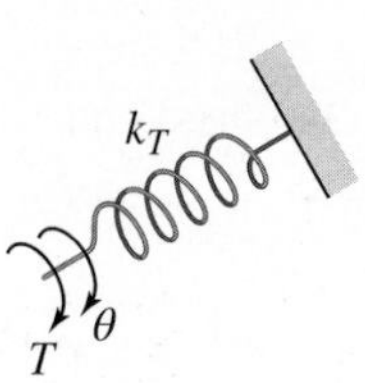

$$T = k_T \theta \tag{4.1.2}$$

where θ is the net angular twist in the element, T is the torque causing the twist, and k_T is the *torsional spring constant*. We assign $\theta = 0$ at the spring position where there is no torque in the spring. This is analogous to the free length position of a translational spring. Note that the units of the torsional and translational spring constants are not the same. FPS units for k_T are lb-ft/rad; in SI the units are N·m/rad.

Figure 4.1.5 Symbol for a torsional spring element.

k_T θ T

Table 4.1.1 Spring constants of common elements.

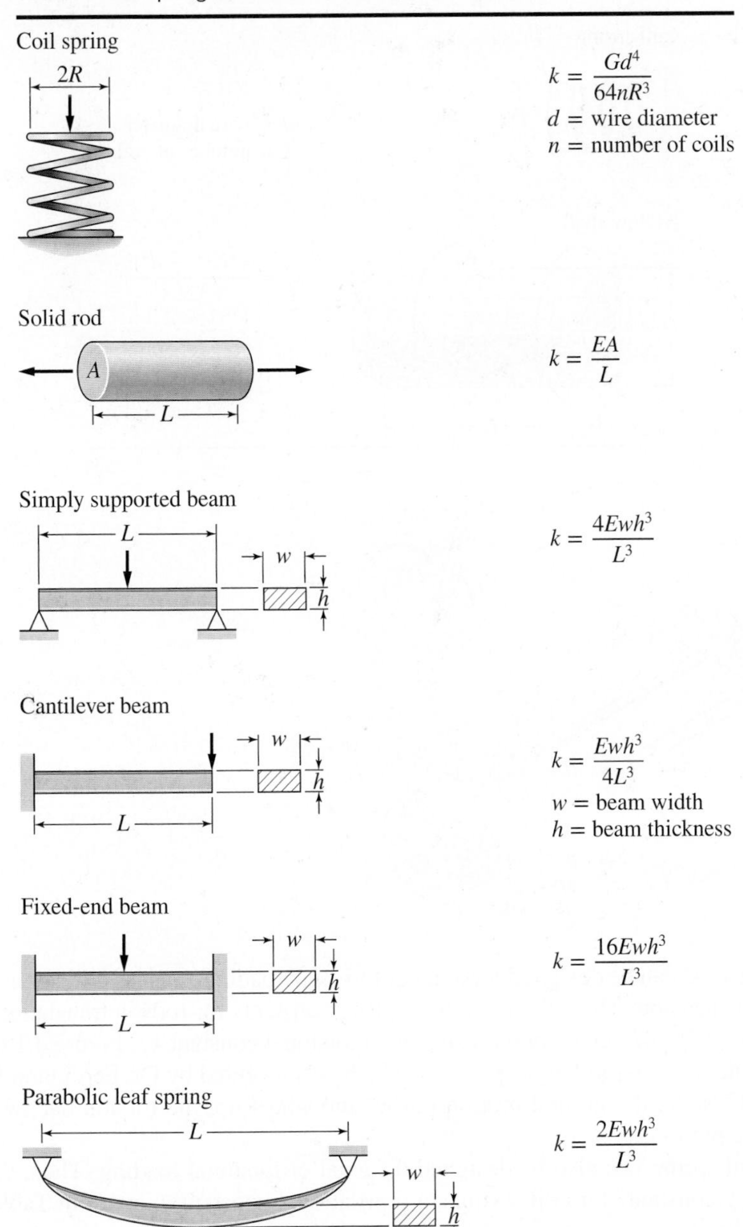

Element	Spring constant
Coil spring	$k = \dfrac{Gd^4}{64nR^3}$ d = wire diameter n = number of coils
Solid rod	$k = \dfrac{EA}{L}$
Simply supported beam	$k = \dfrac{4Ewh^3}{L^3}$
Cantilever beam	$k = \dfrac{Ewh^3}{4L^3}$ w = beam width h = beam thickness
Fixed-end beam	$k = \dfrac{16Ewh^3}{L^3}$
Parabolic leaf spring	$k = \dfrac{2Ewh^3}{L^3}$

Torsional spring constants of two elements are given in Table 4.1.2. They depend on the geometry of the element and its material properties, namely, E and G, the *shear* modulus of elasticity.

If the cylinder is solid, the formula for k_T given in Table 4.1.2 becomes

$$k_T = \frac{\pi G D^4}{32L}$$

Table 4.1.2 Torsional spring constants of common elements.

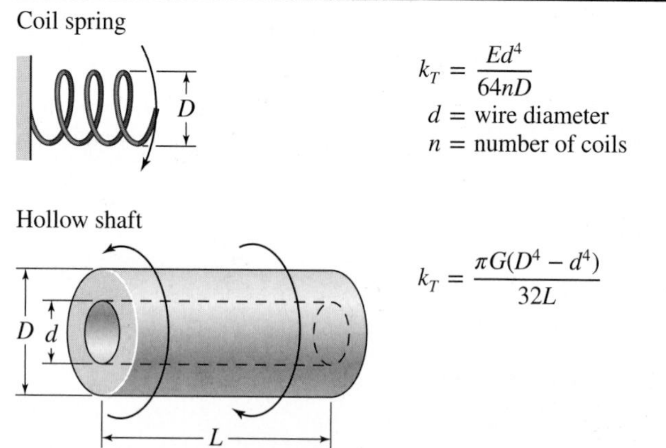

Element	Torsional spring constant
Coil spring	$k_T = \dfrac{Ed^4}{64nD}$ d = wire diameter n = number of coils
Hollow shaft	$k_T = \dfrac{\pi G(D^4 - d^4)}{32L}$

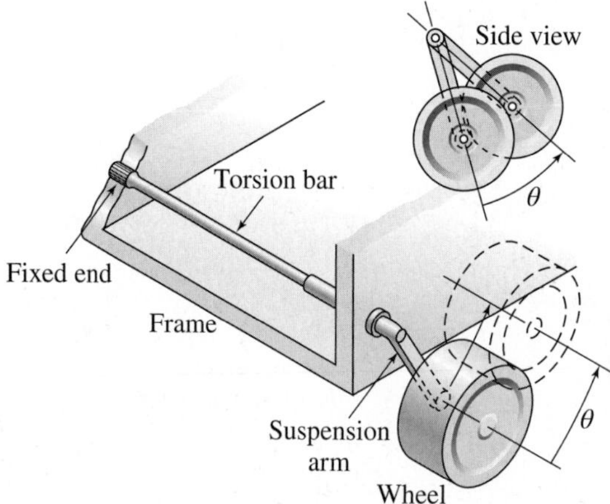

Figure 4.1.6 A torsion-bar suspension.

Note that a rod can be designed for axial or torsional loading, such as with a torsion-bar vehicle suspension. Thus, there are two spring constants for rods, a translational constant $k = \pi ED^2/4L$, given previously, and a torsional constant k_T. Figure 4.1.6 shows an example of a torsion-bar suspension, which was invented by Dr. Ferdinand Porsche in the 1930s. As the ground motion pushes the wheel up, the torsion bar twists and resists the motion.

A coil spring can also be designed for axial or torsional loading. Thus, there are two spring constants for coil springs, a translational constant k, given in Table 4.1.1, and the torsional constant k_T, given in Table 4.1.2.

4.1.5 SERIES AND PARALLEL SPRING ELEMENTS

In many applications, multiple spring elements are used, and in such cases we must obtain the equivalent spring constant of the combined elements. When two springs are connected side-by-side, as in Figure 4.1.7a, we can determine their equivalent spring constant as follows. Assuming that the force f is applied so that both springs have the same deflection x but different forces f_1 and f_2, then

$$x = \frac{f_1}{k_1} = \frac{f_2}{k_2}$$

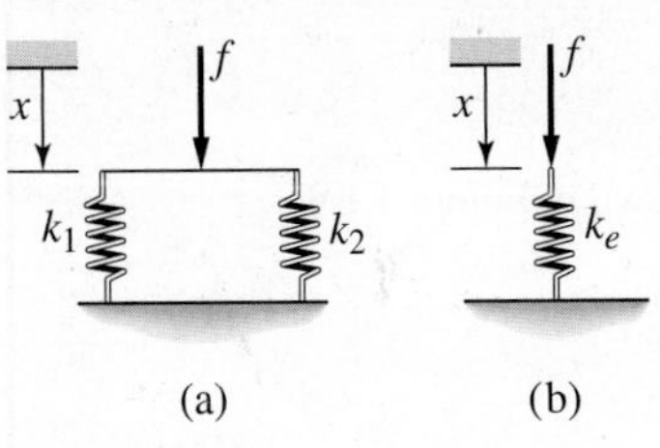

Figure 4.1.7 Parallel spring elements.

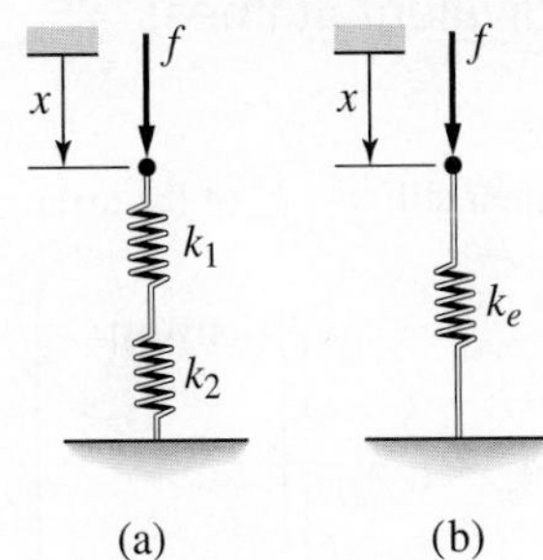

Figure 4.1.8 Series spring elements.

If the system is in static equilibrium, then

$$f = f_1 + f_2 = k_1 x + k_2 x = (k_1 + k_2)x$$

For the equivalent system shown in part (b) of the figure, $f = k_e x$, and thus we see that its equivalent spring constant is given by $k_e = k_1 + k_2$. This formula can be extended to the case of n springs connected side-by-side as follows:

$$k_e = \sum_{i=1}^{n} k_i \tag{4.1.3}$$

When two springs are connected end-to-end, as in Figure 4.1.8a, we can determine their equivalent spring constant as follows. Assuming both springs are in static equilibrium, then both springs are subjected to the same force f, but their deflections f/k_1 and f/k_2 will not be the same unless their spring constants are equal. The total deflection x of the system is obtained from

$$x = \frac{f}{k_1} + \frac{f}{k_2} = \left(\frac{1}{k_1} + \frac{1}{k_2}\right) f$$

For the equivalent system shown in part (b) of the figure, $f = k_e x$, and thus we see that its equivalent spring constant is given by

$$\frac{1}{k_e} = \frac{1}{k_1} + \frac{1}{k_2}$$

This formula can be extended to the case of n springs connected end-to-end as follows:

$$\frac{1}{k_e} = \sum_{i=1}^{n} \frac{1}{k_i} \tag{4.1.4}$$

The derivations of (4.1.3) and (4.1.4) assumed that the product of the spring mass times its acceleration is zero, which means that either the system is in static equilibrium or the spring masses are very small compared to the other masses in the system.

The symbols for springs connected end-to-end look like the symbols for electrical resistors similarly connected. Such resistors are said to be in *series* and therefore, springs connected end-to-end are sometimes said to be in series. However, the equivalent electrical resistance is the sum of the individual resistances, whereas series springs obey the reciprocal rule (4.1.4). This similarity in appearance of the symbols often leads people to mistakenly add the spring constants of springs connected end-to-end, just as series resistances are added. Springs connected side-by-side are sometimes called *parallel* springs, and their spring constants should be added.

EXAMPLE 4.1.3 Determining Equivalent Stiffness

■ **Problem**

Determine the equivalent stiffness k_e of the arrangement shown in Figure 4.1.9a.

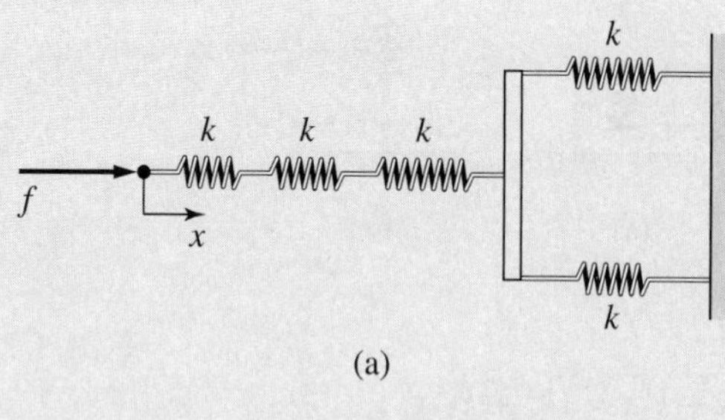

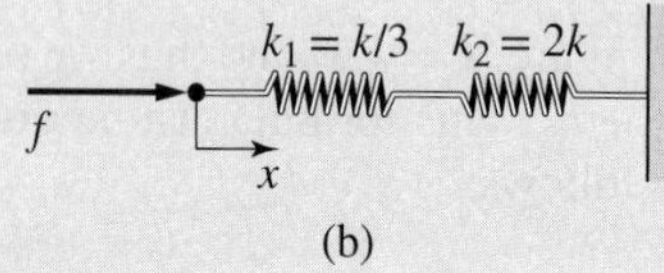

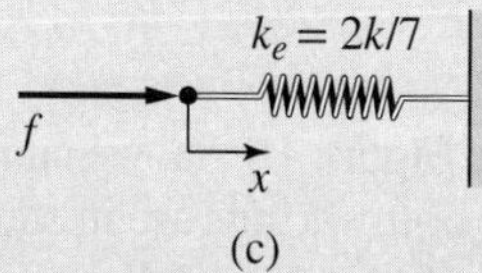

Figure 4.1.9 Determination of an equivalent spring contact.

■ **Solution**

The three springs connected end-to-end have an equivalent stiffness k_1 that is found from

$$\frac{1}{k_1} = \frac{1}{k} + \frac{1}{k} + \frac{1}{k} = \frac{3}{k}$$

or $k_1 = k/3$. The two springs connected side-by-side have an equivalent stiffness k_2 that is found from

$$k_2 = k + k = 2k$$

The equivalent arrangement is shown in Figure 4.1.9b. From this we see that the equivalent stiffness k_e can be found from

$$\frac{1}{k_e} = \frac{1}{k_1} + \frac{1}{k_2} = \frac{3}{k} + \frac{1}{2k} = \frac{7}{2k}$$

which gives $k_e = 2k/7$. The simplest equivalent arrangement is shown in Figure 4.1.9c. Note that the arrangement in Figure 4.1.9a is *less* stiff than a single spring of stiffness k.

Note that springs connected end-to-end result in an arrangement that is *less* stiff than the springs taken individually. Consider two springs with stiffnesses k_1 and k_2 connected end-to-end. Their equivalent stiffness is found from

$$\frac{1}{k_e} = \frac{1}{k_1} + \frac{1}{k_2}$$

Thus

$$k_e = \frac{k_1 k_2}{k_1 + k_2}$$

Now if we write $k_1 = k$ and $k_2 = \alpha k$, then

$$k_e = \frac{\alpha}{1+\alpha}k$$

It can be shown that the ratio $\alpha/(1+\alpha) < 1$, and therefore $k_e < k$.

Similarly, springs connected side-by-side result in an arrangement that is *more* stiff than the springs taken individually. Consider two springs with stiffnesses k and αk connected side-by-side. Their equivalent stiffness is found from

$$k_e = k + \alpha k = (1+\alpha)k$$

Because $\alpha > 0$, then $1 + \alpha > 1$, and therefore $k_e > k$.

Equivalent Spring Constant of Parallel Springs

EXAMPLE 4.1.4

■ Problem

A table with four identical legs supports a vertical force. The solid cylindrical legs are made of metal with $E = 2 \times 10^{11}$ N/m^2. The legs are 1 m in length and 0.03 m in diameter. Compute the equivalent spring constant due to the legs, assuming the table top is rigid.

■ Solution

Because the four legs act like parallel springs, the spring constants add. Thus, from the solid rod formula in Table 4.1.1,

$$k_e = 4\left(\frac{EA}{L}\right) = \frac{4E\pi(d/2)^2}{L} = \frac{4(2\times 10^{11})\pi(0.03)^2}{1} = 17.2\pi \times 10^8 \text{ N/m}$$

Spring Constant of a Stepped Shaft

EXAMPLE 4.1.5

■ Problem

Determine the expression for the equivalent torsional spring constant for the stepped shaft shown in Figure 4.1.10.

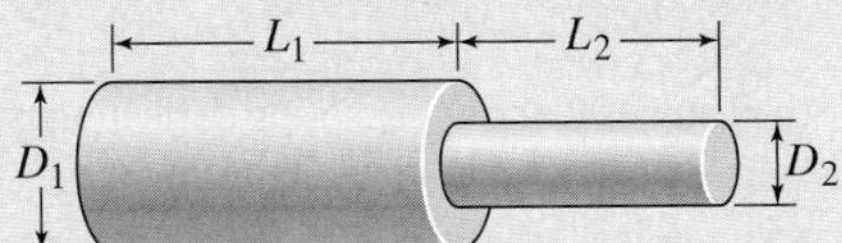

Figure 4.1.10 A stepped shaft.

■ Solution

Each shaft sustains the same torque but has a different twist angle θ. To see why this is true, imagine that shaft 1 is steel and shaft 2 is a soft licorice stick. Which twists more? Therefore, they are in series so that $T = k_{T_1}\theta_1 = k_{T_2}\theta_2$, and the equivalent spring constant is given by

$$\frac{1}{k_{T_e}} = \frac{1}{k_{T_1}} + \frac{1}{k_{T_2}}$$

where k_{T_1} and k_{T_2} are given in Table 4.1.2 as

$$k_{T_i} = \frac{G\pi D_i^4}{32L_i} \qquad i = 1, 2$$

Thus

$$k_{T_e} = \frac{k_{T_1}k_{T_2}}{k_{T_1} + k_{T_2}}$$

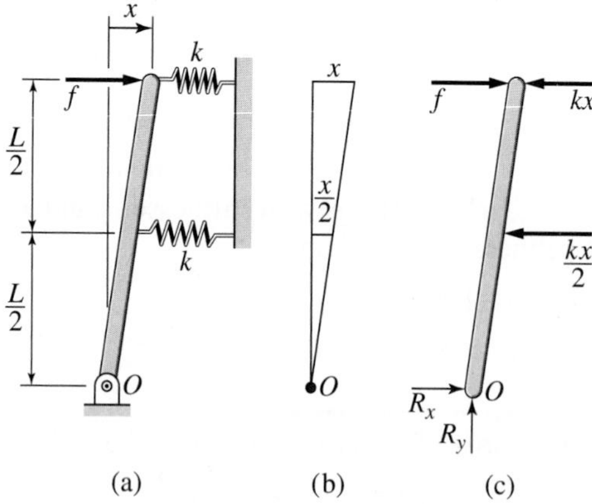

Figure 4.1.11 A lever-spring system.

Although it is frequently useful to recognize when two or more spring elements are connected side-by-side or end-to-end, it is often necessary to rely on the basic definitions of parallel and series elements. For example, are the springs in Figure 4.1.11 in series or in parallel? Note that although these two springs appear to be connected side-by-side, they are not in parallel, because they do not have the same deflection. Thus, their equivalent spring constant is not given by the sum, $2k$. To determine the equivalent spring constant in any situation, we can always go back to the basic principles of statics.

EXAMPLE 4.1.6

Spring Constant of a Lever System

■ Problem

Figure 4.1.11 shows a horizontal force f acting on a lever that is attached to two springs. Assume that the resulting motion is small enough to be only horizontal and determine the expression for the equivalent spring constant that relates the applied force f to the resulting displacement x.

■ Solution

From the triangle shown in part (b) of the figure, for small angles θ, the upper spring deflection is x and the deflection of the lower spring is $x/2$. Thus, the free body diagram is as shown in part (c) of the figure. For static equilibrium, the net moment about point O must be zero. This gives

$$fL - kxL - k\frac{x}{2}\frac{L}{2} = 0$$

Therefore,

$$f = k\left(x + \frac{x}{4}\right) = \frac{5}{4}kx$$

and the equivalent spring constant is $k_e = 5k/4$.

Note that although these two springs appear to be connected side-by-side, they are not in parallel, because they do not have the same deflection. Thus, their equivalent spring constant is not given by the sum, $2k$.

4.1.6 NONLINEAR SPRING ELEMENTS

Up to now we have used the linear spring model $f = kx$. Even though this model is sometimes only an approximation, nevertheless it leads to differential equation models that are linear and therefore relatively easy to solve. Sometimes, however, the use of a nonlinear model is unavoidable. This is the case when a system is designed to utilize two or more spring elements to achieve a spring constant that varies with the applied load. Even if each spring element is linear, the combined system will be nonlinear.

An example of such a system is shown in Figure 4.1.11a. This is a representation of systems used for packaging and in vehicle suspensions, for example. The two side springs provide additional stiffness when the weight W is too heavy for the center spring.

Deflection of a Nonlinear System

EXAMPLE 4.1.7

■ Problem

Obtain the deflection of the system model shown in Figure 4.1.12a, as a function of the weight W. Assume that each spring exerts a force that is proportional to its compression.

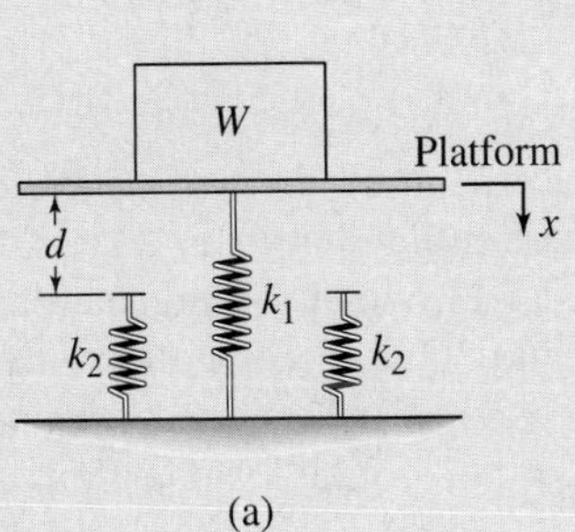

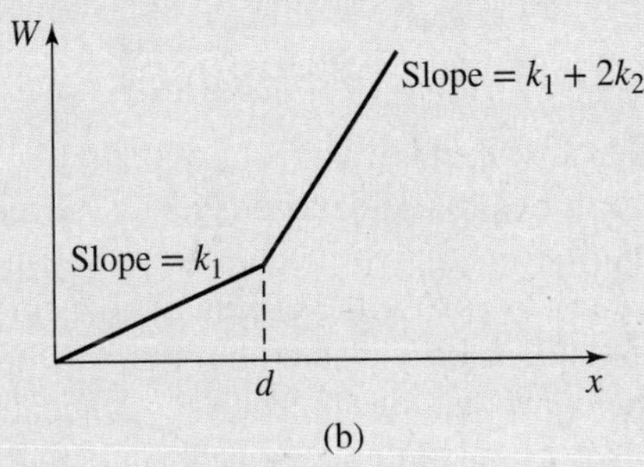

Figure 4.1.12 A nonlinear spring arrangement.

■ Solution

When the weight W is *gently* placed, it moves through a distance x before coming to rest. From statics, we know that the weight force must balance the spring forces at this new position. Thus,

$$W = k_1 x \qquad \text{if } x < d$$

$$W = k_1 x + 2k_2(x - d) \qquad \text{if } x \geq d$$

We can solve these relations for x as follows:

$$x = \frac{W}{k_1} \qquad \text{if } x < d$$

$$x = \frac{W + 2k_2 d}{k_1 + 2k_2} \qquad \text{if } x \geq d$$

These relations can be used to generate the plot of W versus x, shown in part (b) of the figure.

The system in Example 4.1.6 consists of linear spring elements but it has a nonlinear force-deflection relation because of the way the springs are arranged. However, most spring elements display nonlinear behavior if the deflection is large enough. Figure 4.1.13 is a plot of the force-deflection relations for three types of spring elements: the linear spring element, shown by the dashed line in parts (a) and (b) of the figure; a *hardening* spring element; and a *softening* spring element. The stiffness k is the slope of the force-deflection curve and is constant for the linear spring element. A nonlinear spring element does not have a single stiffness value because its slope is variable. For the hardening element, sometimes called a *hard* spring, its slope and thus its stiffness

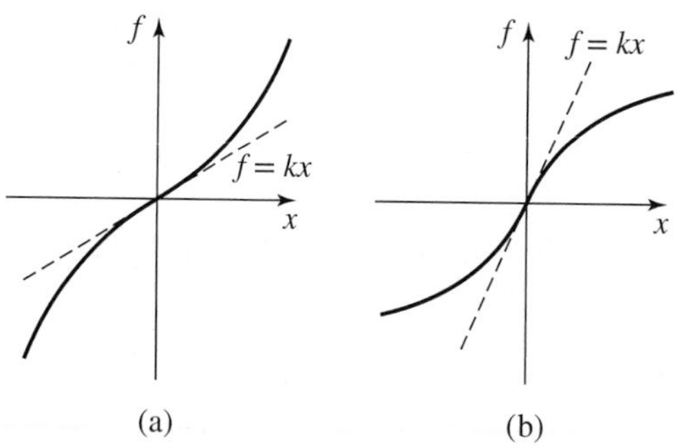

Figure 4.1.13 Force-deflection curves for (a) a hardening spring and (b) a softening spring.

increases with deflection. The stiffness of a softening element, also called a *soft* spring, decreases with deflection.

From the plot in Figure 4.1.12b, we can see that the spring arrangement in part (a) of the figure creates a *hardening* spring.

4.2 MODELING MASS-SPRING SYSTEMS

If we assume that an object is a rigid body and if we neglect the force distribution within the object, we can treat the object as if its mass were concentrated at its mass center. This is the *point-mass* assumption, which makes it easier to obtain the translational equations of motion, because the object's dimensions can be ignored and all external forces can be treated as if they acted through the mass center. If the object can rotate, then the translational equations must be supplemented with the rotational equations of motion, which were treated in Sections 3.2 and 3.4.

If the system cannot be modeled as a rigid body, then we must develop a distributed-parameter model that consists of a *partial* differential equation, which is more difficult to solve.

4.2.1 REAL VERSUS IDEAL SPRING ELEMENTS

By their very nature, all real spring elements have mass and are not rigid bodies. Thus, because it is much easier to derive an equation of motion for a rigid body than for a distributed-mass, flexible system, the basic challenge in modeling mass-spring systems is to first decide whether and how the system can be modeled as one or more rigid bodies.

If the system consists of an object attached to a spring, the simplest way to do this is to neglect the spring mass relative to the mass of the object and take the mass center of the system to be located at the mass center of the object. This assumption is accurate in many practical applications, but to be comfortable with this assumption you should know the numerical values of the masses of the object and the spring element. In some of the homework problems and some of the examples to follow, the numerical values are not given. In such cases, unless otherwise explicitly stated, you should assume that the spring mass can be neglected.

In Section 4.3 we will develop a method to account for the spring mass without the need to develop a partial differential equation model. This method will be useful for applications where the spring mass is neither negligible nor the dominant mass in the system.

An *ideal* spring element is massless. A real spring element can be represented by an ideal element either by neglecting its mass or by including it in another mass in the system.

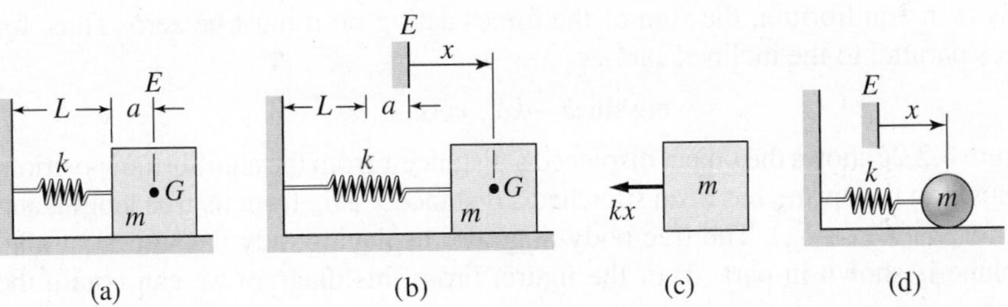

Figure 4.2.1 Models of mass-spring systems.

4.2.2 EFFECT OF SPRING FREE LENGTH AND OBJECT GEOMETRY

Suppose we attach a cube of mass m and side length $2a$ to a linear spring of negligible mass, and we fix the other end of the spring to a rigid support, as shown in Figure 4.2.1a. We assume that the horizontal surface is frictionless. If the mass is homogeneous, its center of mass is at the geometric center G of the cube. The free length of the spring is L and the mass m is in equilibrium when the spring is at its free length. The equilibrium location of G is the point marked E. Part (b) of the figure shows the mass displaced a distance x from the equilibrium position. In this position, the spring has been stretched a distance x from its free length, and thus its force is kx. The free body diagram, displaying only the horizontal force, is shown in part (c) of the figure. From this diagram we can obtain the following equation of motion:

$$m\ddot{x} = -kx \tag{4.2.1}$$

Note that neither the free length L nor the cube dimension a appears in the equation of motion. These two parameters need to be known only to locate the equilibrium position E of the mass center. Therefore, we can represent the object as a point mass, as shown in Figure 4.2.1d.

Unless otherwise specified, you should assume that the objects in our diagrams can be treated as point masses and therefore their geometric dimensions need not be known to obtain the equation of motion. You should also assume that the location of the equilibrium position is known. The shaded-line symbol shown in Figure 4.2.1a is used to indicate a rigid support, such as the horizontal surface and the vertical wall, and also to indicate the location of a fixed coordinate origin, such as the origin of x.

4.2.3 EFFECT OF GRAVITY

Now suppose the object slides on an inclined frictionless surface. In Figure 4.2.2a the spring is at its free length. If we let the object slide until it reaches equilibrium (Figure 4.2.2b), the spring stretches a distance δ_{st}, which is called the *static spring deflection*. Because

Figure 4.2.2 Effect of inclination on a mass-spring model.

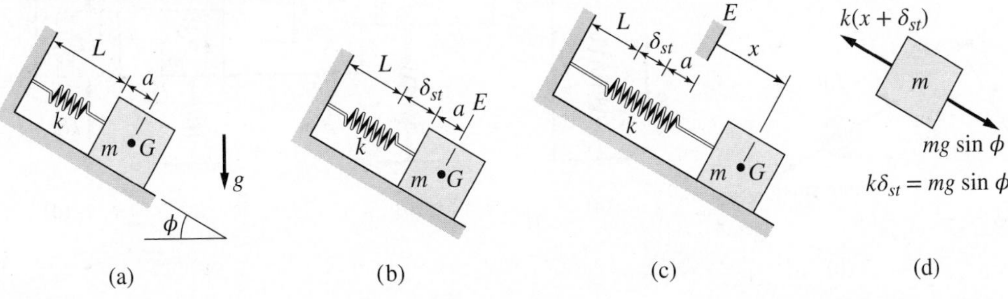

the mass is in equilibrium, the sum of the forces acting on it must be zero. Thus, for the forces parallel to the inclined surface,

$$mg\sin\phi - k\delta_{st} = 0$$

Figure 4.2.2c shows the object displaced a distance x from the equilibrium position. In this position the spring has been stretched a distance $x + \delta_{st}$ from its free length, and thus its force is $k(x + \delta_{st})$. The free body diagram displaying only the forces parallel to the plane is shown in part (d) of the figure. From this diagram we can obtain the following equation of motion:

$$m\ddot{x} = -k(x + \delta_{st}) + mg\sin\phi = -kx + (mg\sin\phi - k\delta_{st})$$

Because $mg\sin\phi = k\delta_{st}$, the term within parentheses is zero and the equation of motion reduces to $m\ddot{x} = -kx$, the same as for the system shown in Figure 4.2.1.

4.2.4 CHOOSING THE EQUILIBRIUM POSITION AS COORDINATE REFERENCE

The example in Figure 4.2.2 shows that for a mass connected to a *linear* spring element, the force due to gravity is canceled out of the equation of motion by the force in the spring due to its static deflection, *as long the displacement of the mass is measured from the equilibrium position*. We will refer to the spring force caused by its static deflection as the *static* spring force and the spring force caused by the variable displacement x as the *dynamic* spring force.

We need not choose the equilibrium location as the coordinate reference. If we choose another coordinate, however, the corresponding equation of motion will contain additional terms that correspond to the static forces in the system. For example, in Figure 4.2.3a, if we choose the coordinate y, the corresponding free body diagram is shown in part (b) of the figure. The resulting equation of motion is

$$m\ddot{y} = -k(y - L) + mg\sin\phi = -ky + kL + mg\sin\phi$$

Note that $kL + mg\sin\phi \neq 0$ so the static force terms do not cancel out of the equation.

The advantages of choosing the equilibrium position as the coordinate origin are (1) that we need not specify the geometric dimensions of the mass and (2) that this choice simplifies the equation of motion by eliminating the static forces.

Now suppose we place the mass m on a spring as shown in Figure 4.2.4a. Assume that the mass is constrained to move in only the vertical direction. If we let the mass

Figure 4.2.3 Choice of coordinate origin for a mass-spring model.

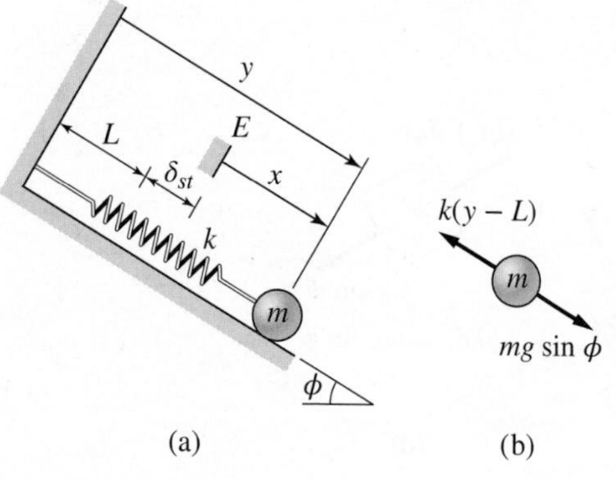

Figure 4.2.4 Static deflection in a mass-spring system.

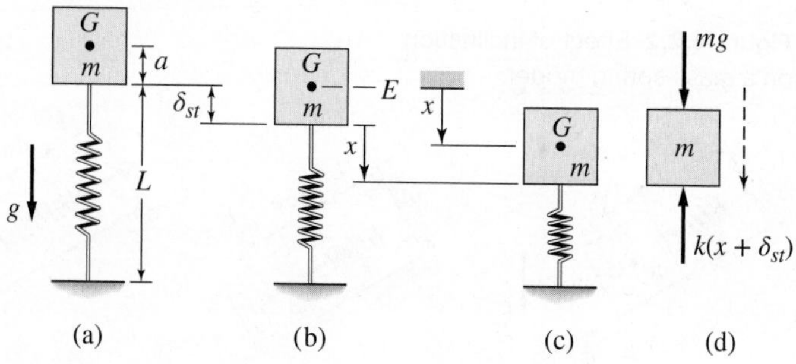

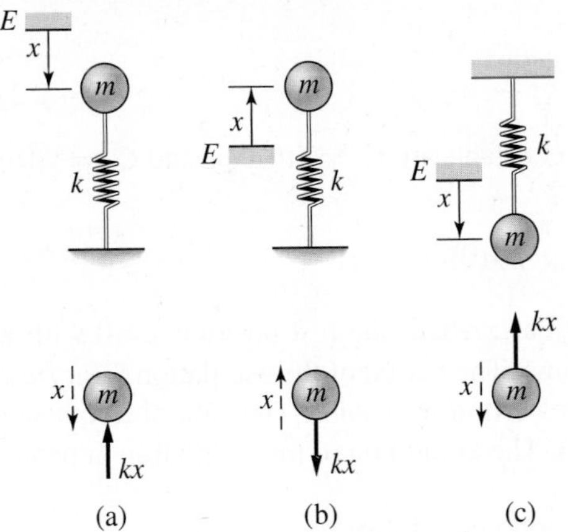

Figure 4.2.5 Choice of coordinate direction for a mass-spring model.

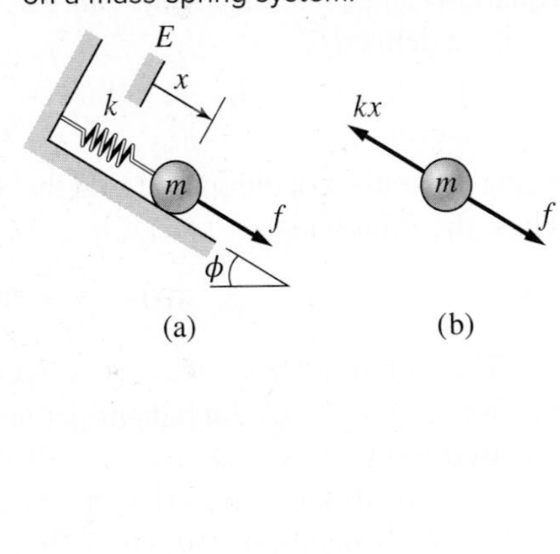

Figure 4.2.6 Modeling an external force on a mass-spring system.

settle down to its equilibrium position at E, the spring compresses an amount δ_{st} and thus $mg = k\delta_{st}$ (Figure 4.2.4b). If the mass is displaced a distance x down from equilibrium (Figure 4.2.4c), the resulting spring force is $k(x + \delta_{st})$ and the resulting free body diagram is shown in part (d) of the figure. Thus, the equation of motion is

$$m\ddot{x} = -k(x + \delta_{st}) + mg = -kx + (mg - k\delta_{st}) = -kx$$

So the equation of motion reduces to $m\ddot{x} = -kx$.

In Figure 4.2.4c, we imagined the mass to be displaced downward from equilibrium and thus we chose the coordinate x to be positive downward. However, we are free to imagine the mass being displaced upward; in this case, we would choose x to be positive upward and would obtain the same equation of motion: $m\ddot{x} = -kx$. You should draw the free body diagram for this case to make sure that you understand the principles.

Figure 4.2.5 shows three situations, and the corresponding free body diagrams, that have the same equation of motion, $m\ddot{x} = -kx$.

It is important to understand that any forces acting on the mass, other than gravity and the spring force, are not to be included when determining the location of the equilibrium position. For example, in Figure 4.2.6a a force f acts on the mass. The equilibrium position E is the location of the mass at which $k\delta_{st} = mg \sin \phi$ when $f = 0$. From the free body diagram shown in part (b), the equation of motion is $m\ddot{x} = f - kx$.

The previous analysis is based on a system model that contains a linear spring and a constant gravity force. For nonlinear spring elements, the gravity forces may or may not appear in the equation of motion. The gravity force acts like a spring in some applications and thus it might appear in the equation of motion. For example, the equation of motion for a pendulum, derived in Chapter 3 for small angles, is $mL^2\ddot{\theta} = -mgL\theta$. The gravity term is not canceled out in this equation because the effect of gravity here is not a constant torque but rather is a torque $mgL\theta$ that is a function of the coordinate θ.

4.2.5 SOLVING THE EQUATION OF MOTION

We have seen that the equation of motion for many mass-spring systems has the form $m\ddot{x} + kx = f$, where f is an applied force other than gravity and the spring force. Suppose that the force f is zero and that we set the mass in motion at time $t = 0$ by pulling it to a position $x(0)$ and releasing it with an initial velocity $\dot{x}(0)$. The solution form of the equation can be obtained from Table 2.3.2 and is $x(t) = C_1 \sin \omega_n t + C_2 \cos \omega_n t$, where we have defined

$$\omega_n = \sqrt{\frac{k}{m}} \tag{4.2.2}$$

Using the initial conditions we find that the constants are $C_1 = \dot{x}(0)/\omega_n$ and $C_2 = x(0)$. Thus, the solution is

$$x(t) = \frac{\dot{x}(0)}{\omega_n} \sin \omega_n t + x(0) \cos \omega_n t \tag{4.2.3}$$

This solution shows that the mass oscillates about the rest position $x = 0$ with a frequency of $\omega_n = \sqrt{k/m}$ radians per unit time. The period of the oscillation is $2\pi/\omega_n$. The frequency of oscillation ω_n is called the *natural frequency*. The natural frequency is greater for stiffer springs (larger k values). The amplitude of the oscillation depends on the initial conditions $x(0)$ and $\dot{x}(0)$.

The solution (4.2.3) can be put into the following form:

$$x(t) = A \sin(\omega_n t + \phi) \tag{4.2.4}$$

where

$$\sin \phi = \frac{x(0)}{A} \qquad \cos \phi = \frac{\dot{x}(0)}{A\omega_n} \tag{4.2.5}$$

$$A = \sqrt{[x(0)]^2 + \left[\frac{\dot{x}(0)}{\omega_n}\right]^2} \tag{4.2.6}$$

If f is a unit-step function, and if the initial displacement $x(0)$ and initial velocity $\dot{x}(0)$ are zero, then you should be able to show that the unit-step response is given by

$$x(t) = \frac{1}{k}(1 - \cos \omega_n t) = \frac{1}{k}\left[1 + \sin\left(\omega_n t - \frac{\pi}{2}\right)\right] \tag{4.2.7}$$

The displacement oscillates about $x = 1/k$ with an amplitude of $1/k$ and a radian frequency ω_n.

EXAMPLE 4.2.1 Calculation of Inertia from Oscillation Frequency

■ Problem

The moment of inertia of an object can be calculated by measuring its oscillation frequency. A connecting rod having a mass of 3.6 kg is shown in Figure 4.2.7. It oscillates with a frequency of 40 cycles per minute when supported on a knife edge, as shown. Its mass center is located 0.15 m below the support. Calculate the moment of inertia about the mass center.

Figure 4.2.7 A connecting rod supported on a knife edge.

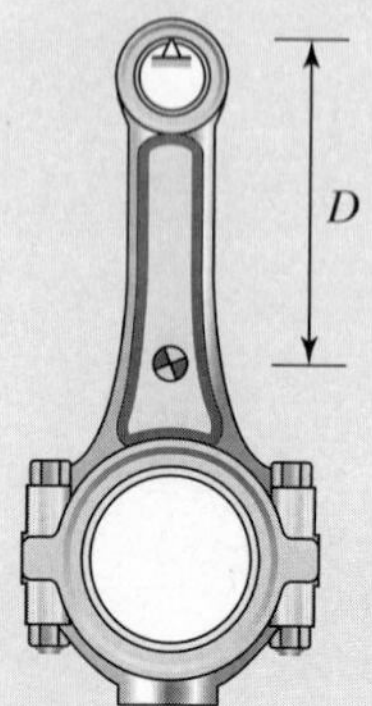

■ Solution

The rod acts like a pendulum rotating about the pivot point O. Its mass is m and its mass center is a distance D from the pivot point. Thus, for small angles θ, its equation of motion is (see Example 3.5.1 in Chapter 3):

$$I_O \ddot{\theta} + mgD\theta = 0 \tag{1}$$

where

$$I_O = I_G + mD^2$$

and $D = 0.15$ and $m = 3.6$. The measured natural frequency is $\omega_n = 2\pi(40/60) = 4\pi/3$ rad/s. From equation (1),

$$\omega_n = \sqrt{\frac{mgD}{I_O}} = \frac{4\pi}{3} \text{ rad/s}$$

Thus

$$I_O = \frac{mgD}{(4\pi/3)^2} = \frac{3.6(9.81)(0.15)}{(4\pi/3)^2} = 0.3019$$

and

$$I_G = I_O - mD^2 = 0.3019 - 3.6(0.15)^2 = 0.2209 \text{ kg·m}^2$$

Beam Vibration — EXAMPLE 4.2.2

■ Problem

The vertical motion of the mass m attached to the beam in Figure 4.2.8a can be modeled as a mass supported by a spring, as shown in part (b) of the figure. Assume that the beam mass is negligible compared to m so that the beam can be modeled as an ideal spring. Determine the system's natural frequency of oscillation.

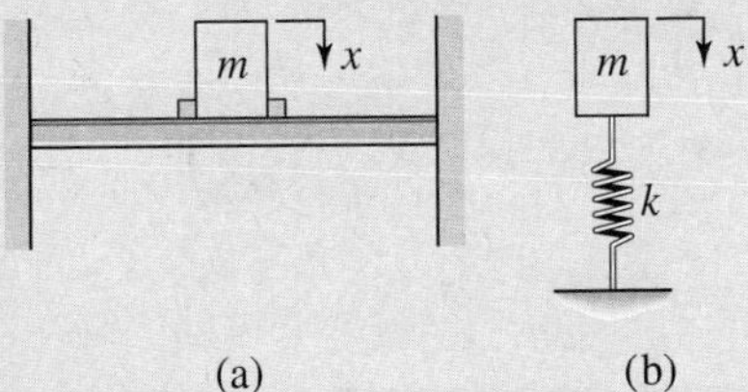

Figure 4.2.8 Model of a mass supported by a fixed-end beam.

■ Solution

The spring constant k is that of the fixed-end beam, and is found from Table 4.1.1 to be $k = 16Ewh^3/L^3$. The mass m has the same equation of motion as (4.2.1), where x is measured from the equilibrium position of the mass. Thus, if the mass m on the beam is initially displaced vertically, it will oscillate about its rest position with a frequency of

$$\omega_n = \sqrt{\frac{k}{m}} = \sqrt{\frac{16Ewh^3}{mL^3}}$$

The source of disturbing forces that initiate such motion will be examined in later chapters. If the beam mass is appreciable, then we must modify the equation of motion. We will see how to do this in Section 4.3.

A Torsional Spring System — EXAMPLE 4.2.3

■ Problem

Consider a torsional system like that shown in Figure 4.2.9a. A cylinder having inertia I is attached to a rod, whose torsional spring constant is k_T. The angle of twist is θ. Assume that the inertia of the rod is negligible compared to the inertia I so that the rod can be modeled as an ideal torsional spring. Obtain the equation of motion in terms of θ and determine the natural frequency.

■ Solution

Because the rod is modeled as an ideal torsional spring, this system is conceptually identical to that shown in Figure 4.2.9b. The free body diagram is shown in part (c) of the figure. From this diagram we obtain the following equation of motion.

$$I\ddot{\theta} = -k_T\theta$$

This has the same form as (4.2.1), and thus we can see immediately that the natural frequency is

$$\omega_n = \sqrt{\frac{k_T}{I}}$$

If the cylinder is twisted and then released, it will oscillate about the equilibrium $\theta = 0$ with a frequency of $\sqrt{k_T/I}$ radians per unit time. This result assumes that the inertia of the rod is very small compared to the inertia I of the attached cylinder. If the rod inertia is appreciable, then we must modify the equation of motion, as will be discussed in Section 4.3.

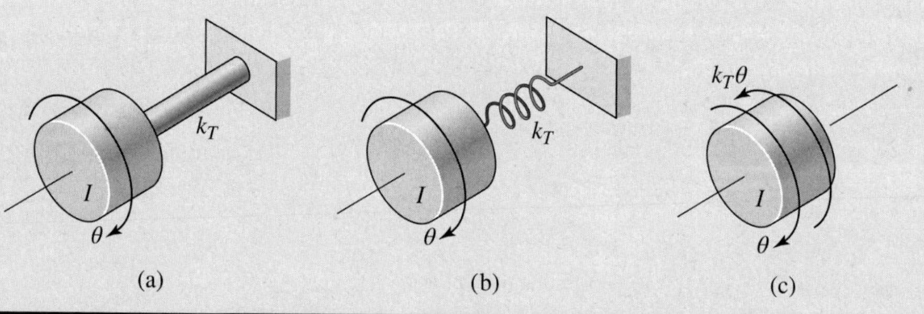

Figure 4.2.9 A torsional spring system.

EXAMPLE 4.2.4

Cylinder on an Incline

■ Problem

Determine the equation of motion of the cylinder shown in Figure 4.2.10a in terms of the coordinate x. The mass moment of inertia about the cylinder's center of mass is I and its mass is m. Assume the cylinder rolls without slipping.

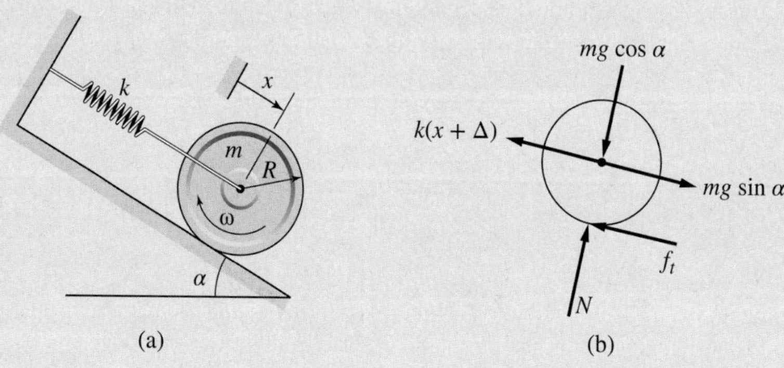

Figure 4.2.10 (a) A cylinder supported by a spring on an incline. (b) Free body diagram.

■ Solution

Taking $x = 0$ to be the equilibrium position, f_t to be the tangential force acting on the cylinder, and Δ to be the static deflection, we see from the free body diagram in part (b) of the figure that

$$m\ddot{x} = mg\sin\alpha - k(x + \Delta) - f_t \tag{1}$$

$$I\dot{\omega} = Rf_t \tag{2}$$

From the geometry of a circle, we know that if the cylinder does not slip, then $R\theta = x$, where θ is the cylinder rotation angle, so that $\omega = \dot{\theta}$. Thus, $R\omega = \dot{x}$ and $R\dot{\omega} = \ddot{x}$. From (2) we find that

$$f_t = I\dot{\omega}/R = I\ddot{x}/R^2$$

Substituting this into (1) and using the fact from statics that

$$mg\sin\alpha = k\Delta$$

we obtain from (1)

$$m\ddot{x} = -kx - \frac{I\ddot{x}}{R^2}$$

or

$$\left(m + \frac{I}{R^2}\right)\ddot{x} + kx = 0$$

The natural frequency is

$$\omega_n = \sqrt{\frac{k}{m + I/R^2}}$$

The cylinder will roll back and forth with this radian frequency.

Note that because the acceleration due of gravity g does not appear in the two previous equations, this system would have the same oscillation frequency in the moon, say, as it does on Earth. Only the static deflection and thus the location of the equilibrium point are affected by g.

Note that if the cylinder is solid, then $I = mR^2/2$, and the previous relations reduce to

$$\frac{3}{2}m\ddot{x} + kx = 0$$

$$\omega_n = \sqrt{\frac{2k}{3m}}$$

Natural Frequency of Pulley System

EXAMPLE 4.2.5

■ Problem

A rigid, stepped pulley is shown in Figure 4.2.11. Assume small motions and assume that the pulley mass is small compared to the mass m. Use the values $R_1 = 0.1$ m, $R_2 = 0.15$ m, $m = 100$ kg, and $k = 10^4$ N/m. Compute the natural frequency of the system.

■ Solution

Assume that x is measured from the equilibrium position. At equilibrium the spring is stretched a length Δ. Let y be the dynamic deflection in the spring (the extra stretch in the spring that occurs when the mass m drops below its equilibrium position). Note that if the cables do not slip when the pulley rotates through an angle θ, then $x = R_2\theta$, $y = R_1\theta$, and thus $y = (R_1/R_2)x = 2x/3$.

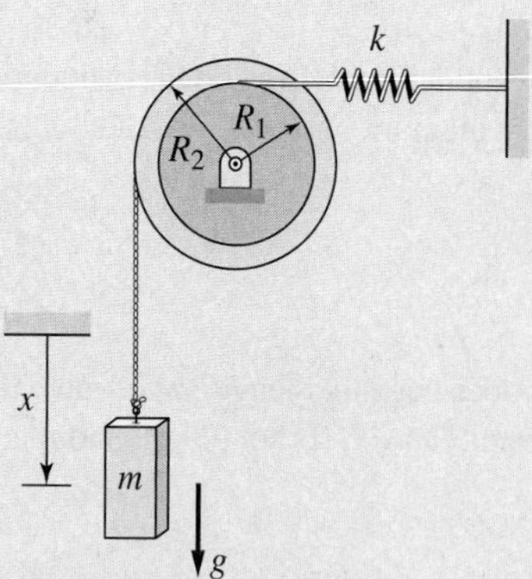

Figure 4.2.11 A system using a rigid, stepped pulley.

Let T be the tension in the cable pulling up on m. Since the effect of the weight mg is canceled by the effect of the static spring force, mg does not appear in the equation of motion for m, which is

$$100\ddot{x} = -T$$

Summing moments about the pulley center gives

$$0.15T - 0.1ky = 0$$

or

$$T = \frac{2}{3}ky = \frac{2}{3}k\left(\frac{2}{3}x\right) = \left(\frac{2}{3}\right)^2 kx$$

Thus

$$100\ddot{x} = -\left(\frac{2}{3}\right)^2 kx = -\left(\frac{2}{3}\right)^2 10^4 x$$

and

$$\omega_n = \frac{2}{3}\sqrt{\frac{10^4}{100}} = \frac{20}{3} \text{ rad/s}$$

4.2.6 DISPLACEMENT INPUTS AND SPRING ELEMENTS

Consider the mass-spring system and its free body diagram shown in Figure 4.2.12a. This gives the equation of motion $m\ddot{x} + kx = f$. To solve this equation for $x(t)$, we must know the force $f(t)$.

Now consider the system shown in Figure 4.2.12b, where we are given the displacement $y(t)$ of the left-hand end of the spring. This represents a practical application in which a rotating cam causes the follower to move the left-hand end of the spring, as in part (c) of the figure. If we know the cam profile and its rotational speed, then we can determine $y(t)$. Suppose that when $x = y = 0$, both springs are at their free lengths. To draw the free body diagram, we must make an assumption about the relative displacements of the endpoints of the spring element. The free body diagram has been drawn with the assumption that $y > x$. Here we are not given an applied force as an input, but nevertheless we must draw the free body diagram showing the forces acting on the mass. The force produced by the given displacement $y(t)$ is the resulting spring force $k_1(y - x)$. The equation of motion is $m\ddot{x} = k_1(y - x) - k_2 x$. Note that we must be given $y(t)$ to solve this equation for $x(t)$. If we need to obtain the force acting on the follower

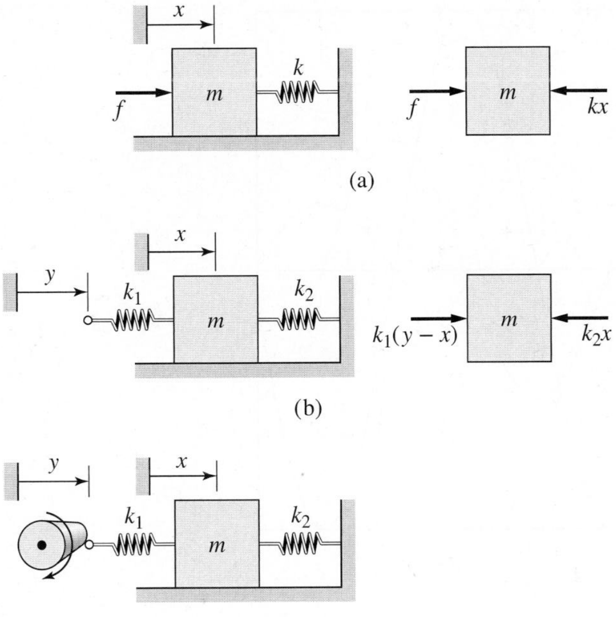

Figure 4.2.12 Force and displacement inputs.

as a result of the motion, we must first solve for $x(t)$ and then compute the follower force from $k_1(y-x)$. This force is of interest to designers because it indicates how much wear will occur on the follower surface.

When displacement inputs are given, it is important to realize that ultimately the displacement is generated by a force (or torque) and that this force must be great enough to generate the specified displacement in the presence of any resisting forces or system inertia. For example, the motor driving the cam must be able to supply enough torque to generate the motion $y(t)$.

4.2.7 SIMPLE HARMONIC MOTION

From the equation of motion $m\ddot{x}=-kx$, we can see that the acceleration is $\ddot{x}=-kx/m = -\omega_n^2 x$. This type of motion, where the acceleration is proportional to the displacement but opposite in sign, is called *simple harmonic motion*. It occurs when the restoring force—here, the spring force—is proportional to the displacement. It is helpful to understand the relation between the displacement, velocity, and acceleration in simple harmonic motion. Expressions for the velocity and acceleration are obtained by differentiating $x(t)$, whose expression is given by (4.2.4):

$$\dot{x}(t) = A\omega_n \cos(\omega_n t + \phi) = A\omega_n \sin\left(\omega_n t + \phi + \frac{\pi}{2}\right)$$

$$\ddot{x}(t) = -A\omega_n^2 \sin(\omega_n t + \phi) = A\omega_n^2 \sin(\omega_n t + \phi + \pi)$$

The displacement, velocity, and acceleration all oscillate with the same frequency ω_n but they have different amplitudes and are shifted in time relative to one another. The velocity is zero when the displacement and acceleration reach their extreme values. The sign of the acceleration is the opposite of that of the displacement, and the magnitude of the acceleration is ω_n^2 times the magnitude of the displacement. These functions are plotted in Figure 4.2.13 for the case where $x(0) = 1$, $\dot{x}(0) = 0$, and $\omega_n = 2$.

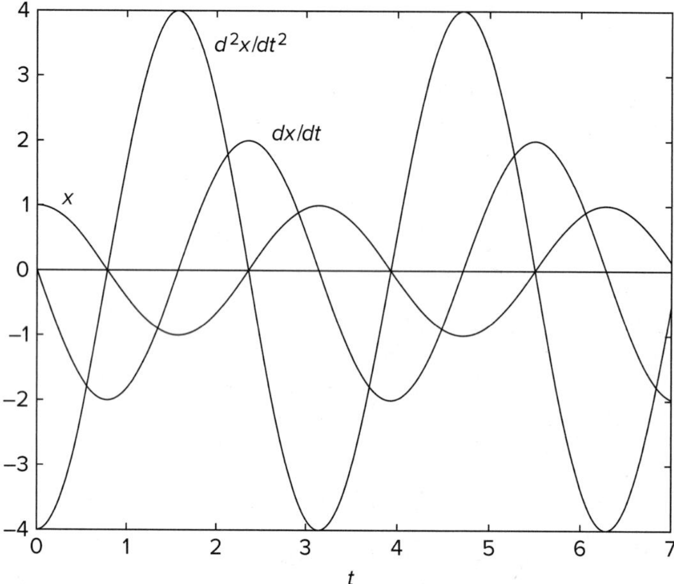

Figure 4.2.13 Plots of displacement, velocity, and acceleration for simple harmonic motion.

4.2.8 SYSTEMS WITH TWO OR MORE MASSES

Up to now our examples have included only a single mass or inertia. If two or more masses or inertias are connected by spring elements, the basic principles remain the same, but we must first make some assumptions regarding the relative motion of the mass or inertia elements. This allows us to assign directions to the force or moment vectors on the free body diagrams. If you are consistent with your assumptions, the resulting equations of motion will correctly describe the dynamics of the system even when the relative motion is different from what you assumed. This is often a difficult concept for beginners, and a common mistake is to make one assumption for one mass and make the opposite assumption for the other mass.

We usually choose the coordinates as the displacements of the masses from their *equilibrium* positions because in equilibrium the static deflection forces in the spring elements cancel the weights of the masses. Thus, the free body diagrams show only the *dynamic* forces, and not the static forces. Thus, for linear spring elements, the equations of motion expressed in such coordinates will not contain the weight forces or the static deflection forces. These concepts are illustrated in the following example.

EXAMPLE 4.2.6

Equation of Motion of a Two-Mass System

■ Problem

Derive the equations of motion of the two-mass system shown in Figure 4.2.14a.

■ Solution

Choose the coordinates x_1 and x_2 as the displacements of the masses from their equilibrium positions. In equilibrium the static forces in the springs cancel the weights of the masses. Thus, the free body diagrams showing the dynamic forces, and not the static forces, are as shown in Figure 4.2.14b. These diagrams have been drawn with the assumption that the displacement x_1 of mass m_1 from its equilibrium position is greater than the displacement of m_2. From these

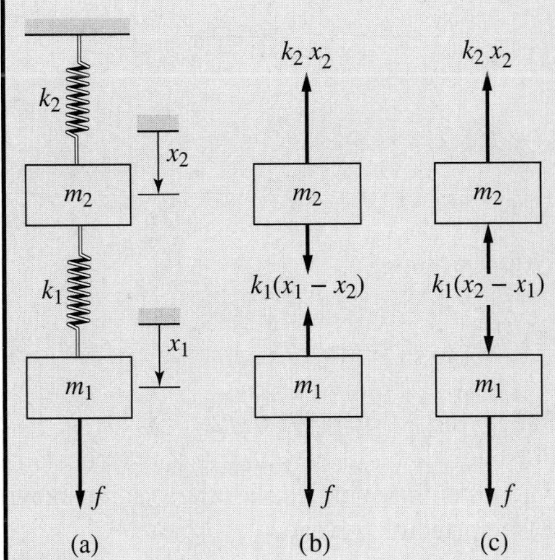

Figure 4.2.14 A system with two masses.

diagrams we obtain the equations of motion:

$$m_1\ddot{x}_1 = f - k_1(x_1 - x_2)$$

$$m_2\ddot{x}_2 = k_1(x_1 - x_2) - k_2x_2$$

If we move all terms to the left side of the equal sign except for the external force f, we obtain

$$m_1\ddot{x}_1 + k_1(x_1 - x_2) = f \tag{1}$$

$$m_2\ddot{x}_2 - k_1(x_1 - x_2) + k_2x_2 = 0 \tag{2}$$

In drawing the free body diagrams of multimass systems, you must make an assumption about the relative motions of each mass. For example, we could have assumed that the displacement x_1 of mass m_1 from its equilibrium position is *less* than the displacement of m_2. Figure 4.2.14c shows the free body diagrams drawn for this assumption. If your assumptions are correct, the forces shown on the diagram must be *positive*. Note that the directions of the forces associated with spring k_1 are the opposite of those in part (b) of the figure. You should confirm that the diagram in part (c) results in equations of motion that are identical to equations (1) and (2).

You must be consistent in your assumptions made about the relative motion when drawing the free body diagrams. A common mistake is to use one assumption to obtain the free body diagram for mass m_1 but another assumption for mass m_2.

4.3 ENERGY METHODS

The force exerted by a spring is a conservative force. If the spring is linear, then its resisting force is given by $f = -kx$ and thus the potential energy of a linear spring is given by

$$V(x) = \frac{1}{2}kx^2 \tag{4.3.1}$$

where x is the deflection from the *free length* of the spring.

A *torsional* spring exerts a moment M if it is twisted. If the spring is linear, the moment is given by $M = k_T\theta$, where θ is the twist angle. The work done by this

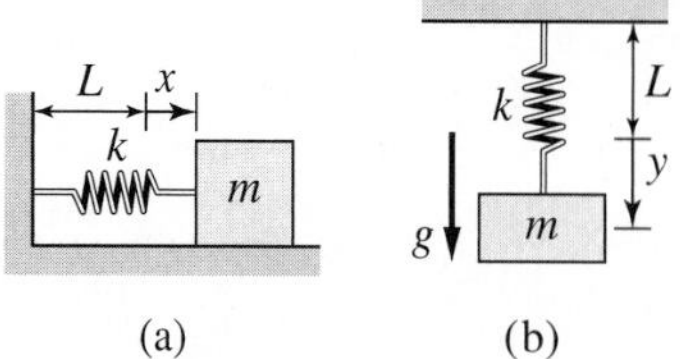

Figure 4.3.1 (a) A system having kinetic and elastic potential energy. (b) A system having kinetic, elastic potential, and gravitational potential energy.

moment and stored as potential energy in the spring is

$$V(\theta) = \int_0^\theta M\, d\theta = \int_0^\theta k_T \theta\, d\theta = \frac{1}{2} k_T \theta^2 \tag{4.3.2}$$

So the potential energy stored in a torsional spring is $V(\theta) = k_T\theta^2/2$.

The conservation of energy principle states that $T + V = T_0 + V_0 =$ constant, where T and V are the system's kinetic and potential energies. For the system shown in Figure 4.3.1a, for a frictionless surface, the principle gives

$$\frac{1}{2}m\dot{x}^2 + \frac{1}{2}kx^2 = \frac{1}{2}m\dot{x}_0^2 + \frac{1}{2}kx_0^2 = \text{constant}$$

This relation can be rearranged as follows:

$$\frac{m}{2}\left(\dot{x}^2 - \dot{x}_0^2\right) + \frac{k}{2}\left(x^2 - x_0^2\right) = 0$$

which states that $\Delta T + \Delta V = 0$.

For the system shown in Figure 4.3.1b, we must include the effect of gravity, and thus the potential energy is the sum of the spring's potential energy V_s and the gravitational potential energy V_g, which we may choose to be zero at $y = 0$. The conservation of energy principle gives

$$T + V_g + V_s = \text{constant}$$

or

$$\Delta T + \Delta V = \Delta T + \Delta V_g + \Delta V_s = 0$$

The spring is at its free length when $y = 0$, so we can write

$$\frac{1}{2}m\dot{y}^2 - mgy + \frac{1}{2}ky^2 = \text{constant}$$

Note that the gravitational potential energy has a negative sign because we have selected y to be positive downward.

The numerical value of the gravitational potential energy depends on the location of the datum and it may be negative. We are free to select the location because only the *change* in gravitational potential energy is significant. Note, however, that a spring's potential energy is always nonnegative and that the potential energy is positive whenever the deflection from the free length is nonzero.

EXAMPLE 4.3.1 A Force Isolation System

■ Problem

Figure 4.3.2 shows a representation of a spring system to isolate the foundation from the force of a falling object. Suppose the weight W is dropped from a height h above the platform attached to the center spring. Determine the maximum spring compression and the maximum force transmitted to the foundation. The given values are $k_1 = 10^4$ N/m, $k_2 = 1.5 \times 10^4$ N/m, $d = 0.1$ m, and $h = 0.5$ m. Consider two cases: (a) $W = 64$ N and (b) $W = 256$ N.

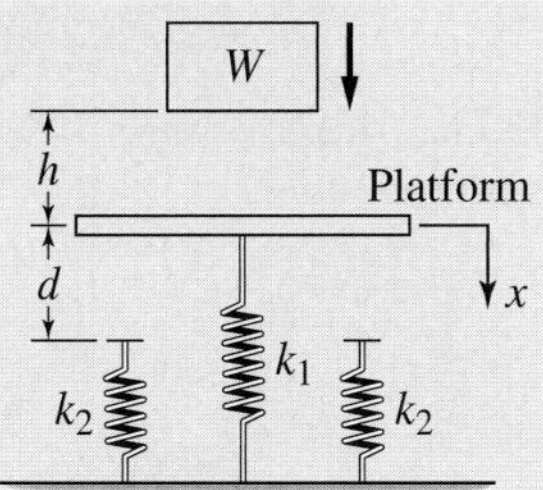

Figure 4.3.2 A force-isolation system.

■ Solution

The velocity of the weight is zero initially and also when the maximum compression is attained. Therefore, $\Delta T = 0$ and we have

$$\Delta T + \Delta V = \Delta T + \Delta V_s + \Delta V_g = 0$$

or

$$\Delta V_s + \Delta V_g = 0$$

That is, if the weight is dropped from a height h above the platform and if we choose the gravitational potential energy to be zero at that height, then the maximum spring compression x can be found by adding the change in the weight's gravitational potential energy $0 - W(h+x) = -W(h + x)$ to the change in potential energy stored in the springs. Thus,

$$\frac{1}{2}k_1(x^2 - 0) + [0 - W(h + x)] = 0 \qquad \text{if } x < d$$

which gives the following quadratic equation to solve for x:

$$\frac{1}{2}k_1x^2 - Wx - Wh = 0 \qquad \text{if } x < d \tag{1}$$

If $x \geq d$, $\Delta V_s + \Delta V_g = 0$ gives

$$\frac{1}{2}k_1(x^2 - 0) + \frac{1}{2}(2k_2)[(x - d)^2 - 0] + [0 - W(h + x)] = 0 \qquad \text{if } x \geq d$$

which gives the following quadratic equation to solve for x:

$$(k_1 + 2k_2)x^2 - (2W + 4k_2d)x + 2k_2d^2 - 2Wh = 0 \qquad \text{if } x \geq d \tag{2}$$

For the given values, equation (1) becomes

$$10^4x^2 - 2Wx - W = 0 \qquad \text{if } x < 0.1 \tag{3}$$

and from equation (2),

$$4 \times 10^4x^2 - (2W + 6000)x + 300 - W = 0 \qquad \text{if } x \geq 0.1 \tag{4}$$

For case (a), the positive root of equation (3) gives $x = 0.0867$, which is less than 0.1. So only the middle spring is compressed, and it is compressed 0.0867 m. The resulting maximum force on the foundation is the spring force $k_1x = 10^4(0.0867) = 867$ N.

For case (b), the positive root of equation (3) gives $x = 0.188$, which is greater than 0.1. So all three springs will be compressed. From equation (4),

$$4 \times 10^4x^2 - (512 + 6000)x + 300 - 256 = 0$$

which has the solutions $x = 0.156$ and $x = 0.007$. We discard the second solution because it is less than 0.1 and thus corresponds to compression in the middle spring only. So the outer springs will be compressed $0.156 - 0.1 = 0.056$ m and the middle spring will be compressed 0.156 m. The resulting maximum force on the foundation is the total spring force $k_1x + 2k_2(x - 0.1) = 10^4(0.156) + 2(1.5 \times 10^4)(0.156 - 0.1) = 3240$ N.

4.3.1 OBTAINING THE EQUATION OF MOTION

In mass-spring systems with negligible friction and damping, we can often use the principle of conservation of energy to obtain the equation of motion and, for simple harmonic motion, to determine the frequency of vibration without obtaining the equation of motion.

EXAMPLE 4.3.2

Equation of Motion of a Mass-Spring System

Figure 4.3.3 A mass-spring system.

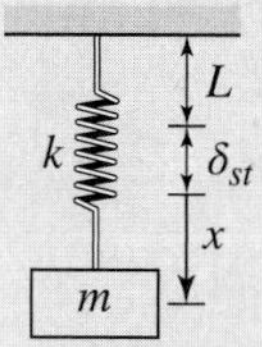

■ Problem

Use the energy method to derive the equation of motion of the mass m attached to a spring and moving in the vertical direction, as shown in Figure 4.3.3.

■ Solution

With the displacement x measured from the equilibrium position, and taking the gravitational potential energy to be zero at $x = 0$, the total potential energy of the system is

$$V = V_s + V_g = \frac{1}{2}k(x + \delta_{st})^2 - mgx = \frac{1}{2}kx^2 + k\delta_{st}x + \frac{1}{2}k\delta_{st}^2 - mgx$$

Because $k\delta_{st} = mg$, the expression for V becomes

$$V = \frac{1}{2}kx^2 + \frac{1}{2}k\delta_{st}^2$$

The total energy of the system is

$$T + V = \frac{1}{2}m\dot{x}^2 + \frac{1}{2}kx^2 + \frac{1}{2}k\delta_{st}^2$$

From conservation of mechanical energy, $T + V$ is constant and thus its time derivative is zero. Therefore,

$$\frac{d}{dt}(T + V) = \frac{d}{dt}\left(\frac{1}{2}m\dot{x}^2\right) + \frac{d}{dt}\left(\frac{1}{2}kx^2\right) + \frac{d}{dt}\left(\frac{1}{2}k\delta_{st}^2\right) = 0$$

Evaluating the derivatives gives

$$m\dot{x}\ddot{x} + kx\dot{x} = 0$$

Canceling $\dot{x}$ gives the equation of motion $m\ddot{x} + kx = 0$.

Example 4.3.2 shows that if we can obtain the expression for the sum of the kinetic and potential energies, $T + V$, the equation of motion can be found by differentiating $T + V$ with respect to time. Although this was a simple example, it illustrates that with this method we need not draw the free body diagrams of every member of a multibody system whose motion can be described by a single coordinate.

4.3.2 RAYLEIGH'S METHOD

We can use the principle of conservation of energy to obtain the natural frequency of a mass-spring system if the spring is linear. This approach is sometimes useful because it does not require that we first obtain the equation of motion.

The method was developed by Lord Rayleigh (John William Strutt) and was presented in his *Theory of Sound* in 1847. A modern reprint is [Rayleigh, 1945]. Rayleigh is considered one of the founders of the study of acoustics and vibration. We illustrate Rayleigh's method here for a second-order system, but it is especially useful for estimating the lowest natural frequency of higher-order systems with several degrees of freedom and distributed parameter systems with an infinite number of natural frequencies.

In simple harmonic motion, the kinetic energy is maximum and the potential energy is minimum at the equilibrium position $x = 0$. When the displacement is maximum, the potential energy is maximum but the kinetic energy is zero. From conservation of energy,

$$T_{\max} + V_{\min} = T_{\min} + V_{\max}$$

Thus

$$T_{\max} + V_{\min} = 0 + V_{\max}$$

or

$$T_{\max} = V_{\max} - V_{\min} \tag{4.3.3}$$

For example, for the mass-spring system oscillating vertically as shown in Figure 4.3.3, $T = m\dot{x}^2/2$ and $V = k(x + \delta_{st})^2/2 - mgx$, and from (4.3.3) we have,

$$T_{\max} = \frac{1}{2}m(\dot{x}_{\max})^2 = V_{\max} - V_{\min} = \frac{1}{2}k(x_{\max} + \delta_{st})^2 - mgx_{\max} - \frac{1}{2}k\delta_{st}^2$$

or

$$\frac{1}{2}m(\dot{x}_{\max})^2 = \frac{1}{2}k(x_{\max})^2$$

where we have used the fact that $k\delta_{st} = mg$. In simple harmonic motion $|\dot{x}_{\max}| = \omega_n|x_{\max}|$, and thus,

$$\frac{1}{2}m(\omega_n|x_{\max}|)^2 = \frac{1}{2}k|x_{\max}|^2$$

Cancel $|x_{\max}|^2$ and solve for ω_n to obtain $\omega_n = \sqrt{k/m}$.

In this simple example, we merely obtained the expression for ω_n that we already knew. However, in other applications the expressions for T and V may be different, but if the motion is simple harmonic, we can directly determine the natural frequency by using the fact that $|\dot{x}_{\max}| = \omega_n|x_{\max}|$ to express $T_{\max}$ as a function of $|x_{\max}|$ and then equating $T_{\max}$ to $V_{\max} - V_{\min}$. This approach is called *Rayleigh's method*.

A common mistake when applying Rayleigh's method is to assume that $V_{\min} = 0$, but this is not always true, as shown by the following example.

A Cylinder and Spring — EXAMPLE 4.3.3

■ Problem

Apply Rayleigh's principle to determine the natural frequency of the cylinder shown in Figure 4.3.4. The mass moment of inertia about the cylinder's center of mass is I and its mass is m. Assume the cylinder rolls without slipping.

Figure 4.3.4 A spring connected to a rolling cylinder.

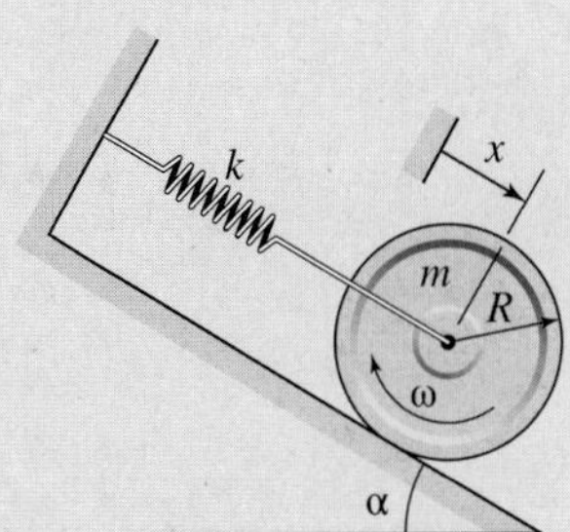

■ Solution

Take $x = 0$ to be the equilibrium position and Δ to be the static deflection. From the geometry of a circle, we know that if the cylinder does not slip, then $R\theta = x$ where θ is the cylinder rotation angle and $\omega = \dot{\theta}$. Take the gravitational potential energy to be zero at the equilibrium position. The elastic potential energy at equilibrium is $k\Delta^2/2$ and the total potential energy is

$$V = \frac{1}{2}k(x + \Delta)^2 - mgh$$

where $h = x \sin\alpha$. Because $mg\sin\alpha = k\Delta$ at equilibrium, the expression for V becomes

$$V = \frac{1}{2}k(x^2 + 2x\Delta + \Delta^2) - mgx\sin\alpha = \frac{1}{2}kx^2 + \frac{1}{2}k\Delta^2$$

Thus, $V_{\min} = k\Delta^2/2$. The kinetic energy is

$$T = \frac{1}{2}m\dot{x}^2 + \frac{1}{2}I\dot{\theta}^2 = \frac{1}{2}m\dot{x}^2 + \frac{1}{2}I\frac{\dot{x}^2}{R^2} = \frac{1}{2}\left(m + \frac{I}{R^2}\right)\dot{x}^2$$

because $\dot{\theta} = \dot{x}/R$.

Assuming simple harmonic motion occurs, we obtain $x(t) = A\sin(\omega_n t + \phi)$ and $\dot{x} = \omega_n A\cos(\omega_n t + \phi)$. Thus, $x_{max} = A$, $\dot{x}_{max} = \omega_n A$, and

$$T_{max} = \frac{1}{2}\left(m + \frac{I}{R^2}\right)(\omega_n A)^2$$

$$V_{max} = \frac{1}{2}kA^2 + \frac{1}{2}k\Delta^2$$

From Rayleigh's method, $T_{max} = V_{max} - V_{min}$, and we have

$$\frac{1}{2}\left(m + \frac{I}{R^2}\right)(\omega_n A)^2 = \frac{1}{2}kA^2 + \frac{1}{2}k\Delta^2 - \frac{1}{2}k\Delta^2 = \frac{1}{2}kA^2$$

This gives

$$\left(m + \frac{I}{R^2}\right)\omega_n^2 = k$$

or

$$\omega_n = \sqrt{\frac{k}{m + I/R^2}}$$

Of course, this is the same answer found in Example 4.2.4, as it should be. The reader must decide which method is easier, the free body diagram method or an energy method such as Rayleigh's.

Rayleigh's method is especially useful when it is difficult to derive the equation of motion or when only the natural frequency must be found. The following example illustrates this point.

EXAMPLE 4.3.4

Natural Frequency of a Suspension System

■ Problem

Figure 4.3.5 shows the suspension of one front wheel of a car in which $L_1 = 0.4$ m and $L_2 = 0.6$ m. The coil spring has a spring constant of $k = 3.6 \times 10^4$ N/m and the car weight associated with that wheel is 3500 N. Determine the suspension's natural frequency for vertical motion.

■ Solution

Imagine that the frame moves down by a distance A_f, while the wheel remains stationary. Then from similar triangles, the amplitude A_s of the spring deflection is related to the amplitude A_f of the frame motion by $A_s = L_1 A_f / L_2 = 0.4A_f/0.6 = 2A_f/3$.

Using the fact that $k\delta_{st} = mg$, the change in potential energy can be written as

$$V_{max} - V_{min} = \frac{1}{2}k(A_s + \delta_{st})^2 - mgA_s - \frac{1}{2}k\delta_{st}^2 = \frac{1}{2}kA_s^2 = \frac{1}{2}k\left(\frac{2}{3}A_f\right)^2$$

The amplitude of the velocity of the mass in simple harmonic motion is $\omega_n A_f$, and thus the maximum kinetic energy is

$$T_{max} = \frac{1}{2}m(\omega_n A_f)^2$$

From Rayleigh's method, $T_{max} = V_{max} - V_{min}$, we obtain

$$\frac{1}{2}m(\omega_n A_f)^2 = \frac{1}{2}k\left(\frac{2}{3}A_f\right)^2$$

Solving this for ω_n we obtain

$$\omega_n = \frac{2}{3}\sqrt{\frac{k}{m}} = \frac{2}{3}\sqrt{\frac{3.6 \times 10^4}{3500/9.8}} = 6.69 \text{ rad/s}$$

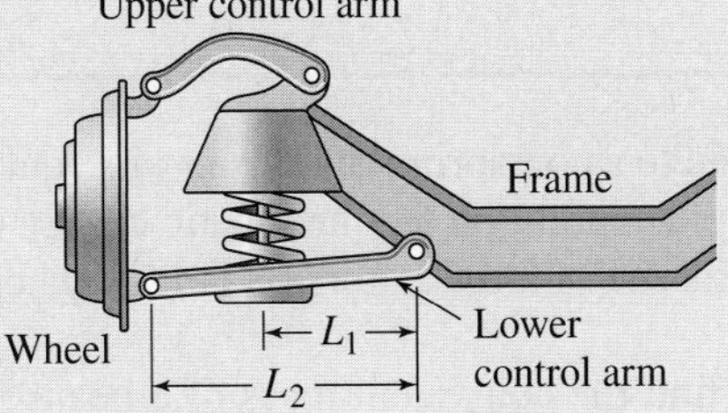

Figure 4.3.5 A vehicle suspension.

Figure 4.3.6 An example of a spring element with distributed mass.

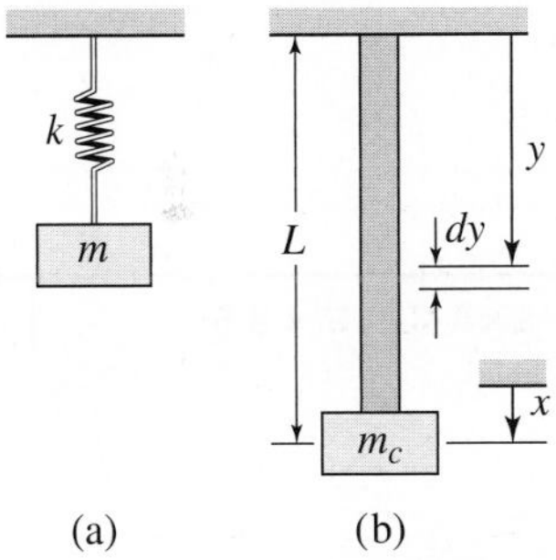

4.3.3 EQUIVALENT MASS OF ELASTIC ELEMENTS

If an elastic element is represented as in Figure 4.3.6a, we assume that the mass of the element either is negligible compared to the rest of the system's mass or has been included in the mass attached to the element. This included mass is called the *equivalent* mass of the element. We do this so that we can obtain a lumped-parameter model of the system. As we did with rigid-body systems in Chapter 3, we compute the equivalent mass by using kinetic energy equivalence, because mass is associated with kinetic energy.

Equivalent Mass of a Rod

EXAMPLE 4.3.5

■ Problem

The rod shown in Figure 4.3.6b acts like a spring when an axially applied force stretches or compresses the rod. Determine the equivalent mass of the rod.

■ Solution

In Figure 4.3.6b, the mass of an infinitesimal element of thickness dy is $dm_r = \rho\, dy$, where ρ is the mass density per unit length of the material. Thus, the kinetic energy of the element is $(dm_r)\dot{y}^2/2$, and the kinetic energy of the entire rod is

$$\text{KE} = \frac{1}{2}\int_0^L \dot{y}^2\, dm_r = \frac{1}{2}\int_0^L \dot{y}^2 \rho\, dy$$

If we assume that the velocity $\dot{y}$ of the element is linearly proportional to its distance from the support, then

$$\dot{y} = \dot{x}\frac{y}{L}$$

where $\dot{x}$ is the velocity of the end of the rod. Thus,

$$\text{KE} = \frac{1}{2}\int_0^L \left(\dot{x}\frac{y}{L}\right)^2 \rho\, dy = \frac{1}{2}\frac{\rho\dot{x}^2}{L^2}\int_0^L y^2\, dy = \frac{1}{2}\frac{\rho\dot{x}^2}{L^2}\frac{y^3}{3}\bigg|_0^L$$

or

$$\text{KE} = \frac{1}{2}\frac{\rho \dot{x}^2}{L^2}\frac{L^3}{3} = \frac{1}{2}\left(\frac{\rho L}{3}\right)\dot{x}^2 = \frac{1}{2}\frac{m_r}{3}\dot{x}^2$$

because $\rho L = m_r$, the mass of the rod. For an equivalent mass m_e concentrated at the end of the rod, its kinetic energy is $m_e\dot{x}^2/2$. Thus, the equivalent mass of the rod is $m_e = m_r/3$. So the mass m in Figure 4.3.6a is $m = m_c + m_e = m_c + m_r/3$.

The approach of this example can be applied to a coil spring of mass m_s; its equivalent mass is $m_s/3$. A similar approach using the expression for the kinetic energy of rotation, $I_r\dot{\theta}^2/2$, will show that the equivalent inertia of a rod in torsion is $I_r/3$, where I_r is the rod inertia.

This type of analysis can also be used to find the equivalent mass of a beam by using the appropriate formula for the beam's static load-deflection curve to obtain an expression for the velocity of a beam element as a function of its distance from a support. The derivations of such formulas are given in basic references on the mechanics of materials and are beyond the scope of this text. The expressions for the equivalent beam masses given in Table 4.3.1 were derived in this manner. Because the static load-deflection curve describes the *static* deflection, it does not account for inertia effects, and therefore the expressions given in Table 4.3.1 are approximations. However, they are accurate enough for many applications.

EXAMPLE 4.3.6

Equivalent Mass of a Spring

■ Problem

A 10-kg mass is attached to a 1.8 kg spring. The mass vibrates at a frequency of 20 Hz when disturbed. Estimate the spring stiffness k.

■ Solution

The equivalent mass is $m_e = 10 + 1.8/3 = 10.6$ kg. The natural frequency is $\omega_n = 20(2\pi) = 40\pi$ rad/s. Since $\omega_n = \sqrt{k/m_e}$, then $k = m_e\omega_n^2 = 10.6(40\pi)^2 = 1.6739 \times 10^5$ N/m.

EXAMPLE 4.3.7

Equivalent Spring Constant of a Fixed-End Beam

■ Problem

The vibration of a motor mounted in the middle of a fixed-end beam can be modeled as a mass-spring system. The motor mass is 40 kg, and the beam mass is 13 kg. When the motor is placed on the beam, it causes an additional static deflection of 3 mm. Find the equivalent mass m_e, the spring constant k, and the natural frequency.

■ Solution

From Table 4.3.1 the equivalent mass is $m_e = 40 + 0.38(13) = 44.94$ kg. Note that the motor causes an *additional* static deflection of 3 mm. From statics,

$$k(0.003) = 40(9.81)$$

or $k = 1.308 \times 10^5$ N/m. The natural frequency is $\omega_n = \sqrt{k/m_e} = 53.95$ rad/s.

Table 4.3.1 Equivalent masses and inertias of common elements.

Translational systems

Nomenclature:
m_c = concentrated mass
m_d = distributed mass
m_e = equivalent lumped mass
System model:
$m_e\ddot{x} + kx = 0$

Equivalent system

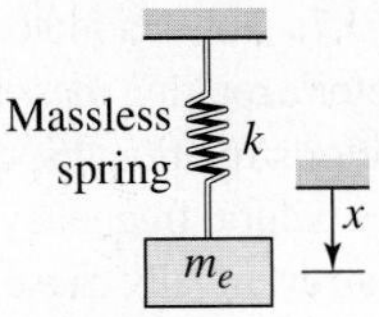

Massless spring k, Rest position, x, m_e

Helical spring, or rod in tension/compression

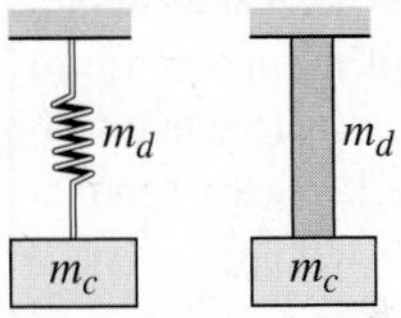

m_d, m_c

$m_e = m_c + m_d/3$

Cantilever beam

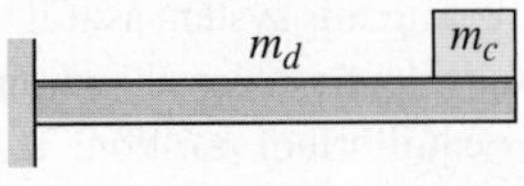

m_d, m_c

$m_e = m_c + 0.23m_d$

Simply supported beam

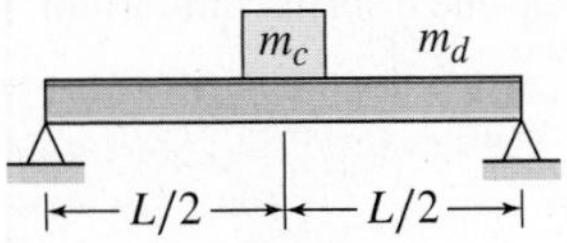

m_c, m_d, $L/2$, $L/2$

$m_e = m_c + 0.50m_d$

Fixed-end beam

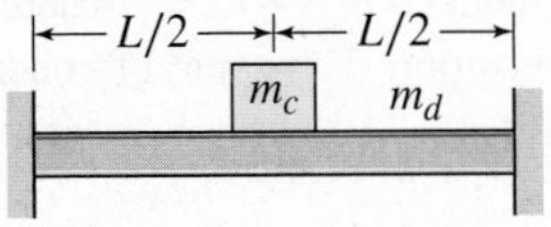

$L/2$, $L/2$, m_c, m_d

$m_e = m_c + 0.38m_d$

Rotational systems

Nomenclature:
I_c = concentrated inertia
I_d = distributed inertia
I_e = equivalent lumped inertia
System model:
$I_e\ddot{\theta} + k\theta = 0$

Equivalent system

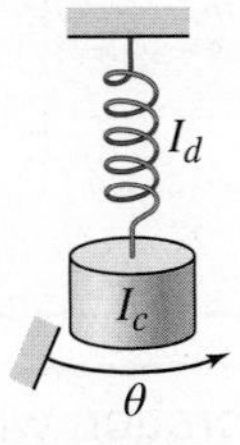

I_d, I_c, θ

Helical spring

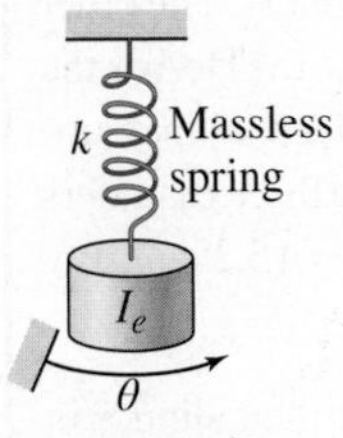

k, Massless spring, I_e, θ

$I_e = I_c + I_d/3$

Rod in torsion

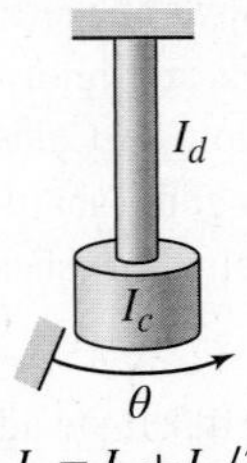

I_d, I_c, θ

$I_e = I_c + I_d/3$

EXAMPLE 4.3.8

Equivalent Mass of a Fixed-End Beam

■ Problem

Figure 4.3.7a shows a motor mounted on a beam with two fixed-end supports. An imbalance in the motor's rotating mass will produce a vertical force f that oscillates at the same frequency as the motor's rotational speed. The resulting beam motion can be excessive if the frequency is near the natural frequency of the beam, as we will see in a later chapter, and excessive beam motion can eventually cause beam failure. Determine the natural frequency of the beam-motor system.

■ Solution

Treating this system as if it were a single mass located at the middle of the beam results in the equivalent system shown in Figure 4.3.7b, where x is the displacement of the motor from its equilibrium position. The equivalent mass of the system is the motor mass (treated as a concentrated mass m_c) plus the equivalent mass of the beam. From Table 4.3.1, the beam's equivalent mass is $0.38m_d$. Thus, the system's equivalent mass is $m_e = m_c + 0.38m_d$.

The equivalent spring constant of the beam is found from Table 4.1.1. It is

$$k = \frac{16Ewh^3}{L^3}$$

where h is the beam's thickness (height) and w is its width (into the page). Thus, the system model is $m_e\ddot{x} + kx = f$ where x is the vertical displacement of the beam end from its equilibrium position. The natural frequency is $\omega_n = \sqrt{k/m_e}$, which gives

$$\omega_n = \sqrt{\frac{k}{m_e}} = \sqrt{\frac{16Ewh^3/L^3}{m_c + 0.38m_d}}$$

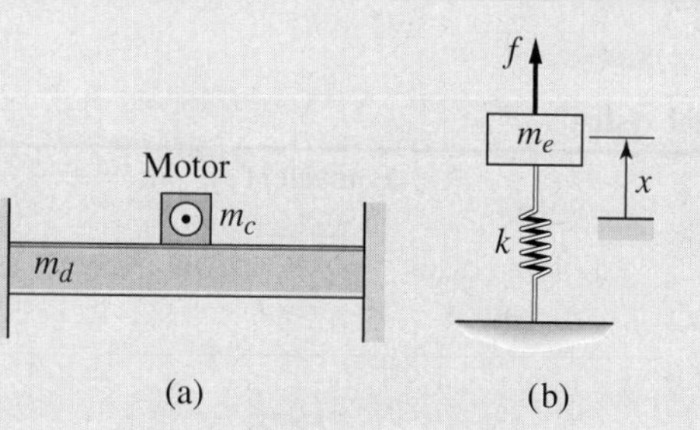

Figure 4.3.7 A motor supported by a fixed-end beam.

EXAMPLE 4.3.9

Torsional Vibration with Fixed Ends

■ Problem

Figure 4.3.8a shows an inertia I_1 rigidly connected to two shafts, each with inertia I_2. The other ends of the shafts are rigidly attached to the supports. The applied torque is T_1. (a) Derive the equation of motion. (b) Calculate the system's natural frequency if I_1 is a cylinder 5 in. in diameter and 3 in. long; the shafts are cylinders 2 in. in diameter and 6 in. long. The three cylinders are made of steel with a shear modulus $G = 1.73 \times 10^9$ lb/ft^2 and a density $\rho = 15.2$ slug/ft^3.

■ Solution

a. As in Table 4.3.1, we add one-third of each shaft's inertia to the inertia of the cylinder in the middle. Thus, the equivalent inertia of the system is

$$I_e = I_1 + 2\left(\frac{1}{3}I_2\right)$$

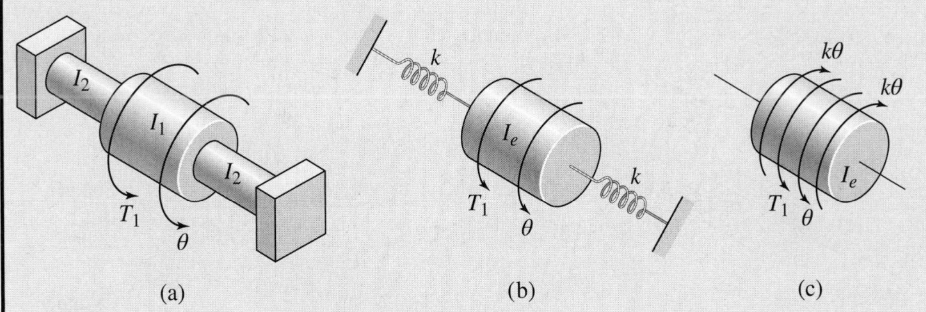

Figure 4.3.8 An inertia fixed to two torsional spring elements.

The equivalent representation is shown in part (b) of the figure, and the free body diagram is shown in part (c) of the figure. From this we obtain the equation of motion:

$$I_e\ddot{\theta} = T - k\theta - k\theta = T - 2k\theta$$

where from Table 4.1.2,

$$k = \frac{G\pi D^4}{32L}$$

b. The value of k is

$$k = \frac{1.73 \times 10^9 \pi\,(2/12)^4}{32(6/12)} = 2.62 \times 10^5 \text{ lb-ft/rad}$$

The moment of inertia of a cylinder of diameter D, length L, and mass density ρ is

$$I = \frac{1}{2}m\left(\frac{D}{2}\right)^2 = \frac{1}{2}\pi\rho L\left(\frac{D}{2}\right)^4$$

The moments of inertia are

$$I_1 = \frac{\pi(15.2)}{2}\frac{3}{12}\left(\frac{5}{24}\right)^4 = 1.12 \times 10^{-2} \text{ slug-ft}^2$$

$$I_2 = \frac{\pi(15.2)}{2}\frac{6}{12}\left(\frac{2}{24}\right)^4 = 5.76 \times 10^{-4} \text{ slug-ft}^2$$

Thus,

$$I_e = 1.12 \times 10^{-2} + 2\left(\frac{1}{3}5.76 \times 10^{-4}\right) = 1.16 \times 10^{-2} \text{ slug/ft}^2$$

This system's natural frequency is $\sqrt{2k/I_e} = 6720$ rad/sec, or $6720/2\pi = 1070$ cycles per second. This gives a period of 9.35×10^{-4} sec.

4.4 DAMPING ELEMENTS

A spring element exerts a reaction force in response to a *displacement*, either compression or extension, of the element. On the other hand, a *damping* element is an element that resists relative *velocity* across it. A common example of a damping element, or *damper*, is a cylinder containing a fluid and a piston with one or more holes (Figure 4.4.1a). If we hold the piston rod in one hand and the cylinder in the other hand, and move the piston and the cylinder at the same velocity, we will feel no reaction force. However, if we move the piston and the cylinder at different velocities, we will feel a resisting force that is caused by the fluid moving through the holes from one side of the piston to the other (Figure 4.4.1b). From this example, we can see that the resisting

Figure 4.4.1 A piston damper.

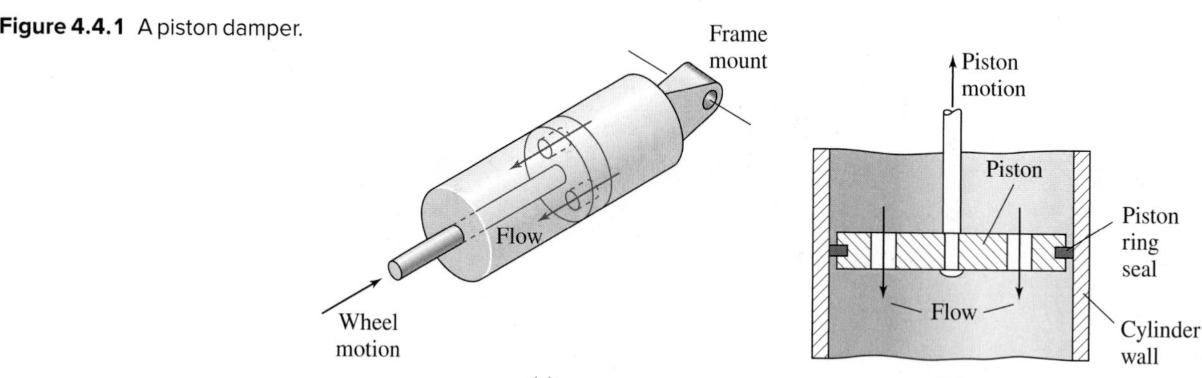

Figure 4.4.2 A pneumatic door closer.

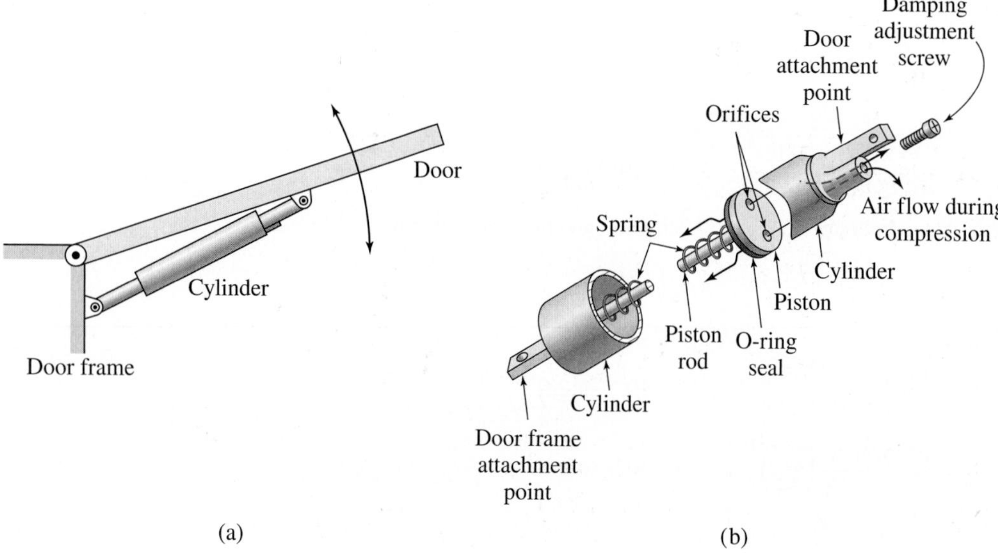

force in a damper is caused by fluid friction (in this case, friction between the fluid and the walls of the piston holes) and that the force depends on the relative velocity of the piston and the cylinder. The faster we move the piston relative to the cylinder, the greater is the resisting force.

4.4.1 A DOOR CLOSER

An example from everyday life of a device that contains a damping element as well as a spring element is the door closer (Figure 4.4.2). In some models, the working fluid is air, while others use a hydraulic fluid. The cylinder is attached to the door and the piston rod is fixed to the door frame. As the door is closed, the air is forced both through the piston holes and out past the adjustment screw, which can be used to adjust the amount of damping resistance (a smaller passageway provides more resistance to the flow and thus more damping force). The purpose of the spring is to close the door; if there were no spring, the door would remain stationary because the damper does not exert any force unless its endpoints are moving relative to each other. The purpose of the damper

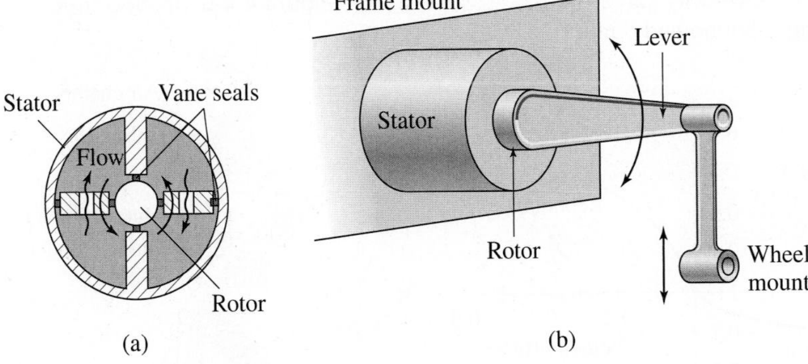

Figure 4.4.3 A rotary damper.

is to exert a force that prevents the door from being opened or closed too quickly (such as may happen due to a gust of wind). If you have such a door closer, try closing the door at different speeds and notice the change in resisting force.

A *rotary* or torsional damper exerts a resisting *torque* in response to an *angular* velocity difference across it. A common example is the vane-type damper shown in Figure 4.4.3a. The rotating part (the rotor) has vanes with holes through which the fluid can flow. The stator is the stationary housing. This device is the basis of some door closers in larger buildings. It is also used to provide damping of wheel motion in some vehicle suspensions [part (b) of the figure].

4.4.2 SHOCK ABSORBERS

The telescopic shock absorber is used in many vehicles. A cutaway view of a typical shock absorber is very complex but the basic principle of its operation is the damper concept illustrated in Figure 4.4.1. The damping resistance can be designed to be dependent on the sign of the relative velocity. For example, Figure 4.4.4 shows a piston containing spring-loaded valves that partially block the piston passageways. If the two spring constants are different or if the two valves have different shapes, then the flow resistance will be dependent on the direction of motion. This design results in a force versus velocity curve like that shown in Figure 4.4.5. During compression (as when the wheel hits a road bump) the resisting force is different than during rebound (when the wheel is forced back to its neutral position by the suspension spring). The resisting force during compression should be small to prevent a large force from being transmitted to the passenger compartment, whereas during rebound the resisting force should be greater to prevent wheel oscillation. An aircraft application of a shock absorber is the *oleo strut*, shown in Figure 4.4.6.

Figure 4.4.4 A damper piston with spring-loaded valves.

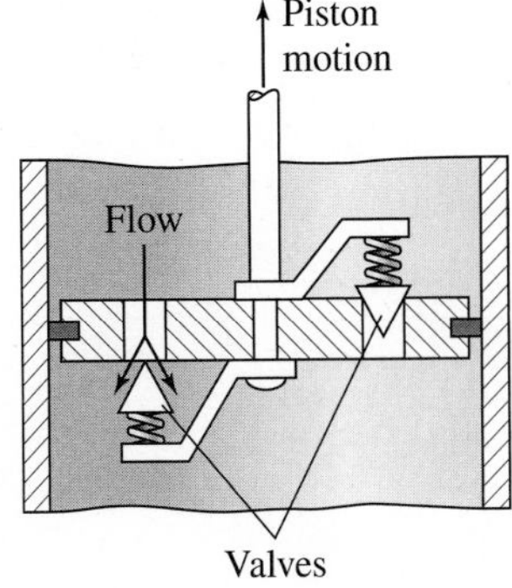

Damping can exist whenever there is a fluid resistance force produced by a fluid layer moving relative to a solid surface. The fluid's viscosity produces a shear stress that exerts a resisting force on the solid surface. Viscosity is an indication of the "stickiness" of the fluid; molasses and oil have greater viscosities than water, for example. Other examples of damping include aerodynamic drag as, discussed in part (b) of Example 1.3.4 in Chapter 1, and hydrodynamic drag. Damping can also be caused by nonfluid effects, such as the energy loss that occurs due to internal friction in solid but flexible materials.

Engineering systems can exhibit damping in bearings and other surfaces lubricated to prevent wear. Damping elements can be deliberately included as part of the design. Such is the case with shock absorbers, fluid couplings, and torque converters.

Figure 4.4.5 Force-velocity curves for a damper during rebound and during compression.

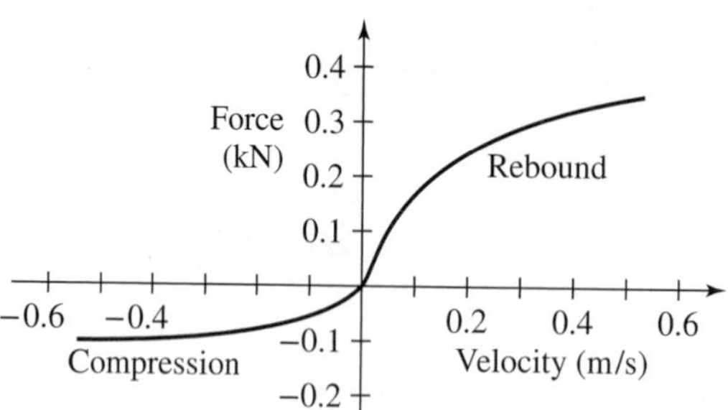

Figure 4.4.6 An oleo strut.

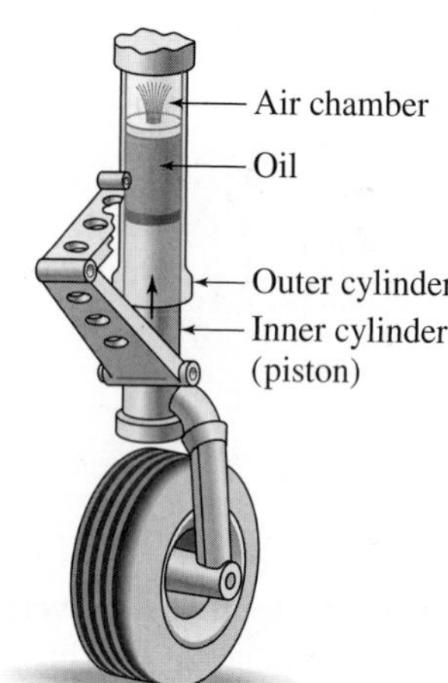

4.4.3 IDEAL DAMPERS

As with spring elements, real damping elements have mass, such as the masses of the piston and the cylinder in a shock absorber. If the system consists of an object attached to a damper (such as a vehicle chassis), a simplifying assumption is to neglect the damper mass relative to the mass of the object and take the mass center of the system to be located at the mass center of the object. This assumption is accurate in many practical applications, but to be comfortable with it you should know the numerical values of the masses of the object and the damper element. In some of the homework problems and some of the examples to follow, the numerical values are not given. In such cases, unless otherwise explicitly stated, you should assume that the damper mass can be neglected. In other cases, where the piston mass and cylinder mass are substantial, for example, the damper must be modeled as two masses, one for the piston and one for the cylinder. An *ideal* damping element is one that is massless.

4.4.4 DAMPER REPRESENTATIONS

The dependence of the damping force on the relative velocity can be quite complicated, and detailed analysis requires application of fluid mechanics principles. Sometimes, to obtain a linear system model, we model the damping as a linear function of the relative velocity. This approach enables us to obtain equations of motion that are easier to solve, without ignoring altogether the effect of the velocity-dependent damping force. The linear model for the damping force f as a function of the relative velocity v is

$$f = cv \tag{4.4.1}$$

where c is the *damping coefficient*. The units of c are force/velocity; for example, N·s/m or lb-sec/ft. In applying this equation to obtain free body diagrams, you must remember that the damping force always opposes the relative velocity.

Using the methods of Chapter 7, Section 7.4, we can derive the following expression for the damping coefficient of a piston-type damper with a single hole.

$$c = 8\pi\mu L\left[\left(\frac{D}{d}\right)^2 - 1\right]^2 \tag{4.4.2}$$

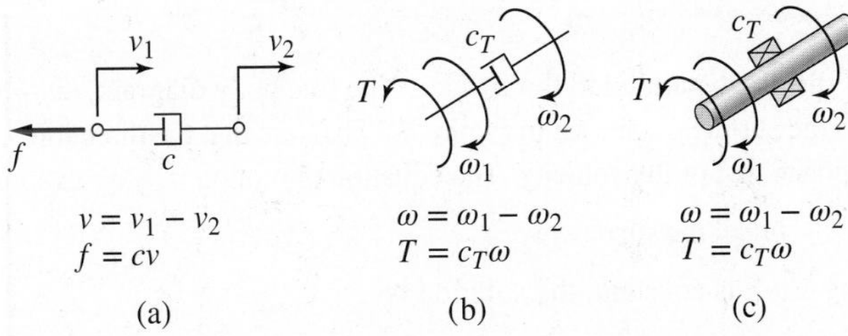

Figure 4.4.7 Symbols for translational and torsional damper elements.

where μ is the viscosity of the fluid, L is the length of the hole through the piston, d is the diameter of the hole, and D is the diameter of the piston. For two holes, as shown in Figure 4.4.1, multiply the result by 2 (this is an example of two damping elements in *series*; their damping coefficients add).

The symbol shown in Figure 4.4.7a is used as the general symbol for a damping element, because it resembles the piston-cylinder device. The symbol is used even when the damping is produced by something other than a piston and cylinder.

The linear model of a torsional damper is

$$T = c_T\omega \tag{4.4.3}$$

where c_T is the *torsional damping coefficient*, ω is the angular velocity, and T is the torque. Torsional dampers are represented by a slightly different symbol, shown in Figure 4.4.7b. When rotational resistance is due to viscous friction in bearings, the bearing symbol shown in Figure 4.4.7c is often used with the symbol c_T to represent the damping. For torsional damping, the units of c_T are torque/angular velocity; for example, N·m·s/rad or lb-ft-sec/rad.

4.4.5 MODELING MASS-DAMPER SYSTEMS

The equations of motion for systems containing damping elements are derived as with spring elements, except that consistent assumptions must also be made about the relative velocities, as well as about the relative displacements, if there is a spring present.

EXAMPLE 4.4.1

Damped Motion on an Inclined Surface

■ Problem

Derive and solve the equation of motion of the block sliding on an inclined, lubricated surface (Figure 4.4.8a). Assume that the damping force is linear. For this application the damping coefficient c depends on the contact area of the block, the viscosity of the lubricating fluid, and the thickness of the fluid layer.

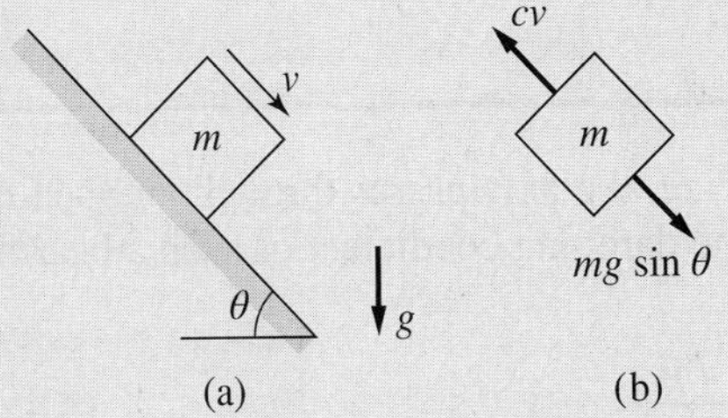

Figure 4.4.8 A mass sliding on a lubricated, inclined surface.

■ **Solution**

We define v to be the velocity of the block parallel to the surface. The free body diagram, displaying only the forces parallel to the surface, is shown in Figure 4.4.8b. Note that the direction of the damping force must be opposite that of the velocity. The equation of motion is

$$m\dot{v} = mg\sin\theta - cv$$

Because the gravity force $mg \sin \theta$ is constant, the solution is

$$v(t) = \left[v(0) - \frac{mg\sin\theta}{c}\right]e^{-ct/m} + \frac{mg\sin\theta}{c}$$

Eventually the block will reach a constant velocity of $mg\sin\theta/c$ regardless of its initial velocity. At this velocity the damping force is great enough to equal the gravity force component. Because $e^{-4} = 0.02$, when $t = 4m/c$, the velocity will be approximately 98% of its final velocity. Thus, a system with a larger mass or less resistance (a smaller damping constant c) will require more time to reach its constant velocity. This apparently unrealistic result is explained by noting that the constant velocity attained is $mg\sin\theta/c$. Thus, a system with a larger mass or a smaller c value will attain a higher velocity, and thus should take longer to reach it.

EXAMPLE 4.4.2

A Wheel-Axle System with Bearing Damping

■ **Problem**

Figure 4.4.9a illustrates a wheel-axle system in which the axle is supported by two sets of bearings that produce damping. Each bearing set has a torsional damping coefficient c_T. The torque T is supplied by a motor. The force F is the friction force due to the road surface. Derive the equation of motion.

■ **Solution**

Part (b) of the figure shows the free body diagram. Note that the damping torque $c_T\omega$ from each bearing set opposes the angular velocity. The inertia I is the combined inertia of the wheel and the two shafts of the axle: $I = I_w + 2I_s$. The equation of motion is

$$I\dot{\omega} = T + RF - 2c_T\omega$$

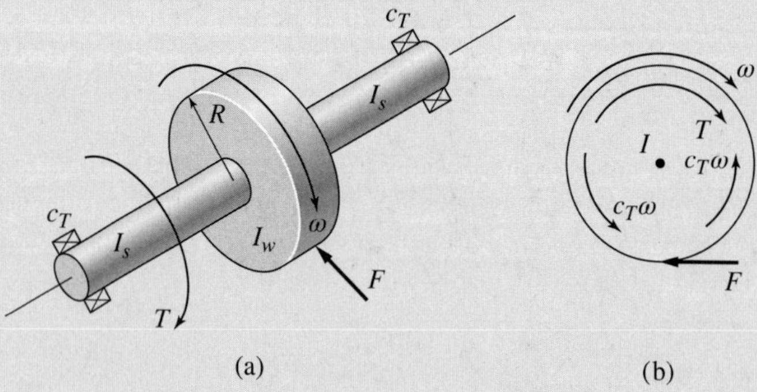

Figure 4.4.9 A wheel-axle system.

In a simple bearing, called a journal bearing, the axle passes through an opening in a support. A commonly used formula for the damping coefficient of such a bearing is *Petrov's* law:

$$c_T = \frac{\pi D^3 L\mu}{4\epsilon} \qquad (4.4.4)$$

where L is the length of the opening, D is the axle diameter, and ϵ is the thickness of the lubricating layer (the radial clearance between the axle and the support).

EXAMPLE 4.4.3

A Generic Mass-Spring-Damper System

■ Problem

Figure 4.4.10a represents a generic mass-spring-damper system with an external force f. Derive its equation of motion and determine its characteristic equation.

■ Solution

The free body diagram is shown in Figure 4.4.10b. Note that because we have defined x, the displacement from equilibrium, to be positive downward, we must take the velocity $\dot{x}$ also to be positive downward. The damper force opposes the velocity. Thus, the equation of motion is

$$m\ddot{x} = -c\dot{x} - k(x + \delta_{st}) + mg + f = -c\dot{x} - kx + f$$

because $k\delta_{st} = mg$. The equation can be rearranged as

$$m\ddot{x} + c\dot{x} + kx = f$$

From this we can recognize the characteristic equation to be $ms^2 + cs + k = 0$.

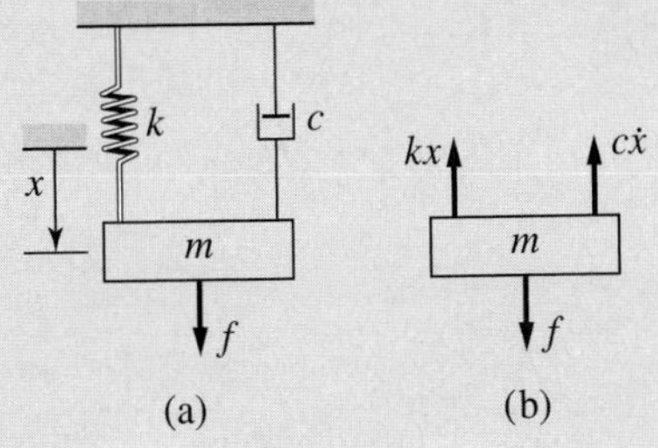

Figure 4.4.10 A mass-spring-damper system.

In Section 4.2 we solved the equation of motion for the case where there is no damping ($c = 0$). Now let us investigate the effects of damping.

EXAMPLE 4.4.4

Effects of Damping

■ Problem

Suppose that for the system shown in Figure 4.4.10a the mass is $m = 1$ and the spring constant is $k = 16$. Investigate the free response as we increase the damping for the four cases: $c = 0$, 4, 8, and 10. Use the initial conditions: $x(0) = 1$ and $\dot{x}(0) = 0$.

■ Solution

The characteristic equation is $s^2 + cs + 16 = 0$. The roots are

$$s = \frac{-c \pm \sqrt{c^2 - 64}}{2}$$

The free response can be obtained with the methods covered in Chapter 2.

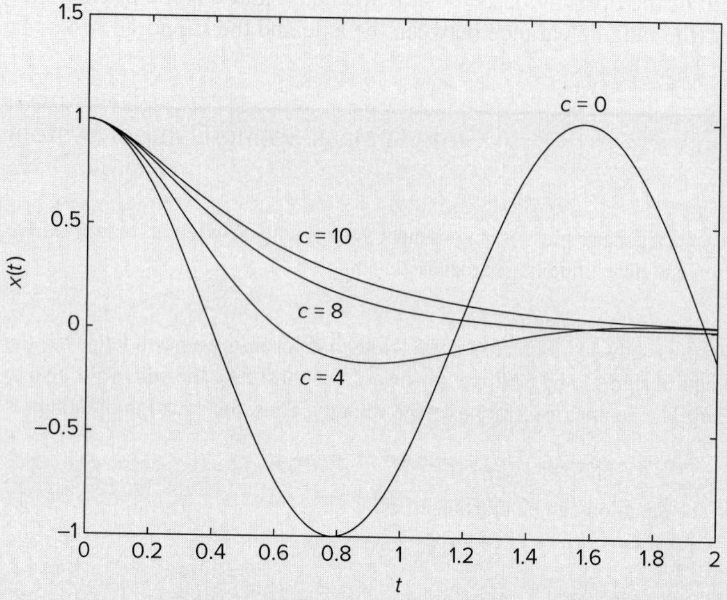

Figure 4.4.11 Free response of a mass-spring-damper system for several values of c.

The free responses for the given initial conditions, and the corresponding values of ζ and τ for each value of c, are as follows:

For $c = 0$ $\quad x(t) = \cos 4t, \zeta = 0, \tau = \infty$

For $c = 4$ $\quad x(t) = 1.155e^{-2t} \sin\left(2\sqrt{3}t + 1.047\right), \zeta = \dfrac{1}{2}, \tau = \dfrac{1}{2}$

For $c = 8$ $\quad x(t) = (1 + 4t)e^{-4t}, \zeta = 1, \tau = \dfrac{1}{4}$

For $c = 10$ $\quad x(t) = \dfrac{4}{3}e^{-2t} - \dfrac{1}{3}e^{-8t}, \zeta = \dfrac{5}{4}, \tau = \dfrac{1}{2}, \dfrac{1}{8}$

The solutions are plotted in Figure 4.4.11. For no damping, the system is neutrally stable and the mass oscillates with a constant amplitude and a radian frequency of $\sqrt{k/m} = 4$, which is the natural frequency ω_n. As the damping is increased slightly to $c = 4$, the system becomes stable and the mass still oscillates but with a smaller radian frequency $\left(\sqrt{12} = 3.464\right)$. The oscillations die out eventually as the mass returns to rest at the equilibrium position $x = 0$. As the damping is increased further to $c = 8$, the mass no longer oscillates because the damping force is large enough to limit the velocity and thus the momentum of the mass to a value that prevents the mass from overshooting the equilibrium position. For a larger value of c ($c = 10$), the mass takes longer to return to equilibrium because the damping force greatly slows down the mass.

When a spring and damper are coupled together, write a force balance equation at the connection point and then apply the Laplace transform to obtain a set of two or more algebraic equations. These equations can then be combined algebraically to obtain a single equation in terms of the given input and the desired output variable.

EXAMPLE 4.4.5

Coupled Spring and Damper

■ **Problem**

Determine the transfer function $X(s)/F(s)$ and the equation of motion for the system shown in Figure 4.4.12a.

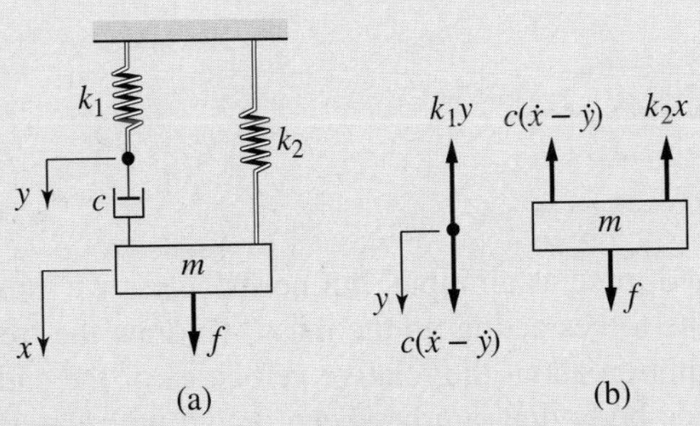

Figure 4.4.12 (a) A coupled spring and damper. (b) Free body diagrams.

■ Solution

Let us assume that $x > 0$, $y > 0$, and $\dot{x} > \dot{y}$. If so, then the damper force pulls up on the mass, and we obtain the free body diagrams shown in part (b) of the figure.

$$m\ddot{x} = f - k_2x - c(\dot{x} - \dot{y})$$

$$c(\dot{x} - \dot{y}) = k_1y$$

Transforming these equations with zero-initial conditions gives

$$(ms^2 + cs + k_2)X(s) = F(s) + csY(s) \tag{1}$$

$$csX(s) = (k_1 + cs)Y(s) \tag{2}$$

Equation (2) gives

$$Y(s) = \frac{cs}{k_1 + cs}X(s)$$

Substituting $Y(s)$ into Equation (2) gives

$$(ms^2 + cs + k_2)X(s) = F(s) + \frac{(cs)^2}{k_1 + ck_1 + cs}X(s)$$

After collecting terms, we obtain

$$\frac{X(s)}{F(s)} = \frac{k_1 + cs}{mcs^3 + mk_1s^2 + (k_1 + k_2)cs + k_1k_2} \tag{3}$$

The corresponding differential equation is

$$mc\dddot{x} + mk_1\ddot{x} + (k_1 + k_2)c\dot{x} + k_1k_2x = k_1f + c\dot{f} \tag{4}$$

Discussion All the equations of motion we have seen up to now have been first- or second-order differential equations. Equation (4) in this example is a third-order equation, and you might wonder as to why this is. The force balance equation at the connection point requires that we introduce an extra coordinate, y, in order to express the damper force. This extra equation, which involves the derivative $\dot{x}$, raises the order of the equation set (1) and (2) by one, and results in a third-order characteristic polynomial.

Note also that equation (3) has numerator dynamics because of the damper.

4.4.6 MOTION INPUTS WITH DAMPING ELEMENTS

Sometimes we are given the motion, either the displacement or the velocity, of the endpoint of a damping element. For example, consider the system shown in Figure 4.4.13, where we are given the displacement $y(t)$ of the left-hand end of the damper. Such a motion could be caused by a cam. Suppose that when $x = 0$ the spring is at its free

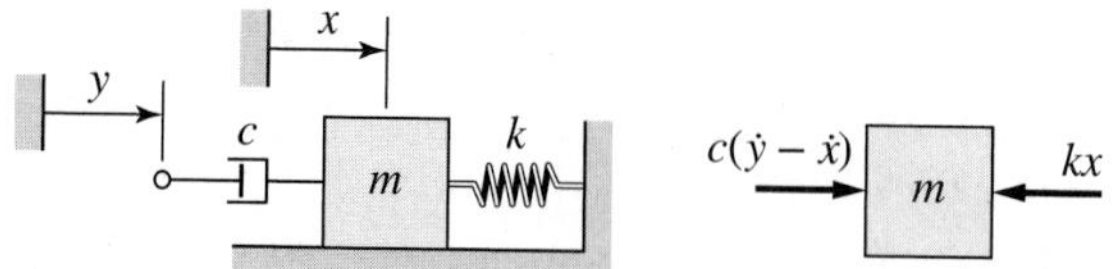

Figure 4.4.13 A system with velocity input.

length. Here we are not given an applied force as an input, but nevertheless we must draw the free body diagram showing the forces acting on the mass. To draw the free body diagram, we must make an assumption about the relative velocities of the end-points of the damping element. The free body diagram has been drawn with the assumption that $\dot{y} > \dot{x}$. The force produced by the given displacement $y(t)$ is the resulting damper force $c(\dot{y} - \dot{x})$. The equation of motion is $m\ddot{x} = c(\dot{y} - \dot{x}) - kx$, or $m\ddot{x} + c\dot{x} + kx = c\dot{y}$. Note that we must be given the velocity $\dot{y}$, or be able to compute $\dot{y}$ from $y(t)$, to solve this equation for $x(t)$.

When motion inputs are given, it is important to realize that ultimately such motion is generated by a force (or torque) and that this force must be great enough to generate the specified motion in the presence of any resisting forces or system inertia. For example, from the principle of action and reaction, we see that the mechanism supplying the velocity $\dot{y}(t)$ must generate a force equal to the damper force $c(\dot{y} - \dot{x})$.

EXAMPLE 4.4.6

A Two-Mass System

■ Problem

Derive the equations of motion of the two-mass system shown in Figure 4.4.14a.

■ Solution

Choose the coordinates x_1 and x_2 as the displacements of the masses from their equilibrium positions. In equilibrium, the static forces in the springs cancel the weights of the masses. Note that the dampers have no effect in equilibrium and thus do not determine the location of the equilibrium position. Therefore, the free body diagrams showing the dynamic forces, and not the static forces, are as shown in Figure 4.4.14b. These diagrams have been drawn with the assumption that the displacement x_2 of mass m_2 from its equilibrium position is greater than the displacement of m_1. Because of the dampers, an additional assumption is required concerning the relative velocities of the masses. The diagrams in part (b) of the figure are based on the assumption that the velocity $\dot{x}_2$ of mass m_2 is greater than the velocity $\dot{x}_1$ of m_1. If your assumptions are correct, the force values shown on the diagram must be *positive*. From

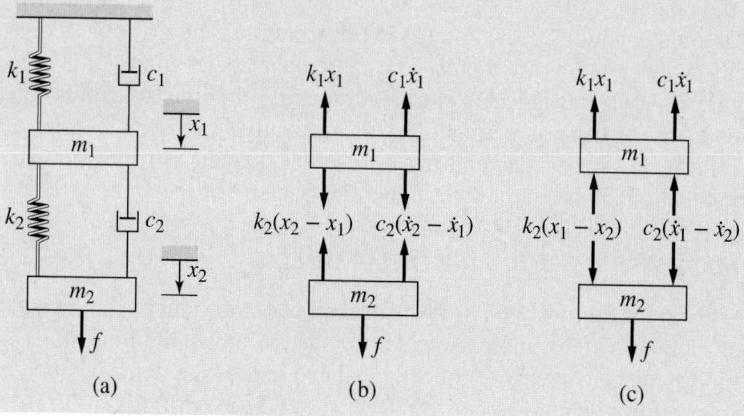

Figure 4.4.14 A two-mass system.

these diagrams we obtain the equations of motion:

$$m_1\ddot{x}_1 = -k_1x_1 + k_2(x_2 - x_1) - c_1\dot{x}_1 + c_2(\dot{x}_2 - \dot{x}_1)$$

$$m_2\ddot{x}_2 = f - k_2(x_2 - x_1) - c_2(\dot{x}_2 - \dot{x}_1)$$

If we move all terms to the left side of the equal sign except for the external force f, and collect terms, we obtain

$$m_1\ddot{x}_1 + (c_1 + c_2)\dot{x}_1 - c_2\dot{x}_2 + (k_1 + k_2)x_1 - k_2x_2 = 0 \quad (1)$$

$$m_2\ddot{x}_2 - k_2x_1 + k_2x_2 - c_2\dot{x}_1 + c_2\dot{x}_2 = f \quad (2)$$

In drawing the free body diagrams of multimass systems having springs and dampers between the masses, you must make assumptions about the relative displacements and relative velocities of each mass. For example, we could have assumed that the displacement x_2 of mass m_2 is *less* than the displacement of m_1, and assumed that the velocity $\dot{x}_2$ of mass m_2 is *less* than the velocity $\dot{x}_1$ of m_1. Figure 4.4.14c shows the free body diagrams drawn for this assumption. Note that the directions of the forces associated with spring k_2 and damper c_2 are the opposite of those in part (b) of the figure. You should confirm that the diagram in part (c) results in equations of motion that are identical to equations (1) and (2).

You must be consistent in your assumptions made to draw the free body diagrams. A common mistake is to use one assumption to obtain the free body diagram for mass m_1 but another assumption for mass m_2.

EXAMPLE 4.4.7

Step Response of a Two-Mass System

■ Problem

Determine the transfer function $X_1(s)/F(s)$ for the system shown in Figure 4.4.14a. Use the parameter values: $m_1 = m_2 = 1, c_1 = 2, c_2 = 3, k_1 = 1$, and $k_2 = 4$. (b) Suppose $f(t)$ is a unit step function. Find the steady-state value of x_1, estimate how long it will take to reach steady state, and discuss any oscillatory response.

■ Solution

a. With the given parameter values, equations (1) and (2) of Example 4.4.6 become

$$\ddot{x}_1 + 5\dot{x}_1 + 5x_1 - 3\dot{x}_2 - 4x_2 = 0$$

$$\ddot{x}_2 + 3\dot{x}_2 + 4x_2 - 3\dot{x}_1 - 4x_1 = f(t)$$

Apply the Laplace transform to these equations, using zero-initial conditions, to obtain

$$(s^2 + 5s + 5)X_1(s) - (3s + 4)X_2(s) = 0 \quad (1)$$

$$-(3s + 4)X_1(s) + (s^2 + 3s + 4)X_2(s) = F(s) \quad (2)$$

To determine the transfer function $X_1(s)/F(s)$, we must eliminate the variable $X_2(s)$. Solve (1) as follows.

$$X_2(s) = \frac{s^2 + 5s + 5}{3s + 4}X_1(s)$$

Substitute this into (2).

$$-(3s + 4)X_1(s) + (s^2 + 3s + 4)\frac{s^2 + 5s + 5}{3s + 4}X_1(s) = F(s)$$

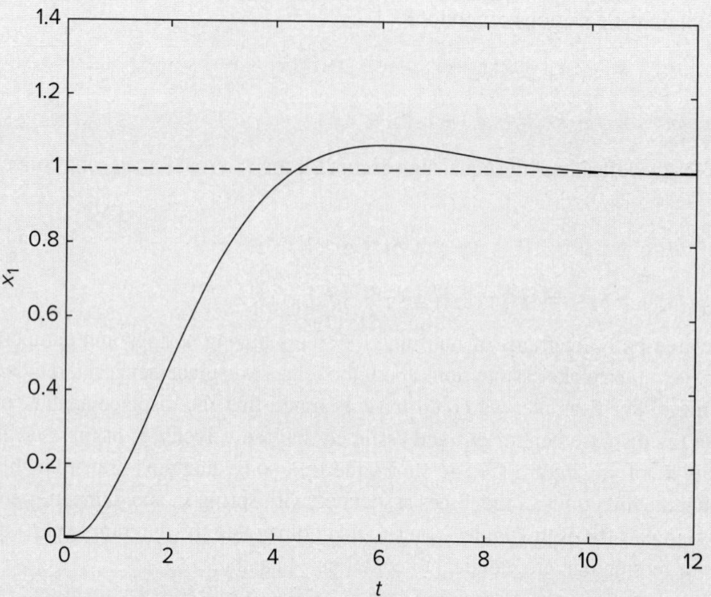

Figure 4.4.15 Unit-step response of a two-mass system.

This simplifies to

$$-(3s+4)^2 X_1(s) + (s^2+3s+4)(s^2+5s+5)X_1(s) = (3s+4)F(s)$$

or

$$\frac{X_1(s)}{F(s)} = \frac{3s+4}{s^4+8s^3+15s^2+11s+4}$$

b. The characteristic roots are found from the roots of the denominator of the transfer function. These can be found with a calculator or with the MATLAB `roots` function, for example. They are

$$s = -5.6773, \qquad -1.3775, \qquad -0.4726 \pm 0.5368j$$

All the roots have negative real parts and thus the Final-Value Theorem can be applied. If $F(s) = 1/s$, this theorem gives

$$x_{1ss} = \lim_{s \to 0} s \left(\frac{3s+4}{s^4+8s^3+15s^2+11s+4} \right) \frac{1}{s} = 1$$

The time constants are the negative reciprocals of the real part of the roots.

$$\tau = 0.1761, \qquad 0.7260, \qquad 2.1161$$

The dominant time constant is 2.1161, and four times that value gives an estimate of how long it will take to reach steady state. This value is 4(2.1161) = 8.4635. The dominant time constant belongs to the complex roots, and so the dominant response is oscillatory with a period of $2\pi/0.5368 = 11.705$. Because the period is longer than four time constants, we will not see a complete oscillation in the response. The response is shown in Figure 4.4.15, which was obtained with the MATLAB `tf` and `step` functions.

4.5 ADDITIONAL MODELING EXAMPLES

This section contains examples for additional practice in deriving equations of motion for systems containing spring and damper elements.

A Translational System with Displacement Input

EXAMPLE 4.5.1

■ Problem

Derive the equation of motion for the system shown in Figure 4.5.1a. The input is the displacement y of the right-end of the spring. The output is the displacement x of the mass. The spring is at its free length when $x = y$.

■ Solution

The free body diagram in Figure 4.5.1b displays only the horizontal forces, and it has been drawn assuming that $y > x$. From this diagram we can obtain the equation of motion.

$$m\ddot{x} = k(y - x) - c\dot{x} \qquad \text{or} \qquad m\ddot{x} + c\dot{x} + kx = ky$$

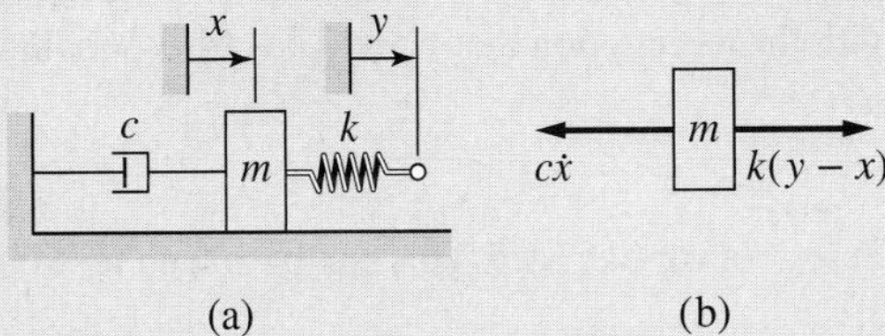

Figure 4.5.1 A translational system with displacement input.

A Rotational System with Displacement Input

EXAMPLE 4.5.2

■ Problem

Derive the equation of motion for the system shown in Figure 4.5.2a. The input is the angular displacement ϕ of the left-end of the rod, which is modeled as a torsional spring. The output is the angular displacement θ of the inertia I. Neglect the inertia of the rod. There is no torque in the rod when $\phi = \theta$.

■ Solution

The free body diagram in Figure 4.5.2b has been drawn assuming that $\phi > \theta$. From this diagram we can obtain the equation of motion.

$$I\ddot{\theta} = k_T(\phi - \theta) - c_T\dot{\theta} \qquad \text{or} \qquad I\ddot{\theta} + c_T\dot{\theta} + k_T\theta = k_T\phi$$

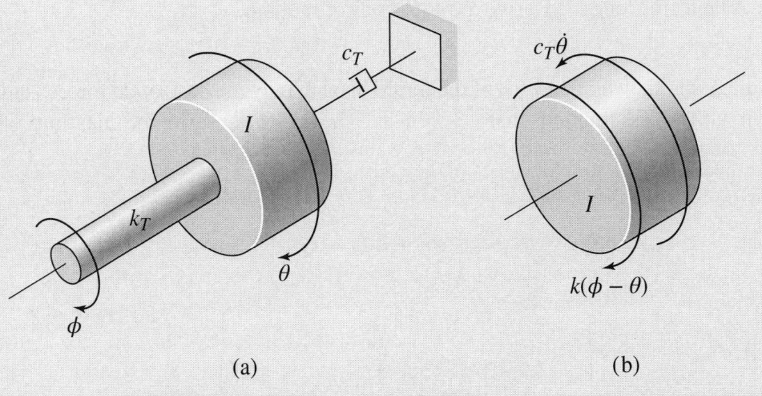

Figure 4.5.2 A rotational system with displacement input.

Sometimes we need to determine the motion of a point in the system where there is no mass. In such problems, it is helpful to place a "fictitious" mass at the point in question, draw the free body diagram of the fictitious mass, and set the mass value to zero in the resulting equation of motion. Although the same result can be obtained by applying the principles of statics to the point in question, this method helps to organize the process.

EXAMPLE 4.5.3

Displacement Input and Negligible System Mass

■ Problem

Obtain the equation of motion of point A for the system shown in Figure 4.5.3a. We are given the displacement $y(t)$. The spring is at its free length when $x = y$.

■ Solution

We place a fictitious mass m_A at point A, and draw its free body diagram, shown in part (b) of the figure. The diagram has been drawn with the assumption that $y > x$. The corresponding equation of motion is

$$m_A \ddot{x} = k(y - x) - c\dot{x}$$

Let $m_A = 0$ to obtain the answer: $0 = k(y - x) - c\dot{x}$ or $c\dot{x} + kx = ky$. This can be solved for $x(t)$ if we know $y(t)$.

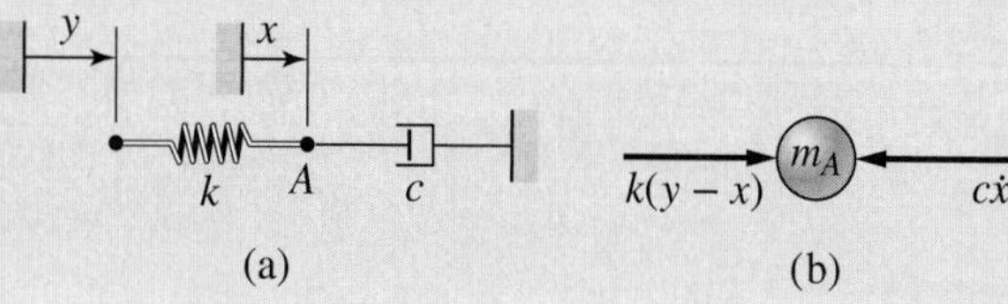

Figure 4.5.3 A system with negligible mass.

EXAMPLE 4.5.4

A Two-Mass System with Displacement Input

■ Problem

Figure 4.5.4a shows a two-mass system where the displacement $y(t)$ of the right-hand end of the spring is a given function. The masses slide on a frictionless surface. When $x_1 = x_2 = y = 0$, the springs are at their free lengths. Derive the equations of motion.

■ Solution

The free body diagrams shown in part (b) of the figure display only the horizontal forces, and they were drawn with the assumptions that $y > x_2 > x_1$. These diagrams give the equations of motion:

$$m_1\ddot{x}_1 = k_1(x_2 - x_1)$$

$$m_2\ddot{x}_2 = -k_1(x_2 - x_1) + k_2(y - x_2)$$

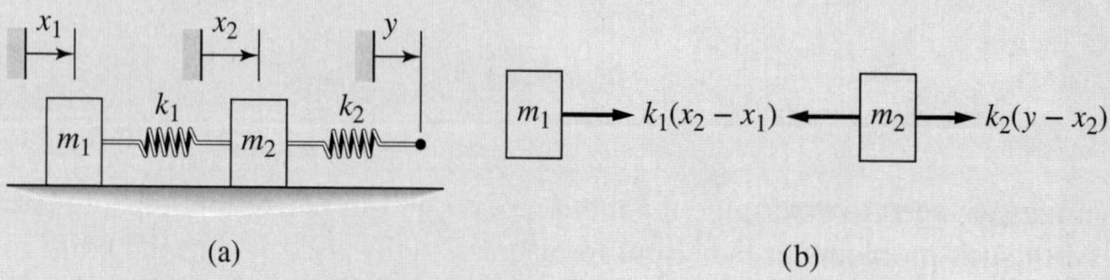

Figure 4.5.4 A two-mass system with displacement input.

EXAMPLE 4.5.5

A Two-Inertia System with Angular Displacement Input

■ Problem

Figure 4.5.5a shows a system with two inertia elements and two torsional dampers. The left-hand end of the shaft is twisted by the angular displacement ϕ, which is a specified function of time. The shaft has a torsional spring constant k_T and negligible inertia. The equilibrium position corresponds to $\phi = \theta_1 = \theta_2 = 0$. Derive the equations of motion.

■ Solution

From the free body diagrams in part (b) of the figure, which are drawn for $\phi > \theta_1 > \theta_2$ and $\dot{\theta}_1 > \dot{\theta}_2$, we obtain

$$I_1\ddot{\theta}_1 = k_T(\phi - \theta_1) - c_{T_1}(\dot{\theta}_1 - \dot{\theta}_2)$$

$$I_2\ddot{\theta}_2 = c_{T_1}(\dot{\theta}_1 - \dot{\theta}_2) - c_{T_2}\dot{\theta}_2$$

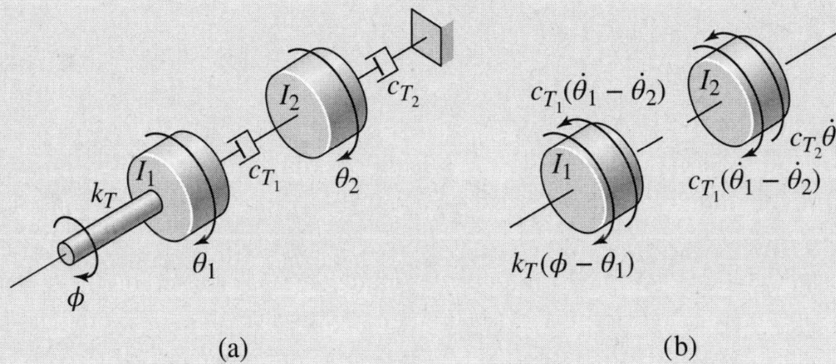

Figure 4.5.5 A system with two inertias.

EXAMPLE 4.5.6

A Single-Inertia Fluid-Clutch Model

■ Problem

Figure 4.5.6a shows a driving disk rotating at a specified speed ω_d. There is a viscous fluid layer between this disk and the driven disk whose inertia plus that of the shaft is I_1. Through the action of the viscous damping, the rotation of the driving disk causes the driven disk to rotate, and this rotation is opposed by the torque T_1, which is due to whatever load is being driven. This model represents the situation in a fluid clutch, which avoids the wear and shock that occurs in friction

Figure 4.5.6 A single-inertia model of a fluid clutch.

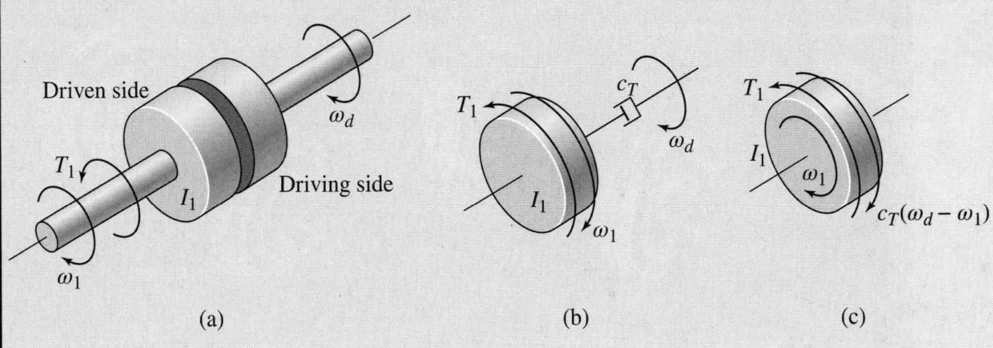

clutches. (a) Derive a model for the speed ω_1. (b) Find the speed $\omega_1(t)$ for the case where the load torque $T_1 = 0$ and the speed ω_d is a step function of magnitude Ω_d.

■ Solution

a. Lacking a more detailed model of the fluid forces in the viscous fluid layer, we will assume that the effect can be modeled as a massless rotational damper obeying the linear damping law. Thus, we will model the system as shown in Figure 4.5.6b. The given input variable is the velocity ω_d.

To draw the free body diagram of I_1, note that the viscous fluid layer opposes any velocity difference across it. The torque that it exerts as a result has a magnitude $c_T|\omega_d - \omega_1|$ and acts in the direction that will reduce the speed difference. Figure 4.5.6c shows the free body diagram, which is drawn assuming that $\omega_d > \omega_1$. The equation of motion is

$$I_1\dot{\omega}_1 = c_T(\omega_d - \omega_1) - T_1$$

b. If ω_d is a step function of magnitude Ω_d, and if $T_1 = 0$, the solution for $\omega(t)$ can be found from Table 2.3.2. It is

$$\omega_1(t) = \Omega_d - [\omega_1(0) - \Omega_d]e^{-c_T t/I_1} \tag{1}$$

This shows that the speed ω_1 will eventually equal Ω_d as $t \to \infty$. If the inertia I_1 is initially rotating in the direction opposite to that of ω_d, it eventually reverses direction. When $t = 4I_1/c_T$, the magnitude of the difference, $|\omega_1(t) - \Omega_d|$, is only 2% of the initial magnitude $|\omega_1(0) - \Omega_d|$.

The model derived in this example is a first-order model, and it is based on the assumption that the torque on the driving shaft is high enough to drive the disk on that side at the specified speed ω_d, regardless of the effects of the damping and the inertia of the driven side. If this is not true, then the model derived in Example 4.5.7 is a better model.

EXAMPLE 4.5.7 A Two-Inertia Fluid-Clutch Model

■ Problem

Figure 4.5.7a is a fluid clutch model that can be used when the torque on the driving shaft is not sufficient to drive the disk on that side at the specified speed ω_d. To account for this situation, we must include the inertia of the driving side in the model. Assume that the torques T_d and T_1

Figure 4.5.7 A two-inertia model of a fluid clutch.

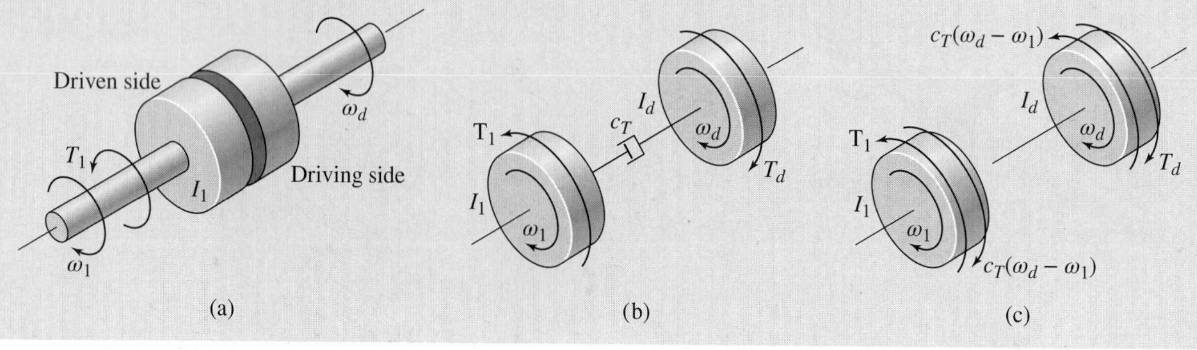

are specified functions of time. (a) Derive the equations of motion for the speeds ω_d and ω_1. (b) Obtain the transfer functions $\Omega_1(s)/T_1(s)$ and $\Omega_1(s)/T_d(s)$. (c) Obtain the expression for the response $\omega_1(t)$ if the initial conditions are zero, $T_1 = 0$, and T_d is a step function of magnitude M.

■ Solution

a. We can model this system with two inertias, as shown in Figure 4.5.7b. The damper torque acts to reduce the difference between the speeds ω_1 and ω_d. The free body diagrams in Figure 4.5.7c are drawn for the case where $\omega_d > \omega_1$, and we can obtain the following equations of motion from them:

$$I_d\dot{\omega}_d = T_d - c_T(\omega_d - \omega_1)$$

$$I_1\dot{\omega}_1 = -T_1 + c_T(\omega_d - \omega_1)$$

b. Applying the Laplace transform to the equations of motion with zero-initial conditions, we obtain

$$(I_d s + c_T)\Omega_d(s) - c_T\Omega_1(s) = T_d(s)$$

$$-c_T\Omega_d(s) + (I_1 s + c_T)\Omega_1(s) = -T_1(s)$$

Eliminating $\Omega_d(s)$ from these equations and solving for $\Omega_1(s)$, we obtain

$$\Omega_1(s) = \frac{c_T}{s(I_1I_ds + c_TI_1 + c_TI_d)}T_d(s) - \frac{I_ds + c_T}{s(I_1I_ds + c_TI_1 + c_TI_d)}T_1(s)$$

Thus, the two transfer functions are

$$\frac{\Omega_1(s)}{T_d(s)} = \frac{c_T}{s(I_1I_ds + c_TI_1 + c_TI_d)} \qquad \frac{\Omega_1(s)}{T_1(s)} = -\frac{I_ds + c_T}{s(I_1I_ds + c_TI_1 + c_TI_d)}$$

c. Setting $T_1(s) = 0$ and $T_d(s) = M/s$ gives

$$\Omega_1(s) = \frac{c_T}{s(I_1I_ds + c_TI_1 + c_TI_d)}\frac{M}{s} = \frac{b}{s(s+a)}\frac{M}{s}$$

where

$$a = \frac{c_TI_1 + c_TI_d}{I_1I_d} \qquad b = \frac{c_T}{I_1I_d}$$

A partial fraction expansion gives

$$\Omega_1(s) = \frac{bM}{a}\left[\frac{1}{s^2} - \frac{1}{as} + \frac{1}{a(s+a)}\right]$$

The corresponding response is

$$\omega_1(t) = \frac{bM}{a}\left(t - \frac{1}{a} + \frac{1}{a}e^{-at}\right)$$

The time constant is $\tau = 1/a$. For $t > 4/a$ approximately, the speed increases linearly with time.

In Chapter 3 we derived the following equation of motion of a pendulum whose mass is concentrated a distance L from the pivot point (see Figure 4.5.8).

$$L\ddot{\theta} = -g\sin\theta$$

We also obtained the following linear model that is approximately correct when the pendulum is hanging nearly vertical at $\theta = 0$.

$$L\ddot{\theta} = -g\theta$$

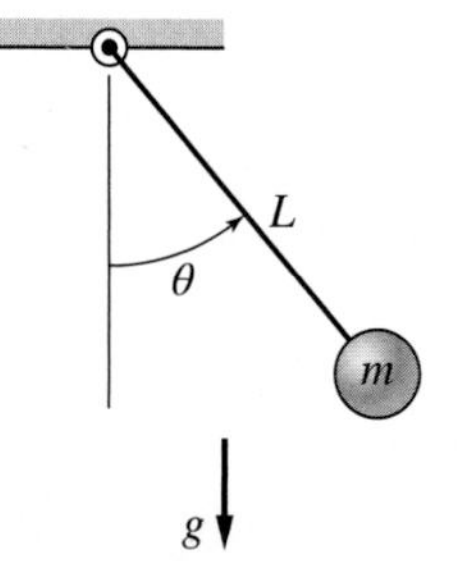

Figure 4.5.8 A pendulum with a concentrated mass.

To obtain this model, we used the small-angle approximation $\sin\theta \approx \theta$. Because the characteristic roots of the model are imaginary ($s = \pm j\sqrt{g/L}$), the equilibrium position at $\theta = 0$ is neutrally stable. Because this conclusion is based on the approximate model, we cannot use this model to predict what will happen if the mass is displaced far from $\theta = 0$. On the other hand, if we move the mass exactly to $\theta = 180°$, common sense tells us that it will stay there, but if it is slightly disturbed, the mass will fall. Thus, we can see that the equilibrium position at $\theta = 180°$ is unstable.

Let us now see how the stability properties are affected if we introduce stiffness and damping into the system.

EXAMPLE 4.5.8

Stability of an Inverted Pendulum

■ Problem

Determine the effects of stiffness and damping on the stability properties of an inverted pendulum (Figure 4.5.9a). Assume that the angle ϕ is small.

■ Solution

Note that for small values of ϕ the motion of the attachment point of the spring and damper is approximately horizontal; its displacement is $L_1\phi$ and its velocity is $L_1\dot{\phi}$. The free body diagram is shown in Figure 4.5.9b. Note that the moment arm of the spring and damper forces is L_1. From the diagram we can write the equation of motion.

$$I_O\ddot{\phi} = M_O \qquad \text{or} \qquad mL^2\ddot{\phi} = mgL\phi - L_1(cL_1\dot{\phi}) - L_1(kL_1\phi)$$

or

$$mL^2\ddot{\phi} + cL_1^2\dot{\phi} + \left(kL_1^2 - mgL\right)\phi = 0$$

which has the form

$$\ddot{\phi} + a\dot{\phi} + b\phi = 0 \qquad a = \frac{cL_1^2}{mL^2} \qquad b = \frac{kL_1^2 - mgL}{mL^2}$$

From the Routh-Hurwitz condition in Chapter 2, the system will be stable if both a and b are positive.

Note that a cannot be negative for physically realistic values of the other parameters, but it can be zero if there is no damping ($c = 0$). Thus, we conclude that some damping is necessary for the system to be stable.

However, b can be positive, negative, or zero depending on the relative values of kL_1^2 and mgL^2. If $b < 0$, the system is unstable ($b < 0$ if $kL_1^2 < mgL$). This indicates that the torque from

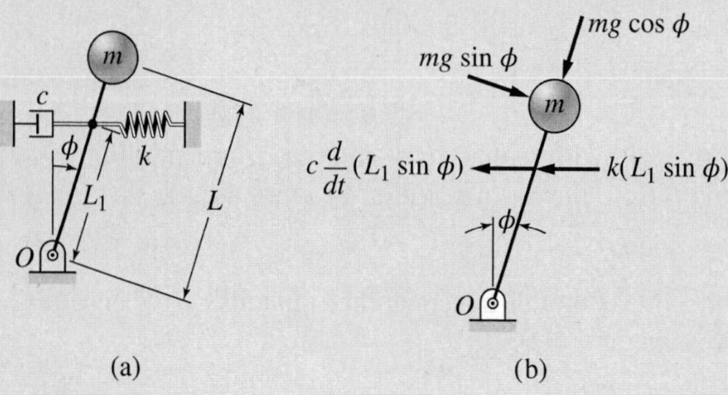

Figure 4.5.9 A pendulum with a damper and spring element.

the spring is not great enough to overcome the torque due to gravity, and thus the mass will fall. We cannot tell how far it will fall because eventually ϕ will become so large that the approximation $\sin\phi \approx \phi$ will no longer be accurate and thus the linear model will be useless.

If $c = 0$ and $b = 0$, the two roots are both zero, which indicates neutral stability. If slightly displaced with zero initial velocity, the mass will remain in that position. If $c = 0$ but $b > 0$, the two roots are imaginary ($s = \pm j\sqrt{b}$) and thus the system is neutrally stable. If slightly displaced, the mass will oscillate about $\phi = 0$ with a constant amplitude.

4.5.1 STEP RESPONSE WITH AN INPUT DERIVATIVE

The following first-order model (4.5.1) contains an input derivative, and so we must be careful in modeling the input and in solving for the response.

$$m\dot{v} + cv = b\dot{f}(t) + f(t) \tag{4.5.1}$$

The transfer function is

$$\frac{V(s)}{F(s)} = \frac{bs + 1}{ms + c} \tag{4.5.2}$$

Thus, we see that the presence of an input derivative is indicated by an s term in the numerator. This presence is called "numerator dynamics." In mechanical systems, numerator dynamics occurs when a displacement input acts directly on a damper.

An example of a device having numerator dynamics is shown in Figure 4.5.10. From the free body diagram,

$$c(\dot{y} - \dot{x}) + k_1(y - x) - k_2 x = 0$$

or

$$c\dot{x} + (k_1 + k_2)x = c\dot{y} + k_1 y$$

Its transfer function is

$$\frac{X(s)}{Y(s)} = \frac{cs + k_1}{cs + k_1 + k_2}$$

Constant Inputs versus Step Inputs Note that a step function changes value at $t = 0$. Thus, it is not a constant. If, however, the input f in (4.5.1) is a constant F for $-\infty \le t \le \infty$, then $\dot{f} = 0$ for $-\infty \le t \le \infty$, and the existence of the input derivative in the model does not affect the response, because the model reduces to $m\dot{v} + cv = f(t)$. Thus, with $v(0) = 0$, $v_{ss} = F/c$ and $\tau = m/c$, we have

$$v(t) = v_{ss}(1 - e^{-t/\tau}) \tag{4.5.3}$$

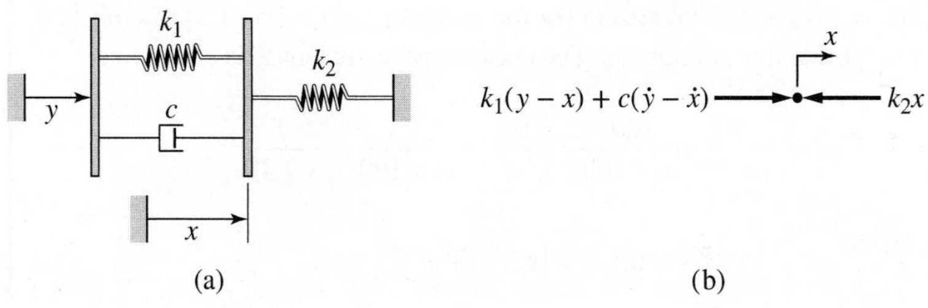

Figure 4.5.10 A device having numerator dynamics.

If, however, the input is a step function, then we cannot use (4.5.3). In this case, we can use the Laplace transform method to derive the response. Assuming that $f(t)$ is a step function of magnitude F, we obtain from (4.5.1)

$$msV(s) + cV(s) = b[sF(s) - f(0-)] + F(s) = b\left[s\frac{F}{s} - f(0-)\right] + \frac{F}{s}$$

If $f(t)$ is a step function, then $f(0-) = 0$ we obtain

$$V(s) = \frac{bF}{ms + c} + \frac{F}{s(ms + c)}$$

The inverse transform gives

$$v(t) = \frac{bF}{m}e^{-ct/m} + \frac{F}{c}(1 - e^{-ct/m}) \tag{4.5.4}$$

Comparing this with (4.5.3) we see that the effect of the term $b\dot{f}(t)$ is to increase the initial value of $v(t)$ by the amount bF/m.

Of course, no physical variable can be discontinuous, and therefore the step function is only an approximate description of an input that changes quickly. For example, the displacement $y(t)$ in Figure 4.5.10a must be continuous. With some models, a step input may produce a physically unrealistic discontinuity in the response. An example is given by (4.5.4), where $v(t)$ is discontinuous at $t = 0$. Taking the limit of $v(t)$ as $t \to 0+$, we find that $v(0+) = bF/m$.

The reason for using a step function is to reduce the complexity of the mathematics required to find the response. The following example illustrates this.

EXAMPLE 4.5.9

An Approximation to the Step Function

■ Problem

Consider the following model:

$$\dot{v} + 10v = \dot{f} + f \tag{1}$$

where $f(t)$ is the input and $v(0) = 0$.

a. Obtain the expression for the unit-step response.
b. Obtain the response to the input $f(t) = 1 - e^{-100t}$ and compare with the results of part (a).

■ Solution

a. Comparing equation (1) with (4.5.1), we see that $m = 1$, $c = 10$, and $b = 1$. Thus, from (4.5.3),

$$v(t) = 0.1 + 0.9e^{-10t} \tag{2}$$

Note that this equation gives $v(0+) = 1$. Thus, for this model, the effect of the step input acting on the input derivative $\dot{f}$ creates an instantaneous jump in the value of v at $t = 0$ from 0 to $v(0+) = 1$.

b. The function $f(t) = 1 - e^{-100t}$ is plotted in the top graph of Figure 4.5.11. It resembles a step function except that it is continuous. The corresponding response is obtained as follows:

$$V(s) = \frac{s+1}{s+10}\frac{100}{s(s+100)} = \frac{0.1}{s} - \frac{1.1}{s+100} + \frac{1}{s+10}$$

This gives

$$v(t) = 0.1 - 1.1e^{-100t} + e^{-10t} \tag{3}$$

This response gives $v(0+) = 0$ and is plotted along with the step response in the bottom graph of Figure 4.5.11. This response resembles the step response except for the latter's unrealistic discontinuity at $t = 0$. The response curves are very close for $t > 0.1$.

Although the response for the approximate step function was easily obtained in this example, this might not be true for higher-order models. In such cases, the approximation introduced by using the step function might be justified to obtain an expression for the response.

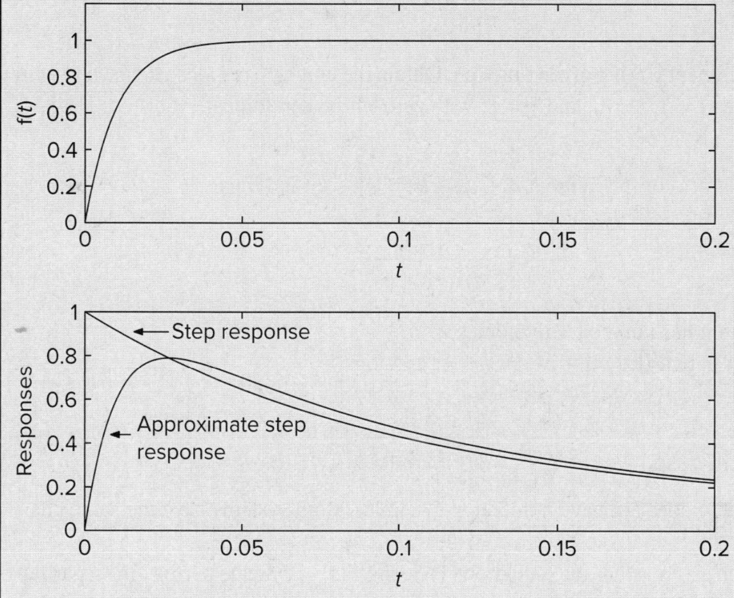

Figure 4.5.11 Response $v(t)$ to the approximate step function $f(t) = 1 - e^{-100t}$ and to a unit-step function.

There will be cases, such as with nonlinear models containing input derivatives, where the Laplace transform method cannot be used to find the response. In such cases we must use an approximate method or a numerical method. Input derivatives with step inputs, however, are difficult to handle numerically. In such cases the approximate step function $1 - e^{-t/\tau}$ and its derivative $e^{-t/\tau}/\tau$ can be used to obtain the response. For this approximate method to work, the value of τ must be chosen to be much smaller than the estimated response time of the system. An example of such an application is the system shown in Figure 4.5.10a if one or both spring elements are nonlinear. For example, if the force-displacement relation of the leftmost spring is a cubic function, then the equation of motion is

$$c(\dot{y} - \dot{x}) + k_{11}(y - x) + k_{12}(y - x)^3 - k_2 x = 0$$

which is nonlinear, and therefore not solvable with the Laplace transform method. This equation must be solved numerically, and so using a pure step function for $y(t)$ gives no advantage and may cause numerical difficulties. The use of MATLAB to solve such an equation is discussed in Section 5.3.

Damper Location and Numerator Dynamics

EXAMPLE 4.5.10

■ **Problem**

By obtaining the equations of motion and the transfer functions of the two systems shown in Figure 4.5.12, investigate the effect of the location of the damper on the step response of the

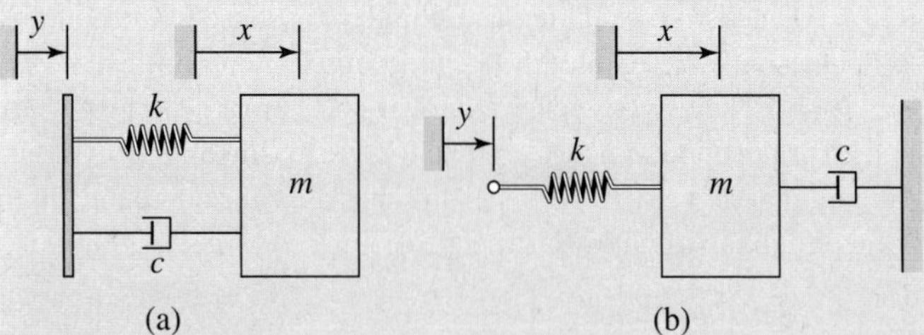

Figure 4.5.12 Effect of damper location on numerator dynamics.

system. The displacement $y(t)$ is a given function. Obtain the unit-step response for each system for the specific case $m = 1$, $c = 6$, and $k = 8$, with zero-initial conditions.

■ Solution

For the system in part (a) of the figure, $m\ddot{x} + c\dot{x} + kx = c\dot{y} + ky$ and thus,

$$\frac{X(s)}{Y(s)} = \frac{cs + k}{ms^2 + cs + k} \tag{1}$$

Therefore, this system has numerator dynamics.

For the system in part (b), $m\ddot{x} + c\dot{x} + kx = ky$ and thus,

$$\frac{X(s)}{Y(s)} = \frac{k}{ms^2 + cs + k} \tag{2}$$

Therefore, this system does not have numerator dynamics. Both systems have the same characteristic equation.

Substituting the given values into equation (1), using $Y(s) = 1/s$ and performing a partial-fraction expansion, we obtain

$$X(s) = \frac{6s + 8}{s(s^2 + 6s + 8)} = \frac{1}{s} + \frac{1}{s + 2} - \frac{2}{s + 4}$$

The response is

$$x(t) = 1 + e^{-2t} - 2e^{-4t}$$

For equation (2),

$$X(s) = \frac{8}{s(s^2 + 6s + 8)} = \frac{1}{s} - \frac{2}{s + 2} + \frac{1}{s + 4}$$

The response is

$$x(t) = 1 - 2e^{-2t} + e^{-4t}$$

The responses are shown in Figure 4.5.13. Curve (a) corresponds to equation (1), which has numerator dynamics. Curve (b) corresponds to the system without numerator dynamics, equation (2). The numerator dynamics causes an overshoot in the response but does not affect the steady-state response. Thus, the damper location can affect the response. But the damper location does not affect the time constants, because both systems have the same characteristic roots. Thus, they take about the same length of time to approach steady state.

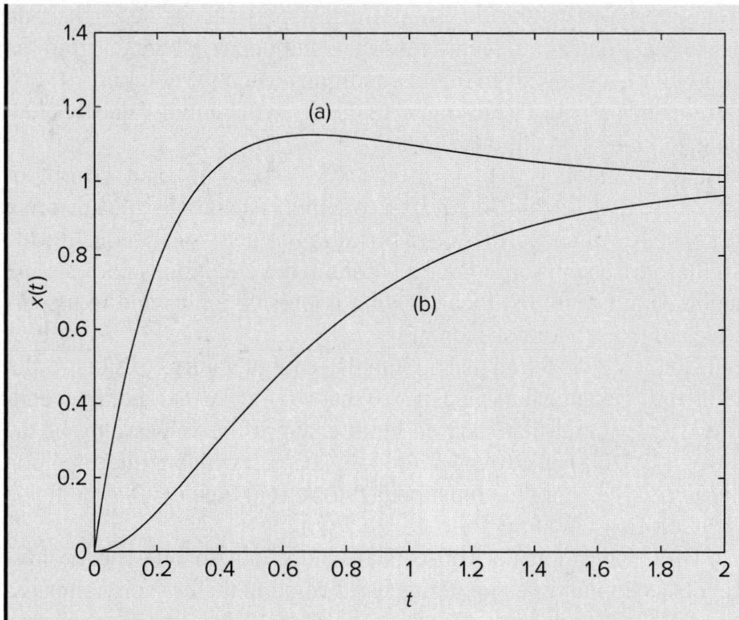

Figure 4.5.13 Effect of numerator dynamics on step response.

4.6 COLLISIONS AND IMPULSE RESPONSE

An input that changes at a constant rate is modeled by the ramp function. The step function models an input that rapidly reaches a constant value, while the rectangular pulse function models a constant input that is suddenly removed. The impulsive function—called an impulse—is similar to the pulse function, but it models an input that is suddenly applied and removed after a very short (infinitesimal) time.

The *strength* of an impulsive input is the area under its curve. The *Dirac delta function* $\delta(t)$ is an impulsive function with a strength equal to unity. Thus,

$$\int_{0-}^{0+} \delta(t)\, dt = 1$$

The Dirac function is an analytically convenient approximation of an input applied for only a very short time, such as the interaction force between two colliding objects. It is also useful for estimating the system's parameters experimentally and for analyzing the effect of differentiating a step or any other discontinuous input function.

The response to an impulsive input is called the *impulse response*. In particular, the response to $\delta(t)$ is called the *unit*-impulse response. Note that the system transfer function $T(s)$ is the Laplace transform of the unit-impulse response, because $\mathcal{L}[\delta(t)] = 1$. That is, if $X(s) = T(s)F(s)$ and if $f(t) = \delta(t)$, then $F(s) = 1$ and $X(s) = T(s)$. Therefore, if we can obtain the transform of the response $x(t)$ to an impulse of strength A, then we can determine the transfer function from $T(s) = X(s)/A$. This relation has some applications in determining the transfer function from the measured response.

4.6.1 INITIAL CONDITIONS AND IMPULSE RESPONSE

The given initial conditions $x(0-), \dot{x}(0-), \ldots,$ for the dependent variable $x(t)$, represent the situation at the start of the process and are the result of any inputs applied prior to $t = 0$. The effects of any inputs starting at $t = 0-$ are not felt by the system until an infinitesimal time later, at $t = 0+$. For some models, $x(t)$ and its derivatives do not change between $t = 0-$ and $t = 0+$, and thus the solution $x(t)$ obtained from the differential equation will match the given initial conditions when the solution $x(t)$ and its derivatives are evaluated at $t = 0$. The results obtained from the initial-value theorem will also match the given initial conditions.

However, for some other models, $x(0-) \neq x(0+)$, or $\dot{x}(0-) \neq \dot{x}(0+)$, and so forth for the higher derivatives of $x(t)$, depending on the type of input. The initial-value theorem gives the value at $t = 0+$, which for these responses is not necessarily equal to the value at $t = 0-$. In these cases the solution of the differential equation is correct only for $t > 0$. This phenomenon occurs in models having impulsive inputs and in models containing derivatives of a discontinuous input.

For example, in Chapter 2 we found that an impulsive input applied to a first-order system will instantaneously change the value of x so that $x(0-) \neq x(0+)$. For a second-order system, however, we found that such an input changes the values only of the derivative, so that $x(0-) = x(0+)$ but $\dot{x}(0-) \neq \dot{x}(0+)$. In keeping with our interpretation of the initial conditions, we consider the function $\delta(t)$ to start at time $t = 0-$ and finish at $t = 0+$, with its effects first felt at $t = 0+$.

Of course, a pure impulsive input does not exist in nature, and, for analytical convenience, we often model as impulsive an input that has a duration that is short compared to the response time of the system.

An example of a force often modeled as impulsive is the force generated when two objects collide. Recall that when Newton's law of motion $m\dot{v} = f(t)$ is integrated over time, we obtain the *impulse-momentum principle* for a system having constant mass:

$$mv(t) - mv(0-) = \int_{0-}^{t} f(u)\, du \tag{4.6.1}$$

This states that the change in momentum mv equals the time integral of the applied force $f(t)$. In the terminology of mechanics, the force integral—the area under the force-time curve—is called the *linear impulse*. The linear impulse is the strength of an impulsive force, but a force need not be impulsive to produce a linear impulse.

If $f(t)$ is an impulsive input of strength A, that is, if $f(t) = A\delta(t)$, then

$$mv(0+) - mv(0-) = \int_{0-}^{0+} A\delta(u)\, du = A \int_{0-}^{0+} \delta(u)\, du = A \tag{4.6.2}$$

since the area under the $\delta(t)$ curve is 1. So the change in momentum equals the strength of the impulsive force.

In practice, the strength is often all we can determine about an input modeled as impulsive. Sometimes we need not determine the input characteristics at all, as the next example illustrates.

EXAMPLE 4.6.1

Inelastic Collision

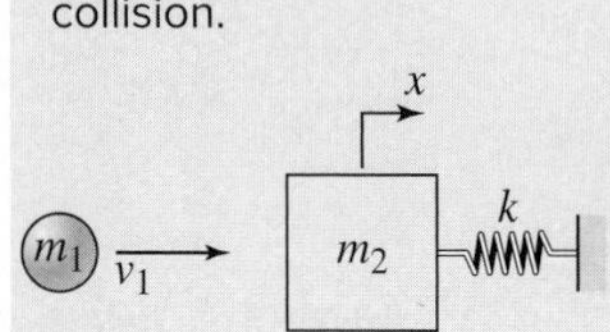

Figure 4.6.1 Inelastic collision.

■ Problem

Suppose a mass $m_1 = m$ moving with a speed v_1 becomes embedded in mass m_2 after striking it (Figure 4.6.1). Suppose $m_2 = 10m$. Determine the expression for the displacement $x(t)$ after the collision.

■ Solution

If we take the entire system to consist of both masses, then the force of collision is *internal* to the system. Because the displacement of m_2 immediately after the collision will be small, we may neglect the spring force initially. Thus, the *external* force $f(t)$ in (4.6.1) is zero, and we have

$$(m + 10m)v(0+) - \left[mv_1 + 10m(0-)\right] = 0$$

or

$$v(0+) = \frac{mv_1}{11m} = \frac{1}{11}v_1$$

The equation of motion for the combined mass is $11m\ddot{x} + kx = 0$. We can solve it for $t \geq 0+$ by using the initial conditions at $t = 0+$; namely, $x(0+) = 0$ and $\dot{x}(0+) = v(0+)$. The solution is

$$x(t) = \frac{v(0+)}{\omega_n}\sin \omega_n t = \frac{v_1}{11}\sqrt{\frac{11m}{k}}\sin\sqrt{\frac{k}{11m}}t$$

Note that it was unnecessary to determine the impulsive collision force. Note also that this force did not change x initially; it changed only $\dot{x}$.

Consider the two colliding masses shown in Figure 4.6.2. Part (a) shows the situation before collision, and part (b) shows the situation after collision. When the two masses are treated as a single system, no external force is applied to the system, and (4.6.1) shows that the momentum is conserved, so that

$$m_1v_1 + m_2v_2 = m_1v_3 + m_2v_4$$

or

$$m_1(v_1 - v_3) = -m_2(v_2 - v_4) \tag{4.6.3}$$

If the collision is *perfectly elastic*, kinetic energy is conserved, so that

$$\frac{1}{2}m_1v_1^2 + \frac{1}{2}m_2v_2^2 = \frac{1}{2}m_1v_3^2 + \frac{1}{2}m_2v_4^2$$

or

$$\frac{1}{2}m_1\left(v_1^2 - v_3^2\right) = -\frac{1}{2}m_2\left(v_2^2 - v_4^2\right) \tag{4.6.4}$$

Using the algebraic identities:

$$v_1^2 - v_3^2 = (v_1 - v_3)(v_1 + v_3)$$
$$v_2^2 - v_4^2 = (v_2 - v_4)(v_2 + v_4)$$

we can write (4.6.4) as

$$\frac{1}{2}m_1(v_1 - v_3)(v_1 + v_3) = -\frac{1}{2}m_2(v_2 - v_4)(v_2 + v_4) \tag{4.6.5}$$

Divide (4.6.5) by (4.6.3) to obtain

$$v_1 + v_3 = v_2 + v_4 \qquad \text{or} \qquad v_1 - v_2 = v_4 - v_3 \tag{4.6.6}$$

This relation says that in a perfectly elastic collision the *relative* velocity of the masses changes sign but its magnitude remains the same.

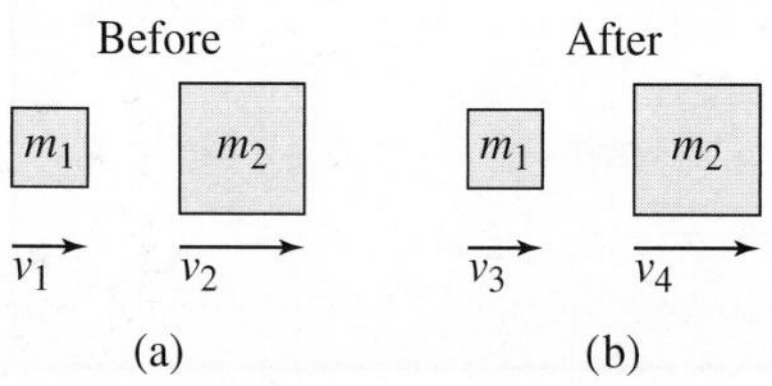

Before

m_1 m_2

v_1 v_2

(a)

After

m_1 m_2

v_3 v_4

(b)

Figure 4.6.2 Two colliding masses.

The most common application is where we know v_1 and mass m_2 is initially stationary, so that $v_2 = 0$. In this case, we can solve (4.6.3) and (4.6.6) for the velocities after collision as follows:

$$v_3 = \frac{m_1 - m_2}{m_1 + m_2} v_1 \qquad v_4 = \frac{2m_1}{m_1 + m_2} v_1 \tag{4.6.7}$$

EXAMPLE 4.6.2 Perfectly Elastic Collision

■ Problem

Consider again the system treated in Example 4.6.1, and shown again in Figure 4.6.3a. Suppose now that the mass $m_1 = m$ moving with a speed v_1 rebounds from the mass $m_2 = 10m$ after striking it. Assume that the collision is perfectly elastic. Determine the expression for the displacement $x(t)$ after the collision.

■ Solution

For a perfectly elastic collision, the velocity v_3 of the mass m after the collision is given by (4.6.7).

$$v_3 = \frac{m_1 - m_2}{m_1 + m_2} v_1 = \frac{m - 10m}{m + 10m} v_1 = -\frac{9}{11} v_1$$

Thus, the change in the momentum of m is

$$m\left(-\frac{9}{11} v_1\right) - mv_1 = \int_{0-}^{0+} f(t)\, dt$$

Thus, the linear impulse applied to the mass m during the collision is

$$\int_{0-}^{0+} f(t)\, dt = -\frac{20}{11} mv_1$$

From Newton's law of action and reaction (Figure 4.6.3b), we see that the linear impulse applied to the $10m$ mass is $+20mv_1/11$, and thus its equation of motion is

$$10m\ddot{x} + kx = \frac{20}{11} mv_1 \delta(t)$$

We can solve this equation for $t > 0$ using the initial conditions $x(0-) = 0$ and $\dot{x}(0-) = 0$. The Laplace transform method gives

$$\left(10ms^2 + k\right) X(s) = \frac{20}{11} mv_1$$

$$X(s) = \frac{20mv_1/11}{10ms^2 + k} = \frac{2v_1}{11} \frac{1}{s^2 + k/10m}$$

Thus

$$x(t) = \frac{2v_1}{11} \sqrt{\frac{10m}{k}} \sin \sqrt{\frac{k}{10m}} t$$

This gives $\dot{x}(0+) = 2v_1/11$, which is identical to the solution for v_4 from (4.6.7), as it should be.

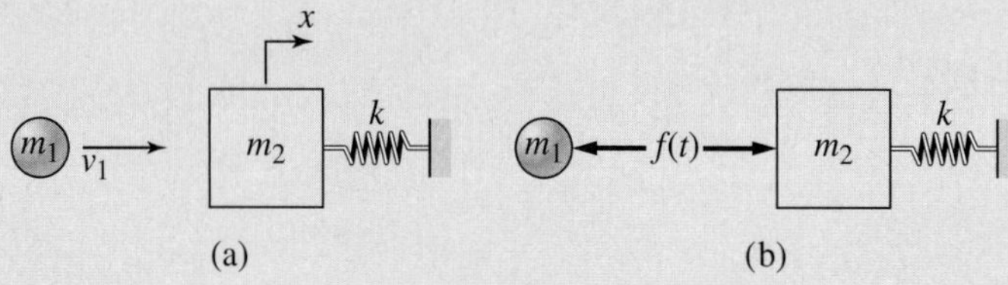

Figure 4.6.3 Perfectly elastic collision.

Table 4.7.1 MATLAB functions used in this chapter.

	Functions Introduced in This Chapter
`floor`	Rounds to the nearest integer toward $-\infty$.
`max`	Computes the algebraically largest element.
`real`	Computes the real part of a complex number.
	Functions Introduced in Earlier Chapters
`conv`	Computes the product of two polynomials (Section 1.6).
`exp`	Computes the exponential function (Section 1.6).
`plot`	Creates a two-dimensional plot on rectilinear axes (Section 1.6).
`residue`	Computes the coefficients of a partial fraction expansion (Section 2.8).
`roots`	Computes the roots of a polynomial (Section 1.6).
`step`	Computes and plots the step response of a linear model (Section 2.9).
`tf`	Creates a model in transfer function form (Section 2.9).
`xlabel`	Puts a label on the abscissa of a plot (Section 1.6).
`ylabel`	Puts a label on the ordinate of a plot (Section 1.6).

4.7 MATLAB APPLICATIONS

We have seen how the `residue` function can be used to obtain the coefficients of a partial-fraction expansion. In addition, with the `conv` function, which multiplies polynomials, you can use MATLAB to perform some of the algebra to obtain a closed-form expression for the response. Table 4.7.1 lists the MATLAB functions used in this section.

MATLAB has the `step` and `impulse` functions to compute the step and impulse responses from the transfer functions. However, MATLAB does not have a function for computing the free response from a transfer-function model. We now show how to use MATLAB to obtain the free response.

Obtaining the Free Response with the `step` Function

EXAMPLE 4.7.1

■ Problem

Use the MATLAB `step` function to obtain a plot of the free response of the following model, where $x(0) = 4$ and $\dot{x}(0) = 2$.

$$5\ddot{x} + 3\dot{x} + 10x = 0$$

■ Solution

Applying the Laplace transform method gives

$$(5s^2 + 3s + 10)X(s) = 20s + 22$$

or

$$X(s) = \frac{20s + 22}{5s^2 + 3s + 10}$$

If we multiply the numerator and denominator by s, we obtain

$$X(s) = \frac{20s^2 + 22s}{5s^2 + 3s + 10}\frac{1}{s} = T(s)\frac{1}{s}$$

Thus, we may compute the free response by using the `step` function with the transfer function

$$T(s) = \frac{20s^2 + 22s}{5s^2 + 3s + 10}$$

The MATLAB program is

```
sys = tf([20, 22, 0],[5, 3, 10]);
step(sys)
```

EXAMPLE 4.7.2 Determining the Free Response with MATLAB

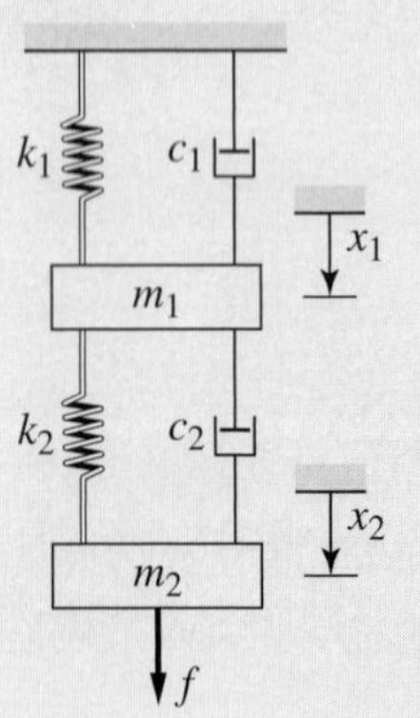

Figure 4.7.1 A two-mass system.

■ Problem

For the system shown in Figure 4.7.1, suppose that $m_1 = m_2 = 1$, $c_1 = 2$, $c_2 = 3$, $k_1 = 1$, and $k_2 = 4$. (a) Obtain the plot of the unit-step response of x_2 for zero-initial conditions. (b) Use two methods with MATLAB to obtain the free response for $x_1(t)$, for the initial conditions $x_1(0) = 3$, $\dot{x}_1(0) = 2$, $x_2(0) = 1$, and $\dot{x}_2(0) = 4$.

■ Solution

The equations of motion are given by equations (1) and (2) of Example 4.4.5. With the given parameter values, they become

$$\ddot{x}_1 + 5\dot{x}_1 + 5x_1 - 3\dot{x}_2 - 4x_2 = 0$$

$$\ddot{x}_2 + 3\dot{x}_2 + 4x_2 - 3\dot{x}_1 - 4x_1 = f(t)$$

Because we need to find the free response, we must now keep the initial conditions in the analysis. So, transforming the equations with the given initial conditions gives

$$s^2X_1(s) - sx_1(0) - \dot{x}_1(0) + 5[sX_1(s) - x_1(0)] + 5X_1(s) - 3[sX_2(s) - x_2(0)] - 4X_2(s) = 0$$

$$s^2X_2(s) - sx_2(0) - \dot{x}_2(0) + 3[sX_2(s) - x_2(0)] + 4X_2(s) - 3[sX_1(s) - x_1(0)] - 4X_1(s) = F(s)$$

Collecting terms results in

$$(s^2 + 5s + 5)X_1(s) - (3s + 4)X_2(s) = x_1(0)s + \dot{x}_1(0) + 5x_1(0) - 3x_2(0) = I_1(s) \quad (1)$$

$$-(3s + 4)X_1(s) + (s^2 + 3s + 4)X_2(s) = x_2(0)s + \dot{x}_2(0) + 3x_2(0) - 3x_1(0) = I_2(s) + F(s) \quad (2)$$

where the terms due to the initial conditions are denoted by $I_1(s)$ and $I_2(s)$.

$$I_1(s) = x_1(0)s + \dot{x}_1(0) + 5x_1(0) - 3x_2(0)$$

$$I_2(s) = x_2(0)s + \dot{x}_2(0) + 3x_2(0) - 3x_1(0)$$

Equations (1) and (2) are two algebraic equations in two unknowns. Their solution can be obtained in several ways, but the most general method for our purposes is to solve them using determinants (this is known as *Cramer's method*).

The determinant of the left-hand side of the equations (1) and (2) is

$$D(s) = \begin{vmatrix} (s^2 + 5s + 5) & -(3s + 4) \\ -(3s + 4) & (s^2 + 3s + 4) \end{vmatrix}$$

$$= (s^2 + 5s + 5)(s^2 + 3s + 4) - (3s + 4)^2$$

The solutions for $X_1(s)$ and $X_2(s)$ can be expressed as

$$X_1(s) = \frac{D_1(s)}{D(s)} \qquad X_2(s) = \frac{D_2(s)}{D(s)}$$

where

$$D_1(s) = \begin{vmatrix} I_1(s) & -(3s+4) \\ I_2(s) + F(s) & (s^2+3s+4) \end{vmatrix}$$

$$= (s^2+3s+4)I_1(s) + (3s+4)I_2(s) + (3s+4)F(s)$$

$$D_2(s) = \begin{vmatrix} (s^2+5s+5) & I_1(s) \\ -(3s+4) & I_2(s) + F(s) \end{vmatrix}$$

$$= (s^2+5s+5)I_2(s) + (3s+4)I_1(s) + (s^2+5s+5)F(s)$$

a. The transfer functions for the input $f(t)$ are found by setting $I_1(s) = I_2(s) = 0$ in $D_1(s)$ and $D_2(s)$:

$$\frac{X_1(s)}{F(s)} = \frac{D_1(s)}{D(s)} = \frac{3s+4}{D(s)} \qquad \frac{X_2(s)}{F(s)} = \frac{D_2(s)}{D(s)} = \frac{s^2+5s+5}{D(s)}$$

The unit-step response for x_2 is plotted with the following MATLAB program.

```
D = conv([1,5,5], [1,3,4])-conv([0,3,4], [0,3,4]);
x2 = tf([1,5,5],D);
step(x2)
```

b. Using the technique introduced in Example 4.7.1, we multiply the numerator of $X_1(s)$ by s. The following MATLAB program performs the calculations. Note that the multiplication by s is accomplished by appending a 0 to the polynomials used to compute `D1` (with $F(s) = 0$).

```
% Specify the initial conditions.
x10 = 3; x1d0 = 2; x20 = 1; x2d0 = 4;
% Form the initial-condition arrays.
I1 = [x10,x1d0+5*x10-3*x20]; I2 = [x20,x2d0+3*x20-3*x10];
% Form the determinant D1.
D1 = conv([1,3,4,0],I1)+conv([0,3,4,0],I2);
% Form the determinant D.
D = conv([1,5,5],[1,3,4])-conv([0,3,4],[0,3,4]);
sys = tf(D1,D);
step(sys)
```

The first method gives the response plot but not the closed-form solution. The following method gives the closed-form solution for the free response.

The characteristic roots can be found with the following MATLAB file.

```
% Find the roots.
D = conv([1,5,5],[1,3,4])-conv([0,3,4],[0,3,4]);
R = roots(D)
```

The roots are -5.6773, -1.3775, and $-0.4726 \pm 0.5368j$.

Note that if all the characteristic roots are distinct, the partial fraction expansion gives the following solution form, even if the roots are complex.

$$x_1(t) = \sum_{i=1}^{4} r_i e^{p_i t}$$

where the r_i are the residues of the partial-fraction expansion and the p_i are the poles (the roots). So if the roots are all distinct, we can use the following MATLAB file to perform the rest of the algebra.

```
% Specify the initial conditions.
x10 = 3; x1d0 = 2; x20 = 1; x2d0 = 4;
% Form the initial-condition vectors.
I1 = [x10,x1d0+5*x10-3*x20]; I2 = [x20,x2d0+3*x20-3*x10];
% Form the determinant D1.
D1 = conv([1,3,4],I1)+conv([0,3,4],I2);
% Compute the partial fraction expansion.
[r,p,K] = residue(D1,D);
% Use 5 time constants to estimate the simulation time.
tmax = floor(-5/max(real(p)));
t = (0:tmax/500:tmax);
% Evaluate the time functions.
x1 = real(r(1)*exp(p(1)*t)+r(2)*exp(p(2)*t)+...
     r(3)*exp(p(3)*t)+r(4)*exp(p(4)*t));
plot(t,x1),xlabel('t'),ylabel('x_1')
```

Although in the expression for `x1` the imaginary parts should cancel, giving a real result, we use the `real` function because numerical round-off errors can produce a small imaginary part. The plot is shown in Figure 4.7.2. We can use a similar approach to plot $x_2(t)$.

If a closed-form expression for the free response is required, you can use the roots (the poles) in the array `p` and the expansion coefficients (the residues) in the array `r`. If the roots are complex, you can use the Euler identities $e^{jx} = \cos x \pm j \sin x$ with the expansion coefficients to obtain the solution, as shown in Section 2.8 in Chapter 2. Here the residues corresponding in order to the four roots given already are -0.3554, 3.7527, and $-0.1987 \pm 4.6238j$.

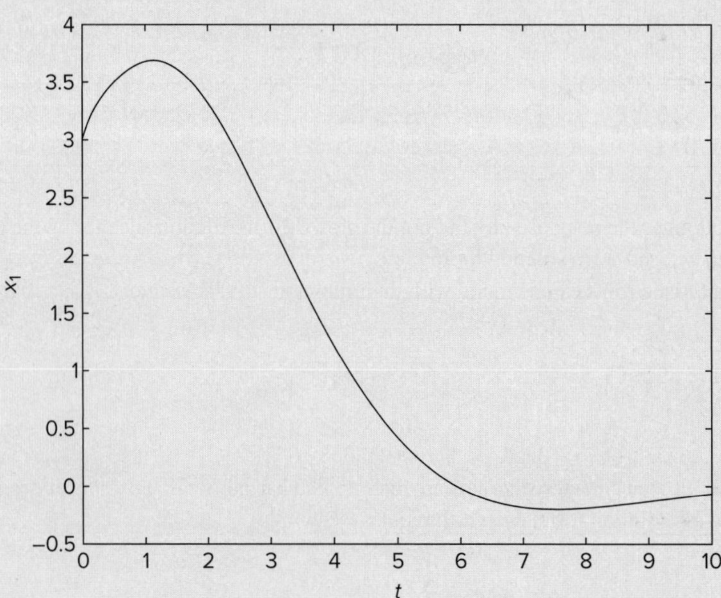

Figure 4.7.2 Plot of the free response for Example 4.7.2.

4.8 CASE STUDY: VEHICLE SUSPENSION DESIGN

A common source of vibration in machines is due to motion of the supporting structure; this is called *base excitation*. A common example of base excitation is caused by vehicle motion along a bumpy road surface. This motion produces a displacement input to the suspension system via the wheels. The primary purpose of a vehicle suspension is to maintain tire contact with the road surface, and the secondary purpose is to minimize the motion and force transmitted to the passenger compartment.

The suspension is an example of a vibration isolation system designed to isolate the source of vibration from other parts of the machine or structure. In addition to metal springs and hydraulic dampers, vibration isolators consisting of highly damped materials like rubber provide stiffness and damping between the source of vibration and the object to be protected. For example, the rubber motor mounts of an automobile engine are used to isolate the automobiles frame from the effects of possible motor unbalance. Cork, felt, and pneumatic springs are also used as isolators.

4.8.1 VEHICLE SUSPENSION MODELS

The quarter-car model of a vehicle suspension is shown in Figure 4.8.1a. In this simplified model, the masses of the wheel, tire, and axle are neglected, and the mass m represents one-fourth of the vehicle mass. The motion of the mass m is produced by the motion $y(t)$ of the base, in this case, the road surface. When we want to reduce the displacement of the mass m caused by the input displacement $y(t)$, we speak of *displacement isolation*. When we want to reduce the force transmitted to the mass m, we speak of *force isolation*.

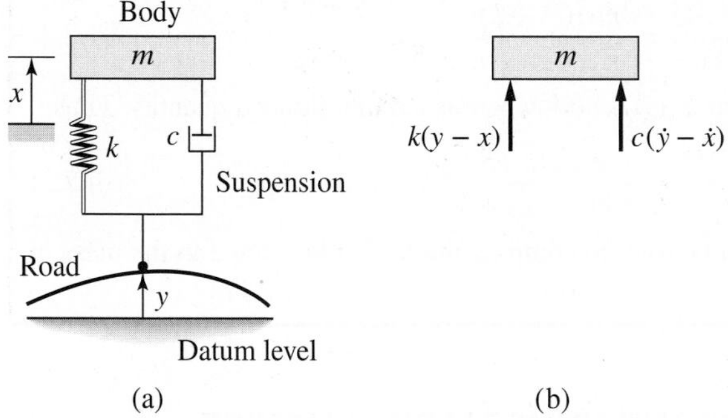

Figure 4.8.1 A quarter-car model with a single mass.

EXAMPLE 4.8.1

The Quarter-Car Model

■ Problem

The model shown in Figure 4.8.1a is a model of many common displacement isolation systems. The spring constant k models the combined elasticity of both the tire and the suspension spring, although usually the tire is much stiffer than the spring and thus k is approximately equal to the spring stiffness. (Example 4.8.2 deals with the tire stiffness.)

The damping constant c models the shock absorber. Assuming that the mass displacement x is measured from the equilibrium position of the mass when $y = 0$, it is important to note that the weight mg is canceled by the static spring force.

The road surface displacement $y(t)$ can be derived from the road surface profile and the vehicle speed. Derive the equation of motion of m with $y(t)$ as the input, and obtain the transfer functions for displacement and force isolation.

■ **Solution**

Figure 4.8.1b shows the free body diagram, which is drawn assuming that $\dot{y} > \dot{x}$ and that $y > x$. Only the dynamic spring force is shown because the static spring force is canceled by the gravity force. From this free body diagram we obtain the equation of motion:

$$m\ddot{x} = c(\dot{y} - \dot{x}) + k(y - x) \quad \text{or} \quad m\ddot{x} + c\dot{x} + kx = c\dot{y} + ky$$

The transfer function $X(s)/Y(s)$ is found by applying the Laplace transform method to the equation, with the initial conditions set to zero.

$$\frac{X(s)}{Y(s)} = \frac{cs + k}{ms^2 + cs + k} \tag{4.8.1}$$

This transfer function is the *displacement transmissibility*.

The force transmitted to the mass by the spring and damper is denoted f_t and is given by

$$f_t = c(\dot{y} - \dot{x}) + k(y - x)$$

The transform is

$$F_t(s) = (cs + k)[Y(s) - X(s)]$$

Substituting for $X(s)$ from (4.8.1) gives

$$F_t(s) = (cs + k)[Y(s) - \frac{cs + k}{ms^2 + cs + k}Y(s)] = (cs + k)\frac{ms^2}{ms^2 + cs + k}Y(s)$$

Thus,

$$\frac{F_t(s)}{Y(s)} = \frac{ms^2(cs + k)}{ms^2 + cs + k} \tag{4.8.2}$$

It is customary to use instead the ratio $F_t(s)/kY(s)$, which is a dimensionless quantity. Thus,

$$\frac{F_t(s)}{kY(s)} = \frac{1}{k}\frac{ms^2(cs + k)}{ms^2 + cs + k} \tag{4.8.3}$$

This the *force transmissibilty*. It can be used to compute the force transmitted to the mass m as a result of the base motion.

4.8.2 TRANSMITTED FORCE AND NUMERATOR DYNAMICS

Note that the order of the numerator of (4.8.3) is greater than the order of the denominator, and thus this transfer function is said to be *improper*. If you attempt to use it with the Laplace transform method, say to find the step response, you will obtain an impulse in the response (a Dirac delta function) because the numerator order is one greater than the denominator order. You cannot use an improper transfer function in MATLAB; if you use the `tf` function, you will be warned that you cannot simulate the time response of improper models in MATLAB.

There are ways to compute the transmitted force in spite of these mathematical difficulties. The solution lies in understanding that numerator dynamics implies that the input is being differentiated, so we can either avoid using a suddenly applied input, like a step or an impulse, or we can replace the derivative with an approximate process or model the step input with an approximate function (see subsection 4.5.1). One way of doing this is to use the ODE solvers or Simulink, both of which are covered in Chapter 5. Another way is explored in one of the problems at the end of this chapter.

Another approach is to realize that a suddenly applied input to a damper is an oversimplification that leads to these mathematical difficulties. A better approach is to apply the input directly to the tire mass. This is the approach of the next example.

A Quarter-Car Model with Two-Masses

EXAMPLE 4.8.2

■ Problem

The suspension model shown in Figure 4.8.2 includes the mass of the wheel-tire-axle assembly. The mass m_1 is one-fourth the mass of the car body, and m_2 is the mass of the wheel-tire-axle assembly. The spring constant k_1 represents the suspension's elasticity, and k_2 represents the tire's elasticity. Derive the equations of motion for m_1 and m_2 in terms of the displacements from equilibrium, x_1 and x_2.

■ Solution

Assuming that $x_2 > x_1$, $\dot{x}_2 > \dot{x}_1$, and $y > x_2$, we obtain the free body diagram shown. The equation of motion for mass m_1 is

$$m_1\ddot{x}_1 = c_1(\dot{x}_2 - \dot{x}_1) + k_1(x_2 - x_1)$$

For mass m_2,

$$m_2\ddot{x}_2 = -c_1(\dot{x}_2 - \dot{x}_1) - k_1(x_2 - x_1) + k_2(y - x_2)$$

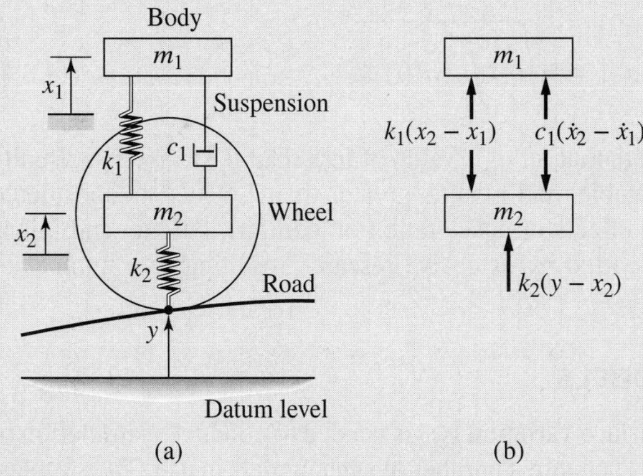

Figure 4.8.2 A quarter-car model with two masses.

4.8.3 RIDE QUALITY AND SUSPENSION DESIGN

The primary purpose of a vehicle suspension is to keep the tires in contact with the road. The secondary purpose is to give the passengers a comfortable ride. Proper design of the suspension provides the primary source of isolation from the road roughness, and the quarter-car model is the starting point for the suspension system design.

We note that the automotive industry often uses the term *spring rate* instead of *spring constant* or *spring stiffness k*. Usually springs are mounted on some pivoting suspension member such as a control arm or swing arm, and the term *wheel rate* is used to distinguish from the spring rate alone. The wheel rate is the effective spring rate when measured at the wheel. Because of the kinematics of the suspension member, the wheel rate is usually less than or equal to the spring rate.

Referring to Figure 4.8.2a, the *ride rate* k_e is the effective stiffness of the suspension spring and the tire stiffness, neglecting the wheel mass m_2. Thus, $k_e = k_1k_2/(k_1 + k_2)$. Note that if the tire is very stiff, so that $k_2 \gg k_1$, then $k_e \approx k_1$.

If $m_1 \gg m_2$, the undamped natural frequency of the suspension system is approximately $\omega_n = \sqrt{k_e/m_1}$ and its damping ratio is $\zeta = c_1/2\sqrt{k_e m_1}$. Most modern cars have a suspension damping ratio ζ between 0.2 and 0.4, and thus their damped natural frequency ω_d is very close to ω_n.

The suspension static deflection due to the weight $m_1 g$ is $\Delta = m_1 g/k_e = g/\omega_n^2$. A natural frequency of 1 Hz ($\omega_n = 2\pi$ rad/sec) is considered to be a design optimum for highway vehicles, and this corresponds to a static deflection of $\Delta = 9.8$ in. or 0.248 m. For a vehicle weighing 3200 lb, this deflection requires an effective stiffness of $k_e = m_1 g/\Delta = 0.25(3200)/9.8 = 81.6$ lb/in., or 14,349 N/m. With a tire stiffness of $k_2 = 1200$ lb/in., the supension stiffness must be $k_1 = k_e k_2/(k_2 - k_e) = 87.6$ lb/in., or 15,341 N/m [Gillespie, 1992].

In addition to specifying damping ratio and oscillation frequency, other performance measures are used to assess ride quality. Some of these are: the maximum displacement of the chassis (and thus the passenger compartment), the maximum force and maximum jerk felt, and the root-mean-square (rms) acceleration of the chassis. You may recall that in Chapters 1 and 3 we applied the root-mean-square measure to assess the average torque required for a motion control system. We can apply the same concept to assess to average acceleration. The rms acceleration a_{rms} over a time T is given by

$$a_{\text{rms}} = \sqrt{\frac{1}{T}\int_0^T a^2(t)\,dt} \tag{4.8.4}$$

In this brief introduction we note that an a_{rms} value of less than 0.315 m/s^2 is usually considered to be not uncomfortable, and a value greater than 1.5 m/s^2 is considered to be very uncomfortable. The effects of acceleration on comfort, fatigue, and injury is a very complex subject supported by extensive research resulting in a number of standards.

4.8.4 ROAD SURFACE MODELS

A suitable model of the road surface variation $y(t)$ is needed to conduct a simulation of a proposed suspension system. There are a number of such models in use. The simplest is the step function, which eases the transient analysis but presents the most severe test of the suspension. The *half-sine function* is sometimes used to model a single bump. If the vehicle speed is v, the bump height is H, and the bump length is L, then the time to cross the bump is $T = L/v$. The half-sine bump shape is $y(z) = H\sin(2\pi z/L)$, where z is the horizontal distance. Thus $z = vt$ and $y(t) = H\sin(2\pi vt/L)$.

The half-sine model is somewhat severe because its slope changes abruptly from 0 to $2\pi H/T$ at $t = 0$, and thus it is a reasonable model for a speed bump. A less severe bump model is the *haversine function* $y(z) = H\sin^2(\pi z/L)$. Thus $y(t) = H\sin^2(\pi vt/L)$. Its slope is 0 at the start and end of the bump, and thus it may be a more reasonable model of a longer road surface variation. The two models are compared in Figure 4.8.3.

The trazepoid is a reasonable model of a longer road surface elevation. The sine function is used to model periodic variations. More advanced simulations sometimes use randomly generated variations based on a statistical analysis of typical road construction techniques.

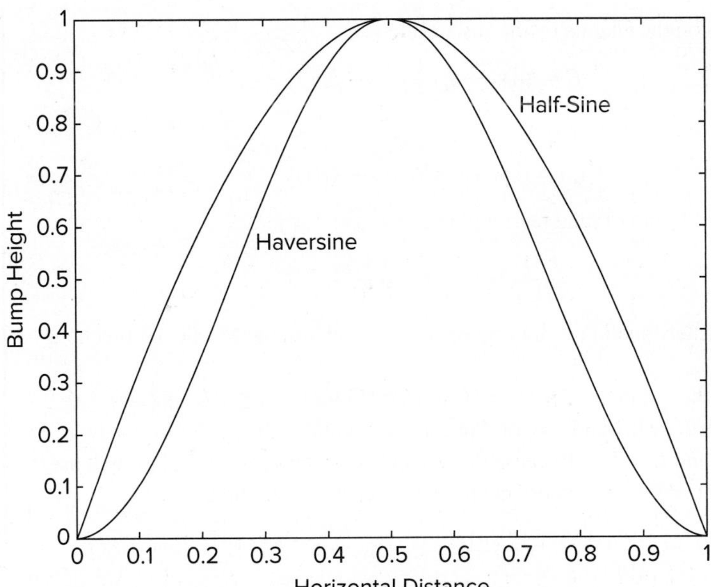

Figure 4.8.3 Comparison of the half-sine and haversine bump models.

Design Example

EXAMPLE 4.8.3

■ Problem

(a) Design a suspension for a vehicle whose quarter mass is 250 kg and whose tire stiffness is 10^5 N/m. The tire and swing arm have a combined mass of 25 kg. (b) Evaluate the design performance when striking a haversine bump 0.1 m high while moving at 18 m/s. Consider two bump lengths: $L = 0.5$ m and $L = 2$ m.

■ Solution

a. The suspension static deflection due to the weight $m_1 g$ is $\Delta = m_1 g/k_e = g/\omega_n^2$. Choosing the natural frequency to be 1 Hz ($\omega_n = 2\pi$ rad/sec), the corresponding static deflection is $\Delta = 0.248$ m. For a quarter-car mass of 250 kg, this deflection requires an effective stiffness of $k_e = m_1 g/\Delta = (250)(9.81)/0.248 = 9889$ N/m. With a tire stiffness of $k_2 = 10^5$ N/m, the supension stiffness must be $k_1 = k_e k_2/(k_2 - k_e) = 10{,}974$ N/m.

   Typical values of the damping ratio ζ for highway vehicles range from $\zeta = 0.2$ to 0.4, so we choose the middle value, $\zeta = 0.3$. This gives $0.3 = c_1/2\sqrt{k_1 m_1} = c_1/2\sqrt{(9889)(250)}$, or $c_1 = 943$ N·s/m.

b. We choose to simulate the design using the transfer functions of the model developed in Example 4.8.2. Transforming the model equations with zero-intial conditions gives

$$(m_1 s^2 + c_1 s + k_1)X_1(s) - (c_1 s + k_1)X_2(s) = 0$$

$$-(c_1 s + k_1)X_1(s) + (m_2 s^2 + c_1 s + k_1)X_2(s) = k_2 Y(s)$$

Let $D(s)$ be the determinant of these equations.

$$D(s) = m_1 m_2 s^4 + (m_1 c_1 + m_2 c_1)s^3 + (m_1 k_1 + m_1 k_2 + m_2 k_1)s^2 + c_1 k_2 s + k_1 k_2 \quad (1)$$

The transfer functions are

$$\frac{X_1(s)}{Y(s)} = \frac{k_2(c_1 s + k_1)}{D(s)} \quad (2)$$

$$\frac{X_2(s)}{Y(s)} = \frac{k_2(m_1 s^2 + c_1 s + k_1)}{D(s)} \quad (3)$$

The force exerted on the chassis by the suspension is

$$f_t = k_1(x_2 - x_1) + c(\dot{x}_2 - \dot{x}_1)$$

Thus,

$$F_t(s) = (c_1 s + k_1)\left[X_2(s) - X_1(s)\right]$$

Substituting from (2) and (3) and rearranging gives

$$\frac{F_t(s)}{Y(s)} = m_1 k_2 s^2 \frac{c_1 s + k_1}{D(s)} \quad (4)$$

Note that this transfer function is not improper, so we can use the MATLAB functions like `tf` and `lsim`.

The input is the bump model $y(t) = H\sin^2(\pi vt/L)$ for $0 \le t \le L/v$ and $y(t) = 0$ for $t > L/v$, where $H = 0.1$ and $v = 18$. We can now simulate the model for the two cases: $L = 0.5$ m and $L = 2$ m. Because the input changes form over time, we will use the `lsim` function. Here is the code fragment that generates the input:

```
T=L/v;
t=linspace(0,8*T,1000);
y=H*(sin(pi*v*t/L)).^2;
for k=1:length(t)
    if t(k) > T
    y(k)=0;
end
```

Figure 4.8.4 shows simulation results for $L = 0.5$ m. The chassis and tire displacements are shown along with the input $y(t)$. Figure 4.8.5 shows simulation results for $L = 2$ m. These show that for a given speed, the bump length has a major effect on the results. Figure 4.8.6 shows the force transmitted to the chassis for both cases. Note that the parameter v/L governs the simulation. So doubling the speed v has the same effect as halving the bump length L.

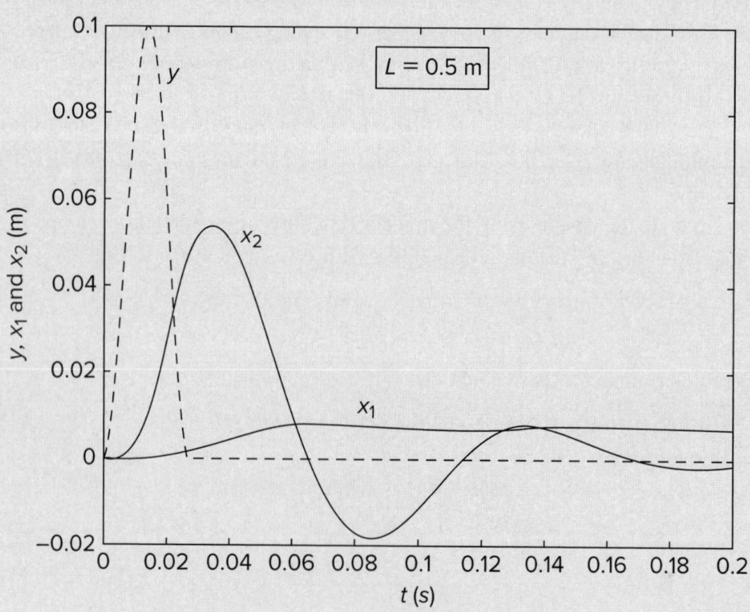

Figure 4.8.4 Simulation results for Example 4.8.3 for $L = 0.5$.

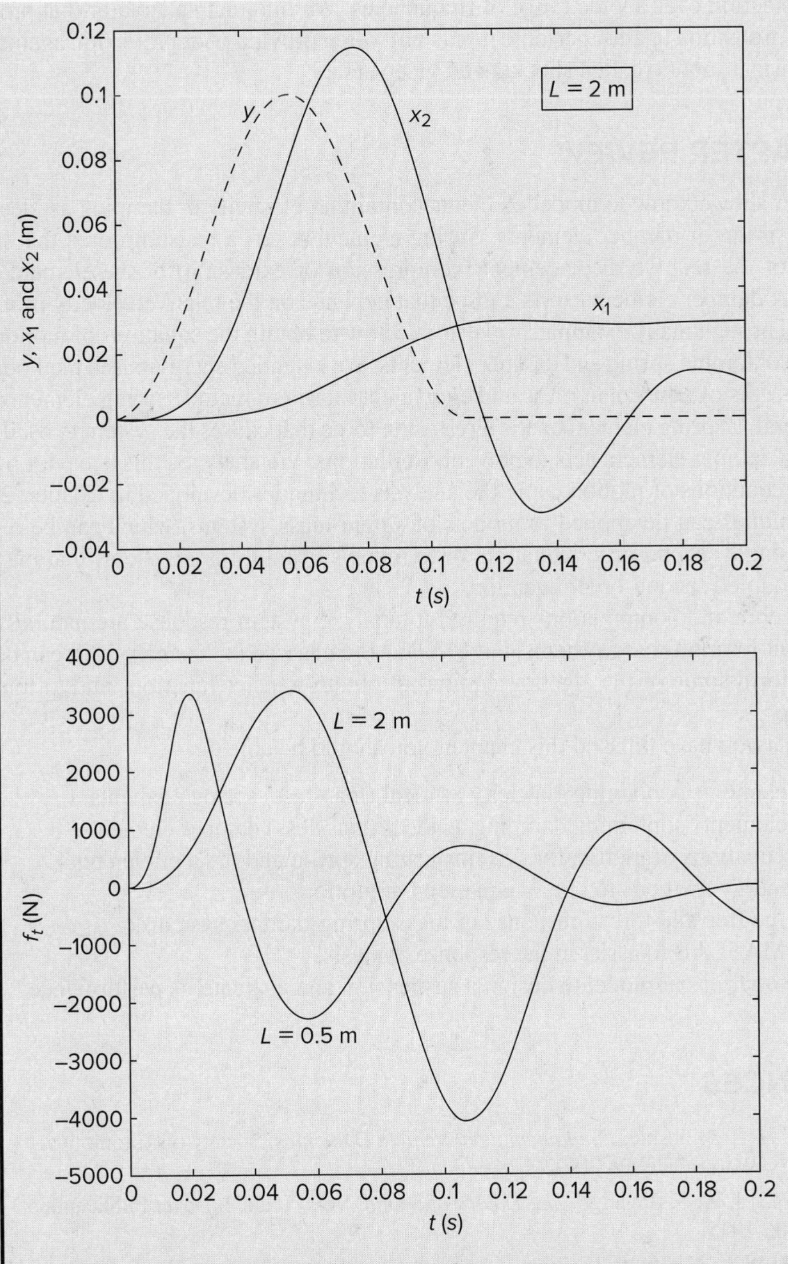

Figure 4.8.5 Simulation results for Example 4.8.3 for $L = 2$.

Figure 4.8.6 Force transmitted to the chassis for Example 4.8.3.

When interpreting these graphs, it is important to remember that the displacements are measured from the respective equilibrium positions of the chassis and tire. So for example, Figure 4.8.5 shows that $x_2 > y$ at some time, but this does not mean that the tire leaves the road. For example, when $L = 2$ m, the maximum of the difference $x_2 - y$ occurs at $t = 0.09$ and is 0.0636 m. This is the stretch in the tire "spring" k_2.

Applications where the input is not random or periodic must be handled with methods based on transient response analysis. Sometimes the transient requirements conflict with the steady-state requirements. For example, with base excitation where the base motion is a step function (a suddenly applied constant displacement), the isolator must provide protection over a wide range of frequencies. We find that an isolator that provides good protection against periodic inputs will often provide poor protection against sudden transient inputs (called shocks) and vice versa.

4.9 CHAPTER REVIEW

This chapter showed how to model elements containing elasticity or damping as ideal (massless) spring or damper elements. Spring elements exert a resisting force that is a function of the relative displacement (compression or extension) of the element's endpoints. A damper element exerts a force that depends on the relative velocity of its endpoints. Through many examples we also saw how to obtain the equations of motion of systems containing spring and damper elements. We extended energy-based methods and the concepts of equivalent mass and equivalent inertia to include spring elements.

In general, a spring element exerts a restoring force that causes the system to oscillate, while a damper element acts to prevent oscillations. We analyzed this behavior by solving the equations of motion using the analysis techniques developed in Chapter 2. These techniques can be applied to models of single-mass systems, which can be reduced to a single second-order equation, or to models of multimass systems that may consist of coupled second-order equations.

The algebra and computations required to analyze system response are naturally more difficult for multimass systems, but MATLAB can assist in the analysis. It can be used to perform some of the algebra required to obtain transfer functions and to find the response.

Now that you have finished this chapter, you should be able to

1. Model elements containing elasticity as ideal (massless) spring elements.
2. Model elements containing damping as ideal (massless) damper elements.
3. Obtain equations of motion for systems having spring and damper elements.
4. Apply energy methods to obtain equations of motion.
5. Obtain the free and forced response of mass-spring-damper systems.
6. Utilize MATLAB to assist in the response analysis.
7. Use the quarter-car model to design a suspension and evaluate its performance.

REFERENCES

[Gillespie, 1992] T. Gillespie, *Fundamentals of Vehicle Dynamics*, Society of Automotive Engineers, Warrendale, PA, 1992.

[Rayleigh, 1945] J. W. S. Rayleigh, *The Theory of Sound*, Vols. 1 and 2, Dover Publications, New York, 1945.

[Young, 2011] W. C. Young, R. G. Budynas, and A. Sadegh, *Roark's Formulas for Stress and Strain*, 8th ed., McGraw-Hill, New York, 2011.

PROBLEMS

Section 4.1 Spring Elements

4.1 Compute the translational spring constant of a particular steel helical coil spring, of the type used in automotive suspensions. The coil has eight turns. The coil diameter is 5 in., and the wire diameter is 0.6 in. For the shear modulus, use $G = 1.7 \times 10^9$ lb/ft^2.

4.2 In the spring arrangement shown in Figure P4.2, the displacement x is caused by the applied force f. Assuming the system is in static equilibrium, sketch the plot of f versus x. Determine the equivalent spring constant k_e for this arrangement, where $f = k_e x$.

4.3 In the arrangement shown in Figure P4.3, a cable is attached to the end of a cantilever beam. We will model the cable as a rod. Denote the translational spring constant of the beam by k_b, and the translational spring constant of the cable by k_c. The displacement x is caused by the applied force f.

a. Are the two springs in series or in parallel?

b. What is the equivalent spring constant for this arrangement?

Figure P4.2

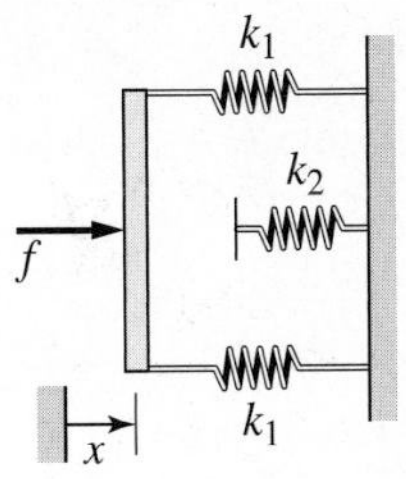

Figure P4.3

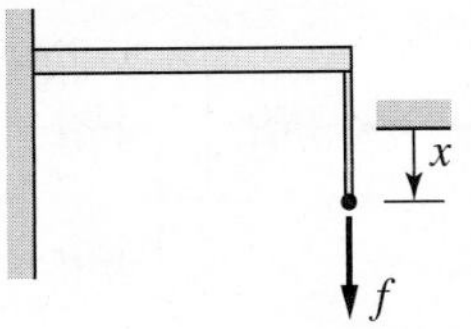

Figure P4.4

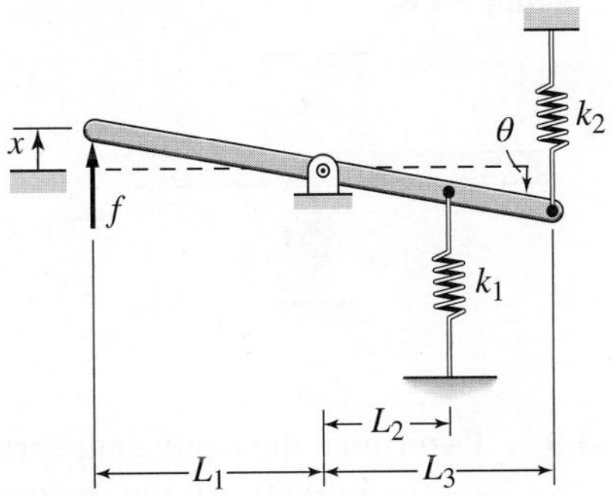

4.4 In the spring arrangement shown in Figure P4.4, the displacement x is caused by the applied force f. Assuming the system is in static equilibrium when $x = 0$ and that the angle θ is small, determine the equivalent spring constant k_e for this arrangement, where $f = k_e x$.

4.5 For the system shown in Figure P4.5, assume that the resulting motion is small enough to be only horizontal, and determine the expression for the equivalent k_e that relates the applied force f to the resulting displacement x.

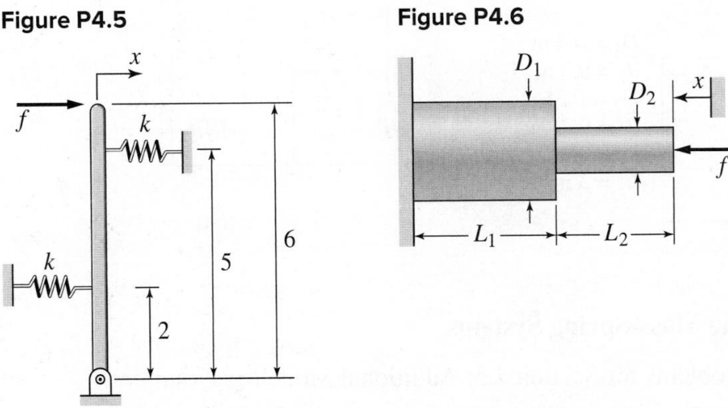

Figure P4.5

Figure P4.6

4.6 The two stepped solid cylinders in Figure P4.6 consist of the same material and have an axial force f applied to them. Determine the equivalent translational spring constant for this arrangement. (Hint: Are the two springs in series or in parallel?)

4.7 A table with four identical legs supports a vertical force. The solid cylindrical legs are made of metal with $E = 2 \times 10^{11}$ N/m^2. The legs are 0.9 m in length and 0.04 m in diameter. Compute the equivalent spring constant due to the legs, assuming the table top is rigid.

4.8 The beam shown in Figure P4.8 has been stiffened by the addition of a spring support. The steel beam is 3 ft long, 1 in. thick, and 1 ft wide, and its mass is 5 slugs. The mass m is 60 slugs. Neglecting the mass of the beam,

a. Compute the spring constant k necessary to reduce the static deflection to one-half its original value before the spring k was added.

b. Compute the natural frequency ω_n of the combined system.

Figure P4.8

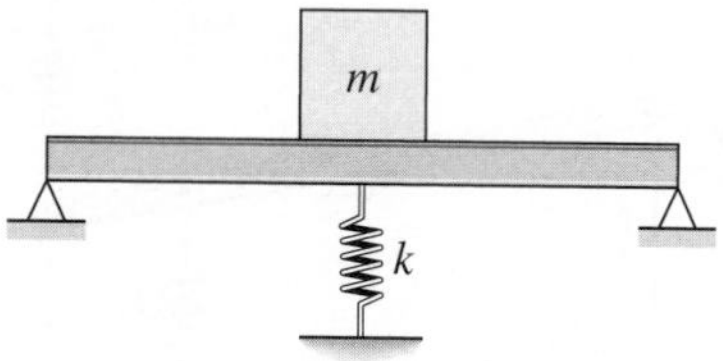

Figure P4.9

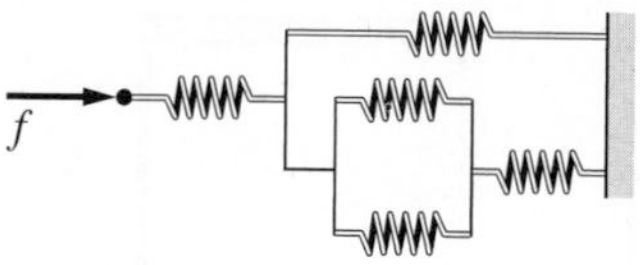

4.9 Determine the equivalent spring constant of the arrangement shown in Figure P4.9. All the springs have the same spring constant k.

4.10 Compute the equivalent torsional spring constant of the stepped shaft arrangement shown in Figure P4.10. For the shaft material, $G = 8 \times 10^{10}$ N/m^2.

4.11 Plot the spring force felt by the mass shown in Figure P4.11 as a function of the displacement x. When $x = 0$, spring 1 is at its free length. Spring 2 is at its free length in the configuration shown.

Figure P4.10

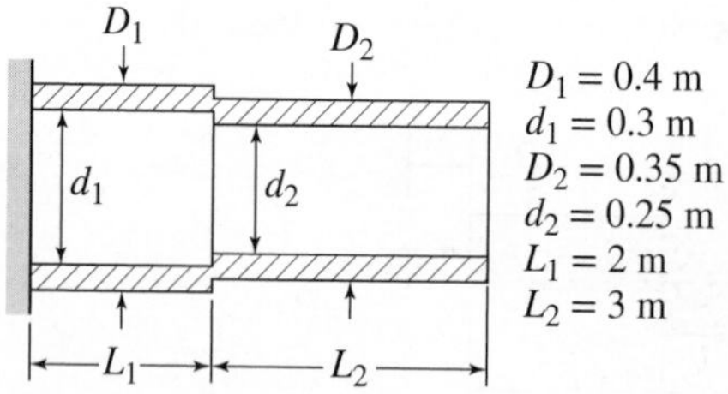

Figure P4.11

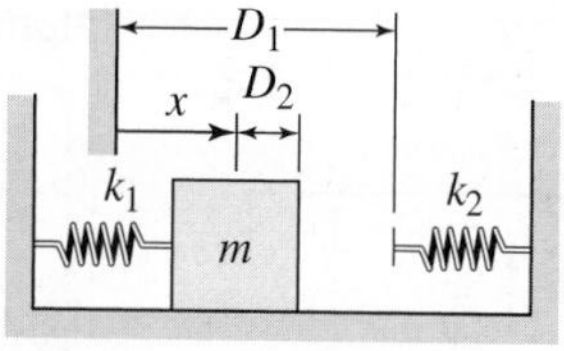

Figure P4.12

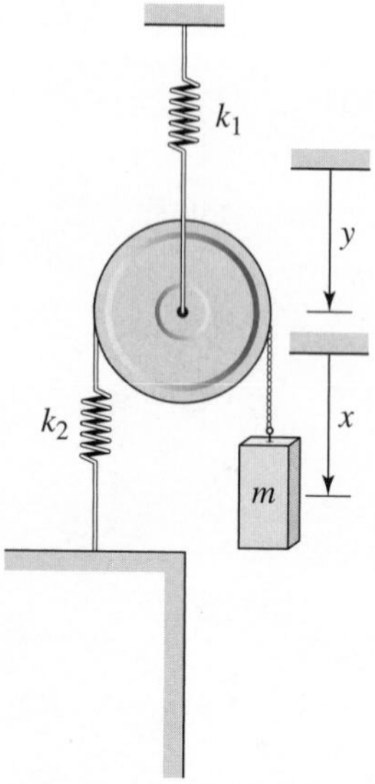

Section 4.2 Modeling Mass-Spring Systems

Note: See also the problems for Section 4.5: Additional Modeling Examples.

4.12 Calculate the expression for the natural frequency of the system shown in Figure P4.12. Disregard the pulley mass.

4.13 Obtain the expression for the natural frequency of the system shown in Figure P4.13. Assume small motions and disregard the pulley mass.

4.14 Obtain the expression for the natural frequency of the system shown in Figure P4.14. Disregard the mass of the L-shaped arm.

Figure P4.13

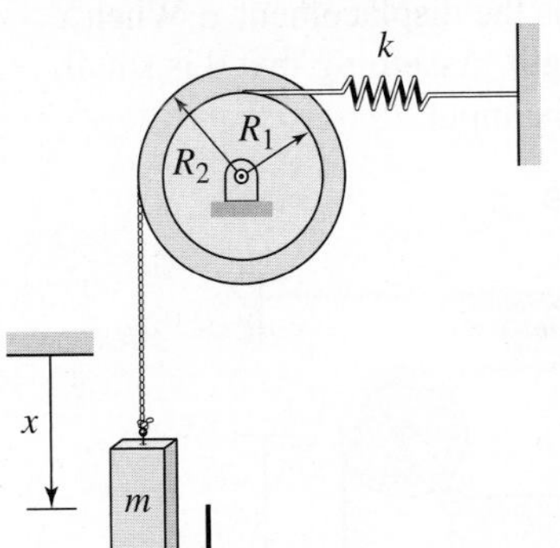

Figure P4.14

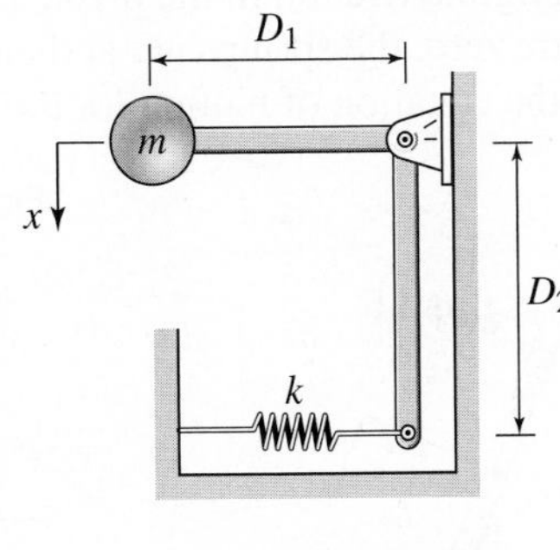

4.15 A connecting rod having a mass of 5 kg is shown in Figure P4.15. It oscillates with a frequency of 30 cycles per minute when supported on a knife edge, as shown. Its mass center is located 0.2 m below the support. Calculate the moment of inertia about the mass center.

Figure P4.15

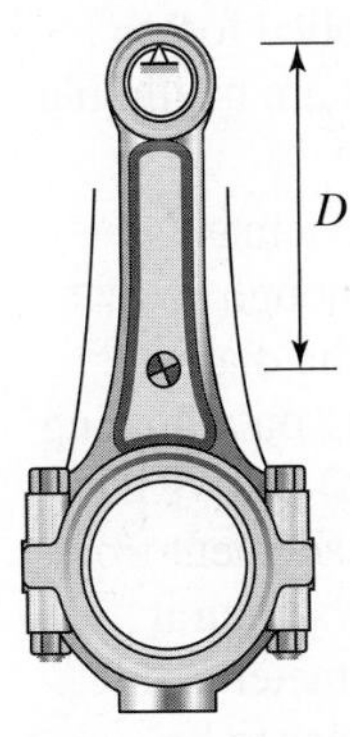

Figure P4.16

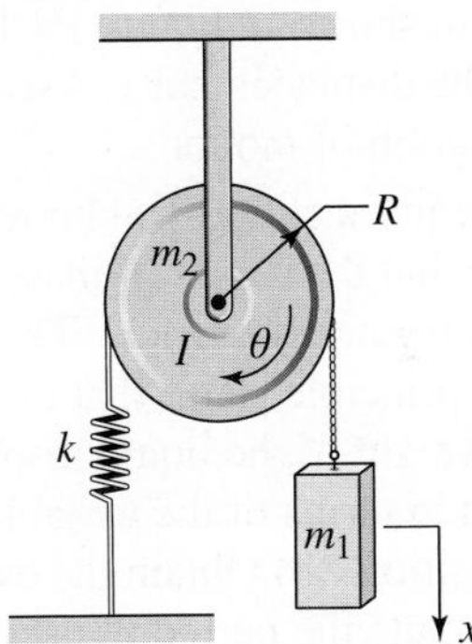

4.16 Calculate the expression for the natural frequency of the system shown in Figure P4.16.

4.17 For each of the systems shown in Figure P4.17, the input is the force f and the outputs are the displacements x_1 and x_2 of the masses. The equilibrium positions with $f = 0$ correspond to $x_1 = x_2 = 0$. Neglect any friction between the masses and the surface. Derive the equations of motion of the systems.

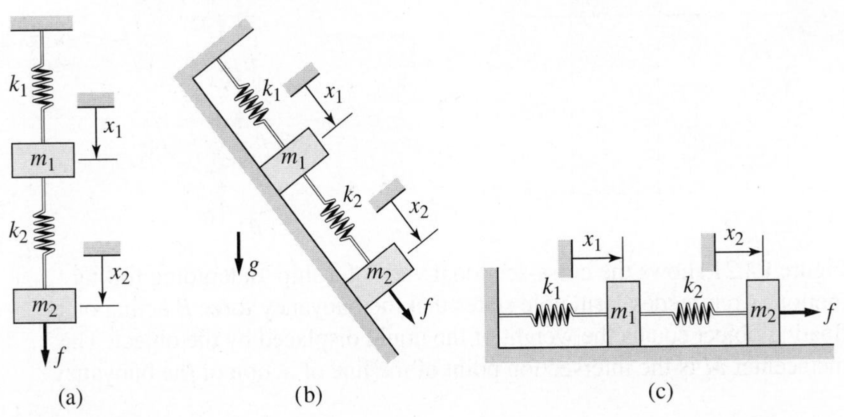

Figure P4.17

4.18 The mass m in Figure P4.18 is attached to a rigid lever having negligible mass and negligible friction in the pivot. The input is the displacement x. When x and θ are zero, the springs are at their free length. Assuming that θ is small, derive the equation of motion for θ with x as the input.

Figure P4.18

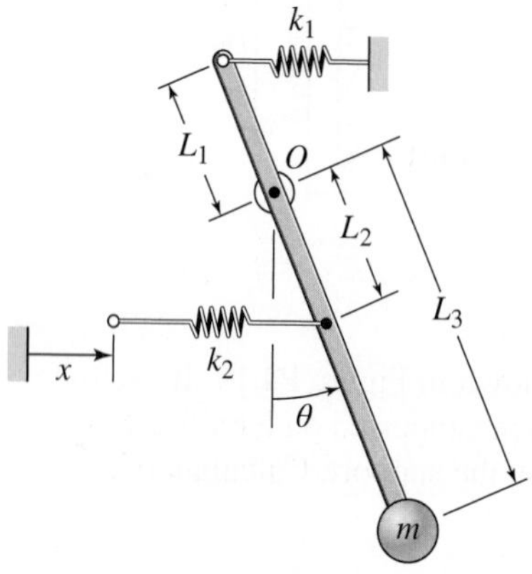

Figure P4.19

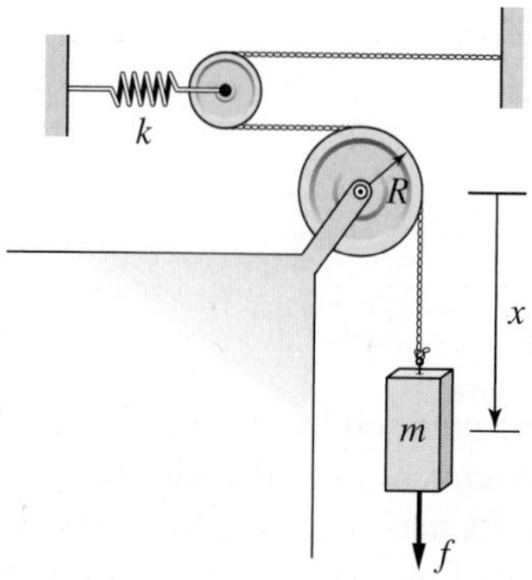

4.19 In the pulley system shown in Figure P4.19, the input is the applied force f, and the output is the displacement x. Assume the pulley masses are negligible and derive the equation of motion.

4.20 Figure P4.20 illustrates a cylindrical buoy floating in water with a mass density ρ. Assume that the center of mass of the buoy is deep enough so that the buoy motion is primarily vertical. The buoy mass is m and the diameter is D. Archimedes' principle states that the buoyancy force acting on a floating object equals the weight of the liquid displaced by the object. (a) Derive the equation of motion in terms of the variable x, which is the displacement from the equilibrium position. (b) Obtain the expression for the buoy's natural frequency. (c) Compute the period of oscillation if the buoy diameter is 4 ft and the buoy weighs 2000 lb. Take the mass density of fresh water to be $\rho = 1.94$ slug/ft^3.

Figure P4.20

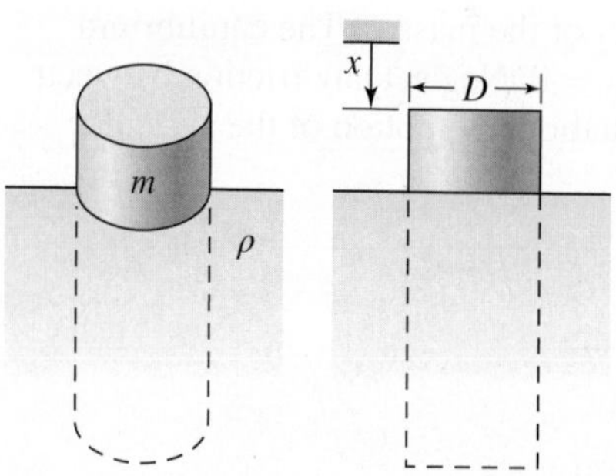

Figure P4.21

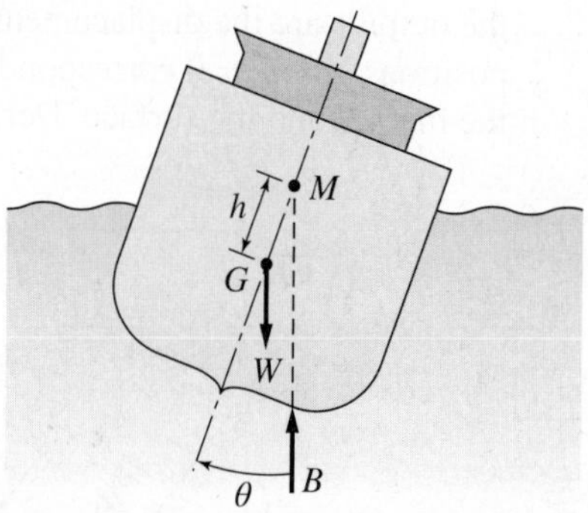

4.21 Figure P4.21 shows the cross-sectional view of a ship undergoing rolling motion. Archimedes' principle states that the buoyancy force B acting on a floating object equals the weight of the liquid displaced by the object. The metacenter M is the intersection point of the line of action of the buoyancy force and the ship's centerline. The distance h of M from the mass center G is called the metacentric height. (a) Obtain the equation of motion describing the ship's rolling motion in terms of the angle θ. (b) The given parameters are the ship's weight W, its metacentric height h, and its moment of inertia I about the center of gravity. Obtain an expression for the natural frequency of the rolling motion.

4.22 In the system shown in Figure P4.22, the input is the angular displacement ϕ of the end of the shaft, and the output is the angular displacement θ of the inertia I. The shafts have torsional stiffnesses k_1 and k_2. The equilibrium position corresponds to $\phi = \theta = 0$. Derive the equation of motion and find the transfer function $\Theta(s)/\Phi(s)$.

4.23 In Figure P4.23, assume that the cylinder rolls without slipping. The spring is at its free length when x and y are zero. (a) Derive the equation of motion in terms of x, with $y(t)$ as the input. (b) Suppose that $m = 20$ kg, $R = 0.4$ m, $k = 1500$ N/m, and that $y(t)$ is a unit-step function. Solve for $x(t)$ if $x(0) = \dot{x}(0) = 0$.

4.24 In Figure P4.24 when $x_1 = x_2 = 0$ the springs are at their free lengths. Derive the equations of motion.

Figure P4.22

Figure P4.23

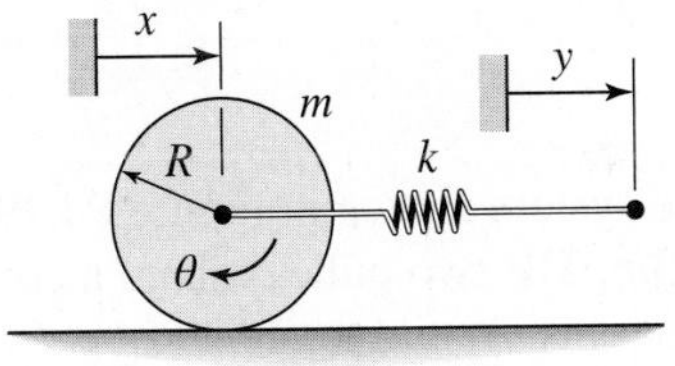

Figure P4.24

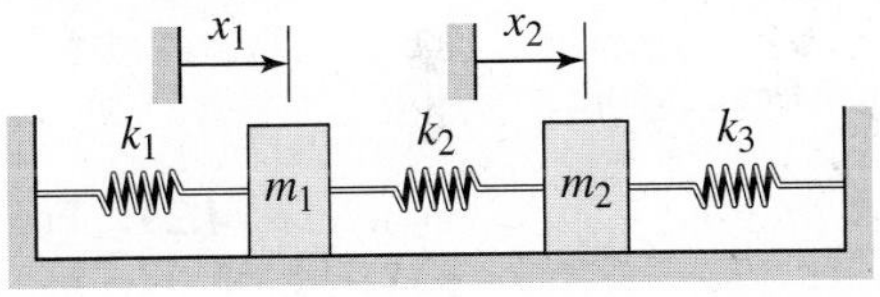

4.25 Figure P4.25 models the three shafts as massless torsional springs. When $\theta_1 = \theta_2 = 0$, the springs are at their free lengths. Derive the equations of motion with the torque T_2 as the input.

4.26 In Figure P4.26, when $\theta_1 = \theta_2 = 0$, the spring is at its free length. Derive the equations of motion, assuming small angles.

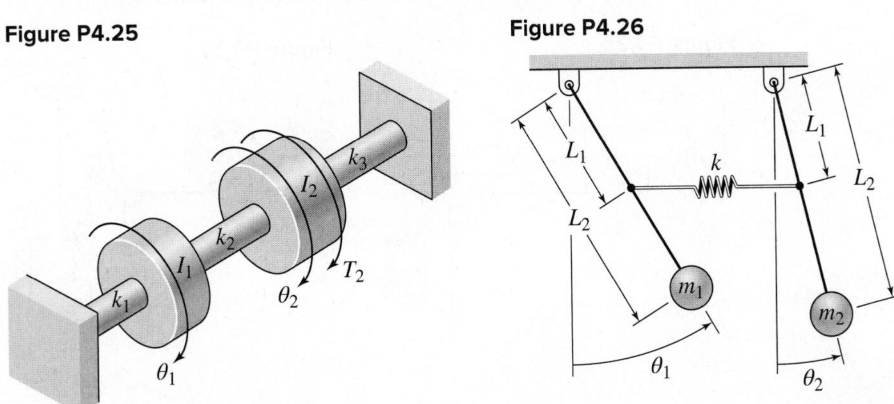

Figure P4.25

Figure P4.26

4.27 Consider the torsion-bar suspension shown in Figure 4.1.6. Assume that the torsion bar is a steel rod with a length of 6 ft and diameter 2 in. The wheel weighs 50 lb and the suspension arm is 2 ft long. Neglect the masses of the torsion bar and the suspension arm, and calculate the natural frequency of the system. Use $G = 1.7 \times 10^9$ lb/ft^2.

4.28 For the system shown in Figure P4.28, suppose that $k_1 = k$, $k_2 = k_3 = 4k$, and $m_1 = m_2 = m$. Obtain the equations of motion in terms of x_1 and x_2.

Figure P4.28

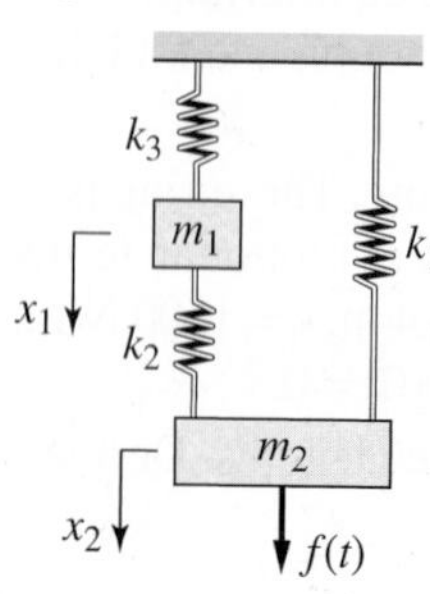

Figure P4.29

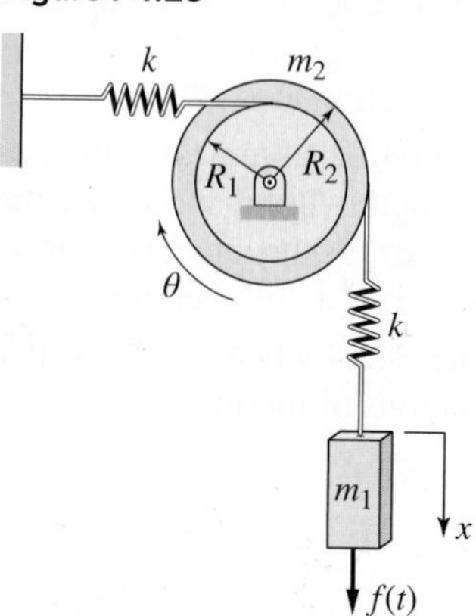

4.29 For the system shown in Figure P4.29, suppose that $R_2 = 3R_1$, $m_1 = m$, and $m_2 = 3m$. The two pulleys share a common hub and are welded together. Their total mass is m_2 and total inertia is I_2. Obtain the equations of motion in terms of x and θ.

Section 4.3 Energy Methods

4.30 For Figure P4.30, assume that the cylinder rolls without slipping and use conservation of energy to derive the equation of motion in terms of x.

4.31 For Figure P4.31, the equilibrium position corresponds to $x = 0$. Neglect the masses of the pulleys and assume that the cable is inextensible, and use conservation of energy to derive the equation of motion in terms of x.

4.32 For Figure P4.32, the equilibrium position corresponds to $x = 0$. Neglect the masses of the pulleys and assume that the cable is inextensible, and use conservation of energy to derive the equation of motion in terms of x.

Figure P4.30

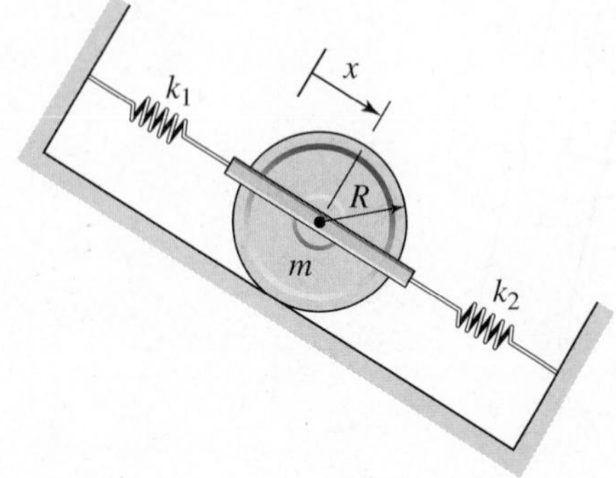

Figure P4.31

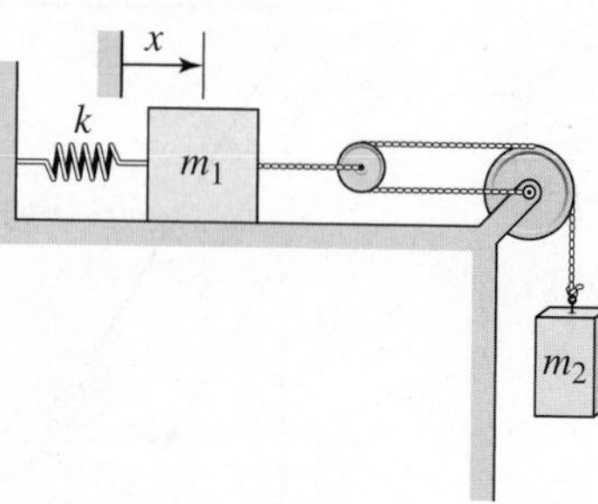

Figure P4.32

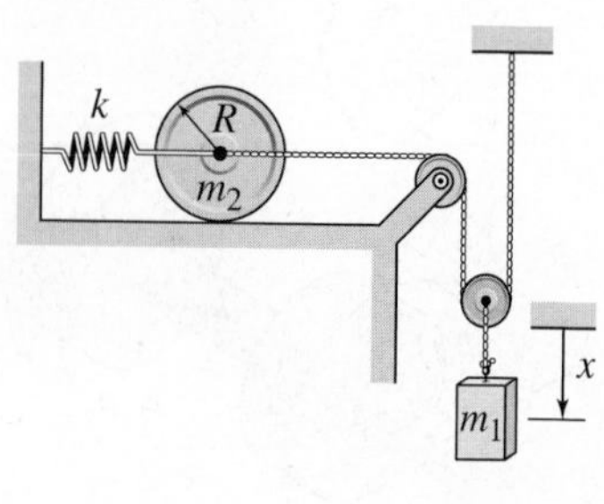

4.33 Use the Rayleigh method to obtain an expression for the natural frequency of the system shown in Figure P4.33. The equilibrium position corresponds to $x = 0$.

4.34 For Figure P4.34, assume that the cylinder rolls without slipping and use the Rayleigh method to obtain an expression for the natural frequency of the system. The equilibrium position corresponds to $x = 0$.

Figure P4.33

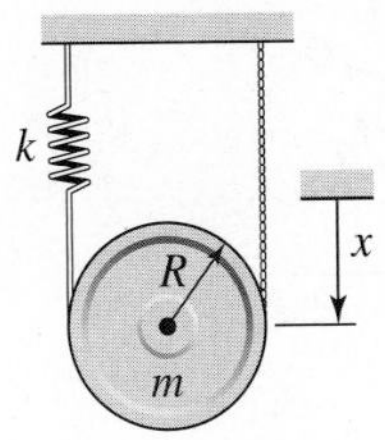

Figure P4.34

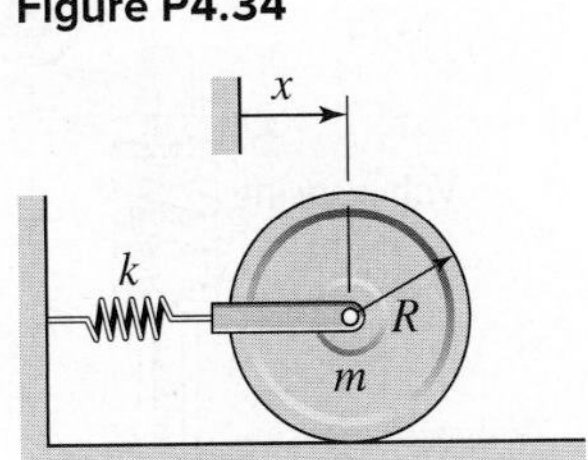

Figure P4.35

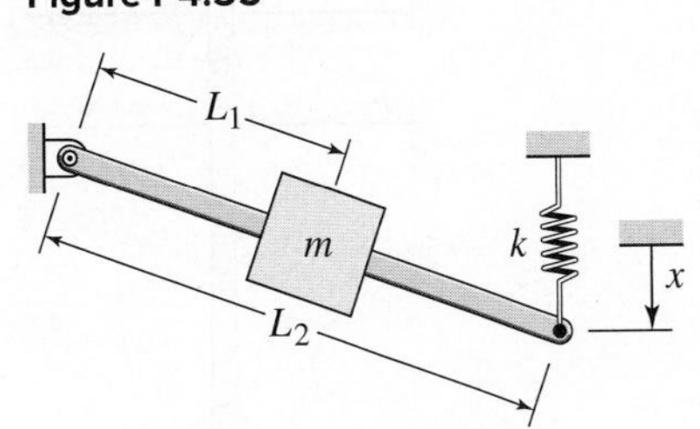

4.35 Use the Rayleigh method to obtain an expression for the natural frequency of the system shown in Figure P4.35. The equilibrium position corresponds to $x = 0$.

4.36 Use an energy method to obtain the expression for the natural frequency of the system shown in Figure P4.36.

4.37 Determine the natural frequency of the system shown in Figure P4.37 using Rayleigh's method. Assume small angles of oscillation.

Figure P4.36

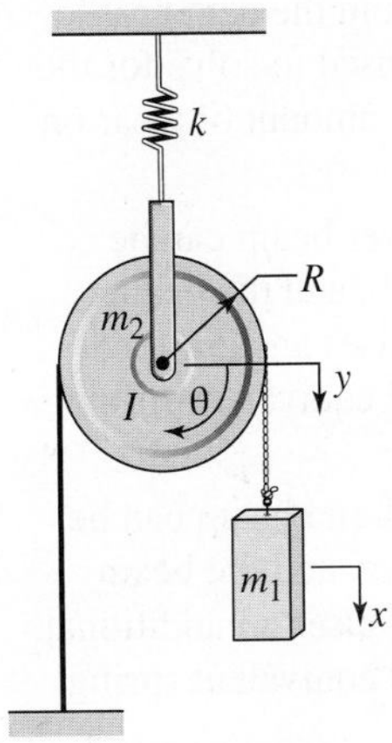

Figure P4.37

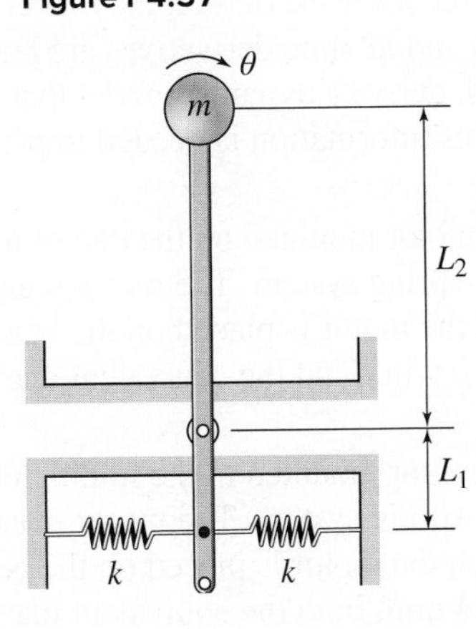

Figure P4.38

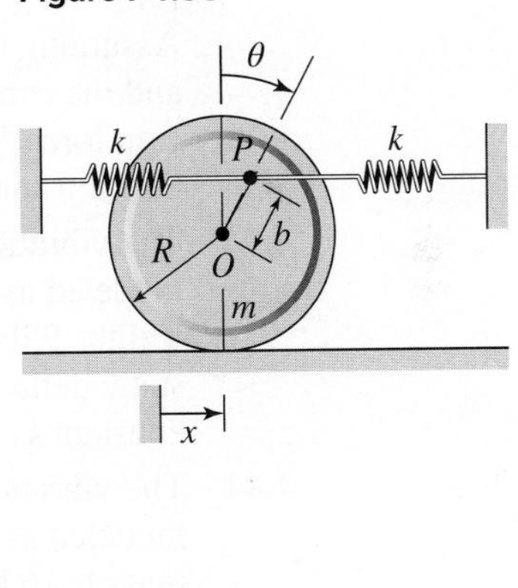

4.38 Determine the natural frequency of the system shown in Figure P4.38 using an energy method. The disk is a solid cylinder. Assume small angles of oscillation.

4.39 Use Rayleigh's method to calculate the expression for the natural frequency of the system shown in Figure P4.13. Assume small motions and neglect the pulley mass.

4.40 Use Rayleigh's method to obtain the expression for the natural frequency of the system shown in Figure P4.14. Disregard the mass of the L-shaped arm.

4.41 Determine the natural frequency of the system shown in Figure P4.41 using Rayleigh's method. Assume small angles of oscillation.

Figure P4.41 **Figure P4.42**

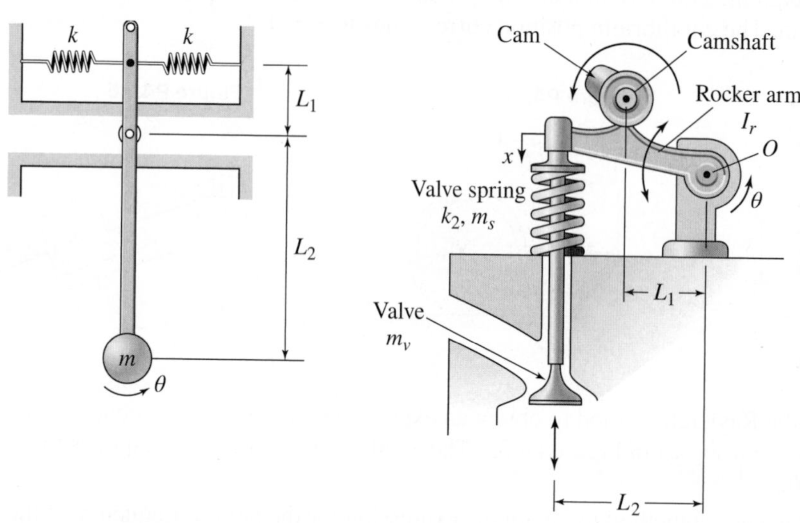

4.42 Figure P4.42 shows an engine valve driven by an overhead camshaft. The rocker arm pivots about the fixed point O and the inertia of the arm about this point is I_r. The valve mass is m_v and the spring mass is m_s; its spring constant is k_s. Let f_c denote the force exerted on the rocker arm by the camshaft. Assuming that $\theta(t)$ and its time derivatives are known (from the cam profile and the cam speed), derive a dynamic model that can be used to solve for the cam force $f_c(t)$. (This information is needed to predict the amount of wear on the cam surface.)

4.43 The vibration of a motor mounted on the end of a cantilever beam can be modeled as a mass-spring system. The motor weighs 40 lb, and the beam weighs 9 lb. When the motor is placed on the beam, it causes an additional static deflection of 0.9 in. Find the equivalent mass m and equivalent spring constant k.

4.44 The vibration of a motor mounted in the middle of a fixed-end beam can be modeled as a mass-spring system. The motor mass is 50 kg, and the beam mass is 16 kg. When the motor is placed on the beam, it causes an additional static deflection of 4 mm. Find the equivalent mass m and equivalent spring constant k.

4.45 The vibration of a motor mounted in the middle of a simply-supported beam can be modeled as a mass-spring system. The motor mass is 40 kg, and the beam mass is 15 kg. When the motor is placed on the beam, it causes an additional static deflection of 3 mm. Find the equivalent mass m and equivalent spring constant k.

4.46 A certain cantilever beam vibrates at a frequency of 6 Hz when a 40 lb motor is placed on the beam. The beam weighs 9 lb. Estimate the beam stiffness k.

4.47 A 12-kg mass is attached to a 3-kg spring. The mass vibrates at a frequency of 30 Hz when disturbed. Estimate the spring stiffness k.

4.48 The static deflection of a cantilever beam is described by

$$x_y = \frac{P}{6EI_A}y^2(3L - y)$$

where P is the load applied at the end of the beam, and x_y is the vertical deflection at a point a distance y from the support (Figure P4.48). Obtain an expression for an equivalent mass located at the end of the beam.

Figure P4.48

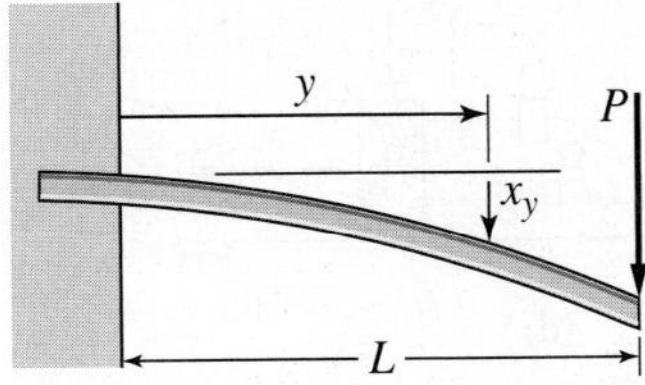

4.49 Figure P4.49 shows a winch supported by a cantilever beam at the stern of a ship. The mass of the winch is m_w, the mass of the beam plus winch bracket and motor is m_b. The object hoisted by the winch has a mass m_h; the wire rope mass m_r is assumed to be negligible compared to the other masses. Find the equation of motion for the vertical motion x_1 of the winch: (a) assuming that the rope does not stretch and (b) assuming that the rope stretches and has a spring constant k_r.

Figure P4.49

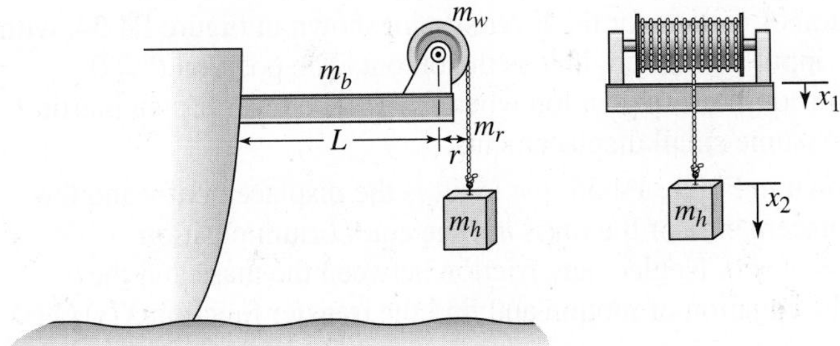

Section 4.4 Damping Elements

4.50 A 60-kg block is placed on an inclined plane whose angle with the horizontal is 35°. The viscous friction coefficient between the block and the plane is $c =$ 8 N·s/m. (a) Derive the equation of motion. (b) Solve the equation of motion for the speed $v(t)$ of the block, assuming that the block is initially given a speed of 7 m/s. (c) Find the steady-state speed of the block and estimate the time required to reach that speed. (d) Discuss the effect of the initial velocity of the steady-state speed.

4.51 A certain mass-spring-damper system has the following equation of motion.

$$60\ddot{x} + c\dot{x} + 1440x = f(t)$$

Suppose that the initial conditions are zero and that the applied force $f(t)$ is a step function of magnitude 3600. Solve for $x(t)$ for the following two cases: (a) $c = 600$ and (b) $c = 240$.

4.52 For each of the systems shown in Figure P4.52, the input is the force f and the outputs are the displacements x_1 and x_2 of the masses. The equilibrium positions with $f = 0$ correspond to $x_1 = x_2 = 0$. Neglect any friction between the masses and the surface. Derive the equations of motion of the systems.

Figure P4.52

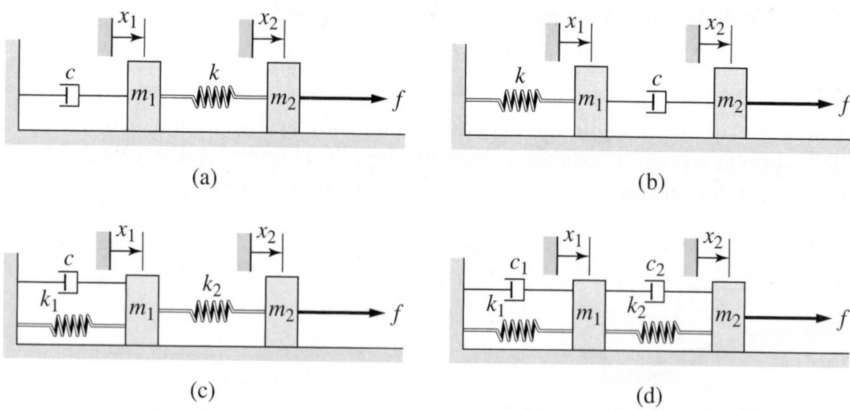

Figure P4.53

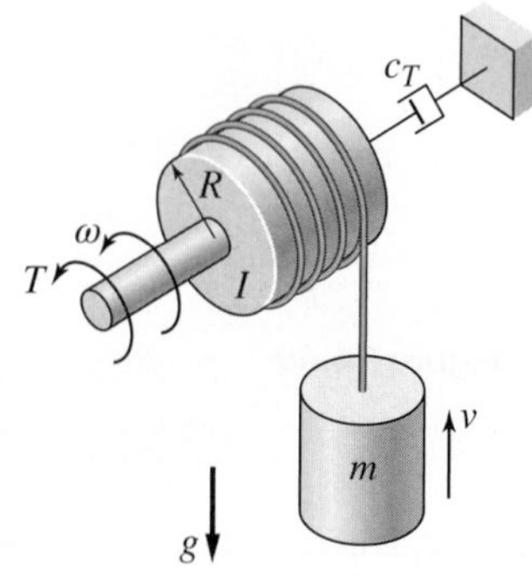

4.53 In Figure P4.53 a motor supplies a torque T to turn a drum of radius R and inertia I about its axis of rotation. The rotating drum lifts a mass m by means of a cable that wraps around the drum. The drum's speed is ω. Viscous torsional damping c_T exists in the drum shaft. Neglect the mass of the cable. (a) Obtain the equation of motion with the torque T as the input and the vertical speed v of the mass as the output. (b) Suppose that $m = 50$ kg, $R = 0.3$ m, $I = 1.1$ kg·m^2, and $c_T = 0.2$ N·m·s. Find the speed $v(t)$ if the system is initially at rest and the torque T is a step function of magnitude 500 N·m.

4.54 Derive the equation of motion for the lever system shown in Figure P4.54, with the force f as the input and the angle θ as the output. The position $\theta = 0$ corresponds to the equilibrium position when $f = 0$. The lever has an inertia I about the pivot. Assume small displacements.

4.55 In the system shown in Figure P4.55, the input is the displacement y and the output is the displacement x of the mass m. The equilibrium position corresponds to $x = y = 0$. Neglect any friction between the mass and the surface. Derive the equation of motion and find the transfer function $X(s)/Y(s)$.

Figure P4.54

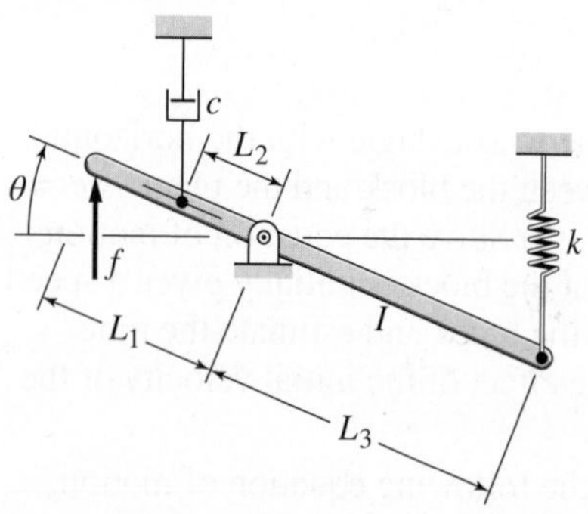

Figure P4.55

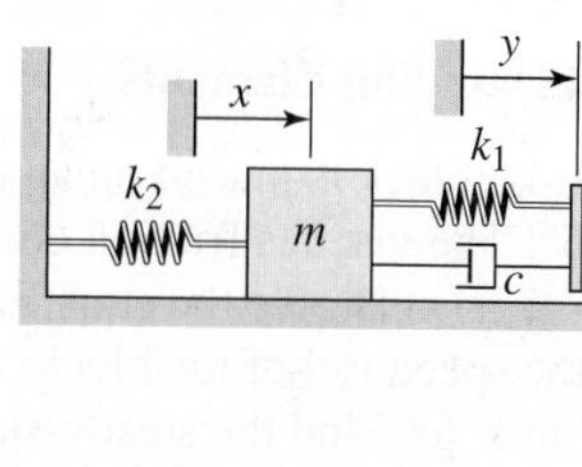

4.56 Figure P4.56a shows a Houdaille damper, which is a device attached to an engine crankshaft to reduce vibrations. The damper has an inertia I_d that is free to rotate within an enclosure filled with viscous fluid. The inertia I_p is the inertia of the fan-belt pulley. Modeling the crankshaft as a torsional spring k_T, the damper system can be modeled as shown in part (b) of the figure. Derive the equation of motion with the angular displacements θ_p and θ_d as the outputs and the crankshaft angular displacement ϕ as the input.

Figure P4.56

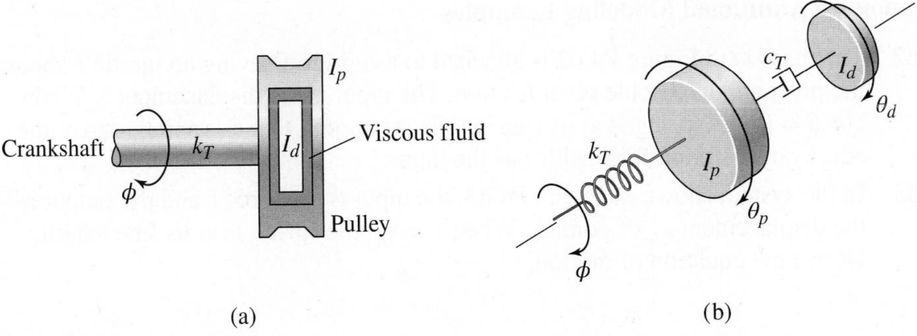

4.57 Refer to Figure P4.57. Determine the relations between c, c_1, and c_2 so that the damper shown in part (c) is equivalent to (a) the arrangement shown in part (a), and the arrangement shown in part (b).

Figure P4.57

Figure P4.58

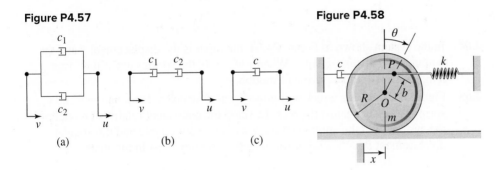

4.58 For the system shown in Figure P4.58, obtain the equation of motion in terms of θ. The disk is a solid cylinder. Assume small angles of oscillation.

4.59 Find the transfer function $Z(s)X(s)$ for the system shown in Figure P4.59.

4.60 Find the transfer function $Y(s)X(s)$ for the system shown in Figure P4.60.

4.61 Find the transfer function $Y(s)X(s)$ for the system shown in Figure P4.61.

Figure P4.59 Figure P4.60 Figure P4.61

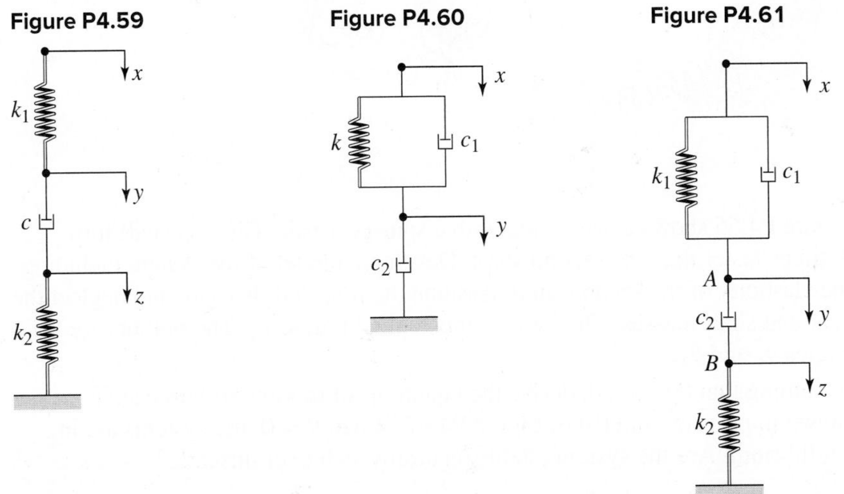

Section 4.5 Additional Modeling Examples

Figure P4.62

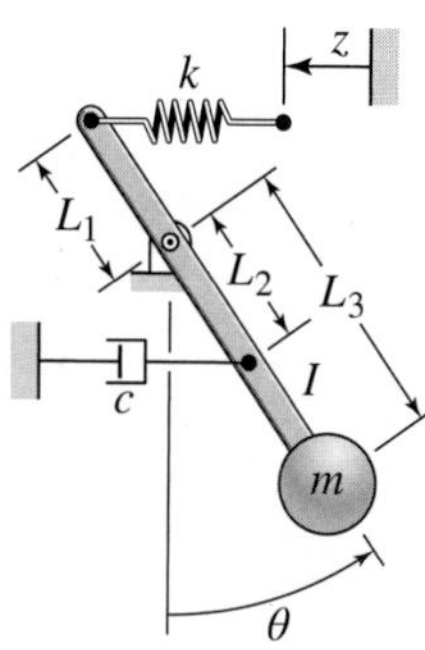

4.62 The mass m in Figure P4.62 is attached to a rigid rod having an inertia I about the pivot and negligible pivot friction. The input is the displacement z. When $z = \theta = 0$, the spring is at its free length. Assuming that θ is small, derive the equation of motion for θ with z as the input.

4.63 In the system shown in Figure P4.63, the input is the force f and the output is the displacement x_A of point A. When $x = x_A$, the spring is at its free length. Derive the equation of motion.

Figure P4.63

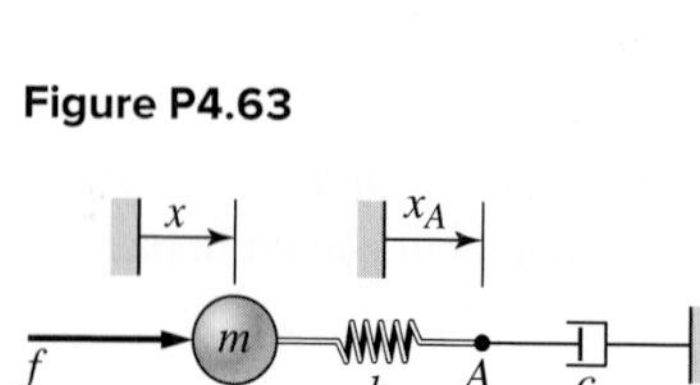

Figure P4.64

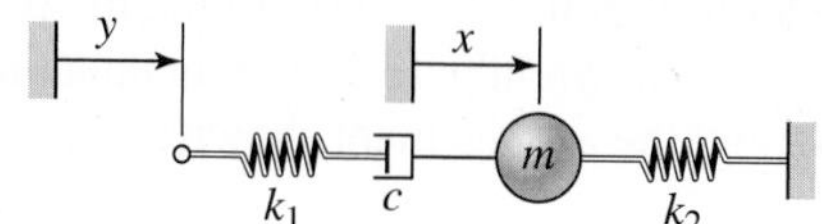

4.64 In the system shown in Figure P4.64, the input is the displacement y and the output is the displacement x. When $x = y = 0$, the springs are at their free lengths. Derive the equation of motion.

4.65 Figure P4.65 shows a rack-and-pinion gear in which a damping force and a spring force act against the rack. Develop the equivalent rotational model of the system with the applied torque T as the input variable and the angular displacement θ is the output variable. Neglect any twist in the shaft.

Figure P4.65

Figure P4.66

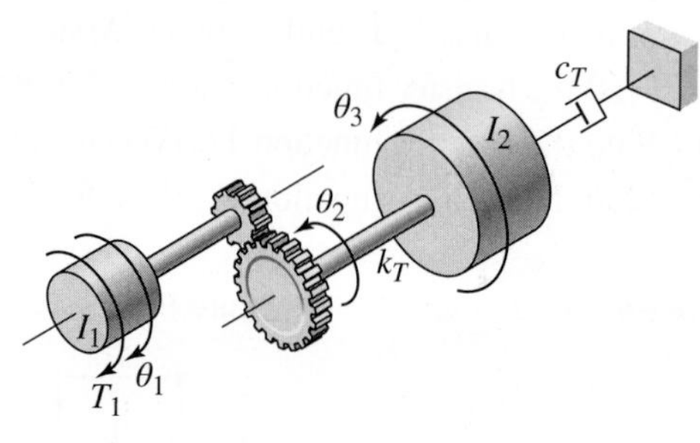

4.66 Figure P4.66 shows a drive train with a spur-gear pair. The first shaft turns N times faster than the second shaft. Develop a model of the system including the elasticity of the second shaft. Assume the first shaft is rigid, and neglect the gear and shaft masses. The input is the applied torque T_1. The outputs are the angles θ_1 and θ_3.

4.67 Assuming that θ is small, derive the equations of motion of the systems shown in parts (a) and (b) of Figure P4.67. When $\theta = 0$, the systems are in equilibrium. Are the systems stable, neutrally stable, or unstable?

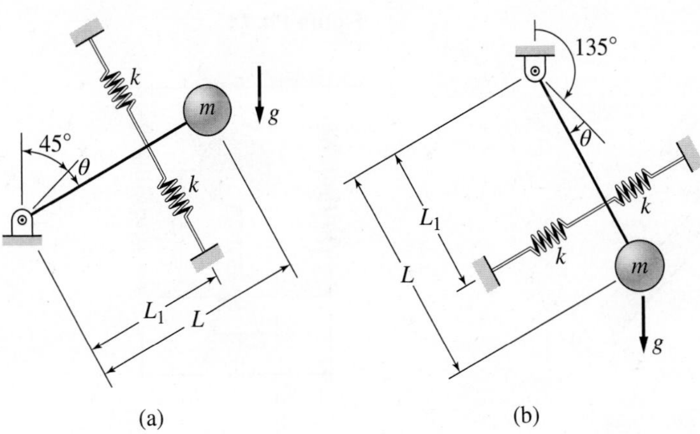

Figure P4.67

4.68 Assuming that θ is small, derive the equation of motion of the pendulum shown in Figure P4.68. The pendulum is in equilibrium when $\theta = 0$. Is the system stable, neutrally stable, or unstable?

Figure P4.68

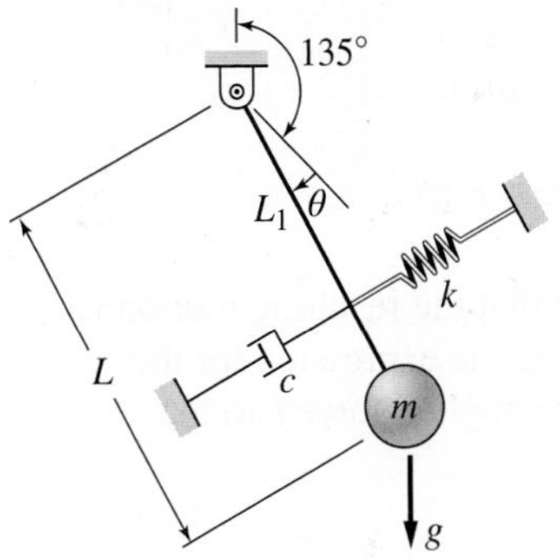

Figure P4.69

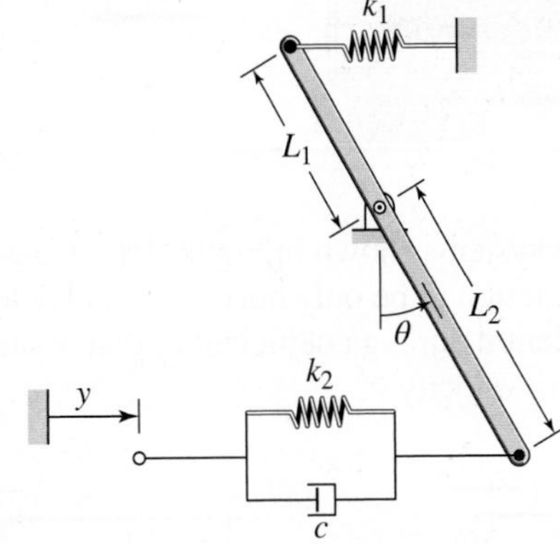

4.69 Assuming that θ is small, derive the equation of motion of the pendulum shown in Figure P4.69. The input is $y(t)$ and the output is θ. The equilibrium corresponds to $y = \theta = 0$, when the springs are at their free lengths. The rod inertia about the pivot is I.

4.70 The top view of a solid door is shown in Figure P4.70. The door has a mass of 50 kg and is 2.3 m high, 1.1 m wide, and 0.1 m thick. Its door closer has a torsional spring constant of 15 N·m/rad. The door will close as fast as possible without oscillating if the torsional damping coefficient c in the closer is set to the critical damping value corresponding to a damping ratio of $\zeta = 1$. Determine this critical value of c.

4.71 Derive the equation of motion for the system shown in Figure P4.71. Assume small angles of oscillation, and neglect the rod mass. What relation among L_1, L_2, m, and k must be satisfied for the system to be stable?

Figure P4.70

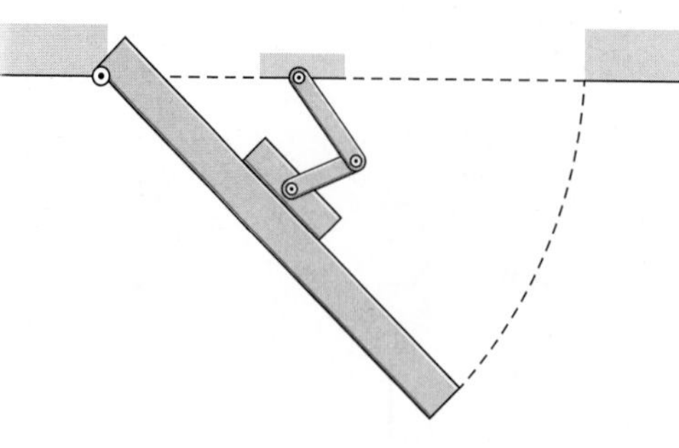

Figure P4.71

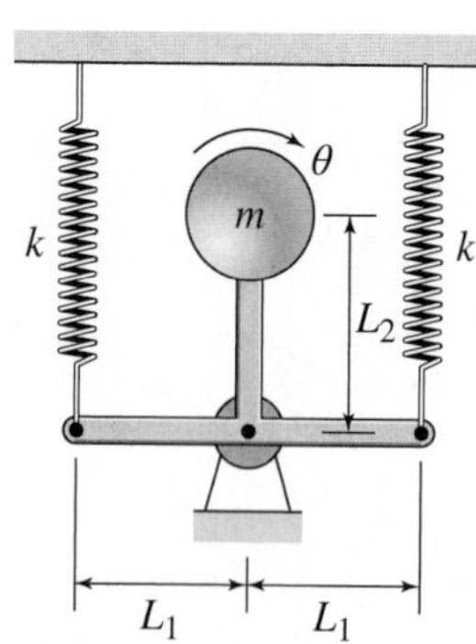

4.72 A boxcar moving at 1.5 m/s hits the shock absorber at the end of the track (Figure P4.72). The boxcar mass is 20,000 kg. The stiffness of the absorber is $k = 80{,}000$ N/m, and the damping coefficient is $c = 90{,}000$ N·s/m. Determine the maximum spring compression and the time for the boxcar to stop.

Figure P4.72

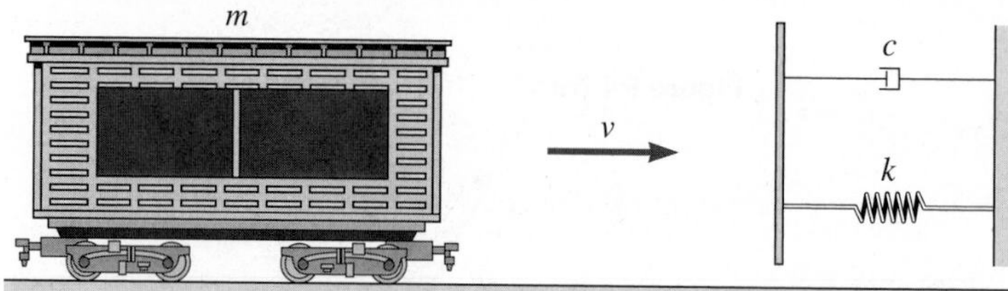

4.73 For the systems shown in Figure P4.73, assume that the resulting motion is small enough to be only horizontal and determine the expression for the equivalent damping coefficient c_e that relates the applied force f to the resulting velocity v.

Figure P4.73

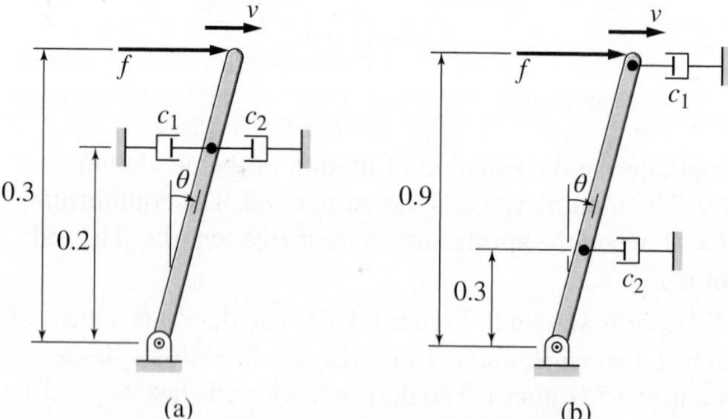

4.74 Refer to Figure P4.74a, which shows a ship's propeller, drive train, engine, and flywheel. The diameter ratio of the gears is $D1/D2 = 2/3$. The inertias in kg·m^2 of gear 1 and gear 2 are 100 and 500, respectively. The flywheel, engine, and propeller inertias are 10^4, 10^3, and 2500, respectively. The torsional stiffness of shaft 1 is 5×10^6 N·m/rad, and that of shaft 2 is 10^6 N·m/rad. Because the flywheel inertia is so much larger than the other inertias, a simpler model of the shaft vibrations can be obtained by assuming the flywheel does not rotate. In addition, because the shaft between the engine and gears is short, we will assume that it is very stiff compared to the other shafts. If we also disregard the shaft inertias, the resulting model consists of two inertias, one obtained by lumping the engine and gear inertias, and one for the propeller (Figure P4.74b). Using these assumptions, obtain the natural frequencies of the system.

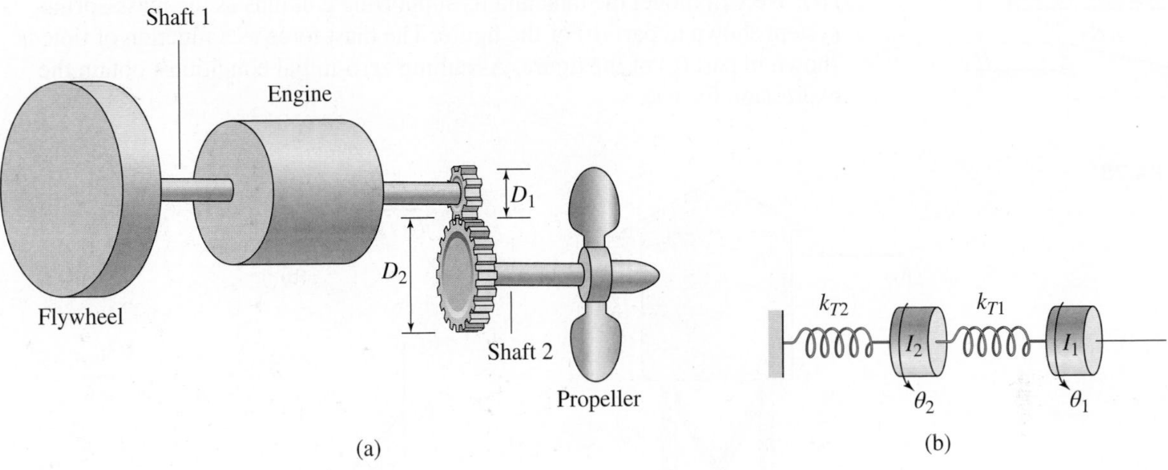

Figure P4.74

4.75 In this problem, we make all the same assumptions as in Problem 4.74, but we do not disregard the flywheel inertia, so our model consists of three inertias, as shown in Figure P4.75. Obtain the natural frequencies of the system.

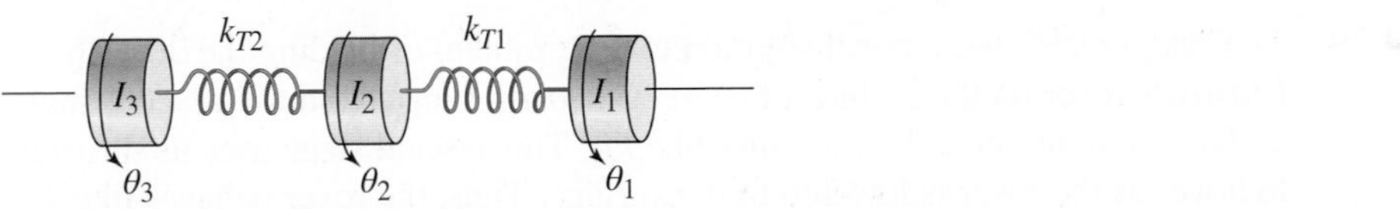

Figure P4.75

4.76 Refer to Figure P4.76, which shows a turbine driving an electrical generator through a gear pair. The diameter ratio of the gears is $D_2/D_1 = 1.5$. The inertias in kg·m^2 of gear 1 and gear 2 are 100 and 500, respectively. The turbine and generator inertias are 2000 and 1000, respectively. The torsion

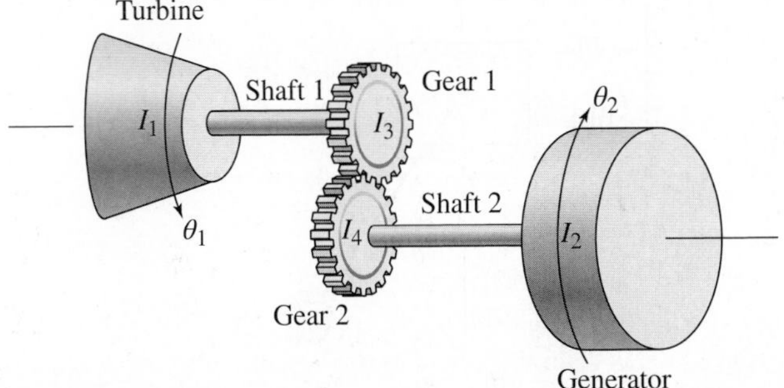

Figure P4.76

Figure P4.77

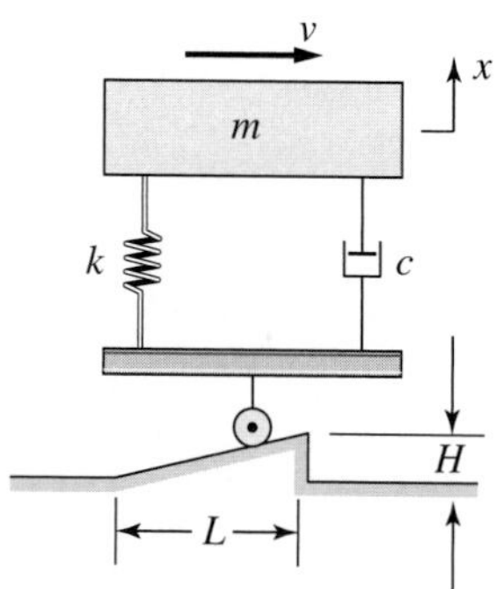

stiffness of shaft 1 is 3×10^5 N·m/rad, and that of shaft 2 is 8×10^4 N·m/rad. Disregard the shaft inertias and obtain the natural frequencies of the system.

4.77 Refer to Figure P4.77, which is a simplified representation of a vehicle striking a bump. The vertical displacement x is 0 when the tire first meets the bump. Assuming that the vehicle's horizontal speed v remains constant and that the system is critically damped, obtain the expression for $x(t)$.

4.78 Refer to Figure P4.78a, which shows a water tank subjected to a blast force $f(t)$. We will model the tank and its supporting columns as the mass-spring system shown in part (b) of the figure. The blast force as a function of time is shown in part (c) of the figure. Assuming zero-initial conditions, obtain the expression for $x(t)$.

Figure P4.78

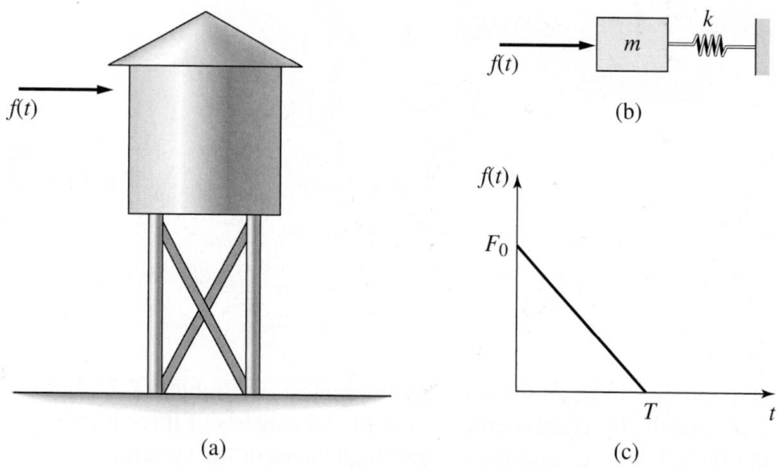

4.79 The "sky crane" was a novel solution to the problem of landing the 2000 lb *Curiosity* rover on the surface of Mars. *Curiosity* hangs from the descent stage by 60-ft long nylon tethers (Figure P4.79a). The descent stage uses its thrusters to hover as the rover is lowered to the surface. Thus, the rover behaves like a pendulum whose base is moving horizontally. The side thruster force is not constant but is controlled to keep the descent stage from deviating left or right from the desired vertical path. As we will see in Chapter 10, such a control

Figure P4.79

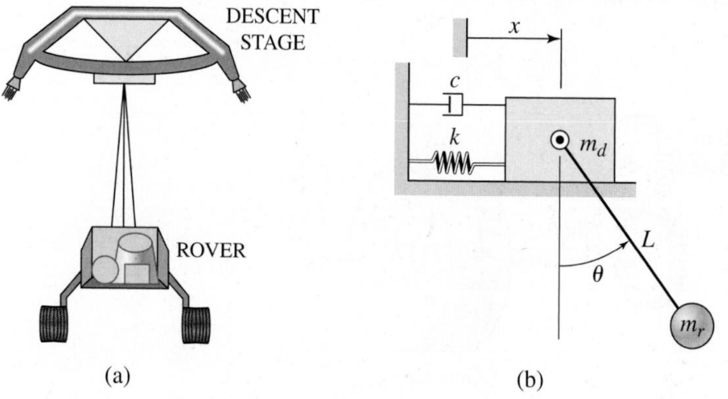

system effectively acts like a spring and a damper, as shown in Figure P4.79b. The rover mass is m_r, the descent stage mass is m_d, and the net horizontal component of the thruster forces is $f = kx + c\dot{x}$. Among other simplifications, this model neglects vertical motion and any rotational motion.

Derive the equations of motion of the system in terms of the angle θ and the displacement x.

4.80 Obtain the equations of motion for the system shown in Figure P4.80 for the case where $m_1 = m$ and $m_2 = 3m$. The cylinder is solid and rolls without slipping. The platform translates without friction on the horizontal surface.

Figure P4.80

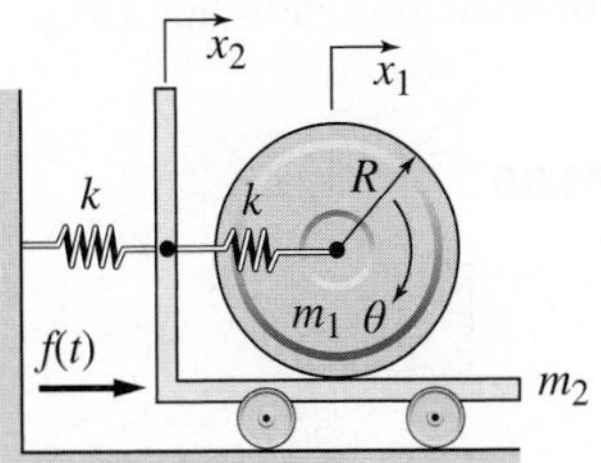

Figure P4.81

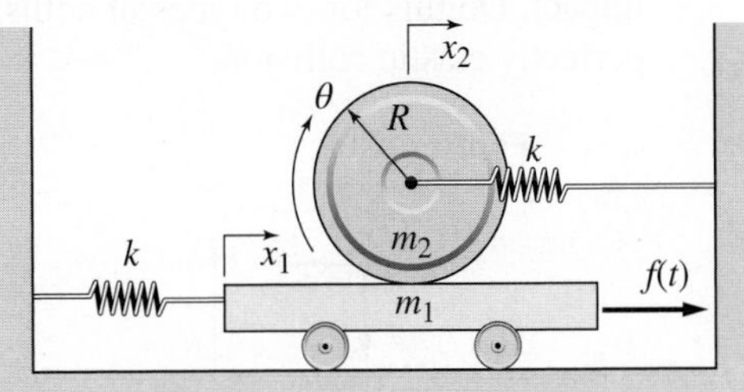

4.81 Obtain the equations of motion for the system shown in Figure P4.81 for the case where $m_1 = m$ and $m_2 = 2m$. The cylinder is solid and rolls without slipping. The platform translates without friction on the horizontal surface.

4.82 In Figure P4.82 a tractor and a trailer is used to carry objects, such as a large paper roll or pipes. Assuming the cylindrical load m_3 rolls without slipping, obtain the equations of motion of the system.

Figure P4.82

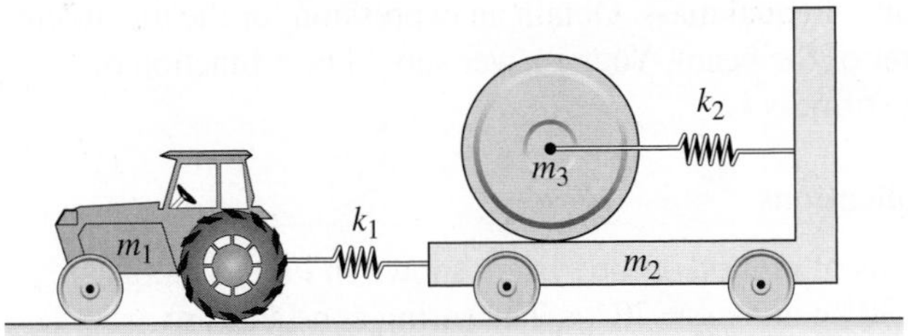

Section 4.6 Collisions and Impulse Response

4.83 Suppose a mass m moving with a speed v_1 becomes embedded in mass m_2 after striking it (Figure 4.6.1). Suppose $m_2 = 4m$. Determine the expression for the displacement $x(t)$ after the collision.

4.84 Consider the system shown in Figure 4.6.3. Suppose that the mass m moving with a speed v_1 rebounds from the mass $m_2 = 4m$ after striking it. Assume that the collision is perfectly elastic. Determine the expression for the displacement $x(t)$ after the collision.

4.85 The mass m_1 is dropped from rest a distance h onto the mass m_2, which is initially resting on the spring support (Figure P4.85). Assume that the impact is inelastic so that m_1 sticks to m_2. Calculate the maximum spring deflection caused by the impact. The given values are $m_1 = 2$ kg, $m_2 = 6$ kg, $k = 500$ N/m, and $h = 3$ m.

Figure P4.85

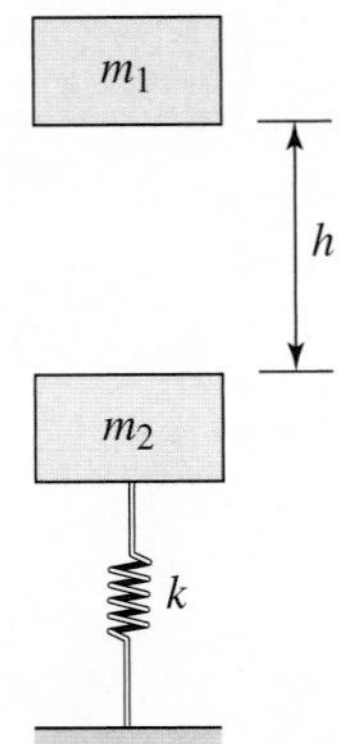

4.86 Figure P4.86 shows a mass m with an attached stiffness, such as that due to protective packaging. The mass drops a distance h, at which time the stiffness element contacts the ground. Let x denote the displacement of m after contact with the ground. Determine the maximum required stiffness of the packaging if a 20 kg package cannot experience a deceleration greater than 7 g when dropped from a height of 3 m.

4.87 Figure P4.87 represents a drop forging process. The anvil mass is $m_1 = 1200$ kg, and the hammer mass is $m_2 = 300$ kg. The support stiffness is $k = 10^7$ N/m, and the damping constant is $c = 2 \times 10^4$ N·s/m. The anvil is at rest when the hammer is dropped from a height of $h = 1$ m. Obtain the expression for the displacement of the anvil as a function of time after the impact. Do this for two types of collisions: (a) an inelastic collision and (b) a perfectly elastic collision.

Figure P4.86

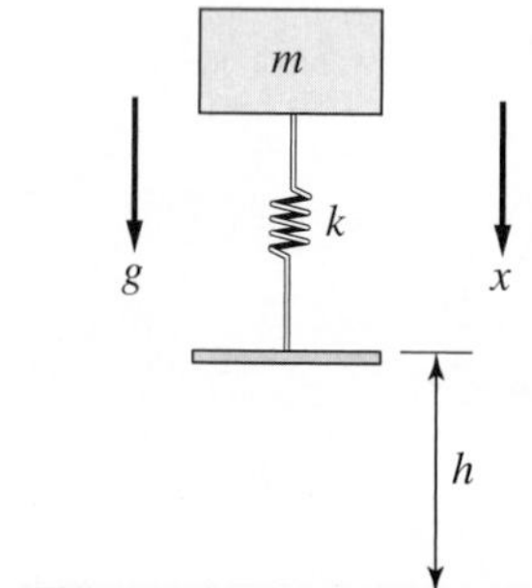

Figure P4.87

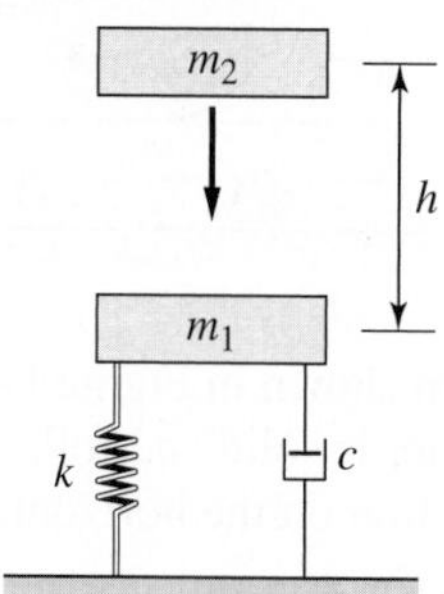

Figure P4.88

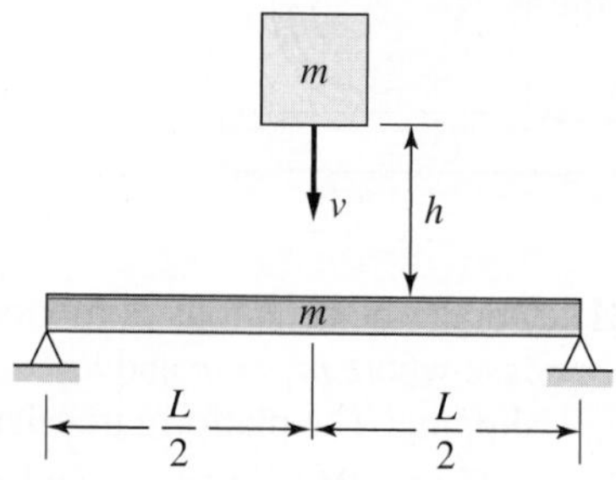

4.88 Refer to Figure P4.88. A mass m drops from a height h and hits and sticks to a simply supported beam of equal mass. Obtain an expression for the maximum deflection of the center of the beam. Your answer should be a function of h, g, m, and the beam stiffness k.

Section 4.7 MATLAB Applications

4.89 (a) Obtain the equations of motion of the system shown in Figure P4.89. The masses are $m_1 = 30$ kg and $m_2 = 70$ kg. The spring constants are $k_1 = 3 \times 10^4$ N/m and $k_2 = 6 \times 10^4$ N/m. (b) Obtain the transfer functions $X_1(s)/F(s)$ and $X_2(s)/F(s)$. (c) Obtain a plot of the unit-step response of x_1 for zero initial conditions.

Figure P4.89

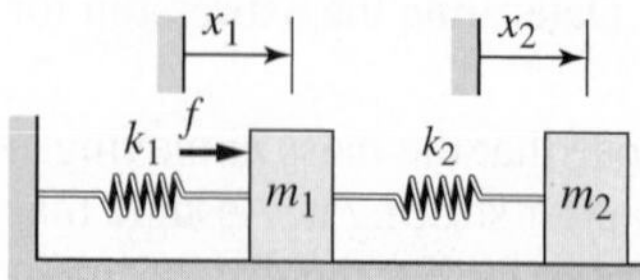

4.90 (a) Obtain the equations of motion of the system shown in Figure P4.25. (b) Suppose the inertias are $I_1 = I$ and $I_2 = 3I$ and the torsional spring constants are $k_1 = k_2 = k_3 = k$. Obtain the transfer functions $\Theta_1(s)/T_2(s)$ and $\Theta_2(s)/T_2(s)$ in terms of I and k. (c) Suppose that $I = 10$ and $k = 60$. Obtain a plot of the unit-impulse response of θ_1 for zero initial conditions.

4.91 Refer to part (a) of Problem 4.89. Use MATLAB to obtain the transfer functions $X_1(s)/F(s)$ and $X_2(s)/F(s)$. Compare your answers with those found in part (b) of Problem 4.90.

4.92 Refer to Problem 4.90. Use MATLAB to obtain the transfer functions $\Theta_1(s)/T_2(s)$ and $\Theta_2(s)/T_2(s)$ for the values $I_1 = 10$, $I_2 = 30$, and $k_1 = k_2 = k_3 = 60$. Compare your answers with those found in part (b) of Problem 4.90.

4.93 (a) Obtain the equations of motion of the system shown in Figure P4.26. Assume small angles. The spring is at its free length when $\theta_1 = \theta_2 = 0$. (b) For the values $m_1 = 1$ kg, $m_2 = 5$ kg, $k = 10$ N/m, $L_1 = 2$ m, and $L_2 = 5$ m, use MATLAB to plot the free response of θ_1 if $\theta_1(0) = 0.1$ rad and $\dot{\theta}_1(0) = \dot{\theta}_2(0) = \theta_2(0) = 0$.

4.94 (a) Obtain the equations of motion of the system shown in Figure P4.94. (b) Suppose that the masses are $m_1 = 1$ kg, $m_2 = 3$ kg, and the spring constants are $k_1 = k_2 = k_3 = 1.6 \times 10^4$ N/m. Use MATLAB to obtain the plot of the free response of x_1. Use $x_1(0) = 0.1$ m, $x_2(0) = \dot{x}_1(0) = \dot{x}_2(0) = 0$.

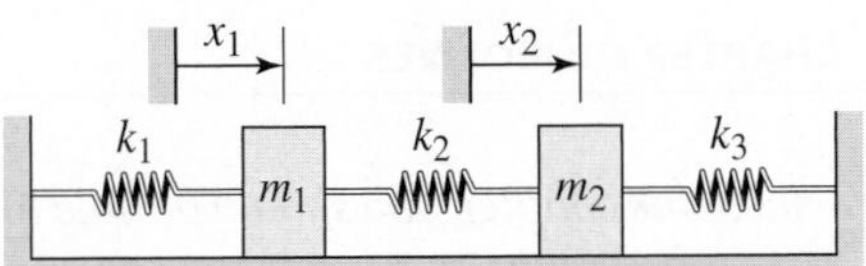

Figure P4.94

4.8 Case Study: Vehicle Suspension Design

4.95 Investigate the performance of the suspension designed in Example 4.8.3 for the longer bump lengths: $L = 5$ and $L = 10$.

4.96 (a) Design a suspension for a vehicle whose quarter mass is 350 kg and whose tire stiffness is 2×10^5 N/m. The tire and swing arm have a combined mass of 35 kg. (b) Evaluate the design performance when striking a haversine bump 0.1 m high while moving at 18 m/s. Consider two bump lengths: $L = 0.5$ m and $L = 2$ m. Compare the results with those of Example 4.8.3.

4.97 Another bump profile is given by $y(z) = 0.5437ze^{-2z}$. This bump is about 2 m in length and 0.1 m high. If the vehicle speed is 18 m/s (about 45 mph), this profile gives $y(t) = 9.7866te^{-36t}$. Use this profile to evaluate the performance of the suspension designed in Example 4.8.3 for the case where $L = 2$.

4.98 Figure P4.98 shows a quarter-car model that includes the mass of the seats (including passengers). The constants k_3 and c_3 represent the stiffness and damping in the seat supports. Derive the equations of motion of this system. The input is the road displacement $y(t)$. The displacements are measured from equilibrium.

Figure P4.98

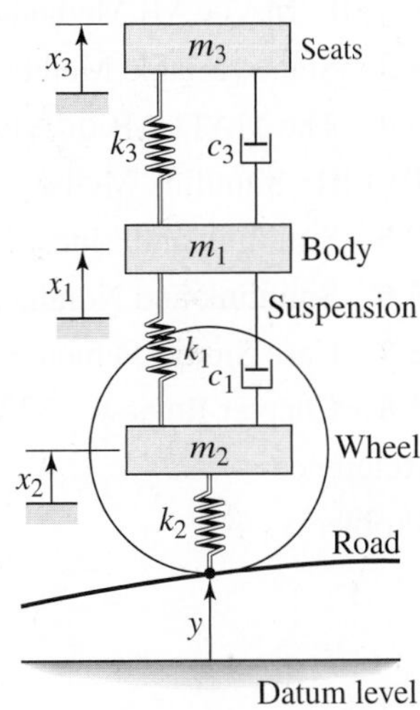

CHAPTER 5

Block Diagrams, State-Variable Models, and Simulation Methods

CHAPTER OUTLINE

CHAPTER OBJECTIVES

When you have finished this chapter, you should be able to

1. Describe a dynamic model as a block diagram.
2. Derive system transfer functions from a block diagram.
3. Convert a differential equation model into state-variable form.
4. Express a linear state-variable model in the standard vector-matrix form.
5. Apply the `ss`, `ssdata`, `tfdata`, `eig`, and `initial` functions to analyze linear models.
6. Use the MATLAB `ode` functions to solve differential equations.
7. Use Simulink to create simulations of dynamic models.

Dynamic models derived from basic physical principles can appear in several forms:

1. As a single equation (which is called the *reduced* form),
2. As a set of coupled first-order equations (which is called the *Cauchy* or *state-variable* form), and
3. As a set of coupled higher-order equations.

In Chapters 2, 3, and 4 we analyzed the response of a single equation, such as $m\ddot{x}+c\dot{x}+kx=f(t)$, and sets of coupled first-order equations, such as $\dot{x}=-5x+7y$, $\dot{y}=3x-9y+f(t)$, by first obtaining the transfer functions and then using the transfer functions to obtain a single, but higher-order equation. We also obtained the response of models that consist of a set of coupled higher-order equations.

Each form has its own advantages and disadvantages. We can convert one form into another, with differing degrees of difficulty. In addition, if the model is *linear*, we can convert any of these forms into the transfer-function form or a vector-matrix form, each of which has its own advantages.

GUIDE TO THE CHAPTER

This chapter has three parts. Part I is required to understand Parts II and III, but Parts II and III are independent of each other.

In Part I, which includes Sections 5.1 and 5.2, we introduce block diagrams, which are based on the transfer-function concept, and the state-variable model form. The block diagram is a way of representing the dynamics of a system in graphical form. Block diagrams will be used often in the subsequent chapters to describe dynamic systems, and they are also the basis for Simulink programming covered in Part III. An advantage of the state-variable form is that it enables us to express a linear model of any order in a standard and compact way that is useful for analysis and software applications. In Chapter 2 we introduced the `tf`, `step`, and `lsim` functions, which can solve models in transfer-function form.

MATLAB has a number of useful functions that are based on the state-variable model form. These functions are covered in Part II, which includes Sections 5.3 and 5.4. Section 5.3 deals with linear models. While the analysis methods of the previous chapters are limited to linear models, the state-variable form is also useful for solving nonlinear equations. It is not always possible or convenient to obtain the closed-form solution of a differential equation, and this is usually true for nonlinear equations. Section 5.4 introduces MATLAB functions that are useful for solving nonlinear differential equations.

Part III includes Sections 5.5 and 5.6 and introduces Simulink, which provides a graphical user interface for solving differential equations. It is especially useful for solving problems containing nonlinear features such as Coulomb friction, saturation, and dead zones, because these features are very difficult to program with traditional programming methods. In addition, its graphical interface might be preferred by some users to the more traditional programming methods offered by the MATLAB solvers covered in Part II. In Section 5.5 we begin with solving linear equations so that we can check the results with the analytical solution. Section 5.6 covers Simulink methods for nonlinear equations. ■

PART I. MODEL FORMS

5.1 TRANSFER FUNCTIONS AND BLOCK DIAGRAM MODELS

We have seen that the complete response of a linear ordinary differential equation (ODE) is the sum of the free and the forced responses. For zero-initial conditions, the free response is zero, and the complete response is the same as the forced response.

Thus, we can focus our analysis on the effects of only the input by taking the initial conditions to be zero temporarily. When we have finished analyzing the effects of the input, we can add to the result the free response due to any nonzero initial conditions.

The transfer function is useful for analyzing the effects of the input. Recall from Chapter 2 that the transfer function $T(s)$ is the transform of the forced response $X(s)$ divided by the transform of the input $F(s)$.

$$T(s) = \frac{X(s)}{F(s)}$$

The transfer function can be used as a multiplier to obtain the forced response transform from the input transform; that is, $X(s) = T(s)F(s)$. The transfer function is a property of the system model only. The transfer function is independent of the input function and the initial conditions.

The transfer function is equivalent to the ODE. If we are given the transfer function, we can reconstruct the corresponding ODE. For example, the transfer function

$$\frac{X(s)}{F(s)} = \frac{5}{s^2 + 7s + 10}$$

corresponds to the equation $\ddot{x} + 7\dot{x} + 10x = 5f(t)$.

Obtaining a transfer function from a single ODE is straightforward, as we have seen. Sometimes, however, models have more than one input or more than one output. It is important to realize that there is one transfer function for each input-output pair. If a model has more than one input, the transfer function for a particular output variable is the ratio of the transform of the forced response of that variable divided by the input transform, with all the remaining inputs ignored (set to zero temporarily). For example, if the variable x is the output for the equation

$$5\ddot{x} + 30\dot{x} + 40x = 6f(t) - 20g(t)$$

then there are two transfer functions, $X(s)/F(s)$ and $X(s)/G(s)$. These are

$$\frac{X(s)}{F(s)} = \frac{6}{5s^2 + 30s + 40} \qquad \frac{X(s)}{G(s)} = -\frac{20}{5s^2 + 30s + 40}$$

We can obtain transfer functions from systems of equations by first transforming the equations and then algebraically eliminating all variables, except for the specified input and output. This technique is especially useful when we want to obtain the response of one or more of the dependent variables in the system of equations. For example, the transfer functions $X(s)/V(s)$ and $Y(s)/V(s)$ of the following system of equations:

$$\dot{x} = -3x + 2y$$

$$\dot{y} = -9y - 4x + 3v(t)$$

are

$$\frac{X(s)}{V(s)} = \frac{6}{s^2 + 12s + 35}$$

and

$$\frac{Y(s)}{V(s)} = \frac{3(s + 3)}{s^2 + 12s + 35}$$

5.1.1 BLOCK DIAGRAMS

We can use the transfer functions of a model to construct a visual representation of the dynamics of the model. Such a representation is a *block diagram*. Block diagrams can be used to describe how system components interact with each other. Unlike a schematic diagram, which shows the physical connections, the block diagram shows the cause-and-effect relations between the components, and thus helps us to understand the system's dynamics.

Block diagrams can also be used to obtain transfer functions for a given system, for cases where the describing differential equations are not given. In addition, as we will see in Section 5.5, block diagrams can be used to develop *simulation diagrams* for use with computer tools such as Simulink.

5.1.2 BLOCK DIAGRAM SYMBOLS

Block diagrams are constructed from the four basic symbols shown in Figure 5.1.1:

1. The arrow, which is used to represent a variable and the direction of the cause-and-effect relation;
2. The block, which is used to represent the input-output relation of a transfer function;
3. The circle, generically called a *summer*, which represents addition as well as subtraction, depending on the sign associated with the variable's arrow; and
4. The *takeoff point*, which is used to obtain the value of a variable from its arrow, for use in another part of the diagram.

The takeoff point does not modify the value of a variable; a variable has the same value along the entire length of an arrow until it is modified by a circle or a block. You may think of a takeoff point as the tip of a voltmeter probe used to measure a voltage at a point on a wire. If the voltmeter is well designed, it will not change the value of the voltage it is measuring.

5.1.3 SOME SIMPLE BLOCK DIAGRAMS

The simplest block diagram is shown in Figure 5.1.1b. Inside the block is the system transfer function $T(s)$. The arrow going into the block represents the transform of the input, $F(s)$; the arrow coming out of the block represents the transform of the output, $X(s)$. Thus, the block diagram is a graphical representation of the cause-and-effect relations operating in a particular system. A specific case is shown in Figure 5.1.2a in which the constant transfer function K represents multiplication and the block is called a *multiplier* or a *gain* block. The corresponding equation in the time domain is $x(t) = Kf(t)$. Another simple case is shown in Figure 5.1.2b in which the transfer function $1/s$ represents integration. The corresponding equation in the time domain is $x(t) = \int f(t)\,dt$. Thus, such a block is called an *integrator*. Note that this relation corresponds to the differential equation $\dot{x} = f(t)$.

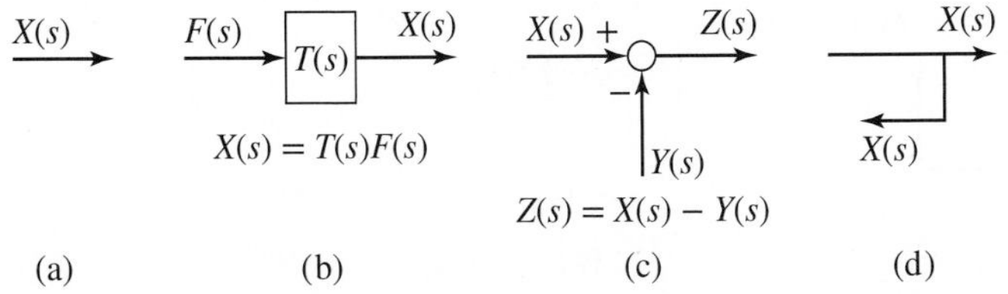

$X(s)$ (a)

$F(s)$ → $T(s)$ → $X(s)$; $X(s) = T(s)F(s)$ (b)

$X(s)$ +, −$Y(s)$, $Z(s)$; $Z(s) = X(s) - Y(s)$ (c)

$X(s)$, $X(s)$ (d)

Figure 5.1.1 The four basic symbols used in block diagrams.

Figure 5.1.2 Two types of blocks. (a) Multiplier. (b) Integrator.

Figure 5.1.3 Diagrams representing the equation $\dot{x} + 7x = f(t)$.

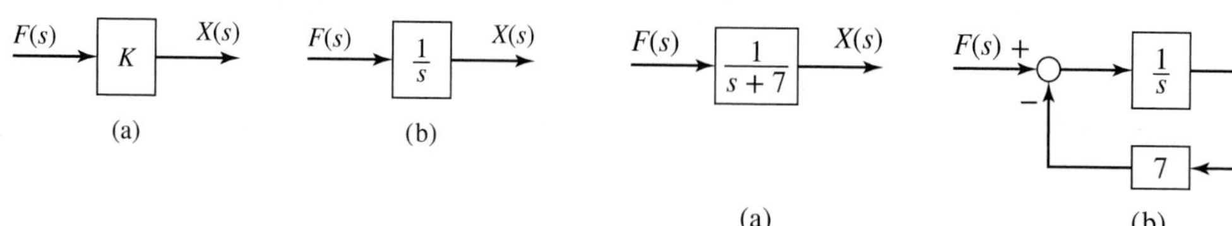

5.1.4 EQUIVALENT BLOCK DIAGRAMS

Figure 5.1.3 shows how more than one diagram can represent the same model, which in this case is $\dot{x} + 7x = f(t)$. The transfer function is $X(s)/F(s) = 1/(s + 7)$, and the corresponding diagram is shown in part (a) of the figure. However, we can rearrange the equation as follows:

$$\dot{x} = f(t) - 7x \qquad \text{or} \qquad x = \int [f(t) - 7x]\, dt$$

which gives

$$X(s) = \frac{1}{s}[F(s) - 7X(s)]$$

In this arrangement the equation corresponds to the diagram shown in part (b) of the figure. Note the use of the takeoff point to feed the variable $X(s)$ to the multiplier. The circle symbol is used to represent addition of the variable $F(s)$ and subtraction of $7X(s)$. The diagram shows how $\dot{x}$, the rate of change of x, is affected by x itself. This is shown by the path from $X(s)$ through the multiplier block to the summer, which changes the sign of $7X(s)$. This path is called a *negative feedback path* or a *negative feedback loop*.

5.1.5 SERIES ELEMENTS AND FEEDBACK LOOPS

Figure 5.1.4 shows two common forms found in block diagrams. In part (a) the two blocks are said to be in *series*. It is equivalent to the diagram in part (b) because we may write

$$B(s) = T_1(s)F(s) \qquad X(s) = T_2(s)B(s)$$

These can be combined algebraically by eliminating $B(s)$ to obtain $X(s) = T_1(s) T_2(s)F(s)$. Note that block diagrams obey the rules of algebra. Therefore, any rearrangement permitted by the rules of algebra is a valid diagram.

Figure 5.1.4 (a) and (b) Simplification of series blocks. (c) and (d) Simplification of a feedback loop.

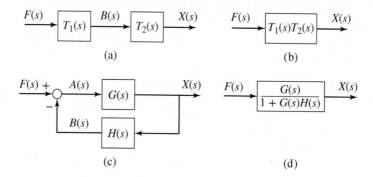

F(s) → $T_1(s)$ → B(s) → $T_2(s)$ → X(s) (a)

F(s) → $T_1(s)T_2(s)$ → X(s) (b)

F(s) + → A(s) → $G(s)$ → X(s); feedback B(s) ← $H(s)$, − (c)

F(s) → $\dfrac{G(s)}{1 + G(s)H(s)}$ → X(s) (d)

Figure 5.1.4c shows a negative feedback loop. From the diagram, we can obtain the following.

$$A(s) = F(s) - B(s) \qquad B(s) = H(s)X(s) \qquad X(s) = G(s)A(s)$$

We can eliminate $A(s)$ and $B(s)$ to obtain

$$X(s) = \frac{G(s)}{1 + G(s)H(s)}F(s) \tag{5.1.1}$$

This is a useful formula for simplifying a feedback loop to a single block.

5.1.6 REARRANGING BLOCK DIAGRAMS

Now consider the second-order model $\ddot{x} + 7\dot{x} + 10x = f(t)$. The transfer function is $X(s)/F(s) = 1/(s^2 + 7s + 10)$, and the simplest diagram for this model is shown in Figure 5.1.5a. However, to show how x and $\dot{x}$ affect the dynamics of the system, we can construct a diagram that contains the appropriate feedback paths for x and $\dot{x}$. To do this, rearrange the equation by solving for the highest derivative.

$$\ddot{x} = f(t) - 7\dot{x} - 10x$$

The transformed equation is

$$X(s) = \frac{1}{s}\left(\frac{1}{s}\{F(s) - 7[sX(s)] - 10X(s)\}\right)$$

With this arrangement we can construct the diagram shown in Figure 5.1.5b. Recall that $sX(s)$ represents $\dot{x}$. The term within the pair of curly braces is the output of the summer and the input to the leftmost integrator. The output of this integrator is shown within the outermost pair of parentheses and is the input to the rightmost integrator.

We may use two summers instead of one, and rearrange the diagram as shown in Figure 5.1.5c. This form shows more clearly the negative feedback loop associated with the derivative $\dot{x}$. Referring to Figure 5.1.3, we see that we may replace this inner loop with its equivalent transfer function $1/(s + 7)$. The result is shown in Figure 5.1.5d, which displays only the feedback loop associated with x.

Figure 5.1.5 Diagrams representing the equation $\ddot{x} + 7\dot{x} + 10x = f(t)$.

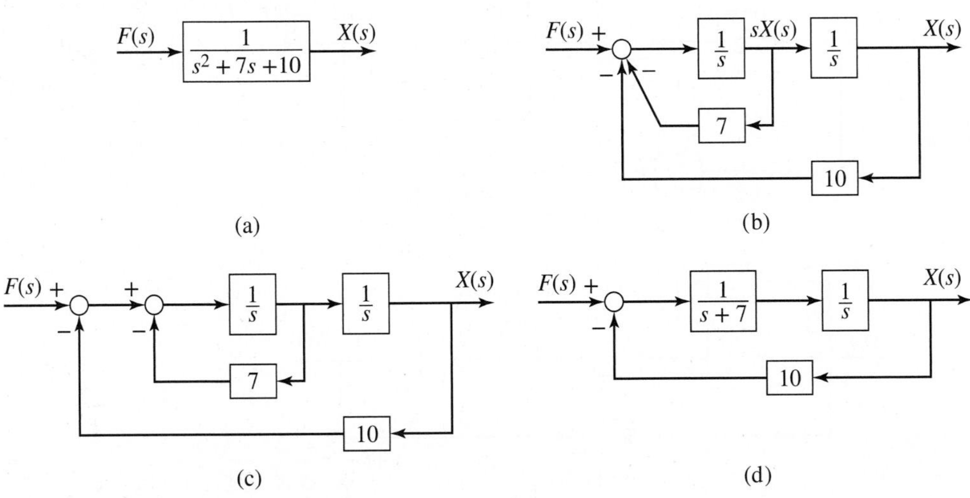

Two important points can be drawn from these examples.

1. More than one correct diagram can be drawn for a given equation; the desired form of the diagram depends on what information we want to display.
2. The form of the resulting diagram depends on how the equation is arranged. A useful procedure for constructing block diagrams is to first solve for the highest derivative of the dependent variable; the terms on the right side of the resulting equation represent the input to an integrator block.

It is important to understand that the block diagram is a "picture" of the algebraic relations obtained by applying the Laplace transform method to the differential equations, assuming that the initial conditions are zero. Therefore, a number of different diagrams can be constructed for a given set of equations and they will all be valid as long as the algebraic relations are correctly represented.

Block diagrams are especially useful when the model consists of more than one differential equation or has more than one input or output. For example, consider the model

$$\dot{x} = -3y + f(t) \qquad \dot{y} = -5y + 4x + g(t)$$

which has two inputs, $f(t)$ and $g(t)$. Suppose we are interested in the variable y as the output. Then the diagram in Figure 5.1.6a is appropriate. Notice that it shows how y affects itself through the feedback loop with the gain of 3, by first affecting x.

Usually we try to place the output variable on the right side of the diagram, with its arrow pointing to the right. We try to place one input on the left side with its arrow point to the right, with a second input, if any, placed at the top of the diagram. The diagram shown in Figure 5.1.6a follows these conventions, which have been established to make it easier for others to interpret your diagrams. Just as in the English language we read from left to right, so the main "flow" of the cause and effect in a diagram (from input to output) should be from left to right if possible.

If instead, we choose the output to be x, then Figure 5.1.6b is more appropriate.

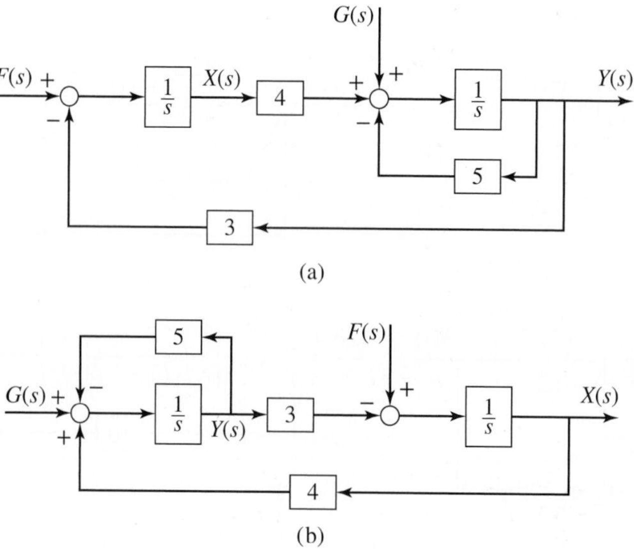

Figure 5.1.6 A diagram with two inputs.

5.1.7 TRANSFER FUNCTIONS FROM BLOCK DIAGRAMS

Sometimes we are given a block diagram and asked to find either the system's transfer function or its differential equation. There are several ways to approach such a problem; the appropriate method depends partly on the form of the diagram and partly on personal preference. The following examples illustrate the process.

EXAMPLE 5.1.1

Series Blocks and Loop Reduction

■ Problem

Determine the transfer function $X(s)/F(s)$ for the system whose diagram is shown in Figure 5.1.7a.

■ Solution

When two blocks are connected by an arrow, they can be combined into a single block that contains the product of their transfer functions. The result is shown in part (b) of the figure. This property, which is called the *series* or *cascade* property, is easily demonstrated. In terms of the variables $X(s)$, $Y(s)$, and $Z(s)$ shown in the diagram, we can write

$$X(s) = \frac{1}{s+12}Y(s) \qquad Y(s) = \frac{1}{s+6}Z(s)$$

Eliminating $Y(s)$ we obtain

$$X(s) = \frac{1}{s+6}\frac{1}{s+12}Z(s)$$

This gives the diagram in part (b) of the figure. So we see that combining two blocks in series is equivalent to eliminating the intermediate variable $Y(s)$ algebraically.

To find the transfer function $X(s)/F(s)$, we can write the following equations based on the diagram in part (b) of the figure:

$$X(s) = \frac{1}{(s+6)(s+12)}Z(s) \qquad Z(s) = F(s) - 8X(s)$$

Eliminating $Z(s)$ from these equations gives the transfer function

$$\frac{X(s)}{F(s)} = \frac{1}{s^2 + 18s + 80}$$

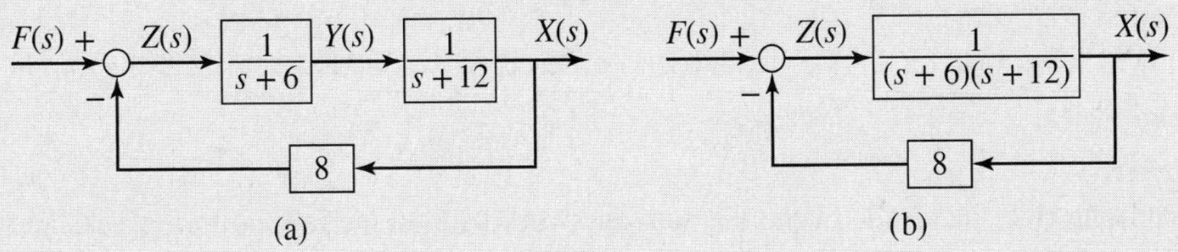

Figure 5.1.7 An example of series combination and loop reduction.

EXAMPLE 5.1.2

Using Integrator Outputs

■ Problem

Determine the model for the output x for the system whose diagram is shown in Figure 5.1.8.

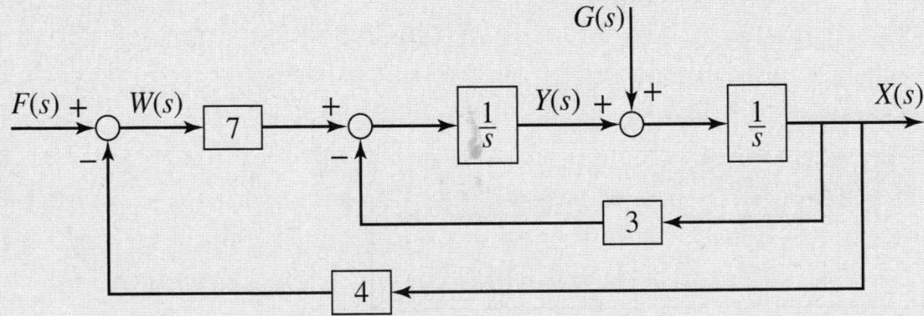

Figure 5.1.8 Diagram for Example 5.1.2.

■ Solution

The input to an integrator block $1/s$ is the time derivative of the output. Thus, by examining the inputs to the two integrators shown in the diagram, we can immediately write the time-domain equations as follows:

$$\dot{x} = g(t) + y \qquad \dot{y} = 7w - 3x \qquad w = f(t) - 4x$$

We can eliminate the variable w from the last two equations to obtain $\dot{y} = 7f(t) - 31x$. Thus, the model in differential equation form is

$$\dot{x} = g(t) + y \qquad \dot{y} = 7f(t) - 31x$$

To obtain the model in transfer-function form, we first transform the equations.

$$sX(s) = G(s) + Y(s) \qquad sY(s) = 7F(s) - 31X(s)$$

Then we eliminate $Y(s)$ algebraically to obtain

$$X(s) = \frac{7}{s^2 + 31}F(s) + \frac{s}{s^2 + 31}G(s)$$

There are two transfer functions, one for each input-output pair. They are

$$\frac{X(s)}{F(s)} = \frac{7}{s^2 + 31} \qquad \frac{X(s)}{G(s)} = \frac{s}{s^2 + 31}$$

Sometimes, we need to obtain the expressions not for just the output variables, but also for some internal variables. The following example illustrates the required method.

EXAMPLE 5.1.3 Deriving Expressions for Internal Variables

■ Problem

Derive the expressions for $C(s)$, $E(s)$, and $M(s)$ in terms of $R(s)$ and $D(s)$ for the diagram in Figure 5.1.9.

■ Solution

Start from the right-hand side of the diagram and work back to the left until *all* blocks and comparators are accounted for. This gives

$$C(s) = \frac{7}{s+3}[M(s) - D(s)] \tag{1}$$

$$M(s) = \frac{K}{4s+1}E(s) \tag{2}$$

$$E(s) = R(s) - C(s) \tag{3}$$

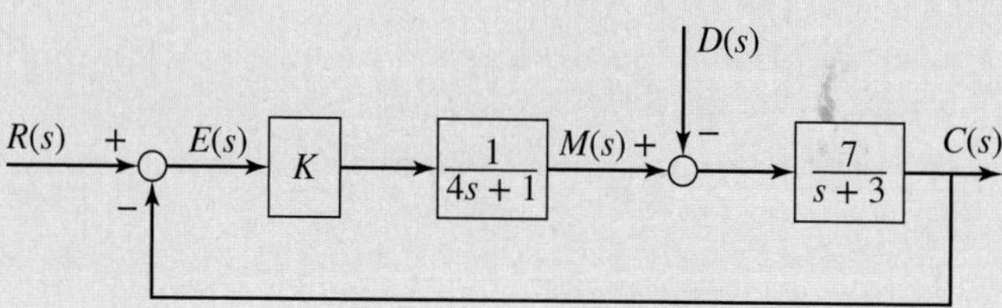

Figure 5.1.9 Block diagram for Example 5.1.3.

Multiply both sides of equation (1) by $s + 3$ to clear fractions, and substitute $M(s)$ and $E(s)$ from equations (2) and (3).

$$(s+3)C(s) = 7M(s) - 7D(s)$$

$$= 7\frac{K}{4s+1}E(s) - 7D(s) = \frac{7K}{4s+1}[R(s) - C(s)] - 7D(s)$$

Multiply both sides by $4s + 1$ to clear fractions, and solve for $C(s)$ to obtain:

$$C(s) = \frac{7K}{4s^2 + 13s + 3 + 7K}R(s) - \frac{7(4s+1)}{4s^2 + 13s + 3 + 7K}D(s) \quad (4)$$

The characteristic polynomial is found from the denominator of either transfer function. It is $4s^2 + 13s + 3 + 7K$.

The equation for $E(s)$ is

$$E(s) = R(s) - C(s)$$

$$= R(s) - \frac{7K}{4s^2 + 13s + 3 + 7K}R(s) + \frac{7(4s+1)}{4s^2 + 13s + 3 + 7K}D(s)$$

$$= \frac{4s^2 + 13s + 3}{4s^2 + 13s + 3 + 7K}R(s) + \frac{7(4s+1)}{4s^2 + 13s + 3 + 7K}D(s)$$

Because $4s^2 + 13s + 3$ can be factored as $(4s + 1)(s + 3)$, the equation for $M(s)$ can be expressed as

$$M(s) = \frac{K}{4s+1}E(s)$$

$$= \frac{K}{4s+1}\left[\frac{(4s+1)(s+3)}{4s^2 + 13s + 3 + 7K}R(s) + \frac{7(4s+1)}{4s^2 + 13s + 3 + 7K}D(s)\right]$$

$$= \frac{K(s+3)}{4s^2 + 13s + 3 + 7K}R(s) + \frac{7K}{4s^2 + 13s + 3 + 7K}D(s)$$

Note the cancellation of the term $4s + 1$. You should always look for such cancellations. Otherwise, the denominator of the transfer functions can appear to be of higher order than the characteristic polynomial.

5.1.8 BLOCK DIAGRAM ALGEBRA USING MATLAB

MATLAB can be used to perform block diagram algebra if all the gains and transfer function coefficients have numerical values. You can combine blocks in series or in feedback loops using the `series` and `feedback` functions to obtain the transfer function and the state-space model.

If the LTI models `sys1` and `sys2` represent blocks in series, their combined transfer function can be obtained by typing `sys3 = series(sys1,sys2)`. A simple gain need not be converted to an LTI model, and does not require the `series` function. For example, if the first system is a simple gain K, use the multiplication symbol * and enter

```
>>sys3 = K*sys2
```

If the LTI model `sys2` is in a *negative* feedback loop around the LTI model `sys1`, then enter

```
>>sys3 = feedback(sys1,sys2)
```

to obtain the LTI model of the closed-loop system.

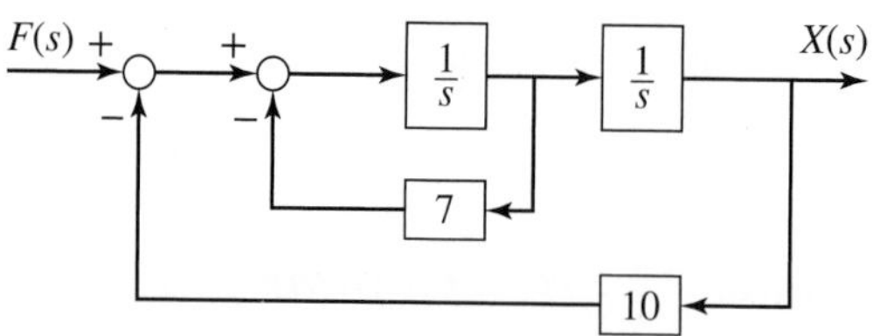

Figure 5.1.10 A typical block diagram.

If the feedback loop is *positive*, use the syntax

```
>>sys3 = feedback(sys1,sys2,+1)
```

If you need to obtain the numerator and denominator of the closed-loop transfer function, you can use the `tfdata` function and enter

```
>>[num,den] = tfdata(sys3,'v')
```

You can then find the characteristic roots by entering

```
>>roots(den)
```

For example, to find the transfer function $X(s)/F(s)$ corresponding to the block diagram shown in Figure 5.1.10, you enter

```
>>sys1=tf(1,[1,0]);
>>sys2=feedback(sys1,7);
>>sys3=series(sys1,sys2);
>>sys4=feedback(sys3,10);
>>[num,den]=tfdata(sys4,'v')
```

The result is num = [0, 0, 1] and den = [1, 7, 10], which corresponds to

$$\frac{X(s)}{F(s)} = \frac{1}{s^2 + 7s + 10}$$

5.2 STATE-VARIABLE MODELS

Models that consist of coupled first-order differential equations are said to be in *state-variable* form. This form, which is also called the *Cauchy* form, has an advantage over the reduced form, which consists of a single, higher-order equation, because it allows a linear model to be expressed in a standard and compact way that is useful for analysis and for software applications. This representation makes use of vector and matrix notation. In this section, we will show how to obtain a model in state-variable form and how to express state-variable models in vector-matrix notation. In Section 5.3 we show how to use this notation with MATLAB.

Consider the second-order equation

$$5\ddot{y} + 7\dot{y} + 4y = f(t)$$

Solve it for the highest derivative:

$$\ddot{y} = \frac{1}{5}f(t) - \frac{4}{5}y - \frac{7}{5}\dot{y}$$

Now define two new variables, x_1 and x_2, as follows: $x_1 = y$ and $x_2 = \dot{y}$. This implies that $\dot{x}_1 = x_2$ and

$$\dot{x}_2 = \frac{1}{5}f(t) - \frac{4}{5}x_1 - \frac{7}{5}x_2$$

These two equations, called the *state equations*, are the state-variable form of the model, and the variables x_1 and x_2 are called the *state variables*.

The general mass-spring-damper model has the following form:

$$m\ddot{x} + c\dot{x} + kx = f \tag{5.2.1}$$

If we define new variables x_1 and x_2 such that

$$x_1 = x \qquad x_2 = \dot{x}$$

these imply that

$$\dot{x}_1 = x_2 \tag{5.2.2}$$

Then we can write the model (5.2.1) as: $m\dot{x}_2 + cx_2 + kx_1 = f$. Next solve for $\dot{x}_2$:

$$\dot{x}_2 = \frac{1}{m}(f - kx_1 - cx_2) \tag{5.2.3}$$

Equations (5.2.2) and (5.2.3) constitute a state-variable model corresponding to the reduced model (5.2.1). The variables x_1 and x_2 are the *state variables*.

If (5.2.1) represents a mass-spring-damper system, the state variable x_1 describes the system's potential energy $kx_1^2/2$, which is due to the spring, and the state variable x_2 describes the system's kinetic energy $mx_2^2/2$, which is due to the mass. Although here we have derived the state-variable model from the reduced form, state-variable models can be derived from basic physical principles. Choosing as state variables those variables that describe the types of energy in the system sometimes helps to derive the model (note that k and m are also needed to describe the energies, but these are parameters, not variables).

The choice of state variables is not unique, but the choice must result in a set of first-order differential equations. For example, we could have chosen the state variables to be $x_1 = x$ and $x_2 = m\dot{x}$, which is the system's momentum. In this case the state-variable model would be

$$\dot{x}_1 = \frac{1}{m}x_2$$

$$\dot{x}_2 = f - \frac{c}{m}x_2 - kx_1$$

State-Variable Model of a Two-Mass System

EXAMPLE 5.2.1

■ Problem

Consider the two-mass system discussed in Chapter 4 (and shown again in Figure 5.2.1). Suppose the parameter values are $m_1 = 5$, $m_2 = 3$, $c_1 = 4$, $c_2 = 8$, $k_1 = 1$, and $k_2 = 4$. The equations of motion are

$$5\ddot{x}_1 + 12\dot{x}_1 + 5x_1 - 8\dot{x}_2 - 4x_2 = 0 \tag{1}$$

$$3\ddot{x}_2 + 8\dot{x}_2 + 4x_2 - 8\dot{x}_1 - 4x_1 = f(t) \tag{2}$$

Put these equations into state-variable form.

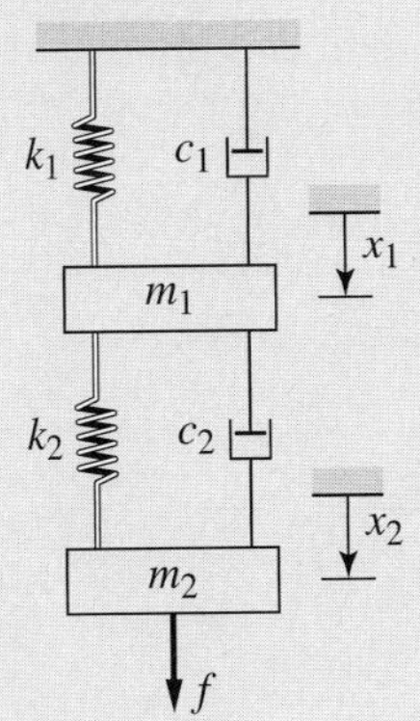

Figure 5.2.1 A two-mass system.

■ Solution

Using the system's potential and kinetic energies as a guide, we see that the displacements x_1 and x_2 describe the system's potential energy and that the velocities $\dot{x}_1$ and $\dot{x}_2$ describe the system's kinetic energy. That is

$$\text{PE} = \frac{1}{2}k_1x_1^2 + \frac{1}{2}k_2(x_1 - x_2)^2$$

and

$$\mathrm{KE} = \frac{1}{2}m_1\dot{x}_1^2 + \frac{1}{2}m_2\dot{x}_2^2$$

This indicates that we need four state variables. (Another way to see that we need four variables is to note that the model consists of two coupled second-order equations, and thus is effectively a fourth-order model.) Thus, we can choose the state variables to be

$$x_1 \qquad x_2 \qquad x_3 = \dot{x}_1 \qquad x_4 = \dot{x}_2 \tag{3}$$

Thus, two of the state equations are $\dot{x}_1 = x_3$ and $\dot{x}_2 = x_4$. The remaining two equations can be found by solving equations (1) and (2) for $\ddot{x}_1$ and $\ddot{x}_2$, noting that $\ddot{x}_1 = \dot{x}_3$ and $\ddot{x}_2 = \dot{x}_4$, and using the substitutions given by equation (3).

$$\dot{x}_3 = \frac{1}{5}(-12x_3 - 5x_1 + 8x_4 + 4x_2)$$

$$\dot{x}_4 = \frac{1}{3}[-8x_4 - 4x_2 + 8x_3 + 4x_1 + f(t)]$$

Note that the left-hand sides of the state equations must contain only the first-order derivative of each state variable. This is why we divided by 5 and 3, respectively. Note also that the right-hand sides must not contain any derivatives of the state variables. Failure to observe this restriction is a common mistake.

Now list the four state equations in ascending order according to their left-hand sides, after rearranging the right-hand sides so that the state variables appear in ascending order from left to right.

$$\dot{x}_1 = x_3 \tag{4}$$

$$\dot{x}_2 = x_4 \tag{5}$$

$$\dot{x}_3 = \frac{1}{5}(-5x_1 + 4x_2 - 12x_3 + 8x_4) \tag{6}$$

$$\dot{x}_4 = \frac{1}{3}[4x_1 - 4x_2 + 8x_3 - 8x_4 + f(t)] \tag{7}$$

These are the state equations in standard form.

5.2.1 VECTOR-MATRIX FORM OF STATE-VARIABLE MODELS

Vector-matrix notation enables us to represent multiple equations as a single matrix equation. For example, consider the following set of linear algebraic equations.

$$2x_1 + 9x_2 = 5 \tag{5.2.4}$$

$$3x_1 - 4x_2 = 7 \tag{5.2.5}$$

The term *matrix* refers to an array with more than one column and more than one row. A *column vector* is an array having only one column. A *row vector* has only one row. A matrix is an arrangement of numbers and is not the same as a determinant, which can be reduced to a single number. Multiplication of a matrix having two rows and two columns (a (2×2) matrix) by a column vector having two rows and one column (a (2 × 1) vector) is defined as follows:

$$\begin{bmatrix} a_{11} & a_{12} \\ a_{21} & a_{22} \end{bmatrix}\begin{bmatrix} x_1 \\ x_2 \end{bmatrix} = \begin{bmatrix} a_{11}x_1 + a_{12}x_2 \\ a_{21}x_1 + a_{22}x_2 \end{bmatrix} \tag{5.2.6}$$

This definition is easily extended to matrices having more than two rows or two columns. In general, the result of multiplying an $(n \times n)$ matrix by an $(n \times 1)$ vector is an $(n \times 1)$ vector. This definition of vector-matrix multiplication requires that the matrix have as many columns as the vector has rows. The order of the multiplication cannot be reversed (vector-matrix multiplication does not have the commutative property).

Two vectors are equal if all their respective elements are equal. Thus, we can represent the set (5.2.4) and (5.2.5) as follows:

$$\begin{bmatrix} 2 & 9 \\ 3 & -4 \end{bmatrix} \begin{bmatrix} x_1 \\ x_2 \end{bmatrix} = \begin{bmatrix} 5 \\ 7 \end{bmatrix} \tag{5.2.7}$$

We usually represent matrices and vectors in boldface type, with matrices usually in uppercase letters and vectors in lowercase, but this is not required. Thus, we can represent the set (5.2.7) in the following compact form.

$$\mathbf{Ax} = \mathbf{b} \tag{5.2.8}$$

where we have defined the following matrices and vectors:

$$\mathbf{A} = \begin{bmatrix} 2 & 9 \\ 3 & -4 \end{bmatrix} \qquad \mathbf{x} = \begin{bmatrix} x_1 \\ x_2 \end{bmatrix} \qquad \mathbf{b} = \begin{bmatrix} 5 \\ 7 \end{bmatrix}$$

The matrix **A** corresponds in an ordered fashion to the coefficients of x_1 and x_2 in (5.2.4) and (5.2.5). Note that the first row in **A** consists of the coefficients of x_1 and x_2 on the left-hand side of (5.2.4), and the second row contains the coefficients on the left-hand side of (5.2.5). The vector **x** contains the variables x_1 and x_2, and the vector **b** contains the right-hand sides of (5.2.4) and (5.2.5).

Vector-Matrix Form of a Single-Mass Model

EXAMPLE 5.2.2

■ **Problem**

Express the mass-spring-damper model (5.2.2) and (5.2.3) as a single vector-matrix equation. These equations are

$$\dot{x}_1 = x_2$$

$$\dot{x}_2 = \frac{1}{m}f(t) - \frac{k}{m}x_1 - \frac{c}{m}x_2$$

■ **Solution**

The equations can be written as one equation as follows:

$$\begin{bmatrix} \dot{x}_1 \\ \dot{x}_2 \end{bmatrix} = \begin{bmatrix} 0 & 1 \\ -\dfrac{k}{m} & -\dfrac{c}{m} \end{bmatrix} \begin{bmatrix} x_1 \\ x_2 \end{bmatrix} + \begin{bmatrix} 0 \\ \dfrac{1}{m} \end{bmatrix} f(t)$$

In compact form this is

$$\dot{\mathbf{x}} = \mathbf{Ax} + \mathbf{B}f(t)$$

where

$$\mathbf{A} = \begin{bmatrix} 0 & 1 \\ -\dfrac{k}{m} & -\dfrac{c}{m} \end{bmatrix} \qquad \mathbf{B} = \begin{bmatrix} 0 \\ \dfrac{1}{m} \end{bmatrix} \qquad \mathbf{x} = \begin{bmatrix} x_1 \\ x_2 \end{bmatrix}$$

EXAMPLE 5.2.3

Vector-Matrix Form of the Two-Mass Model

■ **Problem**

Express the state-variable model of Example 5.2.1 in vector-matrix form. The model is

$$\dot{x}_1 = x_3$$

$$\dot{x}_2 = x_4$$

$$\dot{x}_3 = \frac{1}{5}(-5x_1 + 4x_2 - 12x_3 + 8x_4)$$

$$\dot{x}_4 = \frac{1}{3}[4x_1 - 4x_2 + 8x_3 - 8x_4 + f(t)]$$

■ **Solution**

In vector-matrix form these equations are

$$\dot{\mathbf{x}} = \mathbf{A}\mathbf{x} + \mathbf{B}f(t)$$

where

$$\mathbf{A} = \begin{bmatrix} 0 & 0 & 1 & 0 \\ 0 & 0 & 0 & 1 \\ -1 & \frac{4}{5} & -\frac{12}{5} & \frac{8}{5} \\ \frac{4}{3} & -\frac{4}{3} & \frac{8}{3} & -\frac{8}{3} \end{bmatrix} \qquad \mathbf{B} = \begin{bmatrix} 0 \\ 0 \\ 0 \\ \frac{1}{3} \end{bmatrix}$$

and

$$\mathbf{x} = \begin{bmatrix} x_1 \\ x_2 \\ x_3 \\ x_4 \end{bmatrix} = \begin{bmatrix} x_1 \\ x_2 \\ \dot{x}_1 \\ \dot{x}_2 \end{bmatrix}$$

5.2.2 STANDARD FORM OF THE STATE EQUATION

We may use any symbols we choose for the state variables and the input function, although the common choice is x_i for the state variables and u_i for the input functions. The standard vector-matrix form of the state equations, where the number of state variables is n and the number of inputs is m, is

$$\dot{\mathbf{x}} = \mathbf{A}\mathbf{x} + \mathbf{B}\mathbf{u} \tag{5.2.9}$$

where the vectors **x** and **u** are column vectors containing the state variables and the inputs, if any. The dimensions are as follows:

- The *state vector* **x** is a column vector having n rows.
- The *system matrix* **A** is a square matrix having n rows and n columns.
- The *input vector* **u** is a column vector having m rows.
- The *control* or *input matrix* **B** has n rows and m columns.

In our examples thus far there has been only one input, and for such cases the input vector **u** reduces to a scalar u. The standard form, however, allows for more than one input function. Such would be the case in the two-mass model if external forces f_1 and f_2 are applied to the masses.

5.2.3 THE OUTPUT EQUATION

Some software packages and some design methods require you to define an *output vector*, usually denoted by **y**. The output vector contains the variables that are of interest for the particular problem at hand. These variables are not necessarily the state variables, but might be some combination of the state variables and the inputs. For example, in the mass-spring model, we might be interested in the total force $f - kx - c\dot{x}$ acting on the mass, and in the momentum $m\dot{x}$. In this case, the output vector has two elements. If the state variables are $x_1 = x$ and $x_2 = \dot{x}$, the output vector is

$$\mathbf{y} = \begin{bmatrix} y_1 \\ y_2 \end{bmatrix} = \begin{bmatrix} f - kx - c\dot{x} \\ m\dot{x} \end{bmatrix} = \begin{bmatrix} f - kx_1 - cx_2 \\ mx_2 \end{bmatrix}$$

or

$$\mathbf{y} = \begin{bmatrix} y_1 \\ y_2 \end{bmatrix} = \begin{bmatrix} -k & -c \\ 0 & m \end{bmatrix} \begin{bmatrix} x_1 \\ x_2 \end{bmatrix} + \begin{bmatrix} 1 \\ 0 \end{bmatrix} f = \mathbf{Cx} + \mathbf{D}f$$

where

$$\mathbf{C} = \begin{bmatrix} -k & -c \\ 0 & m \end{bmatrix}$$

and

$$\mathbf{D} = \begin{bmatrix} 1 \\ 0 \end{bmatrix}$$

This is an example of the general form: $\mathbf{y} = \mathbf{Cx} + \mathbf{Du}$.

The standard vector-matrix form of the output equation, where the number of outputs is p, the number of state variables is n, and the number of inputs is m, is

$$\mathbf{y} = \mathbf{Cx} + \mathbf{Du} \tag{5.2.10}$$

where the vector **y** contains the output variables. The dimensions are as follows:

- The *output vector* **y** is a column vector having p rows.
- The *state output matrix* **C** has p rows and n columns.
- The *control output matrix* **D** has p rows and m columns.

The matrices **C** and **D** can always be found whenever the chosen output vector **y** is a linear combination of the state variables and the inputs. However, if the output is a nonlinear function, then the standard form (5.2.10) does not apply. This would be the case, for example, if the output is chosen to be the system's kinetic energy: $\text{KE} = mx_2^2/2$.

The Output Equation for a Two-Mass Model

EXAMPLE 5.2.4

■ Problem

Consider the two-mass model of Example 5.2.1.

(a) Suppose the outputs are x_1 and x_2. Determine the output matrices **C** and **D**. (b) Suppose the outputs are $(x_2 - x_1)$, $\dot{x}_2$, and f. Determine the output matrices **C** and **D**.

■ **Solution**

a. In terms of the **z** vector, $z_1 = x_1$ and $z_3 = x_2$. We can express the output vector **y** as follows:

$$\mathbf{y} = \begin{bmatrix} x_1 \\ x_2 \end{bmatrix} = \begin{bmatrix} 1 & 0 & 0 & 0 \\ 0 & 1 & 0 & 0 \end{bmatrix} \begin{bmatrix} x_1 \\ x_2 \\ x_3 \\ x_4 \end{bmatrix} + \begin{bmatrix} 0 \\ 0 \end{bmatrix} f$$

Thus

$$\mathbf{C} = \begin{bmatrix} 1 & 0 & 0 & 0 \\ 0 & 1 & 0 & 0 \end{bmatrix} \qquad \mathbf{D} = \begin{bmatrix} 0 \\ 0 \end{bmatrix}$$

b. Here the outputs are $y_1 = x_2 - x_1$, $y_2 = \dot{x}_2 = x_4$, and $y_3 = f$. Thus, we can express the output vector as follows:

$$\mathbf{y} = \begin{bmatrix} x_2 - x_1 \\ x_4 \\ f \end{bmatrix} = \begin{bmatrix} -1 & 1 & 0 & 0 \\ 0 & 0 & 0 & 1 \\ 0 & 0 & 0 & 0 \end{bmatrix} \begin{bmatrix} x_1 \\ x_2 \\ x_3 \\ x_4 \end{bmatrix} + \begin{bmatrix} 0 \\ 0 \\ 1 \end{bmatrix} f$$

Thus

$$\mathbf{C} = \begin{bmatrix} -1 & 1 & 0 & 0 \\ 0 & 0 & 0 & 1 \\ 0 & 0 & 0 & 0 \end{bmatrix} \qquad \mathbf{D} = \begin{bmatrix} 0 \\ 0 \\ 1 \end{bmatrix}$$

5.2.4 TRANSFER-FUNCTION VERSUS STATE-VARIABLE MODELS

The decision whether to use a reduced-form model (which is equivalent to a transfer-function model) or a state-variable model depends on many factors, including personal preference. In fact, for many applications both models are equally effective and equally easy to use. Application of basic physical principles sometimes directly results in a state-variable model. An example is the following two-inertia fluid-clutch model derived in Chapter 4.

$$I_d\dot{\omega}_d = T_d - c(\omega_d - \omega_1)$$

$$I_1\dot{\omega}_1 = -T_1 + c(\omega_d - \omega_1)$$

The state and input vectors are

$$\mathbf{x} = \begin{bmatrix} \omega_d \\ \omega_1 \end{bmatrix} \qquad \mathbf{u} = \begin{bmatrix} T_d \\ T_1 \end{bmatrix}$$

The system and input matrices are

$$\mathbf{A} = \begin{bmatrix} -\dfrac{c}{I_d} & \dfrac{c}{I_d} \\ \dfrac{c}{I_1} & -\dfrac{c}{I_1} \end{bmatrix} \qquad \mathbf{B} = \begin{bmatrix} \dfrac{1}{I_d} & 0 \\ 0 & -\dfrac{1}{I_1} \end{bmatrix}$$

For example, this form of the model is easier to use if you need to obtain only numerical values or a plot of the step response, because you can directly use the MATLAB function `step(A,B,C,D)`, to be introduced in Section 5.3. However, if you need to obtain the step response as a function, it might be easier to convert the model to transfer-function form and then use the Laplace transform method to obtain the desired function. To obtain the transfer function from the state-variable model, you may use the MATLAB function `tf(sys)`, as shown in Section 5.3.

The MATLAB functions cited require that all the model parameters have specified numerical values. If, however, you need to examine the effects of a system parameter, say the damping coefficient c in the clutch model, then it is perhaps preferable to convert the model to transfer-function form. In this form, you can examine the effect of c on system response by examining numerator dynamics and the characteristic equation. You can also use the initial- and final-value theorems to investigate the response.

5.2.5 MODEL FORMS HAVING NUMERATOR DYNAMICS

Note that if you only need to obtain the free response, then the presence of input derivatives or numerator dynamics in the model is irrelevant. For example, the free response of the model

$$5\frac{d^3y}{dt^3} + 3\frac{d^2y}{dt^2} + 7\frac{dy}{dt} + 6y = 4\frac{df}{dt} + 9f(t)$$

is identical to the free response of the model

$$5\frac{d^3y}{dt^3} + 3\frac{d^2y}{dt^2} + 7\frac{dy}{dt} + 6y = 0$$

which does not have any inputs. A state-variable model for this equation is easily found to be

$$x_1 = y \qquad x_2 = \dot{y} \qquad x_3 = \ddot{y}$$

$$\dot{x}_1 = x_2 \qquad \dot{x}_2 = x_3 \qquad \dot{x}_3 = -\frac{6}{5}x_1 - \frac{7}{5}x_2 - \frac{3}{5}x_3$$

The free response of this model can be easily found with the MATLAB `initial` function to be introduced in the next section.

For some applications you need to obtain a state-variable model in the standard form. However, in the standard state-variable form $\dot{\mathbf{x}} = \mathbf{A}\mathbf{x} + \mathbf{B}\mathbf{u}$, there is no derivative of the input $\mathbf{u}$. When the model has numerator dynamics or input derivatives, the state variables are not so easy to identify. When there are no numerator dynamics, you can always obtain a state-variable model in standard form from a transfer-function or reduced-form model whose dependent variable is x by defining $x_1 = x$, $x_2 = \dot{x}$, $x_3 = \ddot{x}$, and so forth. This was the procedure followed previously. Note that the initial conditions $x_1(0)$, $x_2(0)$, and $x_3(0)$ are easily obtained from the given conditions $x(0)$, $\dot{x}(0)$, and $\ddot{x}(0)$; that is, $x_1(0) = x(0)$, $x_2(0) = \dot{x}(0)$, and $x_3(0) = \ddot{x}(0)$. However, when numerator dynamics are present, a different technique must be used, and the initial conditions are not as easily related to the state variables.

We now give two examples of how to obtain a state-variable model when numerator dynamics exists.

EXAMPLE 5.2.5

Numerator Dynamics in a First-Order System

■ **Problem**

Consider the transfer-function model

$$\frac{Z(s)}{U(s)} = \frac{5s + 3}{s + 2} \tag{1}$$

This corresponds to the equation

$$\dot{z} + 2z = 5\dot{u} + 3u \tag{2}$$

Note that this equation is not in the standard form $\dot{z} = Az+Bu$ because of the input derivative $\dot{u}$. Demonstrate two ways of converting this model to a state-variable model in standard form.

■ **Solution**

a. One way of obtaining the state-variable model is to divide the numerator and denominator of equation (1) by s.

$$\frac{Z(s)}{U(s)} = \frac{5 + 3/s}{1 + 2/s} \tag{3}$$

The objective is to obtain a 1 in the denominator, which is then used to isolate $Z(s)$ as follows:

$$Z(s) = -\frac{2}{s}Z(s) + 5U(s) + \frac{3}{s}U(s)$$
$$= \frac{1}{s}\left[3U(s) - 2Z(s)\right] + 5U(s)$$

The term within square brackets multiplying $1/s$ is the input to an integrator, and the integrator's output can be selected as a state variable x. Thus,

$$Z(s) = X(s) + 5U(s)$$

where

$$X(s) = \frac{1}{s}\left[3U(s) - 2Z(s)\right] = \frac{1}{s}\left\{3U(s) - 2\left[X(s) + 5U(s)\right]\right\}$$
$$= \frac{1}{s}\left[-2X(s) - 7U(s)\right]$$

This gives

$$\dot{x} = -2x - 7u \tag{4}$$

with the output equation

$$z = x + 5u \tag{5}$$

This fits the standard form (5.2.9) and (5.2.10), with $\mathbf{A} = -2$, $\mathbf{B} = -7$, $\mathbf{y} = z$, $\mathbf{C} = 1$, and $\mathbf{D} = 5$.

Presumably we are given the initial condition $z(0)$, but to solve equation (4) we need $x(0)$. This can be obtained by solving equation (5) for x, $x = z - 5u$, and evaluating it at $t = 0$: $x(0) = z(0) - 5u(0)$. We see that $x(0) = z(0)$ if $u(0) = 0$.

b. Another way is to write equation (1) as

$$Z(s) = (5s + 3)\frac{U(s)}{s + 2} \tag{6}$$

and define the state variable x as follows:

$$X(s) = \frac{U(s)}{s + 2} \tag{7}$$

Thus,

$$sX(s) = -2X(s) + U(s) \tag{8}$$

and the state equation is

$$\dot{x} = -2x + u \tag{9}$$

To find the output equation, note that

$$Z(s) = (5s+3)\frac{U(s)}{s+2} = (5s+3)X(s) = 5sX(s) + 3X(s)$$

Using equation (8) we have

$$Z(s) = 5[-2X(s) + U(s)] + 3X(s) = -7X(s) + 5U(s)$$

and thus the output equation is

$$z = -7x + 5u \tag{10}$$

The initial condition $x(0)$ is found from equation (10) to be $x(0) = [5u(0) - z(0)]/7 = -z(0)/7$ if $u(0) = 0$.

Although the model consisting of equations (9) and (10) looks different from that given by equations (4) and (5), they are both in the standard form and are equivalent, because they were derived from the same transfer function.

This example points out that there is no unique way to derive a state-variable model from a transfer function. It is important to keep this in mind because the state-variable model obtained from the MATLAB `ssdata(sys)` function, to be introduced in the next section, might not be the one you expect. The state-variable model given by MATLAB is $\dot{x} = -2x + 2u$, $z = -3.5x + 5u$. These values correspond to equation (1) being written as

$$Z(s) = \frac{5s+3}{s+2}U(s) = (2.5s+1.5)\left[\frac{2U(s)}{s+2}\right]$$

and defining x as the term within the square brackets; that is,

$$X(s) = \frac{2U(s)}{s+2}$$

The order of the system, and therefore the number of state variables required, can be found by examining the denominator of the transfer function. If the denominator polynomial is of order n, then n state variables are required. Frequently a convenient choice is to select the state variables as the outputs of integrations $(1/s)$, as was done in Example 5.2.5.

Numerator Dynamics in a Second-Order System

EXAMPLE 5.2.6

■ Problem

Obtain a state-variable model for

$$\frac{X(s)}{U(s)} = \frac{4s+7}{5s^2+4s+7} \tag{1}$$

Relate the initial conditions for the state variables to the given initial conditions $x(0)$ and $\dot{x}(0)$.

■ Solution

Divide by $5s^2$ to obtain a 1 in the denominator.

$$\frac{X(s)}{U(s)} = \frac{\frac{7}{5}s^{-2} + \frac{4}{5}s^{-1}}{1 + \frac{4}{5}s^{-1} + \frac{7}{5}s^{-2}}$$

Use the 1 in the denominator to solve for $X(s)$.

$$X(s) = \left(\frac{7}{5}s^{-2} + \frac{4}{5}s^{-1}\right)U(s) - \left(\frac{4}{5}s^{-1} + \frac{7}{5}s^{-2}\right)X(s)$$

$$= \frac{1}{s}\left\{-\frac{4}{5}X(s) + \frac{4}{5}U(s) + \frac{1}{s}\left[\frac{7}{5}U(s) - \frac{7}{5}X(s)\right]\right\} \tag{2}$$

This equation shows that $X(s)$ is the output of an integration. Thus, x can be chosen as a state variable x_1. Thus,

$$X_1(s) = X(s)$$

The term within square brackets in (2) is the input to an integration, and thus the second state variable can be chosen as

$$X_2(s) = \frac{1}{s}\left[\frac{7}{5}U(s) - \frac{7}{5}X(s)\right] = \frac{1}{s}\left[\frac{7}{5}U(s) - \frac{7}{5}X_1(s)\right] \tag{3}$$

Then from equation (2)

$$X_1(s) = \frac{1}{s}\left[-\frac{4}{5}X_1(s) + \frac{4}{5}U(s) + X_2(s)\right] \tag{4}$$

The state equations are found from (3) and (4).

$$\dot{x}_1 = -\frac{4}{5}x_1 + x_2 + \frac{4}{5}u \tag{5}$$

$$\dot{x}_2 = -\frac{7}{5}x_1 + \frac{7}{5}u \tag{6}$$

and the output equation is $x = x_1$. The matrices of the standard form are

$$\mathbf{A} = \begin{bmatrix} -\frac{4}{5} & 1 \\ -\frac{7}{5} & 0 \end{bmatrix} \qquad \mathbf{B} = \begin{bmatrix} \frac{4}{5} \\ \frac{7}{5} \end{bmatrix}$$

$$\mathbf{C} = [1 \quad 0] \qquad \mathbf{D} = [0]$$

Note that the state variables obtained by this technique do not always have straightforward physical interpretations. If the model $m\ddot{x} + c\dot{x} + kx = c\dot{u} + ku$ represents a mass-spring-damper system with a displacement input u with $m = 5$, $c = 4$, $k = 7$, the variable x_2 is the integral of the spring force $k(u - x)$, divided by the mass m. Thus, x_2 is the acceleration of the mass due to this force. Sometimes convenient physical interpretations of the state variables are sacrificed to obtain special forms of the state equations that are useful for analytical purposes.

Using equations (5) and (6), we need to relate the values of $x_1(0)$ and $x_2(0)$ to $x(0)$ and $\dot{x}(0)$. Because x_1 was defined to be $x_1 = x$, we see that $x_1(0) = x(0)$. To find $x_2(0)$, we solve the first state equation, equation (5), for x_2.

$$x_2 = \dot{x}_1 + \frac{4}{5}(x_1 - u)$$

This gives

$$x_2(0) = \dot{x}_1(0) + \frac{4}{5}[x_1(0) - u(0)] = \dot{x}(0) + \frac{4}{5}[x(0) - u(0)]$$

Thus, if $u(0) = 0$,

$$x_2(0) = \dot{x}(0) + \frac{4}{5}x(0)$$

Table 5.2.1 A state-variable form for numerator dynamics.

Transfer-function model:	$\dfrac{Y(s)}{U(s)} = \dfrac{\beta_n s^n + \beta_{n-1}s^{n-1} + \cdots + \beta_1 s + \beta_0}{s^n + \alpha_{n-1}s^{n-1} + \cdots + \alpha_1 s + \alpha_0}$
State-variable model:	$\dot{x}_1 = \gamma_{n-1}u - \alpha_{n-1}x_1 + x_2$ $\dot{x}_2 = \gamma_{n-2}u - \alpha_{n-2}x_1 + x_3$ ⋮ $\dot{x}_j = \gamma_{n-j}u - \alpha_{n-j}x_1 + x_{j+1}, \quad j = 1, 2, \ldots, n-1$ ⋮ $\dot{x}_n = \gamma_0 u - \alpha_0 x_1$ $y = \beta_n u + x_1$
where	$\gamma_i = \beta_i - \alpha_i \beta_n$
Usual case:	If $u(0) = \dot{u}(0) = \cdots = 0$, then $x_i(0) = y^{(i-1)}(0) + \alpha_{n-1}y^{(i-2)}(0) + \cdots + \alpha_{n-i+1}y(0)$ $i = 1, 2, \ldots, n$
where	$y^{(i)}(0) = \left.\dfrac{d^i y}{dt^i}\right\rvert_{t=0}$

The method of the previous example can be extended to the general case where the transfer function is

$$\frac{Y(s)}{U(s)} = \frac{\beta_n s^n + \beta_{n-1}s^{n-1} + \cdots + \beta_1 s + \beta_0}{s^n + \alpha_{n-1}s^{n-1} + \cdots + \alpha_1 s + \alpha_0} \tag{5.2.11}$$

The results are shown in Table 5.2.1. The details of the derivation are given in [Palm, 1986].

PART II. MATLAB METHODS

5.3 STATE-VARIABLE METHODS WITH MATLAB

The MATLAB `step`, `impulse`, and `lsim` functions, treated in Section 2.9, can also be used with state-variable models. However, the `initial` function, which computes the free response, can be used only with a state-variable model. MATLAB also provides functions for converting models between the state-variable and transfer-function forms.

Recall that to create an LTI object from the reduced form

$$5\ddot{x} + 7\dot{x} + 4x = f(t) \tag{5.3.1}$$

or the transfer-function form

$$\frac{X(s)}{F(s)} = \frac{1}{5s^2 + 7s + 4} \tag{5.3.2}$$

you use the `tf(num,den)` function by typing:

```
>>sys1 = tf(1, [5, 7, 4]);
```

The result, `sys1`, is the LTI object that describes the system in the transfer-function form.

The LTI object `sys2` in transfer-function form for the equation

$$8\frac{d^3x}{dt^3} - 3\frac{d^2x}{dt^2} + 5\frac{dx}{dt} + 6x = 4\frac{d^2f}{dt^2} + 3\frac{df}{dt} + 5f \tag{5.3.3}$$

is created by typing

```
>>sys2 = tf([4, 3, 5],[8, -3, 5, 6]);
```

5.3.1 LTI OBJECTS AND THE ss(A,B,C,D) FUNCTION

To create an LTI object from a state model, you use the ss(A,B,C,D) function, where ss stands for *state space*. The matrix arguments of the function are the matrices in the following standard form of a state model:

$$\dot{\mathbf{x}} = \mathbf{A}\mathbf{x} + \mathbf{B}\mathbf{u} \tag{5.3.4}$$

$$\mathbf{y} = \mathbf{C}\mathbf{x} + \mathbf{D}\mathbf{u} \tag{5.3.5}$$

where $\mathbf{x}$ is the vector of state variables, $\mathbf{u}$ is the vector of input functions, and $\mathbf{y}$ is the vector of output variables. For example, to create an LTI object in state-model form for the system described by

$$\dot{x}_1 = x_2$$

$$\dot{x}_2 = \frac{1}{5}f(t) - \frac{4}{5}x_1 - \frac{7}{5}x_2$$

where x_1 is the desired output, the required matrices are

$$\mathbf{A} = \begin{bmatrix} 0 & 1 \\ -\frac{4}{5} & -\frac{7}{5} \end{bmatrix} \qquad \mathbf{B} = \begin{bmatrix} 0 \\ \frac{1}{5} \end{bmatrix}$$

$$\mathbf{C} = [1 \quad 0] \qquad \mathbf{D} = 0$$

In MATLAB you type

```
>>A = [0, 1; -4/5, -7/5];
>>B = [0; 1/5];
>>C = [1, 0];
>>D = 0;
>>sys3 = ss(A,B,C,D);
```

5.3.2 THE ss(sys) AND ssdata(sys) FUNCTIONS

An LTI object defined using the tf function can be used to obtain an equivalent state model description of the system. To create a state model for the system described by the LTI object sys1 created previously in transfer-function form, you type ss(sys1). You will then see the resulting **A**, **B**, **C**, and **D** matrices on the screen. To extract and save the matrices as A1, B1, C1, and D1 (to avoid overwriting the matrices from the second example here), use the ssdata function as follows:

```
>>[A1, B1, C1, D1] = ssdata(sys1);
```

The results are

$$\mathbf{A1} = \begin{bmatrix} -1.4 & -0.8 \\ 1 & 0 \end{bmatrix}$$

$$\mathbf{B1} = \begin{bmatrix} 0.5 \\ 0 \end{bmatrix}$$

$$\mathbf{C1} = [0 \quad 0.4]$$

$$\mathbf{D1} = [0]$$

which correspond to the state equations:

$$\dot{x}_1 = -1.4x_1 - 0.8x_2 + 0.5f(t)$$

$$\dot{x}_2 = x_1$$

and the output equation $y = 0.4x_2$.

Note that these state equations are different than those obtained in Example 5.2.2. This shows that the definition of state variables is not unique. You can use the `ssdata` function to avoid algebra like that used in Examples 5.2.5 and 5.2.6, but you will obtain different definitions for the state variables and different state equations.

5.3.3 RELATING STATE VARIABLES TO THE ORIGINAL VARIABLES

When using `ssdata` to convert a transfer-function form into a state model, note that the output y will be a scalar that is identical to the solution variable of the reduced form; in this case the solution variable of (5.3.1) is the variable x. To interpret the state model, we need to relate its state variables x_1 and x_2 to x. The values of the matrices **C1** and **D1** tell us that the output variable is $y = 0.4x_2$. Because the output y is the same as x, we then see that $x_2 = x/0.4 = 2.5x$. The other state-variable x_1 is related to x_2 by the second state equation $\dot{x}_2 = x_1$. Thus, $x_1 = 2.5\dot{x}$.

5.3.4 THE `tfdata` FUNCTION

To create a transfer-function description of the system `sys3`, previously created from the state model, you type `tfsys3 = tf(sys3)`. However, there can be situations where we are given the model `tfsys3` in transfer-function form and we need to obtain the numerator and denominator. To extract and save the coefficients of the transfer function, use the `tfdata` function as follows:

```
>>[num, den] = tfdata(tfsys3, 'v');
```

The optional parameter `'v'` tells MATLAB to return the coefficients as vectors if there is only one transfer function; otherwise, they are returned as cell arrays.

For this example, the vectors returned are `num = [0, 0, 0.2]` and `den = [1, 1.4, 0.8]`. This corresponds to the transfer function

$$\frac{X(s)}{F(s)} = \frac{0.2}{s^2 + 1.4s + 0.8} = \frac{1}{5s^2 + 7s + 4}$$

which is the correct transfer function, as seen from (5.2.2).

EXAMPLE 5.3.1

Transfer Functions of a Two-Mass System

■ Problem

Obtain the transfer functions $X_1(s)/F(s)$ and $X_2(s)/F(s)$ of the state-variable model obtained in Example 5.2.3. The matrices and state vector of the model are

$$\mathbf{A} = \begin{bmatrix} 0 & 0 & 1 & 0 \\ 0 & 0 & 0 & 1 \\ -1 & \frac{4}{5} & -\frac{12}{5} & \frac{8}{5} \\ \frac{4}{3} & -\frac{4}{3} & \frac{8}{3} & -\frac{8}{3} \end{bmatrix} \qquad \mathbf{B} = \begin{bmatrix} 0 \\ 0 \\ 0 \\ \frac{1}{3} \end{bmatrix}$$

and

$$\mathbf{z} = \begin{bmatrix} x_1 \\ x_2 \\ x_3 \\ x_4 \end{bmatrix} = \begin{bmatrix} x_1 \\ x_2 \\ \dot{x}_1 \\ \dot{x}_2 \end{bmatrix}$$

■ **Solution**

Because we want the transfer functions for x_1 and x_2, we must define the **C** and **D** matrices to indicate that z_1 and z_3 are the output variables y_1 and y_2. Thus,

$$\mathbf{C} = \begin{bmatrix} 1 & 0 & 0 & 0 \\ 0 & 1 & 0 & 0 \end{bmatrix} \qquad \mathbf{D} = \begin{bmatrix} 0 \\ 0 \end{bmatrix}$$

The MATLAB program is as follows:

```
A = [0, 0, 1, 0; 0, 0, 0, 1;...
     -1, 4/5, -12/5, 8/5; 4/3, -4/3, 8/3, -8/3];
B = [0; 0; 0; 1/3];
C = [1, 0, 0, 0; 0, 1, 0, 0]; D = [0; 0]
sys4 = ss(A, B, C, D);
tfsys4 = tf(sys4)
```

The results displayed on the screen are labeled #1 and #2. These correspond to the first and second transfer functions in order. The answers are

$$\frac{X_1(s)}{F(s)} = \frac{0.5333s + 0.2667}{s^4 + 5.067s^3 + 4.467s^2 + 1.6s + 0.2667}$$

$$\frac{X_2(s)}{F(s)} = \frac{0.3333s^2 + 0.8s + 0.3333}{s^4 + 5.067s^3 + 4.467s^2 + 1.6s + 0.2667}$$

Table 5.3.1 summarizes these functions.

Table 5.3.1 LTI object functions.

Command	Description
`sys = ss(A, B, C, D)`	Creates an LTI object in state-space form, where the matrices A, B, C, and D correspond to those in the model $\dot{\mathbf{x}} = \mathbf{Ax} + \mathbf{Bu}$, $\mathbf{y} = \mathbf{Cx} + \mathbf{Du}$.
`[A, B, C, D] = ssdata(sys)`	Extracts the matrices A, B, C, and D of the LTI object `sys`, corresponding to those in the model $\dot{\mathbf{x}} = \mathbf{Ax} + \mathbf{Bu}$, $\mathbf{y} = \mathbf{Cx} + \mathbf{Du}$.
`sys = tf(num,den)`	Creates an LTI object in transfer-function form, where the vector `num` is the vector of coefficients of the transfer-function numerator, arranged in descending order, and `den` is the vector of coefficients of the denominator, also arranged in descending order.
`sys2=tf(sys1)`	Creates the transfer-function model `sys2` from the state model `sys1`.
`sys1=ss(sys2)`	Creates the state model `sys1` from the transfer-function model `sys2`.
`[num, den] = tfdata(sys, 'v')`	Extracts the coefficients of the numerator and denominator of the transfer-function model `sys`. When the optional parameter `'v'` is used, if there is only one transfer function, the coefficients are returned as vectors rather than as cell arrays.

5.3.5 LINEAR ODE SOLVERS

The Control System Toolbox provides several solvers for linear models. These solvers are categorized by the type of input function they can accept: zero input, impulse input, step input, and a general input function.

5.3.6 THE `initial` FUNCTION

The `initial` function computes and plots the free response of a state model. This is sometimes called the *initial condition response* or the *undriven response* in the MATLAB documentation. The basic syntax is `initial(sys,x0)`, where `sys` is the LTI object in state-variable form, and `x0` is the initial condition vector. The time span and number of solution points are chosen automatically.

EXAMPLE 5.3.2

Free Response of the Two-Mass Model

■ Problem

Compute the free response $x_1(t)$ and $x_2(t)$ of the state model derived in Example 5.2.3, for $x_1(0) = 5$, $\dot{x}_1(0) = -3$, $x_2(0) = 1$, and $\dot{x}_2(0) = 2$. The model is

$$\dot{x}_1 = x_3$$

$$\dot{x}_2 = x_4$$

$$\dot{x}_3 = \frac{1}{5}(-5x_1 + 4x_2 - 12x_3 + 8x_4)$$

$$\dot{x}_4 = \frac{1}{3}[4x_1 - 4x_2 + 8x_3 - 8x_4 + f(t)]$$

or

$$\dot{\mathbf{x}} = \mathbf{A}\mathbf{x} + \mathbf{B}f(t)$$

where

$$\mathbf{A} = \begin{bmatrix} 0 & 0 & 1 & 0 \\ 0 & 0 & 0 & 1 \\ -1 & \frac{4}{5} & -\frac{12}{5} & \frac{8}{5} \\ \frac{4}{3} & -\frac{4}{3} & \frac{8}{3} & -\frac{8}{3} \end{bmatrix} \qquad \mathbf{B} = \begin{bmatrix} 0 \\ 0 \\ 0 \\ \frac{1}{3} \end{bmatrix}$$

and

$$\mathbf{x} = \begin{bmatrix} x_1 \\ x_2 \\ x_3 \\ x_4 \end{bmatrix} = \begin{bmatrix} x_1 \\ x_2 \\ \dot{x}_1 \\ \dot{x}_2 \end{bmatrix}$$

■ Solution

We must first relate the initial conditions given in terms of the original variables to the state variables. From the definition of the state vector **x**, we see that $x_1(0) = 5$, $x_2(0) = 1$, $x_3(0) = -3$, $x_4(0) = 2$. Next we must define the model in state-variable form. The system `sys4` created in Example 5.3.1 specified two outputs, x_1 and x_2. Because we want to obtain only one output here (x_1), we must create a new state model using the same values for the **A** and **B** matrices, but now using

$$\mathbf{C} = \begin{bmatrix} 1 & 0 & 0 & 0 \\ 0 & 1 & 0 & 0 \end{bmatrix} \qquad \mathbf{D} = \begin{bmatrix} 0 \\ 0 \end{bmatrix}$$

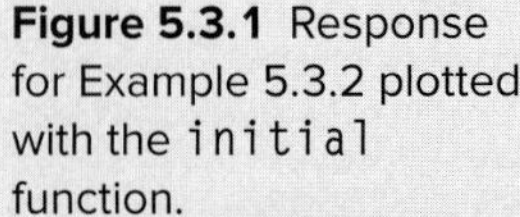

Figure 5.3.1 Response for Example 5.3.2 plotted with the `initial` function.

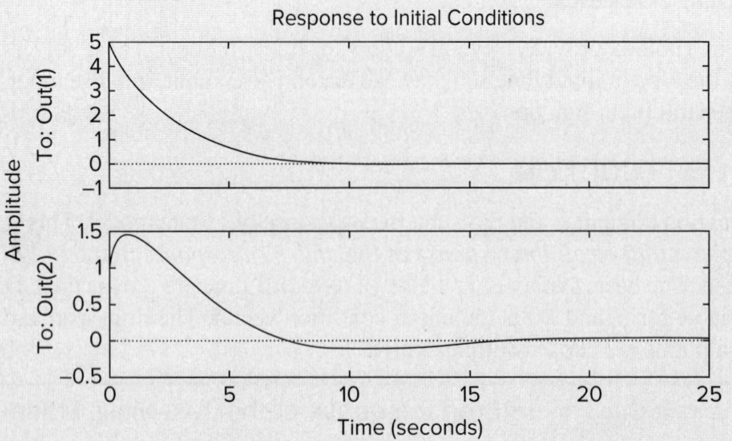

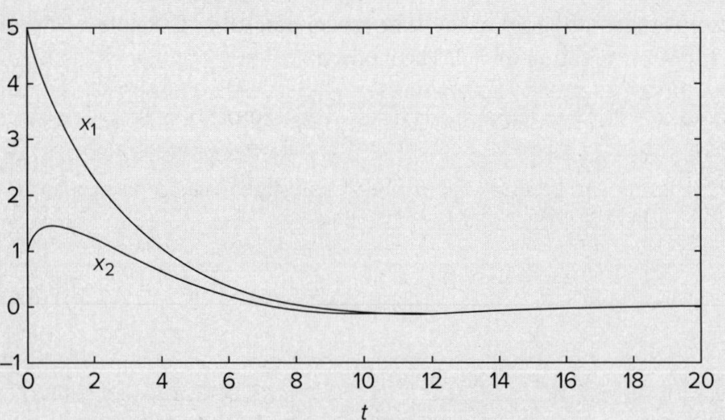

Figure 5.3.2 Response for Example 5.3.2 plotted with the `plot` function.

The MATLAB program is as follows:

```
A = [0, 0, 1, 0; 0, 0, 0, 1;...
     -1, 4/5, -12/5, 8/5; 4/3, -4/3, 8/3, -8/3];
B = [0; 0; 0; 1/3];
C = [1, 0, 0, 0; 0, 1, 0, 0]; D = [0; 0]
sys5 = ss(A, B, C, D);
initial(sys5, [5, 1, -3, 2])
```

The plot of $x_1(t)$ and $x_2(t)$ will be displayed on the screen (see Figure 5.3.1).

To plot x_1 and x_2 on the same plot, you can replace the last line with the following two lines.

```
[y,t] = initial(sys,[5,1,-3,2]);
plot(t,y),gtext('x_1'),gtext('x_2'),xlabel('t')
```

The resulting plot is shown in Figure 5.3.2.

To specify the final time `tfinal`, use the syntax `initial(sys,x0,tfinal)`. To specify a vector of times of the form `t = (0:dt:tfinal)`, at which to obtain the solution, use the syntax `initial(sys,x0,t)`. When called with left-hand arguments, as `[y, t, x]=initial(sys,x0, ...)`, the function returns the output response `y`, the time vector `t` used for the simulation, and the state vector `x` evaluated at those times. The columns of the matrices `y` and `x` are the outputs and the states, respectively. The number of rows in `y` and `x` equals `length(t)`. No plot is drawn. The syntax `initial(sys1,sys2, ...,x0,t)` plots the free response of multiple LTI systems on a single plot. The time vector `t` is optional. You can specify line color, line style, and marker for each system; for example, `initial(sys1,'r',sys2, 'y--',sys3,'gx',x0)`.

5.3.7 THE `impulse`, `step`, AND `lsim` FUNCTIONS

You may use the `impulse`, `step`, and `lsim` functions with state-variable models the same way they are used with transfer-function models. However, when used with state-variable models, there are some additional features available, which we illustrate with the `step` function. When called with left-hand arguments, as `[y, t] = step(sys, ...)`, the function returns the output response `y` and the time vector `t` used for the simulation. No plot is drawn. The array `y` is $(p \times q \times m)$, where p is `length(t)`, q is the number of outputs, and m is the number of inputs. To obtain the state vector solution for state-space models, use the syntax `[y, t, x] = step(sys, ...)`.

To use the `lsim` function for nonzero initial conditions with a state-space model, use the syntax `lsim(sys,u,t,x0)`. The initial condition vector `x0` is needed only if the initial conditions are nonzero.

These functions are summarized in Table 5.3.2.

Table 5.3.2 Basic syntax of linear solvers for state-variable models.

`initial(sys,x0,tfinal)`	Generates a plot of the free response of the state-variable model `sys`, for the initial conditions specified in the array `x0`. The final time `tfinal` is optional.
`initial(sys,x0,t)`	Generates the free response plot using the user-supplied array of regularly spaced time values `t`.
`[y,t,x]=initial(sys,x0,...)`	Generates and saves the free response in the array `y` of the output variables, and in the array `x` of the state variables. No plot is produced.
`step(sys)`	Generates a plot of the unit-step response of the LTI model `sys`.
`step(sys,t)`	Generates a plot of the unit-step response using the user-supplied array of regularly spaced time values `t`.
`[y,t]= step(sys)`	Generates and saves the unit-step response in the arrays `y` and `t`. No plot is produced.
`[y,t,x]=step(sys,...)`	Generates and saves the free response in the array `y` of the output variables, and in the array `x` of the state variables, which is optional. No plot is produced.
`impulse(sys)`	Generates and plots the unit-impulse response of the LTI model `sys`. The extended syntax is identical to that of the `step` function.
`lsim(sys,u,t,x0)`	Generates a plot of the total response of the state-variable model `sys`. The array `u` contains the values of the forcing function, which must have the same number of values as the regularly spaced time values in the array `t`. The initial conditions are specified in the array `x0`, which is optional if the initial conditions are zero.
`[y,x]= lsim(sys,u,t,x0)`	Generates and saves the total response in the array `y` of the output variables, and in the array `x` of the state variables, which is optional. No plot is produced.

Recall that the total response of a linear model is the sum of the free response and the forced response. The free response can be obtained with the `initial` function, and the forced response obtained with the `step` function, if the forcing function is a step. Keep in mind that to add the responses, the free and forced solutions must have the same time span and must have the same time spacing. The following example shows how this is done.

EXAMPLE 5.3.3

Total Response of a Two-Mass Model

■ Problem

Obtain the total response $x_1(t)$ and $x_2(t)$ of the two-mass model given in Example 5.3.2, using the same initial conditions but now subjected to a step input of magnitude 3.

■ Solution

We first define the **A**, **B**, **C**, and **D** matrices and then create the LTI model `sys`. Then we compute the step response, saving it in the arrays `ystep` and `t`. Note that the `step` function automatically selects a time span and a time spacing for the array `t`. We then use this array to compute the free response `yfree`. Finally we add the two arrays `ystep` and `yfree` to obtain the total response. Note that we could not add these two arrays if they did not have the same number of points. Note also, that if they had different time increments, we could add them, but the sum would be meaningless. The following script file shows the procedure. The resulting plot is shown in Figure 5.3.3.

```
% InitialPlusStep.m
A = [0,0,1,0;0,0,0,1;-1,4/5,-12/5,8/5;4/3,-4/3,8/3,-8/3];
B = [0;0;0;1/3];
C = [1,0,0,0;0,1,0,0];
D = [0;0];
sys = ss(A,B,C,D);
[ystep,t] = step(3*sys);
yfree = initial(sys,[5,1,-3,2],t);
y = yfree + ystep;
plot(t,y),xlabel('t'),gtext('x_1'),gtext('x_2')
```

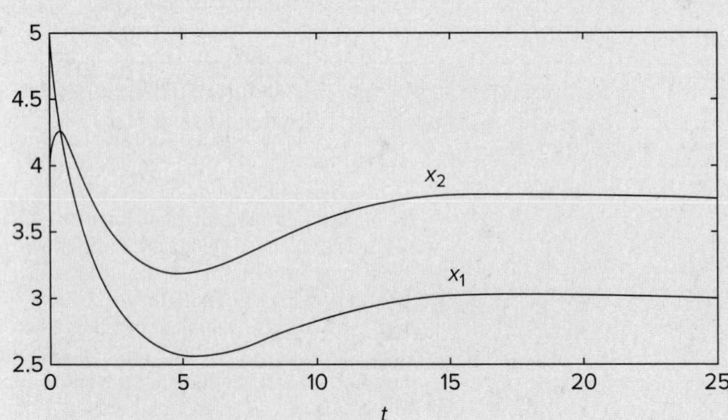

Figure 5.3.3 Step plus free response for Example 5.3.3.

5.3.8 OBTAINING THE CHARACTERISTIC POLYNOMIAL

The MATLAB command `poly` can find the characteristic polynomial that corresponds to a specified state matrix **A**. For example, the matrix **A** given in Example 5.2.2 is, for $m = 1$, $c = 5$, $k = 6$,

$$\mathbf{A} = \begin{bmatrix} 0 & 1 \\ -6 & -5 \end{bmatrix} \tag{5.3.6}$$

This matrix corresponds to the equation $\ddot{x} + 5\dot{x} + 6x = 0$, whose characteristic polynomial is $s^2 + 5s + 6 = 0$.

The coefficients of its characteristic polynomial are found by typing

```
>>A = [0, 1; -6, -5];
>>poly(A)
```

MATLAB returns the answer `[1, 5, 6]`, which corresponds to the polynomial $s^2 + 5s + 6$. The `roots` function can be used to compute the characteristic roots; for example, `roots(poly(A))`, which gives the roots `[-2, -3]`.

MATLAB provides the `eig` function to compute the characteristic roots without obtaining the characteristic polynomial, when the model is given in the state-variable form. Its syntax is `eig(A)`. (The function's name is an abbreviation of *eigenvalue*, which is another name for characteristic root.) For example, typing `eig([0, 1; -6, -5])` returns the roots `[-2, -3]`.

5.4 THE MATLAB ode FUNCTIONS

Recall that we can categorize differential equations as *linear* or *nonlinear*. Linear differential equations are recognized by the fact that they contain only linear functions of the dependent variable and its derivatives. Nonlinear functions of the *independent* variable do *not* make a differential equation nonlinear. For example, the following equations are linear.

$$\dot{y} + 3y = 5 + t^2 \qquad \dot{y} + 3t^2 y = 5 \qquad \ddot{y} + 7\dot{y} + t^2 x = \sin t$$

whereas the following equations are nonlinear:

$$\dot{y} = -y^2 \sin t \qquad \text{because of } y^2$$

$$\ddot{y} + 5\dot{y}^2 + 6y = 4 \qquad \text{because of } \dot{y}^2$$

$$\ddot{y} + 4y\dot{y} + 3y = 10 \qquad \text{because of } y\dot{y}$$

The equation $\ddot{y} + 7\dot{y} + t^2 x = \sin t$ is a *variable-coefficient* differential equation, so named because one of its coefficients (t^2) is a function of the independent variable t.

5.4.1 SELECTING A SOLUTION METHOD

Recall that the Laplace transform method and the state-variable MATLAB methods presented in Section 5.3.1 cannot be used to solve variable-coefficient equations and nonlinear equations.

It is possible sometimes, mainly for first order equations, to obtain the closed-form solution of a nonlinear differential equation. If not, then we must solve the equation numerically. In this section, we introduce numerical methods for solving differential equations. First we will treat first-order equations, and then we will show how to extend the techniques to higher-order equations.

The essence of a numerical method is to convert the differential equation into a *difference equation* that can be programmed on a digital computer. Numerical algorithms differ partly as a result of the specific procedure used to obtain the difference equations. In general, as the accuracy of the approximation is increased, so is the complexity of the programming involved. It is important to understand the concept of *step size* and its effects on solution accuracy. In order to provide a simple introduction to these issues, we begin with the simplest numerical method, the *Euler method.*

5.4.2 THE EULER METHOD

The Euler method is the simplest algorithm for numerical solution of a differential equation. It usually gives the least accurate results, but provides a basis for understanding more sophisticated methods. Consider the equation

$$\frac{dy}{dt} = r(t)y(t) \tag{5.4.1}$$

where $r(t)$ is a known function. From the definition of the derivative,

$$\frac{dy}{dt} = \lim_{\Delta t \to 0} \frac{y(t + \Delta t) - y(t)}{\Delta t}$$

If the time increment Δt is chosen small enough, the derivative can be replaced by the approximate expression

$$\frac{dy}{dt} \approx \frac{y(t + \Delta t) - y(t)}{\Delta t}$$

Assume that the right-hand side of (5.4.1) remains constant over the time interval $[t, t + \Delta t]$, and replace (5.4.1) by the following approximation:

$$\frac{y(t + \Delta t) - y(t)}{\Delta t} = r(t)y(t)$$

or

$$y(t + \Delta t) = y(t) + r(t)y(t)\Delta t \tag{5.4.2}$$

The smaller Δt is, the more accurate are our two assumptions leading to (5.4.2). This technique for replacing a differential equation with a difference equation is the Euler method. The increment Δt is called the *step size.*

Equation (5.4.2) can be written in more convenient form by replacing t with t_k as follows:

$$y(t_k + \Delta t) = y(t_{k+1}) = y(t_k) + r(t_k)y(t_k)\Delta t \tag{5.4.3}$$

where $t_{k+1} = t_k + \Delta t$.

The Euler method for the general first-order equation $\dot{y} = f(t, y)$ is

$$y(t_{k+1}) = y(t_k) + \Delta t f[t_k, y(t_k)] \tag{5.4.4}$$

This equation can be applied successively at the times t_k by putting it in a `for` loop. (See Table 5.4.1 for definitions of MATLAB functions used in this section.)

For example, the following script file solves the differential equation $\dot{y} = ry$ and plots the solution over the range $0 \le t \le 0.5$ for the case where $r = -10$ and the initial condition is $y(0) = 2$. The time constant is $\tau = -1/r = 0.1$, and the true solution is $y(t) = 2e^{-10t}$. To illustrate the effect of the step size on the solution's accuracy, we use a step size $\Delta t = 0.02$, which is 20% of the time constant.

Table 5.4.1 MATLAB functions used in this section.

	Functions Introduced in This Section
`axis[x1 x2 y1 y2]`	Sets the minimum and maximum values for the x and y axes.
`cos(x)`	Computes $\cos x$.
`end`	Terminates a loop, such as a `for` loop.
`exp(x)`	Computes e^x.
`for`	Denotes the beginning of a `for` loop.
`function`	Indicates the beginning of a function file.
`gtext('text')`	Enables placement of text on a plot using the cursor.
`ode45`	Implements the fourth and fifth order Runge-Kutta algorithm for solving differential equations.
`sin(x)`	Computes $\sin x$.
`sqrt(x)`	Computes $\sqrt{x}$.
`\theta`	Puts the Greek letter θ in a plot label.
	Functions Introduced in Earlier Sections
`eig(A)`	Computes the eigenvalues of a matrix **A** (Section 5.3).
`plot(x,y)`	Creates a two-dimensional plot on rectilinear axes (Section 1.6).
`xlabel('text')`	Puts a label on the abscissa of a plot (Section 1.6).
`ylabel('text')`	Puts a label on the ordinate of a plot (Section 1.6).

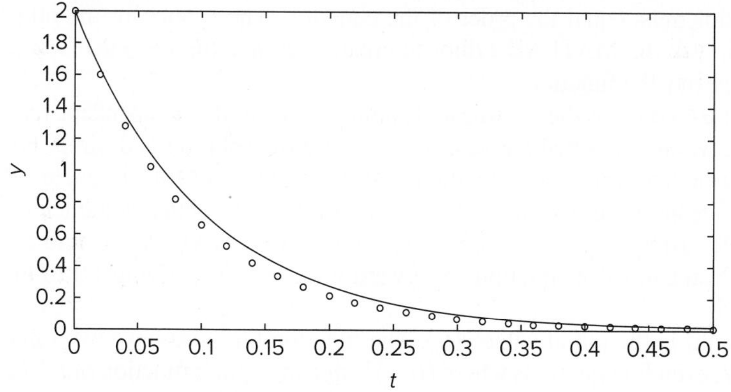

Figure 5.4.1 Euler-method solution for the free response of $\dot{y} = -10y$, $y(0) = 2$.

```
r = -10; delta = 0.02; y(1) = 2;
k = 0;
for time = [delta:delta:.5]
    k = k + 1;
    y(k+1) = y(k) + r*y(k)*delta;
end
t = (0:delta:0.5);
y_exact=2*exp(-10*t);
plot(t,y, 'o',t,y_exact),xlabel('t'),ylabel('y')
```

Figure 5.4.1 shows the results. The numerical solution is shown by the small circles. The exact solution is shown by the solid line. There is some noticeable error. If we had used a step size equal to 5% of the time constant, the error would not be noticeable on the plot.

MATLAB provides functions called *solvers* that implement several numerical solution methods. The `ode45` solver is sufficient to solve the problems encountered in this text.[1]

[1] The `ode45` solver is based on the fourth and fifth order Runge-Kutta algorithms. For an introduction to Runge-Kutta algorithms, see Appendix E on the text website.

Table 5.4.2 Basic Syntax of the `ode45` solver.

`[t, y] = ode45(@ydot, tspan, y0)`   Solves the vector differential equation $\dot{\mathbf{y}} = \mathbf{f}(t, \mathbf{y})$ specified in the function file `ydot`, whose inputs must be t and $\mathbf{y}$, and whose output must be a *column* vector representing $d\mathbf{y}/dt$; that is, $\mathbf{f}(t, \mathbf{y})$. The number of rows in this column vector must equal the order of the equation. The vector `tspan` contains the starting and ending values of the independent variable t, and optionally, any intermediate values of t where the solution is desired. The vector `y0` contains the initial values $\mathbf{y}(t_0)$. The function file must have the two input arguments, `t` and `y`, even for equations where $\mathbf{f}(t, \mathbf{y})$ is not an explicit function of t.

5.4.3 SOLVER SYNTAX

We begin our coverage with several examples of solving first-order equations. Solution of higher-order equations is covered later in this section. When used to solve the equation $\dot{y} = f(t, y)$, the basic syntax is:

```
[t,y] = ode45(@ydot, tspan, y0)
```

where `ydot` is the name of the function file whose inputs must be t and y, and whose output must be a *column* vector representing dy/dt; that is, $f(t, y)$. The number of rows in this column vector must equal the order of the equation. The syntax for the other solvers is identical. Use the MATLAB Editor to create and save the file `ydot`. Note that we use @ to specify the function.

The vector `tspan` contains the starting and ending values of the independent variable t, and optionally, any intermediate values of t where the solution is desired. For example, if no intermediate values are specified, `tspan` is `[t0, tfinal]`, where `t0` and `tfinal` are the desired starting and ending values of the independent parameter t. As another example, using `tspan = [0, 5, 10]` forces MATLAB to find the solution at $t = 5$. You can solve equations backward in time by specifying `t0` to the greater than `tfinal`.

The parameter `y0` is the initial value $y(t_0)$. The function file must have two input arguments, `t` and `y`, even for equations where $f(t, y)$ is *not* an explicit function of t. You need not use array operations in the function file because the ODE solvers call the file with scalar values for the arguments.

Table 5.4.1 summarizes the MATLAB functions used in this section. Table 5.4.2 summarizes the basic syntax of the ODE solvers.

As a first example of using a solver, let us solve an equation whose solution is known in closed form, so that we can make sure we are using the method correctly. We can also assess the performance of the `ode45` solver when applied to find an oscillating solution, which can be difficult to obtain numerically.

EXAMPLE 5.4.1

MATLAB Solution of $\dot{y} = -y^2 \sin t$

■ Problem

Equations with oscillating solutions provide a good test of the accuracy of a numerical solver. To check our results, we choose an equation whose closed-form solution can be found. We can use separation of variables to solve the following equation, which is nonlinear because of the y^2 term.

$$\dot{y} = -y^2 \sin t \qquad y(0) = 1 \tag{1}$$

$$\int_{y(0)}^{y(t)} \frac{dy}{y^2} = -\int_0^t \sin t \, dt$$

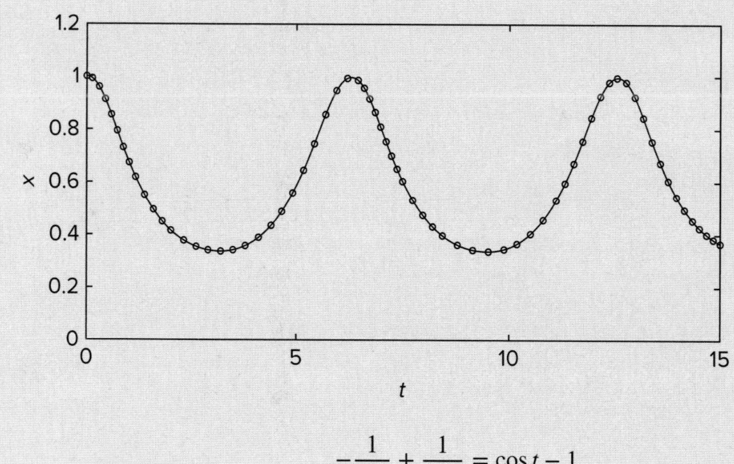

Figure 5.4.2 MATLAB (`ode45`) and exact solutions of $\dot{y} = -y^2 \sin t, y(0) = 1$.

$$-\frac{1}{y(t)} + \frac{1}{y(0)} = \cos t - 1$$

or, since $y(0) = 1$,

$$y(t) = \frac{1}{2 - \cos t}$$

Use the `ode45` solver to solve equation (1) for $0 \leq t \leq 15$.

■ **Solution**

Create and save the following function file. Give it a unique name of your choice. Here we will name it after the example number 5.4.1, abbreviated as `Ex5p4p1`.

```
function ydot=Ex5p4p1(t,y)
ydot=-sin(t)*y^2;
```

Note that we need not use array arithmetic (.* and .^) in the expression `sin(t)*y^2`, because the solver does not call the function `ydot` with array inputs. Type the following lines in the MATLAB Command window. The solution from `ode45` is a discrete set of points stored in the arrays `y` and `t`, which we show on the plot as small circles.

```
>>[t,y] = ode45(@Ex5p4p1,[0,15],1);
>>y_exact = 1./(2-cos(t));
>>plot(t,y,'o',t,y_exact),xlabel('t'),ylabel('y'),axis([0 15 0 1.2])
```

Note that we use the t values generated by the `ode45` function to compute the exact solution at the same values of t. Note also that in practice we do not know the range of the solution $y(t)$ ahead of time, so we must first run the program using guessed values for the range of $y(t)$. The final choice of 0 and 1.2 gives the plot shown in Figure 5.4.2. The solid line is the exact solution and the circular data markers are the numerical solution. This plot shows that the numerical solution is accurate.

The main application of numerical methods is to solve equations for which a closed-form solution cannot be obtained. The next example shows such an application.

A Rocket-Propelled Sled — EXAMPLE 5.4.2

■ **Problem**

A rocket-propelled sled on a track is represented in Figure 5.4.3 as a mass m with an applied force f that represents the rocket thrust. The rocket thrust initially is horizontal, but the engine

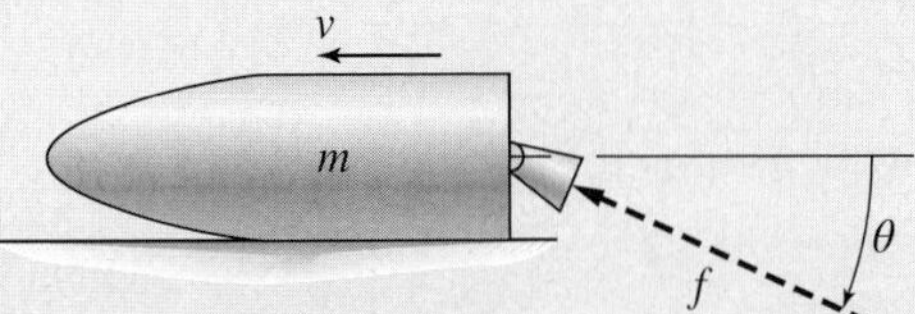

Figure 5.4.3 A rocket-propelled sled.

accidentally pivots during firing and rotates with an angular acceleration of $\ddot{\theta} = \pi/50$ rad/s. Compute the sled's velocity v for $0 \le t \le 6$ if $v(0) = 0$. The rocket thrust is 4000 N and the sled mass is 450 kg.

■ Solution

The sled's equation of motion is $450\dot{v} = 4000 \cos \theta(t)$. To obtain $\theta(t)$, note that

$$\dot{\theta} = \int_0^t \ddot{\theta}\, dt = \frac{\pi}{50} t$$

and

$$\theta = \int_0^t \dot{\theta}\, dt = \int_0^t \frac{\pi}{50} t\, dt = \frac{\pi}{100} t^2$$

Thus, the equation of motion becomes

$$\dot{v} = \frac{80}{9} \cos\left(\frac{\pi}{100} t^2\right) \tag{1}$$

The solution is formally given by

$$v(t) = \frac{80}{9} \int_0^t \cos\left(\frac{\pi}{100} t^2\right) dt$$

Unfortunately, no closed-form solution is available for the integral, which is called *Fresnel's cosine integral*. The value of the integral has been tabulated numerically, but we will use a MATLAB ODE solver to obtain the solution.

First create the following user-defined function file, which is based on equation (1).

```
function vdot = sled(t,v)
vdot = 80*cos(pi*t^2/100)/9;
```

As a check on our results, we will use the solution for $\theta = 0$ as a comparison. The equation of motion for this case is $\dot{v} = 80/9$, which gives $v(t) = 80t/9$. The following session solves the equation and plots the two solutions, which are shown in Figure 5.4.4.

```
[t,v] = ode45(@sled,[0 6],0);
plot(t,v,t,(80*t/9)),xlabel('t (s)'),...
   ylabel('v (m/s)'),gtext('\theta = 0'),gtext('\theta \neq 0')
```

We can make two observations that help us determine whether or not our numerical solution is correct. From the plot we see that the solution for $\theta \neq 0$ is almost identical to the solution for $\theta = 0$, for small values of θ. This is correct because $\cos\theta \approx 1$ for small values of θ. As θ increases, we would expect the velocity to be smaller than the velocity for $\theta = 0$ because the horizontal component of the thrust is smaller. The plot confirms this.

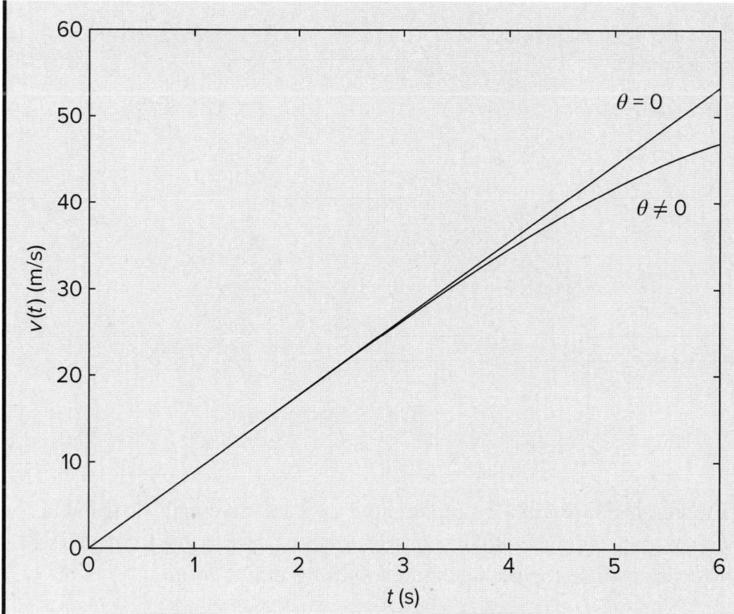

Figure 5.4.4 Speed response of the sled for $\theta = 0$ and $\theta \neq 0$.

5.4.4 EXTENSION TO HIGHER-ORDER EQUATIONS

To use the ODE solvers to solve an equation of order two or greater, you must first write the equation in state-variable form, as a set of first-order equations, and then create a function file that computes the derivatives of the state variables. One advantage of using the `ode45` solver, even for *linear* equations, is that it can handle nonzero initial conditions and variable coefficients, whereas the `step` and `impulse` functions cannot.

EXAMPLE 5.4.3

Sinusoidal Response of a Mass-Spring-Damper System

■ Problem

Consider the second-order equation $5\ddot{y} + 7\dot{y} + 4y = f(t)$. Define the variables, $x_1 = y$ and $x_2 = \dot{y}$. The state-variable form is

$$\dot{x}_1 = x_2 \tag{1}$$

$$\dot{x}_2 = \frac{1}{5}f(t) - \frac{4}{5}x_1 - \frac{7}{5}x_2 \tag{2}$$

Now write a function file that computes the values of $\dot{x}_1$ and $\dot{x}_2$ and stores them in a *column* vector. To do this, we must first have a function specified for $f(t)$. Suppose that $f(t) = \sin t$. Solve the equation over the interval $0 \leq t \leq 12$ using the initial conditions $y(0) = 3$ and $\dot{y}(0) = 9$.

■ Solution

The required file is

```
function xdot = Ex5p4p3(t,x)
% Computes derivatives of two equations
xdot(1) = x(2);
xdot(2) = (1/5)*(sin(t)-4*x(1)-7*x(2));
xdot = [xdot(1); xdot(2)];
```

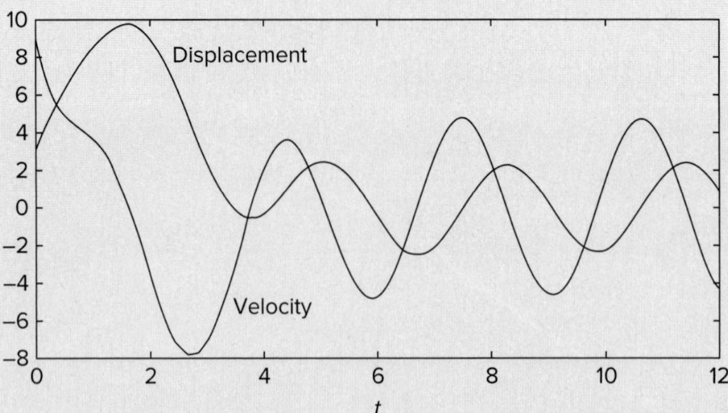

Figure 5.4.5 Sinusoidal response of a mass-spring-damper system.

Note that `xdot(1)` represents $\dot{x}_1$, `xdot(2)` represents $\dot{x}_2$, `x(1)` represents x_1, and `x(2)` represents x_2. Once you become familiar with the notation for the state-variable form, you will see that the preceding code could be replaced with the following shorter form.

```
function xdot = example1(t,x)
% Computes derivatives of two equations
xdot = [x(2); (1/5)*(sin(t)-4*x(1)-7*x(2))];
```

Suppose we want to solve (1) and (2) for $0 \leq t \leq 12$ with the initial conditions $y(0) = x_1(0) = 3$ and $\dot{y}(0) = x_2(0) = 9$. Then the initial condition for the *vector* `x` is `[3, 9]`. To use `ode45`, you type

```
[t, x] = ode45(@Ex5p4p3, [0, 12], [3, 9]);
```

Each row in the vector `x` corresponds to a time returned in the column vector `t`. If you type `plot(t,x)`, you will obtain a plot of both x_1 and x_2 versus t. Note that `x` is a matrix with two columns; the first column contains the values of x_1 at the various times generated by the solver. The second column contains the values of x_2. Thus, to plot only x_1, type `plot(t,x(:,1))`. Figure 5.4.5 shows the results after labeling with the MATLAB plot Editor.

Most variable coefficient equations do not have closed form solutions, and so we must usually solve these numerically.

EXAMPLE 5.4.4

Response Due to an Aging Spring

■ Problem

Consider a mass-spring-damper system in which the spring element gets weaker with time due to metal fatigue. Suppose the spring constant varies with time as follows:

$$k = 20(1 + e^{-t/10})$$

The equation of motion is

$$m\ddot{x} + c\dot{x} + 20(1 + e^{-t/10})x = f(t)$$

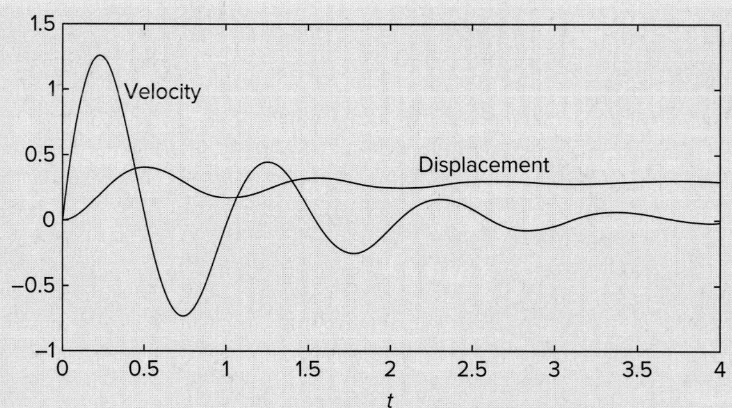

Figure 5.4.6 Displacement and velocity of a mass with a damper and aging spring.

Use the values $m = 1$, $c = 2$, and $f = 10$, and solve the equations for zero-initial conditions over the interval $0 \le t \le 4$.

■ Solution

Defining the state variables to be $x_1 = x$ and $x_2 = \dot{x}$, the state-variable equations are

$$\dot{x}_1 = x_2$$

$$\dot{x}_2 = \frac{1}{m}f(t) - \frac{20(1 + e^{-t/10})}{m}x_1 - \frac{c}{m}x_2$$

If $m = 1$, $c = 2$, and $f = 10$, these equations become

$$\dot{x}_1 = x_2$$

$$\dot{x}_2 = 10 - 20(1 + e^{-t/10})x_1 - 2x_2$$

Create a function file that computes the values of $\dot{x}_1$ and $\dot{x}_2$ and stores them in a column vector. This file is shown below.

```
function xdot = aging_spring(t,x)
% Mass-spring-damper with aging spring
k = 20*(1+exp(-t/10));
xdot = [x(2);10-k*x(1)-2*x(2)];
```

To solve these equations for zero-initial conditions over the interval $0 \le t \le 4$, enter the following in the Command window.

```
>>[t,x] = ode45(@aging_spring,[0,4],[0,0]);
>>plot(t,x),xlabel('t'),gtext('Velocity'),gtext('Displacement')
```

The result is shown in Figure 5.4.6.

When solving nonlinear equations, sometimes it is possible to check the numerical results by using an approximation that reduces the equation to a linear one. Example 5.4.5 illustrates such an approach with a second-order equation.

EXAMPLE 5.4.5

A Nonlinear Pendulum Model

Figure 5.4.7 A pendulum.

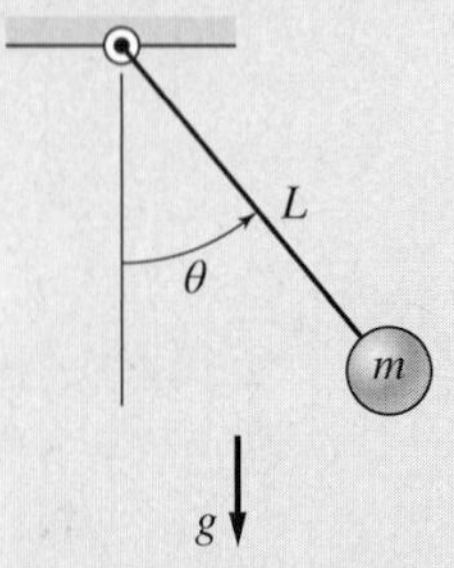

■ Problem

By studying the dynamics of a pendulum like that shown in Figure 5.4.7, we can better understand the dynamics of machines such as a robot arm. The pendulum shown consists of a concentrated mass m attached to a rod whose mass is small compared to m. The rod's length is L. The equation of motion for this pendulum is

$$\ddot{\theta} + \frac{g}{L}\sin\theta = 0 \tag{1}$$

Suppose that $L = 1$ m and $g = 9.81$ m/s^2. Use MATLAB to solve this equation for $\theta(t)$ for two cases: $\theta(0) = 0.5$ rad, and $\theta(0) = 0.8\pi$ rad. In both cases $\dot{\theta}(0) = 0$. Discuss how to check the accuracy of the results.

■ Solution

If we use the small-angle approximation sin $\approx \theta$, the equation becomes

$$\ddot{\theta} + \frac{g}{L}\theta = 0 \tag{2}$$

which is linear and has the solution:

$$\theta(t) = \theta(0)\cos\sqrt{\frac{g}{L}}t \tag{3}$$

Thus, the amplitude of oscillation is $\theta(0)$ and the period is $P = 2\pi\sqrt{L/g} = 2$ s. We can use this information to select a final time, and to check our numerical results.

First, rewrite the pendulum equation (1) as two first-order equations. To do this, let $x_1 = \theta$ and $x_2 = \dot{\theta}$. Thus

$$\dot{x}_1 = \dot{\theta} = x_2$$
$$\dot{x}_2 = \ddot{\theta} = -\frac{g}{L}\sin x_1 = -9.81\sin x_1$$

The following function file is based on the last two equations. Remember that the output `xdot` must be a *column* vector.

```
function xdot = pendulum(t,x)
xdot = [x(2); -9.81*sin(x(1))];
```

The function file is called as follows. The vectors `ta` and `xa` contain the results for the case where $\theta(0) = 0.5$. The vectors `tb` and `xb` contain the results for $\theta(0) = 0.8\pi$.

```
[ta, xa] = ode45(@pendulum, [0, 5], [0.5, 0]);
[tb, xb] = ode45(@pendulum, [0, 5], [0.8*pi, 0]);
plot(ta,xa(:,1),tb,xb(:,1)),xlabel('Time (s)'),...
   ylabel('Angle (rad)'),gtext('Case 1'),gtext('Case 2')
```

The results are shown in Figure 5.4.8. The amplitude remains constant, as predicted by the small-angle analysis, and the period for the case where $\theta(0) = 0.5$ is a little larger than 2 s, the value predicted by the small-angle analysis. So we can place some confidence in the numerical procedure. For the case where $\theta(0) = 0.8\pi$, the period of the numerical solution is about 3.3 s. This illustrates an important property of nonlinear differential equations. The free response of a linear equation has the same period for any initial conditions; however, the form of the

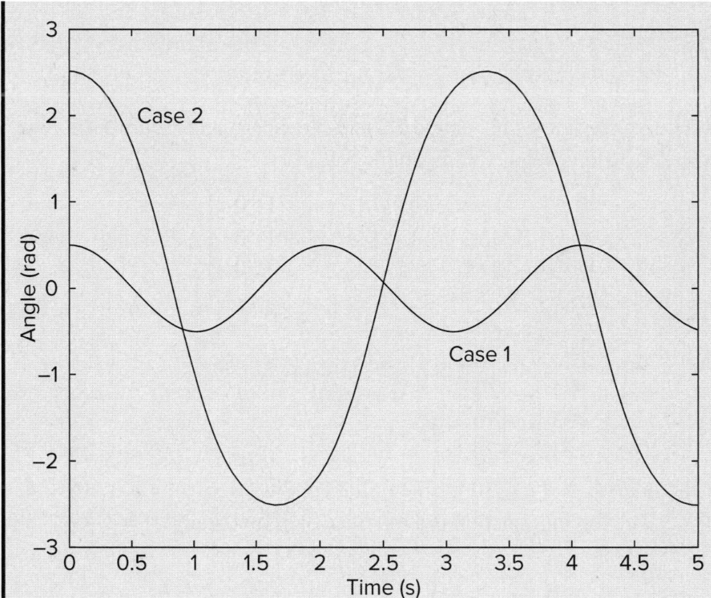

Figure 5.4.8 The pendulum angle as a function of time for two starting positions.

free response of a nonlinear equation often depends on the particular values of the initial conditions.

In the previous example, the values of g and L did not appear in the function `pendulum(t,x)`. Now suppose you want to obtain the pendulum response for different lengths L or different gravitational accelerations g. Starting with MATLAB 7, the preferred method is to use a nested function. Nested functions are discussed in [Palm, 2009] and [Palm, 2019]. The following program shows how this is done.

```
function pendula
g = 9.81; L = 0.75; % First case.
tfinal = 6*pi*sqrt(L/g); % Approximately 3 periods.
[t1, x1] = ode45(@pendulum, [0,tfinal], [0.4, 0]);
%
g = 1.63; L = 2.5; % Second case.
tfinal = 6*pi*sqrt(L/g); % Approximately 3 periods.
[t2, x2] = ode45(@pendulum, [0,tfinal], [0.2, 0]);
plot(t1, x1(:,1), t2, x2(:,1)), ...
    xlabel ('time (s)'), ylabel ('\theta (rad)')
% Nested function.
    function xdot = pendulum(t,x)
    xdot = [x(2);-(g/L)*sin(x(1))];
    end
end
```

5.4.5 MATRIX METHODS

We can use matrix operations to reduce the number of lines to be typed in the derivative function file.

EXAMPLE 5.4.6

A Fourth-Order System

■ Problem

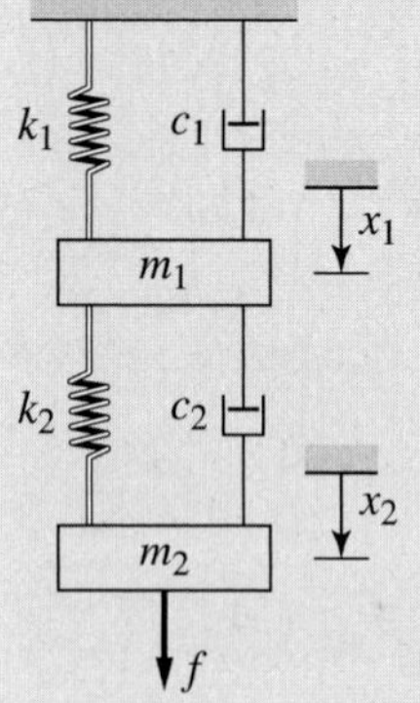

Figure 5.4.9 A two-mass system.

Consider the two-mass system considered in Section 5.2 and shown again in Figure 5.4.9. The state-variable model $\dot{\mathbf{x}} = \mathbf{Ax} + \mathbf{Bu}$ developed in Example 5.2.3 is

$$\mathbf{A} = \begin{bmatrix} 0 & 0 & 1 & 0 \\ 0 & 0 & 0 & 1 \\ -1 & 4/5 & -12/5 & 8/5 \\ 4/3 & -4/3 & 8/3 & -8/3 \end{bmatrix} \qquad \mathbf{B} = \begin{bmatrix} 0 \\ 0 \\ 0 \\ 1/3 \end{bmatrix}$$

$$\mathbf{x} = \begin{bmatrix} x_1 \\ x_2 \\ x_3 \\ x_4 \end{bmatrix} = \begin{bmatrix} x_1 \\ x_2 \\ \dot{x}_1 \\ \dot{x}_2 \end{bmatrix} \qquad \mathbf{u} = [f(t)]$$

Solve this model over the interval $0 \leq t \leq 30$ for the initial conditions $x_1(0) = 5$, $x_2(0) = 4$, $\dot{x}_1(0) = -3$, and $\dot{x}_2(0) = 2$. The forcing function is a modified step function: $f(t) = 1 - e^{-t/8}$.

■ Solution

The following function file shows how to use matrix operations with the `ode45` function.

```
function xdot = Ex5p4p6(t,x)
%Two-mass system with non-zero initial conditions and time
%varying input.
f = 1-exp(-t/8);
A = [0,0,1,0;0,0,0,1;-1,4/5,-12/5,8/5;4/3,-4/3,8/3,-8/3];
B = [0;0;0;1/3];
xdot = A*x+B*f;
```

Note that `xdot` will be a column vector because of the definition of matrix-vector multiplication.

The solution for x_1 and x_2 can be plotted by typing the following in the Command window.

```
>>plot(t,x(:,1),t,x(:,2),'-'),xlabel('t'),...
   gtext('Mass 1'),gtext('Mass 2')
```

The result is shown in Figure 5.4.10.

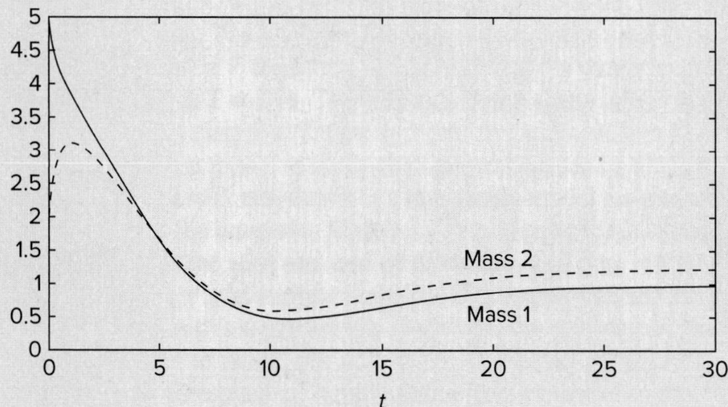

Figure 5.4.10 Response of a two-mass system to modified step input.

The characteristic roots and the time constants may be found by typing

```
>>A = [0,0,1,0;0,0,0,1;-1,4/5,-12/5,8/5;4/3,-4/3,8/3,-8/3];
>>r = eig(A);
>>tau = -1./real(r)
```

The results for the eigenvalues are −4.0595, −0.5261, $-0.2405 \pm 0.2588j$. The time constants are $\tau = 0.2463, 1.9006, 4.1575, 4.1575$. The dominant time constant is 4.1575, and four times that value is 16.63, but steady state will be reached later than this time because the time constant of the input function $f(t)$ is 8. Thus, steady state should be reached after about $t = 4(8) = 32$. This is verified by the plot on Figure 5.4.10.

Table 5.4.2 summarizes the basic syntax of the `ode45` solver.

PART III. SIMULINK METHODS

5.5 SIMULINK AND LINEAR MODELS

Simulink is built on top of MATLAB, so you must have MATLAB to use Simulink. It is included in the Student Edition of MATLAB and is also available separately from The MathWorks, Inc. It provides a graphical user interface that uses various types of elements called *blocks* to create a simulation of a dynamic system; that is, a system that can be modeled with differential or difference equations whose independent variable is time. For example, one block type is a multiplier, another performs a sum, and another is an integrator. Its graphical interface enables you to position the blocks, resize them, label them, specify block parameters, and interconnect the blocks to describe complicated systems for simulation.

Type `simulink` in the MATLAB Command window to start Simulink. The Simulink window opens. To create a new model, double-click on Blank Model. A new untitled window opens for you to create the model. Then open the Library Browser by selecting it in the drop-down menu under the View menu. To select a block from the Library Browser, double-click on the appropriate library category and a list of blocks within that category then appears, as shown in Figure 5.5.1. Figure 5.5.1 shows the result of double-clicking on the Continuous library. Click on the block name or icon, hold the mouse button down, drag the block to the new model window, and release the button. You can access help for that block by right-clicking on its name or icon, and selecting **Help** from the drop-down menu.

Simulink model files have the default extension `.slx` as of Release 2019a. The `.slx` format is a new compressed format that provides smaller file sizes. The `.mdl` format is still available as a save option. Use the **File** menu in the model window to Open, Close, and Save model files. To print the block diagram of the model, select **Print** on the **File** menu. Use the **Edit** menu to copy, cut, and paste blocks. You can also use the mouse for these operations. For example, to delete a block, click on it, and press the **Delete** key.

Getting started with Simulink is best done through examples, which we now present.

5.5.1 SIMULATION DIAGRAMS

You construct Simulink models by constructing a diagram that shows the elements of the problem to be solved. Such diagrams are called *simulation diagrams*. Consider the

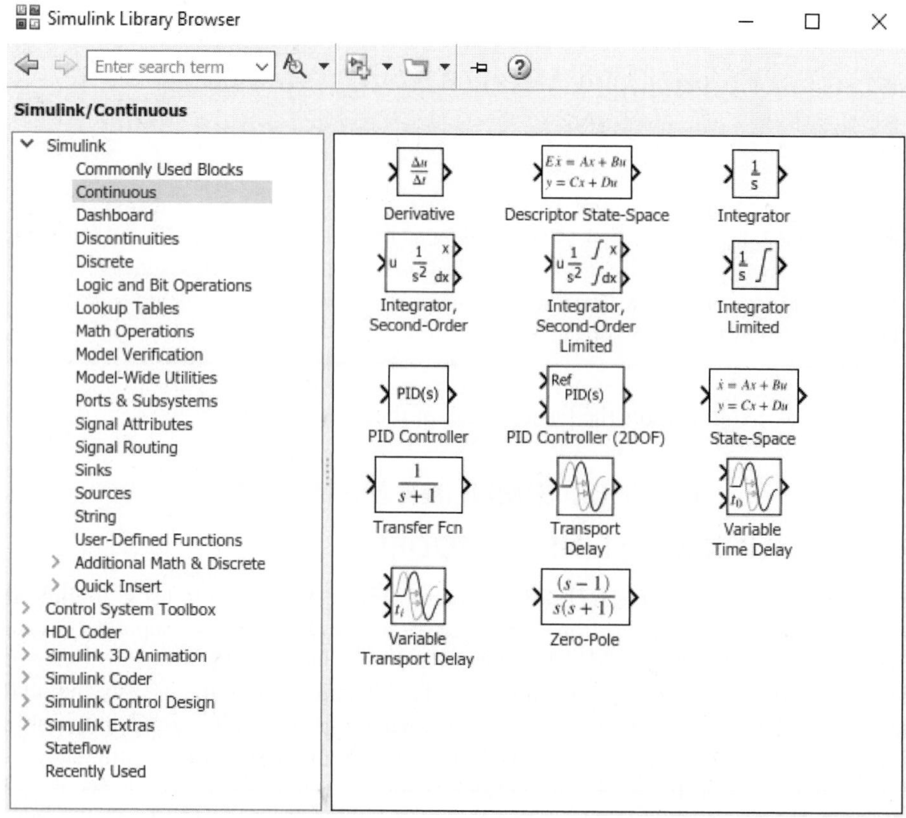

Figure 5.5.1 Simulink Library Browser showing the Continuous library being selected.
(*The MathWorks, Inc.*)

equation $\dot{y} = 10f(t)$. Its solution can be represented symbolically as

$$y(t) = \int 10f(t)\,dt$$

which can be thought of as two steps, using an intermediate variable x:

$$x(t) = 10f(t) \qquad \text{and} \qquad y(t) = \int x(t)\,dt$$

This solution can be represented graphically by the simulation diagram shown in Figure 5.5.2a. The arrows represent the variables y, x, and f. The blocks represent cause-and-effect processes. Thus, the block containing the number 10 represents the process $x(t) = 10f(t)$, where $f(t)$ is the cause (the *input*) and $x(t)$ represents the effect (the *output*). This type of block is called a *multiplier* or *gain* block.

The block containing the integral sign $\int$ represents the integration process $y(t) = \int x(t)\,dt$, where $x(t)$ is the cause (the *input*) and $y(t)$ represents the effect (the *output*). This type of block is called an *integrator* block.

There is some variation in the notation and symbols used in simulation diagrams. Part (b) of Figure 5.5.2 shows one variation. Instead of being represented by a box, the

Figure 5.5.2 Simulation diagrams for $\dot{y} = 10f(t)$.

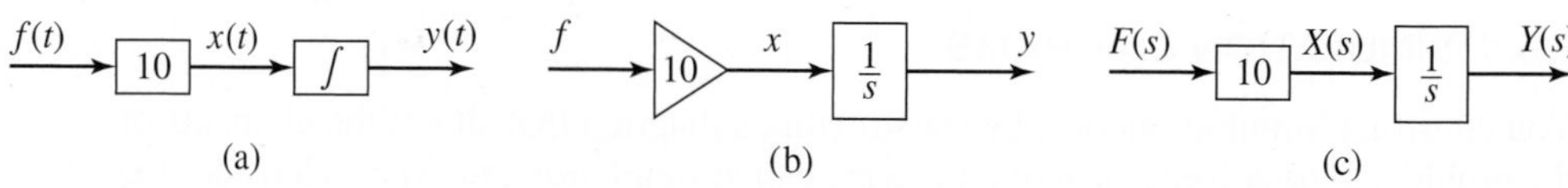

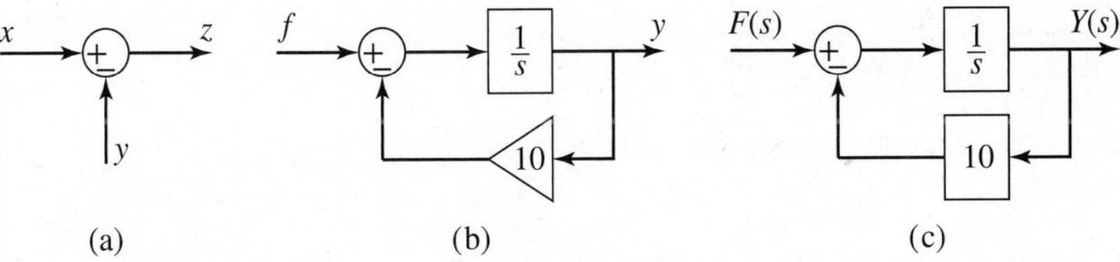

Figure 5.5.3 (a) The summer element. (b) Simulation diagram for $\dot{y} = f(t) - 10y$. (c) Block diagram.

multiplication process is now represented by a triangle like that used to represent an electrical amplifier, hence the name *gain* block.

Part (c) of Figure 5.5.2 shows the block diagram representation. If you are familiar with block diagrams, you can immediately create a Simulink model from the block diagram.

In addition, the integration symbol in the integrator block has been replaced by the symbol $1/s$, which represents integration in Laplace transform notation. Thus, the equation $\dot{y} = 10f(t)$ is represented by $sy = 10f$, and the solution is represented as

$$y = \frac{10f}{s}$$

or as the two equations

$$x = 10f \qquad \text{and} \qquad y = \frac{1}{s}x$$

Another element used in simulation diagrams is the *summer* that, despite its name, is used to subtract as well as to sum variables. Its symbol is shown in Figure 5.5.3a. The symbol represents the equation $z = x - y$. Note that a plus or minus sign is required for each input arrow.

The summer symbol can be used to represent the equation $\dot{y} = f(t) - 10y$, which can be expressed as

$$y(t) = \int [f(t) - 10y]\, dt$$

or as

$$y = \frac{1}{s}(f - 10y)$$

You should study the simulation diagram shown in part (b) of Figure 5.5.3 to confirm that it represents this equation.

Part (c) of Figure 5.5.3 shows the corresponding block diagram. Once again, note the similarity with the simulation diagram.

Figure 5.5.2b forms the basis for developing a Simulink model to solve the equation $\dot{y} = f(t)$.

Simulink Solution of $\dot{y} = 10 \sin t$ — EXAMPLE 5.5.1

■ Problem

Let us use Simulink to solve the following problem for $0 \le t \le 13$.

$$\frac{dy}{dt} = 10 \sin t \qquad y(0) = 0$$

The exact solution is $y(t) = 10(1 - \cos t)$.

■ Solution

To construct the simulation, do the following steps. Refer to Figure 5.5.4.

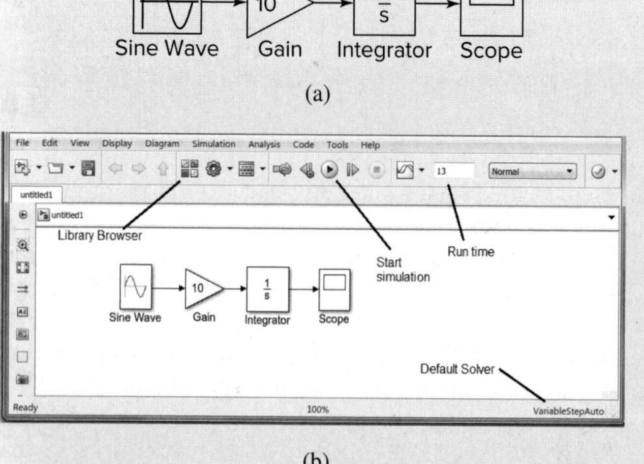

Figure 5.5.4 (a) Simulink model for $\dot{y} = 10 \sin t$. (b) Simulink model window. (*The MathWorks, Inc.*)

1. Start Simulink and open a new model window as described previously.
2. Select and place in the new window the Sine Wave block from the Sources category. Double-click on it to open the Block Parameters window, and make sure the Amplitude is set to 1, the Bias to 0, the Frequency to 1, the Phase to 0, and the Sample time to 0. Then click **OK.**
3. Select and place the Gain block from the Math category, double-click on it, and set the Gain Value to 10 in the Block Parameters window. Then click **OK.** Note that the value 10 then appears in the triangle. To make the number more visible, click on the block, and drag one of the corners to expand the block so that all the text is visible.
4. Select and place the Integrator block from the Continuous category, double-click on it to obtain the Block Parameters window, and set the Initial Condition to 0 [this is because $y(0) = 0$]. Then click **OK.**
5. Select and place the Scope block from the Sinks category.
6. Once the blocks have been placed as shown in Figure 5.5.4a, connect the input port on each block to the ouput port on the preceding block. To do this, click on the preceding block, hold down the **Ctrl** key, and then click on the destination block. Simulink will connect the blocks with an arrow pointing at the input port. Your model should now look like that shown in Figure 5.5.4b.
7. Enter 13 for the Stop time in the window to the right of the Start Simulation icon (the black triangle). See Figure 5.5.4b. The default value is 10, which can be deleted and replaced with 13.
8. Run the simulation by clicking on the **Start Simulation** icon on the toolbar.
9. You will hear a bell sound when the simulation is finished. You should see an oscillating curve with an amplitude of 10 and a period of 2π. The independent variable in the Scope block is time t; the input to the block is the dependent variable y. This completes the simulation.

Note that the default solver is `ode45`, a variable step solver that is automatically selected by Simulink. By clicking on this item, you can specify the solver to be the same solver discussed in Section 5.4.

Note that blocks have a Block Parameters window that opens when you double-click on the block. This window contains several items, the number and nature of which depend on the specific type of block. In general, you can use the default values of these parameters, except where we have explicitly indicated that they should be changed. You can always click on Help within the Block Parameters window to obtain more information.

When you click Apply in the Block Parameters window, any parameter changes immediately take effect and the window remains open. If you click Close, the changes take effect and the window closes.

Note that most blocks have default labels. You can edit text associated with a block by clicking on the text and making the changes.

The Scope block is useful for examining the solution, but if you want to obtain a labeled and printed plot you should use the To Workspace block, which is described in the next example.

Exporting to the MATLAB Workspace

EXAMPLE 5.5.2

■ **Problem**

We now demonstrate how to export the results of the simulation to the MATLAB workspace, where they can be plotted or analyzed with any of the MATLAB functions.

■ **Solution**

Modify the Simulink model constructed in Example 5.5.1 as follows. Refer to Figure 5.5.5.

1. Delete the arrow connecting the Scope block by clicking on it and pressing the Delete key. Delete the Scope block in the same way.
2. Select and place the To Workspace block from the Sinks category. Your model should now look like that shown in Figure 5.5.5.
3. Double-click on the To Workspace block. You can specify any variable name you want as the output; the default is `simout`. Change its name to `y`. The output variable `y` will have as many rows as there are simulation time steps, and as many columns as there are inputs to the block. Specify the Save Format as Array. Use the default values for the other parameters (these should be `inf`, 1, and `-1` for Limit data points to last: Decimation, and Sample Time, respectively). Click on **OK.**
4. After running the simulation, you can use the MATLAB plotting commands from the Command window to plot the columns of `y` (or `simout` in general). Simulink puts the time variable `tout` into the MATLAB workspace automatically when using the To Workspace block. To plot $y(t)$, type in the MATLAB Command window:

```
>>plot(tout,y),xlabel('t'),ylabel('y')
```

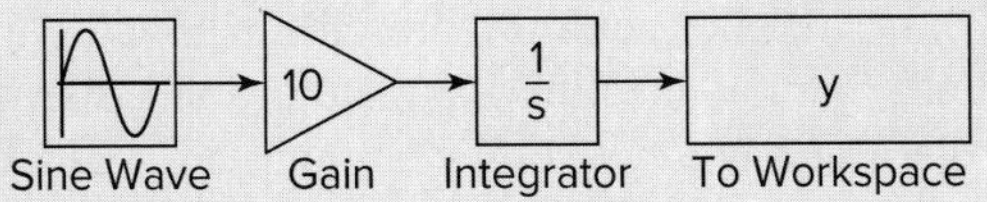

Figure 5.5.5 Simulink model for $\dot{y} = 10 \sin t$ using the To Workspace block.

Simulink Model for $\dot{y} = -10y + f(t)$

EXAMPLE 5.5.3

■ **Problem**

Construct a Simulink model to solve

$$\dot{y} = -10y + f(t) \qquad y(0) = 1$$

where $f(t) = 2 \sin 4t$, for $0 \le t \le 3$.

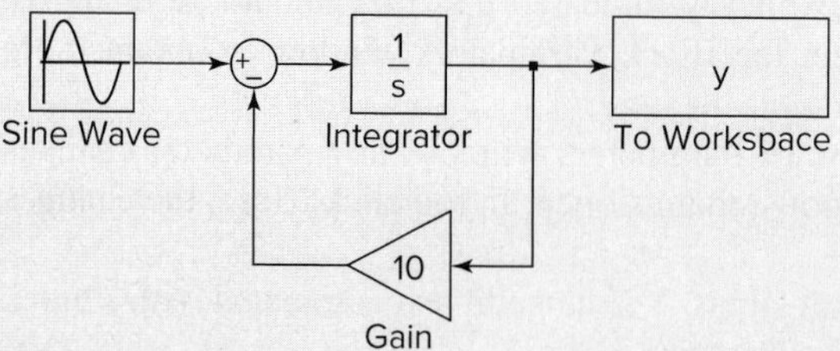

Figure 5.5.6 Simulink model for $\dot{y} = -10y + f(t)$.

■ **Solution**

To construct the simulation, do the following steps.

1. You can use the model shown in Figure 5.5.4 by rearranging the blocks as shown in Figure 5.5.6. You will need to add a Sum block.
2. Select the Sum block from the Math Operations library and place it as shown in the simulation diagram. Its default setting adds two input signals. To change this, double-click on the block, and in the List of Signs window, type | + −. The signs are ordered counterclockwise from the top. The symbol | is a spacer indicating here that the top port is to be empty.
3. To reverse the direction of the gain block, right-click on the block, select **Rotate/Flip** from the pop-up menu, and select **Flip Block.**
4. When you connect the negative input port of the Sum block to the output port of the Gain block, Simulink will attempt to draw the shortest line. To obtain the more standard appearance shown in Figure 5.5.6, first extend the line vertically down from the Sum input port. Release the mouse button and then click on the end of the line and attach it to the Gain block. The result will be a line with a right angle. Do the same to connect the input of the Gain to the arrow connecting the Integrator and the To Workspace block. A small dot appears to indicate that the lines have been successfully connected.
5. Remember to specify Save Format as Array, and to change the name to `y`. Set the Stop time to 3.
6. Run the simulation as before; you can plot the results by typing the following:

```
>>plot(tout,y)
```

5.5.2 SIMULATING STATE-VARIABLE MODELS

State-variable models, unlike transfer-function models, can have more than one input and more than one output. Simulink has the State-Space block that represents the linear state-variable model $\dot{\mathbf{x}} = \mathbf{A}\mathbf{x} + \mathbf{B}\mathbf{u}$, $\mathbf{y} = \mathbf{C}\mathbf{x} + \mathbf{D}\mathbf{u}$. The vector $\mathbf{u}$ represents the inputs, and the vector $\mathbf{y}$ represents the outputs. Thus, when connecting inputs to the State-Space block, care must be taken to connect them in the proper order. Similar care must be taken when connecting the block's outputs to another block. The following example illustrates how this is done.

EXAMPLE 5.5.4 Simulink Model of a Two-Mass System

■ **Problem**

The state-variable model of the two-mass system discussed in Example 5.2.3 is

$$\dot{\mathbf{x}} = \mathbf{A}\mathbf{x} + \mathbf{B}f(t)$$

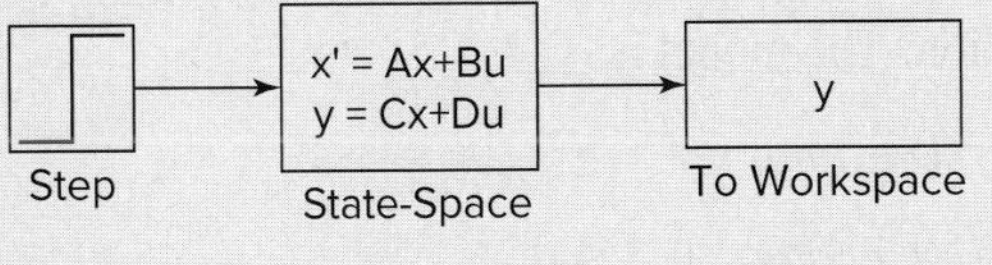

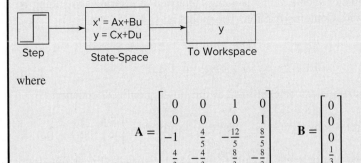

Figure 5.5.7 Simulink model containing the State-Space block and the Step block.

where

$$\mathbf{A} = \begin{bmatrix} 0 & 0 & 1 & 0 \\ 0 & 0 & 0 & 1 \\ -1 & \frac{4}{5} & -\frac{12}{5} & \frac{8}{5} \\ \frac{4}{3} & -\frac{4}{3} & \frac{8}{3} & -\frac{8}{3} \end{bmatrix} \qquad \mathbf{B} = \begin{bmatrix} 0 \\ 0 \\ 0 \\ \frac{1}{3} \end{bmatrix}$$

and

$$\mathbf{x} = \begin{bmatrix} x_1 \\ x_2 \\ x_3 \\ x_4 \end{bmatrix} = \begin{bmatrix} x_1 \\ x_2 \\ \dot{x}_1 \\ \dot{x}_2 \end{bmatrix}$$

Develop a Simulink model to plot the unit-step response of the variables x_1 and x_2 with the initial conditions $x_1(0) = 5$, $\dot{x}_1(0) = -3$, $x_2(0) = 1$, and $\dot{x}_2(0) = 2$.

■ **Solution**

First select appropriate values for the matrices in the output equation $\mathbf{y} = \mathbf{Cz} + \mathbf{D}f(t)$. Since we want to plot x_1 and x_2, which are z_1 and z_3, we choose **C** and **D** as follows.

$$\mathbf{C} = \begin{bmatrix} 1 & 0 & 0 & 0 \\ 0 & 1 & 0 & 0 \end{bmatrix} \qquad \mathbf{D} = \begin{bmatrix} 0 \\ 0 \end{bmatrix}$$

To create this simulation, obtain a new model window. Then do the following to create the model shown in Figure 5.5.7.

1. Select and place in the new window the Step block from the Sources category. Double-click on it to obtain the Block Parameters window, and set the Step time to 0, the Initial and Final values to 0 and 1, and the Sample time to 0. Click **OK**.
2. Select and place the State-Space block from the Continuous Library. Double-click on it, and enter `[0, 0, 1, 0; 0, 0, 0, 1; -1, 4/5, -12/5, 8/5; 4/3, -4/3, 8/3, -8/3]` for **A**, `[0; 0; 0; 1/3]` for **B**, `[1, 0, 0, 0; 0, 1, 0, 0]` for **C**, and `[0; 0]` for **D**. Then enter `[5, 1, -3, 2]` for the initial conditions. Click **OK**. Note that the dimension of the matrix **B** tells Simulink that there is one input. The dimensions of the matrices **C** and **D** tell Simulink that there are two outputs.
3. Select and place the To Workspace block. Change the name to `y` and the Save As format to Array.
4. Once the blocks have been placed, connect the input port on each block to the output port on the preceding block as shown in the figure.
5. For this application, a Stop time of 20 is satisfactory.

5.6 SIMULINK AND NONLINEAR MODELS

Unlike linear models, closed-form solutions are not available for most nonlinear differential equations, and we must therefore solve such equations numerically. *Piecewise-linear* models are actually nonlinear, although they may appear to be linear. They are composed of linear models that take effect when certain conditions are satisfied. The effect of switching back and forth between these linear models makes the overall model nonlinear. An example of such a model is a mass attached to a spring and sliding on a horizontal surface with Coulomb friction. The model is

$$m\ddot{x} + kx = f(t) - \mu mg \quad \text{if } \dot{x} \geq 0$$

$$m\ddot{x} + kx = f(t) + \mu mg \quad \text{if } \dot{x} < 0$$

These two linear equations can be expressed as the single, nonlinear equation

$$m\ddot{x} + kx = f(t) - \mu mg \,\text{sign}(\dot{x}) \qquad \text{where} \qquad \text{sign}(\dot{x}) = \begin{cases} +1 & \text{if } \dot{x} \geq 0 \\ -1 & \text{if } \dot{x} < 0 \end{cases}$$

Solution of linear or nonlinear models that contain piecewise-linear functions is very tedious to program. However, Simulink has built-in blocks that represent many of the commonly found functions such as Coulomb friction. Therefore, Simulink is especially useful for such applications.

EXAMPLE 5.6.1

Simulink Model of a Rocket-Propelled Sled

■ Problem

A rocket-propelled sled on a track is represented in Figure 5.6.1 as a mass m with an applied force f that represents the rocket thrust. The rocket thrust initially is horizontal, but the engine accidentally pivots during firing and rotates with an angular acceleration of $\ddot{\theta} = \pi/50$ rad/s. Compute the sled's velocity v for $0 \leq t \leq 10$ if $v(0) = 0$. The rocket thrust is 4000 N and the sled mass is 450 kg.

The sled's equation of motion was derived in Example 5.4.2 and is

$$\dot{v} = \frac{80}{9} \cos\left(\frac{\pi}{100} t^2\right) \tag{1}$$

The solution is formally given by

$$v(t) = \frac{80}{9} \int_0^t \cos\left(\frac{\pi}{100} t^2\right) dt$$

Unfortunately, no closed-form solution is available for the integral, which is called *Fresnel's cosine integral*. The value of the integral has been tabulated numerically, but we will use Simulink to obtain the solution.

a. Create a Simulink model to solve this problem for $0 \leq t \leq 10$ s.
b. Now suppose that the engine angle is limited by a mechanical stop to 60°, which is $60\pi/180$ rad. Create a Simulink model to solve the problem.

■ Solution

a. There are several ways to create the input function $\theta = (\pi/100)t^2$. Here we note that $\ddot{\theta} = \pi/50$ rad/s and that

$$\dot{\theta} = \int_0^t \ddot{\theta}\, dt = \frac{\pi}{50} t$$

and

$$\theta = \int_0^t \dot{\theta}\, dt = \frac{\pi}{100} t^2$$

Thus, we can create $\theta(t)$ by integrating the constant $\ddot{\theta} = \pi/50$ twice. The simulation diagram is shown in Figure 5.6.2. This diagram is used to create the corresponding Simulink model shown in Figure 5.6.3.

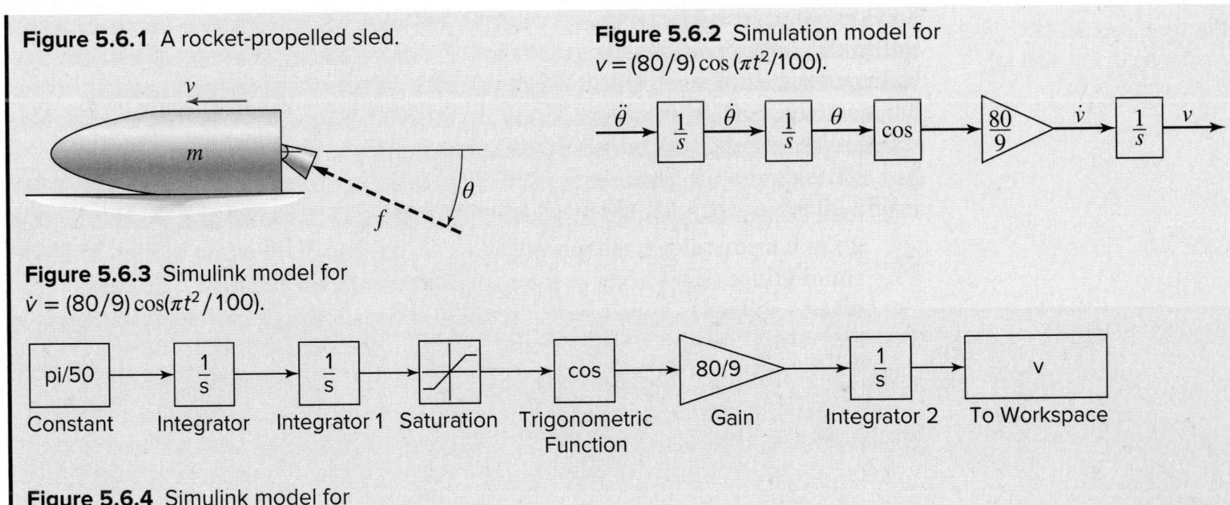

Figure 5.6.1 A rocket-propelled sled.

Figure 5.6.2 Simulation model for $\dot{v} = (80/9)\cos(\pi t^2/100)$.

Figure 5.6.3 Simulink model for $\dot{v} = (80/9)\cos(\pi t^2/100)$.

Figure 5.6.4 Simulink model for $\dot{v} = (80/9)\cos(\pi t^2/100)$ with a Saturation block.

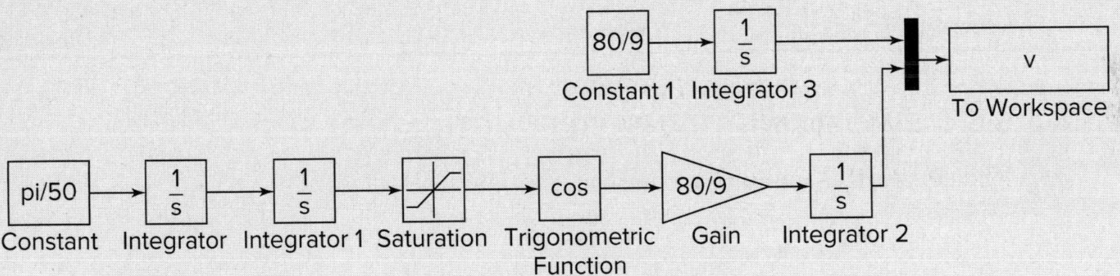

There are two new blocks in this model. The Constant block is in the Sources library. After placing it, double-click on it and type `pi/50` in its Constant Value window.

The Trigonometric function block is in the Math Operations library. After placing it, double-click on it and select `cos` in its Function window.

Set the Stop Time to 10, run the simulation, and examine the results in the Scope.

b. Modify the model in Figure 5.6.3 as follows to obtain the model shown in Figure 5.6.4. We use the Saturation block in the Discontinuities library to limit the range of θ to $60\pi/180$ rad. After placing the block as shown in Figure 5.6.4, double-click on it and type `60*pi/180` in its Upper Limit window. Then type 0 or any negative value in its Lower Limit window.

Select and place the Mux block from the Signal Routing category, double-click on it, and set the Number of inputs to 2. Click **OK**. (The name "Mux" is an abbreviation for "multiplexer," which is an electrical device for transmitting several signals.)

Enter and connect the remaining elements as shown, and run the simulation. The upper Constant block and Integrator block are used to generate the solution when the engine angle is $\theta = 0$, as a check on our results. [The equation of motion for $\theta = 0$ is $\dot{v} = 80/9$, which gives $v(t) = 80t/9$.]

You can plot the results in MATLAB by typing the following:

```
>>plot(tout,v)
```

The array `v(:,1)` appears at the topmost (first) connection on the Mux; array `v(:,2)` appears at the second (bottom) connection. The resulting plot is shown in Figure 5.6.5.

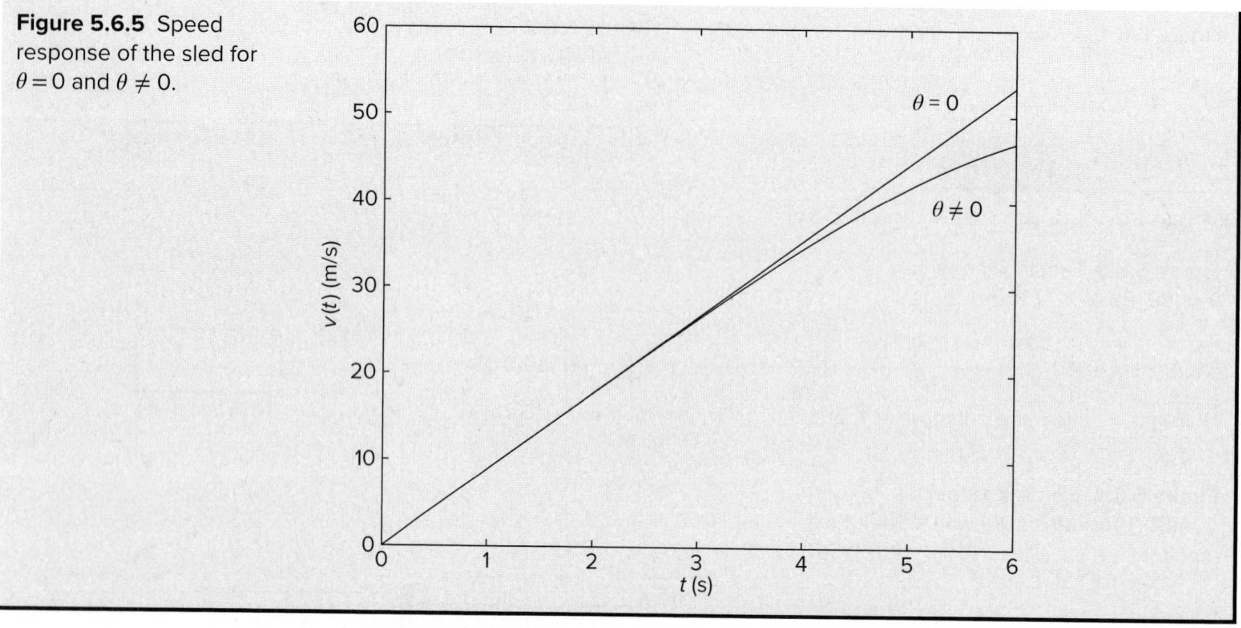

Figure 5.6.5 Speed response of the sled for $\theta = 0$ and $\theta \neq 0$.

5.6.1 SIMULATING TRANSFER-FUNCTION MODELS

The equation of motion of a mass-spring-damper system is

$$m\ddot{y} + c\dot{y} + ky = f(t) \tag{5.6.1}$$

Simulink can accept a system description in transfer-function form and in state-variable form. If the mass-spring system is subjected to a sinusoidal forcing function $f(t)$, it is easy to use the MATLAB commands presented thus far to solve and plot the response $y(t)$. However, suppose that the force $f(t)$ is created by applying a sinusoidal input voltage to a hydraulic piston that has a *dead-zone* nonlinearity. This means that the piston does not generate a force until the input voltage exceeds a certain magnitude, and thus the system model is piecewise linear. A graph of a particular dead-zone nonlinearity is shown in Figure 5.6.6. When the input (the independent variable on the graph) is

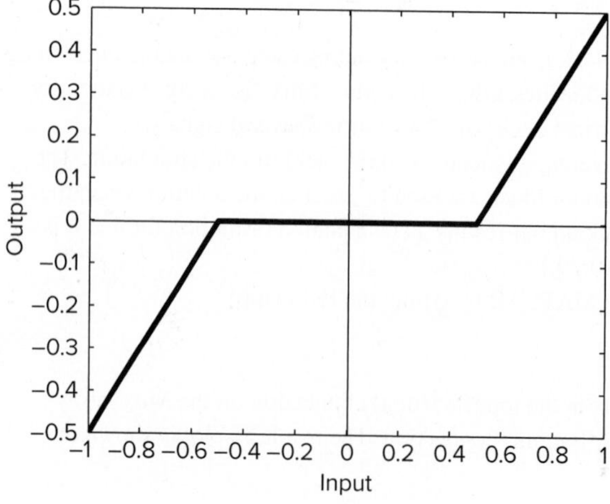

Figure 5.6.6 A dead-zone nonlinearity.

between −0.5 and 0.5, the output is zero. When the input is greater than or equal to the upper limit of 0.5, the output is the input minus the upper limit. When the input is less than or equal to the lower limit of −0.5, the output is the input minus the lower limit. In this example, the dead zone is symmetric about 0, but it need not be in general.

Simulations with dead-zone nonlinearities are somewhat tedious to program in MATLAB, but are easily done in Simulink. Example 5.6.2 illustrates how it is done.

EXAMPLE 5.6.2

A Simulink Model of Response with a Dead Zone

■ Problem

Create and run a Simulink simulation of a mass-spring-damper system (5.6.1) using the parameter values $m = 1$, $c = 2$, and $k = 4$. The forcing function is the function $f(t) = \sin 1.4t$. The system has the dead-zone nonlinearity shown in Figure 5.6.6.

■ Solution

To construct the simulation, do the following steps.

1. Start Simulink and open a new model window as described previously.
2. Select and place in the new window the Sine Wave block from the Sources category. Double-click on it, and set the Amplitude to 1, the Bias to 0, the Frequency to 1.4, the Phase to 0, and the Sample time to 0. Click **OK**.
3. Select and place the Dead Zone block from the Discontinuities category, double-click on it, and set the Start of dead zone to −0.5 and the End of dead zone to 0.5. Click **OK**.
4. Select and place the Transfer Fcn block from the Continuous category, double-click on it, and set the Numerator to `[1]` and the Denominator to `[1, 2, 4]`. Click **OK**.
5. Select and place the To Workspace block from the Sinks category. Change the name to `y` and the Save As Format to Array.
6. Once the blocks have been placed, connect the input port on each block to the output port on the preceding block. Your model should now look like that shown in Figure 5.6.7.
7. Enter 10 for the Stop time.
8. Run the simulation.
9. You should see an oscillating curve if you plot the output.

It is informative to plot both the input and the output of the Transfer Fcn block versus time on the same graph. To do this,

1. Delete the arrow connecting the To Workspace block to the Transfer Fcn block. Do this by clicking on the arrow line and then pressing the **Delete** key.
2. Select and place the Mux block from the Signal Routing category, double-click on it, and set the Number of inputs to 2. Click **Apply**, then **OK**.
3. Connect the top input port of the Mux block to the output port of the Transfer Fcn block. Then use the same technique to connect the bottom input port of the Mux block to the arrow from the output port of the Dead Zone block. Just remember to start with the input port. Simulink will sense the arrow automatically and make the connection. Your model should now look like that shown in Figure 5.6.8.
4. Set the Stop time to 10, run the simulation as before. Plot the results by typing:

```
>>plot(tout,y(:,1),tout,y(:,2))
```

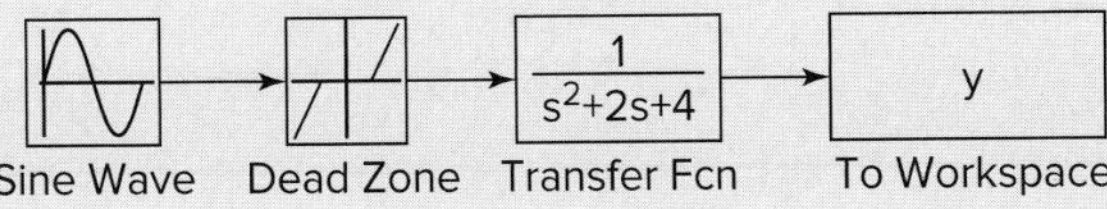

Figure 5.6.7 The Simulink model of dead-zone response.

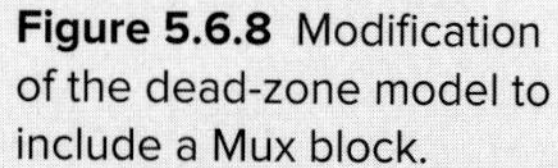
Figure 5.6.8 Modification of the dead-zone model to include a Mux block.

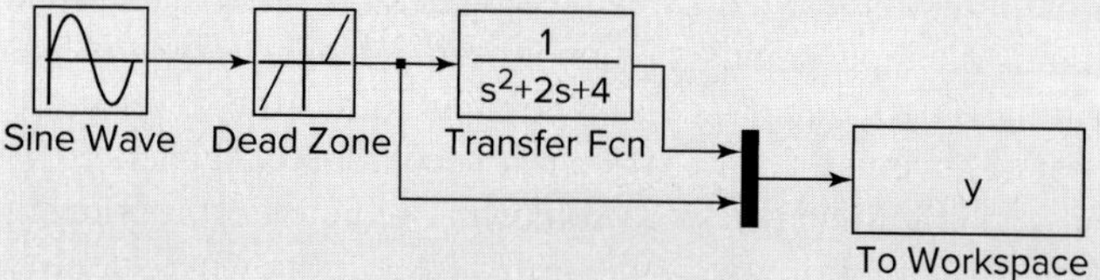

Figure 5.6.9 Plot of the response of the dead-zone model shown in Figure 5.6.8.

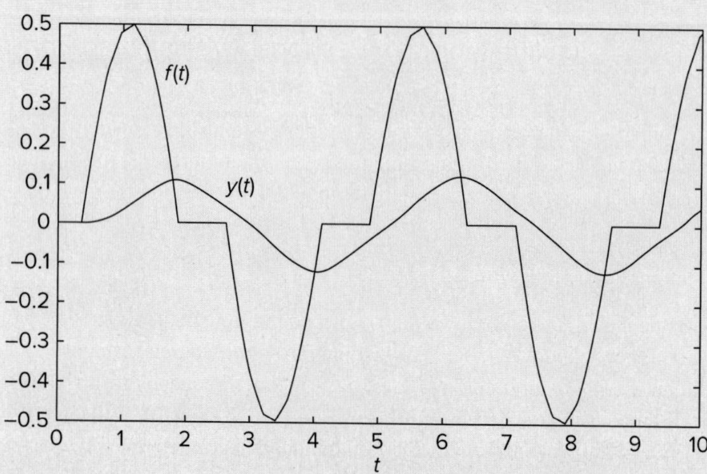

Figure 5.6.10 Modification of the dead-zone model to export variables to the MATLAB workspace.

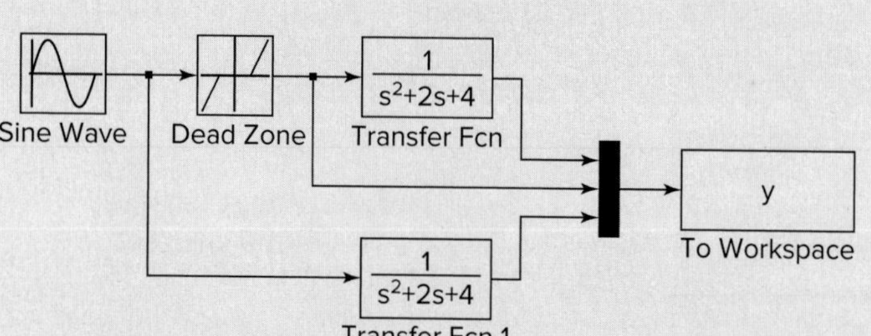

You should see what is shown in Figure 5.6.9. This plot shows the effect of the dead zone on the sine wave.

Suppose we want to examine the effects of the dead zone by comparing the response of the system with and without a dead zone. We can do this with the model shown in Figure 5.6.10. To create this model,

1. Copy the Transfer Fcn block by right-clicking on it, holding down the mouse button, and dragging the block copy to a new location. Then release the button.
2. Double-click on the Mux block and change the number of its inputs to 3.
3. The output variable `y` will have as many rows as there are simulation time steps, and as many columns as there are inputs to the Mux block.
4. Connect the blocks as shown, and run the simulation.
5. You can use the MATLAB plotting commands from the Command window to plot the columns of `y`; for example, to plot the response of the two systems and the output of the Dead Zone block versus time, type

```
>>plot(tout,y)
```

Nonlinear models cannot be put into transfer-function form or the state-variable form $\dot{\mathbf{x}} = \mathbf{Ax} + \mathbf{Bu}$. However, they can be solved in Simulink. Example 5.6.3 shows how this can be done.

EXAMPLE 5.6.3

Simulink Model of a Nonlinear Pendulum

■ **Problem**

The pendulum shown in Figure 5.6.11 has the following nonlinear equation of motion, if there is viscous friction in the pivot and if there is an applied moment $M(t)$ about the pivot.

$$I\ddot{\theta} + c\dot{\theta} + mgL\sin\theta = M(t) \tag{1}$$

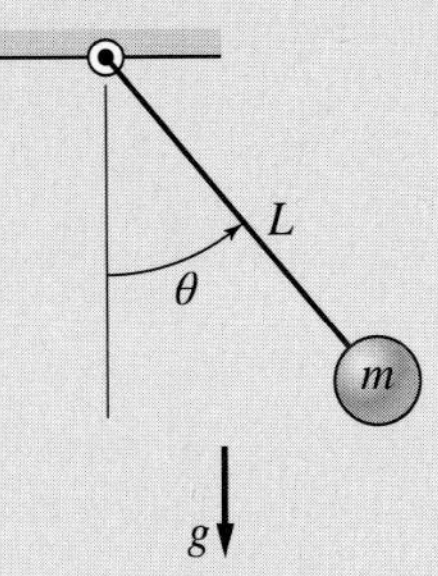

Figure 5.6.11 A pendulum.

where I is the mass moment of inertia about the pivot. Create a Simulink model for this system for the case where $I = 4$, $mgL = 10$, $c = 0.8$, and $M(t)$ is a square wave with an amplitude of 3 and a frequency of 0.5 Hz. Assume that the initial conditions are $\theta(0) = \pi/4$ rad and $\dot{\theta}(0) = 0$.

■ **Solution**

To simulate this model in Simulink, define a set of variables that lets you rewrite the equation as two first-order equations. Thus, let $\omega = \dot{\theta}$. Then the model can be written as

$$\dot{\theta} = \omega$$

$$\dot{\omega} = \frac{1}{I}[-c\omega - mgL\sin\theta + M(t)] = 0.25\,[-0.8\omega - 10\sin\theta + M(t)]$$

Integrate both sides of each equation over time to obtain

$$\theta = \int \omega\,dt$$

$$\omega = 0.25 \int [-0.8\omega - 10\sin\theta + M(t)]\,dt$$

We will introduce four new blocks to create this simulation. Obtain a new model window and do the following.

1. Select and place in the new window the Integrator block from the Continuous category, and change its label to Integrator 1 as shown in Figure 5.6.12. You can edit text associated with a block by clicking on the text and making the changes. Double-click on

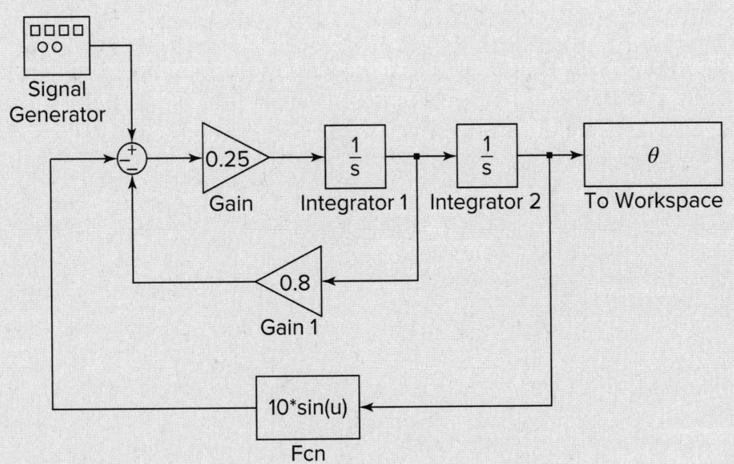

Figure 5.6.12 Simulink model of a nonlinear pendulum.

the block to obtain the Block Parameters window, and set the Initial condition to 0 [this is the initial condition $\dot{\theta}(0) = 0$]. Click **OK**.

2. Copy the Integrator block to the location shown and change its label to Integrator 2. Set its initial condition to $\pi/4$ by typing `pi/4` in the Block Parameters window. This is the initial condition $\theta(0) = \pi/4$.
3. Select and place a Gain block from the Math Operations category, double-click on it, and set the Gain value to 0.25. Click **OK**. Change its label to `1/I`. Then click on the block, and drag one of the corners to expand the box so that all the text is visible.
4. Copy the Gain box, change its label to `c`, and place it as shown in Figure 5.6.12. Double-click on it, and set the Gain value to 0.8. Click **OK**. To flip the box left to right, right-click on it, select **Format**, and select **Flip Block**.
5. Select and place the To Workspace block from the Sinks category. Edit the name and Save As format.
6. For the term $10 \sin \theta$, we cannot use the Trig function block in the Math Operations category without using a separate gain block to multiply the $\sin \theta$ by 10. Instead we will use the Fcn block under the User-Defined Functions category (Fcn stands for function). Select and place this block as shown. Double-click on it, and type `10*sin(u)` in the expression window. This block uses the variable `u` to represent the input to the block. Click **OK**. Then flip the block.
7. Select and place the Sum block from the Math Operations category. Double-click on it, and select round for the Icon shape. In the List of signs window, type $+--$. Click **OK**.
8. Select and place the Signal Generator block from the Sources category. Double-click on it, select square wave for the Wave form, 3 for the Amplitude, and 0.5 for the Frequency, and Hertz for the Units. Click **OK**.
9. Once the blocks have been placed, connect arrows as shown in the figure.
10. Set the Stop time to 10, run the simulation, and plot the results by typing

```
>>plot(tout,y)
```

5.7 CASE STUDY: VEHICLE SUSPENSION SIMULATION

In this section, we introduce four additional Simulink elements that enable us to model a wide range of nonlinearities and input functions.

As our example, we will use the single-mass suspension model shown in Figure 5.7.1, where the spring and damper elements have the nonlinear models shown in Figures 5.7.2 and 5.7.3. These models represent a hardening spring and a degressive damper. In addition, the damper model is asymmetric. It represents a damper whose

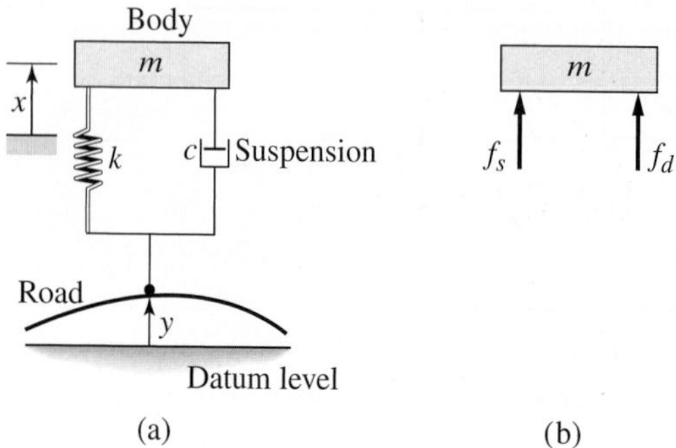

Figure 5.7.1 Single-mass suspension model.

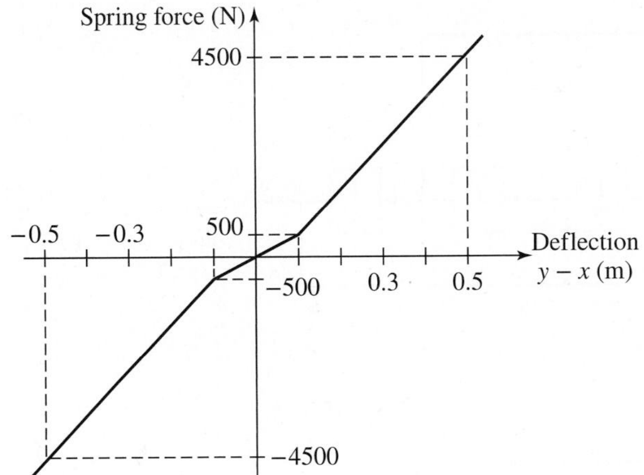

Figure 5.7.2 Hardening spring model.

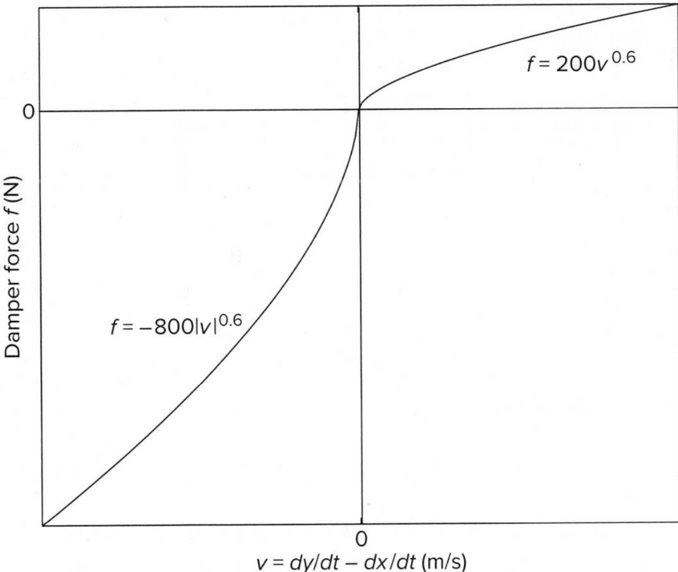

Figure 5.7.3 Degressive damper model.

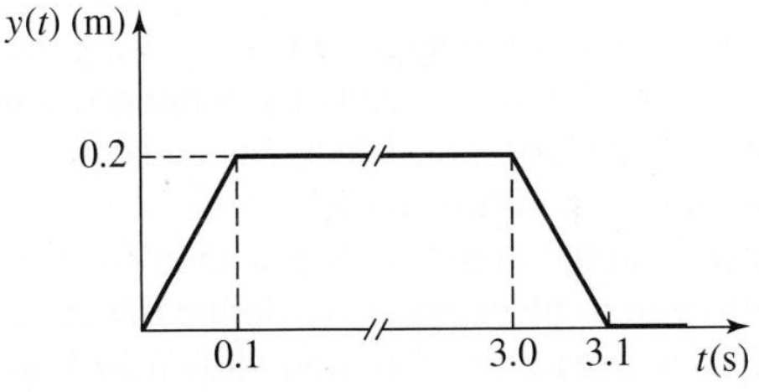

Figure 5.7.4 Bump profile.

damping during rebound is higher than during jounce (to minimize the force transmitted to the passenger compartment when the vehicle strikes a bump). The bump is represented by the trapezoidal function $y(t)$ shown in Figure 5.7.4. This function corresponds approximately to a vehicle traveling at 30 mi/h over a 0.2-m-high road surface elevation 48 m long.

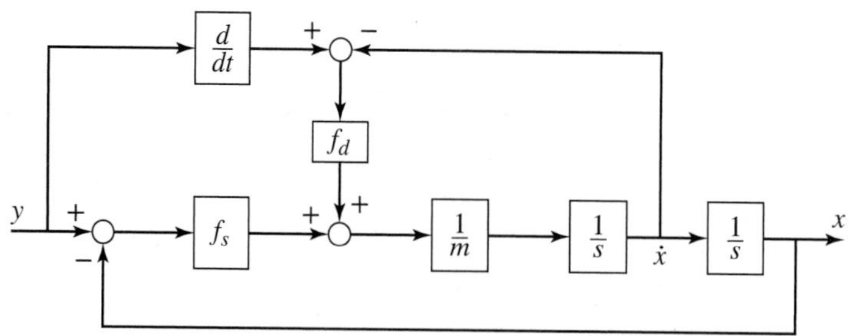

Figure 5.7.5 Simulation diagram for the suspension model.

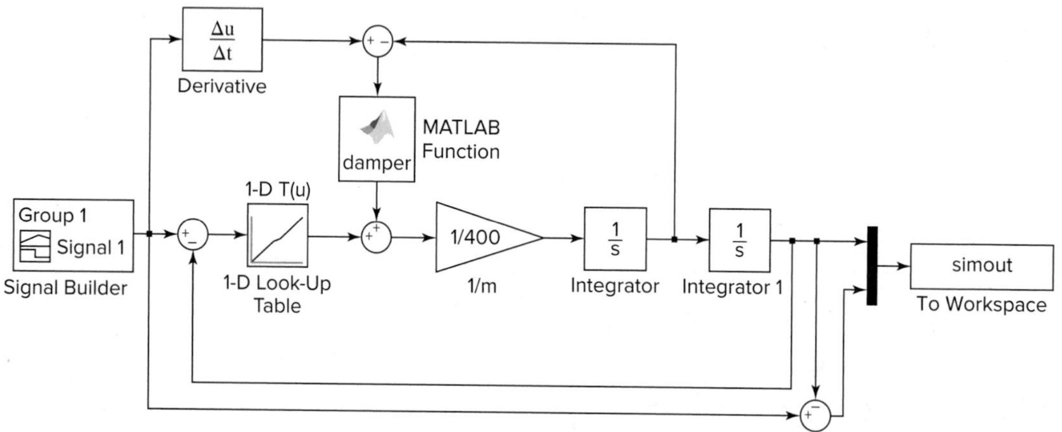

Figure 5.7.6 Simulink diagram for the suspension model.

The system model from Newton's law is

$$m\ddot{x} = f_s + f_d \tag{5.7.1}$$

where $m = 400$ kg, f_s is the nonlinear spring function shown in Figure 5.7.2, and f_d is the nonlinear damper function shown in Figure 5.7.3. The corresponding simulation diagram is shown in Figure 5.7.5.

This diagram shows that we need to compute $\dot{y}$. Because Simulink uses numerical and not analytical methods, it computes derivatives only approximately, using the Derivative block, which is in the Continuous library. We must keep this in mind when using rapidly changing or discontinuous inputs. The Derivative block has no settings, so merely place it in the Simulink diagram as shown in Figure 5.7.6.

Next place the Signal Builder block, which is in the Sources library, then double-click on it. A plot window appears that enables you to place points to define the input function. Follow the directions in the window to create the function shown in Figure 5.7.4. The spring function f_s is created with the Lookup Table block, which is in the Lookup Tables library. After placing it, double-click on it and enter `[-0.5, -0.1, 0, 0.1, 0.5]` for the `Vector of input values` and `[-4500, -500, 0, 500, 4500]` for the `Vector of output values`. Use the default settings for the remaining parameters.

Place the two integrators as shown, and make sure the initial values are set to 0. Then place the Gain block and set its gain to 1/400. The To Workspace block will enable us to plot $x(t)$ and $y(t) - x(t)$ versus t in the MATLAB Command window.

CREATING FUNCTIONS

For the damper function shown in Figure 5.7.3, we write a user-defined function to describe it. This function is as follows:

```
function f = damper(v)
if v <= 0
   f = -800*(abs(v)).^(0.6);
else
   f = 200*v.^(0.6);
end
```

Create and save this function file. After placing the MATLAB Function block, double-click on it and the MATLAB Function editor opens. When you close this editor, the function is saved. Enter its name `damper` and enter the code.

The Simulink model when completed should look like Figure 5.7.6. After running it, you can plot the response $x(t)$ in the Command window as follows:

```
>>plot(tout,xy(:,1)),grid,xlabel('t (s)'),ylabel('x (m)')
```

The result is shown in Figure 5.7.7. The maximum overshoot is seen to be $(0.26-0.2) = 0.06$ m, but the maximum *under*shoot is seen to be much greater, -0.168 m.

You can plot the difference $y - x$ by typing `plot(tout,xy(:,2))`.

The force transmitted to the chassis can be displayed by placing a Scope block just before the Gain block containing the gain $1/400$.

Problems involving piecewise-linear functions such as the saturation block are much easier to solve with Simulink. In later chapters we will discover other blocks and other advantages to using Simulink. There are menu items in the model window we have not discussed. However, the ones we have discussed are the most important ones for getting started. We have introduced just a few of the blocks available within Simulink and we will introduce more in later chapters. In addition, some blocks have additional properties that we have not mentioned. However, the examples given here will help you get started in exploring the other features of Simulink. Consult the online help for information about these items.

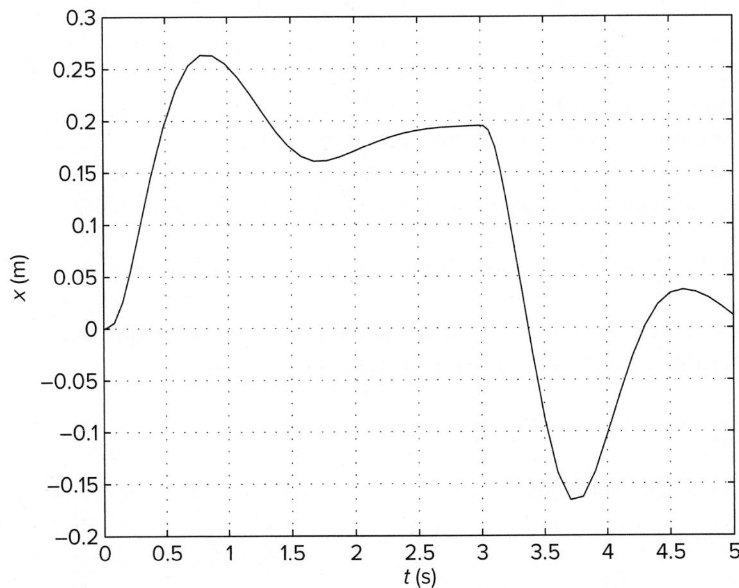

Figure 5.7.7 Response plot for the suspension model.

5.8 CHAPTER REVIEW

Part I of the chapter covers block diagrams and state-variable models. The Laplace transform enables an algebraic description of a dynamic system model to be developed. This model form, the transfer function, is the basis of a graphical description of the system, called the block diagram (Section 5.1). Thus, a block diagram is a "picture" of the algebraic description of the system that shows how the subsystem elements interact with the other subsystems. Block diagrams can be developed either from the differential equation model or from a schematic diagram that shows how the system components are connected.

The state-variable model form, which can be expressed as a vector-matrix equation, is a concise representation that is useful for analytical purposes and for writing general-purpose computer programs. Section 5.2 shows how to convert models into state-variable form.

In theory it is possible to use the Laplace transform to obtain the closed-form solution of a linear constant-coefficient differential equation if the input function is not too complicated. However, the Laplace transform method cannot be used when the Laplace transform or inverse transform either does not exist or cannot be found easily, as is the case with higher-order models. The reasons include the large amount of algebra required and the need to solve the characteristic equation numerically (closed-form solutions for polynomial roots do not exist for polynomials of order five and higher). Therefore, in Part II of this chapter we introduced several types of numerical methods for solving differential equations.

Section 5.3 focuses on MATLAB functions for solving linear state-variable models. These are the `ss`, `ssdata`, `tfdata`, `step`, `impulse`, `lsim`, `initial`, and `eig` functions. Section 5.4 treats the MATLAB `ode` functions, which are useful for solving both linear and nonlinear equations.

Part III covers Simulink, which provides a graphical user interface for solving differential equations. It includes many program blocks and features that enable you to create simulations that are otherwise difficult to program in MATLAB. Section 5.5 treats applications to linear systems, while Section 5.6 treats nonlinear system applications.

Now that you have finished this chapter, you should be able to

1. Draw a block diagram either from the differential equation model or from a schematic diagram.
2. Obtain the differential equation model from a block diagram.
3. Convert a differential equation model into state-variable form.
4. Express a linear state-variable model in the standard vector-matrix form.
5. Apply the `ss`, `ssdata`, `tfdata`, `eig`, and `initial` functions to analyze linear models.
6. Use the MATLAB `ode` functions to solve linear and nonlinear differential equations.
7. Use Simulink to create simulations of linear and nonlinear models expressed either as differential equations or, if linear, as transfer functions.

REFERENCES

[Palm, 1986] Palm, W. J. III, *Control Systems Engineering*, John Wiley & Sons, New York, 1986.

[Palm, 2009] Palm, W. J. III, *A Concise Introduction to MATLAB*, McGraw-Hill, New York, 2009.

[Palm, 2019] Palm, W. J. III, *Introduction to MATLAB for Engineers*, 4th ed., McGraw-Hill, New York, 2019.

PROBLEMS

Section 5.1 Transfer Functions and Block Diagram Models

5.1 Obtain the transfer function $X(s)/F(s)$ from the block diagram shown in Figure P5.1.

Figure P5.1

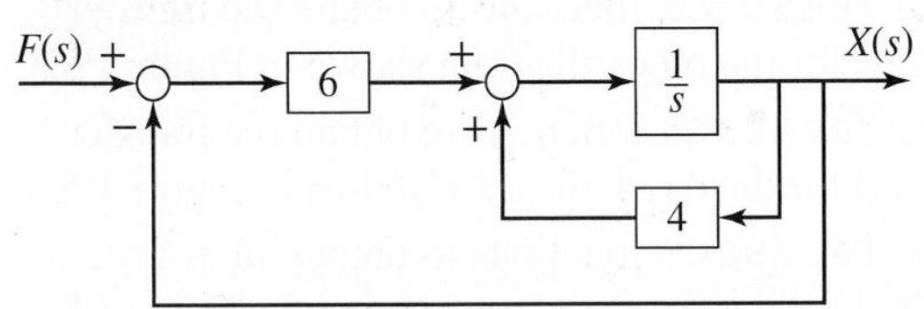

Figure P5.2

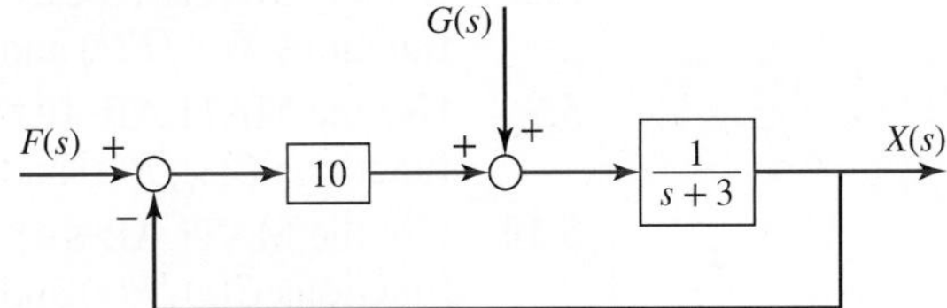

5.2 Obtain the transfer function $X(s)/F(s)$ from the block diagram shown in Figure P5.2.

5.3 Obtain the transfer function $X(s)/F(s)$ from the block diagram shown in Figure P5.3.

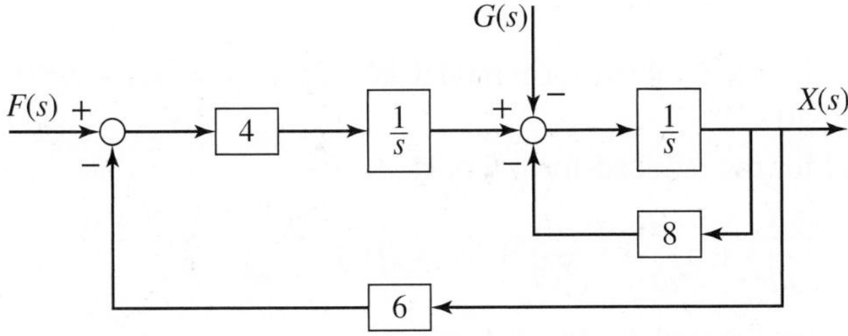

Figure P5.3

5.4 Draw a block diagram for the following equation. The output is $X(s)$; the inputs are $F(s)$ and $G(s)$.

$$5\ddot{x} + 3\dot{x} + 7x = 10f(t) - 4g(t)$$

5.5 Draw a block diagram for the following model. The output is $X(s)$; the inputs are $F(s)$ and $G(s)$. Indicate the location of $Y(s)$ on the diagram.

$$\dot{x} = y - 5x + g(t) \qquad \dot{y} = 10f(t) - 30x$$

5.6 Referring to Figure P5.6, derive the expressions for the variables $C(s)$, $E(s)$, and $M(s)$ in terms of $R(s)$ and $D(s)$.

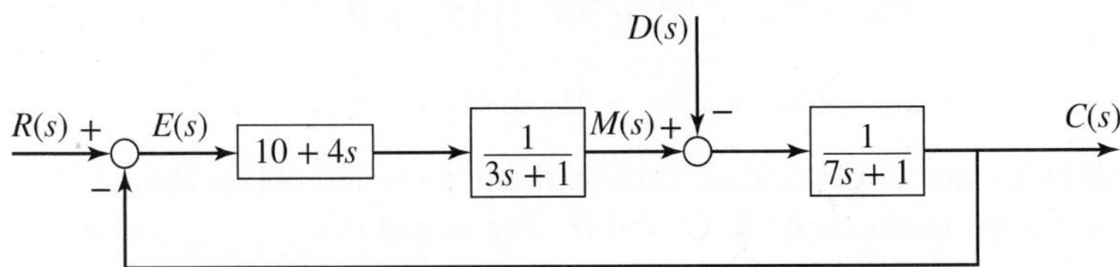

Figure P5.6

5.7 Referring to Figure P5.7, derive the expressions for the variables $C(s)$, $E(s)$, and $M(s)$ in terms of $R(s)$ and $D(s)$.

Figure P5.7

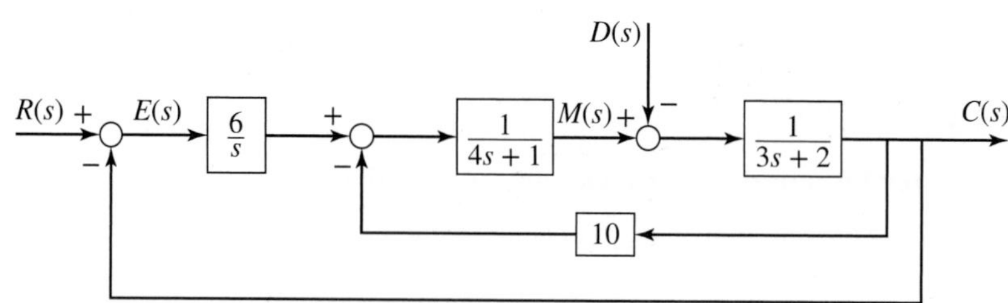

5.8 Use the MATLAB `series` and `feedback` functions to obtain the transfer functions $X(s)/F(s)$ and $X(s)/G(s)$ for the block diagram shown in Figure P5.3.

5.9 Use the MATLAB `series` and `feedback` functions to obtain the transfer functions $C(s)/R(s)$ and $C(s)/D(s)$ for the block diagram shown in Figure P5.6.

5.10 Use the MATLAB `series` and `feedback` functions to obtain the transfer functions $C(s)/R(s)$ and $C(s)/D(s)$ for the block diagram shown in Figure P5.7.

Section 5.2 State-Variable Models

5.11 Obtain the state model for the reduced-form model $6\ddot{x} + 4\dot{x} + 9x = 7f(t)$. The input is $f(t)$.

5.12 Obtain the state model for the reduced-form model $3\ddot{x} + 5\dot{x} + 2x = 7y(t)$. The input is $y(t)$.

5.13 Obtain the state model for the reduced-form model $4\ddot{x} + 2\dot{x} + 9x = f(t) + 7g(t)$. The inputs are $f(t)$ and $g(t)$.

5.14 Obtain the state model for the reduced-form model

$$4\frac{d^3y}{dt^3} + 6\frac{d^2y}{dt^2} + 3\frac{dy}{dt} + 8y = 2f(t)$$

5.15 Obtain the state model for the transfer-function model

$$\frac{Y(s)}{F(s)} = \frac{4}{9s^2 + 2s + 7}$$

5.16 Obtain the state model for the two-mass system whose equations of motion are

$$m_1\ddot{x}_1 + k_1(x_1 - x_2) = f(t)$$

$$m_2\ddot{x}_2 - k_1(x_1 - x_2) + k_2x_2 = 0$$

5.17 Obtain the state model for the two-mass system whose equations of motion for specific values of the spring and damping constants are

$$15\ddot{x}_1 + 7\dot{x}_1 - 4\dot{x}_2 + 30x_1 - 15x_2 = 0$$

$$6\ddot{x}_2 - 15x_1 + 15x_2 - 4\dot{x}_1 + 4\dot{x}_2 = f(t)$$

5.18 Put the following model in standard state-variable form and obtain the expressions for the matrices **A**, **B**, **C**, and **D**. The output is x.

$$4\ddot{x} + 7\dot{x} + 5x = 5y(t)$$

5.19 Given the state-variable model

$$\dot{x}_1 = -6x_1 + 4x_2 + 7u_1$$

$$\dot{x}_2 = -5x_2 + 9u_2$$

and the output equations

$$y_1 = x_1 + 4x_2 + 7u_1$$

$$y_2 = x_2$$

obtain the expressions for the matrices **A**, **B**, **C**, and **D**.

5.20 Given the following state-variable models, obtain the expressions for the matrices **A**, **B**, **C**, and **D** for the given inputs and outputs.

a. The outputs are x_1 and x_2; the input is u.

$$\dot{x}_1 = -7x_1 + 4x_2$$

$$\dot{x}_2 = -3x_2 + 8u$$

b. The output is x_1; the inputs are u_1 and u_2.

$$\dot{x}_1 = -7x_1 + 5x_2 + 3u_1$$

$$\dot{x}_2 = -9x_2 + 2u_2$$

5.21 Obtain the expressions for the matrices **A**, **B**, **C**, and **D** for the state-variable model you obtained in Problem 5.17. The outputs are x_1 and x_2.

5.22 The transfer function of a certain system is

$$\frac{Y(s)}{F(s)} = \frac{4s + 7}{s + 5}$$

Use two methods to obtain a state-variable model in standard form. For each model, relate the initial value of the state variable to the given initial value $y(0)$.

5.23 The transfer function of a certain system is

$$\frac{Y(s)}{F(s)} = \frac{s + 9}{s^2 + 6s + 5}$$

Use two methods to obtain a state-variable model in standard form. For each model, relate the initial values of the state variables to the given initial values $y(0)$ and $\dot{y}(0)$.

Section 5.3 State-Variable Methods with MATLAB

5.24 Use MATLAB to create a state-variable model; obtain the expressions for the matrices **A**, **B**, **C**, and **D**, and then find the transfer functions of the following models, for the given inputs and outputs.

a. The outputs are x_1 and x_2; input is u.

$$\dot{x}_1 = -6x_1 + 7x_2$$

$$\dot{x}_2 = x_1 - 5x_2 + 6u$$

b. The output is x_1; the inputs are u_1 and u_2.

$$\dot{x}_1 = -6x_1 + 7x_2 + 4u_1$$

$$\dot{x}_2 = x_1 - 5x_2 + 3u_2$$

5.25 Use MATLAB to obtain a state model for the following equations; obtain the expressions for the matrices **A**, **B**, **C**, and **D**. In both cases, the input is $f(t)$; the output is y.

a.

$$5\frac{d^3y}{dt^3} + 7\frac{d^2y}{dt^2} + 3\frac{dy}{dt} + 6y = f(t)$$

b.

$$\frac{Y(s)}{F(s)} = \frac{5}{s^2 + 7s + 4}$$

5.26 Use MATLAB to obtain a state-variable model for the following transfer functions.

a.

$$\frac{Y(s)}{F(s)} = \frac{4s + 7}{s + 6}$$

b.

$$\frac{Y(s)}{F(s)} = \frac{s + 2}{s^2 + 4s + 3}$$

5.27 For the following model the output is x_1 and the input is $f(t)$.

$$\dot{x}_1 = -4x_1 + 2x_2$$

$$\dot{x}_2 = -5x_2 + 9f(t)$$

a. Use MATLAB to compute and plot the free response for $x_1(0) = 3$, and $x_2(0) = 5$.

b. Use MATLAB to compute and plot the unit-step response for zero-initial conditions.

c. Use MATLAB to compute and plot the response for zero-initial conditions with the input $f(t) = 3 \sin 10\pi t$, for $0 \le t \le 2$.

d. Use MATLAB to compute and plot the total response using the initial conditions given in part (a) and the forcing function given in part (c).

5.28 Given the state-variable model

$$\dot{x}_1 = -7x_1 + 9x_2 + 3u_1$$

$$\dot{x}_2 = -5x_2 + 2u_2$$

and the output equations

$$y_1 = x_1 + 7x_2 + 4u_1$$

$$y_2 = x_2$$

use MATLAB to find the characteristic polynomial and the characteristic roots.

5.29 The equations of motion for a two-mass, quarter-car model of a suspension system are

$$m_1\ddot{x}_1 = c_1(\dot{x}_2 - \dot{x}_1) + k_1(x_2 - x_1)$$

$$m_2\ddot{x}_2 = -c_1(\dot{x}_2 - \dot{x}_1) - k_1(x_2 - x_1) + k_2(y - x_2)$$

Suppose the coefficient values are: $m_1 = 240$ kg, $m_2 = 36$ kg, $k_1 = 1.6 \times 10^4$ N/m, $k_2 = 1.6 \times 10^5$ N/m, $c_1 = 98$ N · s/m.

a. Use MATLAB to create a state model. The input is $y(t)$; the outputs are x_1 and x_2.

b. Use MATLAB to compute and plot the response of x_1 and x_2 if the input $y(t)$ is a unit impulse and the initial conditions are zero.

c. Use MATLAB to find the characteristic polynomial and the characteristic roots.

d. Use MATLAB to obtain the transfer functions $X_1(s)/Y(s)$ and $X_2(s)/Y(s)$.

5.30 A representation of a car's suspension suitable for modeling the bounce and pitch motions is shown in Figure P5.30, which is a side view of the vehicle's body showing the front and rear suspensions. Assume that the car's motion is constrained to a vertical translation x of the mass center and rotation θ about a single axis, which is perpendicular to the page. The body's mass is m and its moment of inertia about the mass center is I_G. As usual, x and θ are the displacements from the equilibrium position corresponding to $y_1 = y_2 = 0$. The displacements $y_1(t)$ and $y_2(t)$ can be found knowing the vehicle's speed and the road surface profile.

a. Assume that x and θ are small, and derive the equations of motion for the bounce motion x and pitch motion θ.

b. For the values $k_1 = 1100$ lb/ft, $k_2 = 1525$ lb/ft, $c_1 = c_2 = 4$ lb-sec/ft, $L_1 = 4.8$ ft, $L_2 = 3.6$ ft, $m = 50$ slugs, and $I_G = 1000$ slug-ft^2, use MATLAB to obtain a state-variable model in standard form.

c. Use MATLAB to obtain and plot the solution for $x(t)$ and $\theta(t)$ when $y_1 = 0$ and y_2 is a unit impulse. The initial conditions are zero.

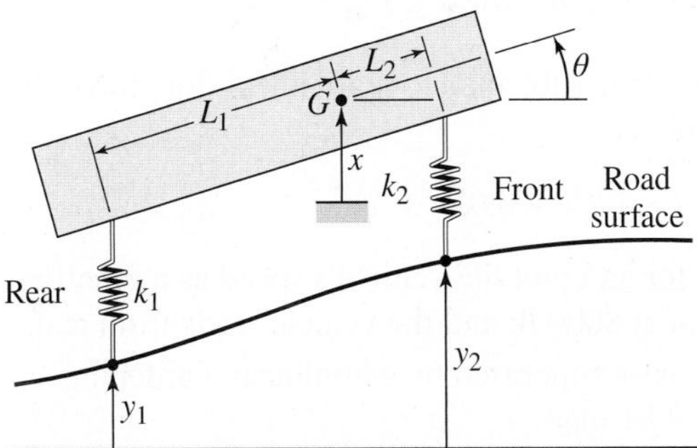

Figure P5.30

Section 5.4 The MATLAB ode Functions

5.31 a. Use a MATLAB `ode` function to solve the following equation for $0 \le t \le 12$. Plot the solution.

$$\dot{y} = \cos t \qquad y(0) = 6$$

b. Use the closed-form solution to check the accuracy of the numerical method.

5.32 a. Use a MATLAB `ode` function to solve the following equation for $0 \le t \le 1$. Plot the solution.

$$\dot{y} = 5e^{-4t} \qquad y(0) = 2$$

b. Use the closed-form solution to check the accuracy of the numerical method.

5.33 a. Use a MATLAB `ode` function to solve the following equation for $0 \le t \le 1$. Plot the solution.

$$\dot{y} + 3y = 5e^{4t} \qquad y(0) = 10$$

b. Use the closed-form solution to check the accuracy of the numerical method.

5.34 a. Use a MATLAB `ode` function to solve the following nonlinear equation for $0 \le t \le 4$. Plot the solution.

$$\dot{y} + \sin y = 0 \qquad y(0) = 0.1 \tag{1}$$

b. For small angles, $\sin y \approx y$. Use this fact to obtain a linear equation that approximates equation (1). Use the closed-form solution of this linear equation to check the output of your program.

5.35 Sometimes it is tedious to obtain a solution of a linear equation, especially if all we need is a plot of the solution. In such cases, a numerical method might be preferred. Use a MATLAB `ode` function to solve the following equation for $0 \le t \le 7$. Plot the solution.

$$\dot{y} + 2y = f(t) \qquad y(0) = 2$$

where

$$f(t) = \begin{cases} 3t & \text{for } 0 \le t \le 2 \\ 6 & \text{for } 2 \le t \le 5 \\ -3(t-5)+6 & \text{for } 5 \le t \le 7 \end{cases}$$

5.36 A certain jet-powered ground vehicle is subjected to a nonlinear drag force. Its equation of motion, in British units, is

$$50\dot{v} = f - (20v + 0.05v^2)$$

Use a numerical method to solve for and plot the vehicle's speed as a function of time if the jet's force is constant at 8000 lb and the vehicle starts from rest.

5.37 The following model describes a mass supported by a nonlinear, hardening spring. The units are SI. Use $g = 9.81$ m/s^2.

$$5\ddot{y} = 5g - (900y + 1700y^3)$$

Suppose that $\dot{y}(0) = 0$. Use a numerical method to solve for and plot the solution for two different initial conditions: (1) $y(0) = 0.06$ and (2) $y(0) = 0.1$.

5.38 Van der Pol's equation is a nonlinear model for some oscillatory processes. It is

$$\ddot{y} - b(1 - y^2)\dot{y} + y = 0$$

Use a numerical method to solve for and plot the solution for the following cases:

1. $b = 0.1, y(0) = \dot{y}(0) = 1, 0 \leq t \leq 25$
2. $b = 0.1, y(0) = \dot{y}(0) = 3, 0 \leq t \leq 25$
3. $b = 3, y(0) = \dot{y}(0) = 1, 0 \leq t \leq 25$

5.39 Van der Pol's equation is

$$\ddot{y} - b(1 - y^2)\dot{y} + y = 0$$

This equation can be difficult to solve for large values of the parameter b. Use $b = 1000$ and $0 \leq t \leq 3000$, with the initial conditions $y(0) = 2$ and $\dot{y}(0) = 0$. Use `ode45` to plot the response.

5.40 The equation of motion for a pendulum whose base is accelerating horizontally with an acceleration $a(t)$ is

$$L\ddot{\theta} + g \sin \theta = a(t) \cos \theta$$

Suppose that $g = 9.81$ m/s^2, $L = 1$ m, and $\dot{\theta}(0) = 0$. Solve for and plot $\theta(t)$ for $0 \leq t \leq 10$ s for the following three cases.

a. The acceleration is constant: $a = 5$ m/s^2, and $\theta(0) = 0.5$ rad.
b. The acceleration is constant: $a = 5$ m/s^2, and $\theta(0) = 3$ rad.
c. The acceleration is linear with time: $a = 0.5t$ m/s^2, and $\theta(0) = 3$ rad.

5.41 Suppose the spring in Figure P5.41 is nonlinear and is described by the cubic force-displacement relation. The equation of motion is

$$m\ddot{x} = c(\dot{y} - \dot{x}) + k_1(y - x) + k_2(y - x)^3$$

where $m = 100$, $c = 600$, $k_1 = 8000$, and $k_2 = 24{,}000$. Approximate the unit-step input $y(t)$ with $y(t) = 1 - e^{-t/\tau}$, where τ is chosen to be small compared to the period and time constant of the model when the cubic term is neglected. Use MATLAB to plot the forced response $x(t)$.

Figure P5.41

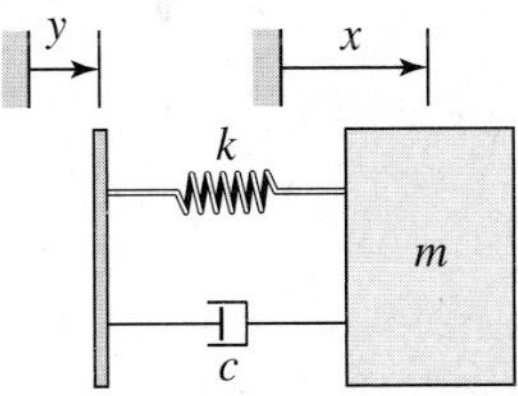

Section 5.5 Simulink and Linear Models

5.42 Create a Simulink model to plot the solution of the following equation for $0 \leq t \leq 6$.

$$10\ddot{y} = 7 \sin 4t + 5 \cos 3t \qquad y(0) = 4 \qquad \dot{y}(0) = 1$$

5.43 A projectile is launched with a velocity of 100 m/s at an angle of 30° above the horizontal. Create a Simulink model to solve the projectile's equations of motion, where x and y are the horizontal and vertical displacements of the projectile.

$$\ddot{x} = 0 \qquad x(0) = 0 \qquad \dot{x}(0) = 100 \cos 30°$$

$$\ddot{y} = -g \qquad y(0) = 0 \qquad \dot{y}(0) = 100 \sin 30°$$

Use the model to plot the projectile's trajectory y versus x for $0 \leq t \leq 10$ s.

5.44 In Chapter 2 we obtained an approximate solution of the following problem, which has no analytical solution even though it is linear.

$$\dot{x} + x = \tan t \qquad x(0) = 0$$

The approximate solution, which is less accurate for large values of t, is

$$x(t) = \frac{1}{3}t^3 - t^2 + 3t - 3 + 3e^{-t}$$

Create a Simulink model to solve this problem and compare its solution with the approximate solution over the range $0 \le t \le 1$.

5.45 Construct a Simulink model to plot the solution of the following equation for $0 \le t \le 10$.

$$15\dot{x} + 5x = 4u_s(t) - 4u_s(t-2) \qquad x(0) = 2$$

5.46 Use Simulink to solve Problem 5.19 for zero initial conditions, u_1 a unit-step input, and $u_2 = 0$.

5.47 Use Simulink to solve Problem 5.19 for the initial conditions $x_1(0) = 4$, $x_2(0) = 3$, and $u_1 = u_2 = 0$.

5.48 Use Simulink to solve Problem 5.20a for zero initial conditions and $u = 3 \sin 2t$.

5.49 Use Simulink to solve Problem 5.27c.

Section 5.6 Simulink and Nonlinear Models

5.50 Use the Transfer Function block to construct a Simulink model to plot the solution of the following equation for $0 \le t \le 4$.

$$2\ddot{x} + 12\dot{x} + 10x^2 = 5u_s(t) - 5u_s(t-2) \qquad x(0) = \dot{x}(0) = 0$$

5.51 Construct a Simulink model to plot the solution of the following equation for $0 \le t \le 4$.

$$2\ddot{x} + 12\dot{x} + 10x^2 = 5 \sin 0.8t \qquad x(0) = \dot{x}(0) = 0$$

5.52 Use the Saturation block to create a Simulink model to plot the solution of the following equation for $0 \le t \le 6$.

$$3\dot{y} + y = f(t) \qquad y(0) = 2$$

where

$$f(t) = \begin{cases} 8 & \text{if } 10 \sin 3t > 8 \\ -8 & \text{if } 10 \sin 3t < -8 \\ 10 \sin 3t & \text{otherwise} \end{cases}$$

5.53 Construct a Simulink model of the following problem.

$$5\dot{x} + \sin x = f(t) \qquad x(0) = 0$$

The forcing function is

$$f(t) = \begin{cases} -5 & \text{if } g(t) \le -5 \\ g(t) & \text{if } -5 \le g(t) \le 5 \\ 5 & \text{if } g(t) \ge 5 \end{cases}$$

where $g(t) = 10 \sin 4t$.

5.54 Create a Simulink model to plot the solution of the following equation for $0 \le t \le 3$.

$$\dot{x} + 10x^2 = 2 \sin 4t \qquad x(0) = 1$$

5.55 Construct a Simulink model of the following problem.

$$10\dot{x} + \sin x = f(t) \qquad x(0) = 0$$

The forcing function is $f(t) = \sin 2t$. The system has the dead-zone nonlinearity shown in Figure 5.6.6.

5.56 The following model describes a mass supported by a nonlinear, hardening spring. The units are SI. Use $g = 9.81$ m/s^2.

$$5\ddot{y} = 5g - (900y + 1700y^3) \qquad y(0) = 0.5 \qquad \dot{y}(0) = 0.$$

Create a Simulink model to plot the solution for $0 \le t \le 2$.

5.57 Consider the system for lifting a mast, discussed in Chapter 3 and shown again in Figure P5.57. The 70-ft-long mast weighs 500 lb. The winch applies a force $f = 380$ lb to the cable. The mast is supported initially at an angle of $\theta = 30°$, and the cable at A is initially horizontal. The equation of motion of the mast is

$$25{,}400\,\ddot{\theta} = -17{,}500\cos\theta + \frac{626{,}000}{Q}\sin(1.33 + \theta)$$

where

$$Q = \sqrt{2020 + 1650\cos(1.33 + \theta)}$$

Create and run a Simulink model to solve for and plot $\theta(t)$ for $\theta(t) \le \pi/2$ rad.

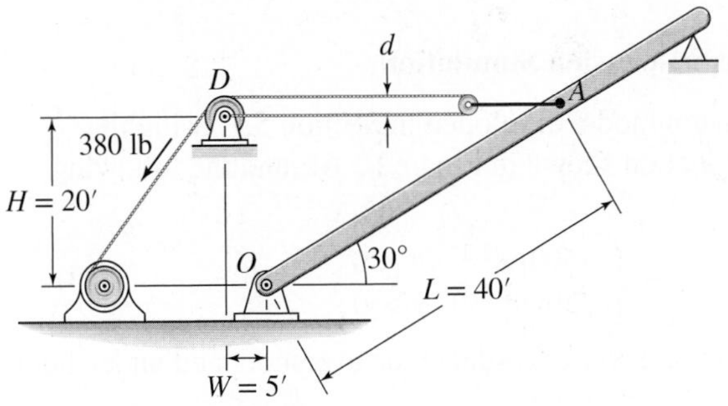

Figure P5.57

5.58 A certain mass, $m = 2$ kg, moves on a surface inclined at an angle $\phi = 30°$ above the horizontal. Its initial velocity is $v(0) = 3$ m/s up the incline. An external force of $f_1 = 5$ N acts on it parallel to and up the incline. The coefficient of dynamic friction is $\mu_d = 0.5$. Use the Coulomb Friction block or the Sign block and create a Simulink model to solve for the velocity of the mass until the mass comes to rest. Use the model to determine the time at which the mass comes to rest.

5.59 If a mass-spring system has Coulomb friction on the horizontal surface rather than viscous friction, its equation of motion is

$$m\ddot{y} = -ky + f(t) - \mu mg \quad \text{if } \dot{y} \ge 0$$

$$m\ddot{y} = -ky + f(t) + \mu mg \quad \text{if } \dot{y} < 0$$

where μ is the coefficient of friction. Develop a Simulink model for the case where $m = 1$ kg, $k = 5$ N/m, $\mu = 0.4$, and $g = 9.8$ m/s^2. Run the simulation for two cases: (a) the applied force $f(t)$ is a step function with a magnitude of 10 N, and (b) the applied force is sinusoidal: $f(t) = 10\sin 2.5t$.

5.60 Consider the system shown in Figure P5.60. The equations of motion are

$$m_1\ddot{x}_1 + (c_1 + c_2)\dot{x}_1 + (k_1 + k_2)x_1 - c_2\dot{x}_2 - k_2x_2 = 0$$

$$m_2\ddot{x}_2 + c_2\dot{x}_2 + k_2x_2 - c_2\dot{x}_1 - k_2x_1 = f(t)$$

Suppose that $m_1 = m_2 = 1$, $c_1 = 3$, $c_2 = 1$, $k_1 = 1$, and $k_2 = 4$.

a. Develop a Simulink simulation of this system. In doing this, consider whether to use a state-variable representation or a transfer-function representation of the model.

b. Use the Simulink program to plot the response $x_1(t)$ for the following input. The initial conditions are zero.

$$f(t) = \begin{cases} t & 0 \le t \le 1 \\ 2 - t & 1 < t < 2 \\ 0 & t \ge 2 \end{cases}$$

Figure P5.60

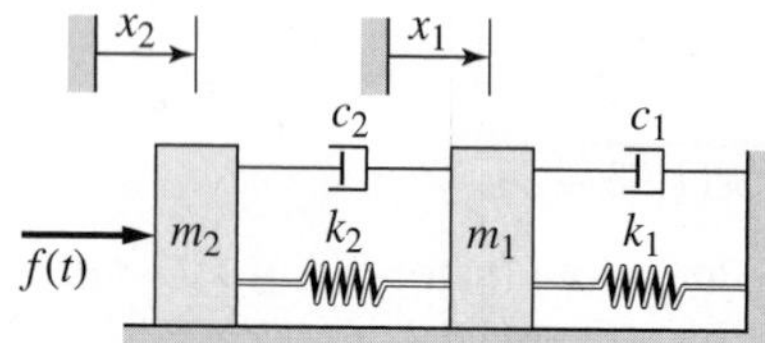

Section 5.7 Case Study: Vehicle Suspension Simulation

5.61 Redo the Simulink suspension model developed in Section 5.7, using the spring relation and input function shown in Figure P5.61, and the following damper relation.

$$f_d(v) = \begin{cases} -500|v|^{1.2} & v \le 0 \\ 50v^{1.2} & v > 0 \end{cases}$$

Use the simulation to plot the response. Evaluate the overshoot and undershoot.

Figure P5.61

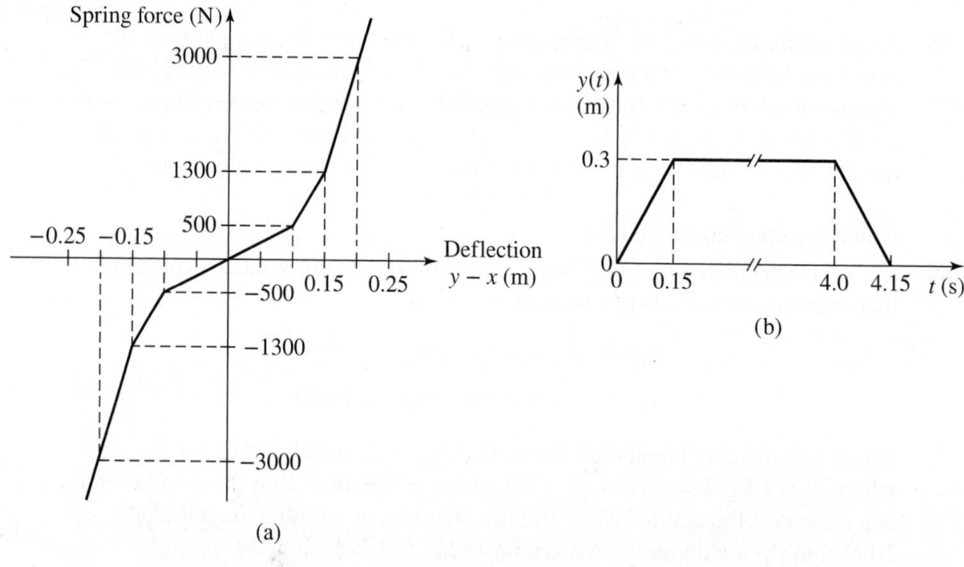

5.62 Modify the Simulink model given in Section 5.7 to replace the road input profile (the Signal Builder block) with a haversine function (see Section 4.8). The bump length is 3 m and the height is 0.2 m. The vehicle speed is 30 mph. Plot $x(t)$.

CHAPTER 6

Electrical and Electromechanical Systems

CHAPTER OUTLINE

CHAPTER OBJECTIVES

When you have finished this chapter, you should be able to

1. Develop models of electrical circuits. This includes
   - Application of Kirchhoff's voltage and current laws,
   - Application of the voltage- and current-divider rules,
   - Use of loop analysis,
   - Selection of appropriate state variables, and
   - Use of transformed equations and impedance.
2. Obtain circuit models in transfer-function, state-variable, and block-diagram form.
3. Apply impedance methods to obtain models of circuits and op-amp systems.
4. Apply Newton's laws, electrical circuit laws, and electromagnetic principles to develop models of electromechanical systems, including dc motors and sensors.
5. Assess the performance of motors and amplifiers when used in motion-control systems.
6. Apply MATLAB and Simulink to analyze models of circuits and electromechanical systems in state-variable, transfer-function, and block-diagram form.

The majority of engineering systems now have at least one electrical subsystem. This may be a power supply, sensor, motor, controller, or an acoustic device such as a speaker. So an understanding of electrical systems is essential to understanding the behavior of many systems.

Section 6.1 introduces the basic physics, common elements, and terminology of electrical circuits and treats the two main physical laws needed to develop circuit models. These are Kirchhoff's current and voltage laws. Section 6.2 is an extensive collection of circuit examples that emphasize resistance networks and circuits having one or two capacitors or inductors.

Impedance, a generalization of the electrical resistance concept, is covered in Section 6.3. This concept enables you to derive circuit models more easily, especially for more complex circuits, and is especially useful for obtaining models of circuits containing operational amplifiers, the subject of Section 6.4.

The principles of direct-current (dc) motors are established in Section 6.5, and these principles are used to develop transfer-function and state-variable models of motors. Section 6.6 examines some practical considerations in motor modeling, and Section 6.7 introduces methods for assessing the performance of motors and amplifiers in motion-control systems. In Section 6.8, these principles are extended to other electromechanical devices such as sensors and speakers. Many of the systems treated in this chapter are more easily designed with computer simulation, and Sections 6.9 and 6.10 show how to apply MATLAB and Simulink in electromechanical systems analysis. ■

6.1 ELECTRICAL ELEMENTS

Voltage and current are the primary variables used to describe a circuit's behavior. *Current* is the flow of electrons. It is the time rate of change of electrons passing through a defined area, such as the cross section of a wire. Because electrons are negatively charged, the positive direction of current flow is opposite to that of the electron flow. The mathematical description of the relation between the number of electrons (called *charge* Q) and current i is

$$i = \frac{dQ}{dt} \qquad \text{or} \qquad Q(t) = \int i\,dt$$

The unit of charge is the *coulomb* (C), and the unit of current is the *ampere* (A), which is one coulomb per second. These units, and the others we will use for electrical systems, are the same in both the SI and FPS systems.

Energy is required to move a charge between two points in a circuit. The work per unit charge required to do this is called *voltage*. The unit of voltage is the *volt* (V), which is defined to be one joule per coulomb. The voltage difference between two points in a circuit is a measure of the energy required to move charge from one point to the other.

The sign of voltage difference is important. Figure 6.1.1a shows a battery connected to a light bulb. The electrons in the wire are attracted to the battery's positive terminal; thus the positive direction of current is clockwise, as indicated by the arrow. Because the battery supplies energy to move electrons and the light bulb dissipates energy (through light and heat), the sign of voltage difference across the battery is the opposite of the sign of the voltage difference across the light bulb. This is indicated by

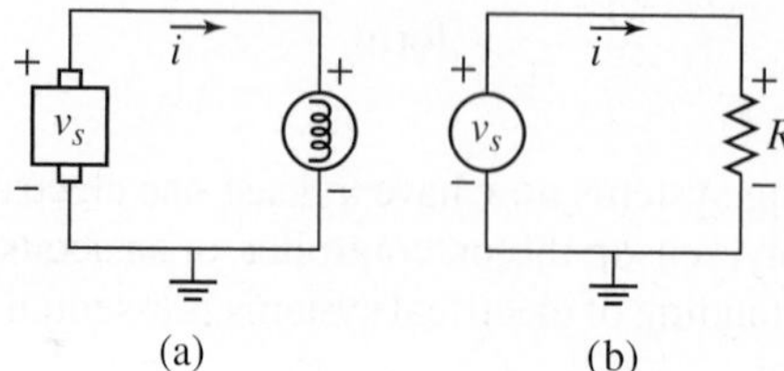

Figure 6.1.1 (a) A battery-light bulb circuit. (b) Circuit diagram representation of the battery-light bulb circuit.

the + and – signs in the diagram. Although the charge flows counterclockwise, we can think of a positive current flowing clockwise. This current picks up energy as it passes through the battery from the negative to the positive terminal. As it flows through the light bulb, the current loses energy; this is indicated by the + and – signs above and below the bulb.

The bulb current depends on the voltage difference and the bulb's material properties, which resist the current and cause the energy loss. When a current flows through wire or other circuit elements, it encounters resistance. Sometimes this resistance is desirable and intentionally introduced; sometimes not. A *resistor* is an element designed to provide resistance. Most resistors are designed to have a linear relation between the current passing through them and the voltage difference across them. This linear relation is *Ohm's law*. It states that

$$v = iR$$

where i is the current, v is the voltage difference, and R is the *resistance*. The unit of resistance is the *ohm* (Ω), which is one volt per ampere.

Figure 6.1.1b shows a voltage source, such as a battery, connected to a resistor. Because of conservation of energy, the voltage increase v_s supplied by the source must equal the voltage drop iR across the resistor. Thus, the model of this circuit is $v_s = iR$. If we know the voltage and the resistance, we can calculate the current that must be supplied by the source as follows: $i = v_s/R$.

6.1.1 ACTIVE AND PASSIVE ELEMENTS

Circuit elements may be classified as *active* or *passive*. Passive elements such as resistors, capacitors, and inductors are not sources of energy, although the latter two can store it temporarily. Elements that provide energy are *sources*, and elements that dissipate energy are *loads*.

The active elements are energy sources that drive the system. There are several types available; for example, chemical (batteries), mechanical (generators), thermal (thermocouples), and optical (solar cells). Active elements are modeled as either *ideal voltage sources* or *ideal current sources*. Their circuit symbols are given in Table 6.1.1. An ideal voltage source supplies the desired voltage no matter how much current is drawn by the loading circuit. An ideal current source supplies whatever current is needed by the loading circuit. Obviously no real source behaves exactly this way. For example, battery voltage drops because of heat produced as current is drawn from the battery. If the current is small, the battery can be treated as an ideal voltage source. If not, the battery is frequently modeled as an ideal voltage source plus an internal resistance whose value can be obtained from the voltage-current curve for the battery.

Power P is work done per unit time, so the power generated by an active element, or the power dissipated or stored by a passive element, can be calculated as follows:

$$\text{power} = \frac{\text{work}}{\text{time}} = \frac{\text{work}}{\text{unit charge}}\frac{\text{charge}}{\text{time}} = \text{voltage} \times \text{current}$$

Thus, the power generated, dissipated, or stored by a circuit element equals the product of the voltage difference across the element and the current flowing through the element. That is, $P = iv$. The unit of power in the SI system is one joule per second, which is defined to be a *watt* (W).

If the element is a linear resistor, the power dissipated is given by

$$P = iv = i^2R = \frac{v^2}{R}$$

Table 6.1.1 Electrical quantities.

Quantity	Units	Circuit symbol
Voltage	volt (V)	Voltage source
Charge	coulomb (C) = N · m/V	
Current	ampere (A) = C/s	Current source
Resistance	ohm (Ω) = V/A	R
Capacitance	farad (F) = C/V	C
Inductance	henry (H) = V · s/A	L
Battery	—	
Ground	—	
Terminals (input or output)	—	

The appropriate form to use depends on which two of the three quantities i, v, and R are known.

6.1.2 MODELING CIRCUITS

The dynamics of physical systems result from the transfer, loss, and storage of mass or energy. A basic law used to model electrical systems is conservation of charge, also known as *Kirchhoff's current law*. Another basic law is conservation of energy. In electrical systems, conservation of energy is commonly known as *Kirchhoff's voltage law*, which states that the algebraic sum of the voltages around a closed circuit or *loop* must be zero. These physical laws alone do not provide enough information to write the equations that describe the system. Three more types of information must be provided; the four requirements are

1. The appropriate physical laws, such as conservation of charge and energy.
2. Empirically based descriptions, called *constitutive relations*, for some or all of the system elements.
3. The specific way the system elements are arranged or connected.
4. Any relationships due to integral causality, such as the relation between charge and current, $Q = \int i\,dt$.

The voltage-current relation for a resistor, $v = iR$, is an example of an empirically based description of a system element, the third type of required information. This type of description is an algebraic relation not derivable from a basic physical law. Rather, the relation is obtained from a series of measurements. For example, if we apply a range of currents to a resistor, then measure the resulting voltage difference for each current, we would find that the voltage is directly proportional to the applied current. The constant of proportionality is the resistance R, and we can determine its value from the test results.

Knowledge of the elements' arrangement is important because it is possible to connect two elements in more than one way. For example, two resistors can be connected differently to form two different circuits (Figures 6.1.2 and 6.1.3); the models are different for each circuit. We can use the voltage-current relation for a resistor along with conservation of charge and conservation of energy to obtain the models.

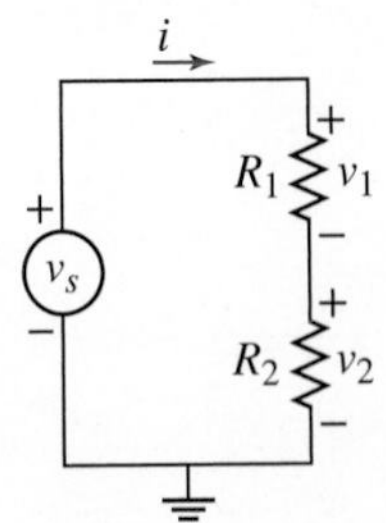

Figure 6.1.2 Series resistors.

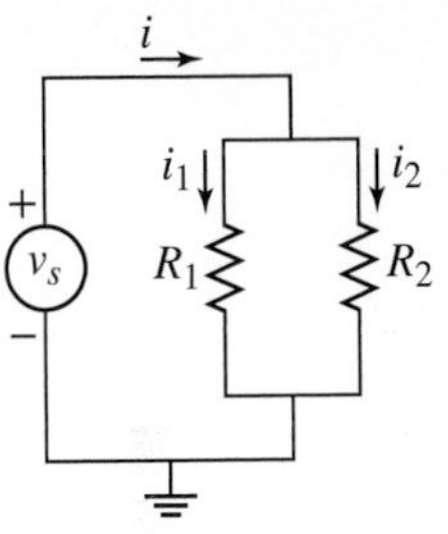

Figure 6.1.3 Parallel resistors.

6.1.3 SERIES RESISTANCES

For Figure 6.1.2, conservation of charge implies that the current is the same through each resistor. When traversing the loop clockwise, a voltage *increase* occurs when we traverse the source v_s, and a voltage *decrease* occurs when we traverse each resistor. Assign a positive sign to a voltage increase, and a negative sign to a voltage decrease. Then Kirchhoff's voltage law gives

$$v_s - v_1 - v_2 = v_s - iR_1 - iR_2 = 0$$

or

$$v_s = (R_1 + R_2)i \tag{6.1.1}$$

Thus, the supply voltage v_s must equal the sum of the voltages across the two resistors.

Equation (6.1.1) is an illustration of the *series resistance law*. If the same current passes through two or more electrical elements, those elements are said to be in *series*, and the series resistance law states that they are equivalent to a single resistance R that is the sum of the individual resistances. Thus, the circuit in Figure 6.1.2 can be modeled as the simpler circuit in Figure 6.1.1b.

Because $i = v_s/(R_1 + R_2)$ in Figure 6.1.2,

$$v_1 = R_1 i = \left(\frac{R_1}{R_1 + R_2}\right)v_s \tag{6.1.2}$$

$$v_2 = R_2 i = \left(\frac{R_2}{R_1 + R_2}\right)v_s \tag{6.1.3}$$

These equations express the *voltage-divider rule*. They show that

$$\frac{v_1}{v_2} = \frac{R_1}{R_2} \tag{6.1.4}$$

This rule is useful for reducing a circuit with many resistors to an equivalent circuit containing one resistor.

6.1.4 PARALLEL RESISTANCES

For Figure 6.1.3, conservation of charge gives $i = i_1 + i_2$. From Kirchhoff's voltage law we see that the voltage across each resistor must be v_s, and thus, $i_1 = v_s/R_1$ and $i_2 = v_s/R_2$. Combining these three relations gives

$$i = i_1 + i_2 = \frac{v_s}{R_1} + \frac{v_s}{R_2}$$

Solve for i to obtain

$$i = \left(\frac{1}{R_1} + \frac{1}{R_2}\right)v_s \tag{6.1.5}$$

Equation (6.1.5) is an illustration of the *parallel resistance law*. If the same voltage difference exists across two or more electrical elements, those elements are said to be in *parallel*. For two resistors in parallel, the parallel resistance law states that they are equivalent to a single resistance R given by

$$\frac{1}{R} = \frac{1}{R_1} + \frac{1}{R_2} \tag{6.1.6}$$

Thus, the circuit in Figure 6.1.3 can be modeled as the simpler circuit in Figure 6.1.1b. This formula can be extended to more than two resistors.

Note that

$$i_1 = \left(\frac{R_2}{R_1 + R_2}\right) i \tag{6.1.7}$$

$$i_2 = \left(\frac{R_1}{R_1 + R_2}\right) i \tag{6.1.8}$$

These equations express the *current-divider rule*. They show that

$$\frac{i_1}{i_2} = \frac{R_2}{R_1} \tag{6.1.9}$$

The current-divider rule can be used to find the equivalent resistance of a circuit with many resistors.

6.1.5 NONLINEAR RESISTANCES

Not all resistance elements have the linear voltage-current relation $v = iR$. An example of a specific diode's voltage-current relationship found from experiments is

$$i = 0.16(e^{0.12v} - 1)$$

For low voltages, we can approximate this curve with a straight line whose slope equals the derivative di/dv at $v = 0$.

$$\left.\frac{di}{dv}\right|_{v=0} = 0.16\left(0.12e^{0.12v}\right)\big|_{v=0} = 0.0192$$

Thus, for small voltages, $i = 0.0192v$, and the resistance is $R = 1/0.0192 = 52\ \Omega$.

6.1.6 CAPACITANCE

A *capacitor* is designed to store charge. The *capacitance* C of a capacitor is a measure of how much charge can be stored for a given voltage difference across the element. Capacitance thus has the units of charge per volt. This unit is named the *farad* (F).

For a capacitor, $Q = \int i\, dt$, where Q is the charge on the capacitor, and i is the current passing through the capacitor. The constitutive relation for a capacitor is $v = Q/C$, where v is the voltage across the capacitor. Combining these two relations gives

$$v = \frac{1}{C}\int i\, dt = \frac{1}{C}\int_0^t i\, dt + \frac{Q_0}{C}$$

where Q_0 is the initial charge on the capacitor at time $t = 0$. In derivative form, this relation is expressed as

$$i = C\frac{dv}{dt}$$

6.1.7 INDUCTANCE

A magnetic field (a *flux*) surrounds a moving charge or current. If the conductor of the current is coiled, the flux created by the current in one loop affects the adjacent loops. This flux is proportional to the time integral of the applied voltage, and the current is proportional to the flux. The constitutive relation for an inductor is $\phi = Li$, where L is the *inductance* and ϕ is the flux across the inductor.

The integral causality relation between flux and voltage is

$$\phi = \int v\,dt$$

Combining the two preceding expressions for ϕ gives the voltage-current relation for the inductor

$$i = \frac{1}{L}\int v\,dt$$

which is equivalent to

$$v = L\frac{di}{dt}$$

The unit of inductance is the *henry* (H), which is one volt-second per ampere.

6.1.8 POWER AND ENERGY

The power dissipated by or stored by an electrical element is the product of its current and the voltage across it: $P = iv$. Capacitors and inductors store electrical energy as stored charge and in a magnetic field, respectively. The energy E stored in a capacitor can be found as follows:

$$E = \int P\,dt = \int iv\,dt = \int \left(C\frac{dv}{dt}\right) v\,dt = C\int v\,dv = \frac{1}{2}Cv^2$$

Similarly, the energy E stored in an inductor is

$$E = \int P\,dt = \int iv\,dt = \int i\left(L\frac{di}{dt}\right) dt = L\int i\,di = \frac{1}{2}Li^2$$

Table 6.1.2 summarizes the voltage-current relations and the energy expressions for resistance, capacitance, and inductance elements.

Table 6.1.2 Voltage-current and energy relations for circuit elements.

Resistance:	$v = iR$	$P = Ri^2 = \dfrac{v^2}{R}$
Capacitance:	$v = \dfrac{1}{C}\displaystyle\int_0^t i\,dt + \dfrac{Q_0}{C}$	$E = \dfrac{1}{2}Cv^2$
Inductance:	$v = L\dfrac{di}{dt}$	$E = \dfrac{1}{2}Li^2$

6.2 CIRCUIT EXAMPLES

The examples in this section illustrate how to apply the basic circuit principles introduced in Section 6.1.

EXAMPLE 6.2.1 Current and Power in a Resistance Circuit

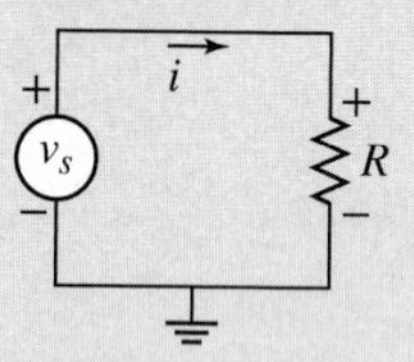

Figure 6.2.1 A simple resistance circuit.

■ Problem

For the circuit shown in Figure 6.2.1, the applied voltage is $v_s = 6$ V and the resistance is $R = 10\ \Omega$. Determine the current and the power that must be produced by the power supply.

■ Solution

The current is found from $i = v_s/R = 6/10 = 0.6$ A. The power is computed from $P = v_s^2/R = 6^2/10 = 3.6$ W. Note that we can also compute the power from $P = iv_s$.

EXAMPLE 6.2.2 A Summing Circuit

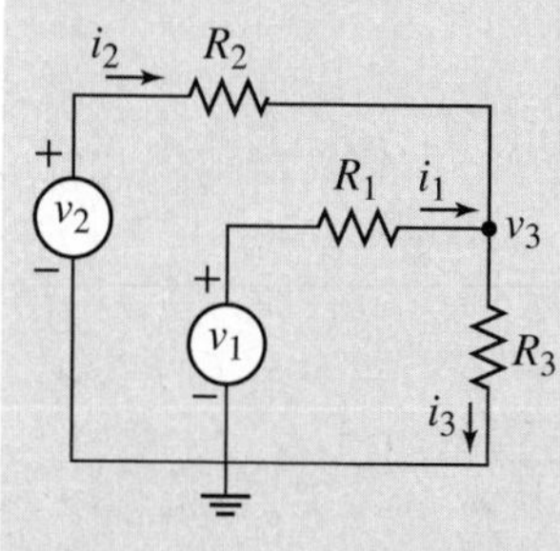

Figure 6.2.2 A summing circuit.

■ Problem

Figure 6.2.2 shows a circuit for summing the voltages v_1 and v_2 to produce v_3. Derive the expression for v_3 as a function of v_1 and v_2, for the case where $R_1 = R_2 = 10\ \Omega$, and $R_3 = 20\ \Omega$.

■ Solution

Define the currents shown in the diagram. The voltage-current relation for each resistor gives

$$i_1 = \frac{v_1 - v_3}{R_1}$$

$$i_2 = \frac{v_2 - v_3}{R_2}$$

$$i_3 = \frac{v_3}{R_3}$$

There is no capacitance in the circuit. Therefore, no charge can be stored anywhere in the circuit. Thus, conservation of charge gives $i_3 = i_1 + i_2$. Substituting the expressions for the currents into this equation, we obtain

$$\frac{v_3}{R_3} = \frac{v_1 - v_3}{R_1} + \frac{v_2 - v_3}{R_2}$$

which gives

$$v_3 = 0.4(v_1 + v_2)$$

Thus, v_3 is proportional to the sum of v_1 and v_2. This is true in general only if $R_1 = R_2$.

EXAMPLE 6.2.3 Application of the Voltage-Divider Rule

■ Problem

Consider the circuit shown in Figure 6.2.3. Obtain the voltage v_o as a function of the applied voltage v_s by applying the voltage-divider rule. Use the values $R_1 = 5\ \Omega$, $R_2 = 10\ \Omega$, $R_3 = 6\ \Omega$, and $R_4 = 2\ \Omega$.

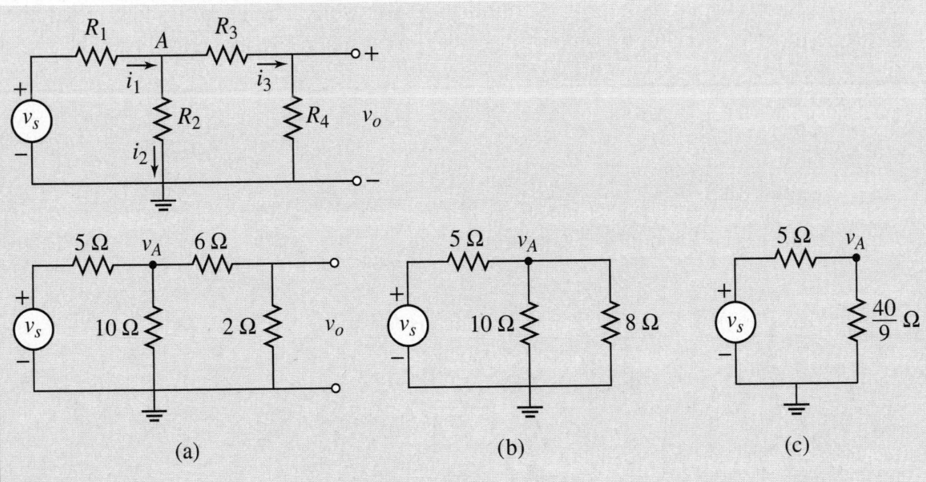

Figure 6.2.3 A resistance network with two loops.

Figure 6.2.4 Application of the voltage-divider rule.

■ Solution

Let v_A be the voltage at the node shown in Figure 6.2.4a. The voltage-divider rule applied to resistors R_3 and R_4 gives

$$v_o = \frac{R_4}{R_3 + R_4} v_A = \frac{2}{6+2} v_A = \frac{1}{4} v_A \tag{1}$$

Because resistors R_3 and R_4 are in series, we can add their values to obtain their equivalent resistance $R_s = 2 + 6 = 8\ \Omega$. The equivalent circuit is shown in Figure 6.2.4b.

Resistors R_s and R_2 are parallel, so we can combine their values to obtain their equivalent resistance R_p as follows:

$$\frac{1}{R_p} = \frac{1}{10} + \frac{1}{8} = \frac{9}{40}$$

Thus, $R_p = 40/9$. The equivalent circuit is shown in Figure 6.2.4c.

Finally, we apply the voltage-divider rule again to obtain

$$v_A = \frac{40/9}{5 + 40/9} v_s = \frac{8}{17} v_s \tag{2}$$

Using (1) and (2) we obtain

$$v_o = \frac{1}{4} v_A = \frac{1}{4}\left(\frac{8}{17}\right) v_s = \frac{2}{17} v_s$$

A *potentiometer* is a resistance with a sliding electrical pick-off (Figure 6.2.5a). Thus, the resistance R_1 between the sliding contact and ground is a function of the distance x of the contact from the end of the potentiometer. Potentiometers, commonly

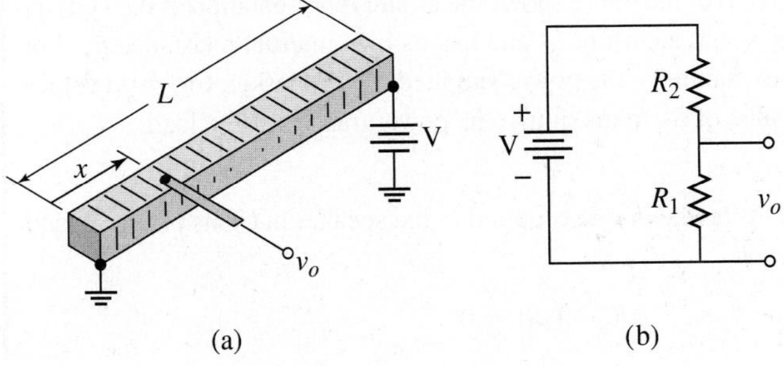

Figure 6.2.5 A translational (linear) potentiometer.

called *pots*, are used as linear and angular position sensors. The volume control knob on some radios is a potentiometer that is used to adjust the voltage to the speakers.

EXAMPLE 6.2.4

Potentiometers

■ Problem

Assuming the potentiometer resistance R_1 is proportional to x, derive the expression for the output voltage v_o as a function of x.

■ Solution

The length of the pot is L and its total resistance is $R_1 + R_2$. Figure 6.2.5b shows the circuit diagram of the system. From the voltage-divider rule,

$$v_o = \frac{R_1}{R_1 + R_2} V \tag{1}$$

Because the resistance R_1 proportional to x,

$$R_1 = (R_1 + R_2)\left(\frac{x}{L}\right)$$

Substituting this into equation (1) gives

$$v_o = \frac{x}{L} V = Kx$$

where $K = V/L$ is the gain of the pot.

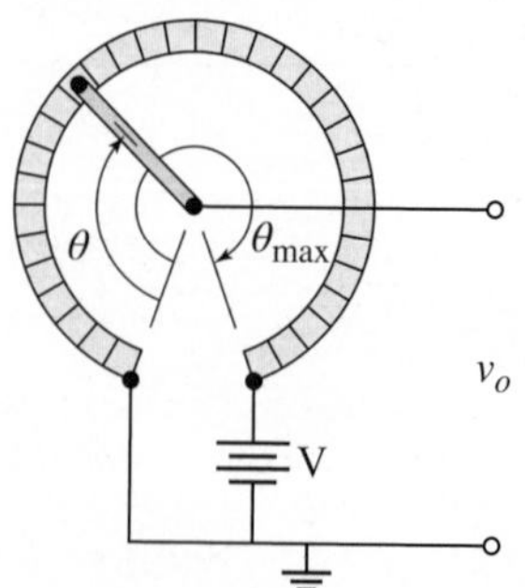

Figure 6.2.6 A rotational potentiometer.

The voltage source for the pot can be a battery or it can be a power supply.

A *rotational potentiometer* is shown in Figure 6.2.6. Following a similar procedure we can show that if the resistance is proportional to θ, then

$$v_o = \frac{\theta}{\theta_{\max}} V = K\theta$$

where $K = V/\theta_{\max}$.

EXAMPLE 6.2.5

Maximum Power Transfer in a Speaker-Amplifier System

■ Problem

A common example of an electrical system is an amplifier and a speaker. The load is the speaker, which requires current from the amplifier in order to produce sound. In Figure 6.2.7a the resistance R_L is that of the load. Part (b) of the figure shows the circuit representation of the system. The source supplies a voltage v_S and a current i_S, and has its own internal resistance R_S. For optimum efficiency, we want to maximize the power supplied to the speaker, for given values of v_S and R_S. Determine the value of R_L to maximize the power transfer to the load.

■ Solution

The required model should describe the power supplied to the speaker in terms of v_S, R_S, and R_L. From Kirchhoff's voltage law,

$$v_s - i_S R_S - i_S R_L = 0$$

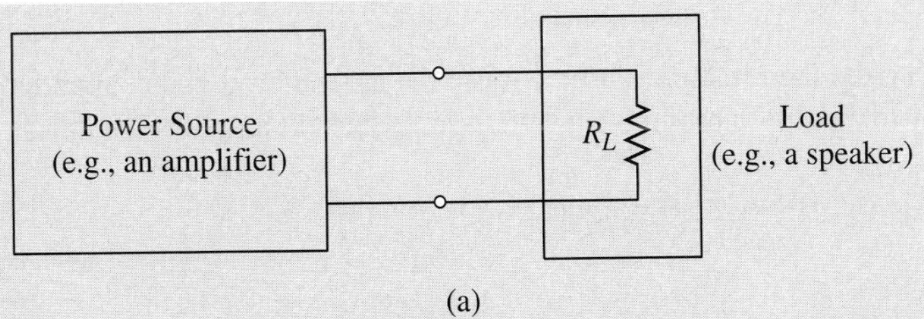

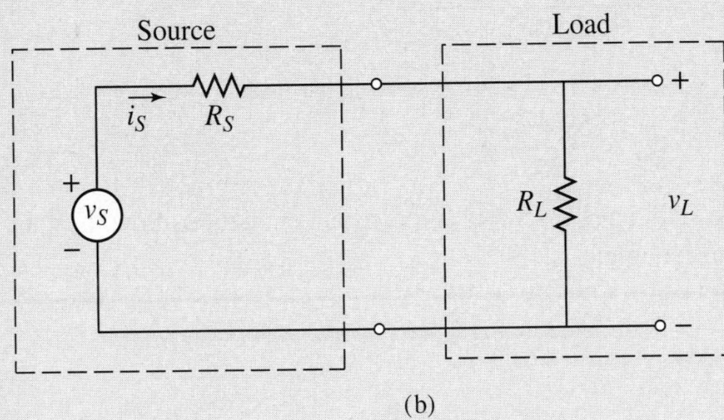

Figure 6.2.7 (a) An amplifier-speaker system. (b) Circuit representation with a voltage source and a resistive load.

We want to find v_L in terms of v_S. From the voltage-divider rule,

$$v_L = \frac{R_L}{R_S + R_L} v_S$$

The power consumed by the load is $P_L = i_S^2 R_L = v_L^2 / R_L$. Using the relation between v_L and v_S, we can express P_L in terms of v_S as

$$P_L = \frac{R_L}{(R_S + R_L)^2} v_S^2$$

To maximize P_L for a fixed value of v_S, we must choose R_L to maximize the ratio

$$r = \frac{R_L}{(R_S + R_L)^2}$$

The maximum of r occurs where $dr/dR_L = 0$.

$$\frac{dr}{dR_L} = \frac{(R_S + R_L)^2 - 2R_L(R_S + R_L)}{(R_S + R_L)^4} = 0$$

This is true if $R_L = R_S$. Thus, to maximize the power to the load we should choose the load resistance R_L to be equal to the source resistance R_S. This result for a resistance circuit is a special case of the more general result known as *impedance matching*.

A Feedback Amplifier

EXAMPLE 6.2.6

■ Problem

Early in the twentieth century engineers struggled to design vacuum-tube amplifiers whose gain remained constant at a predictable value. The gain is the ratio of the output voltage to the input voltage. The vacuum-tube gain G can be made large but is somewhat unpredictable and unreliable due to heat effects and manufacturing variations. A solution to the problem is shown in Figure 6.2.8. Derive the input-output relation for v_o as a function of v_i. Investigate the case where the gain G is very large.

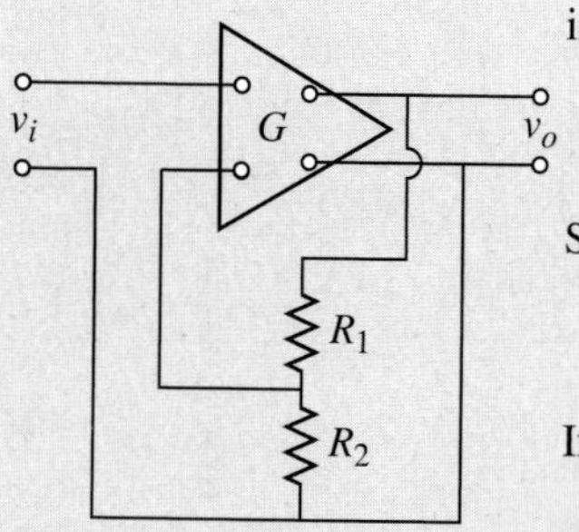

Figure 6.2.8 A feedback amplifier.

■ Solution

Part of the voltage drop across the resistors is used to raise the ground level at the amplifier input, so the input voltage to the amplifier is $v_i - R_2 v_o/(R_1 + R_2)$. Thus, the amplifier's output is

$$v_o = G\left(v_i - \frac{R_2}{R_1 + R_2} v_o\right)$$

Solve for v_o:

$$v_o = \frac{G}{1 + GR_2/(R_1 + R_2)} v_i$$

If $GR_2/(R_1 + R_2) \gg 1$, then

$$v_o \approx \frac{R_1 + R_2}{R_2} v_i$$

Presumably, the resistor values are sufficiently accurate and constant enough to allow the gain $(R_1 + R_2)/R_2$ to be predictable and reliable.

6.2.1 LOOP CURRENTS

Sometimes the circuit equations can be simplified by using the concept of a *loop current*, which is a current identified with a specific loop in the circuit. A loop current is not necessarily an actual current. If an element is part of two or more loops, the actual current through the element is the algebraic sum of the loop currents. Use of loop currents usually reduces the number of unknowns to be found, although when deriving the circuit equations, you must be careful to use the proper algebraic sum for each element.

EXAMPLE 6.2.7 Analysis with Loop Currents

■ Problem

We are given the values of the voltages and the resistances for the circuit in Figure 6.2.9a. (a) Solve for the currents i_1, i_2, and i_3 passing through the three resistors. (b) Use the loop-current method to solve for the currents.

■ Solution

a. Applying Kirchhoff's voltage law to the left-hand loop gives

$$v_1 - R_1 i_1 - R_2 i_2 = 0$$

For the right-hand loop,

$$v_2 - R_2 i_2 + i_3 R_3 = 0$$

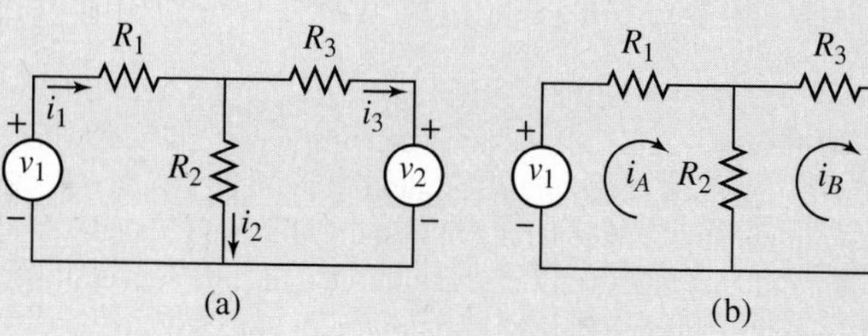

Figure 6.2.9 Example of loop analysis.

From conservation of charge, $i_1 = i_2 + i_3$. These are three equations in three unknowns. Their solution is

$$
\begin{aligned}
i_1 &= \frac{(R_2 + R_3)v_1 - R_2 v_2}{R_1 R_2 + R_1 R_3 + R_2 R_3} \\
i_2 &= \frac{R_3 v_1 + R_1 v_2}{R_1 R_2 + R_1 R_3 + R_2 R_3} \qquad (1) \\
i_3 &= \frac{R_2 v_1 - (R_1 + R_2)v_2}{R_1 R_2 + R_1 R_3 + R_2 R_3}
\end{aligned}
$$

b. Define the loop currents i_A and i_B positive clockwise, as shown in Figure 6.2.9b. Note that there is a voltage *drop* $R_2 i_A$ across R_2 due to i_A and a voltage *increase* $R_2 i_B$ due to i_B. Applying Kirchhoff's voltage law to the left-hand loop gives

$$v_1 - R_1 i_A - R_2 i_A + R_2 i_B = 0$$

For the right-hand loop,

$$v_2 + R_3 i_B + R_2 i_B - R_2 i_A = 0$$

Now we have only two equations to solve. Their solution is

$$i_A = \frac{(R_2 + R_3)v_1 - R_2 v_2}{R_1 R_2 + R_1 R_3 + R_2 R_3} \qquad (2)$$

$$i_B = \frac{R_2 v_1 - (R_1 + R_2)v_2}{R_1 R_2 + R_1 R_3 + R_2 R_3} \qquad (3)$$

The current i_1 through R_1 is the same as i_A, and the current i_3 through R_3 is the same as i_B. The current i_2 through R_2 is $i_2 = i_A - i_B$, which gives expression (1) when expressions (2) and (3) are substituted.

6.2.2 CAPACITANCE AND INDUCTANCE IN CIRCUITS

Examples 6.2.8 through 6.2.13 illustrate how models are developed for circuits containing capacitors or inductors.

Series *RC* Circuit — EXAMPLE 6.2.8

■ Problem

The resistor and capacitor in the circuit shown in Figure 6.2.10 are said to be in series because the same current flows through them. Obtain the model of the capacitor voltage v_1. Assume that the supply voltage v_s is known.

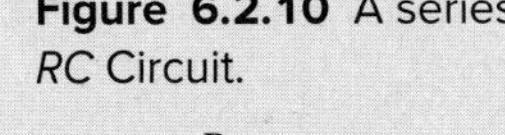

Figure 6.2.10 A series *RC* Circuit.

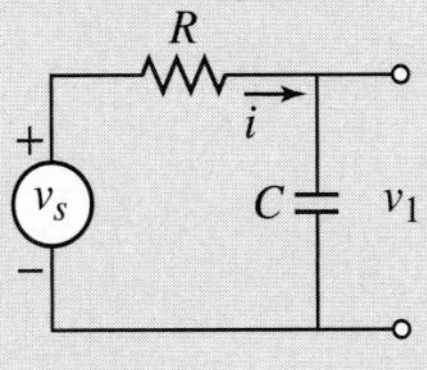

■ Solution

From Kirchhoff's voltage law, $v_s - Ri - v_1 = 0$. This gives

$$i = \frac{1}{R}(v_s - v_1) \qquad (1)$$

For the capacitor,

$$v_1 = \frac{1}{C}\int_0^t i\,dt + \frac{Q_0}{C}$$

Differentiate this with respect to t to obtain

$$\frac{dv_1}{dt} = \frac{1}{C}i$$

Then substitute for i from (1):

$$\frac{dv_1}{dt} = \frac{1}{RC}(v_s - v_1)$$

This the required model. It is often expressed in the following rearranged form:

$$RC\frac{dv_1}{dt} + v_1 = v_s \tag{2}$$

EXAMPLE 6.2.9

Series *RCL* Circuit

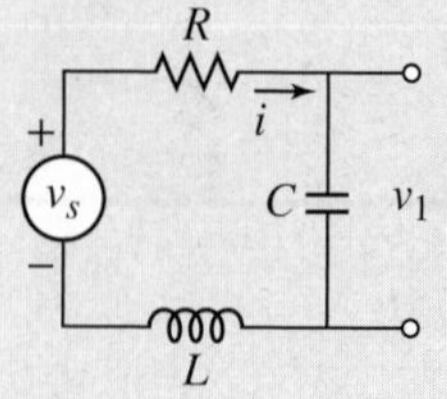

Figure 6.2.11 A series *RCL* circuit.

■ Problem

The resistor, inductor, and capacitor in the circuit shown in Figure 6.2.11 are in series because the same current flows through them. Obtain the model of the capacitor voltage v_1 with the supply voltage v_s as the input.

■ Solution

From Kirchhoff's voltage law,

$$v_s - Ri - L\frac{di}{dt} - v_1 = 0 \tag{1}$$

For the capacitor,

$$v_1 = \frac{1}{C}\int_0^t i\,dt$$

Differentiate this with respect to t to obtain

$$i = C\frac{dv_1}{dt}$$

and substitute this for i in (1):

$$v_s - RC\frac{dv_1}{dt} - LC\frac{d^2v_1}{dt^2} - v_1 = 0$$

This is the required model. It can be expressed in the following form:

$$LC\frac{d^2v_1}{dt^2} + RC\frac{dv_1}{dt} + v_1 = v_s \tag{2}$$

Parallel *RL* Circuit

EXAMPLE 6.2.10

■ **Problem**

The resistor and inductor in the circuit shown in Figure 6.2.12 are said to be in parallel because they have the same voltage v_1 across them. Obtain the model of the current i_2 passing through the inductor. Assume that the supply current i_s is known.

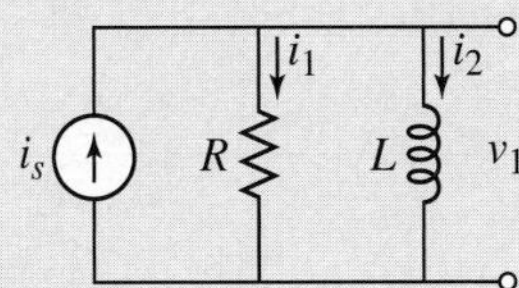

Figure 6.2.12 A parallel *RL* circuit.

■ **Solution**

The currents i_1 and i_2 are defined in the figure. Then,

$$v_1 = L\frac{di_2}{dt} = Ri_1 \tag{1}$$

From conservation of charge, $i_1 + i_2 = i_s$. Thus, $i_1 = i_s - i_2$. Substitute this expression into (1) to obtain

$$L\frac{di_2}{dt} = R(i_s - i_2)$$

This is the required model. It can be rearranged as follows:

$$\frac{L}{R}\frac{di_2}{dt} + i_2 = i_s \tag{2}$$

Analysis of a Telegraph Line

EXAMPLE 6.2.11

■ **Problem**

Figure 6.2.13 shows a circuit representation of a telegraph line. The resistance R is the line resistance and L is the inductance of the solenoid that activates the receiver's clicker. The switch represents the operator's key. Assume that when sending a "dot," the key is closed for 0.1 s. Using the values $R = 20\ \Omega$ and $L = 4$ H, obtain the expression for the current $i(t)$ passing through the solenoid.

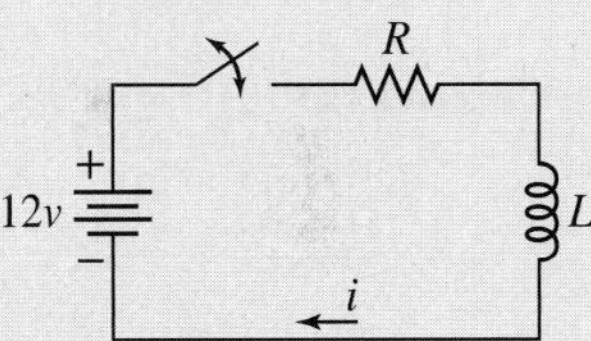

Figure 6.2.13 Circuit representation of a telegraph line.

■ **Solution**

From the voltage law we have

$$L\frac{di}{dt} + Ri = v_i(t) \tag{1}$$

where $v_i(t)$ represents the input voltage due to the switch and the 12-V supply. We could model $v_i(t)$ as a rectangular pulse of height 12 V and duration 0.1 s, but the differential equation (1) is easier to solve if we model $v_i(t)$ as an impulsive input of strength $12(0.1) = 1.2$ V · s. This model can be justified by the fact that the circuit time constant, $L/R = 4/20 = 0.2$, is greater than the duration of $v_i(t)$. Thus, we model $v_i(t)$ as $v_i(t) = 1.2\delta(t)$. The Laplace transform of equation (1) with $i(0) = 0$ gives

$$(4s + 20)I(s) = 1.2 \qquad \text{or} \qquad I(s) = \frac{1.2}{4s + 20} = \frac{0.3}{s + 5}$$

This gives the solution

$$i(t) = 0.3e^{-5t} \text{ A}$$

Note that this solution gives $i(0+) = 0.3$, whereas $i(0) = 0$. The difference is due to the impulsive input.

EXAMPLE 6.2.12

An *RLC* Circuit with Two Input Voltages

Figure 6.2.14 An *RLC* circuit with two voltage sources.

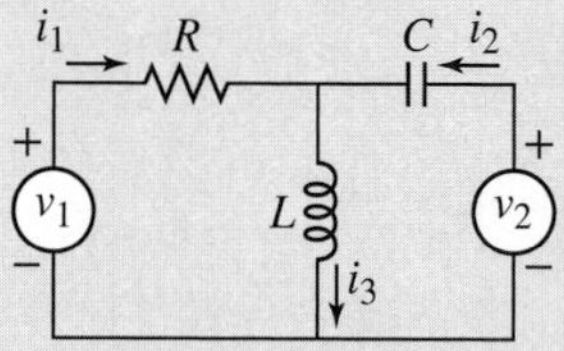

■ Problem

The *RLC* circuit shown in Figure 6.2.14 has two input voltages. Obtain the differential equation model for the current i_3.

■ Solution

Applying Kirchhoff's voltage law to the left-hand loop gives

$$v_1 - Ri_1 - L\frac{di_3}{dt} = 0 \tag{1}$$

For the right-hand loop,

$$v_2 - \frac{1}{C}\int i_2\, dt - L\frac{di_3}{dt} = 0$$

Differentiate this equation with respect to t:

$$\frac{dv_2}{dt} - \frac{1}{C}i_2 - L\frac{d^2i_3}{dt^2} = 0 \tag{2}$$

From conservation of charge,

$$i_3 = i_1 + i_2 \tag{3}$$

These are three equations in the three unknowns i_1, i_2, and i_3. To eliminate i_1 and i_2, solve equation (1) for i_1

$$i_1 = \frac{1}{R}\left(v_1 - L\frac{di_3}{dt}\right) \tag{4}$$

In equation (2), substitute for i_2 from equation (3):

$$\frac{dv_2}{dt} - \frac{1}{C}(i_3 - i_1) - L\frac{d^2i_3}{dt^2} = 0$$

Now substitute for i_1 from equation (4):

$$\frac{dv_2}{dt} - \frac{1}{C}i_3 + \frac{1}{C}\left[\frac{1}{R}\left(v_1 - L\frac{di_3}{dt}\right)\right] - L\frac{d^2i_3}{dt^2} = 0$$

Rearrange this equation to obtain the answer:

$$LRC\frac{d^2i_3}{dt^2} + L\frac{di_3}{dt} + Ri_3 = v_1 + RC\frac{dv_2}{dt} \tag{5}$$

6.2.3 STATE-VARIABLE MODELS OF CIRCUITS

The presence of several current and voltage variables in a circuit can sometimes lead to difficulty in identifying the appropriate variables to use for expressing the circuit model. Use of state variables can often reduce this confusion. To choose a proper set of state variables, identify where the energy is stored in the system. The variables that describe the stored energy are appropriate choices for state variables.

EXAMPLE 6.2.13

State-Variable Model of a Series *RLC* Circuit

■ Problem

Consider the series *RLC* circuit shown in Figure 6.2.15. Choose a suitable set of state variables, and obtain the state-variable model of the circuit in matrix form. The input is the voltage v_s.

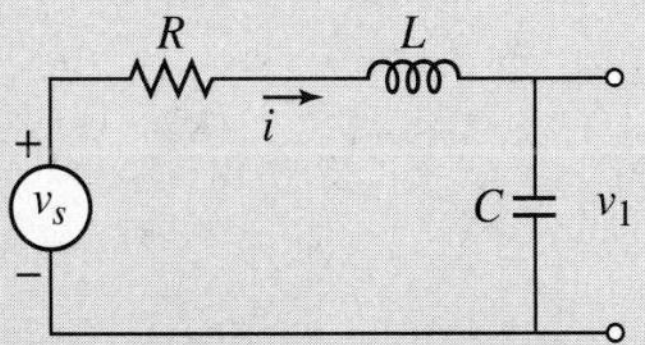

Figure 6.2.15 Series *RLC* circuit.

■ **Solution**

In this circuit the energy is stored in the capacitor and in the inductor. The energy stored in the capacitor is $Cv_1^2/2$ and the energy stored in the inductor is $Li^2/2$. Thus, a suitable choice of state variables is v_1 and i.

From Kirchhoff's voltage law,

$$v_s - Ri - L\frac{di}{dt} - v_1 = 0$$

Solve this for di/dt:

$$\frac{di}{dt} = \frac{1}{L}v_s - \frac{1}{L}v_1 - \frac{R}{L}i$$

This is the first state equation.

Now find an equation for dv_1/dt. From the capacitor relation,

$$v_1 = \frac{1}{C}\int i\,dt$$

Differentiating gives the second state equation.

$$\frac{dv_1}{dt} = \frac{1}{C}i$$

The two state equations can be expressed in matrix form as follows.

$$\begin{bmatrix} \dfrac{di}{dt} \\ \dfrac{dv_1}{dt} \end{bmatrix} = \begin{bmatrix} -\dfrac{R}{L} & -\dfrac{1}{L} \\ \dfrac{1}{C} & 0 \end{bmatrix} \begin{bmatrix} i \\ v_1 \end{bmatrix} + \begin{bmatrix} \dfrac{1}{L} \\ 0 \end{bmatrix} v_s$$

6.3 TRANSFER FUNCTIONS AND IMPEDANCE

The Laplace transform method and the transfer-function concept enable us to deal with algebraic equations rather than differential equations, and thus ease the task of analyzing circuit models. This section illustrates why this is so, and introduces a related concept, impedance, which is simply the transfer function between voltage and current.

6.3.1 USE OF TRANSFORMED EQUATIONS

Applying the Laplace transform method to the circuit equations often helps in eliminating intermediate variables, because the transformed equations are algebraic and therefore are easier to solve. This is especially true when the circuit contains dynamic elements (capacitances or inductances).

An *RLC* Circuit with Two Inputs — EXAMPLE 6.3.1

■ **Problem**

Consider the *RLC* circuit treated in Example 6.2.14 and shown in Figure 6.2.17. Use the Laplace transform method to obtain the differential equation model for the current i_3.

■ **Solution**

Applying Kirchhoff's voltage law to the left-hand loop gives

$$v_1 - Ri_1 - L\frac{di_3}{dt} = 0$$

which, when transformed for zero-initial conditions, gives

$$V_1(s) - RI_1(s) - LsI_3(s) = 0 \tag{1}$$

Similarly for the right-hand loop,

$$v_2 - \frac{1}{C}\int i_2\, dt - L\frac{di_3}{dt} = 0$$

or

$$V_2(s) - \frac{1}{Cs}I_2(s) - LsI_3(s) = 0 \tag{2}$$

From conservation of charge, $i_3 = i_1 + i_2$, or

$$I_3(s) = I_1(s) + I_2(s) \tag{3}$$

These are three equations in the three unknowns $I_1(s)$, $I_2(s)$, and $I_3(s)$. To eliminate $I_1(s)$ and $I_2(s)$, solve equation (1) for $I_1(s)$:

$$I_1(s) = \frac{1}{R}[V_1(s) - LsI_3(s)] \tag{4}$$

From equation (2),

$$I_2(s) = Cs[V_2(s) - LsI_3(s)] \tag{5}$$

Substitute equations (4) and (5) into equation (3):

$$I_3(s) = I_1(s) + I_2(s) = \frac{1}{R}[V_1(s) - LsI_3(s)] + Cs[V_2(s) - LsI_3(s)]$$

and collect the $I_3(s)$ terms:

$$(LRCs^2 + Ls + R)I_3(s) = V_1(s) + RCsV_2(s)$$

This transformed equation corresponds to the differential equation:

$$LRC\frac{d^2i_3}{dt^2} + L\frac{di_3}{dt} + Ri_3 = v_1 + RC\frac{dv_2}{dt} \tag{6}$$

EXAMPLE 6.3.2

Coupled *RC* Loops

■ **Problem**

(a) Determine the transfer function $V_o(s)/V_s(s)$ of the circuit shown in Figure 6.3.1. (b) Use a block-diagram representation of the circuit to show how the output voltage v_o affects the internal voltage v_1.

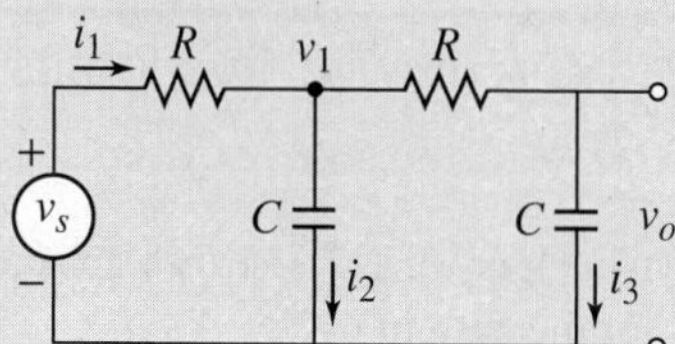

Figure 6.3.1 Coupled *RC* loops.

■ Solution

a. The energy in this circuit is stored in the two capacitors. Because the energy stored in a capacitor is expressed by $Cv^2/2$, appropriate choices for the state variables are the voltages v_1 and v_o. The capacitance relations are

$$v_o = \frac{1}{C}\int i_3 \qquad v_1 = \frac{1}{C}\int i_2$$

which give the state equations

$$\frac{dv_o}{dt} = \frac{i_3}{C} \tag{1}$$

$$\frac{dv_1}{dt} = \frac{i_2}{C} \tag{2}$$

For the right-hand loop,

$$v_1 - Ri_3 - v_o = 0$$

Thus

$$i_3 = \frac{v_1 - v_o}{R} \tag{3}$$

and equation (1) becomes

$$\frac{dv_o}{dt} = \frac{1}{RC}(v_1 - v_o) \tag{4}$$

For the left-hand loop,

$$v_s - Ri_1 - v_1 = 0$$

which gives

$$i_1 = \frac{v_s - v_1}{R} \tag{5}$$

From conservation of charge and equations (3) and (5),

$$i_2 = i_1 - i_3 = \frac{v_s - v_1}{R} - \frac{v_1 - v_o}{R} = \frac{1}{R}(v_s - 2v_1 + v_o) \tag{6}$$

Substituting this into equation (2) gives

$$\frac{dv_1}{dt} = \frac{1}{RC}(v_s - 2v_1 + v_o) \tag{7}$$

Equations (4) and (7) are the state equations. To obtain the transfer function $V_o(s)/V_s(s)$, transform these equations for zero-initial conditions to obtain

$$sV_o(s) = \frac{1}{RC}[V_1(s) - V_o(s)]$$

$$sV_1(s) = \frac{1}{RC}[V_s(s) - 2V_1(s) + V_o(s)]$$

Eliminating $V_1(s)$ from these two equations gives

$$V_o(s) = \frac{1}{R^2C^2s^2 + 3RCs + 1}V_s(s)$$

So the transfer function is

$$\frac{V_o(s)}{V_s(s)} = \frac{1}{R^2C^2s^2 + 3RCs + 1} \tag{8}$$

b. In the block diagram in Figure 6.3.2a, the left-hand inner loop is based on equations (2) and (6). The right-hand inner loop is based on equations (1) and (3). The diagram can be simplified by reducing each inner loop to a single block, as shown in part (b) of the figure.

Figure 6.3.2 Block diagrams of coupled *RC* loops.

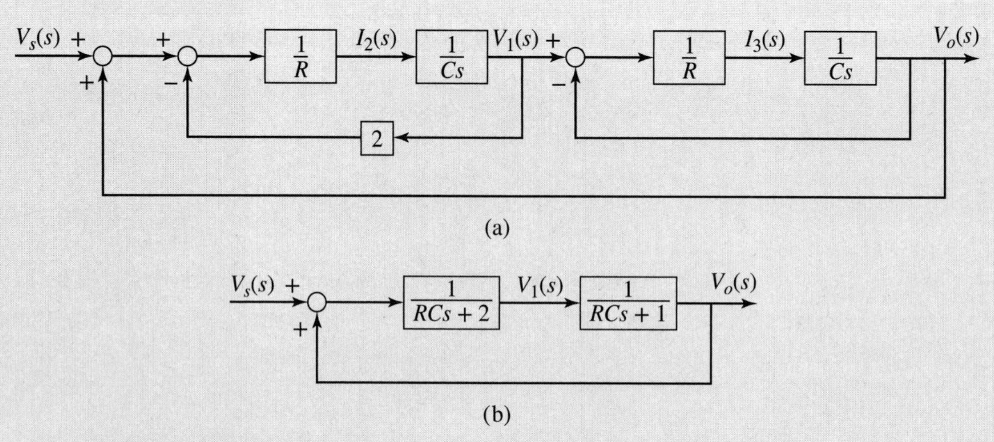

This simpler diagram shows how the output voltage v_o affects the inner voltage v_1 through the outer feedback loop.

Some models have more than one input or more than one output. In such cases, there is one transfer function for each input-output pair. If a model has more than one input, a particular transfer function is the ratio of the output transform over the input transform, with all the remaining inputs ignored (set to zero temporarily).

EXAMPLE 6.3.3

Transfer Functions for Two Inputs

■ **Problem**

For the circuit of Example 6.3.1, the model is

$$LRC\frac{d^2i_3}{dt^2} + L\frac{di_3}{dt} + Ri_3 = v_1 + RC\frac{dv_2}{dt} \tag{1}$$

where the two inputs are v_1 and v_2 and the current i_3 has been selected as the output. Determine the transfer functions.

■ **Solution**

Transforming equation (1) for zero-initial conditions and solving for $I_3(s)$ gives

$$I_3(s) = \frac{V_1(s) + RCsV_2(s)}{LRCs^2 + Ls + R} = \frac{1}{LRCs^2 + Ls + R}V_1(s) + \frac{RCs}{LRCs^2 + Ls + R}V_2(s) \tag{2}$$

Temporarily setting $V_2(s) = 0$, we have

$$\frac{I_3(s)}{V_1(s)} = \frac{1}{LRCs^2 + Ls + R}$$

This is the transfer function for the input v_1. To find the transfer function for the input v_2, set $V_1(s) = 0$ in (2) to obtain

$$\frac{I_3(s)}{V_2(s)} = \frac{RCs}{LRCs^2 + Ls + R}$$

The characteristic polynomial for the system is $LRCs^2 + Ls + R$, which can be obtained from the denominator of either transfer function. The numerators are different, however. Note that the second transfer function has numerator dynamics, which indicates that i_3 responds to the derivative $\dot{v}_2$.

6.3.2 IMPEDANCE

We have seen that a resistance resists or "impedes" the flow of current. The corresponding relation is $v/i = R$. Capacitance and inductance elements also impede the flow of current. In electrical systems, an *impedance* is a generalization of the resistance concept and is defined as the ratio of a voltage transform $V(s)$ to a current transform $I(s)$ and thus implies a current source. A standard symbol for impedance is $Z(s)$. Thus,

$$Z(s) = \frac{V(s)}{I(s)} \tag{6.3.3}$$

The impedance of a resistor is its resistance R. The impedances of the other two common electrical elements are found as follows. For a capacitor,

$$v(t) = \frac{1}{C}\int_0^t i\,dt$$

or $V(s) = I(s)/(Cs)$. The impedance of a capacitor is thus

$$Z(s) = \frac{1}{Cs} \tag{6.3.4}$$

For an inductor,

$$v(t) = L\frac{di}{dt}$$

or $V(s) = LsI(s)$. Thus, the impedance of an inductor is

$$Z(s) = Ls \tag{6.3.5}$$

6.3.3 SERIES AND PARALLEL IMPEDANCES

The concept of impedance is useful because the impedances of individual elements can be combined with series and parallel laws to find the impedance at any point in the system. The laws for combining series or parallel impedances are extensions to the dynamic case of the laws governing series and parallel resistance elements. Two impedances are in *series* if they have the *same current*. If so, the total impedance is the sum of the individual impedances.

$$Z(s) = Z_1(s) + Z_2(s)$$

For example, a resistor R and capacitor C in series, as shown in Figure 6.3.3a, have the equivalent impedance

$$Z(s) = R + \frac{1}{Cs} = \frac{RCs + 1}{Cs}$$

Thus, the relation between the current i flowing through them and the total voltage drop v across them is

$$\frac{V(s)}{I(s)} = Z(s) = \frac{RCs + 1}{Cs}$$

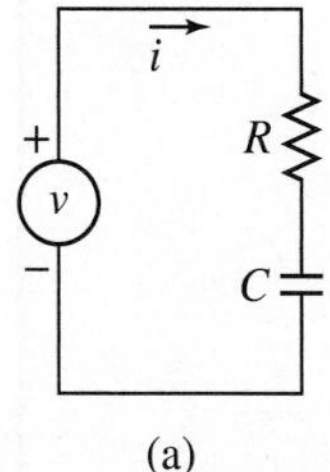

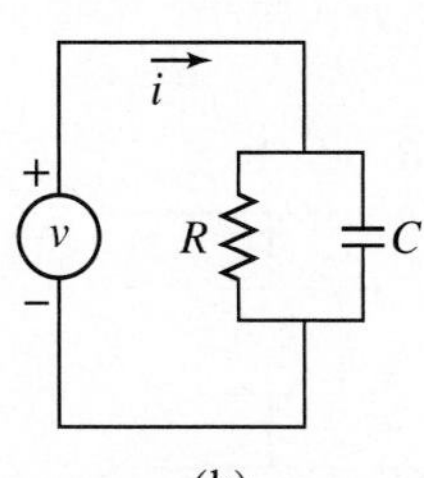

Figure 6.3.3 Series and parallel *RC* circuits.

and the differential equation model is

$$C\frac{dv}{dt} = RC\frac{di}{dt} + i(t)$$

If the impedances have the *same voltage difference* across them, they are in *parallel*, and their impedances combine by the reciprocal rule

$$\frac{1}{Z(s)} = \frac{1}{Z_1(s)} + \frac{1}{Z_2(s)}$$

where $Z(s)$ is the total equivalent impedance. If a resistor R and capacitor C are in parallel, as shown in Figure 6.3.3b, their equivalent total impedance $Z(s)$ is found from

$$\frac{1}{Z(s)} = \frac{1}{1/Cs} + \frac{1}{R}$$

or

$$Z(s) = \frac{R}{RCs + 1}$$

Thus, the relation between the total current i and the voltage drop v across them is

$$\frac{V(s)}{I(s)} = Z(s) = \frac{R}{RCs + 1}$$

and the differential equation model is

$$RC\frac{dv}{dt} + v = Ri(t)$$

EXAMPLE 6.3.4

Circuit Analysis Using Impedance

■ Problem

For the circuit shown in Figure 6.3.4a, determine the transfer function between the input voltage v_s and the output voltage v_o.

■ Solution

Note that R and C are in parallel. Therefore, their equivalent impedance $Z(s)$ is found from

$$\frac{1}{Z(s)} = \frac{1}{1/Cs} + \frac{1}{R}$$

or

$$Z(s) = \frac{R}{RCs + 1}$$

An impedance representation of the equivalent circuit is shown in Figure 6.3.4b. In this representation, we may think of the impedance as a simple resistance, provided we express

Figure 6.3.4 Circuit analysis using impedance.

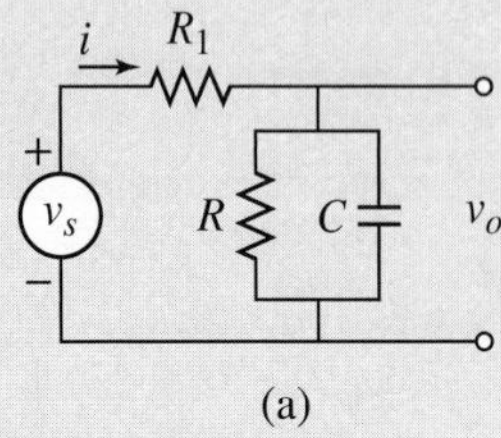

(a)

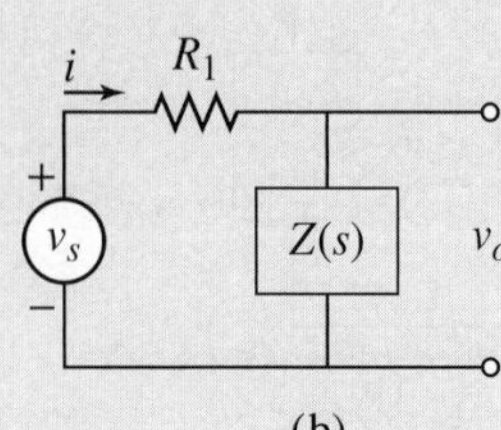

(b)

the relations in Laplace transform notation. Kirchhoff's voltage law gives

$$V_s(s) - R_1 I(s) - Z(s)I(s) = 0$$

The output voltage is related to the current by $V_o(s) = Z(s)I(s)$. Eliminating $I(s)$ from these two relations gives

$$V_s(s) - R_1 \frac{V_o(s)}{Z(s)} - V_o(s) = 0$$

which yields the desired transfer function:

$$\frac{V_o(s)}{V_s(s)} = \frac{Z(s)}{R_1 + Z(s)} = \frac{R}{RR_1Cs + R + R_1}$$

This network is a first-order system whose time constant is

$$\tau = \frac{RR_1}{R + R_1}$$

If the voltage output is measured at the terminals to which the driving current is applied, the impedance so obtained is the *driving-point* or *input* impedance. If the voltage is measured at another place in the circuit, the impedance obtained is a *transfer* impedance (because the effect of the input current has been transferred to another point). Sometimes the term *admittance* is used. This is the reciprocal of impedance, and it is an indication of to what extent a circuit "admits" current flow.

6.3.4 ISOLATION AMPLIFIERS

A *voltage-isolation amplifier* is designed to produce an output voltage that is proportional to the input voltage. It is intended to boost the electrical signal from a low-power source. Therefore, the amplifier requires an external power source, which is not usually shown in the circuit diagrams. We will not be concerned with the internal design details of amplifiers, but we need to understand their effects on any circuit in which they are used.

Such an amplifier may be considered to be a voltage source if it does not affect the behavior of the source circuit that is attached to the amplifier input terminals, and if the amplifier is capable of providing the voltage independently of the particular circuit (the "load") attached to amplifier output terminals.

Consider the system in Figure 6.3.5. The internal impedances of the amplifier at its input and output terminals are $Z_i(s)$ and $Z_o(s)$, respectively. The impedance of the source circuit is $Z_s(s)$ and the impedance of the load is $Z_L(s)$. A simple circuit analysis will reveal that

$$V_s(s) - I_1(s)Z_s(s) - V_i(s) = 0$$

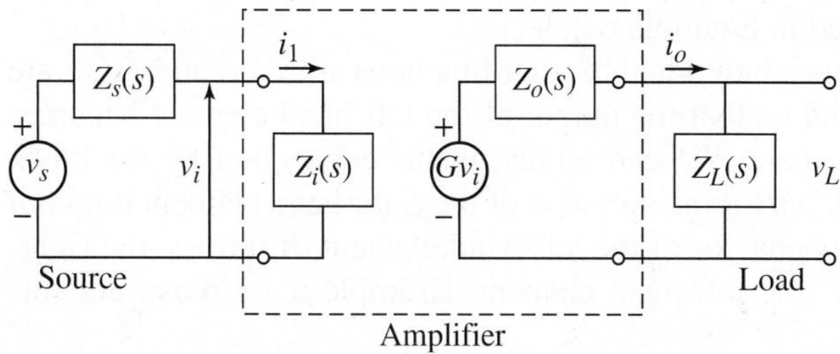

Figure 6.3.5 Input and output impedance in an amplifier.

and $I_1(s) = V_i(s)/Z_i(s)$. Thus, if $Z_i(s)$ is large, the current i_1 drawn by the amplifier will be small. Therefore, if the input impedance $Z_i(s)$ is large, the amplifier does not affect the current i_1, and thus the amplifier does not affect the behavior of the source circuit. In addition, if $Z_i(s)$ is large,

$$V_i(s) = \frac{Z_i(s)}{Z_i(s) + Z_s(s)} V_s(s) \approx V_s(s)$$

So we conclude that a voltage-isolation amplifier must have a high input impedance.

Denote the amplifier's voltage gain as G. This means that the amplifier's output voltage v_o is $v_o = Gv_i$. For the load circuit,

$$GV_i(s) - I_o(s)Z_o(s) - I_o(s)Z_L(s) = 0$$

Thus

$$I_o(s) = \frac{GV_i(s)}{Z_o(s) + Z_L(s)}$$

and

$$V_L(s) = Z_L(s)I_o(s) = \frac{Z_L(s)}{Z_o(s) + Z_L(s)} GV_o(s)$$

Thus, if $Z_o(s)$ is small, $v_L \approx GV_o$. So if the amplifier output impedance is small, the voltage v_L delivered to the load is independent of the load.

A similar analysis applies to a *current amplifier*, which provides a current proportional to its input signal regardless of the load. Thus, such an amplifier acts as a current source.

6.3.5 LOADING EFFECTS AND BLOCK DIAGRAMS

One application of transfer functions is to provide a graphical representation of the system's dynamics in the form of a block diagram, which illustrates the cause-and-effect relations operating in a particular system. The algebraic representation of the system's equations in terms of transfer functions enables easier manipulation for analysis and design purposes, and the graphical representation allows the engineer to see the interaction between the system's components. For example, the equation for the feedback amplifier discussed in Example 6.2.6 and shown in Figure 6.2.8 is

$$v_o = G\left(v_i - \frac{R_2 v_o}{R_1}\right)$$

The block diagram of this device is given in Figure 6.3.6. It shows the presence and action of the negative feedback loop, by which the output voltage v_o is used to modify the behavior of the device. A block-diagram example with two inputs is given in Figure 6.3.7, which shows the block diagram of the circuit in Figure 6.2.17, whose transfer functions were obtained in Example 6.3.3.

Suppose two elements, whose individual transfer functions are $T_1(s)$ and $T_2(s)$, are physically connected end-to-end so that the output of the left-hand element becomes the input to the right-hand element. We can represent this connection by the block diagram shown in Figure 6.3.8, *only if* the output w of the right-hand element does *not* affect the inputs u and v or the behavior of the left-hand element. If it does, the right-hand element is said to "load" the left-hand element. Example 6.3.5 illustrates this point.

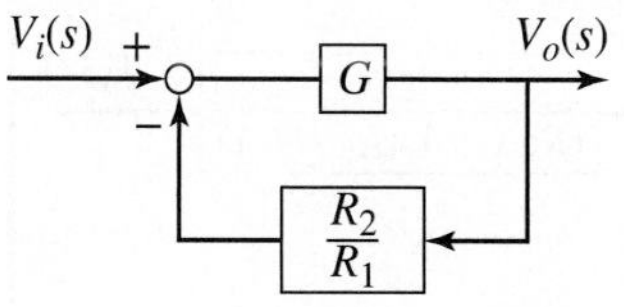

Figure 6.3.6 Block diagram of a feedback amplifier.

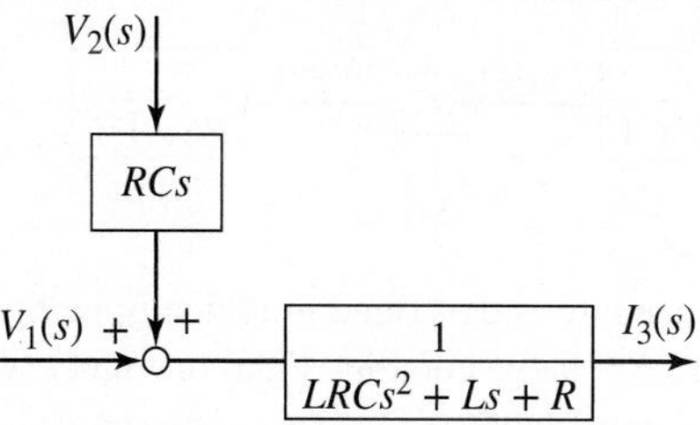

Figure 6.3.7 Block diagram of an *RLC* circuit with two voltage sources.

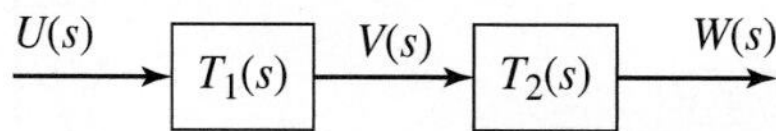

Figure 6.3.8 Block diagram of two series elements.

Loading Effects, Transfer Functions, and Block Diagrams

EXAMPLE 6.3.5

■ Problem

The circuit shown in Figure 6.3.9a consists of two series *RC* circuits wired so that the output voltage of the first circuit is the input voltage to an isolation amplifier. The output voltage of the amplifier is the input voltage to the second *RC* circuit. The amplifier has a voltage gain G; that is, $v_2(t) = Gv_1(t)$. Derive the transfer function $V_o(s)/V_s(s)$ for this circuit, and for the case $G = 1$ compare it with the transfer function of the circuit shown in Figure 6.3.1.

■ Solution

The amplifier isolates the first *RC* loop from the effects of the second loop; that is, the amplifier prevents the voltage v_1 from being affected by the second *RC* loop. This in effect creates two separate loops with an intermediate voltage source $v_2 = Gv_1$, as shown in Figure 6.3.9b. Thus, for the left-hand loop, we obtain

$$\frac{V_1(s)}{V_s(s)} = \frac{1}{RCs + 1}$$

For the right-hand *RC* loop,

$$\frac{V_o(s)}{V_2(s)} = \frac{1}{RCs + 1}$$

For the amplifier with gain G,

$$V_2(s) = GV_1(s)$$

To obtain the transfer function $V_o(s)/V_s(s)$, eliminate the variables $V_1(s)$ and $V_2(s)$ from these equations as follows:

$$\frac{V_o(s)}{V_s(s)} = \frac{V_o(s)}{V_2(s)}\frac{V_2(s)}{V_1(s)}\frac{V_1(s)}{V_s(s)} = \frac{1}{RCs + 1}G\frac{1}{RCs + 1} = \frac{G}{R^2C^2s^2 + 2RCs + 1} \tag{1}$$

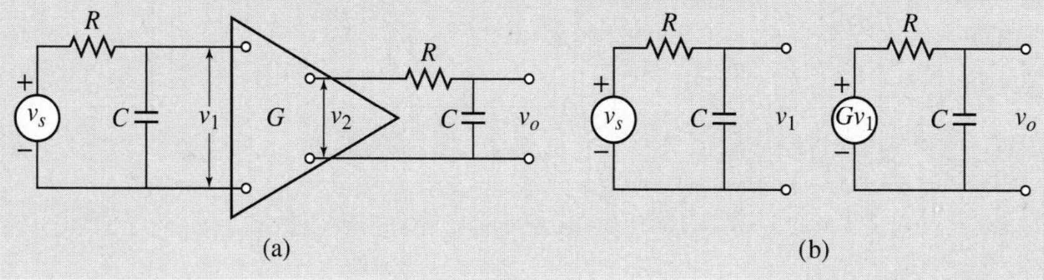

Figure 6.3.9 Two *RC* loops with an isolation amplifier.

Figure 6.3.10 Block diagrams of two RC loops with an isolation amplifier.

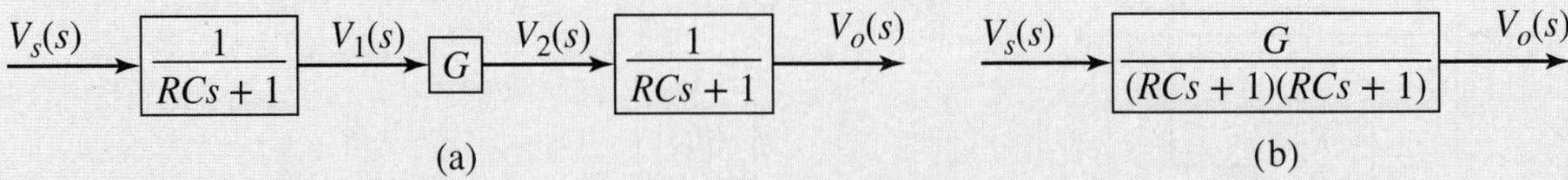

This procedure is described graphically by the block diagram shown in Figure 6.3.10a. The three blocks can be combined into one block as shown in part (b) of the figure.

The transfer function of the circuit shown in Figure 6.3.1 was derived in Example 6.3.2. It is

$$\frac{V_o(s)}{V_s(s)} = \frac{1}{R^2C^2s^2 + 3RCs + 1} \tag{2}$$

Note that it is *not* the same as the transfer function given by equation (1) with $G = 1$.

A common mistake is to obtain the transfer function of loops connected end-to-end by multiplying their transfer functions. This is equivalent to treating them as independent loops, which they are not, because each loop "loads" the adjacent loops and thus changes the currents and voltages in those loops. An isolation amplifier prevents a loop from loading an adjacent loop, and when such amplifiers are used, we can multiply the loop transfer functions to obtain the overall transfer function.

This mistake is sometimes made when drawing block diagrams. The circuit of Figure 6.3.9 can be represented by the block diagram of Figure 6.3.10, where the transfer function of each block can be multiplied to obtain the overall transfer function $V_o(s)/V_i(s)$. However, the circuit of Figure 6.3.1 *cannot* be represented by a simple series of blocks because the output voltage v_o affects the voltage v_1. To show this effect requires a feedback loop, as shown previously in Figure 6.3.2.

In general, even though elements are physically connected end-to-end, we cannot represent them by a series of blocks if the output of one element affects its input or the behavior of any preceding elements.

6.4 OPERATIONAL AMPLIFIERS

A modern version of the feedback amplifier discussed in Example 6.2.6 is the *operational amplifier* (op amp), which is a voltage amplifier with a very large gain G (greater than 10^5). The op amp has a large input impedance so it draws a negligible current. The op amp is an integrated circuit chip that contains many transistors, capacitors, and resistors and has several external terminals. We can attach two resistors in series with and parallel to the op amp, as shown in Figure 6.4.1a. This circuit diagram does not show

Figure 6.4.1 An op-amp multiplier.

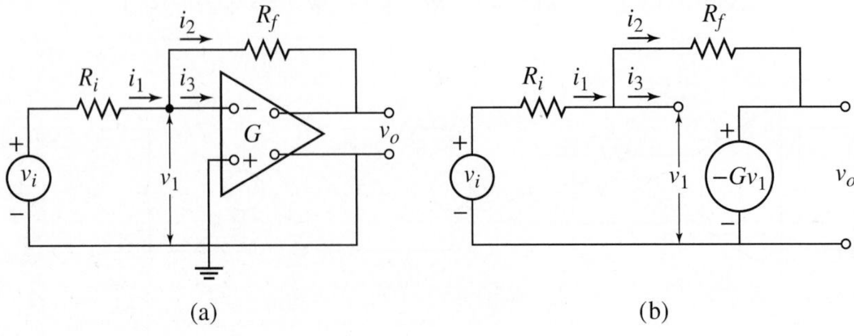

all of the op amp's external terminals; for example, some terminals are needed to power the device and to provide constant bias voltages. Our diagram shows only two pairs of terminals: the input terminals intended for time-varying input signals and the output terminals. A plus sign or a minus sign on an input terminal denotes it as a *noninverting* terminal or an *inverting* terminal, respectively.

Op amps are widely used in instruments and control systems for multiplying, integrating, and differentiating signals.

Op-Amp Multiplier

EXAMPLE 6.4.1

■ Problem

Determine the relation between the input voltage v_i and the output voltage v_o of the op-amp circuit shown in Figure 6.4.1a. Assume that the op amp has the following properties:

1. The op-amp gain G is very large,
2. $v_o = -Gv_1$; and
3. The op-amp input impedance is very large, and thus the current i_3 drawn by the op amp is very small.

■ Solution

Because the current i_3 drawn by the op amp is very small, the input terminal pair can be represented as an open circuit, as in Figure 6.4.1b. The voltage-current relation for each resistor gives

$$i_1 = \frac{v_i - v_1}{R_1}$$

and

$$i_2 = \frac{v_1 - v_o}{R_2}$$

From conservation of charge, $i_1 = i_2 + i_3$. However, from property 3, $i_3 \approx 0$, which implies that $i_1 \approx i_2$. Thus,

$$\frac{v_i - v_1}{R_1} = \frac{v_1 - v_o}{R_2}$$

From property 1, $v_1 = -v_o/G$. Substitute this into the preceeding equation:

$$\frac{v_i}{R_1} + \frac{v_o}{R_1 G} = -\frac{v_o}{R_1 G} - \frac{v_o}{R_2}$$

Because G is very large, the terms containing G in the denominator are very small, and we obtain

$$\frac{v_i}{R_1} = -\frac{v_o}{R_2}$$

Solve for v_o:

$$v_o = -\frac{R_2}{R_1} v_i \qquad (1)$$

This circuit can be used to multiply a voltage by the factor R_2/R_1, and is called an op-amp *multiplier*. Note that this circuit inverts the sign of the input voltage.

Resistors usually can be made so that their resistance values are known with sufficient accuracy and are constant enough to allow the gain R_2/R_1 to be predictable and reliable.

6.4.1 GENERAL OP-AMP INPUT-OUTPUT RELATION

We can use the impedance concept to simplify the process of obtaining a model of an op-amp circuit. A circuit diagram of the op amp with general feedback and input elements $Z_f(s)$ and $Z_i(s)$ is shown in Figure 6.4.2a. A similar but simplified form is given in part (b). The impedance $Z_i(s)$ of the input elements is defined such that

$$V_i(s) - V_1(s) = Z_i(s)I_1(s)$$

For the feedback elements,

$$V_1(s) - V_o(s) = Z_f(s)I_2(s)$$

The high internal impedance of the op amp implies that $i_1 \approx 0$, and thus $i_1 \approx i_2$. The final relation we need is the amplifier relation $v_o = -Gv_1$ or

$$V_o(s) = -GV_1(s) \tag{6.4.1}$$

When the preceding relations are used to eliminate $I_1(s)$ and $I_2(s)$, the result is

$$V_1(s) = \frac{Z_f(s)}{Z_f(s) + Z_i(s)}V_i(s) + \frac{Z_i(s)}{Z_f(s) + Z_i(s)}V_o(s)$$

Using (6.4.1) to eliminate $V_1(s)$, the transfer function between $V_i(s)$ and $V_o(s)$ is found to be

$$\frac{V_o(s)}{V_i(s)} = -\frac{Z_f(s)}{Z_f(s) + Z_i(s)}\frac{G}{1 + GH(s)}$$

where

$$H(s) = \frac{Z_i(s)}{Z_f(s) + Z_i(s)}$$

Because G is a very large number, $|GH(s)| \gg 1$, and we obtain

$$\frac{V_o(s)}{V_i(s)} \approx -\frac{Z_f(s)}{Z_i(s)} \tag{6.4.2}$$

This is the transfer-function model for op-amp circuits.

An op-amp multiplier is created by using two resistors as shown in Figure 6.4.1a, where $Z_i(s) = R_i$ and $Z_f(s) = R_f$. Thus,

$$\frac{V_o(s)}{V_i(s)} \approx -\frac{R_f}{R_i}$$

The gain of this multiplier is R_f/R_i, with a sign reversal. In some applications, we want to eliminate the sign reversal. We can do this by using an *inverter*, which is a multiplier

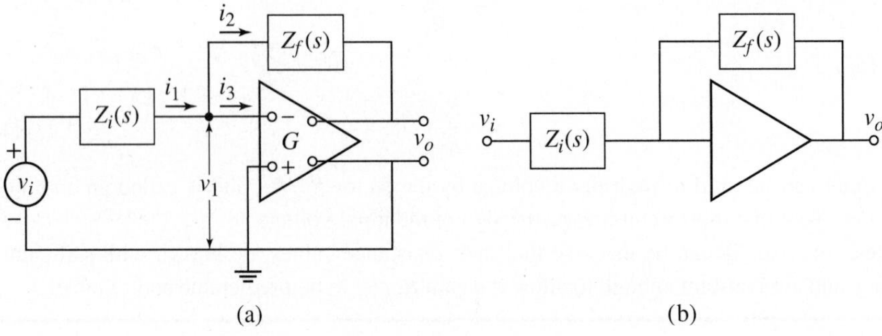

Figure 6.4.2 General circuit representation of an op-amp system.

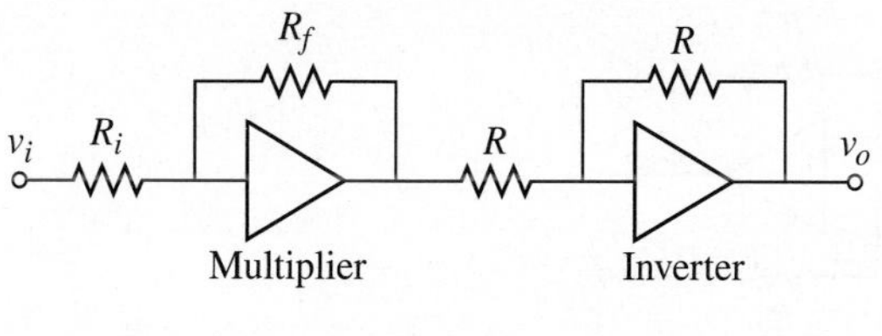

Figure 6.4.3 An op-amp multiplier with an inverter.

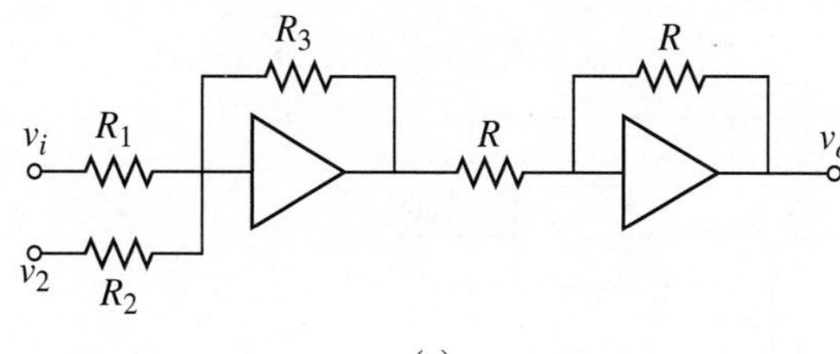

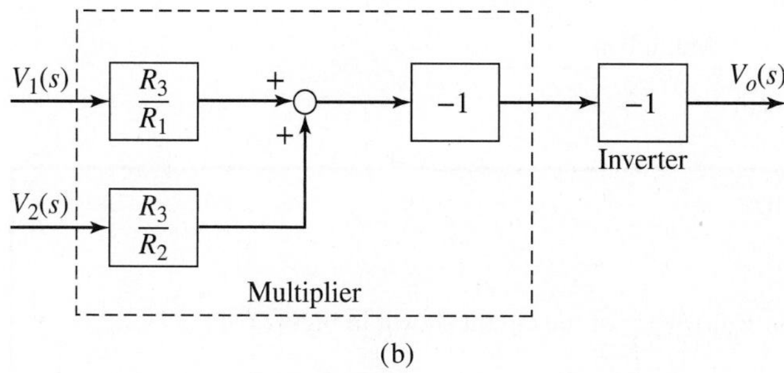

Figure 6.4.4 An op-amp adder. (a) Circuit. (b) Block diagram.

having equal resistances. Using an inverter in series with the multiplier, as shown in Figure 6.4.3, eliminates the overall sign reversal.

6.4.2 OP-AMP COMPARATOR

The multiplier circuit can be modified to act as an *adder* (Figure 6.4.4) or as a *subtractor* if another inverter is included (Figure 6.4.5). The output relation for the adder is

$$v_o = \frac{R_3}{R_1}v_1 + \frac{R_3}{R_2}v_2$$

The output relation for the subtractor is

$$v_o = \frac{R_3}{R_1}v_1 - \frac{R_3}{R_2}v_2$$

The block diagrams are shown in parts (b) of the figures.

We can make a *comparator* by selecting $R_1 = R_2 = R_3$ in Figure 6.4.5, in which case

$$v_o = v_1 - v_2$$

In Chapter 10 we will see how this is used in a control system.

6.4.3 INTEGRATION AND DIFFERENTIATION WITH OP AMPS

Control systems and many types of instrumentation utilize op amps to integrate and differentiate electrical signals.

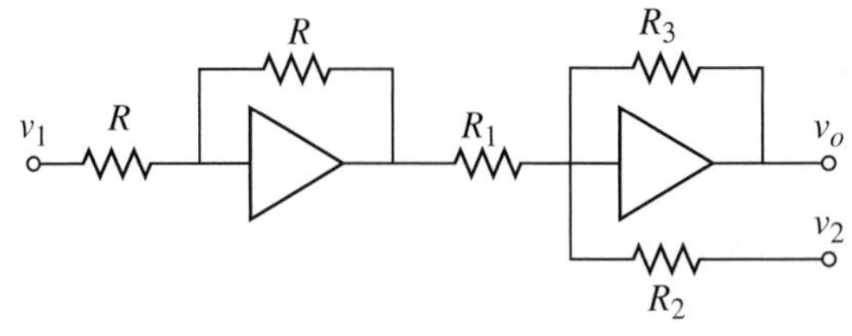

(a)

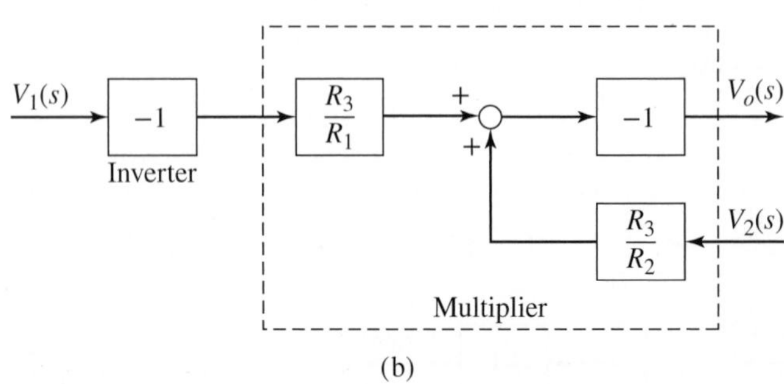

(b)

Figure 6.4.5 An op-amp subtractor. (a) Circuit. (b) Block diagram.

EXAMPLE 6.4.2

Integration with Op Amps

■ Problem

Determine the transfer function $V_o(s)/V_i(s)$ of the circuit shown in Figure 6.4.6.

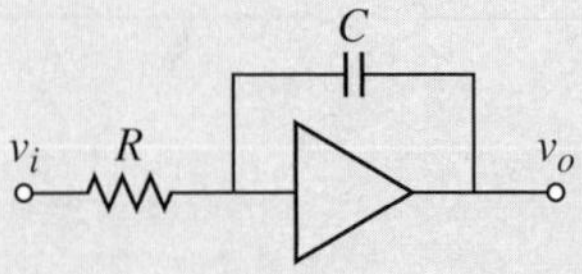

Figure 6.4.6 An op-amp integrator.

■ Solution

The impedance of a capacitor is $1/Cs$. Thus, the transfer function of the circuit is found from (6.4.2) with $Z_i(s) = R$ and $Z_f(s) = 1/Cs$. It is

$$\frac{V_o(s)}{V_i(s)} = -\frac{Z_f(s)}{Z_i(s)} = -\frac{1}{RCs}$$

Thus, in the time domain, the circuit model is

$$v_o = -\frac{1}{RC}\int_0^t v_i\,dt \tag{1}$$

Thus, the circuit integrates the input voltage, reverses its sign, and divides it by RC. It is called an op-amp *integrator*, and is used in many devices for computing, signal generation, and control, some of which will be analyzed in later chapters.

EXAMPLE 6.4.3

Differentiation with Op Amps

■ Problem

Design an op-amp circuit that differentiates the input voltage.

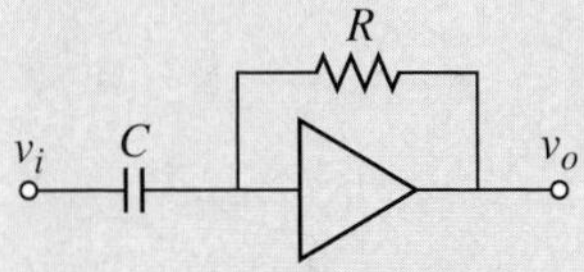

Figure 6.4.7 An op-amp differentiator.

■ Solution

In theory, a differentiator can be created by interchanging the resistance and capacitance in the integrator circuit. The result is shown in Figure 6.4.7, where $Z_i(s) = 1/Cs$ and $Z_f(s) = R$. The input-output relation for this ideal differentiator is

$$\frac{V_o(s)}{V_i(s)} = -\frac{Z_f(s)}{Z_i(s)} = -RCs$$

The model in the time domain is

$$v_o(t) = -RC\frac{dv_i(t)}{dt}$$

The difficulty with this design is that no electrical signal is "pure." Contamination always exists as a result of voltage spikes, ripple, and other transients generally categorized as "noise." These high-frequency signals have large slopes compared with the more slowly varying primary signal, and thus they will dominate the output of the differentiator. Example 6.4.4 shows an improved differentiator design that does not have this limitation.

Design of an Improved Differentiator

EXAMPLE 6.4.4

■ Problem

The differentiator analyzed in Example 6.4.3 is susceptible to high-frequency noise. In practice, this problem is often solved by using a redesigned differentiator, such as the one shown in Figure 6.4.8. We will analyze its performance in Chapter 9 when we study the response of systems to sinusoidal inputs. Derive its transfer function.

■ Solution

Using the op-amp equation (6.4.2), we have

$$\frac{V_o(s)}{V_i(s)} = -\frac{Z_f(s)}{Z_i(s)}$$

where $Z_i(s) = R_1 + 1/Cs$ and $Z_f(s) = R$. The circuit's transfer function is

$$\frac{V_o(s)}{V_i(s)} = -\frac{Z_f(s)}{Z_i(s)} = -\frac{RCs}{R_1Cs + 1} \tag{1}$$

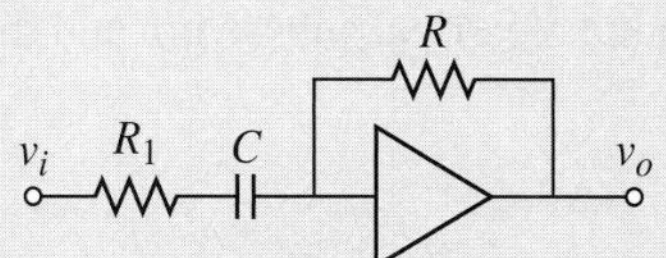

Figure 6.4.8 A modified op-amp differentiator.

6.5 ELECTRIC MOTORS

Electromechanical systems consist of an electrical subsystem and a mechanical subsystem with mass and possibly elasticity and damping. In some devices, such as motors and speakers, the mass is driven by a force generated by the electrical subsystem. In other devices, such as microphones, the motion of the mass generates a voltage or current in the electrical subsystem. Thus, we must apply electrical principles and Newton's laws to develop a model of an electromechanical system. Often the forces and torques are generated electromagnetically, but other methods are used as well; for example, *piezoelectric* devices contain crystals that generate forces when a voltage is applied to them.

In this section and Section 6.6 we treat electric motors. In Section 6.8 we treat other electromechanical devices.

6.5.1 MAGNETIC COUPLING

The majority of electromechanical devices utilize a magnetic field. Two basic principles apply to a conductor, such as a wire, carrying a current within a magnetic field: (1) a force is exerted on the conductor by the field and (2) if the conductor moves relative to the field, the field induces a voltage in the conductor that opposes the voltage producing the current. In many applications, the following model relates the force f to the current i:

$$f = BLi \tag{6.5.1}$$

where B is the flux density of the field and L is the length of the conductor. In SI the units of B are webers per square meter (Wb/m^2). The direction of the force, which is perpendicular to the conductor and the field, can be found with the right-hand rule. Sweep the fingers from the positive current direction to the positive field direction; the thumb will point in the positive force direction. Equation (6.5.1) is a special case of the more general physical principle. It applies to two commonly found situations: (1) straight conductors that are perpendicular to a uniform magnetic field and (2) circular conductors in a radial field.

When the directions of the field, the conductor, and its velocity are mutually perpendicular, the second principle can be expressed as

$$v_b = BLv \tag{6.5.2}$$

where v_b is the voltage induced in the conductor by its velocity v in the field. Again using the right-hand rule, we find the positive direction of the current induced by v_b by sweeping the fingers from the positive direction of v to the positive direction of the field.

The two principles and the expressions (6.5.1) and (6.5.2) can be represented graphically as in Figure 6.5.1. The circuit represents the electrical behavior of the conductor and the mass m represents the mass of the conductor and any attached mass. The power generated by the circuit is $v_b i$. The power applied to the mass m by the force f is fv. Neglecting any energy loss due to resistance in the conductor or friction or damping acting on the mass, we see that no power will be lost between the electrical subsystem and the mechanical subsystem, and thus,

Figure 6.5.1 Electromagnetic interaction between electrical and mechanical subsystems.

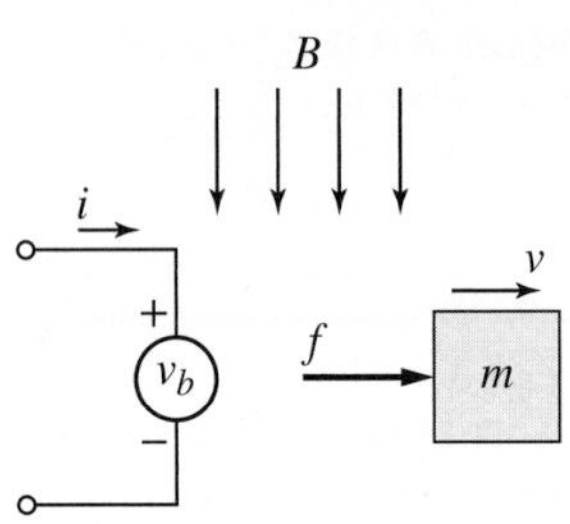

$$v_b i = fv = BLiv$$

from which we obtain (6.5.2). In addition, from Newton's law,

$$m\dot{v} = f = BLi$$

The basic principles underlying the operation of electric motors can be best understood by first considering a simpler device called the *D'Arsonval meter*, named after its inventor. The device is also known as a *galvanometer*.

6.5.2 THE D'ARSONVAL METER

A D'Arsonval meter can be used to measure current (Figure 6.5.2a). The current to be measured is passed through a coil to which a pointer is attached. The coil is positioned within a magnetic field and is wrapped around an iron core to strengthen the effects of the field. The core thus acts like an inductor. The interaction between the current and the field produces a torque that tends to rotate the coil and pointer. This rotation is opposed by a torsional spring of stiffness k_T.

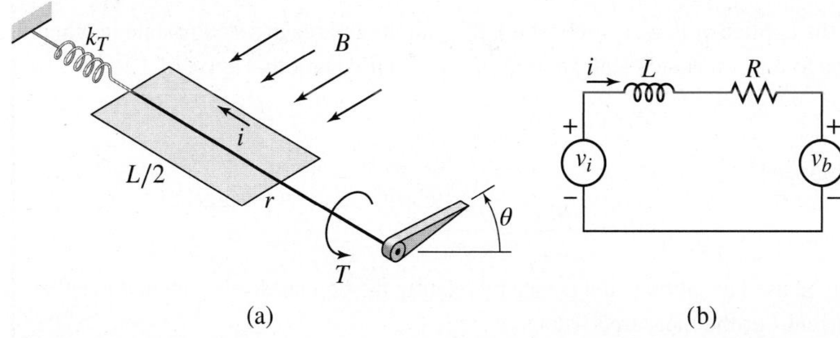

Figure 6.5.2 (a) D'Arsonval meter. (b) Circuit model.

EXAMPLE 6.5.1

A Model of the D'Arsonval Meter

■ Problem

Derive a model of a D'Arsonval meter in terms of the coil angular displacement θ and the coil current i. The input is the applied voltage v_i. Discuss the case where there are n coils around the core.

■ Solution

Let the length of one side of the coil be $L/2$ and its radius be r. Then the torque T acting on both sides of the coil due to the magnetic field B is

$$T = fr = \left(2B\frac{L}{2}i\right)r = (BLr)i$$

If a torsional viscous damping torque $c\dot{\theta}$, for example, due to air resistance or damping in the bearings, also acts on the core shaft as it rotates, the equation of motion of the core/coil unit is

$$I\frac{d^2\theta}{dt^2} + c\frac{d\theta}{dt} + k_T\theta = T = (BLr)i \tag{1}$$

where I is the inertia of the core/coil unit.

The rotation of the coil induces a voltage v_b in the coil that is proportional to the coil's linear velocity v such that $v_b = BLv$. The linear velocity is related to the coil's angular velocity $\dot{\theta}$ by $v = r\dot{\theta}$. Thus,

$$v_b = BLv = BLr\frac{d\theta}{dt}$$

The coil circuit is represented in part (b) of Figure 6.5.2, where R represents the resistance of the wire in the coil. Kirchhoff's voltage law applied to the coil circuit gives

$$v_i - L\frac{di}{dt} - Ri - v_b = 0$$

or

$$L\frac{di}{dt} + Ri + BLr\frac{d\theta}{dt} = v_i \tag{2}$$

The model consists of equations (1) and (2). Note that the system model is third order.

If there are n coils, the resulting torque expression is $T = n(BLr)i$ and the induced voltage expression is $v_b = nBLr\dot{\theta}$. Thus, equations (1) and (2) become

$$I\frac{d^2\theta}{dt^2} + c\frac{d\theta}{dt} + k_T\theta = n(BLr)i \tag{3}$$

$$L\frac{di}{dt} + Ri + nBLr\frac{d\theta}{dt} = v_i \tag{4}$$

Note that if the applied voltage v_i is constant, the system will reach a steady state in which the pointer comes to rest. At steady state, $\dot{\theta} = di/dt = 0$, and equation (4) gives

$$i = \frac{v_i}{R}$$

and equation (3) gives

$$\theta = \frac{nBLri}{k_T} = \frac{nBLrv_i}{Rk_T}$$

This equation can be used to calibrate the device by relating the pointer displacement θ to either the measured current i or the measured voltage v_i.

6.5.3 DC MOTORS

There are many types of electric motors, but the two main categories are *direct current (dc)* motors and *alternating current (ac)* motors. Within the dc motor category there are the armature-controlled motor and the field-controlled motor.

The basic elements of a motor, like that shown in Figure 6.5.3, are the *stator*, the *rotor*, the *armature*, and the *commutator*. The stator is stationary and provides the magnetic field. The rotor is an iron core that is supported by bearings and is free to rotate. The coils are attached to the rotor, and the combined unit is called the armature. A dc motor operates on the same principles as a D'Arsonval meter, but the design of a practical dc motor requires the solution of the problems caused by the fact that the coils must be free to rotate continually. As a coil rotates through 180°, the torque will reverse direction unless the current can be made to reverse direction also. In addition, a means must be found to maintain electrical contact between the rotating coil and the power supply leads. A solution is provided by the commutator, which is a pair of electrically conducting, spring-loaded carbon sticks (called *brushes*) that slide on the armature and transfer power to the coil contacts.

The stator may be a permanent magnet or an electromagnet with its own separate power supply, which creates additional cost. It is now possible to manufacture permanent magnets of high field intensity and armatures of low inertia so that permanent-magnet motors with a high torque-to-inertia ratio are now available.

6.5.4 MODEL OF AN ARMATURE-CONTROLLED DC MOTOR

We now develop a model for the armature-controlled motor shown in Figure 6.5.4. The armature voltage v_a is the input, and the armature current i_a and motor speed ω are the outputs.

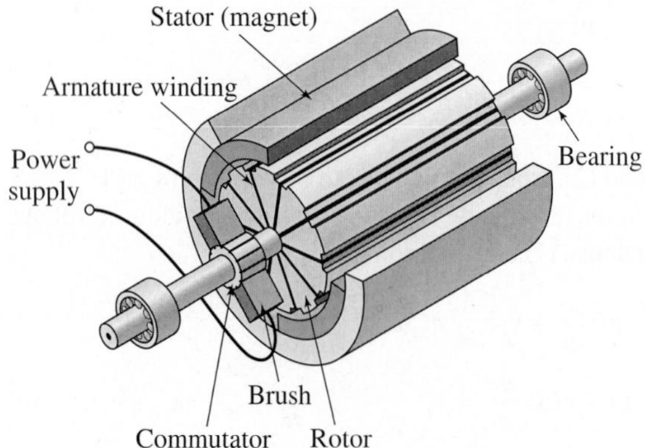

Figure 6.5.3 Cutaway view of a permanent magnet motor.

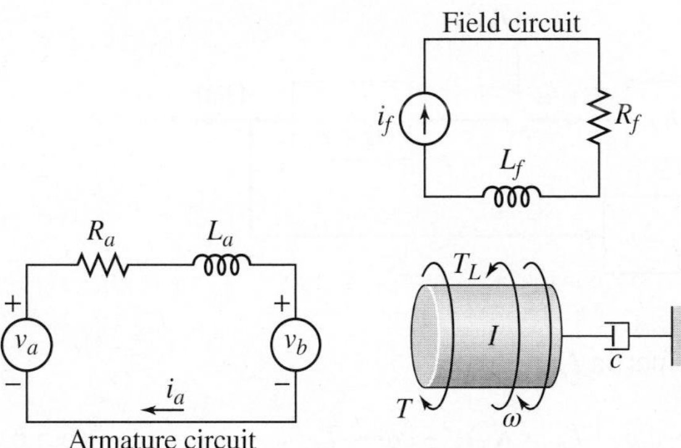

Figure 6.5.4 Diagram of an armature-controlled dc motor.

The electrical subsystems of the motor can be represented by the armature circuit and the field circuit in Figure 6.5.4. In a permanent-magnet motor, the field circuit is replaced by the magnet. The mechanical subsystem consists of the inertia I and the damping c. The inertia is due to the load inertia as well as the armature inertia. Damping can be present because of shaft bearings or load damping, such as with a fan or pump. The external torque T_L represents an additional torque acting on the load, other than the damping torque. The load torque T_L opposes the motor torque in most applications, so we have shown it acting in the direction opposite that of T. However, sometimes the load torque assists the motor. For example, if the load is the wheel of a vehicle, then T_L could be the torque produced by gravity as the vehicle ascends or descends a hill. When descending, the load torque assists the motor, and in such a case we would reverse the direction of T_L shown in Figure 6.5.4.

The motor produces a torque T that is proportional to the armature current i_a. This relation can be derived by noting that the force on the armature due to the magnetic field is, from (6.5.1), $f = nBLi_a$, where n is the number of armature coils. If the armature radius is r, then the torque on the armature is

$$T = (nBLi_a)r = (nBLr)i_a = K_T i_a \tag{6.5.3}$$

where $K_T = nBLr$ is the motor's *torque constant*. This relation can be used by motor designers to determine the effect of changing the number of coils, the field strength, or the armature geometry. The user of such motors (as opposed to the motor's designer) can obtain values of K_T for a specific motor from the manufacturer's literature.

As we have seen, the motion of a current-carrying conductor in a field produces a voltage in the conductor that opposes the current. This voltage in the armature is called the *back emf* (for *electromotive force*, an older term for voltage). Its magnitude is proportional to the speed. The coils' linear velocity v is related to their angular velocity by $v = r\omega$. Thus, from (6.5.2),

$$v_b = nBLv = (nBLr)\omega = K_b\omega \tag{6.5.4}$$

where $K_b = nBLr$ is the motor's *back emf constant*, and is sometimes called the *voltage constant*. Note that the expressions for K_T and K_b are identical and thus, K_T and K_b have the same numerical value if expressed in the same units. For this reason, motor manufacturers usually do not give values for K_b.

The back emf is a voltage drop in the armature circuit. Thus, Kirchhoff's voltage law gives

$$v_a - R_a i_a - L_a\frac{di_a}{dt} - K_b\omega = 0 \tag{6.5.5}$$

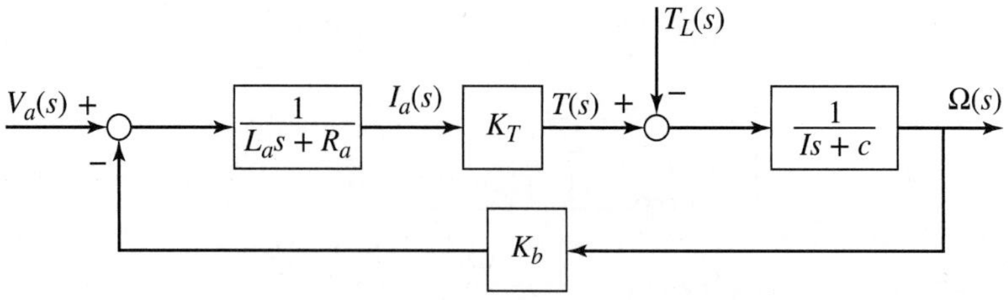

Figure 6.5.5 Block diagram of an armature-controlled dc motor.

From Newton's law applied to the inertia I,

$$I\frac{d\omega}{dt} = T - c\omega - T_L = K_T i_a - c\omega - T_L \tag{6.5.6}$$

Equations (6.5.5) and (6.5.6) constitute the system model.

Motor Block Diagram Figure 6.5.5 is a block diagram of an armature-controlled motor, with motor speed as the output. The equations used to obtain the diagram can be obtained by transforming the motor equations using zero-initial conditions. Solving (6.5.5) for $I_a(s)$,

$$I_a(s) = \frac{1}{L_a s + R_a}\left[V_a(s) - K_b\Omega(s)\right] \tag{6.5.7}$$

This equation is the basis for the left-half of the diagram. Now solve (6.5.6) for $\Omega(s)$:

$$\Omega(s) = \frac{1}{Is + c}\left[K_T I_a(s) - T_L(s)\right] \tag{6.5.8}$$

This equation is the basis for the right-half of the diagram.

The diagram shows how the back emf acts as a negative feedback loop to slow down the motor's speed. It also shows that the dynamics of the motor are affected by the dynamics of the armature circuit, whose time constant is L_a/R_a, and by the dynamics of the mechanical subsystem, whose time constant is I/c.

Motor Transfer Functions Normally we are interested in both the motor speed ω and the current i_a. The two inputs are the applied voltage v_a and the load torque T_L. Thus, there are four transfer functions for the motor, one transfer function for each input-output pair. We can obtain these transfer functions either by reducing the block diagram shown in Figure 6.5.5, or by solving (6.5.7) and (6.5.8) for $I_a(s)$ and $\Omega(s)$. The result for the output $I_a(s)$ is

$$\frac{I_a(s)}{V_a(s)} = \frac{Is + c}{L_a I s^2 + \left(R_a I + cL_a\right)s + cR_a + K_b K_T} \tag{6.5.9}$$

$$\frac{I_a(s)}{T_L(s)} = \frac{K_b}{L_a I s^2 + \left(R_a I + cL_a\right)s + cR_a + K_b K_T} \tag{6.5.10}$$

For the output $\Omega(s)$,

$$\frac{\Omega(s)}{V_a(s)} = \frac{K_T}{L_a I s^2 + \left(R_a I + cL_a\right)s + cR_a + K_b K_T} \tag{6.5.11}$$

$$\frac{\Omega(s)}{T_L(s)} = -\frac{L_a s + R_a}{L_a I s^2 + \left(R_a I + cL_a\right)s + cR_a + K_b K_T} \tag{6.5.12}$$

The denominator is the same in each of the motor's four transfer functions. It is the characteristic polynomial and it gives the characteristic equation:

$$L_a I s^2 + (R_a I + cL_a)\, s + cR_a + K_b K_T = 0 \tag{6.5.13}$$

Note that $I_a(s)/V_a(s)$ and $\Omega(s)/T_L(s)$ have numerator dynamics. This can cause a large overshoot in i_a if v_a is a step function, and a large overshoot in ω if T_L is a step function.

State-Variable Form of the Motor Model Equations (6.5.5) and (6.5.6) can be put into state-variable form by isolating the derivatives of the state variables i_a and ω. The state equations thus obtained are the following.

$$\frac{di_a}{dt} = \frac{1}{L_a}\left(v_a - R_a i_a - K_b\omega\right) \tag{6.5.14}$$

$$\frac{d\omega}{dt} = \frac{1}{I}\left(K_T i_a - c\omega - T_L\right) \tag{6.5.15}$$

Note that these state variables describe the energies $Li_a^2/2$ and $I\omega^2/2$ stored in the system.

Letting $x_1 = i_a$ and $x_2 = \omega$, the state equations become

$$\frac{dx_1}{dt} = \frac{1}{L_a}\left(v_a - R_a x_1 - K_b x_2\right) = \begin{bmatrix} -\frac{R_a}{L_a} & -\frac{K_b}{L_a} \end{bmatrix} \begin{bmatrix} x_1 \\ x_2 \end{bmatrix} + \begin{bmatrix} \frac{1}{L_a} & 0 \end{bmatrix} \begin{bmatrix} v_a \\ T_L \end{bmatrix}$$

$$\frac{dx_2}{dt} = \frac{1}{I}\left(K_T x_1 - c x_2 - T_L\right) = \begin{bmatrix} \frac{K_T}{I} & -\frac{c}{I} \end{bmatrix} \begin{bmatrix} x_1 \\ x_2 \end{bmatrix} + \begin{bmatrix} 0 & -\frac{1}{I} \end{bmatrix} \begin{bmatrix} v_a \\ T_L \end{bmatrix}$$

To handle models having multiple inputs, the general vector-matrix form of the state equations allows for a column vector of inputs, usually denoted **u**. A column vector—called the input *vector* **u**—is then formed from the two inputs v_a and T_L, in any order. Define **u** as

$$\mathbf{u} = \begin{bmatrix} v_a \\ T_L \end{bmatrix}$$

Then the vector-matrix form is

$$\dot{\mathbf{x}} = \mathbf{A}\mathbf{x} + \mathbf{B}\mathbf{u} \tag{6.5.16}$$

where **x** is defined as before, and **A** and **B** are now defined as

$$\mathbf{A} = \begin{bmatrix} -\frac{R_a}{L_a} & -\frac{K_b}{L_a} \\ \frac{K_T}{I} & -\frac{c}{I} \end{bmatrix}$$

$$\mathbf{B} = \begin{bmatrix} \frac{1}{L_a} & 0 \\ 0 & -\frac{1}{I} \end{bmatrix}$$

The equation for dx_1/dt contains the first rows of the matrices **A** and **B**, and the equation for dx_2/dt contains the second rows.

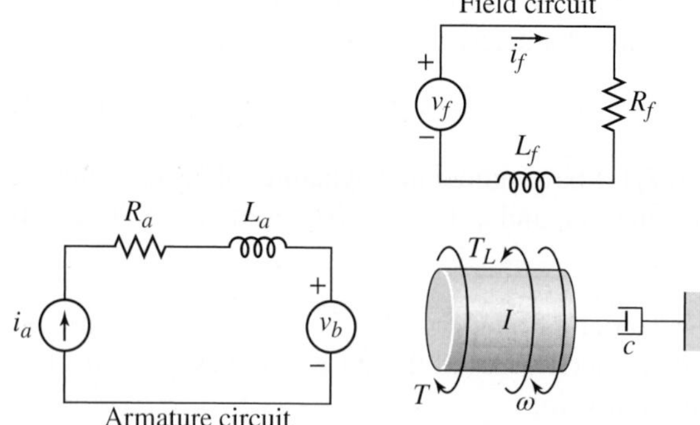

Figure 6.5.6 Diagram of a field-controlled dc motor.

6.5.5 FIELD-CONTROLLED MOTORS

Another way to control a dc motor is to keep the armature current constant while adjusting the voltage applied to the field windings to vary the intensity of the magnetic field surrounding the armature (see Figure 6.5.6). Thus, unlike permanent-magnet motors, field-controlled motors require two power supplies, one for the armature circuit and one for the field circuit. They also require a control circuit to maintain a constant armature current in the presence of the back emf, which varies with motor speed and field strength.

In general, the field strength B is a nonlinear function of the field current i_f and can be expressed as $B(i_f)$. Thus, if the armature radius is r, the torque on the armature is

$$T = nB(i_f)Li_a r = (nLri_a)B(i_f) = T(i_f)$$

and we see that the motor torque is also a nonlinear function of i_f. Often the linear approximation $T - T_r = K_T(i_f - i_{fr})$ is used, where T_r and i_{fr} are the torque and current values at a reference operating equilibrium, and the torque constant K_T is the slope of the $T(i_f)$ curve at the reference condition. In the rest of our development, we will assume that $T_r = i_{fr} = 0$ to simplify the discussion. Thus, we will use the relation $T = K_T i_f$.

EXAMPLE 6.5.2

Model of a Field-Controlled dc Motor

■ Problem

Develop a model of the field-controlled motor shown in Figure 6.5.6.

■ Solution

The voltage v_f is applied to the field circuit, whose inductance and resistance are L_f and R_f. No back emf exists in the field circuit, and Kirchhoff's voltage law applied to the field circuit gives

$$v_f = R_f i_f + L_f \frac{di_f}{dt} \tag{1}$$

For the inertia I,

$$I\frac{d\omega}{dt} = T - c\omega - T_L = K_T i_f - c\omega - T_L \tag{2}$$

where T_L is the load torque. These two equations form the motor model.

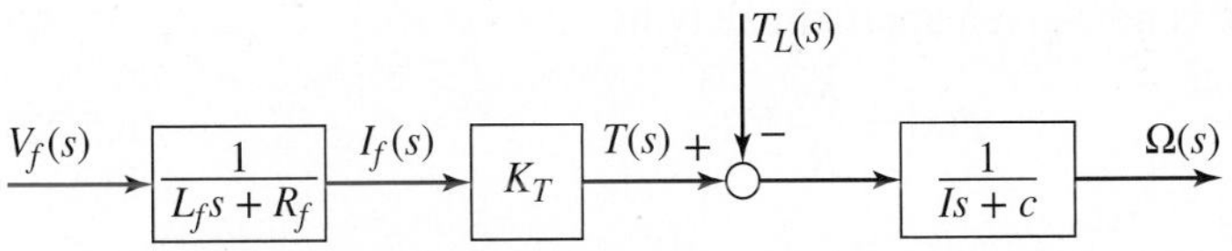

Figure 6.5.7 Block diagram of a field-controlled dc motor.

The block diagram of a field-controlled motor is shown in Figure 6.5.7. To see how it was obtained, take the transform of equation (1) of Example 6.5.2 with zero-initial conditions, and solve for $I_f(s)$ to obtain

$$I_f(s) = \frac{1}{L_f s + R_f} V_f(s) \tag{6.5.17}$$

This equation was used to obtain the left side of the block diagram. Take the transform of (2) of Example 6.5.2 with zero-initial conditions, and solve for $\Omega(s)$ to obtain

$$\Omega(s) = \frac{1}{Is + c}[K_T I_f(s) - T_L(s)]$$

This equation was used to obtain the right side of the block diagram. Note that this motor has no feedback loop because it does not have back emf.

Dynamic Response of a Field-Controlled Motor From (6.5.17) we see that the transfer function for the field current is

$$\frac{I_f(s)}{V_f(s)} = \frac{1}{L_f s + R_f} \tag{6.5.18}$$

The characteristic root is $s = -R_f/L_f$ and the time constant is L_f/R_f. Thus, if $v_f(t)$ is a step function, the field current will take approximately $4L_f/R_f$ to reach the constant value of $i_f = v_f/R_f$. Note that, unlike the armature-controlled motor, the current in the field-controlled motor is not affected by the load torque T_L, because there is no feedback loop due to back emf.

From the block diagram, we can easily obtain the transfer functions for the speed. Setting $T_L(s) = 0$ temporarily, and using the series law for diagram reduction, we find that

$$\frac{\Omega(s)}{V_f(s)} = \frac{K_T/R_f c}{[(L_f/R_f)s + 1][(I/c)s + 1]} \tag{6.5.19}$$

The characteristic roots are $s = -R_f/L_f$ and $s = -c/I$, which are real. So the speed ω will not oscillate if the applied voltage v_f is a step function. Its response time is governed by the larger of the two time constants, I/c and L_f/R_f, which are due to the mechanical and electrical subsystems, respectively. In most cases, the largest time constant is I/c. The real roots are due to the fact that the current equation (1) of Example 6.5.2 does not contain the speed ω, and thus is not coupled to the speed equation (2) of that example.

Setting $V_f(s) = 0$ in the block diagram, we obtain

$$\frac{\Omega(s)}{T_L(s)} = -\frac{1}{Is + c} \tag{6.5.20}$$

The time constant is I/c. Note that the minus sign on the right side indicates that the speed will decrease for a positive torque. This is because of the chosen direction of positive load torque in Figure 6.5.7.

If the time constant of the electrical subsystem is small compared to that of the mechanical subsystem, the speed response can be approximately described by the first-order model

$$\frac{\Omega(s)}{V_f(s)} = \frac{K_T/R_f c}{(I/c)s + 1} = \frac{K_T/R_f}{Is + c}, \qquad \frac{L_f}{R_f} \ll \frac{I}{c} \tag{6.5.21}$$

where the motor torque is now given approximately by

$$T(s) = \frac{K_T}{R_f} V_f(s) \tag{6.5.22}$$

6.6 ANALYSIS OF MOTOR PERFORMANCE

We now use the transfer-function model of an armature-controlled dc motor to investigate the performance of such motors. The transfer functions given by (6.5.9) through (6.5.12) are repeated here.

$$\frac{I_a(s)}{V_a(s)} = \frac{Is + c}{L_a I s^2 + \left(R_a I + cL_a\right)s + cR_a + K_b K_T} \tag{6.6.1}$$

$$\frac{I_a(s)}{T_L(s)} = \frac{K_b}{L_a I s^2 + \left(R_a I + cL_a\right)s + cR_a + K_b K_T} \tag{6.6.2}$$

$$\frac{\Omega(s)}{V_a(s)} = \frac{K_T}{L_a I s^2 + \left(R_a I + cL_a\right)s + cR_a + K_b K_T} \tag{6.6.3}$$

$$\frac{\Omega(s)}{T_L(s)} = -\frac{L_a s + R_a}{L_a I s^2 + \left(R_a I + cL_a\right)s + cR_a + K_b K_T} \tag{6.6.4}$$

STEADY-STATE MOTOR RESPONSE

The steady-state operating conditions can be obtained by applying the final value theorem to the transfer functions. If v_a and T_L are step functions of magnitude V_a and T_L, respectively, then the steady-state current and speed are

$$i_a = \frac{cV_a + K_b T_L}{cR_a + K_b K_T} \tag{6.6.5}$$

$$\omega = \frac{K_T V_a - R_a T_L}{cR_a + K_b K_T} \tag{6.6.6}$$

Thus, an increased load torque leads to an increased current and a decreased speed, as would be expected. From equation (6.6.6) the steady-state speed is often plotted versus T_L for different values of the applied voltage V_a. This plot is known as the load-speed curve of the motor. For a given value of V_a, it gives the maximum load torque the motor can handle at a specified speed.

The *no-load speed* is the motor speed when the load torque is zero. Setting $T_L = 0$ in (6.6.6) gives $\omega = K_T V_a/(cR_a + K_b K_T)$. This is the highest motor speed for a given applied voltage. The corresponding no-load current required can be found by setting $T_L = 0$ in (6.6.5). It is $i_a = cV_a/(cR_a + K_b K_T)$.

The *stall torque* is the value of the load torque that produces zero motor speed. Setting $\omega = 0$ in (6.6.6) gives the stall torque: $T_L = K_T V_a/R_a$. The corresponding *stall current* can be found by substituting this value into (6.6.5).

EXAMPLE 6.6.1 No-Load Speed and Stall Torque

■ **Problem**

The parameter values for a certain motor are

$$K_T = K_b = 0.05 \text{ N} \cdot \text{m/A}$$

$$c = 10^{-4} \text{ N} \cdot \text{m} \cdot \text{s/rad} \qquad R_a = 0.5\ \Omega$$

The manufacturer's data states that the motor's maximum speed is 3000 rpm, and the maximum armature current it can withstand without demagnetizing is 30 A.

Compute the no-load speed, the no-load current, and the stall torque. Determine whether the motor can be used with an applied voltage of $v_a = 10$ V.

■ **Solution**

For $v_a = 10$ V, (6.6.5) and (6.6.6) give

$$i_a = 0.392 + 19.61T_L \text{ A} \qquad \omega = 196.1 - 196.1T_L \text{ rad/s}$$

The no-load speed is found from the second equation with $T_L = 0$. It is 196.1 rad/s, or 1872 rpm, which is less than the maximum speed of 3000 rpm. The corresponding no-load current is $i_a = 0.392$ A, which is less than the maximum allowable current of 30 A. The no-load current is required to provide a motor torque $K_T i_a$ to cancel the damping torque $c\omega$.

The stall torque is found by setting $\omega = 0$. It is $T_L = 1$ N · m. The corresponding stall current is $i_a = 20$ A, which is less than the maximum allowable current.

6.6.1 MOTOR DYNAMIC RESPONSE

The steady-state relations are often used because they are algebraic relations and thus are easier to use than the motor differential equations. However, they can be misleading. Because $I_a(s)/V_a(s)$ and $\Omega(s)/T_L(s)$ have numerator dynamics, the actual maximum current required and the actual maximum speed attained might be quite different than their steady-state values. Example 6.6.2 illustrates this effect.

Response of an Armature-Controlled dc Motor

EXAMPLE 6.6.2

■ **Problem**

The parameter values for a certain motor are

$$K_T = K_b = 0.05 \text{ N} \cdot \text{m/A}$$

$$c = 10^{-4} \text{ N} \cdot \text{m} \cdot \text{s/rad} \qquad R_a = 0.5\ \Omega$$

$$L_a = 2 \times 10^{-3} \text{ H} \qquad I = 9 \times 10^{-5} \text{ kg} \cdot \text{m}^2$$

where I includes the inertia of the armature and that of the load. The load torque T_L is zero.

Obtain the step response of $i_a(t)$ and $\omega(t)$ if the applied voltage is $v_a = 10$ V.

■ **Solution**

Substituting the given parameter values into (6.6.1) and (6.6.3), gives

$$\frac{I_a(s)}{V_a(s)} = \frac{9 \times 10^{-5}s + 10^{-4}}{18 \times 10^{-8}s^2 + 4.52 \times 10^{-5}s + 2.55 \times 10^{-3}}$$

$$\frac{\Omega(s)}{V_a(s)} = \frac{0.05}{18 \times 10^{-8}s^2 + 4.52 \times 10^{-5}s + 2.55 \times 10^{-3}}$$

If v_a is a step function of magnitude 10 V,

$$I_a(s) = \frac{5 \times 10^3 s + 5.555 \times 10^4}{s(s + 165.52)(s + 85.59)} = \frac{C_1}{s} + \frac{C_2}{s + 165.52} + \frac{C_3}{s + 85.59}$$

$$\Omega(s) = \frac{2.777 \times 10^6}{s(s + 165.52)(s + 85.59)} = \frac{D_1}{s} + \frac{D_2}{s + 165.52} + \frac{D_3}{s + 85.59}$$

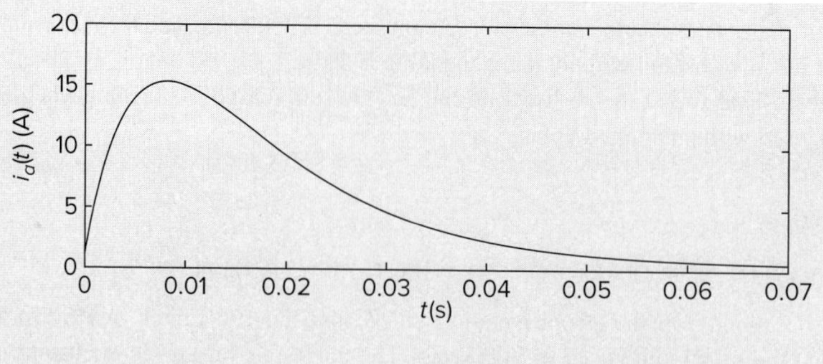

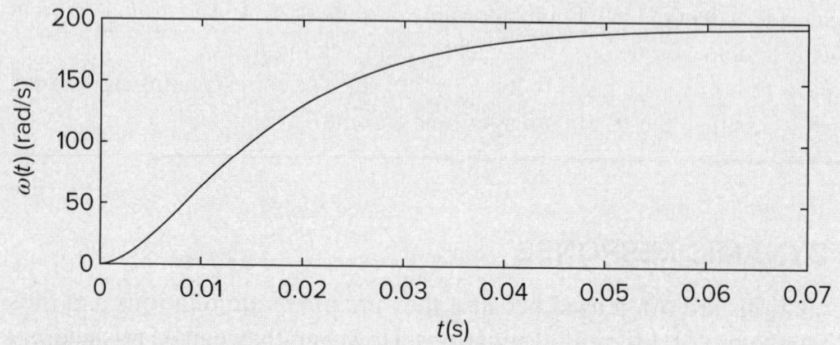

Figure 6.6.1 Step response of an armature-controlled dc motor.

Evaluating the partial-fraction coefficients by hand or with MATLAB, as described in Chapter 2, we obtain

$$i_a(t) = 0.39 - 61e^{-165.52t} + 61.74e^{-85.59t}$$

$$\omega(t) = 196.1 + 210^{-165.52t} - 406e^{-85.59t}$$

The plots are shown in Figure 6.6.1. Note the large overshoot in i_a, which is caused by the numerator dynamics. The plot shows that the steady-state calculation of $i_a = 0.39$ A greatly underestimates the maximum required current, which is approximately 15 A.

In practice, of course, a pure step input is impossible, and thus the required current will not be as high as 15 A. The real input would take some time to reach 10 V. The response to such an input is more easily investigated by computer simulation, so we will return to this topic in Section 6.7.

6.6.2 THE EFFECT OF ARMATURE INDUCTANCE

If we set $L_a = 0$, the second-order motor model reduces to a first-order model, which is easier to use. For this reason, even though L_a must be nonzero for physical reasons, you often see L_a treated as negligible.

Consider the motor of Example 6.6.2. Suppose the combined inertia of the armature and load is $I = 3 \times 10^{-5}$ kg · m^2. The roots of the characteristic equation are the complex pair $s = -126.7 \pm 162.3j$, which correspond to an *oscillatory* response that reaches steady state after approximately $4/126.7 = 0.032$ s. If we neglect the inductance and set $L_a = 0$ in (6.5.13), we obtain the first-order equation $1.5 \times 10^{-5}s + 0.00255 = 0$, which has the single root $s = -170$. Thus, the $L_a = 0$ approximation incorrectly predicts a *nonoscillatory* response that reaches steady state after approximately $4/170 = 0.024$ s, which differs by 25% from the correct value.

If instead the inertia is larger, say $I = 9 \times 10^{-5}$, (6.5.13) gives the real roots $s = -165.5$ and -85.6, which corresponds to a nonoscillatory response that reaches steady state after approximately $4/85.6 = 0.047$ s. Setting $L_a = 0$ in (6.5.13) gives the single root $s = -56.7$, which correctly predicts a nonoscillatory response but implies that steady state is reached after approximately $4/56.7 = 0.071$ s, which differs by 51% from the correct value.

We conclude from this example that you should be careful in using the approximation $L_a = 0$, although one sees it in common use. With $L_a \neq 0$ the characteristic equation and the motor differential equations are still only second order and thus are manageable. So this approximation really is not needed here. However, models of some types of control systems are third order or higher if the $L_a = 0$ approximation is not used, as we will see in Chapter 10. In such cases the mathematics becomes much more difficult, and so the approximation is used to reduce the order of the equations. In such cases, the correct, nonzero value of L_a is used in computer simulation studies to assess accuracy of the predictions obtained from the lower-order model.

6.6.3 DETERMINING MOTOR PARAMETERS

Motor parameter values can often be obtained from the manufacturer. If not, they must be either calculated or measured. Calculating K_T and K_b for an existing motor from the formula $nBLr$ is not always practical because the value of the magnetic field parameter B might be difficult to determine. An approximate value of the armature inertia I_a can be calculated from the formula for the inertia of a cylinder using the density of iron, assuming that the length and radius are available. The inertia can be measured by suspending it with a metal wire and measuring the torsional oscillation frequency f_n Hz as the armature twists on the wire. The inertia can be calculated from $I_a = k_T/(2\pi f_n)^2$, where k_T is the torsional spring constant of the wire.

Some parameters can be measured with static (steady-state) tests. By slowly increasing the load torque T_L until the motor stalls and measuring the resulting stall current, we can compute K_T from $K_T = T_L/i_a$. Knowing the voltage V_a, we can compute the armature resistance from $R_a = V_a/i_a$. By measuring the no-load speed ω and the resulting current i_a, and knowing V_a, R_a, and K_T, we can compute c from the steady-state relations (6.6.5) and (6.6.6) with $T_L = 0$.

Much of the viscous damping in the motor is due to air drag as the armature rotates. Drag force is a nonlinear function of speed, and so the linear relation $c\omega$ is an approximation. Therefore, the value of c might be different at lower speeds. However, most of the damping in a given application might be due to whatever load the motor is driving (examples include pumps and fans), and so the motor's damping might be small enough to be ignored. Then the damping constant c will need to be measured at the load or calculated from a model of the load.

Because c is difficult to determine precisely, its value is rarely reported by motor manufacturers. However, in motors with good bearings the damping can be slight and is often taken to be zero (perhaps this is why its value is rarely reported!).

The inductance L_a can be difficult to determine because it involves the rate di_a/dt and thus requires a dynamic test. Special instruments such as an *impedance meter* can be used to measure L_a. As we have discussed, the inductance L_a is often assumed to be very small and therefore is often taken to be zero. This is sometimes a good approximation, but not always.

Units for K_T and K_b The units of K_T and K_b can be a source of confusion. In manufacturer's data, the units of K_T are often different from those of K_b. We have seen that the formula for K_T ($K_T = nBLr$) is identical to that for K_b. Therefore, the units and numerical values will be identical if a consistent set of units is used in the formula. From the definition of K_T, we see the units in SI to be:

$$[K_T] = \frac{[T]}{[i_a]} = \frac{\text{N}\cdot\text{m}}{\text{A}}$$

Similarly, from the definition of K_b, we see the units in SI to be:

$$[K_b] = \frac{[v_b]}{[\omega]} = \frac{\text{V}}{1/\text{s}} = \text{V}\cdot\text{s}$$

These units are equivalent to one another because 1 V = 1 W/A. Thus, 1 V · s = 1 (W/A) · s = 1 N · m/A. So K_T and K_b have the same units and numerical values when SI is used.

However, while values of K_b are usually reported in SI as V · s, values of K_T are sometimes reported in non-SI units. For example, K_T is often reported in FPS units as lb-ft/A. If so, the numerical values of K_b and K_T will not be the same. The difference corresponds to using SI units in the formula for K_b and non-SI units in the formula for K_T.

If no value for K_b is reported and K_T is reported in FPS units as lb-ft/A, the value of K_b in SI units can be obtained from the relation $K_b = 1.3558K_T$. This relation is derived in one of the homework problems.

6.7 CASE STUDY: DESIGN OF A MOTION-CONTROL SYSTEM

In many motion-control applications we want to accelerate the system to a desired speed and run it at that speed for some time before decelerating to a stop, as illustrated by the *trapezoidal* speed profile shown in Figure 6.7.1. We studied this profile in Chapters 1 and 3 (Sections 1.5 and 3.6), and we derived formulas to compute the maximum torque and rms torque required to achieve the given profile. These are repeated in Table 6.7.1.

We also derived the principal of inertia matching to select a reducer ratio. Now we want to develop a means to properly select a motor and amplifier to drive the system. We will also investigate the energy consumption of such a system, including the effect of the reducer ratio and inertia.

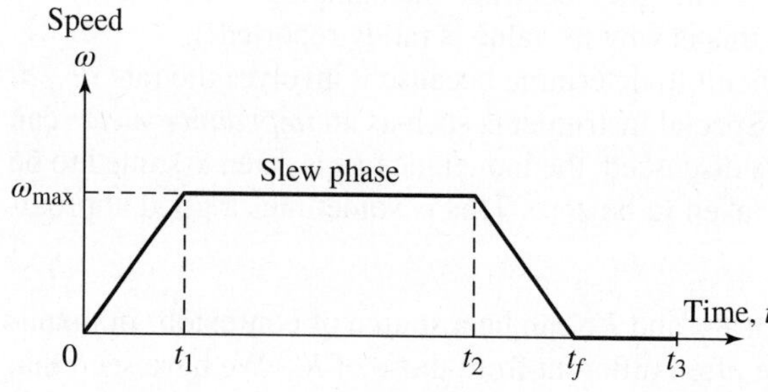

Figure 6.7.1 Trapezoidal speed profile.

6.7.1 MOTOR AND AMPLIFIER PERFORMANCE

We will treat motor control in Chapters 10, 11, and 12, but for now we develop methods to determine whether or not a specified motor and amplifier are capable of delivering the required performance. In evaluating the performance of a motion-control system, the following are important:

- Energy consumption per cycle, E. This is the sum of the energy loss in the motor resistance and in the damping. Thus, the energy loss per cycle is

$$E = \int_0^{t_f} Ri^2(t)\,dt + \int_0^{t_f} c\omega^2(t)\,dt \tag{6.7.1}$$

  where t_f is the duration of the cycle and ω is the speed of the system at the location of the damping.
- Maximum required current and motor torque, i_{max} and T_{max}. Motor current must be limited to prevent damage to the motor. In addition, the amplifier has an upper limit on the current it can supply. Because $T = i/K_T$, the motor torque is limited by the available current. The profile cannot require more current or torque than is available.
- Maximum required motor speed, ω_{max}. Motor manufacturers state a limit on motor speed to prevent damage to the motor.
- Maximum required voltage, v_{max}. Amplifiers have an upper limit on the voltage they can deliver.
- Average required current and motor torque, i_{rms} and T_{rms}. Most amplifiers and motors have a *rated continuous current* and a *rated continuous torque*. These specify the current and torque that can be supplied for a long period of time. If the profile requires an average current or average torque greater than the rated values, the amplifier or motor can be damaged, often by overheating. The rated values are *less* than the maximum values. The average usually stated is the *rms average*, which stands for *root mean square*. For torque, it is calculated as follows:

$$T_{\text{rms}} = \sqrt{\frac{1}{t_f}\int_0^{t_f} T^2(t)\,dt} \tag{6.7.2}$$

  with a similar expression used for i_{rms}. Thus, the rms average gives equal weight to positive and negative values. Note that $i_{\text{rms}} = T_{\text{rms}}/K_T$.
- Maximum speed error: This is the maximum difference between the desired speed given by the profile and the actual speed.
- Average speed error: The average speed error is commonly computed as the rms value.
- Displacement error: For applications requiring motion through a specified displacement (distance or angle), this error is the difference between the specified displacement and the actual displacement.
- System response time: The system must be able to respond fast enough to follow the profile. For a quick check, if the system's largest time constant is greater than one-fourth of the profile's ramp time t_1, then the system is not fast enough.

Not all these criteria are important for every application. For example, the speed errors probably are not relevant for applications requiring motion through a specified distance or angle. Energy consumption may not be important where a single cycle is to be performed, but for applications where the cycle is performed repeatedly, the energy consumption per cycle will probably be important.

In the following we assume that the damping constant c is zero. This is often a good assumption for well-designed electromechanical systems used for motion control. By modifying (6.5.5) and (6.5.6) to account for a speed reduction ratio of N, we obtain the following motor model. See Figure 6.5.4.

$$v = Ri + L\frac{di}{dt} + K_b\omega \tag{6.7.3}$$

$$I\frac{d\omega}{dt} = K_T i - T_d \tag{6.7.4}$$

We have omitted the subscripts on v, i, R, and L to simplify the notation. The speed ω is the *motor* speed. The *load* speed is $\omega_L = \omega/N$. The torque T_d opposing the motor torque T is due to the torque T_L acting on the load, where $T_d = T_L/N$. The inertia I includes the motor inertia I_m and the load inertia I_L reflected to the motor shaft. Thus, $I = I_m + I_L/N^2$.

Energy Loss With $c = 0$, the expression for the energy loss per cycle becomes

$$E = \int_0^{t_f} Ri^2(t)\,dt = R\int_0^{t_f}\left(\frac{I\dot{\omega} + T_d}{K_T}\right)^2 dt$$

where (6.7.4) has been used to substitute for i. Assuming that T_d is constant, and expanding this expression we obtain

$$E = \frac{RI^2}{K_T^2}\int_0^{t_f}\dot{\omega}^2\,dt + \frac{2RIT_d}{K_T^2}\int_0^{t_f}\dot{\omega}\,dt + \frac{RT_d^2}{K_T^2}\int_0^{t_f}dt$$

With the assumption that the motion begins and ends at rest so that $\omega(0) = \omega(t_f)$, the second integral is zero. Thus

$$E = \frac{RI^2}{K_T^2}\int_0^{t_f}\dot{\omega}^2\,dt + \frac{RT_d^2 t_f}{K_T^2} \tag{6.7.5}$$

Note that the first term on the right depends on the profile, while the second term depends on the disturbance torque T_d.

Maximum Motor Speed We now evaluate the integral in (6.7.5) for the trapezoidal profile, which is specified in terms of the *motor* speed $\omega(t)$. If the load speed $\omega_L(t)$ is specified instead, we can find $\omega(t)$ from $N\omega_L(t)$. The total angular displacement is the area under the trapezoidal profile shown in Figure 6.7.1. Assuming that $\theta(0) = 0$, and because $t_f - t_2 = t_1$, we have

$$\theta(t_f) = \theta_f = \int_0^{t_f}\omega\,dt = 2\left(\frac{1}{2}\omega_{\max}t_1\right) + \omega_{\max}(t_2 - t_1) = \omega_{\max}t_2$$

Thus

$$\omega_{\max} = \frac{\theta_f}{t_2} \tag{6.7.6}$$

So, if we specify the displacement θ_f and the time t_2, the maximum required speed is determined from this equation.

For $0 \le t \le t_1$, the acceleration $\dot{\omega}$ is $\omega_{max}/t_1 = \theta_f/t_1 t_2$, after using (6.7.6). For $t_1 < t < t_2$, $\dot{\omega} = 0$, and for $t_2 \le t \le t_f$, $\dot{\omega} = -\omega_{max}/t_1 = -\theta_f/t_1 t_2$. Therefore,

$$\int_0^{t_f} \dot{\omega}^2\, dt = \int_0^{t_1} \left(\frac{\theta_f}{t_1 t_2}\right)^2 dt + \int_{t_1}^{t_2} 0\, dt + \int_{t_2}^{t_f} \left(-\frac{\theta_f}{t_1 t_2}\right)^2 dt$$

$$= \left(\frac{\theta_f}{t_1 t_2}\right)^2 (t_1 + t_f - t_2) = \left(\frac{\theta_f}{t_1 t_2}\right)^2 2t_1$$

since $t_f - t_2 = t_1$. Therefore, expression (6.7.5) becomes

$$E = \frac{RI^2}{K_T^2}\left[\left(\frac{\theta_f}{t_1 t_2}\right)^2 2t_1\right] + \frac{RT_d^2 t_f}{K_T^2}$$

or

$$E = \frac{R}{K_T^2}\left(\frac{2I^2\theta_f^2}{t_1 t_2^2} + T_d^2 t_f\right) \tag{6.7.7}$$

Maximum Motor Torque The maximum required acceleration for the trapezoidal profile is

$$\alpha_{max} = \frac{\omega_{max}}{t_1} = \frac{\theta_f}{t_1 t_2} \tag{6.7.8}$$

Using (6.7.4) and the fact that the motor torque is $T = K_T i$, we have

$$T = K_T i = I\frac{d\omega}{dt} + T_d = I\alpha + T_d \tag{6.7.9}$$

Thus, the maximum required motor torque is

$$T_{max} = I\alpha_{max} + T_d = I\frac{\omega_{max}}{t_1} + T_d = I\frac{\theta_f}{t_1 t_2} + T_d \tag{6.7.10}$$

RMS Motor Torque The rms torque is calculated from (6.7.2), using (6.7.9): As shown in Section 1.5, the result is

$$T_{rms} = \sqrt{\frac{2I^2\theta_f^2}{t_f t_1 t_2^2} + T_d^2} \tag{6.7.11}$$

These equations are summarized in Table 6.7.1. They are used to compute the motor requirements for the trapezoidal speed profile. To determine whether a given motor will be satisfactory, the calculated values of ω_{max}, T_{max}, and T_{rms} are compared with the motor manufacturer's data on maximum speed, peak torque, and rated continuous torque.

Table 6.7.1 Motor/amplifier requirements for trapezoidal speed profile.

Profile times t_1, t_2, t_f	See Figure 6.7.1
Motor displacement	θ_f = area under speed profile
Load torque felt at motor	T_d
Motor requirements	
Energy consumption/cycle	$E = \dfrac{R}{K_T^2}\left(\dfrac{2I^2\theta_f^2}{t_1 t_2^2} + T_d^2 t_f\right)$
Maximum speed	$\omega_{\text{max}} = \dfrac{\theta_f}{t_2}$
Maximum torque	$T_{\text{max}} = I\dfrac{\theta_f}{t_1 t_2} + T_d$
rms torque	$T_{\text{rms}} = \sqrt{\dfrac{2I^2\theta_f^2}{t_f t_1 t_2^2} + T_d^2}$
Amplifier requirements	
Maximum current	$i_{\text{max}} = \dfrac{T_{\text{max}}}{K_T}$
rms current	$i_{\text{rms}} = \dfrac{T_{\text{rms}}}{K_T}$
Maximum voltage	$v_{\text{max}} = Ri_{\text{max}} + K_b\omega_{\text{max}}$

Amplifier Requirements We now derive expressions for the amplifier requirements to drive a specific motor through a given profile. Using the motor current equation $i = T/K_T$, we see that the maximum current and the rms current required are

$$i_{\text{max}} = \frac{T_{\text{max}}}{K_T} \tag{6.7.12}$$

$$i_{\text{rms}} = \frac{T_{\text{rms}}}{K_T} \tag{6.7.13}$$

The motor voltage equation is

$$v = Ri + L\frac{di}{dt} + K_b\omega$$

Divide by R:

$$\frac{v}{R} = i + \frac{L}{R}\frac{di}{dt} + \frac{K_b}{R}\omega$$

If the electrical time constant L/R is very small, we can neglect the second term on the right to obtain

$$\frac{v}{R} = i + \frac{K_b}{R}\omega$$

which gives $v = Ri + K_b\omega$. Thus, the maximum voltage required is given approximately by

$$v_{\text{max}} = Ri_{\text{max}} + K_b\omega_{\text{max}} \tag{6.7.14}$$

These equations are summarized in Table 6.7.1. They are used to compute the amplifier requirements for the trapezoidal speed profile. To determine whether a given amplifier will be satisfactory, the calculated values of i_{max}, i_{rms}, and v_{max} are compared with the amplifier manufacturer's data on peak current, rated continuous current, and maximum voltage.

EXAMPLE 6.7.1

Calculating Motor-Amplifier Requirements

■ Problem

The trapezoidal profile requirements for a specific application are given in the following table, along with the load and motor data. Determine the motor and amplifier requirements.

Profile data		
	Cycle times	$t_1 = 0.2$ s, $t_2 = 0.4$ s, $t_f = 0.6$ s
Load data		
	Inertia $I_L = 4 \times 10^{-3}$ kg·m^2	Displacement $\theta_{Lf} = 10\pi$ rad
	Torque $T_L = 0.1$ N·m	Reduction ratio $N = 2$
Motor data		
	Resistance $R = 2\ \Omega$	Torque constant $K_T = 0.3$ N·m/A
	Inductance $L = 3 \times 10^{-3}$ H	Damping $c = 0$
	Inertia $I_m = 10^{-3}$ kg·m^2	
	Time constants	1.56×10^{-3} s and 0.043 s

■ Solution

The total inertia I is the sum of the motor inertia and the reflected load inertia. Thus,

$$I = I_m + \frac{I_L}{N^2} = 10^{-3} + \frac{4 \times 10^{-3}}{2^2} = 2 \times 10^{-3} \text{ kg} \cdot \text{m}^2$$

Because the reduction ratio is $N = 2$, the required motor displacement is $N\theta_{Lf} = 2(10\pi) = 20\pi$ rad, and the load torque as felt at the motor shaft is $T_d = T_L/N = 0.1/2 = 0.05$ N·m.

The motor's energy consumption per cycle is found from (6.7.7).

$$E = \frac{2}{(0.3)^2}\left[\frac{2(4 \times 10^{-6})(20\pi)^2}{0.2(0.4)^2} + (0.05)^2 0.6\right] = 22 \text{ J/cycle}$$

The power consumption is $22/t_f = 37$ J/s, or 37 W.

Equation (6.7.6) shows that the maximum speed for the trapezoidal profile is

$$\omega_{\max} = \frac{\theta_f}{t_2} = 50\pi \text{ rad/s}$$

which is $50\pi(60)/(2\pi) = 1500$ rpm. So the motor's maximum permissible speed must be greater than 1500 rpm.

The rms torque is found from (6.7.11) to be

$$T_{\text{rms}} = \sqrt{\frac{2(4 \times 10^{-6})(20\pi)^2}{0.6(0.2)(0.4)^2} + (0.05)^2} = 1.28 \text{ N} \cdot \text{m}$$

Use (6.7.10) to compute the maximum required torque.

$$T_{\max} = \frac{2 \times 10^{-3}(20\pi)}{0.2(0.4)} + 0.05 = 1.57 + 0.05 = 1.62 \text{ N} \cdot \text{m}$$

Note that the load torque contributes little to $T_{\max}$. Most of the required torque is needed to accelerate the inertia.

The system time constants are obtained from the roots of (6.5.13), which are $s = -643$ and $s = -23.3$. The system must be fast enough to respond to the profile command. Its largest time constant, $1/23.3 = 0.043$ s, is less than one-fourth of the ramp time $t_1 = 0.2$, so the system is fast enough.

The amplifier requirements are calculated as follows. Note that because the motor data is given in SI units, $K_b = K_T = 0.3$. From (6.7.12), (6.7.13), and (6.7.14),

$$i_{\text{max}} = \frac{1.62}{0.3} = 5.4 \text{ A}$$

$$i_{\text{rms}} = \frac{1.28}{0.3} = 4.27 \text{ A}$$

$$v_{\text{max}} = 2(5.4) + 0.3(50\pi) = 10.8 + 47.1 = 57.9 \text{ V}$$

Note that most of the required voltage is needed to oppose the back emf.

The preceding analysis neglected the damping constant c and assumed that the term $(L/R)di/dt$ is very small. If these conditions are not satisfied in a given application, the performance evaluation is best done by computer solution. A MATLAB example is given in Section 6.9.

EXAMPLE 6.7.2 Effect of Reducer Ratio on Energy Consumption

■ Problem

In Chapter 3, Section 3.6, we showed that the reducer ratio N selected by inertia matching maximizes the load acceleration. Now we want to see if that value of N is optimal with respect to energy consumption. We will use the conveyor system shown in Figure 6.7.2 as our example. This system was analyzed in Example 3.6.1.

■ Solution

Using the results from Example 3.6.1, for a specific load and motor, the total inertia felt at the motor shaft is

$$I = I_m + I_r + \frac{I_L}{N^2} = 0.003 + I_r + \frac{0.085}{N^2} \text{ kg} \cdot \text{m}^2 \tag{1}$$

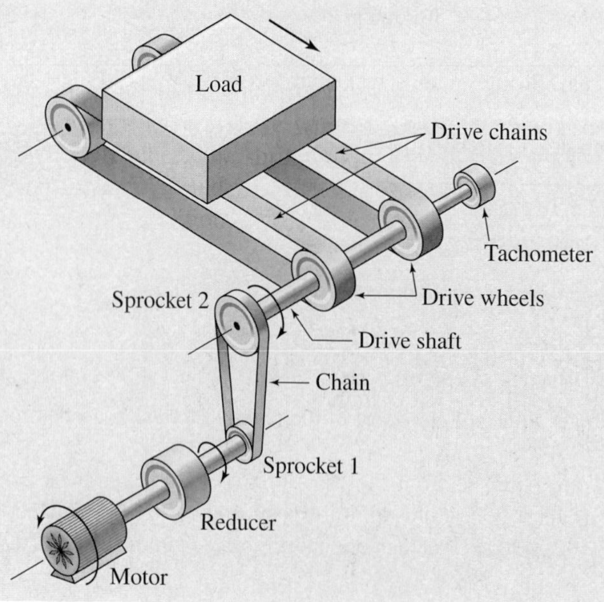

Figure 6.7.2 Mechanical drive for a conveyor system.

where I_m is the motor inertia, I_r is the reducer inertia felt at the motor shaft, and I_L is the load inertia felt on the load side of the reducer. We will use the formulas in Table 6.7.1. The pertinent values are

$$\theta_f = N\theta_{fL} = 50N \text{ rad} \qquad T_d = \frac{T_L}{N} = \frac{0.6}{N} \text{ N}\cdot\text{m}$$

$$t_1 = 0.5 \qquad t_2 = 2.5 \qquad t_f = 3 \text{ s}$$

Substituting these values into (6.7.7) gives

$$E = \frac{R}{K_T^2}\left(1600N^2I^2 + \frac{1.08}{N^2}\right) = \frac{R}{K_T^2}F \tag{2}$$

where F is the factor within the parentheses.

Unfortunately, further analysis will not be straightforward because the reflected reducer inertia I_r depends on its ratio N in an involved way that depends on the reducer type (spur gear, worm gear, etc.). Instead, we will use a basic engineering approach by first assuming that the effect of N on I_r is small (using the same value $I_r = 0.002$ used in Example 3.6.1), and then investigating how sensitive the result is to that assumption.

Setting $I_r = 0.002$ in (1) gives $I = 0.005 + 0.085/N^2$. Substituting this into (2) gives

$$E = \frac{R}{K_T^2}\left(0.04N^2 + 1.36 + \frac{12.64}{N^2}\right) \tag{3}$$

Setting $dE/dN = 0$ and solving gives the value $N = 4.216$. (MATLAB users will notice that this process could be done with the `min` function.) This value gives the minimum of E and is very close to the value computed by inertia matching ($N = 4.123$).

For reference, if $I_r = 0$, E has a minimum at $N = 5.44$. Section 3.6 also derived a formula for N that accounts for the load torque T_2 and the motor torque T_1. From (3.6.1),

$$N = -\frac{T_2}{T_1} + \sqrt{\left(\frac{T_2}{T_1}\right)^2 + \frac{I_2}{I_1}}$$

Here $T_1 = T_{\max} = 1.798$ and $T_2 = -0.6$ N·m. Also, $I_1 = I_m + I_r = 0.003 + 0.002 = 0.005$ and $I_2 = I_L = 0.085$. These give $N = 4.47$. In Example 3.6.1, the available reducer has the ratio $N = 4$. The following table summarizes these results.

N	I_r	Energy factor F	Comment
4.47	0.002	2.78	Maximizes $\dot{\omega}_{max}$. Includes load and motor torques.
4	0.002	2.79	Uses available reducer.
4.123	0.002	2.78	Uses inertia matching.
4.216	0.002	2.78	Minimizes energy.
5.44	0	1.669	Inertia matching with negligible I_r.

Next we look at the effect of the reducer inertia I_r. Figure 6.7.3 shows the energy factor F as a function of N for values of I_r ranging from 0 up to the motor inertia. Note how the minimum point shifts to smaller values of N as the inertia increases. Note also how the energy consumption rises dramatically for values of N not close to the optimum.

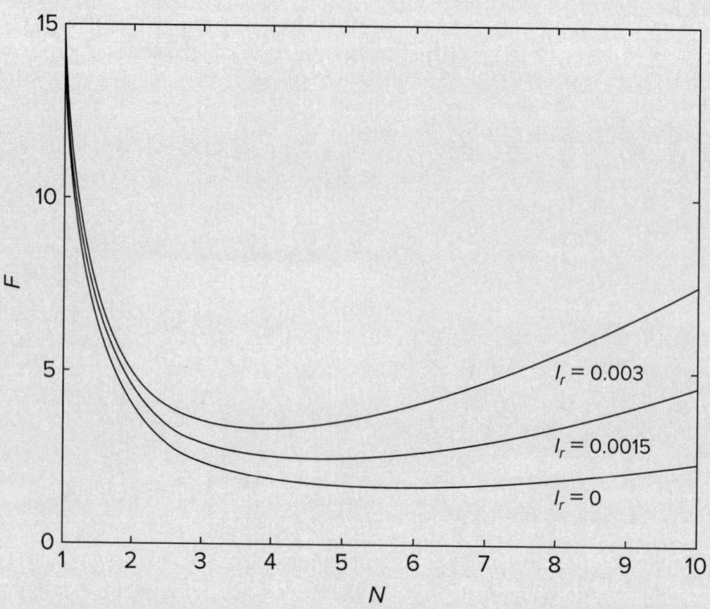

Figure 6.7.3 Energy consumption factor versus reducer ratio for several reducer inertia values.

The following table shows how the energy consumption is affected as the reducer inertia increases. It shows that it can significantly increase energy consumption.

I_r	Optimum N	Energy Factor F	Percent Increase
0.001	4.71	2.23	-
0.0015	4.44	2.500	12%
0.003	3.85	3.339	50%

Of course, these results are for a specific load inertia, load torque, and motor inertia. You may obtain different conclusions for other applications. Sometimes, in a practial setting the engineer may not have much choice over available motors and reducers, and may need to order from a specific and limited catalog. But, if there is some flexibility, this example shows how to optimize the design.

6.8 SENSORS AND ELECTROACOUSTIC DEVICES

In this section, we consider two common sensors: the tachometer for measuring velocity and the accelerometer, which can be used to measure either acceleration or displacement. In addition, some electroacoustic devices, such as microphones and speakers, are based on the electromagnetic principles explained in Section 6.5.

6.8.1 A TACHOMETER

There are many devices available to measure linear and rotational velocity. One such device can be constructed in a manner similar to a motor. However, instead of applying an input voltage, we use the load torque as the input. Consider the circuit equation for an armature-controlled motor:

$$v_a - i_a R_a - L_a \frac{di_a}{dt} - K_b \omega = 0$$

With the tachometer there is no applied voltage v_a. Thus, with $v_a = 0$, at steady state, when the derivative $di_a/dt = 0$, this equation becomes

$$-i_a R_a - K_b \omega = 0$$

The voltage $i_a R_a$ across the resistor is thus given by $i_a R_a = K_b \omega$. If we denote this voltage by v_t, we see that

$$v_t = K_b \omega$$

If we measure the voltage v_t, we can use it to determine the velocity ω.

6.8.2 AN ACCELEROMETER

Figure 6.8.1 illustrates the construction of an electromechanical accelerometer or a *seismograph*. A mass m, often called the *seismic mass* or the *proof mass*, is supported in a case by two springs. Its motion, which is damped by a fluid within the case, is measured by a potentiometer and amplifier. The displacement z of the case is its displacement relative to an inertial reference. With proper selection of m, c, and k, the device can be used either as a *vibrometer* to measure the amplitude of a sinusoidal displacement $z = A \sin \omega t$, or as an *accelerometer* to measure the amplitude of the acceleration $\ddot{z} = -A\omega^2 \sin \omega t$. When used to measure ground motion from an earthquake, for example, the instrument is commonly referred to as a *seismograph*.

The mass displacement x is defined relative to an inertial reference, with $x = 0$ corresponding to the equilibrium position of m when $z = 0$. With the potentiometer arrangement shown, the voltage v is proportional to the relative displacement y between the case and the mass m, where $y = x - z$. So the measured voltage is $v = Ky$.

We model the system as a mass-spring-damper system. Newton's law gives

$$m\ddot{x} = -c(\dot{x} - \dot{z}) - \frac{k}{2}(x - z) - \frac{k}{2}(x - z)$$

Substituting y for $x - z$, we obtain

$$m\ddot{y} + c\dot{y} + ky = -m\ddot{z} \tag{6.8.1}$$

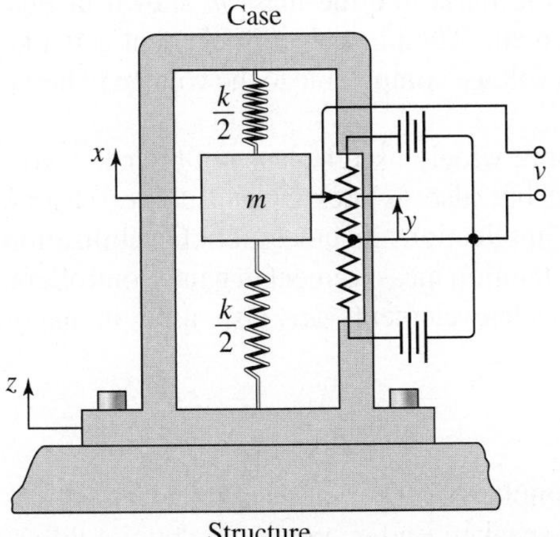

Figure 6.8.1
An accelerometer.

The transfer function between the input z and the output y is

$$\frac{Y(s)}{Z(s)} = \frac{-ms^2}{ms^2 + cs + k} \tag{6.8.2}$$

In Chapter 9 we will investigate the response of such a system to a sinusoidal input. We will see that depending on the selection of m, c, and k, the displacement y, and therefore the output voltage v, can be used to indicate either the acceleration $\ddot{z}$ or the displacement z. A relatively smaller mass is used for an accelerometer, and relatively stiffer springs are used for a seismograph.

Other principles can be used to construct an accelerometer. For example, some accelerometers utilize piezoelectric crystals, whose electrical output is a function of pressure. A coil and magnet can be used instead of the potentiometer to measure the displacement of the mass.

6.8.3 STRAIN GAGE ACCELEROMETERS

Figure 6.8.1 illustrates the general principle of an accelerometer, but there are many types of accelerometers based on that principle that use different technology. An example is an accelerometer based on a strain gage. In this case, the spring element in Figure 6.8.1 is replaced by a cantilever beam. The strain gage measures the beam strain, which is proportional to the inertia force, which depends on the acceleration. Since the voltage output from the gage is proportional to the strain, we see that the voltage is also proportional to the acceleration.

6.8.4 PIEZOELECTRIC DEVICES

The piezoelectric effect refers to the property of crystals and some ceramics to generate an electric potential in response to an applied force. In crystals, the force causes an electric charge across the crystal lattice. This charge produces a voltage across the crystal. This effect is reversible so that a force is produced when a voltage is applied. Thus, a piezoelectric element can be used as a force *sensor* or as an *actuator*.

Many sensors and actuators are based on the piezoelectric effect. They are used to measure force, pressure, acceleration, or strain. Their large modulus of elasticity means that they show almost no deflection under load, are very rugged, and are very linear over a wide range of input values. They are also not affected by electromagnetic fields and radiation. However, their small deflection means they cannot be used to measure forces that are truly static.

In piezoelectric accelerometers, a seismic mass like the mass m shown in Figure 6.8.1 is attached to the piezoelectric element. The piezoelectric element acts like the spring element in Figure 6.8.1, exerting a voltage comparable to the voltage v shown in that figure.

Piezoelectric elements are becoming more widely used because of their low cost and small size, not only as accelerometers, but also as pressure and force sensors, gyroscopes, and positioning systems. Some applications include: motion stabilization systems for cameras and orientation and acceleration measurement in game controllers. Because of their small deflection, piezoelectric elements are also used in nanopositioning systems.

6.8.5 ELECTROACOUSTIC DEVICES

Speakers and microphones are common examples of a class of electromechanical devices called *electroacoustic*. There are other speakers and microphones that use different principles, such as a capacitance microphone, but we will focus on those devices

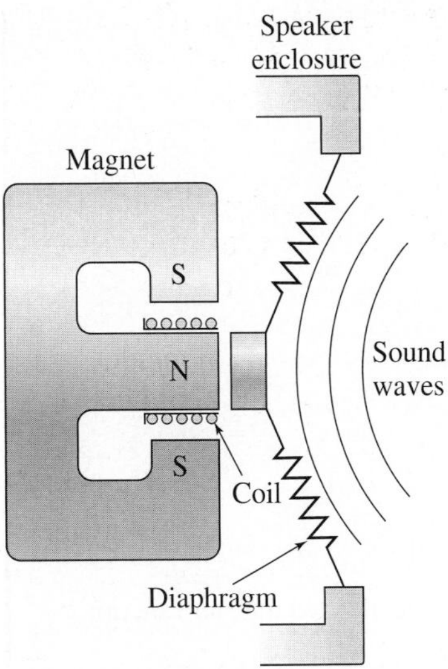

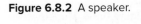

Figure 6.8.2 A speaker.

that utilize a magnet, a coil, and a cone. A speaker converts electrical energy into mechanical energy (sound waves) by causing the coil to move the cone. On the other hand, a microphone converts the mechanical energy in sound into electrical energy by moving the cone, thus producing a voltage and current in the coil. Here we consider the operation of a speaker. The model for a microphone is treated in the problems at the end of the chapter.

The operation of a speaker is illustrated by Figure 6.8.2. A stereo or radio amplifier produces a current in a coil that is attached to a diaphragm in the cone. This causes the coil and diaphragm to move relative to the permanent magnet. The motion of the diaphragm produces air pressure waves, which is sound.

An Electromagnetic Speaker

EXAMPLE 6.8.1

■ Problem

Develop a model of the electromagnetic speaker shown in Figure 6.8.2, and obtain the transfer function relating the diaphragm displacement x to the applied voltage v.

■ Solution

Figure 6.8.3a shows a simplified model of the mechanical subsystem, along with its free body diagram. The mass m represents the combined mass of the diaphragm and the coil. The spring constant k and damping constant c depend on the material properties of the diaphragm. The force f is the magnetic force, which is related to the coil current i by (6.5.1), $f = nBLi$, where n is the number of turns in the coil. Let $K_f = nBL$. From Newton's law

$$m\frac{d^2x}{dt^2} = -c\frac{dx}{dt} - kx + K_f i \tag{1}$$

Figure 6.8.3b shows the electrical subsystem. The coil's inductance and resistance are L and R. The coil experiences a back emf because it is a current conductor moving in a magnetic field. This back emf is given by $K_b\dot{x}$. The voltage v is the signal from the amplifier. From Kirchhoff's voltage law,

$$v = L\frac{di}{dt} + Ri + K_b\frac{dx}{dt} \tag{2}$$

The speaker model consists of equations (1) and (2).

Figure 6.8.3 Models of the mechanical and electrical subsystems of a speaker.

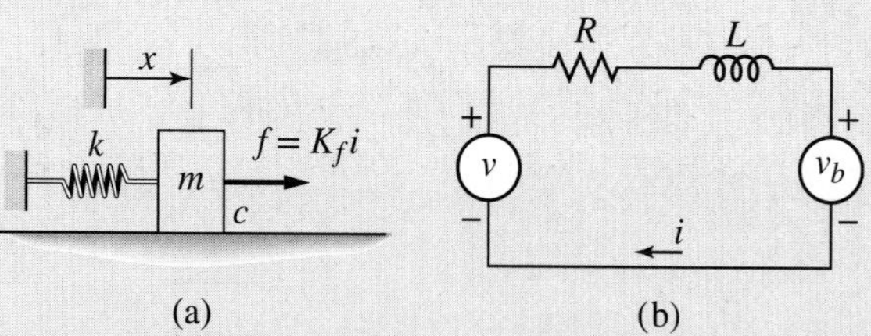

Transforming equation (1) and solving for $X(s)$ gives

$$X(s) = \frac{K_f}{ms^2 + cs + k} I(s)$$

Transforming equation (2) and solving for $I(s)$ gives

$$I(s) = \frac{1}{Ls + R}\left[V(s) - K_b s X(s)\right]$$

Eliminating $I(s)$ from the previous two equations, we obtain the desired transfer function.

$$\frac{X(s)}{V(s)} = \frac{K_f}{mLs^3 + (cL + mR)s^2 + (kL + cR + K_f K_b)s + kR} \tag{3}$$

6.9 MATLAB APPLICATIONS

In this section, we illustrate how to use the `lsim` and `step` functions with motor models in transfer-function form and state-variable form. We also show how to use the `ode` solvers to obtain the response of a nonlinear system.

6.9.1 STEP RESPONSE FROM TRANSFER FUNCTIONS

The following transfer functions of an armature-controlled motor were developed in Section 6.5.

$$\frac{I_a(s)}{V(s)} = \frac{Is + c}{L_a I s^2 + (R_a I + cL_a)s + cR_a + K_b K_T}$$

$$\frac{\Omega(s)}{V(s)} = \frac{K_T}{L_a I s^2 + (R_a I + cL_a)s + cR_a + K_b K_T}$$

where the input is the armature voltage $v(t)$. We will use three MATLAB programs to plot the step response of the motor speed and current, for an applied voltage of 10 volts.

MATLAB Program `motor_par.m` contains the motor parameters. When this is executed, the values of the parameters will be available in the MATLAB workspace, for use by our other programs. The use of such a program is good practice, because it enables you to develop modular programs that do not depend on a specific set of parameters. To investigate another motor having different parameter values, simply edit the program and run it. We declare the parameters to be global so they can be accessed by the user-defined function `nlmotor`, which is used in our second example. The parameter values are those of Example 6.6.2. We use `Im` and `cm` to represent the inertia and damping of the motor, so that we can later distinguish them from the inertia and damping of the load.

MATLAB Program motor_par.m

```
% Program motor_par.m (Motor parameters in SI units)
global KT Kb La Ra Im cm
```

```
KT = 0.05;Kb = KT;
La = 2e-3;Ra = 0.5;
Im = 9e-5;cm = 1e-4;
```

MATLAB Program `motor_tf.m` creates the LTI models for the current and speed based on the motor-transfer functions.

MATLAB Program motor_tf.m

```
% Program motor_tf.m (Transfer functions for voltage input)
I = Im;
c = cm;
% current transfer function:
current_tf = tf([I,c],[La*I,Ra*I+c*La,c*Ra+Kb*KT]);
% speed transfer function:
speed_tf = tf(KT,[La*I,Ra*I+c*La,c*Ra+Kb*KT]);
```

The MATLAB Program `motor_step.m` computes and plots the step response for an input of 10 V. Note that since `step` gives the *unit*-step response, we must multiply the response by 10. This is done in the `plot` functions.

MATLAB Program motor_step.m

```
% Program motor_step.m (Motor step response)
motor_par
motor_tf
[current, tc] = step(current_tf);
[speed, ts] = step(speed_tf);
subplot(2,1,1),plot(tc,10*current),...
   xlabel('t (s)'),ylabel('Current (A)')
subplot(2,1,2),plot(ts,10*speed),...
   xlabel('t (s)'),ylabel('Speed (rad/s)')
```

The result is shown in Figure 6.9.1.

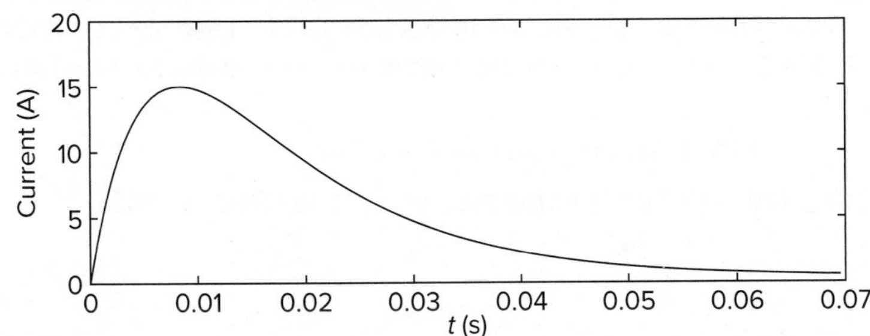

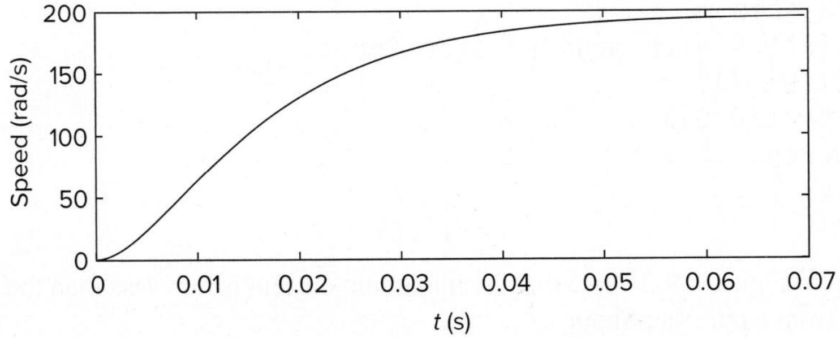

Figure 6.9.1 Motor step response.

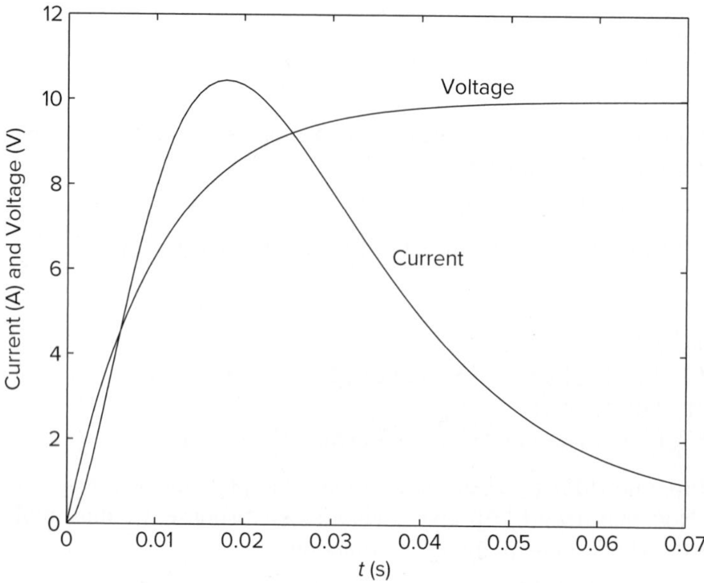

Figure 6.9.2 Motor response to a modified step input.

6.9.2 MODIFIED STEP RESPONSE

In Example 6.6.2 we noted that the large current response is due partly to the step input acting on the numerator dynamics of the transfer function $I_a(s)/V(s)$. A more realistic model of a suddenly applied voltage input is $v(t) = 10(1 - e^{-t/\tau})$, where τ is no larger than the system's dominant time constant. MATLAB Program `motor_mod.m` computes motor response to this modified step input for $\tau = 0.01$ s. This program first calls `motor_par` and `motor_tf`, which establish the motor parameters and transfer functions these programs are shown in Section 6.9.1. Then `motor_mod` calls program `mod_step`, which computes the current response to the modified step input.

MATLAB Program motor_mod.m

```
% Program motor_mod (Motor response with modified step)
motor_par
motor_tf
mod_step
```

MATLAB Program mod_step.m

```
% Program mod_step.m
% Motor simulation with modified step input
t = (0:0.0001:0.07);
v = 10*(1-exp(-t/0.01));
ia = lsim(current_tf,v,t);
plot(t,ia,t,v)
```

The plot is shown in Figure 6.9.2. Note that the maximum current is now less than the 15 A that results from a pure step input.

6.9.3 STEP RESPONSE FROM STATE-VARIABLE MODEL

We now use the state-variable model of the motor to plot the step response. The following model was developed in Section 6.5.

$$\frac{di_a}{dt} = \frac{v - i_a R_a - K_b \omega}{L_a} \tag{6.9.1}$$

$$\frac{d\omega}{dt} = \frac{K_T i_a - c\omega - T_d}{I} \tag{6.9.2}$$

This model is in state-variable form, where the state variables are the armature current i_a and the speed ω. The inputs are the applied voltage v and the load torque T_L, which is reflected through the gear ratio N to produce $T_d = T_L/N$. In this case the appropriate state and input matrices are the following (see Section 5.2 for a discussion of the standard matrix-vector form of a state-variable model).

$$\mathbf{A} = \begin{bmatrix} -R_a/L_a & -K_b/L_a \\ K_T/I & -c/I \end{bmatrix} \qquad \mathbf{B} = \begin{bmatrix} 1/L_a & 0 \\ 0 & -1/(NI) \end{bmatrix}$$

where the state vector and input vector are

$$\mathbf{x} = \begin{bmatrix} i_a \\ \omega \end{bmatrix} \qquad \mathbf{u} = \begin{bmatrix} v \\ T_L \end{bmatrix}$$

Choosing the outputs to be i_a and ω, we use the following output matrices.

$$\mathbf{C} = \begin{bmatrix} 1 & 0 \\ 0 & 1 \end{bmatrix} \qquad \mathbf{D} = \begin{bmatrix} 0 & 0 \\ 0 & 0 \end{bmatrix}$$

When the program `load_par` is run, it places the load parameters into the workspace.

MATLAB Program load_par.m

```
% load_par.m Load parameters.
IL = 0;
cL = 0;
N = 1;
TL = 0;
```

These particular values match those of Example 6.6.2, in which there is no reducer, no load torque, no load damping, and no load inertia. This program, however, is useful for solving the general problem where a reducer and a load are present.

The following program `motor_mat.m` computes reflected inertia and damping, the matrices for the motor model, and the state space model `sysmotor`. When this is executed, the matrices and state space model will be available in the MATLAB workspace. Note that the inertia I and damping c depend on both the motor and the load.

MATLAB Program motor_mat.m

```
% motor_mat.m Motor state matrices.
I = Im +IL/N^2;
c = cm + cL/N^2;
A = [-Ra/La,-Kb/La;KT/I,-c/I;];
B = [1/La,0;0, -1/(N*I)];
C = [1,0;0,1];
D = [0,0;0,0];
sysmotor = ss(A,B,C,D);
```

The following program state_step.m computes and plots the response due to a step voltage of magnitude 10 v, with the load torque T_L equal to zero. Note that it also calls programs motor_par and load_par from Sections 6.9.1 and 6.9.3.

MATLAB Program state_step.m

```
% state_step.m (Motor step response with state model)
motor_par
load_par
motor_mat
[y, t] = step(sysmotor);
subplot(2,1,1),plot(t,10*y(:,1)),...
   xlabel('t (s)'),ylabel('Current (A)')
subplot(2,1,2),plot(t,10*y(:,2)),...
   xlabel('t (s)'),ylabel('Speed (rad/s)')
```

The resulting plot looks like Figure 6.9.1.

6.9.4 TRAPEZOIDAL RESPONSE

Suppose the applied voltage is the following trapezoidal function, which is shown in Figure 6.9.3 for the case where $v_{max} = 20$ V, $t_1 = 0.3$ s, $t_2 = 0.9$ s, $t_f = 1.2$ s, and $t_3 = 1.5$ s.

$$v(t) = \begin{cases} \dfrac{v_{max}}{t_1}t & 0 \le t \le t_1 \\ v_{max} & t_1 < t < t_2 \\ \dfrac{v_{max}}{t_1}(t_f - t) & t_2 \le t \le t_f \\ 0 & t_f < t \le t_3 \end{cases}$$

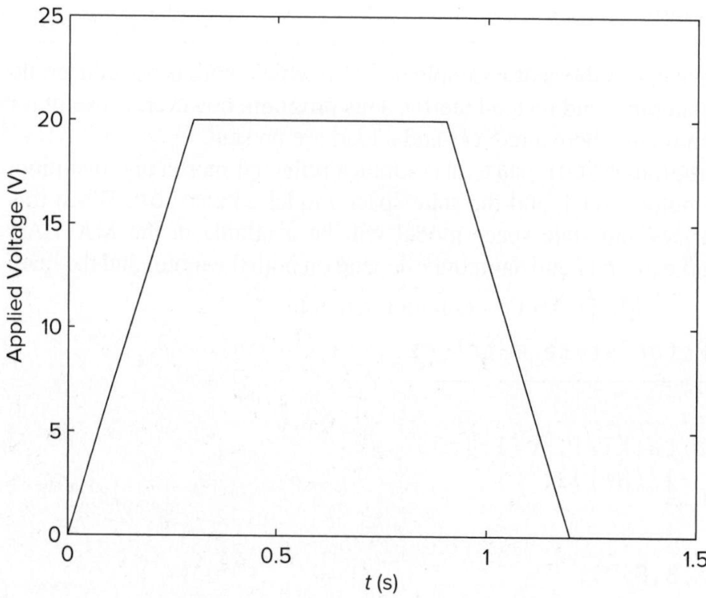

Figure 6.9.3 Trapezoidal voltage profile.

The program `trapezoid.m` creates the voltage array `v`.

MATLAB Program trapezoid.m

```
% Program trapezoid.m (Trapezoidal voltage profile)
t1 = 0.3; t2 = 0.9; tfinal = 1.2; t3 = 1.5;
v_max = 20;
dt = t3/1000;
t = (0:dt:t3);
for k = 1:1001
  if t(k) <= t1
     v(k) = (v_max/t1)*t(k);
  elseif t(k) <= t2
     v(k) = v_max;
  elseif t(k) <= tfinal
     v(k) = (v_max/t1)*(tfinal-t(k));
  else
     v(k) = 0;
  end
end
```

The next program, `performance.m`, computes the performance measures relating to energy consumption, maximum current, voltage, and torque required and rms current and torque. It uses the `trapz` function, which computes an integral with the trapezoidal rule. Note that this program does not use the formulas in Table 6.7.1, which represent a simplified case, but rather it computes the measures directly from the computed response.

MATLAB Program performance.m

```
% Program performance.m
% Computes motor performance measures.
ia = y(:,1);
speed = y(:,2);
E = trapz(t,Ra*ia.^2)+trapz(t,c*speed.^2)
i_max = max(ia)
i_rms = sqrt(trapz(t,ia.^2)/t3)
T_max = KT*i_max
T_rms = KT*i_rms
speed_max = max(speed);
v_max = Ra*i_max+Kb*speed_max
```

The next program, `trapresp.m`, uses `performance` and `trapezoid` to compute the response.

MATLAB Program trapresp.m

```
% Program trapresp.m (Motor trapezoidal response)
motor_par
load_par
motor_mat
trapezoid
u = [v', TL*ones(size(v'))];
y = lsim(sysmotor,u,t);
subplot(2,1,1),plot(t,y(:,1)),...
```

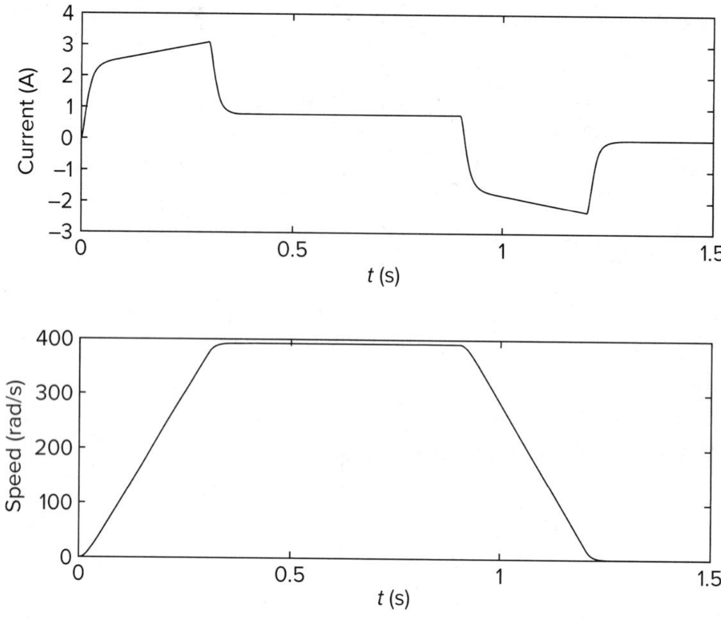

Figure 6.9.4 Motor response to a trapezoidal voltage profile.

```
   xlabel('t (s)'),ylabel('Current (A)')
subplot(2,1,2),plot(t,y(:,2)),...
   xlabel('t (s)'),ylabel('Speed (rad/s)')
performance
```

The resulting plot is shown in Figure 6.9.4. The computed performance measures are

$$E = 14.12 \text{ J/cycle} \qquad i_{max} = 3.09 \text{ A} \qquad i_{rms} = 1.56 \text{ A}$$
$$T_{max} = 0.15 \text{ N}\cdot\text{m} \qquad T_{rms} = 0.08 \text{ N}\cdot\text{m} \qquad v_{max} = 21.15 \text{ V}$$
$$\omega_{max} = 392 \text{ rad/s}$$

Note that the computed value of v_{max} is different from that computed from the formula $v_{max} = Ri_{max} + K_b\omega_{max}$ because this formula neglects the electrical time constant L/R.

6.9.5 NONLINEAR DAMPING

In Chapter 4 we saw that the force or torque due to viscous friction is sometimes a nonlinear function of the speed. Let us investigate the effects of a nonlinear damping torque $c_N\omega^2$ on the speed and current, for a step input voltage. Suppose that $c_N = 5 \times 10^{-6}$ $\text{N}\cdot\text{m}\cdot\text{s}^2/\text{rad}^2$ and suppose that we are interested in the speed range from 0 to 2000 rpm. The plot of the nonlinear damping torque is shown in Figure 6.9.5. As discussed in Chapter 1, a linear approximation that gives an overestimate of the damping torque is the straight line shown on the plot. The coefficient for the linear model is $c = 10^{-3}$ $\text{N}\cdot\text{m}\cdot\text{s/rad}$.

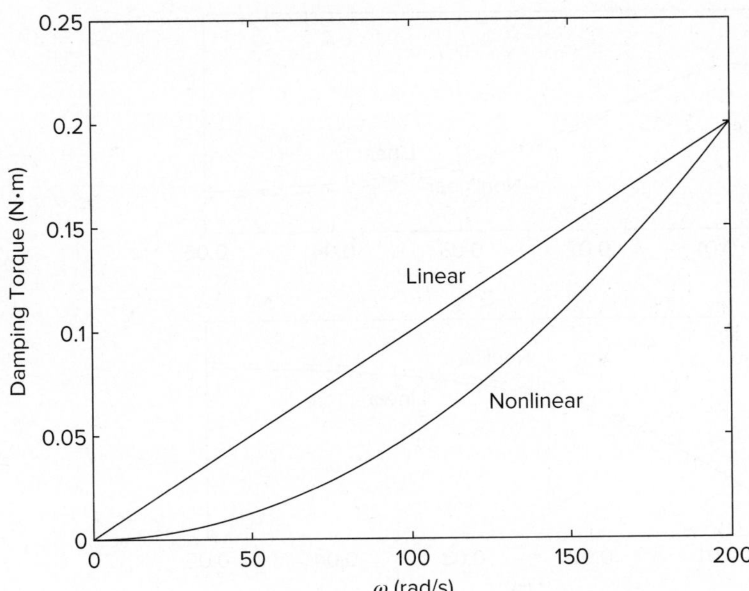

Figure 6.9.5 Linear and nonlinear models of damping torque.

The current equation is given by (6.9.1). The nonlinear motor equation is a modification of (6.9.2) with $T_d = 0$:

$$\frac{d\omega}{dt} = \frac{K_T i_a - c_N \omega^2}{I} \tag{6.9.3}$$

To use one of the `ode` solvers, we must first create a function file that computes the derivatives di_a/dt and $d\omega/dt$. This file is based on (6.9.1) and (6.9.3). We use the program `nlmotor.m`. The current is `x(1)` and the speed is `x(2)`.

MATLAB Program nlmotor.m

```
function xdot = nlmotor(t,x)
% nonlinear damping in motor.
global KT Kb La Ra I cN v
xdot = [(-Ra*x(1)-Kb*x(2)+v)/La;(KT*x(1)-cN*x(2).^2)/I];
```

Note that the use of `global` in the program `motor_par` enables the parameters to be accessed by the function `nlmotor`.

Program `motor.m` computes the step response of the linear and nonlinear models for an applied voltage of 10 V. Note that because we have changed the value of the motor damping c_m from that originally used in `motor_par`, we must type `cm=1e-3` *after* running `motor_par` but *before* running `motor_mat`. Because `step` gives the *unit*-step response, we must multiply its result by $v = 10$ V. Note that the output order of the linear solvers `lsim` and `step` is `[y, t]`, whereas for the `ode` solvers the order is `[t, y]`.

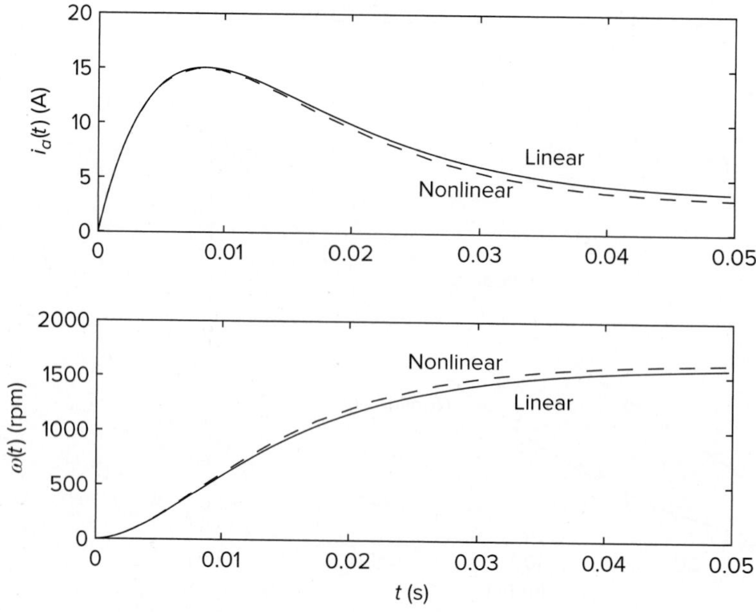

Figure 6.9.6 Motor response with linear and nonlinear damping.

MATLAB Program motor.m

```
Program motor.m (Computes response for nonlinear damping)
clear
global I cN v
motor_par
load_par
cm = 1e-3;
motor_mat
cN = 5e-6; v = 10;
[y1, t1]= step(sysmotor);
y1 = v*y1;
[t2, y2] = ode45(@nlmotor, [0, 0.05], [0, 0]);
subplot(2,1,1), plot(t2,y2(:,1),'--',t1,y1(:,1)),...
   ylabel('i_a(t) (A)'),...
   axis([0 0.05 0 20]), gtext('Linear'),gtext('Nonlinear'),
subplot(2,1,2), ...
   plot(t2,y2(:,2)*60/(2*pi),'--',t1,y1(:,2)*60/(2*pi)),...
   xlabel('t (s)'),axis([0 0.05 0 2000]),...
   ylabel('\omega(t) (rpm)'),gtext('Linear'),...
   gtext('Nonlinear')
```

The resulting plots are shown in Figure 6.9.6. Note that the current of the linear model is greater than that of the nonlinear model but its speed is less. This is because the linear damping model used here overestimates the damping torque.

Table 6.9.1 summarizes the MATLAB functions used in this section.

Table 6.9.1 MATLAB functions used in this chapter.

	Functions Introduced in This Chapter
`clear`	Clear all variables and functions from memory.
`global x`	Defines a global variable `x`.
`max(x)`	Computes the algebraically largest element in the array `x`.
`subplot`	Creates axes in tiled positions.
`trapz(y)`	Computes an approximation of the integral of the array `y` via the trapezoidal method (with unit spacing).
	Functions Introduced in Earlier Chapters
`axis ([x1 x2 y1 y2]`	Sets the minimum and maximum values for the *x* and *y* axes.
`else`	Specifies an alternative action following an `if` statement.
`elseif`	Specifies an alternative `if` statement.
`end`	Terminates a loop, such as a `for` loop, or an `if-else` block.
`for`	Denotes the beginning of a `for` loop.
`gtext('text')`	Enables placement of text on a plot using the cursor.
`lsim`	Computes the response of an LTI object to a user-defined input function.
`plot(x,y)`	Creates a two-dimensional plot on rectilinear axes.
`sqrt(x)`	Computes $\sqrt{x}$.
`step(sys)`	Computes and plots the step response of a linear model `sys`.
`sys=tf(num,den)`	Creates an LTI object `sys` in transfer-function form, having `num` as the numerator and `den` as the denominator.
`xlabel`	Puts a label on the abscissa of a plot.
`ylabel`	Puts a label on the ordinate of a plot.

6.10 SIMULINK APPLICATIONS

Simulink is especially useful for obtaining the response of systems to input functions that are more complicated than step, impulse, ramp, or sine functions. Simulink is also helpful for computing the response of systems that contain nonlinear elements whose behavior is difficult to analyze by hand and tedious to program in MATLAB. In this section, we use several electrical systems to illustrate how to accomplish this.

6.10.1 SIMULATION WITH A PULSE INPUT

It is relatively easy to derive an expression for the pulse response of a series *RC* circuit for a single pulse. To derive the response expression when the input is a series of pulses is, however, much more difficult. We can use Simulink to find the response rather easily. We will introduce another block, the Pulse Generator block, to accomplish this. The Pulse Generator block is in the Sources library.

We will first use a single pulse of amplitude 12 V and duration 0.02 s as an input to check the results, using the values $R = 10^4\ \Omega$ and $C = 10^{-6}$ F. Thus, the transfer function for the circuit is

$$\frac{V_o(s)}{V_i(s)} = \frac{1}{RCs + 1} = \frac{1}{0.01s + 1}$$

After placing the blocks as shown in Figure 6.10.1 and entering the coefficients of the transfer function, double-click on the Pulse Generator block and set the Amplitude to 12, the Period to 1, the Pulse Width to 2%, and the Phase Delay to 0. Use the default value, Time-based, for the Pulse type. Set the Save format for the Simout block to Array. Set the Stop time to 0.1. The output of the program when plotted should look like Figure 6.10.2 after being enhanced with the Plot Editor.

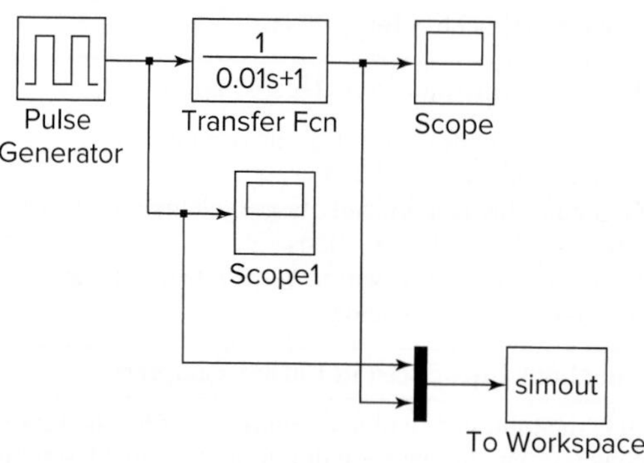

Figure 6.10.1 Simulink model using the Pulse Generator block.

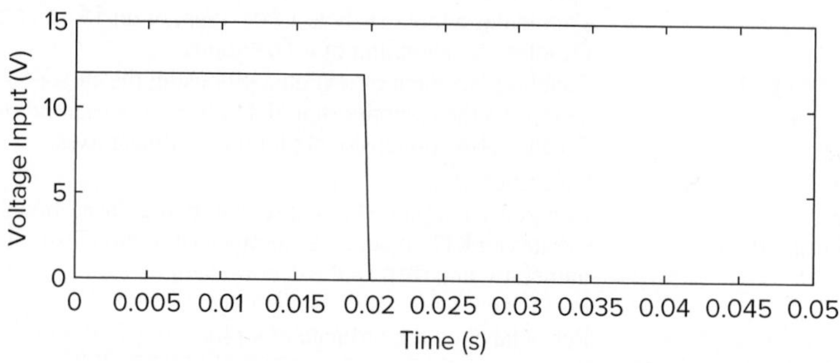

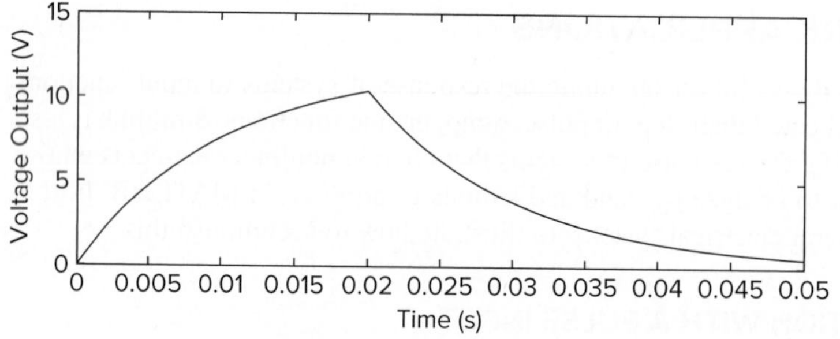

Figure 6.10.2 *RC* circuit response to a pulse input.

To obtain the response for a series of pulses of amplitude 12 V, of duration 0.005 s, and period 0.02 s, double-click on the Pulse Generator block and change the Period to 0.02, and the Pulse width to 25%. The output of the program when plotted should look like Figure 6.10.3 after being enhanced with the Plot Editor.

6.10.2 TORQUE LIMITATION IN MOTORS

In Section 6.6 we saw that motors have a maximum available torque. We can use Simulink to examine the effects of torque limits on the step response. The Simulink diagram in Figure 6.10.4 is based on the following equations.

$$\Omega(s) = \frac{1}{I_m s + c_m}[T(s) - T_d(s)]$$

$$T(s) = K_T I_a(s) = K_T \frac{1}{L_a s + R_a} V_a(s)$$

$$V_a(s) = V(s) - K_b \Omega(s)$$

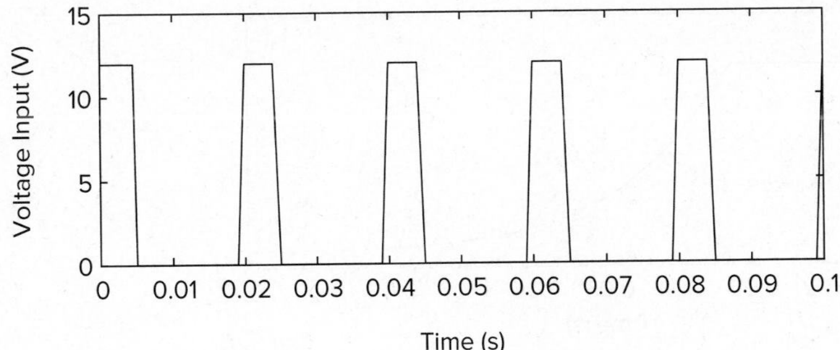

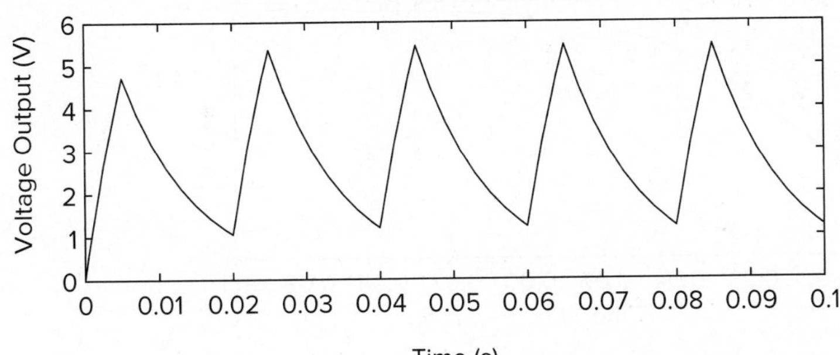

Figure 6.10.3 *RC* circuit response to a series of pulses.

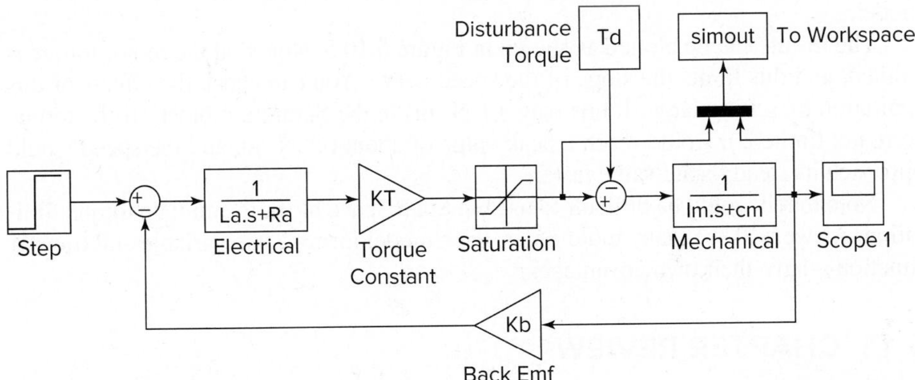

Figure 6.10.4 Simulink model of a torque-limited motor.

It uses two features we have not yet seen. These are the Constant block, labeled Disturbance Torque, and the use of variables as coefficients in a block. The Constant block gives a fixed constant input for all time, whereas the Step block can be set to switch on at a specified time.

The blocks labeled Electrical and Mechanical are Transfer Fcn blocks. The Torque Constant and Back Emf blocks are Gain blocks. After placing the blocks as shown, enter the denominator of the Electrical block as `[La, Ra]` and the denominator of the Mechanical block as `[Im, cm]`. Similarly, set the gains to `KT` and `Kb`, and set the constant in the Disturbance Torque block to `Td`. Note that these parameters do not yet have values.

Next set the lower and upper limits in the Saturation block to −0.4 and 0.4, respectively, to limit the motor torque to ±0.4 N · m. Set the Step Time of the Step block to 0 and the Final Value to 10, which corresponds to a 10 V input. Then, in the MATLAB Command window, run the program `motor_par` described in Section 6.9. This will set the values of all the parameters except for the torque T_d, which can be set to

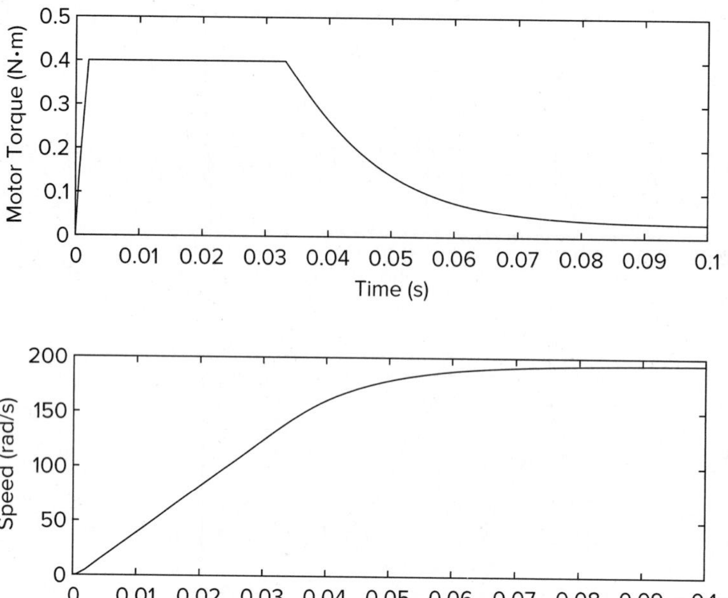

Figure 6.10.5 Response of a torque-limited motor.

0.01 N · m by typing `Td = 0.01` in the Command window. Then run the Simulink model.

The results can be plotted as shown in Figure 6.10.5. Note that the motor torque is limited, and this limits the slope of the speed curve. You can check the effects of this limitation by setting larger limits (say ±1 N · m) in the Saturation block. If the torque were not limited, it would reach a peak value of about 0.8 N · m, and the speed would approach its steady-state value faster.

Note that it would be difficult to use the State Space block to simulate torque limitation. As we will see many more times, each model form—state variable and transfer function—have their own advantages.

6.11 CHAPTER REVIEW

This chapter introduced the basic physics, common elements, and terminology of electric circuits, and treated the two main physical laws needed to develop circuit models. These are Kirchhoff's current and voltage laws. Impedance, which is a generalization of the electrical resistance concept, enables you to derive circuit models more easily, especially for more complex circuits, and is especially useful for obtaining models of circuits containing operational amplifiers. The principles of electromechanical devices, of which direct-current (dc) motors are the most important example, were extended to other electromechanical devices such as sensors and speakers.

Some of the systems treated in this chapter are more easily analyzed with computer simulation, and the chapter showed how to apply MATLAB and Simulink to analyze electromechanical systems.

Now that you have finished this chapter, you should be able to

- Develop models of electrical circuits. This includes

  Application of Kirchhoff's voltage and current laws,

  Application of the voltage- and current-divider rules,

Use of loop analysis,
Selection of appropriate state variables, and
Use of impedance.

- Obtain circuit models in transfer-function and block-diagram form.
- Apply impedance methods to obtain models of op-amp systems.
- Apply Newton's laws, electrical circuit laws, and electromagnetic principles to develop models of electromechanical systems.
- Analyze the performance of motors, amplifiers, and speed reducers in motion-control systems.
- Apply MATLAB and Simulink to analyze models of circuits and electromechanical systems in state-variable and transfer-function form.

PROBLEMS

Section 6.1 Electrical Elements

6.1 Determine the equivalent resistance R_e of the circuit shown in Figure P6.1, such that $v_s = R_e i$. All the resistors are identical and have the resistance R.

6.2 Determine the voltage v_1 in terms of the supply voltage v_s for the circuit shown in Figure P6.2.

6.3 The Wheatstone bridge, like that shown in Figure P6.3, is used for various measurements. For example, a strain gage sensor utilizes the fact that the resistance of wire changes when deformed. If the sensor is one resistance leg of the bridge, then the deformation can be determined from the voltage v_1. Determine the relation between the voltage v_1 and the supply voltage v_s.

Figure P6.1

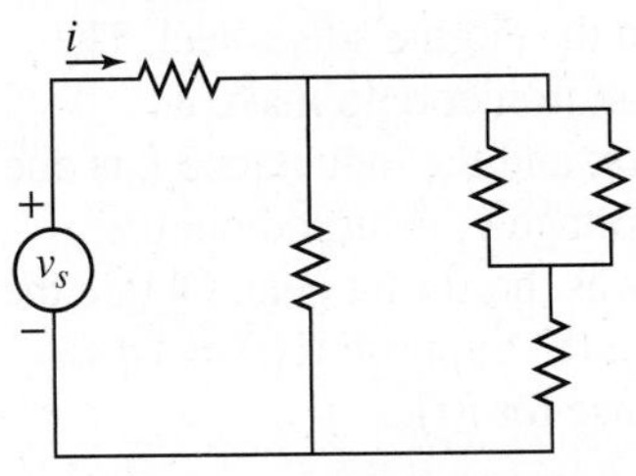

Figure P6.2

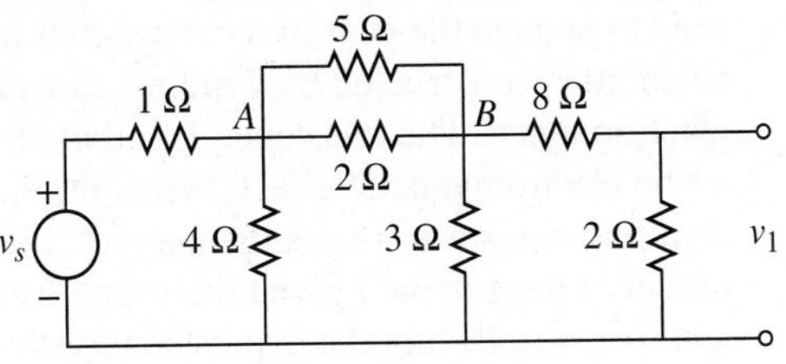

Figure P6.3

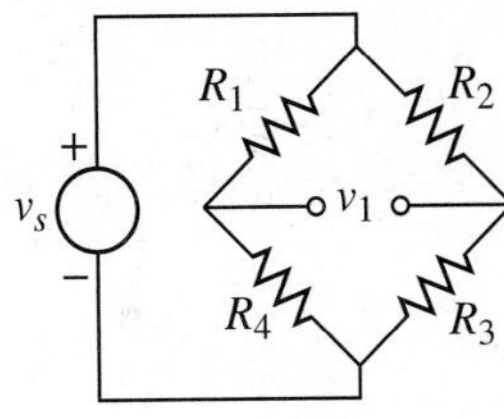

Section 6.2 Circuit Examples

6.4 The power supply of the circuit shown in Figure P6.4 supplies a voltage of 9 V. Compute the current i and the power P that must be supplied.

6.5 Obtain the model of the voltage v_1, given the current i_s, for the circuit shown in Figure P6.5.

Figure P6.4

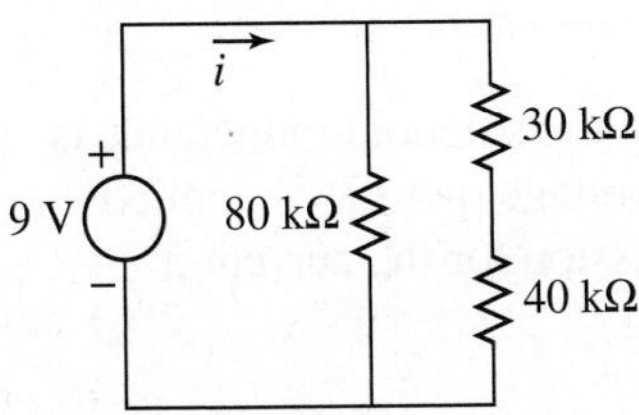

Figure P6.5

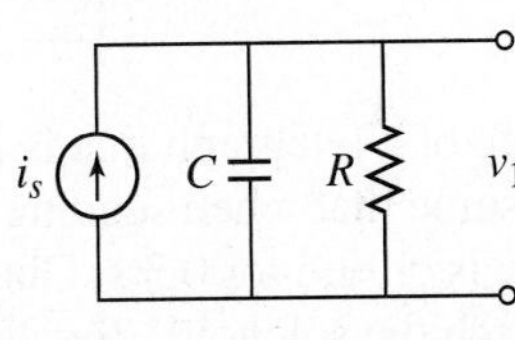

Figure P6.6

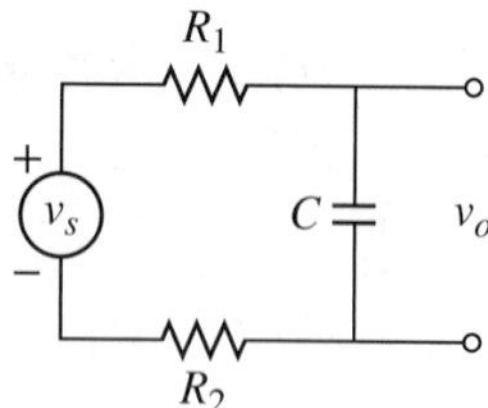

6.6 (a) Obtain the model of the voltage v_o, given the supply voltage v_s, for the circuit shown in Figure P6.6. (b) Suppose $v_s(t) = Vu_s(t)$. Obtain the expressions for the free and forced responses for $v_o(t)$.

6.7 A rectangular pulse input is a positive step function that lasts a duration D. One way of producing a step voltage input is to use a switch like that shown in Figure P6.7. The battery voltage V is constant and the switch is initially closed at point B. At $t = 0$ the switch is suddenly moved from point B to point A. Then at $t = D$ the switch is suddenly moved back to point B. Obtain the expression for the capacitor voltage $v_1(t)$ assuming that $v_1(0) = 0$.

Figure P6.7

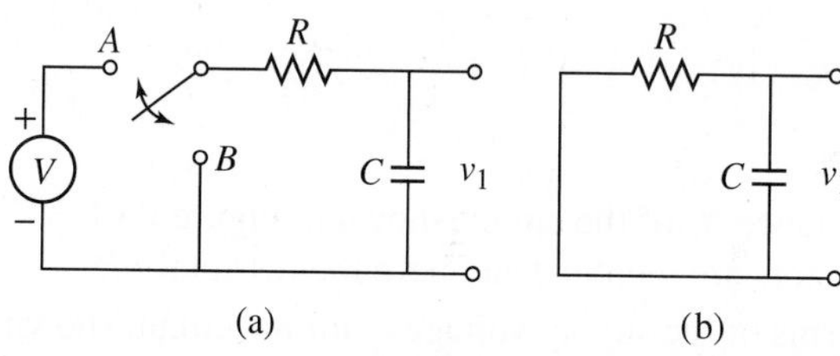

6.8 (a) Obtain the model of the voltage v_o, given the supply voltage v_s, for the circuit shown in Figure P6.8. (b) Suppose $v_s(t) = Vu_s(t)$. Obtain the expressions for the free and forced responses for $v_o(t)$.

6.9 (a) Obtain the model of the voltage v_o, given the supply voltage v_s, for the circuit shown in Figure P6.9. (b) Suppose $v_s(t) = Vu_s(t)$. Obtain the expressions for the free and forced responses for $v_o(t)$.

6.10 (a) The circuit shown in Figure P6.10 is a model of a solenoid, such as that used to engage the gear of a car's starter motor to the engine's flywheel. The solenoid is constructed by winding wire around an iron core to make an electromagnet. The resistance R is that of the wire, and the inductance L is due to the electromagnetic effect. When the supply voltage v_s is turned on, the resulting current activates the magnet, which moves the starter gear. Obtain the model of the current i given the supply voltage v_s. (b) Suppose $v_s(t) = Vu_s(t)$ and $i(0) = 0$. Obtain the expression for the response for $i(t)$.

Figure P6.8

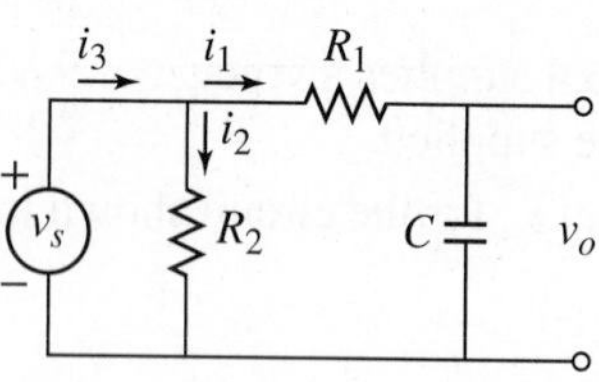

Figure P6.9

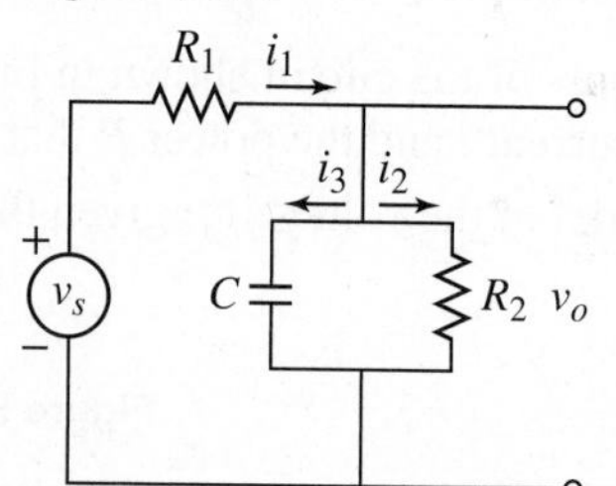

Figure P6.10

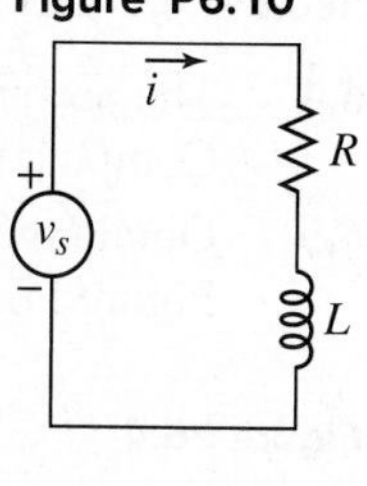

6.11 The resistance of a telegraph line is $R = 10\ \Omega$, and the solenoid inductance is $L = 5$ H. Assume that when sending a "dash," a voltage of 12 V is applied while the key is closed for 0.3 s. Obtain the expression for the current $i(t)$ passing through the solenoid. (See Figure 6.2.15.)

6.12 Obtain the model of the voltage v_o, given the supply voltage v_s, for the circuit shown in Figure P6.12.

6.13 Obtain the model of the voltage v_o, given the supply voltage v_s, for the circuit shown in Figure P6.13.

6.14 Obtain the model of the current i, given the supply voltage v_s, for the circuit shown in Figure P6.14.

Figure P6.12

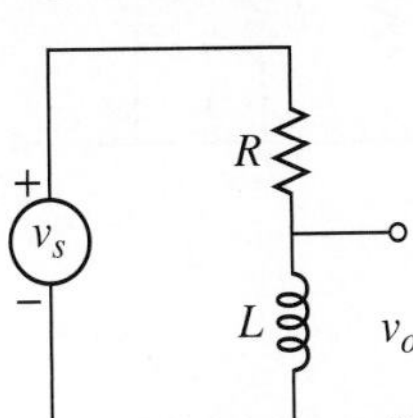

Figure P6.13

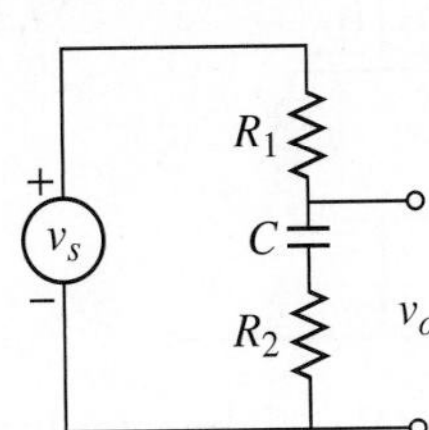

Figure P6.14

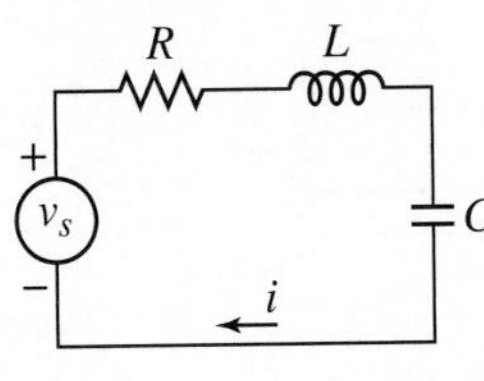

6.15 Obtain the model of the voltage v_o, given the supply current i_s, for the circuit shown in Figure P6.15.

Figure P6.15

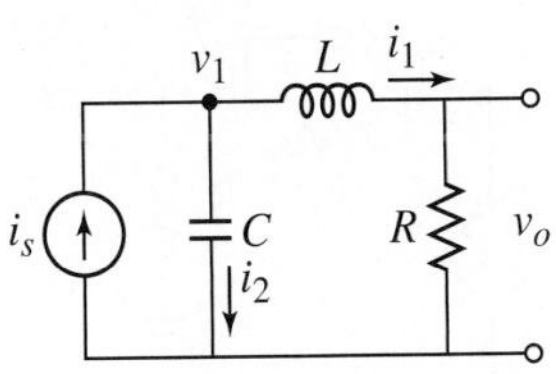

Figure P6.16

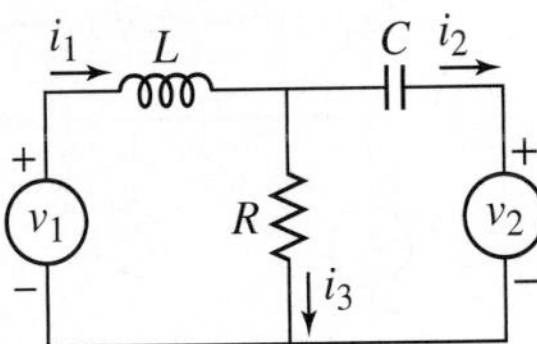

6.16 Obtain the model of the currents i_1, i_2, and i_3, given the input voltages v_1 and v_2, for the circuit shown in Figure P6.16.

6.17 Obtain the model of the currents i_1, i_2, and the voltage v_3, given the input voltages v_1 and v_2, for the circuit shown in Figure P6.17.

Figure P6.17

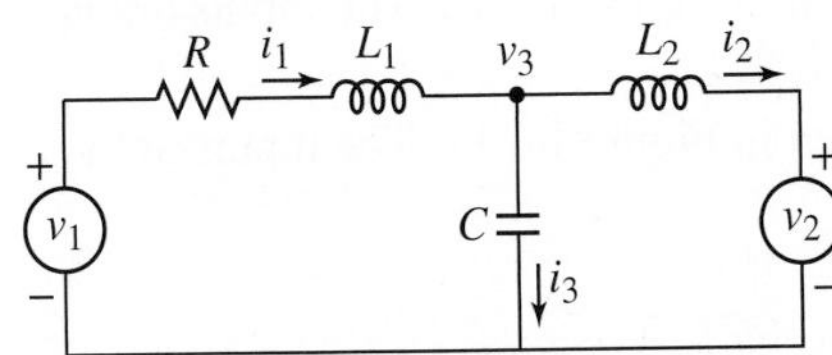

6.18 For the circuit shown in Figure P6.15, determine a suitable set of state variables, and obtain the state equations.

6.19 For the circuit shown in Figure P6.16, determine a suitable set of state variables, and obtain the state equations.

6.20 For the circuit shown in Figure P6.17, determine a suitable set of state variables, and obtain the state equations.

Section 6.3 Transfer Functions and Impedance

6.21 Use the impedance method to obtain the transfer function $V_o(s)/V_s(s)$ for the circuit shown in Figure P6.21.

6.22 Use the impedance method to obtain the transfer function $I(s)/V_s(s)$ for the circuit shown in Figure P6.22.

6.23 Use the impedance method to obtain the transfer function $V_o(s)/V_s(s)$ for the circuit shown in Figure P6.23.

6.24 Use the impedance method to obtain the transfer function $V_o(s)/I_s(s)$ for the circuit shown in Figure P6.24.

Figure P6.21

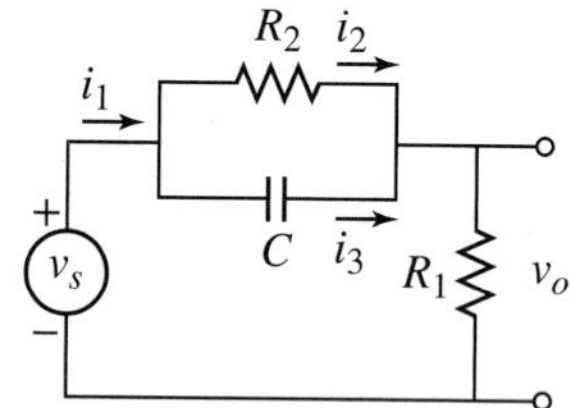

Figure P6.22

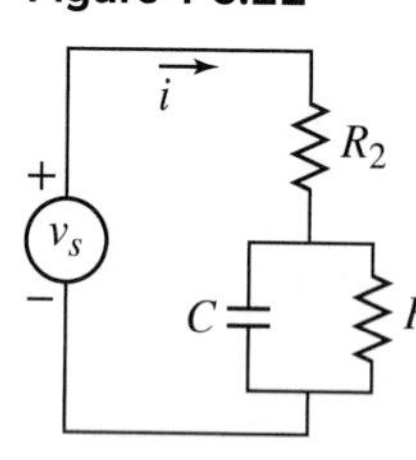

Figure P6.23

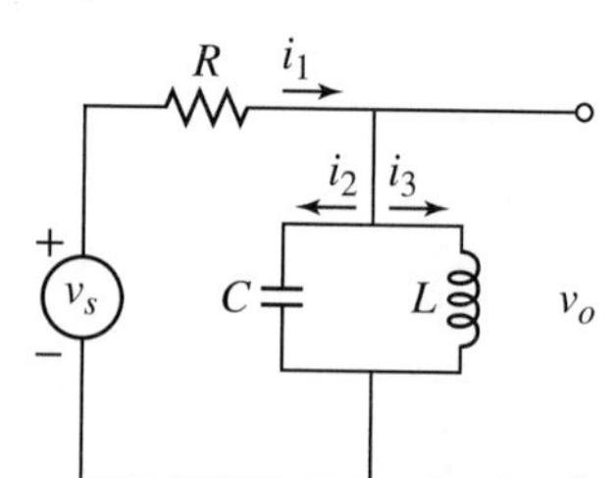

Figure P6.24

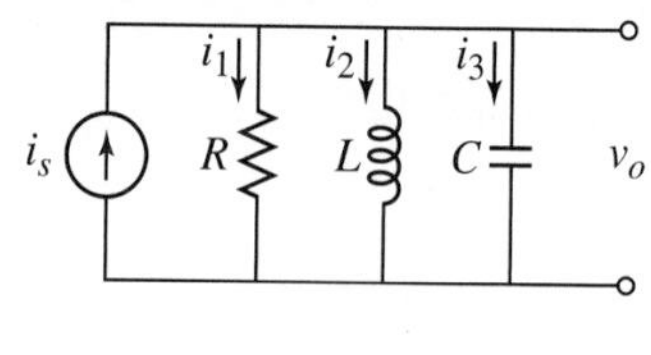

6.25 Use the impedance method to obtain the transfer function $V_o(s)/V_s(s)$ for the circuit shown in Figure P6.25.

Figure P6.25

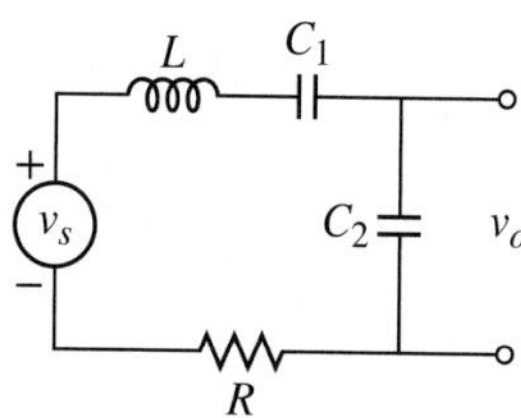

Figure P6.26

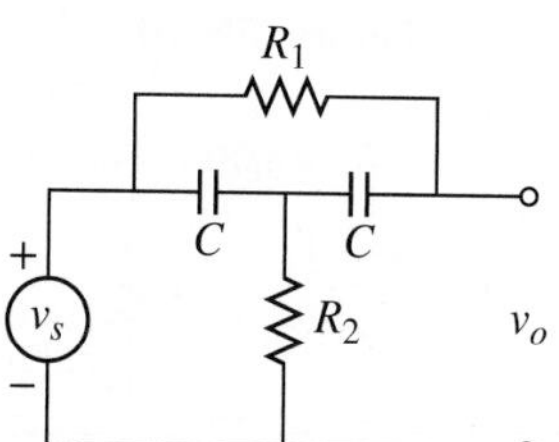

Figure P6.27

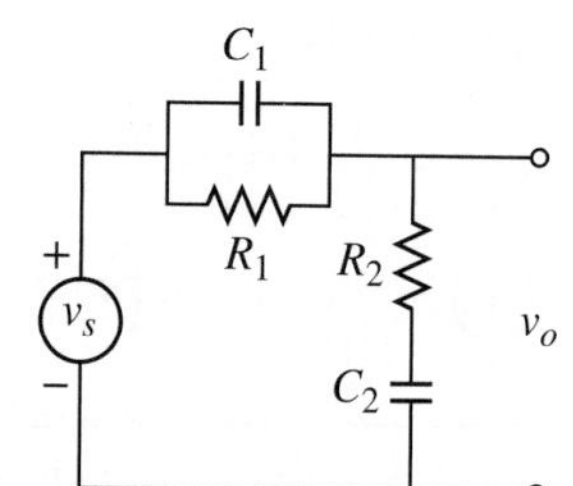

6.26 Use the impedance method to obtain the transfer function $V_o(s)/V_s(s)$ for the circuit shown in Figure P6.26.

6.27 Use the impedance method to obtain the transfer function $V_o(s)/V_s(s)$ for the circuit shown in Figure P6.27.

6.28 Draw a block diagram of the circuit shown in Figure P6.16. The inputs are v_1 and v_2. The output is i_2.

6.29 Draw a block diagram of the circuit shown in Figure P6.17. The inputs are v_1 and v_2. The output is v_3.

Section 6.4 Operational Amplifiers

6.30 Obtain the transfer function $V_o(s)/V_i(s)$ for the op-amp system shown in Figure P6.30.

6.31 Obtain the transfer function $V_o(s)/V_i(s)$ for the op-amp system shown in Figure P6.31.

Figure P6.30

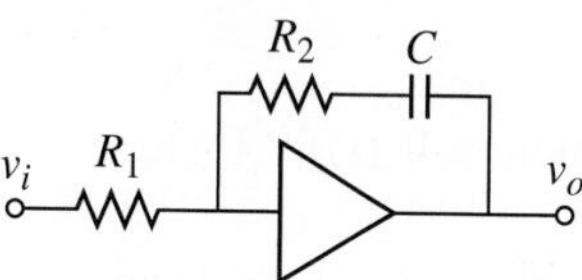

Figure P6.31

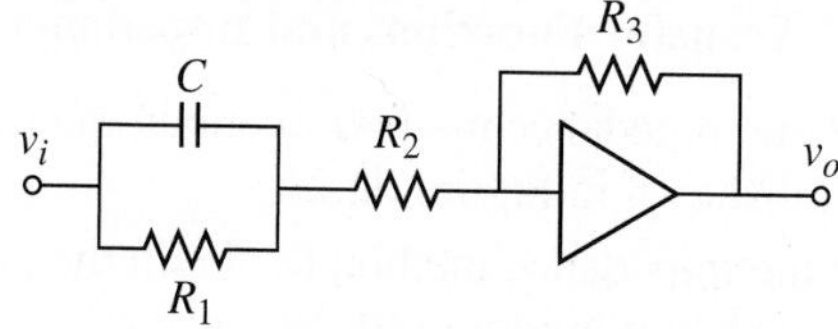

6.32 Obtain the transfer function $V_o(s)/V_i(s)$ for the op-amp system shown in Figure P6.32.

6.33 Obtain the transfer function $V_o(s)/V_i(s)$ for the op-amp system shown in Figure P6.33.

Figure P6.32

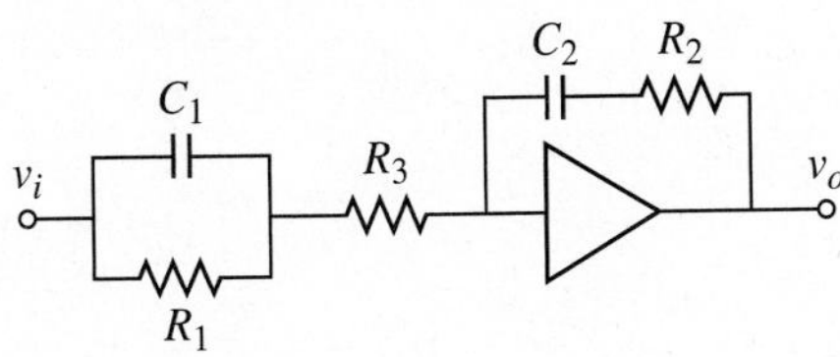

Figure P6.33

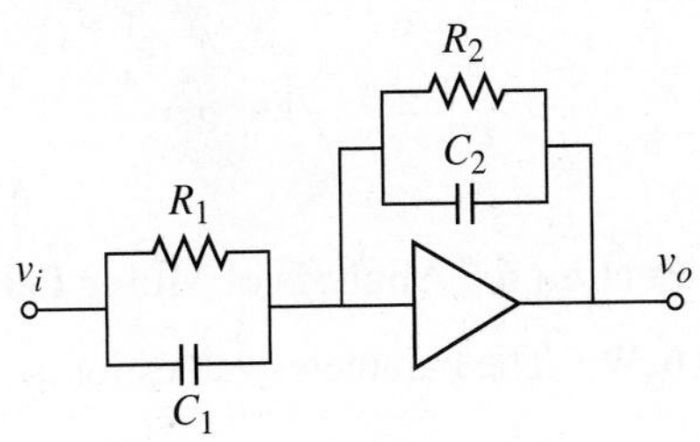

6.34 Obtain the transfer function $V_o(s)/V_i(s)$ for the op-amp system shown in Figure P6.34.

Figure P6.34

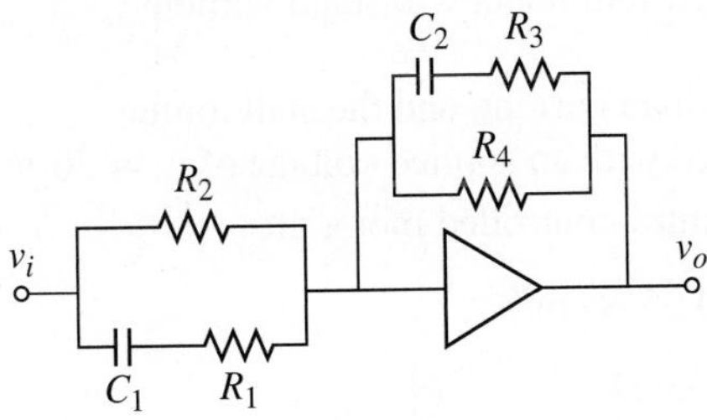

Section 6.5 Electric Motors

6.35 (a) Obtain the transfer function $\Theta(s)/V_i(s)$ for the D'Arsonval meter. (b) Use the final value theorem to obtain the expression for the steady-state value of the angle θ if the applied voltage v_i is a step function.

6.36 (a) Obtain the transfer function $\Omega(s)/T_L(s)$ for the field-controlled motor of Example 6.5.2. (b) Modify the field-controlled motor model in Example 6.5.2 so that the output is the angular displacement θ, rather than the speed ω, where $\omega = \dot{\theta}$. Obtain the transfer functions $\Theta(s)/V_f(s)$ and $\Theta(s)/T_L(s)$.

6.37 Modify the motor model given in Example 6.5.2 to account for a gear pair between the motor shaft and the load. The ratio of motor speed to load speed ω_L is N. The motor inertia is I_m and the motor damping is c_m. The load inertia is I_L and the load damping is c_L. The load torque T_L acts directly on the load inertia. Obtain the transfer functions $\Omega_L(s)/V_f(s)$ and $\Omega_L(s)/T_L(s)$.

6.38 Figure P6.38 is the circuit diagram of a speed-control system in which the dc motor voltage v_a is supplied by a generator driven by an engine. This system has been used on locomotives whose diesel engine operates most efficiently at one speed. The efficiency of the electric motor is not as sensitive to speed and thus can be used to drive the locomotive at various speeds. The motor voltage v_a is varied by changing the generator input voltage v_f. The voltage v_a is related to the generator field current i_f by $v_a = K_f i_f$.

Derive the system model relating the output speed ω to the voltage v_f, and obtain the transfer function $\Omega(s)/V_f(s)$.

Figure P6.38

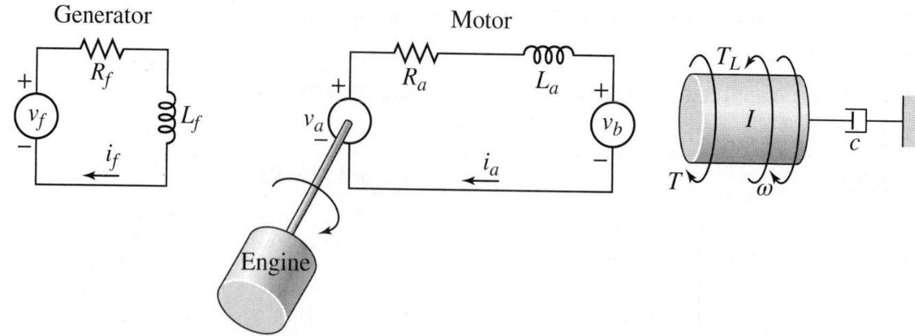

Section 6.6 Analysis of Motor Performance

6.39 The parameter values for a certain armature-controlled motor are

$$K_T = K_b = 0.3 \text{ N} \cdot \text{m/A}$$

$$c = 5 \times 10^{-4} \text{ N} \cdot \text{m} \cdot \text{s/rad} \qquad R_a = 1.2\ \Omega$$

The manufacturer's data states that the motor's maximum speed is 3500 rpm, and the maximum armature current it can withstand without demagnetizing is 40 A.

Compute the no-load speed, the no-load current, and the stall torque. Determine whether the motor can be used with an applied voltage of $v_a = 20$ V.

6.40 The parameter values for a certain armature-controlled motor are

$$K_T = K_b = 0.05 \text{ N} \cdot \text{m/A}$$

$$R_a = 0.6\ \Omega$$

$$L_a = 3 \times 10^{-3} \text{ H} \qquad I = 5 \times 10^{-5} \text{ kg} \cdot \text{m}^2$$

where I includes the inertia of the armature and that of the load. Investigate the effect of the damping constant c on the motor's characteristic roots and on its response to a step voltage input. Use the following values of c (in N · m · s/rad): $c = 0$, $c = 0.01$, and $c = 0.1$. For each case, estimate how long the motor's speed will take to become constant, and discuss whether or not the speed will oscillate before it becomes constant.

6.41 The following measurements were performed on a permanent magnet motor when the applied voltage was $v_a = 10$ V. The measured stall current was 15 A. The no-load speed was 300 rpm and the no-load current was 0.8 A. Estimate the values of K_b, K_T, R_a, and c.

Section 6.7 Case Study: Design of a Motion-Control System

6.42 A single link of a robot arm is shown in Figure P6.42. The arm mass is m and its center of mass is located a distance L from the joint, which is driven by a motor torque T_m through spur gears. Suppose that the equivalent inertia felt at the motor shaft is 0.215 kg · m^2. As the arm rotates, the effect of the arm weight generates an opposing torque that depends on the arm angle, and is therefore nonlinear. For this problem, however, assume that the effect of the opposing torque is a constant 4.2 N · m at the motor shaft. Neglect damping in the system. It is desired to have the motor shaft rotate through $3\pi/4$ rad in a total time of 2 s, using a trapezoidal speed profile with $t_1 = 0.3$ s and $t_2 = 1.7$ s.

The given motor parameters are $R_a = 4\ \Omega$, $L_a = 3 \times 10^{-3}$ H, and $K_T = 0.3$ N · m/A. Compute the energy consumption per cycle; the maximum required torque, current, and voltage; the rms torque; and the rms current.

6.43 A conveyor drive system to produce translation of the load is shown in Figure P6.43. Suppose that the equivalent inertia felt at the motor shaft is 0.05 kg · m^2, and that the effect of Coulomb friction in the system produces an opposing torque of 3.6 N · m at the motor shaft. Neglect damping in the system. It is desired to have the motor shaft rotate through 11 revolutions in a total time of 3 s, using a trapezoidal speed profile with $t_1 = 0.5$ s and $t_2 = 2.5$ s.

The given motor parameters are $R_a = 3\ \Omega$, $L_a = 4 \times 10^{-3}$ H, and $K_T = 0.4$N · m/A. Compute the energy consumption per cycle; the maximum required torque, current, and voltage; the rms torque; and the rms current.

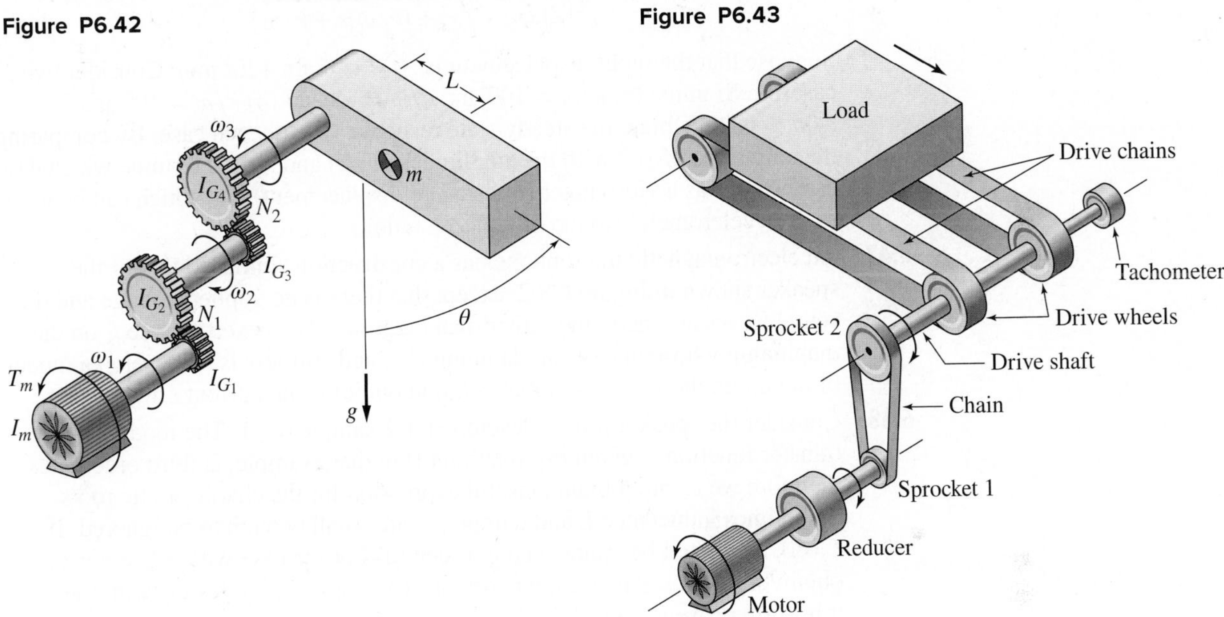

Figure P6.42

Figure P6.43

Section 6.8 Sensors and Electroacoustic Devices

6.44 Consider a reducer consisting of two spur gears whose diameters are D_1 and D_2. The gear whose diameter is D_1 is attached to the motor and its inertia is I_1. The reducer ratio is $N = D_2/D_1$. Model the gears as simple cylinders and derive a formula for the reflected inertia in terms of I_1 and N.

6.45 Use the values in the table in Example 6.7.1 except for the reducer ratio N, which we will treat as a variable.

a. Use inertia matching to determine a suitable value for N, (1) neglecting the reflected reducer inertia, and (2) assuming that the reflected reducer inertia I_r is one-half the motor inertia.

b. Taking into account load torque and the maximum required motor torque, compute the value of N to minimize the maximum acceleration.

c. Express the energy consumption in the form

$$E = \frac{R}{K_T^2}F$$

and obtain the expression for F as a function of N.

d. Assume that the reflected reducer inertia I_r is 0 and compute the value of N to minimize F.

e. Assume that the reflected reducer inertia I_r is one-half the motor inertia, and compute the value of N to minimize F.

6.46 Consider the accelerometer model in Section 6.8. Its transfer function can be expressed as

$$\frac{Y(s)}{Z(s)} = -\frac{s^2}{s^2 + (c/m)s + k/m}$$

Suppose that the input displacement is $z(t) = 10 \sin 120t$ mm. Consider two cases, in SI units: (a) $k/m = 100$ and $c/m = 18$ and (b) $k/m = 10^6$ and $c/m = 1800$. Obtain the steady-state response $y(t)$ for each case. By comparing the amplitude of $y(t)$ with the amplitudes of $z(t)$ and $\ddot{z}(t)$, determine which case can be used as a vibrometer (to measure displacement) and which can be used as an accelerometer (to measure acceleration).

6.47 An electromagnetic microphone has a construction similar to that of the speaker shown in Figure 6.8.2, except that there is no applied voltage and the sound waves are incoming rather than outgoing. They exert a force f_s on the diaphragm whose mass is m, damping is c, and stiffness is k. Develop a model of the microphone, whose input is f_s and output is the current i in the coil.

6.48 Consider the speaker model developed in Example 6.8.1. The model, whose transfer function is given by equation (3) in that example, is third order and therefore we cannot obtain a useful expression for the characteristic roots. Sometimes inductance L and damping c are small enough to be ignored. If $L = 0$, the model becomes second order. (a) For the case where $L = c = 0$, obtain the expressions for the two roots. (b) Compare the results with the third-order case where

$$m = 0.002 \text{ kg} \qquad k = 4 \times 10^5 \text{ N/m}$$
$$K_f = 16 \text{ N/A} \qquad K_b = 13 \text{ V} \cdot \text{s/m}$$
$$R = 12\ \Omega \qquad L = 10^{-3} \text{ H}$$
$$c = 0$$

Section 6.9 MATLAB Applications

6.49 The parameter values for a certain armature-controlled motor are

$$K_T = K_b = 0.3 \text{ N} \cdot \text{m/A}$$
$$c = 5 \times 10^{-4} \text{ N} \cdot \text{m} \cdot \text{s/rad} \qquad R_a = 0.8\ \Omega$$
$$L_a = 4 \times 10^{-3} \text{ H} \qquad I = 5 \times 10^{-4} \text{ kg} \cdot \text{m}^2$$

where c and I include the effect of the load. The load torque is zero.

Use MATLAB to obtain a plot of the step response of $i_a(t)$ and $\omega(t)$ if the applied voltage is $v_a = 10$ V. Determine the peak value of $i_a(t)$.

6.50 Consider the motor whose parameters are given in Problem 6.49. Use MATLAB to obtain a plot of the response of $i_a(t)$ and $\omega(t)$ if the applied voltage is the modified step $v_a(t) = 10(1 - e^{-100t})$ V. Determine the peak value of $i_a(t)$.

6.51 Consider the circuit shown in Figure P6.51. The parameter values are $R = 10^3\ \Omega$, $C = 2 \times 10^{-6}$ F, and $L = 2 \times 10^{-3}$ H. The voltage v_1 is a step input of magnitude 5 V, and the voltage v_2 is sinusoidal with frequency of 60 Hz and an amplitude of 4 V. The initial conditions are zero. Use MATLAB to obtain a plot of the current response $i_3(t)$.

Figure P6.51

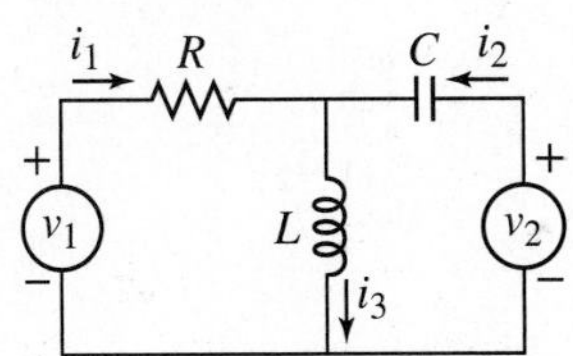

6.52 The parameter values for a certain armature-controlled motor are

$$K_T = K_b = 0.2\ \text{N} \cdot \text{m/A}$$
$$c = 3 \times 10^{-4}\ \text{N} \cdot \text{m} \cdot \text{s/rad} \qquad R_a = 0.8\ \Omega$$
$$L_a = 4 \times 10^{-3}\ \text{H} \qquad I = 4 \times 10^{-4}\ \text{kg} \cdot \text{m}^2$$

The system uses a gear reducer with a reduction ratio of 3:1. The load inertia is 10^{-3} kg · m^2, the load torque is 0.04 N · m, and the load-damping constant is 1.8×10^{-3} N · m · s/rad.

Use MATLAB to obtain a plot of the step response of $i_a(t)$ and $\omega(t)$ if the applied voltage is $v_a = 20$ V. Determine the peak value of $i_a(t)$.

6.53 The parameter values for a certain armature-controlled motor are

$$K_T = K_b = 0.05\ \text{N} \cdot \text{m/A}$$
$$c = 0 \qquad R_a = 0.8\ \Omega$$
$$L_a = 3 \times 10^{-3}\ \text{H} \qquad I = 8 \times 10^{-5}\ \text{kg} \cdot \text{m}^2$$

where I includes the inertia of the armature and that of the load. The load torque is zero. The applied voltage is a trapezoidal function defined as follows.

$$v(t) = \begin{cases} 60t & 0 \le t \le 0.5 \\ 30 & 0.5 < t < 2 \\ 60(2.5 - t) & 2 \le t \le 2.5 \\ 0 & 2.5 < t \le 4 \end{cases}$$

a. Use MATLAB to obtain of plot of the response of $i_a(t)$ and $\omega(t)$.
b. Compute the energy consumption per cycle; the maximum required torque, current, and voltage; the rms torque; and the rms current.

6.54 A single link of a robot arm is shown in Figure P6.42. The arm mass is m and its center of mass is located a distance L from the joint, which is driven by a motor torque T_m through spur gears. Suppose that the equivalent inertia felt at the motor shaft is 0.215 kg · m^2. As the arm rotates, the effect of the arm weight generates an opposing torque that depends on the arm angle, and is therefore nonlinear. The effect of the opposing torque at the motor shaft is $4.2 \sin \theta$ N · m. Neglect damping in the system. It is desired to have the motor shaft rotate through $3\pi/4$ rad in a total time of 2 s, using a trapezoidal speed profile with $t_1 = 0.3$ s and $t_2 = 1.7$ s.

The given motor parameters are $R_a = 4\ \Omega$, $L_a = 3 \times 10^{-3}$ H, and $K_T = 0.3$ N · m/A. Use MATLAB to obtain a plot of the response of the motor current and the motor speed.

Section 6.10 Simulink Applications

6.55 Consider the circuit shown in Figure P6.51. The parameter values are $R = 10^4\ \Omega$, $C = 2 \times 10^{-6}$ F, and $L = 2 \times 10^{-3}$ H. The voltage v_1 is a single pulse of magnitude 5 V and duration 0.05 s, and the voltage v_2 is sinusoidal with frequency of 60 Hz and an amplitude of 4 V. The initial conditions are zero. Use Simulink to obtain a plot of the current response $i_3(t)$.

6.56 Consider the circuit shown in Figure P6.56. The parameter values are $R = 2 \times 10^4\ \Omega$ and $C = 3 \times 10^{-6}$ F. The voltage v_s is $v_s(t) = 12u_s(t) + 3 \sin 120\pi t$ V. The initial conditions are zero. Use Simulink to obtain a plot of the responses $v_o(t)$ and $v_1(t)$.

Figure P6.56

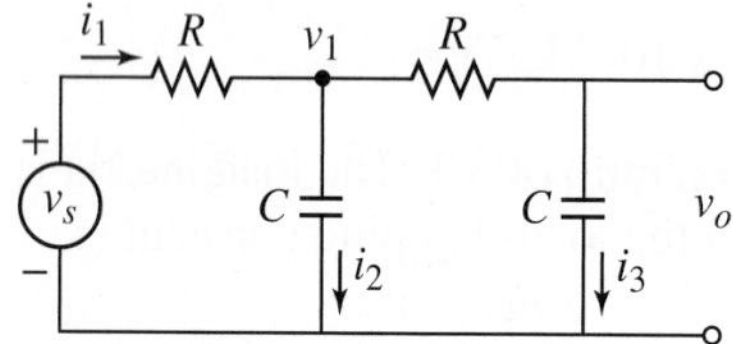

6.57 The parameter values for a certain armature-controlled motor are

$$K_T = K_b = 0.2\ \text{N} \cdot \text{m/A}$$
$$c = 5 \times 10^{-4}\ \text{N} \cdot \text{m} \cdot \text{s/rad} \qquad R_a = 0.8\ \Omega$$
$$L_a = 4 \times 10^{-3}\ \text{H} \qquad I = 5 \times 10^{-4}\ \text{kg} \cdot \text{m}^2$$

where c and I include the effect of the load. The load torque is zero.

a. Use Simulink to obtain a plot of the step response of the motor torque and speed if the applied voltage is $v_a = 10$ V. Determine the peak value of the motor torque.

b. Now suppose that the motor torque is limited to one-half the peak value found in part (a). Use Simulink to obtain a plot of the step response of the motor torque and speed if the applied voltage is $v_a = 10$ V.

6.58 The parameter values for a certain armature-controlled motor are

$$K_T = K_b = 0.05\ \text{N} \cdot \text{m/A}$$
$$c = 0 \qquad R_a = 0.8\ \Omega$$
$$L_a = 3 \times 10^{-3}\ \text{H} \qquad I = 8 \times 10^{-5}\ \text{kg} \cdot \text{m}^2$$

where I includes the inertia of the armature and that of the load. The load torque is zero. The applied voltage is a trapezoidal function defined as follows.

$$v(t) = \begin{cases} 60t & 0 \le t \le 0.5 \\ 30 & 0.5 < t < 2 \\ 60(2.5 - t) & 2 \le t \le 2.5 \\ 0 & 2.5 < t \le 4 \end{cases}$$

A trapezoidal profile can be created by adding and subtracting ramp functions starting at different times. Use several Ramp source blocks and Sum blocks in Simulink to create the trapezoidal input. Obtain a plot of the response of $i_a(t)$ and $\omega(t)$.

CHAPTER 7

Fluid and Thermal Systems

CHAPTER OUTLINE

CHAPTER OBJECTIVES

When you have finished this chapter, you should be able to

1. Apply the conservation of mass principle to model simple hydraulic and pneumatic systems.
2. Determine the appropriate resistance relation to use for laminar, turbulent, and orifice flow.
3. Develop a dynamic model of hydraulic and pneumatic systems containing one or more fluid containers.
4. Determine the appropriate thermal resistance relation to use for conduction, convection, and radiation heat transfer.
5. Develop a model of a thermal process having one or more thermal storage compartments.
6. Apply MATLAB and Simulink to solve fluid and thermal system models.

A fluid system uses one or more fluids to achieve its purpose. The dampers, shock absorbers, and door closer we saw in Chapter 4 are examples of fluid systems because they depend on the viscous nature of a fluid to provide damping. A fluid might be either a liquid or a gas. Part I of this chapter concerns the study of fluid systems, which can be divided into *hydraulics* and *pneumatics*. Hydraulics is the study of systems in which the fluid is *incompressible*, that is, its density stays approximately constant over a range of pressures. Pneumatics is the study of systems in which the fluid is *compressible*. Hydraulics and pneumatics share a common modeling principle: conservation of mass. Modeling pneumatic systems also requires application of thermodynamics, because the temperature and density of a gas can change when its pressure changes.

Thus, pneumatics provides a bridge to the treatment of *thermal* systems, which is the subject of Part II of the chapter. Thermal systems are systems that operate due to temperature differences. They thus involve the flow and storage of thermal energy, or *heat*, and conservation of heat energy forms the basis of our thermal models.

Part III illustrates applications of MATLAB and Simulink to fluid and thermal systems.

Fluid and thermal systems are more complicated than most electrical and mechanical systems. While, for example, there are formulas available to compute the spring constant of typical elastic elements, few formulas are available for the coefficients that will appear in our fluid and thermal models, and the coefficients' values often must be determined experimentally. For this reason, the methods for developing models from data, covered in Chapter 8 and Appendix C, are most important for modeling fluid and thermal systems. ■

PART I. FLUID SYSTEMS

Hydraulics and pneumatics share a common modeling principle: conservation of mass. It will form the basis of all our models of such systems.

7.1 CONSERVATION OF MASS

For incompressible fluids, conservation of mass is equivalent to conservation of volume, because the fluid density is constant. If we know the mass density ρ and the volume flow rate, we can compute the mass flow rate. That is, $q_m = \rho q_v$, where q_m and q_v are the mass and volume flow rates. The FPS and SI units for mass flow rate are slug/sec and kg/s, respectively. The units for volume rates are ft^3/sec and m^3/s, respectively. Other common units for volume are the *U.S. gallon*, which is 0.13368 ft^3, and the *liter*, which is 0.001 m^3.

The units for mass density are slug/ft^3 and kg/m^3. Sometimes one encounters *weight density*, whose common symbol is γ. Its units are lb/ft^3 or N/m^3, and it is related to the mass density as $\gamma = \rho g$, where g is the acceleration due to gravity. The mass density of fresh water near room temperature is 1.94 slug/ft^3, or 1000 kg/m^3. The mass density of air at sea level and near room temperature is approximately 0.0023 slug/ft^3 or 1.185 kg/m^3.

Pressure is the force per unit area that is exerted by the fluid. The FPS and SI units of pressure are lb/ft^2 and the Pascal (1 Pa = 1 N/m^2), respectively. Another common unit is psi (lb/in.^{2}). At sea level near room temperature, atmospheric pressure, usually abbreviated p_a, is 14.7 psi (2117 lb/ft^2) or 1.0133×10^5 Pa. *Gage pressure* is the pressure difference between the *absolute pressure* and atmospheric pressure, and is often abbreviated as psig. For example, 3 psig is 17.7 psi absolute (which is abbreviated as psia).

Hydrostatic pressure is the pressure that exists in a fluid at rest. It is caused by the weight of the fluid. For example, the hydrostatic pressure at the bottom of a column of fluid of height h is $\rho g h$. If the atmospheric pressure above the column of liquid is p_a, then the total pressure at the bottom of the column is $\rho g h + p_a$.

Conservation of mass can be stated as follows. For a container holding a mass of fluid m, the time rate of change $\dot{m}$ of mass in the container must equal the total mass inflow rate minus the total mass outflow rate. That is,

$$\dot{m} = q_{mi} - q_{mo} \tag{7.1.1}$$

where q_{mi} is the mass inflow rate and q_{mo} is the mass outflow rate.

The fluid mass m is related to the container volume V by $m = \rho V$. For an incompressible fluid, ρ is constant, and thus $\dot{m} = \rho \dot{V}$. Let q_{vi} and q_{vo} be the total *volume* inflow and outflow rates. Thus, $q_{mi} = \rho q_{vi}$, and $q_{mo} = \rho q_{vo}$. Substituting these relationships into (7.1.1) gives

$$\rho \dot{V} = \rho q_{vi} - \rho q_{vo}$$

Cancel ρ to obtain

$$\dot{V} = q_{vi} - q_{vo} \tag{7.1.2}$$

This is a statement of conservation of *volume* for the fluid, and it is equivalent to conservation of mass, equation (7.1.1), when the fluid is incompressible.

A common hydraulic actuator is the piston-and-cylinder actuator used on many types of heavy equipment, such as the backhoe shown in Figure 7.1.1. When the operator moves a handle, hydraulic fluid under high pressure is sent through the line to the cylinder. The fluid acts on the piston within the cylinder and produces a force that is equal to the pressure times the piston area. This large force moves the linkage. Example 7.1.1 develops a simple model of such a device.

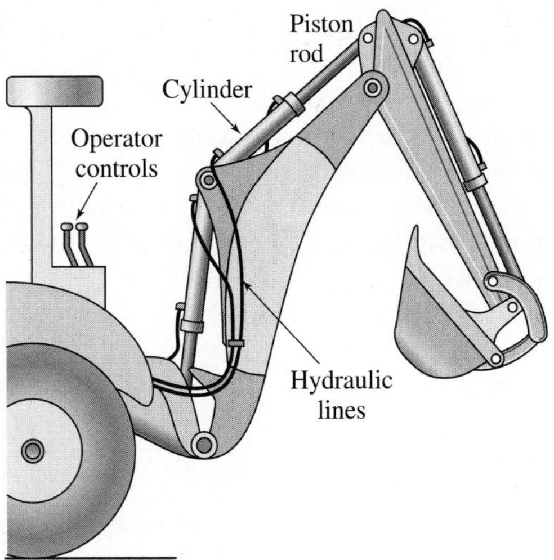

Figure 7.1.1 A backhoe.

A Hydraulic Cylinder

EXAMPLE 7.1.1

■ Problem

Figure 7.1.2a shows a cylinder and piston connected to a load mass m, which slides on a frictionless surface. Part (b) of the figure shows the piston rod connected to a rack-and-pinion gear. The pressures p_1 and p_2 are applied to each side of the piston by two pumps. Assume the piston rod diameter is small compared to the piston area, so the effective piston area A is the same on both sides of the piston. Assume also that the piston and rod mass have been lumped into m and that any friction is negligible. (a) Develop a model of the motion of the displacement x of the

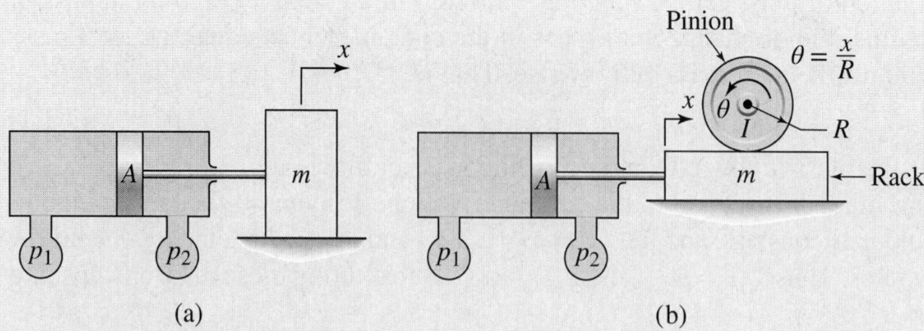

Figure 7.1.2 A hydraulic cylinder for (a) translating a mass and for (b) rotating a pinion gear.

mass in part (a) of the figure, assuming that p_1 and p_2 are given functions of time. Also, obtain the expression for the mass flow rate that must be delivered or absorbed by the two pumps. (b) Develop a model of the displacement x in part (b) of the figure. The inertia of the pinion and the load connected to the pinion is I.

■ Solution

a. Assuming that $p_1 > p_2$, the net force acting on the piston and mass m is $(p_1 - p_2)A$, and thus from Newton's law,

$$m\ddot{x} = (p_1 - p_2)A$$

Because p_1 and p_2 are given functions of time, we can integrate this equation once to obtain the velocity:

$$\dot{x}(t) = \dot{x}(0) + \frac{A}{m}\int_0^t [p_1(u) - p_2(u)]\, du$$

The rate at which fluid volume is swept out by the piston is $A\dot{x}$, and thus if $\dot{x} > 0$, the pump providing pressure p_1 must supply fluid at the mass rate $\rho A\dot{x}$, and the pump providing pressure p_2 must absorb fluid at the same mass rate.

b. Because we want an expression for the displacement x, we obtain an expression for the equivalent mass of the rack, pinion, and load. The kinetic energy of the system is

$$\text{KE} = \frac{1}{2}m\dot{x}^2 + \frac{1}{2}I\dot{\theta}^2 = \frac{1}{2}\left(m + \frac{I}{R^2}\right)\dot{x}^2$$

because $R\dot{\theta} = \dot{x}$.

Thus, the equivalent mass is

$$m_e = m + \frac{I}{R^2}$$

The required model can now be obtained by replacing m with m_e in the model developed in part (a).

7.2 FLUID CAPACITANCE

Sometimes it is very useful to think of fluid systems in terms of electrical circuits. Table 7.2.1 gives the fluid quantity, its common nomenclature, its linear relation, and its analogous electrical property. *Fluid resistance* is the relation between pressure and mass flow rate. *Fluid capacitance* is the relation between pressure and stored mass. Fluid resistance relates to energy dissipation while fluid capacitance relates to potential energy. *Fluid inertance* relates to fluid acceleration and kinetic energy.

Table 7.2.1 Analogous fluid and electrical quantities.

Fluid quantity	Electrical quantity
Fluid mass, m	Charge, Q
Mass flow rate, q_m	Current, i
Pressure, p	Voltage, v
Fluid linear resistance, R	Electrical resistance, R
$R = p/q_m$	$R = v/i$
Fluid capacitance, C	Electrical capacitance, C
$C = m/p$	$C = Q/v$
Fluid inertance, I	Electrical inductance, L
$I = p/(dq_m/dt)$	$L = v/(di/dt)$

Fluid systems obey two laws that are analogous to Kirchhoff's current and voltage laws; these laws are the *continuity* and the *compatibility* laws. The continuity law is simply a statement of conservation of fluid mass. This says that the total mass flow into a junction must equal the total flow out of the junction. This is analogous to Kirchhoff's current law. Flow through two rigid pipes joined together to make one pipe is an example where this applies. If, however, the flow is through *flexible* tubes that can expand and contract under pressure, then the outflow rate is not the sum of the inflow rates. This is an example where fluid mass can accumulate within the system and is analogous to having a capacitor in an electrical circuit.

The compatibility law is analogous to Kirchhoff's voltage law, which states that the sum of signed voltage differences around a closed loop must be zero. It is an expression of conservation of energy. The compatibility law states that the sum of signed pressure differences around a closed loop must be zero.

Figure 7.2.1 shows the commonly used symbols for fluid system elements. The resistance symbol is used to represent *fixed* resistances, for example, due to pipe flow, orifice flow, or a restriction. A valve that can be manually adjusted, such as a faucet, is a *variable* resistance and has a slightly different symbol. An actuated valve, driven, for example, by an electric motor or a pneumatic device, has a different symbol. Such valves are usually operated under computer control.

Just as there are ideal voltage and current sources in electrical systems, so we use ideal pressure and flow sources in our fluid system models. An *ideal pressure source* is capable of supplying the specified pressure at any flow rate. An *ideal flow source* is capable of supplying the specified flow. These ideal sources are approximations to real devices such as pumps.

Fluid capacitance is the relation between stored fluid mass m and the resulting pressure p caused by the stored mass. Figure 7.2.2 illustrates this relation, which holds

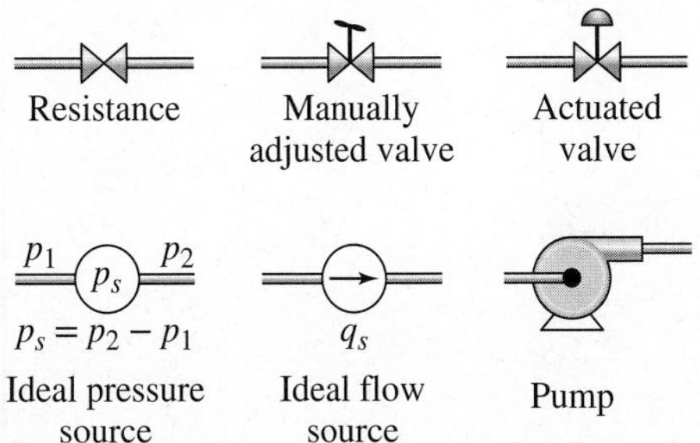

Figure 7.2.1 Fluid system symbols.

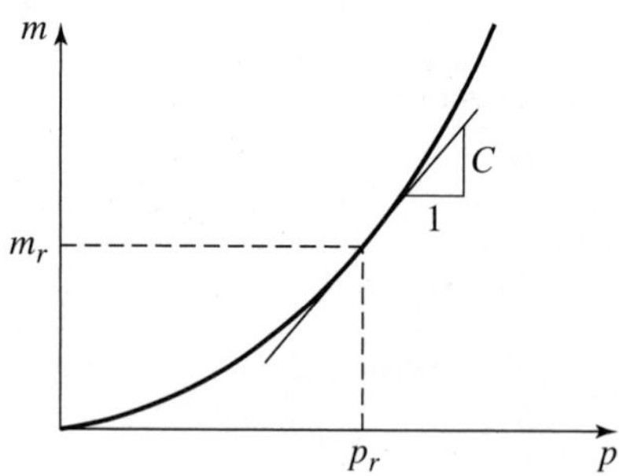

Figure 7.2.2 General fluid capacitance relation and its linear approximation.

for both pneumatic and hydraulic systems. At a particular reference point (p_r, m_r) the slope is C, where

$$C = \left.\frac{dm}{dp}\right|_{p=p_r} \tag{7.2.1}$$

Thus, fluid capacitance C is the ratio of the change in stored mass to the change in pressure.

EXAMPLE 7.2.1

Capacitance of a Storage Tank

■ Problem

Consider the tank shown in Figure 7.2.3. Assume that the sides are vertical so that the cross-sectional area A is constant. This is the case, for example, with a cylindrical tank whose horizontal cross section is circular, or with a tank having vertical sides and a rectangular horizontal cross section. (a) Derive the expression for the tank's capacitance. (b) Derive the differential equation model for the pressure p and the height h if fluid flows into the top of the tank at the mass flow rate $q_{mi}(t)$.

■ Solution

a. Because the tank's sides are vertical, the liquid height h is related to m, the liquid mass in the tank, by $m = \rho A h$. The total pressure at the bottom of the tank is $\rho g h + p_a$, but the pressure due only to the stored fluid mass is $p = \rho g h$. We can therefore express the pressure as a function of the mass m stored in the tank as $p = mg/A$.

Thus,

$$m = \frac{pA}{g}$$

and the capacitance of the tank is given by

$$C = \frac{dm}{dp} = \frac{A}{g}$$

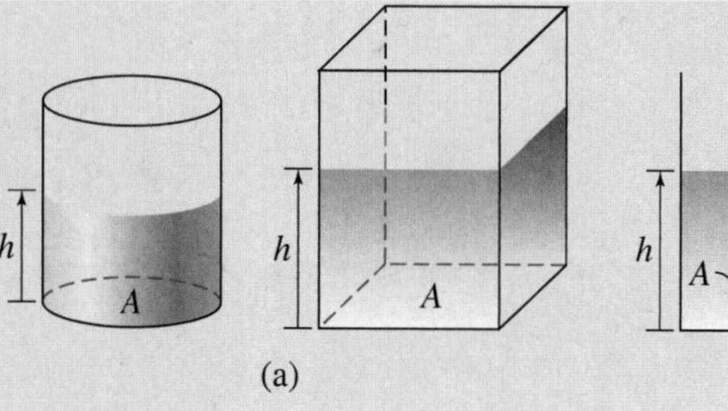

Figure 7.2.3 Capacitance of a storage tank having vertical sides.

b. Conservation of mass gives

$$\frac{dm}{dt} = q_{mi} \tag{1}$$

But

$$\frac{dm}{dt} = \frac{dm}{dp}\frac{dp}{dt} = C\frac{dp}{dt}$$

Thus,

$$C\frac{dp}{dt} = q_{mi} \tag{2}$$

since $p = \rho g h + p_a$.

$$C\frac{dp}{dt} = C\rho g\frac{dh}{dt} = q_{mi}$$

Since $Cg = A$, this equation simplifies to

$$\rho A\frac{dh}{dt} = q_{mi} \tag{3}$$

Equations (1), (2), and (3) are alternative, but equivalent, hydraulic models of a container of fluid. They suggest that either mass m, pressure p, or height h can be chosen as the model variable. These variables are all indicators of the system potential energy, and as such any one can be chosen as a state variable. If the container cross-sectional area is constant, then $V = Ah$, and thus the liquid volume V can also be used as the model variable.

When the container does not have vertical sides, the cross-sectional area A is a function of the liquid height h, and the relations between m and h and between p and m are nonlinear. In such cases, there is no single value for the container's capacitance.

Capacitance of a V-Shaped Trough

EXAMPLE 7.2.2

■ Problem

A V-shaped trough is shown in Figure 7.2.4a. Derive the dynamic models for the bottom pressure p and the height h. The mass inflow rate is $q_{mi}(t)$, and there is no outflow.

■ Solution

From part (b) of the figure, $D = 2h\tan\theta$, and the vertical cross-sectional area of the liquid is $hD/2$. Thus, the fluid mass is given by

$$m = \rho V = \rho\left(\frac{1}{2}hD\right)L = (\rho L\tan\theta)h^2$$

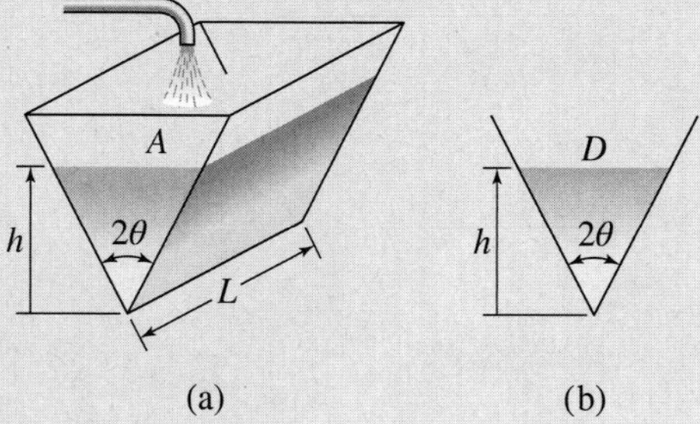

Figure 7.2.4 A V-shaped trough.

But $p = \rho g h$ and thus,

$$m = (\rho L \tan\theta)\left(\frac{p}{\rho g}\right)^2 = \left(\frac{L\tan\theta}{\rho g^2}\right)p^2$$

$$\frac{dm}{dt} = \frac{2L\tan\,\theta}{\rho g^2}p\frac{dp}{dt} = q_{mi}$$

which is a nonlinear equation because of the product $p\dot{p}$. We can obtain the model for the height by substituting $h = p/\rho g$. The result is

$$(2\rho L \tan\theta)h\frac{dh}{dt} = q_{mi}$$

7.3 FLUID RESISTANCE

Fluid meets resistance when flowing through a conduit such as a pipe, through a component such as a valve, or even through a simple opening or *orifice*, such as a hole. We now consider appropriate models for each type of resistance.

The mass flow rate $\hat{q}_m$ through a resistance is related to the pressure difference $\hat{p}$ across the resistance. This relation, $\hat{p} = f(\hat{q}_m)$, is illustrated in general by Figure 7.3.1. We define the *fluid resistance* R_r as the slope of $f(\hat{q}_m)$ evaluated at a reference equilibrium condition (p_r, q_{mr}). That is,

$$R_r = \left.\frac{d\hat{p}}{d\hat{q}_m}\right|_{\hat{q}_m=\hat{q}_{mr}} \overset{\Delta}{=} \left(\frac{d\hat{p}}{d\hat{q}_m}\right)_r \tag{7.3.1}$$

If we need to obtain an approximate linear model of the pressure-flow rate relation, we can use a Taylor series expansion to linearize the expression $\hat{p}_r = f(\hat{q}_m)$ near a reference operating point (p_r, q_{mr}) as follows (after dropping the second order and higher terms in the expansion):

$$\hat{p} = p_r + \left(\frac{d\hat{p}}{d\hat{q}_m}\right)_r (q_m - q_{mr}) = p_r + R_r(q_m - q_{mr}) \tag{7.3.2}$$

where R_r is the linearized resistance defined by (7.3.1).

Referring to Figure 7.3.1, we define a new set of variables p and q_m, called *deviation variables*, that represent small but finite changes in $\hat{p}$ and $\hat{q}_m$ from their reference values p_r and q_{mr}. From Figure 7.3.1, we see that

$$p = \hat{p} - p_r \tag{7.3.3}$$

$$q_m = \hat{q}_m - q_{mr} \tag{7.3.4}$$

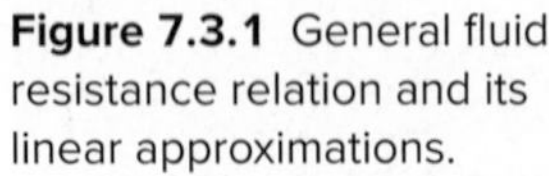

Figure 7.3.1 General fluid resistance relation and its linear approximations.

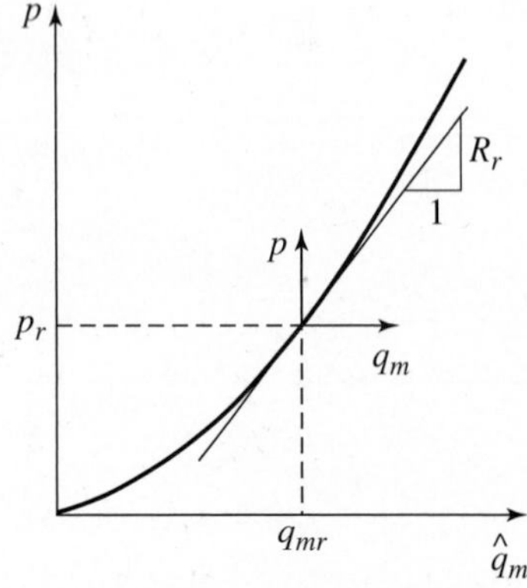

In terms of these deviation variables, we can rewrite (7.3.2) as

$$\hat{p} - p_r = R_r(\hat{q}_m - q_{mr})$$

or $p = R_r q_m$, or

$$q_m = \frac{p}{R_r} \tag{7.3.5}$$

which is a linear relation. Thus, the resistance R_r is called the linearized resistance. Thus, we may think of the linearized relation (7.3.5) as

$$\frac{\text{small pressure change}}{R_r} = \text{resulting small change in mass flow rate}$$

The values of p_r and q_{mr} depend on the particular application. So the resistance R_r depends on these values, as well as the functional form of $f(\hat{q}_m)$, which depends on the application.

In a limited number of cases, such as pipe flow under certain conditions, the relation of $\hat{p}$ versus $\hat{q}_m$ is linear so that $\hat{p} = R\hat{q}_m$, or

$$\hat{q}_m = \frac{\hat{p}}{R} \tag{7.3.6}$$

where R is the linear resistance.

In some other applications the relation is a square-root relation.

$$\hat{q}_m = \sqrt{\frac{\hat{p}}{B}} \tag{7.3.7}$$

where B is a constant that often must be determined empirically.

The relation (7.3.7) gives $\hat{p} = B\hat{q}_m^2$ and thus

$$R_r = \left(\frac{d\hat{p}}{d\hat{q}_m}\right)_r = 2Bq_{mr}$$

and

$$q_m = \frac{p}{2Bq_{mr}} \tag{7.3.8}$$

This is the linearized model corresponding to the relation (7.3.7). This relation can be expressed instead in terms of p_r as follows. Because $q_{mr} = \sqrt{p_r/B}$, we obtain

$$R_r = 2B\sqrt{p_r/B} = 2\sqrt{Bp_r}$$

and thus

$$q_m = \left(\frac{1}{2\sqrt{Bp_r}}\right)p \tag{7.3.9}$$

When only the curve of $\hat{p}$ versus $\hat{q}_m$ is available, we can obtain a linearized model by graphically computing the slope S of the tangent line that passes through the reference point (p_r, q_{mr}). The equivalent, linearized resistance R_r is the slope S.

The resistance symbol shown in Figure 7.3.2 represents all types of fluid resistance, whether linear or not. Although the symbol looks like a valve, it can represent fluid resistance due to other causes, such as pipe wall friction and orifices.

As with electrical resistances, linear fluid resistance elements obey the series and parallel combination rules. These are illustrated in Figure 7.3.2. Series fluid resistances carry the same flow rate; parallel fluid resistances have the same pressure difference across them.

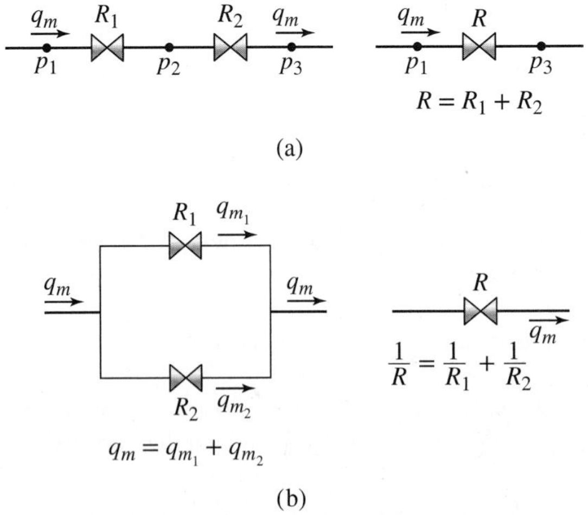

Figure 7.3.2 Combination of (a) series resistances and (b) parallel resistances.

$$R = R_1 + R_2$$

(a)

$$q_m = q_{m_1} + q_{m_2}$$

$$\frac{1}{R} = \frac{1}{R_1} + \frac{1}{R_2}$$

(b)

7.3.1 LAMINAR PIPE RESISTANCE

Fluid motion is generally divided into two types: *laminar flow* and *turbulent flow*. Laminar flow can be described as "smooth" in the sense that the average fluid particle velocity is the same as the actual particle velocity. If the flow is "rough," the *average* particle velocity will be less than the *actual* particle velocity, because the fluid particles meander while moving downstream. This is turbulent flow. You can see the difference between laminar and turbulent flow by slightly opening a faucet; the flow will be smooth. As you open the faucet more, eventually the flow becomes rough.

If the pipe flow is laminar, the following linear relation applies.

$$\hat{q}_m = \frac{\hat{p}}{R} \tag{7.3.10}$$

or equivalently

$$q_m = \frac{p}{R} \tag{7.3.11}$$

The laminar resistance for a level pipe of diameter D and length L is given by the *Hagen-Poiseuille* formula

$$R = \frac{128\mu L}{\pi \rho D^4} \tag{7.3.12}$$

where μ is the fluid viscosity. The viscosity is a measure of the "stickiness" of the fluid. Thus, molasses has a higher value of μ than that of water.

Not all pipe flow is laminar. A useful criterion for predicting the existence of laminar flow is the *Reynolds number* N_e, the ratio of the fluid's inertial forces to the viscosity forces. For a circular pipe,

$$N_e = \frac{\rho \bar{v} D}{\mu} \tag{7.3.13}$$

where $\bar{v} = q_v/(\pi D^2/4)$, the average fluid velocity. For $N_e > 2300$ the flow is often turbulent, while for $N_e < 2300$ laminar flow usually exists. The precise value of N_e above which the flow becomes turbulent depends on, for example, the flow conditions at the pipe inlet. However, the criterion is useful as a rule of thumb.

The resistance formula (7.3.12) applies only if the so-called "entrance length" L_e, which is the distance from the pipe entrance beyond which the velocity profile no longer changes with increasing distance, is much less than $0.06DN_e$. Because laminar flow can be expected only if $N_e < 2300$, L_e might be as long as 138 pipe diameters. Of course, for small Reynolds numbers, L_e is shorter. The smaller L_e is relative to the pipe length, the more reliable will be our resistance calculations.

7.3.2 TORRICELLI'S PRINCIPLE

An orifice can simply be a hole in the side of a tank or it can be a passage in a valve. An example of orifice flow is given in Figure 7.3.3, which shows the flow rate of water through a small hole in the side of a plastic milk bottle. The data is plotted in Figure 7.3.4. In Example C.1.2 in Appendix C we show that the fitted function is $f = 9.4h^{0.558}$, where f is the outflow rate in ml/s and the water height h is in centimeters. It turns out that the empirically determined exponent 0.558 is close to its theoretical value of 0.5, as we will now demonstrate.

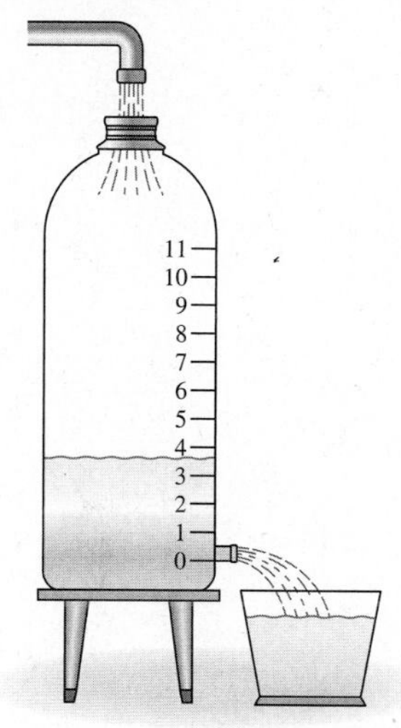

Figure 7.3.3 An experiment to determine flow rate versus liquid height.

Around 1640 Torricelli discovered that the flow rate through an orifice is proportional to the square root of the pressure difference. This observation can be simply derived by considering a mass m of fluid a height h above the orifice (see Figure 7.3.5). The potential energy of the mass is mgh. As the mass falls toward the orifice, its potential energy is converted to kinetic energy $mv^2/2$. If all the potential energy is converted to kinetic energy at the orifice, then $mgh = mv^2/2$, and the maximum speed the fluid mass can attain through the orifice is $v = \sqrt{2gh}$. Because the pressure drop across the orifice is $p = \rho gh$, we can express the maximum speed as $v = \sqrt{2p/\rho}$. Thus, the mass flow rate q_m through the orifice of area A_o can be no greater than $A_o \rho v = A_o \rho \sqrt{2p/\rho} = A_o \sqrt{2p\rho}$. The actual flow rate will be less than this value because of friction effects. To account for these frictional effects, we introduce a factor C_d in the flow rate equation as follows:

$$q_m = C_d A_o \sqrt{2p\rho} \tag{7.3.14}$$

The factor C_d is the *discharge coefficient*, which must lie in the range $0 < C_d \leq 1$. A typical value for water is 0.6.

Figure 7.3.4 Plot of flow rate data.

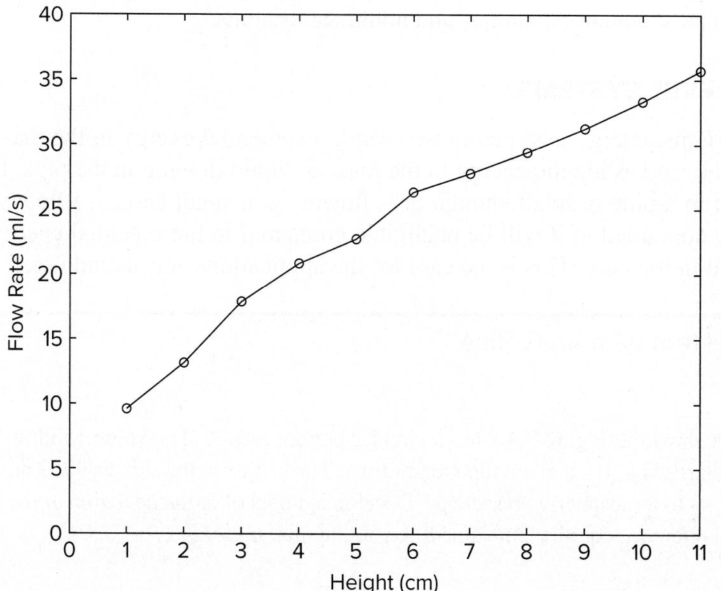

Figure 7.3.5 Derivation of Torricelli's principle.

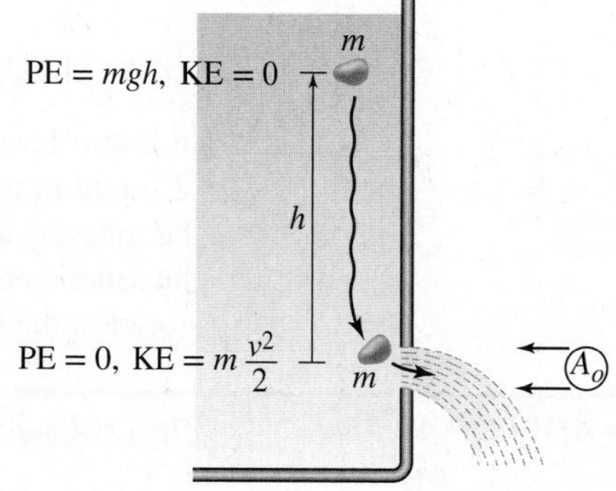

Because $p = \rho gh$, (7.3.14) can be expressed in terms of the volume flow rate q_v and the height h as follows:

$$q_v = \left(C_d A_o \sqrt{2g}\right) h^{0.5}$$

Thus, the theoretical value of the exponent (0.5) is close to the value obtained in the bottle experiment.

Equation (7.3.14) depends on the orifice area being small enough so that the pressure variation over the orifice area is negligible compared to the average pressure at the orifice. For a liquid-level system with a circular orifice, this implies that the liquid height above the orifice must be large compared to the orifice diameter.

The orifice relation (7.3.14) can be rearranged as:

$$q_m = C_d A_o \sqrt{2\rho}\sqrt{p} = \sqrt{\frac{p}{R_o}} \tag{7.3.15}$$

where the *orifice resistance* is defined as

$$R_o = \frac{1}{2\rho C_d^2 A_o^2} \tag{7.3.16}$$

7.3.3 TURBULENT AND COMPONENT RESISTANCE

For us, the practical importance of the difference between laminar and turbulent flow lies in the fact that laminar flow can be described by the *linear* relation (7.3.5), while turbulent flow is described by the *nonlinear* relation (7.3.7). Components, such as valves, elbow bends, couplings, porous plugs, and changes in flow area resist flow and usually induce turbulent flow at typical pressures, and (7.3.7) is often used to model them. Experimentally determined values of B are available for common types of components.

7.4 DYNAMIC MODELS OF HYDRAULIC SYSTEMS

In this section we consider a number of hydraulic system examples dealing with liquid-level systems, dampers, actuators, pumps, and nonlinear systems.

7.4.1 LIQUID-LEVEL SYSTEMS

In liquid-level systems, energy is stored in two ways: as potential energy in the mass of liquid in the tank, and as kinetic energy in the mass of liquid flowing in the pipe. If the mass of liquid in a pipe is small enough or is flowing at a small enough velocity, the kinetic energy contained in it will be negligible compared to the potential energy stored in the liquid in the tank. This is the case for the applications considered here.

EXAMPLE 7.4.1 Liquid-Level System with an Orifice

■ Problem

The cylindrical tank shown in Figure 7.4.1 has a circular bottom area A. The volume inflow rate from the flow source is $\hat{q}_{vi}(t)$, a given function of time. The orifice in the side wall has an area A_o and discharges to atmospheric pressure p_a. Develop a model of h, the deviation of the liquid height from a reference equilibrium value h_r, assuming that $h_1 > L$.

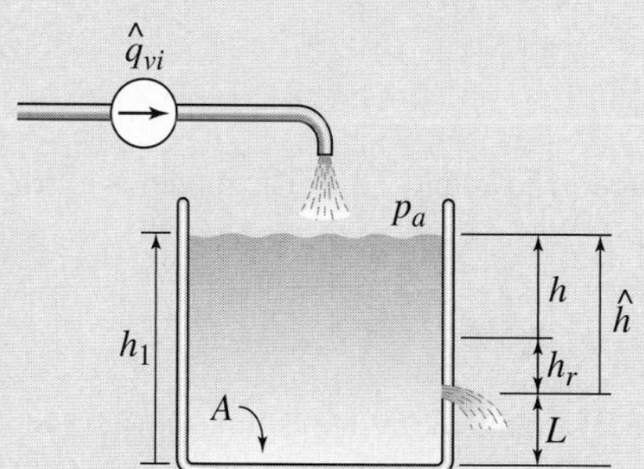

Figure 7.4.1 A liquid-level system with an orifice.

■ **Solution**

If the inflow rate $\hat{q}_{vi}$ is held constant at the rate q_{vir}, the liquid level eventually becomes constant at the height h_r. Using the orifice flow relation (7.3.14), this height can be found from

$$\rho q_{vir} = C_d A_o \sqrt{2\rho(\rho g h_r)}$$

or

$$h_r = \frac{1}{2g}\left(\frac{q_{vir}}{C_d A_o}\right)^2$$

So if we are given q_{vir}, we can determine h_r, or vice versa.

Noting that $h_1 = h + h_r + L$, the rate of change of liquid mass in the tank is

$$\frac{d(\rho A h_1)}{dt} = \rho A \frac{d(h + h_r + L)}{dt} = \rho A \frac{dh}{dt}$$

Conservation of mass implies that

$$\rho A \frac{dh}{dt} = \rho \hat{q}_{vi} - C_d A_o \sqrt{2\hat{p}\rho} \tag{1}$$

where the pressure drop across the orifice is

$$\hat{p} = p_a + \rho g(h + h_r) - p_a = \rho g(h + h_r)$$

Therefore, equation (1) becomes

$$\rho A \frac{dh}{dt} = \rho \hat{q}_{vi} - C_d A_o \sqrt{2g\rho^2(h + h_r)}$$

Canceling ρ gives the desired model.

$$A \frac{dh}{dt} = \hat{q}_{vi} - C_d A_o \sqrt{2g(h + h_r)}$$

or, because $\hat{h} = h + h_r$,

$$A \frac{d\hat{h}}{dt} = \hat{q}_{vi} - C_d A_o \sqrt{2g\hat{h}} \tag{2}$$

Note that the height L does not appear in the model because the liquid below the orifice does not affect the pressure at the orifice.

EXAMPLE 7.4.2 Linearizing a Model

■ Problem

Consider the liquid-level system with an orifice, treated in Example 7.4.1. The model is given by equation (2) of that example.

$$A\frac{d\hat{h}}{dt} = \hat{q}_{vi} - \hat{q}_{vo} = \hat{q}_{vi} - C_d A_o \sqrt{2g\hat{h}}$$

Consider the case where $A = 2$ ft^2 and $C_d A_o \sqrt{2g} = 6$. Estimate the system's time constant for two cases: (i) the inflow rate is held constant at $\hat{q}_{vi} = 12$ ft^3/sec and (ii) the inflow rate is held constant at $\hat{q}_{vi} = 24$ ft^3/sec.

■ Solution

Substituting the given values, we obtain

$$2\frac{d\hat{h}}{dt} = \hat{q}_{vi} - \hat{q}_{vo} = \hat{q}_{vi} - 6\sqrt{\hat{h}} \tag{1}$$

When the inflow rate is held constant at the value $\hat{q}_{vir}$, the liquid height $\hat{h}$ reaches an equilibrium value h_r that can be found from the preceding equation by setting $\hat{h} = h_r$ and $d\hat{h}/dt$ equal to zero. This gives $36h_r = q_{vir}^2$.

The two cases of interest to us are (i) $h_r = (12)^2/36 = 4$ ft and (ii) $h_r = (24)^2/36 = 16$ ft. Figure 7.4.2 is a plot of the outflow flow rate $\hat{q}_{vo} = 6\sqrt{\hat{h}}$ through the orifice as a function of the height $\hat{h}$. The two points corresponding to $\hat{h} = 4$ and $\hat{h} = 16$ are indicated on the plot.

In Figure 7.4.2 two straight lines are shown, each passing through one of the points of interest ($\hat{h} = 4$ and $\hat{h} = 16$) and having a slope equal to the slope of the curve at that point. The general equation for these lines is

$$\hat{q}_{vo} = 6\sqrt{\hat{h}} = 6\sqrt{\hat{h}_r} + \left(\frac{d\hat{q}_{vo}}{d\hat{h}}\right)_r (\hat{h} - \hat{h}_r) = 6\sqrt{\hat{h}_r} + 3\hat{h}_r^{-1/2}(\hat{h} - \hat{h}_r)$$

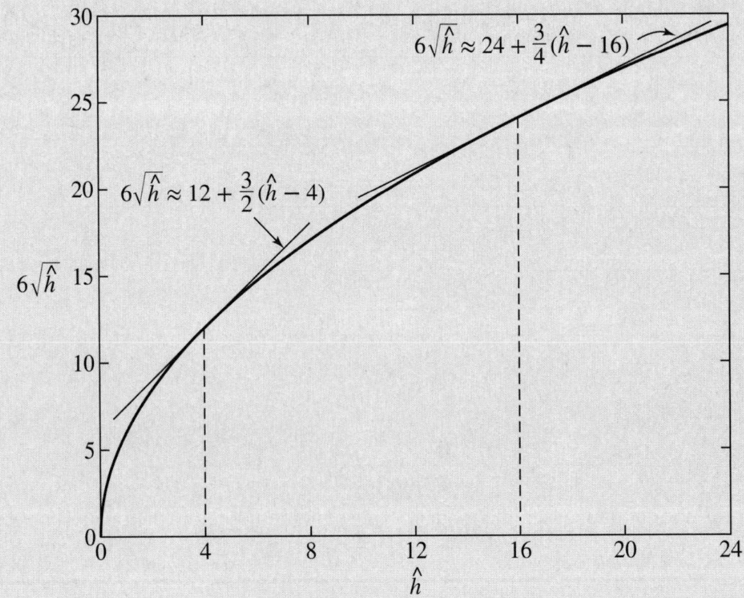

Figure 7.4.2 Linearized approximations of the resistance relation.

and is the same as a Taylor series expansion truncated after the first-order term. Noting that $\hat{h}_r = h_r$, this equation becomes

$$\hat{q}_{vo} = 6\sqrt{h_r} + 3h_r^{-1/2}(\hat{h} - h_r) \tag{2}$$

For Case (i), this equation becomes

$$\hat{q}_{vo} = 12 + \frac{3}{2}(\hat{h} - 4)$$

and for Case (ii),

$$\hat{q}_{vo} = 24 + \frac{3}{4}(\hat{h} - 16)$$

These are the equations of the straight lines shown in the figure.

Noting that $\hat{h} - h_r = h$, equation (2) can be expressed in the simpler form

$$\hat{q}_{vo} = 6\sqrt{h_r} + (3h_r^{-1/2})h$$

Substitute this into equation (1), and note that $d\hat{h}/dt = dh/dt$ and $\hat{q}_{vi} - 6\sqrt{h_r} = q_{vi}$, to obtain

$$2\frac{dh}{dt} = q_{vi} + (3h_r^{-1/2})h \tag{3}$$

This is the linearized model that is a good approximation of the nonlinear model (1) near the reference height h_r.

The time constant of the linearized model (3) is $2\sqrt{h_r}/3$, and is $4/3$ sec for $h_r = 4$ and $8/3$ sec for $h_r = 16$. Thus, for Case (i), if the input flow rate is changed slightly from its equilibrium value of 12, the liquid height will take about 4(4/3), or 16/3, sec to reach its new height. For Case (ii), if the input flow rate is changed slightly from its value of 24, the liquid height will take about 4(8/3), or 32/3, seconds to reach its new height.

Note that the model's time constant depends on the particular equilibrium solution chosen for the linearization. Because the straight line is an approximation to the $\hat{q}_{vo} = 6\sqrt{\hat{h}}$ curve, we cannot use the linearized models to make predictions about the system's behavior far from the equilibrium point. However, despite this limitation, a linearized model is useful for designing a flow control system to keep the height near some desired value. If the control system works properly, the height will stay near the equilibrium value, and the linearized model will be accurate.

Liquid-Level System with a Flow Source

EXAMPLE 7.4.3

■ Problem

The cylindrical tank shown in Figure 7.4.3 has a bottom area A. The total mass inflow rate from the flow source is $\hat{q}_{mi}(t)$, a given function of time. The total mass outflow rate $\hat{q}_{mo}$ is not given and must be determined. The outlet resistance R is the linearized resistance about the reference condition (h_r, q_{mir}). Develop a model of h, the deviation of the liquid height from the constant reference height h_r, where $\hat{h} = h_r + h$.

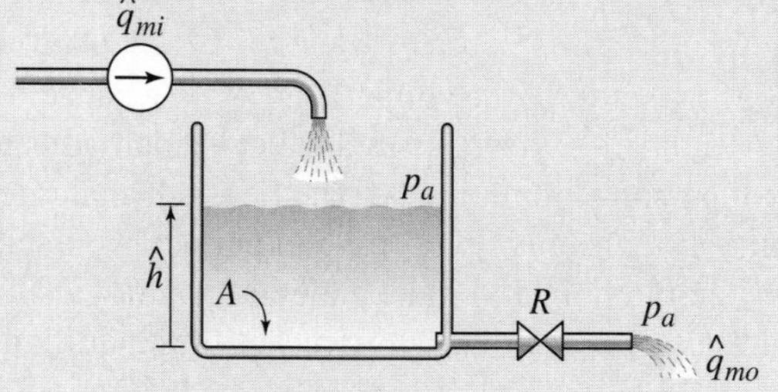

Figure 7.4.3 A liquid-level system with a flow source.

■ **Solution**

The total mass in the tank is $m = \rho A\hat{h} = \rho A(h + h_r)$ and from conservation of mass

$$\frac{dm}{dt} = \frac{d[\rho A(h + h_r)]}{dt} = \rho A\frac{dh}{dt} = \hat{q}_{mi} - \hat{q}_{mo}$$

because ρ, h_r, and A are constants. Expressing $\hat{q}_{mi}$ and $\hat{q}_{mo}$ in terms of the deviation variables q_{mi} and q_{mo}, we have

$$\rho A\frac{dh}{dt} = (q_{mi} + q_{mir}) - (q_{mo} + q_{mor}) = (q_{mir} - q_{mor}) + (q_{mi} - q_{mo})$$

Because the reference height h_r is a constant, the outflow rate at equilibrium must equal the inflow rate. Thus $q_{mir} - q_{mor} = 0$, and the model becomes

$$\rho A\frac{dh}{dt} = q_{mi} - q_{mo} \tag{1}$$

Because R is a linearized resistance, then for small changes h in the height,

$$q_{mo} = \frac{1}{R}[(\rho gh + p_a) - p_a] = \frac{1}{R}\rho gh$$

Substituting this into equation (1) gives the desired model:

$$\rho A\frac{dh}{dt} = q_{mi} - \frac{1}{R}\rho gh$$

which can be rearranged as

$$\frac{RA}{g}\frac{dh}{dt} + h = \frac{R}{\rho g}q_{mi}$$

So the time constant is $\tau = RA/g$. If $q_{mi} = 0$ (so the inflow rate remains constant at q_{mir}), then h will be essentially zero for $t > 4\tau$, which means that the height will return to nearly the equilibrium height h_r at that time.

Some engineers are helped by thinking of a fluid system in terms of an analogous electric circuit, in which pressure difference plays the role of voltage difference, and mass flow rate is analogous to current. A fluid resistance resists flow just as an electrical resistor resists current. A fluid capacitance stores fluid mass just as an electrical capacitor stores charge. Figure 7.4.4 shows an electric circuit that is analogous to the tank system of Figure 7.4.3. The circuit model is

Figure 7.4.4 Electric circuit analogous to the hydraulic system shown in Figure 7.4.3.

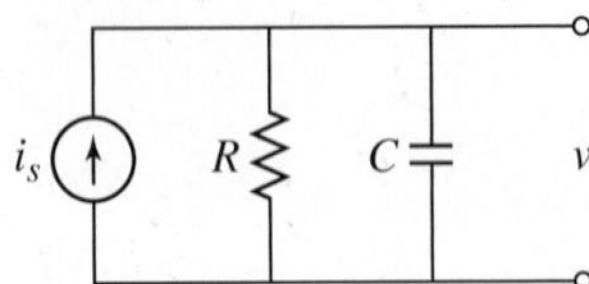

$$C\frac{dv}{dt} = i_s - \frac{1}{R}v$$

The input current i_s is analogous to the inflow rate q_{mi}, the voltage v across the capacitor is analogous to the fluid pressure ρgh, and the electrical capacitance C is analogous to the fluid capacitance A/g. It is a matter of personal opinion as to whether such analogies help to understand the dynamics of fluid systems, and you should decide for yourself. Always keep in mind, however, that we should not get too dependent on analogies for developing models, because they might not always properly represent the underlying physics of the original system.

Liquid-Level System with a Pressure Source

EXAMPLE 7.4.4

■ Problem

The tank shown in cross section in Figure 7.4.5 has a bottom area A. A pressure source $\hat{p}_s = p_s(t) + p_{sr}$ is connected through a resistance to the bottom of the tank, where $p_s(t)$ is a given function of time. The resistances R_1 and R_2 are linearized resistances about the reference condition (p_{sr}, h_r). Develop a model of h, the deviation of the liquid height from the constant reference height h_r, where $\hat{h} = h_r + h$.

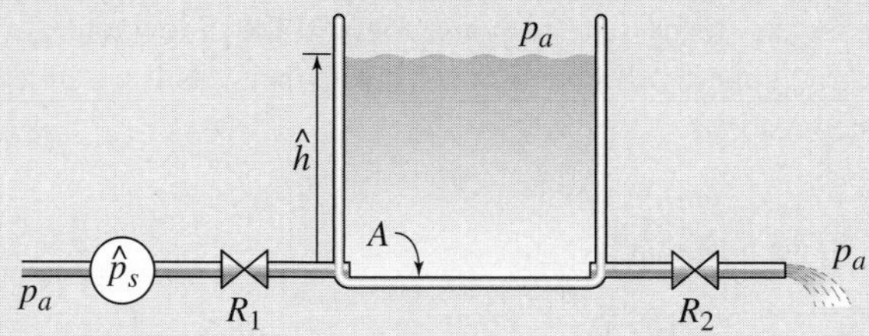

Figure 7.4.5 A liquid-level system with a pressure source.

■ Solution

The total mass in the tank is $m = \rho A\hat{h} = \rho A(h + h_r)$, and from conservation of mass

$$\frac{dm}{dt} = \frac{d[\rho A(h + h_r)]}{dt} = \rho A\frac{dh}{dt} = \hat{q}_{mi} - \hat{q}_{mo}$$

or

$$\rho A\frac{dh}{dt} = (q_{mi} + q_{mir}) - (q_{mo} + q_{mor}) = (q_{mi} - q_{mo}) + (q_{mir} - q_{mor})$$

Because at the reference equilibrium, the outflow rate equals the inflow rate, $q_{mir} - q_{mor} = 0$, and we have

$$\rho A\frac{dh}{dt} = q_{mi} - q_{mo} \tag{1}$$

This is a linearized model that is valid for small changes around the equilibrium state.

Because the outlet resistance has been linearized,

$$q_{mo} = \frac{1}{R_2}[(\rho gh + p_a) - p_a] = \frac{\rho gh}{R_2}$$

Similarly for the mass inflow rate, we have

$$q_{mi} = \frac{1}{R_1}[(p_s + p_a) - (\rho gh + p_a)] = \frac{1}{R_1}(p_s - \rho gh)$$

Substituting into equation (1) gives

$$\rho A\frac{dh}{dt} = \frac{1}{R_1}(p_s - \rho gh) - \frac{\rho gh}{R_2} = \frac{1}{R_1}p_s - \rho g\frac{R_1 + R_2}{R_1R_2}h$$

This can be rearranged as

$$\frac{R_1R_2A}{g(R_1 + R_2)}\frac{dh}{dt} + h = \frac{R_2}{\rho g(R_1 + R_2)}p_s$$

The coefficient of dh/dt gives the time constant, which is $\tau = R_1R_2A/g(R_1 + R_2)$.

When a fluid system contains more than one capacitance, you should apply the conservation of mass principle to each capacitance, and then use the appropriate resistance relations to couple the resulting equations. To do this, you must assume that some pressures or liquid heights are greater than others and assign the positive-flow directions accordingly. If you are consistent, the mathematics will handle the reversals of flow direction automatically.

EXAMPLE 7.4.5

Two Connected Tanks

■ Problem

The cylindrical tanks shown in Figure 7.4.6a have bottom areas A_1 and A_2. The total mass inflow rate from the flow source is $\hat{q}_{mi}(t)$, a given function of time. The resistances are linearized resistances about the reference condition h_{1r}, h_{2r}, q_{mir}. (a) Develop a model of the liquid heights in terms of the deviation variables h_1 and h_2. (b) Suppose the resistances are equal: $R_1 = R_2 = R$, and the areas are $A_1 = A$ and $A_2 = 3A$. Obtain the transfer function $H_1(s)/Q_{mi}(s)$. (c) Use the transfer function to solve for the steady-state response for h_1 if the inflow rate q_{mi} is a unit-step function, and estimate how long it will take to reach steady state. Is it possible for liquid heights to oscillate in the step response?

■ Solution

a. Using deviation variables as usual, we note that

$$\hat{h}_1 = h_{1r} + h_1$$
$$\hat{h}_2 = h_{2r} + h_2$$
$$\hat{q}_{mi} = q_{mir} + q_{mi}$$

For convenience, assume that $\hat{h}_1 > \hat{h}_2$. This is equivalent to assuming that the flow rate is from tank 1 to tank 2. From conservation of mass applied to tank 1, we obtain

$$\frac{d(\rho A_1 \hat{h}_1)}{dt} = \frac{d[\rho A_1(h_1 + h_{1r})]}{dt} = \rho A_1 \frac{dh_1}{dt} = -\hat{q}_{1m} = -(q_{1m} + q_{1mr})$$

From physical reasoning, we can see that the two heights must be equal at equilibrium, and thus $q_{1mr} = 0$. Therefore,

$$\rho A_1 \frac{dh_1}{dt} = -q_{1m}$$

Because R_1 is a linearized resistance,

$$q_{1m} = \frac{\rho g}{R_1}(h_1 - h_2)$$

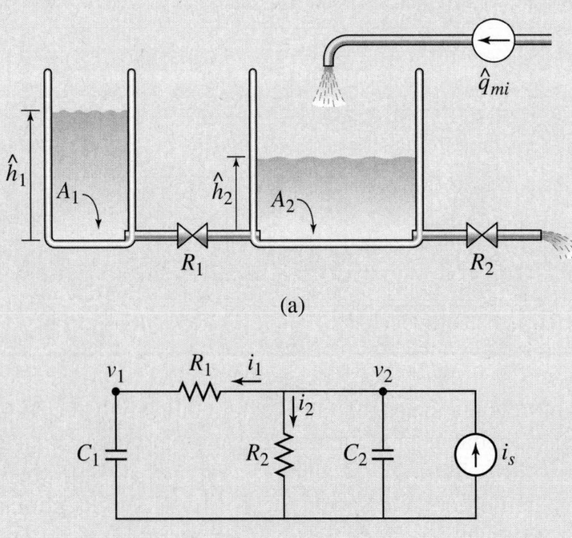

Figure 7.4.6 (a) Two connected tanks. (b) Analogous electric circuit.

So, after canceling ρ on both sides, the model for tank 1 is

$$A_1 \frac{dh_1}{dt} = -\frac{g}{R_1}(h_1 - h_2) \tag{1}$$

Similarly for tank 2,

$$\frac{d(\rho A_2 \hat{h}_2)}{dt} = \frac{d[\rho A_2(h_2 + h_{2r})]}{dt} = \rho A_2 \frac{dh_2}{dt}$$

Conservation of mass gives

$$\rho A_2 \frac{dh_2}{dt} = \hat{q}_{mi} + \hat{q}_{1m} - \hat{q}_{mo} = (q_{mi} + q_{mir}) + (q_{1m} + q_{1mr}) - (q_{mo} + q_{mor})$$

Recalling that $q_{1mr} = 0$, we note that this implies that $q_{mir} = q_{mor}$, and thus

$$\rho A_2 \frac{dh_2}{dt} = q_{mi} + q_{1m} - q_{mo}$$

Because the resistances are linearized, we have

$$\rho A_2 \frac{dh_2}{dt} = q_{mi} + \frac{\rho g}{R_1}(h_1 - h_2) - \frac{\rho g}{R_2} h_2 \tag{2}$$

The desired model consists of equations (1) and (2).

b. Substituting $R_1 = R_2 = R$, $A_1 = A$, and $A_2 = 3A$ into the differential equations and dividing by A, and letting $B = g/RA$ we obtain

$$\dot{h}_1 = -B(h_1 - h_2)$$

$$3\dot{h}_2 = \frac{q_{mi}}{\rho A} + B(h_1 - h_2) - Bh_2 = \frac{q_{mi}}{\rho A} + Bh_1 - 2Bh_2$$

Apply the Laplace transform method of each equation, assuming zero-initial conditions, and collect terms to obtain

$$(s + B)H_1(s) - BH_2(s) = 0 \tag{3}$$

$$-BH_1(s) + (3s + 2B)H_2(s) = \frac{1}{\rho A} Q_{mi}(s) \tag{4}$$

Solve equation (3) for $H_2(s)$, substitute the expression into equation (4), and solve for $H_1(s)$ to obtain

$$\frac{H_1(s)}{Q_{mi}(s)} = \frac{RB^2/\rho g}{3s^2 + 5Bs + B^2} \tag{5}$$

c. The characteristic equation is $3s^2 + 5Bs + B^2 = 0$ and has the two real roots

$$s = \frac{-5 \pm \sqrt{13}}{6} B = -1.43B, -0.232B$$

Thus, the system is stable, and there will be a constant steady-state response to a step input. The step response cannot oscillate because both roots are real. The steady-state height can be obtained by applying the final-value theorem to equation (5) with $Q_{mi}(s) = 1/s$.

$$h_{1ss} = \lim_{s \to 0} sH_1(s) = \lim_{s \to 0} s \frac{RB^2/\rho g}{3s^2 + 5Bs + B^2} \frac{1}{s} = \frac{R}{\rho g}$$

The time constants are

$$\tau_1 = \frac{1}{1.43B} = \frac{0.699}{B} \qquad \tau_2 = \frac{1}{0.232B} = \frac{4.32}{B}$$

The largest time constant is τ_2 and thus it will take a time equal to approximately $4\tau_2 = 17.2/B$ to reach steady state.

Figure 7.4.6b shows an electrical circuit that is analogous to the hydraulic system shown in part (a) of the figure. The currents i_s, i_1, and i_2 are analogous to the mass flow rates q_{mi}, q_{m_1}, and q_{mo}. The voltages v_1 and v_2 are analogous to the pressures $\rho g h_1$ and $\rho g h_2$, and the capacitances C_1 and C_2 are analogous to the fluid capacitances A_1/g and A_2/g.

7.4.2 HYDRAULIC DAMPERS

Dampers oppose a velocity difference across them, and thus they are used to limit velocities. A common application of dampers is in vehicle shock absorbers.

EXAMPLE 7.4.6

Linear Damper

■ Problem

A damper exerts a force as a result of a velocity difference across it. Figure 7.4.7 shows the principle used in automotive shock absorbers. A piston of diameter W and thickness L has a cylindrical hole of diameter D. The piston rod extends out of the housing, which is sealed and filled with a viscous incompressible fluid. Assuming that the flow through the hole is laminar and that the entrance length L_e is small compared to L, develop a model of the relation between the applied force f and $\dot{x}$, the relative velocity between the piston and the cylinder.

■ Solution

Assume that the rod's cross-sectional area and the hole area $\pi(D/2)^2$ are small compared to the piston area A. Let m be the combined mass of the piston and rod. Then the force f acting on the piston rod creates a pressure difference $(p_1 - p_2)$ across the piston such that

$$m\ddot{y} = f - A(p_1 - p_2) \tag{1}$$

If the mass m or the acceleration $\ddot{y}$ is small, then $m\ddot{y} \approx 0$, and we obtain

$$f = A(p_1 - p_2) \tag{2}$$

For laminar flow through the hole,

$$q_v = \frac{1}{\rho} q_m = \frac{1}{\rho R}(p_1 - p_2) \tag{3}$$

The volume flow rate q_v is the rate at which the piston sweeps out volume as it moves, and can be expressed as

$$q_v = A(\dot{y} - \dot{z}) = A\dot{x} \tag{4}$$

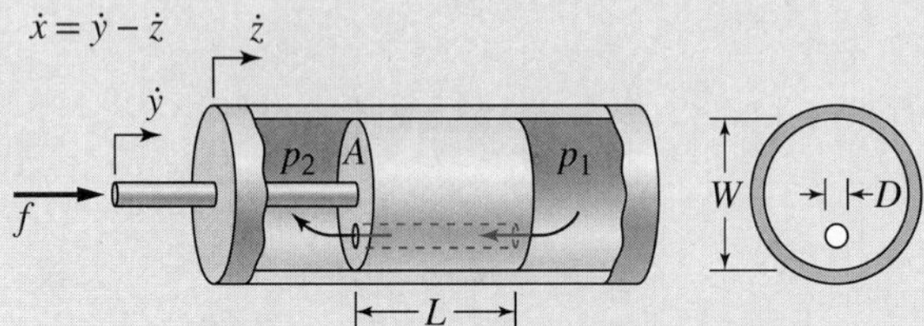

Figure 7.4.7 A damper.

because the fluid is incompressible. Combining equations (2), (3), and (4), we obtain

$$f = A(\rho R A\dot{x}) = \rho R A^2 \dot{x} = c\dot{x}$$

where the damping coefficient c is given by $c = \rho R A^2$. From the Hagen-Poiseuille formula (7.3.12) for a cylindrical conduit,

$$R = \frac{128\mu L}{\pi \rho D^4}$$

and thus the damping coefficient can be expressed as

$$c = \frac{128\mu L A^2}{\pi D^4}$$

The approximation $m\ddot{y} \approx 0$ is commonly used for hydraulic systems to simplify the resulting model. To see the effect of this approximation, use instead equation (1) with equations (3) and (4) to obtain

$$q_v = \frac{1}{\rho R}(p_1 - p_2) = \frac{1}{\rho R A}(f - m\ddot{y}) = A\dot{x}$$

Thus,

$$f = m\ddot{y} + \rho R A^2 \dot{x}$$

Therefore, if $m\ddot{y}$ cannot be neglected, the damper force is a function of the absolute acceleration as well as the relative velocity.

7.4.3 HYDRAULIC ACTUATORS

Hydraulic actuators are widely used with high pressures to obtain high forces for moving large loads or achieving high accelerations. The working fluid may be liquid, as is commonly found with construction machinery, or it may be air, as with the air cylinder-piston units frequently used in manufacturing and parts-handling equipment.

Hydraulic Piston and Load — EXAMPLE 7.4.7

■ Problem

Figure 7.4.8 shows a double-acting piston and cylinder. The device moves the load mass m in response to the pressure sources p_1 and p_2. Assume the fluid is incompressible, the resistances are linear, and the piston mass is included in m. Derive the equation of motion for m.

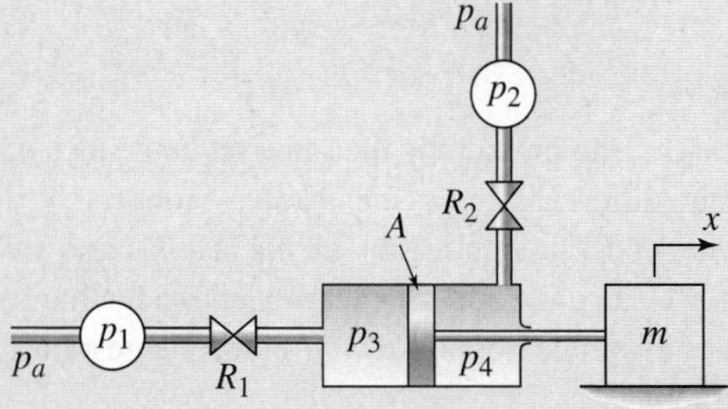

Figure 7.4.8 A double-acting piston and cylinder.

■ Solution

Define the pressures p_3 and p_4 to be the pressures on the left- and right-hand sides of the piston. The mass flow rates through the resistances are

$$q_{m_1} = \frac{1}{R_1}(p_1 + p_a - p_3) \tag{1}$$

$$q_{m_2} = \frac{1}{R_2}(p_4 - p_2 - p_a) \tag{2}$$

From conservation of mass, $q_{m_1} = q_{m_2}$ and $q_{m_1} = \rho A\dot{x}$. Combining these four equations we obtain

$$p_1 + p_a - p_3 = R_1 \rho A\dot{x} \tag{3}$$

$$p_4 - p_2 - p_a = R_2 \rho A\dot{x} \tag{4}$$

Adding equations (3) and (4) gives

$$p_4 - p_3 = p_2 - p_1 + (R_1 + R_2)\rho A\dot{x} \tag{5}$$

From Newton's law,

$$m\ddot{x} = A(p_3 - p_4) \tag{6}$$

Substitute equation (5) into (6) to obtain the desired model:

$$m\ddot{x} + (R_1 + R_2)\rho A^2\dot{x} = A(p_1 - p_2) \tag{7}$$

Note that if the resistances are zero, the $\dot{x}$ term disappears, and we obtain

$$m\ddot{x} = A(p_1 - p_2)$$

which is identical to the model derived in part (a) of Example 7.1.1.

EXAMPLE 7.4.8

Hydraulic Piston with Negligible Load

■ Problem

Develop a model for the motion of the load mass m in Figure 7.4.8, assuming that the product of the load mass m and the load acceleration $\ddot{x}$ is very small.

■ Solution

If $m\ddot{x}$ is very small, from equation (7) of Example 7.4.7, we obtain the model

$$(R_1 + R_2)\rho A^2\dot{x} = A(p_1 - p_2)$$

which can be expressed as

$$\dot{x} = \frac{p_1 - p_2}{(R_1 + R_2)\rho A} \tag{1}$$

From this we see that if $p_1 - p_2$ is constant, the mass velocity $\dot{x}$ will also be constant.

The implications of the approximation $m\ddot{x} = 0$ can be seen from Newton's law:

$$m\ddot{x} = A(p_3 - p_4) \tag{2}$$

If $m\ddot{x} = 0$, equation (2) implies that $p_3 = p_4$; that is, the pressure is the same on both sides of the piston. From this we can see that the pressure difference across the piston is produced by a large load mass or a large load acceleration. The modeling implication of this fact is that if we neglect the load mass or the load acceleration, we can develop a simpler model of a hydraulic system—a model based only on conservation of mass and not on Newton's law. The resulting model will be first order rather than second order.

Hydraulic Motor

EXAMPLE 7.4.9

■ Problem

A *hydraulic motor* is shown in Figure 7.4.9. The pilot valve controls the flow rate of the hydraulic fluid from the supply to the cylinder. When the pilot valve is moved to the right of its neutral position, the fluid enters the right-hand piston chamber and pushes the piston to the left. The fluid displaced by this motion exits through the left-hand drain port. The action is reversed for a pilot valve displacement to the left. Both return lines are connected to a sump from which a pump draws fluid to deliver to the supply line. Derive a model of the system assuming that $m\ddot{x}$ is negligible.

■ Solution

Let y denote the displacement of the pilot valve from its neutral position, and x the displacement of the load from its last position before the start of the motion. Note that a positive value of x (to the left) results from a positive value of y (to the right). The flow through the cylinder port uncovered by the pilot valve can be treated as flow through an orifice. Let Δp be the pressure drop across the orifice. Thus, from (7.3.14) with p replaced by Δp, the volume flow rate through the cylinder port is given by

$$q_v = \frac{1}{\rho}q_m = \frac{1}{\rho}C_d A_o \sqrt{2\Delta p \rho} = C_d A_o \sqrt{2\Delta p/\rho} \tag{1}$$

where A_o is the uncovered area of the port, C_d is the discharge coefficient, and ρ is the mass density of the fluid. The area A_o is approximately equal to yD, where D is the port depth (into the page). If C_d, ρ, Δp, and D are taken to be constant, equation (1) can be written as

$$q_v = C_d D y \sqrt{2\Delta p/\rho} = By \tag{2}$$

where $B = C_d D\sqrt{2\Delta p/\rho}$.

Assuming that the rod's area is small compared to the piston area, the piston areas on the left and right sides are equal to A. The rate at which the piston pushes fluid out of the cylinder is $A\,dx/dt$. Conservation of volume requires the volume flow rate into the cylinder be equal to the volume flow rate out. Therefore,

$$q_v = A\frac{dx}{dt} \tag{3}$$

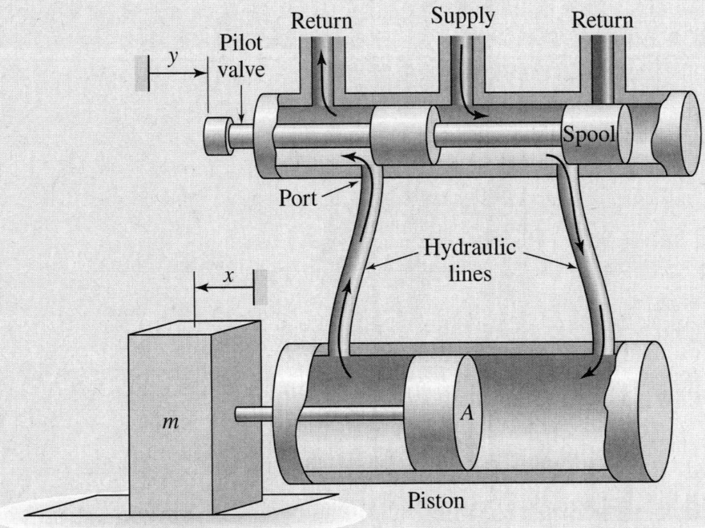

Figure 7.4.9 A hydraulic motor.

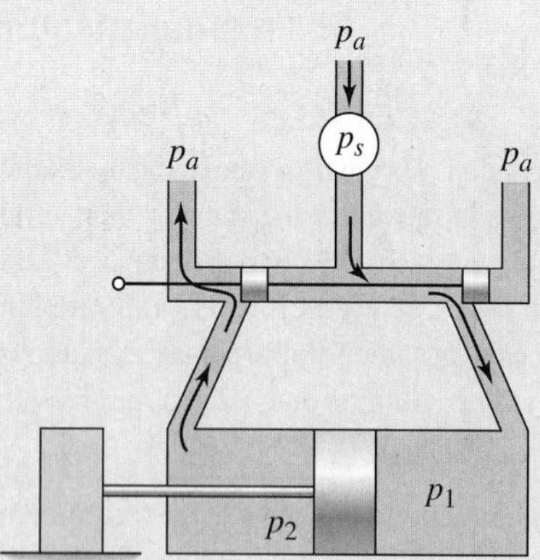

Figure 7.4.10 Pressures in a hydraulic motor.

Combining the last two equations gives the model for the servomotor:

$$\frac{dx}{dt} = \frac{B}{A} y \tag{4}$$

This model predicts a constant piston velocity dx/dt if the pilot valve position y is held fixed.

The pressure drop Δp can be determined as follows. We assume that because of geometric symmetry, the pressure drop is the same across both the inlet and outlet valves. From Figure 7.4.10 we see that

$$\Delta p = (p_s + p_a) - p_1 = p_2 - p_a$$

and thus

$$p_1 - p_2 = p_s - 2\Delta p \tag{5}$$

where p_1 and p_2 are the pressures on either side of the piston. The force on the piston is $A(p_1 - p_2)$, and from Newton's law, $m\ddot{x} = A(p_1 - p_2)$. Using the approximation $m\ddot{x} = 0$, we see that $p_1 = p_2$, and thus equation (5) shows that $\Delta p = p_s/2$. Therefore, $B = C_d D\sqrt{p_s/\rho}$.

Equation (4) is accurate for many applications, but for high accelerations or large loads, this model must be modified because it neglects the effects of the inertia of the load and piston on the pressures on each side of the piston.

7.4.4 FLUID INERTANCE

We have defined fluid inertance I as

$$I = \frac{p}{dq_m/dt}$$

which is the ratio of the pressure difference over the rate of change of the mass flow rate. The inertance is the change in pressure required to produce a unit rate of change in mass flow rate. Thus, inertance relates to fluid acceleration and kinetic energy, which are often negligible either because the moving fluid mass is small or because it is moving at a steady rate. There are, however, some cases where inertance may be significant. The effect known as *water hammer* is due partly to inertance. Also, as we will see in the following example, inertance can be significant in conduits that are long or that have small cross sections.

EXAMPLE 7.4.10

Calculation of Inertance

■ **Problem**

Consider fluid flow (either liquid or gas) in a nonaccelerating pipe (Figure 7.4.11). Derive the expression for the inertance of a slug of fluid of length L.

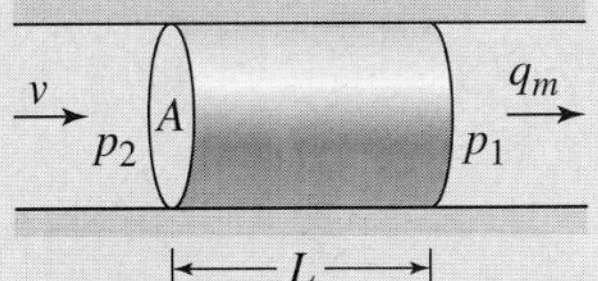

Figure 7.4.11 Derivation of pipe inertance expression.

■ **Solution**

The mass of the slug is ρAL, where ρ is the fluid mass density. The net force acting on the slug due to the pressures p_1 and p_2 is $A(p_2 - p_1)$. Applying Newton's law to the slug, we have

$$\rho AL\frac{dv}{dt} = A(p_2 - p_1)$$

where v is the fluid velocity. The velocity v is related to the mass flow rate q_m by $\rho Av = q_m$. Using this to substitute for v, we obtain

$$L\frac{dq_m}{dt} = A(p_2 - p_1)$$

or

$$\frac{L}{A}\frac{dq_m}{dt} = p_2 - p_1$$

With $p = p_2 - p_1$, we obtain

$$\frac{L}{A} = \frac{p}{dq_m/dt}$$

Thus, from the definition of inertance I,

$$I = \frac{L}{A}$$

Note that the inertance is larger for longer pipes and for pipes with smaller cross section.

The significance of inertance in a given application is often difficult to assess. Often a model is developed by ignoring inertance effects at first, and then, if possible, the model is verified by experiment to see if the neglected inertance is significant.

7.5 PNEUMATIC SYSTEMS

The working medium in a pneumatic device is a compressible fluid, most commonly air. The availability of air is an advantage for pneumatic devices, because it can be exhausted to the atmosphere at the end of the device's work cycle, thus eliminating the need for return lines. On the other hand, because of the compressibility of the working fluid, the response of pneumatic systems can be slower and more oscillatory than that of hydraulic systems.

Because the kinetic energy of a gas is usually negligible, the inertance relation is not usually needed to develop a model. Instead, capacitance and resistance elements form the basis of most pneumatic system models.

Temperature, pressure, volume, and mass are functionally related for a gas. The model most often used to describe this relationship is the *perfect gas law*, which is a good model of gas behavior under normal pressures and temperatures. The law states that

$$pV = mR_gT \tag{7.5.1}$$

where p is the absolute pressure of the gas with volume V, m is the mass, T is its absolute temperature, and R_g is the gas constant that depends on the particular type of gas. The values of R_g for air are 1715 ft-lb/slug-°R and 287 N · m/kg · K.

If heat is added to a gas from its surroundings, some of this heat can do external work on its surroundings, and the rest can increase the internal energy of the gas by raising its temperature. The *specific heat* of a substance at a specified temperature is the ratio of two heat values, the amount of heat needed to raise the temperature of a unit mass of the substance by 1° divided by the heat needed to raise a unit mass of water 1° at the specified temperature. Because the pressure and volume of a gas can change as its temperature changes, two specific heats are defined for a gas: one at constant pressure (c_p), and one at constant volume (c_v).

The amount of energy needed to raise the temperature of 1 kg of water from 14.5°C to 15.5°C is 4186 J. In the British Engineering system, the British thermal unit (BTU) can be considered to be the energy needed to raise 1 pound of water (1/32.174 of a slug in mass units) 1 degree Fahrenheit.

The perfect gas law enables us to solve for one of the variables p, V, m, or T if the other three are given. Additional information is usually available in the form of a pressure-volume or "process" relation. The following process models are commonly used, where the subscripts 1 and 2 refer to the start and the end of the process, respectively. We assume the mass m of the gas is constant.

1. *Constant-Pressure Process* ($p_1 = p_2$). The perfect gas law thus implies that $V_2/V_1 = T_2/T_1$. If the gas receives heat from the surroundings, some of it raises the temperature and some expands the volume.
2. *Constant-Volume Process* ($V_1 = V_2$). Here, $p_2/p_1 = T_2/T_1$. When heat is added to the gas, it merely raises the temperature because no external work is done in a constant-volume process.
3. *Constant-Temperature Process (an* isothermal *process)* ($T_1 = T_2$). Thus, $p_2/p_1 = V_1/V_2$. Any added heat does not increase the internal energy of the gas because of the constant temperature. It only does external work.
4. *Reversible Adiabatic (Isentropic) Process.* This process is described by the relation

$$p_1 V_1^\gamma = p_2 V_2^\gamma \tag{7.5.2}$$

where $\gamma = c_p/c_v$. *Adiabatic* means that no heat is transferred to or from the gas. *Reversible* means the gas and its surroundings can be returned to their original thermodynamic conditions. Because no heat is transferred, any external work done by the gas changes its internal energy by the same amount. Thus, its temperature changes; that is,

$$W = mc_v(T_1 - T_2)$$

where W is the external work. The work W is positive if work is done on the surroundings.

5. *Polytropic Process.* A process can be more accurately modeled by properly choosing the exponent n in the polytropic process

$$p\left(\frac{V}{m}\right)^n = \text{constant}$$

If the mass m is constant, this general process reduces to the previous processes if n is chosen as 0, ∞, 1, and γ, respectively, and if the perfect gas law is used.

7.5.1 PNEUMATIC CAPACITANCE

Fluid capacitance is the relation between stored mass and pressure. Specifically, fluid capacitance C is the ratio of the change in stored mass to the change in pressure, or

$$C = \frac{dm}{dp} \tag{7.5.3}$$

For a container of constant volume V with a gas density ρ, $m = \rho V$, and the capacitance equation may be written as

$$C = \frac{d(\rho V)}{dp} = V\frac{d\rho}{dp} \tag{7.5.4}$$

If the gas undergoes a polytropic process,

$$p\left(\frac{V}{m}\right)^n = \frac{p}{\rho^n} = \text{constant} \tag{7.5.5}$$

Differentiating this expression gives

$$\frac{d\rho}{dp} = \frac{\rho}{np} = \frac{m}{npV}$$

For a perfect gas, this shows the capacitance of the container to be

$$C = \frac{mV}{npV} = \frac{V}{nR_gT} \tag{7.5.6}$$

Note that the same container can have a different capacitance for different expansion processes, temperatures, and gases, because C depends on n, T, and R_g.

EXAMPLE 7.5.1

Capacitance of an Air Cylinder

■ **Problem**

Obtain the capacitance of air in a rigid cylinder of volume 0.03 m^3, if the cylinder is filled by an isothermal process. Assume the air is initially at room temperature, 293 K.

■ **Solution**

The filling of the cylinder can be modeled as an isothermal process if it occurs slowly enough to allow heat transfer to occur between the air and its surroundings. In this case, $n = 1$ in the polytropic process equation, and from (7.5.6) we obtain,

$$C = \frac{0.03}{1(287)(293)} = 3.57 \times 10^{-7} \text{ kg} \cdot \text{m}^2/\text{N}$$

EXAMPLE 7.5.2

Pressurizing an Air Cylinder

■ **Problem**

Air at temperature T passes through a valve into a rigid cylinder of volume V, as shown in Figure 7.5.1. The mass flow rate through the valve depends on the pressure difference $\Delta p = p_i - p$, and is given by an experimentally determined function:

$$q_{mi} = f(\Delta p) \tag{1}$$

Develop a dynamic model of the gage pressure p in the container as a function of the input pressure p_i. Assume the filling process is isothermal.

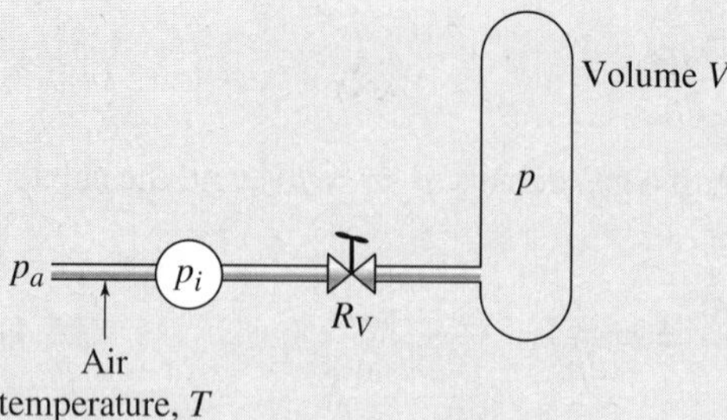

Figure 7.5.1 Pressurizing an air cylinder.

■ Solution

From conservation of mass, the rate of mass increase in the container equals the mass flow rate through the valve. Thus, if $p_i - p > 0$, from equation (1)

$$\frac{dm}{dt} = q_{mi} = f(\Delta p)$$

But

$$\frac{dm}{dt} = \frac{dm}{dp}\frac{dp}{dt} = C\frac{dp}{dt}$$

and thus,

$$C\frac{dp}{dt} = f(\Delta p) = f(p_i - p) \tag{2}$$

where the capacitance C is given by (7.5.6) with $n = 1$.

$$C = \frac{V}{R_g T}$$

If the function f is nonlinear, then the dynamic model is nonlinear.

PART II. THERMAL SYSTEMS

A *thermal system* is one in which energy is stored and transferred as thermal energy, commonly called heat. Although heat energy can play a role in pneumatic systems, we chose to treat pneumatic systems primarily as fluid systems because their dynamics is largely governed by fluid capacitance and fluid resistance. Examples of thermal systems include heating and cooling systems in buildings and mixing processes where heat must be added or removed to maintain an optimal reaction temperature.

Thermal systems operate because of temperature differences, as heat energy flows from an object with the higher temperature to an object with the lower temperature. Just as conservation of mass, fluid resistance, and fluid capacitance form the basis of fluid system models, so conservation of heat energy forms the basis of thermal system models, along with the concepts of thermal resistance and thermal capacitance. Thus, it is natural to study thermal systems along with fluid systems. We will see that thermal systems are also analogous to electric circuits, where conservation of charge plays the same role as conservation of heat, and voltage difference plays the same role as temperature difference.

7.6 THERMAL CAPACITANCE

For a mass m whose specific heat is c_p, the amount of heat energy E stored in the object at a temperature T is

$$E = mc_p(T - T_r) \tag{7.6.1}$$

where T_r is an arbitrarily selected reference temperature. As with gravitational potential energy, which is computed relative to an arbitrary height h_r as $mg(h - h_r)$, we can select the reference temperature T_r for convenience. This is because only the *change* in stored heat energy affects the dynamics of a thermal system, just as only the change in gravitational potential energy affects the dynamics of a mechanical system.

Thermal capacitance relates an object's temperature to the amount of heat energy stored. It is defined as

$$C = \frac{dE}{dT} \tag{7.6.2}$$

where E is the stored heat energy. If the temperature range is wide enough, c_p might vary considerably with temperature. However, if c_p is not a function of temperature, from (7.6.1) we have

$$C = mc_p = \rho V c_p \tag{7.6.3}$$

where ρ and V are the density and the volume of the mass m. The thermal capacitance C can be interpreted as the amount of heat energy required to raise the temperature of the object by 1°. Thus, in SI, the units of C are J/°C or J/K. The FPS units of C are ft-lb/°F. Another common unit for C is BTU/°F.

The concept of thermal capacitance applies to fluids as well as solids. For example, at room temperature and atmospheric pressure, the ratio of the specific heat of water to that of air is 4.16. Thus, for the same *mass* of air and water, to raise the water temperature by 1° requires 4.16 times more energy than for air.

Temperature Dynamics of a Mixing Process

EXAMPLE 7.6.1

■ Problem

Liquid at a temperature T_i is pumped into a mixing tank at a constant volume flow rate q_v (Figure 7.6.1). The container walls are perfectly insulated so that no heat escapes through them. The container volume is V, and the liquid within is well mixed so that its temperature throughout is T. The liquid's specific heat and mass density are c_p and ρ. Develop a model for the temperature T as a function of time, with T_i as the input.

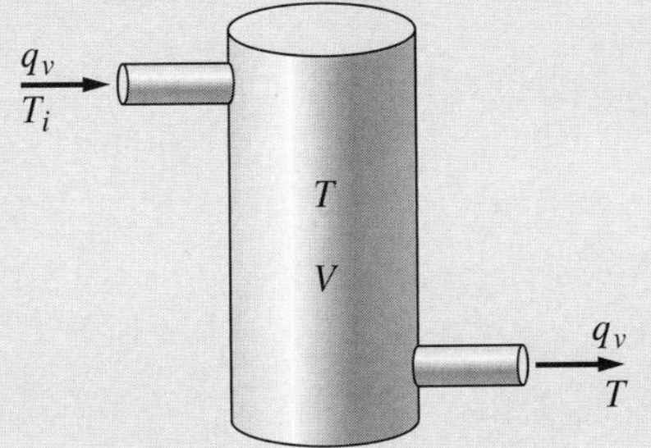

Figure 7.6.1 Temperature dynamics of a mixing process.

■ **Solution**

The amount of heat energy in the tank liquid is $\rho c_p V(T - T_r)$, where T_r is an arbitrarily selected reference temperature. From conservation of energy,

$$\frac{d[\rho c_p V(T - T_r)]}{dt} = \text{heat rate in} - \text{heat rate out} \tag{1}$$

Liquid mass is flowing into the tank at the rate $\dot{m} = \rho q_v$. Thus, heat energy is flowing into the tank at the rate

$$\text{heat rate in} = \dot{m} c_p (T_i - T_r) = \rho q_v c_p (T_i - T_r)$$

Similarly,

$$\text{heat rate out} = \rho q_v c_p (T - T_r)$$

Therefore, from equation (1), since ρ, c_p, V, and T_r are constants,

$$\rho c_p V \frac{dT}{dt} = \rho q_v c_p (T_i - T_r) - \rho q_v c_p (T - T_r) = \rho q_v c_p (T_i - T)$$

Cancel ρc_p and rearrange to obtain

$$\frac{V}{q_v}\frac{dT}{dt} + T = T_i$$

Note that T_r, ρ, and c_p do not appear in the final model form, so their specific numerical values are irrelevant to the problem. The system's time constant is V/q_v, and thus the liquid temperature T changes slowly if the tank volume V is large or if the inflow rate q_v is small, which agrees with our intuition.

7.7 THERMAL RESISTANCE

Heat energy is conserved, and thus heat in thermal system analysis plays the same role as charge in electrical systems. The flow of heat, called *heat transfer*, causes a change in an object's temperature. Heat transfer between two objects is caused by a difference in their temperatures. Thus, temperature difference in thermal systems plays the same role as voltage difference in electrical systems, and so we utilize the concept of *thermal resistance* in a manner similar to electrical resistance.

7.7.1 CONDUCTION, CONVECTION, AND RADIATION

Heat transfer can occur by one or more modes: *conduction*, *convection*, and *radiation*, as illustrated by Figure 7.7.1. Temperature is a measure of the amount of heat energy in an object. Heat energy and thus temperature can be thought of as due to the kinetic energy of vibrating molecules. A higher temperature is an indication of higher molecule

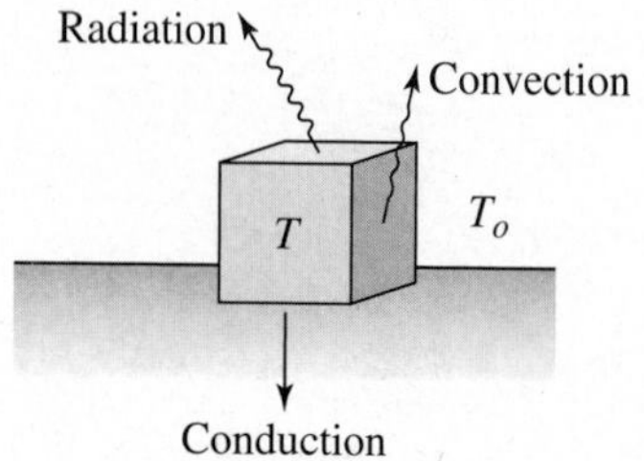

Figure 7.7.1 Modes of heat transfer.

vibration velocity. Heat transfer by conduction occurs by diffusion of heat through a substance. This diffusion occurs by molecules transferring some of their kinetic energy to adjacent, slower molecules.

The mechanism for convection is due to fluid transport. This effect can be seen in boiling water and in thermal air currents. Convection also occurs within a fluid at the boundary of the fluid and a solid surface whose temperature is different from that of the fluid. Convective heat transfer might be due to forced convection, such as when a fluid is pumped past a surface, or natural (free) convection, which is caused by motion produced by density differences in the fluid.

Heat transfer by radiation occurs through infrared waves. Heat lamps are common examples of this type of transfer. Heating by solar radiation is another example.

7.7.2 NEWTON'S LAW OF COOLING

Newton's law of cooling is a linear model for heat flow rate as a function of temperature difference. The law, which is used for both convection and conduction models, is expressed as

$$q_h = \frac{1}{R}\Delta T \tag{7.7.1}$$

where q_h is the heat flow rate, R is the thermal resistance, and ΔT is the temperature difference. In SI, q_h has the units of J/s, which is a watt (W). In the FPS system, the units of q_h are ft-lb/sec, but BTU/hr is also commonly used. For thermal resistance R, the SI units are °C/W, and the FPS units are °F-sec/ft-lb.

For conduction through material of thickness L, an approximate formula for the conductive resistance is

$$R = \frac{L}{kA} \tag{7.7.2}$$

so that

$$q_h = \frac{kA}{L}\Delta T \tag{7.7.3}$$

where k is the *thermal* conductivity of the material and A is the surface area.

If convection occurs, we might need to analyze the system as a fluid as well as a thermal system. Fortunately many analytical and empirical results are available for common situations, and we can use them to obtain the necessary coefficients for our models. The thermal resistance for convection occurring at the boundary of a fluid and a solid is given by

$$R = \frac{1}{hA} \tag{7.7.4}$$

so that

$$q_h = hA\Delta T \tag{7.7.5}$$

where h is the so-called *film coefficient* or *convection coefficient* of the fluid-solid interface and A is the involved surface area. The film coefficient might be a complicated function of the fluid-flow characteristics. For many cases of practical importance, the coefficient has been determined to acceptable accuracy, but a presentation of the results is lengthy and beyond the scope of this work. Standard references on heat transfer contain this information for many cases [Çengel, 2001].

When two bodies are in visual contact, radiation heat transfer occurs through a mutual exchange of heat energy by emission and absorption. Thermal radiation, such as solar energy, produces heat when it strikes a surface capable of absorbing it. The radiation can also be reflected or refracted, and all three mechanisms can occur at a single surface. A net exchange of heat energy occurs from the warmer to the colder body. The rate of this exchange depends on material properties and geometric factors affecting the relative visibility and the amount of surface area involved. The net heat transfer rate depends on the difference of the body temperatures raised to the fourth power (a consequence of the so-called *Stefan-Boltzmann* law).

$$q_h = \beta\left(T_1^4 - T_2^4\right) \tag{7.7.6}$$

The absolute body temperatures are T_1 and T_2, and β is a factor incorporating the other effects. Determining β, like the convection coefficient, is too involved to consider here, but many results are available in the literature.

The radiation model is nonlinear, and therefore we cannot define a specific thermal resistance. However, we can use a linearized model if the temperature change is not too large. Note that linear thermal resistance is a special case of the more general definition of thermal resistance:

$$R = \frac{1}{dq_h/dT} \tag{7.7.7}$$

For example, suppose that T_2 is constant. Then from (7.7.6) and (7.7.7),

$$R = \frac{1}{dq_h/dT_1} = \frac{1}{4\beta T_1^3}$$

When this is evaluated at a specific temperature T_1, we can obtain a specific value for the linearized radiation resistance.

7.7.3 HEAT TRANSFER THROUGH A PLATE

Consider a solid plate or wall of thickness L, as shown in cross section in Figure 7.7.2a. The temperatures of the objects (either solid or fluid) on each side of the plate are T_1 and T_2. If $T_1 > T_2$, heat will flow from the left side to the right side. The temperatures T_1 and T_2 of the adjacent objects will remain constant if the objects are large enough. (One can easily visualize this with a building; the outside air temperature is not affected significantly by heat transfer through the building walls because the mass of the atmosphere is so large.)

If the plate material is homogeneous, eventually the temperature distribution in the plate will look like that shown in part (b) of the figure. This is the steady-state temperature distribution. *Fourier's law of heat conduction* states that the heat transfer rate per unit area within a homogeneous substance is directly proportional to the negative temperature gradient. The proportionality constant is the thermal conductivity k. For the case shown in Figure 7.7.2b, the negative gradient is $(T_1 - T_2)/L$, and the heat transfer rate is thus

$$q_h = \frac{kA(T_1 - T_2)}{L}$$

where A is the plate area in question. Comparing this with (7.7.1) shows that the thermal resistance is given by (7.7.2), with $\Delta T = T_1 - T_2$.

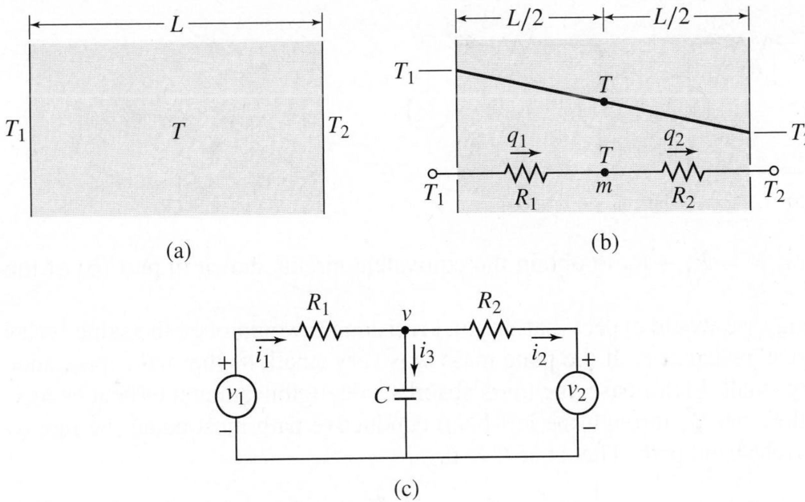

Figure 7.7.2 (a) Conductive heat transfer through a plate. (b) Thermal model. (c) Analogous electric circuit.

Under transient conditions the temperature profile is not linear and must be obtained by solving a partial differential equation (a so-called *distributed-parameter model*). To obtain an ordinary differential equation (a *lumped-parameter model*), which is easier to solve, we must select a point in the plate and use its temperature as the representative temperature of the object. Under steady-state conditions, the average temperature is at the center, and so for this reason we select as an educated guess the center temperature as the representative temperature for the transient calculations. We therefore consider the entire mass m of the plate to be concentrated ("lumped") at the plate centerline, and consider conductive heat transfer to occur over a path of length $L/2$ between temperature T_1 and temperature T. Thus, the thermal resistance for this path is

$$R_1 = \frac{L/2}{kA}$$

Similarly, for the path from T to T_2, the thermal resistance is $R_2 = (L/2)/(kA)$. These resistances and the lumped mass m are represented in Figure 7.7.2b.

From Figure 7.7.2b we can derive the following model by applying conservation of heat energy. Assuming that $T_1 > T > T_2$, we obtain

$$mc_p\frac{dT}{dt} = q_1 - q_2 = \frac{1}{R_1}(T_1 - T) - \frac{1}{R_2}(T - T_2) \tag{7.7.8}$$

The thermal capacitance is $C = mc_p$.

This system is analogous to the circuit shown in Figure 7.7.2c, where the voltages v, v_1, and v_2 are analogous to the temperatures T, T_1, and T_2. Note that the current i_1 is analogous to the heat flow rate into the mass m through the left-hand conductive path, and that current i_2 is analogous to the heat flow rate out of the mass m through the right-hand conductive path. The current i_3 is the net current into the capacitance $(i_3 = i_1 - i_2)$ and increases the voltage v. The current i_3 is analogous to the net heat flow rate into the mass m, which increases the mass temperature T.

7.7.4 SERIES AND PARALLEL THERMAL RESISTANCES

Suppose the capacitance C in the circuit in Figure 7.7.2c is zero. This is equivalent to removing the capacitance to obtain the circuit shown in Figure 7.7.3a, and we can see immediately that the two resistances are in series. Therefore, they can be combined by

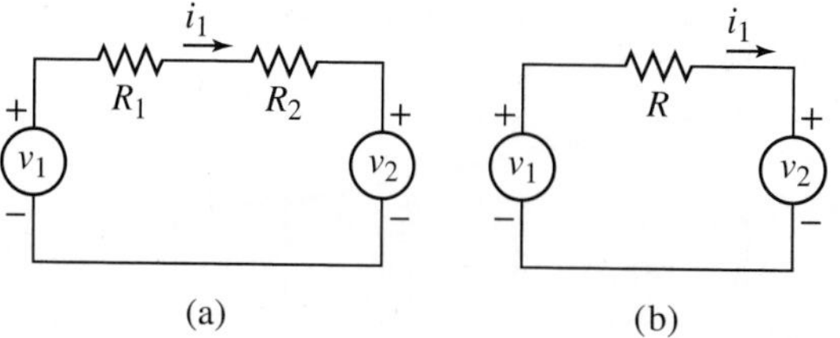

Figure 7.7.3 (a) Circuit analogous to conductive heat transfer through a plate with negligible capacitance. (b) Equivalent circuit.

the series law: $R = R_1 + R_2$ to obtain the equivalent circuit shown in part (b) of the figure.

By analogy we would expect that thermal resistances would obey the same series law as electrical resistances. If the plate mass m is very small, its thermal capacitance C is also very small. In this case, the mass absorbs a negligible amount of heat energy, so the heat flow rate q_1 through the left-hand conductive path must equal the rate q_2 through the right-hand path. That is, if $C = 0$,

$$q_1 = \frac{1}{R_1}(T_1 - T) = q_2 = \frac{1}{R_2}(T - T_2) \tag{7.7.9}$$

The solution of these equations is

$$T = \frac{R_2 T_1 + R_1 T_2}{R_1 + R_2}$$

$$q_1 = q_2 = \frac{T_1 - T_2}{R_1 + R_2} = \frac{T_1 - T_2}{R}$$

The latter solution shows that the resistances R_1 and R_2 are equivalent to the single resistance $R = R_1 + R_2$, which is the series resistance law.

Thus, thermal resistances are in *series* if they pass the same heat flow rate; if so, they are equivalent to a single resistance equal to the sum of the individual resistances. It can also be shown that thermal resistances are in *parallel* if they have the temperature difference; if so, they are equivalent to a single resistance calculated by the reciprocal formula

$$\frac{1}{R} = \frac{1}{R_1} + \frac{1}{R_2} + \cdots$$

We have seen that the resistances R_1 and R_2 in Figure 7.7.2b are in series if $C = 0$. This can occur if the mass m is negligible or if the value of c_p is so small that the product mc_p is negligible. However, by examining the left-hand side of (7.7.8), we see that $q_1 = q_2$ if either $C = 0$ or $dT/dt = 0$. Thus, we conclude that the series resistance law also applies under steady-state conditions, where $dT/dt = 0$. So if $C = 0$ or $dT/dt = 0$ for the plate shown in Figure 7.7.2a, it can be represented as a pure conductive resistance of zero mass, as shown in Figure 7.7.4a, where $R = R_1 + R_2$.

If convection occurs on both sides of the plate, the convective resistances R_{c_1} and R_{c_2} are in series with the conductive resistance R, and the total resistance is given by $R + R_{c_1} + R_{c_2}$, as shown in part (b) of the figure.

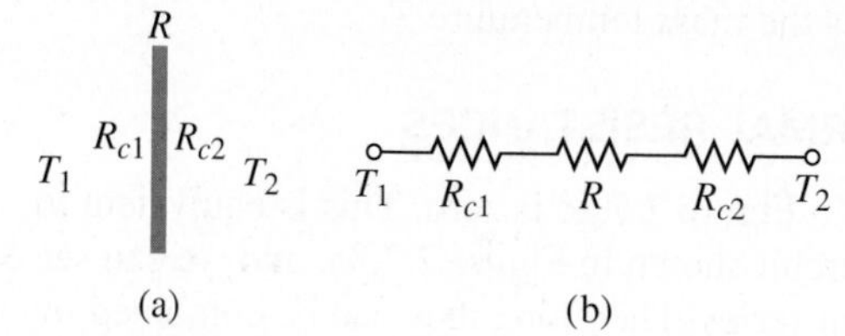

Figure 7.7.4 (a) Convective and conductive heat transfer through a plate with negligible capacitance. (b) Thermal model.

In practice, to obtain a simpler model, the series resistance formula is sometimes used even in applications where the thermal capacitance is not small or where transient conditions exist, but you must be aware that the formula is an approximation in those situations.

Thermal Resistance of Building Wall

EXAMPLE 7.7.1

■ Problem

Engineers must be able to predict the rate of heat loss through a building wall to determine the heating system's requirements. The wall cross section shown in Figure 7.7.5 consists of four layers: an inner layer of plaster/lathe 10 mm thick, a layer of fiberglass insulation 125 mm thick, a layer of wood 60 mm thick, and an outer layer of brick 50 mm thick. For the given materials, the resistances for a wall area of 1 m^2 are $R_1 = 0.036$, $R_2 = 4.01$, $R_3 = 0.408$, and $R_4 = 0.038$°C/W.

Suppose that $T_i = 20$°C and $T_o = -10$°C. (a) Compute the total wall resistance for 1 m^2 of wall area, and compute the heat loss rate if the wall's area is 3 m by 5 m. (b) Find the temperatures T_1, T_2, and T_3, assuming steady-state conditions.

■ Solution

a. The series resistance law gives

$$R = R_1 + R_2 + R_3 + R_4 = 0.036 + 4.01 + 0.408 + 0.038 = 4.492\text{°C/W}$$

which is the total resistance for 1 m^2 of wall area. The wall area is 3(5) = 15 m^2, and thus the total heat loss is

$$q_h = 15\frac{1}{R}(T_i - T_o) = 15\frac{1}{4.492}(20 + 10) = 100.2 \text{ W}$$

This is the heat rate that must be supplied by the building's heating system to maintain the inside temperature at 20°C, if the outside temperature is −10°C.

b. If we assume that the inner and outer temperatures T_i and T_o have remained constant for some time, then the heat flow rate through each layer is the same, q_h. Applying conservation of energy gives the following equations.

$$q_h = \frac{1}{R_1}(T_i - T_1) = \frac{1}{R_2}(T_1 - T_2) = \frac{1}{R_3}(T_2 - T_3) = \frac{1}{R_4}(T_3 - T_o)$$

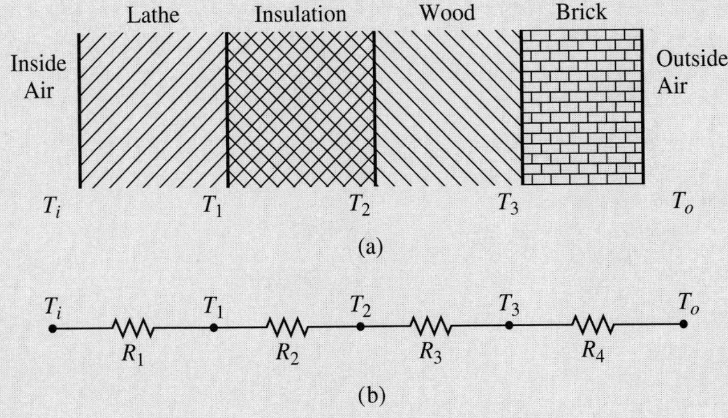

Figure 7.7.5 Heat transfer through a wall with four layers.

The last three equations can be rearranged as follows:

$$(R_1 + R_2)T_1 - R_1 T_2 = R_2 T_i$$

$$R_3 T_1 - (R_2 + R_3)T_2 + R_2 T_3 = 0$$

$$-R_4 T_2 + (R_3 + R_4)T_3 = R_3 T_o$$

For the given values of T_i and T_o, the solution to these equations is $T_1 = 19.7596$, $T_2 = -7.0214$, and $T_3 = -9.7462°\text{C}$.

EXAMPLE 7.7.2

Parallel Resistances

■ Problem

A certain wall section is composed of a 15 cm by 15 cm glass block 8 cm thick. Surrounding the block is a 50 cm by 50 cm brick section, which is also 8 cm thick (see Figure 7.7.6). The thermal conductivity of the glass is $k = 0.81$ W/m·°C. For the brick, $k = 0.45$ W/m·°C. (a) Determine the thermal resistance of the wall section. (b) Compute the heat flow rate through (1) the glass, (2) the brick, and (3) the wall if the temperature difference across the wall is 30°C.

■ Solution

a. The resistances are found from (7.7.2):

$$R = \frac{L}{kA}$$

For the glass,

$$R_1 = \frac{0.08}{0.81(0.15)^2} = 4.39$$

For the brick,

$$R_2 = \frac{0.08}{0.45\left[(0.5)^2 - (0.15)^2\right]} = 0.781$$

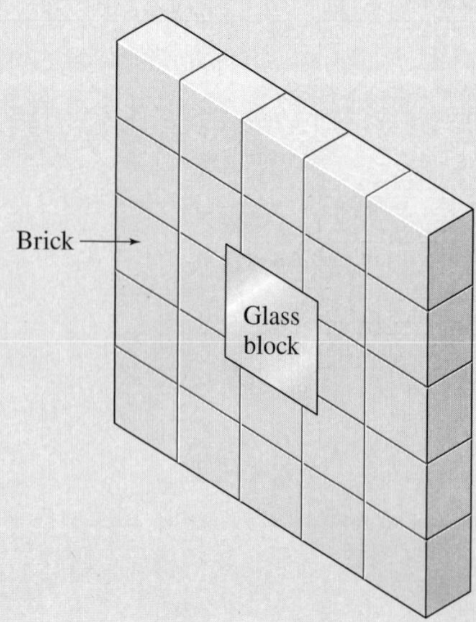

Figure 7.7.6 An example of parallel thermal resistances.

Because the temperature difference is the same across both the glass and the brick, the resistances are in parallel, and thus their total resistance is given by

$$\frac{1}{R} = \frac{1}{R_1} + \frac{1}{R_2} = 0.228 + 1.28 = 1.51$$

or $R = 0.633°\text{C/W}$.

b. The heat flow through the glass is

$$q_1 = \frac{1}{R_1}\Delta T = \frac{1}{4.39}30 = 6.83 \text{ W}$$

The heat flow through the brick is

$$q_2 = \frac{1}{R_2}\Delta T = \frac{1}{0.781}30 = 38.4 \text{ W}$$

Thus, the total heat flow through the wall section is

$$q_h = q_1 + q_2 = 45.2 \text{ W}$$

This rate could also have been calculated from the total resistance as follows:

$$q_h = \frac{1}{R}\Delta T = \frac{1}{0.663}30 = 45.2 \text{ W}$$

Radial Conductive Resistance

EXAMPLE 7.7.3

■ Problem

Consider a cylindrical tube whose inner and outer radii are r_i and r_o. Heat flow in the tube wall can occur in the axial direction along the length of the tube and in the radial direction. If the tube surface is insulated, there will be no radial heat flow, and the heat flow in the axial direction is given by

$$q_h = \frac{kA}{L}\Delta T$$

where L is the length of the tube, ΔT is the temperature difference between the ends a distance L apart, and A is area of the solid cross section (see Figure 7.7.7a).

If only the ends of the tube are insulated, then the heat flow will be entirely radial. Derive an expression for the conductive resistance in the radial direction.

■ Solution

As shown in Figure 7.7.7b the inner and outer temperatures are T_i and T_o, and are assumed to be constant along the length L of the tube. As shown in part (c) of the figure, from Fourier's law, the heat flow rate per unit area through an element of thickness dr is proportional to the negative of the temperature gradient dT/dr. Thus, assuming that the temperature inside the tube wall does not change with time, the heat flow rate q_h out of the section of thickness dr is

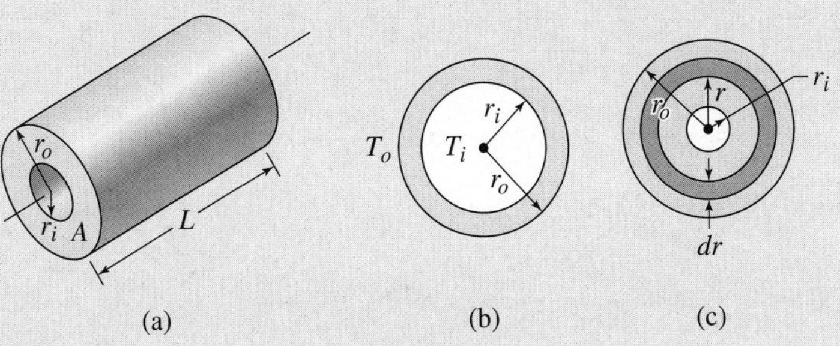

Figure 7.7.7 Radial conductive heat transfer.

the same as the heat flow into the section. Therefore,

$$\frac{q_h}{2\pi rL} = -k\frac{dT}{dr}$$

Thus,

$$q_h = -k\frac{dT}{dr}2\pi rL = -2\pi Lk\frac{dT}{dr/r}$$

or

$$\int_{r_i}^{r_o} q_h\frac{dr}{r} = -2\pi Lk\int_{T_i}^{T_o} dT$$

Because q_h is constant, the integration yields

$$q_h \ln\frac{r_o}{r_i} = -2\pi Lk(T_o - T_i)$$

or

$$q_h = \frac{2\pi Lk}{\ln(r_o/r_i)}(T_i - T_o)$$

The radial resistance is thus given by

$$R = \frac{\ln(r_o/r_i)}{2\pi Lk} \tag{1}$$

EXAMPLE 7.7.4

Heat Loss from Water in a Pipe

■ Problem

Water at 120°F flows in a copper pipe 6 ft long, whose inner and outer radii are 1/4 in. and 3/8 in. The temperature of the surrounding air is 70°F. Compute the heat loss rate from the water to the air in the radial direction. Use the following values. For copper, $k = 50$ lb/sec-°F. The convection coefficient at the inner surface between the water and the copper is $h_i = 16$ lb/sec-ft-°F. The convection coefficient at the outer surface between the air and the copper is $h_o = 1.1$ lb/sec-ft-°F.

■ Solution

Assuming that the temperature inside the pipe wall does not change with time, then the same heat flow rate occurs in the inner and outer convection layers and in the pipe wall. Thus, the three resistances are in series and we can add them to obtain the total resistance. The inner and outer surface areas are

$$A_i = 2\pi r_i L = 2\pi\left(\frac{1}{4}\right)\left(\frac{1}{12}\right)6 = 0.785 \text{ ft}^2$$

$$A_o = 2\pi r_o L = 2\pi\left(\frac{3}{8}\right)\left(\frac{1}{12}\right)6 = 1.178 \text{ ft}^2$$

The inner convective resistance is

$$R_i = \frac{1}{h_i A_i} = \frac{1}{16(0.785)} = 0.08\ \frac{\text{sec-°F}}{\text{ft-lb}}$$

$$R_o = \frac{1}{h_o A_o} = \frac{1}{1.1(1.178)} = 0.77\ \frac{\text{sec-°F}}{\text{ft-lb}}$$

The conductive resistance of the pipe wall is

$$R_c = \frac{\ln\left(\frac{r_o}{r_i}\right)}{2\pi Lk} = \frac{\ln\left(\frac{3/8}{1/4}\right)}{2\pi(6)(50)} = 2.15\times10^{-4}\ \frac{\text{sec-°F}}{\text{ft-lb}}$$

Thus, the total resistance is

$$R = R_i + R_c + R_o = 0.08 + 2.15 \times 10^{-4} + 0.77 = 0.85 \ \frac{\text{sec-°F}}{\text{ft-lb}}$$

The heat loss from the pipe, assuming that the water temperature is a constant 120° along the length of the pipe, is

$$q_h = \frac{1}{R}\Delta T = \frac{1}{0.85}(120 - 70) = 59 \ \frac{\text{ft-lb}}{\text{sec}}$$

To investigate the assumption that the water temperature is constant, compute the thermal energy E of the water in the pipe, using the mass density $\rho = 1.94$ slug/ft^3 and $c_p = 25,000$ ft-lb/slug-°F:

$$E = mc_pT_i = \left(\pi r_i^2 L\rho\right)c_pT_i = 47,624 \text{ ft-lb}$$

Assuming that the water flows at 1 ft/sec, a slug of water will be in the pipe for 6 sec. During that time it will lose 59(6) = 354 ft-lb of heat. Because this amount is very small compared to E, our assumption that the water temperature is constant is confirmed.

7.8 DYNAMIC MODELS OF THERMAL SYSTEMS

In this section, we discuss how to obtain lumped-parameter (ordinary differential equation) models of thermal systems containing one or more thermal capacitances, and how to obtain the value of thermal resistance experimentally.

Heat transfer occurs between two objects, but it will also occur *within* an object if the temperature varies with location in the object. To obtain an ordinary differential equation model of the temperature dynamics of an object, we must be able to assign a single temperature that is representative of the object. Sometimes it is difficult to assign such a representative temperature to a body or fluid because of its complex shape or motion and the resulting complex distribution of temperature throughout the object. When a single representative temperature cannot be assigned, several coupled lumped-parameter models or even a distributed-parameter model will be required.

7.8.1 THE BIOT CRITERION

For solid bodies immersed in a fluid, a useful criterion for determining the validity of the uniform-temperature assumption is based on the *Biot* number, defined as

$$N_B = \frac{hL}{k} \tag{7.8.1}$$

where L is a representative dimension of the object, which is usually taken to be the ratio of the volume to the surface area of the body. For example, the ratio L for a sphere of radius r is $(4/3)\pi r^3/4\pi r^2 = r/3$. If the shape of the body resembles a plate, cylinder, or sphere, it is common practice to consider the object to have a single uniform temperature if N_B is small. Often, if $N_B < 0.1$, the temperature is taken to be uniform. The accuracy of this approximation improves if the inputs vary slowly.

The Biot number is the ratio of the convective heat transfer rate to the conductive rate. This can be seen by expressing N_B for a plate of thickness L as follows, using (7.7.1) through (7.7.3):

$$N_B = \frac{q_{\text{convection}}}{q_{\text{conduction}}} = \frac{hA\Delta T}{kA\Delta T/L} = \frac{hL}{k}$$

So the Biot criterion reflects the fact that if the conductive heat transfer rate is large relative to the convective rate, any temperature changes due to conduction within the object will occur relatively rapidly, and thus the object's temperature will become uniform relatively quickly.

Calculation of the ratio L depends on the surface area that is exposed to convection. For example, a cube of side length d has a value of $L = d^3/(6d^2) = d/6$ if all six sides are exposed to convection, whereas if four sides are insulated, the value is $L = d^3/(2d^2) = d/2$.

EXAMPLE 7.8.1

Quenching with Constant Bath Temperature

■ Problem

Hardness and other properties of metal can be improved by the rapid cooling that occurs during *quenching*, a process in which a heated object is placed into a liquid bath (see Figure 7.8.1). Consider a lead cube with a side length of $d = 20$ mm. The cube is immersed in an oil bath for which $h = 200$ W/(m$^2 \cdot$ °C). The oil temperature is T_b.

Thermal conductivity varies as function of temperature, but for lead the variation is relatively small (k for lead varies from 35.5 W/m · °C at 0°C to 31.2 W/m · °C at 327°C). The density of lead is 1.134×10^4 kg/m^3. Take the specific heat of lead to be 129 J/kg · °C.

(a) Show that temperature of the cube can be considered uniform; and (b) develop a model of the cube's temperature as a function of the liquid temperature T_b, which is assumed to be known.

■ Solution

a. The ratio of volume of the cube to its surface area is

$$L = \frac{d^3}{6d^2} = \frac{d}{6} = \frac{0.02}{6}$$

and using an average value of 34 W/m · °C for k, we compute the Biot number to be

$$N_B = \frac{200(0.02)}{34(6)} = 0.02$$

which is much less than 0.1. According to the Biot criterion, we may treat the cube as a lumped-parameter system with a single uniform temperature, denoted T.

b. If we assume that $T > T_b$, then the heat flows from the cube to the liquid, and from conservation of energy we obtain

$$C\frac{dT}{dt} = -\frac{1}{R}(T - T_b) \qquad (1)$$

When deriving thermal system models, you must make an assumption about the relative values of the temperatures, and assign the heat flow direction consistent with that assumption. The specific assumption does not matter as long as you are consistent. Thus, although the bath temperature will be less than the cube temperature in the quenching application, you may still assume that $T_b > T$ and arrive at the correct model as long as you assign the heat flow direction to be *into* the cube. However, when making such

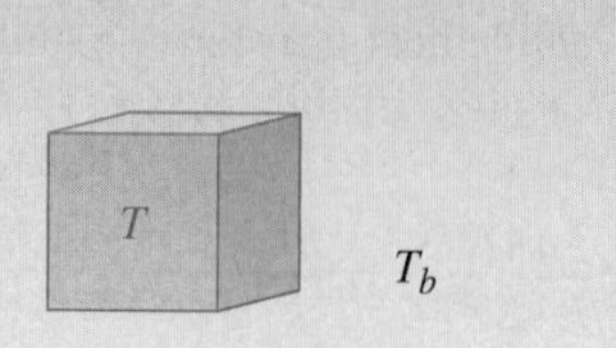

Figure 7.8.1 Quenching with a constant bath temperature.

assumptions, your physical insight is improved if you assume the most likely situation; the nature of the quenching process means that $T > T_b$, so this is the logical assumption to use.

The thermal capacitance of the cube is computed as

$$C = mc_p = \rho V c_p = 1.134 \times 10^4 (0.02)^3 (129) = 11.7\ \text{J/°C}$$

The thermal resistance R is due to convection, and is

$$R = \frac{1}{hA} = \frac{1}{200(6)(0.02)^2} = 2.08\text{°C} \cdot \text{s/J}$$

Thus, the model is

$$11.7\frac{dT}{dt} = -\frac{1}{2.08}(T - T_b)$$

or

$$24.4\frac{dT}{dt} + T = T_b$$

The time constant is $\tau = RC = 24.4$ s. If the bath is large enough so that the cube's energy does not appreciably affect the bath temperature T_b, then when the cube is dropped into the bath, the temperature T_b acts like a step input. The cube's temperature will reach the temperature T_b in approximately $4\tau = 98$ s.

Figure 7.8.2 shows a circuit that is analogous to the thermal model of the quenching process. The voltages v and v_b are analogous to the temperatures T and T_b. The circuit model is

$$C\frac{dv}{dt} = \frac{1}{R}(v_b - v)$$

Figure 7.8.2 Electric circuit analogous to quenching with a constant bath temperature.

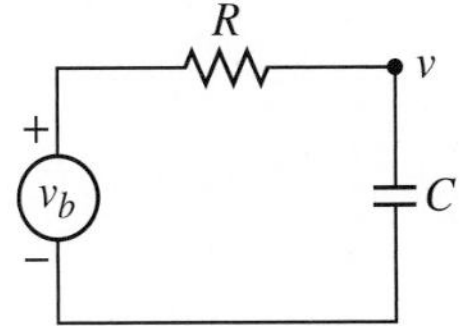

7.8.2 MULTIPLE THERMAL CAPACITANCES

When it is not possible to identify one representative temperature for a system, you must identify a representative temperature for each distinct thermal capacitance. Then, after identifying the resistance paths between each capacitance, apply conservation of heat energy to each capacitance. In doing so, you must arbitrarily but consistently assume that some temperatures are greater than others, to assign directions to the resulting heat flows. The order of the resulting model equals the number of representative temperatures.

Quenching with Variable Bath Temperature

EXAMPLE 7.8.2

■ Problem

Consider the quenching process treated in Example 7.8.1. If the thermal capacitance of the liquid bath is not large, the heat energy transferred from the cube will change the bath temperature, and we will need a model to describe its dynamics. Consider the representation shown in Figure 7.8.3. The temperature outside the bath is T_o, which is assumed to be known. The convective resistance between the cube and the bath is R_1, and the combined convective/conductive resistance of the container wall and the liquid surface is R_2. The capacitances of the cube and the liquid bath are C and C_b, respectively.

Figure 7.8.3 Quenching with a variable bath temperature.

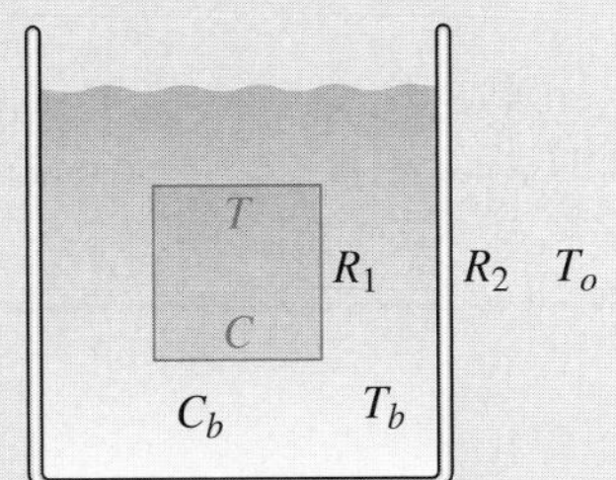

a. Derive a model of the cube temperature and the bath temperature assuming that the bath loses no heat to the surroundings (that is, $R_2 = \infty$).
b. Obtain the model's characteristic roots and the form of the response.

■ **Solution**

a. Assume that $T > T_b$. Then the heat flow is out of the cube and into the bath. From conservation of energy for the cube,

$$C\frac{dT}{dt} = -\frac{1}{R_1}(T - T_b) \tag{1}$$

and for the bath,

$$C_b\frac{dT_b}{dt} = \frac{1}{R_1}(T - T_b) \tag{2}$$

Equations (1) and (2) are the desired model. Note that the heat flow rate in equation (2) must have a sign opposite to that in equation (1) because the heat flow *out* of the cube must be the same as the heat flow *into* the bath.

b. Applying the Laplace transform method to equations (1) and (2) with zero initial conditions, we obtain

$$(R_1Cs + 1)T(s) - T_b(s) = 0 \tag{3}$$

$$(R_1C_bs + 1)T_b(s) - T(s) = 0 \tag{4}$$

Solving equation (3) for $T_b(s)$ and substituting into equation (4) gives

$$[(R_1C_bs + 1)(R_1Cs + 1) - 1]T(s) = 0$$

from which we obtain

$$R_1^2C_bCs^2 + R_1(C + C_b)s = 0$$

So the characteristic roots are

$$s = 0, \quad s = -\frac{C + C_b}{R_1CC_b}$$

Because equations (3) and (4) are homogeneous, the form of the response is

$$T(t) = A_1e^{0t} + B_1e^{-t/\tau} = A_1 + B_1e^{-t/\tau} \qquad \tau = \frac{R_1CC_b}{C + C_b}$$

$$T_b(t) = A_2e^{0t} + B_2e^{-t/\tau} = A_2 + B_2e^{-t/\tau}$$

where the constants A_1, A_2, B_1, and B_2 depend on the initial conditions. The two temperatures become constant after approximately 4τ. Note that $T(t) \to A_1$ and $T_b(t) \to A_2$ as $t \to \infty$. From physical insight we know that T and T_b will become equal as $t \to \infty$. Therefore, $A_2 = A_1$.

The final value of the temperatures, A_1, can be easily found from physical reasoning using conservation of energy. The initial energy in the system consisting of the cube and the bath is the thermal energy in both; namely, $CT(0) + C_bT_b(0)$. The final energy is expressed as $CA_1 + C_bA_1$, and is the same as the initial energy. Thus,

$$CT(0) + C_bT_b(0) = CA_1 + C_bA_1 \qquad \text{or} \qquad A_1 = \frac{CT(0) + C_bT_b(0)}{C + C_b} = A_2$$

Note also that $T(0) = A_1 + B_1$, and $T_b(0) = A_2 + B_2$. Thus, $B_1 = T(0) - A_1$ and $B_2 = T_b(0) - A_2$.

Figure 7.8.4 shows an electric circuit that is analogous to the quenching system of Figure 7.8.3 with $R_2 = \infty$. The voltages v and v_b are analogous to the temperatures T and T_b.

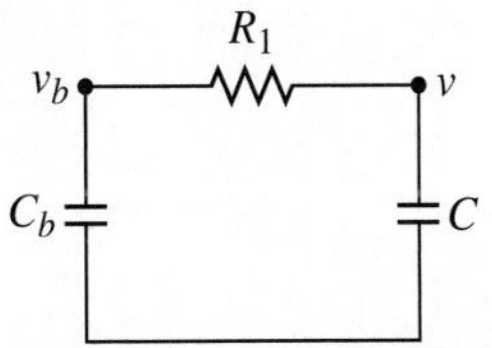

Figure 7.8.4 Electric circuit analogous to quenching with a variable bath temperature and infinite container resistance.

EXAMPLE 7.8.3

Quenching with Heat Loss to the Surroundings

■ Problem

Consider the quenching process treated in the previous example (Figure 7.8.3). (a) Derive a model of the cube temperature and the bath temperature assuming R_2 is finite. (b) Obtain the model's characteristic roots and the form of the response of $T(t)$, assuming that the surrounding temperature T_o is constant.

■ Solution

a. If R_2 is finite, then we must now account for the heat flow into or out of the container. Assume that $T > T_b > T_o$. Then the heat flows from the cube into the bath and then into the surroundings. From conservation of energy,

$$C\frac{dT}{dt} = -\frac{1}{R_1}(T - T_b) \tag{1}$$

and

$$C_b\frac{dT_b}{dt} = \frac{1}{R_1}(T - T_b) - \frac{1}{R_2}(T_b - T_o) \tag{2}$$

Equations (1) and (2) are the desired model.

b. Applying the Laplace transform method with zero-initial conditions, we obtain

$$(R_1Cs + 1)T(s) - T_b(s) = 0 \tag{3}$$

$$(R_1R_2C_bs + R_1 + R_2)T_b(s) - R_2T(s) = R_1T_o \tag{4}$$

Solving equation (3) for $T_b(s)$ and substituting into equation (4) gives the transfer function

$$\frac{T(s)}{T_o(s)} = \frac{1}{R_1R_2C_bCs^2 + \left[(R_1 + R_2)C + R_2C_b\right]s + 1} \tag{5}$$

The denominator gives the characteristic equation

$$R_1R_2C_bCs^2 + \left[(R_1 + R_2)C + R_2C_b\right]s + 1 = 0$$

So there will be two nonzero characteristic roots. If these roots are real, say $s = -1/\tau_1$ and $s = -1/\tau_2$, and if T_o is constant, the response will have the form

$$T(t) = Ae^{-t/\tau_1} + Be^{-t/\tau_2} + D$$

where the constants A and B depend on the initial conditions. Note that $T(t) \to D$ as $t \to \infty$. Applying the final-value theorem to equation (5) gives $T(\infty) = T_o$ and thus $D = T_o$. We could have also obtained this result through physical reasoning.

Figure 7.8.5 shows an electric circuit that is analogous to the quenching system of Figure 7.8.3 with R_2 finite.

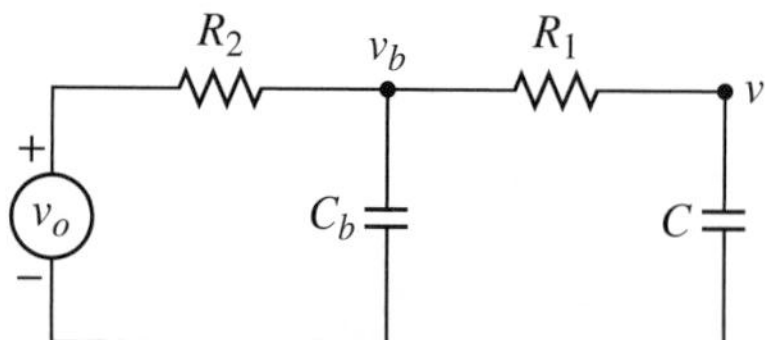

Figure 7.8.5 Electric circuit analogous to quenching with a variable bath temperature and finite container resistance.

7.8.3 EXPERIMENTAL DETERMINATION OF THERMAL RESISTANCE

The mass density ρ, the specific heat c_p, and the thermal conductivity k are accurately known for most materials, and thus we can obtain accurate values of the thermal capacitance C, and also of the conductive resistance L/kA, especially if the thermal capacitance of the conducting element is small. However, determination of the convective resistance is difficult to do analytically, and we must usually resort to experimentally determined values. In some cases, we may not be able to distinguish between the effects of conduction, convection, and radiation heat transfer, and the resulting model will contain a thermal resistance that expresses the aggregated effects.

EXAMPLE 7.8.4

Temperature Dynamics of a Cooling Object

■ Problem

Consider the experiment with a cooling cup of water described in Example 1.5.1 in Chapter 1. Water of volume 250 ml in a glass measuring cup was allowed to cool after being heated to 204°F. The surrounding air temperature was 70°F. The measured water temperature at various times is given in the table in Example 1.5.1. From that data we derived the following model of the water temperature as a function of time.

$$T = 129e^{-0.0007t} + 70 \tag{1}$$

where T is in °F and time t is in seconds. Estimate the thermal resistance of this system.

■ Solution

We model the cup and water as the object shown in Figure 7.8.6. We assume that convection has mixed the water well so that the water has the same temperature throughout. Let R be the aggregated thermal resistance due to the combined effects of (1) conduction through the sides and bottom of the cup, (2) convection from the water surface and from the sides of the cup into the air, and (3) radiation from the water to the surroundings. Assume that the air temperature T_o is constant and select it as the reference temperature. The heat energy in the water is

Figure 7.8.6 Generic representation of a thermal system having a single capacitance and a single resistance.

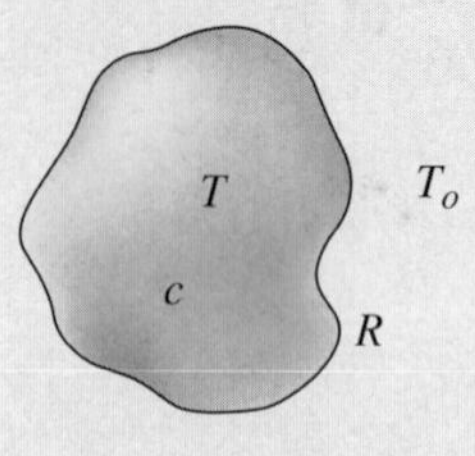

$$E = \rho V c_p (T - T_o)$$

From conservation of heat energy

$$\frac{dE}{dt} = -\frac{1}{R}(T - T_o)$$

or, since ρ, V, c_p, and T_o are constant,

$$\rho V c_p \frac{dT}{dt} = -\frac{1}{R}(T - T_o)$$

The water's thermal capacitance is $C = \rho V c_p$ and the model can be expressed as

$$RC\frac{dT}{dt} + T = T_o \tag{2}$$

The model's complete response is

$$T(t) = T(0)e^{-t/RC} + \left(1 - e^{-t/RC}\right)T_o = [T(0) - T_o]e^{-t/RC} + T_o$$

Comparing this with equation (1), we see that

$$RC = \frac{1}{0.0007} = 1429 \text{ sec}$$

or

$$R = \frac{1429}{C}\frac{°\text{F}}{\text{ft-lb}}$$

where $C = \rho V c_p$. Because the temperature data was given in °F, we must convert the volume of 250 ml to ft^3. Note that $V = 250$ ml $= 2.5 \times 10^{-4}$ m^3, so that $V = 2.5 \times 10^{-4}(3.28 \text{ ft/m})^3\text{m}^3 =$ 8.82×10^{-3} ft^3. Using the values at room temperature for water, we have $\rho = 1.942$ slug/ft^3, $c_p = 25,000$ ft-lb/slug-°F, and thus, $C = 423$ ft-lb-sec/°F. Therefore, the aggregated thermal resistance is

$$R = \frac{1429}{C} = 3.37\frac{°\text{F}}{\text{ft-lb}}$$

The usefulness of this result is that this value of R can be used to predict the temperature dynamics of the water/cup system under somewhat different conditions. For example, we can use it to estimate the temperature of a different amount of water if we also know how much the surface area changes. This is because the thermal resistance is inversely proportional to area, as can be seen from (7.7.2) and (7.7.4). Suppose we double the water volume to 500 ml, so that the new value of the thermal capacitance becomes $C = 2(423) = 846$ ft-lb-sec/°F. Suppose the surface area of the new volume is $5/3$ that of the smaller volume. Then we estimate the new value of R to be $R = (3/5)(3.37) = 2.02$°F/ft-lb. Thus, the model of the new water mass is given by equation (2) with $RC = (2.02)(846) = 1709$ sec.

Temperature Sensor Response

EXAMPLE 7.8.5

■ **Problem**

A thermocouple can be used to measure temperature. The electrical resistance of the device is a function of the temperature of the surrounding fluid. By calibrating the thermocouple and measuring its resistance, we can determine the temperature. Because the thermocouple has mass, it has thermal capacitance, and thus its temperature change (and electrical resistance change) will lag behind any change in the fluid temperature.

Estimate the response time of a thermocouple suddenly immersed in a fluid. Model the device as a sphere of copper constantin alloy, whose diameter is 2 mm, and whose properties are $\rho = 8920$ kg/m^3, $k = 19$ W/m·°C, and $c_p = 362$ J/kg·°C. Take the convection coefficient to be $h = 200$ W/m^2·°C.

■ **Solution**

First compute the Biot number N_B to see if a lumped-parameter model is sufficient. For a sphere of radius r,

$$L = \frac{V}{A} = \frac{(4/3)\pi r^3}{4\pi r^2} = \frac{r}{3} = \frac{0.001}{3} = 3.33 \times 10^{-4}$$

The Biot number is

$$N_B = \frac{hL}{k} = \frac{200(3.33 \times 10^{-4})}{19} = 0.0035$$

which is much less than 0.1. So we can use a lumped-parameter model.

Applying conservation of heat energy to the sphere, we obtain the model:

$$c_p \rho V \frac{dT}{dt} = hA(T_o - T)$$

where T is the temperature of the sphere and T_o is the fluid temperature. The time constant of this model is

$$\tau = \frac{c_p \rho}{h}\frac{V}{A} = \frac{362(8920)}{200} 3.33 \times 10^{-4} = 5.38 \text{ s}$$

The thermocouple temperature will reach 98% of the fluid temperature within $4\tau = 21.5$ s.

EXAMPLE 7.8.6

State-Variable Model of Wall Temperature Dynamics

■ Problem

Consider the wall shown in cross section in Figure 7.7.5 and treated in Example 7.7.1. In that example the thermal capacitances of the layers were neglected. We now want to develop a model that includes their effects. Neglect any convective resistance on the inside and outside surfaces.

■ Solution

We lump each thermal mass at the centerline of its respective layer and assign half of the layer's thermal resistance to the heat flow path on the left and half to the path on the right side of the lumped mass. The representation is shown in Figure 7.8.7a. Let

$$R_a = \frac{R_1}{2} \qquad R_b = \frac{R_1}{2} + \frac{R_2}{2}$$

$$R_c = \frac{R_2}{2} + \frac{R_3}{2} \qquad R_d = \frac{R_3}{2} + \frac{R_4}{2} \qquad R_e = \frac{R_4}{2}$$

An equivalent electrical circuit is shown in part (b) for those who benefit from such an analogy.

For thermal capacitance C_1, conservation of energy gives

$$C_1 \frac{dT_1}{dt} = \frac{T_i - T_1}{R_a} - \frac{T_1 - T_2}{R_b}$$

For C_2,

$$C_2 \frac{dT_2}{dt} = \frac{T_1 - T_2}{R_b} - \frac{T_2 - T_3}{R_c}$$

For C_3,

$$C_3 \frac{dT_3}{dt} = \frac{T_2 - T_3}{R_c} - \frac{T_3 - T_4}{R_d}$$

Figure 7.8.7 (a) A wall model with four capacitances. (b) Analogous electric circuit.

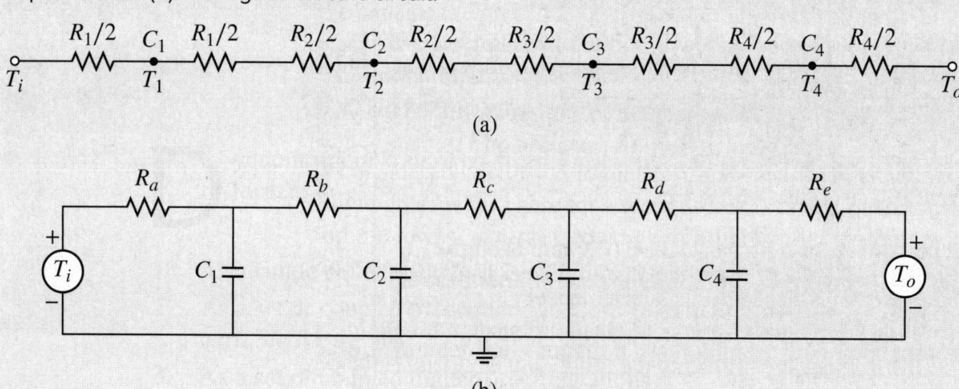

Finally, for C_4,

$$C_4\frac{dT_4}{dt} = \frac{T_3 - T_4}{R_d} - \frac{T_4 - T_0}{R_e}$$

These four equations may be put into state-variance form as follows.

$$\frac{d\mathbf{T}}{dt} = \mathbf{AT} + \mathbf{Bu}$$

where

$$\mathbf{T} = \begin{bmatrix} T_1 \\ T_2 \\ T_3 \\ T_4 \end{bmatrix} \qquad \mathbf{u} = \begin{bmatrix} T_i \\ T_o \end{bmatrix}$$

$$\mathbf{A} = \begin{bmatrix} a_{11} & a_{12} & 0 & 0 \\ a_{21} & a_{22} & a_{23} & 0 \\ 0 & a_{32} & a_{33} & a_{34} \\ 0 & 0 & a_{43} & a_{44} \end{bmatrix} \qquad \mathbf{B} = \begin{bmatrix} b_{11} & 0 \\ 0 & 0 \\ 0 & 0 \\ 0 & b_{42} \end{bmatrix}$$

where

$$a_{11} = -\frac{R_a + R_b}{C_1 R_a R_b} \qquad a_{12} = \frac{1}{C_1 R_b}$$

$$a_{21} = \frac{1}{C_2 R_b} \qquad a_{22} = -\frac{R_b + R_c}{C_2 R_b R_c} \qquad a_{23} = \frac{1}{C_2 R_c}$$

$$a_{32} = \frac{1}{C_3 R_c} \qquad a_{33} = -\frac{R_c + R_d}{C_3 R_c R_d} \qquad a_{34} = \frac{1}{C_3 R_d}$$

$$a_{43} = \frac{1}{C_4 R_d} \qquad a_{44} = -\frac{R_d + R_e}{C_4 R_d R_e}$$

$$b_{11} = \frac{1}{C_1 R_a} \qquad b_{42} = \frac{1}{C_4 R_e}$$

In Section 7.9 we will use MATLAB to solve these equations.

PART III. MATLAB AND SIMULINK APPLICATIONS

Sections 7.9 and 7.10 show how MATLAB and Simulink can be used to solve problems involving fluid and thermal systems.

7.9 MATLAB APPLICATIONS

Fluid and thermal system models are often nonlinear. When a differential equation is nonlinear, we often have no analytical solution to use for checking our numerical results. In such cases, we can use our physical insight to guard against grossly incorrect results. We can also check the equation singularities that might affect the numerical procedure. Finally, we can sometimes use an approximation to replace the nonlinear equation with a linear one that can be solved analytically. Although the linear approximation does not give the exact answer, it can be used to see if our numerical answer is "in the ball park." Example 7.9.1 illustrates this approach.

Lumped-parameter models of thermal systems often need several lumped masses in order to represent the dynamics well. This requires a higher-order model that is difficult to solve in closed form and must be solved numerically. Such models naturally occur in state-variable form and are therefore easily handled with the MATLAB state-variable functions. Example 7.9.2 deals with heat transfer through a multilayered wall. It shows how to obtain the response of a state-variable model that is subjected to two different inputs and whose initial conditions are nonzero.

EXAMPLE 7.9.1

Liquid Height in a Spherical Tank

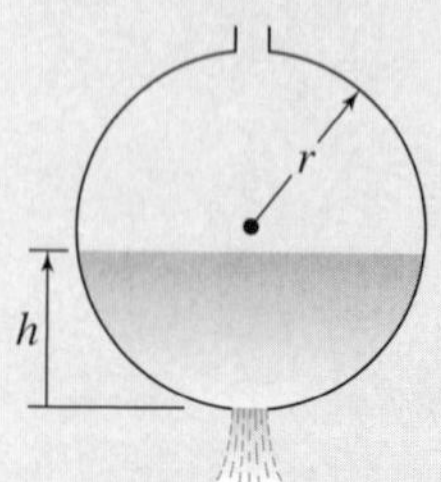

Figure 7.9.1 Draining of a spherical tank.

■ Problem

Figure 7.9.1 shows a spherical tank for storing water. The tank is filled through a hole in the top and drained through a hole in the bottom. The following model for the liquid height h is developed in Problems 7.37 and 7.38.

$$\pi(2Rh - h^2)\frac{dh}{dt} = -C_d A_o \sqrt{2gh} \tag{1}$$

For water, $C_d = 0.6$ is a common value.

Use MATLAB to solve this equation to determine how long it will take for the tank to empty if the initial height is 9 ft. The tank has a radius of $R = 5$ ft and has a 1-in.-diameter hole in the bottom. Use $g = 32.2$ ft/sec^2. Discuss how to check the solution.

■ Solution

With $C_d = 0.6$, $R = 5$, $g = 32.2$, and $A_o = \pi(1/24)^2$, equation (1) becomes

$$\frac{dh}{dt} = -\frac{0.0334\sqrt{h}}{10h - h^2} \tag{2}$$

We can use our physical insight to guard against grossly incorrect results. We can also check the preceding expression for dh/dt for singularities. The denominator does not become zero unless $h=0$ or $h=10$, which correspond to a completely empty and a completely full tank. So we will avoid singularities if $0 < h(0) < 10$.

We can use the following approximation to estimate the time to empty. Replace h on the right side of equation (2) with its average value, namely, $(9-0)/2=4.5$ feet. This gives $dh/dt=-0.00286$, whose solution is $h(t)=h(0)-0.00286t=9-0.00286t$. According to this equation, if $h(0)=9$, the tank will be empty at $t=9/0.00286=3147$ sec, or 52 min. We will use this value as a "reality check" on our answer.

The function file based on this equation is

```
function hdot = height(t,h)
hdot = -(0.0334*sqrt(h))/(10*h-h^2);
```

The file is called as follows, using the `ode45` solver.

```
[t, h] = ode45(@height, [0, 2475], 9);
plot(t,h),xlabel('Time (sec)'),ylabel('Height (ft)')
```

The resulting plot is shown in Figure 7.9.2. Note how the height changes more rapidly when the tank is nearly full or nearly empty. This is to be expected because of the effects of the tank's shape. The tank empties in 2475 sec, or 41 min. This value is not grossly different from our rough estimate of 52 min, so we should feel comfortable accepting the numerical results.

The value of the final time of 2475 sec was found by increasing the final time until the plot showed that the height became zero. You could use a `while` loop to do this, by increasing the final time in the loop while calling `ode45` repeatedly.

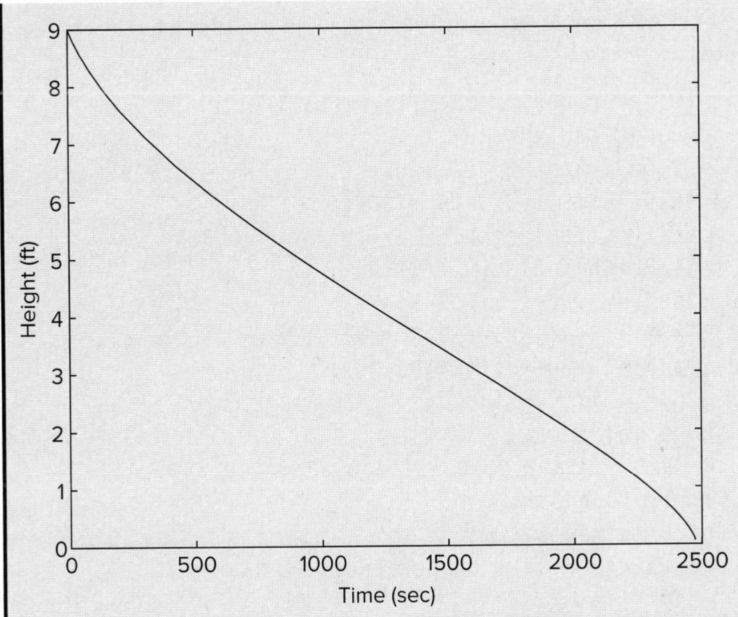

Figure 7.9.2 Plot of liquid height in a draining spherical tank.

EXAMPLE 7.9.2

Heat Transfer Through a Wall

■ **Problem**

Consider the wall cross section shown in Figure 7.9.3. The temperature model was developed in Example 7.8.6. Use the following values and plot the temperatures versus time for the case where the inside temperature is constant at $T_i = 20°\text{C}$ and the outside temperature T_o decreases linearly from 5°C to −10°C in 1 h. The initial wall temperatures are 10°C.

The resistance values in °C/W are

$$R_a = 0.018 \qquad R_b = 2.023 \qquad R_c = 2.204$$
$$R_d = 0.223 \qquad R_e = 0.019$$

The capacitance values in J/°C are

$$C_1 = 8720 \qquad C_2 = 6210$$
$$C_3 = 6637 \qquad C_4 = 2.08 \times 10^4$$

■ **Solution**

The model was developed in Example 7.8.6.

The given information shows that the outside temperature is described by

$$T_o(t) = 5 - 15t \quad 0 \le t \le 3600 \text{ s}$$

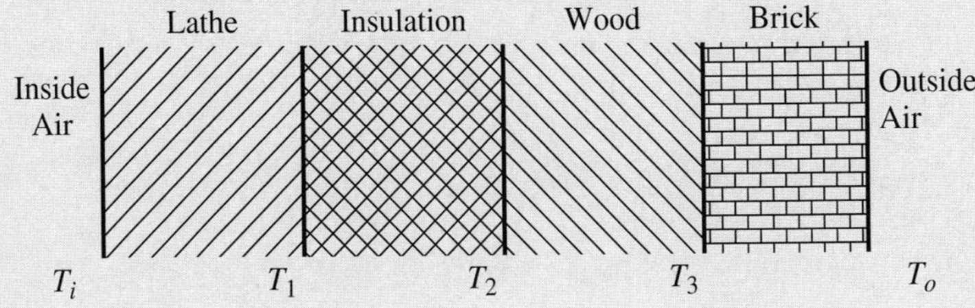

Figure 7.9.3 Heat transfer through a wall with four layers.

The following MATLAB program creates the required plots. Recall that the total response is the sum of the forced and the free responses.

```
% htwall.m Heat transfer thru a multilayer wall.
% Resistance and capacitance data.
Ra = 0.018; Rb = 2.023; Rc = 2.204; Rd = 0.223; Re = 0.019;
C1 = 8720; C2 = 6210; C3 = 6637; C4 = 20800;
% Compute the matrix coefficients.
a11 = -(Ra+Rb)/(C1*Ra*Rb); a12 = 1/(C1*Rb);
a21 = 1/(C2*Rb); a22 = -(Rb+Rc)/(C2*Rb*Rc); a23 = 1/(C2*Rc);
a32 = 1/(C3*Rc); a33 = -(Rc+Rd)/(C3*Rc*Rd); a34 = 1/(C3*Rd);
a43 = 1/(C4*Rd); a44 = -(Rd+Re)/(C4*Rd*Re);
b11 = 1/(C1*Ra); b42 = 1/(C4*Re);
% Define the A and B matrices.
A = [a11,a12,0,0; a21,a22,a23,0; 0,a32,a33,a34; 0,0,a43,a44];
B = [b11,0; 0,0; 0,0; 0,b42];
% Define the C and D matrices.
% The outputs are the four wall temperatures.
C = eye(4);
D = zeros(size(B));
% Create the LTI model.
sys = ss(A,B,C,D);
% Create the time vector for 1 hour (3600 seconds).
t = (0:1:3600);
% Create the input vector.
u = [20*ones(size(t));(5-15*ones(size(t)).*t/3600)];
% Compute the forced response.
[yforced,t] = lsim(sys,u,t);
% Compute the free response.
[yfree,t] = initial(sys,[10,10,10,10],t);
% Plot the response along with the outside temperature.
plot(t,yforced+yfree,t,u(2,:))
% Compute the time constants.
tau =(-1./real(eig(A)))/60
```

The plot is shown in Figure 7.9.4. Note how T_1 follows the inside temperature, while T_4 follows the outside temperature, as expected. The time constants are 2.6, 6, 24, and 117 min.

From the plot note that T_1 reaches steady state in less than 1000 s, but the dominant time constant is 117 min, or 7020 s. The discrepancy is explained by examining the transfer functions for numerator dynamics. After running `htwall`, which puts the matrices `A` and `B` in the workspace, the following MATLAB session can be used to obtain the transfer functions for T_1 and T_2. The `C` and `D` matrices need to be replaced to obtain only T_1 and T_2 as the outputs.

```
>>C1 = [1,0,0,0]; D1=[0,0];
>>sys1 = ss(A,B,C1,D1);
>>C2 = [0,1,0,0]; D2 = [0,0];
>>sys2 = ss(A,B,C2,D2);
>>tf1 = tf(sys1)
>>tf2 = tf(sys2)
```

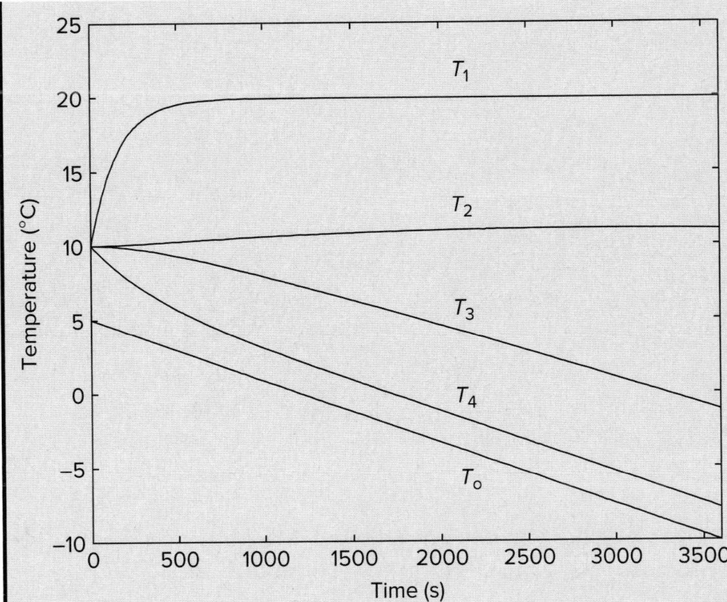

Figure 7.9.4 Wall temperatures as functions of time.

This session produces four transfer functions. Two of them are

$$\frac{T_1(s)}{T_i(s)} = \frac{6.371 \times 10^{-3}s^3 + 2.321 \times 10^{-5}s^2 + 1.545 \times 10^{-8}s + 1.758 \times 10^{-12}}{P(s)}$$

$$\frac{T_2(s)}{T_i(s)} = \frac{5.071 \times 10^{-7}s^2 + 1.77 \times 10^{-9}s + 9.622 \times 10^{-13}}{P(s)}$$

where the denominator is

$$P(s) = s^4 + 0.01007s^3 + 2.583 \times 10^{-5}s^2 + 1.585 \times 10^{-8}s + 1.765 \times 10^{-12}$$

The transfer function $T_1(s)/T_i(s)$ has higher-order numerator dynamics than $T_2(s)/T_i(s)$. This accounts for the faster response of T_1.

7.10 SIMULINK APPLICATIONS

One potential disadvantage of a graphical interface such as Simulink is that to simulate a complex system, the diagram can become rather large, and therefore somewhat cumbersome. Simulink, however, provides for the creation of *subsystem blocks*, which play a role analogous to subprograms in a programming language. A subsystem block is actually a Simulink program represented by a single block. A subsystem block, once created, can be used in other Simulink programs. In this section, we introduce subsystem blocks, and also show how to use the Relay block, which is an example of something that is tedious to program in MATLAB. We also introduce the Fcn block.

7.10.1 SUBSYSTEM BLOCKS

You can create a subsystem block in one of two ways, by dragging the Subsystem block from the block library to the model window, or by first creating a Simulink model and then "encapsulating" it within a bounding box. We will illustrate the latter method.

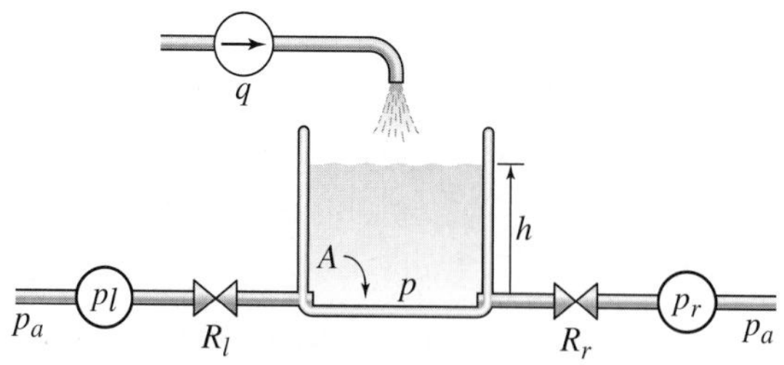

Figure 7.10.1 A liquid-level system.

We will create a subsystem block for the liquid-level system shown in Figure 7.10.1, in which the resistances are nonlinear and obey the following signed-square-root relation:

$$q = \frac{1}{R}\text{SSR}(\Delta p)$$

where q is the mass flow rate, R is the resistance, Δp is the pressure difference across the resistance, and

$$\text{SSR}(u) = \begin{cases} \sqrt{u} & \text{if } u \geq 0 \\ -\sqrt{|u|} & \text{if } u < 0 \end{cases}$$

Note that we can express the SSR(u) function in MATLAB as follows: `sgn(u)*sqrt(abs(u))`.

The model of the system in Figure 7.10.1 is the following:

$$\rho A\frac{dh}{dt} = q + \frac{1}{R_l}\text{SSR}(p_l - p) - \frac{1}{R_r}\text{SSR}(p - p_r)$$

where p_l and p_r are the *gage* pressures at the left- and right-hand sides, A is the bottom area, q is a mass flow rate, and $p = \rho gh$. Note that the atmospheric pressure p_a cancels out of the model because of the use of gage pressure.

First construct the Simulink model shown in Figure 7.10.2. The oval blocks are input and outport ports (In 1 and Out 1), which are available in the Ports and Subsystems library. When entering the gains in each of the four Gain blocks, note that you can use MATLAB variables and expressions. Before running the program we will assign values to these variables in the MATLAB Command window. Enter the gains for the four Gain

Figure 7.10.2 Simulink model of the system shown in Figure 7.10.1.

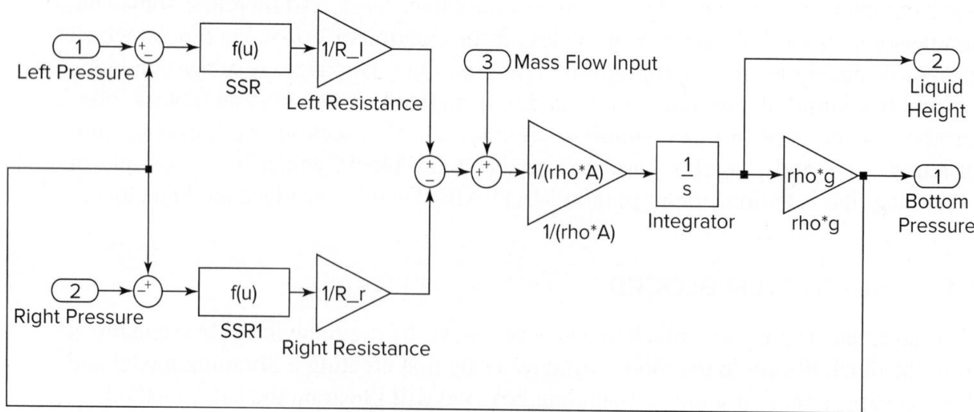

blocks using the expressions shown in the block. You may also use a variable as the Initial condition of the Integrator block. Name this variable h0.

The SSR blocks are examples of the Fcn block in the User-Defined Functions Library. Double-click on the block and enter the MATLAB expression sgn(u)*sqrt(abs(u)). Note that the Fcn block requires you to use the variable u. The output of the Fcn block must be a scalar, as is the case here, and you cannot perform matrix operations in the Fcn block, but these also are not needed here. (Alternatives to the Fcn block are the Math Function block and the MATLAB Fcn block, which are discussed in Section 8.8.) Save the model and give it a name, such as Tank.

Now create a "bounding box" surrounding the diagram. Do this by placing the mouse cursor in the upper left, holding the mouse button down, and dragging the expanding box to the lower right to enclose the entire diagram. Then choose **Diagram > Subsystem & Model Reference > Create Subsystem from Selection**. Simulink will then replace the diagram with a single block having as many input and output ports as required, and will assign default names. You can resize the block to make the labels readable (see Figure 7.10.3). You can view or edit the subsystem by double-clicking on it.

Suppose we want to create a simulation of the system shown in Figure 7.10.4, where the mass inflow rate q_1 is a step function. To do this, create the Simulink model shown in Figure 7.10.5. The square blocks are Constant blocks from the Sources library. These give constant inputs. The larger rectangular blocks are two subsystem blocks of the type just created. To insert them into the model, first open the Tank subsystem model, select **Copy** from the **Edit** menu, then paste it twice into the new model window. Connect the input and output ports and edit the labels as shown. Then double-click on the Tank 1 subsystem block, set the gain 1/R_l equal to 0, the gain 1/R_r equal to 1/R_1, and the gain 1/rho*A equal to 1/rho*A_1. Set the Initial condition of the integrator to h10. Note that setting the gain 1/R_l equal to 0 is equivalent to $R\_l = \infty$, which represents the fact that there is no inlet on the left-hand side.

Then double-click on the Tank 2 subsystem block, set the gain 1/R_l equal to 1/R_1, the gain 1/R_r equal to 1/R_2, and the gain 1/rho*A equal to 1/rho*A_2. Set the Initial condition of the integrator to h20. For the Step block, set the Step time to 0, the Initial value to 0, the Final value to the variable q_1, and the Sample time to 0. Save the model using a name other than Tank.

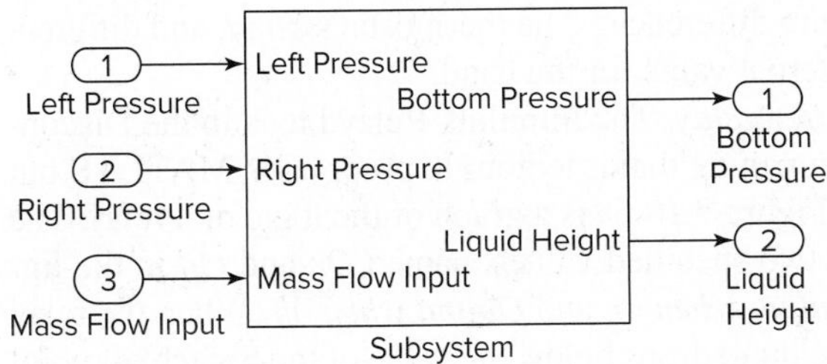

Figure 7.10.3 The Subsystem block.

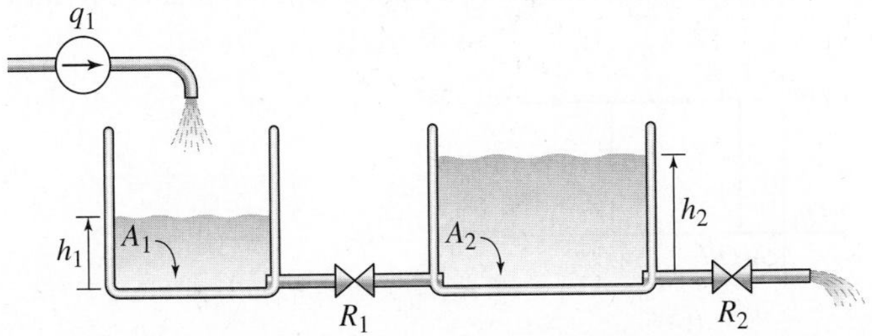

Figure 7.10.4 A liquid-level system with two tanks.

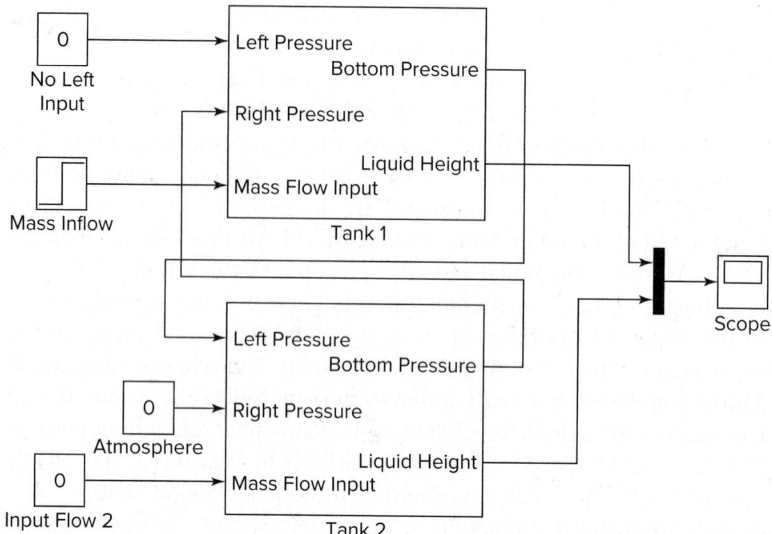

Figure 7.10.5 Simulink model of the system shown in Figure 7.10.4.

Before running the model, in the Command window assign numerical values to the variables. As an example, you may type the following values for water, in U.S. Customary units, in the Command window.

```
>>A_1 = 2;A_2 = 5;rho = 1.94;g = 32.2;
>>R_1 = 20;R_2 = 50;q_1 = 0.3;h10 = 1;h20 = 10;
```

After selecting a simulation Stop time, you may run the simulation. The Scope will display the plots of the heights h_1 and h_2 versus time.

7.10.2 SIMULATION OF THERMAL SYSTEMS

Home heating systems are controlled by a thermostat, which measures the room temperature and compares it with the desired temperature set by the user. Suppose the user selects 70°F. A typical thermostat would switch the heating system on whenever the inside temperature drops below 69° and switch the system off whenever the temperature is above 71°. The 2° temperature difference is the thermostat's *band*, and different thermostat models might use a different value for the band.

The thermostat is an example of a *relay*. The Simulink Relay block in the Discontinuities library is an example of something that is tedious to program in MATLAB but is easy to implement in Simulink. Figure 7.10.6a is a graph of the logic of a relay. The relay switches the output between two specified values, named *On* and *Off* in the figure. Simulink calls these values *Output when on* and *Output when off*. When the relay output is *On*, it remains *On* until the input drops below the value of the Switch off point parameter, named *SwOff* in the figure. When the relay output is *Off*, it remains *Off* until

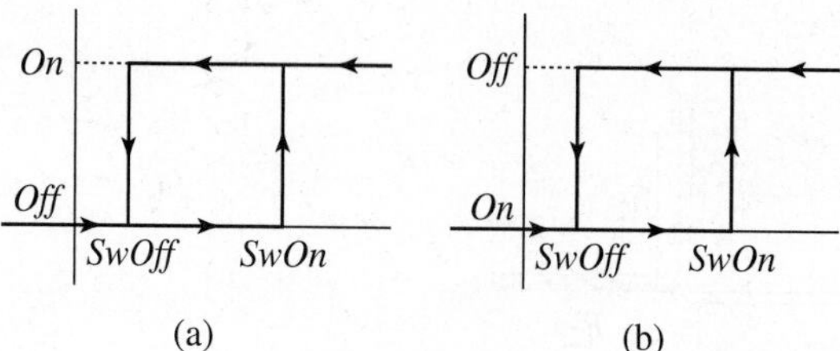

Figure 7.10.6 The relay function. (a) The case where *On* > *Off*. (b) The case where *On* < *Off*.

the input exceeds the value of the Switch on point parameter, named *SwOn* in the figure. The Switch on point parameter value must be greater than or equal to the Switch off point value. Note that the value of *Off* need not be zero. Note also that the value of *Off* need not be less than the value of *On*. As we will see in Example 7.10.1, it is sometimes necessary to use this case. The case where *Off* > *On* is shown in Figure 7.10.6b.

EXAMPLE 7.10.1

Thermostatic Control of Temperature

■ Problem

(a) Develop a Simulink model of a thermostatic control system in which the temperature model is

$$RC\frac{dT}{dt} + T = Rq + T_a(t)$$

where T is the room air temperature in °F, T_a is the ambient (outside) air temperature in °F, time t is measured in hours, q is the input from the heating system in ft-lb/hr, R is the thermal resistance, and C is the thermal capacitance. The thermostat switches q on at the value $q_{\max}$ whenever the temperature drops below 69°, and switches q to $q = 0$ whenever the temperature is above 71°. The value of $q_{\max}$ indicates the heat output of the heating system.

Run the simulation for the case where $T(0) = 70°$ and $T_a(t) = 50 + 10\sin(\pi t/12)$. Use the values $R = 5 \times 10^{-5}$ °F-hr/lb-ft and $C = 4 \times 10^4$ lb-ft/°F. Plot the temperatures T and T_a versus t on the same graph, for $0 \le t \le 24$ hr. Do this for two cases: $q_{\max} = 4 \times 10^5$ and $q_{\max} = 8 \times 10^5$ lb-ft/hr. Investigate the effectiveness of each case. (b) The integral of q over time is the energy used. Plot $\int q\,dt$ versus t and determine how much energy is used in 24 hr for the case where $q_{\max} = 8 \times 10^5$.

■ Solution

The model can be arranged as follows:

$$\frac{dT}{dt} = \frac{1}{RC}\left[Rq + T_a(t) - T\right]$$

The Simulink model is shown in Figure 7.10.7, where $1/RC = 0.5$ and the gain labeled R has the value $R = 5 \times 10^{-5}$, which is entered as `5e-5`. The output of the Relay block is $q(t)$. The input $T_a(t)$ is produced with the Sine block. For the Sine block, set the `Amplitude` to 10, the `Bias` to 50, the `Frequency` to pi/12, the `Phase` to 0, and the `Sample time` to 0. The second Integrator block and Scope 2 were included to compute and display $\int q\,dt$ versus t.

For the Relay block, set the Switch on point to 71, the Switch off point to 69, the Output when on to 0, and the Output when off to the variable `qmax`. This corresponds to Figure 7.10.6a. Set the simulation stop time to 24.

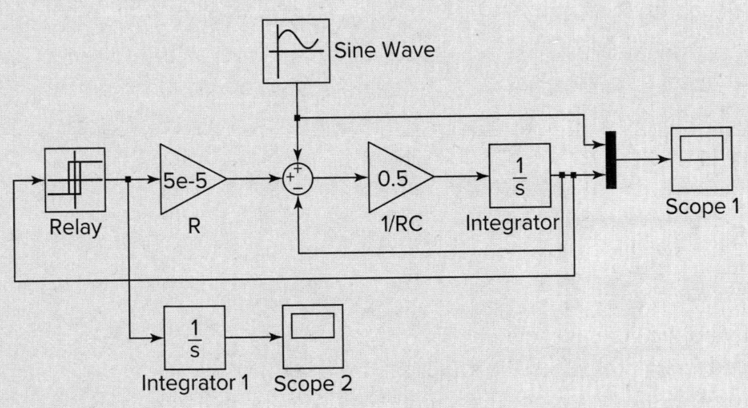

Figure 7.10.7 Simulink model of a temperature control system.

You can set the value of `qmax` by editing the Relay block or by setting its value in the MATLAB Command window before running the simulation, for example, by typing `qmax = 4e+5` before the first simulation.

The simulation results show that when $q_{max} = 4 \times 10^5$, the system is unable to keep the temperature from falling below 69°. When $q_{max} = 8 \times 10^5$, the temperature stays within the desired band. The plot of $\int q\,dt$ versus t for this case shows that the energy used at the end of 24 hr is 9.6158×10^6 ft-lb. This value can be obtained by exporting the output of the second integrator to the workspace.

7.11 CHAPTER REVIEW

Part I of this chapter treated fluid systems, which can be divided into hydraulics and pneumatics. Hydraulics is the study of systems in which the fluid is incompressible; that is, its density stays approximately constant over a range of pressures. Pneumatics is the study of systems in which the fluid is compressible. Hydraulics and pneumatics share a common modeling principle: conservation of mass. It forms the basis of all our models of such systems.

Modeling pneumatic systems also requires application of thermodynamics, because the temperature of a gas can change when its pressure changes. Thus, pneumatics provides a bridge to the treatment of thermal systems, which is the subject of Part II of the chapter. Thermal systems are systems that operate due to temperature differences. They thus involve the flow and storage of heat energy, and conservation of heat energy forms the basis of our thermal models.

Now that you have finished this chapter, you should be able to

1. Apply conservation of mass to model simple hydraulic and pneumatic systems.
2. Derive expressions for the capacitance of simple hydraulic and pneumatic systems.
3. Determine the appropriate resistance relation to use for laminar, turbulent, and orifice flow.
4. Develop a dynamic model of hydraulic and pneumatic systems containing one or more capacitances.
5. Determine the appropriate thermal resistance relation to use for conduction, convection, and radiation heat transfer.
6. Develop a model of a thermal process having one or more thermal storage compartments.
7. Apply MATLAB and Simulink to solve fluid and thermal system models.

REFERENCE

[Çengel, 2001] Y. A. Çengel and R. H. Turner, *Fundamentals of Thermal-Fluid Sciences*, McGraw-Hill, NY, 2001.

PROBLEMS

Section 7.1 Conservation of Mass

7.1 For the hydraulic system shown in Figure P7.1, given $A_1 = 15$ in.^{2}, $A_2 = 45$ in.^{2}, and $mg = 80$ lb, find the force f_1 required to lift the mass m a distance $x_2 = 8$ in. Also find the distance x_1 and the work done by the force f_1.

Figure P7.1

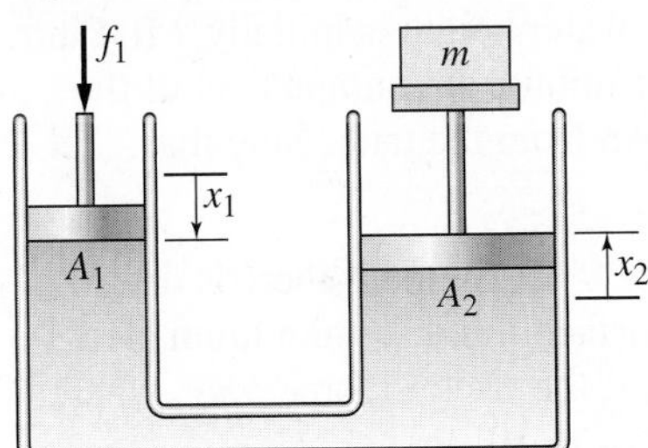

Figure P7.2

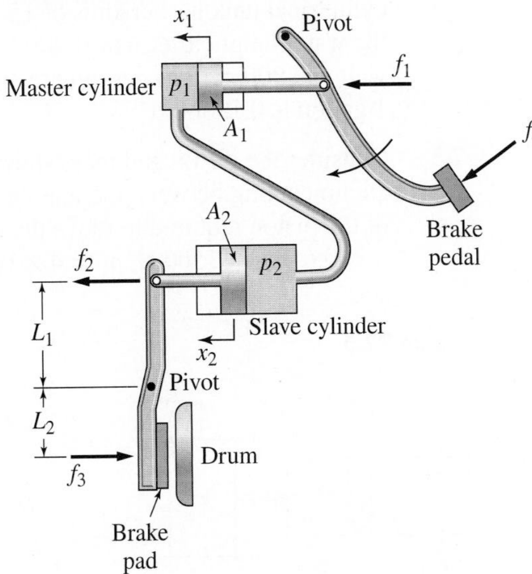

7.2 Figure P7.2 is a representation of a hydraulic brake system. The piston in the master cylinder moves in response to the foot pedal. The resulting motion of the piston in the slave cylinder causes the brake pad to be pressed against the brake drum with a force f_3. Obtain the expression for the force f_3 with the force f_1 as the input. The force f_1 depends on the force f_4 applied by the driver's foot. The precise relation between f_1 and f_4 depends on the geometry of the pedal arm.

7.3 Water is pumped as needed at the mass flow rate $q_{mo}(t)$ from the tank shown in Figure P7.3a. Replacement water is pumped from a well at the mass flow rate $q_{mi}(t)$. A simple representation is shown in Figure P7.3b. Determine the water height $h(t)$, assuming that the tank is cylindrical with a cross section A.

Figure P7.3

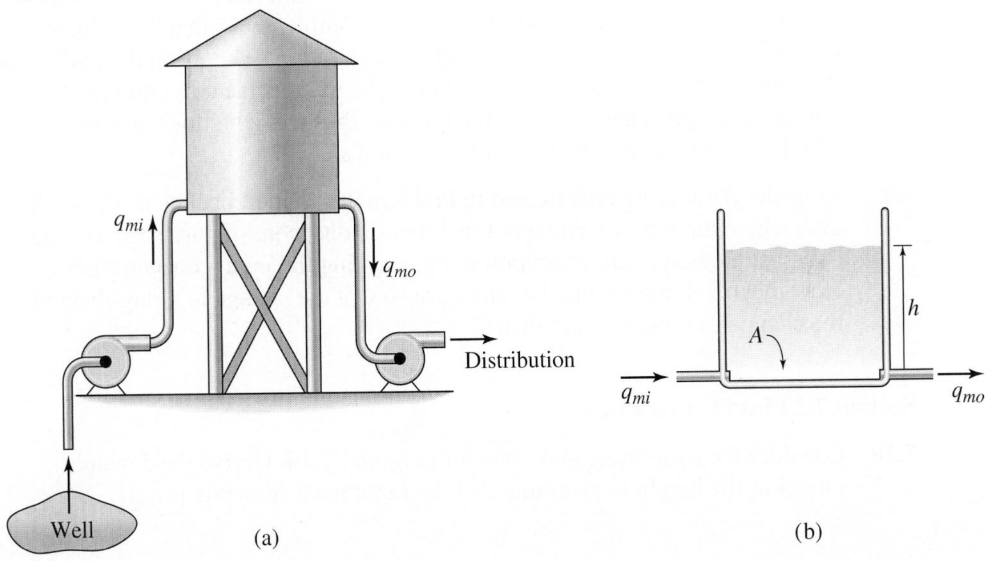

7.4 Refer to the water storage and supply system shown in Figure P7.3. The cylindrical tank has a radius of 15 ft, and the water height is initially 7 ft. Find the water height after 6 hr if 2000 gallons per minute are pumped out of the well and 900 gallons per minute are withdrawn from the tank. Note that 1 gallon is 0.13368 ft^3.

7.5 Consider the piston and mass shown in Figure P7.5. Suppose there is dry friction acting between the mass m and the surface. Find the minimum area A of the piston required to move the mass against the friction force μmg, where $\mu = 0.6$, $mg = 1500$ N, $p_1 = 4 \times 10^5$ Pa, and $p_2 = 10^5$ Pa.

Figure P7.5

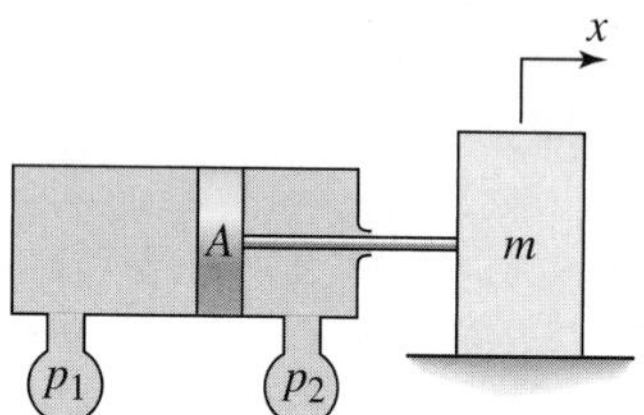

Figure P7.6

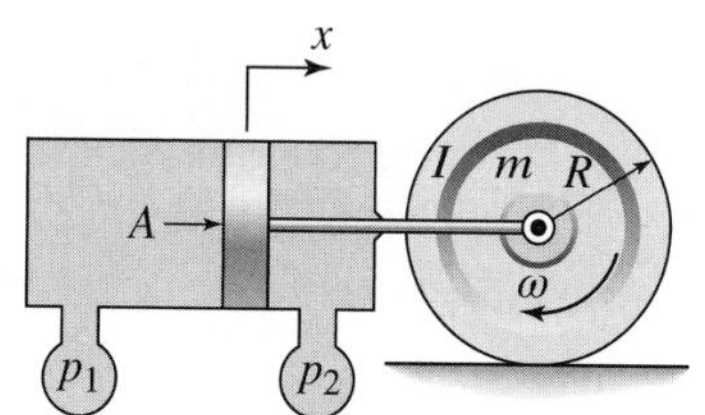

7.6 In Figure P7.6 the piston of area A is connected to the axle of the cylinder of radius R, mass m, and inertia I about its center. Given $p_1 - p_2 = 3 \times 10^5$ Pa, $A = 0.008\ \text{m}^2$, $R = 0.6$ m, $m = 150$ kg, and $I = 9\ \text{kg}\cdot\text{m}^2$, determine the angular velocity $\omega(t)$ of the cylinder assuming that it starts from rest.

7.7 Refer to Figure P7.5, and suppose that $p_1 - p_2 = 10\ \text{lb/in.}^2$, $A = 4\ \text{in.}^2$, and $mg = 700$ lb. If the mass starts from rest at $x(0) = 0$, how far will it move in 0.5 sec, and how much hydraulic fluid will be displaced?

7.8 Pure water flows into a mixing tank of volume $V = 400\ \text{m}^3$ at the constant-volume rate of $12\ \text{m}^3/\text{s}$. A solution with a salt concentration of $s_i\ \text{kg/m}^3$ flows into the tank at a constant-volume rate of $3\ \text{m}^3/\text{s}$. Assume that the solution in the tank is well mixed so that the salt concentration in the tank is uniform. Assume also that the salt dissolves completely so that the volume of the mixture remains the same. The salt concentration $s_o\ \text{kg/m}^3$ in the outflow is the same as the concentration in the tank. The input is the concentration $s_i(t)$, whose value may change during the process, thus changing the value of s_o. Obtain a dynamic model of the concentration s_o.

7.9 Consider the mixing tank treated in Problem 7.8. Generalize the model to the case where the tank's volume is $V\ \text{m}^3$. For quality-control purposes, we want to adjust the output concentration s_o by adjusting the input concentration s_i. How much volume should the tank have so that the change in s_o lags behind the change in s_i by no more than 30 s?

Section 7.2 Fluid Capacitance

7.10 Consider the cylindrical tank shown in Figure P7.10. Derive the dynamic model of the height h, assuming that the input mass flow rate is $q_m(t)$.

Figure P7.10

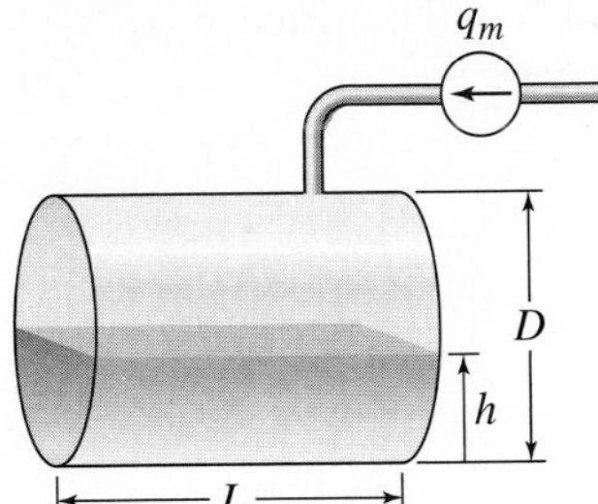

Figure P7.11

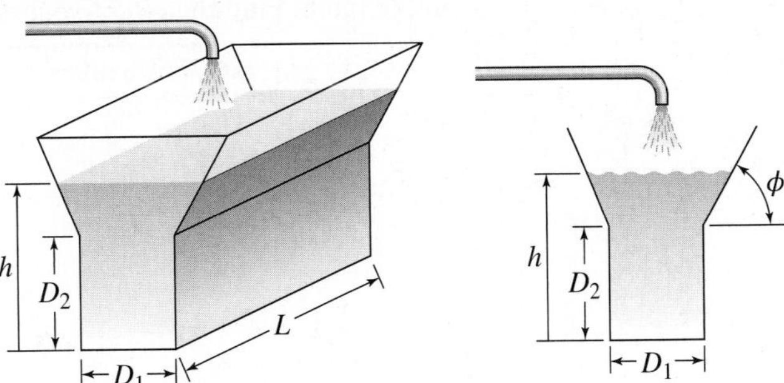

7.11 Consider the tank shown in Figure P7.11. Derive the dynamic model of the height h, assuming that the input mass flow rate is $q_m(t)$.

Section 7.3 Fluid Resistance

7.12 Air flows in a certain cylindrical pipe 1 m long with an inside diameter of 1 mm. The pressure difference between the ends of the pipe is 0.1 atm. Compute the laminar resistance, the Reynolds number, the entrance length, and the mass flow rate. Comment on the accuracy of the resistance calculation. For air use $\mu = 1.58 \times 10^{-5}$ N · s/m^2 and $\rho = 1.2885$ kg/m^3.

7.13 Derive the expression for the linearized resistance due to orifice flow near a reference height h_r.

Section 7.4 Dynamic Models of Hydraulic Systems

7.14 Consider the cylindrical container treated in Example 7.4.3. Suppose the outlet flow is turbulent. Derive the dynamic model of the system (a) in terms of the gage pressure p at the bottom of the tank and (b) in terms of the height h.

7.15 A certain tank has a bottom area $A = 30$ m^2. The liquid level in the tank is initially 6 m. When the outlet is opened, it takes 280 s to empty by 98%.

a. Estimate the value of the linear resistance R.

b. Find the steady-state height if the inflow is $q = 4$ m^3/s.

7.16 A certain tank has a circular bottom area $A = 30$ ft^2. It is drained by a pipe whose linear resistance is $R = 200$ m^{-1}sec^{-1}. The tank contains water whose mass density is 1.94 slug/ft^3.

a. Estimate how long it will take for the tank to empty if the water height is initially 20 ft.

b. Suppose we dump water into the tank at a rate of 0.2 ft^3/sec. If the tank is initially empty and the outlet pipe remains open, find the steady-state height and the time to reach one-third that height, and estimate how long it will take to reach the steady-state height.

7.17 The water inflow rate to a certain tank was kept constant until the water height above the orifice outlet reached a constant level. The inflow rate was then measured, and the process repeated for a larger inflow rate. The data are given in the table. Find the effective area $C_d A_o$ for the tank's outlet orifice.

Inflow rate (liters/min)	Liquid height (cm)
98	30
93	27
91	24
86	21
81	18
75	15
68	12
63	9
56	6
49	3

7.18 In the system shown in Figure P7.18, a component such as a valve has been inserted between the two lengths of pipe. Assume that turbulent flow exists throughout the system. Use the resistance relation 7.3.7. (a) Find the total turbulent resistance. (b) Develop a model for the behavior of the liquid height h, with the mass flow rate q_{mi} as the input.

Figure P7.18

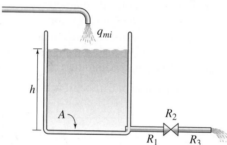

7.19 The cylindrical tank shown in Figure 7.4.3 has a circular bottom area A. The mass inflow rate from the flow source is $\hat{q}_{mi}(t)$, a given function of time. The flow through the outlet is *turbulent*, and the outlet discharges to atmospheric pressure p_a. Develop a model of the liquid height h.

7.20 In the liquid-level system shown in Figure P7.20, the resistances R_1 and R_2 are linear, and the input is the pressure source p_s. Obtain the differential equation model for the height h, assuming that $\hat{h} > D$.

Figure P7.20

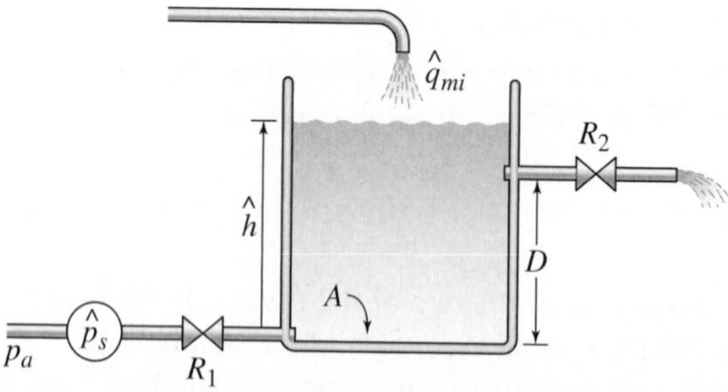

7.21 The water height in a certain tank was measured at several times with no inflow applied. See Figure 7.4.3. The resistance R is a linearized resistance. The data are given in the table. The tank's bottom area is $A = 6$ ft^2.

a. Estimate the resistance R.

b. Suppose the initial height is known to be exactly 20.2 ft. How does this change the results of part (a)?

Time (sec)	Height (ft)
0	20.2
300	17.26
600	14.6
900	12.4
1200	10.4
1500	9.0
1800	7.6
2100	6.4
2400	5.4

7.22 Derive the model for the system shown in Figure P7.22. The flow rate $\hat{q}_{mi}$ is a mass flow rate and the resistances are linearized.

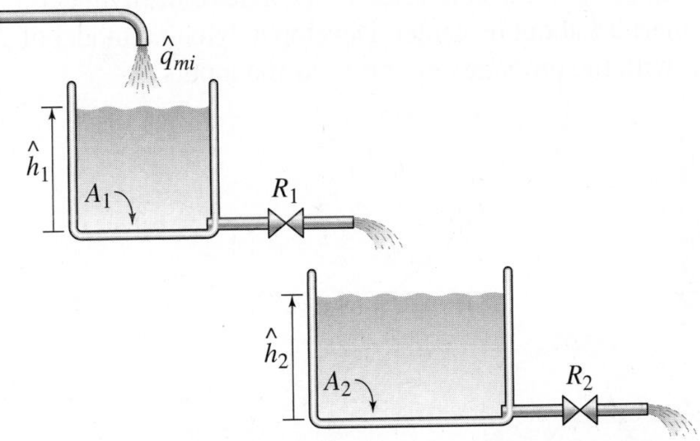

Figure P7.22

7.23 (a) Develop a model of the two liquid heights in the system shown in Figure P7.23. The inflow rate $q_{mi}(t)$ is a mass flow rate. (b) Using the values $R_1 = R$, $R_2 = 3R$, $A_1 = A$, and $A_2 = 4A$, find the transfer function $H_2(s)/Q_{mi}(s)$.

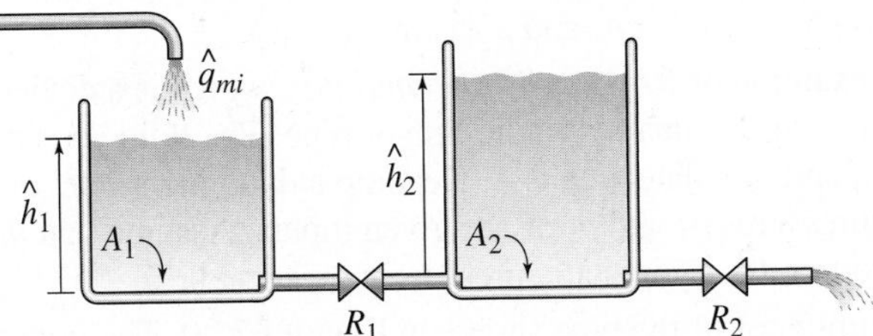

Figure P7.23

7.24 Consider Example 7.4.6. Suppose that $R_1 = R$, $R_2 = 4R$, $A_1 = A$, and $A_2 = 3A$. Find the transfer function $H_1(s)/Q_{mi}(s)$ and the characteristic roots.

7.25 Design a piston-type damper using an oil with a viscosity at 20°C of $\mu = 0.9\ \text{kg/(m} \cdot \text{s)}$. The desired damping coefficient is 2000 N · s/m. See Figure 7.4.4.

7.26 For the damper shown in Figure 7.4.7, assume that the flow through the hole is turbulent, and neglect the term $m\ddot{y}$. Develop a model of the relation between the force f and $\dot{x}$, the relative velocity between the piston and the cylinder.

7.27 An electric motor is sometimes used to move the spool valve of a hydraulic motor. In Figure P7.27 the force f is due to an electric motor acting through a rack-and-pinion gear. Develop a model of the system with the load displacement y as the output and the force f as the input. Consider two cases: (a) $m_1 = 0$ and (b) $m_1 \neq 0$.

Figure P7.27

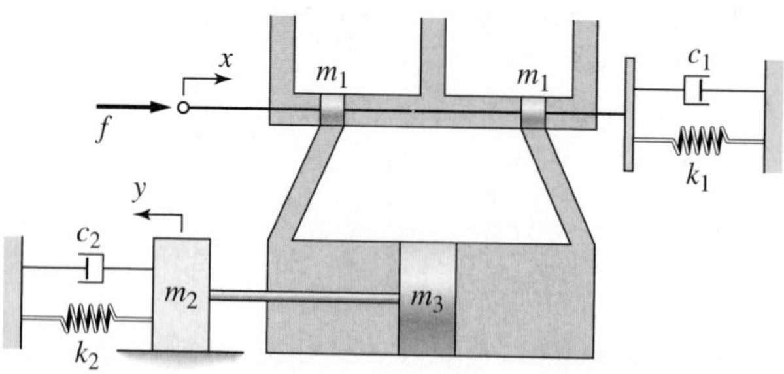

7.28 In Figure P7.28 the piston of area A is connected to the axle of the cylinder of radius R, mass m, and inertia I about its center. Develop a dynamic model of the axle's translation x, with the pressures p_1 and p_2 as the inputs.

Figure P7.28

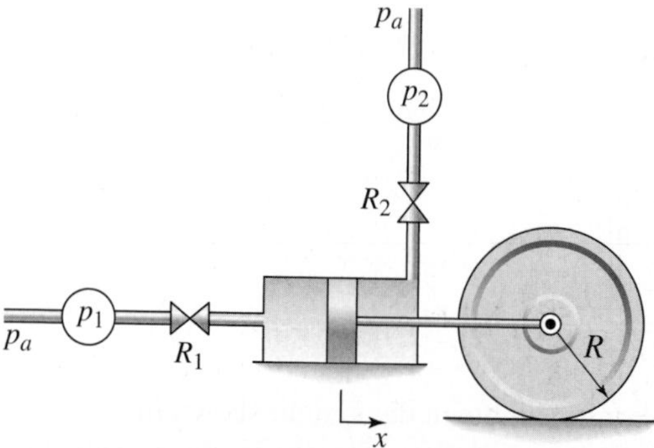

7.29 Figure P7.29 shows a pendulum driven by a hydraulic piston. Assuming small angles θ and a concentrated mass m a distance L_1 from the pivot, derive the equation of motion with the pressures p_1 and p_2 as inputs.

7.30 Figure P7.30 shows an example of a *hydraulic accumulator*, which is a device for reducing pressure fluctuations in a hydraulic line or pipe. The fluid density is ρ, the plate mass is m, and the plate area is A. Develop a dynamic model of the pressure p with the pressures p_1 and p_2 as the given inputs. Assume that $m\ddot{x}$ of the plate is small, and that the hydrostatic pressure ρgh is small.

7.31 Design a hydraulic accumulator of the type shown in Figure P7.30. The liquid volume in the accumulator should increase by 30 in.^{3} when the pressure p increases by 1.5 lb/in.^{2}. Determine suitable values for the plate area A and the spring constant k.

7.32 Consider the V-shaped container treated in Example 7.2.2, whose cross section is shown in Figure P7.32. The outlet resistance is linear. Derive the dynamic model of the height h.

7.33 Consider the V-shaped container treated in Example 7.2.2, whose cross section is shown in Figure P7.33. The outlet is an orifice of area A_o and discharge coefficient C_d. Derive the dynamic model of the height h.

Figure P7.29 **Figure P7.30**

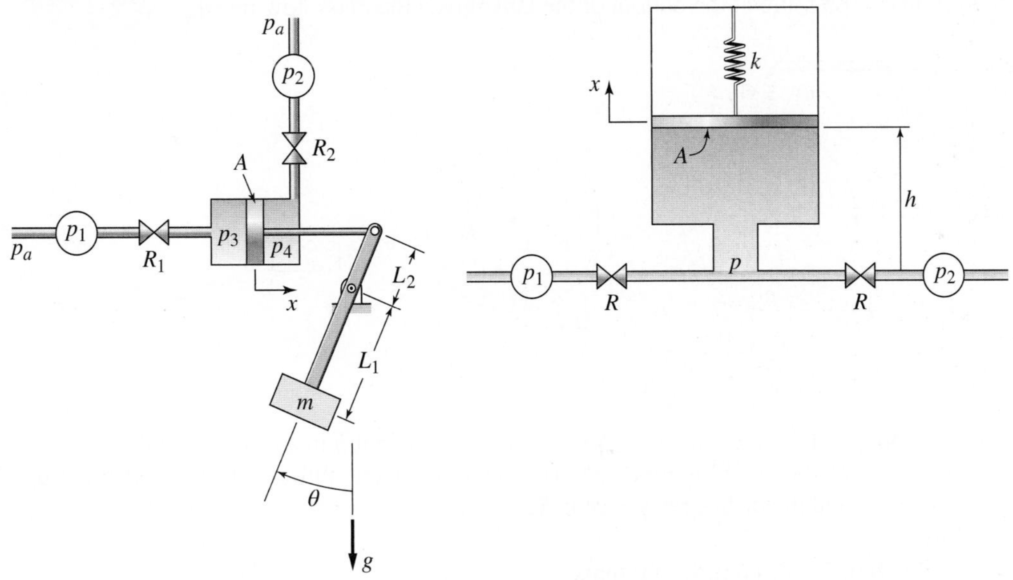

Figure P7.32

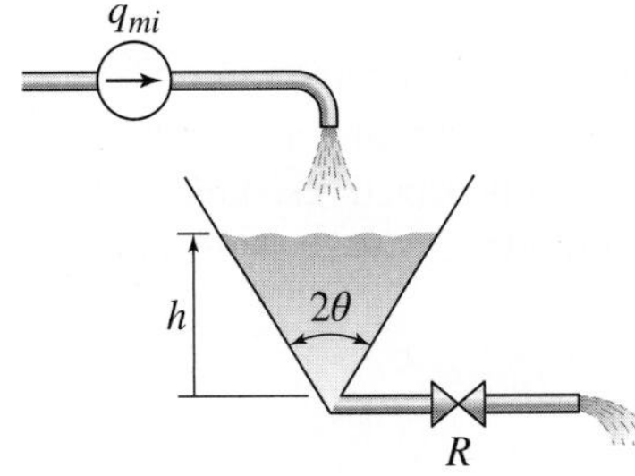

Figure P7.33

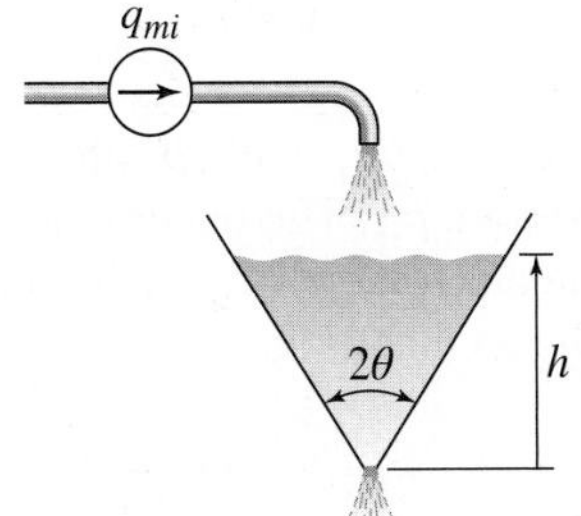

Figure P7.34

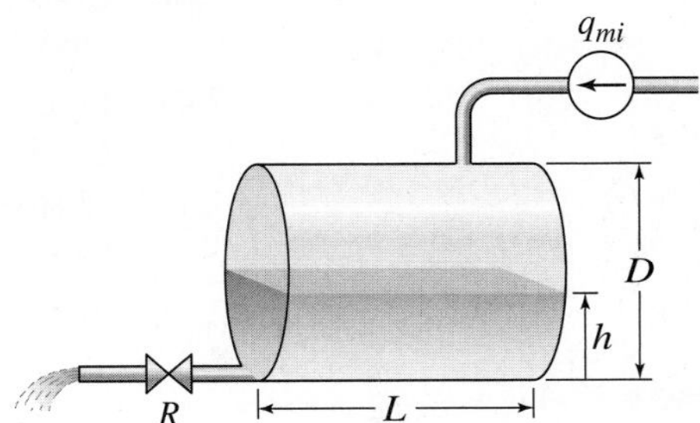

7.34 Consider the cylindrical container treated in Problem 7.10. In Figure P7.34 the tank is shown with a valve outlet at the bottom of the tank. Assume that the flow through the valve is turbulent with a resistance R. Derive the dynamic model of the height h.

7.35 A certain tank contains water whose mass density is 1.94 slug/ft^3. The tank's circular bottom area is $A = 100$ ft^2. It is drained by an orifice in the bottom. The effective cross-sectional area of the orifice is $C_d A_o = 0.5$ ft^2. A pipe dumps water into the tank at the *volume* flow rate q_v.

a. Derive the model for the tank's height h with the input q_v.
b. Compute the steady-state height if the input flow rate is $q_v = 5$ ft^3/sec.
c. Estimate the tank's time constant when the height is near the steady-state height.

7.36 A conical tank is shown in Figure P7.36. The cone angle θ is a constant and should appear in your answer as a parameter. Derive the dynamic model of the liquid height h. The mass inflow rate is $q_{mi}(t)$. The resistance R is linear.

Figure P7.36

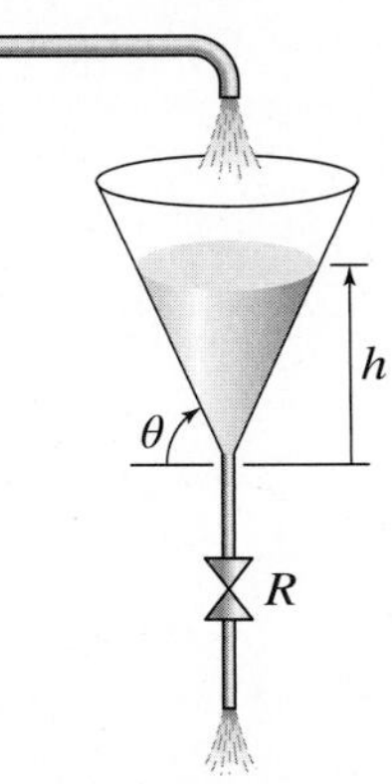

7.37 A spherical tank of radius R is shown in Figure P7.37. Obtain a model of the pressure at the bottom of the tank, given the mass flow rate q_{mi}.

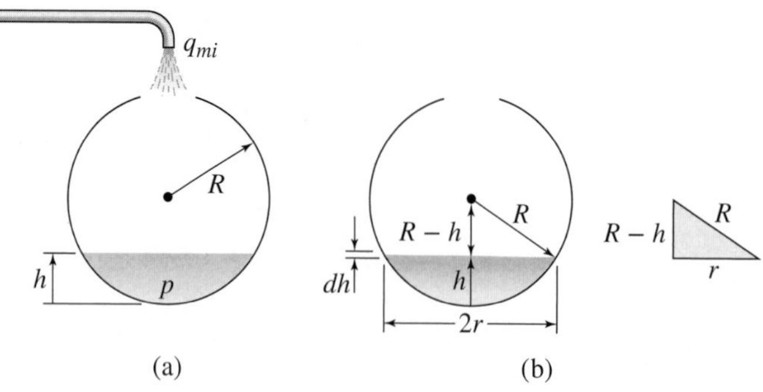

Figure P7.37 A spherical tank.

7.38 Obtain the dynamic model of the liquid height h in a spherical tank of radius R, shown in Figure P7.38. The mass inflow rate through the top opening is q_{mi} and the orifice resistance is R_o.

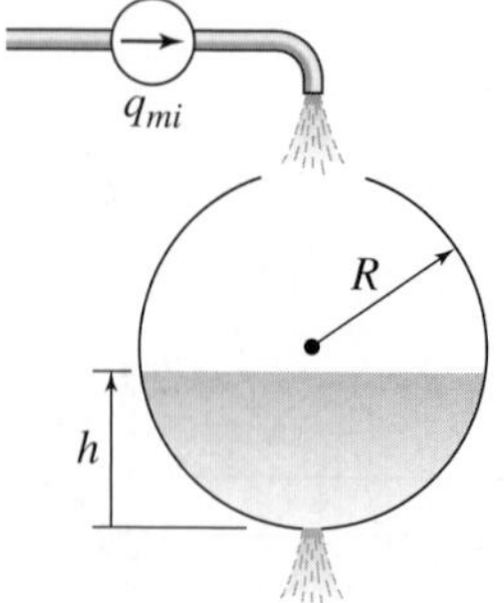

Figure P7.38 A spherical tank with an orifice resistance.

Section 7.5 Pneumatic Systems

7.39 A rigid container has a volume of 30 ft^3. The air inside is initially at 75°F. Find the pneumatic capacitance of the container for an isothermal process.

7.40 Consider the pneumatic system treated in Example 7.5.2. Derive the linearized model for the case where $p_i < p$.

7.41 Figure P7.41 shows two rigid tanks whose pneumatic capacitances are C_1 and C_2. The variables δp_i, δp_1, and δp_2 are small deviations around a reference steady-state pressure p_{ss}. The pneumatic lines have linearized resistances R_1 and R_2. Assume an isothermal process. Derive a model of the pressures δp_1 and δp_2 with δp_i as the input.

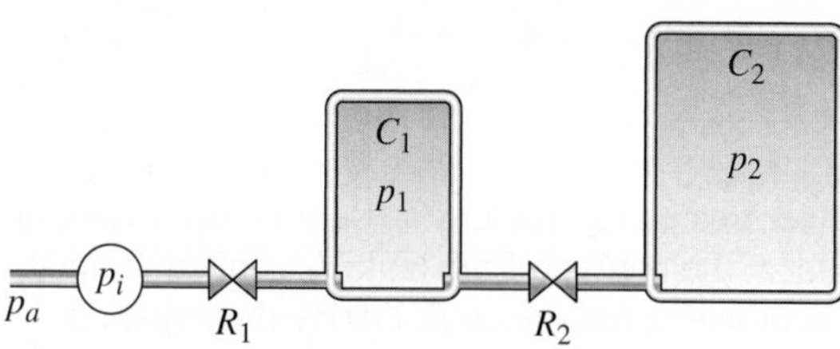

Figure P7.41

Section 7.6 Thermal Capacitance

7.42 (a) Compute the thermal capacitance of 300 ml of water, for which $\rho = 1000$ kg/m^3 and $c_p = 4.18 \times 10^3$ J/kg · °C. Note that 1 ml = 10^{-6} m^3. (b) How much energy does it take to raise the water temperature from room temperature (20°C) to 99°C (just below boiling).

7.43 A certain room measures 20 ft by 12 ft by 8 ft. (a) Compute the thermal capacitance of the room air using $c_p = 6.012 \times 10^3$ ft-lb/slug-°F and $\rho = 0.0023$ slug/ft^3. (b) How much energy is required to raise the air temperature from 68°F to 72°F, neglecting heat transfer to the walls, floor, and ceiling?

7.44 Liquid initially at 30°C is pumped into a mixing tank at a constant volume flow rate of 0.2 m^3/s. See Figure 7.6.1. At time $t = 0$ the temperature of the incoming liquid suddenly is changed to 90°C. The tank walls are perfectly insulated. The tank volume is 20 m^3, and the liquid within is well-mixed so that its temperature is uniform throughout, and denoted by T. The liquid's specific heat and mass density are c_p and ρ. Given that $T(0) = 30$°C, develop and solve a dynamic model for the temperature T as a function of time.

Section 7.7 Thermal Resistance

7.45 The copper shaft shown in Figure P7.45 consists of two cylinders with the following dimensions: $L_1 = 15$ mm, $L_2 = 10$ mm, $D_1 = 3$ mm, and $D_2 =$ 2 mm. The shaft is insulated around its circumference so that heat transfer occurs only in the axial direction. (a) Compute the thermal resistance of each section of the shaft and of the total shaft. Use the following value for the conductivity of copper: $k = 400$ W/m · °C. (b) Compute the heat flow rate in the axial direction if the temperature difference across the endpoints of the shaft is 30°C.

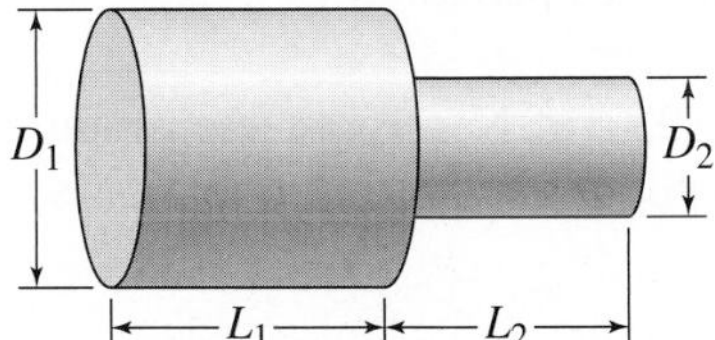

Figure P7.45

7.46 A certain radiator wall is made of copper with a conductivity $k = 47$ lb/sec-°F at 212°F. The wall is 5/16 in. thick and has circulating water on one side with a convection coefficient $h_1 = 100$ lb/sec-ft-°F. A fan blows air over the other side, which has a convection coefficient $h_2 = 20$ lb/sec-ft-°F. Find the thermal resistance of the radiator on a square foot basis.

7.47 A particular house wall consists of three layers and has a surface area of 30 m^2. The inside layer is 10 mm thick and made of plaster board with a thermal conductivity of $k = 0.2$ W/(m · °C). The middle layer is made of fiberglass insulation with $k = 0.04$ W/(m · °C). The outside layer is 20 mm thick and made of wood siding with $k = 0.1$ W/(m · °C). The inside temperature is 20°C, and the convection coefficient for the inside wall surface is $h_i = 40$ W/(m^2 · °C). The convection coefficient for the outside wall surface is $h_o =$ 70 W/(m^2 · °C). How thick must the insulation layer be so that the heat loss is no greater than 400 W if the outside temperature is −20°C?

7.48 A certain wall section is composed of a 12 in. by 12 in. brick area 4 in. thick. Surrounding the brick is a 36 in. by 36 in. concrete section, which is also 4 in. thick. The thermal conductivity of the brick is $k = 0.086$ lb/sec-°F. For the concrete, $k = 0.02$ lb/sec-°F. (a) Determine the thermal resistance of the wall section. (b) Compute the heat flow rate through (1) the concrete, (2) the brick, and (3) the wall section if the temperature difference across the wall is 40°F.

7.49 Water at 120°F flows in an iron pipe 10 ft long, whose inner and outer radii are 1/2 in. and 3/4 in. The temperature of the surrounding air is 70°F. (a) Assuming that the water temperature remains constant along the length of the pipe, compute the heat loss rate from the water to the air in the radial direction, using the following values. For iron, $k = 10.1$ lb/sec-°F. The convection coefficient at the inner surface between the water and the iron is $h_i = 16$ lb/sec-ft-°F. The convection coefficient at the outer surface between the air and the iron is $h_o = 1.1$ lb/sec-ft-°F. (b) Suppose the water is flowing at 0.5 ft/sec. Check the validity of the constant-temperature assumption. For water, $\rho = 1.94$ slug/ft^3 and $c_p = 25,000$ ft-lb/slug-°F.

Section 7.8 Dynamic Models of Thermal Systems

7.50 Consider the water pipe treated in Example 7.7.4. Suppose now that the water is not flowing. The water is initially at 120°F. The copper pipe is 6 ft long, with inner and outer radii of 1/4 in. and 3/8 in. The temperature of the surrounding air is constant at 70°F. Neglect heat loss from the ends of the pipe, and use the following values. For copper, $k = 50$ lb/sec-°F. The convection coefficient at the inner surface between the water and the copper is now different because the water is standing. Use $h_i = 6$ lb/sec-ft-°F. The convection coefficient at the outer surface between the air and the copper is $h_o = 1.1$ lb/sec-ft-°F. Develop and solve a dynamic model of the water temperature $T(t)$ as a function of time.

7.51 A steel tank filled with water has a volume of 1000 ft^3. Near room temperature, the specific heat for water is $c = 25,000$ ft-lb/slug-°F, and its mass density is $\rho = 1.94$ slug/ft^3.

a. Compute the thermal capacitance C_1 of the water in the tank.
b. Denote the total thermal resistance (convective and conductive) of the tank's steel wall by R_1. The temperature of the air surrounding the tank is T_o. The tank's water temperature is T_1. Assume that the thermal capacitance of the steel wall is negligible. Derive the differential equation model for the water's temperature, with T_o as the input.

7.52 Consider the tank of water discussed in Problem 7.51. A test was performed in which the surrounding air temperature T_o was held constant at 70°F. The tank's water temperature was heated to 90° and then allowed to cool. The following data show the tank's water temperature as a function of time. Use these data to estimate the value of the thermal resistance R_1.

Time t (sec)	Water temperature T_1 (°F)
0	90
500	82
1000	77
1500	75
2000	73
2500	72
3000	71
4000	70

7.53 The oven shown in Figure P7.53 has a heating element with appreciable capacitance C_1. The other capacitance is that of the oven air C_2. The corresponding temperatures are T_1 and T_2, and the outside temperature is T_o. The thermal resistance of the heater-air interface is R_1; that of the oven wall is

R_2. Develop a model for T_1 and T_2, with input q_i, the heat flow rate delivered to the heater mass.

Figure P7.53

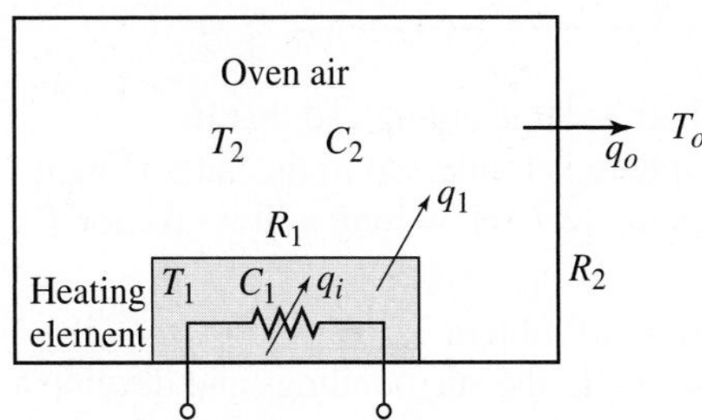

Figure P7.54

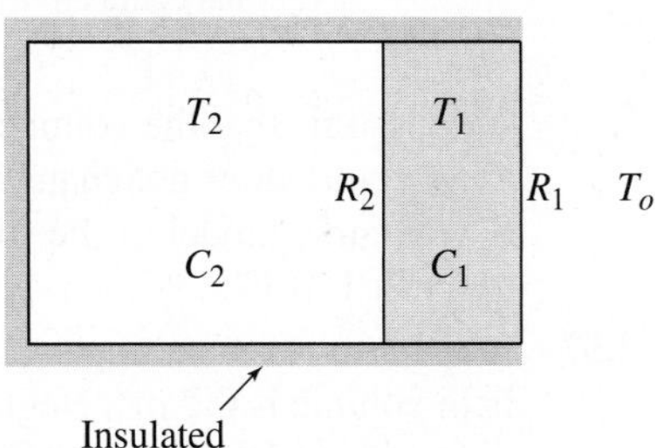

7.54 A simplified representation of the temperature dynamics of two adjacent masses is shown in Figure P7.54. The mass with capacitance C_2 is perfectly insulated on all sides except one, which has a convective resistance R_2. The thermal capacitances of the masses are C_1 and C_2, and their representative uniform temperatures are T_1 and T_2. The thermal capacitance of the surroundings is very large and the temperature is T_o. (a) Develop a model of the behavior of T_1 and T_2. (b) Discuss what happens if the thermal capacitance C_2 is very small.

7.55 A metal sphere 25 mm in diameter was heated to 95°C, and then suspended in air at 22°C. The mass density of the metal is 7920 kg/m^3, its specific heat at 30°C is $c_p = 500$ J/(kg · °C), and its thermal conductivity at 30°C is 400 W/(m · °C). The following sphere temperature data were measured as the sphere cooled.

t (s)	T (°C)	t (s)	T (°C)	t (s)	T(°C)
0	95	120	85	540	67
15	93	135	84	600	65
30	92	180	82	660	62
45	90	240	79	720	61
60	89	300	76	780	59
75	88	360	73	840	57
90	87	420	71	900	56
105	86	480	69	960	54

a. Assume that the sphere's heat loss rate is due entirely to convection. Estimate the convection coefficient h.

b. Compute the Biot number and discuss the accuracy of the lumped-parameter model used in part (a).

c. Discuss whether some of the heat loss rate could be due to radiation. Give a numerical reason for your answer.

7.56 A copper sphere is to be quenched in an oil bath whose temperature is 60°C. The sphere's radius is 40 mm, and the convection coefficient is $h =$ 300 W/(m^2 · °C). Assume the sphere and the oil properties are constant. These properties are given in the following table. The sphere's initial temperature is 450°C.

Property	Sphere	Oil
Density ρ (kg/m^3)	8900	7900
Specific heat c_p J/(kg · °C)	385	400
Thermal conductivity k [W/(m · °C)]	400	—

Assume that the volume of the oil bath is large enough so that its temperature does not change when the sphere is immersed in the bath. Obtain the dynamic model of the sphere's temperature T. How long will it take for T to reach 150°C?

7.57 Consider the quenching process discussed in Problem 7.56. Suppose the oil bath volume is 0.2 m^3. Neglect any heat loss to the surroundings and develop a dynamic model of the sphere's temperature and the bath temperature. How long will it take for the sphere temperature to reach 150°C?

Section 7.9 MATLAB Applications

7.58 Consider Example 7.7.1. The MATLAB left division operator can be used to solve the set of linear algebraic equations $\mathbf{AT} = \mathbf{b}$ as follows: `T = A\b`. Use this method to write a script file to solve for the three steady-state temperatures T_1, T_2, and T_3, given values for the resistances and the temperatures T_i and T_o. Use the results of Example 7.7.1 to test your file.

Figure P7.59

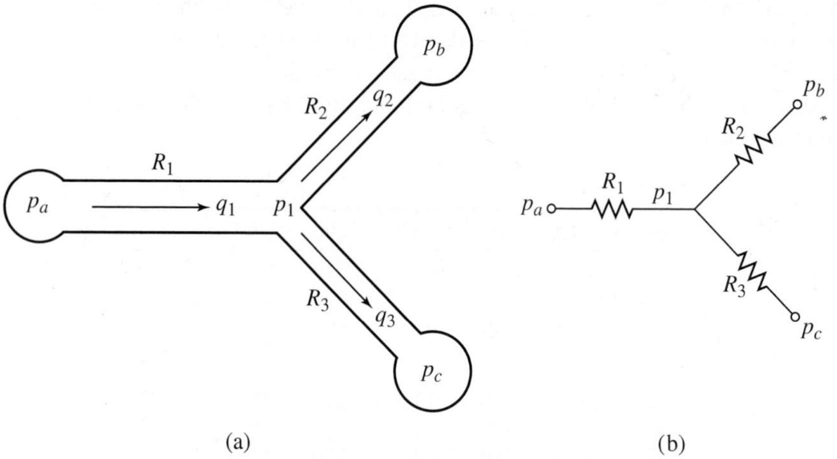

7.59 Fluid flows in pipe networks can be analyzed in a manner similar to that used for electric resistance networks. Figure P7.59a shows a network with three pipes, which is analogous to the electrical network shown in part (b) of the figure. The volume flow rates in the pipes are q_1, q_2, and q_3. The pressures at the pipe ends are p_a, p_b, and p_c. The pressure at the junction is p_1.

a. Assuming that the linear resistance relation applies, we have

$$q_1 = \frac{1}{R_1}(p_a - p_1)$$

Obtain the equations for q_2 and q_3.

b. Note that conservation of mass gives $q_1 = q_2 + q_3$. Set up the equations in a matrix form $\mathbf{Aq} = \mathbf{b}$ suitable for solving for the three flow rates q_1, q_2, and q_3, and the pressure p_1, given the values of the pressures p_a, p_b, and p_c, and the values of the resistances R_1, R_2, and R_3. Find the expressions for matrix **A** and the column vector **b**.

c. Use MATLAB to solve the matrix equations obtained in part (b) for the case: $p_a = 30$ psi, $p_b = 25$ psi, and $p_c = 20$ psi. Use the resistance values $R_1 = 10{,}000$, $R_2 = R_3 = 14{,}000$ 1/(ft-sec). These values correspond to fuel oil flowing through pipes 2 ft long, with 2 in. and 1.4 in. diameters, respectively. The units of the answers should be ft^3/sec for the flow rates, and lb/ft^2 for pressure.

7.60 The equation describing the water height h in a spherical tank with a drain at the bottom is

$$\pi(2rh - h^2)\frac{dh}{dt} = -C_d A_o \sqrt{2gh}$$

Suppose the tank's radius is $r = 3$ m and that the circular drain hole has a radius of 2 cm. Assume that $C_d = 0.5$, and that the initial water height is $h(0) = 5$ m. Use $g = 9.81$ m/s^2.

a. Use an approximation to estimate how long it takes for the tank to empty.

b. Use MATLAB to solve the nonlinear equation and plot the water height as a function of time until $h(t)$ is not quite zero.

7.61 The following equation describes a certain dilution process, where $y(t)$ is the concentration of salt in a tank of fresh water to which salt brine is being added.

$$\frac{dy}{dt} + \frac{2}{10 + 2t}y = 4$$

Suppose that $y(0) = 0$.

a. Use MATLAB to solve this equation for $y(t)$ and to plot $y(t)$ for $0 \le t \le 10$.

b. Check your results by using an approximation that converts the differential equation into one having constant coefficients.

7.62 A tank having vertical sides and a bottom area of 100 ft^2 is used to store water. To fill the tank, water is pumped into the top at the rate given in the following table. Use MATLAB to solve for and plot the water height $h(t)$ for $0 \le t \le 10$ min.

Time (min)	0	1	2	3	4	5	6	7	8	9	10
Flow Rate (ft^3/min)	0	80	130	150	150	160	165	170	160	140	120

7.63 A cone-shaped paper drinking cup (like the kind used at water fountains) has a radius R and a height H. If the water height in the cup is h, the water volume is given by

$$V = \frac{1}{3}\pi\left(\frac{R}{H}\right)^2 h^3$$

Suppose that the cup's dimensions are $R = 1.5$ in. and $H = 4$ in.

a. If the flow rate from the fountain into the cup is 2 in.^{3}/sec, use MATLAB to determine how long will it take to fill the cup to the brim.

b. If the flow rate from the fountain into the cup is given by $2(1 - e^{-2t})$ in.^{3}/sec, use MATLAB to determine how long will it take to fill the cup to the brim.

Section 7.10 Simulink Applications

7.64 Refer to Figure 7.10.1. Assume that the resistances obey the linear relation, so that the mass flow q_l through the left-hand resistance is $q_l = (p_l - p)/R_l$, with a similar linear relation for the right-hand resistance.

a. Create a Simulink subsystem block for this element.

b. Use the subsystem block to create a Simulink model of the system discussed in Example 7.4.3 and shown in Figure 7.4.3a. Assume that the mass inflow rate q_{mi} is a step function.

c. Use the Simulink model to obtain plots of $h_1(t)$ and $h_2(t)$ for the following parameter values: $A_1 = 2\ \text{m}^2$, $A_2 = 5\ \text{m}^2$, $R_1 = 400\ 1/(\text{m} \cdot \text{s})$, $R_2 = 600\ 1/(\text{m} \cdot \text{s})$, $\rho = 1000\ \text{kg/m}^3$, $q_{mi} = 50$ kg/s, $h_1(0) = 1.5$ m, and $h_2(0) = 0.5$ m.

7.65 Use Simulink to solve Problem 7.60(b).

7.66 Use Simulink to solve Problem 7.62.

7.67 Use Simulink to solve Problem 7.63. Plot $h(t)$ for both parts (a) and (b).

7.68 Refer to Example 7.10.1. Use the simulation with $q = 8 \times 10^5$ to compare the energy consumption and the thermostat cycling frequency for the two temperature bands (69°, 71°) and (68°, 72°).

7.69 Consider the liquid-level system shown in Figure 7.3.3. Suppose that the height h is controlled by using a relay to switch the flow rate q_{mi} between the values 0 and 50 kg/s. The flow rate is switched on when the height is less than 4.5 m and is switched off when the height reaches 5.5 m. Create a Simulink model for this application using the values $A = 2\ \text{m}^2$, $R = 400\ 1/(\text{m} \cdot \text{s})$, $\rho = 1000\ \text{kg/m}^3$, and $h(0) = 1$ m. Obtain a plot of $h(t)$.

CHAPTER 8

System Analysis in the Time Domain

CHAPTER OUTLINE

CHAPTER OBJECTIVES

When you have finished this chapter, you should be able to

1. Obtain and interpret the free, step, ramp, and impulse response of linear models.
2. Compute and use the time constant τ, the undamped natural frequency ω_n, and other parameters to describe and assess system response.
3. Use time-domain response data to estimate coefficient values in dynamic models.
4. Use Simulink to simulate nonlinear systems and systems with inputs more complicated than the impulse, step, and ramp functions.

Now that we have seen how to model the various types of physical systems (mechanical, electrical, fluid, and thermal), it is appropriate at this point to pull together our modeling knowledge and our analytical and computer solution methods. The purpose of this chapter is to integrate this knowledge with emphasis on understanding system behavior in the *time domain*. The forcing functions commonly used to model real inputs or to test a system's response in the time domain are the *impulse*, the *step*, and the *ramp* functions. The impulse models a suddenly applied and suddenly removed input. The step function models a suddenly applied input that remains constant. The ramp models an input that is changing at a constant rate. In this chapter we will show how to analyze systems subjected to these inputs. In Chapter 9, we will treat system response in the *frequency domain* by analyzing the response to the other commonly used input, the sinusoid.

Sections 8.1 and 8.2 treat the response of first- and second-order systems. We have already introduced the concepts of the *time constant* τ and the *undamped natural frequency* ω_n. These parameters help to describe the speed and the oscillation characteristics of the response. In Section 8.3 we will introduce additional parameters that are useful for describing and for specifying the response.

Response prediction cannot be precisely made unless we have numerical values for the model's coefficients, and Section 8.4 shows how measurements of the response as a function of time can be used to obtain such values.

Section 8.5 shows how to use MATLAB to analyze the step response. Section 8.6 introduces several Simulink blocks that are useful for simulating systems with complicated inputs. ■

8.1 RESPONSE OF FIRST-ORDER SYSTEMS

We now examine the response of first-order systems to step and ramp inputs. The free response of a model is its solution in the absence of an input. We have seen that the *free* response of the model

$$m\dot{v} + cv = f(t) \tag{8.1.1}$$

is given by

$$v(t) = v(0)e^{-ct/m} \tag{8.1.2}$$

where $v(0)$ is the initial value of the response $v(t)$, and m and c are constants. When $c/m < 0$, the solution grows exponentially; this is the *unstable* case. If $c/m = 0$, the model is *neutrally stable*, and $v(t) = v(0)$. If c/m is positive, the model is *stable*, and the solution decays exponentially.

8.1.1 THE TIME CONSTANT

For the stable case, we introduced a new parameter τ for (8.1.1) with the definition

$$\tau = \frac{m}{c} \tag{8.1.3}$$

The free response of (8.1.1) can be written as

$$v(t) = v(0)e^{-t/\tau} \tag{8.1.4}$$

and is illustrated in Figure 8.1.1.

Figure 8.1.1 The free response $v(t) = v(0)e^{-t/\tau}$.

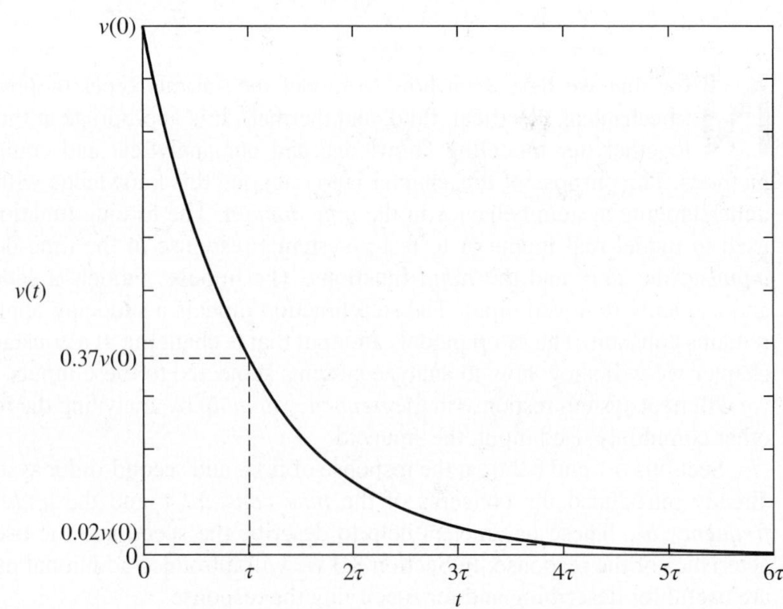

(a)

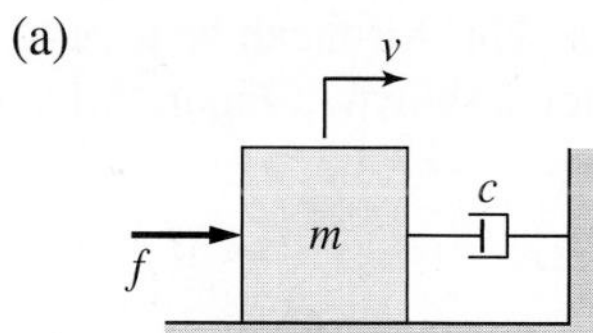

$$m\frac{dv}{dt} + cv = f$$

$$\tau = \frac{m}{c}$$

(b)

$$I\frac{d\omega}{dt} + c\omega = T$$

$$\tau = \frac{I}{c}$$

(c)

$$RC\frac{dv}{dt} + v = v_s$$

$$\tau = RC$$

(d)

$$AR\frac{dh}{dt} + gh = Rq_v$$

$$\tau = \frac{AR}{g}$$

(e)

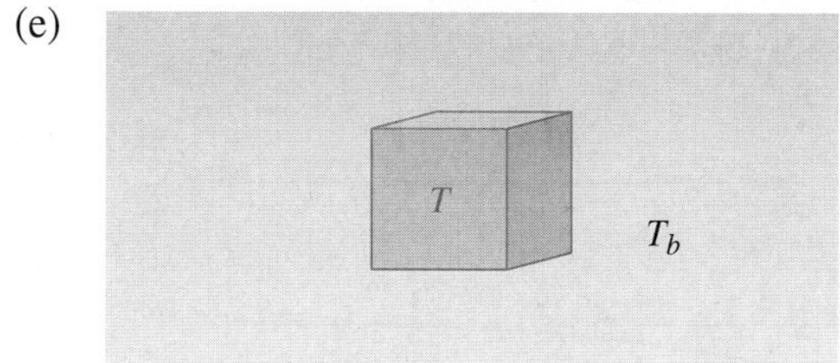

$$mc_pR\frac{dT}{dt} + T = T_b$$

$$\tau = mc_pR$$

Figure 8.1.2 First-order systems having the model form $a\dot{y} + by = f(t)$.

The new parameter τ has units of time and is the model's *time constant*. It gives a convenient measure of the exponential decay curve. After a time equal to one time constant has elapsed, v has decayed to 37% of its initial value. We can also say that v has decayed *by* 63%. At $t = 4\tau$, $v(t)$ has decayed to 2% of its initial value. At $t = 5\tau$, $v(t)$ has decayed to 1% of its initial value.

Figure 8.1.2 shows some systems whose models have the same form as that of (8.1.1); namely,

$$a\dot{y} + by = f(t)$$

This can be rewritten as

$$\frac{a}{b}\dot{y} + y = \frac{1}{b}f(t)$$

In this form, the time constant may be identified as $\tau = a/b$. This form can be used to determine the time constant expressions for each of the systems shown in Figure 8.1.2.

8.1.2 STEP RESPONSE OF A FIRST-ORDER MODEL

The step response of the model $m\dot{v} + cv = f$, which has no input derivative, can be obtained with the substitution method or the Laplace transform method. The result is

$$v(t) = \underbrace{v(0)e^{-ct/m}}_{\text{Free response}} + \underbrace{\frac{F}{c}\left(1 - e^{-ct/m}\right)}_{\text{Forced response}} \tag{8.1.5}$$

Note that the response is the sum of the free and the forced responses.

The time constant is useful also for analyzing the response when the forcing function is a step. We can express the solution in terms of the time constant τ by substituting $c/m = 1/\tau$ to obtain

$$v(t) = v(0)e^{-t/\tau} + \frac{F}{c}\left(1 - e^{-t/\tau}\right) = v(0)e^{-t/\tau} + v_{ss}\left(1 - e^{-t/\tau}\right) \tag{8.1.6}$$

where $v_{ss} = F/c$. The solution approaches the constant value F/c as $t \to \infty$. This is called the *steady-state* response, denoted v_{ss}.

The response is plotted in Figure 8.1.3 for $v(0) = 0$. At $t = \tau$, the response is 63% of the steady-state value. At $t = 4\tau$, the response is 98% of the steady-state value, and at $t = 5\tau$, it is 99% of steady state. Thus, because the difference between 98% and 99% is so small, for most engineering purposes we can say that $v(t)$ reaches steady state at $t = 4\tau$, although mathematically, steady state is not reached until $t = \infty$.

If $v(0) \neq 0$, the response is shifted by $v(0)e^{-t/\tau}$. At $t = \tau$, 37% of the difference between the initial value $v(0)$ and the steady-state value remains. At $t = 4\tau$, only 2% of this difference remains. Figure 8.1.4 shows the complete response.

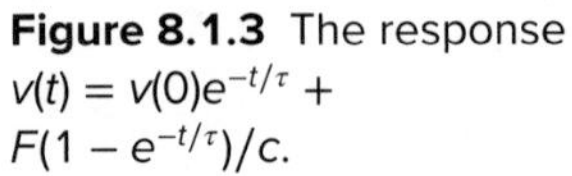

Figure 8.1.3 The response $v(t) = v(0)e^{-t/\tau} + F(1 - e^{-t/\tau})/c$.

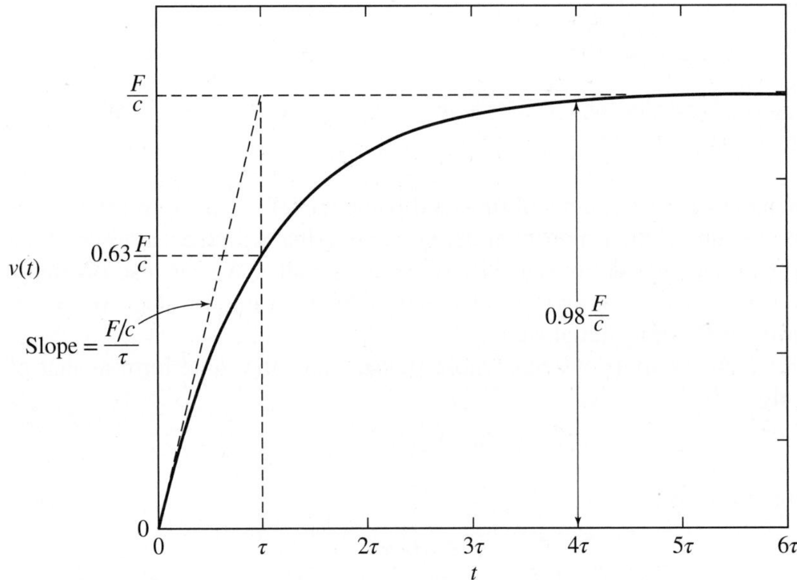

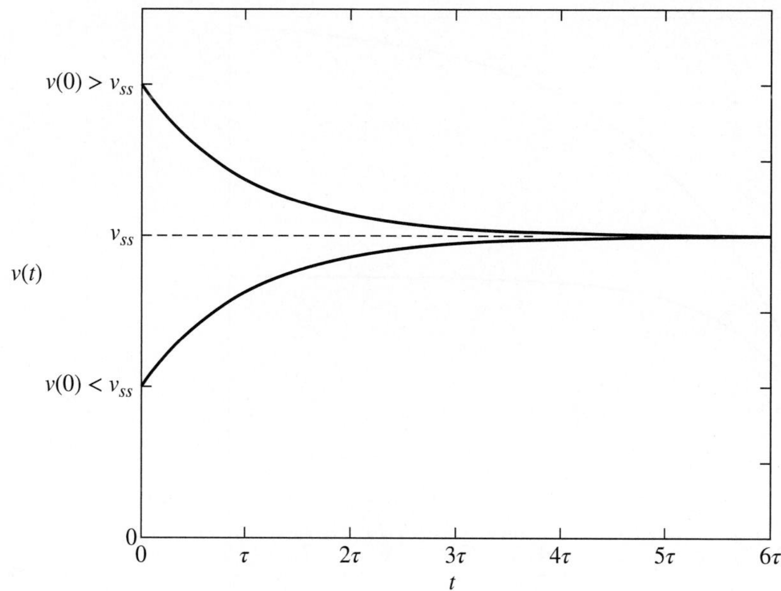

Figure 8.1.4 The response $v(t) = v(0)e^{-t/\tau} + v_{ss}(1 - e^{-t/\tau})$.

Table 8.1.1 summarizes the response of a first-order linear model.

8.1.3 TRANSIENT AND STEADY-STATE RESPONSES

For the stable case ($c/m > 0$), we can separate the step response into the sum of a term that eventually disappears (the *transient response*) and a term that remains (the *steady-state response*). Referring to (8.1.6), we see that

$$v(t) = \underbrace{\left[v(0) - \frac{F}{c}\right] e^{-t/\tau}}_{\text{Transient response}} + \underbrace{\frac{F}{c}}_{\text{Steady-state response}}$$

Table 8.1.1 Free, step, and ramp response of $\tau\dot{y} + y = r(t)$.

Free response [$r(t) = 0$]
$y(t) = y(0)e^{-t/\tau}$
$y(\tau) \approx 0.37y(0)$
$y(4\tau) \approx 0.02y(0)$
Step response [$r(t) = Ru_s(t)$, $y(0) = 0$]
$y(t) = R(1 - e^{-t/\tau})$
$y(\infty) = y_{ss} = R$
$y(\tau) \approx 0.63y_{ss}$
$y(4\tau) \approx 0.98y_{ss}$
Ramp response [$r(t) = mt$, $y(0) = 0$]
$y(t) = m(t - \tau + \tau e^{-t/\tau})$

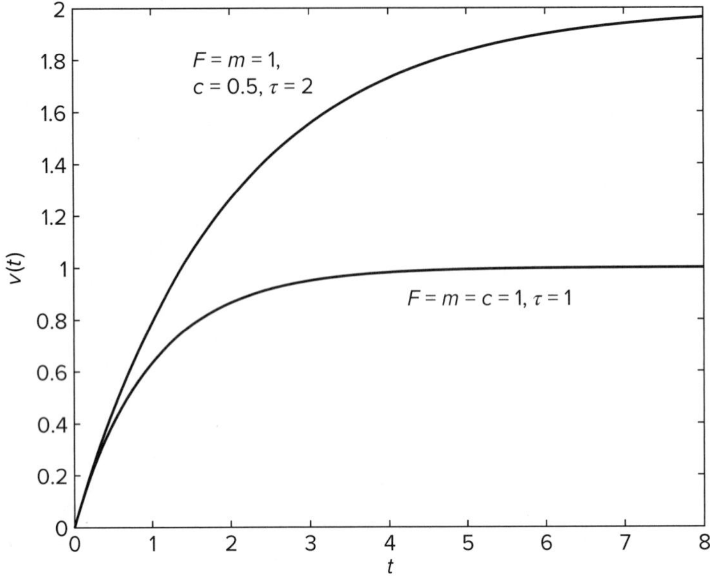

Figure 8.1.5 Step response for two values of c.

8.1.4 RESPONSE TIME AND THE TIME CONSTANT

When using the time constant to compare the response times of two systems, the comparison should be done only for systems whose steady-state responses are identical. Note that for the model $m\dot{v} + cv = f(t)$ because $\tau = m/c$, a small value of the damping constant c produces a large time constant, which indicates a sluggish system. The result is counter to our intuition, which tells us that a small amount of damping should correspond to a system with a fast response. This counterintuitive result is explained by noting that a larger steady-state response F/c results from a smaller damping constant. Thus, for a large value of c, even though the time constant is smaller, the steady-state response is smaller, and thus it takes less time to reach a specific steady-state value. This is illustrated in Figure 8.1.5 for $v(0) = 0$ with two values of c.

8.1.5 THE STEP FUNCTION APPROXIMATION

The step function is an approximate description of an input that can be switched on in a time interval that is very short compared to the time constant of the system. A good example of a step input is the voltage applied to a circuit due to the sudden closure of a switch.

In some applications, however, it may not be clear whether a step function is a good model of the input. For example, we might model rocket thrust as a step function if it reaches a constant value quickly compared to the vehicle's time constant. Solid-rocket thrust as a function of time depends on the cross-sectional shape of the solid fuel. For a propellant grain having a tubular cross section, the fuel burns from the inside to the outside. The thrust increases with time because the propellant's burning area increases. A step function usually would not be used to model the thrust in this case, unless we needed a quick, very approximate answer, in which case we would take the average thrust as the step magnitude. Another shape, called rod-and-tube, was designed to give a constant thrust by keeping the total burning area approximately constant. The step function is a good model for this case if the time to reach constant thrust is short compared to the vehicle time constant.

Speed Response of a Rational System — **EXAMPLE 8.1.1**

■ **Problem**

A certain rotational system has an inertia $I = 50\ \text{kg}\cdot\text{m}^2$ and a viscous damping constant $c = 10\ \text{N}\cdot\text{m}\cdot\text{s/rad}$. The torque $T(t)$ is applied by an electric motor (Figure 8.1.6a). From the free body diagram shown in part (b) of the figure, the equation of motion is

$$50\frac{d\omega}{dt} + 10\omega = T(t) \tag{1}$$

The model of the motor's field current i_f in amperes is

$$0.001\frac{di_f}{dt} + 5i_f = v(t) \tag{2}$$

where $v(t)$ is the voltage applied to the motor. The motor torque constant is $K_T = 25\ \text{N}\cdot\text{m/A}$.

Suppose the applied voltage is 10 V. Determine the steady-state speed of the inertia and estimate the time required to reach that speed.

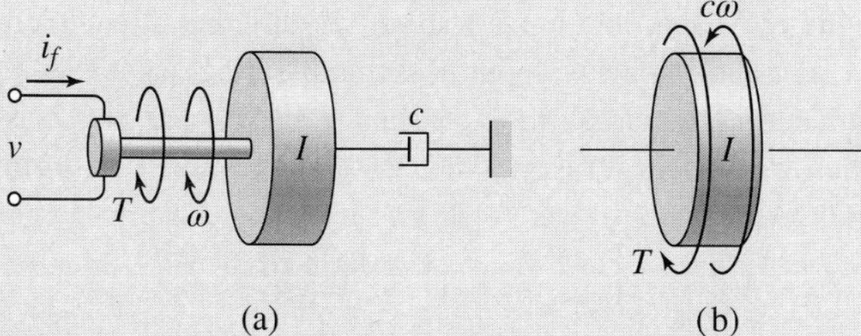

Figure 8.1.6 A rotational system.

■ **Solution**

From equation (2) we see that the time constant of the motor circuit is $0.001/5 = 2 \times 10^{-4}$ s. Thus, the current will reach a steady-state value of $10/5 = 2$ A in approximately $4(2 \times 10^{-4}) = 8 \times 10^{-4}$ s. The resulting steady-state torque is $K_T(2) = 25(2) = 50\ \text{N}\cdot\text{m}$.

From equation (1) we find the time constant of the rotational system to be $50/10 = 5$ s. Since this is much larger than the time constant of the circuit (2×10^{-4} s), we conclude that the motor torque may be modeled as a step function. The magnitude of the step function is $50\ \text{N}\cdot\text{m}$. The steady-state speed is $\omega = 50/10 = 5$ rad/s, and therefore it will take approximately $4(5) = 20$ s to reach this speed.

8.1.6 IMPULSE RESPONSE

The impulse response of $m\dot{v} + cv = f(t)$ is found as follows, where A is the impulse *strength*, or the area under the impulse versus time curve.

$$msV(s) - mv(0-) + cV(s) = A$$

$$V(s) = \frac{mv(0-) + A}{ms + c} = \frac{v(0-) + A/m}{s + c/m}$$

$$v(t) = \left[v(0-) + \frac{A}{m}\right] e^{-ct/m}$$

We can see that the effect of the impulse is to increase the effective initial condition by A/m.

Because the derivative $\dot{f}(t)$ is undefined if $f(t)$ is an impulse, we do not treat the impulse response of the equation $m\dot{v} + cv = f(t) + b\dot{f}(t)$.

EXAMPLE 8.1.2

Analysis of a Telegraph Line

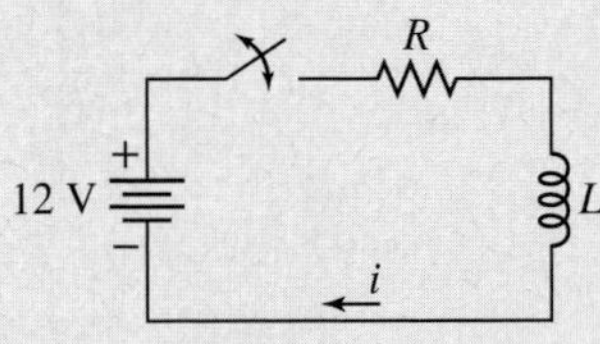

Figure 8.1.7 Circuit representation of a telegraph line.

■ **Problem**

Figure 8.1.7 shows a circuit representation of a telegraph line. The resistance R is the line resistance and L is the inductance of the solenoid that activates the receiver's clicker. The switch represents the operator's key. Assume that when sending a "dot," the key is closed for 0.1 s. Using the values $R = 20\ \Omega$ and $L = 4$ H, obtain the expression for the current $i(t)$ passing through the solenoid.

■ **Solution**

From the voltage law we have

$$L\frac{di}{dt} + Ri = v_i(t) \tag{1}$$

where $v_i(t)$ represents the input voltage due to the switch and the 12-V supply. We could model $v_i(t)$ as a rectangular pulse of height 12 V and duration 0.1 s, but the differential equation (1) is easier to solve if we model $v_i(t)$ as an impulsive input of strength $12(0.1) = 1.2\ \text{V}\cdot\text{s}$. This model can be justified by the fact that the circuit time constant, $L/R = 4/20 = 0.2$, is greater than the duration of $v_i(t)$. Thus, we model $v_i(t)$ as $v_i(t) = 1.2\delta(t)$. The Laplace transform method of equation (1) with $i(0) = 0$ gives

$$(4s + 20)I(s) = 1.2 \qquad \text{or} \qquad I(s) = \frac{1.2}{4s + 20} = \frac{0.3}{s + 5}$$

This gives the solution

$$i(t) = 0.3e^{-5t}\ \text{A}$$

Note that this solution gives $i(0+) = 0.3$, whereas $i(0) = 0$.

8.1.7 RAMP RESPONSE

The ramp function models an input that is changing at a constant rate. Thus, a ramp input function has the form $f(t) = mt$, where m is the slope of the ramp. We can obtain the response to the ramp using the Laplace transform method.

EXAMPLE 8.1.3

Ramp Response of a First-Order Model

■ **Problem**

Obtain the ramp response of the model $5\dot{v} + v = f(t)$, where $f(t) = 3t$ and $v(0) = 0$.

■ **Solution**

Transforming the equation with $F(s) = 3/s^2$ gives

$$5sV(s) + V(s) = F(s) = \frac{3}{s^2}$$

or

$$V(s) = \frac{3}{s^2(5s + 1)} = \frac{0.6}{s^2(s + 0.2)} = \frac{3}{s^2} - \frac{15}{s} + \frac{15}{s + 0.2}$$

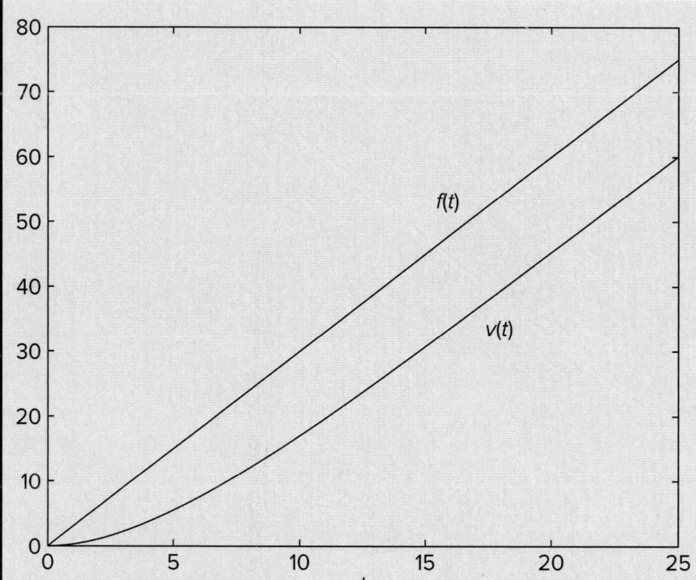

Figure 8.1.8 Ramp input and ramp response of the model $5\dot{v} + v = 3t$.

The inverse transforms give

$$v(t) = \underbrace{3t - 15}_{\text{Steady state}} + \underbrace{15e^{-0.2t}}_{\text{Transient}}$$

Note that $v(0) = 0$, as it should.

The response is shown in Figure 8.1.8 along with the ramp input. The time constant indicates how long it takes for the transient response to disappear. Here $\tau = 5$, so the transient response disappears after approximately $t = 4\tau = 20$, and the steady-state response is a straight line with a slope of 3 and a v intercept of -15.

We can obtain the response of the general equation $\tau\dot{v} + v = f(t)$ to the ramp input $f(t) = mt$ as follows. Setting $v(0) = 0$ and transforming the equation with $F(s) = m/s^2$ gives

$$V(s) = \frac{m}{s^2(\tau s + 1)} = \frac{m}{s^2} - \frac{m\tau}{s} + \frac{m\tau}{s + 1/\tau}$$

The inverse transforms give

$$v(t) = m(t - \tau) + m\tau e^{-t/\tau}$$

The response is in steady state after approximately $t = 4\tau$. At steady state, $v(t) = m(t - \tau)$, so the response is parallel to the input but lags behind it by a time τ (Figure 8.1.9).

Obtaining the ramp response can be tedious for higher-order systems, but sometimes we only need to find the steady-state *difference* between the input and the output. This can be done easily with the final value theorem, as illustrated by Example 8.1.4.

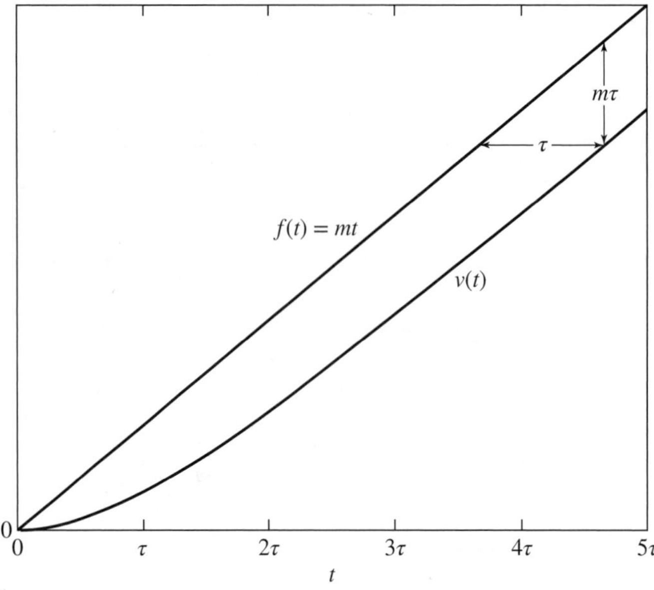

Figure 8.1.9 Ramp input and ramp response of the model $\tau\dot{v} + v = f(t)$.

EXAMPLE 8.1.4

The Final Value Theorem and Ramp Response

■ Problem

Obtain the steady-state difference $f(\infty) - v(\infty)$ between the input and output of the following model: $\tau\dot{v} + v = bf(t)$, where b is a constant and $f(t) = mt$. Assume that $v(0) = 0$ and that the model is stable ($\tau > 0$).

■ Solution

The transform of the response is

$$V(s) = \frac{b}{\tau s + 1} F(s) = \frac{b}{\tau s + 1} \frac{m}{s^2}$$

Use this with the final-value theorem to find the steady-state difference:

$$\begin{aligned} f(\infty) - v(\infty) &= \lim_{s \to 0} [sF(s)] - \lim_{s \to 0} [sV(s)] = \lim_{s \to 0} s[F(s) - V(s)] \\ &= \lim_{s \to 0} s\left(\frac{m}{s^2} - \frac{b}{\tau s + 1}\frac{m}{s^2}\right) \\ &= \lim_{s \to 0} \frac{m}{s}\left(\frac{\tau s + 1 - b}{\tau s + 1}\right) \\ &= \begin{cases} \infty & b \neq 1 \\ m\tau & b = 1 \end{cases} \end{aligned}$$

Thus, the steady-state difference is infinite unless $b = 1$. Both the input and output approach straight lines at steady state. The preceding result shows that the lines diverge unless $b = 1$. If $b = 1$, the lines are a vertical distance $m\tau$ apart.

8.2 RESPONSE OF SECOND-ORDER SYSTEMS

The equations of motion of many systems containing mass, spring, and damping elements have the form

$$m\ddot{x} + c\dot{x} + kx = f(t) \tag{8.2.1}$$

where $f(t)$ is the input. Its transfer function is

$$\frac{X(s)}{F(s)} = \frac{1}{ms^2 + cs + k} \tag{8.2.2}$$

Figure 8.2.1 shows examples of other types of systems that have the same model form. The solution of this equation, and therefore the form of the free and forced responses, depends on the values of the two characteristic roots, obtained from the characteristic equation $ms^2 + cs + k = 0$. Recall from the discussion in Chapter 2 that this model is

Figure 8.2.1 Some second-order systems.

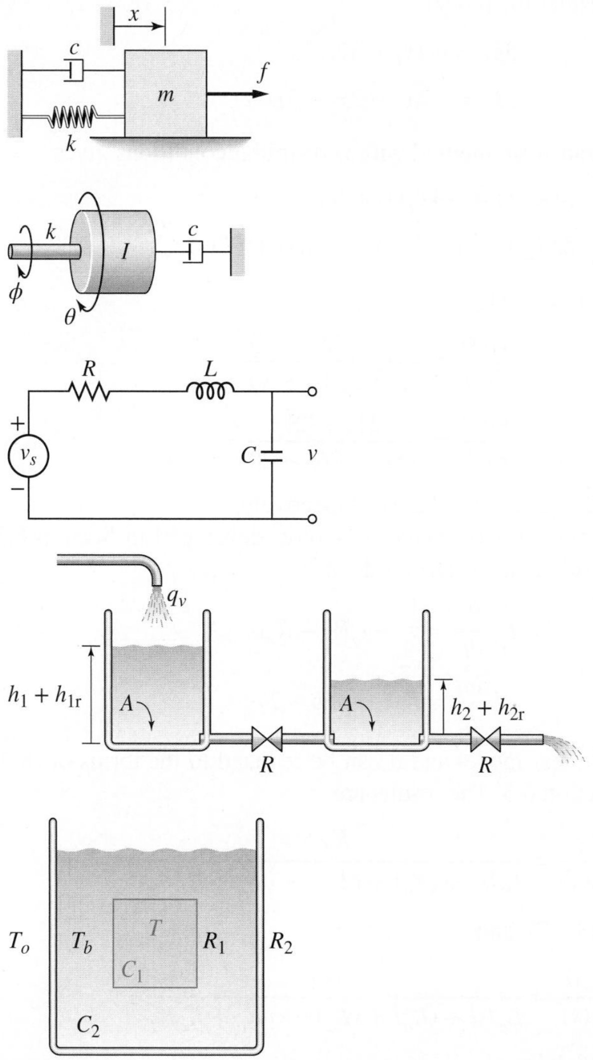

(a)

$$m\frac{d^2x}{dt^2} + c\frac{dx}{dt} + kx = f$$

(b)

$$I\frac{d^2\theta}{dt^2} + c\frac{d\theta}{dt} + k\theta = k\phi$$

(c)

$$LC\frac{d^2v}{dt^2} + RC\frac{dv}{dt} + v = v_s$$

(d)

$$RA\frac{dh_1}{dt} + g(h_1 - h_2) = Rq_v$$

$$RA\frac{dh_2}{dt} + g(h_2 - h_1) + gh_2 = 0$$

(e)

$$R_1C_1\frac{dT}{dt} + T = T_b$$

$$R_1R_2C_2\frac{dT_b}{dt} + (R_1 + R_2)T_b = R_2T + R_1T_o$$

stable if both of its roots are real and negative or if the roots are complex with negative real parts. This is true if m, c, and k have the same sign.

A related model form is

$$m\ddot{x} + c\dot{x} + kx = a\dot{g}(t) + bg(t) \tag{8.2.3}$$

where $g(t)$ is the input. Its transfer function is

$$\frac{X(s)}{G(s)} = \frac{as + b}{ms^2 + cs + k} \tag{8.2.4}$$

So this model has numerator dynamics. It is important to understand that the input does not affect the characteristic equation, and therefore does not affect the stability of the model or its free response. Thus, (8.2.1) and (8.2.3) have the same stability characteristics and the same free response, because they have the same characteristic equation, $ms^2 + cs + k = 0$.

The formulas to be developed in this section are based on the transfer-function model form. Models in state variable form can always be reduced to transfer-function form. For example, consider the model

$$3\dot{x}_1 = -4x_1 + 6x_2$$

$$5\dot{x}_2 = -2x_1 - 2x_2 + 7f(t)$$

Applying the Laplace transform method with zero-initial conditions gives

$$3sX_1(s) = -4X_1(s) + 6X_2(s)$$

$$5sX_2(s) = -2X_1(s) - 2X_2(s) + 7F(s)$$

Solving these for $X_1(s)$ and $X_2(s)$ gives

$$\frac{X_1(s)}{F(s)} = \frac{42}{15s^2 + 26s + 20}$$

$$\frac{X_2(s)}{F(s)} = \frac{21s + 28}{15s^2 + 26s + 20}$$

These have the forms of (8.2.2) and (8.2.4), respectively.

The model of an armature-controlled dc motor, developed in Section 6.5, is an example of the state-variable form. The model is

$$L_a\frac{di_a}{dt} = v_a - i_aR_a - K_b\omega \tag{8.2.5}$$

$$I\frac{d\omega}{dt} = K_Ti_a - c\omega - T_L \tag{8.2.6}$$

This is a second-order linear model and it can be reduced to the forms of (8.2.2) and (8.2.4), as shown in Section 6.5. The results are

$$\frac{\Omega(s)}{V_a(s)} = \frac{K_T}{L_aIs^2 + (R_aI + cL_a)s + cR_a + K_bK_T} \tag{8.2.7}$$

which has the form of (8.2.2), and

$$\frac{I_a(s)}{V_a(s)} = \frac{Is + c}{L_aIs^2 + (R_aI + cL_a)s + cR_a + K_bK_T} \tag{8.2.8}$$

which has the form of (8.2.4).

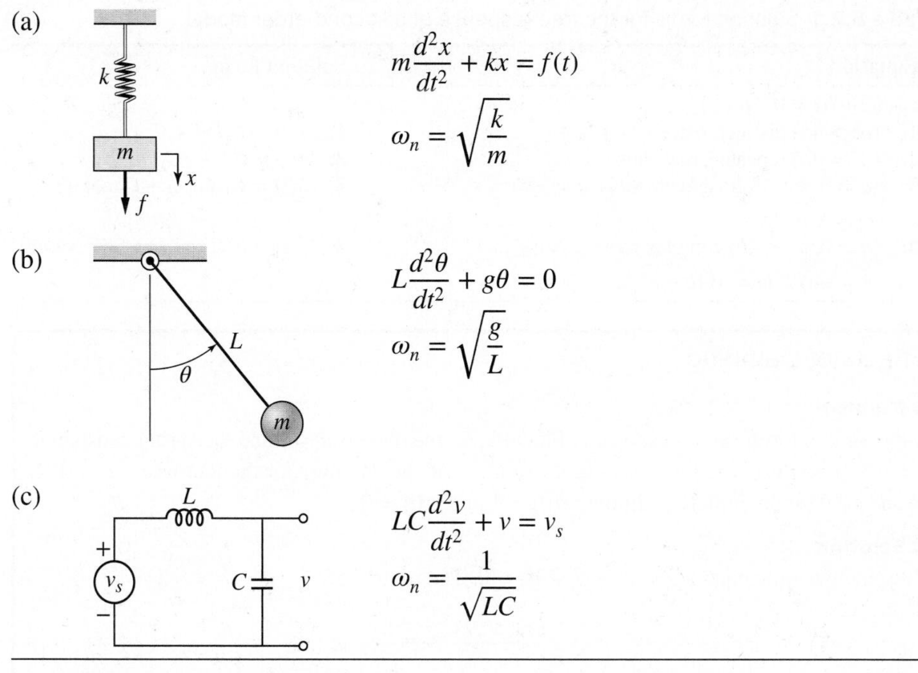

Figure 8.2.2 Examples of undamped systems.

8.2.1 UNDAMPED RESPONSE

Consider the undamped systems shown in Figure 8.2.2. They all have the same model form: $m\ddot{x} + kx = f(t)$. In the first system, suppose $f(t) = 0$ and we set the mass in motion at time $t = 0$ by pulling it to a position $x(0)$ and releasing it with an initial velocity $\dot{x}(0)$. The solution can be obtained by using the Laplace transform method or by using the trial-solution method discussed in Chapter 2. The form is $x(t) = C_1 \sin \omega_n t + C_2 \cos \omega_n t$, where we have defined

$$\omega_n = \sqrt{\frac{k}{m}} \tag{8.2.9}$$

Using the initial conditions we find that the constants are $C_1 = \dot{x}(0)/\omega_n$ and $C_2 = x(0)$. Thus, the solution is

$$x(t) = \frac{\dot{x}(0)}{\omega_n} \sin \omega_n t + x(0) \cos \omega_n t \tag{8.2.10}$$

This solution shows that the mass oscillates about the rest position $x = 0$ with a frequency of $\omega_n = \sqrt{k/m}$ radians per unit time. The period of the oscillation is $2\pi/\omega_n$. The frequency of oscillation ω_n is called the *natural frequency*. The natural frequency is greater for stiffer springs (larger k values). The amplitude of the oscillation depends on the initial conditions $x(0)$ and $\dot{x}(0)$.

It is important to realize that (8.2.9) gives the oscillation frequency of *any* model of the *form* $m\ddot{x} + kx = f(t)$. So, for example, the free response of the voltage $v(t)$ in the circuit in Figure 8.2.2c will oscillate with the radian frequency $\omega_n = \sqrt{1/LC}$.

8.2.2 RESPONSE WITH DAMPING

Now let us investigate the effects of damping. Table 8.2.1 summarizes the free response forms obtained in Chapter 2.

Table 8.2.1 Solution forms for the free response of a second-order model.

Equation	Solution form
$\ddot{x} + a\dot{x} + bx = 0 \quad b \neq 0$	
1. $(a^2 > 4b)$ distinct, real roots: s_1, s_2	**1.** $x(t) = C_1 e^{s_1 t} + C_2 e^{s_2 t}$
2. $(a^2 = 4b)$ repeated, real roots: s_1, s_1	**2.** $x(t) = (C_1 + tC_2)e^{s_1 t}$
3. $(a = 0, b > 0)$ imaginary roots: $s = \pm j\omega$, $\omega = \sqrt{b}$	**3.** $x(t) = C_1 \sin \omega t + C_2 \cos \omega t$
4. $(a \neq 0, a^2 < 4b)$ complex roots: $s = r \pm j\omega$, $r = -a/2, \omega = \sqrt{4b - a^2}/2$	**4.** $x(t) = e^{rt}(C_1 \sin \omega t + C_2 \cos \omega t)$

EXAMPLE 8.2.1

Effects of Damping

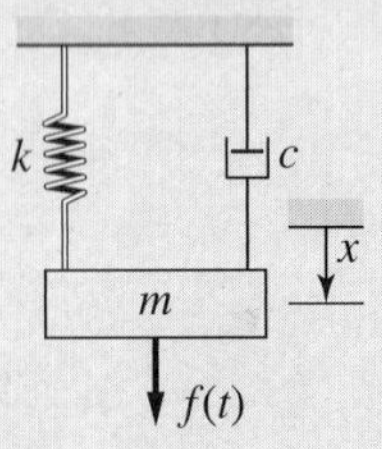

Figure 8.2.3 The system used in Example 8.2.1.

■ Problem

Suppose that for the system shown in Figure 8.2.3 the mass is $m = 1$ and the spring constant is $k = 16$. Investigate the free response as we increase the damping for the four cases: $c = 0$, 4, 8, and 10. Use the initial conditions: $x(0) = 1$ and $\dot{x}(0) = 0$.

■ Solution

The characteristic equation is $s^2 + cs + 16 = 0$. The roots are

$$s = \frac{-c \pm \sqrt{c^2 - 64}}{2}$$

The free response can be obtained with the trial-solution method or with the Laplace transform method. The free responses for the given initial conditions, for each value of c, are as follows:

$$\text{For } c = 0 \qquad x(t) = \cos 4t$$

$$\text{For } c = 4 \qquad x(t) = 1.155e^{-2t} \sin\left(\sqrt{12}t + 1.047\right)$$

$$\text{For } c = 8 \qquad x(t) = (1 + 4t)e^{-4t}$$

$$\text{For } c = 10 \qquad x(t) = \frac{4}{3}e^{-2t} - \frac{1}{3}e^{-8t}$$

The solutions are plotted in Figure 8.2.4. For no damping, the system is neutrally stable and the mass oscillates with a constant amplitude and a radian frequency of $\sqrt{k/m} = 4$, which is the

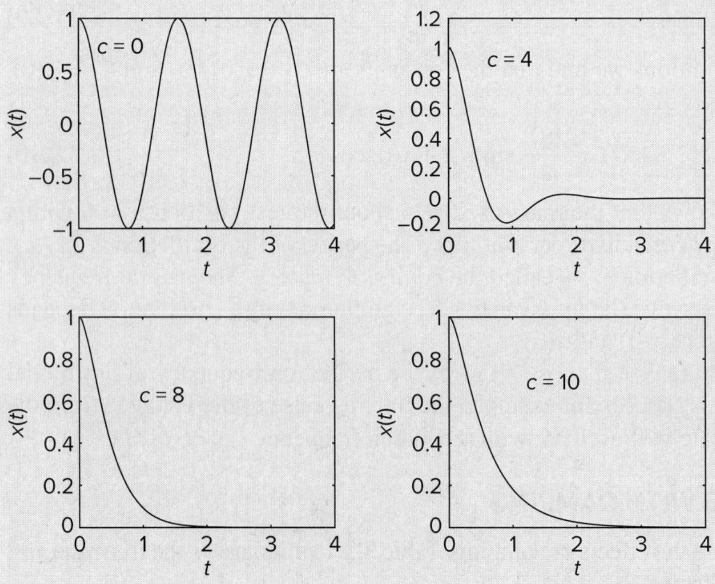

Figure 8.2.4 Responses for four values of c.

natural frequency ω_n. As the damping is increased slightly to $c = 4$, the system becomes stable and the mass still oscillates but with a smaller radian frequency ($\sqrt{12} = 3.464$). The oscillations die out eventually as the mass returns to rest at the equilibrium position $x = 0$. As the damping is increased further to $c = 8$, the mass no longer oscillates because the damping force is large enough to limit the velocity and thus the momentum of the mass to a value that prevents the mass from overshooting the equilibrium position. For a larger value of c ($c = 10$), the mass takes longer to return to equilibrium because the damping force greatly slows down the mass.

8.2.3 EFFECT OF ROOT LOCATION

Figure 8.2.5 shows how the location of the characteristic roots in the complex plane affects the free response. The real part of the root is plotted on the horizontal axis, and the imaginary part is plotted on the vertical axis. Because the roots are conjugate, we show only the upper root. Using the results we have found in earlier chapters, we see that

1. Unstable behavior occurs if any root lies to the right of the imaginary axis.
2. The response oscillates only when a root has a nonzero imaginary part.
3. The greater the imaginary part, the higher the frequency of the oscillation.
4. The farther to the left the root lies, the faster the response due to that root decays.

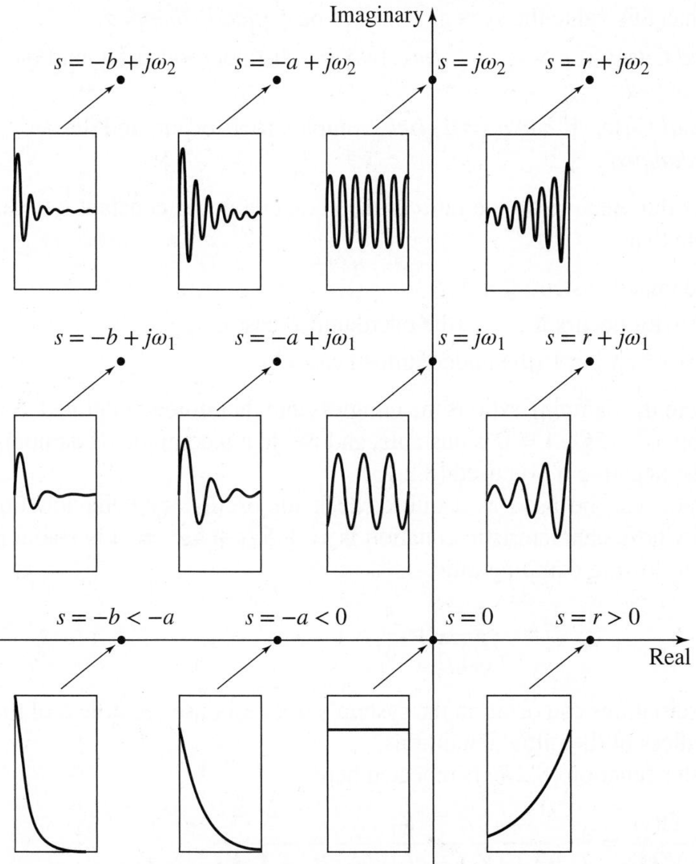

Figure 8.2.5 Effect of root location on the free response.

The characteristic equation of the model (8.2.1) is

$$ms^2 + cs + k = 0 \tag{8.2.11}$$

Its roots are

$$s = \frac{-c \pm \sqrt{c^2 - 4mk}}{2m} = r \pm j\omega \tag{8.2.12}$$

where r and ω denote the real and imaginary parts of the roots.

8.2.4 THE DAMPING RATIO

A second-order system's free response for the stable case can be conveniently characterized by the *damping ratio* ζ (sometimes called the *damping factor*). For the characteristic equation (8.2.11), the damping ratio is defined as

$$\zeta = \frac{c}{2\sqrt{mk}} \tag{8.2.13}$$

This definition is not arbitrary but is based on the way the roots change from real to complex as the value of c is changed; that is, from (8.2.12) we see that three cases can occur:

1. *The Critically Damped Case:* Repeated roots occur if $c^2 - 4mk = 0$; that is, if $c = 2\sqrt{mk}$. This value of the damping constant is the *critical damping constant* c_c, and when c has this value the system is said to be *critically damped.*
2. *The Overdamped Case:* If $c > c_c = 2\sqrt{mk}$, two real distinct roots exist, and the system is *overdamped.*
3. *The Underdamped Case:* If $c < c_c = 2\sqrt{mk}$, complex roots occur, and the system is *underdamped.*

The damping ratio is thus seen to be the ratio of the actual damping constant c to the critical value c_c. Note that

1. For a critically damped system, $\zeta = 1$.
2. Exponential behavior occurs if $\zeta > 1$ (the overdamped case).
3. Oscillations exist when $\zeta < 1$ (the underdamped case).

For an unstable system the damping ratio is meaningless and therefore not defined. For example, the equation $3s^2 - 5s + 4 = 0$ is unstable, and we do not compute its damping ratio, which would be negative if you used (8.2.13).

The damping ratio can be used as a quick check for oscillatory behavior. For example, the model whose characteristic equation is $s^2 + 5ds + 4d^2 = 0$ is stable if $d > 0$ and it has the following damping ratio.

$$\zeta = \frac{5d}{2\sqrt{4d^2}} = \frac{5}{4} > 1$$

Because $\zeta > 1$, no oscillations can occur in the system's free response regardless of the value of d and regardless of the initial conditions.

The motor transfer function (8.2.7) is repeated here.

$$\frac{\Omega(s)}{V_a(s)} = \frac{K_T}{L_a I s^2 + (R_a I + cL_a)s + cR_a + K_b K_T}$$

The denominator of this model has the standard form $ms^2 + cs + k$, and thus the damping ratio of the motor model is, from (8.2.13),

$$\zeta = \frac{R_a I + cL_a}{2\sqrt{L_a I(R_a c + K_b K_T)}}$$

Even if the damping constant c is zero, the damping ratio is still nonzero because of the term $R_a I$. We can see this by setting $c = 0$ in the previous equation to obtain

$$\zeta = \frac{R_a I}{2\sqrt{L_a I K_b K_T}}$$

In SI units, $K_T = K_b$, and the previous expression becomes

$$\zeta = \frac{R_a}{2K_T}\sqrt{\frac{I}{L_a}}$$

Thus, increasing either R_a or I increases the damping ratio, while increasing either K_T or L_a decreases the damping ratio.

8.2.5 NATURAL AND DAMPED FREQUENCIES OF OSCILLATION

The roots of (8.2.11) are purely imaginary when there is no damping. The imaginary part and the frequency of oscillation for this case is the *undamped natural frequency* $\omega_n = \sqrt{k/m}$.

We can write the characteristic equation in terms of the parameters ζ and ω_n. First divide (8.2.11) by m and use the fact that $\omega_n^2 = k/m$ and that

$$2\zeta\omega_n = 2\left(\frac{c}{2\sqrt{mk}}\right)\left(\sqrt{\frac{k}{m}}\right) = \frac{c}{m}$$

The characteristic equation becomes

$$s^2 + 2\zeta\omega_n s + \omega_n^2 = 0 \tag{8.2.14}$$

and the roots are

$$s = -\zeta\omega_n \pm j\omega_n\sqrt{1-\zeta^2} \tag{8.2.15}$$

The frequency of oscillation is $\omega_n\sqrt{1-\zeta^2}$ and is called the *damped natural frequency*, or simply the damped frequency, to distinguish it from the undamped natural frequency ω_n. We will denote the damped frequency by ω_d.

$$\omega_d = \omega_n\sqrt{1-\zeta^2} \tag{8.2.16}$$

The frequencies ω_n and ω_d have meaning only for the underdamped case ($\zeta < 1$). For this case (8.2.16) shows that $\omega_d < \omega_n$. Thus, the damped frequency is always less than the undamped frequency.

8.2.6 TIME CONSTANTS OF SECOND-ORDER SYSTEMS

Comparison of (8.2.12) with (8.2.15) shows that $r = -\zeta\omega_n$ and $\omega = \omega_n\sqrt{1-\zeta^2}$. Because the time constant τ is $-1/r$, we have

$$\tau = \frac{1}{\zeta\omega_n} \tag{8.2.17}$$

Remember that this formula applies only if $\zeta \le 1$ (otherwise, $\sqrt{1-\zeta^2}$ is imaginary).

Table 8.2.2 Free response of $m\ddot{x} + c\dot{x} + kx = f(t)$ for the stable case.

Characteristic roots	$s = \dfrac{-c \pm \sqrt{c^2 - 4mk}}{2m}$
Undamped natural frequency	$\omega_n = \sqrt{\dfrac{k}{m}}$
Damping ratio	$\zeta = \dfrac{c}{2\sqrt{km}}$
Damped natural frequency	$\omega_d = \omega_n\sqrt{1-\zeta^2}$
Overdamped case ($\zeta > 1$)	Distinct, real roots: $s = -r_1, s = -r_2 \quad (r_1 \neq r_2)$ $x(t) = A_1 e^{-r_1 t} + A_2 e^{-r_2 t}$ $A_1 = \dfrac{\dot{x}(0) + r_2 x(0)}{r_2 - r_1}$ $A_2 = \dfrac{-r_1 x(0) - \dot{x}(0)}{r_2 - r_1} = x(0) - A_1$
Critically damped case ($\zeta = 1$)	Repeated roots: $s = -r_1, s = -r_1$ $x(t) = (A_1 + A_2 t)e^{-r_1 t}$ $A_1 = x(0)$ $A_2 = \dot{x}(0) + r_1 x(0)$
Underdamped case ($0 \leq \zeta < 1$)	Complex conjugate roots: $s = -a \pm jb, b > 0$ $x(t) = De^{-at}\sin(bt + \phi)$ $D = \dfrac{1}{b}\sqrt{[bx(0)]^2 + [\dot{x}(0) + ax(0)]^2}, D > 0$ $\sin\phi = \dfrac{x(0)}{D} \quad \cos\phi = \dfrac{\dot{x}(0) + ax(0)}{bD}$
Alternative form for $0 \leq \zeta < 1$	$x(t) = e^{-\zeta\omega_n t}[A\sin\omega_d t + x(0)\cos\omega_d t]$ $A = \dfrac{\zeta}{\sqrt{1-\zeta^2}}x(0) + \dfrac{1}{\omega_d}\dot{x}(0)$

Table 8.2.2 summarizes the free response of the stable, linear, second-order model in terms of the parameters ζ, ω_n, and ω_d.

8.2.7 APPLICATIONS

Transfer functions of second-order models have denominators of the form $ms^2 + cs + k$, on which the expressions for ω_n, ζ, ω_d, and τ are based. These expressions are given by (8.2.9), (8.2.13), (8.2.16), (8.2.17), and are summarized in Table 8.2.3. You need to be able to apply these expressions to a given transfer function. The characteristic equation $ms^2 + cs + k = 0$ can be expressed in terms of ζ and ω_n by dividing the polynomial by m, as follows:

$$s^2 + \frac{c}{m}s + \frac{k}{m} = s^2 + 2\zeta\omega_n s + \omega_n^2 = 0$$

For example, the motor transfer function (8.2.7) is repeated here.

$$\frac{\Omega(s)}{V_a(s)} = \frac{K_T}{L_a I s^2 + (R_a I + cL_a)s + cR_a + K_b K_T}$$

Table 8.2.3 Response parameters for second-order systems.

Model	$m\ddot{x} + c\dot{x} + kx = f(t)$ m, c, k constant
Characteristic Equation	$ms^2 + cs + k = 0$
1. Roots	$s = \dfrac{-c \pm \sqrt{c^2 - 4mk}}{2m}$
2. Stability Property	Stable if and only if both roots have negative real parts. This occurs if and only if m, c, and k have the same sign.
3. Alternative forms for underdamped systems Characteristic Equation:	$s^2 + 2\zeta\omega_n s + \omega_n^2 = 0$
Roots:	$s = -\zeta\omega_n \pm j\omega_n\sqrt{1-\zeta^2}$
4. Damping ratio or damping factor	$\zeta = \dfrac{c}{2\sqrt{mk}}$
5. Undamped natural frequency	$\omega_n = \sqrt{\dfrac{k}{m}}$
6. Damped natural frequency	$\omega_d = \omega_n\sqrt{1-\zeta^2}$
7. Time constant	$\tau = 2m/c = 1/\zeta\omega_n$ if $\zeta \leq 1$

The damping ratio of this model is

$$\zeta = \frac{R_a I + cL_a}{2\sqrt{L_a I(cR_a + K_b K_T)}}$$

and its undamped natural frequency is

$$\omega_n = \sqrt{\frac{cR_a + K_b K_T}{L_a I}}$$

Thus, if $\zeta \leq 1$, the motor time constant is

$$\tau = \frac{1}{\zeta\omega_n} = \frac{2L_a I}{R_a I + cL_a}$$

If the damping c is slight and if $\zeta \leq 1$, then the time constant is given approximately by

$$\tau = \frac{2L_a}{R_a}$$

The *RLC* circuit shown in Figure 8.2.1c has the characteristic equation $LCs^2 + RCs + 1 = 0$. Its undamped natural frequency and damping ratio are

$$\omega_n = \sqrt{\frac{1}{LC}}$$

$$\zeta = \frac{RC}{2\sqrt{LC}} = \frac{R}{2}\sqrt{\frac{C}{L}}$$

From this we see that larger values of R and C will increase the damping ratio, while a larger value of L will decrease the damping ratio. If $\zeta \leq 1$, the expression for the circuit time constant is

$$\tau = \frac{1}{\zeta\omega_n} = \frac{2L}{R}$$

which gives the interesting result that τ is independent of C as long as $\zeta \leq 1$.

8.2.8 THE DOMINANT-ROOT APPROXIMATION

If $\zeta > 1$, the two characteristic roots are real and distinct. Denote these roots by $s = -r_1$ and $s = -r_2$. In the response there will be the exponentials $e^{-r_1 t}$ and $e^{-r_2 t}$ associated with each of these roots. If both roots are negative, the system is stable and a time constant is associated with each root. Denote these time constants by τ_1 and τ_2, where $\tau_1 = 1/r_1$ and $\tau_2 = 1/r_2$.

A time constant τ is a measure of the decay rate of an exponential $e^{-t/\tau}$. Thus, the two exponentials in the response will decay at different rates because their time constants are different. The exponential having the *largest* time constant decays the slowest and thus dominates the response. Its corresponding root is thus called the *dominant* root, and its time constant is the *dominant* time constant. If the roots of a stable system are plotted on the complex plane, the dominant root is the one lying the farthest to the *right*.

Consider a second-order model whose roots are $s = -3, -15$. Its free response has the form

$$x(t) = A_1 e^{-3t} + A_2 e^{-15t}$$

The time constants are $\tau = 1/3$ and $\tau = 1/15$. Thus, the second exponential will disappear five times faster than the first exponential. The dominant-time constant is thus $\tau = 1/3$, and for $t > 4/3$ the response will essentially look like $A_1 e^{-3t}$.

We cannot make exact predictions based on the dominant root, because the initial conditions, which determine the values of A_1 and A_2, may be such that $A_2 \gg A_1$, so that the second exponential influences the response for longer than expected. The dominant-root approximation, however, is often used to obtain a quick estimate of response time. The farther away the dominant root is from the other roots (the "secondary" roots), the better the approximation.

EXAMPLE 8.2.2

Response of a Liquid-Level System

■ Problem

For the system shown in Figure 8.2.1d, $A = 50$ ft^2, $R = 60$ ft^{-1} sec^{-1}, and $g = 32.2$ ft/sec^2.

a. Determine the time constants.
b. Suppose the inflow rate is $q_v = u_s(t)$ ft^3/sec. Determine the steady-state liquid heights, and estimate how long it will take to reach those heights if both tanks are initially empty.

■ Solution

a. From Figure 8.2.1d, the equations are

$$3000\frac{dh_1}{dt} + 32.2(h_1 - h_2) = 60u_s(t) \qquad (1)$$

$$3000\frac{dh_2}{dt} + 32.2(h_2 - h_1) + 32.2h_2 = 0 \qquad (2)$$

Applying the Laplace transform method with zero-initial conditions and collecting terms give

$$(3000s + 32.2)H_1(s) - 32.2H_2(s) = \frac{60}{s}$$

$$-32.2H_1(s) + (3000s + 64.4)H_2(s) = 0$$

The determinant of the left side gives the characteristic equation.

$$\begin{vmatrix} (3000s + 32.2) & -32.2 \\ -32.2 & (3000s + 64.4) \end{vmatrix} = (3000s + 32.2)(3000s + 64.4) - (32.2)^2 = 0$$

or

$$9 \times 10^6 s^2 + 2.898 \times 10^5 s + 1036.84 = 0$$

The roots are $s = -0.0041$ and $s = -0.0281$, so the time constants are

$$\tau_1 = \frac{1}{0.0041} = 244 \text{ sec} \qquad \tau_2 = \frac{1}{0.0281} = 36 \text{ sec}$$

The dominant-time constant is $\tau_1 = 244$ sec.

b. Since the system is stable and the input is a step, the steady-state heights will be constant. Thus, we may set the derivatives equal to zero in equations (1) and (2). With $u_s(t) = 1$, this gives

$$32.2(h_1 - h_2) = 60$$

$$32.2(h_2 - h_1) + 32.2h_2 = 0$$

and the solution for the steady-state heights is $h_1 = 3.73$ ft and $h_2 = 1.86$ ft. It will take approximately $4\tau = 976$ sec to reach the steady-state heights.

8.2.9 GRAPHICAL INTERPRETATION

The preceding relations can be represented graphically by plotting the location of the roots (8.2.15) in the complex plane (Figure 8.2.6). The parameters ζ, ω_n, ω_d, and τ are normally used to describe stable systems only, and so we will assume for now that all the roots lie to the left of the imaginary axis (in the "left half-plane").

The lengths of two sides of the right triangle shown in Figure 8.2.6 are $\zeta\omega_n$ and $\omega_n\sqrt{1-\zeta^2}$. Thus, the hypotenuse is of length ω_n. It makes an angle θ with the *negative* real axis, and

$$\cos\theta = \zeta \tag{8.2.18}$$

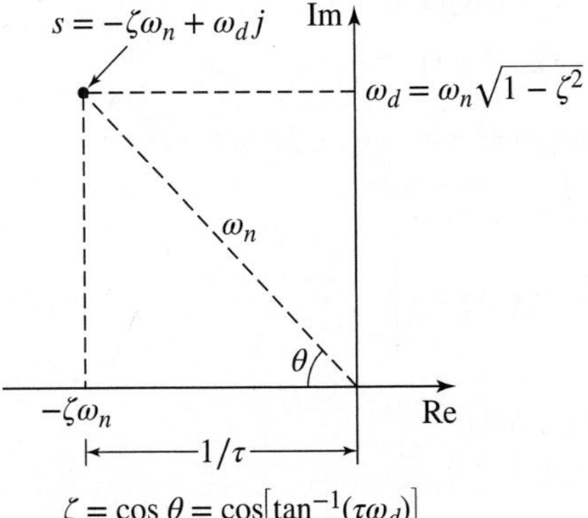

Figure 8.2.6 Graphical interpretation of the parameters ζ, τ, ω_n, and ω_d.

Therefore, all roots lying on the circumference of a given circle centered on the origin are associated with the same undamped natural frequency ω_n. From (8.2.18) we see that all roots lying on the same line passing through the origin are associated with the same damping ratio. The limiting values of ζ correspond to the imaginary axis ($\zeta = 0$) and the negative real axis ($\zeta = 1$). Roots lying on a given line parallel to the real axis all give the same damped natural frequency. All roots lying on a line parallel to the imaginary axis have the same time constant. The farther to the left this line is, the smaller the time constant.

8.3 DESCRIPTION AND SPECIFICATION OF STEP RESPONSE

We can express the free response and the step response of the underdamped second-order model $m\ddot{x} + c\dot{x} + kx = f(t)$ in terms of the parameters ζ and ω_n as follows. The form of the free response is

$$x(t) = Be^{-\zeta\omega_n t}\sin(\omega_d t + \phi) \tag{8.3.1}$$

where B and ϕ depend on the initial conditions $x(0)$ and $\dot{x}(0)$. The unit-step response for zero-initial conditions is

$$x(t) = \frac{1}{k}\left[\frac{1}{\sqrt{1-\zeta^2}}e^{-\zeta\omega_n t}\sin\left(\omega_n\sqrt{1-\zeta^2}t + \phi\right) + 1\right] \tag{8.3.2}$$

where

$$\phi = \tan^{-1}\left(\frac{\sqrt{1-\zeta^2}}{\zeta}\right) + \pi \tag{8.3.3}$$

Because $\zeta > 0$, ϕ lies in the third quadrant. Table 8.3.1 gives the step response in terms of ζ and ω_n for the underdamped, critically damped, and overdamped cases.

Table 8.3.1 Unit-step response of a stable second-order model.

Model: $m\ddot{x} + c\dot{x} + kx = u_s(t)$

Initial conditions: $x(0) = \dot{x}(0) = 0$

Characteristic roots: $s = \dfrac{-c \pm \sqrt{c^2 - 4mk}}{2m} = -r_1, -r_2$

1. Overdamped case ($\zeta > 1$): distinct, real roots: $r_1 \neq r_2$

$$x(t) = A_1e^{-r_1 t} + A_2e^{-r_2 t} + \frac{1}{k} = \frac{1}{k}\left(\frac{r_2}{r_1 - r_2}e^{-r_1 t} - \frac{r_1}{r_1 - r_2}e^{-r_2 t} + 1\right)$$

2. Critically damped case ($\zeta = 1$): repeated, real roots: $r_1 = r_2$

$$x(t) = (A_1 + A_2 t)e^{-r_1 t} + \frac{1}{k} = \frac{1}{k}[(-r_1 t - 1)e^{-r_1 t} + 1]$$

3. Underdamped case ($0 \leq \zeta < 1$): complex roots: $s = -\zeta\omega_n \pm j\omega_n\sqrt{1-\zeta^2}$

$$x(t) = Be^{-t/\tau}\sin\left(\omega_n\sqrt{1-\zeta^2}t + \phi\right) + \frac{1}{k}$$

$$= \frac{1}{k}\left[\frac{1}{\sqrt{1-\zeta^2}}e^{-\zeta\omega_n t}\sin\left(\omega_n\sqrt{1-\zeta^2}t + \phi\right) + 1\right]$$

$$\phi = \tan^{-1}\left(\frac{\sqrt{1-\zeta^2}}{\zeta}\right) + \pi \quad \text{(third quadrant)}$$

Time constant: $\tau = 1/\zeta\omega_n$

The free response for $\dot{x}(0) = 0$ and the step response for $x(0) = \dot{x}(0) = 0$ of the second-order model $m\ddot{x} + c\dot{x} + kx = f(t)$ are illustrated in Figures 8.3.1 and 8.3.2 for several values of the damping ratio ζ. Note that the response axis has been normalized by k and the time axis has been normalized by $\omega_n = \sqrt{k/m}$. Thus, a variation of ζ should be interpreted as a variation of c, and not of k or m. When $\zeta > 1$, the response is sluggish and does not overshoot the steady-state value. As ζ is decreased, the speed of response increases. The critically damped case $\zeta = 1$ is the case in which the steady-state value is reached most quickly but without oscillation. As ζ is decreased below 1, the response overshoots and oscillates about the final value. The smaller ζ is, the larger

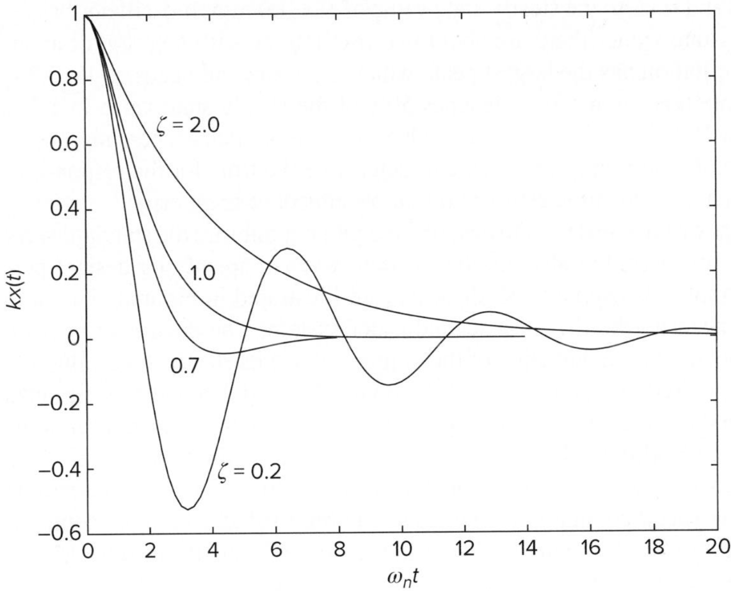

Figure 8.3.1 Free response of second-order systems for various values of ζ.

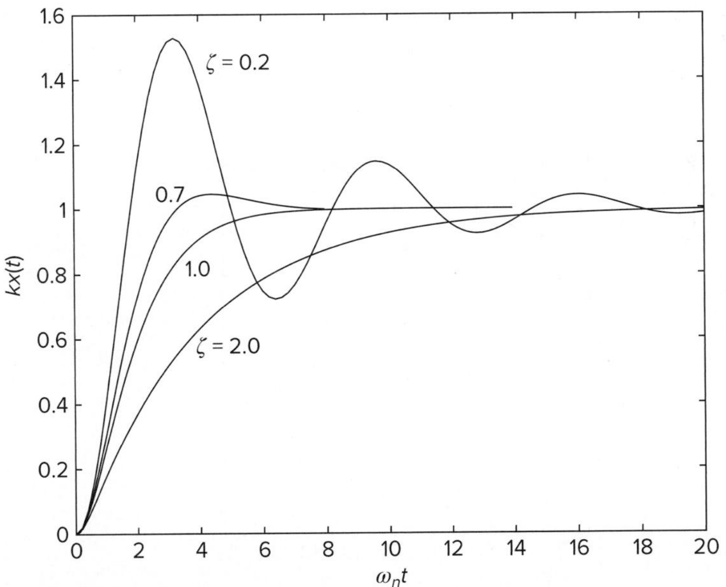

Figure 8.3.2 Step response of second-order systems for various values of ζ.

the overshoot, and the longer it takes for the oscillations to die out. There are design applications in which we wish the response to be near its final value as quickly as possible, with some oscillation tolerated. This corresponds to a value of ζ slightly less than 1. The value $\zeta = 0.707$ is a common choice for such applications. As ζ is decreased to zero (the undamped case), the oscillations never die out.

8.3.1 GRAPHICAL DESCRIPTION OF STEP RESPONSE

Suppose you obtained the response plot shown in Figure 8.3.3 either from a measured response or from a computer simulation. Suppose also that you want to describe the plot to someone over the phone (assuming you cannot send the plot!). You would say that $x(t)$ starts at 0 and rises to the steady-state value of $x = 100$ mm. It oscillates briefly around the steady-state value (there are about two oscillations with a period of about 6.6 s). The first oscillation has the largest peak, which is 37 mm and occurs at $t = 3.3$ s. You might also note that the response reaches 50% of the steady-state value in 1.2 s, and it first reaches 100% of the steady-state value in 2 s. Note that, depending on the resolution of the plot, you might not be able to determine the time for the response to reach the steady state, so this time estimate might be subject to great error.

The parameters we have just used to describe the plot not only are the standard ways of describing step response, but also are the standard ways of specifying desired performance. These transient-response specifications are illustrated in Figure 8.3.4. Note that the response need not be that of a second-order system. The *maximum or peak overshoot* M_p is the maximum deviation of the output x above its steady-state value x_{ss}. It is sometimes expressed as a percentage of the final value and denoted $M_{\%}$. Because the maximum overshoot increases with decreasing ζ, it is sometimes used as an indicator of the relative stability of the system. The *peak time* t_p is the time at which the maximum overshoot occurs. The *settling time* t_s is the time required for the oscillations to stay within some specified small percentage of the final value. The most common values used are 2% and 5%. If the final value of the response differs from some desired value, a steady-state error exists.

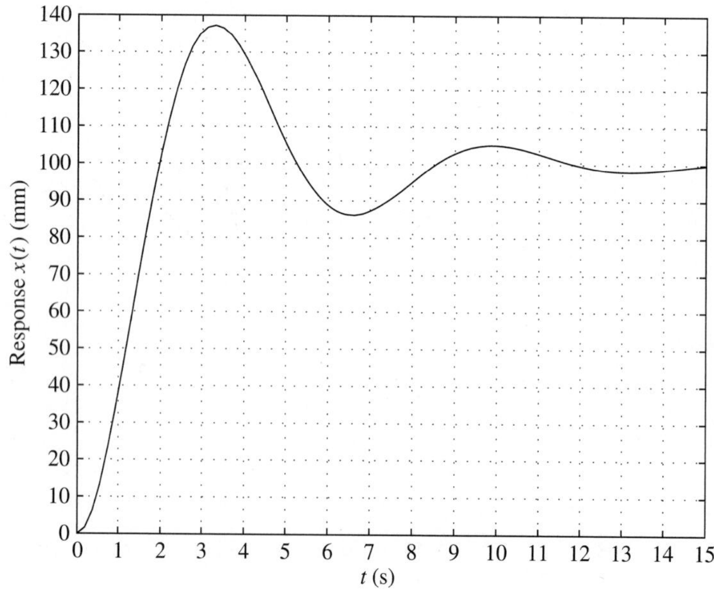

Figure 8.3.3 Underdamped step response.

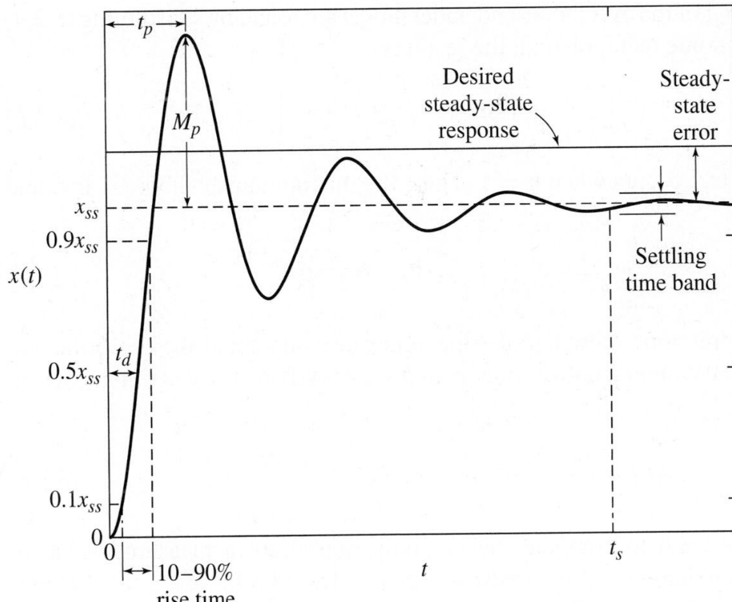

Figure 8.3.4 Transient-response specifications.

The *rise time* t_r can be defined as the time required for the output to rise from 0% to 100% of its final value. However, no agreement exists on this definition. Sometimes, the rise time is taken to be the time required for the response to rise from 10% to 90% of the final value. Finally, the *delay time* t_d is the time required for the response to reach 50% of its final value.

Except for the settling time, these parameters are relatively easy to obtain from an experimentally determined step-response plot. They can also be determined in analytical form for a second-order model, whose underdamped step response is given by (8.3.2) and (8.3.3).

8.3.2 MAXIMUM OVERSHOOT

Setting the derivative of (8.3.2) equal to zero gives expressions for both the maximum overshoot and the peak time t_p. After some trigonometric manipulation, the result is

$$\frac{dx}{dt} = \frac{1}{k}\left(\frac{\omega_n}{\sqrt{1-\zeta^2}}e^{-\zeta\omega_n t}\sin\omega_n\sqrt{1-\zeta^2}t\right) = 0$$

For $t < \infty$, this gives

$$\omega_n\sqrt{1-\zeta^2}t = n\pi \quad n = 0, 1, 2, \dots$$

The times at which extreme values of the oscillations occur are thus

$$t = \frac{n\pi}{\omega_n\sqrt{1-\zeta^2}} \tag{8.3.4}$$

The odd values of n give the times of overshoots, and the even values correspond to the times of undershoots. The maximum overshoot occurs when $n = 1$. Thus,

$$t_p = \frac{\pi}{\omega_n\sqrt{1-\zeta^2}} \tag{8.3.5}$$

The magnitudes $|x_n|$ of the overshoots and undershoots are found by substituting (8.3.4) into (8.3.2). After some manipulation, the result is

$$|x_n| = \frac{1}{k}\left[1 + (-1)^{n-1} e^{-n\pi\zeta/\sqrt{1-\zeta^2}}\right] \tag{8.3.6}$$

The largest value $|x_n|$ occurs when $n = 1$. Thus, the maximum overshoot M_p is found when $n = 1$. It is

$$M_p = x_{\max} - x_{ss} = \frac{1}{k} e^{-\pi\zeta/\sqrt{1-\zeta^2}} \tag{8.3.7}$$

The preceding expressions show that the maximum overshoot and the peak time are functions of only the damping ratio ζ for a second-order system. The *percent* maximum overshoot $M_{\%}$ is

$$M_{\%} = \frac{x_{\max} - x_{ss}}{x_{ss}} 100 = 100 e^{-\pi\zeta/\sqrt{1-\zeta^2}} \tag{8.3.8}$$

Frequently we need to compute the damping ratio from a measured value of the maximum overshoot. In this case, we can solve (8.3.8) for ζ as follows. Let $R = \ln 100/M_{\%}$. Then (8.3.8) gives

$$\zeta = \frac{R}{\sqrt{\pi^2 + R^2}} \qquad R = \ln \frac{100}{M_{\%}} \tag{8.3.9}$$

8.3.3 RISE TIME

To obtain the expression for the 100% rise time t_r, set $x = x_{ss} = 1/k$ in (8.3.2) to obtain

$$e^{-\zeta\omega_n t} \sin\left(\omega_n\sqrt{1-\zeta^2}\, t + \phi\right) = 0$$

This implies that for $t < \infty$,

$$\omega_n\sqrt{1-\zeta^2}\, t + \phi = n\pi \qquad n = 0, 1, 2, \ldots \tag{8.3.10}$$

For $t_r > 0$, $n = 2$, because ϕ is in the third quadrant. Thus,

$$t_r = \frac{2\pi - \phi}{\omega_n\sqrt{1-\zeta^2}} \tag{8.3.11}$$

where ϕ is given by (8.3.3). The rise time is inversely proportional to the natural frequency ω_n for a given value of ζ.

No closed form expression exists for the 10%–90% rise time.

8.3.4 SETTLING TIME

To express the settling time in terms of the parameters ζ and ω_n, we can use the fact that the exponential term in the solution (8.3.2) provides the envelopes of the oscillations. The time constant of these envelopes is $1/\zeta\omega_n$, and thus the 2% settling time t_s is

$$t_s = \frac{4}{\zeta\omega_n} \tag{8.3.12}$$

Table 8.3.2 Step-response specifications for the underdamped model $m\ddot{x} + c\dot{x} + kx = f$.

Maximum percent overshoot	$M_\% = 100e^{-\pi\zeta/\sqrt{1-\zeta^2}}$
	$\zeta = \dfrac{R}{\sqrt{\pi^2 + R^2}}, \quad R = \ln \dfrac{100}{M_\%}$
Peak time	$t_p = \dfrac{\pi}{\omega_n\sqrt{1-\zeta^2}}$
Delay time	$t_d \approx \dfrac{1+0.7\zeta}{\omega_n}$
100% rise time	$t_r = \dfrac{2\pi - \phi}{\omega_n\sqrt{1-\zeta^2}}$
	$\phi = \tan^{-1}\left(\dfrac{\sqrt{1-\zeta^2}}{\zeta}\right) + \pi$

8.3.5 DELAY TIME

An exact analytical expression for the delay time is difficult to obtain, but we can obtain an approximate expression as follows. Set $x = 0.5x_{ss} = 0.5/k$ in (8.3.2) to obtain

$$e^{-\zeta\omega_n t} \sin\left(\omega_n\sqrt{1-\zeta^2}t + \phi\right) = -0.5\sqrt{1-\zeta^2} \tag{8.3.13}$$

where ϕ is given by (8.3.3). For a given ζ and ω_n, t_d can be obtained by a numerical procedure, using the following straight-line approximation as a starting guess:

$$t_d \approx \frac{1+0.7\zeta}{\omega_n} \tag{8.3.14}$$

Table 8.3.2 summarizes these formulas. Figure 8.3.5 shows the plots of the maximum percent overshoot, the peak time, and the 100% rise time as functions of ζ. In

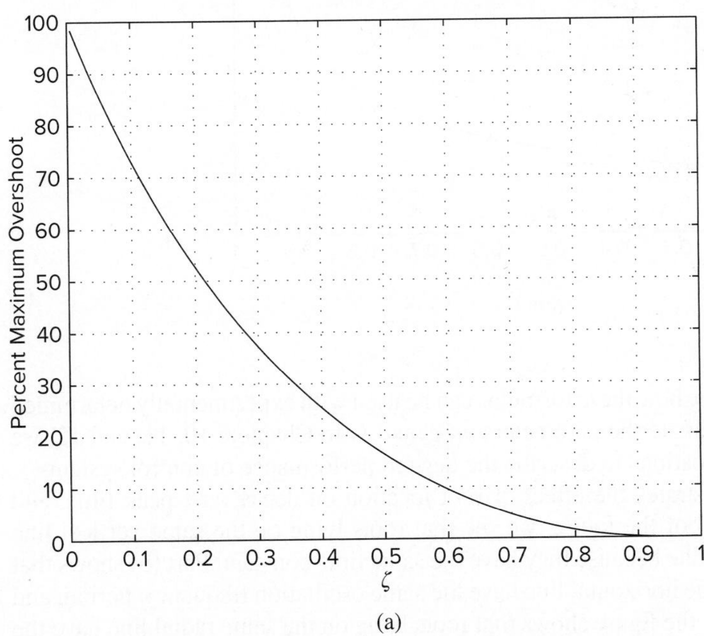

Figure 8.3.5 Percent maximum overshoot, peak time, and rise time as functions of the damping ratio ζ. *(Continued)*

Figure 8.3.5 *(Continued)*

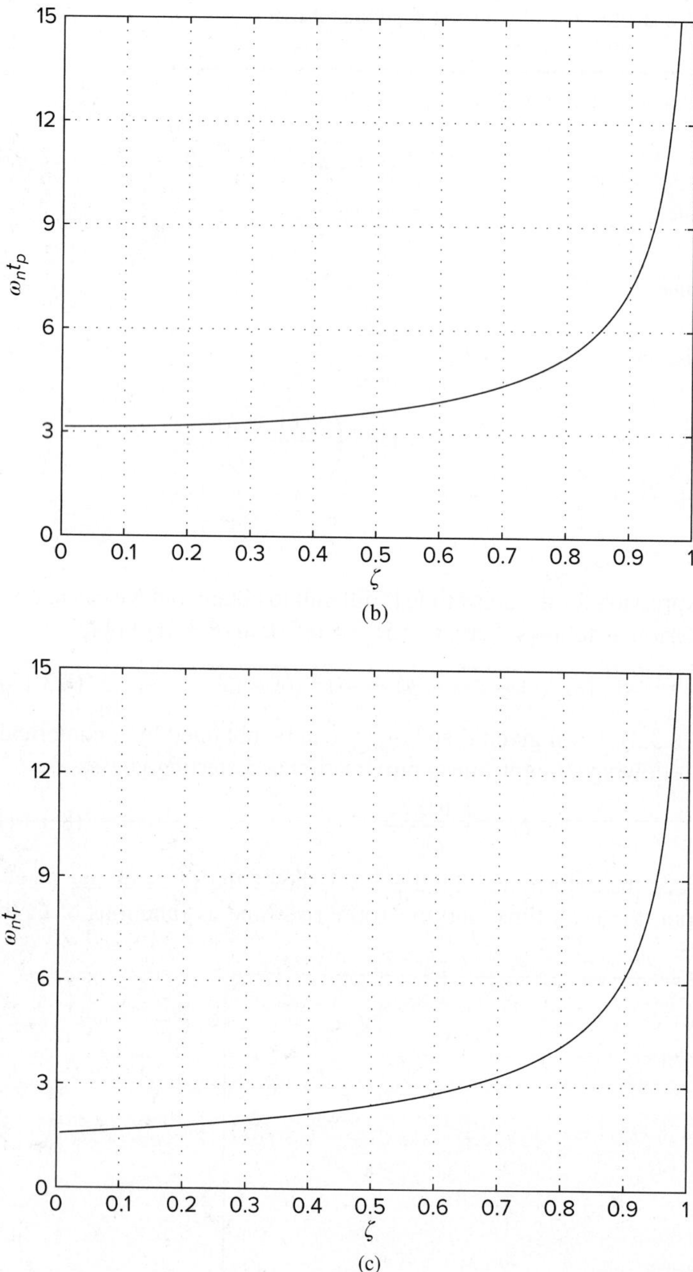

Section 8.5 we will see how these formulas can be used with experimentally determined response plots to estimate the parameters m, c, and k. In Chapters 10, 11, and 12, we will use these specifications to describe the desired performance of control systems.

Figure 8.3.6 illustrates the effect of root location on decay rate, peak time, and overshoot. In part (a) of the figure, we see that roots lying on the same vertical line have the same decay rate because they have the same time constant. Part (b) shows that roots lying on the same horizontal line have the same oscillation frequency, period, and peak time. Part (c) of the figure shows that roots lying on the same radial line have the

Models A and B have the same real part, the same time constant, and the same decay time.

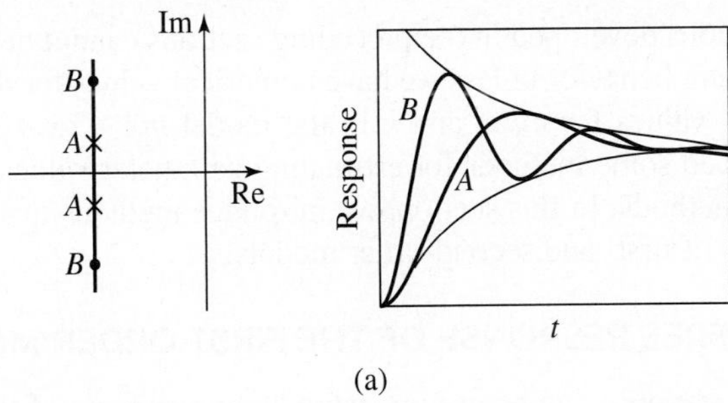

(a)

Models A and B have the same imaginary part, the same period, and the same peak time.

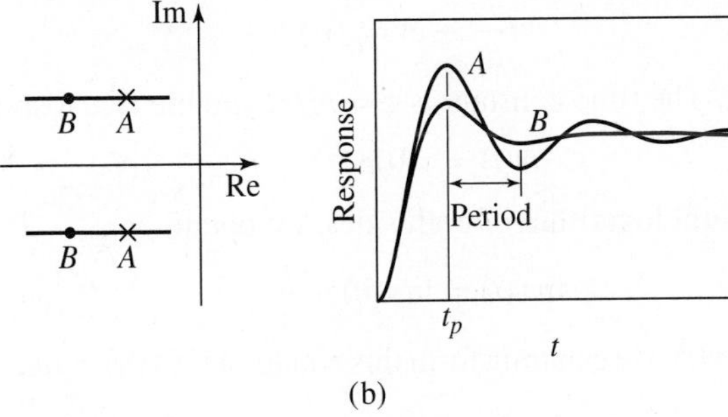

(b)

Models A, B, and C have the same damping ratio and the same overshoot.

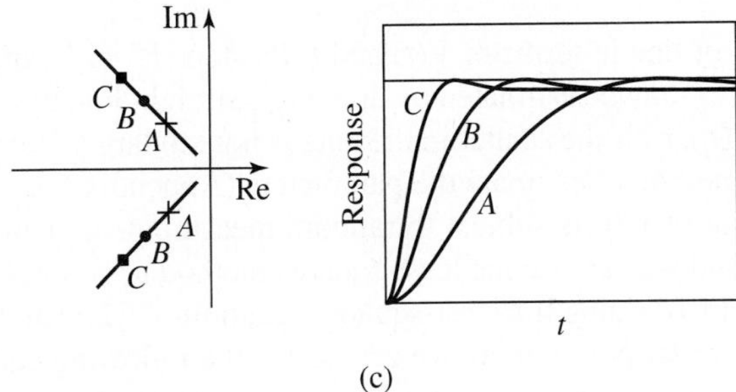

(c)

Figure 8.3.6 Effect of root location on decay rate, peak time, and overshoot.

same damping ratio and therefore the same maximum percent overshoot $M_{\%}$. You can see this from the $M_{\%}$ formula, (8.3.8), which is a function of ζ only. Roots lying on the same horizontal line have the same damped frequency ω_d, and therefore, have the same peak time t_p. This is true because $t_p = \pi/\omega_n\sqrt{1-\zeta^2}$, $\omega_d = \omega_n\sqrt{1-\zeta^2}$, and therefore, $t_p = \pi/\omega_d$. Thus, if ω_d is constant, so is t_p.

8.3.6 NUMERATOR DYNAMICS AND SECOND-ORDER SYSTEM RESPONSE

Be aware that the preceding analysis is based on the model $m\ddot{x} + c\dot{x} + kx = f$, which does not have numerator dynamics. The presence of numerator dynamics can change the system behavior quite a bit, decreasing the response time and producing overshoots or undershoots in the response. Thus, the response of the model $m\ddot{x} + c\dot{x} + kx = f + b\dot{f}$ can be quite different. In particular, the rise time and maximum overshoot will be different from what is predicted by our formulas.

8.4 PARAMETER ESTIMATION IN THE TIME DOMAIN

The response calculations developed in the preceding sections cannot be used to make predictions about system behavior unless we have numerical values for the parameters in the model, such as values for m, c, and k in the model $m\ddot{x} + c\dot{x} + kx = f(t)$. In Chapter 1 we introduced some methods for estimating parameter values. Appendix C has some additional methods. In this section, we introduce methods that are based on the response solutions of first- and second-order models.

8.4.1 USING THE FREE RESPONSE OF THE FIRST-ORDER MODEL

The free and the step response can be used to estimate one or more of the parameters of a dynamic model. For example, consider the first-order model

$$m\frac{dv}{dt} + cv = f(t) \tag{8.4.1}$$

where $f(t)$ is the input. The time constant is $\tau = m/c$, and the free response is

$$v(t) = v(0)e^{-t/\tau} \tag{8.4.2}$$

If we take the natural logarithm of both sides, we obtain

$$\ln v(t) = \ln v(0) - \frac{t}{\tau}$$

By defining $V(t) = \ln\, v(t)$, we can transform this equation into the equation of a straight line, as follows:

$$V(t) = V(0) - \frac{t}{\tau} \tag{8.4.3}$$

This describes a straight line in terms of $V(t)$ and t. Its slope is $-1/\tau$ and its intercept is $V(0)$. These quantities may be estimated by drawing a straight line through the transformed data points $[V(t_i), t_i]$ if the scatter in the data is not too large. Otherwise we can use the least-squares method to estimate the parameters (Appendix C).

If the measurement of $v(0)$ is subject to random measurement error, then $V(0)$ is not known precisely, and we can use the least-squares method to compute estimates of the coefficients τ and $V(0)$. Using the least-squares equations (C.2.1) and (C.2.2) from Appendix C for a first-order polynomial, we can derive the following equations:

$$-\frac{1}{\tau}\sum_{i=1}^{n} t_i^2 + V(0)\sum_{i=1}^{n} t_i = \sum_{i=1}^{n} V_i t_i \tag{8.4.4}$$

$$-\frac{1}{\tau}\sum_{i=1}^{n} t_i + V(0)n = \sum_{i=1}^{n} V_i \tag{8.4.5}$$

These are two linear algebraic equations, which can be solved for τ and $V(0)$.

On the other hand, in many applications the starting value $v(0)$ can be measured without significant error. In this case, we can transform the data by using $z(t) = V(t) - V(0)$ so that the model (8.4.3) becomes $z(t) = -t/\tau$, which is a linear equation constrained to pass through the origin. We can then use (C.1.3) from Appendix C, expressed here as

$$-\frac{1}{\tau}\sum_{i=1}^{n} t_i^2 = \sum_{i=1}^{n} t_i z_i \tag{8.4.6}$$

to find the time constant τ.

EXAMPLE 8.4.1

Estimating Capacitance from the Free Response

■ Problem

Commercially available resistors are marked with a color code that indicates their resistance value. Suppose that the resistance in the circuit of Figure 8.4.1a is $10^5\ \Omega$. A voltage is applied to the circuit for $t < 0$ and then is suddenly removed at time $t = 0$. The voltage across the capacitor as measured by a data acquisition system is plotted in part (b) of the figure and is given in the following table. Use the data to estimate the value of the capacitance C.

Time t (s)	Voltage v_C (V)
0	5
0.25	3.3
0.5	2.2
0.75	1.4
1	0.9
1.25	0.6
1.5	0.4
1.75	0.3
2	0.2

■ Solution

The circuit model may be derived from Kirchhoff's voltage law, which gives

$$v_s = Ri + v_C \qquad \text{or} \qquad i = \frac{v_s - v_C}{R} \tag{1}$$

For the capacitor, we have

$$v_C = \frac{1}{C}\int i\,dt \qquad \text{or} \qquad \frac{dv_C}{dt} = \frac{i}{C}$$

Substituting from equation (1) gives the circuit model.

$$\frac{dv_C}{dt} = \frac{v_s - v_C}{RC} \qquad \text{or} \qquad RC\frac{dv_C}{dt} + v_C = v_s$$

The free response has the form

$$v_C(t) = v_C(0)e^{-t/RC} = v_C(0)e^{-t/\tau}$$

where the time constant is $\tau = RC$. Taking the natural logarithm of both sides gives

$$\ln v_C(t) = \ln v_C(0) - \frac{t}{\tau} \tag{2}$$

When the logarithmic transformation is applied to the original data, we obtain the following table.

Time t (s)	$\ln v_C$
0	1.609
0.25	1.194
0.5	0.789
0.75	0.337
1	−0.105
1.25	−0.511
1.5	−0.916
1.75	−1.204
2	−1.609

Figure 8.4.1 An *RC* circuit.

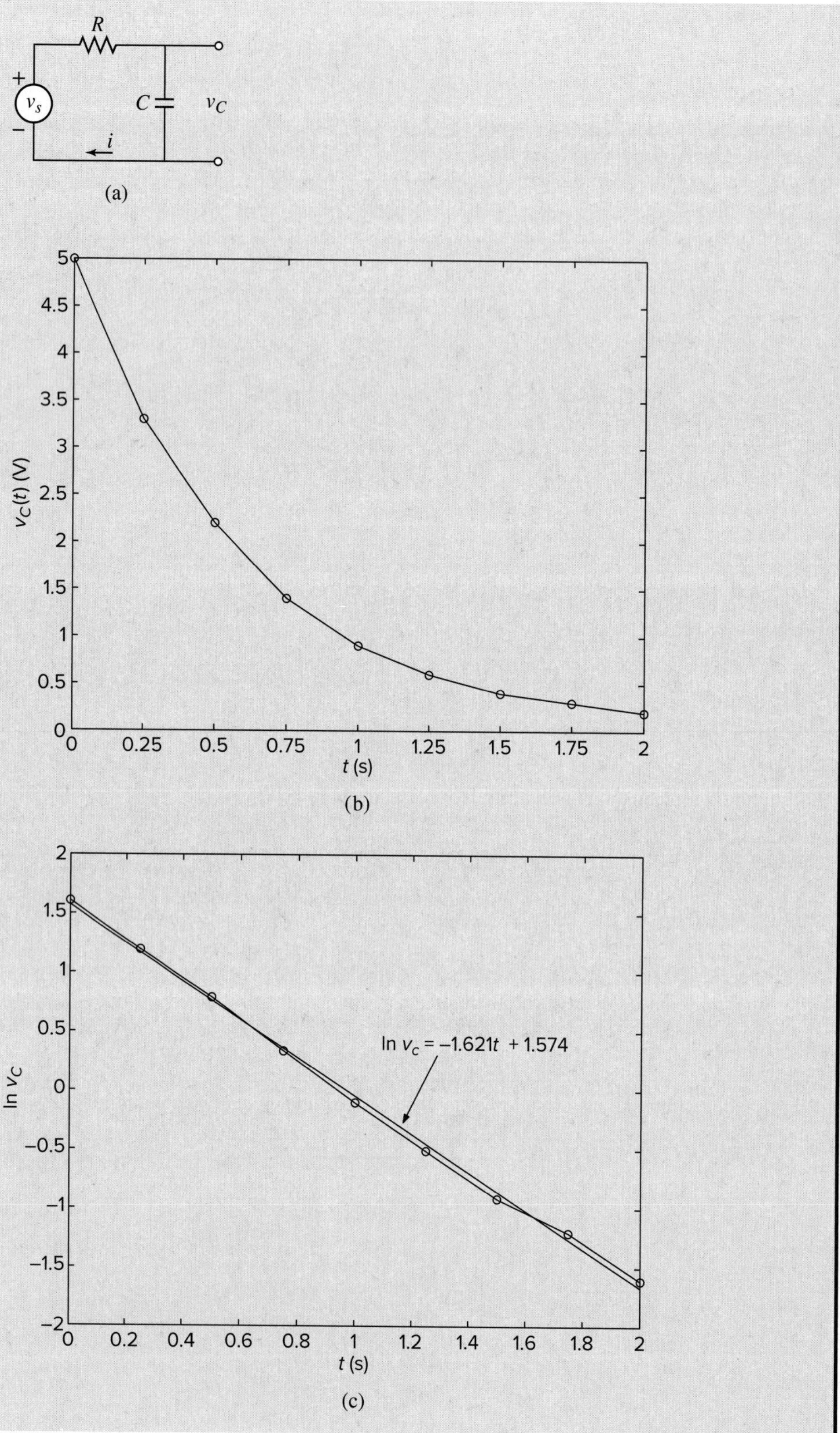

These transformed data are plotted in Figure 8.4.1c. Note that the data lie close to a straight line, which is given by $\ln v_C = -1.621t + 1.574$. This line was found with the least-squares method, but a similar line could have been obtained by using a straightedge to draw a line through the data. The least-squares method is required when there is considerable scatter in the data.

Comparing the equation for the line with equation (2), we obtain $\ln v_C(0) = 1.5737$, which gives $v_C(0) = 4.825$, and $1/\tau = 1.6206$, which gives $\tau = 0.617$. Because we know that $R = 10^5\ \Omega$, we obtain $C = \tau/R = 0.617/10^5 = 6.17 \times 10^{-6}$ F.

Temperature Dynamics

EXAMPLE 8.4.2

■ Problem

The temperature of liquid cooling in a porcelain mug at room temperature (68°F) was measured at various times. The data are given below.

Time t (sec)	Temperature T (°F)
0	145
620	130
2266	103
3482	90

Develop a model of the liquid temperature as a function of time, and use it to estimate how long it takes the temperature to reach 120°F.

■ Solution

We will model the liquid as a single lumped thermal mass with a representative temperature T. From conservation of heat energy, we have

$$\frac{dE}{dt} = -\frac{T - T_o}{R}$$

where E is the heat energy in the liquid, $T_o = 68$°F is the temperature of the air surrounding the cup, and R is the total thermal resistance of the cup. We have $E = mc_p(T - T_o) = C(T - T_o)$, where m is the liquid mass, c_p is its specific heat, and $C = mc_p$ is the thermal capacitance. Assuming that m, c_p, and T_o are constant, we obtain

$$C\frac{dT}{dt} = -\frac{T - T_o}{R}$$

If we let $\Delta T = T - T_o$ and note that

$$\frac{d(\Delta T)}{dt} = \frac{dT}{dt}$$

we obtain

$$RC\frac{d(\Delta T)}{dt} + \Delta T = 0 \qquad (1)$$

The time constant is $\tau = RC$, and the solution has the form

$$\Delta T(t) = \Delta T(0)e^{-t/\tau}$$

Thus,

$$\ln \Delta T(t) = \ln \Delta T(0) - \frac{t}{\tau} \qquad (2)$$

The transformed data $\ln \Delta T(t)$ are plotted in Figure 8.4.2a. Because they fall near a straight line, we can use equation (2) to fit the data. The values obtained are $\ln \Delta T(0) = 4.35$ and

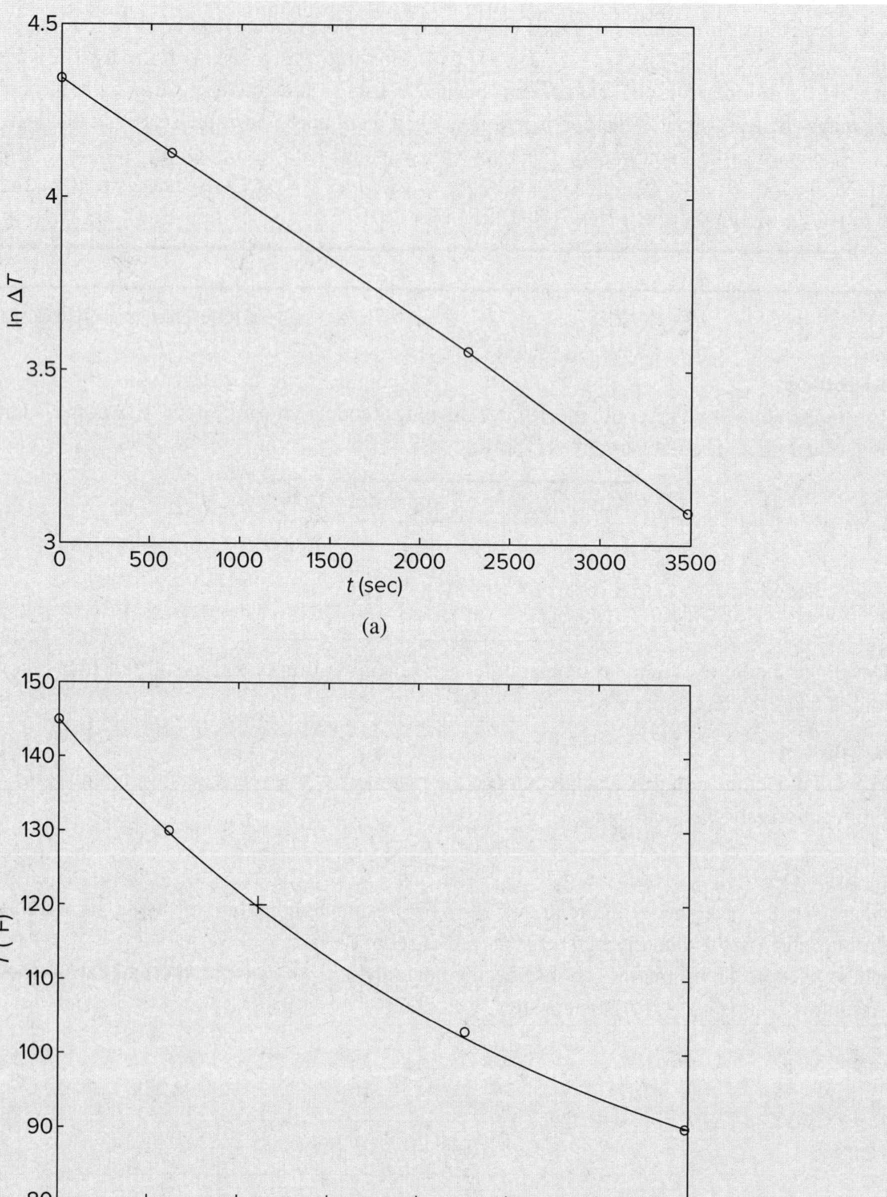

Figure 8.4.2 Temperature data. (a) Plot of transformed data. (b) Plot of the fitted equation.

$\tau = 2792$ sec. This gives $\Delta T(0) = 77°\text{F}$. Thus, the model is

$$T(t) = 68 + 77e^{-t/2792} \tag{3}$$

The computed time to reach 120°F is

$$t = -2792 \ln \frac{120 - 68}{77} = 1112 \text{ sec}$$

The plot of equation (3), along with the data and the estimated point (1112, 120) marked with a "+" sign, is shown in part (b) of Figure 8.4.2. Because the graph of our model lies near the data points, we can treat its prediction of 1112 sec with some confidence.

8.4.2 USING THE STEP RESPONSE OF THE FIRST-ORDER MODEL

The free response of the model $m\dot{v} + cv = f(t)$ enables us to estimate τ, but does not give enough information to find both m and c separately. However, we may use the step response if available. The step response of (8.5.1) for a step input of magnitude F is

$$v(t) = \left[v(0) - \frac{F}{c}\right] e^{-t/\tau} + \frac{F}{c} \tag{8.4.7}$$

Assume that we know F and $v(0)$ accurately and that we can measure the step response long enough to estimate accurately the steady-state response v_{ss}. Then we can compute c from the steady-state response $v_{ss} = F/c$; that is, $c = F/v_{ss}$. To estimate m, we rearrange the step response (8.4.7) as follows, using the fact that $F/c = v_{ss}$.

$$\ln\left[\frac{v(t) - v_{ss}}{v(0) - v_{ss}}\right] = -\frac{t}{\tau}$$

Transform the data $v(t)$ using

$$z(t) = \ln\left[\frac{v(t) - v_{ss}}{v(0) - v_{ss}}\right] \tag{8.4.8}$$

This gives the zero-intercept, linear model: $z(t) = -t/\tau$, and we can use (8.4.6) to find τ. Assuming we have calculated c from the steady-state response, we can find m from $m = c\tau$.

8.4.3 PARAMETER ESTIMATION FOR THE SECOND-ORDER MODEL

Depending on the available experimental method, we can use either the free or the step-response formulas to estimate the parameters of a second-order model.

Estimating Mass, Stiffness, and Damping from the Step Response

EXAMPLE 8.4.3

■ Problem

Figure 8.4.3 shows the response of a system to a step input of magnitude 6×10^3 N. The equation of motion is

$$m\ddot{x} + c\dot{x} + kx = f(t)$$

Estimate the values of m, c, and k.

■ Solution

From the graph we see that the steady-state response is $x_{ss} = 6$ cm. At steady state, $x_{ss} = f_{ss}/k$, and thus $k = 6 \times 10^3/6 \times 10^{-2} = 10^5$ N/m.

The peak value from the plot is $x = 8.1$ cm, so the maximum percent overshoot is $M_{\%} = [(8.1 - 6)/6]100 = 35\%$. From Table 8.3.2, we can compute the damping ratio as follows:

$$R = \ln\frac{100}{35} = 1.0498 \qquad \zeta = \frac{R}{\sqrt{\pi^2 + R^2}} = 0.32$$

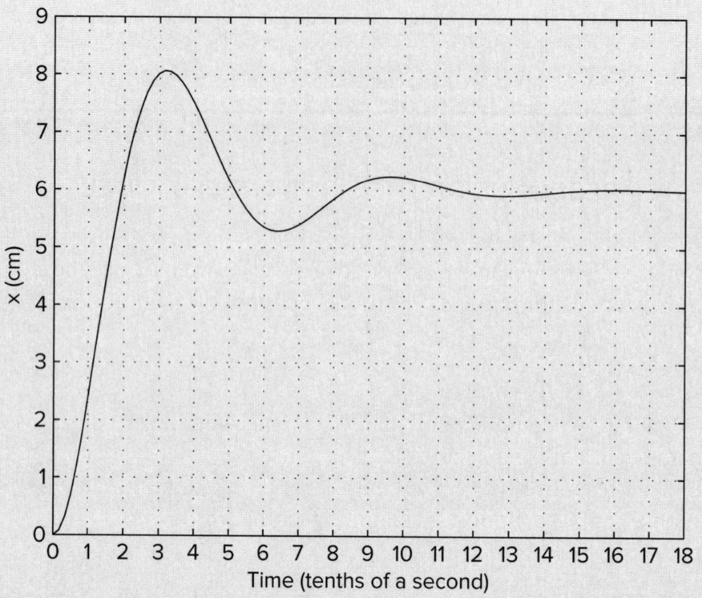

Figure 8.4.3 Measured step response.

The peak occurs at $t_p = 0.32$ s. From Table 8.3.2,

$$t_p = 0.32 = \frac{\pi}{\omega_n\sqrt{1-\zeta^2}} = \frac{3.316}{\omega_n}$$

Thus, $\omega_n^2 = 107$ and

$$m = \frac{k}{\omega_n^2} = \frac{10^5}{107} = 930 \text{ kg}$$

From the expression for the damping ratio,

$$\zeta = \frac{c}{2\sqrt{mk}} = \frac{c}{2\sqrt{930(10^5)}} = 0.32$$

Thus, $c = 6170$ N · s/m, and the model is

$$930\ddot{x} + 6170\dot{x} + 10^5 x = f(t)$$

8.4.4 THE LOGARITHMIC DECREMENT

Usually the damping coefficient c is the parameter most difficult to estimate. Mass m and stiffness k can be measured with static tests, but measuring damping requires a dynamic test. If the system exists and dynamic testing can be done with it, then the *logarithmic decrement* provides a good way to estimate the damping ratio ζ, from which we can compute c ($c = 2\zeta\sqrt{mk}$). To see how this is done, use the form $s = -\zeta\omega_n \pm \omega_d j$ for the characteristic roots, and write the free response for the underdamped case as follows:

$$x(t) = Be^{-\zeta\omega_n t}\sin(\omega_d t + \phi) \tag{8.4.9}$$

The frequency of the oscillation is ω_d, and thus the period P is $P = 2\pi/\omega_d$. The logarithmic decrement δ is defined as the natural logarithm of the ratio of two successive amplitudes; that is,

$$\delta = \ln \frac{x(t)}{x(t+P)} \tag{8.4.10}$$

Using (8.4.9) this becomes

$$\delta = \ln \frac{Be^{-\zeta\omega_n t}\sin(\omega_d t + \phi)}{Be^{-\zeta\omega_n (t+P)}\sin(\omega_d t + \omega_d P + \phi)} \tag{8.4.11}$$

Note that $e^{-\zeta\omega_n(t+P)} = e^{-\zeta\omega_n t}e^{-\zeta\omega_n P}$. In addition, because $\omega_d P = 2\pi$ and $\sin(\theta + 2\pi) = \sin\theta$, $\sin(\omega_d t + \omega_d P + \phi) = \sin(\omega_d t + \phi)$, and (8.4.11) becomes

$$\delta = \ln e^{\zeta\omega_n P} = \zeta\omega_n P$$

Because $P = 2\pi/\omega_d = 2\pi/\omega_n\sqrt{1-\zeta^2}$, we have

$$\delta = \frac{2\pi\zeta}{\sqrt{1-\zeta^2}} \tag{8.4.12}$$

We can solve this for ζ to obtain

$$\zeta = \frac{\delta}{\sqrt{4\pi^2 + \delta^2}} \tag{8.4.13}$$

If we have a plot of $x(t)$ from a test, we can measure two values x at two times t and $t + P$. These values can be measured at two successive peaks in x. The x values are then substituted into (8.4.10) to compute δ. Equation (8.4.13) gives the value of ζ, from which we compute $c = 2\zeta\sqrt{mk}$.

The plot of $x(t)$ will contain some measurement error, and for this reason, the preceding method is usually modified to use measurements of two peaks n cycles apart (Figure 8.4.4). Let the peak values be denoted B_1, B_2, etc. and note that

$$\ln\left(\frac{B_1}{B_2}\frac{B_2}{B_3}\frac{B_3}{B_4}\cdots\frac{B_n}{B_{n+1}}\right) = \ln\left(\frac{B_1}{B_{n+1}}\right)$$

or

$$\ln\frac{B_1}{B_2} + \ln\frac{B_2}{B_3} + \ln\frac{B_3}{B_4}\cdots\ln\frac{B_n}{B_{n+1}} = \ln\frac{B_1}{B_{n+1}}$$

Thus,

$$\delta + \delta + \delta + \cdots + \delta = n\delta = \ln\frac{B_1}{B_{n+1}}$$

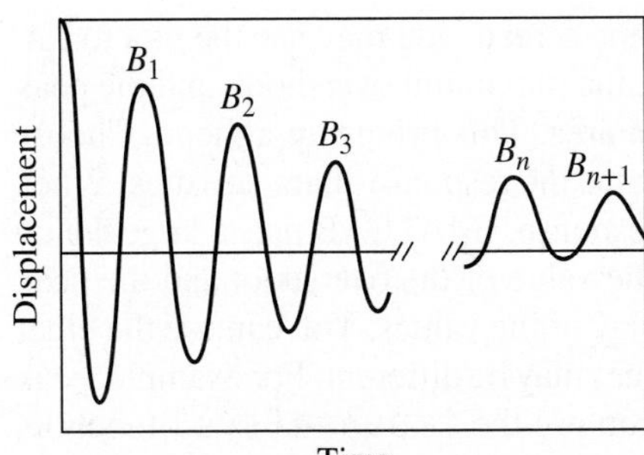

Figure 8.4.4 Terminology for logarithmic decrement.

or

$$\delta = \frac{1}{n} \ln \frac{B_1}{B_{n+1}} \tag{8.4.14}$$

We normally take the first peak to be B_1, because this is the highest peak and least subject to measurement error, but this is not required. The above formula applies to any two points n cycles apart.

EXAMPLE 8.4.4

Estimating Damping and Stiffness

■ Problem

Measurement of the free response of a certain system whose mass is 500 kg shows that after six cycles the amplitude of the displacement is 10% of the first amplitude. Also, the time for these six cycles to occur was measured to be 30 s. Estimate the system's damping c and stiffness k.

■ Solution

From the given data, $n = 6$ and $B_7/B_1 = 0.1$. Thus, from (8.5.14),

$$\delta = \frac{1}{6} \ln \left(\frac{B_1}{B_7}\right) = \frac{1}{6} \ln 10 = \frac{2.302}{6} = 0.384$$

From (8.4.13),

$$\zeta = \frac{0.384}{\sqrt{4\pi^2 + (0.384)^2}} = 0.066$$

Because the measured time for six cycles was 30 s, the period P is $P = 30/6 = 5$ s. Thus, $\omega_d = 2\pi/P = 2\pi/5$. The damped frequency is related to the undamped frequency as

$$\omega_d = \frac{2\pi}{5} = \omega_n \sqrt{1 - \zeta^2} = \omega_n \sqrt{1 - (0.066)^2}$$

Thus, $\omega_n = 1.26$ and

$$k = m\omega_n^2 = 500(1.26)^2 = 794 \text{ N/m}$$

The damping constant is calculated as follows:

$$c = 2\zeta\sqrt{mk} = 2(0.066)\sqrt{500(794)} = 83.2 \text{ N} \cdot \text{s/m}$$

8.4.5 COMPUTING RESPONSE CHARACTERISTICS WITH MATLAB

When the `step(sys)` function puts a plot on the screen, you may use the plot to calculate the settling time, the 10%–90% rise time, the maximum overshoot, and the peak time by right-clicking anywhere within the plot area. This brings up a menu. Choose "Characteristics" to obtain a submenu that contains the response characteristics. When you select a specific characteristic, say "Peak Response," MATLAB puts a large dot on the peak and displays dashed lines indicating the value of the overshoot and the peak time. Move the cursor over this dot to see a display of the values. You can use the other solvers in the same way, although the menu choices may be different. For example, peak response and settling time are available when you use the `impulse(sys)` function, but not the rise time.

You can read values off the curve by placing the cursor on the curve at the desired point. You can also move the cursor along the curve and read the values as they change.

8.5 MATLAB APPLICATIONS

You can obtain a list of all the performance parameters with the `stepinfo` function. After you have created the model in either transfer-function form or state-space form, you type `stepinfo(sys)`, where `sys` is the model name. For example, the model $\ddot{x} + 2\dot{x} + 5x = f(t)$ has the transfer function

$$\frac{X(s)}{F(s)} = \frac{1}{s^2 + 2s + 5}$$

In MATLAB, if you enter

```
>>sys = tf(1,[1,2,5]);
>>stepinfo(sys)
```

you will see the following on the screen:

```
ans =
        RiseTime: 0.6903
    SettlingTime: 3.7352
     SettlingMin: 0.1830
     SettlingMax: 0.2416
       Overshoot: 20.7866
      Undershoot: 0
            Peak: 0.2416
        PeakTime: 1.5658
```

Entering `step(sys)` produces the plot shown in Figure 8.5.1. Note that `stepinfo` gives the overshoot as a percent, the settling time using the 2% criterion, and the rise time as the 10%–90% rise time. You can change the definition of the rise time to the

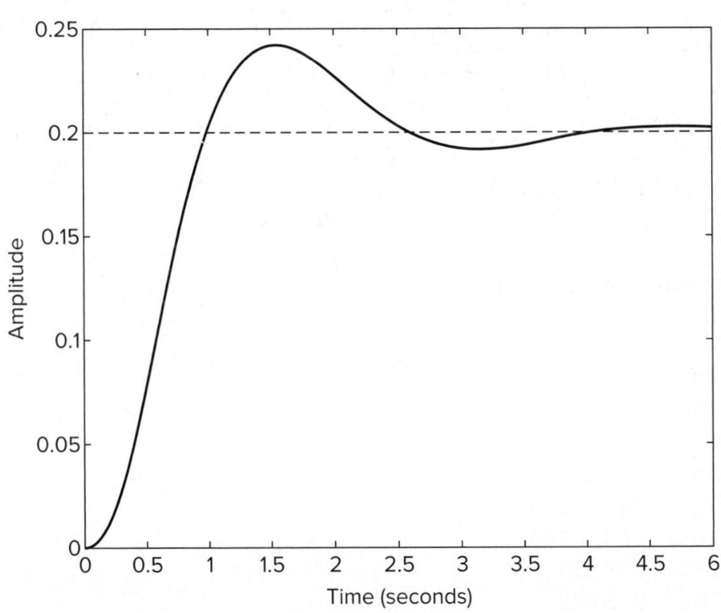

Figure 8.5.1 Unit-step response of the model $\ddot{x} + 2\dot{x} + 5x = f(t)$.

0%–100% value by typing

```
>>stepinfo(sys,'RiseTimeLimits',[0,1])
ans =
        RiseTime: 1.0174
    SettlingTime: 3.7352
     SettlingMin: 0.1914
     SettlingMax: 0.2416
       Overshoot: 20.7866
      Undershoot: 0
            Peak: 0.2416
        PeakTime: 1.5658
```

You can also change the definition of the settling time from the default value of 2% to 5% by entering `stepinfo(sys,'SettlingTimeThreshold',0.05)`. The `stepinfo` function assumes zero-initial conditions.

The `Overshoot` is calculated using the steady-state response as the reference value, just as explained in Section 8.3. Here it is calculated as

$$\frac{0.2416 - 0.2}{0.2} 100 = 20.8\%$$

The quantities `SettlingMin` and `SettlingMax` are the minimum and maximum values of the response $x(t)$ after it has entered the $\pm 2\%$ band around the steady-state response (which here is $x = 0.25$.)

The `Undershoot` is the opposite of `Overshoot`. It is the smallest value of the response before it enters the settling time band and is also expressed as a percent, but it uses the initial value as a reference instead of the steady-state value. So the response shown in Figure 8.5.1 has no undershoot.

An example of a model with undershoot is the model $\ddot{x} + 2\dot{x} + 64x = 64f(t) - 32\dot{f}(t)$, which has the transfer function

$$\frac{X(s)}{F(s)} = \frac{-32s + 64}{s^2 + 2s + 64}$$

In MATLAB, if you enter

```
>>sys = tf([-32,64],[1,2,64]);
>>stepinfo(sys)
```

You will see the following on the screen:

```
ans =
        RiseTime: 0.0335
    SettlingTime: 3.8161
     SettlingMin: -0.6506
     SettlingMax: 3.4521
       Overshoot: 245.2056
      Undershoot: 264.0444
            Peak: 3.4521
        PeakTime: 0.5498
```

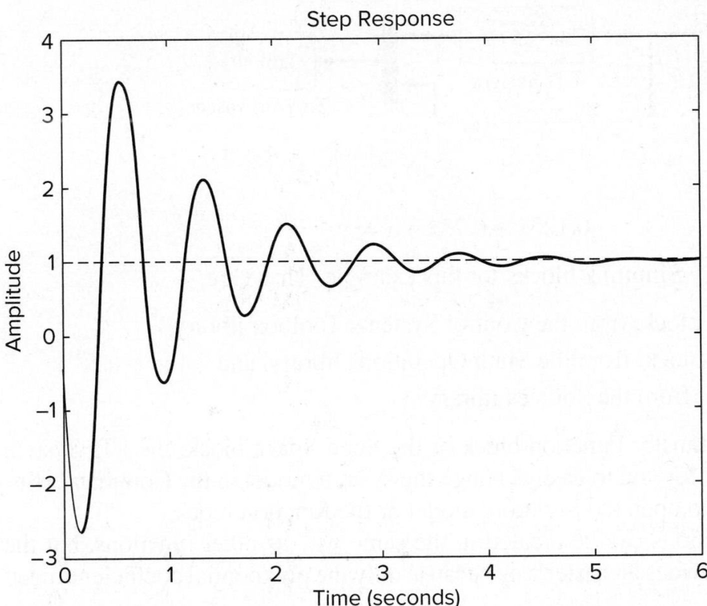

Figure 8.5.2 Unit-step response of the model $\ddot{x} + 2\dot{x} + 64x = 64f(t) - 32\dot{f}(t)$.

Entering `step(sys)` produces the plot shown in Figure 8.5.2. The overshoot is calculated as

$$\frac{3.45 - 1}{1} 100 = 245\%$$

while the undershoot is calculated differently, using: $(2.64 - 0)100\% = 264\%$.

8.6 SIMULINK APPLICATIONS

There are many engineering applications where the input is described by a polynomial. Applications include computer-controlled machines that are used to cut and form metal and other materials during manufacture. Robots for welding and gluing must follow a path that is described by polynomials.

If the model is linear with constant coefficients, the response can be found with the Laplace transform method, but if the polynomial contains many terms, the solution process would be tedious. In such cases, MATLAB or Simulink can be used to obtain the response.

8.6.1 A CAM-DRIVEN SYSTEM

Consider the cam-driven system shown in Figure 8.6.1. The motion y produced by the cam profile is often described by a polynomial. The engineer wants to predict the mass displacement $x(t)$ for chosen values of the stiffness k and damping c. To take a simple example, suppose the displacement y is described by the simple polynomial $y(t) = -t^2 + 5t$ for $0 \leq t \leq 5$. Suppose also that $m = 0.125$, $c = 0.75$, and $k = 1$.

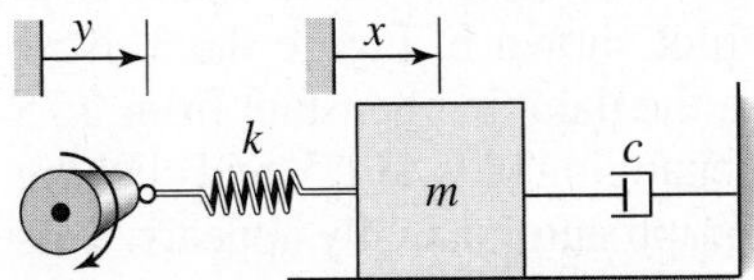

Figure 8.6.1 The cam-driven system discussed in the text.

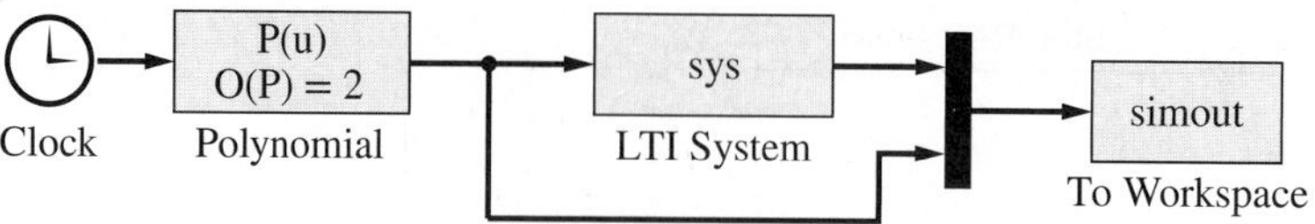

Figure 8.6.2 Simulink diagram for the text example.

Then the model is

$$0.125\ddot{x} + 0.75\dot{x} + x = y$$

We will use three new Simulink blocks for this example. These are:

1. The LTI System block, from the Control Systems Toolbox library,
2. The Polynomial block, from the Math Operations library, and
3. The Clock block, from the Sources library.

As opposed to the Transfer Function block or the State Space block, the LTI System block provides a way for you to easily change the system model in the Command window without having to open the Simulink model or the function block.

Polynomial functions can be created in the same way as other functions, but the Polynomial block provides an easier way because only the polynomial coefficients need to be entered.

The Simulink model is shown in Figure 8.6.2. The Clock block requires no parameters to be set. It provides the simulation time values t to the Polynomial block. Open this block and enter `[-1, 5, 0]` for the polynomial coefficients. Do not forget to change the Save format to Array in the To Workspace block.

Before opening the LTI System block, in the Command window enter the model's transfer function, as

```
>>sys = tf(1,[0.125,0.75,1]);
```

Then open the LTI System block and enter `sys` in the LTI system variable window. This method lets you easily change the system model in the Command window without reopening the LTI Systems block. The alternative is to type `tf(1,[0.125,0.75,1])` in the LTI system variable window. Do not specify any initial states, because this option is available only if the model is in state-space form.

You can now run the model, which we have named `CamSystem`.

8.6.2 RUNNING MODELS FROM SCRIPTS

However, to illustrate the usefulness of the LTI System block, we instead create the following script file.

```
% Program CamDemo.m
% Polynomial input for a cam-driven system.
sys = tf(1,[0.125,0.75,1]);
sim('CamSystem')
plot(tout,simout(:,1),tout,simout(:,2),'--'),legend('x','y')
```

This file uses the `sim` function, which runs the Simulink model whose name is in quotes. When this program is run, it produces the plot shown in Figure 8.6.3. Now suppose we want to investigate the effect of changing the damping constant from 0.75 to 1.5. You simply edit the third line of the script file, changing 0.75 to 1.5, and clicking on the Run icon in the MATLAB toolbar. The new graph automatically appears. This provides a way to quickly check the effects of different parameter values.

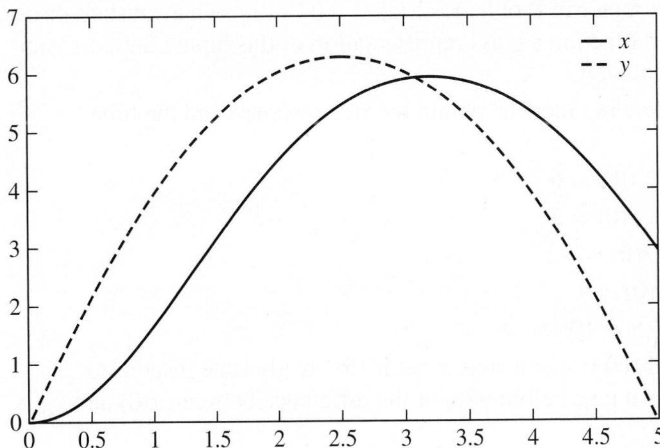

Figure 8.6.3 Plot of the desired trajectory (y) and the actual trajectory (x).

8.7 CHAPTER REVIEW

This chapter emphasizes understanding system behavior in the time domain. The forcing functions commonly used to model real inputs or to test a system's response in the time domain are the impulse, the step, and the ramp functions. The impulse models a suddenly applied and suddenly removed input. The step function models a suddenly applied input that remains constant. The ramp models an input that is changing at a constant rate.

Sections 8.1 and 8.2 treated the response of first- and second-order systems. The time constant τ, the damping ratio ζ, and the undamped natural frequency ω_n are important for assessing system response. In Section 8.3 we introduced the concepts of maximum overshoot M_p, peak time t_p, delay time t_d, rise time t_r, and settling time t_s. These are useful for describing and for specifying the step response. Section 8.4 treated parameter estimation in the time domain.

Section 8.5 showed how to use MATLAB to compute performance specifications, such as overshoot and rise time. Finally, Section 8.6 introduced several Simulink blocks that are useful for simulating nonlinear systems or systems with complicated inputs.

Now that you have completed Chapter 8, you should be able to do the following:

1. Obtain and interpret the free, step, ramp, and impulse response of linear models.
2. Compute and apply the time constant τ, the damping ratio ζ, and the undamped natural frequency ω_n to assess system response.
3. Compute and apply maximum overshoot M_p, peak time t_p, delay time t_d, rise time t_r, and settling time t_s to describe and assess system response.
4. Use time-domain response data to estimate coefficients in dynamic models.
5. Use MATLAB to compute performance specifications.
6. Use Simulink to simulate systems with complicated inputs.

PROBLEMS

Section 8.1 Response of First-Order Systems

8.1 A rocket sled has the following equation of motion: $6\dot{v} = 2700 - 24v$. How long must the rocket fire before the sled travels 2000 m? The sled starts from rest.

8.2 Suppose the rocket motor in Problem 8.1 takes 0.04 s to reach a constant thrust of 2700 N. Is a step function a good representation of this input? Support your answer with a calculation.

8.3 For each of the following models, obtain the free response and the time constant, if any.

a. $16\dot{x} + 14x = 0,\ x(0) = 6$

b. $12\dot{x} + 5x = 15,\ x(0) = 3$

c. $13\dot{x} + 6x = 0,\ x(0) = -2$

d. $7\dot{x} - 5x = 0,\ x(0) = 9$

8.4 For the model $2\dot{x} + x = 10f(t)$,

a. If $x(0) = 0$ and $f(t)$ is a unit step, what is the steady-state response x_{ss}? How long does it take before 98% of the difference between $x(0)$ and x_{ss} is eliminated?

b. Repeat part (a) with $x(0) = 5$.

c. Repeat part (a) with $x(0) = 0$ and $f(t) = 20u_s(t)$.

8.5 Obtain the steady-state response of each of the following models, and estimate how long it will take the response to reach steady state.

a. $6\dot{x} + 5x = 20u_s(t),\ x(0) = 0$

b. $6\dot{x} + 5x = 20u_s(t),\ x(0) = 1$

c. $13\dot{x} - 6x = 18u_s(t),\ x(0) = -2$

8.6 Obtain the total response of the following models.

a. $6\dot{x} + 5x = 20u_s(t),\ x(0) = 0$

b. $6\dot{x} + 5x = 20u_s(t),\ x(0) = 1$

c. $13\dot{x} - 6x = 18u_s(t),\ x(0) = -2$

8.7 A certain rotational system has the equation of motion

$$100\frac{d\omega}{dt} + 5\omega = T(t)$$

where $T(t)$ is the torque applied by an electric motor, as shown in Figure 8.1.8. The model of the motor's field current i_f in amperes is

$$0.002\frac{di_f}{dt} + 4i_f = v(t)$$

where $v(t)$ is the voltage applied to the motor. The motor torque constant is $K_T = 15\ \text{N} \cdot \text{m/A}$. Suppose the applied voltage is $12u_s(t)$ V. Determine the steady-state speed of the inertia and estimate the time required to reach that speed.

8.8 The *RC* circuit shown in Figure 8.1.2c has the parameter values $R = 3 \times 10^6\ \Omega$ and $C = 1\ \mu\text{F}$. If the inital capacitor voltage is 6 V and the applied voltage is $v_s(t) = 12u_s(t)$, obtain the expression for the capacitor voltage response $v(t)$.

8.9 The liquid-level system shown in Figure 8.1.2d has the parameter values $A = 50\ \text{ft}^2$ and $R = 60\ \text{ft}^{-1}\text{sec}^{-1}$. If the inflow rate is $q_v(t) = 10u_s(t)\ \text{ft}^3/\text{sec}$, and the initial height is 2 ft, how long will it take for the height to reach 15 ft?

8.10 The immersed object shown in Figure 8.1.2e is steel and has a mass of 100 kg and a specific heat of $c_p = 500\ \text{J/kg} \cdot {}^\circ\text{C}$. Assume the thermal resistance is $R = 0.09^\circ\text{C} \cdot \text{s/J}$. The inital temperature of the object is 20° when it is dropped into a large bath of temperature 80°C. Obtain the expression for the temperature $T(t)$ of the object after it is dropped into the bath.

8.11 Compare the responses of $2\dot{v} + v = \dot{g}(t) + g(t)$ and $2\dot{v} + v = g(t)$ if $g(t) = 10u_s(t)$ and $v(0) = 5$.

8.12 Compare the responses of $5\dot{v} + v = \dot{g} + g$ and $5\dot{v} + v = g$ if $v(0) = 5$ and $g = 10$ for $-\infty \leq t \leq \infty$.

8.13 Consider the following model:

$$6\dot{v} + 3v = \dot{g}(t) + g(t)$$

where $v(0) = 0$.

a. Obtain the response $v(t)$ if $g(t) = u_s(t)$.

b. Obtain the response $v(t)$ to the approximate step input $g(t) = 1 - e^{-5t}$ and compare with the results of part (a).

8.14 Obtain the response of the model $2\dot{v} + v = f(t)$, where $f(t) = 5t$ and $v(0) = 0$. Identify the transient and steady-state responses.

8.15 Obtain the response of the model $9\dot{v} + 3v = f(t)$, where $f(t) = 7t$ and $v(0) = 0$. Is steady-state response parallel to $f(t)$?

8.16 The resistance of a telegraph line is $R = 10\ \Omega$, and the solenoid inductance is $L = 5$ H. Assume that when sending a "dash," a voltage of 12 V is applied while the key is closed for 0.3 s. Obtain the expression for the current $i(t)$ passing through the solenoid.

Section 8.2 Response of Second-Order Systems

8.17 Obtain the oscillation frequency and amplitude of the response of the model $3\ddot{x} + 12x = 0$ for (a) $x(0) = 5$ and $\dot{x}(0) = 0$ and (b) $x(0) = 0$ and $\dot{x}(0) = 5$.

8.18 Find the response for the following models. The initial conditions are zero.

a. $3\ddot{x} + 21\dot{x} + 30x = 4t$

b. $5\ddot{x} + 20\dot{x} + 20x = 7t$

c. $2\ddot{x} + 8\dot{x} + 58x = 5t$

8.19 Suppose the input $f(t)$ to the following model is a ramp function: $f(t) = at$. Assuming that the model is stable, for what values of a, m, c, and k will the steady-state response be parallel to the input? For this case, what is the steady-state difference between the input and the response?

$$m\ddot{x} + c\dot{x} + kx = f(t)$$

8.20 If applicable, compute ζ, τ, ω_n, and ω_d for the following roots, and find the corresponding characteristic polynomial.

1. $s = -2 \pm 6j$
2. $s = 1 \pm 5j$
3. $s = -10, -10$
4. $s = -10$

8.21 If applicable, compute ζ, τ, ω_n, and ω_d for the dominant root in each of the following sets of characteristic roots.

1. $s = -2, -3 \pm j$
2. $s = -3, -2 \pm 2j$

8.22 A certain fourth-order model has the roots

$$s = -2 \pm 4j, -10 \pm 7j$$

Identify the dominant roots and use them to estimate the system's time constant, damping ratio, and oscillation frequency.

8.23 Given the model

$$\ddot{x} - (\mu + 2)\dot{x} + (2\mu + 5)x = 0$$

a. Find the values of the parameter μ for which the system is
   1. Stable
   2. Neutrally stable
   3. Unstable

b. For the stable case, for what values of μ is the system
   1. Underdamped?
   2. Overdamped?

8.24 The characteristic equation of the system shown in Figure 8.2.3 for $m = 3$ and $k = 27$ is $3s^2 + cs + 27 = 0$. Obtain the free response for the following values of damping: $c = 0$, 9, 18, and 22. Use the initial conditions $x(0) = 1$ and $\dot{x}(0) = 0$.

8.25 The characteristic equation of a certain system is $4s^2 + 6ds + 25d^2 = 0$, where d is a constant. (a) For what values of d is the system stable? (b) Is there a value of d for which the free response will consist of decaying oscillations?

8.26 The characteristic equation of a certain system is $s^2 + 6bs + 5b - 10 = 0$, where b is a constant. (a) For what values of b is the system stable? (b) Is there a value of b for which the free response will consist of decaying oscillations?

8.27 A certain system has two coupled subsystems. One subsystem is a rotational system with the equation of motion:

$$50\frac{d\omega}{dt} + 10\omega = T(t)$$

where $T(t)$ is the torque applied by an electric motor, Figure 8.1.8. The second subsystem is a field-controlled motor. The model of the motor's field current i_f in amperes is

$$0.001\frac{di_f}{dt} + 5i_f = v(t)$$

where $v(t)$ is the voltage applied to the motor. The motor torque constant is $K_T = 25$ N · m/A. Obtain the damping ratio ζ, time constants, and undamped natural frequency ω_n of the combined system.

8.28 A certain armature-controlled dc motor has the characteristic equation

$$L_a I s^2 + (R_a I + cL_a)s + cR_a + K_b K_T = 0$$

Using the following parameter values:

$$K_b = K_T = 0.1 \text{ N} \cdot \text{m/A} \qquad I = 6 \times 10^{-5} \text{ kg} \cdot \text{m}^2$$

$$R_a = 0.6\ \Omega \qquad L_a = 4 \times 10^{-3} \text{ H}$$

obtain the expressions for the damping ratio ζ and undamped natural frequency ω_n in terms of the damping c. Assuming that $\zeta < 1$, obtain the expression for the time constant.

Section 8.3 Description and Specification of Step Response

8.29 Compute the maximum percent overshoot, the maximum overshoot, the peak time, the 100% rise time, the delay time, and the 2% settling time for the following model. The initial conditions are zero. Time is measured in seconds.

$$\ddot{x} + 4\dot{x} + 8x = 2u_s(t)$$

8.30 A certain system is described by the model

$$\ddot{x} + c\dot{x} + 4x = u_s(t)$$

Set the value of the damping constant c so that both of the following specifications are satisfied. Give priority to the overshoot specification. If both cannot be satisfied, state the reason. Time is measured in seconds.

1. Maximum percent overshoot should be as small as possible and no greater than 20%.
2. 100% rise time should be as small as possible and no greater than 3 s.

8.31 A certain system is described by the model

$$9\ddot{x} + c\dot{x} + 4x = u_s(t)$$

Set the value of the damping constant c so that both of the following specifications are satisfied. Give priority to the overshoot specification. If both cannot be satisfied, state the reason. Time is measured in seconds.

1. Maximum percent overshoot should be as small as possible and no greater than 20%.
2. 100% rise time should be as small as possible and no greater than 3 s.

8.32 Derive the fact that the peak time is the same for all characteristic roots having the same imaginary part.

Section 8.4 Parameter Estimation in the Time Domain

8.33 Suppose that the resistance in the circuit of Figure 8.4.1a is 3×10^6 Ω. A voltage is applied to the circuit and then is suddenly removed at time $t = 0$. The measured voltage across the capacitor is given in the following table. Use the data to estimate the value of the capacitance C.

Time t (s)	Voltage v_C (V)
0	12.0
2	11.2
4	10.5
6	9.8
8	9.2
10	8.6
12	8.0
14	7.5
16	7.0
18	6.6
20	6.2

8.34 The temperature of liquid cooling in a cup at room temperature (68°F) was measured at various times. The data are given next.

Time t (sec)	Temperature T (°F)
0	178
500	150
1000	124
1500	110
2000	97
2500	90
3000	82

Develop a model of the liquid temperature as a function of time, and use it to estimate how long it will take the temperature to reach 135°F.

8.35 Figure P8.35 shows the response of a system to a step input of magnitude 1000 N. The equation of motion is

$$m\ddot{x} + c\dot{x} + kx = f(t)$$

Estimate the values of m, c, and k.

Figure P8.35

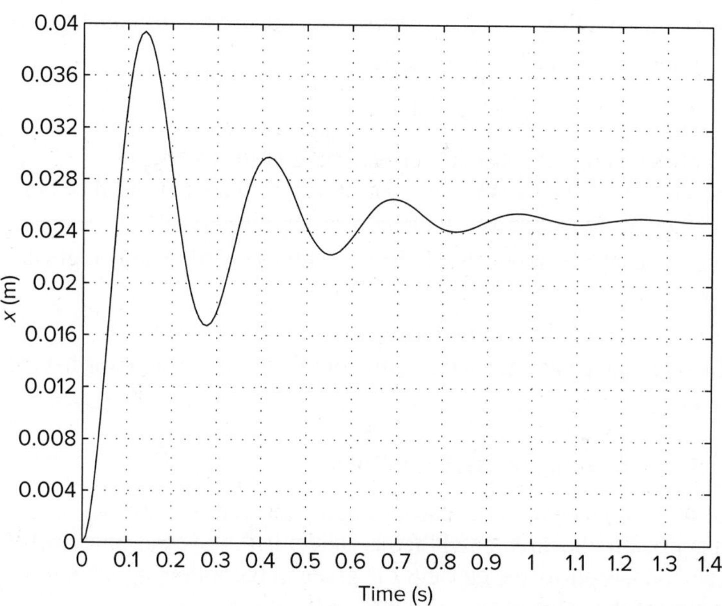

8.36 A mass-spring-damper system has a mass of 100 kg. Its free response amplitude decays such that the amplitude of the 30th cycle is 20% of the amplitude of the 1st cycle. It takes 60 s to complete 30 cycles. Estimate the damping constant c and the spring constant k.

8.37 a. For the following model, find the steady-state response and use the dominant-root approximation to find the dominant response (how long will it take to reach steady state? does it oscillate?). The initial conditions are zero.

$$\frac{d^3x}{dt^3} + 22\frac{d^2x}{dt^2} + 131\frac{dx}{dt} + 110x = u_s(t)$$

b. Obtain the exact solution for the response of the above model, and use it to check the prediction based on the dominant-root approximation.

8.38 The following model has a dominant root of $s = -3 \pm 5j$ as long as the parameter μ is no less than 3.

$$\frac{d^3y}{dt^3} + (6+\mu)\frac{d^2y}{dt^2} + (34+6\mu)\frac{dy}{dt} + 34\mu y = u_s(t)$$

Investigate the accuracy of the estimate of the maximum overshoot, the peak time, the 100% rise time, and the 2% settling time based on the dominant root, for the following three cases: (a) $\mu = 30$, (b) $\mu = 6$, and (c) $\mu = 3$. Discuss the effect of μ on these predictions.

8.39 Estimate the maximum overshoot, the peak time, and the rise time of the unit-step response of the following model if $f(t) = 5000u_s(t)$ and the initial conditions are zero.

$$\frac{d^4y}{dt^4} + 26\frac{d^3y}{dt^3} + 269\frac{d^2y}{dt^2} + 1524\frac{dy}{dt} + 4680y = f(t)$$

Its roots are

$$s = -3 \pm 6j, \qquad -10 \pm 2j$$

8.40 What is the form of the unit-step response of the following model? Find the steady-state response. How long does the response take to reach steady state?

$$2\frac{d^4y}{dt^4} + 52\frac{d^3y}{dt^3} + 6250\frac{d^2y}{dt^2} + 4108\frac{dy}{dt} + 1.1202 \times 10^4 y = 5 \times 10^4 f(t)$$

8.41 Use a software package such as MATLAB to plot the step response of the following model for three cases: $a = 0.2$, $a = 1$, and $a = 10$. The step input has a magnitude of 2500.

$$\frac{d^4y}{dt^4} + 24\frac{d^3y}{dt^3} + 225\frac{d^2y}{dt^2} + 900\frac{dy}{dt} + 2500y = f + a\frac{df}{dt}$$

Compare the response to that predicted by the maximum overshoot, peak time, 100% rise time, and 2% settling time calculated from the dominant roots.

Section 8.5 MATLAB Applications

8.42 Use MATLAB to find the maximum percent overshoot, peak time, 2% settling time, and 100% rise time for the following equation. The initial conditions are zero.

$$\ddot{x} + 4\dot{x} + 8x = 2u_s(t)$$

8.43 Use MATLAB to compare the maximum percent overshoot, peak time, and 100% rise time of the following models where the input $f(t)$ is a unit-step function. The initial conditions are zero.

a. $3\ddot{x} + 18\dot{x} + 10x = 10f(t)$

b. $3\ddot{x} + 18\dot{x} + 10x = 10f(t) + 10\dot{f}(t)$

8.44 a. Use MATLAB to find the maximum percent overshoot, peak time, and 100% rise time for the following equation. The initial conditions are zero.

$$\frac{d^3x}{dt^3} + 22\frac{d^2x}{dt^2} + 113\frac{dx}{dt} + 110x = u_s(t)$$

b. Use the dominant-root pair to compute the maximum percent overshoot, peak time, and 100% rise time, and compare the results with those found in part (a).

8.45 a. Use MATLAB to find the maximum percent overshoot, peak time, and 100% rise time for the following equation. The initial conditions are zero.

$$\frac{d^4y}{dt^4} + 26\frac{d^3y}{dt^3} + 269\frac{d^2y}{dt^2} + 1524\frac{dy}{dt} + 4680y = 5000u_s(t)$$

b. The characteristic roots are $s = -3 \pm 6j, -10 \pm 2j$. Use the dominant-root pair to compute the maximum percent overshoot, peak time, and 100% rise time, and compare the results with those found in part (a).

8.46 a. Use MATLAB to find the maximum percent overshoot, peak time, and 100% rise time for the following equation. The initial conditions are zero.

$$\frac{d^4y}{dt^4} + 14\frac{d^3y}{dt^3} + 127\frac{d^2y}{dt^2} + 426\frac{dy}{dt} + 962y = 926u_s(t)$$

b. Use the dominant-root pair to compute the overshoot, peak time, and 100% rise time, and compare the results with those found in part (a).

Section 8.6 Simulink Applications

8.47 The following equation has a polynomial input.

$$0.125\dddot{x} + 0.75\ddot{x} + x = y(t) = -\frac{27}{800}t^3 + \frac{270}{800}t^2$$

Use Simulink to plot $x(t)$ and $y(t)$ on the same graph. The initial conditions are zero.

8.48 The following model has a polynomial input.

$$\dot{x}_1 = -3x_1 + 4x_2$$

$$\dot{x}_2 = -5x_1 - x_2 + f(t)$$

$$f(t) = -\frac{5}{3}t^2 + \frac{25}{3}t$$

The initial conditions are $x_1(0) = 3$ and $x_2(0) = 7$. Use Simulink to plot $x_1(t)$, $x_2(t)$, and $f(t)$ on the same graph.

8.49 First create a Simulink model containing an LTI System block to plot the unit-step response of the following equation for $k = 4$. The initial conditions are zero.

$$5\frac{d^3x}{dt^3} + 3\frac{d^2x}{dt^2} + 7\frac{dx}{dt} + kx = u_s(t)$$

Then create a script file to run the Simulink model. Use the file to experiment with the value of k to find the largest possible value of k such that the system remains stable.

8.50 Figure P8.50 shows an engine valve driven by an overhead camshaft. The rocker arm pivots about the fixed point O. For particular values of the parameters shown, the valve displacement $x(t)$ satisfies the following equation.

$$10^{-6}\ddot{x} + 0.3x = 5\theta(t)$$

where $\theta(t)$ is determined by the cam shaft speed and the cam profile. A particular profile is

$$\theta(t) = 16 \times 10^3(10^4t^4 - 200t^3 + t^2) \qquad 0 \le t \le 0.01$$

Use Simulink to plot $x(t)$. The initial conditions are zero.

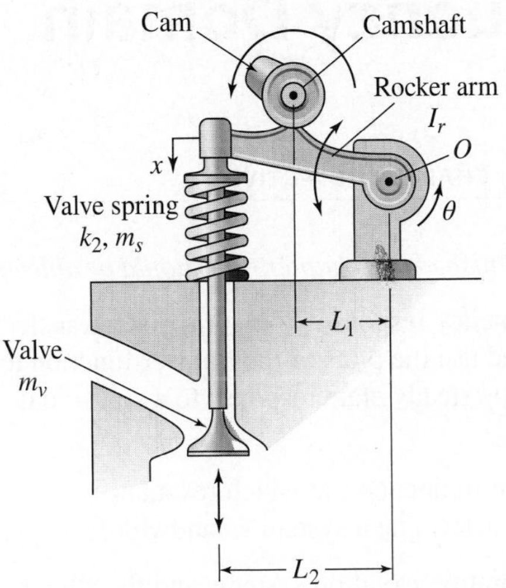

Figure P8.50

CHAPTER 9

System Analysis in the Frequency Domain

CHAPTER OUTLINE

CHAPTER OBJECTIVES

When you have finished this chapter, you should be able to

1. Sketch frequency response plots for a given transfer function, and use the plots or the transfer function to determine the steady-state response to a sinusoidal input.
2. Compute the frequencies at which resonance occurs, and determine a system's bandwidth.
3. Analyze vibration isolation systems and the effects of base motion and rotating unbalance.
4. Determine the steady-state response to a periodic input, given the Fourier series description of the input.
5. Estimate the form of a transfer function and its parameter values, given frequency response data.
6. Use frequency response methods to evaluate vehicle suspension performance.
7. Use MATLAB as an aid in the preceding tasks.

The term *frequency response* refers to how a system responds to a periodic input, such as a sinusoid. An input $f(t)$ is periodic with a period P if $f(t + P) = f(t)$ for all values of time t, where P is a constant called the *period*. Periodic inputs are commonly found in many applications. The most common perhaps is ac voltage, which is sinusoidal. For the common ac frequency of 60 Hz, the period is $P = 1/60$ s. Rotating unbalanced machinery produces periodic forces on the supporting structures, internal combustion engines produce a periodic torque, and reciprocating pumps produce hydraulic and pneumatic pressures that are periodic.

Frequency response analysis focuses on harmonic inputs, such as sines and cosines. A sine function has the form $A \sin \omega t$, where A is its *amplitude* and ω is its frequency in radians per unit time. Note that a cosine is simply a sine shifted by 90° or $\pi/2$ rad, as $\cos \omega t = \sin(\omega t + \pi/2)$. Not all periodic inputs are sinusoidal or cosinusoidal, but an important result of mathematics, called the *Fourier series*, enables us to represent a periodic function as the sum of a constant term plus a series of cosine and sine terms of different frequencies and amplitudes. Thus, we will be able to apply the results of this chapter to any such periodic input.

The transfer function is central to understanding and applying frequency response methods. The transfer function enables us to obtain a concise graphical description of a system's frequency response. This graphical description is called a *frequency response plot*. We analyze the response of first-order systems to sinusoidal inputs in Section 9.1. In Section 9.2 we generalize the method to higher-order systems. Section 9.3 discusses several phenomena and applications of frequency response, including *beating*, *resonance*, the effect of base motion and rotating unbalance, and instrument design.

In Section 9.4 we introduce the important concept of *bandwidth*, which enables us to develop a concise quantitative description of a system's frequency response, much like the time constant characterizes the step response. Section 9.5 shows how to analyze the response due to general periodic inputs that can be described with a Fourier series.

Experiments involving frequency response can often be used to determine the form of the transfer function of a system and to estimate the numerical values of the parameters in the transfer function. Section 9.6 gives some examples of this process. MATLAB has several useful functions for obtaining and for analyzing frequency response. These are treated in Section 9.7. ■

9.1 FREQUENCY RESPONSE OF FIRST-ORDER SYSTEMS

Consider the model of a mass m subjected to a damping force cv and another force $f(t)$. The model for the velocity v is $m\dot{v} + cv = f(t)$. Suppose that $m = 0.2$ kg, $c = 1$ N · s/m, and that the force $f(t)$ is the sum of two sinusoidal terms of different frequencies: $f(t) = f_1(t) + f_2(t) = \sin 15t + \sin 60t$. These two forces are shown separately in Figure 9.1.1. If we were to solve the model using one of our analytical or computer techniques, we would obtain the plot of the steady-state response $v_{ss}(t)$ shown in Figure 9.1.2. The plot is quite messy and somewhat difficult to interpret. However, if we obtain the steady-state response by considering each sine term separately, we would better understand the dynamics of the system. This is done in Example 9.1.1.

So the themes of this chapter are (1) how to obtain the steady-state response to sine and cosine inputs, and (2) how to use that knowledge to analyze the effects of general periodic inputs. We will see that the transfer function is the most important tool for doing this, and we will see that applying this tool requires facility with complex numbers, so that is where we will begin.

9.1.1 REPRESENTATION OF COMPLEX NUMBERS

In this chapter we will frequently need to obtain the magnitude and angle of complex numbers. A complex number N can be represented in *rectangular* form as $N = x + jy$, where x is the real part and y is the imaginary part. The number can be plotted in two dimensions with x as the abscissa (on the horizontal axis) and y as the ordinate (on the vertical axis). We can think of the number as a two-dimensional vector whose head is at the point (x, y) and whose tail is at the origin. The vector length is the magnitude M of the number and the vector angle ϕ is measured counterclockwise from the positive

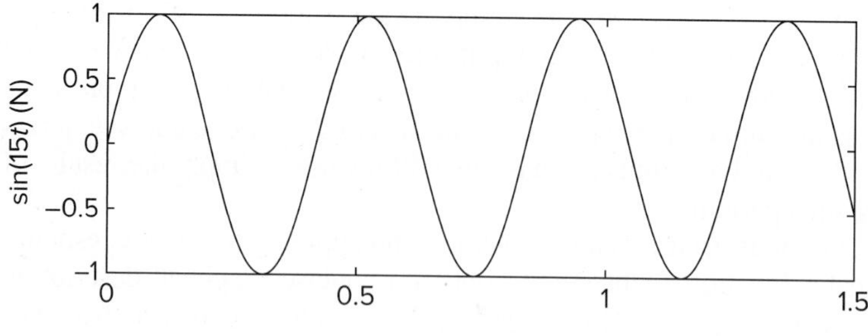

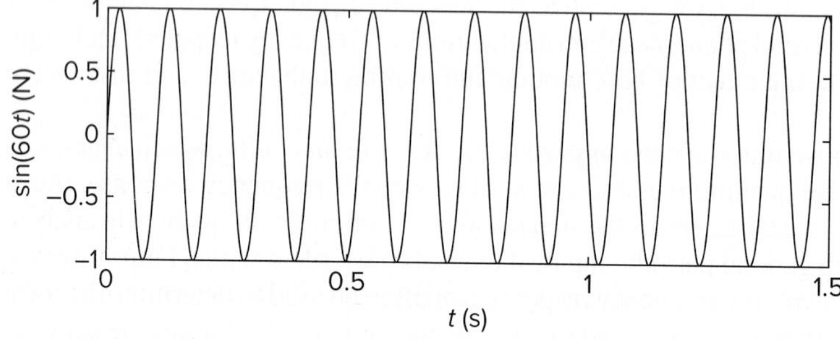

Figure 9.1.1 Plots of two forces having the same amplitude but different frequencies.

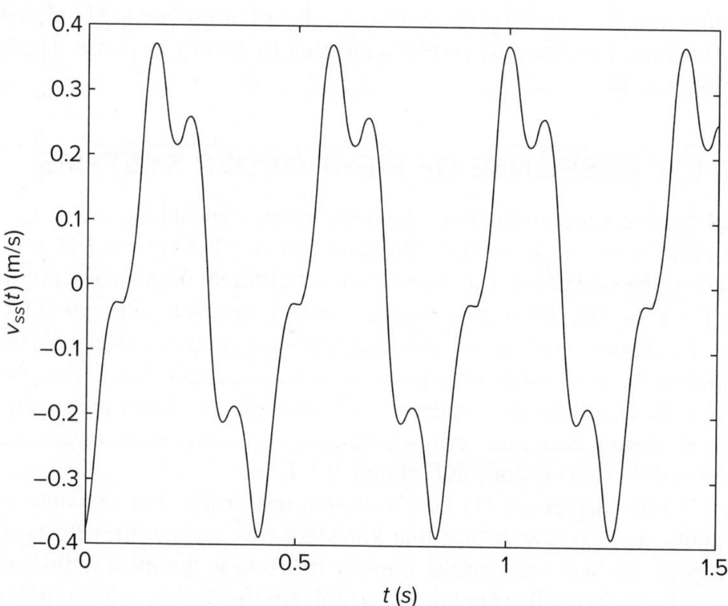

Figure 9.1.2 Steady-state response when the input is the sum of the two forces shown in Figure 9.1.1.

real axis. In this form, the magnitude and angle of the number can be calculated from $|N| = \sqrt{x^2 + y^2}$. The complex conjugate of N is $x - yj$. See Figure 9.1.3a.

Another form is the *complex exponential* form:

$$N = Me^{j\phi} = M(\cos\phi + j\sin\phi)$$

Note that the complex conjugate of N is $Me^{-j\phi}$.

9.1.2 PRODUCTS AND RATIOS OF COMPLEX NUMBERS

The complex exponential form can be used to show that the magnitude and angle of a number consisting of products and ratios of complex numbers $N_1, N_2, \ldots,$

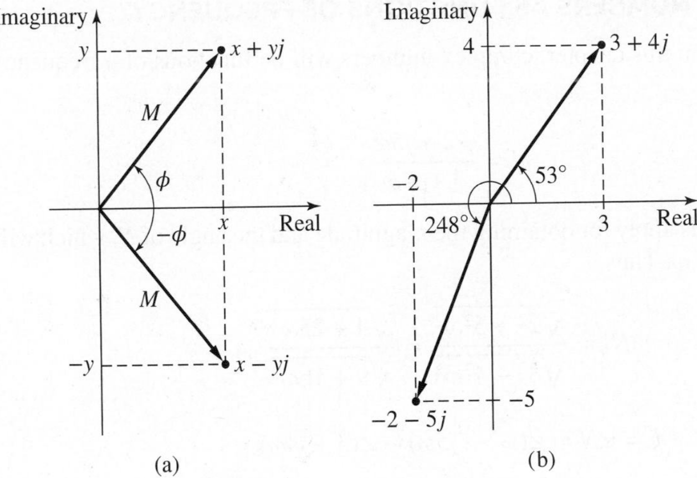

Figure 9.1.3 Vector representation of complex numbers.

can be calculated as follows.

$$N = \frac{N_1 N_2}{N_3 N_4} = \frac{M_1 e^{j\phi_1} M_2 e^{j\phi_2}}{M_3 e^{j\phi_3} M_4 e^{j\phi_4}} = \frac{M_1 M_2}{M_3 M_4} e^{j(\phi_1+\phi_2-\phi_3-\phi_4)}$$

Thus, $N = |N|e^{j\phi}$, where

$$|N| = \frac{M_1 M_2}{M_3 M_4} \qquad \phi = \phi_1 + \phi_2 - \phi_3 - \phi_4$$

So the magnitudes combine as products and ratios, and the angles combine as sums and differences.

For example, consider the number

$$N = \frac{-2-5j}{3+4j}$$

The magnitude of N can be calculated by computing the ratio of the magnitudes of the numerator and denominator, as follows:

$$|N| = \left|\frac{-2-5j}{3+4j}\right| = \frac{|-2-5j|}{|3+4j|} = \frac{\sqrt{(-2)^2+(-5)^2}}{\sqrt{3^2+4^2}} = \frac{\sqrt{29}}{5}$$

The angle of N, denoted by $\angle N$, is the difference between the angle of the numerator and the angle of the denominator. These angles are shown in Figure 9.1.1b, which is a vector representation of the complex numbers in the numerator and denominator. From this diagram we can see that the angles are given by

$$\angle(-2-5j) = 180° + \tan^{-1}\frac{5}{2} = 180° + 68° = 248°$$

$$\angle(3+4j) = \tan^{-1}\frac{4}{3} = 53°$$

Thus,

$$\angle N = 248° - 53° = 195°$$

and

$$N = \frac{\sqrt{29}}{5}\angle 195° = \frac{\sqrt{29}}{5}(\cos 195° + j\sin 195°) = -1.04 + 0.28j$$

9.1.3 COMPLEX NUMBERS AS FUNCTIONS OF FREQUENCY

In our applications in this chapter, complex numbers will be functions of a frequency ω, as for example,

$$N = \frac{-2 - j5\omega}{3 + j4\omega}$$

but the same methods apply for obtaining the magnitude and the angle of N, which will then be functions of ω. Thus,

$$|N| = \frac{\sqrt{2^2 + 5^2\omega^2}}{\sqrt{3^2 + 4^2\omega^2}} = \frac{\sqrt{4 + 25\omega^2}}{\sqrt{9 + 16\omega^2}}$$

$$\phi = \angle N = \angle(-2 - j5\omega) - \angle(3 + j4\omega)$$

$$= 180^\circ + \tan^{-1}\left(\frac{5\omega}{2}\right) - \tan^{-1}\left(\frac{4\omega}{3}\right)$$

Once a value is given for ω, we can compute $|N|$ and ϕ.

9.1.4 FREQUENCY RESPONSE PROPERTIES

The methods of this chapter pertain to *any stable, linear, time-invariant (LTI)* system. The basic frequency response property of such systems is summarized in Table 9.1.1 and Figure 9.1.4. Any linear, time-invariant system, stable or not, has a transfer function, say $T(s)$. If a sinusoidal input of frequency ω is applied to such a system, and if the system is *stable*, the transient response eventually disappears, leaving a steady-state response that is sinusoidal with the same frequency ω as the input, but with a different amplitude and shifted in time relative to the input.

To prove this result, suppose the system transfer function $T(s)$ is of order n, the output is $x(t)$, and the input is $f(t) = A \sin \omega t$. Then

$$X(s) = F(s)T(s) = \frac{A\omega}{s^2 + \omega^2}T(s)$$

Table 9.1.1 Frequency response of a stable LTI system.

The transfer function $T(s)$ with s replaced by $j\omega$ is called the *frequency transfer function.* Therefore, the frequency transfer function is a complex function of ω, and it has a magnitude and an angle, just as any complex number. If the system is *stable*, the magnitude M of the frequency transfer function $T(j\omega)$ is the ratio of the sinusoidal steady-state output amplitude over a sinusoidal input amplitude. The phase shift of the steady-state output relative to the input is the angle of $T(j\omega)$. Thus, denoting the input by $A \sin \omega t$ and the steady-state output by $B \sin(\omega t + \phi)$, we have

$$|T(j\omega)| = \frac{B}{A} \equiv M(\omega) \tag{1}$$

$$\phi(\omega) = \angle T(j\omega) \tag{2}$$

$$B = AM(\omega) \tag{3}$$

where $|T(j\omega)|$ and $\angle T(j\omega)$ denote the magnitude and angle of the complex number $T(j\omega)$. Because of equation (1), M is called the *amplitude ratio* or the *magnitude ratio.*

These results are illustrated in Figure 9.1.4.

Sinusoidal input — Stable, linear system — Steady-state response

$A \sin \omega t \longrightarrow \boxed{T(s)} \longrightarrow B \sin(\omega t + \phi)$

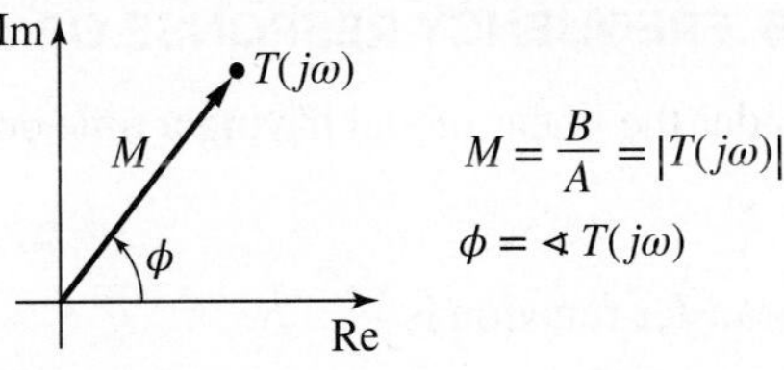

Figure 9.1.4 Frequency response of a stable linear system.

which can be expressed as a partial-fraction expansion.

$$\frac{A\omega}{s^2+\omega^2}T(s) = \frac{C_1}{s+j\omega} + \frac{C_2}{s-j\omega} + \cdots$$

The factors containing $s + j\omega$ and $s - j\omega$ correspond to the term $s^2 + \omega^2$, while the remaining terms in the expansion correspond to the factors introduced by the denominator of $T(s)$. If the system is *stable*, all these factors will be negative or have negative real parts. Thus, the response will have the form

$$x(t) = C_1 e^{-j\omega t} + C_2 e^{j\omega t} + \sum_{i=1}^{m} D_i e^{-\alpha_i t} \sin(\beta_i t + \phi_i) + \sum_{i=m+1}^{n} D_i e^{-\alpha_i t}$$

where the roots of $T(s)$ are $s_i = -\alpha_i \pm \beta_i j$ for $i = 1, \ldots, m$ and $s_i = -\alpha_i$ for $i = 1 + m, \ldots, n$. The steady-state response is thus given by

$$x_{ss}(t) = C_1 e^{-j\omega t} + C_2 e^{j\omega t}$$

We can evaluate C_1 and C_2 as follows. Note that $T(j\omega) = |T(j\omega)|e^{j\phi}$, where $\phi = \angle T(j\omega)$.

$$C_1 = T(s)\frac{A\omega}{s^2+\omega^2}(s+j\omega)\bigg|_{s=-j\omega} = -\frac{A}{2j}T(-j\omega) = -\frac{A}{2j}|T(j\omega)|e^{-j\phi}$$

$$C_2 = T(s)\frac{A\omega}{s^2+\omega^2}(s-j\omega)\bigg|_{s=j\omega} = \frac{A}{2j}T(j\omega) = \frac{A}{2j}|T(j\omega)|e^{j\phi}$$

The steady-state response can thus be expressed as

$$x_{ss}(t) = -\frac{A}{2j}|T(j\omega)|e^{-j\phi}e^{-j\omega t} + \frac{A}{2j}|T(j\omega)|e^{j\phi}e^{j\omega t}$$

$$= |T(j\omega)|A\frac{e^{j(\omega t+\phi)} - e^{-j(\omega t+\phi)}}{2j}$$

or

$$x_{ss}(t) = |T(j\omega)|A \sin(\omega t + \phi)$$

because

$$\frac{e^{j(\omega t+\phi)} - e^{-j(\omega t+\phi)}}{2j} = \sin(\omega t + \phi)$$

Thus, at steady state, the output amplitude is $|T(j\omega)|A$ and the phase shift is $\phi = \angle T(j\omega)$.

A specific example will help to clarify these concepts.

9.1.5 FREQUENCY RESPONSE OF $\tau\dot{y}+y=f(t)$

Consider the linear model having a time constant τ.

$$\tau\dot{y}+y=f(t) \tag{9.1.1}$$

The transfer function is

$$T(s)=\frac{Y(s)}{F(s)}=\frac{1}{\tau s+1} \tag{9.1.2}$$

If the input is sinusoidal, $f(t)=A\sin\omega t$, we can use the Laplace transform method to show that the *forced* response is

$$y(t)=\frac{A\omega\tau}{1+\omega^2\tau^2}\left(e^{-t/\tau}-\cos\omega t+\frac{1}{\omega\tau}\sin\omega t\right) \tag{9.1.3}$$

For $t\geq 4\tau$, the transient response has essentially disappeared, and we can express the steady-state response as

$$y_{ss}(t)=\frac{A}{1+\omega^2\tau^2}(\sin\omega t-\omega\tau\cos\omega t)=\frac{A}{\sqrt{1+\omega^2\tau^2}}\sin(\omega t+\phi) \tag{9.1.4}$$

where we have used the identity

$$\sin(\omega t+\phi)=\cos\phi\sin\omega t+\sin\phi\cos\omega t$$

with $\phi=-\tan^{-1}\omega\tau$.

If we substitute $s=j\omega$ into the transfer function (9.1.2), we obtain

$$T(j\omega)=\frac{1}{1+j\omega\tau}$$

The magnitude is

$$|T(j\omega)|=\left|\frac{1}{1+j\omega\tau}\right|=\frac{1}{\sqrt{1+\omega^2\tau^2}} \tag{9.1.5}$$

and the angle is

$$\phi=\angle T(j\omega)=-\angle(1+j\omega\tau)=-\tan^{-1}(\omega\tau) \tag{9.1.6}$$

These results are summarized in Table 9.1.2.

So, if $f(t)=A\sin\omega t$, the steady state response of $\tau\dot{y}+y=f(t)$ obtained from (9.1.5) and (9.1.6) is identical to (9.1.4), as it should be. The point is that the transfer function provides an easier way to obtain the steady-state response compared to the Laplace transform solution method, if the forcing function is sinusoidal.

Table 9.1.2 Frequency response of the model $\tau\dot{y}+y=f(t)$.

$$M=\frac{|Y|}{|F|}=\frac{1}{\sqrt{1+\omega^2\tau^2}} \tag{1}$$

$$\phi=-\tan^{-1}(\omega\tau) \tag{2}$$

$$f(t)=A\sin\omega t \tag{3}$$

$$B=AM \tag{4}$$

$$y_{ss}(t)=B\sin(\omega t+\phi) \tag{5}$$

Note that for a stable system, the free response disappears eventually, so the steady-state response is independent of the initial conditions.

Frequency Response of a Mass with Damping

EXAMPLE 9.1.1

■ Problem

Consider a mass subjected to a sinusoidal applied force $f(t)$. The mass is $m = 0.2$ kg and the damping constant is $c = 1$ N · s/m. If v is the speed of the mass, then the equation of motion is $0.2\dot{v} + v = f(t)$ where $f(t) = \sin \omega t$ and ω is the oscillation frequency of the applied force. The initial speed is $v(0) = 0$. Find the total response for two cases: (a) $\omega = 15$ rad/s and (b) $\omega = 60$ rad/s.

■ Solution

This equation is of the form of (9.1.1) with $\tau = 0.2$ and $y = v$. From (9.1.3) the response is

$$v(t) = \frac{0.2\omega}{1 + (0.2)^2\omega^2}\left(e^{-5t} - \cos \omega t + \frac{1}{0.2\omega} \sin \omega t\right)$$

The transient response contains the exponential e^{-5t}, which is essentially zero for $t > 4/5$ s. The steady-state response from (9.1.4) with $A = 1$ and $y = v$, is

$$v_{ss}(t) = \frac{1}{\sqrt{1 + (0.2)^2\omega^2}} \sin(\omega t + \phi)$$

where from (9.1.6), $\phi = -\tan^{-1}(0.2\omega)$.

The steady-state response can be seen to oscillate at the input frequency ω with an amplitude of $1/\sqrt{1 + 0.04\omega^2}$, and a phase shift of ϕ relative to the input. Since the phase shift is negative, a peak in the input is followed by a peak in the response a time $|\phi|/\omega$ later. If ϕ is in radians and ω is in radians per second, the time shift $|\phi|/\omega$ will be in seconds.

Figure 9.1.5a shows the total response for the case where $\omega = 15$. Part (b) shows the case where $\omega = 60$. Note that the transient response has in both cases essentially disappeared by $t = 4/5$ as predicted. Note also that the amplitude of the response is much smaller at the higher frequency. This is because the inertia of the mass prevents it from reacting to a rapidly changing input. Although the time shift is also smaller at the higher frequency, this does not indicate a faster response, because the peak attained is much smaller.

Circuit Response to a Step-Plus-Cosine Input

EXAMPLE 9.1.2

■ Problem

The model of the voltage v_2 across the capacitor in a series RC circuit having an input voltage v_1 is

$$RC\frac{dv_2}{dt} + v_2 = v_1(t)$$

Suppose that $RC = 0.02$ s and that the applied voltage consists of a step function plus a cosine function: $v_1(t) = 5u_s(t) + 3\cos 80t$. Obtain the circuit's steady-state response.

■ Solution

We can apply the principle of superposition here to separate the effects of the step input from those of the cosine input. Note that if only a step voltage of 5 were applied, the steady-state response to the step would be 5 (this can be shown by setting $dv_2/dt = 0$ in the model to see that $v_2 = v_1 = 5$ at steady state). Next find the steady-state response due to the *cosine* function $3\cos 80t$. To do this, we can use the same formulas developed for the amplitude and phase shift

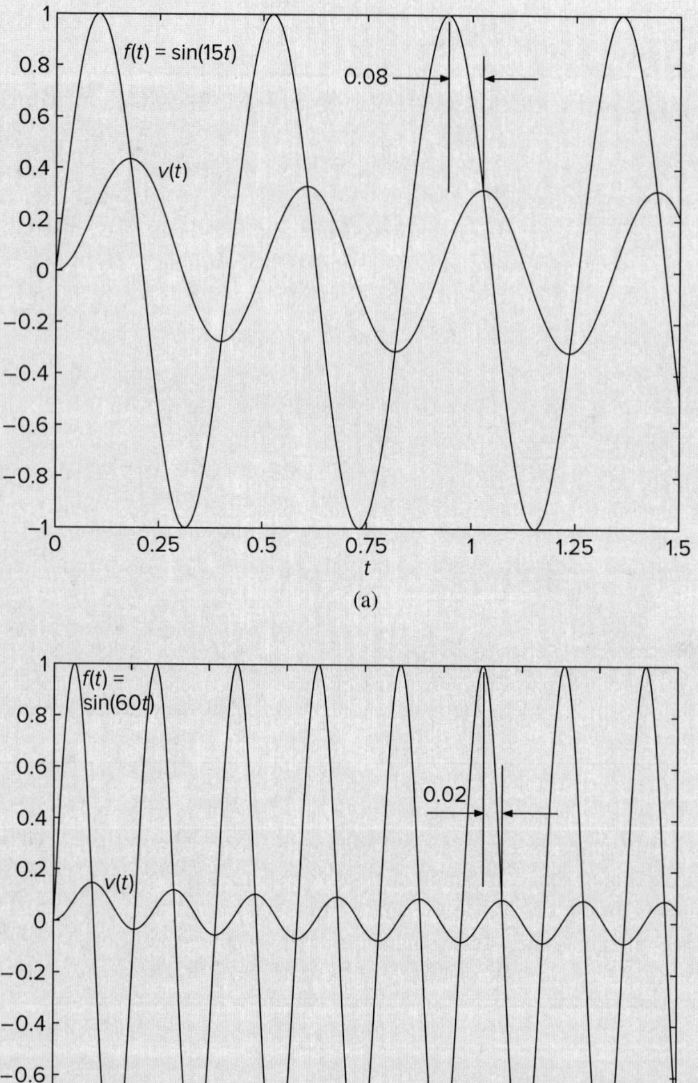

Figure 9.1.5 Response of the model $0.2\dot{v} + v = \sin \omega t$: (a) $\omega = 15$ and (b) $\omega = 60$.

due to a sine function, but express the response in terms of a cosine function. From (9.1.5) and (9.1.6) with $A = 3$,

$$B = AM = \frac{3}{\sqrt{1 + (80)^2(0.02)^2}} = 1.59$$

$$\phi = -\tan^{-1}[(80)(0.02)] = -\tan^{-1}(1.6) = -1.012 \text{ rad}$$

The steady-state response is $v_2(t) = 5 + 1.59\cos(80t - 1.012)$. Thus, at steady state, the voltage oscillates about the mean value 5 V with an amplitude of 1.59 and a frequency of 80 rad/s.

9.1.6 THE LOGARITHMIC PLOTS

Inspecting (9.1.5) reveals that the steady-state amplitude of the output decreases as the frequency of the input increases. The larger τ is, the faster the output amplitude decreases with frequency. At a high frequency, the system's "inertia" prevents it from closely following the input. The larger τ is, the more sluggish is the system response. This also produces an increasing phase lag as ω increases. Figure 9.1.6 shows the magnitude ratio and phase curves for two values of τ. In both cases, the magnitude ratio is close to 1 at low frequencies and approaches 0 as the frequency increases. The rate of decrease is greater for systems having larger time constants. The phase angle is close to 0 at low frequencies and approaches $-90°$ as the frequency increases.

Logarithmic scales are usually used to plot the frequency response curves. There are two reasons for using logarithmic scales. The curves of M and ϕ versus ω can be sketched more easily if logarithmic axes are used because they enable us to add or subtract the magnitude plots of simple transfer functions to sketch the plot for a transfer function composed of the product and ratio of simpler ones. Logarithmic scales also can display a wider variation in numerical values. When plotted using logarithmic scales, the frequency response plots are frequently called *Bode* plots, after H. W. Bode, who applied these techniques to the design of amplifiers.

Keep in mind the following basic properties of logarithms:

$$\log(xy) = \log x + \log y \qquad \log\left(\frac{x}{y}\right) = \log x - \log y$$

$$\log x^n = n \log x$$

When using logarithmic scales, the amplitude ratio M is specified in decibel units, denoted dB. (The decibel is named for Alexander Graham Bell.) The relationship between a number M and its decibel equivalent m is

$$m = 10 \log M^2 = 20 \log M \text{ dB} \tag{9.1.7}$$

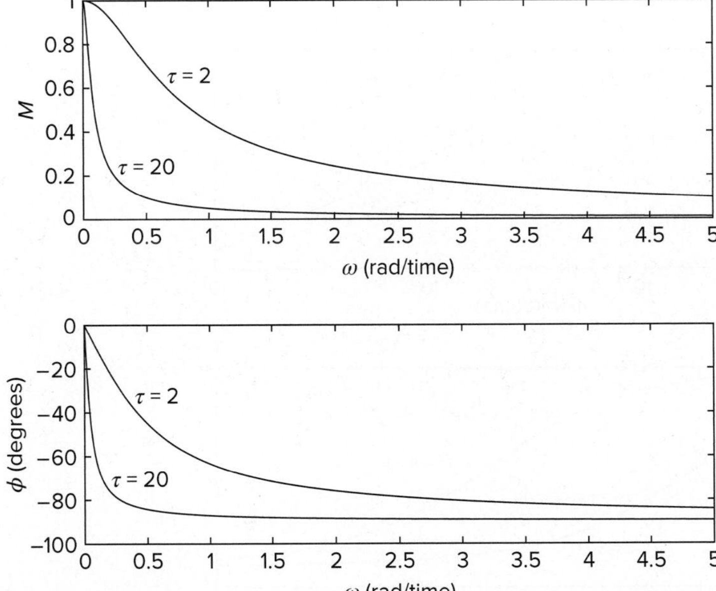

Figure 9.1.6 Magnitude ratio and phase angle of the model $\tau\dot{y} + y = f(t)$ for $\tau = 2$ and $\tau = 20$.

where the logarithm is to the base 10. For example, the $M = 10$ corresponds to 20 dB; the $M = 1$ corresponds to 0 dB; numbers less than 1 have negative decibel values. It is common practice to plot $m(\omega)$ in decibels versus $\log \omega$. For easy reference, $\phi(\omega)$ is also plotted versus $\log \omega$, and it is common practice to plot ϕ in degrees.

9.1.7 LOGARITHMIC PLOTS FOR $\tau\dot{y} + y = f(t)$

From equation (1) of Table 9.1.2,

$$M = \frac{1}{\sqrt{1 + \omega^2\tau^2}} \tag{9.1.8}$$

So we have

$$\begin{aligned} m(\omega) &= 20 \log \frac{1}{\sqrt{1 + \tau^2\omega^2}} \\ &= 20 \log(1) - 10 \log(1 + \tau^2\omega^2) \\ &= -10 \log(1 + \tau^2\omega^2) \end{aligned} \tag{9.1.9}$$

Figure 9.1.7 shows the logarithmic plots for the two systems whose plots with rectilinear axes were given in Figure 9.1.6. Note that the shape of the m versus $\log \omega$ curve is very different from the shape of the M versus ω curve. The logarithmic plots can be confusing for beginners; take time to study them. Remember that $m = 0$ corresponds to a magnitude ratio of $M = 1$. Positive values of m correspond to $M > 1$, which means that the system *amplifies* the input. Negative values of m correspond to $M < 1$, which means the system *attenuates* the input. When you need to obtain values of M from a plot of m versus $\log \omega$, use the relation

$$M = 10^{m/20} \tag{9.1.10}$$

For example, $m = 50$ corresponds to $M = 316.2$; $m = -15$ corresponds to $M = 0.1778$; and $m = -3.01$ corresponds to $M = 1/\sqrt{2} = 0.7071$.

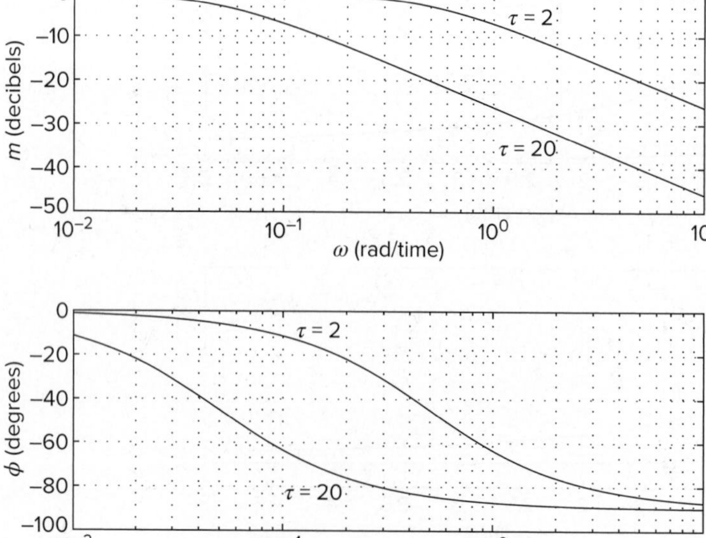

Figure 9.1.7 Semilog plots of log magnitude ratio and phase angle of the model $\tau\dot{y} + y = f(t)$ for $\tau = 2$ and $\tau = 20$.

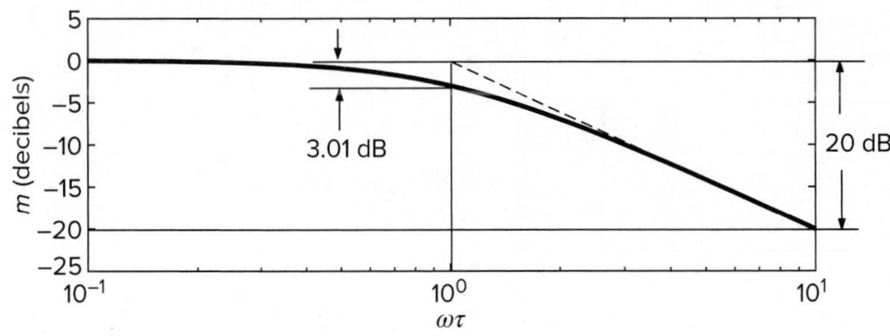

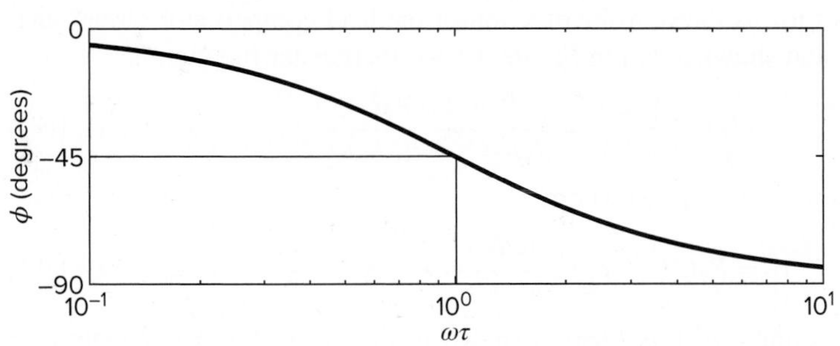

Figure 9.1.8 Asymptotes and corner frequency $\omega = 1/\tau$ of the model $1/(\tau s + 1)$.

To sketch the logarithmic plot of m versus $\log \omega$, we approximate $m(\omega)$ in three frequency ranges. For $\tau\omega \ll 1$, $1 + \tau^2\omega^2 \approx 1$, and (9.1.9) gives

$$m(\omega) \approx -10 \log 1 = 0 \tag{9.1.11}$$

For $\tau\omega \gg 1$, $1 + \tau^2\omega^2 \approx \tau^2\omega^2$, and (9.1.9) gives

$$m(\omega) \approx -10 \log \tau^2\omega^2 = -20 \log \tau\omega = -20 \log \tau - 20 \log \omega \tag{9.1.12}$$

This gives a straight line versus $\log \omega$. This line is the high-frequency asymptote. Its slope is −20 dB/decade, where a *decade* is any 10 : 1 frequency range. At $\omega = 1/\tau$, (9.1.12) gives $m(\omega) = 0$. This is useful for plotting purposes but does not represent the true value of m at that point because (9.1.12) was derived assuming $\tau\omega \gg 1$. For $\omega = 1/\tau$, (9.1.9) gives $m(\omega) = -10 \log 2 = -3.01$. Thus, at $\omega = 1/\tau$, $m(\omega)$ is 3.01 dB below the low-frequency asymptote given by (9.1.11). The low-frequency and high-frequency asymptotes meet at $\omega = 1/\tau$, which is the *breakpoint frequency*. It is also called the *corner frequency*. The plot is shown in Figure 9.1.8 (upper plot).

The phase angle is given by $\phi = -\tan^{-1}(\omega\tau)$, and the curve of ϕ versus ω is constructed as follows. For $\omega\tau \ll 1$, this equation gives $\phi(\omega) \approx \tan^{-1}(0) = 0°$. For $\omega\tau = 1$, $\phi(\omega) = -\tan^{-1}(1) = -45°$, and for $\omega\tau \gg 1$, $\phi(\omega) \approx -\tan^{-1}(\infty) = -90°$. Because the phase angle is negative, the output "lags" behind the input sine wave. Such a system is called a *lag* system. The $\phi(\omega)$ curves are easily sketched using these facts and are shown in Figure 9.1.8 (lower plot).

9.1.8 A COMMON FORM HAVING NUMERATOR DYNAMICS

An example of a first-order system having numerator dynamics is the transfer function

$$T(s) = K\frac{\tau_1 s + 1}{\tau_2 s + 1} \tag{9.1.13}$$

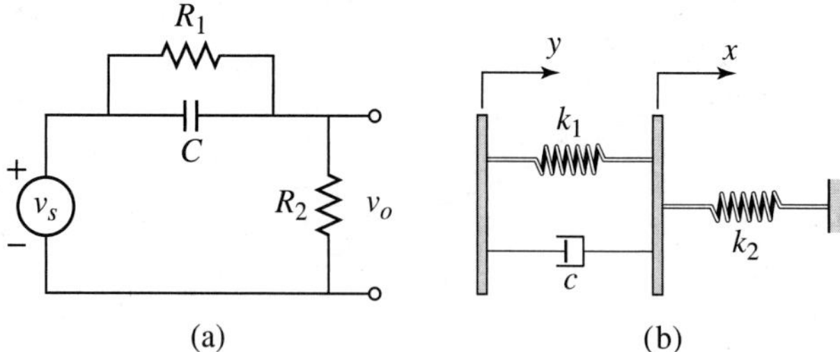

Figure 9.1.9 Electrical and mechanical examples having the transfer-function form $K(\tau_1 s + 1)/(\tau_2 s + 1)$.

An example of this form is the transfer function of the lead compensator circuit analyzed in Chapter 6, and shown again in Figure 9.1.9a. Its transfer function is

$$T(s) = \frac{V_o(s)}{V_s(s)} = \frac{R_1 R_2 Cs + R_2}{R_1 R_2 Cs + R_1 + R_2} \tag{9.1.14}$$

which can be rearranged as (9.1.13) where

$$\tau_1 = R_1 C \qquad \tau_2 = \frac{R_1 R_2 C}{R_1 + R_2} \qquad K = \frac{\tau_2}{\tau_1} \tag{9.1.15}$$

A mechanical example of this form is shown in Figure 9.1.9b. The equation of motion is

$$c\dot{x} + (k_1 + k_2)x = c\dot{y} + k_1 y$$

With y as the input, the transfer function is

$$T(s) = \frac{X(s)}{Y(s)} = \frac{cs + k_1}{cs + k_1 + k_2} \tag{9.1.16}$$

which can be rearranged as (9.1.13), where

$$\tau_1 = \frac{c}{k_1} \qquad \tau_2 = \frac{c}{k_1 + k_2} \qquad K = \frac{\tau_2}{\tau_1} \tag{9.1.17}$$

EXAMPLE 9.1.3

A Model with Numerator Dynamics

■ Problem

Find the steady-state response of the following system:

$$\dot{y} + 5y = 4\dot{g} + 12g$$

if the input is $g(t) = 20 \sin 4t$.

■ Solution

First, obtain the transfer function.

$$T(s) = \frac{Y(s)}{G(s)} = \frac{4s + 12}{s + 5} = 4\frac{s + 3}{s + 5}$$

Here $\omega = 4$, so we substitute $s = 4j$ to obtain

$$T(j\omega) = 4\frac{3 + j\omega}{5 + j\omega} = 4\frac{3 + 4j}{5 + 4j}$$

Then,

$$M = |T(j\omega)| = 4\frac{|3 + 4j|}{|5 + 4j|} = 4\frac{\sqrt{3^2 + 4^2}}{\sqrt{5^2 + 4^2}} = 3.123$$

The phase angle is found as follows:

$$\phi = \angle T(j\omega) = \angle\left(4\frac{3+j\omega}{5+j\omega}\right) = \angle 4 + \angle(3+j\omega) - \angle(5+j\omega)$$

$$= 0° + \tan^{-1}\frac{\omega}{3} - \tan^{-1}\frac{\omega}{5}$$

Substitute $\omega = 4$ to obtain

$$\phi = \tan^{-1}\frac{4}{3} - \tan^{-1}\frac{4}{5} = 0.253 \text{ rad}$$

Thus, the steady-state response is

$$y_{ss}(t) = 20M\sin(4t+\phi) = 62.46\sin(4t+0.253)$$

9.1.9 SKETCHING PLOTS USING ASYMPTOTES

Substituting $s = j\omega$ into the transfer function (9.1.13) gives

$$T(j\omega) = K\frac{\tau_1\omega j + 1}{\tau_2\omega j + 1} \tag{9.1.18}$$

Thus,

$$M(\omega) = |K|\frac{\sqrt{(\tau_1\omega)^2+1}}{\sqrt{(\tau_2\omega)^2+1}}$$

$$m(\omega) = 20\log|K| + 10\log[(\tau_1\omega)^2+1] - 10\log[(\tau_2\omega)^2+1] \tag{9.1.19}$$

Thus, the plot of $m(\omega)$ can be obtained by subtracting the plot of $\tau_2 s + 1$ from that of $\tau_1 s + 1$. The scale is then adjusted by $20\log|K|$. The sketches in Figure 9.1.10 are for $K = 1$, so that $20\log K = 0$.

The term $\tau_1 s + 1$ causes the curve to break upward at $\omega = 1/\tau_1$. The term $\tau_2 s + 1$ causes the curve to break downward at $\omega = 1/\tau_2$. If $1/\tau_1 > 1/\tau_2$, the composite curve looks like Figure 9.1.10a. If $1/\tau_1 < 1/\tau_2$, the plot is given by Figure 9.1.10b. The plots of $m(\omega)$ were obtained by using only the asymptotes of the terms $\tau_1 s + 1$ and $\tau_2 s + 1$ without using the 3-dB corrections at the corner frequencies $1/\tau_1$, and $1/\tau_2$. This sketching technique enables the designer to understand the system's general behavior quickly.

From (9.1.18), the phase angle is

$$\phi(\omega) = \angle K + \angle(\tau_1\omega j + 1) - \angle(\tau_2\omega j + 1)$$
$$= \angle K + \tan^{-1}(\tau_1\omega) - \tan^{-1}(\tau_2\omega) \tag{9.1.20}$$

where $\angle K = 0$ if $K > 0$. The plot can be found by combining the plots of $\phi(\omega)$ for K, $\tau_1 s + 1$, and $\tau_2 s + 1$, using the low-frequency, corner frequency, and high-frequency values of 0°, 45°, and 90° for the terms of the form $\tau s + 1$, supplemented by evaluations of (9.1.20) in the region between the corner frequencies. This sketching technique is more accurate when the corner frequencies are far apart.

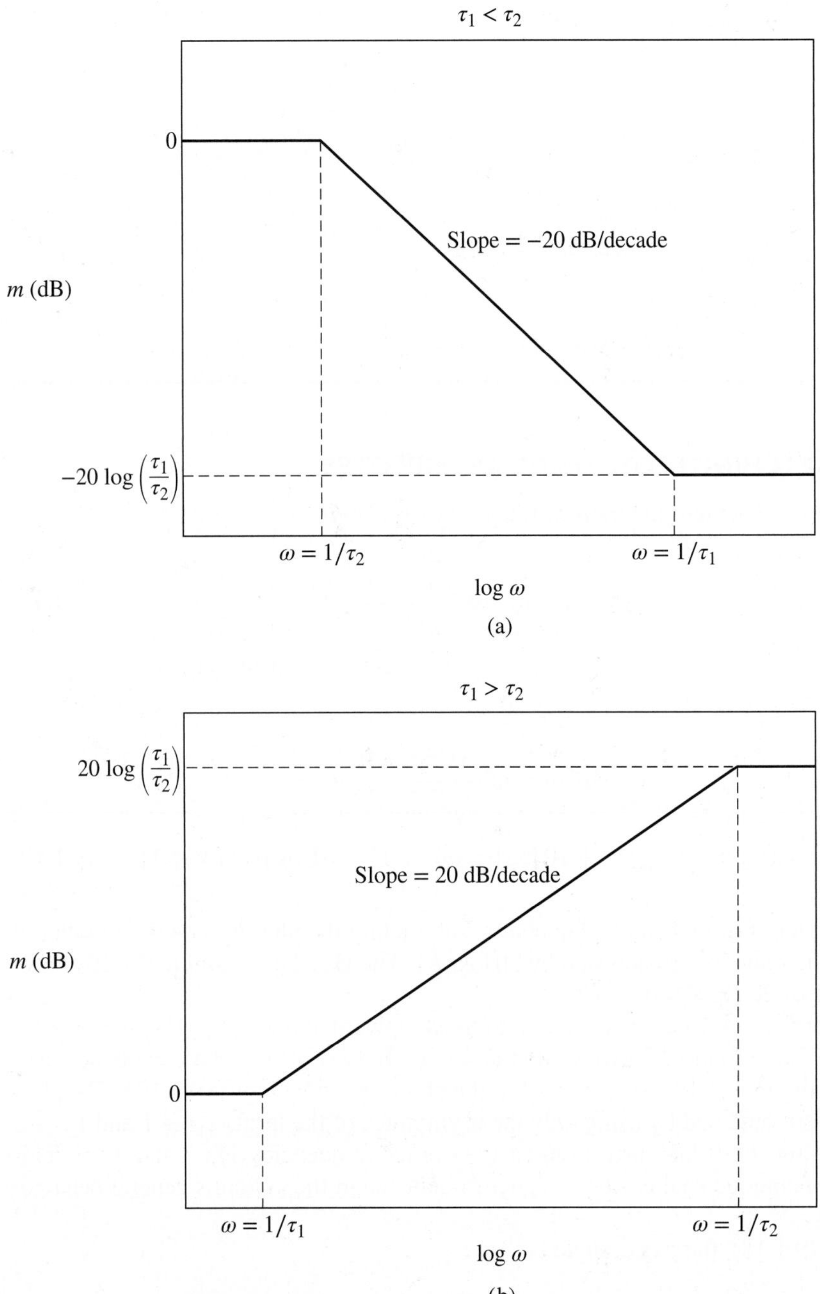

Figure 9.1.10 The log magnitude plots for the transfer-function form $(\tau_1 s + 1)/(\tau_2 s + 1)$ for (a) $\tau_1 < \tau_2$ and (b) $\tau_1 > \tau_2$.

9.2 FREQUENCY RESPONSE OF HIGHER-ORDER SYSTEMS

The results of Section 9.1 can be easily extended to a stable, invariant, linear system of any order. The general form of a transfer function is

$$T(s) = K\frac{N_1(s)N_2(s)\ldots}{D_1(s)D_2(s)\ldots} \tag{9.2.1}$$

where K is a constant real number. In general, if a complex number $T(j\omega)$ consists of products and ratios of complex factors, such that

$$T(j\omega) = K\frac{N_1(j\omega)N_2(j\omega)\ldots}{D_1(j\omega)D_2(j\omega)\ldots} \tag{9.2.2}$$

where K is constant and real, then from the properties of complex numbers

$$|T(j\omega)| = \frac{|K||N_1(j\omega)||N_2(j\omega)|\ldots}{|D_1(j\omega)||D_2(j\omega)|\ldots} \tag{9.2.3}$$

In decibel units, this implies that

$$\begin{aligned} m(\omega) &= 20\log|T(j\omega)| \\ &= 20\log|K| + 20\log|N_1(j\omega)| + 20\log|N_2(j\omega)| + \cdots \\ &\quad - 20\log|D_1(j\omega)| - 20\log|D_2(j\omega)| - \cdots \end{aligned} \tag{9.2.4}$$

That is, when expressed in logarithmic units, multiplicative factors in the numerator of the transfer function are summed, while those in the denominator are subtracted. We can use this principle graphically to add or subtract the contribution of each term in the transfer function to obtain the plot for the overall system transfer function.

For the form (9.2.1), the phase angle is

$$\begin{aligned} \phi(j\omega) = \angle T(j\omega) &= \angle K + \angle N_1(j\omega) + \angle N_2(j\omega) + \cdots \\ &\quad -\angle D_1(j\omega) - D_2(j\omega) - \cdots \end{aligned} \tag{9.2.5}$$

The phase angles of multiplicative factors in the numerator are summed, while those in the denominator are subtracted. This enables us to build the composite phase angle plot from the plots for each factor.

9.2.1 COMMON TRANSFER-FUNCTION FACTORS

Most transfer functions occur in the form given by (9.2.1). In addition, the factors $N_i(s)$ and $D_i(s)$ usually take the forms shown in Table 9.2.1. We have already obtained the frequency response plots for form 3. The effect of a multiplicative constant K, which is form 1, is to shift the m curve up by $20\log|K|$. If $K > 0$, the phase plot is unchanged because the angle of a positive number is 0°. If $K < 0$, the phase plot is shifted down by 180° because the angle of a negative number is −180°.

Table 9.2.1 Common factors in the transfer-function form: $T(s) = K\frac{N_1(s)N_2(s)\ldots}{D_1(s)D_2(s)\ldots}$.

Factor $N_i(s)$ or $D_i(s)$
1. Constant, K
2. s^n
3. $\tau s + 1$
4. $s^2 + 2\zeta\omega_n s + \omega_n^2 = \left[\left(\frac{s}{\omega_n}\right)^2 + 2\zeta\frac{s}{\omega_n} + 1\right]\omega_n^2, \quad \zeta < 1$

9.2.2 OVERDAMPED CASE

The denominator of a second-order transfer function with two real roots can be written as the product of two first-order factors like form 3. For example,

$$T(s) = \frac{1}{2s^2 + 14s + 20} = \frac{1}{2(s+2)(s+5)} = \frac{0.05}{(0.5s+1)(0.2s+1)} \tag{9.2.6}$$

where the time constants are $\tau_1 = 0.5$ and $\tau_2 = 0.2$.

EXAMPLE 9.2.1

Steady-State Response of an Underdamped System

■ Problem

a. Obtain the expressions for $m(\omega)$ and $\phi(\omega)$ for the following transfer function.

$$T(s) = \frac{X(s)}{F(s)} = \frac{0.05}{(0.5s+1)(0.2s+1)}$$

b. Obtain the steady-state response if $f(t) = 14 \sin 3t$.

■ Solution

a. First obtain M.

$$M(\omega) = |T(j\omega)| = \left| \frac{0.05}{(0.5\omega j + 1)(0.2\omega j + 1)} \right| = \frac{0.05}{\sqrt{(0.5\omega)^2 + 1}\sqrt{(0.2\omega)^2 + 1}} \tag{1}$$

Then

$$m(\omega) = 20 \log M = 20 \log 0.05 - 20 \log \sqrt{(0.5\omega)^2 + 1} - 20 \log \sqrt{(0.2\omega)^2 + 1}$$

or

$$m(\omega) = -26.0206 - 10 \log[(0.5\omega)^2 + 1] - 10 \log[(0.2\omega)^2 + 1] \tag{2}$$

The phase angle is

$$\phi(\omega) = \angle 0.05 - \angle(0.5\omega j + 1) - \angle(0.2\omega j + 1) = 0 - \tan^{-1} \frac{0.5\omega}{1} - \tan^{-1} \frac{0.2\omega}{1}$$

or

$$\phi(\omega) = -\tan^{-1} 0.5\omega - \tan^{-1} 0.2\omega \tag{3}$$

Equations (2) and (3) can be plotted versus ω, but the easiest way is to obtain the plots with MATLAB. How this is done is the subject of Section 9.7. The plots are shown in Figure 9.2.1.

b. Because the input is given as $f(t) = 14 \sin 3t$, we substitute $\omega = 3$ into (1) and (3) to obtain

$$M = \frac{0.05}{\sqrt{(1.5)^2 + 1}\sqrt{(0.6)^2 + 1}} = 0.0238$$

$$\phi = -\tan^{-1} 1.5 - \tan^{-1} 0.6 = -1.5232 \text{ rad}$$

Thus, the steady-state response is

$$x(t) = 14M \sin(3t + \phi) = 14(0.0238) \sin(3t - 1.5232) = 0.3332 \sin(3t - 1.5232)$$

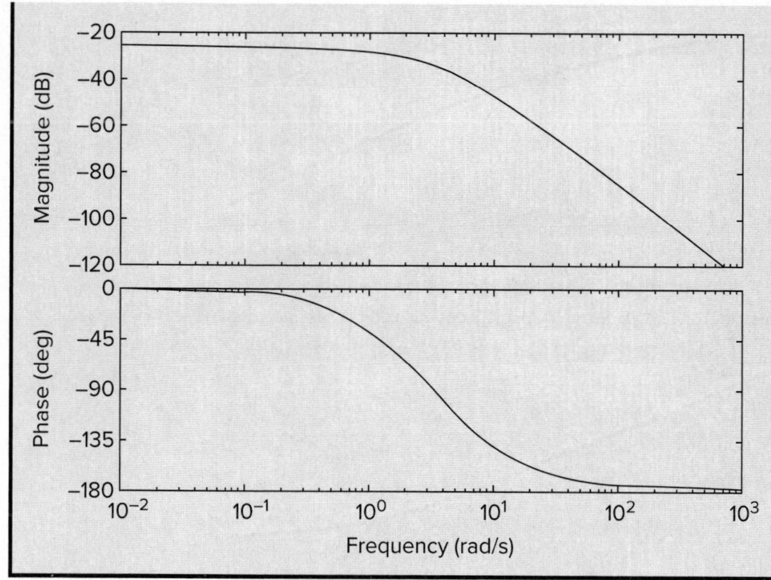

Figure 9.2.1 Frequency response plots for the transfer function $0.05/(0.5s+1)(0.2s+1)$.

Consider the second-order model

$$m\ddot{x} + c\dot{x} + kx = f(t)$$

Its transfer function is

$$T(s) = \frac{X(s)}{F(s)} = \frac{1}{ms^2 + cs + k} \tag{9.2.7}$$

If the system is *overdamped*, both roots are real and distinct, and we can write $T(s)$ as

$$T(s) = \frac{1/k}{(m/k)s^2 + (c/k)s + 1} = \frac{1/k}{(\tau_1 s + 1)(\tau_2 s + 1)} \tag{9.2.8}$$

where τ_1 and τ_2 are the time constants of the roots.

Substitute $s = j\omega$ into (9.2.8).

$$T(j\omega) = \frac{1/k}{(\tau_1 j\omega + 1)(\tau_2 j\omega + 1)}$$

The magnitude ratio is

$$M(\omega) = |T(j\omega)| = \frac{|1/k|}{|\tau_1 j\omega + 1||\tau_2 j\omega + 1|}$$

Thus,

$$m(\omega) = 20\log M(\omega) = 20\log\left|\frac{1}{k}\right| - 20\log|\tau_1\omega j + 1| - 20\log|\tau_2\omega j + 1| \tag{9.2.9}$$

The phase angle is

$$\phi(\omega) = \angle\frac{1}{k} - \angle(\tau_1\omega j + 1) - \angle(\tau_2\omega j + 1) \tag{9.2.10}$$

The magnitude ratio plot in decibels consists of a constant term, $20\log|1/k|$, minus the sum of the plots for two first-order lead terms. Assume that $\tau_1 > \tau_2$. Then for $1/\tau_1 < \omega < 1/\tau_2$, the slope is approximately −20 dB/decade. For $\omega > 1/\tau_2$, the

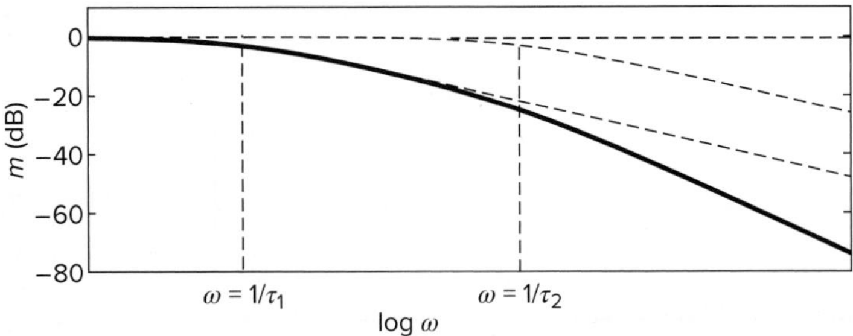

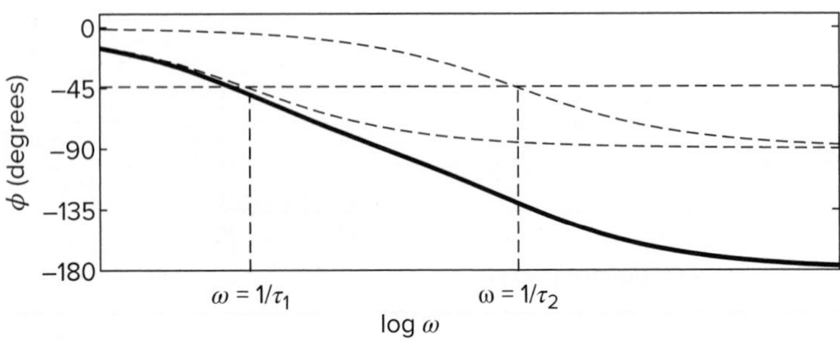

Figure 9.2.2 Semilog plots of log magnitude ratio and phase angle of the model $1/(\tau_1 s + 1)(\tau_2 s + 1)$.

contribution of the term $(\tau_2 \omega j + 1)$ becomes significant. This causes the slope to decrease by an additional 20 dB/decade, to produce a net slope of −40 dB/decade for $\omega > 1/\tau_2$. The rest of the plot can be sketched as before. The result is shown in Figure 9.2.2 (upper plot) for $k = 1$. The phase angle plot shown in Figure 9.2.2 (lower plot) is produced in a similar manner by using (9.2.10). Note that if $k > 0$, $\angle(1/k) = 0°$.

9.2.3 UNDERDAMPED CASE

We now consider the *underdamped* case of a second-order system that has two complex roots.

EXAMPLE 9.2.2

Response with Two Complex Roots

■ Problem

The model of a certain system is

$$6\ddot{x} + 12\dot{x} + 174x = 15f(t)$$

a. Obtain its steady-state response for $f(t) = 5 \sin 7t$.
b. Obtain the expressions for $m(\omega)$ and $\phi(\omega)$.

■ Solution

a. The system's transfer function is

$$T(s) = \frac{X(s)}{F(s)} = \frac{15}{6s^2 + 12s + 174}$$

Substitute $s = 7j$.

$$T(7j) = \frac{15}{-6(7)^2 + 12(7j) + 174} = \frac{15}{-120 + 84j}$$

Then

$$M = |T(7j)| = \frac{15}{\sqrt{(120)^2 + (84)^2}} = 0.1024$$

The phase angle is

$$\phi = \angle T(7j) = \angle(15) - \angle(-120 + 84j) = 0^\circ - \angle(-120 + 84j)$$

Noting that $\angle(-120 + 84j)$ is in the second quadrant, we obtain

$$\phi = -\left[\pi + \tan^{-1}\left(\frac{-84}{120}\right)\right] = -2.531 \text{ rad}$$

Thus, the steady-state response is

$$x_{ss}(t) = 5M\sin(7t + \phi) = 0.512\sin(7t - 2.531)$$

b. Replacing s with $j\omega$ gives

$$T(j\omega) = \frac{15}{-6\omega^2 + 12\omega j + 174}$$

Thus,

$$M(\omega) = \frac{15}{\sqrt{(174 - 6\omega^2)^2 + 144\omega^2}}$$

and

$$m(\omega) = 20\log 15 - 10\log\left[\left(174 - 6\omega^2\right)^2 + 144\omega^2\right]$$

$$\phi(\omega) = \angle(15) - \angle\left[\left(174 - 6\omega^2\right) + 12\omega j\right]$$

$$= \begin{cases} -\tan^{-1}\dfrac{12\omega}{174 - 6\omega^2} & \text{if } 174 - 6\omega^2 > 0 \\ \tan^{-1}\dfrac{12\omega}{6\omega^2 - 174} - 180^\circ & \text{if } 174 - 6\omega^2 < 0 \end{cases}$$

The plots are shown in Figure 9.2.3.

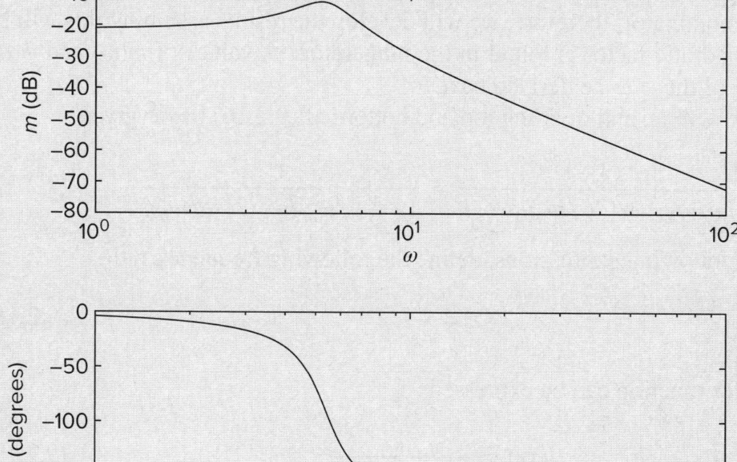

Figure 9.2.3 Semilog plots of log magnitude ratio and phase angle of the model $15/(6s^2 + 12s + 174)$.

If the transfer function

$$\frac{X(s)}{F(s)} = \frac{1}{ms^2 + cs + k} \tag{9.2.11}$$

has complex conjugate roots, it can be expressed as form 4 in Table 9.2.1 as follows:

$$T(s) = \frac{kX(s)}{F(s)} = \frac{1}{(m/k)s^2 + (c/k)s + 1} = \frac{1}{(s/\omega_n)^2 + 2\zeta(s/\omega_n) + 1} \tag{9.2.12}$$

where we have defined the natural frequency and damping as usual to be

$$\omega_n = \sqrt{\frac{k}{m}} \tag{9.2.13}$$

$$\zeta = \frac{c}{2\sqrt{mk}} \tag{9.2.14}$$

As shown in Chapter 2, the roots can be expressed in terms of these parameters as

$$s = -\zeta\omega_n \pm j\omega_n\sqrt{1-\zeta^2} \tag{9.2.15}$$

The roots are complex if $\zeta < 1$. Thus, $T(s)$ becomes

$$T(s) = \frac{kX(s)}{F(s)} = \frac{\omega_n^2}{s^2 + 2\zeta\omega_n s + \omega_n^2} \tag{9.2.16}$$

It is important to note that the reason for multiplying by k is that the quantity $kX(s)$ represents a force, as does $F(s)$. Therefore, the ratio $kX(s)/F(s)$ represents a dimensionless quantity that is a function of only two parameters, ζ and ω_n, instead of the three parameters m, c, and k. This allows us to obtain a more general understanding of the frequency response.

For most applications of interest here, the quadratic factor given by form 4 in Table 9.2.1 occurs in the denominator; therefore, we will develop the results assuming this will be the case. If a quadratic factor is found in the numerator, its values of $m(\omega)$ and $\phi(\omega)$ are the negative of those to be derived next.

Replacing s with $j\omega$ and dividing top and bottom of (9.2.16) by ω_n^2 gives

$$T(j\omega) = \frac{1}{(j\omega/\omega_n)^2 + (2\zeta/\omega_n)j\omega + 1} = \frac{1}{1 - (\omega/\omega_n)^2 + (2\zeta\omega/\omega_n)j} \tag{9.2.17}$$

To simplify the following expressions, define the following frequency ratio:

$$r = \frac{\omega}{\omega_n} \tag{9.2.18}$$

Thus, the transfer function can be expressed as

$$T(r) = \frac{1}{1 - r^2 + 2\zeta rj} \tag{9.2.19}$$

The magnitude ratio is

$$M = \frac{1}{|1 - r^2 + 2\zeta rj|} = \frac{1}{\sqrt{(1-r^2)^2 + (2\zeta r)^2}} \tag{9.2.20}$$

The log magnitude ratio is

$$m = 20\log\left|\frac{1}{1 - r^2 + 2\zeta rj}\right|$$
$$= -20\log\sqrt{(1 - r^2)^2 + (2\zeta r)^2}$$
$$= -10\log\left[(1 - r^2)^2 + (2\zeta r)^2\right] \tag{9.2.21}$$

The asymptotic approximations are as follows. For $r \ll 1$ (that is, for $\omega \ll \omega_n$), $m \approx -20\log 1 = 0$. For $r \gg 1$ (that is, for $\omega \gg \omega_n$),

$$m \approx -20\log\sqrt{r^4 + 4\zeta^2 r^2}$$
$$\approx -20\log\sqrt{r^4}$$
$$= -40\log r$$

Thus, at low frequencies where $\omega \ll \omega_n$, the curve is horizontal at $m = 0$, while for high frequencies $\omega \gg \omega_n$ where $r \gg 1$, the curve has a slope of −40 dB/decade, just as in the overdamped case (Figure 9.2.4a). The high-frequency and low-frequency asymptotes intersect at the corner frequency $\omega = \omega_n$.

The phase angle plot can be obtained in a similar manner (Figure 9.4.2b). From the additive property for angles, we see that for (9.2.19),

$$\phi = -\angle(1 - r^2 + 2\zeta rj)$$

Thus,

$$\phi = -\tan^{-1}\left(\frac{2\zeta r}{1 - r^2}\right)$$

where ϕ is in the third or fourth quadrant. For $r \ll 1$, $\phi \approx 0°$. For $r \gg 1$, $\phi \approx -180°$. At $\omega = \omega_n$, $\phi = -90°$ independently of ζ. The curve is skew-symmetric about the inflection point at $\phi = -90°$ for all values of ζ.

9.2.4 RESONANCE

The complex roots case differs from the real roots case in the vicinity of the corner frequency. To see this, note that M given by (9.2.20) has a maximum value when the denominator has a minimum. Setting the derivative of the denominator with respect to r equal to zero shows that the maximum M occurs at $r = \sqrt{1 - 2\zeta^2}$, which corresponds to the frequency $\omega = \omega_n\sqrt{1 - 2\zeta^2}$. This frequency is the *resonant frequency* ω_r. The peak of M exists only when the term under the radical is positive; that is, when $\zeta \le 0.707$. Thus, the resonant frequency is given by

$$\omega_r = \omega_n\sqrt{1 - 2\zeta^2} \qquad 0 \le \zeta \le 0.707 \tag{9.2.22}$$

The peak, or resonant, value of M, denoted by M_r, is found by substituting $r = \sqrt{1 - 2\zeta^2}$ into the expression (9.2.20) for M. This gives

$$M_r = \frac{1}{2\zeta\sqrt{1 - \zeta^2}} \qquad 0 \le \zeta \le 0.707 \tag{9.2.23}$$

If $\zeta > 0.707$, no peak exists, and the maximum value of M occurs at $\omega = 0$ where $M = 1$. Note that as $\zeta \to 0$, $\omega_r \to \omega_n$, and $M_r \to \infty$. For an undamped system, the roots are purely imaginary, $\zeta = 0$, and the resonant frequency is the natural frequency ω_n. These formulas are summarized in Table 9.2.2.

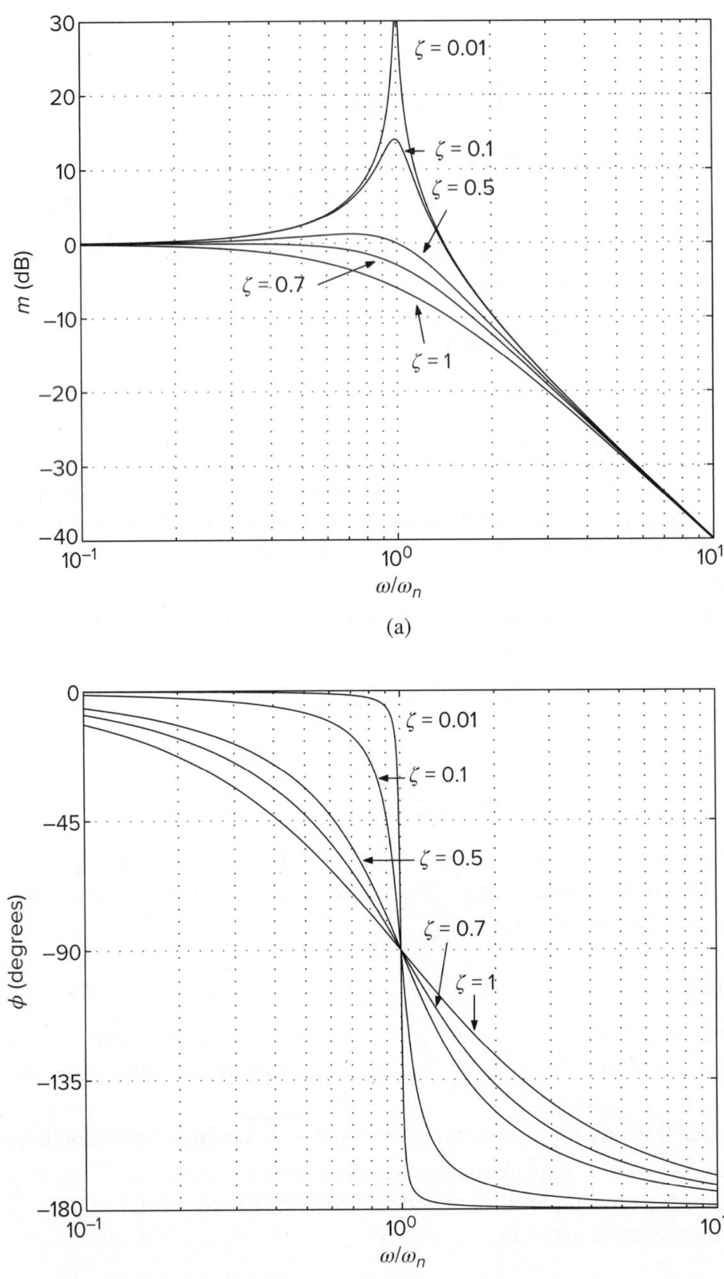

Figure 9.2.4 Semilog plots of log magnitude ratio and phase angle of the model $\omega_n^2/(s^2 + 2\zeta\omega_n s + \omega_n^2)$.

Resonance occurs when the input frequency is close to the resonant frequency of the system. If the damping is small, the output amplitude will continue to increase until either the linear model is no longer accurate or the system fails. When ϕ is near $-90°$, the *velocity* $\dot{x}$ is *in phase* with the input. This causes the large amplitude. Circuit designers take advantage of resonance by designing amplification circuits whose natural frequency is close to the frequency of a signal they want to amplify (such as the signal from a radio station). Designers of structural systems and suspensions try to avoid resonance because of the damage or discomfort that large motions can produce.

Table 9.2.2 Frequency response of a second-order system.

Transfer Function:	$\dfrac{k}{ms^2+cs+k} = \dfrac{\omega_n^2}{s^2+2\zeta\omega_n s+\omega_n^2}$
Natural Frequency:	$\omega_n = \sqrt{\dfrac{k}{m}}$
Damping ratio:	$\zeta = \dfrac{c}{2\sqrt{mk}}$
Resonant frequency:	$\omega_r = \omega_n\sqrt{1-2\zeta^2} \quad 0 \le \zeta \le 0.707$
Resonant response:	$M_r = \dfrac{1}{2\zeta\sqrt{1-\zeta^2}} \quad 0 \le \zeta \le 0.707$
	$\phi_r = -\tan^{-1}\dfrac{\sqrt{1-2\zeta^2}}{\zeta} \quad 0 \le \zeta \le 0.707$

In Figure 9.2.4a, the correction to the asymptotic approximations in the vicinity of the corner frequency depends on the value of ζ. The peak value in decibels is

$$m_r = 20\log M_r = -20\log\left(2\zeta\sqrt{1-\zeta^2}\right) \tag{9.2.24}$$

At the resonant frequency ω_r, the phase angle is

$$\phi|_{r=\sqrt{1-2\zeta^2}} = -\tan^{-1}\frac{\sqrt{1-2\zeta^2}}{\zeta} \tag{9.2.25}$$

When $r = 1$ (at $\omega = \omega_n$),

$$m|_{r=1} = -20\log 2\zeta \tag{9.2.26}$$

The transfer function of a mass-spring-damper model $m\ddot{x} + c\dot{x} + kx = f(t)$ with an applied force $f(t)$ has the form

$$\frac{X(s)}{F(s)} = \frac{1}{ms^2+cs+k} \tag{9.2.27}$$

This can be put into the form required by Table 9.2.2 by multiplying the numerator and denominator by k to obtain

$$\frac{X(s)}{F(s)} = \frac{1}{k}\frac{k}{ms^2+cs+k} \tag{9.2.28}$$

Thus, we can use the results of Table 9.2.1 for (9.2.27), but we must divide the table formula for M_r by k. The formula for ϕ_r is unchanged.

Determining Maximum Response — EXAMPLE 9.2.3

■ **Problem**

The model of a certain mass-spring-damper system is

$$2\ddot{x} + c\dot{x} + 18x = f(t) = 14\sin\omega t$$

Determine its resonant frequency ω_r and its peak response x_{peak} at resonance if $\zeta = 0.4$.

■ **Solution**

The transfer function is

$$\frac{X(s)}{F(s)} = \frac{1}{2s^2+cs+18}$$

To put this in the form required by Table 9.2.2, we multiply and divide by 18 to obtain

$$\frac{X(s)}{F(s)} = \frac{1}{18}\frac{18}{2s^2 + cs + 18}$$

So to use the expression for M_r from Table 9.2.2, we must divide the result by 18. This gives

$$M_r = \frac{1}{18}\frac{1}{2\zeta\sqrt{1-\zeta^2}}$$

With $\zeta = 0.4$,

$$M_r = \frac{1}{18}\frac{1}{2(0.4)\sqrt{1-(0.4)^2}} = 0.0758$$

Thus, the peak response is

$$x_{\text{peak}} = 14M_r = 14(0.0758) = 1.061$$

Note that $\omega_n = \sqrt{18/2} = 3$. Thus, the resonant frequency is given by

$$\omega_r = \omega_n\sqrt{1-2\zeta^2} = 3\sqrt{1-2(0.4)^2} = 2.4739$$

Thus, if $f(t) = 14\sin 2.4739t$, the amplitude of the steady-state response will be 1.061. Note that we did not need the value of the damping coefficient c.

EXAMPLE 9.2.4

Limits on Damping

■ Problem

The model of a certain mass-spring-damper system is

$$5\ddot{x} + c\dot{x} + 80x = f(t) = 7\sin\omega t$$

How large must the damping constant c be so that the maximum steady-state amplitude of x is no greater than 2, for an arbitrary value of ω?

■ Solution

The transfer function is

$$\frac{X(s)}{F(s)} = \frac{1}{5s^2 + cs + 80}$$

To put this in the form required by Table 9.2.2, we multiply and divide by 80 to obtain

$$\frac{X(s)}{F(s)} = \frac{1}{80}\frac{80}{5s^2 + cs + 80}$$

So to use the expression for M_r from the table, we must divide the result by 80. This gives

$$M_r = \frac{1}{80}\frac{1}{2\zeta\sqrt{1-\zeta^2}}$$

Because the amplitude of the forcing function is 7, the maximum response amplitude, which occurs at $\omega = \omega_r$, is $7M_r$, and it can be no greater than 2. Therefore,

$$7M_r = \frac{1}{80}\frac{7}{2\zeta\sqrt{1-\zeta^2}} = 2$$

Solve for ζ by squaring each side to obtain

$$\zeta^4 - \zeta^2 + \frac{49}{102{,}400} = 0$$

This is a quadratic in ζ^2, and the solutions are $\zeta^2 = 0.9995, 0.00048$. Because ζ must be positive, we have two possible solutions: $\zeta = 0.9997, 0.0219$. The first solution is not acceptable because our formula for M_r is valid only for $\zeta \le 0.707$. So the solution is $\zeta = 0.0219$. We now find c.

From the definition of the damping ratio,

$$\zeta = 0.0219 = \frac{c}{2\sqrt{5(80)}}$$

Thus gives $c = 0.876$. Thus, if $c = 0.876$, there is no value of the forcing frequency ω that will produce a response amplitude greater than 2.

Once you understand how each of the forms shown in Table 9.2.1 contributes to the magnitude and phase plots, you can quickly determine the effects of each term in a more complicated transfer function.

A Fourth-Order Model

EXAMPLE 9.2.5

■ **Problem**

Determine the effect of the parameter τ of the magnitude plot of the following transfer function. Assume that time is measured in seconds.

$$T(s) = \frac{\tau s + 1}{s^4 + 40.8s^3 + 8337s^2 + 4.1184 \times 10^4 + 5.184 \times 10^5}$$

Is it possible to choose a value for τ to increase the gain at low frequencies?

■ **Solution**

First determine the roots of the denominator. They are

$$s = -2.4 \pm 7.632j \qquad -18 \pm 88.10j$$

For the first root pair,

$$\zeta = \cos\left[\tan^{-1}\left(\frac{7.632}{2.4}\right)\right] = 0.3 \qquad \omega_n = \sqrt{2.4^2 + 7.632^2} = 8$$

Similarly, for the second pair,

$$\zeta = \cos\left[\tan^{-1}\left(\frac{88.1}{18}\right)\right] = 0.2 \qquad \omega_n = \sqrt{18^2 + 88.1^2} = 90$$

Thus, $T(s)$ can be expressed in factored form as

$$T(s) = \frac{\tau s + 1}{(s^2 + 4.8s + 64)(s^2 + 36s + 8100)}$$

The model has two resonant frequencies corresponding to the two root pairs. These frequencies are

$$\omega_{r_1} = \omega_n\sqrt{1 - 2\zeta^2} = 8\sqrt{1 - 0.18} = 7.24 \text{ rad/s}$$

$$\omega_{r_2} = \omega_n\sqrt{1 - 2\zeta^2} = 90\sqrt{1 - 0.08} = 86.3 \text{ rad/s}$$

Because ζ is small for each factor, we can expect a resonant peak at these frequencies. However, we cannot use the formula (9.2.24) to calculate m_r at each peak because it applies only to a second-order system. Each quadratic term contributes −40 dB/decade to the composite slope at high frequencies. The numerator term contributes a slope of +20 dB/decade at frequencies above $1/\tau$. So the composite curve will have a slope of $20 - 2(40) = -60$ dB/decade at high frequencies.

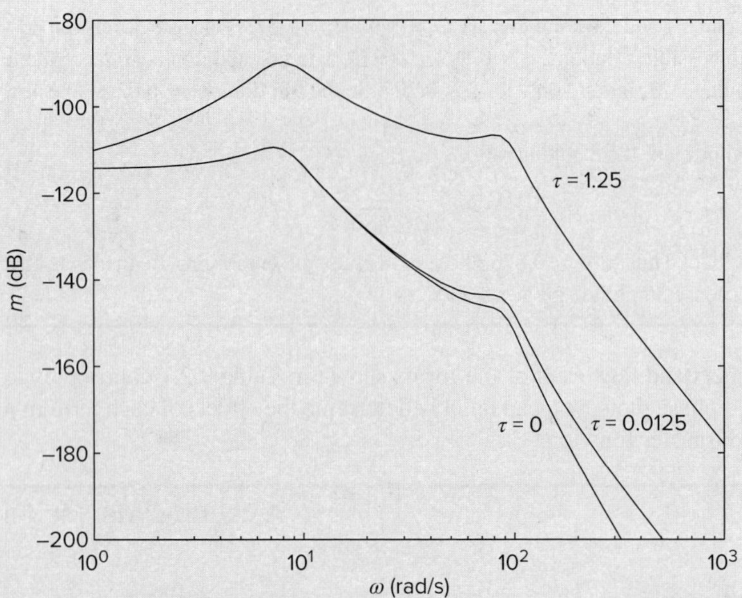

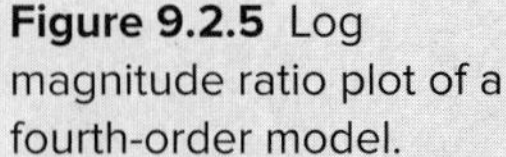
Figure 9.2.5 Log magnitude ratio plot of a fourth-order model.

The numerator term causes the m curve to break upward at $\omega = 1/\tau$. If $1/\tau$ is less than ω_{r_1}, the m curve will break upward before the −40 dB/decade slope of the first quadratic term takes effect. Figure 9.2.5 shows the m plot for three cases, including the case where $\tau = 1.25$, which corresponds to a corner frequency of $\omega = 1/\tau = 0.8$. As compared with the case having no numerator dynamics ($\tau = 0$), the choice of $\tau = 1.25$ can be seen to raise the magnitude. For example, at $\omega = 7.24$ rad/s, the resonant peaks for the two cases are −109 dB ($M = 3.55 \times 10^{-6}$) for $\tau = 0$, and −89.6 ($M = 3.31 \times 10^{-5}$) for $\tau = 1.25$. The amplitude ratio is $3.31 \times 10^{-5}/3.55 \times 10^{-6} = 9.34$ times larger for $\tau = 1.25$.

Using a value of $1/\tau$ that is larger than the smallest resonant frequency will not increase the amplitude ratio because the −40 dB/decade slope from the first quadratic term will take effect before the +20 dB/decade slope of the numerator term makes its contribution. An example of this is shown in Figure 9.2.5 for $\tau = 0.0125$ s, which corresponds to a corner frequency of $1/\tau = 80$ rad/s.

9.3 FREQUENCY RESPONSE APPLICATIONS

In this section we present more examples illustrating the concepts and applications of frequency response.

9.3.1 A NEUTRALLY STABLE CASE

While all real systems will have some damping, it is instructive and useful to obtain some results for the undamped case, where $c = \zeta = 0$. This is because the mathematical results for the undamped case are more easily derived and analyzed, and they give insight into the behavior of many real systems that have a small amount of damping.

If the model is stable, the free response term disappears in time. The results derived in Section 9.2 therefore must be reexamined for the case where there is no damping because this is not a stable case (it is neutrally stable). In this case, the magnitude and phase angle of the transfer function do not give the entire steady-state response. The free response for the undamped equation $m\ddot{x} + kx = F \sin \omega t$ is found with the

Laplace transform method as follows:

$$(ms^2 + k)X_{\text{free}}(s) - m\dot{x}(0) - msx(0) = 0$$

$$X_{\text{free}}(s) = \frac{m\dot{x}(0) + msx(0)}{ms^2 + k} = \frac{\dot{x}(0) + sx(0)}{s^2 + \omega_n^2}$$

Thus,

$$x_{\text{free}}(t) = \frac{\dot{x}(0)}{\omega_n} \sin \omega_n t + x(0) \cos \omega_n t$$

The forced response is found as follows:

$$(ms^2 + k)X_{\text{forced}}(s) = \frac{F\omega}{s^2 + \omega^2}$$

$$X_{\text{forced}}(s) = \frac{1}{m} \frac{F\omega}{\left(s^2 + \omega_n^2\right)(s^2 + \omega^2)}$$

Assuming that $\omega \neq \omega_n$, the partial-fraction expansion is

$$X_{\text{forced}}(s) = \frac{F\omega}{m\left(\omega^2 - \omega_n^2\right)} \left(\frac{\omega_n}{s^2 + \omega_n^2} - \frac{\omega_n}{\omega} \frac{\omega}{s^2 + \omega^2} \right)$$

Thus, the forced response is

$$x_{\text{forced}}(t) = \frac{F\omega}{m\left(\omega^2 - \omega_n^2\right)} \left(\sin \omega_n t - \frac{\omega_n}{\omega} \sin \omega t \right)$$

or, with $r = \omega/\omega_n$,

$$x_{\text{forced}}(t) = -\frac{Fr}{k(1 - r^2)} \sin \omega_n t + \frac{F}{k(1 - r^2)} \sin \omega t$$

Thus, the total response is

$$x(t) = \left[\frac{\dot{x}(0)}{\omega_n} - \frac{F}{k} \frac{r}{1 - r^2} \right] \sin \omega_n t + x(0) \cos \omega_n t + \frac{F}{k} \frac{1}{1 - r^2} \sin \omega t \tag{9.3.1}$$

There is no transient response here; the entire solution is the steady-state response.

9.3.2 BEATING

We may examine the effects of the forcing function independently of the effects of the initial conditions by setting $x(0) = \dot{x}(0) = 0$ in (9.3.1). The result is the forced response:

$$x(t) = \frac{F}{k} \frac{1}{1 - r^2} \left(\sin \omega t - r \sin \omega_n t \right) \tag{9.3.2}$$

When the forcing frequency ω is substantially different from the natural frequency ω_n, the forced response looks somewhat like Figure 9.3.1 and consists of a higher-frequency oscillation superimposed on a lower-frequency oscillation.

If the forcing frequency ω is close to the natural frequency ω_n, then $r \approx 1$ and the forced response (9.3.2) can be expressed as follows:

$$x(t) = \frac{F}{k} \frac{1}{1 - r^2} \left(\sin \omega t - \sin \omega_n t \right) = \frac{F}{m} \frac{1}{\omega_n^2 - \omega^2} \left(\sin \omega t - \sin \omega_n t \right)$$

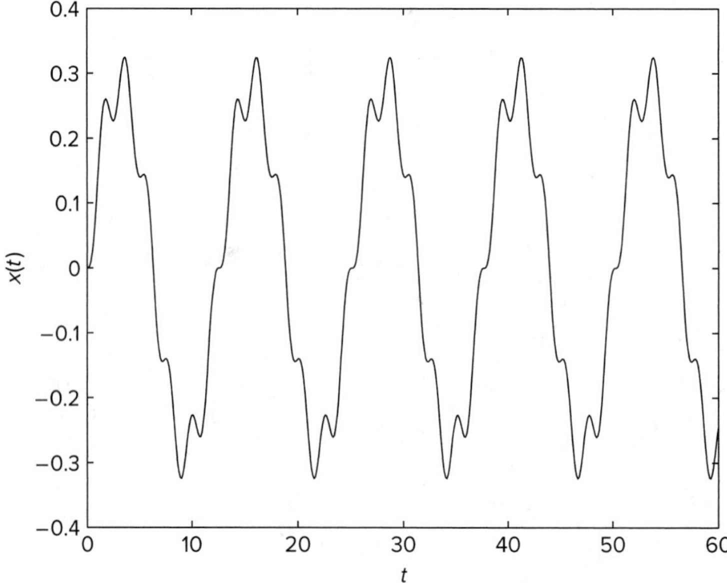

Figure 9.3.1 Response when the forcing frequency is substantially different from the natural frequency.

Using the identity

$$2 \sin \frac{1}{2}(A - B) \cos \frac{1}{2}(A + B) = \sin A - \sin B$$

we see that

$$2 \sin \frac{\omega - \omega_n}{2} t \cos \frac{\omega + \omega_n}{2} t = \sin \omega t - \sin \omega_n t$$

Thus, the forced response is given by

$$x(t) = \left(\frac{2F}{m} \frac{1}{\omega_n^2 - \omega^2} \sin \frac{\omega - \omega_n}{2} t \right) \cos \frac{\omega + \omega_n}{2} t \tag{9.3.3}$$

This can be interpreted as a cosine with a frequency $(\omega + \omega_n)/2$ and a time-varying amplitude of

$$\frac{2F}{m} \frac{1}{\omega_n^2 - \omega^2} \sin \frac{\omega - \omega_n}{2} t$$

The amplitude varies sinusoidally with the frequency $(\omega - \omega_n)/2$, which is lower than the frequency of the cosine. The response thus looks like Figure 9.3.2. This behavior, in which the amplitude increases and decreases periodically, is called *beating*. The *beat period* is the time between the occurrence of zeros in $x(t)$ and thus is given by the half-period of the sine wave, which is $2\pi/|\omega - \omega_n|$. The *vibration period* is the period of the cosine wave, $4\pi/(\omega + \omega_n)$.

The concept of beating is important for tuning musical instruments. When a piano tuner strikes a piano wire and a tuning fork emitting a standard frequency, he or she listens for the beats caused by the difference between the two frequencies and then tightens or loosens the piano wire until the beating disappears.

9.3.3 RESPONSE AT RESONANCE

Equation (9.2.20) shows that when $\zeta = 0$ the amplitude of the steady-state response becomes infinite when $r = 1$; that is, when the forcing frequency ω equals the natural frequency ω_n. The phase shift ϕ is exactly −90 degrees at this frequency.

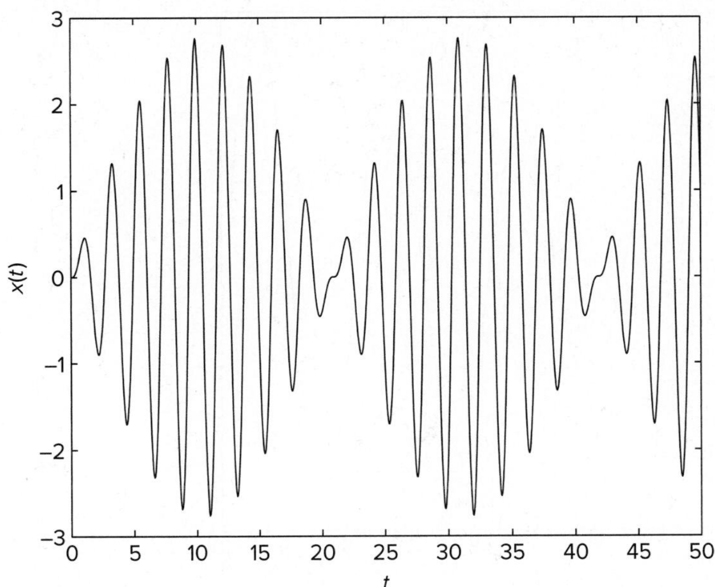

Figure 9.3.2 Beating response when the forcing frequency is close to the natural frequency.

To obtain the expression for $x(t)$ at resonance, we use (9.3.2) to compute the limit of $x(t)$ as $r \to 1$ after replacing ω in $x(t)$ using the relation $\omega = \omega_n r$. We must use l'Hôpital's rule to compute the limit.

$$\begin{aligned} x(t) &= \lim_{r\to 1} \frac{F}{k}\frac{1}{1-r^2}(\sin \omega_n rt - r \sin \omega_n t) \\ &= \frac{F}{k}\lim_{r\to 1}\frac{d(\sin \omega_n rt - r\sin \omega_n t)/dr}{d(1-r^2)/dr} \\ &= \frac{F}{k}\lim_{r\to 1}\frac{\omega_n t \cos \omega_n rt - \sin \omega_n t}{-2r} \\ &= \frac{F\omega_n}{2k}\left(\frac{\sin \omega_n t}{\omega_n} - t\cos \omega_n t\right) \end{aligned} \tag{9.3.4}$$

The plot is shown in Figure 9.3.3 for the case $m = 4$, $c = 0$, $k = 36$, and $F = 10$. The amplitude increases linearly with time. The behavior for lightly damped systems is similar except that the amplitude does not become infinite. Figure 9.3.4 shows the damped case: $m = 4$, $c = 4$, $k = 36$, and $F = 10$.

For large amplitudes, the linear model on which this analysis is based will no longer be accurate. In addition, all physical systems have some damping, so c will never be exactly zero, and the response amplitude will never be infinite. The important point, however, is that the amplitude might be large enough to damage the system or cause some other undesirable result. At resonance the output amplitude will continue to increase until either the linear model is no longer accurate or the system fails.

9.3.4 RESONANCE AND TRANSIENT RESPONSE

Although the analysis in Section 9.2 was a steady-state analysis and assumed that the forcing frequency ω is constant, resonance can still occur in transient processes if the input varies slowly enough to allow the steady-state response to begin to appear. For example, resonance can be a problem even if a machine's operating speed is well above its

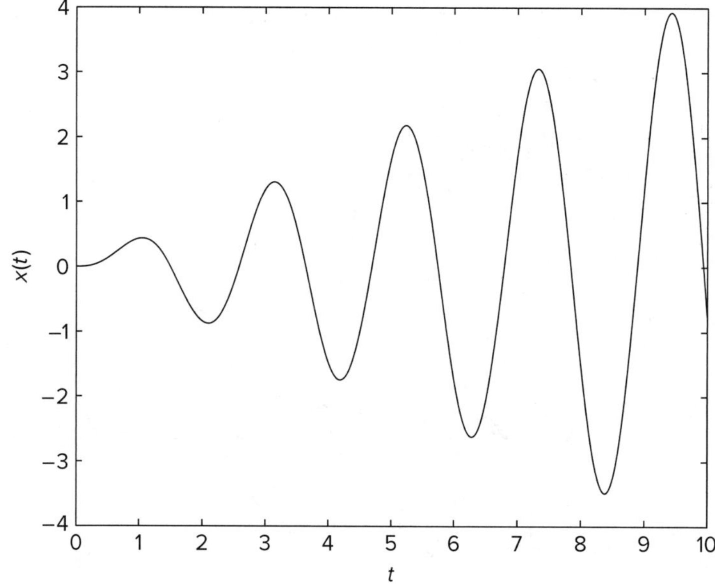

Figure 9.3.3 Response near resonance for an undamped system.

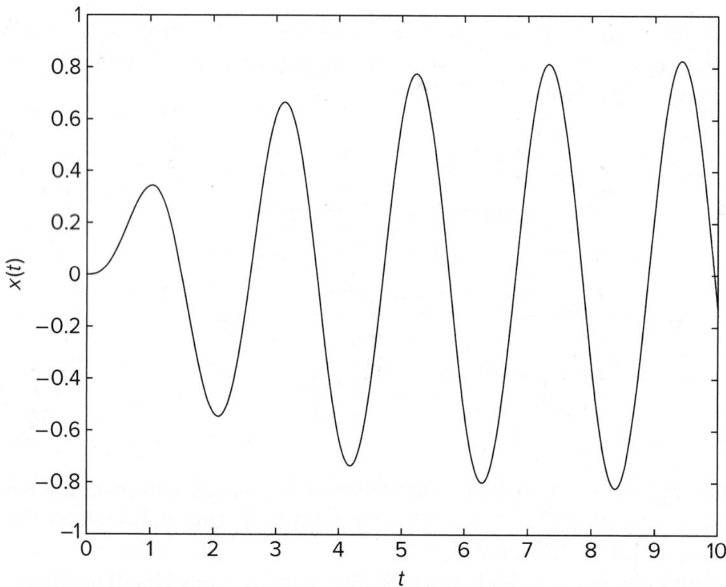

Figure 9.3.4 Response near resonance for a damped system.

resonant frequency, because the machine's speed must pass through the resonant frequency at startup. If the machine's speed does not pass through the resonant frequency quickly enough, high amplitude oscillations will result. Figure 9.3.5 shows the transient response of the model

$$4\ddot{x} + 3\dot{x} + 100x = 295\sin[\omega(t)t]$$

where the frequency $\omega(t)$ increases linearly with time, as $\omega(t) = 0.7t$. Thus, the frequency passes through the resonant frequency of this model, which is $\omega_r = 4.97$, at $t = 7.1$ s. The plot shows the large oscillations that occur as the input frequency passes close to the resonant frequency.

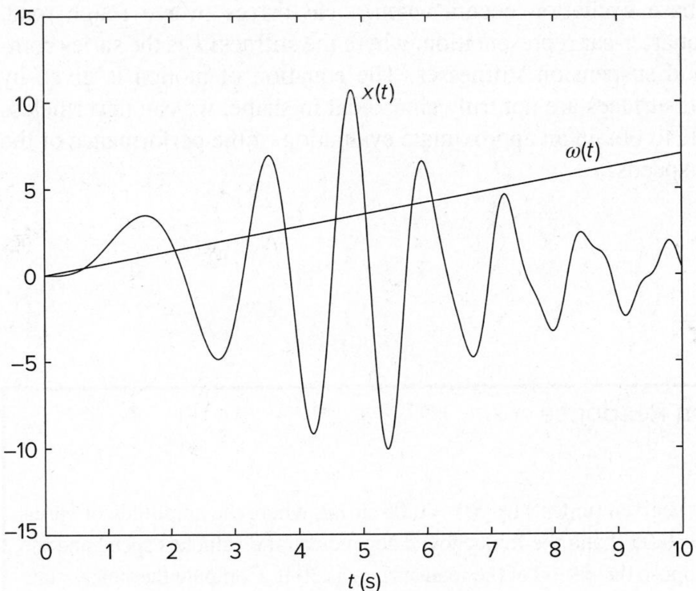

Figure 9.3.5 Transient response when an increasing forcing frequency passes through the resonant frequency.

9.3.5 BASE MOTION AND TRANSMISSIBILITY

The motion of the mass shown in Figure 9.3.6 is produced by the motion $y(t)$ of the base. This system is a model of many common displacement isolation systems. Assuming that the mass displacement x is measured from the rest position of the mass when $y = 0$, the weight mg is canceled by the static spring force. The force transmitted to the mass by the spring and damper is denoted f_t and is given by

$$f_t = c(\dot{y} - \dot{x}) + k(y - x) \tag{9.3.5}$$

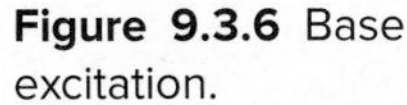
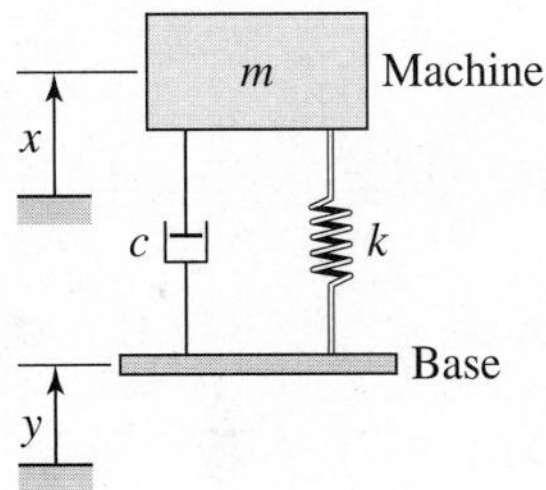

Figure 9.3.6 Base excitation.

This gives the following equation of motion:

$$m\ddot{x} + c\dot{x} + kx = c\dot{y} + ky \tag{9.3.6}$$

The transfer function is

$$\frac{X(s)}{Y(s)} = \frac{cs + k}{ms^2 + cs + k} \tag{9.3.7}$$

From (9.3.5)

$$F_t(s) = (cs + k)\,[Y(s) - X(s)] \tag{9.3.8}$$

Substituting for $X(s)$ from (9.3.7) gives

$$F_t(s) = (cs + k)\left[Y(s) - \frac{cs + k}{ms^2 + cs + k}Y(s)\right] = (cs + k)\frac{ms^2}{ms^2 + cs + k}Y(s)$$

Thus, the second desired ratio is

$$\frac{F_t(s)}{Y(s)} = (cs + k)\frac{ms^2}{ms^2 + cs + k} \tag{9.3.9}$$

This is the ratio of the transmitted force to the base motion.

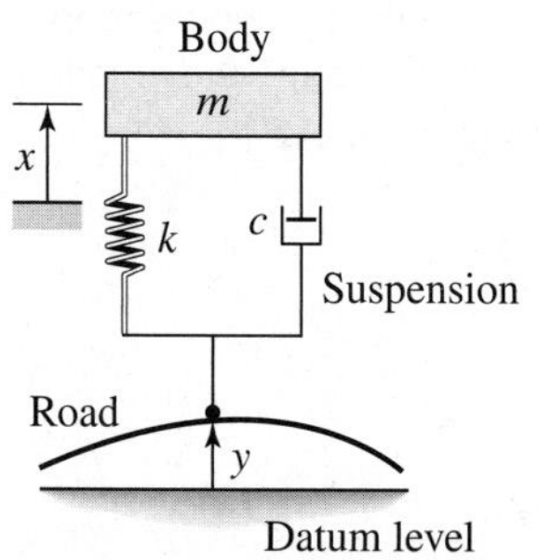

Figure 9.3.7 Single-mass suspension model.

An example of base excitation occurs when a car drives over a rough road. Figure 9.3.7 shows a quarter-car representation, where the stiffness k is the series combination of the tire and suspension stiffnesses. The equation of motion is given by (9.3.6). Although road surfaces are not truly sinusoidal in shape, we can nevertheless use a sinusoidal profile to obtain an approximate evaluation of the performance of the suspension at various speeds.

EXAMPLE 9.3.1

Vehicle Suspension Response

■ Problem

Suppose the road profile is given (in feet) by $y(t) = 0.05 \sin \omega t$, where the amplitude of variation of the road surface is 0.05 ft and the frequency ω depends on the vehicle's speed and the road profile's period. Suppose the period of the road surface is 30 ft. Compute the steady-state motion amplitude and the force transmitted to the chassis, for a car traveling at a speed of 30 mi/hr. The car weighs 3200 lb. The effective stiffness, which is a series combination of the tire stiffness and the suspension stiffness, is $k = 3000$ lb/ft. The damping is $c = 300$ lb-sec/ft.

■ Solution

For a period of 30 ft and a vehicle speed of 30 mi/hr, the frequency ω is

$$\omega = \left(\frac{5280}{30}\right)\left(\frac{1}{3600}\right)(2\pi)30 = 9.215 \text{ rad/sec}$$

For the car weighing 3200 lb, the quarter-car mass is $m = 800/32.2$ slug. From (9.3.7) with $s = j\omega = 9.215j$,

$$\frac{|X|}{|Y|} = \frac{\sqrt{k^2 + (c\omega)^2}}{\sqrt{(k - m\omega^2)^2 + (c\omega)^2}} = 1.405$$

Thus, the displacement amplitude is $|X| = 0.05(1.405) = 0.07$ ft.

The transmitted force is obtained from (9.3.9).

$$\frac{|F_t|}{|Y|} = \sqrt{k^2 + (c\omega)^2}\frac{m\omega^2}{\sqrt{(k - m\omega^2)^2 + (c\omega)^2}} = 2960$$

Thus, the magnitude of the transmitted force is $|F_t| = 0.05(2960) = 148$ lb.

9.3.6 INSTRUMENT DESIGN

When no s term occurs in the numerator of a transfer function, its magnitude ratio is small at high frequencies. On the other hand, introducing numerator dynamics can produce a large magnitude ratio at high frequencies. This effect can be used to advantage in instrument design, for example. The instrument shown in Figure 9.3.8 illustrates this point. With proper selection of the natural frequency of the device, it can be used either as a *vibrometer* to measure the amplitude of a sinusoidal input displacement $z = A_z \sin \omega t$, or an *accelerometer* to measure the amplitude of the acceleration

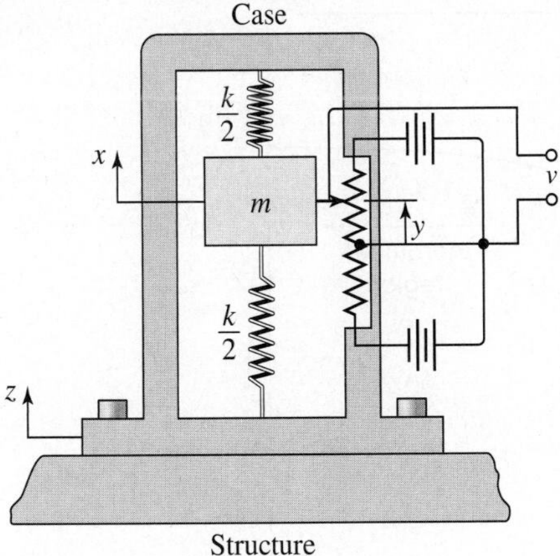

Figure 9.3.8
An accelerometer.

$\ddot{z} = -A_z\omega^2 \sin \omega t$. When used to measure ground motion from an earthquake, for example, the instrument is commonly referred to as a *seismograph*.

The model was obtained in Section 6.6 and is

$$T(s) = \frac{Y(s)}{Z(s)} = \frac{-ms^2}{ms^2 + cs + k} = \frac{-s^2/\omega_n^2}{s^2/\omega_n^2 + 2\zeta s/\omega_n + 1} \tag{9.3.10}$$

where $y(t)$ is the output displacement, which is measured from the voltage $v(t)$. At steady state, $y(t) = A_y \sin(\omega t + \phi)$.

The numerator $N(s) = -s^2/\omega_n^2$ gives the following contribution to the log magnitude ratio:

$$20 \log |N(j\omega)| = 20 \log \left| \left(\frac{j\omega}{\omega_n} \right)^2 \right| = 40 \log \frac{\omega}{\omega_n} \tag{9.3.11}$$

This term contributes 0 dB to the net curve at the corner frequency $\omega = \omega_n$, and it increases the slope by 40 dB/decade over all frequencies. Thus, at low frequencies, the slope of $m(\omega)$ corresponding to (9.3.11) is 40 dB/decade, and at high frequencies, the slope is zero. The plot is sketched in Figure 9.3.9.

For the device to act like a vibrometer, this plot shows that the device's natural frequency ω_n must be selected so that $\omega \gg \omega_n$, where ω is the oscillation frequency of the displacement to be measured. For $\omega \gg \omega_n$,

$$|T(j\omega)| \approx 40 \log \frac{\omega}{\omega_n} - 40 \log \frac{\omega}{\omega_n} = 0 \text{ dB}$$

and thus $A_y \approx A_z$ as desired. This is because the mass m cannot respond to high-frequency input displacements. Its displacement x therefore remains approximately constant, and the motion z directly indicates the motion y. To design a vibrometer having specific characteristics, we must know the lower bound of the input displacement frequency ω. The frequency $\omega_n = \sqrt{k/m}$ is then made much smaller than this bound by selecting a large mass and a soft spring (small k). However, these choices are governed by constraints on the allowable deflection. For example, a very soft spring will have a large distance between the free length and the equilibrium positions.

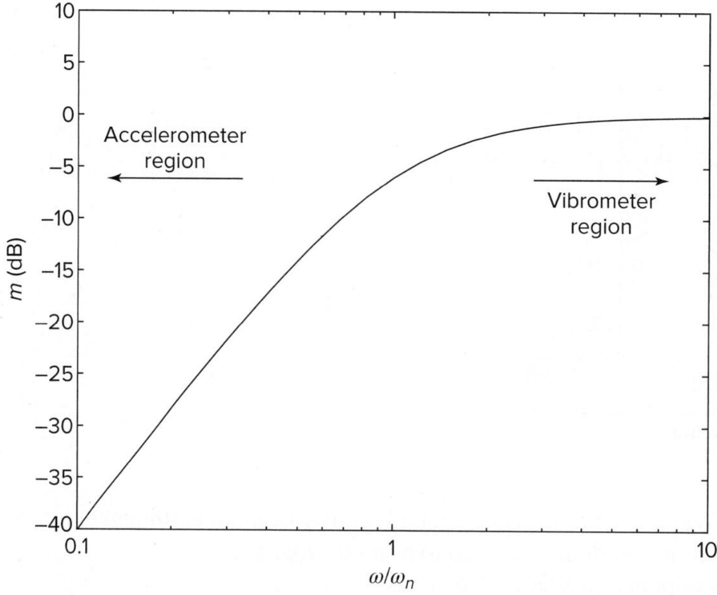

Figure 9.3.9 Frequency response of an accelerometer.

An accelerometer can be obtained by using the lower end of the frequency range; that is, selecting $\omega_n \gg \omega$, or equivalently, for s near zero, (9.3.10) gives

$$T(s) \approx -\frac{s^2}{\omega_n^2} = \frac{Y(s)}{Z(s)}$$

or

$$Y(s) \approx -\frac{1}{\omega_n^2} s^2 Z(s)$$

The term $s^2 Z(s)$ represents the transform of $\ddot{z}$ so the output of the accelerometer is

$$A_y \approx \frac{1}{\omega_n^2}|\ddot{z}| = \frac{\omega^2}{\omega_n^2} A_z$$

With ω_n chosen large (using a small mass and a stiff spring), the input acceleration amplitude $\omega^2 A_z$ can be determined from A_y.

9.3.7 VIBRATION ABSORBERS

A *vibration absorber* is used to reduce vibration amplitude in situations where a disturbing force has a constant frequency. A common example of a disturbing force is vibration from rotating machinery. Vibration absorbers are often found on devices that run at constant speed. These include saws, sanders, shavers, and devices powered by ac motors, because such motors are usually designed to operate at constant speed. The absorber is a device consisting of another mass and stiffness element, which are attached to the main mass to be protected from vibration (Figure 9.3.10). The new system consisting of the main mass and the absorber mass has a fourth-order model, and thus the new system has two natural frequencies.

Figure 9.3.10 A vibration absorber (m_2, k_2) attached to the main system (m_1, k_1, c).

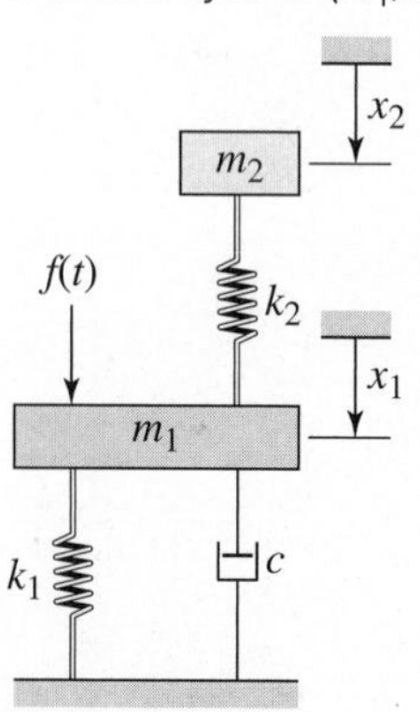

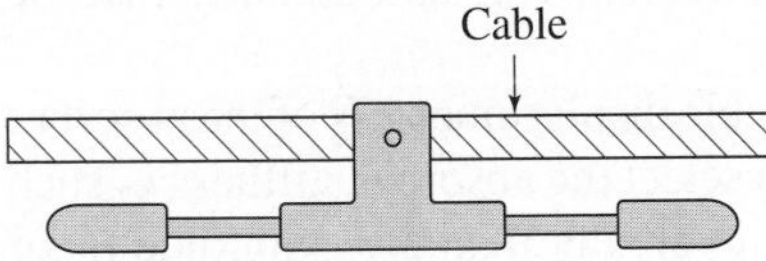

Figure 9.3.11 A Stockbridge damper.

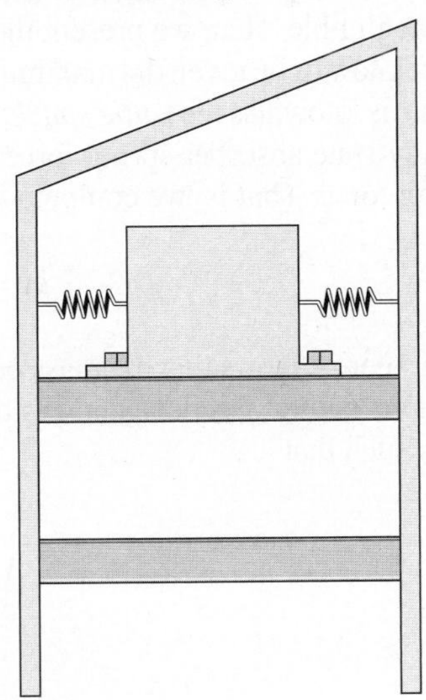

Figure 9.3.12 Vibration absorber for a tall building.

Power lines and structural cables often have vibration absorbers, called Stockbridge dampers, that are shaped somewhat like elongated dumbbells (Figure 9.3.11). These protect the lines and cables from excessive vibration caused by the wind. Wind blowing across the cables produces shedding vortices that cause the cables to vibrate.

Many modern buildings and bridges have vibration absorbers to counteract the effects of wind and earthquakes. These are quite large; for example, the one in the Citicorp building in New York City uses a 400-ton concrete block that is allowed to slide horizontally on an oil bearing (see Figure 9.3.12). The tower has a natural frequency 0.15 Hz. The springs were designed to allow the mass to move up to 55 inches. Some bridges and buildings use a pendulum damper. The designer must properly select the pendulum inertia and the pendulum length. The Taipei 101, located in Taipei, Taiwan, has 101 floors and a tuned mass damper weighing 728 tons. It consists of a steel sphere suspended as a pendulum from the 92nd to the 88th floor.

If the disturbing force is sinusoidal, $f(t) = F_o \sin \omega t$, and if we know its frequency ω and the natural frequency ω_{n1} of the original system, we can select values for the absorber's mass and stiffness so that the motion of the original mass is very small, which means that its kinetic and potential energies will be small. To achieve this small motion, the energy delivered to the system by the disturbing input must be "absorbed" by the absorber's mass and stiffness. Thus, the resulting absorber motion may be large. Because the principle of the absorber depends on the absorber's motion, such devices are sometimes called *dynamic* vibration absorbers. Another term for vibration absorber is a *tuned mass damper*.

Figure 9.3.10 illustrates a simple vibration absorber. The absorber consists of a mass m_2 and stiffness element k_2 that are connected to the main mass m_1. The disturbing input is the applied force $f(t)$. The equations of motion for the system are

$$m_1\ddot{x}_1 = -k_1x_1 - k_2(x_1 - x_2) - c\dot{x}_1 + f(t) \tag{9.3.12}$$

$$m_2\ddot{x}_2 = k_2(x_1 - x_2) \tag{9.3.13}$$

where x_1 and x_2 are measured from the rest positions of the masses. The theory for designing the absorber is presented in Chapter 13, Section 13.3 and assumes that the damping is negligible. Here we present the results.

We must know or be given the maximum allowable displacement X_2 of the absorber mass m_2. This is known as the *rattle space*. We then select the absorber stiffness k_2 such that the steady-state absorber spring force k_2X_2 is no greater than the amplitude F_o of the disturbing force. That is, we compute k_2 as follows:

$$k_2 = \frac{F_o}{X_2} \tag{9.3.14}$$

The absorber theory shows that the mass m_1 will be motionless if we select an absorber having the same natural frequency ω_{n2} as the frequency ω of the applied force. That is, we select m_2 such that

$$\omega_{n2} = \sqrt{\frac{k_2}{m_2}} = \omega$$

or

$$m_2 = \frac{k_2}{\omega^2} \tag{9.3.15}$$

Equations (9.3.14) and (9.3.15) are the design equations for the absorber. Note that we need not know the values of k_1 or m_1. Note, however, that the amplitude F_o of the disturbing force can be difficult to determine.

EXAMPLE 9.3.2

Design of a Vibration Absorber

■ Problem

A certain machine with supports will be subjected to a disturbing force having an amplitude of 2 lb and a frequency of 4 Hz. Design a dynamic vibration absorber for this machine. The available clearance for the absorber's motion is 0.5 in.

■ Solution

The maximum allowable clearance is $X_2 = 0.5$ in., or $1/24$ ft. Using (9.3.14), we obtain

$$k_2 = \frac{2}{1/24} = 48 \text{ lb/ft}$$

The frequency of the applied force is $\omega = 2\pi(4) = 8\pi$ rad/sec. The absorber's design requires that $\omega_{n2} = \omega = 8\pi$. Thus,

$$\omega_{n2} = \sqrt{\frac{k_2}{m_2}} = 8\pi$$

or

$$m_2 = \frac{k_2}{64\pi^2} = \frac{48}{64\pi^2} = 0.076 \text{ slug}$$

To simulate the performance of the absorber, we need to know the values of m_1, k_1, and F_o, and we add a little bit of damping. Let us suppose that $c = 10$ lb-sec/ft, $m_1 = 0.8$ slug, $k_1 = 790$ lb/ft, and $F_o = 2$ lb, so that $f(t) = 2\sin 8\pi t$. Substituting these values into (9.3.12) and (9.3.13) enables us to generate the plots shown in Figure 9.3.13

Figure 9.3.13 Response of a vibration absorber.

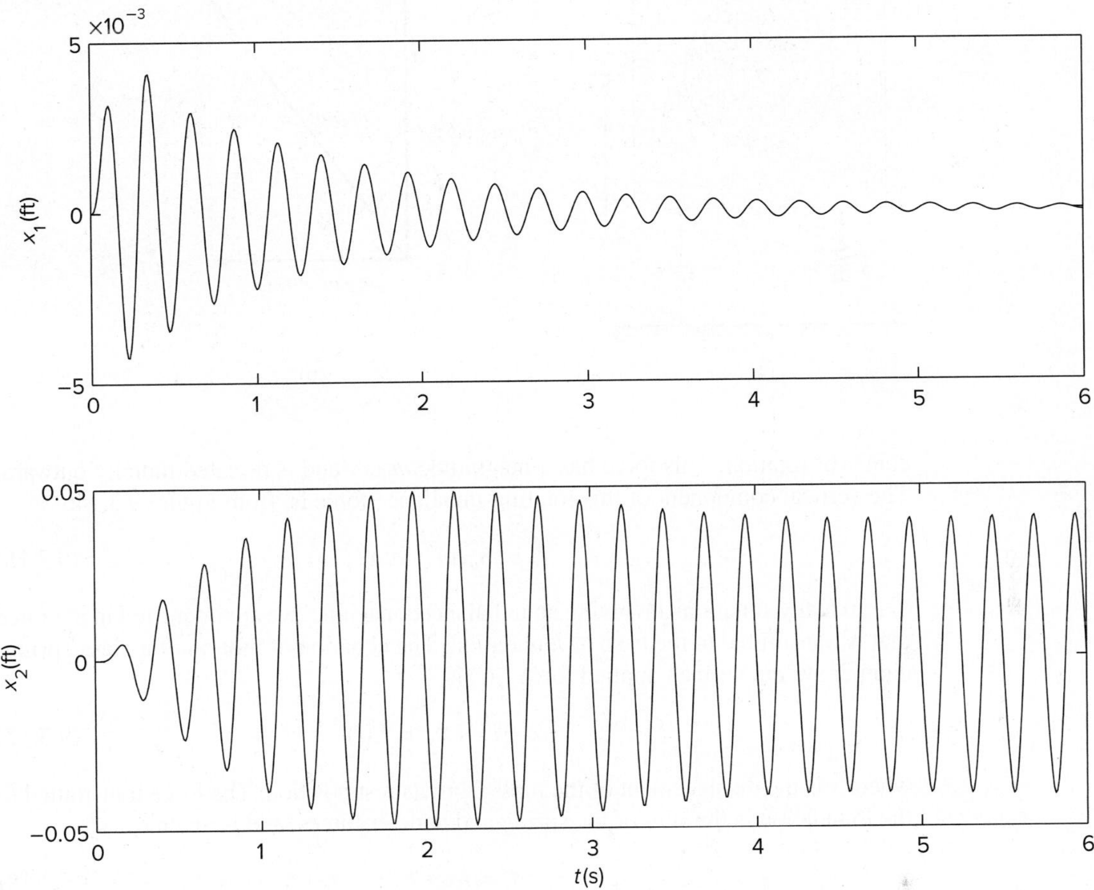

(using for example, the MATLAB `ss` and `lsim` functions). Note that the absorber design equations apply to the steady-state response; the transient response will be different. We see that the absorber mass is oscillating at the forcing frequency, and its amplitude at steady state is 0.042 ft, as predicted. The main mass is motionless at steady state, also as predicted.

9.3.8 ROTATING UNBALANCE

A common cause of sinusoidal forcing in machines is the unbalance that exists to some extent in every rotating machine. The unbalance is caused by the fact that the center of mass of the rotating part does not coincide with the center of rotation.

Let m be the total mass of the machine and m_u the rotating mass causing the unbalance. Consider the entire unbalanced mass m_u to be lumped at its center of mass, a distance ϵ from the center of rotation. This distance is the *eccentricity*. Figure 9.3.14a shows this situation. The main mass is thus $(m - m_u)$ and is assumed to be constrained to allow only vertical motion. The motion of the unbalanced mass m will consist of the vector combination of its motion relative to the main mass $(m - m_u)$ and the motion of the main mass. For a constant speed of rotation ω, the rotation produces a radial acceleration of m_u equal $\epsilon\omega^2$. This causes a force to be exerted on the bearings at the

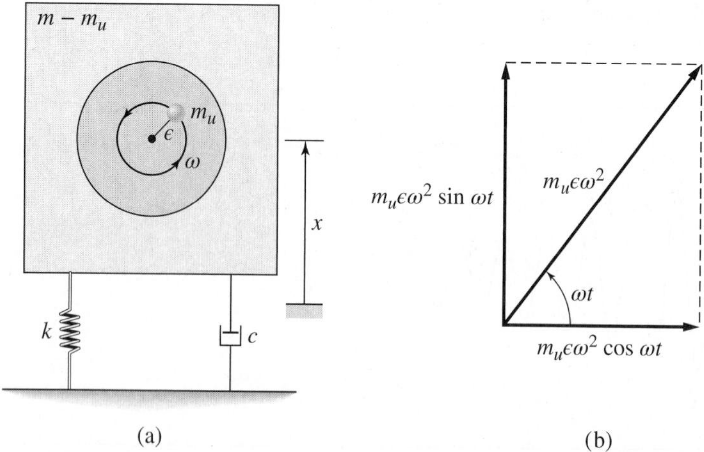

Figure 9.3.14 A machine having rotating unbalance.

center of rotation. This force has a magnitude $m_u\epsilon\omega^2$and is directed radially outward. The vertical component of this rotating-unbalance force is, from Figure 9.3.14b,

$$f_r = m_u\epsilon\omega^2 \sin \omega t \tag{9.3.16}$$

In many situations involving an unbalanced machine, we are interested in the force that is transmitted to the base or foundation. The equation of motion of a mass-spring-damper system with an applied force $f_r(t)$ is

$$m\ddot{x} + c\dot{x} + kx = f_r(t) \tag{9.3.17}$$

where x is the displacement of the mass from its rest position. The force transmitted to the foundation is the sum of the spring and damper forces, and is given by

$$f_t = kx + c\dot{x} \tag{9.3.18}$$

The force transmissibility of this system is the ratio $F_t(s)/F_r(s)$, which represents the ratio of the force f_t transmitted to the foundation to the applied force f_r. The most common case of such an applied force is the rotating-unbalance force. From (9.3.16), we see that the amplitude F_r of the rotating-unbalance force is

$$F_r = m_u\epsilon\omega^2 \tag{9.3.19}$$

The transfer function is

$$\frac{X(s)}{F_r(s)} = \frac{1}{ms^2 + cs + k} \tag{9.3.20}$$

From (9.3.18),

$$F_t(s) = (k + cs)X(s) \tag{9.3.21}$$

Substituting $X(s)$ from (9.3.20) into (9.3.21) gives the force-transmissibility transfer function:

$$\frac{F_t(s)}{F_r(s)} = \frac{k + cs}{ms^2 + cs + k} \tag{9.3.22}$$

EXAMPLE 9.3.3

Foundation Force Due to Rotating Unbalance

■ **Problem**

A system having a rotating unbalance, like that shown in Figure 9.3.14, has a total mass of $m = 20$ kg, an unbalanced mass of $m_u = 0.05$ kg, and an eccentricity of $\epsilon = 0.01$ m. The machine rotates at 1150 rpm. Its vibration isolator has a stiffness of $k = 2\times10^4$ N/m. Compute the force transmitted to the foundation if the isolator damping ratio is $\zeta = 0.5$.

■ **Solution**

First convert the machine speed to radians per second.

$$\omega = \frac{1150(2\pi)}{60} = 120 \text{ rad/s}$$

Then

$$m_u \epsilon \omega^2 = 0.05(0.01)(120)^2 = 7.25 \text{ N}$$

The damping constant can be obtained from the damping ratio: $c = 2\zeta\sqrt{mk} = 632$ N · s/m. The amplitude of the steady-state transmitted force can be calculated from (9.3.22).

$$|F_t| = |F_r| \left| \frac{k + c\omega j}{k - m\omega^2 + c\omega j} \right| = m_u \epsilon \omega^2 \frac{\sqrt{k^2 + (c\omega)^2}}{\sqrt{(k - m\omega^2)^2 + (c\omega)^2}} = 2 \text{ N}$$

EXAMPLE 9.3.4

Isolation of a Motor

■ **Problem**

Often motors are mounted to a base with an isolator consisting of an elastic pad. The pad serves to reduce the rotating-unbalance force transmitted to the base. A particular motor weighs 10 lb and runs at 3200 rpm. Neglect damping in the pad and calculate the pad stiffness required to provide a 90% reduction in the force transmitted from the motor to the base.

■ **Solution**

A 90% force reduction corresponds to a transmissibility ratio of $|F_t|/|F_r| = 0.1$. Since we are neglecting damping, (9.3.22) becomes

$$\frac{F_t(s)}{F_r(s)} = \frac{k}{ms^2 + k}$$

and from the frequency transfer function:

$$\left| \frac{F_t(j\omega)}{F_r(j\omega)} \right| = \left| \frac{k}{k - m\omega^2} \right| = \left| \frac{1}{1 - \frac{\omega^2}{\omega_n^2}} \right| = 0.1$$

This has two solutions: $\omega^2/\omega_n^2 = -9$, which is impossible, and $\omega^2/\omega_n^2 = 11$. Thus,

$$\omega_n^2 = \frac{k}{m} = \frac{\omega^2}{11} = \frac{[3200(2\pi/60]^2)}{11} = 1.0209 \times 10^4$$

and $k = (10/32.2)1.0209 \times 10^4 = 3170$ lb/ft.

9.4 FILTERING PROPERTIES OF DYNAMIC SYSTEMS

We call a system a *low-pass filter* if it responds more to sinusoidal inputs having low frequencies. Similarly, a *high-pass filter* responds more to high-frequency inputs, while a *band-pass filter* responds to inputs having a frequency in the midrange. While the term "filter" is normally used to refer to electrical circuits, we may think of other system types as filters. For example, a mass-spring-damper system having a large damping ratio will act like a low-pass filter. If such a system represents a vehicle suspension, the passenger compartment will undergo negligible motion when the vehicle moves with high speed over closely spaced road surface variations.

9.4.1 FREQUENCY RESPONSE OF ELECTRICAL CIRCUITS

Simple electrical circuits composed of resistors, capacitors and op amps can be designed to act as low-pass, high-pass, and band-pass filters, and are found in many applications.

EXAMPLE 9.4.1

A Low-Pass Filter

■ Problem

A series RC circuit is shown in Figure 9.4.1, where the input is the voltage v_s and the output is the voltage v_o. The two amplifiers serve to isolate the circuit from the loading effects of adjacent elements. Describe its frequency response characteristics.

■ Solution

The circuit model obtained in Chapter 6 has the following transfer function.

$$T(s) = \frac{V_o(s)}{V_s(s)} = \frac{1}{RCs + 1}$$

Comparing this with the transfer function $1/(\tau s + 1)$, we see that $\tau = RC$. Referring to the frequency response plots shown in Figure 9.1.8, we see that the breakpoint frequency is $1/\tau = 1/RC$. This means that the circuit filters out sinusoidal voltage inputs whose frequency is higher than $1/RC$, and the higher the frequency, the greater is the filtering. Inputs having frequencies lower than $1/RC$ pass through the circuit with almost no loss of amplitude and with little phase shift. This is because $T(j\omega) \approx 1$ and $\phi \approx 0$ for $\omega < 1/RC$. The circuit is called a low-pass filter for this reason. It can be used to filter out unwanted high-frequency components of the input voltage, such as 60-Hz interference from nearby ac equipment.

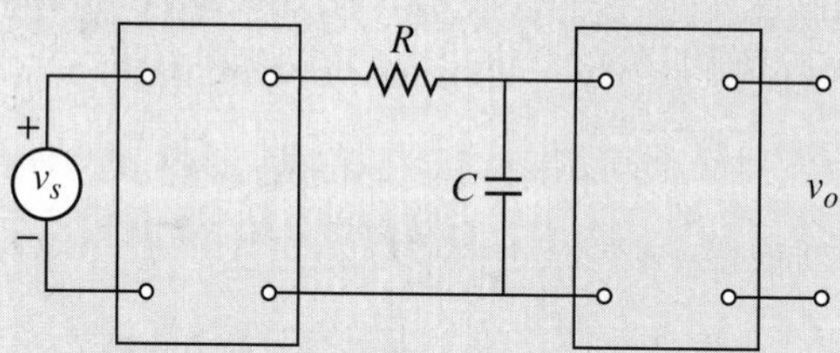

Figure 9.4.1 Series RC circuit configured as a low-pass filter.

EXAMPLE 9.4.2

A High-Pass Filter

■ Problem

Consider the series RC circuit shown in Figure 9.4.2, where the output voltage is taken to be across the resistor. The input is the voltage v_s and the output is the voltage v_o. Obtain the circuit's frequency response plots, and interpret the circuit's effect on the input.

■ Solution

The input impedance between the voltage v_s and the current i is found from the series law.

$$\frac{V_s(s)}{I(s)} = R + \frac{1}{Cs}$$

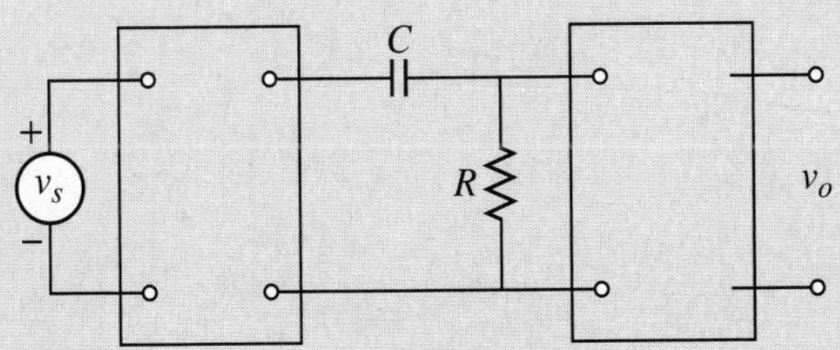

Figure 9.4.2 Series *RC* circuit configured as a high-pass filter.

Thus,

$$V_o(s) = I(s)R = \frac{V_s(s)}{R + 1/Cs}R = \frac{RCs}{RCs + 1}V_s(s)$$

The transfer function is

$$T(s) = \frac{V_o(s)}{V_s(s)} = \frac{RCs}{RCs + 1} = \frac{\tau s}{\tau s + 1}$$

where $\tau = RC$.

The frequency transfer function and magnitude ratio are

$$T(j\omega) = \frac{\tau\omega j}{1 + \tau\omega j}$$

$$M = \left|\frac{\tau\omega j}{1 + \tau\omega j}\right| = \frac{\tau\omega}{\sqrt{1 + (\tau\omega)^2}}$$

$$\begin{aligned} m(\omega) &= 20\log|\tau| + 20\log|j\omega| - 20\log|1 + \tau\omega j| \\ &= 20\log|\tau| + 20\log\omega - 20\log\sqrt{1 + (\tau\omega)^2} \\ &= 20\log|\tau| + 20\log\omega - 10\log\left[1 + (\tau\omega)^2\right] \end{aligned}$$

At frequencies where $\omega \ll 1/\tau$, the last term on the right is negligible, and thus the low-frequency asymptote is described by

$$m(\omega) = 20\log|\tau| + 20\log\omega, \qquad \omega \ll \frac{1}{\tau}$$

The plot of this equation can be sketched by noting that it has a slope of 20 dB/decade and it passes through the point $m = 0$, $\omega = 1/\tau$. The asymptote is shown by the dashed line in Figure 9.4.3 (upper plot). At high frequencies where $\omega \gg 1/\tau$, the slope of −20 due to the term $\tau s + 1$ in the denominator cancels the slope of +20 due to the term s in the numerator. Therefore, at high frequencies, $m(\omega)$ has a slope of approximately zero, and

$$\begin{aligned} m(\omega) &\approx 20\log|\tau| + 20\log\omega - 20\log|\tau\omega| \\ &= 20\log|\tau| + 20\log\omega - 20\log|\tau| - 20\log\omega = 0 \quad \text{for } \omega \gg \frac{1}{\tau} \end{aligned}$$

At $\omega = 1/\tau$, the denominator term $\tau s + 1$ contributes −3 dB. The composite curve for $m(\omega)$ is obtained by "blending" the low-frequency and high-frequency asymptotes through this point, as shown in Figure 9.4.3 (upper plot). A similar technique can be used to sketch $\phi(\omega)$.

The phase angle is

$$\phi(\omega) = \angle\tau + \angle(j\omega) - \angle(1 + \tau\omega j) = 0° + 90° - \tan^{-1}(\omega\tau)$$

For $\omega \ll 1/\tau$, the low-frequency asymptote is $\phi(\omega) \approx 90° - \tan^{-1}(0) = 90°$. For $\omega = 1/\tau$, $\phi(\omega) = 90° - \tan^{-1}(1) = 45°$. For $\omega \gg 1/\tau$, the high-frequency asymptote is $\phi(\omega) \approx 90° - \tan^{-1}(\infty) = 90° - 90° = 0°$. The result is sketched in Figure 9.4.3 (lower plot).

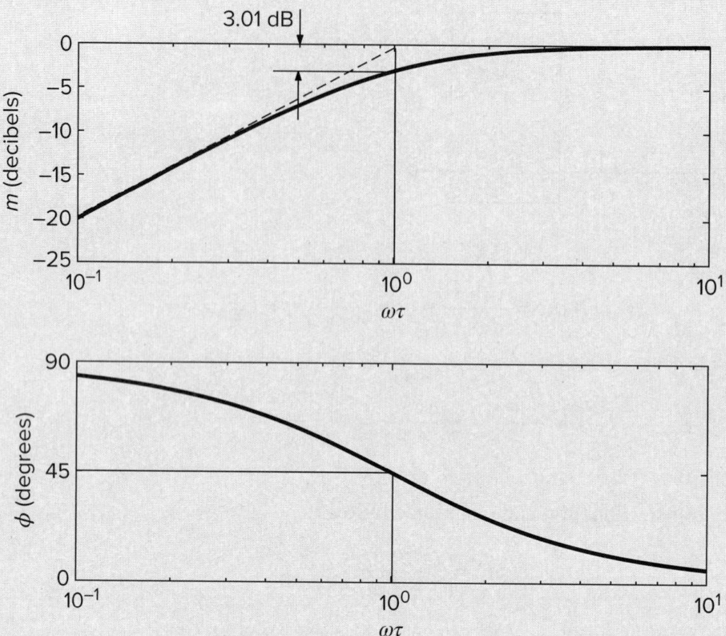

Figure 9.4.3 Asymptotes and corner frequency $\omega = 1/\tau$ of the model $\tau s/(\tau s + 1)$.

The log magnitude plot shows that the circuit passes signals with frequencies above $\omega = 1/\tau$ with little attenuation, and thus it is called a high-pass filter. It is used to remove dc and low-frequency components from a signal. This is desirable when we wish to study high-frequency components whose small amplitudes would be indiscernible in the presence of large-amplitude, low-frequency components. Such a circuit is incorporated into oscilloscopes for this reason. The voltage scale can then be selected so that the signal components to be studied will fill the screen.

EXAMPLE 9.4.3

Frequency Response of a Differentiating Circuit

■ Problem

The op-amp differentiator analyzed in Chapter 6 and shown in Figure 9.4.4a has the transfer function

$$\frac{V_o(s)}{V_s(s)} = -RCs$$

The circuit is susceptible to high-frequency noise, because the derivative of a rapidly changing signal is difficult to compute and produces an exaggerated output. In practice, this problem is often solved by filtering out high-frequency signals either with a low-pass filter inserted in series with the differentiator or by using a redesigned differentiator, such as the one shown in Figure 9.4.4b. Its transfer function is

$$\frac{V_o(s)}{V_s(s)} = -\frac{RCs}{R_1 Cs + 1}$$

Analyze its frequency response characteristics.

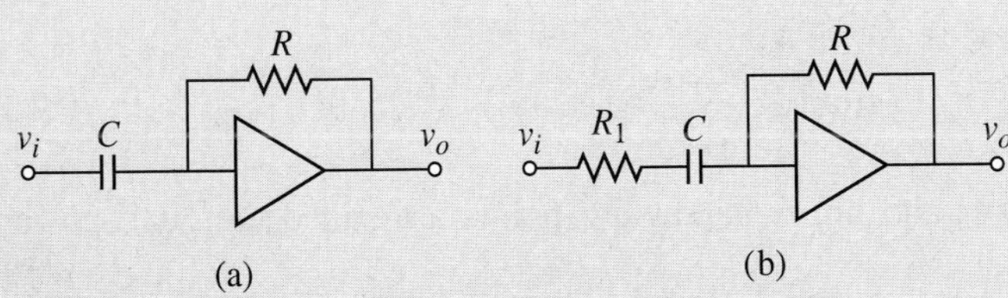

Figure 9.4.4 (a) Op-amp differentiator. (b) Modified op-amp differentiator.

■ Solution

Note that if a circuit is a pure differentiator, then $v_o = \dot{v}_s$, or

$$\frac{V_o(s)}{V_s(s)} = s$$

So the slope of $m(\omega)$ curve for a pure differentiator is 20 dB/decade. The transfer function of the redesigned differentiator can be rearranged as follows:

$$\frac{V_o(s)}{V_s(s)} = -\frac{RCs}{R_1Cs + 1} = -\frac{R}{R_1}\frac{R_1Cs}{R_1Cs + 1}$$

Thus, the frequency response plot of this transfer function is similar to that of the high-pass filter shown in Figure 9.4.3 (upper plot) with $\tau = R_1C$, except for the factor $-R/R_1$. We will neglect the factor -1, which can be eliminated by using a series op amp inverter. The circuit's corner frequency is $\omega_c = 1/R_1C$, and the circuit acts like the ideal differentiator for frequencies up to about $\omega = 1/R_1C$. For higher frequencies, the magnitude ratio curve has zero slope rather than the 20 dB/decade slope required for differentiation, and thus the circuit does not differentiate the high-frequency signals, but merely amplifies them with a gain of R/R_1. Because the amplitudes of noisy high-frequency signals are generally small, this amplification effect is negligible. For $\omega < 1/R_1C$, the circuit's gain is RC. Thus, after choosing a convenient value for C, the two resistors are selected as follows: R_1 is used to set the cutoff frequency $1/R_1C$, and R is used to set the gain RC.

9.4.2 BANDWIDTH

It is useful to have a specific measure of the range of forcing frequencies to which a system is especially responsive. This single measure may be used as a design specification. The most common such measure is the *bandwidth*. It is commonly defined as the range of frequencies over which the power transmitted or dissipated by the system is no less than one-half of the peak power.* For many systems the power transmitted or dissipated is proportional to M^2, where M is the magnitude of the frequency transfer function. For example, the power dissipated by a resistor is proportional to the square of the current, and the power dissipated by a damper is proportional to the square of the amplitude of the velocity difference across the damper endpoints.

Thus, the bandwidth is commonly defined as the range of frequencies (ω_1, ω_2) over which

$$M^2(\omega_1) \leq \frac{M^2_{\text{peak}}}{2} \geq M^2(\omega_2)$$

*The term *bandwidth* has a different meaning in the communications and computer industries, where it is used to describe the rate of data transfer.

or, equivalently,

$$M(\omega_1) \le \frac{M_{\text{peak}}}{\sqrt{2}} \ge M(\omega_2) \tag{9.4.1}$$

For this reason, the lower and upper bandwidth points are called the "half-power" points.

EXAMPLE 9.4.4

Bandwidth of a First-Order Model

■ Problem

Consider the model $\tau\dot{v} + v = f(t)$, for which

$$M = \frac{V}{F} = \frac{1}{\sqrt{1 + \omega^2\tau^2}}$$

Obtain an expression for the bandwidth in terms of τ and interpret its significance.

■ Solution

The peak in M occurs at $\omega = 0$ and is $M_{\text{peak}} = 1$. Thus, $\omega_1 = 0$ and ω_2 is found from (9.4.1) as follows:

$$\frac{M_{\text{peak}}}{\sqrt{2}} = \frac{1}{\sqrt{2}} = M(\omega_2) = \frac{1}{\sqrt{1 + \omega_2^2\tau^2}}$$

This gives $\omega_2 = 1/\tau$. Thus, the bandwidth of the model $\tau\dot{v}+v = f(t)$ is the range of frequencies $(0, 1/\tau)$.

Because a small time constant indicates a fast system, the bandwidth is another measure of the speed of response. So the faster the system, the larger the bandwidth, and some engineers describe a system's response speed in terms of its bandwidth rather than in terms of its time constant. However, for other models, the relation between the time constant and the bandwidth is not always so simple, as we will see.

When expressed in decibel units, $m = 20\log M$, the bandwidth points are those points on the m curve that lie 3 dB below the peak value of m (because $20\log(1/\sqrt{2}) \approx -3$ dB). For this reason, the bandwidth points are sometimes called the "3 dB" points.

EXAMPLE 9.4.5

Bandwidth of a Second-Order Model

■ Problem

Determine the expression for the bandwidth of the second-order model $m\ddot{x} + c\dot{x} + kx = f(t)$.

■ Solution

For this model, the amplitude ratio $M = X/F$ is

$$M = \frac{1}{k}\frac{1}{\sqrt{(1 - r^2)^2 + (2\zeta r)^2}} \tag{1}$$

where $r = \omega/\omega_n = \omega\sqrt{m/k}$. The peak value of M, denoted M_r, occurs when $0 \le \zeta \le 0.707$ and $r = \sqrt{1 - 2\zeta^2}$, and is

$$M_r = \frac{1}{k}\frac{1}{2\zeta\sqrt{1 - \zeta^2}}$$

The value of r that makes $M(\omega) = M_r/\sqrt{2}$ is found from

$$\frac{1}{k}\frac{1}{\sqrt{(1-r^2)^2+(2\zeta r)^2}} = \frac{1}{\sqrt{2}}\left(\frac{1}{k}\frac{1}{2\zeta\sqrt{1-\zeta^2}}\right)$$

This can be solved for r by squaring both sides and rearranging to obtain

$$r^4 + (4\zeta^2 - 2)r^2 + 1 - 8\zeta^2 + 8\zeta^4 = 0$$

The solution is

$$r = \sqrt{1 - 2\zeta^2 \pm 2\zeta\sqrt{1-\zeta^2}} \tag{2}$$

If two positive, real solutions r_1 and r_2 of this equation exist, we can obtain the lower and upper bandwidth frequencies from $\omega_1 = r_1\omega_n$ and $\omega_2 = r_2\omega_n$. If one solution for r is imaginary and the other solution is positive, the lower bandwidth frequency is $\omega_1 = 0$ and the upper frequency ω_2 is obtained from the positive solution. This occurs when $\zeta > 0.382$. So, two half-power points exist only if $0 \leq \zeta \leq 0.382$.

Figure 9.4.5 shows several cases that can occur with various other models. In part (a) of the figure, the peak M_r is large enough so that $M_r/\sqrt{2} > M(0)$, and thus the lower bandwidth frequency ω_1 exists and is greater than zero. Such a system is called a *band-pass* system. Part (b) shows the plot of a *low-pass* system, for which $M(0) > M_r/\sqrt{2}$, so that $\omega_1 = 0$. This occurs for the model $m\ddot{x} + c\dot{x} + kx = f(t)$ when $\zeta > 0.382$. Part (c) shows a case where the upper bandwidth frequency is infinite. Such a system is called a *high-pass* system because it responds more to high-frequency inputs. Part (d) shows a magnitude ratio plot with two peaks. This can occur only with a model of fourth or higher order. Depending on the exact shape of the plot, such a system can have two bandwidths, one for each peak.

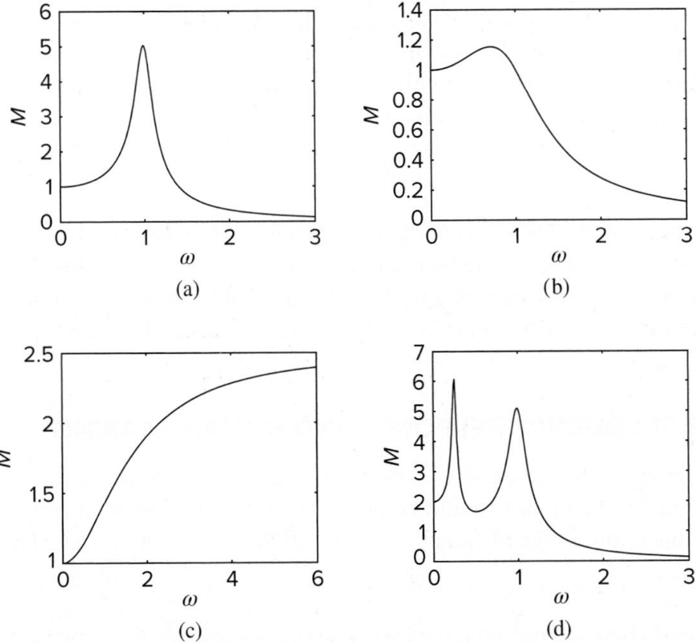

Figure 9.4.5 Some possible frequency response plots.

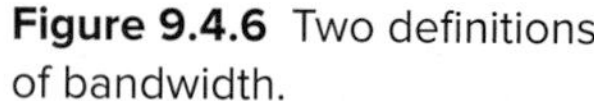
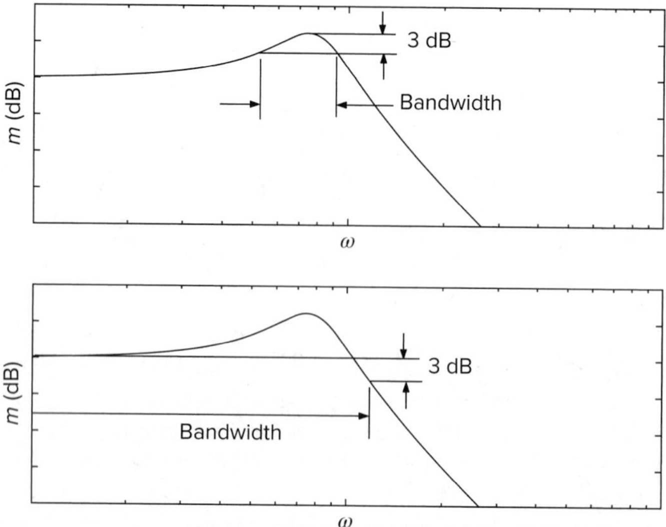

Figure 9.4.6 Two definitions of bandwidth.

9.4.3 ALTERNATIVE DEFINITION OF BANDWIDTH

In the definition of bandwidth specified by (9.4.1), the power transmitted by an input having a frequency outside the bandwidth is less than one-half the power transmitted by an input whose frequency corresponds to M_{peak}, *assuming* that both inputs have the *same* amplitude. However, often the amplitudes of low-frequency forcing functions are larger than those of high-frequency inputs, so the low-frequency inputs might account for more power. Thus, inputs whose frequencies lie below the lower bandwidth frequency ω_1 may contribute significantly to the response, and cannot always be neglected. Because of this, a modified definition of bandwidth is sometimes used. With this alternative definition of bandwidth, the lower bandwidth frequency ω_1 is assumed to be zero and the upper bandwidth frequency ω_2 is defined to be that frequency at which the power is one-half of the power at *zero* frequency (see Figure 9.4.6).

The two definitions give the same bandwidth for those systems whose peak value of M is $M(0)$. For some systems, however, this alternative definition of bandwidth cannot be applied.[1] For example, with the transfer function of a high-pass filter,

$$T(s) = \frac{\tau s}{\tau s + 1}$$

the value of $M(0)$ is zero and thus $M(0)$ is not the peak value of M. In fact, $m(0) = -\infty$ (see Figure 9.4.3). On the other hand, the bandwidth definition (9.4.1) gives the filter bandwidth to be $1/\tau \leq \omega \leq \infty$, which is a useful and physically meaningful result. Therefore, unless explicitly stated otherwise, we will use the bandwidth definition specified by (9.4.1).

9.4.4 EARPHONE FREQUENCY RESPONSE AND HUMAN HEARING

The peak in the frequency response curve is not always used as a measure of system performance. For example, earphones should be designed so that the magnitude ratio is 1 ($m = 0$ dB) over the entire range of frequencies detectable by the human ear. This

[1]The MATLAB function `bandwidth` uses this alternative definition and thus can give meaningless results.

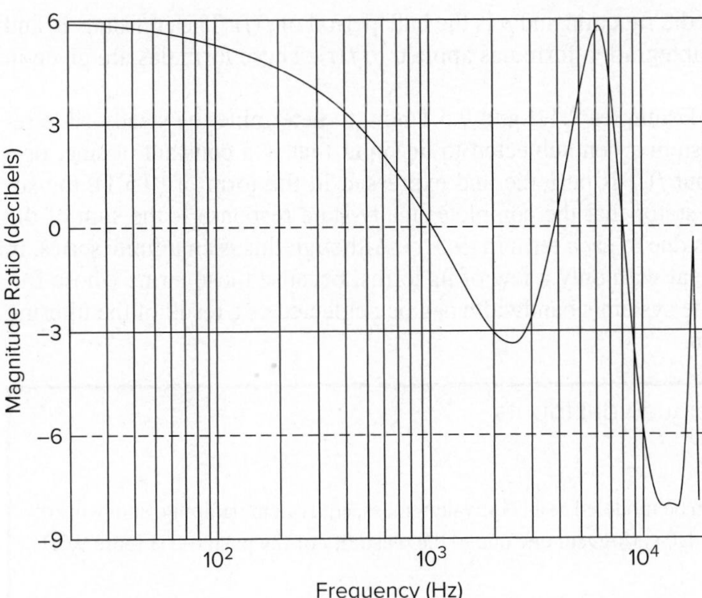

Figure 9.4.7 Frequency response of an earphone.

range is normally from about 20 Hz to 20,000 Hz. Consequently, a measure of earphone performance is the amount of deviation of its m curve from 0 dB over this frequency range.

Often earphone response is considered acceptable if its m curve deviates from the 0 db line by no more than ±3 dB. Figure 9.4.7 shows an experimentally determined frequency response curve of a particular earphone supplied with a handheld music player. We see that the earphone amplifies signals in the 20–400 Hz and the 5,000–8,000 Hz ranges, whereas it attenuates most signals above 9,000 Hz.

Thus, this earphone distorts bass and treble tones but reproduces well those tones in the mid-range of frequencies.

9.5 RESPONSE TO GENERAL PERIODIC INPUTS

The application of the sinusoidal input response is not limited to cases involving a single sinusoidal input. If the input consists of a sum of functions, the total forced response can be found from the sum of the forced responses of each term. This is one result of the superposition property of linear differential equations.

A basic theorem of analysis states that under some assumptions, which are generally satisfied in most practical applications, any periodic function can be expressed by a constant term plus an infinite series of sines and cosines with increasing frequencies. This theorem is the *Fourier theorem*, and its associated series is the *Fourier series*. The series has the form

$$
\begin{aligned}
f(t) = \frac{a_0}{2} &+ a_1 \cos\left(\frac{\pi t}{p}\right) + a_2 \cos\left(\frac{2\pi t}{p}\right) + \cdots \\
&+ b_1 \sin\left(\frac{\pi t}{p}\right) + b_2 \sin\left(\frac{2\pi t}{p}\right) + \cdots
\end{aligned}
\tag{9.5.1}
$$

where $f(t)$ is the periodic function and p is the half-period of $f(t)$. The constants a_i and b_i are determined by integration formulas applied to $f(t)$. These formulas are given in Appendix B.

We have seen in Examples 9.1.2 and 9.5.1 how to determine the steady-state response of a linear system when subjected to an input that is a constant, a sine, or a cosine. When the input $f(t)$ is periodic and expressed in the form of (9.5.1), the superposition principle states that the complete steady-state response is the sum of the steady-state responses due to each term in (9.5.1). Although this is an infinite series, in practice we have to deal with only a few of its terms, because those terms whose frequencies lie outside the system's bandwidth can be neglected as a result of the filtering property of the system.

EXAMPLE 9.5.1 Response to Nonsinusoidal Inputs

Problem

An engine valve train can be modeled as an equivalent mass, equivalent damping, and two stiffnesses, one due to the valve spring and one due to the elasticity of the push rod (Figure 9.5.1). The equation of motion is

$$\frac{I_o}{a^2}\ddot{x} + c_e\dot{x} + \left(k_1 + k_2\frac{a^2}{b^2}\right)x = \frac{a}{b}k_2 y(t)$$

The input displacement $y(t)$ is determined by the shape and rotational speed of the cam. Suppose the input is as shown in Figure 9.5.2a. Determine the steady-state response of the model for the following parameter values:

$$\ddot{x} + 20\dot{x} + 625x = 600y(t)$$

Solution

The transfer function is

$$T(s) = \frac{X(s)}{Y(s)} = \frac{600}{s^2 + 20s + 625}$$

and

$$T(j\omega) = \frac{600}{625 - \omega^2 + 20\omega j}$$

$$M(\omega) = \frac{600}{\sqrt{\left(625 - \omega^2\right)^2 + 400\omega^2}} \tag{1}$$

$$\phi(\omega) = -\tan^{-1}\frac{20\omega}{625 - \omega^2} \tag{2}$$

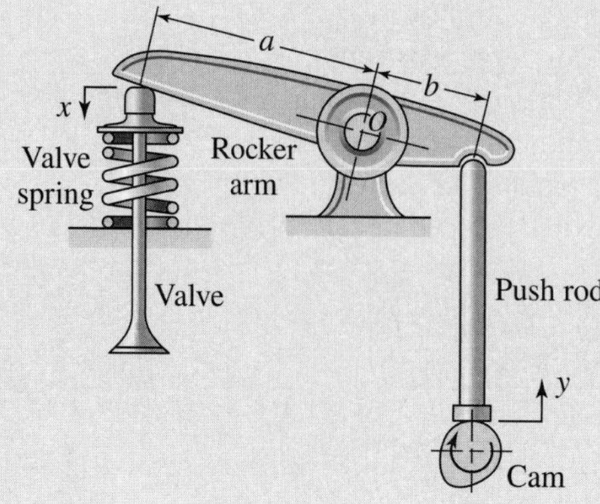

Figure 9.5.1 A valve system.

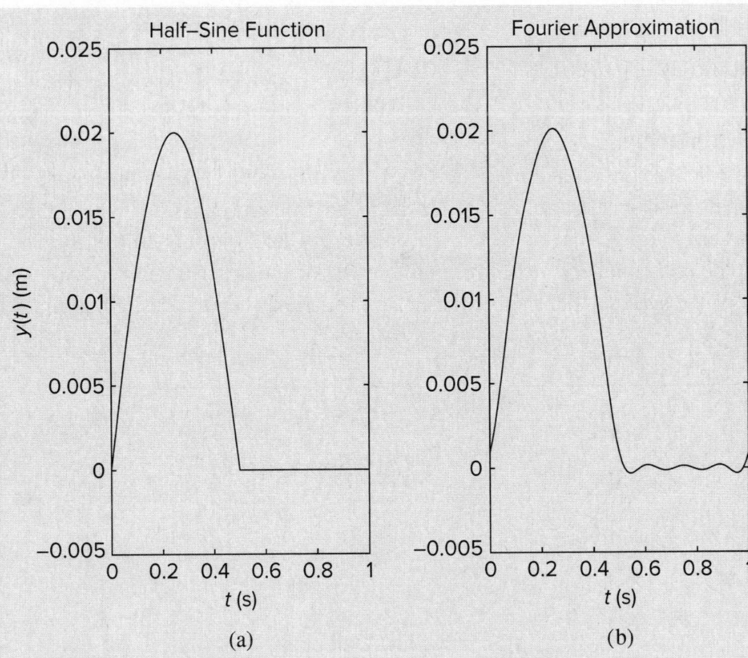

Figure 9.5.2 (a) Half-sine function. (b) Fourier series approximation.

The Fourier series' representation of $y(t)$ can be determined from the formulas in Appendix B, and is

$$y(t) = A_0 + A_1 \sin \omega_1 t + A_2 \cos \omega_2 t + A_3 \cos \omega_3 t + A_4 \cos \omega_4 t + \cdots$$

$$= 0.02\left[\frac{1}{\pi} + \frac{1}{2}\sin 2\pi t - \frac{2}{\pi}\left(\frac{\cos 4\pi t}{1(3)} + \frac{\cos 8\pi t}{3(5)} + \frac{\cos 12\pi t}{5(7)} + \cdots\right)\right] \tag{3}$$

Figure 9.5.2b is a plot of equation (3) including only those series terms shown. The plot illustrates how well the Fourier series represents the input function.

The steady-state response will have the form

$$x(t) = B_0 + B_1 \sin(\omega_1 t + \phi_1) + B_2 \cos(\omega_2 t + \phi_2) + B_3 \cos(\omega_3 t + \phi_3) + B_4 \cos(\omega_4 t + \phi_4) + \cdots$$

where

$$B_i = A_i M_i \tag{4}$$

Note that the amplitudes M_i and phases ϕ_i of the response due to cosine inputs are computed just as for a sine input.

Using equation (2) of Example 9.4.5, we obtain one positive solution, $r = 1.187$, and one imaginary solution. Thus, the lower bandwidth frequency is 0, and the upper frequency is $\omega_2 = 1.187\sqrt{625} = 29.7$ rad/sec. Because the bandwidth is from 0 to 29.7 rad/sec, the only series terms in $y(t)$ lying within the bandwidth are the constant term (whose frequency is 0), and the $\sin 2\pi t$, $\cos 4\pi t$, and $\cos 8\pi t$ terms, whose frequencies are 6.28, 12.6, and 25.1 rad/sec, respectively. To demonstrate the filtering property of the system, however, we also compute the effect of the first term lying outside the bandwidth. This is the $\cos 12\pi t$ term, whose frequency is 37.7 rad/sec. The following table was computed using equations (1), (2), (3), and (4).

i	ω_i	A_i	M_i	B_i	ϕ_i
0	0	0.006366	0.96	0.006112	0
1	2π	0.01	1.001913	0.010019	−0.211411
2	4π	−0.004244	1.1312	−0.004801	−0.493642
3	8π	−0.000849	1.193557	−0.001013	−1.584035
4	12π	−0.000364	0.547162	−0.000199	−2.383436

Using this table we can express the steady-state response as follows:

$$x(t) = 0.006112 + 0.010019\sin(2\pi t - 0.211411) - 0.004801\cos(4\pi t - 0.493642) - 0.001013\cos(8\pi t - 1.584035) - 0.000199\cos(12\pi t - 2.383436)$$

The input and the response are plotted in Figure 9.5.3. The difference between the input and output results from the resistive or lag effect of the system, not from the omission of the higher-order terms in the series. To see this, note that A_4 is 43% of A_3 but B_4 is only 20% of B_3. The decreasing amplitude of the higher-order terms in the series for $y(t)$, when combined with the filtering property of the system, enables us to truncate the series when the desired accuracy has been achieved.

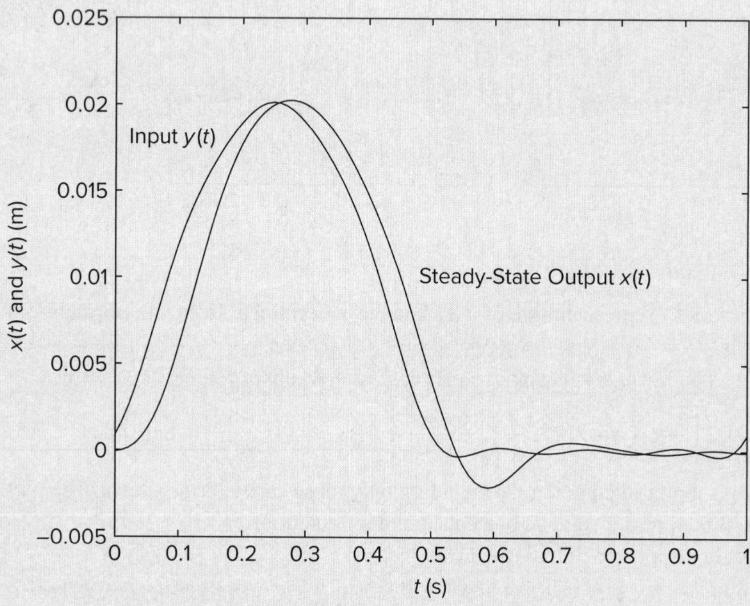

Figure 9.5.3 Steady-state response to a half-sine input.

9.6 SYSTEM IDENTIFICATION FROM FREQUENCY RESPONSE

In cases where a transfer-function or differential equation model is difficult to derive from general principles, or where the model's coefficient values are unknown, often an experimentally obtained frequency response plot can be used to determine the form of an appropriate model and the values of the model's coefficients.

9.6.1 TEST PROCEDURES

Often a sinusoidal input is easier to apply to a system than a step input, because many devices, such as ac circuits and rotating machines, naturally produce a sinusoidal signal or motion. If a suitable apparatus can be devised to provide a sinusoidal input of adjustable frequency, then the system's output amplitude and phase shift relative to the input can be measured for various input frequencies. When these data are plotted on the logarithmic plot for a sufficient frequency range, the form of the model can often be determined. This procedure is easiest for systems with electrical inputs and outputs, because for these systems, variable-frequency oscillators and frequency response analyzers are commonly available. Some of these can automatically sweep through a range of frequencies and plot the decibel and phase angle data. For nonelectrical outputs, such as a displacement, a suitable transducer can be used to produce an electrical measurement that can be analyzed.

Another advantage of frequency response tests is that they can often be applied to a system or process without interrupting its normal operations. A small-amplitude sinusoidal test signal is superimposed on the operating inputs, and the sinusoidal component of the output having the input's frequency is subtracted from the output measurements. Specialized computer algorithms and test equipment are available for this purpose.

9.6.2 USE OF ASYMPTOTES AND CORNER FREQUENCIES

When using frequency response data for identification, it is important to understand how to reconstruct a transfer-function form from the asymptotes and corner frequencies. The slopes of the asymptotes can be used to determine the orders of the numerator and the denominator. The corner frequencies can be used to estimate time constants and natural frequencies.

All real data will have some "scatter," and thus a perfect model fit will not be practical. However, to illustrate the methods clearly, in the following examples we use data that have very little scatter, and thus the derived models are unambiguous.

Identifying a First-Order System

EXAMPLE 9.6.1

■ Problem

An input $v_s(t) = 5 \sin \omega t$ V was applied to a certain electrical system for various values of the frequency ω, and the amplitude $|v_o|$ of the steady-state output was recorded. The data are shown in the first two columns of the following table. Determine the transfer function.

ω (rad/s)	$\lvert v_o \rvert$ (V)	$\lvert v_o \rvert/5$	$20 \log (\lvert v_o \rvert/5)$
1	10.95	2.19	6.81
2	10.67	2.13	6.57
3	10.3	2.06	6.28
4	9.84	1.97	5.89
5	9.33	1.87	5.44
6	8.80	1.76	4.91
7	8.28	1.66	4.40
8	7.782	1.56	3.86
9	7.31	1.46	3.29
10	6.87	1.37	2.73
15	5.18	1.04	0.34
20	4.09	0.82	−1.72
30	2.83	0.57	−4.88
40	2.16	0.43	−7.33
50	1.74	0.35	−9.12
60	1.45	0.29	−10.75
70	1.25	0.25	−12.04
80	1.10	0.22	−13.15
90	0.97	0.19	−14.43
100	0.88	0.18	−14.89

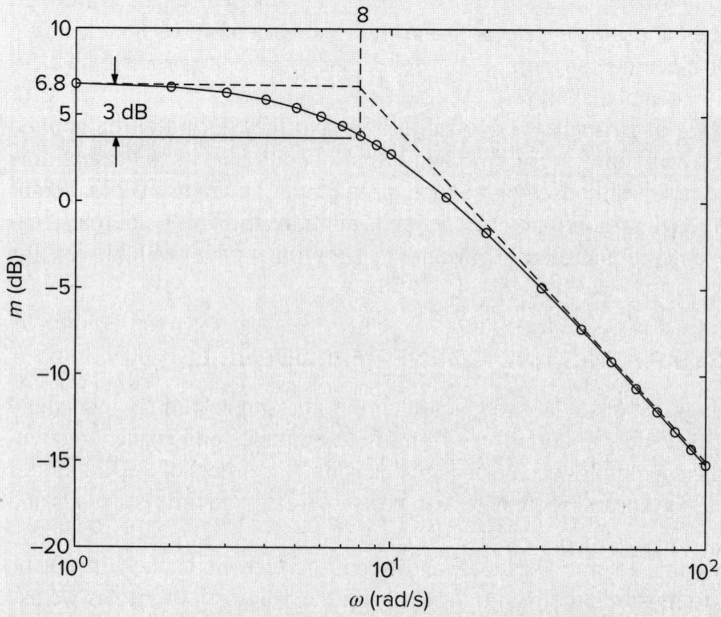

Figure 9.6.1 Frequency response data for Example 9.6.1.

■ Solution

First divide the output amplitude $|v_o|$ by the amplitude of the input. The result is given in the third column. This is the amplitude ratio M. Next convert this data to decibels using the conversion $m = 20 \log M$. This is the fourth column. The plot of m versus ω is shown by the small circles in Figure 9.6.1.

After drawing the asymptotes shown by the dashed lines, we first note that the data has a low-frequency asymptote of zero slope, and a high-frequency asymptote of slope −20 dB/decade.

This suggests a model of the form

$$T(s) = \frac{K}{\tau s + 1}$$

In decibel units,

$$m = 20 \log K - 10 \log(\tau^2 \omega^2 + 1)$$

The corner frequency $\omega = 1/\tau$ occurs where m is 3 dB below the peak value of 6.81. From the plot or the data, we can see that the corner frequency is $\omega = 8$ rad/s. Thus, $\tau = 1/8$ s.

At low frequencies, $\omega \ll 1/\tau$ and $m \approx 20 \log K$. From the plot, at low frequency, $m =$ 6.81 dB. Thus, $6.81 = 20 \log K$, which gives

$$K = 10^{6.81/20} = 2.19$$

Thus, the estimated model is

$$\frac{V_o(s)}{V_s(s)} = \frac{2.19}{\frac{1}{8}s + 1}$$

Another first-order model form is

$$\frac{V_o(s)}{V_s(s)} = K\frac{\tau_1 s + 1}{\tau_2 s + 1}$$

Because the high-frequency asymptote of this model has zero slope, it cannot describe the given data.

Identifying a Second-Order System

EXAMPLE 9.6.2

■ Problem

Measured response data are shown by the small circles in Figure 9.6.2. Determine the transfer function.

■ Solution

After drawing the asymptotes shown by the dashed lines, we first note that the data have a low-frequency asymptote of zero slope, and a high-frequency asymptote of slope −40 dB/decade. This suggests a second-order model without numerator dynamics, either of the form having real roots:

$$T(s) = \frac{K}{(\tau_1 s + 1)(\tau_2 s + 1)}$$

or the form having complex roots:

$$T(s) = \frac{K}{s^2 + 2\zeta\omega_n s + \omega_n^2}$$

However, the peak in the data eliminates the form having real roots.

At low frequencies, $m \approx 20 \log K$. From the plot, at low frequencies, $m = 75$ dB. Thus, $75 = 20 \log K$, which gives

$$K = 10^{75/20} = 5623$$

The peak is estimated to be 83 dB. From Table 9.2.2 the peak when $K = 1$ is given by $m_r = -20 \log(2\zeta\sqrt{1-\zeta^2})$. Thus, with $K = 5623$, the formula for the peak becomes

$$m_r = 20 \log 5623 - 20 \log \left(2\zeta\sqrt{1-\zeta^2}\right)$$

or

$$83 = 75 - 20 \log \left(2\zeta\sqrt{1-\zeta^2}\right)$$

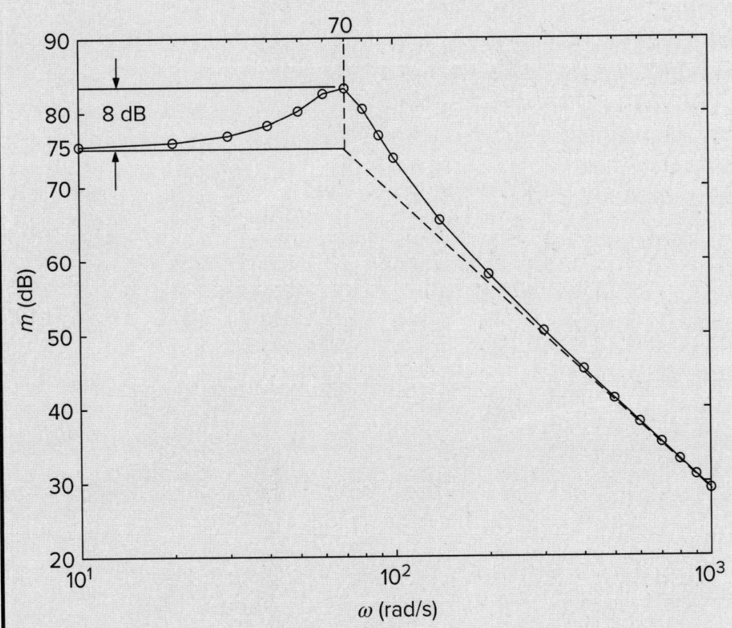

Figure 9.6.2 Frequency response data for Example 9.6.2.

Thus,

$$\log\left(2\zeta\sqrt{1-\zeta^2}\right) = \frac{75-83}{20} = -0.4$$

and

$$2\zeta\sqrt{1-\zeta^2} = 10^{-0.4}$$

Solve for ζ by squaring both sides.

$$4\zeta^2\left(1-\zeta^2\right) = 10^{-0.8}$$

$$4\zeta^4 - 4\zeta^2 + 10^{-0.8} = 0$$

This gives $\zeta^2 = 0.9587$ and 0.0413. The positive solutions are $\zeta = 0.98$ and 0.2. Because there is a resonance peak in the data, the first solution is not valid, and we obtain $\zeta = 0.2$.

Knowing ζ, we can now estimate ω_n from the peak frequency, which is estimated to be $\omega_r = 70$ rad/s. Thus from Table 9.2.1, $\omega_r = \omega_n\sqrt{1-2\zeta^2}$, or

$$70 = \omega_n\sqrt{1-2(0.2)^2}$$

This gives $\omega_n = 73$ rad/s.

Thus, the estimated model is

$$T(s) = \frac{5623}{s^2 + 29.2s + 5329}$$

EXAMPLE 9.6.3

Application of the Phase Plot

■ Problem

Consider the experimentally determined plots shown in Figure 9.6.3. Determine the forms of the transfer function.

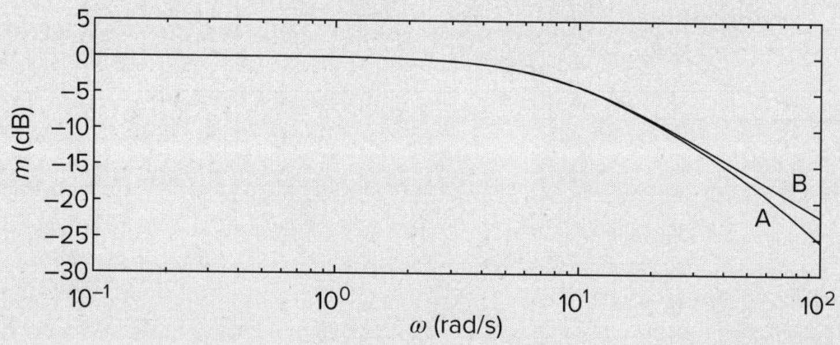

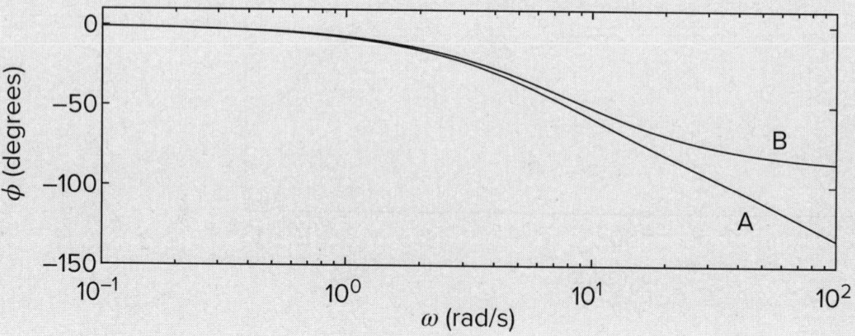

Figure 9.6.3 Frequency response data for Example 9.6.3.

■ **Solution**

At low frequencies both magnitude curves start at 0 dB and have zero slope. Therefore, the numerator of both transfer functions is 1 at low frequencies. Both magnitudes drop by 3 dB near $\omega = 8$ rad/s, and both phase curves pass near $-45°$ near $\omega = 8$ rad/s. Thus, we conclude that each transfer function has a denominator term $\tau s + 1$ where $\tau = 1/8$ s, approximately.

The high-frequency slope of curve B is -20 dB/decade, and thus, we conclude that its transfer function is

$$T_B(s) = \frac{1}{\frac{1}{8}s + 1}$$

The high-frequency slope of curve A is steeper, and thus we conclude that its transfer function has another denominator term that contributes to the slope at high frequencies. This is more apparent in the phase plot, where ϕ_B becomes horizontal but ϕ_A appears to be heading toward $-180°$. Thus, we suspect that its transfer function has the form

$$T_A(s) = \frac{1}{\left(\frac{1}{8}s + 1\right)(\tau_2 s + 1)}$$

but we cannot confirm this or determine the value of τ_2 without data at frequencies higher than 10^2 rad/s.

9.7 CASE STUDY: VEHICLE SUSPENSION DESIGN

As we saw in Chapter 4, Section 4.7, a common example of base excitation is caused by vehicle motion along a bumpy road surface. The primary purpose of a vehicle suspension is to maintain tire contact with the road surface, and the secondary purpose is to minimize the motion and force transmitted to the passenger compartment. In that chapter we analyzed the performance of a suspension system when encountering a *single* bump. Here we will analyze the effects of a road surface that is *periodic*.

We will see that a recurring theme in the design of vibration isolators is the conflict between design to protect against a single short-lived disturbance, such as a short bump, versus an irregular disturbance that lasts for some time. Sometimes the transient requirements conflict with the steady-state requirements. For example, with base excitation where the base motion is a step function (a suddenly applied constant displacement), the isolator must provide protection over a wide range of frequencies. We find that an isolator that provides good protection against periodic inputs will often provide poor protection against sudden transient inputs (called *shocks*) and vice versa.

There are advanced methods used by vehicle manufacturers and the U.S. Department of Transportation to assess and specify road surface roughness. Some of these methods use Fourier series analysis to identify the amplitudes and wavelengths associated with certain construction techniques. This information can be applied with Bode plot analysis to suspension system design.

The plots in this section are best generated by computer. Section 9.8 shows how to do this with MATLAB.

9.7.1 QUARTER-CAR MODEL

The quarter-car model of a vehicle suspension was treated in Section 4.8 and is shown again in Figure 9.7.1a. In this simplified model, the masses of the wheel, tire, and axle are neglected, and the mass m represents one-fourth of the vehicle mass. From the free

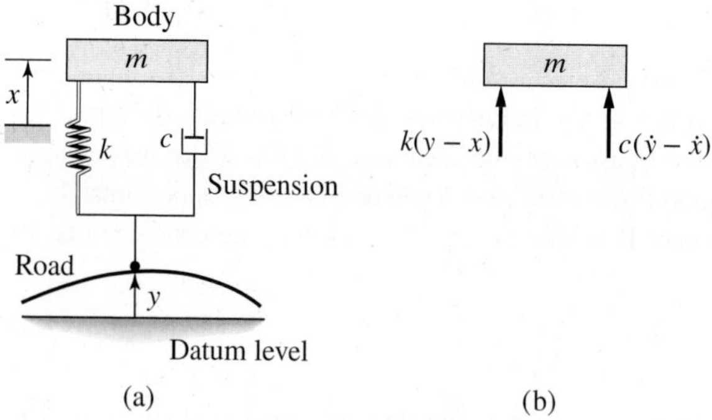

Figure 9.7.1 A quarter-car model with a single mass.

body diagram shown in Figure 9.7.1b we obtain the equation of motion:

$$m\ddot{x} + c\dot{x} + kx = c\dot{y} + ky$$

The displacement transfer function is

$$\frac{X(s)}{Y(s)} = \frac{cs + k}{ms^2 + cs + k} \tag{9.7.1}$$

The transfer function for the force f_t transmitted to the mass by the spring and damper is given by

$$\frac{F_t(s)}{Y(s)} = \frac{ms^2(cs + k)}{ms^2 + cs + k} \tag{9.7.2}$$

In Example 4.8.3 we used the single-mass model to design a suspension for a vehicle whose quarter mass is 250 kg and whose tire stiffness is 10^5 N/m. The input $y(t)$ was the haversine bump model, $y(t) = H \sin^2(\pi vt/L)$ for $0 \leq t \leq L/v$ and $y(t) = 0$ for $t > L/v$, where the bump height is $H = 0.1$ m and the vehicle velocity is $v = 18$ m/s (or about 40 mph). We considered two bump lengths: $L = 0.5$ m and $L = 2$ m. Choosing the natural frequency to be 1 Hz, we determined that the supension stiffness must be $k = 1.0974 \times 10^4$ N/m. Typical values of the damping ratio ζ for highway vehicles range from $\zeta = 0.2$ to 0.4, so we chose the middle value, $\zeta = 0.3$. This gives $c = 943$ N · s/m.

We will simulate the design using the transfer functions of the quarter-car model that includes the wheel-tire mass m_2 and the tire stiffness k_2, developed in Example 4.8.2 (see Figure 9.7.2). The displacement transfer functions are

$$\frac{X_1(s)}{Y(s)} = \frac{k_2(c_1 s + k_1)}{D(s)} \tag{9.7.3}$$

$$\frac{X_2(s)}{Y(s)} = \frac{k_2(m_1 s^2 + c_1 s + k_1)}{D(s)} \tag{9.7.4}$$

where

$$D(s) = m_1 m_2 s^4 + (m_1 c_1 + m_2 c_1)s^3 + (m_1 k_1 + m_1 k_2 + m_2 k_1)s^2 + c_1 k_2 s + k_1 k_2 \tag{9.7.5}$$

The force exerted on the chassis by the suspension is given by

$$\frac{F_t(s)}{Y(s)} = m_1 k_2 s^2 \frac{c_1 s + k_1}{D(s)} \tag{9.7.6}$$

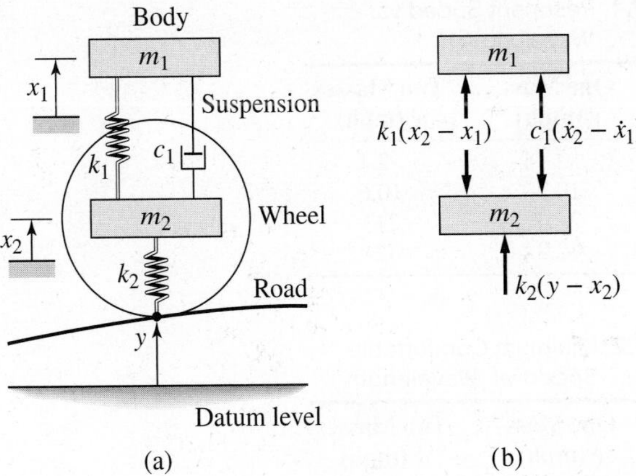

Figure 9.7.2 A quarter-car model with two masses.

The input $y(t)$ was the same haversine bump model used with the single-mass model. For both bump lengths the chassis motion was well isolated from the bump (less than 10% of the bump height for $L = 0.5$ m, and 30% for $L = 2$ m). However, the wheel displacement was a little more than half the bump height for $L = 0.5$ m, and was about 17% *more* than the bump displacement for $L = 2$ m.

9.7.2 FREQUENCY RESPONSE OF SINGLE MASS MODEL

We now analyze the Bode plots for the two models, using the design parameters given previously. For the single-mass model,

$$\frac{X(s)}{Y(s)} = \frac{943s + 1.0974 \times 10^4}{250s^2 + 943s + 1.0974 \times 10^4} \tag{9.7.7}$$

Its magnitude and phase angle plots are shown in Figure 9.7.3. The peak value of M is 2 and the peak occurs at $\omega = 6$ rad/s. If the road surface is sinusoidal with a wavelength

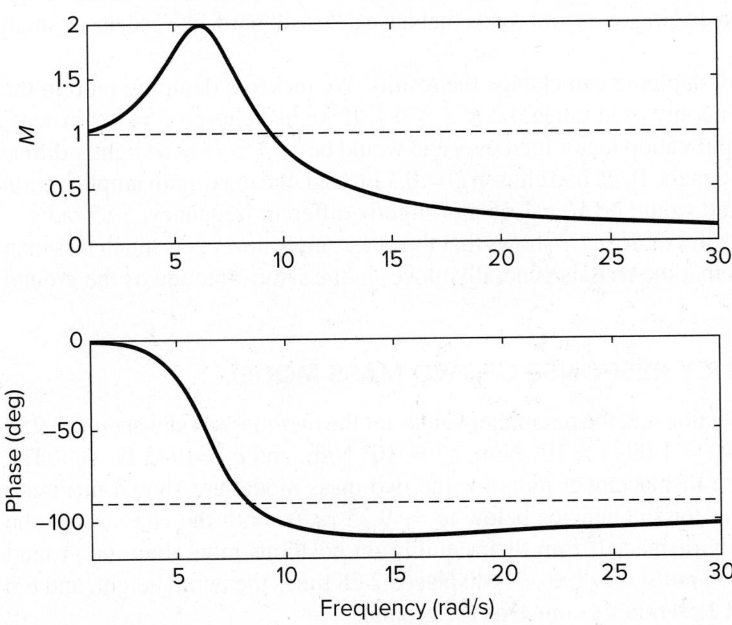

Figure 9.7.3 Magnitude and phase angle plots for the single-mass model (9.7.7).

Table 9.7.1 Resonant Speed vs. Wavelength

L (m)	One Mass v (mph)	Two Masses v (mph)
1	2.14	2.1
5	10.7	10.6
10	21.4	21
20	42.8	42

Table 9.7.2 Minimum Comfortable Speed vs. Wavelength

L (m)	One Mass v (mph)	Two Masses v (mph)
1	3.3	3.3
5	13.7	16.5
10	33.4	33
20	66.8	66

L and a height A, the input displacement is $y(t) = A \sin \omega t = A \sin(2\pi vt/L)$, where v is the vehicle speed. Thus, the maximum chassis displacement will be twice as high as the bump (a bad result!) and will occur if $\omega = 2\pi v/L = 6$, or $v = 0.9549L$. Table 9.7.1 shows the vehicle speed that produces the maximum chassis displacement, for a given wavelength L. So if you drive at 21.4 mph over a road whose wavelength is 10 m, the chassis displacement will be twice the road displacement.

The M value is less than 1 for $\omega > 9.38$ rad/s. Thus, for speeds greater than $v = (9.38/2\pi)L = 1.493L$, the chassis displacement will be less than the road surface height. Table 9.7.2 shows the minimum vehicle speed that produces a chassis displacement less than the bump height, for a given wavelength L. So if the wavelength is 10 m, and if you drive slower than 33.4 mph, the chassis displacement will be more than the road displacement. If these results seem counterintuitive, observe the rough ride one gets when driving on gravel roads that have a "washboard-like" surface (small wavelength).

The amount of damping can change the results. We picked a damping ratio in the middle of the commonly used range $0.2 \leq \zeta \leq 0.4$. If we had chosen $\zeta = 0.2$ instead, the maximum amplification factor increases and would be $M = 2.74$ at a slightly different frequency, 6.36 rad/s. If we had chosen $\zeta = 0.4$ instead, the maximum amplification factor decreases and would be $M = 1.66$ at a slightly different frequency, 5.85 rad/s.

The phase plot in Figure 9.7.3 shows that the phase angle never gets much less than $-90°$. This means that the chassis generally moves in the same direction as the ground motion.

9.7.3 FREQUENCY RESPONSE OF TWO-MASS MODEL

From Chapter 4, Section 4.8, the parameter values for the two-mass model are $m_1 = 250$ kg, $m_2 = 25$ kg, $k_1 = 1.0975 \times 10^4$ N/m, $k_2 = 10^5$ N/m, and $c_1 = 943$ N $\cdot$ s/m. The frequency response displacement plots for the two-mass model are shown in Figure 9.7.4. It shows that for frequencies below $\omega = 9.25$ rad/s, both the chassis and the wheel are vertically displaced from their equilibrum positions more than the ground motion. At $\omega = 5.94$ rad/s, the chassis is displaced 2.28 times the bump height, and the wheel is displaced 1.59 times as much as the ground.

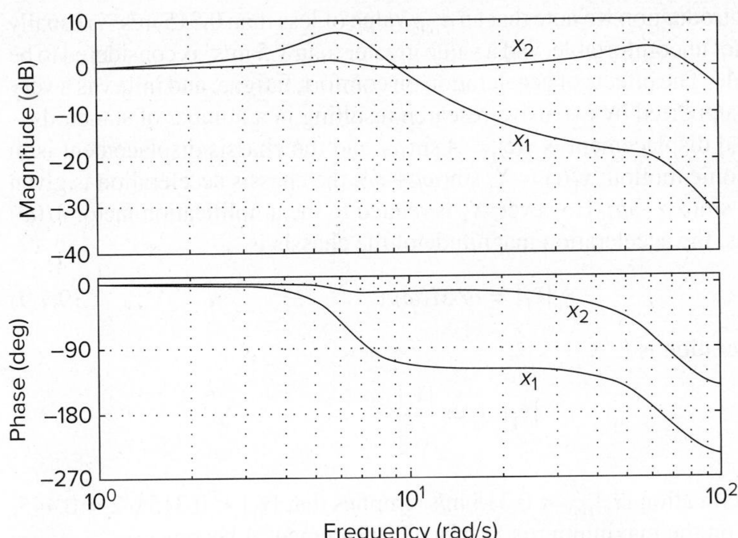

Figure 9.7.4 Magnitude and phase angle plots of the chassis and wheel displacement for the two-mass model.

Thus, the comfortable speed range for the chassis is $v > (9.25/2\pi)L = 1.472L$. Table 9.7.2 shows the comfortable speed range for chassis motion versus wavelength. Thus, we see that the predictions given by the single-mass model are fairly accurate.

Figure 9.7.4 shows that the wheel motion is more pronounced at higher frequencies, which means at smaller wavelengths. The peak in M for the wheel is 1.59 at $\omega = 59.1$ rad/s. These results make sense because the chassis is so much more massive than the wheel, and thus less sensitive to rapidly changing inputs.

The range of frequencies for which the wheel amplification factor M is less than 1 is $8.08 \leq \omega \leq 20.2$ rad/s. The chassis factor is also less than 1 in this range, so this range defines the speeds where the ride is most comfortable.

The relative motion of the chassis, wheel, and road can be determined from the phase plot. For example, near $\omega = 60$ rad/s, the wheel motion and the chassis motion lag behind the ground sinusoidal motion by 90° and 180°, respectively. Thus near this frequency, the chassis motion is in the opposite direction of the ground motion and it lags the wheel motion by 90°. Such information is useful to designers for understanding the suspension behavior.

9.7.4 ASSESSING RIDE QUALITY

In addition to oscillation frequency and the maximum displacement of the chassis (and thus the passenger compartment), other performance measures are used to assess ride quality. Some of these are: the maximum or root-mean-square (rms) acceleration of the chassis, and the maximum force and maximum jerk felt. Recall that in Chapters 1 and 3 we applied the root-mean-square measure to assess the average torque required for a motion-control system. We can apply the same concept to assess to average acceleration. The rms acceleration a_{rms} over a time T is given by

$$a_{rms} = \sqrt{\frac{1}{T}\int_0^T a^2(t)\,dt} \tag{9.7.8}$$

If $a(t)$ is sinusoidal, $a(t) = A\sin(\omega t + \phi)$, then $a_{rms} = A/\sqrt{2}$.

In this brief introduction we note that an a_{rms} value of less than 0.315 m/s^2 is usually considered to be not uncomfortable, and a value greater than 1.5 m/s^2 is considered to be very uncomfortable. The effects of acceleration on comfort, fatigue, and injury is a very complex subject supported by extensive research resulting in a number of standards.

When the road displacement is $y(t) = A \sin \omega t$ and the chassis displacement is in steady-state harmonic motion: $x_1(t) = X_1 \sin(\omega t + \phi)$, the chassis acceleration is given by $\ddot{x}_1(t) = -\omega^2 X_1 \sin(\omega t + \phi)$. However, X_1 is related to the amplification factor $M(\omega)$ by $X_1 = MA$. Thus, the acceleration magnitude of the chassis is

$$|\ddot{x}_1| = \omega^2 M(\omega) A \tag{9.7.9}$$

and the rms acceleration is

$$|\ddot{x}_1|_{rms} = \frac{|\ddot{x}_1|}{\sqrt{2}}$$

Thus, the specification $|\ddot{x}_1|_{rms} < 0.315$ m/s^2 implies that $|\ddot{x}_1| < 0.315\sqrt{2} = 0.445$, and the condition on the maximum road surface displacement A becomes

$$A \leq \frac{0.445}{\omega^2 M(\omega)} \text{ m} \tag{9.7.10}$$

This can be a complicated relation, depending on how M varies with ω, and is best analyzed by plotting it (Figure 9.7.5). Road displacements below the curve give a comfortable ride (an acceleration less than 0.315 m/s^2).

For example, the chassis peak M value is 2.28 at $\omega = 5.94$ rad/s. Thus, the road amplitude A should be less than $0.445/[2.28(5.94)^2] = 0.0055$ m, or 0.55 cm. At the "comfort" limit where $M = 1$, $\omega = 9.25$ and A should be less than $A = 0.445/[1(9.25)^2] = 0.0052$ m, or 0.52 cm. This result for $M = 1$ is not much different from the peak value of $M = 2.28$, because $M(\omega)$ decreases with frequency, tending to cancel the effect of ω^2.

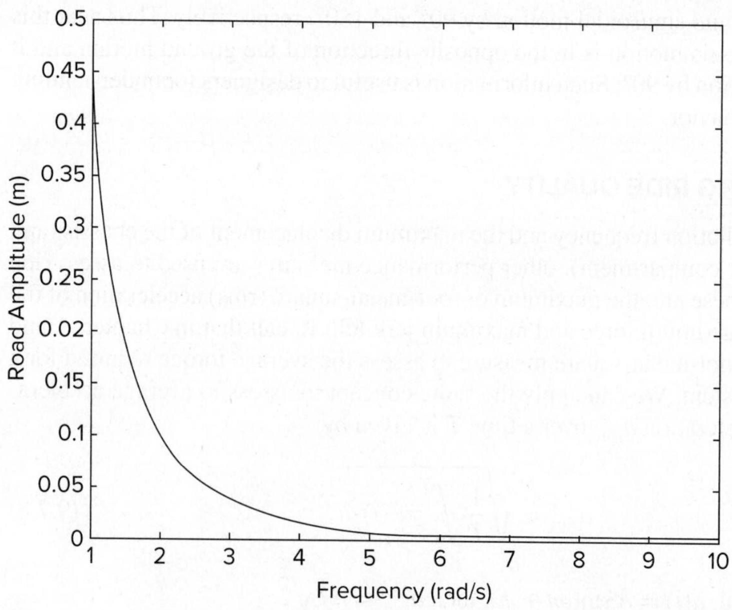

Figure 9.7.5 Maximum road surface displacement for a comfortable ride.

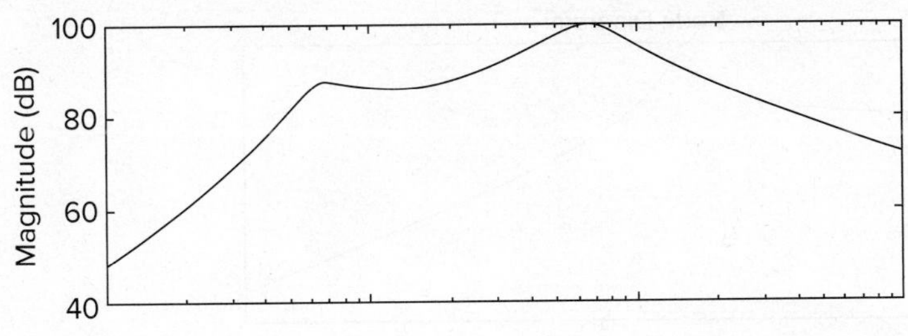

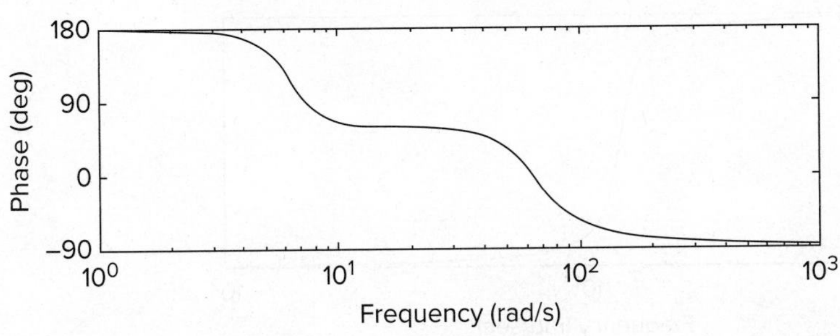

Figure 9.7.6 Bode plot of force transmitted to the chassis.

Our final indicator of ride quality is the force transmitted to the chassis. Figure 9.7.6 is the frequency response plot for the two-mass model (9.7.6). The peak value of m is 100 dB ($M = 10^5$) and would generate a force of 1000 N if the road amplitude is 1 cm. But the peak force occurs at $\omega = 67.1$ rad/s, which is way above most practical speeds (the road wavelength needs to be less than 2.93 m to generate this frequency at 70 mph).

A secondary peak in transmitted force is $m = 87.6$ dB ($M = 2.4 \times 10^4$), and it occurs at $\omega = 6.95$. A 1-cm road displacement will produce a force of 240 N. This could be the more troublesome force because it occurs at reasonable speeds.

9.8 FREQUENCY RESPONSE ANALYSIS USING MATLAB

The logarithmic frequency response plots are sometimes called "Bode" plots, and are named after H. W. Bode, who developed frequency response methods for electronic circuit design. MATLAB provides the `bode`, `bodemag`, `evalfr`, and `freresp` functions to compute and plot frequency response. These functions are available in the Control Systems Toolbox.

9.8.1 THE `bode` FUNCTION

The `bode` function generates frequency response plots. The `bodemag(sys)` function plots the magnitude only. The basic syntax of the `bode` function is `bode(sys)`, where `sys` is an LTI system model. Such a model can be created with the `tf` function or the `ss` function. In this basic syntax the `bode` function produces a screen plot of the logarithmic magnitude ratio m and the phase angle ϕ versus ω. Note that the phase angle is plotted in *degrees*, not radians. MATLAB uses the corner frequencies of the numerator and denominator of the transfer function to automatically select an appropriate frequency range for the plots. For example, to obtain the plots for the transfer function

$$T(s) = \frac{70}{3s^2 + 4s + 192} \tag{9.8.1}$$

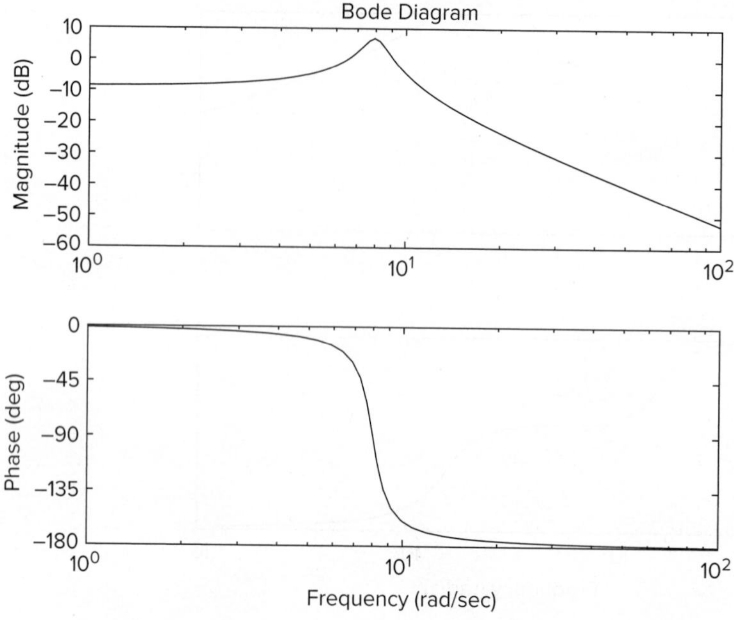

Figure 9.8.1 Plots produced with the MATLAB `bode` function.

the session is

```
>>sys = tf(70,[3, 4, 192]);
>>bode(sys)
```

Figure 9.8.1 shows the plots you will see on the screen.

If you created the LTI model from a state variable model that has more than one input or more than one output, then the `bode` function will display a series of plots, one pair for each input-output combination.

9.8.2 SPECIFYING THE FREQUENCY RANGE

To specify a set of frequencies rather than letting MATLAB select them, you can use the syntax `bode(sys,w)`, where `w` is a vector containing the desired frequencies. The frequencies must be specified in *radians* per unit time (for example, radians/second). For example, you might want to examine more closely the peak in the m curve near the resonant frequency of (9.8.1). To see 401 regularly spaced points on the curve for $7 \leq \omega \leq 9$, the session is

```
>>sys = tf(70,[3, 4, 192]);
>>w = (7:0.005:9);
>>bode(sys,w)
```

To obtain 401 points in the range $7 \leq \omega \leq 9$ that are logarithmically spaced, type instead `w = logspace(log10(7),log10(9),401)`.

If you want to specify just the lower and upper frequencies `wmin` and `wmax`, and let MATLAB select the spacing, generate the vector `w` by typing `w = {7, 9}`.

A variation is the following: `[mag, phase, w] = bode(sys)`. This form returns the magnitude ratio M, the phase angle ϕ, and the frequencies ω in the arrays `mag`, `phase`, and `w`, but does not display a plot. It is very important to note that the array `mag` contains the ratio M, and *not* the logarithmic ratio m. Note also that the array `phase` contains the phase angle in *degrees*, not radians. However, the vector `w` contains the frequencies in *radians* per unit time. The frequencies in the vector `w` are the frequencies automatically selected by MATLAB. If you want to use your own set of frequencies, use the syntax: `[mag, phase] = bode(sys,w)`. The extended syntax of the `bodemag` function is similar.

These two forms are useful when you want to save the frequency response calculations generated by the `bode` function. You can use the arrays `mag`, `phase`, and `w` for further analysis, or to generate plots that you can format as you want. If you do this note that the variables `mag` and `phase` are returned as $1 \times 1 \times n$ arrays. To collapse the values into a $1 \times n$ array suitable for plotting, type `mag = mag(:)` and `phase = phase(:)`.

For example, to plot M versus ω for (9.8.1) using 401 regularly spaced points over the range $7 \leq \omega \leq 9$, the session is

```
>>sys = tf(70,[3, 4, 192]);
>>w = (7:0.005:9);
>>[mag] = bode(sys,w); mag = mag(:);
>>plot(w,mag),xlabel('\omega (rad/sec)'),ylabel('M')
```

Note that the `phase` array need not be generated if you are not going to use it.

To plot the logarithmic magnitude ratio m versus $\log \omega$, using the frequencies generated by MATLAB, the session is

```
>>sys = tf(70,[3, 4, 192]);
>>[mag, phase, w] = bode(sys); mag = mag(:);
>>semilogx(w,20*log10(mag)),grid,...
>>xlabel('\omega (rad/sec)'),ylabel('m (dB)')
```

Note that you must use the decibel conversion $m = 20 \log M$.

Both the `bodemag` and the `bode` functions can be used to obtain the curves for several transfer functions on the same plot by using the syntax `bode(sys1, sys2, ...)` or `bodemag(sys1, sys2, ...)`.

9.8.3 OBTAINING INFORMATION FROM THE PLOT

Once the plot is displayed on the screen, right-clicking on it brings up a menu. One choice on the menu is "Characteristics." Left-clicking on this choice brings up a submenu, one of whose choices is "Peak Response." When this is selected, the peak of the magnitude curve is identified with a dot, and the corresponding frequency is displayed. Left-clicking on the curve enables you to trace the curve with the cursor while the coordinates are displayed. With this method you can determine the bandwidth. For the transfer function (9.8.1), with the plot covering the frequency range $7 \leq \omega \leq 9$, the peak is found to be $m = 6.83$ dB at $\omega = 7.94$ rad/sec, and the lower and upper bandwidth frequencies are found to be 7.25 and 8.59 rad/sec.

9.8.4 THE `evalfr` FUNCTION

The `evalfr` function stands for "evaluation of frequency response." It computes the complex value of a transfer function at a specified value of s. For example, to compute the value of $T(s)$ given by (8.6.1) at $s = 6j$, and find its magnitude and phase angle, the session is

```
>>fr = evalfr(sys,6j);
>>mag = abs(fr);
>>ang = angle(fr);
```

The results are `fr` $= 0.7704 - 0.2201j$, `mag` $= 0.8013$, and `phase` $= -0.2783$ rad.

9.8.5 THE `freqresp` FUNCTION

The `evalfr` function can be used to evaluate `sys` at one frequency only. For a vector of frequencies, use the `freqresp` function, whose syntax is `fr = freqresp(sys,w)`, where `w` is a vector of frequencies, which must be in radians/unit time.

The following MATLAB script file calculates the coefficients and phase angles of the steady-state response described in Example 9.5.1.

```
sys1 = tf(600, [1, 20, 625]);
w = [0, 2*pi, 4*pi, 8*pi, 12*pi];
A = 0.02*[1/pi, 1/2, -2/(3*pi), -2/(3*5*pi), -2/(5*7*pi)];
fr = freqresp(sys1,w);
M = abs(fr);
B = A.*M(:)'
ph = angle(fr(:))
```

The results are shown in the table in Example 9.5.1.

9.9 CHAPTER REVIEW

The frequency response of a system describes its steady-state behavior resulting from periodic inputs. This chapter demonstrates the usefulness of the transfer function for understanding and analyzing a system's frequency response.

A sinusoidal input applied to a stable linear system produces a steady-state sinusoidal output of the same frequency, but with a different amplitude and a phase shift. The frequency transfer function is the transfer function with the Laplace variable s replaced by $j\omega$, where ω is the input frequency. The magnitude M of the frequency transfer function is the amplitude ratio between the input and output, and its argument is the phase shift ϕ. Both are functions of ω and usually plotted against $\log \omega$, with M expressed in decibels as $m = 20 \log M$. The logarithmic plots can be sketched by using low- and high-frequency asymptotes intersecting at the corner frequencies.

Most transfer functions consist of a combination of the following terms: K, s, $\tau s + 1$, and $s^2 + 2\zeta\omega_n s + \omega_n^2$, so we analyzed the frequency response of each term. The composite frequency response plot consists of the sum of the plots of the numerator terms minus the sum of the plots of the denominator terms.

Important engineering applications of frequency response covered in this chapter are an understanding of resonance and bandwidth. The bandwidth measures the filtering property of the system. This concept can be used with the Fourier series representation of a periodic function, which is an infinite series, by eliminating those terms that lie outside the bandwidth. The truncated series can be used to compute the response to a general periodic input.

The form of the transfer function and an estimate of the numerical values of the parameters can often be obtained from experiments involving frequency response.

The MATLAB Control Systems toolbox provides the `bode`, `bodemag`, `evalfr`, and `freresp` functions, which are useful for frequency response analysis.

Now that you have finished this chapter, you should be able to

1. Sketch frequency response plots using asymptotes, and use the plots or the frequency transfer function to determine the steady-state output amplitude and phase that results from a sinusoidal input.
2. Compute resonance frequencies and bandwidth.
3. Analyze vibration isolation and absorber systems, and the effects of base motion.
4. Determine the steady-state response to a periodic input, given the Fourier series description of the input.
5. Estimate the form of a transfer function and its parameter values, given frequency response data.
6. Use MATLAB as an aid in the preceding tasks.

PROBLEMS

Section 9.1 Frequency Response of First-Order Systems

9.1 Evaluate the following products and ratios. Put the answer in the form $a + jb$.

a. $(2 + 4j)(7 + 9j)$ b. $\dfrac{2 + 4j}{7 + 9j}$

c. $\dfrac{(2 + 4j)(7 + 9j)}{(3 + 7j)(5 + 2j)}$ d. $(-2 + 4j)(-7 + 9j)$

e. $\dfrac{-2 + 4j}{-7 + 9j}$

9.2 Find the magnitude and angle of the following numbers.

a. $2 + 4j$ b. $-2 + 4j$ c. $2 - 4j$ d. $-2 - 4j$

9.3 Find the magnitude $|N(\omega)|$ and angle $\phi(\omega)$ of the following expressions. Assume that $\omega \geq 0$.

a. $N(\omega) = \dfrac{2 + 4\omega j}{7 + 9\omega j}$ b. $N(\omega) = \dfrac{-2 + 4\omega j}{-7 + 9\omega j}$

9.4 A certain mass oscillates with the displacement given by $x(t) = 5\sin(3t - 1.2)$ for $t \geq 0$. (a) What is the peak displacement and when does the first peak occur? (b) What is the peak velocity and when does the first peak occur?

9.5 Use the following transfer functions to find the steady-state response $y_{ss}(t)$ to the given input function $f(t)$.

a. $$T(s) = \frac{Y(s)}{F(s)} = \frac{75}{14s + 18}, \quad f(t) = 10\sin 1.5t$$

b. $$T(s) = \frac{Y(s)}{F(s)} = \frac{5s}{3s + 4}, \quad f(t) = 30\sin 2t$$

c. $$T(s) = \frac{Y(s)}{F(s)} = \frac{s + 50}{s + 150}, \quad f(t) = 15\sin 100t$$

d. $$T(s) = \frac{Y(s)}{F(s)} = \frac{33}{100}\frac{s + 100}{s + 33}, \quad f(t) = 8\sin 50t$$

9.6 Use asymptotic approximations to sketch the frequency response plots for the following transfer functions.

a. $$T(s) = \frac{15}{6s+2}$$

b. $$T(s) = \frac{9s}{8s+4}$$

c. $$T(s) = 6\frac{14s+7}{10s+2}$$

9.7 Figure P9.7 is a representation of the effects of the tide on a small body of water connected to the ocean by a channel. Assume that the ocean height h_i varies sinusoidally with a period of 12 hr with an amplitude of 3 ft about a mean height of 10 ft. If the observed amplitude of variation of $\hat{h}$ is 2 ft, determine the time constant of the system and the time lag between a peak in h_i and a peak in $\hat{h}$.

Figure P9.7

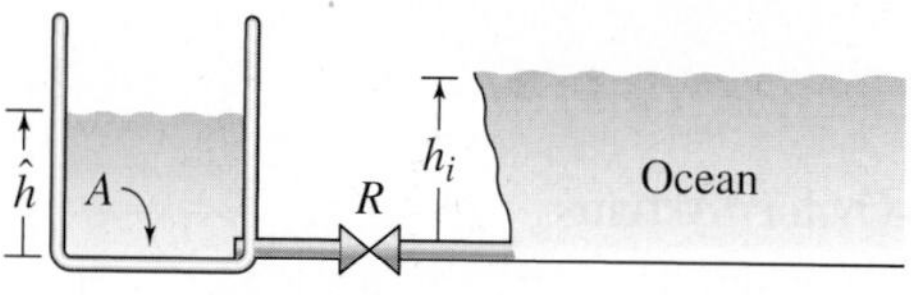

9.8 A single-room building has four identical exterior walls, 5 m wide by 3 m high, with a perfectly insulated roof and floor. The thermal resistance of the walls is $R = 4.5 \times 10^{-3}$ K/W · m^2. Taking the only significant thermal capacitance to be the room air, obtain the expression for the steady-state room air temperature if the outside air temperature varies sinusoidally about 15°C with an amplitude of 5° and a period of 24 h. The specific heat and density of air at these conditions are $c_p = 1004$ J/kg · K and $\rho = 1.289$ kg/m^3.

9.9 For the rotational system shown in Figure P9.9, $I = 4$ kg · m^2 and $c = 8$ N · m · s/rad. Obtain the transfer function $\Omega(s)/T(s)$, and derive the expression for the steady-state speed $\omega_{ss}(t)$ if the applied torque in N · m is given by

$$T(t) = 20 + 15 \sin 3t + 6 \cos 5t$$

9.10 For the system shown in Figure P9.10, the bottom area is $A = 2\pi$ ft^2 and the linear resistance is $R = 3000$ sec^{-1}ft^{-1}. Suppose the volume inflow rate is

$$q_{vi}(t) = 0.1 + 0.2 \sin 0.002t \text{ ft}^3\text{/sec}$$

Obtain the expression for the steady-state height $\hat{h}_{ss}(t)$. Compute the lag in seconds between a peak in $q_{vi}(t)$ and a peak in $\hat{h}_{ss}(t)$.

Figure P9.9

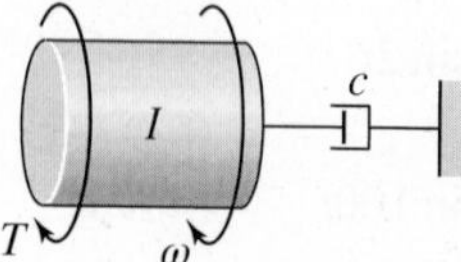

Figure P9.10

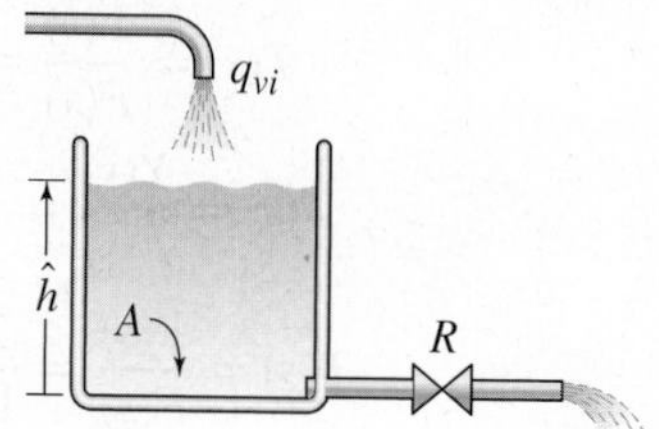

Section 9.2 Frequency Response of Higher-Order Systems

9.11 Use the following transfer functions to find the steady-state response $y_{ss}(t)$ to the given input function $f(t)$.

a. $$T(s) = \frac{Y(s)}{F(s)} = \frac{5}{(5s+1)(2s+1)}, \quad f(t) = 10 \sin 0.2t$$

b. $$T(s) = \frac{Y(s)}{F(s)} = \frac{1}{s^2 + 10s + 100}, \quad f(t) = 16 \sin 5t$$

9.12 Use the following transfer functions to find the steady-state response $y_{ss}(t)$ to the given input function $f(t)$.

a. $$T(s) = \frac{Y(s)}{F(s)} = \frac{8}{s(s^2 + 10s + 100)}, \quad f(t) = 6 \sin 9t$$

b. $$T(s) = \frac{Y(s)}{F(s)} = \frac{10}{s^2(s+1)}, \quad f(t) = 9 \sin 2t$$

c. $$T(s) = \frac{Y(s)}{F(s)} = \frac{s}{(2s+1)(5s+1)}, \quad f(t) = 9 \sin 0.7t$$

d. $$T(s) = \frac{Y(s)}{F(s)} = \frac{s^2}{(2s+1)(5s+1)}, \quad f(t) = 9 \sin 0.7t$$

9.13 The model of a certain mass-spring-damper system is

$$10\ddot{x} + c\dot{x} + 20x = f(t)$$

Determine its resonant frequency ω_r and its peak magnitude M_r if (a) $\zeta = 0.1$ and (b) $\zeta = 0.3$.

9.14 The model of a certain mass-spring-damper system is

$$10\ddot{x} + c\dot{x} + 20x = f(t)$$

How large must the damping constant c be so that the maximum steady-state amplitude of x is no greater than 3, if the input is $f(t) = 11 \sin \omega t$, for an arbitrary value of ω?

9.15 The model of a certain mass-spring-damper system is

$$13\ddot{x} + 2\dot{x} + kx = 10 \sin \omega t$$

Determine the value of k required so that the maximum response occurs at $\omega = 4$ rad/sec. Obtain the steady-state response at that frequency.

9.16 Determine the resonant frequencies of the following models.

a. $$T(s) = \frac{7}{s(s^2 + 6s + 58)}$$

b. $$T(s) = \frac{7}{(3s^2 + 18s + 174)(2s^2 + 8s + 58)}$$

9.17 For the circuit shown in Figure P9.17, $L = 0.1$ H, $C = 10^{-6}$ F, and $R = 100\ \Omega$. Obtain the transfer functions $I_3(s)/V_1(s)$ and $I_3(s)/V_2(s)$. Using asymptotic approximations, sketch the m curves for each transfer function and discuss how the circuit acts on each input voltage (does it act like a low-pass filter, a high-pass filter, or other?).

Figure P9.17

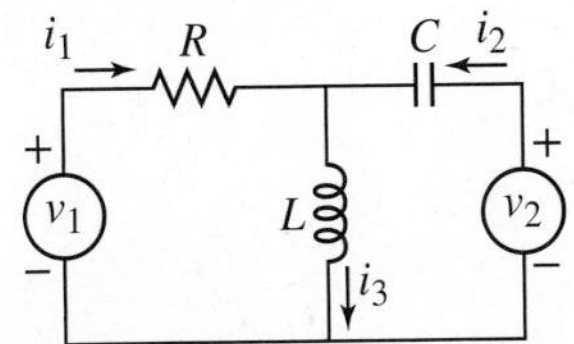

Figure P9.18

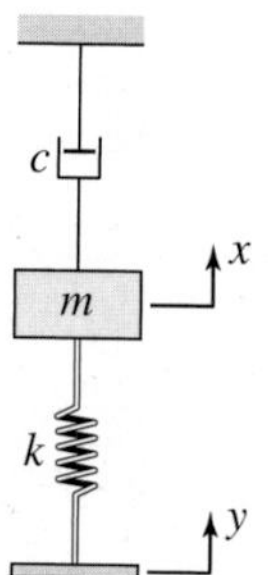

9.18 (a) For the system shown in Figure P9.18, $m = 1$ kg and $k = 600$ N/m. Derive the expression for the peak amplitude ratio M_r and resonant frequency ω_r, and discuss the effect of the damping c on M_r and on ω_r. (b) Extend the derivation of the expressions for M_r and ω_r to the case where the values of m, c, and k are arbitrary.

9.19 For the RLC circuit shown in Figure P9.19, $C = 10^{-5}$ F and $L = 5 \times 10^{-3}$ H. Consider two cases: (a) $R = 10\ \Omega$ and (b) $R = 1000\ \Omega$. Obtain the transfer function $V_o(s)/V_s(s)$ and the log magnitude plot for each case. Discuss how the value of R affects the filtering characteristics of the system.

9.20 A model of a fluid clutch is shown in Figure P9.20. Using the values $I_1 = I_2 = 0.02\ \text{kg}\cdot\text{m}^2$, $c_1 = 0.04\ \text{N}\cdot\text{m}\cdot\text{s/rad}$, and $c_2 = 0.02\ \text{N}\cdot\text{m}\cdot\text{s/rad}$, obtain the transfer function $\Omega_2(s)/T_1(s)$, and derive the expression for the steady-state speed $\omega_2(t)$ if the applied torque in N · m is given by

$$T_1(t) = 4 + 2\sin 1.5t + 0.9\sin 2t$$

Figure P9.19

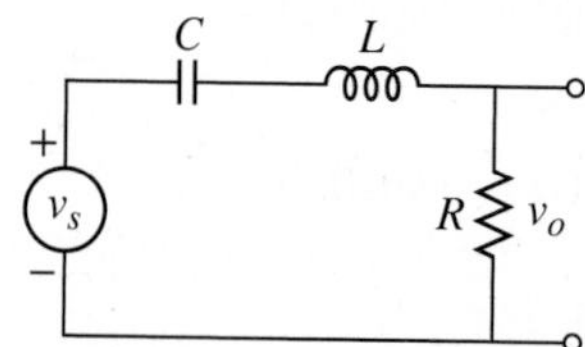

Figure P9.20

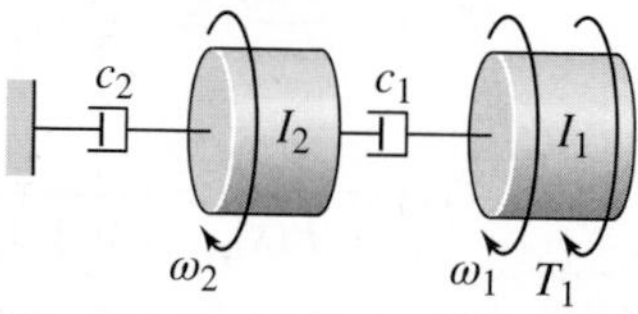

Section 9.3 Frequency Response Applications

9.21 Determine the beat period and the vibration period of the model

$$3\ddot{x} + 75x = 7\sin 5.2t$$

9.22 Resonance will produce large vibration amplitudes, which can lead to system failure. For a system described by the model

$$\ddot{x} + 64x = 0.2\sin \omega t$$

where x is in feet, how long will it take before $|x|$ exceeds 0.1 ft, if the forcing frequency ω is close to the natural frequency?

9.23 A certain factory contains a heavy rotating machine that causes the factory floor to vibrate. We want to operate another piece of equipment nearby and we measure the amplitude of the floor's motion at that point to be 0.01 m. The mass of the equipment is 1500 kg and its support has a stiffness of $k = 2 \times 10^4$ N/m and a damping ratio of $\zeta = 0.04$. Calculate the maximum force that will be transmitted to the equipment at resonance.

9.24 An electronics module inside an aircraft must be mounted on an elastic pad to protect it from vibration of the airframe. The largest amplitude vibration produced by the airframe's motion has a frequency of 40 cycles per second. The module weighs 200 N, and its amplitude of motion is limited to 0.003 m because of space. Neglect damping and calculate the percent of the airframe's motion transmitted to the module.

9.25 An electronics module used to control a large crane must be isolated from the crane's motion. The module weighs 2 lb. (a) Design an isolator so that no more than 10% of the crane's motion amplitude is transmitted to the module. The crane's vibration frequency is 3000 rpm. (b) What percentage of the crane's motion will be transmitted to the module if the crane's frequency can be anywhere between 2500 and 3500 rpm?

9.26 Design a vibrometer having a mass of 0.1 kg, to measure displacements having a frequency near 200 Hz.

9.27 A motor mounted on a beam vibrates too much when it runs at a speed of 6000 rpm. At that speed the measured force produced on the beam is 40 lb. Design a vibration absorber to attach to the beam. Because of space limitations, the absorber's mass cannot have an amplitude of motion greater than 0.06 in.

9.28 The supporting table of a radial saw weighs 180 lb. When the saw operates at 300 rpm, it transmits a force of 6 lb to the table. Design a vibration absorber to be attached underneath the table. The absorber's mass cannot vibrate with an amplitude greater than 0.5 in.

Section 9.4 Filtering Properties of Dynamic Systems

9.29 A certain mass is driven by base excitation through a spring (see Figure P9.29). Its parameter values are $m = 200$ kg, $c = 2000$ N · s/m, and $k = 2 \times 10^4$ N/m. Determine its resonant frequency ω_r, its resonance peak M_r, and its bandwidth.

Figure P9.29

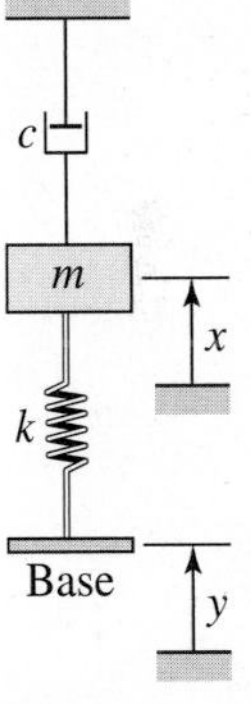

9.30 A certain series *RLC* circuit has the following transfer function.

$$T(s) = \frac{I(s)}{V(s)} = \frac{Cs}{LCs^2 + RCs + 1}$$

Suppose that $L = 300$ H, $R = 10^4$ Ω, and $C = 10^{-6}$ F. Find the bandwidth of this system.

9.31 Obtain the expressions for the bandwidths of the two circuits shown in Figure P9.31.

Figure P9.31

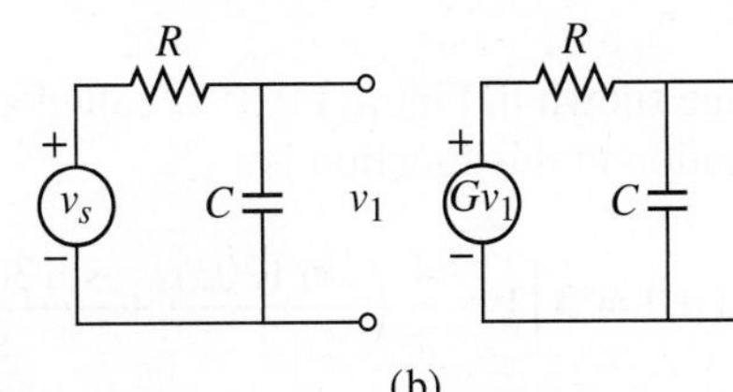

Figure P9.32

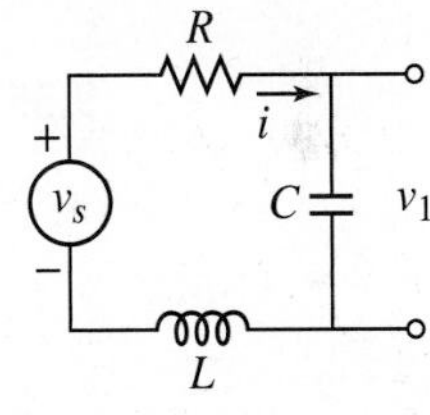

9.32 For the circuit shown in Figure P9.32, $L = 0.1$ H and $C = 10^{-6}$ F. Investigate the effect of the resistance R on the bandwidth, resonant frequency, and resonant peak over the range $100 \le R \le 1000$ Ω.

9.33 For the circuit shown in Figure 9.1.7a, can values be found for R_1, R_2, and C to make a low-pass filter? Prove your answer mathematically.

Section 9.5 Response to Nonperiodic Inputs

9.34 The voltage shown in Figure P9.34 is produced by applying a sinusoidal voltage to a *full wave rectifier*. The Fourier series approximation to this function is

$$v_s(t) = \frac{20}{\pi} - \frac{40}{\pi}\left[\frac{\cos 240\pi t}{1(3)} + \frac{\cos 480\pi t}{3(5)} + \frac{\cos 720\pi t}{5(7)} + \cdots\right]$$

Suppose this voltage is applied to a series RC circuit whose transfer function is

$$\frac{V_o(s)}{V_s(s)} = \frac{1}{RCs + 1}$$

where $R = 600\ \Omega$ and $C = 10^{-6}$ F. Keeping only those terms in the Fourier series whose frequencies lie within the circuit's bandwidth, obtain the expression for the steady-state voltage $v_o(t)$.

Figure P9.34

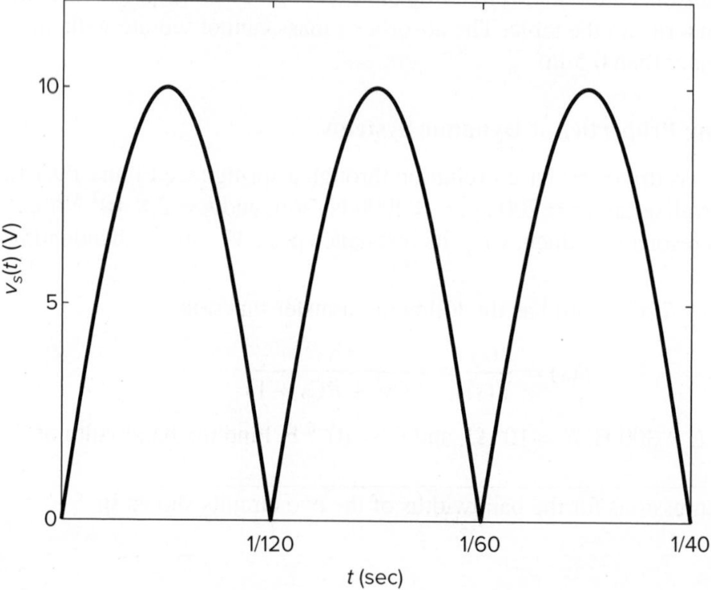

9.35 The voltage shown in Figure P9.35 is called a square wave. The Fourier series approximation to this function is

$$v_s(t) = 5\left[1 + \frac{4}{\pi}\left(\frac{\sin 120\pi t}{1} + \frac{\sin 360\pi t}{3} + \frac{\sin 600\pi t}{5} + \cdots\right)\right]$$

Suppose this voltage is applied to a series RC circuit whose transfer function is

$$\frac{V_o(s)}{V_s(s)} = \frac{1}{RCs + 1}$$

where $R = 10^3\ \Omega$ and $C = 10^{-6}$ F. Keeping only those terms in the Fourier series whose frequencies lie within the circuit's bandwidth, obtain the expression for the steady-state voltage $v_o(t)$.

9.36 The displacement shown in Figure P9.36a is produced by the cam shown in part (b) of the figure. The Fourier series approximation to this function is

$$y(t) = \frac{1}{20\pi}\left[\pi - 2\left(\frac{\sin 10\pi t}{1} + \frac{\sin 20\pi t}{2} + \frac{\sin 30\pi t}{3} + \cdots\right)\right]$$

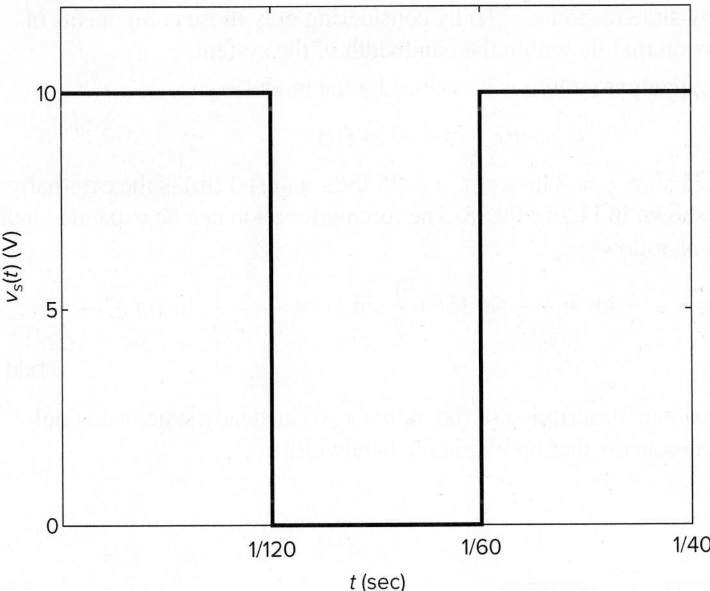

Figure P9.35

Figure P9.36

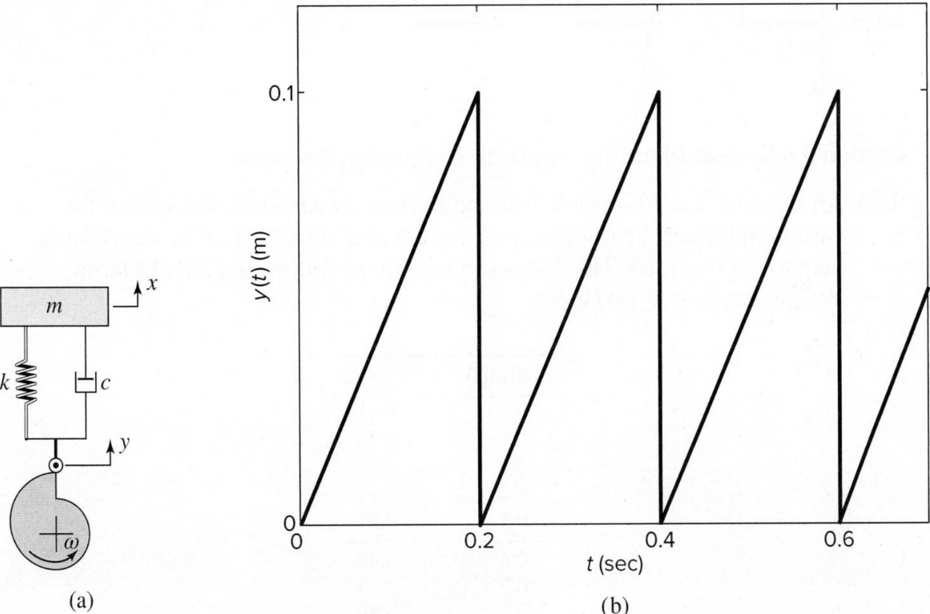

For the values $m = 1$ kg, $c = 98$ N · s/m, and $k = 4900$ N/m, keeping only those terms in the Fourier series whose frequencies lie within the system's bandwidth, obtain the expression for the steady-state displacement $x(t)$.

9.37 Given the model

$$2\dot{y} + 20y = f(t)$$

with the following Fourier series representation of the input

$$f(t) = 2 \sin 4t + 8 \sin 8t + 0.08 \sin 12t + 0.12 \sin 16t + \cdots$$

Find the steady-state response $y_{ss}(t)$ by considering only those components of the $f(t)$ expansion that lie within the bandwidth of the system.

9.38 A mass-spring-damper system is described by the model

$$m\ddot{x} + c\dot{x} + kx = f(t)$$

where $m = 0.25$ slug, $c = 2$ lb-sec/ft, $k = 25$ lb/ft, and $f(t)$ (lb) is the externally applied force shown in Figure P9.38. The forcing function can be expanded in a Fourier series as follows:

$$f(t) = -\left(\sin 3t + \frac{1}{3}\sin 9t + \frac{1}{5}\sin 15t + \frac{1}{7}\sin 21t + \cdots + \frac{1}{n}\sin 3nt \pm \cdots \right)$$

n odd

Find an approximate description of the output $x_{ss}(t)$ at steady state, using only those input components that lie within the bandwidth.

Figure P9.38

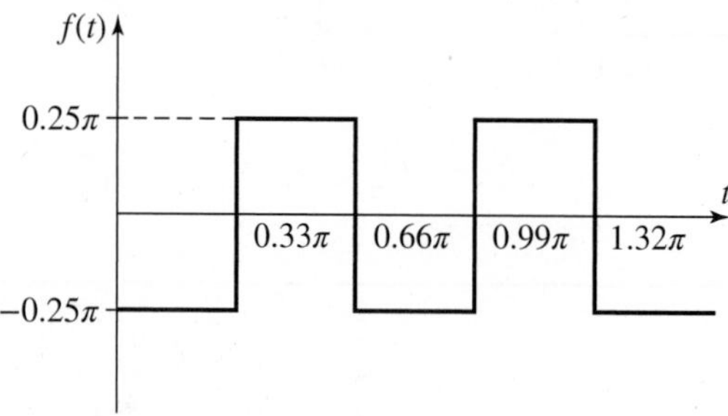

Section 9.6 System Identification from Frequency Response

9.39 An input $v_s(t) = 20 \sin \omega t$ V was applied to a certain electrical system for various values of the frequency ω, and the amplitude $|v_o|$ of the steady-state output was recorded. The data are shown in the following table. Determine the transfer function $V_o(s)/V_s(s)$.

ω (rad/s)	$\lvert v_o \rvert$ (V)
0.1	5.48
0.2	5.34
0.3	5.15
0.4	4.92
0.5	4.67
0.6	4.40
0.7	4.14
0.8	3.89
0.9	3.67
1	3.20
1.5	2.59
2	2.05
3	1.42
4	1.08
5	0.87
6	0.73
7	0.63

9.40 An input $f(t) = 15 \sin \omega t$ N was applied to a certain mechanical system for various values of the frequency ω, and the amplitude $|x|$ of the steady-state output was recorded. The data are shown in the following table. Determine the transfer function $X(s)/F(s)$.

ω (rad/s)	$\lvert x \rvert$ (mm)
0.1	209
0.4	52
0.7	28
1	19
2	7
4	2
6	1

9.41 The following data were taken by driving a machine on its support with a rotating unbalance force at various frequencies. The machine's mass is 50 kg, but the stiffness and damping in the support are unknown. The frequency of the driving force is f Hz. The measured steady-state displacement of the machine is $|x|$ mm. Estimate the stiffness and damping in the support.

f (Hz)	$\lvert x \rvert$ (mm)	f (Hz)	$\lvert x \rvert$ (mm)
0.2	2	3.8	26
1	4	4	22
2	8	5	16
2.6	24	6	14
2.8	36	7	12
3	50	8	12
3.4	36	9	12
3.6	30	10	10

Section 9.7 Case Study: Vehicle Suspension Design

9.42 The quarter-car weight of a certain vehicle is 625 lb and the weight of the associated wheel and axle is 190 lb. The suspension stiffness is 8000 lb/ft and the tire stiffness is 10,000 lb/ft. If the amplitude of variation of the road surface is 0.25 ft with a period of 20 ft, determine the critical (resonant) speeds of this vehicle.

9.43 Write a program to generate the frequency response plot shown in Figure 9.7.3, based on the transfer function given by (9.7.7).

9.44 Write a program to generate the frequency response plot shown in Figure 9.7.4, based on the transfer functions given by (9.7.3) and (9.7.4). Use the values: $m_1 = 250$ kg, $m_2 = 25$ kg, $k_1 = 1.0975 \times 10^4$ N/m, $k_2 = 10^5$ N/m, and $c_1 = 943$ N $\cdot$ s/m.

9.45 Write a program to generate the frequency response plot shown in Figure 9.7.6, based on the transfer function given by (9.7.6). Use the values: $m_1 = 250$ kg, $m_2 = 25$ kg, $k_1 = 1.0975 \times 10^4$ N/m, $k_2 = 10^5$ N/m, and $c_1 = 943$ N $\cdot$ s/m.

9.46 Problem 4.96 in Chapter 4 called for the design of a suspension for a vehicle whose quarter mass is 350 kg and whose tire stiffness is 2×10^5 N/m. The tire and swing arm have a combined mass of 35 kg. Evaluate the displacement frequency response of the design using the two-mass model. Use the following criteria: (1) The frequency range over which M for the chassis is less than 1. (Create a table like Table 9.7.2 for the two-mass case.) (2) Create a plot like Figure 9.7.5, showing the maximum comfortable road surface displacement as a function of frequency, and (3) Create a plot like Figure 9.7.6 for the transmitted force. How large can we expect the force to be if the road surface amplitude is 1 cm?

9.47 A certain vehicle has the following parameter values:

$$m_1 = 250 \text{ kg} \qquad k_1 = 1.5 \times 10^4 \text{ N/m}$$
$$m_2 = 40 \text{ kg} \qquad k_2 = 1.5 \times 10^5 \text{ N/m}$$
$$c_1 = 1000 \text{ N} \cdot \text{m/s}$$

(a) Obtain the suspension's frequency response plots. Determine the bandwidth and resonant frequencies, if any. (b) If the road surface is approximately sinusoidal with a period of 10 m, at what speeds will the car mass experience the greatest oscillation amplitude?

Section 9.8 Frequency Response Analysis Using MATLAB

9.48 For the system shown in Figure P9.48, $m_1 = m_2$, $k_1 = k_2$, and $k_1/m_1 = 64$ N/(m · kg). Obtain the transfer function $X_1(s)/Y(s)$ and its Bode plots. Identify the resonant frequencies and bandwidth.

9.49 For the system shown in Figure P9.49, $I_1 = I_2$, $c_{T_1} = c_{T_2}$, $c_{T_1}/I_1 = 0.1$ rad$^{-1}\cdot$ s^{-1}, and $k_T/I_1 = 1$ s^{-2}. Obtain the transfer function $\Theta_1(s)/\Phi(s)$ and its Bode plots. Identify the resonant frequencies and bandwidth.

Figure P9.48

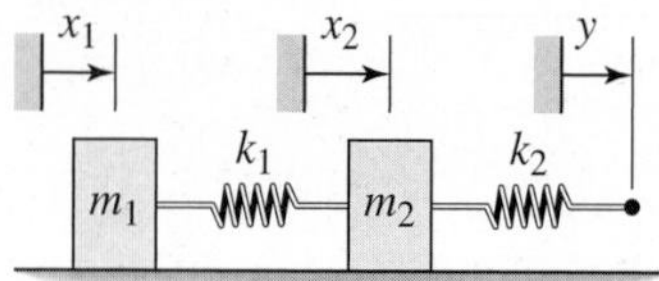

Figure P9.49

9.50 A certain mass is driven by base excitation through a spring (see Figure P9.29). Its parameter values are $m = 50$ kg, $c = 200$ N · s/m, and $k = 5000$ N/m. Determine its resonant frequency ω_r, its resonance peak M_r, and the lower and upper bandwidth frequencies.

9.51 The transfer functions for an armature-controlled dc motor with the speed as the output are

$$\frac{\Omega(s)}{V(s)} = \frac{K_T}{(Is + c)(Ls + R) + K_b K_T}$$

$$\frac{\Omega(s)}{T_d(s)} = -\frac{Ls + R}{(Is + c)(Ls + R) + K_b K_T}$$

A certain motor has the following parameter values:

$$K_T = 0.04 \text{ N} \cdot \text{m/A} \qquad K_b = 0.04 \text{ V} \cdot \text{s/rad}$$
$$c = 7 \times 10^{-5} \text{ N} \cdot \text{m} \cdot \text{s/rad} \qquad R = 0.6\ \Omega$$
$$L = 0.1 \text{ H} \qquad I_m = 2 \times 10^{-5} \text{ kg} \cdot \text{m}^2$$
$$I_L = 4 \times 10^{-5} \text{ kg} \cdot \text{m}^2$$

where I_m is the motor's inertia and I_L is the load inertia. Thus, $I = I_m + I_L$. Obtain the frequency response plots for both transfer functions. Determine the bandwidth and resonant frequency, if any.

9.52 The transfer function of the speaker model derived in Chapter 6 is, for $c = 0$,

$$\frac{X(s)}{V(s)} = \frac{K_f}{mLs^3 + mRs^2 + (kL + K_f K_b)s + kR}$$

where x is the diaphragm's displacement and v is the applied voltage. A certain speaker has the following parameter values:

$$m = 0.002 \text{ kg} \qquad k = 10^6 \text{ N/m}$$
$$K_f = 20 \text{ N/A} \qquad K_b = 15 \text{ V} \cdot \text{s/m}$$
$$R = 10\ \Omega \qquad L = 10^{-3} \text{ H}$$

Obtain the speaker's frequency response plots. Determine the speaker's bandwidth and resonant frequency, if any.

CHAPTER 10

Introduction to Feedback Control Systems

CHAPTER OUTLINE

CHAPTER OBJECTIVES

When you have finished this chapter, you should be able to

1. Model common control system components.
2. Select an appropriate control algorithm of the PID type or one of its variations, for a given application and for given steady-state and transient performance specifications.
3. Analyze the performance of a control algorithm using transfer functions, block diagrams, and computer methods, in light of given performance specifications.
4. Use MATLAB and Simulink to analyze and simulate control systems.

A major application of the methods of system dynamics is the design of control systems. In this chapter we introduce the basic concepts of feedback control and show how to model control systems so that we can analyze their performance. To this end we will use many of the modeling and analysis techniques developed in earlier chapters.

Section 10.1 introduces the concept of feedback control, also called closed-loop control. Control system terminology is covered in Section 10.2, and Section 10.3 shows how the various components in control systems are modeled. The most common control algorithm is the so-called PID algorithm, which is introduced in Section 10.4. A general control system analysis procedure is presented in Section 10.5.

The performance of PID control and variations of it are analyzed in Sections 10.6 and 10.7 for a variety of system models. Additional examples showing specific applications are covered in Section 10.8. Computer analysis and simulation of control systems are the topics of Sections 10.9 and 10.10, which use MATLAB and Simulink, respectively. Section 10.11 summarizes the major topics of the chapter. ■

10.1 CLOSED-LOOP CONTROL

A common example of a control system is a timer used to turn room lights on and off. Such a control system is called an *open-loop* system because it does not use a sensor to alter its behavior. If a light sensor is used to adjust the timing, the system would be *closed-loop*. The term derives from the block diagram of such a system, in which the sensor measurement is represented by a feedback loop.

While open-loop systems have their applications and can be used to supplement closed-loop systems, the emphasis in this text is on closed-loop control, because of the numerous and great advantages afforded by the use of a sensor to measure the output of the object or process to be controlled.

10.1.1 BLOCK-DIAGRAM REVIEW

The block diagrams of closed-loop systems have feedback loops (hence, the name "closed loop"!). Thus we will be using many block diagrams with feedback loops throughout this chapter, so now is a good time to review their properties (see Section 5.1 in Chapter 5 for a thorough introduction to block diagrams). Block diagrams are useful for several reasons: They provide one type of visual representation of the dynamics of a system; they help in building a Simulink model of the system; and they provide a way of obtaining the transfer functions of the system. We will use all three applications in this chapter, but for now let us review how to find transfer functions.

Consider the equation

$$\frac{d^2x}{dt^2} + 10\frac{dx}{dt} + 30x = 3r(t) - 3\frac{df}{dt} - 18f(t) \tag{10.1.1}$$

where $x(t)$ is the output, and the inputs are $r(t)$ and $f(t)$. Remember that transfer functions describe only the *forced* response and so are independent of the initial conditions. Thus, we find the transfer functions using the Laplace transform method with all initial conditions set to zero. Transforming (10.1.1) with zero-initial conditions gives

$$(s^2 + 10s + 30)X(s) = 3R(s) - (3s + 18)F(s) \tag{10.1.2}$$

There is one transfer function for each input-output pair. Each transfer function is found by setting all the other inputs to zero. Thus (10.1.1) has two transfer functions. The first is found by setting $F(s) = 0$ and solving for the ratio of the output $X(s)$ over the input $R(s)$.

$$\frac{X(s)}{R(s)} = \frac{3}{s^2 + 10s + 30} \tag{10.1.3}$$

The second transfer function is found in a similar way by setting $R(s) = 0$ in (10.1.2).

$$\frac{X(s)}{F(s)} = \frac{-(3s + 18)}{s^2 + 10s + 30} \tag{10.1.4}$$

The characteristic roots are the roots of $s^2 + 10s + 30 = 0$. They are $s = -5 \pm j\sqrt{5}$. Thus, the time constant is $\tau = 1/5$ and the damped natural frequency is $\omega_d = \sqrt{5}$.

We frequently will not have the differential equations but only a block-diagram description of the system. There are two ways to find the transfer functions from the block diagram. One way is block-diagram reduction; the other way is to write the transformed equations directly from the diagram. All variables except the output and the input of interest are eliminated algebraically to obtain the transfer function for that input.

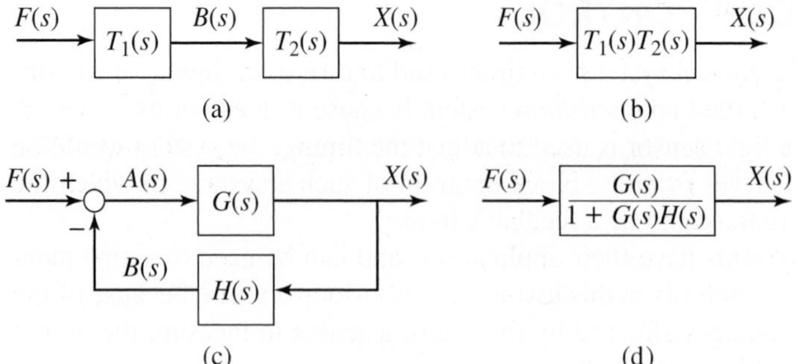

Figure 10.1.1 (a) and (b) Simplification of series blocks. (c) and (d) Simplification of a feedback loop.

Figure 10.1.1 is repeated from Section 5.1 in Chapter 5. It shows two common forms found in block diagrams. In part (a) the two blocks are said to be in *series*. It is equivalent to the diagram in part (b) because we may write

$$B(s) = T_1(s)F(s) \qquad \text{and} \qquad X(s) = T_2(s)B(s)$$

These can be combined algebraically by eliminating $B(s)$ to obtain $X(s) = T_1(s)T_2(s)F(s)$. Note that block diagrams obey the rules of algebra. Therefore, any rearrangement permitted by the rules of algebra is a valid diagram.

Figure 10.1.1c shows a negative feedback loop. From the diagram, we can obtain the following.

$$A(s) = F(s) - B(s), \qquad B(s) = H(s)X(s), \qquad X(s) = G(s)A(s)$$

We can eliminate $A(s)$ and $B(s)$ to obtain

$$X(s) = \frac{G(s)}{1 + G(s)H(s)}F(s) \tag{10.1.5}$$

This is a useful formula for reducing a feedback loop to a single block, from which we can easily see the transfer function.

The following example illustrates the block-diagram reduction method.

EXAMPLE 10.1.1

Block-Diagram Reduction

■ Problem

Obtain the transfer functions $X(s)/R(s)$ and $X(s)/F(s)$ from the block diagram shown in Figure 10.1.2.

■ Solution

To obtain $X(s)/R(s)$, set $F(s) = 0$ and redraw the diagram as shown in part b of the figure. Applying the series rule and the feedback loop rule, we obtain the first transfer function.

$$\frac{X(s)}{R(s)} = \frac{\dfrac{3}{(s+6)(s+4)}}{1 + \dfrac{3(2)}{(s+6)(s+4)}} = \frac{3}{(s+6)(s+4)+6} = \frac{3}{s^2 + 10s + 30}$$

Now return to part (a) of the figure, set $R(s) = 0$, and redraw it so that the input $F(s)$ enters from the left. The result is shown in part (c) of the figure. Note that even though we set $R(s) = 0$, we

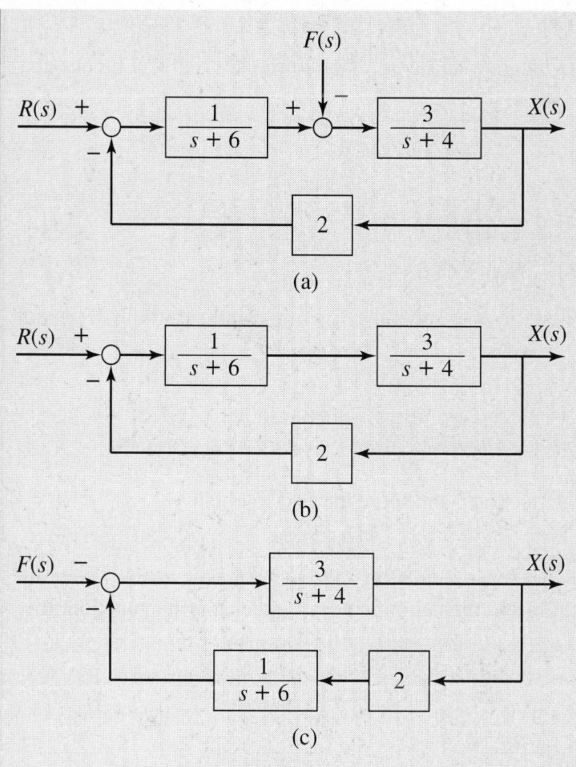

Figure 10.1.2 Systematic reduction of a block diagram.

must still retain the effect of the negative sign on the leftmost comparator. Using the series rule and the feedback loop rule on part (c) of the figure, we obtain

$$\frac{X(s)}{F(s)} = -\frac{\dfrac{3}{(s+4)}}{1 + \dfrac{3(2)}{(s+6)(s+4)}} = -\frac{3(s+6)}{(s+6)(s+4)+6} = -\frac{3(s+6)}{s^2 + 10s + 30}$$

Sometimes, we need to obtain the expressions not for just the output variables, but also for some internal variables. The following example illustrates the required method.

Deriving Expressions for Internal Variables

EXAMPLE 10.1.2

■ **Problem**

Derive the expressions for $C(s)$, $E(s)$, and $M(s)$ in terms of $R(s)$ and $D(s)$ for the diagram in Figure 10.1.3.

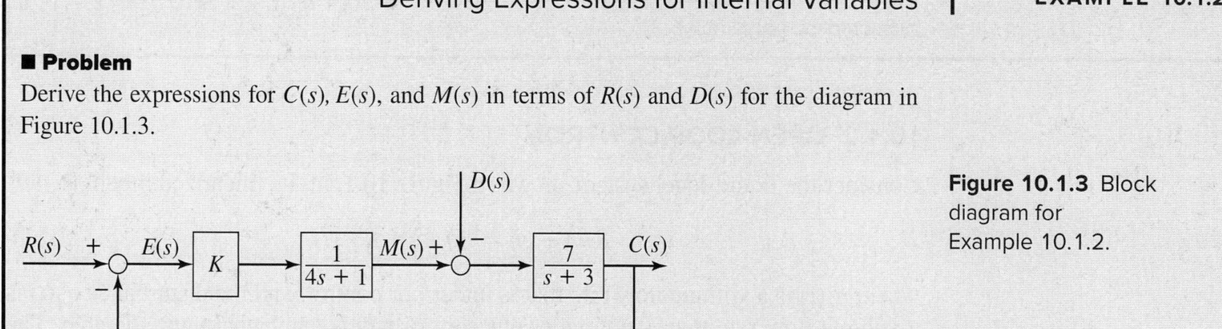

Figure 10.1.3 Block diagram for Example 10.1.2.

■ Solution

Start from the right-hand side of the diagram and work back to the left until *all* blocks and comparators are accounted for. This gives

$$C(s) = \frac{7}{s+3}[M(s) - D(s)] \tag{1}$$

$$M(s) = \frac{K}{4s+1}E(s) \tag{2}$$

$$E(s) = R(s) - C(s) \tag{3}$$

Multiply both sides of equation (1) by $s + 3$ to clear fractions, and substitute $M(s)$ and $E(s)$ from equations (2) and (3).

$$(s+3)C(s) = 7M(s) - 7D(s)$$

$$= 7\frac{K}{4s+1}E(s) - 7D(s) = \frac{7K}{4s+1}[R(s) - C(s)] - 7D(s)$$

Multiply both sides by $4s + 1$ to clear fractions, and solve for $C(s)$ to obtain:

$$C(s) = \frac{7K}{4s^2 + 13s + 3 + 7K}R(s) - \frac{7(4s+1)}{4s^2 + 13s + 3 + 7K}D(s) \tag{4}$$

The characteristic polynomial is found from the denominator of either transfer function. It is $4s^2 + 13s + 3 + 7K$.

The equation for $E(s)$ is

$$E(s) = R(s) - C(s)$$

$$= R(s) - \frac{7K}{4s^2 + 13s + 3 + 7K}R(s) + \frac{7(4s+1)}{4s^2 + 13s + 3 + 7K}D(s)$$

$$= \frac{4s^2 + 13s + 3}{4s^2 + 13s + 3 + 7K}R(s) + \frac{7(4s+1)}{4s^2 + 13s + 3 + 7K}D(s)$$

Because $4s^2 + 13s + 3$ can be factored as $(4s + 1)(s + 3)$, the equation for $M(s)$ can be expressed as

$$M(s) = \frac{K}{4s+1}E(s)$$

$$= \frac{K}{4s+1}\left[\frac{(4s+1)(s+3)}{4s^2 + 13s + 3 + 7K}R(s) + \frac{7(4s+1)}{4s^2 + 13s + 3 + 7K}D(s)\right]$$

$$= \frac{K(s+3)}{4s^2 + 13s + 3 + 7K}R(s) + \frac{7K}{4s^2 + 13s + 3 + 7K}D(s)$$

Note the cancellation of the term $4s + 1$. You should always look for such cancellations. Otherwise, the denominator of the transfer functions can appear to be of higher order than the characteristic polynomial.

10.1.2 OPEN-LOOP CONTROL

Consider the liquid-level system shown in Figure 10.1.4a. Its linearized model is

$$RA\dot{h} + gh = Rq_c(t) + Rq_d(t)$$

where $q_c(t)$ is a volume flow rate that is under our control, and the disturbance $q_d(t)$ is a volume flow rate that is not under our control and is somewhat unpredictable. The tank's bottom area is A and the outlet resistance is R.

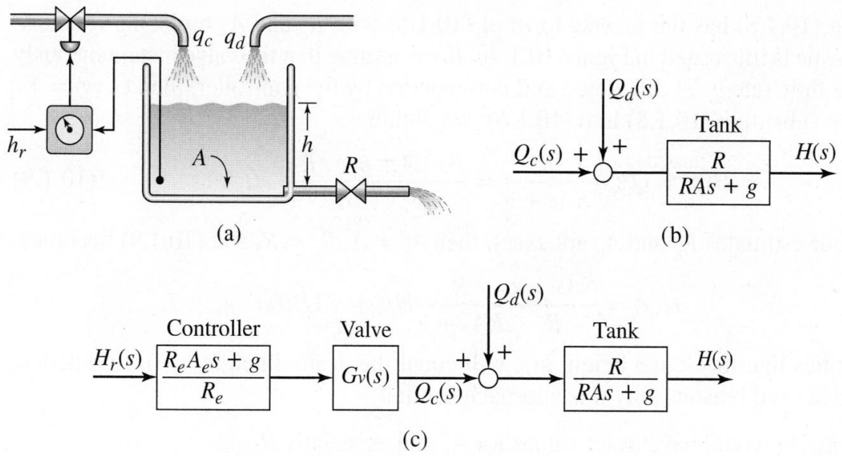

Figure 10.1.4 Open-loop control of a liquid-level system.

The block diagram is shown in Figure 10.1.4b. Note that

$$\frac{H(s)}{Q_c(s)} = \frac{R}{RAs + g} \tag{10.1.6}$$

Suppose we want to control the rate q_c so that the liquid height $h(t)$ is some specified or requested function of time, $h_r(t)$. A control valve used for this purpose is illustrated in Figure 10.1.5. The valve opening and therefore the flow rate are controlled by the current-to-pressure transducer that regulates the air pressure applied to the valve.

Assuming that (1) we know the value of g accurately, (2) $q_d = 0$, and (3) we have estimates A_e and R_e of the area A and resistance R, we can calculate the required flow rate q_c from the model as follows:

$$q_c(t) = A_e\dot{h}_r + \frac{g}{R_e}h_r \tag{10.1.7}$$

The transfer function of this algorithm is

$$\frac{Q_c(s)}{H_r(s)} = \frac{R_eA_es + g}{R_e} \tag{10.1.8}$$

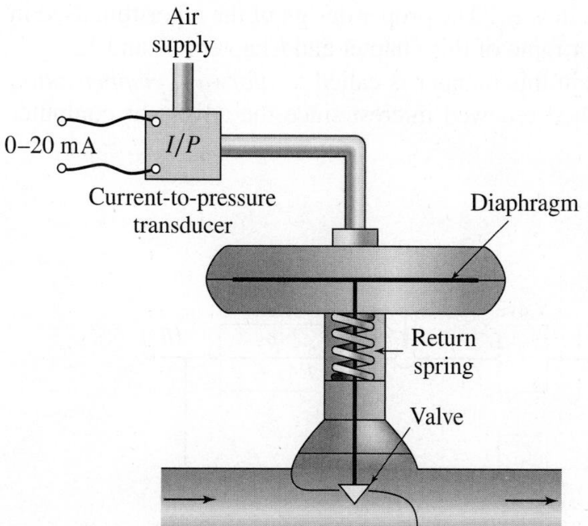

Figure 10.1.5 A pneumatic control valve.

Note that (10.1.8) has the inverse form of (10.1.6) with R_e and A_e replacing R and A. This scheme is illustrated in Figure 10.1.4c. If we assume that the valve instantaneously gives the flow rate $q_c(t)$ calculated and commanded by the controller, then $G_v(s) = 1$.

If we substitute (10.1.8) into (10.1.6), we obtain

$$H(s) = Q_c(s)\frac{R}{RAs + g} = \frac{R_eA_es + g}{R_e}\frac{R}{RAs + g}H_r(s) \tag{10.1.9}$$

Now if our estimates R_e and A_e are exact, then $A_e = A$, $R_e = R$, and (10.1.9) becomes

$$H(s) = \frac{RAs + g}{R}\frac{R}{RAs + g}H_r(s) = 1H_r(s)$$

This implies that the actual height $h(t)$ will equal the desired height $h_r(t)$. In practice, there are several reasons why this approach can fail:

1. We rarely would have exact values for A_e and especially R_e.
2. The controller, which is the physical device making the calculation required by (10.1.7) cannot compute the derivative $\dot{h}_r$ exactly. This is true for any type of controller, whether it be an analog controller, composed of op amps, for example, or a digital computer.
3. The model (10.1.6) only approximately describes the real system, which may be nonlinear, for example.
4. This scheme will not work whenever the disturbance flow q_d is present.

This example illustrates the difficulties with using open-loop control.

10.1.3 CLOSED-LOOP CONTROL

Suppose we use a sensor to measure the liquid height h and compare the measured height with the desired height h_r, as shown in Figure 10.1.6. The *error signal*, $e = h_r - h$, is used to adjust the signal going to the valve that produces the flow rate q_c. This part of the controller is the *feedback*, or *closed-loop* controller. The total signal to the valve is the sum of this signal and the signal produced by the open-loop part of the controller. We may thus think of the feedback controller as providing a correction to the open-loop calculation. This correction, which is based on the measured height, enables the controller to adjust for inaccuracies in the model or in the estimates A_e and R_e, and for changes caused by the disturbance flow q_d. The proper design of the algorithm used in the feedback controller is the major topic of this chapter and Chapters 11 and 12.

The use of open-loop control in this manner is called *feedforward compensation of the command input*. It has gained renewed interest since the advent of computer

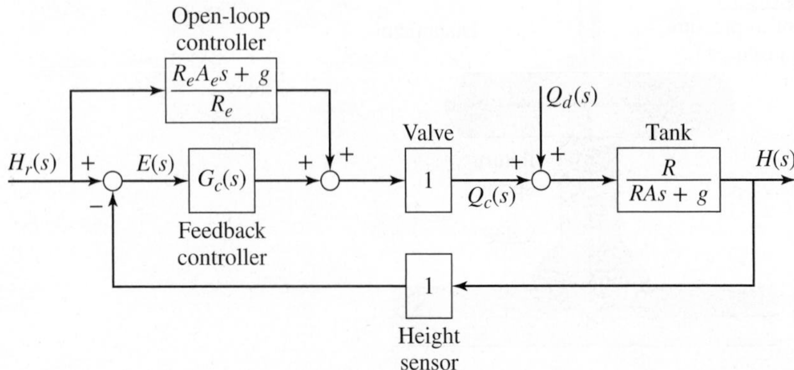

Figure 10.1.6 Feedforward compensation for a liquid-level controller.

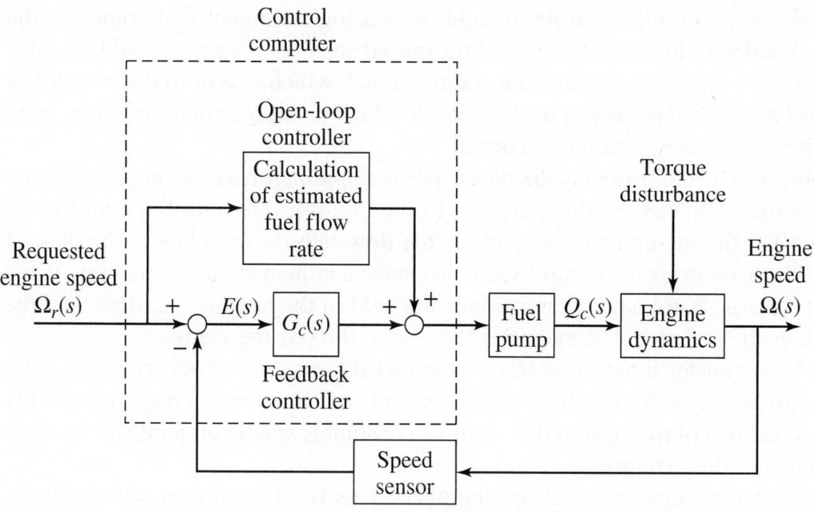

Figure 10.1.7 Use of feedforward compensation in a fuel control system.

control systems. It is based on the principle that if there are no disturbances, you can effectively control a system if you have an accurate enough model of it. Some modern systems are so complex that a simple differential equation model cannot be obtained that completely describes their behavior. Computer simulation models, however, can often be developed to describe such plants. These models can then be used as command compensation in control systems having digital computers.

For example, computers can store lookup tables that can be used to control systems that are difficult to model entirely with differential equations and analytical functions. Figure 10.1.7 shows a fuel control system designed for an internal combustion engine. The fuel flow rate required to achieve a desired speed depends in a complicated way on the desired speed and other variables. This dependence was measured in engine tests and was summarized in a table stored in the control computer. This table is used to estimate the required fuel flow rate. A feedback control algorithm is used to adjust the flow rate estimate based on the speed error (because the lookup-table model will not be exact) and to counteract the effects of disturbances such as a load torque on the engine.

For now, we will concentrate on how to analyze the performance of a feedback control algorithm. Section 10.8 discusses feedforward compensation in more detail.

10.2 CONTROL SYSTEM TERMINOLOGY

Figure 10.2.1 is a block diagram that represents the structure of many, but not all, feedback control systems. To simplify the example, we have not included feedforward compensation.

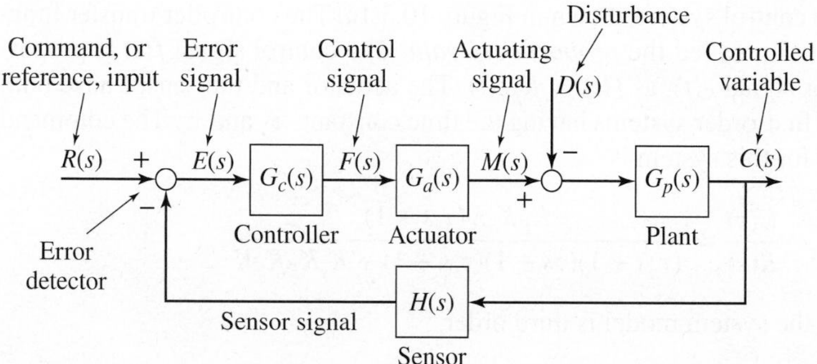

Figure 10.2.1 Control system terminology.

The feedback controller can be thought of as a logic element that compares the command signal with the measurement of the output and decides what should be done. The control logic elements produce the control signal, which is sent to the actuator or *final control element*. This is the device that develops enough torque, pressure, heat, etc., to influence the elements under control.

The object to be controlled is the *plant*. The *manipulated variable* is generated by the final control elements for this purpose. In the liquid-level control system treated in Section 10.1, the manipulated variable is the flow rate q_c, the plant is the tank of liquid, and the actuator is the control valve. To make a human analogy, the brain is the controller; the arm, hand, and whatever is being held in the hand is the plant; and the arm muscle is the actuator. The eye plays the role of the sensing element.

The sensor transfer function is $H(s)$. Note that if the sensor gives error-free measurements instantaneously, then $H(s) = 1$. This implies that the sensor responds quickly compared to the rest of the system (i.e., its time constant is small compared to the other time constants of the system).

The disturbance input also acts on the plant. This is an input over which the designer has no control, and perhaps for which little information is available about the magnitude, functional form, or time of occurrence. The disturbance can be a random input, such as wind gusts on a radar antenna, or deterministic, such as Coulomb friction effects. In the latter case, we can include the friction force in the system model by using a nominal value for the coefficient of friction. The disturbance input would then be the deviation of the friction force from this estimated value and would represent the uncertainty in our estimate. The *command* transfer function is $C(s)/R(s)$. The *disturbance* transfer function is $C(s)/D(s)$.

10.3 MODELING CONTROL SYSTEMS

Modeling the control system with all its components is the first step in the design process. The key to successful design often lies in the engineer's ability to develop a model that describes the important behavior of the system without including so much detail that the required mathematical analysis is cumbersome. In this section, we show some examples of how such modeling is accomplished.

10.3.1 MODELING ACTUATOR AND SENSOR RESPONSE

Every actuator and every sensor requires time to respond to its input. However, if actuator and sensor response times are small compared to the response times of the controller and the plant, then we usually model the actuator and the sensor with constant gains. The purpose of this approximation is to reduce the order of the system and thus simplify the required mathematics.

Consider the control system shown in Figure 10.3.1a. The controller transfer function is a constant, K_p, called the *proportional gain*. The control signal $f(t)$ is proportional to the error signal $e(t)$, as $f(t) = K_p e(t)$. The actuator and the sensor have both been modeled as first-order systems having the time constants τ_a and τ_s. The command transfer function for this system is

$$\frac{C(s)}{R(s)} = \frac{K_P K_a K(\tau_s s + 1)}{(\tau_a s + 1)(\tau s + 1)(\tau_s s + 1) + K_s K_P K_a K}$$

We thus see that the system model is third order.

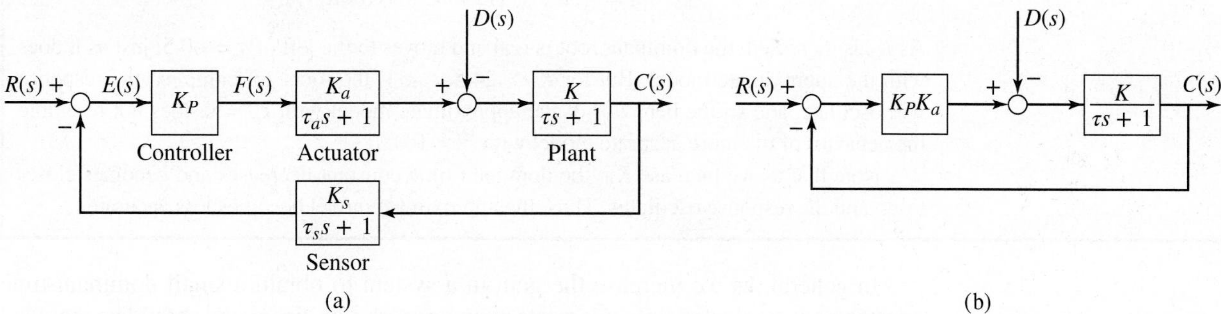

Figure 10.3.1 (a) A feedback control system. (b) Approximate model obtained by neglecting the time constants of the sensor and actuator.

The sensor gain K_s will be unity if the sensor gives an accurate measurement. If the sensor time constant τ_s is very small compared to that of the plant, τ, then we can set $\tau_s = 0$ and approximate the sensor transfer function as a simple gain of 1. We will usually use this sensor model.

Similarly, if the actuator time constant τ_a is very small compared to that of the plant, then we may set $\tau_a = 0$ and approximate the actuator transfer function as a simple gain K_a, which may be absorbed into the proportional gain, as shown in Figure 10.3.1(b). The command transfer function of this simplified model is

$$\frac{C(s)}{R(s)} = \frac{K_P K_a K}{\tau s + 1 + K_P K_a K}$$

which is first order and therefore much easier to analyze.

In most cases, the sensor time constant is the smallest of the three, so we will usually neglect it in our examples. The actuator time constant may not always be negligible, however. Quite often in our examples we assume that the actuator gain K_a is 1 and do not show its block to simplify the examples. In these cases, the hidden block has units that convert the controller output (e.g., volts) to the actuator output (e.g., force). If the actuator gain is not 1 in practice, it can be absorbed into the controller gain.

Effect of Gain on Modeling Approximations

EXAMPLE 10.3.1

■ Problem

For the system shown in Figure 10.3.2 the plant time constant is 2 and the nominal value of the actuator time constant is $\tau_a = 0.02$. Investigate the effects of neglecting this time constant as the gain K_P is increased.

■ Solution

If we set $\tau_a = 0$, there is one characteristic root, which is $s = -(1 + K_P)/2$. As K_P is increased, the root remains real and moves to the left of $s = -0.5$. The predicted step response will not oscillate.

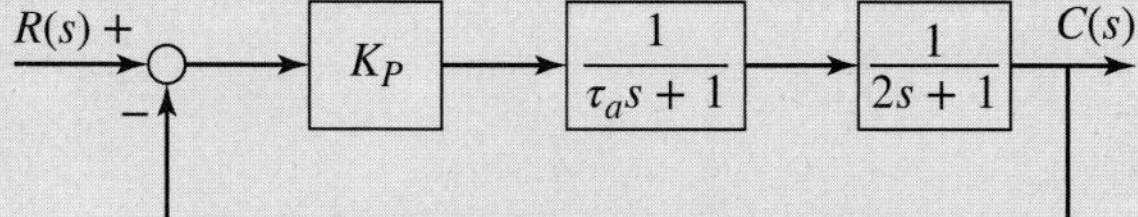

Figure 10.3.2 Effects of neglecting the actuator time constant.

If we set $\tau_a = 0.02$, there are two characteristic roots and the characteristic equation is $0.04s^2 + 2.02s + 1 + K_P = 0$.

$$s = -50.5 \pm \sqrt{2450.25 - 100K_p}$$

As K_P is increased, the dominant root is real and moves to the left of $s = -0.5$, just as it does with the approximate model. But for $K_P > 24.5$, $\zeta < 1$, the roots are complex, the response will oscillate, and so the behavior of the approximate model with $\tau_a = 0$ does not resemble the behavior of the more accurate model with $\tau_a = 0.02$.

Note that as we increase K_P, the dominant-time constant decreases and becomes closer to τ_a, and the response oscillates. Thus, the approximate model becomes less accurate.

In general, as we increase the gain in a system to obtain a small dominant-time constant, any neglected time constants become more significant and should be included in the model. Doing this, however, increases the order of the model and the complexity of the required mathematics. So, in practice, the control engineer begins the design of a system by neglecting the smaller time constants, just as we neglected the actuator time constant τ_a in Example 10.3.1. Once a preliminary design has been made, the more accurate model is used to refine the design and check the effect of the approximations, including the effect of high gain, often by simulation if the model is highly detailed.

10.3.2 MODELING DISTURBANCES

The disturbance inputs experienced by real systems are usually not simple or deterministic functions of time like steps, ramps, and sine waves. In fact, we often do not know the exact functional form of the disturbance. It is often a random function, such as the disturbance torque due to wind gusts. The analysis of random inputs is beyond the scope of this text. However, the frequency response plot of the disturbance transfer function is often very useful for evaluating the effects of a disturbance. Although such a plot describes only the steady-state response for sinusoidal and cosinusoidal inputs, periodic disturbance inputs can be represented as a sum of sines and cosines, as we saw in Chapter 9. In addition, measurements of an actual disturbance can be analyzed to determine its frequency content. If these frequencies lie outside the bandwidth of the disturbance transfer function, then we may conclude that the control system adequately handles the disturbance.

Some system inputs may or may not be treated as a disturbance, depending on whether or not they are known. The following example illustrates this.

EXAMPLE 10.3.2 Controlling a Thermal Process

■ Problem

The inside temperature T of the oven shown in Figure 10.3.3a is to be controlled. The oven wall has a thermal resistance R. The only significant thermal capacitance is that of the oven contents, C. The air temperature outside the oven is T_o. The heater supplies a heat flow rate q.

A power amplifier supplies current to the resistance-type heater. The amplifier inputs are the voltage from the command potentiometer, which represents the desired value T_r of the temperature T, and the voltage from the temperature sensor. These voltages are $v_r = K_r T_r$ and $v_s = K_s T$, respectively.

Develop a model of the system.

Figure 10.3.3 (a) An oven temperature control system. (b) Model of the control system if T_o is a known constant temperature. (c) Outside temperature T_o modeled as a disturbance.

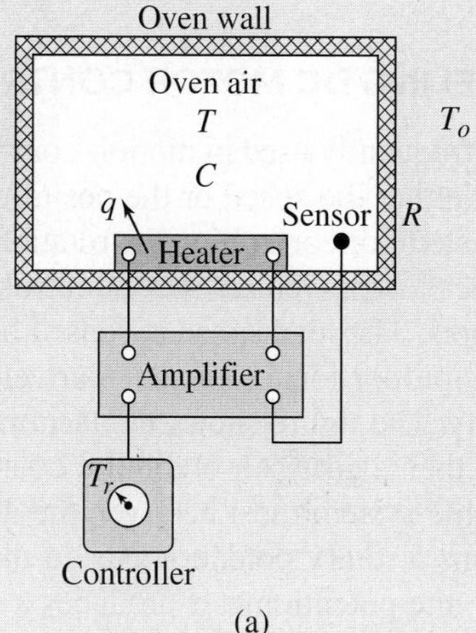

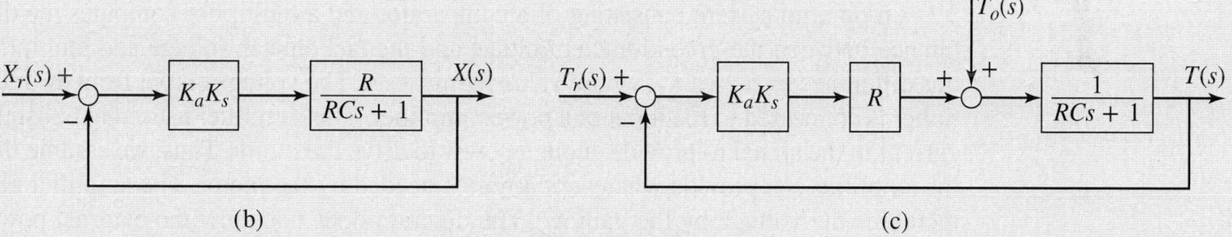

■ **Solution**

Using the methods of Section 7.8, the model of the plant is found to be

$$C\frac{dT}{dt} = q - \frac{1}{R}(T - T_o)$$

We should choose the scale factor of the potentiometer so that $K_r = K_s$. This means that the voltage difference $v_r - v_s$ will be proportional to the temperature difference $T_r - T$, as $v_r - v_s = K_s(T_r - T)$. The heat flow rate q from the heater is equal to $K_a(v_r - v_s)$, where K_a is the amplifier gain. Thus, the system model is

$$C\frac{dT}{dt} = q - \frac{1}{R}(T - T_o) = K_aK_s(T_r - T) - \frac{1}{R}(T - T_o) \tag{1}$$

If T_o is a known constant value, then we can define two new variables x and x_r such that

$$x = T - T_o \qquad x_r = T_r - T_o$$

Then $T = x + T_o$ and $T_r = x_r + T_o$. Substitution of these into the system model gives

$$C\frac{dx}{dt} = K_aK_s(x_r - x) - \frac{1}{R}x$$

The corresponding block diagram is shown in part (b) of the figure. The effect of the new variables x and x_r is to remove T_o from the equation model and the block diagram, thus simplifying both.

If, however, the outside temperature T_o fluctuates in an unknown manner, then it is a disturbance. In this case, the variable transformation $x = T - T_o$ is useless because we would not know the value of T_o. The system model is equation (1) and its block diagram is shown in part (c) of the figure. We will analyze the performance of this system later in this chapter.

10.3.3 MODELING DC MOTOR CONTROL

DC motors are frequently used in motion-control applications. In such applications, we need to control either the speed or the position of a load. Figure 10.3.4 is a schematic diagram of a system for controlling the rotational speed of a load. Applications of such a system include electric-vehicle speed control, tape drives, conveyor belts, robot arms, and mobile robots. The load speed is sensed by a tachometer. The tachometer outputs a voltage that is proportional to its angular velocity; thus, $v_{\text{tach}} = K_{\text{tach}}\omega_L$, where ω_L is the load velocity. The figure shows the tachometer connected to the load shaft; if this is not possible, then a gear pair or similar connection can be used.

Somehow the system must be given the desired load speed. Here we assume that an operator turns a rotary potentiometer to the desired speed marked on a calibrated dial face. Thus, the potentiometer produces a voltage $K_d K_{\text{pot}}\omega_r$ that is proportional to the desired load speed ω_r, where K_d is the dial factor and K_{pot} is the potentiometer gain.

An op-amp system consisting of a comparator and a multiplier computes the difference between the potentiometer voltage and the tachometer voltage and multiplies this difference by a gain K_1, whose value we must set. The voltage output from the multiplier is connected to the input of a power amplifier. This amplifier is used to boost the current in the signal to provide enough power to drive the motor. Thus, we assume that the amplifier can provide whatever current is needed by the motor. The amplifier also increases the voltage by the gain K_a. The diagram does not show the external power connections required to drive the op amp and the potentiometer.

The voltage difference v_e is given by

$$v_e = K_d K_{\text{pot}}\omega_r - K_{\text{tach}}\omega_L$$

where ω_r is the desired value of the load speed. This is sometimes called the *requested* or *commanded* speed. It is important to understand that the voltage difference v_e must be proportional to the speed error $\omega_r - \omega_L$. Therefore, we must set the dial factor K_d

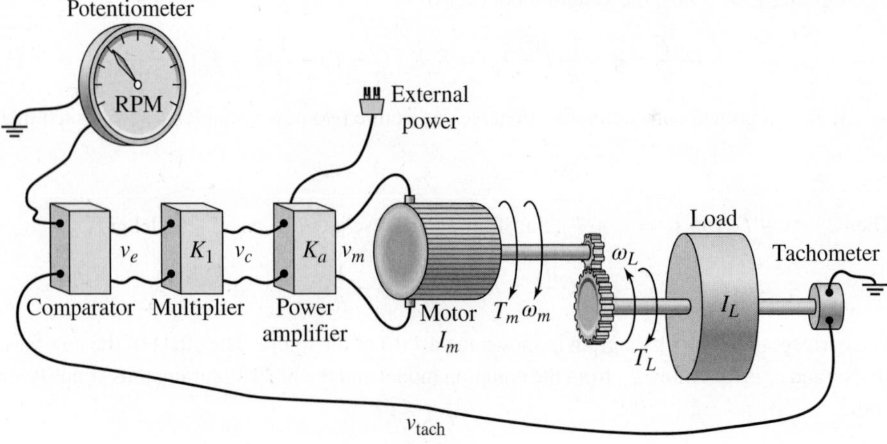

Figure 10.3.4 A speed-control system.

so that $K_d K_{pot} = K_{tach}$. If we do this, then $v_e = K_{tach}(\omega_r - \omega_L)$ as required. (Note that it is easier to set K_d by calibrating the dial face than it is to find a potentiometer or tachometer with the necessary gain.)

Reflecting the load and tachometer inertias and the load damping back to the motor shaft using the concepts of Chapter 3, we obtain the following equivalent inertia and damping values:

$$I_e = I_m + \frac{I_L + I_t}{N^2} \qquad c_e = \frac{c}{N^2}$$

where I_m is the motor inertia, I_L is the load inertia, I_t is the tachometer inertia, and c is the torsional viscous damping constant acting on the load shaft. Here, N is the gear ratio ($N = \omega_m/\omega_L$, where ω_m is the motor speed). Thus, the dynamics of the mechanical subsystem are described by

$$I_e \frac{d\omega_m}{dt} = T_m - \frac{T_L}{N} - c_e \omega_m$$

Thus,

$$\Omega_m(s) = \frac{1}{I_e s + c_e}\left[T_m(s) - \frac{T_L(s)}{N}\right]$$

Use of a Field-Controlled Motor

EXAMPLE 10.3.3

■ **Problem**

Assuming that the motor is field-controlled, draw the block diagram of the speed-control system shown in Figure 10.3.4 and obtain its command and disturbance transfer functions.

■ **Solution**

A field-controlled motor is controlled by varying the field current i_f while keeping the armature current constant. Figure 10.3.5 shows the block diagram of a proportional control system for controlling the speed of such a motor and was obtained by modifying the motor diagram in Figure 6.5.7 in Chapter 6.

The transfer functions can be obtained either by reducing the block diagram or by transforming the equations and eliminating all variables except the inputs and output. Recalling that $K_d K_{pot} = K_{tach}$, we can write the following equation from the diagram

$$V_m(s) = K_1 K_a K_{tach}[\Omega_r(s) - \Omega_L(s)]$$

Figure 10.3.5 Block diagram of a speed-control system using a field-controlled motor.

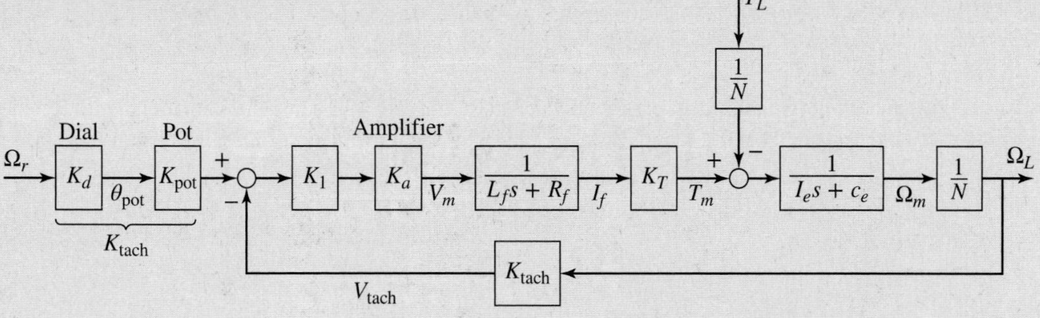

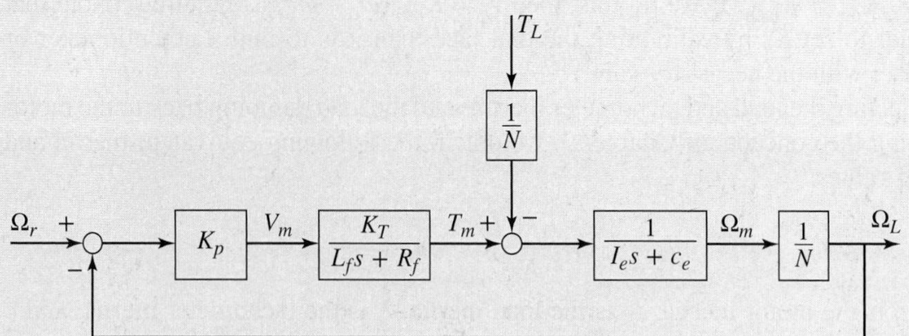

Figure 10.3.6 Simplified block diagram of a speed-control system using a field-controlled motor.

From the diagram we can see that the motor torque is given by

$$T_m(s) = \frac{K_T}{L_f s + R_f} V_m(s)$$

Finally, we have $\Omega_L(s) = \Omega_m(s)/N$. We obtain the transfer functions by eliminating $V_m(s)$, $T_m(s)$, and $\Omega_m(s)$ from these equations. Let

$$K_P = K_{\text{tach}} K_1 K_a \tag{1}$$

Using the parameter K_P, the diagram can be simplified as shown in Figure 10.3.6. The transfer functions are

$$\frac{\Omega_L(s)}{\Omega_r(s)} = \frac{K_P K_T}{N L_f I_e s^2 + N(R_f I_e + c_e L_f)s + N c_e R_f + K_P K_T} \tag{2}$$

$$\frac{\Omega_L(s)}{T_L(s)} = -\frac{(L_f s + R_f)/N}{N L_f I_e s^2 + N(R_f I_e + c_e L_f)s + N c_e R_f + K_P K_T} \tag{3}$$

EXAMPLE 10.3.4

Approximate First-Order Model

■ Problem

Derive the transfer functions of the system shown in Figure 10.3.6 for the case where the field time constant L_f/R_f is small.

■ Solution

If $T_L = 0$, Figure 10.3.6 shows that the transfer function $\Omega_m(s)/V_m(s)$ is

$$\frac{\Omega_m(s)}{V_m(s)} = \frac{K_T}{(L_f s + R_f)(I_e s + c_e)} = \frac{K_T/R_f c_e}{(\tau_f s + 1)(\tau_m s + 1)} \tag{1}$$

where $\tau_f = L_f/R_f$, which is called the field time constant, and $\tau_m = I_e/c_e$, which is the mechanical time constant. Quite often $\tau_f \ll \tau_m$, in which case equation (1) reduces to a first-order model (by setting $\tau_f = 0$ in equation (1)):

$$\frac{\Omega_m(s)}{V_m(s)} = \frac{K_T/R_f c_e}{\tau_m s + 1} = \frac{K_T/R_f}{I_e s + c_e}$$

The approximation $\tau_f \ll \tau_m$ is equivalent to setting $L_f = 0$. If this approximation is valid, equations (1) and (2) of Example 10.3.3 reduce to the following first-order models:

$$\frac{\Omega_L(s)}{\Omega_r(s)} = \frac{K_P K_T}{N R_f I_e s + N c_e R_f + K_P K_T}$$

$$\frac{\Omega_L(s)}{T_L(s)} = -\frac{R_f/N}{N R_f I_e s + N c_e R_f + K_P K_T}$$

Figure 10.3.7 Model simplification for a speed-control system with negligible field time constant.

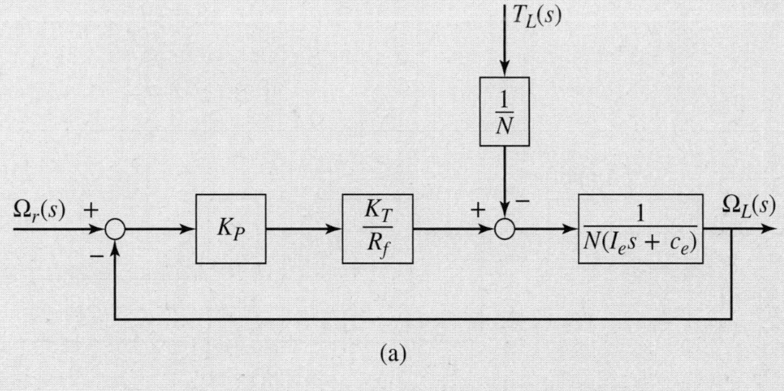

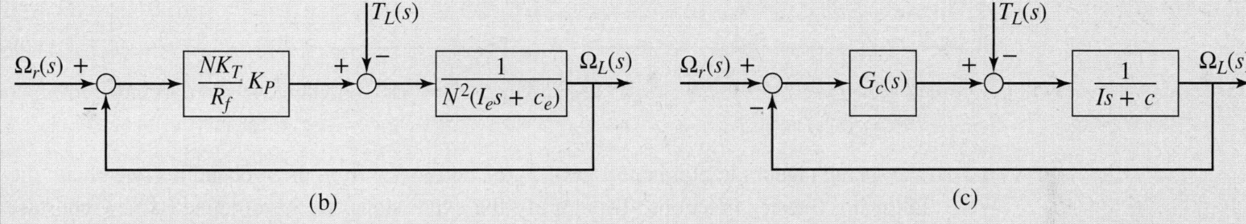

These correspond to the block diagram shown in Figure 10.3.7a. Moving the factor $1/N$ gives the diagram shown in part (b). Note that we may thus absorb the factor NK_T/R_f into the controller gain K_P. Part (c) shows the case where the control algorithm is some arbitrary transfer function $G_c(s)$, where $I = N^2 I_e$ and $c = N^2 c_e$. We will use this simplified diagram quite often in discussing the use of various control algorithms in Sections 10.6 and 10.7.

Use of an Armature-Controlled Motor

EXAMPLE 10.3.5

■ Problem

Assuming that the motor is armature-controlled, draw the block diagram of the speed-control system shown in Figure 10.3.4 and obtain its transfer functions.

■ Solution

The system inputs are the desired or required value ω_r of the load speed ω_L and the load torque T_L. The block diagram is shown in Figure 10.3.8 and was obtained by modifying the motor diagram in Figure 6.5.5 in Chapter 6.

The transfer functions can be obtained either by reducing the block diagram or by transforming the equations and eliminating all variables except the inputs and the output. The latter method is perhaps the easier one when the diagram has multiple inputs and loops containing comparators within them. Recalling that $K_d K_{\text{pot}} = K_{\text{tach}}$, we can write the following equation from the diagram

$$V_m(s) = K_{\text{tach}} K_1 K_a [\Omega_r(s) - \Omega_L(s)]$$

From the diagram we see that the motor torque is given by

$$T_m(s) = \frac{K_T}{L_a s + R_a}[V_m(s) - K_b \Omega_m(s)]$$

Finally, we have $\Omega_L(s) = \Omega_m(s)/N$.

Figure 10.3.8 Block diagram of a speed-control system using an armature-controlled motor.

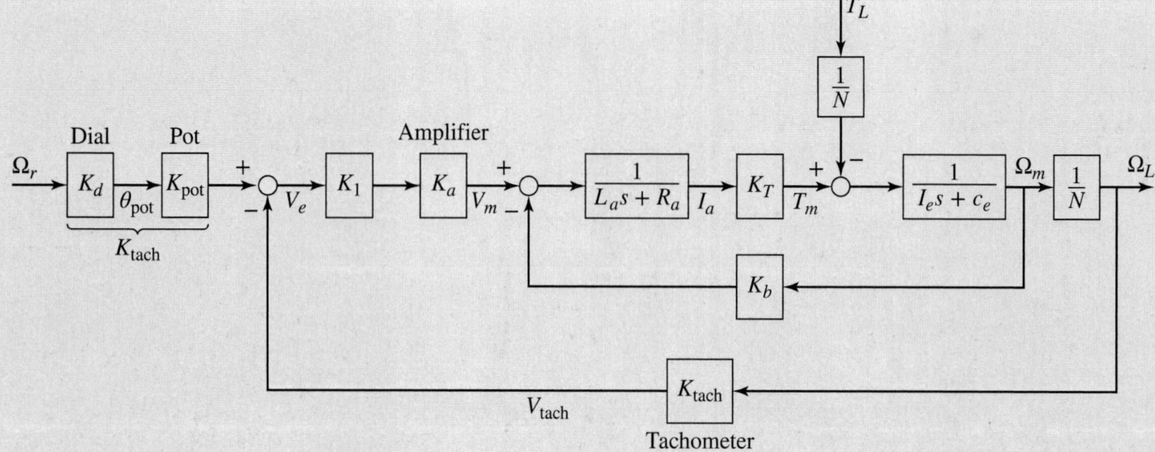

Using algebra to eliminate $V_m(s)$, $T_m(s)$, and $\Omega_m(s)$ from these equations, we obtain the following transfer functions. To simplify the expressions, we have defined a new constant $K_P = K_1 K_a K_{tach}$.

$$\frac{\Omega_L(s)}{\Omega_r(s)} = \frac{K_P K_T}{D(s)} \tag{1}$$

$$\frac{\Omega_L(s)}{T_L(s)} = -\frac{(L_a s + R_a)/N}{D(s)} \tag{2}$$

where the denominator is

$$D(s) = N L_a I_e s^2 + N(R_a I_e + c_e L_a)s + N R_a c_e + N K_T K_b + K_P K_T \tag{3}$$

Using the parameter K_P, the block diagram can be simplified as shown in Figure 10.3.9. These transfer functions can be analyzed to determine whether or not the system is able to control the speed successfully and to determine an appropriate value for the gain K_1. Such analysis would consider the system's stability, steady-state error, and transient response. We will illustrate this analysis in Sections 10.6, 10.7, and 10.8.

Figure 10.3.9 Simplified block diagram of a speed-control system using an armature-controlled motor.

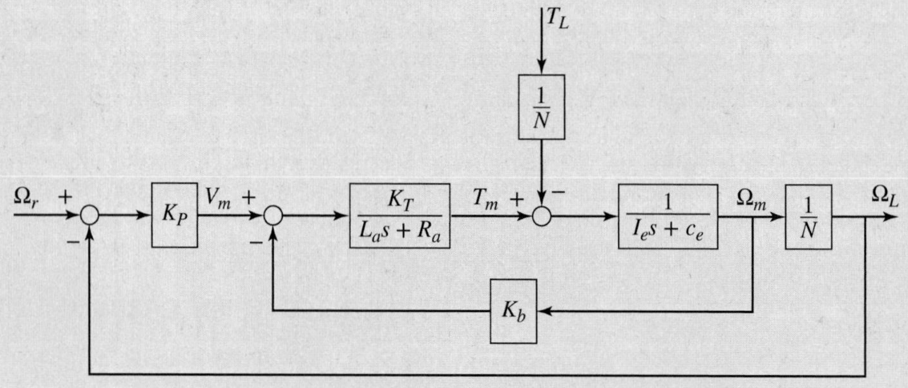

Effect of Negligible Armature Time Constant

EXAMPLE 10.3.6

■ Problem

Derive the transfer functions for the case where the armature time constant L_a/R_a is small.

■ Solution

If $T_L = 0$, we obtain Figure 10.3.10a. We can reduce the inner loop containing K_b to obtain part (b) of the figure. Setting $L_a = 0$ we obtain the diagram in part (c). Part (d) is obtained by replacing K_P with the arbitrary controller transfer function $G_c(s)$ and defining

$$I = \frac{NR_a I_e}{K_T}$$

$$c = N\frac{R_a c_e + K_T K_b}{K_T}$$

The block diagram in part (d) is obviously a simpler representation than the diagram in part (a). This simplified diagram will help us to demonstrate the use of various control algorithms in Sections 10.6 and 10.7. Note that it does not include the disturbance. To analyze the disturbance response, we need to use the diagram in Figure 10.3.9 or the disturbance transfer function given by equations (2) and (3) in Example 10.3.5.

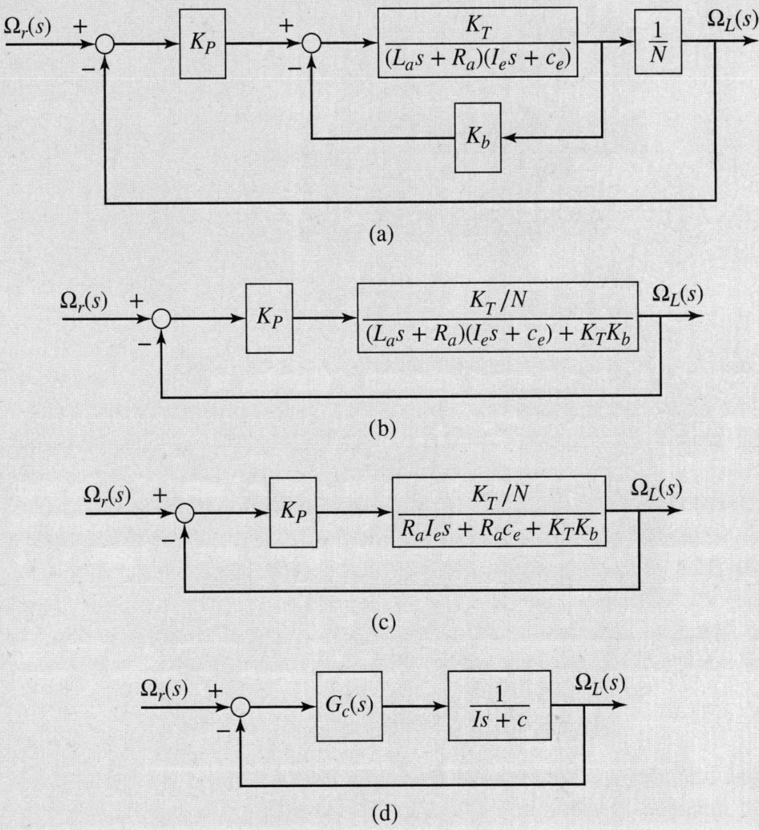

Figure 10.3.10 Model simplification for a speed-control system with negligible armature time constant.

EXAMPLE 10.3.7

Controlling the Speed of a Conveyor

■ Problem

A conveyor drive system to produce translation of the load is shown in Figure 10.3.11. To translate the load a specified distance, the drive wheels must rotate through a required angle, and this can be accomplished by controlling the speed, often with a trapezoidal speed profile. The equivalent inertia of the load and all the drive components felt at the motor shaft is I_e. The effect of Coulomb friction in the system produces an opposing torque T_{Fe} at the motor shaft, and the damping in the system is negligible. Develop the block diagram of a proportional control system using an armature-controlled motor for this application. Assume that the drive wheel speed ω_L is measured by a tachometer and that the motor speed ω_m is related to the drive wheel speed by $\omega_m = N\omega_L$, where N is the speed ratio due to the reducer and the chain drive.

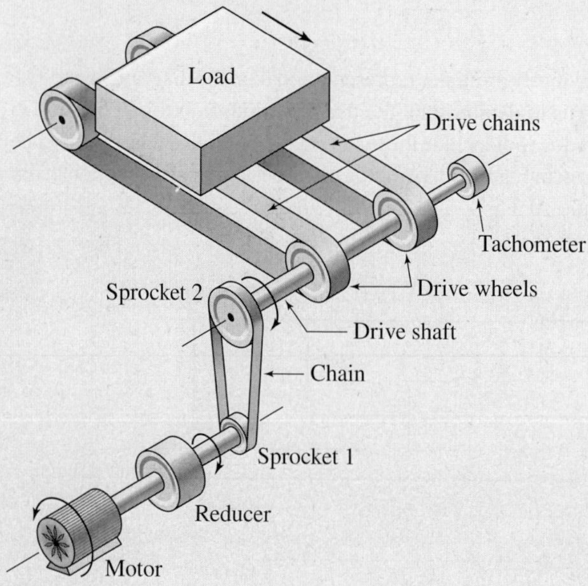

Figure 10.3.11 A conveyor system.

■ Solution

As shown in Example 3.3.6, t the mechanical subsystem is described by

$$I_e \frac{d\omega_m}{dt} = T - T_{Fe} \tag{1}$$

where the motor torque is $T = K_T i_a$. The system is like that shown in Figure 10.3.4. The block diagram can be obtained by modifying Figure 10.3.8 using equation (1) and collecting the various gains into one gain: $K_P = K_{\text{tach}} K_1 K_a$. The resulting diagram is shown in Figure 10.3.12.

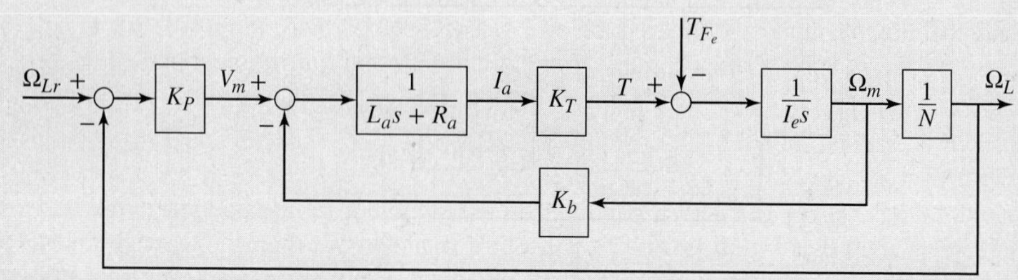

Figure 10.3.12 Block diagram of a speed-control system for a conveyor.

EXAMPLE 10.3.8

Controlling the Position of a Robot-Arm Link

■ Problem

The drive system for one link of a robot arm is illustrated in Figure 10.3.13. The equivalent inertia of the link and all the drive components felt at the motor shaft is I_e. Gravity produces an opposing torque that is proportional to sin θ but which we model as a constant torque T_d felt at the motor shaft (this is a good approximation if the change in θ is small). Neglect friction and damping in the system. Develop the block diagram of a proportional control system using an armature-controlled motor for this application. Assume that motor rotation angle θ_m is measured by a sensor and is related to the arm rotation angle θ by $\theta_m = N\theta$, where N is the gear ratio.

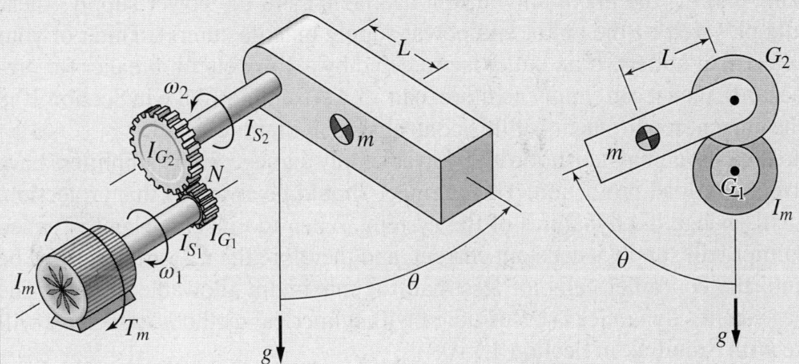

Figure 10.3.13 A robot-arm link.

■ Solution

As shown in Example 3.5.5, t the mechanical subsystem is described by

$$I_e \frac{d^2\theta_m}{dt^2} = T - \frac{1}{N}T_d \qquad (1)$$

where the motor torque is $T = K_T i_a$. The system is like that shown in Figure 10.3.4. The block diagram can be obtained by modifying Figure 10.3.8 using equation (1) and collecting the various gains into one gain: $K_P = K_{\text{tach}} K_1 K_a$. The resulting diagram is shown in Figure 10.3.14.

Figure 10.3.14 Block diagram of a position control system for a robot-arm link.

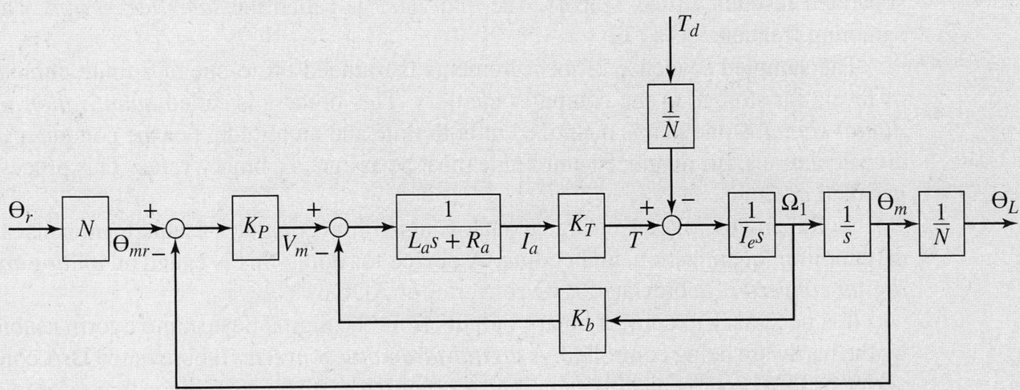

Note that the system actually controls the *motor* rotation angle, so the arm angle command θ_r must be converted to the motor angle command θ_{mr}.

10.3.4 CURRENT AND TORQUE LIMITATIONS

Obviously there is a limit to how large we can make the gain K_P. The limit depends on the power available in the actuator system, which consists of the motor and the power amplifier.

In practice, because the motor torque depends on the armature current, as $T_m = K_T i_a$, the maximum torque available is limited by the armature current. This current is limited by two factors: (1) the maximum current the armature can withstand without demagnetizing and (2) the maximum current available from the power supply (these values should be given by the motor and power supply manufacturers). Think of your household electrical outlets. If an outlet is protected by a 15-A circuit breaker (to prevent a fire hazard), then it can supply no more than 15 A. Example 10.8.6 in Section 10.8 examines the current requirements of this control system in more detail.

To guard against demagnetization and power supply damage, some amplifiers have built-in current overload protection. The engineer should be aware of this protection, because it will change the dynamics of the system. When too much current is called for, the amplifier will supply a constant current, and therefore the motor torque will be constant, until the controller calls for less than the maximum allowable current. The effect on the system's dynamics is best studied with numerical methods, which we will demonstrate with Simulink in Section 10.10.

10.3.5 DIGITAL CONTROL SYSTEMS

In many modern control systems, a digital computer, rather than an op amp, performs the calculations required to implement the control algorithm.

An *analog signal* is one that is continuous in both time and amplitude. Given a large enough plot of a signal $y(t)$, we can theoretically obtain a value of y from the plot for any value of t over its plotted range. Thus, both y and t have an infinite number of values. Digital computers, however, can handle mathematical relations and operations only when expressed as a finite set of numbers rather than as infinite-valued functions. Thus, any continuous-measurement signal must be converted into a set of pulses by *sampling*—the process by which a continuous-time variable is measured at distinct, separated instants a time T apart. The interval T is called the *sampling period*. The sampling frequency is $1/T$.

The sampled sequence of measurements is rounded off to one of a finite number of levels for storage in the computer memory. This process is called *quantization*. A *digital signal* is one that is quantized in both time and amplitude. For the computer to do calculations, the quantized amplitude must be assigned a binary value. This process is called *coding*.

So the process of converting an analog signal into computer-usable form consists of sampling, quantization, and coding. A device for doing this is called an *analog-to-digital converter* (abbreviated A/D converter or ADC).

It is necessary to convert binary outputs from the digital device into a form usable by the hardware being controlled. A *digital-to-analog converter* (abbreviated D/A converter or DAC) performs this function. A common mathematical algorithm used in a D/A converter is the *zero-order hold (ZOH)*. The binary form of the output from the digital device is first converted to a sequence of short-duration voltage pulses. However, it is usually not possible to drive a load, such as a motor, with short pulses. To deliver sufficient energy, the pulse amplitude might have to be so large that it is infeasible to be generated. Also, large-voltage pulses might saturate or even damage the system being driven. The solution to this problem is to smooth the output pulses to produce a signal in analog form. The simplest way of converting a pulse sequence into a continuous signal is to hold the value of the pulse until the next one arrives. A device called a *zero-order hold* accomplishes this task.

10.3.6 DIGITAL CONTROL STRUCTURES

There are two types of applications of digital computers to control problems. The first is *supervisory control*, where analog controllers are directly involved with the plant to be controlled, while the digital computer provides command signals to the controllers. If this were the case for the system shown in Figure 10.3.4, the op-amp comparator and multiplier would still be present, but the command input ω_r would be provided by the computer through a D/A converter, instead of by the potentiometer. In the second application, *direct digital control (DDC)*, the digital device acts at the lowest levels of the system, replacing the op-amp comparator and multiplier, in direct control of the amplifier-motor system.

Figure 10.3.15 shows two possible structures of a single-loop DDC controller. In both structures, the computer with its internal clock drives the D/A and A/D converters. It compares the command signals with the feedback signals and generates the control signals to be sent to the D/A and eventually to the actuator. These control signals are computed from the control algorithm stored in the memory. In part (a) of the figure, the feedback signal is an analog signal (from an analog sensor such as a tachometer), which must be fed to an ADC. To convert the controller shown in Figure 10.3.4 to such a system, we must replace the op-amp comparator and multiplier with a computer

Figure 10.3.15 Possible structures for digital control systems.

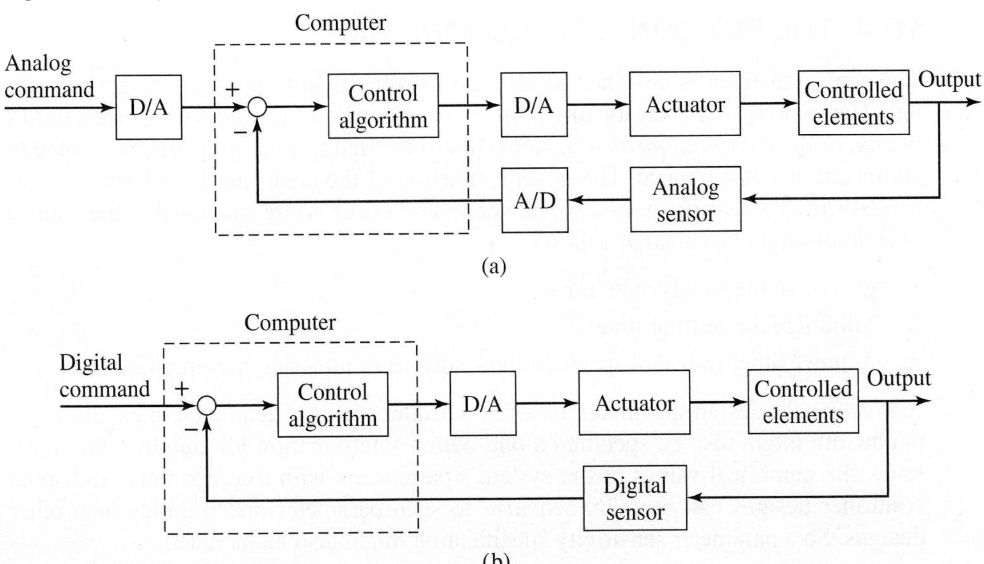

and DAC, and use an ADC on the tachometer signal. An alternative is shown in Figure 10.3.15b, which illustrates the use of a digital sensor (such as an optical encoder), whose signal can be communicated directly to the computer without using an ADC. If we retain the potentiometer, we must use an ADC on its voltage signal. An alternative is to replace the potentiometer with a command entered into the computer, perhaps via a keyboard.

10.3.7 PULSE-WIDTH MODULATION

The voltage output of the zero-order hold is constant over the sampling interval T but varies in amplitude over successive sampling intervals. The task of providing a variable-amplitude voltage can be somewhat difficult because of the dynamics caused by the motor inductance. Instead of varying the amplitude, an alternative to the zero-order hold is to output a constant voltage but vary the amount of time the voltage is applied over the sampling interval. This is the basis of *pulse-width modulation (PWM)*. Because it requires only a voltage source that can be switched between two constant values, say $v_{min} = 0$ and $v_{max} = V$, pulse-width modulation is particularly well suited for use with digital computers. If we consider the voltage v_c computed from the control algorithm to be the average required voltage over the interval T, then the pulse time t_i can be calculated from

$$t_i = \frac{v_c}{V}T$$

For the modulated voltage to be fully equivalent to v_c, t_i should be calculated so that the pulse of duration t_i delivers the same power as v_c over the interval T. However, this is difficult to calculate because of the dynamics of the circuit. The formula for t_i can be extended to include the case where $v_{min} < 0$ (negative voltage is required to slow down or reverse the motor's speed). Note that it is possible for the computed value of t_i to be greater than T. This unacceptable condition may occur if the control gains are too large.

10.4 THE PID CONTROL ALGORITHM

The control element is designed to act on the error signal to produce the control signal. The logic that is used for this purpose is also called the *control law*, the *control action*, or the *control algorithm*. A nonzero error signal results from either a change in command or a disturbance. The general function of the controller is to keep the controlled variable near its desired value when these occur. More specifically, the control objectives might be stated as follows:

1. Minimize the steady-state error.
2. Minimize the settling time.
3. Achieve other transient specifications, such as minimizing the overshoot.

In practice, the design specifications for a controller are more detailed. For example, the bandwidth might also be specified along with a safety margin for stability. We never know the numerical values of the system's parameters with true certainty, and some controller designs can be more sensitive to such parameter uncertainties than other designs. So a parameter sensitivity specification might also be included.

10.4.1 TWO-POSITION CONTROL

Two-position control action is the most familiar type perhaps because of its use in home thermostats. The control output takes on one of two values. A special case is an *on-off controller*, in which the controller output is either on or off. Such is the case with a thermostatically controlled heating system, in which the heater or furnace is turned on and off. If the outside conditions (temperature, wind, solar insolation, etc.) were predictable, we could design a building heating system that would operate continuously to supply heat at a predetermined rate just large enough to replace the heat lost to the outside environment. Of course, the real world does not behave so predictably, so we must use a feedback controller to adjust the heat output according to the actual room temperature.

A common example illustrates how the capacitance of the plant affects the suitability of on-off control. Regulation of the temperature of shower water by using on-off control of the hot water valve will obviously be unsuitable, but it is acceptable for a bath, because the thermal capacitance of the bath water is greater than that of the shower water.

Two-position control is acceptable for many applications in which the requirements are not too severe. In the home heating application, two-position control is acceptable because the typical 2°F temperature gap is hardly detectable by the occupants. In situations requiring finer control, however, use of two-position control results in frequent switching of the actuator, which shortens its life span and might waste energy.

10.4.2 PROPORTIONAL CONTROL

We have already seen proportional control, in which the control signal is proportional to the error. Referring to Figure 10.2.1, the transfer function of proportional control action is $G_c(s) = K_P$, and the controller output is given by

$$F(s) = G_c(s)E(s) = K_P E(s) \tag{10.4.1}$$

Sometimes the term "proportional band" is used in industrial controllers, rather than "proportional gain." The percent change in error needed to move the actuator full scale is the proportional band. It is related to the gain as follows:

$$\text{proportional band } (\%) = \frac{100}{K_P} \tag{10.4.2}$$

Thus, a gain $K_P = 1$ corresponds to a band of 100%. A gain $K_P = 2$ corresponds to a band of 50%.

Figure 10.4.1 illustrates a proportional controller using an op amp, where the proportional gain is $K_P = R_f/R_i$. Figure 10.4.2a is a diagram of a system using proportional control, and part (b) of the figure is a circuit diagram showing how op amps can be used to create the controller. An op-amp adder with an inverter in the feedback loop implements the comparator. Each op amp inverts its input signal. So the voltage v_r is inverted twice before entering the power amplifier, and there is no net change of sign. Similarly, the sensor voltage is inverted three times, for one net sign change, as required by the minus sign of the comparator.

Figure 10.4.1 Op-amp implementation of proportional action.

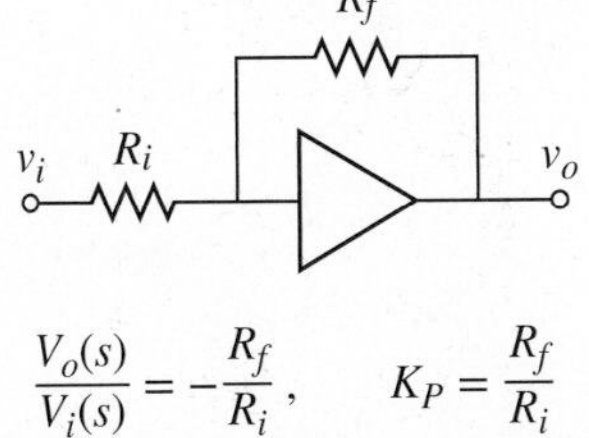

$$\frac{V_o(s)}{V_i(s)} = -\frac{R_f}{R_i}, \qquad K_P = \frac{R_f}{R_i}$$

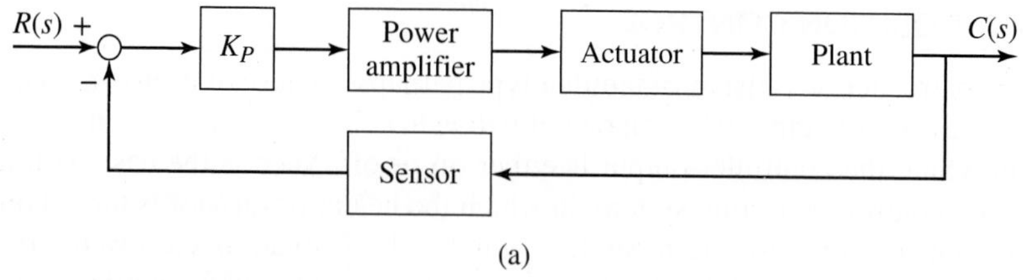

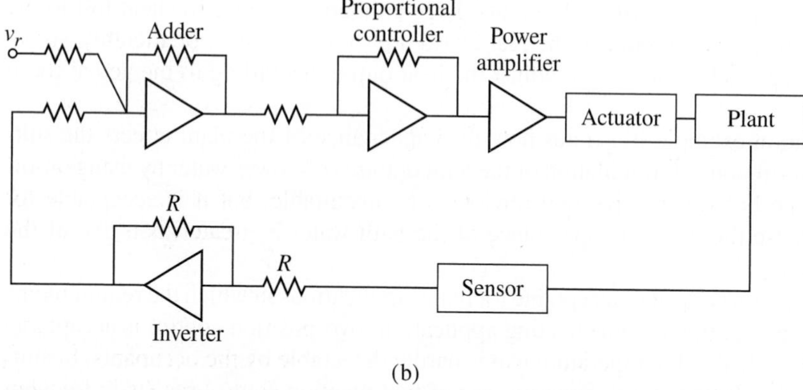

Figure 10.4.2 (a) Block-diagram representation of a proportional control system. (b) Op-amp implementation of a proportional controller.

EXAMPLE 10.4.1 Hydraulic Implementation of Proportional Control

■ Problem

Figure 10.4.3 shows a hydraulic implementation of proportional action to control the angle of an aircraft rudder, elevator, or aileron (see [Cannon, 1967]). The input motion y is produced by the motion of the pilot's control stick acting through cables. Analyze its motion assuming that the rudder inertia is small, and show that the system gives proportional action.

■ Solution

When the input motion y occurs, the beam pivots about its lower end. For small motions $x = L_3\theta$ and the beam geometry is such that

$$z = \frac{L_1 + L_2}{L_2} y - \frac{L_1}{L_2} x \tag{1}$$

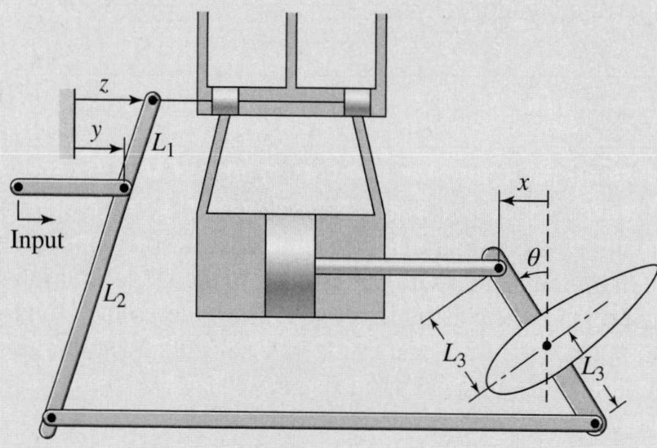

Figure 10.4.3 Hydraulic implementation of a proportional controller for an aerodynamic surface.

The proof of this equation for small angular motions is as follows. First, suppose that x is fixed to be 0. Then from similar triangles,

$$\frac{z}{L_1 + L_2} = \frac{y}{L_2} \quad \text{or} \quad z = (L_1 + L_2)\frac{y}{L_2}$$

Now suppose that y is fixed to be 0. From similar triangles,

$$\frac{z}{L_1} = -\frac{x}{L_2} \quad \text{or} \quad z = -L_1\frac{x}{L_2}$$

Since the general motion is the superposition of the two individual motions, adding these two relations gives equation (1).

For the servomotor, as shown in Example 7.4.7, if the rudder inertia is small,

$$\frac{X(s)}{Z(s)} = \frac{C_1}{As}$$

These equations give the following transfer function.

$$\frac{\Theta(s)}{Y(s)} = \frac{L_1 + L_2}{L_1 L_3}\frac{1}{\tau s + 1}$$

where

$$\tau = \frac{L_2 A}{L_1 C_1}$$

If the servomotor gain C_1/A is large, then $\tau \approx 0$, and the transfer function becomes

$$\frac{\Theta(s)}{Y(s)} = \frac{L_1 + L_2}{L_1 L_3} \equiv K_P$$

Thus, the system implements proportional control with proportional gain K_P if the motion y represents the error signal and θ represents the controller output.

10.4.3 INTEGRAL CONTROL ACTION

We will see that sometimes proportional control results in the system reaching an equilibrium in which the control signal no longer changes, thus allowing a constant error to exist. If the controller is modified to produce an increasing signal as long as the error is nonzero, the error might be eliminated. This is the principle of *integral control action*, also called *reset action*. In this mode, the control signal is proportional to the integral of the error. In the terminology of Figure 10.2.1, this corresponds to a controller transfer function of $G_c(s) = K_I/s$, or

$$F(s) = G_c(s)E(s) = \frac{K_I}{s}E(s)$$

where K_I is called the *integral gain*. In the time domain, the relation is

$$f(t) = K_I \int_0^t e(t)\,dt \tag{10.4.3}$$

if $f(0) = 0$. In this form, it can be seen that if the error signal $e(t)$ remains positive, the integration will theoretically produce an infinite value of $f(t)$. This implies that special care must be taken to reinitialize the controller.

The integral term does not react instantaneously to the error signal but continues to correct, which tends to cause oscillations if the designer does not take this effect into account.

When integral and proportional actions are combined, we obtain the PI control algorithm:

$$f(t) = K_P e(t) + K_I \int_0^t e(t)\,dt \tag{10.4.4}$$

or

$$F(s) = \left(K_P + \frac{K_I}{s}\right)E(s) \tag{10.4.5}$$

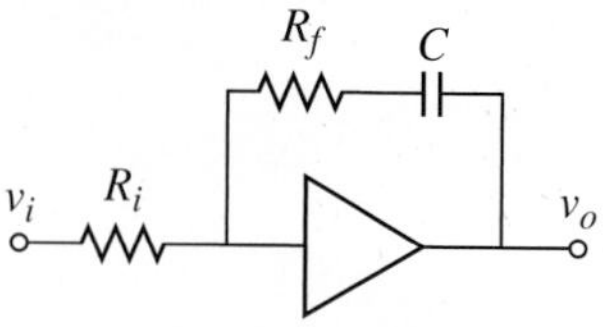

Figure 10.4.4 Op-amp implementation of PI action.

$$\frac{V_o(s)}{V_i(s)} = -\frac{R_f}{R_i} - \frac{1}{R_i Cs}$$

$$K_P = \frac{R_f}{R_i} \qquad K_I = \frac{1}{R_i C}$$

An op-amp implementation of PI action is shown in Figure 10.4.4. With three component values to be chosen, normally the capacitance value is selected first, and the two resistances computed from the desired gain values, using the equations given in the figure.

EXAMPLE 10.4.2

Hydraulic Implementation of PI Control

■ **Problem**

Figure 10.4.5 shows a hydraulic system whose input is the motion y. Show that this system implements PI action, if the inertia of the load is small.

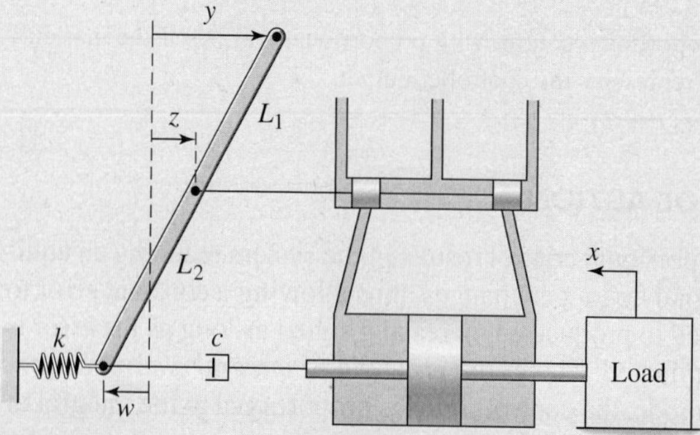

Figure 10.4.5 Hydraulic implementation of PI action.

■ **Solution**

For small motions,

$$z = \frac{L_2}{L_1 + L_2}y - \frac{L_1}{L_1 + L_2}w$$

A force balance at the bottom of the beam gives

$$c\dot{w} + kw = c\dot{x}$$

For the servomotor, if the load inertia is small

$$\frac{X(s)}{Z(s)} = \frac{C_1}{As}$$

If

$$\frac{C_1}{A} \gg \left|\frac{(L_1 + L_2)(c\omega j + k)}{cL_1}\right|$$

for ω lying within the operating range of frequencies, then these equations may be combined to yield

$$\frac{X(s)}{Y(s)} = \frac{L_2}{L_1}\left(1 + \frac{k}{cs}\right) = K_P + \frac{K_I}{s}$$

where

$$K_P = \frac{L_2}{L_1} \qquad K_I = \frac{L_2}{L_1}\frac{k}{c}$$

10.4.4 DERIVATIVE CONTROL ACTION

Integral action produces a control signal even after the error has vanished, and this tends to cause oscillations. This suggests that the controller should be made aware that the error is approaching zero. One way to accomplish this is to design the controller to react to the rate of change of the error. This is the basis of *derivative control* action, in which the controller transfer function is $G_c(s) = K_D s$ and

$$F(s) = G_c(s)E(s) = K_D s E(s) \tag{10.4.6}$$

where K_D is the *derivative gain*. This algorithm is also called *rate action*. It is used to damp out oscillations.

The output of derivative control action depends on the error rate, and thus should never be used alone because it will produce zero output even if the error is large but constant. When used with proportional action, the following PD control algorithm results.

$$f(t) = K_P e(t) + K_D \frac{de}{dt} \tag{10.4.7}$$

or

$$F(s) = (K_P + K_D s)E(s) \tag{10.4.8}$$

An op-amp implementation of PD action is shown in Figure 10.4.6, where αT_D should be chosen to be small.

10.4.5 PID ACTION

When integral and derivative action are used with proportional action, we obtain the following proportional-plus-integral-plus-derivative control algorithm, abbreviated PID.

$$F(s) = \left(K_P + \frac{K_I}{s} + K_D s\right) E(s) = K_P\left(1 + \frac{1}{T_I s} + T_D s\right) E(s) \tag{10.4.9}$$

This is called a *three-mode controller*. Some industrial controllers are programmed not in terms of K_I and K_D, but in terms of the *reset time* T_I and the *rate time* T_D, defined as follows:

$$T_I = \frac{K_P}{K_I} \tag{10.4.10}$$

$$T_D = \frac{K_D}{K_P} \tag{10.4.11}$$

The reset time is the time required for the integral action signal to equal that of the proportional term if a constant error exists (a hypothetical situation). The reciprocal of reset time is expressed as repeats per minute and is the frequency with which the integral action repeats the proportional correction signal.

Figure 10.4.6 Op-amp implementation of PD action.

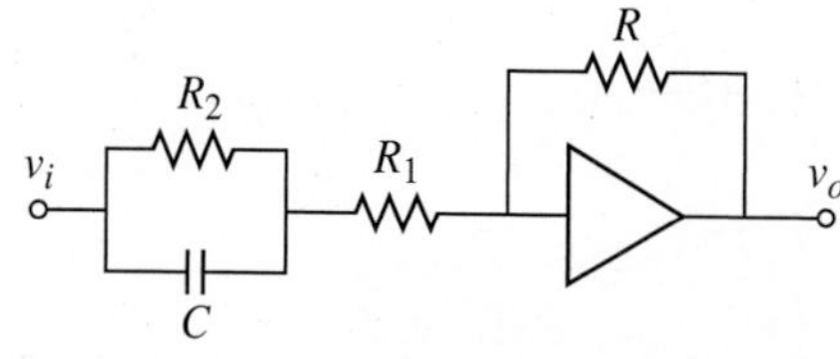

$$\frac{V_o(s)}{V_i(s)} = -\frac{R(R_2CS+1)}{R_1+R_2+R_1R_2Cs} = -\frac{K_P(1+T_Ds)}{1+\alpha T_Ds}$$

$$K_P = \frac{R}{R_1+R_2} \qquad T_D = R_2C \qquad \alpha = \frac{R_1}{R_1+R_2}$$

Figure 10.4.7 Op-amp implementation of PID action.

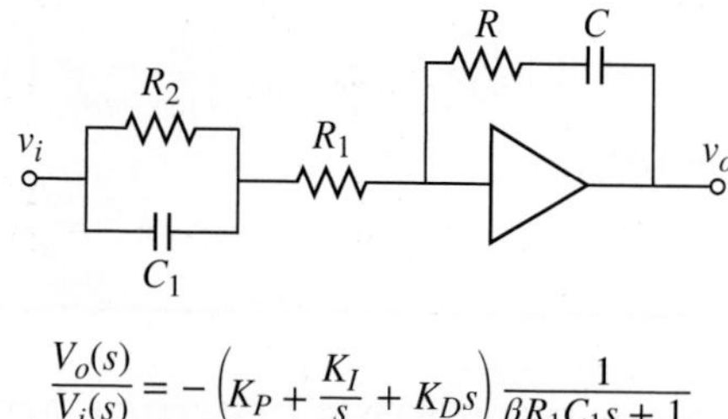

$$\frac{V_o(s)}{V_i(s)} = -\left(K_P + \frac{K_I}{s} + K_Ds\right)\frac{1}{\beta R_1C_1s+1}$$

$$K_P = \beta\frac{RC+R_2C_1}{R_2C} \qquad K_I = \frac{\beta}{R_2C}$$

$$K_D = \beta RC_1 \qquad \beta = \frac{R_2}{R_1+R_2}$$

Proportional action is frequently used, but integral or derivative action might or might not be used, depending on the application. An op-amp implementation of PID action is shown in Figure 10.4.7, where βR_1C_1 should be chosen to be small.

The typical effects of P, I, and D action on a system's step response are illustrated in Figure 10.4.8, which shows the response to a unit-step command (with no disturbance) of three systems having the same plant but using P, PI, and PID control, respectively. The response with P control is non-oscillatory and reasonably fast, but it allows a nonzero steady-state error to exist. When I action is added, the response under PI control becomes oscillatory with a large overshoot, but the steady-state error is now zero. When D action is added, the response under PID control has fewer oscillations and a smaller overshoot, and the steady-state error remains zero. In general, but not always, as control action is changed from P to PI to PID, first, the steady-state error is eliminated, and then the oscillations and settling time are reduced. Note that D action can never change a constant steady-state error, because it produces no output for a constant error.

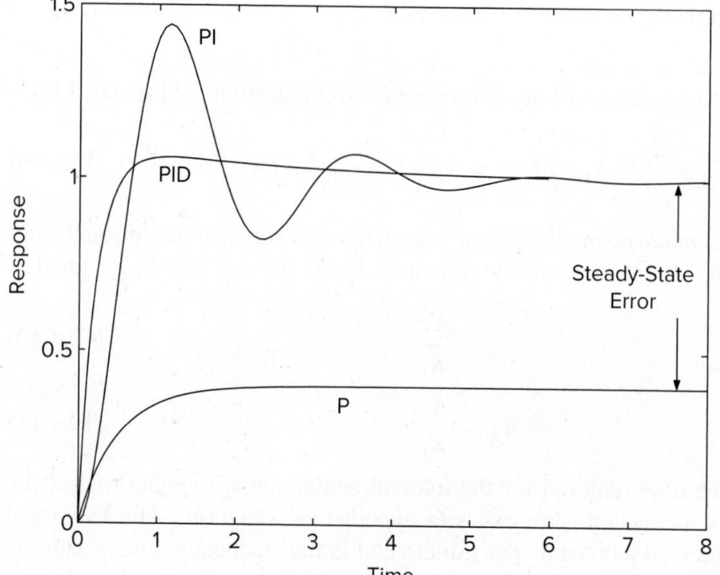

Figure 10.4.8 Typical step response of P, PI, and PID control systems.

Table 10.4.1 Comparison of control actions.

Control action	Advantages	Disadvantages
On-Off	Simple to implement.	Cannot achieve small error without excessive oscillation.
Proportional	Simple to implement. Improves response time.	Allows steady-state error.
Integral	Reduces steady-state error.	Increases system order, creating potential for overshoot, oscillations, and instability.
Derivative	Reduces oscillations. Improves response time.	Amplifies noise in measurement signals.
Feedforward compensation	Reduces command input error.	Depends on accurate system model.
Disturbance compensation	Reduces disturbance error.	Depends on accurate system model. Depends on usable measurement of the disturbance.

Table 10.4.1 gives a comparison of the various control actions.

Proportional, integral, and derivative actions and their various combinations are not the only control laws possible, but they are the most common. It has been estimated that most controllers have PI or PID as their basic control action, and these actions will remain for some time the standard against which any new designs must compete.

10.4.6 DIGITAL CONTROL ALGORITHMS

The analog PID control algorithm

$$f(t) = K_P e(t) + K_I \int_0^t e(t)\, dt + K_D \frac{de}{dt} \tag{10.4.12}$$

cannot be implemented directly in a digital controller because digital hardware, which is limited to arithmetic operations, cannot perform differentiation and integration. Rather, the algorithm must first be converted to a difference equation. There are several finite difference equivalents used for integral and derivative action. A commonly used form is

$$f(t_k) = K_P e(t_k) + K_I T \sum_{i=0}^{k} e(t_i) + \frac{K_D}{T}[e(t_k) - e(t_{k-1})] \tag{10.4.13}$$

where $f(t_k)$ and $e(t_k)$ are the control and error signals at time $t_k = kT$, and T is the sampling period. The integral has been replaced with a sum of the areas of rectangles of width T. The derivative has been replaced with a simple difference expression.

The algorithm (10.4.13) is the *position* version of the PID control law. The *incremental* or *velocity* version of the algorithm determines the *change* in the control signal: $f(t_k) - f(t_{k-1})$. To obtain it, decrement k by 1 in (10.4.13) and subtract the resulting equation from (10.4.13). This gives

$$\begin{aligned} f(t_k) - f(t_{k-1}) &= K_P[e(t_k) - e(t_{k-1})] + K_I T e(t_k) \\ &\quad + \frac{K_D}{T}[e(t_k) - 2e(t_{k-1}) + e(t_{k-2})] \end{aligned} \tag{10.4.14}$$

Suppose that the control signal $f(k)$ is a valve position. The position version of the algorithm is so named because it specifies the valve position directly as a function of the error signal. The incremental algorithm, on the other hand, specifies the change in valve position. The incremental version has the advantage that the valve will maintain its last position in the event of failure or shutdown of the control computer. Also, the valve will not "saturate" at start-up if the controller is not matched to the current valve position. In addition to having these safety features, the incremental algorithm is also well suited for use with incremental output devices, such as stepper motors.

These approximations for derivative and integral actions are the simplest, but are useful for introducing the concept of a digital control algorithm. Many other, more accurate approximations have been implemented in controllers. These are discussed in more advanced references dealing with digital control.

Throughout this chapter and the next we will concentrate on the analog version of the PID algorithm, but this does not prevent these results from being applied in a digital controller. The important point to keep in mind when doing so is the sampling rate of the controller. For a very fast sampling rate, the sampling time T is very small, and the finite difference approximations for the derivative and the integral approach their analog counterparts. That is,

$$\lim_{T\to 0} T \sum_{i=0}^{k} e(t_i) \to \int_0^t e(t)\,dt$$

and

$$\lim_{T\to 0} \frac{e(t_k) - e(t_{k-1})}{T} \to \frac{de}{dt}$$

Thus, if we assume that the sampling time is small enough, the gains computed for the analog PID algorithm can also be used for digital control.

10.5 CONTROL SYSTEM ANALYSIS

In this section we outline the approach to selecting an appropriate control algorithm and computing its gain values. Central to this process is an understanding of the system's transient and steady-state behavior, especially with regard to the effect of the command and disturbance inputs on the error.

10.5.1 SOURCES OF SYSTEM ERRORS

We have seen that the error $e(t)$ in a control system is the difference between the command input $r(t)$ and the controlled variable $c(t)$; that is, $e(t) = r(t) - c(t)$. The error is obviously a function of time, and when designing a control system we might specify and examine the peak error, the settling time for the error, and the steady-state error, for example. For a system having a command input and a disturbance input, the error is the sum of the errors caused by the command and the disturbance. These errors are called the *command error* and the *disturbance error*. Figure 10.5.1 illustrates the two types of error for the case where both the command and the disturbance are step functions, with the command input applied at $t = 0$ and the disturbance acting at time $t = t_1$. We wish the controlled variable to be equal to the command; that is, we want $c = r$. In this figure, it is assumed that the effect of the disturbance is to decrease the controlled variable, whereas in some systems it may increase the controlled variable.

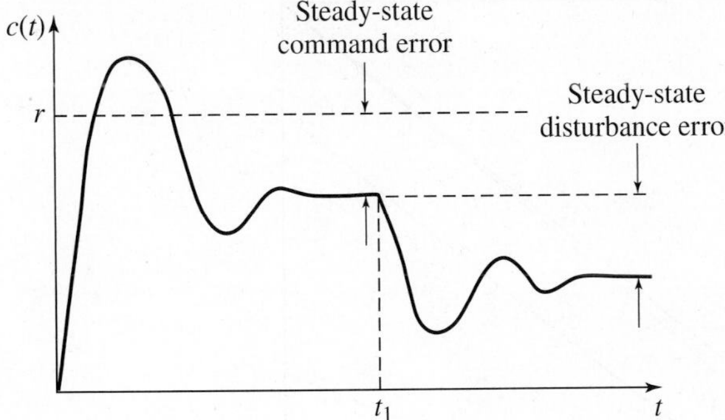

Figure 10.5.1 Illustration of steady-state command error and disturbance error.

10.5.2 STEP INPUTS AND THE FINAL-VALUE THEOREM

Calculation of the steady-state response due to a step input is simple if the system is stable. Let $T(s)$ denote the command transfer function, so that $C(s) = T(s)R(s)$. The steady-state output is obtained from the final-value theorem.

$$c_{ss} = \lim_{s\to 0} sC(s) = \lim_{s\to 0} sT(s)R(s) \tag{10.5.1}$$

If $R(s)$ represents a step of magnitude M, then $R(s) = M/s$ and

$$c_{ss} = \lim_{s\to 0} s\frac{M}{s}T(s) = M\lim_{s\to 0} T(s) \tag{10.5.2}$$

The theorem applies only if all the roots of the denominator of $T(s)$ have negative real parts; that is, if $T(s)$ represents a stable system. Stated simply, this result says the steady-state response of a stable system to a step input of magnitude M is $MT(0)$. A similar procedure can be used for the disturbance input. This result is summarized in Table 10.5.1.

Table 10.5.1 The final-value theorem.

1. **General Case:** If $T(s)$ is the transfer function and $R(s)$ is the input, the output is $C(s) = T(s)R(s)$. The steady-state value c_{ss} is given by

$$c_{ss} = \lim_{s\to 0} sC(s) = \lim_{s\to 0} sT(s)R(s)$$

if all the roots of the denominator of $sT(s)R(s)$ have negative real parts.

2. **Step Inputs:** If $R(s)$ represents a step of magnitude M, then $R(s) = M/s$ and

$$c_{ss} = \lim_{s\to 0} s\frac{M}{s}T(s) = M\lim_{s\to 0} T(s) = MT(0)$$

only if all the roots of the denominator of $T(s)$ have negative real parts; that is, if $T(s)$ represents a stable system.

3. **Ramp Inputs:** The transformed error given by $E(s) = R(s) - C(s)$ has the steady-state value given by

$$e_{ss} = \lim_{s\to 0} s[R(s) - C(s)] = \lim_{s\to 0} sR(s)[1 - T(s)]$$

if all the roots of the denominator of $sR(s)[1 - T(s)]$ have negative real parts. If $R(s)$ represents a ramp of slope m, then $R(s) = m/s^2$ and

$$e_{ss} = \lim_{s\to 0} s\frac{m}{s^2}[1 - T(s)] = \lim_{s\to 0} \frac{m}{s}[1 - T(s)]$$

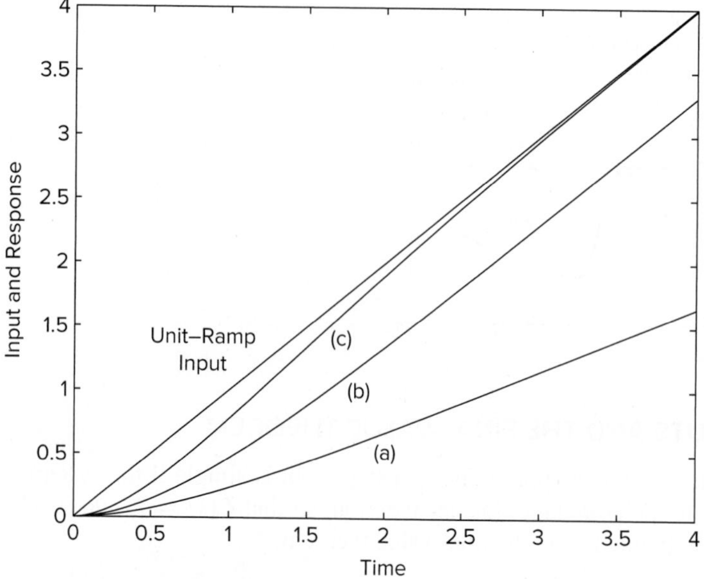

Figure 10.5.2 Ramp responses of three control systems.

10.5.3 RAMP INPUTS

Step inputs are not the only input types to be considered when designing control systems. Ramps, parabolas, and sine functions are also commonly used. Figure 10.5.2 shows some possible responses to a ramp input. For curve (a) the controlled variable continues to deviate from the ramp command, and thus the steady-state command error is infinite. For curve (b) the controlled variable eventually becomes parallel to the ramp command, and thus the steady-state command error is constant but nonzero. For curve (c) the controlled variable eventually becomes equal to the ramp command, and thus the steady-state command error is zero.

Because the output due to a ramp input eventually becomes infinite, the final-value theorem gives no useful information when applied to the output equation. Rather, the theorem is applied to the *error* equation to obtain the steady-state error. Let $T(s)$ denote the command transfer function, so that $C(s) = T(s)R(s)$. The transformed error is given by $E(s) = R(s) - C(s)$, and thus the steady-state error is obtained from

$$e_{ss} = \lim_{s\to 0} s[R(s) - C(s)] = \lim_{s\to 0} sR(s)[1 - T(s)] \tag{10.5.3}$$

if all the roots of the denominator of $sR(s)[1 - T(s)]$ have negative real parts. If $R(s)$ represents a ramp of slope m, then $R(s) = m/s^2$ and

$$e_{ss} = \lim_{s\to 0} s\frac{m}{s^2}[1 - T(s)] = \lim_{s\to 0} \frac{m}{s}[1 - T(s)] \tag{10.5.4}$$

A similar procedure can be used for the disturbance input. This result is summarized in Table 10.5.1.

Control systems are never expected to respond to a pure ramp command or ramp disturbance, because no real input increases indefinitely. A ramp command function, however, is a good test of a control system's ability to follow a constantly changing command. Although the ramp function continues to increase without limit, in practice we are interested in the system's response only up to the time the transient response disappears. The reason for this is illustrated by Figure 10.5.3, which shows a command input called the *trapezoidal profile*. This is a typical command to speed-control systems; it represents the desired speed in many positioning system applications such as tape drives, conveyor drives, and robotic systems. For example, when controlling

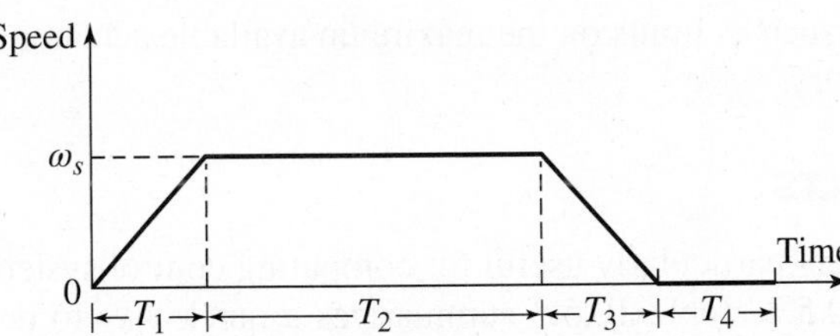

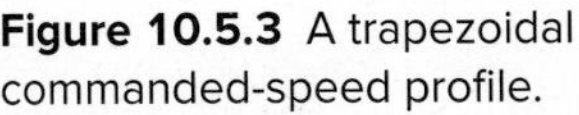
Figure 10.5.3 A trapezoidal commanded-speed profile.

the rotational speed of a motor, in order to bring the load up to some desired speed ω_s, called the *slew speed*, as quickly as possible, the drive motor should accelerate at its maximum acceleration α_{max}, which is limited by the maximum available motor torque. The motor then decelerates to bring the load to rest. To follow such a command input, the controller must be able to deal with both step and ramp commands.

10.5.4 DESIGN INFORMATION

When designing a control system, the control systems engineer is usually given the following information.

1. Usually the plant is given, and cannot be changed. The engineer might be given a model of the plant and the actuator, or might be expected to develop a model.
2. Usually the physical type of controller (electronic, pneumatic, hydraulic, or digital) and the actuator type (electric, pneumatic, or hydraulic) is specified, or the choice will be obvious from the application.
3. The command and disturbance inputs might be specified, or the engineer might be expected to develop suitable test inputs based on the application. Step functions are the principal test inputs, because they are the most common and perhaps represent the severest test of system performance. Ramp, trapezoidal, and sinusoidal test inputs are also employed. The type to use should be made clear in the design specifications.

10.5.5 CONTROL SYSTEM DESIGN PROCEDURE

In general, the design steps are as follows:

1. Based on the system model and the performance specifications, choose a control action and obtain the output, error, and actuator equations.
2. Analyze the system for stability. If it cannot be made stable, stop and try a different control action.
3. If it can be made stable, determine the constraints on the control gains to achieve stability.
4. Using the given command and disturbance input functions (step, ramp, etc.), evaluate the steady-state response. This usually can be done quickly with the final-value theorem. Determine the constraints on the gains to satisfy the steady-state specifications.
5. Evaluate the transient performance in light of the transient specifications, using the given input functions. These specifications often are stated in terms of the desired dominant-time constant and damping ratio, but they can be given in terms of overshoot, rise time, settling time, or bandwidth, for example. If the model of the closed-loop system is third order or higher, then we have no formulas to use for time constant, damping ratio, rise time, settling time, or bandwidth, and a trial-and-error approach must be used. This situation is explored in Chapter 11.

6. Evaluate other specifications, such as limits on the maximum available actuator output, and redesign if necessary.

10.5.6 SOME USEFUL RESULTS

Some formulas from earlier chapters particularly useful for computing control system gains are summarized in Table 10.5.2. Table 10.5.3 summarizes a quick way to determine the stability of a model without the need to solve for the roots. These results are obtained with the method developed by Routh and Hurwitz, which is discussed in

Table 10.5.2 Useful results for second-order systems.

1. Model: $m\ddot{x} + c\dot{x} + kx = f(t)$
2. Transfer function:

$$\frac{X(s)}{F(s)} = \frac{1}{ms^2 + cs + k}$$

3. Characteristic equation: $ms^2 + cs + k = 0$
4. Characteristic roots:

$$s = \frac{-c \pm \sqrt{c^2 - 4mk}}{2m}$$

5. Damping ratio and undamped natural frequency:

$$\zeta = \frac{c}{2\sqrt{mk}} \qquad \omega_n = \sqrt{\frac{k}{m}}$$

6. Time constant: If $\zeta \le 1$,

$$\tau = \frac{2m}{c}$$

If $\zeta > 1$, the dominant (larger) time constant is

$$\tau_1 = \frac{2m}{c - \sqrt{c^2 - 4mk}}$$

and the secondary (smaller) time constant is

$$\tau_2 = \frac{2m}{c + \sqrt{c^2 - 4mk}}$$

7. Maximum percent overshoot and peak time:

$$M_{\%} = 100e^{-\pi\zeta/\sqrt{1-\zeta^2}} \qquad t_p = \frac{\pi}{\omega_n\sqrt{1-\zeta^2}}$$

8. The complex root pair $s = -a \pm bj$ corresponds to the characteristic equation

$$(s + a)^2 + b^2 = 0$$

9. The value $\zeta = 0.707$ corresponds to a root pair having equal real and imaginary parts: $s = -a \pm aj$.

Table 10.5.3 Routh-Hurwitz stability conditions.

1. **Second-Order:** $a_2s^2 + a_1s + a_0 = 0$
   Stable if and only if a_2, a_1, and a_0 all have the same sign.
2. **Third-Order:** $a_3s^3 + a_2s^2 + a_1s + a_0 = 0$
   Assuming $a_3 > 0$, stable if and only if a_2, a_1, and a_0 are all positive and $a_2a_1 > a_3a_0$.
3. **Fourth-Order:** $a_4s^4 + a_3s^3 + a_2s^2 + a_1s + a_0 = 0$
   Assuming $a_4 > 0$, stable if and only if a_3, a_2, a_1, and a_0 are all positive, $a_2a_3 > a_1a_4$, and

$$a_1(a_2a_3 - a_1a_4) - a_0a_3^2 > 0$$

detail in texts devoted to control theory and is applicable to a model of any order. We state the results for second-, third-, and fourth-order models only, because we will not use higher-order models in this text.

10.6 CONTROLLING FIRST-ORDER PLANTS

The success of a chosen control action depends partly on the form of the plant transfer function. In this section and Sections 10.7 and 10.8, we show how to apply the design steps listed in Section 10.5 to various plant models. It is important to realize that the mathematical results of these examples can be used to design a control system for any plant having the same transfer-function form. For example, the plant transfer function for the liquid-level system considered in Section 10.1 (Figure 10.1.1c) is

$$G_p(s) = \frac{R}{RAs + g} = \frac{1}{As + g/R}$$

while the plant transfer function for the oven heating system (Figure 10.3.3b) is

$$G_p(s) = \frac{R}{RCs + 1} = \frac{1}{Cs + 1/R}$$

Both transfer functions are first order with no numerator dynamics and have the form

$$G_p(s) = \frac{1}{Is + c}$$

which is the plant transfer function for a speed-control system (Figures 10.3.7b and 10.3.10d). Therefore the results of the following example may be applied to design the liquid-level system, the heating system, and the speed-control system.

Proportional Control of a First-Order Plant

EXAMPLE 10.6.1

■ Problem

Consider proportional control of the first-order plant whose transfer function is $1/(Is + c)$. This can represent a rotational system whose model is $I\dot{\omega} + c\omega = T - T_d$, where the controlled variable is the speed ω, the actuator torque is T, and the disturbance torque is T_d. To illustrate the methods simply, we will assume that the actuator has a gain of 1 and has an instantaneous response. This implies that the actuator transfer function is $G_a(s) = 1$. The block diagram is shown in Figure 10.6.1. Discuss the effect of the value of K_P on the system performance when the inputs are step functions.

■ Solution

The transfer functions are

$$\frac{\Omega(s)}{\Omega_r(s)} = \frac{K_P}{Is + c + K_P} \tag{1}$$

$$\frac{\Omega(s)}{T_d(s)} = \frac{-1}{Is + c + K_P} \tag{2}$$

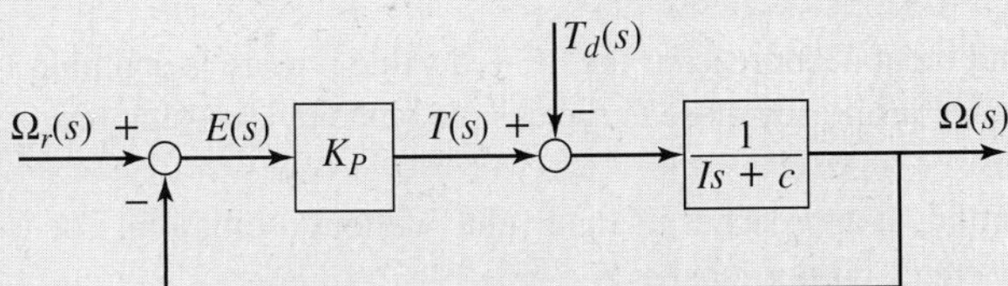

Figure 10.6.1 P control of a first-order plant.

Steady-State Errors: For a unit-step command, $\Omega_r(s) = 1/s$, the speed approaches the steady-state value

$$\omega_{ss} = \lim_{s\to 0} s\frac{K_P}{Is + c + K_P}\frac{1}{s} = \frac{K_P}{c + K_P} < 1$$

Thus, the steady-state speed is less than the desired value of 1, but it might be close enough if the damping c is small. The time required to reach this value is approximately four time constants, or $4\tau = 4I/(c + K_P)$.

It is impossible for any control system to have a disturbance response that is zero for all time, because it takes time for any system to respond to the disturbance. In many applications, we are satisfied if the disturbance response is zero at steady state. In other applications, we might tolerate a small nonzero disturbance response. The performance specifications should indicate what is required for the design.

The response due to a unit-step disturbance, $T_d(s) = 1/s$, is found from equation (2).

$$\Omega(s) = \frac{-1}{Is + c + K_P}\frac{1}{s}$$

The steady-state response to the disturbance is found with the final-value theorem to be $-1/(c + K_P)$. If $(c + K_P)$ is large, this response will be small. Note that this is the speed *change* caused by the disturbance, and is thus the disturbance error.

Frequency Response: The disturbance transfer function has the form of a low-pass filter whose low-frequency gain is $1/(c+K_P)$, and whose bandwidth is $1/\tau = (c+K_P)/I$. Increasing the gain K_P will increase the bandwidth, thus making the system sensitive to higher-frequency disturbances.

Actuator Requirements: It is important to predict what the actuator requirements will be. In the absence of a disturbance, the expression for the motor torque T can be found from the block diagram and equation (1). It is

$$T(s) = K_P[\Omega_r(s) - \Omega(s)] = K_P\left(1 - \frac{K_P}{Is + c + K_P}\right)\Omega_r(s)$$

or

$$T(s) = K_P\frac{Is + c}{Is + c + K_P}\Omega_r(s)$$

If the commanded speed is a unit step, the torque response can be found from a partial fraction expansion to be

$$T(t) = \frac{K_P}{c + K_P}\left[c + K_P e^{-(c+K_P)t/I}\right]$$

Thus, the torque decays exponentially to the value $cK_P/(c+K_P)$. The maximum torque occurs at $t = 0$ and is $T(0) = K_P$. Thus, the larger the gain K_P, the more torque is required. To speed up the system (by reducing τ) or to reduce the steady-state error, we must increase K_P and therefore must increase the available torque. This has implications for the size and cost of the motor that must be used.

Summary: The performance of proportional control action thus far can be summarized as follows. For the first-order plant $1/(Is + c)$, whose inputs are step functions,

1. The response to the command input never reaches the desired value if there is damping $(c \neq 0)$, although it can be made arbitrarily close to the desired value by choosing the gain K_P large enough.
2. The response to the command input approaches its final value without oscillation. The time to reach this value is inversely proportional to K_P.

3. The steady-state disturbance error is inversely proportional to the gain K_P even if there is no damping.
4. For a step command input of magnitude M, the maximum required motor torque occurs at $t = 0$ and equals MK_P. To reduce the time constant or the errors, we must increase K_P and thus the maximum available torque.

The chief disadvantage of proportional control action is that it results in steady-state errors. It can be used only when the gain can be made large enough to reduce errors and reduce the time constant without requiring too large an actuator output. It can be augmented with feedforward command compensation to reduce the command error. An advantage to proportional control action is that the control signal responds to the error instantaneously (in theory at least) and does not cause oscillations.

Response of a Proportional Control System

EXAMPLE 10.6.2

■ Problem

Suppose the plant shown in Figure 10.6.1 has the parameter values $I = 5$ and $c = 2$ in a consistent set of units in which time is measured in seconds. (a) Determine the smallest value of the gain K_P required so that the steady-state command error will be no greater than 0.05 rad/s if ω_r is a unit-step input. Evaluate the resulting time constant and the steady-state disturbance error for a unit-step disturbance. (b) Obtain the steady-state error due to a unit-ramp command.

■ Solution

a. The error equation is

$$E(s) = \Omega_r(s) - \Omega(s)$$

$$= \left(1 - \frac{K_P}{5s + 2 + K_P}\right)\Omega_r(s) + \frac{1}{5s + 2 + K_P}T_d(s)$$

$$= \left(\frac{5s + 2}{5s + 2 + K_P}\right)\Omega_r(s) + \frac{1}{5s + 2 + K_P}T_d(s)$$

The steady-state command error for a unit-step command with no disturbance is

$$e_{ss} = \lim_{s \to 0} s\left(\frac{5s + 2}{5s + 2 + K_P}\right)\frac{1}{s} = \frac{2}{2 + K_P}$$

The error will be 0.05 rad/s if $K_P = 38$. The time constant for this value of K_P is

$$\tau = \frac{I}{c + K_P} = \frac{5}{2 + K_P} = \frac{1}{8}\text{ s}$$

The steady-state error due to a unit-step disturbance is

$$e_{ss} = \lim_{s \to 0} s\left(\frac{1}{5s + 2 + K_P}\right)\frac{1}{s} = \frac{1}{2 + K_P} = \frac{1}{40} = 0.025\text{ rad/s}$$

If both inputs are applied, the total steady-state error will be

$$e_{ss} = 0.05 + 0.025 = 0.075\text{ rad/s}$$

and the steady-state speed will be

$$\omega_{ss} = 1 - 0.075 = 0.925\text{ rad/s}$$

Note that we can make the command error, the disturbance error, and the time constant smaller only by making K_P larger than 38, but this increases the maximum required torque and probably the cost of the system.

b. Using $\Omega_r(s) = 1/s^2$ with the final-value theorem, we find the steady-state command error is

$$e_{ss} = \lim_{s\to 0} s\left(\frac{5s+2}{5s+40}\right)\frac{1}{s^2} = \infty$$

The speed $\omega(t)$ never catches up with the command input $\omega_r(t)$, and the steady-state error is infinite.

The speed-control system with P action is a first-order system and thus cannot oscillate with step inputs. Integral control action often eliminates steady-state error but it always raises the order of the system by one, thus resulting in a higher-order system more likely to oscillate. The logical choice is to combine the two actions and use PI control.

EXAMPLE 10.6.3

Computing PI Gains

■ Problem

A PI control system is shown in Figure 10.6.2. Suppose the plant has the parameter values $I = 5$ and $c = 2$. The dominant-time constant is specified to be 0.5 s.

a. Compute the required values for K_P and K_I for each of the following three cases: (1) $\zeta = 0.707$, (2) $\zeta = 1$, and (3) the secondary time constant must be 0.05 s.

Evaluate the steady-state command error and the steady-state disturbance error for each case given that both the command input $\omega_r(t)$ and the disturbance $T_d(t)$ are unit-step functions.

b. Plot the output response $\omega(t)$ and the actuator response $T(t)$ for each case, given that the command input $\omega_r(t)$ is a unit-step function and the disturbance $T_d(t)$ is zero.

c. Evaluate the steady-state command error for each case given that the command input $\omega_r(t)$ is a unit-ramp function and the disturbance $T_d(t)$ is zero.

d. Evaluate the frequency response characteristics of the disturbance transfer function.

■ Solution

From the block diagram we obtain the following output, error, and actuator equations.

$$\Omega(s) = \frac{K_P s + K_I}{5s^2 + (2 + K_P)s + K_I}\Omega_r(s) - \frac{s}{5s^2 + (2 + K_P)s + K_I}T_d(s) \tag{1}$$

$$E(s) = \frac{5s^2 + 2s}{5s^2 + (2 + K_P)s + K_I}\Omega_r(s) + \frac{s}{5s^2 + (2 + K_P)s + K_I}T_d(s) \tag{2}$$

$$T(s) = \frac{(K_P s + K_I)(5s + 2)}{5s^2 + (2 + K_P)s + K_I}\Omega_r(s) + \frac{K_P s + K_I}{5s^2 + (2 + K_P)s + K_I}T_d(s) \tag{3}$$

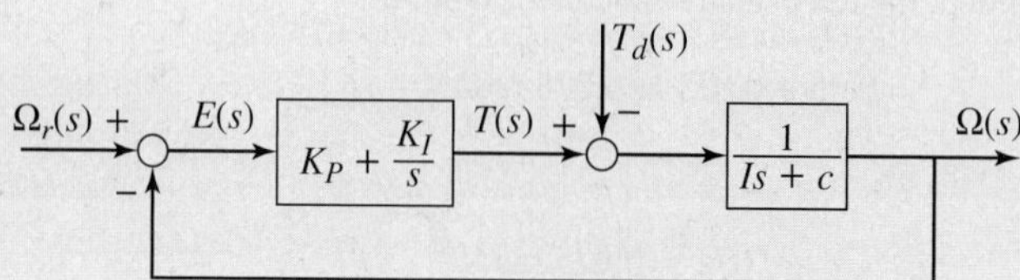

Figure 10.6.2 PI control of a first-order plant.

The characteristic equation is

$$5s^2 + (2 + K_P)s + K_I = 0 \tag{4}$$

and its roots are

$$s = \frac{-(2 + K_P) \pm \sqrt{(2 + K_P)^2 - 20K_I}}{10}$$

The Routh-Hurwitz condition shows the system is stable if $2 + K_P > 0$ and $K_I > 0$. Applying the final-value theorem to the error equation (2) with $\Omega_r(s) = 1/s$ and $T_d(s) = 0$ gives

$$e_{ss} = \lim_{s \to 0} sE(s) = \lim_{s \to 0} s \frac{5s^2 + 2s}{5s^2 + (2 + K_P)s + K_I} \frac{1}{s} = 0$$

if the system is stable. Thus, the system has zero command error for a step command.

Applying the final-value theorem to the error equation (2) with $\Omega_r(s) = 0$ and $T_d(s) = 1/s$ gives

$$e_{ss} = \lim_{s \to 0} sE(s) = \lim_{s \to 0} s \frac{s}{5s^2 + (2 + K_P)s + K_I} \frac{1}{s} = 0$$

if the system is stable. Thus, the system has zero disturbance error for a step disturbance.

The damping ratio is

$$\zeta = \frac{2 + K_P}{2\sqrt{5K_I}}$$

The presence of K_I enables the damping ratio to be selected without fixing the value of the dominant-time constant. For example, if the system is underdamped or critically damped, the time constant is

$$\tau = \frac{10}{2 + K_P} \quad (\text{if } \zeta \le 1)$$

The gain K_P can be picked to obtain the desired time constant, while K_I is used to set the damping ratio. Complete description of the transient response requires the numerator dynamics present in the transfer functions to be accounted for.

a. *Case 1:* If $\zeta < 1$, the real part of the roots is $-(2 + K_P)/10$, and thus the time constant is

$$\tau = \frac{10}{2 + K_P} = 0.5 \text{ sec}$$

which gives $K_P = 18$. The damping ratio of equation (4) is

$$\zeta = \frac{2 + K_P}{2\sqrt{5K_I}} = 0.707$$

Using $K_P = 18$ and squaring both sides gives

$$\frac{20^2}{4(5K_I)} = (0.707)^2 = 0.5$$

Solve for K_I to obtain $K_I = 40$.

Case 2: If $\zeta = 1$, the roots of equation (4) are both equal to $-(2 + K_P)/10$, and thus the time constant is

$$\tau = \frac{10}{2 + K_P} = 0.5$$

which gives $K_P = 18$. Using this value in the damping ratio of equation (4), we have

$$\zeta = \frac{2 + K_P}{2\sqrt{5K_I}} = \frac{10}{\sqrt{5K_I}} = 1$$

Squaring both sides gives

$$\frac{10^2}{5K_I} = 1$$

Solve for K_I to obtain $K_I = 20$.

Case 3: Since $\zeta > 1$ in this case, the expressions for the time constants of equation (4) are too complicated to be useful. The easier approach is to express the desired characteristic equation in factored form. The desired time constants are 0.5 s and 0.05 s, which correspond to the roots $s = -2$ and $s = -20$. The desired characteristic equation may thus be expressed as

$$(s+2)(s+20) = s^2 + 22s + 40 = 0$$

To compare this with equation (4), we must multiply by 5 to obtain

$$5s^2 + 110s + 200 = 0$$

Comparison of this with equation (4) shows that $K_I = 200$ and $2 + K_P = 110$, which gives $K_P = 108$. These gain values give a damping ratio of $\zeta = 110/2\sqrt{1000} = 1.74$.

b. The plots can be easily obtained with MATLAB using the `tf` and `step` functions applied to equations (1) and (3). The plots are shown in Figures 10.6.3 and 10.6.4. The design having the shortest rise time and shortest settling time is case 3 with $\zeta = 1.74$. It also has the smallest overshoot. That case, however, requires the greatest output from the actuator, as seen in Figure 10.6.4. The maximum torque values for each case are 18, 18, and 108, respectively.

c. Applying the final-value theorem to the error equation (2) with $\Omega_r(s) = 1/s^2$ and $T_d(s) = 0$ gives

$$e_{ss} = \lim_{s\to 0} sE(s) = \lim_{s\to 0} s\frac{5s^2 + 2s}{5s^2 + (2+K_P)s + K_I}\frac{1}{s^2} = \frac{2}{K_I}$$

Thus, the steady-state ramp command error is nonzero and is different for each of the three cases. It is the smallest for case 3 because that case has the largest value of K_I.

Figure 10.6.3 Step command response of a PI control system.

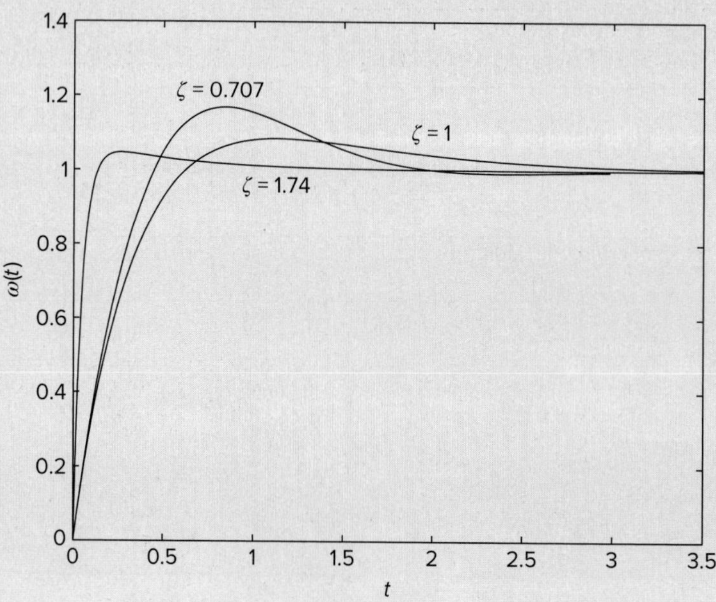

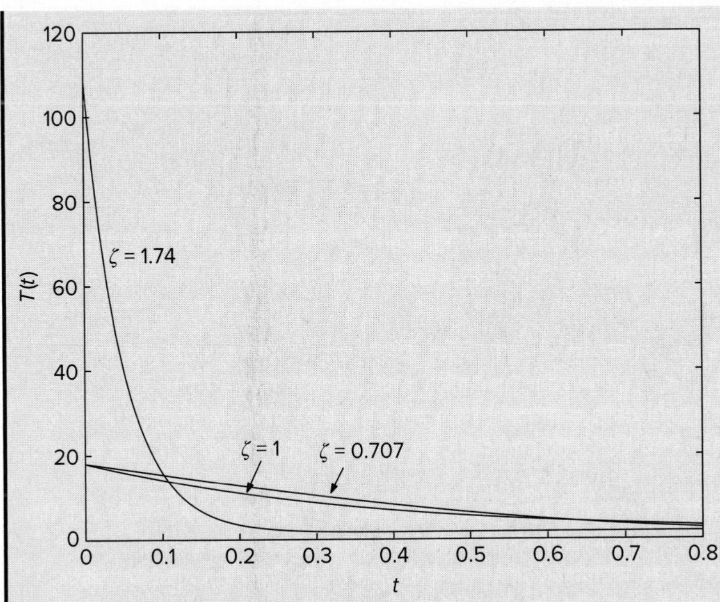

Figure 10.6.4 Actuator response of a PI control system.

The following table summarizes the results.

Case	K_P	K_I	ζ	Maximum torque	Ramp command error
1	18	40	0.707	18	0.05
2	18	20	1	18	0.1
3	108	200	1.74	108	0.005

d. The disturbance frequency response plot for each case is shown in Figure 10.6.5. The bandwidths for each case are slightly different, but the disturbance amplification for case 3 is much smaller than for the other two cases. The peak values are −26.02, −26.02, and −40.85 dB, respectively. These correspond to magnitude ratios of 0.05, 0.05, and 0.0091.

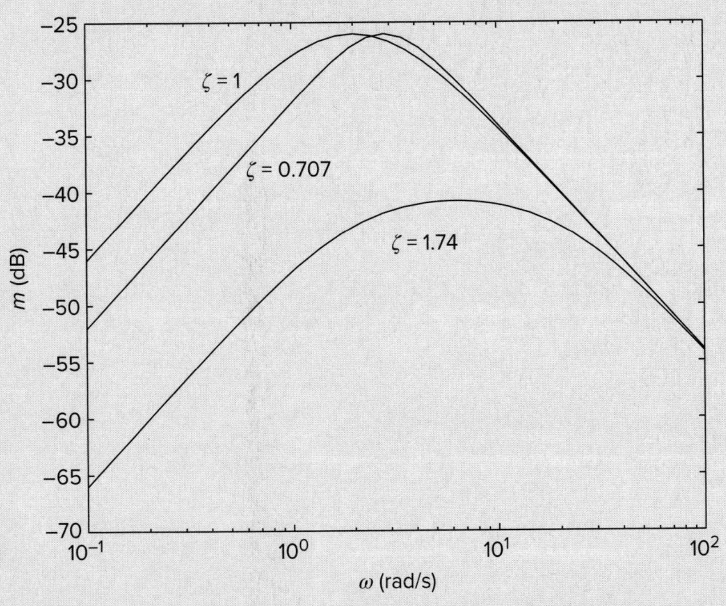

Figure 10.6.5 Frequency response of a PI control system.

EXAMPLE 10.6.4

Use of Internal Feedback

■ **Problem**

Suppose the plant to be controlled is

$$G_p(s) = \frac{1}{Is + c}$$

where $I = 5$ and $c = 2$. Figure 10.6.6 shows a proposed alternative to PI control for this plant. It uses an internal feedback loop to adjust the output of the controller. The dominant-time constant is specified to be 0.5 s.

a. Compute the required values for K_2 and K_I for each of the following three cases: (1) $\zeta = 0.707$, (2) $\zeta = 1$, and (3) the secondary time constant must be 0.05 s.

   Evaluate the steady-state command error and the steady-state disturbance error for each case given that both the command input $\omega_r(t)$ and the disturbance $T_d(t)$ are unit-step functions.

b. Plot the output response $\omega(t)$ and the actuator response $T(t)$ for each case, given that the command input $\omega_r(t)$ is a unit-step function and the disturbance $T_d(t)$ is zero.

c. Evaluate the steady-state command error for each case given that the command input $\omega_r(t)$ is a unit-ramp function and the disturbance $T_d(t)$ is zero.

d. Evaluate the frequency response characteristics of the disturbance transfer function.

■ **Solution**

a. From the figure, we obtain the output equation as follows:

$$\Omega(s) = \frac{1}{5s + 2}[T(s) - T_d(s)] \tag{1}$$

$$T(s) = \frac{K_I}{s}E(s) - K_2\Omega(s) \tag{2}$$

$$E(s) = \Omega_r(s) - \Omega(s) \tag{3}$$

Substituting equations (2) and (3) into equation (1) and multiplying both sides by $5s + 2$ gives

$$(5s + 2)\Omega(s) = T(s) - T_d(s) = \frac{K_I}{s}[\Omega_r(s) - \Omega(s)] - K_2\Omega(s) - T_d(s)$$

Solve for $\Omega(s)$:

$$\Omega(s) = \frac{K_I}{5s^2 + (2 + K_2)s + K_I}\Omega_r(s) - \frac{s}{5s^2 + (2 + K_P)s + K_I}T_d(s) \tag{4}$$

For the error equation,

$$E(s) = \Omega_r(s) - \Omega(s)$$

$$= \Omega_r(s) - \frac{K_I}{5s^2 + (2 + K_2)s + K_I}\Omega_r(s) + \frac{s}{5s^2 + (2 + K_P)s + K_I}T_d(s)$$

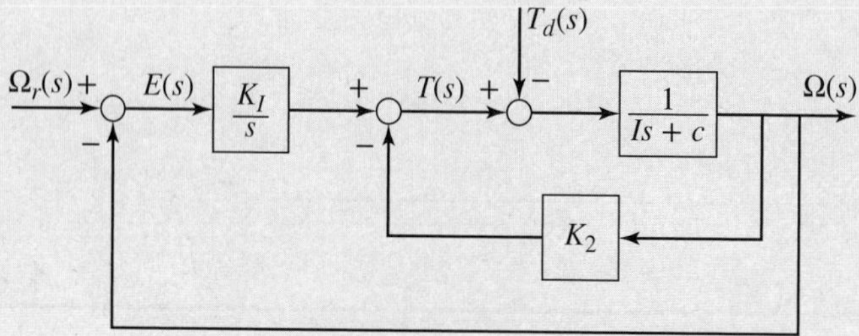

Figure 10.6.6 I action with an internal feedback loop.

Solve for $E(s)$:

$$E(s) = \frac{5s^2 + (2 + K_2)s}{5s^2 + (2 + K_2)s + K_I}\Omega_r(s) + \frac{s}{5s^2 + (2 + K_2)s + K_I}T_d(s) \qquad (5)$$

For the actuator equation,

$$T(s) = \frac{K_I}{s}E(s) - K_2\Omega(s) = \frac{K_I}{s}[\Omega_r(s) - \Omega(s)] - K_2\Omega(s)$$

Solve for $T(s)$:

$$T(s) = \frac{(5s + 2)K_I}{5s^2 + (2 + K_2)s + K_I}\Omega_r(s) + \frac{K_2 s + K_I}{5s^2 + (2 + K_2)s + K_I}T_d(s) \qquad (6)$$

The characteristic equation is

$$5s^2 + (2 + K_2)s + K_I = 0 \qquad (4)$$

This has the same form as equation (4) of Example 10.6.3 with K_2 replacing K_P. Therefore, the gain values obtained in that example can be used here. This gives the gain values shown in the following table.

Case	K_2	K_I	ζ	Maximum torque	Ramp command error
1	18	40	0.707	7.23	0.5
2	18	20	1	4.29	1
3	108	200	1.74	8.06	5.5

Although the denominators in this example are the same as those in Example 10.6.3, the numerators are different. Applying the final-value theorem to the error equation (5) with $\Omega_r(s) = 1/s$ and $T_d(s) = 0$ gives

$$e_{ss} = \lim_{s\to 0} sE(s) = \lim_{s\to 0} s\frac{5s^2 + (2 + K_2)s}{5s^2 + (2 + K_2)s + K_I}\frac{1}{s} = 0$$

for all three cases. Thus, the system has zero command error for a step command.

Applying the final-value theorem to the error equation (5) with $\Omega_r(s) = 0$ and $T_d(s) = 1/s$ gives

$$e_{ss} = \lim_{s\to 0} sE(s) = \lim_{s\to 0} s\frac{s}{5s^2 + (2 + K_2)s + K_I}\frac{1}{s} = 0$$

for all three cases. Thus, the system has zero disturbance error for a step disturbance.

b. The plots can be easily obtained with MATLAB using the `tf` and `step` functions applied to equations (1) and (3). The plots are shown in Figures 10.6.7 and 10.6.8. The rise times are much longer than for PI control, but the overshoots are much smaller. The peak torque required is 7.23, 4.29, and 8.06 for cases 1, 2, and 3, respectively. These are much less than the peak torque required with PI control.

c. Applying the final-value theorem to the error equation (5) with $\Omega_r(s) = 1/s^2$ and $T_d(s) = 0$ gives

$$e_{ss} = \lim_{s\to 0} sE(s) = \lim_{s\to 0} s\frac{5s^2 + (2 + K_2)s}{5s^2 + (2 + K_2)s + K_I}\frac{1}{s^2} = \frac{2 + K_2}{K_I}$$

Thus, the ramp command error is different for each of the three cases and much larger than for PI action.

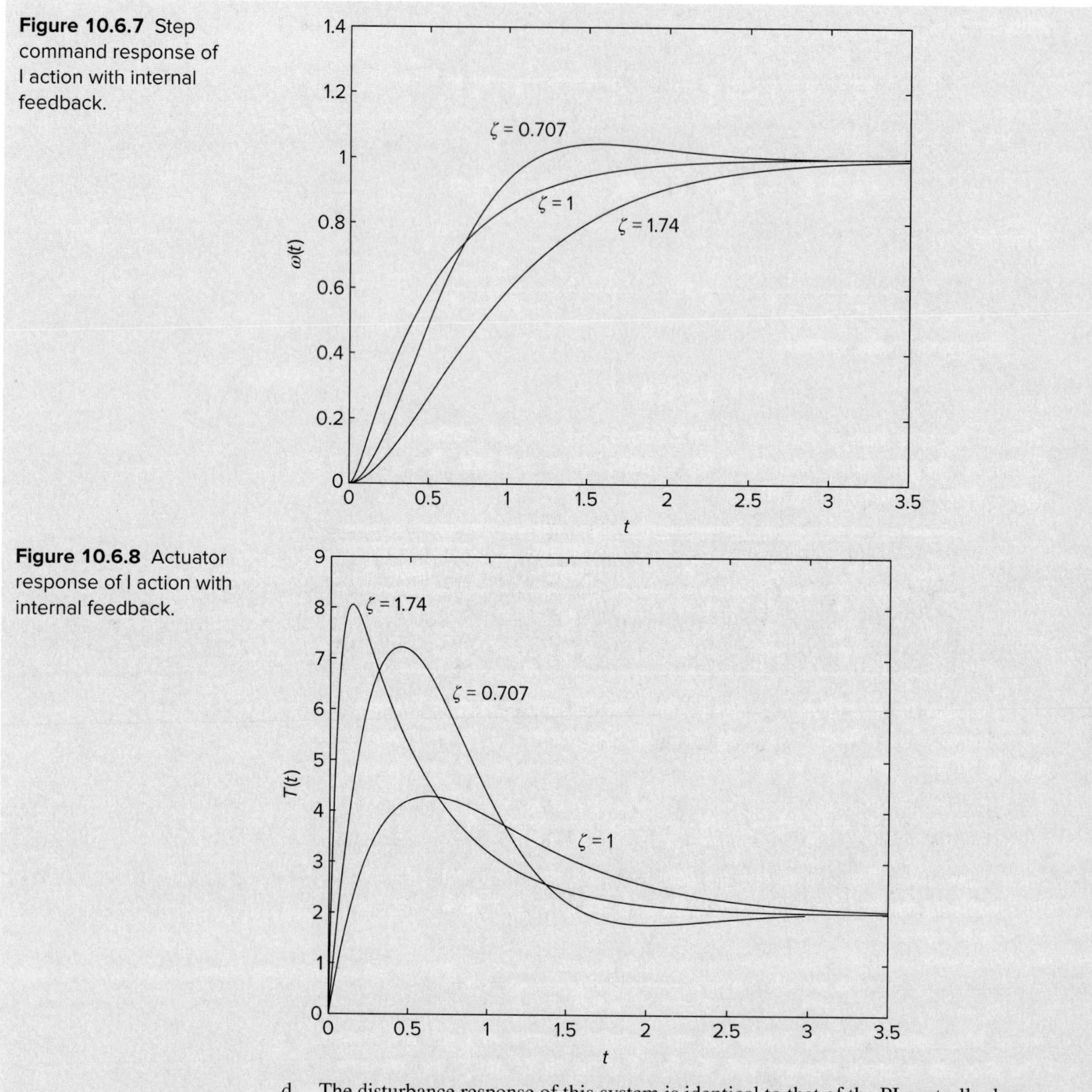

Figure 10.6.7 Step command response of I action with internal feedback.

Figure 10.6.8 Actuator response of I action with internal feedback.

d. The disturbance response of this system is identical to that of the PI controller because they both have the same disturbance transfer function.

10.7 CONTROLLING SECOND-ORDER PLANTS

We now investigate how to control a second-order plant. An example of a general form of a second-order plant without numerator dynamics is given by a rotational system of inertia I, damping c, and torsional elasticity k, whose output is angular position θ and whose inputs are the actuator torque T and a disturbance torque T_d. The equation of motion is

$$I\ddot{\theta} + c\dot{\theta} + k\theta = T(t) - T_d(t)$$

The plant transfer function is

$$G_p(s) = \frac{1}{Is^2 + cs + k}$$

This plant is stable if I, c, and k have the same sign. If there is no elasticity, the plant's transfer function is $1/s(Is + c)$, which is neutrally stable because of the root at $s = 0$.

The following examples deal with a neutrally stable plant. Applications involving a plant of the form $1/(Is^2 + cs + k)$ are treated in the chapter problems.

PD Control of a Neutrally Stable Second-Order Plant

EXAMPLE 10.7.1

■ **Problem**

PD control of a neutrally stable second-order plant is shown in Figure 10.7.1. Investigate its performance for step and ramp inputs for $c \geq 0$.

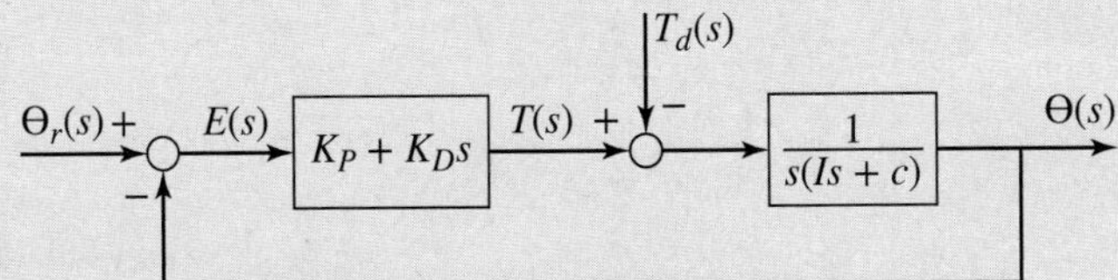

Figure 10.7.1 PD control of a second-order plant.

■ **Solution**

The output equation is

$$\Theta(s) = \frac{K_P + K_D s}{Is^2 + (c + K_D)s + K_P}\Theta_r(s) - \frac{1}{Is^2 + (c + K_D)s + K_P}T_d(s)$$

The characteristic polynomial is $Is^2 + (c + K_D)s + K_P$, and the system is stable if $c + K_D > 0$ and if $K_P > 0$. For unit-step inputs, the steady-state command response is $K_P/K_P = 1$, which is perfect, and the disturbance response is $-1/K_P$. The damping ratio is

$$\zeta = \frac{c + K_D}{2\sqrt{IK_P}}$$

If $\zeta \leq 1$, the time constant is given by

$$\tau = \frac{2I}{c + K_D}$$

For P control (with $K_D = 0$), $\zeta = c/2\sqrt{IK_P}$. Thus, introducing rate action allows the proportional gain K_P to be selected large to reduce the steady-state disturbance response, while K_D can be used to achieve an acceptable damping ratio. The rate action also helps to stabilize the system by adding damping (if $c = 0$, the system with P control is not stable).

Design of a PD Control System

EXAMPLE 10.7.2

■ **Problem**

For the system shown in Figure 10.7.1, we are given that $I = 10$ and $c = 2$. The dominant-time constant τ is specified to be 2 s, and the damping ratio is specified to be $\zeta = 1$.

a. Compute the required values for K_P and K_D. Evaluate the steady-state command error and the steady-state disturbance error given that both the command input $\Theta_r(t)$ and the disturbance $T_d(t)$ are unit-step functions.

b. Evaluate the steady-state command error for each case given that the command input $\Theta_r(t)$ is a unit-ramp function and the disturbance $T_d(t)$ is zero.
c. Evaluate the frequency response characteristics of the disturbance transfer function.
d. Discuss the actuator output as a function of time when the command input $\Theta_r(t)$ is a unit-step function and the disturbance $T_d(t)$ is zero.

■ Solution

a. From the figure we obtain the following output, error, and actuator equations.

$$\Theta(s) = \frac{K_P + K_D s}{10s^2 + (2 + K_D)s + K_P}\Theta_r(s) - \frac{1}{10s^2 + (2 + K_D)s + K_P}T_d(s) \tag{1}$$

$$E(s) = \frac{10s^2 + 2s}{10s^2 + (2 + K_D)s + K_P}\Theta_r(s) + \frac{1}{10s^2 + (2 + K_D)s + K_P}T_d(s) \tag{2}$$

$$T(s) = \frac{(10s^2 + 2s)(K_P + K_D s)}{10s^2 + (2 + K_D)s + K_P}\Theta_r(s) + \frac{K_D s + K_P}{10s^2 + (2 + K_D)s + K_P}T_d(s) \tag{3}$$

The characteristic equation is

$$10s^2 + (2 + K_D)s + K_P = 0$$

and the damping ratio is

$$\zeta = \frac{2 + K_D}{2\sqrt{10K_P}} = 1 \tag{4}$$

Since $\zeta = 1$, the expression for the time constant is

$$\tau = \frac{20}{2 + K_D} = 2 \text{ sec}$$

which gives $K_D = 8$. Using this value in equation (4) gives

$$\frac{2 + 8}{2\sqrt{10K_P}} = 1$$

which gives $K_P = 2.5$.

Applying the final-value theorem to the error equation (2) with $\Theta_r(s) = 1/s$ and $T_d(s) = 0$ gives

$$e_{ss} = \lim_{s\to 0} sE(s) = \lim_{s\to 0} s\frac{10s^2 + 2s}{10s^2 + (2 + K_D)s + K_P}\frac{1}{s} = 0$$

Thus, the system has zero command error for a step command.

Applying the final-value theorem to the error equation (2) with $\Theta_r(s) = 0$ and $T_d(s) = 1/s$ gives

$$e_{ss} = \lim_{s\to 0} sE(s) = \lim_{s\to 0} s\frac{1}{10s^2 + (2 + K_D)s + K_P}\frac{1}{s} = \frac{1}{K_P} = \frac{1}{2.5} = 0.4 \text{ rad}$$

Thus, the system has a nonzero disturbance error for a step disturbance.

b. Applying the final-value theorem to the error equation (2) with $\Theta_r(s) = 1/s^2$ and $T_d(s) = 0$ gives

$$e_{ss} = \lim_{s\to 0} sE(s) = \lim_{s\to 0} s\frac{10s^2 + 2s}{10s^2 + (2 + K_D)s + K_P}\frac{1}{s^2} = \frac{2}{K_P} = 0.8 \text{ rad}$$

Thus, the ramp command error is nonzero.

c. The disturbance transfer function is

$$\frac{\Theta_r(s)}{T_d(s)} = -\frac{1}{10s^2 + (2 + K_D)s + K_P} = -\frac{1}{10s^2 + 10s + 2.5}$$

Its bandwidth is the frequency range $0 \le \omega \le 0.32$ rad/s. Its low-frequency gain is $1/2.5 = 0.4$ or $20 \log 0.4 = -7.86$ dB. Thus, the system will not respond very much to disturbances that have a frequency content higher than 0.32 rad/s.

d. Substituting the values of K_P and K_D into the actuator equation (3) and using $\Theta_r(s) = 1/s$ and $T_d(s) = 0$, we obtain

$$T(s) = \frac{(10s^2 + 2s)(8s + 2.5)}{10s^2 + 10s + 2.5}\frac{1}{s} = \frac{80s^2 + 41s + 5}{10s^2 + 10s + 2.5} \tag{5}$$

Because the orders of the numerator and denominator are equal, this can be expressed as

$$T(s) = C_1 + \frac{C_2 s + C_3}{10s^2 + 10s + 2.5}$$

which may be arranged as a single fraction as follows.

$$T(s) = \frac{10C_1 s^2 + (10C_1 + C_2)s + 2.5C_1 + C_3}{10s^2 + 10s + 2.5} \tag{6}$$

Comparing the numerators of equations (5) and (6), we find that $C_1 = 8$, $C_2 = -39$, and $C_3 = -15$, so that

$$T(s) = 8 - \frac{39s + 15}{10s^2 + 10s + 2.5} = 8 - \frac{3.9s + 1.5}{(s + 0.5)^2}$$

This could also have been obtained with synthetic division. Using the inverse transform we obtain

$$T(t) = 8\delta(t) - 3.9e^{-0.5t} + 0.45te^{-0.5t}$$

where $\delta(t)$ is the unit-impulse function. Therefore, the actuator output predicted by the model will contain an impulse of strength 8 at $t = 0$. Of course, this is impossible physically, and so we must view these results with skepticism. The impulse is caused by the derivative term $K_D s$ in the control algorithm.

10.7.1 D ACTION AND STEP INPUTS

D action can be used to improve the system's speed of response, but it is best to avoid using it in the main controller if an equivalent effect can be obtained with a compensator elsewhere in the system. When a change in command input occurs, the error signal and therefore the controller are instantaneously affected. If the change is sudden, as with a step input, the physical limitations of the differentiating device mean that an accurate derivative is not computed, and the actual performance of the system will be degraded relative to the ideal performance predicted by the mathematical model.

In addition, as seen in Example 10.7.2, with a step command, the D action calls for an impulse in the actuator output, which is physically impossible. Note that in Example 10.7.2 a step *disturbance* does not produce an impulse in the actuator output because the disturbance passes through the plant's transfer function before entering the control action block via the feedback loop. Because the plant's transfer function contains an integration process, its output will not include an impulse. In conclusion, we must be careful in interpreting the transient response of a system containing a perfect model of a physical or numerical differentiator, whenever a step function is applied to such a differentiator model.

We have seen some devices, such as op-amp circuits, for differentiating the error signal, but it is impossible to construct a differentiating device that performs perfectly. Therefore, the interpretation of the step response must be considered with care. Note that this caution is not necessary in interpreting the step response of a controller with numerator dynamics due to I action, because in that case a step input is not applied to a differentiator.

If D action is required, a good design practice is to place the differentiator at a point in the loop where the signals are more slowly varying. Then the behavior of the differentiator will more closely correspond to its behavior as predicted by the mathematical model. A frequent choice for a differentiator location is in a feedback loop. The output is usually the result of several integrations and therefore will exhibit behavior that is smoothed and slowly varying with respect to the other signals in the system. A common example of this approach is in position control. In addition to the position sensor, a velocity sensor such as a tachometer is used to provide an internal feedback loop. In Example 10.7.3 we will see how this scheme gives the equivalent of derivative action without producing numerator dynamics or an impulse in the actuator output.

EXAMPLE 10.7.3

Velocity Feedback Compensation

■ Problem

Consider the system shown in Figure 10.7.2 where $I = 10$ and $c = 2$. The output of the feedback sensor, which measures the rate of change $\omega = \dot{\theta}$ of the output, is multiplied by the gain K_2, whose value must be selected. Analyze the system and compare its performance with that of PD control for the specifications $\tau = 2$ s and $\zeta = 1$.

■ Solution

The output equation is

$$\Theta(s) = \frac{K_P}{10s^2 + (2 + K_2)s + K_P}\Theta_r(s) - \frac{1}{10s^2 + (2 + K_2)s + K_P}T_d(s)$$

Note that this command transfer function does not possess numerator dynamics, unlike the PD system. The disturbance transfer function is identical to that of the PD system.

The error equation is

$$E(s) = \frac{10s^2 + (2 + K_2)s}{10s^2 + (2 + K_2)s + K_P}\Theta_r(s) + \frac{1}{10s^2 + (2 + K_2)s + K_P}T_d(s)$$

The actuator equation is

$$T(s) = \frac{K_P(10s^2 + 2s)}{10s^2 + (2 + K_2)s + K_P}\Theta_r(s) + \frac{K_2 s + K_P}{10s^2 + (2 + K_2)s + K_P}T_d(s)$$

The characteristic equation is the same as that for PD control with K_2 replacing K_D, so we may use the results of Example 10.7.2 to obtain $K_P = 2.5$ and $K_2 = 8$. The numerators of some of the transfer functions differ from those of PD control, however.

Applying the final-value theorem to the error equation with $\Theta_r(s) = 1/s$ and $T_d(s) = 0$ gives $e_{ss} = 0$. Thus, the system has zero command error for a step command. With $\Theta_r(s) = 0$

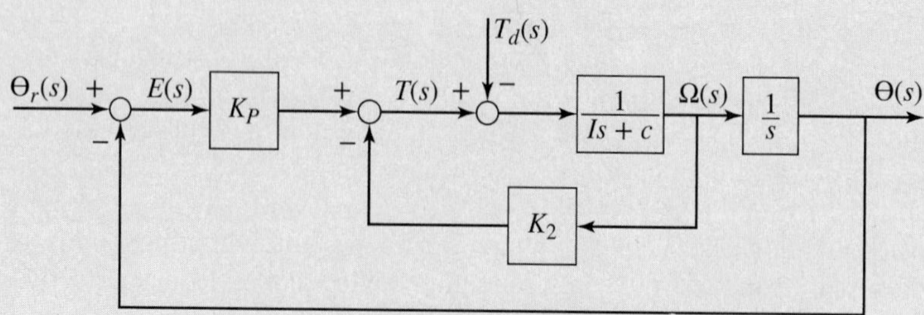

Figure 10.7.2 P action with rate feedback.

and $T_d(s) = 1/s$ we obtain

$$e_{ss} = \lim_{s\to 0} sE(s) = \lim_{s\to 0} s\frac{1}{10s^2 + (2+K_2)s + K_P}\frac{1}{s} = \frac{1}{K_P} = 0.4 \text{ rad}$$

Thus, the system has the same nonzero disturbance error as PD control for a step disturbance.

For a ramp command, the error equation with $\Theta_r(s) = 1/s^2$ and $T_d(s) = 0$ gives

$$e_{ss} = \lim_{s\to 0} sE(s) = \lim_{s\to 0} s\frac{10s^2 + (2+K_2)s}{10s^2 + (2+K_2)s + K_P}\frac{1}{s^2} = \frac{2+K_2}{K_P} = 4 \text{ rad}$$

Thus, the ramp command error is greater than with the PD control system.

The actuator equation with $\Theta_r(s) = 1/s$ and $T_d(s) = 0$ gives

$$T(s) = \frac{K_P(10s+2)}{10s^2 + (2+K_2)s + K_P} = \frac{2.5s + 0.5}{(s+0.5)^2}$$

Following the same procedure used in Example 10.7.2, we obtain

$$T(t) = (2.5 - 0.75t)e^{-0.5t}$$

The actuator output does not contain an impulse, as opposed to the result for PD control.

I action might be included when we need to reduce or eliminate steady-state error. However, this results in a third-order model to analyze. For such models we do not have convenient formulas to use for overshoot, and for the time constant, damping ratio, and natural frequency of the dominant roots.

PID Control of a Neutrally Stable Second-Order Plant

EXAMPLE 10.7.4

■ Problem

The designs in Examples 10.7.1 and 10.7.3 have a non-zero-error for a step disturbance. We now add integral action to the design in Example 10.7.1. The resulting PID control of a neutrally stable second-order plant is shown in Figure 10.7.3. Suppose that $I = 10$ and $c = 2$. The performance specifications require that $\tau = 2$ s and $\zeta = 0.707$. Compute the gain values, and discuss the system's frequency response to a disturbance.

■ Solution

The output equation is

$$\Theta(s) = \frac{K_D s^2 + K_P s + K_I}{10s^3 + (2+K_D)s^2 + K_P s + K_I}\Theta_r(s) - \frac{s}{10s^3 + (2+K_D)s^2 + K_P s + K_I}T_d(s)$$

The characteristic polynomial is $10s^3 + (2+K_D)s^2 + K_P s + K_I$ and system is stable if $2+K_D > 0$, $K_P > 0$, and $K_I > 0$, and if

$$(2+K_D)K_P - 10K_I > 0$$

Recall that for PI control, K_P needs to be large to achieve stability. The presence of K_D relaxes this requirement somewhat.

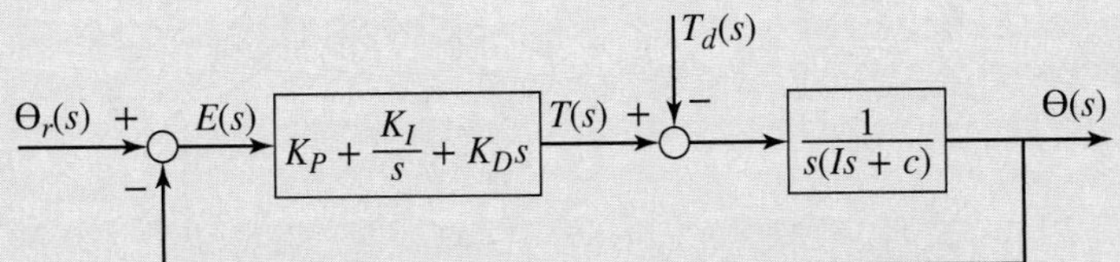

Figure 10.7.3 PID control of a second-order plant.

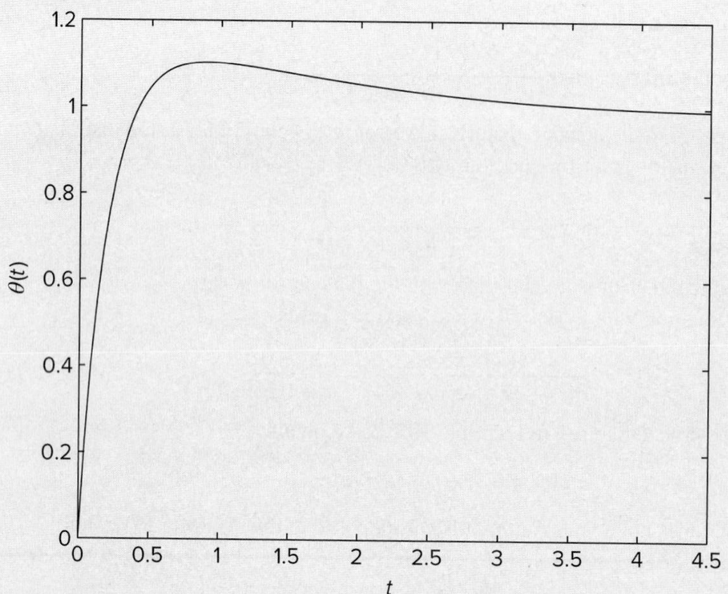

Figure 10.7.4 Step command response of a PID control system.

The steady-state errors are zero for step inputs, and the transient response can be improved relative to that of PD action, because three of the coefficients of the characteristic equation can be selected.

Because this system is third order, we do not have formulas to use for the damping ratio and the time constant. In addition, we must interpret the specifications to apply to the *dominant* root or root pair, and must choose the secondary root somewhat arbitrarily. The values $\zeta = 0.707$ and $\tau = 2$ s correspond to the dominant-root pair $s = -0.5 \pm 0.5j$. The third root must be less than -0.5 so that it will not be dominant. We arbitrarily select the third root to be $s = -5$. We can later investigate other choices (this is done in one of the chapter problems). These three roots correspond to the polynomial equation

$$(s+5)\left[(s+0.5)^2+(0.5)^2\right] = (s+5)\left(s^2+s+0.5\right) = s^3+6s^2+5.5s+2.5 = 0$$

To compare this with the system's characteristic equation, we multiply it by 10.

$$10s^3+60s^2+55s+25 = 0$$

Compare this with the system's characteristic equation:

$$10s^3+(2+K_D)s^2+K_Ps+K_I$$

Thus, $K_P = 55$, $K_I = 25$, and $2+K_D = 60$, or $K_D = 58$. The response to a unit-step command is shown in Figure 10.7.4. The overshoot is 10.3% and the 2% settling time is 2.89 s. This is quite less than the settling time of 8 s predicted from the dominant-time constant of 2 s. The difference is due to the first- and second-order numerator dynamics of the transfer function.

The resulting disturbance transfer function for this system is

$$\frac{\Theta(s)}{T_d(s)} = \frac{-s}{10s^3+60s^2+55s+25}$$

The frequency response plot is shown in Figure 10.7.5. It shows that the system attenuates disturbance inputs by a factor of $m = -34.1$ dB or more. This corresponds to an amplitude reduction of $M = 10^{-34.1/20} = 0.02$. The system rejects by an even greater amount any disturbances whose frequencies are outside the bandwidth of $0.363 \leq \omega \leq 1.33$ rad/s. Compare this performance with the PD control system of Example 10.7.2, which does not have as great an attenuation.

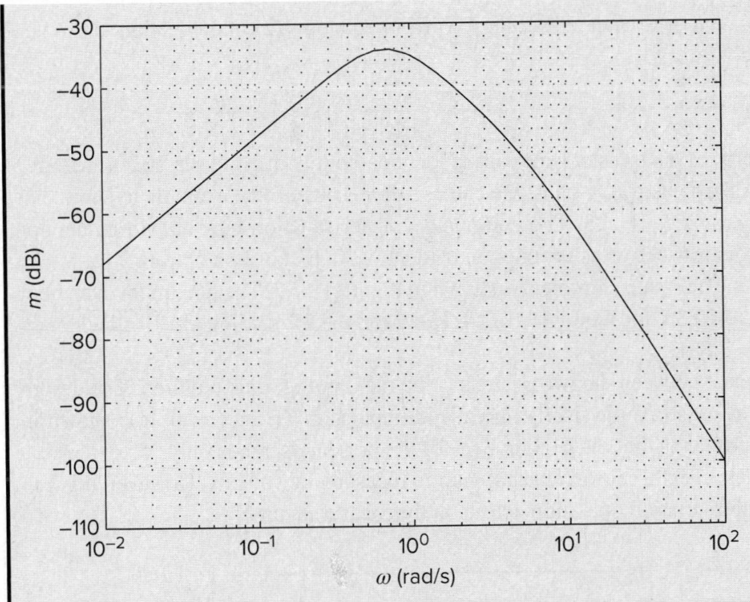

Figure 10.7.5 Frequency response of a PID control system.

Because of the D action, a step command input to a PID controller will cause an impulse in the actuator response, and we must therefore exercise caution in interpreting the transient response due to step commands. The use of an internal feedback loop (velocity feedback compensation) avoids this difficulty.

EXAMPLE 10.7.5

Velocity Feedback with a Second-Order Plant

■ Problem

Note that the gains K_P and K_D produce numerator dynamics in the command transfer function of Example 10.7.4 and that the K_D term produces an unrealistic impulse in the torque $T(t)$ when the command is a step function. We can eliminate the numerator dynamics by measuring the angular velocity ω and using this measurement to modify the output of the controller. This replaces the derivative action. The proportional action is preserved by using an additional internal-feedback loop with the measurement of θ. The diagram is shown in Figure 10.7.6. Compare its performance with the PID system of Example 10.7.4 using the same performance specifications.

■ Solution

Starting from the output $\Theta(s)$ and working to the left, we can write

$$\Theta(s) = \frac{1}{s(10s+2)}\left[T(s) - T_d(s)\right] \tag{1}$$

$$T(s) = \frac{K_I}{s}\left[\Theta_r(s) - \Theta(s)\right] - K_1\Theta(s) - K_2\Omega(s) \tag{2}$$

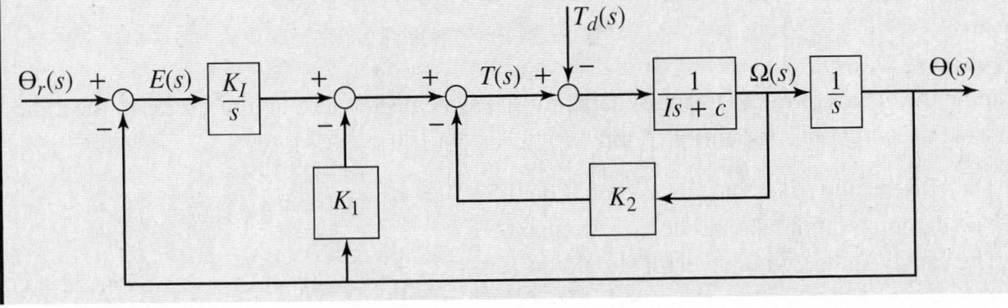

Figure 10.7.6 Velocity feedback with integral control for a second-order plant.

Noting that $\Omega(s) = s\Theta(s)$, we can substitute $T(s)$ from equation (2) into equation (1). After some algebra, we obtain

$$\Theta(s) = \frac{K_I}{10s^3 + (2 + K_2)s^2 + K_1 s + K_I}\Theta_r(s) - \frac{s}{10s^3 + (2 + K_2)s^2 + K_1 s + K_I}T_d(s)$$

The denominator of the transfer functions is identical to that of Example 10.7.4, with K_1 replacing K_P and K_2 replacing K_D. Thus, we can use the results of that example to show that $K_1 = 55$, $K_2 = 58$, and $K_I = 25$. The command transfer function now has no numerator dynamics, and the disturbance transfer function is identical to that of Example 10.7.4.

The response to a unit-step command is shown in Figure 10.7.7. The maximum overshoot is 4% as compared to 10.3% for Example 10.7.4. However, the 2% settling time is now longer ($t = 8.64$ versus $t = 2.89$).

The big difference between the two designs is the maximum torque required. The design of Example 10.7.4 results in a physically unrealistic impulse in $T(t)$ at $t = 0$. It is shown in one of the chapter problems that the peak in $T(t)$ for the present design is only 25.

This design uses an extra sensor, a tachometer to measure the velocity. However, this can be replaced by a differentiation operation acting on the measurement of θ.

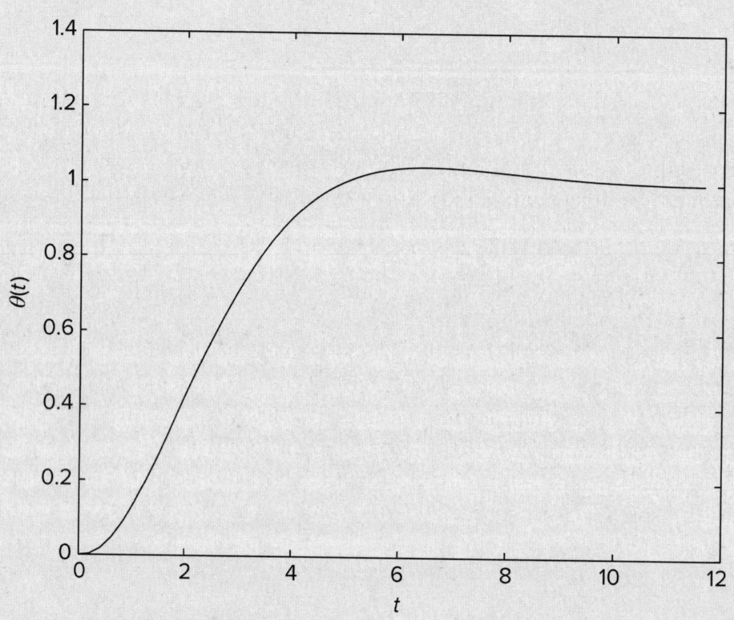

Figure 10.7.7 Unit-step command response for Example 10.7.5.

10.8 ADDITIONAL EXAMPLES

EXAMPLE 10.8.1 PD Control

■ Problem

Compute the values of the PD control gains for the system shown in Figure 10.8.1 to meet the following specifications for unit-step inputs:

1. The closed-loop time constant τ should be 0.5 s.
2. The damping ratio ζ should be no less than 0.707.
3. The steady-state command error should be zero.

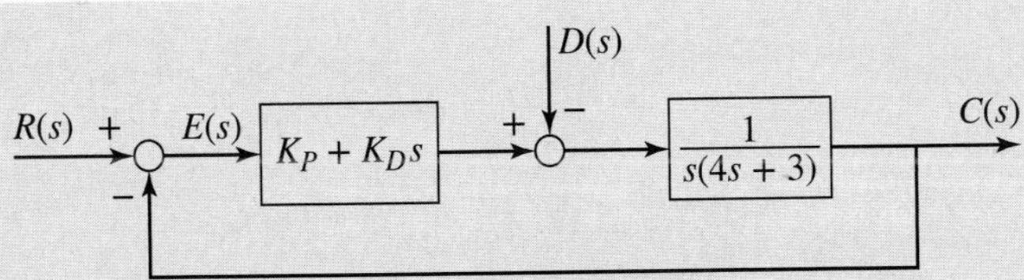

Figure 10.8.1 PD control of a second-order plant.

4. The magnitude of the steady-state disturbance error should be as small as possible, as long as the first three specifications are satisfied.

■ Solution

The error equation is

$$E(s) = \frac{4s^2 + 3s}{4s^2 + (3 + K_D)s + K_P} R(s) + \frac{1}{4s^2 + (3 + K_D)s + K_P} D(s)$$

The characteristic equation is $4s^2 + (3 + K_D)s + K_P = 0$. Since $\zeta < 1$, we may express τ as

$$\tau = \frac{8}{3 + K_D} = 0.5 \text{ s}$$

which gives $K_D = 13$. With this value for K_D, the expression for ζ becomes

$$\zeta = \frac{3 + K_D}{2\sqrt{4K_P}} = \frac{16}{2\sqrt{4K_P}} = \frac{4}{\sqrt{K_P}}$$

From the error equation we see that the steady-state command error is 0, and the steady-state disturbance error is $1/K_P$. Thus, to minimize the error, we should make K_P as large as possible, as long as we keep $\zeta \geq 0.707$. Thus, we set $\zeta = 0.707$ to obtain

$$\zeta = \frac{4}{\sqrt{K_P}} = 0.707$$

This gives $K_P = 32$. So the solution is $K_D = 13$ and $K_P = 32$.

Velocity Feedback

EXAMPLE 10.8.2

■ Problem

Determine the values of the gains in the system shown in Figure 10.8.2 to satisfy the following specifications for unit-step inputs, and compute the maximum percent overshoot.

1. The closed-loop time constant τ should be no greater than 0.05 s.
2. The steady-state command error should be zero.
3. The steady-state disturbance error should be no greater than 0.001.
4. To minimize cost, the gains should be as small as possible as long as the first three specifications are satisfied.

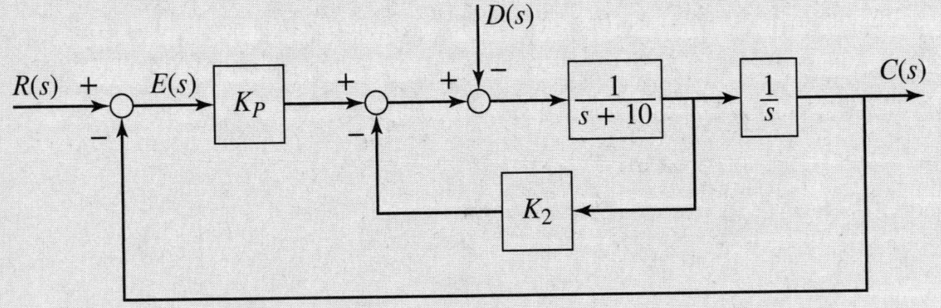

Figure 10.8.2 P action with internal feedback.

■ Solution

The error equation is

$$E(s) = \frac{s^2 + (10 + K_2)s}{s^2 + (10 + K_2)s + K_P} R(s) + \frac{1}{s^2 + (10 + K_2)s + K_P} D(s)$$

For unit-step inputs, the steady-state command error is 0 and the steady-state disturbance error is $1/K_P$. Thus, to satisfy the third specification, K_P may be no less than 10^3.

Assuming that $\zeta \leq 1$, the expression for the time constant is

$$\tau = \frac{2}{10 + K_2}$$

To minimize K_2, we must set τ equal to its largest permissible value, 0.05 s. This gives $K_2 = 30$. Thus, the tentative solution is $K_P = 1000$ and $K_2 = 30$, which is valid if $\zeta \leq 1$. So we must check ζ:

$$\zeta = \frac{10 + K_2}{2\sqrt{K_P}} = 0.632$$

Thus, the solution is valid.

The resulting overshoot is

$$M_{\%} = 100e^{-\pi\zeta/\sqrt{1-\zeta^2}} = 7.7\%$$

EXAMPLE 10.8.3

Control of an Unstable System

■ Problem

It is desired to stabilize the unstable plant shown in Figure 10.8.3 so that the time constant and damping ratio of the dominant root are $\tau = 0.1$ sec and $\zeta = 0.707$.

a. It is desired to have zero steady-state command error for a unit-step input. Determine the required values of the PID gains.

b. Since D action and I action increase the overshoot, can we eliminate them and still satisfy the specifications?

■ Solution

a. With PID action the error equation is

$$E(s) = \frac{s^3 - 4s}{s^3 + K_D s^2 + (K_P - 4)s + K_I} R(s)$$

This shows that if the system is stable, the steady-state command error is zero for a unit-step input. The characteristic equation is

$$s^3 + K_D s^2 + (K_P - 4)s + K_I = 0 \tag{1}$$

The Routh-Hurwitz criterion shows that the system is stable if and only if $K_D > 0$, $K_I > 0$, $K_P > 4$, and $K_D(K_P - 4) > K_I$. Note that we can place all three roots where we want because the last three coefficients are independent functions of the gains. The

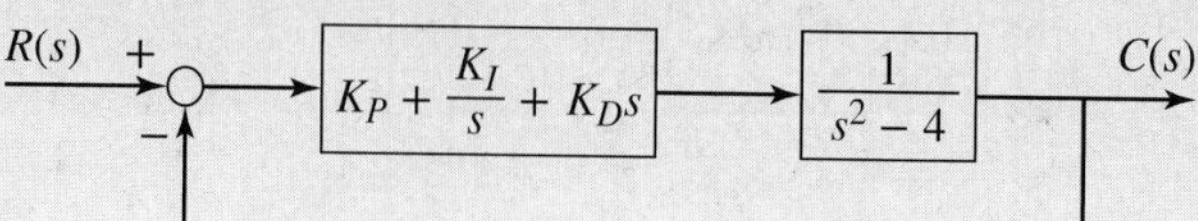

Figure 10.8.3 PID control of an unstable plant.

specifications require that the dominant roots be $s = -10 \pm 10j$. Choosing the third root to lie to the left, say at $s = -20$, the characteristic equation must be

$$(s+20)\left[(s+10)^2 + 10^2\right] = s^3 + 40s^2 + 600s + 4000 = 0$$

Comparing this with equation (1) shows that $K_D = 40$, $K_P = 604$, and $K_I = 4000$.

b. If $K_D = 0$ there will be no s^2 term in the characteristic polynomial, and thus the system will be unstable. So D action is needed for stability.

If we use P and D action but not I action, then setting $K_I = 0$ gives the error equation

$$E(s) = \frac{s^2 - 4}{s^2 + K_D s + K_P - 4} R(s)$$

The system is stable if and only if $K_D > 0$ and $K_P > 4$. If so, the steady-state command error for a unit-step input is

$$e_{ss} = -\frac{4}{K_P - 4} = \frac{4}{4 - K_P}, \quad K_P > 4$$

The error is nonzero for finite values of K_P, so I action is needed.

To achieve the desired roots $s = -10 \pm 10j$ with PD action requires that $K_D = 20$ and $K_P = 204$. The resulting steady-state error is $-1/150$. The negative error means that the steady-state response is $1 + 1/150$, which is *above* the desired value of 1.

Simulation of the PID and PD designs shows that the PID system has an overshoot of 24.6% while the overshoot of the PD system is 20%, but it has nonzero steady-state error.

Vehicle Stabilization

EXAMPLE 10.8.4

■ Problem

Figure 10.8.4 illustrates a vehicle moving through the atmosphere in a horizontal plane. The vehicle attitude θ is inherently unstable because the aerodynamic forces do not act through the mass center G. Instead they act through point P, the center of pressure, thus creating a net moment that tends to rotate the vehicle in the positive θ direction.

We want to control the attitude by controlling the elevator angle ϕ, which exerts a torque $T = B\phi$ about G, where B is a known positive constant. The equation of motion is

$$I\ddot{\theta} - LC_n\theta = T$$

where I is the inertia, C_n is the normal-force coefficient, and L is the distance between G and P.

Develop a control algorithm to control ϕ to stabilize the vehicle.

■ Solution

This problem is different than Example 10.8.3 because the desired value of the output θ is zero. Thus, the command transfer function has no use here because the command input is zero. It is

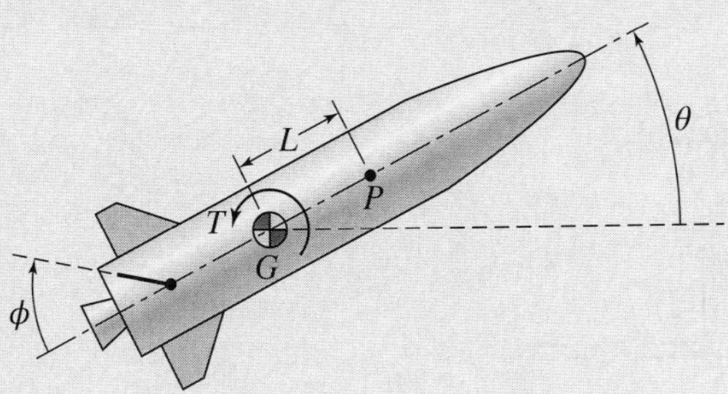

Figure 10.8.4 Vehicle attitude stabilization.

typical of many control applications where we simply need to stabilize the system to make the output zero. Instead of using the transfer function, we will work directly with the equation of motion.

Substituting $T = B\phi$ into the equation of motion gives

$$I\ddot{\theta} - LC_n\theta = B\phi \tag{1}$$

Stability requires an equation of the form

$$I\ddot{\theta} + c\dot{\theta} + k\theta = 0$$

where c and k must be positive. From this we see that stability requires that a $\dot{\theta}$ term be present. This suggests that the control algorithm must include D action, $K_D\dot{\theta}$. For the coefficient of θ to be positive, we need P action, $K_P\theta$. Thus, the control algorithm should be

$$\phi = -K_P\theta - K_D\dot{\theta}$$

Substituting this into equation (1) and rearranging gives

$$I\ddot{\theta} + BK_D\dot{\theta} + (BK_P - LC_n)\theta = 0$$

which will be stable if $K_D > 0$ and $BK_P > LC_n$. The gains K_P and K_D can be selected to give the desired time constant and damping ratio.

EXAMPLE 10.8.5 A Liquid-Level Control System

■ Problem

Design a control system to keep the height h_2 in Figure 10.8.5 near a desired value by controlling the volume inflow rate q_1. The volume flow rate q_d is a disturbance. Assume that sensors are available to measure both heights h_1 and h_2. Use the following numerical values:

$$A_1 = 1 \qquad A_2 = 2$$

$$\alpha_1 = \frac{R_1}{g} = 4 \qquad \alpha_2 = \frac{R_2}{g} = 1$$

We require zero steady-state command error for a step input. The dominant-time constant must be 2 sec to achieve a 2% settling time of approximately 8 sec, and the damping ratio of the dominant roots must be $\zeta = 0.707$.

Propose a control structure, compute the required gain values, and evaluate the resulting steady-state disturbance error if q_d is a unit step.

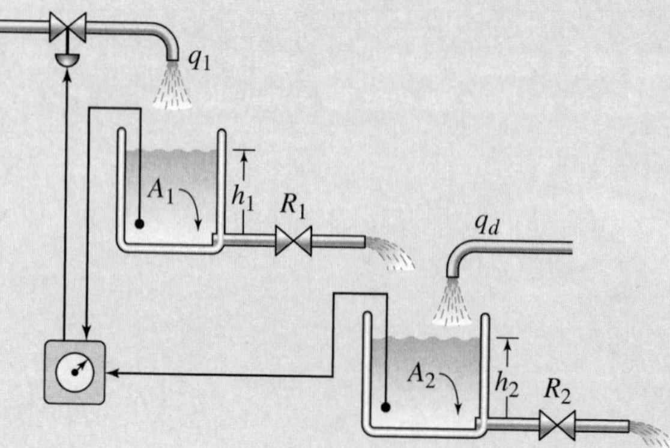

Figure 10.8.5 Liquid-level control.

■ **Solution**

Using a method similar to that of Example 7.4.3, we find that the model for tank 1 is

$$\rho A_1 \dot{h}_1 = \rho q_1 - \frac{\rho g}{R_1} h_1$$

The model for tank 2 is

$$\rho A_2 \dot{h}_2 = \rho q_d + \frac{\rho g}{R_1} h_1 - \frac{\rho g}{R_2} h_2$$

The mass density ρ cancels out of the equations, and they can be rearranged as

$$\frac{R_1}{g} A_1 \dot{h}_1 = \frac{R_1}{g} q_1 - h_1$$

$$\frac{R_1}{g}\frac{R_2}{g} A_2 \dot{h}_2 = \frac{R_1}{g}\frac{R_2}{g} q_d + \frac{R_2}{g} h_1 - \frac{R_1}{g} h_2$$

Using the definitions of α_1 and α_2, these equations can be expressed as

$$\alpha_1 A_1 \dot{h}_1 = \alpha_1 q_1 - h_1$$

$$\alpha_1 \alpha_2 A_2 \dot{h}_2 = \alpha_1 \alpha_2 q_d + \alpha_2 h_1 - \alpha_1 h_2$$

Figure 10.8.6a shows the block diagram of these equations. Combining the two inner loops gives the diagram shown in part (b).

More than one control structure can be used to solve this problem, but because measurements of h_1 and h_2 are available, we choose to use two feedback loops based on these measurements. In addition, the requirement for zero steady-state error suggests that the main controller use integral action. The result is shown in Figure 10.8.7.

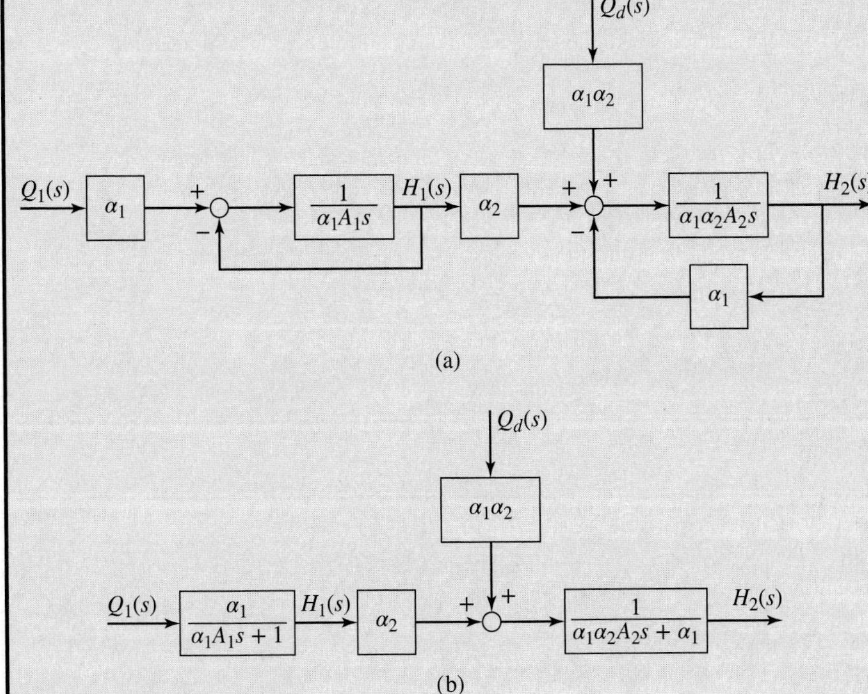

Figure 10.8.6 Model simplification for a liquid-level system.

Figure 10.8.7 Block diagram of a liquid-level control system using two feedback loops.

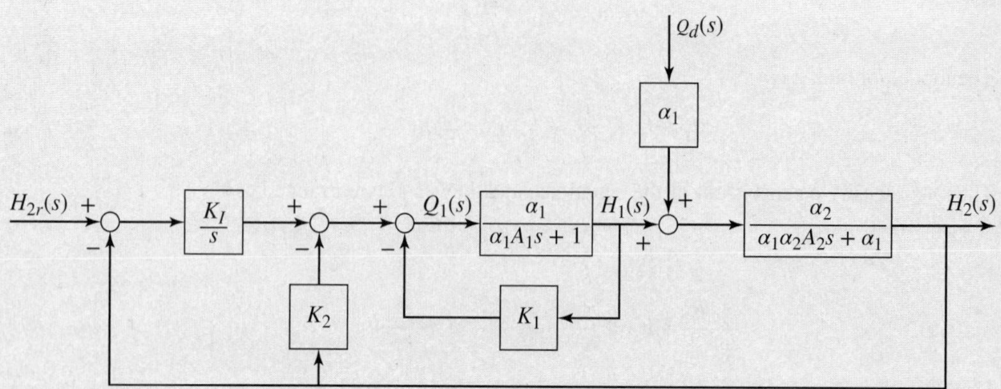

Using the given parameter values, the command transfer function is found to be

$$\frac{H_2(s)}{H_{2r}(s)} = \frac{K_I}{8s^3 + (6 + 8K_1)s^2 + (1 + 4K_1 + K_2)s + K_I}$$

For a step command input, the final-value theorem shows that $h_{2ss} = h_{2r}$ for any stable values of the gains. To satisfy the transient-response requirement, the dominant roots must be $s = -0.5 \pm 0.5j$. The third root should be chosen to lie to the left of these roots. Arbitrarily choosing $s = -5$ we can express the characteristic equation as

$$8\{[(s + 0.5)^2 + (0.5)^2](s + 5)\} = 8s^3 + 48s^2 + 44s + 20 = 0$$

Comparing the coefficients with those of the denominator of the transfer function, we obtain

$$6 + 8K_1 = 48 \qquad 1 + 4K_1 + K_2 = 44 \qquad K_I = 20$$

which gives $K_I = 20$, $K_1 = 21/4$, and $K_2 = 22$. Simulation shows that the unit-step response has a 2% settling time of 8.64 sec, which is close to the value of 8 sec predicted from the dominant roots. The difference is due to the existence of a third root.

The disturbance transfer function is

$$\frac{H_2(s)}{Q_d(s)} = \frac{4s^2 + (1 + 4K_1)s}{8s^3 + (6 + 8K_1)s^2 + (1 + 4K_1 + K_2)s + K_I}$$

and so the steady-state disturbance error for a unit step is 0.

EXAMPLE 10.8.6 The Effect of Gain on Current Demand

■ **Problem**

The block diagram of a speed-control system is shown in Figure 10.8.8. Determine (a) the transfer functions of the current and (b) obtain the current response to a step input of magnitude 104.7 rad/s (1000 rpm), using the parameter values given below.

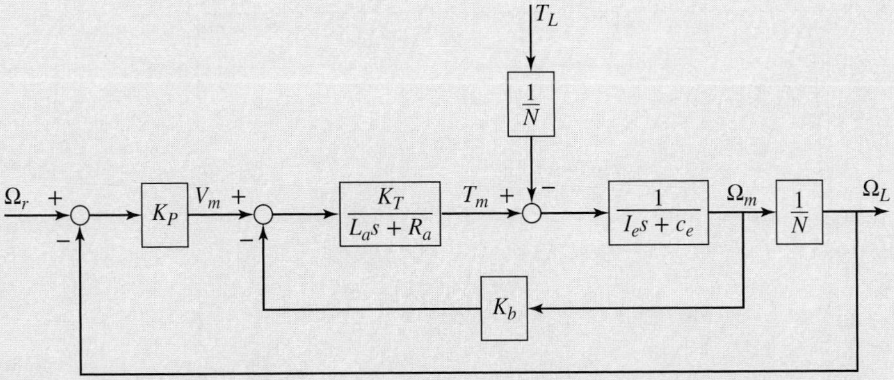

Figure 10.8.8 A speed-control system.

■ Solution

(a) After some algebra we obtain the equation for the motor torque from the block diagram.

$$K_T = K_b = 0.04\ \mathrm{N \cdot m/A} \qquad N = 1.5 \qquad K_P = 0.63$$

$$I_e = 1.802 \times 10^{-3}\ \mathrm{kg \cdot m^2} \qquad c_e = 4.444 \times 10^{-4}\ \mathrm{N \cdot m \cdot s/rad}$$

$$R_a = 0.6\ \Omega \qquad L_a = 2 \times 10^{-3}\ \mathrm{H}$$

$$T_m(s) = NK_PK_T \frac{L_aI_es^2 + (R_aI_e + c_eL_a)s + R_ac_e}{(L_as + R_a)D(s)}\Omega_r(s) + \frac{K_T(K_P + NK_b)/N}{D(s)}T_L(s) \qquad (1)$$

where

$$D(s) = NL_aI_es^2 + N(R_aI_e + c_eL_a)s + NR_ac_e + NK_TK_b + K_PK_T$$

The current is found from

$$I_a(s) = \frac{1}{K_T}T_m(s)$$

Equation (1) should raise our suspicions because the denominators of the two transfer functions $T_m(s)/\Omega_r(s)$ and $T_m(s)/T_L(s)$ differ by the factor $L_as + R_a$. Always keeping in mind the physics of the problem, we observe that there are only two ways energy can be stored in this system (as electromagnetic energy in the inductor L_a and as kinetic energy in the inertia I_e), so the system model should be second order. However, the denominator of the transfer function $T_m(s)/\Omega_r(s)$ appears to be third order. The only way the denominators could be the same is if the term $L_as + R_a$ is canceled by an identical term in the numerator. If this is the case, we could express the numerator as

$$(as + b)(L_as + R_a) = aL_as^2 + (bL_a + aR_a)s + bR_a$$

Comparing the coefficients with those of the numerator of equation (1), we see that $a = I_e$ and $b = c_e$. Thus, our suspicion is confirmed, and we may write equation (1) as

$$T_m(s) = \frac{NK_PK_T(I_es + c_e)}{D(s)}\Omega_r(s) + \frac{K_T(K_P + NK_b)/N}{D(s)}T_L(s)$$

Another way to detect the cancellation of the factor $L_as + R_a$ is to obtain the partial-fraction expansion for either the free or forced response. The expansion coefficient (the residue) corresponding to the factor $s + R_a/L_a$ would be zero and thus this factor has no influence on the response. A zero residue indicates that a denominator factor has been canceled by a numerator factor.

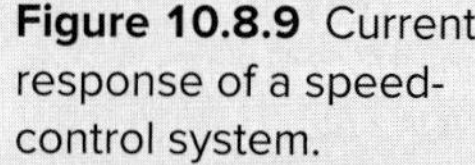

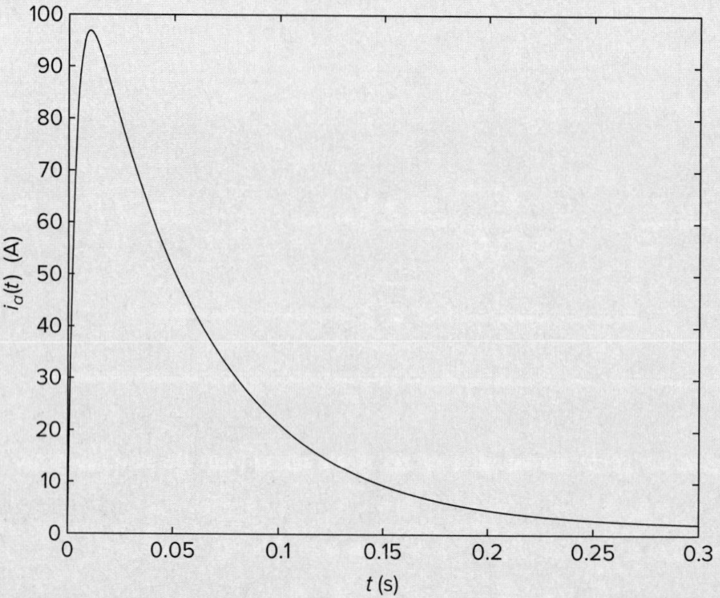

Figure 10.8.9 Current response of a speed-control system.

(b) Use the transfer function $I_a(s)/\Omega_r(s)$ with $\Omega_r(s) = 104.7/s$ to compute the required armature current as a function of time. This can be done analytically, using the methods of Chapter 2, or numerically, using the methods of Chapter 5. The result is shown in Figure 10.8.9. The plot shows that the motor will require 96 A! This is a large value, as most motors with similar parameter values have a demagnetization current of less than 50 A. We can reduce the maximum current by accepting a steady-state speed error greater than 10%, by using a smaller value for K_P. For example, a speed error of 20% with $K_P = 0.28$ results in a maximum current of 45 A.

One way to reduce the maximum required current is to use a speed command that increases slowly, such as a ramp function. The step and impulse functions are the most severe inputs that can be applied to a system, because they change their value instantaneously. In many systems it is physically difficult to apply such an input, but these functions are often used in analysis because they result in simpler mathematics.

Suppose $K_P = 0.63$, which corresponds to a 10% speed error. Consider the command input ω_r that is a ramp for $0 \leq t \leq t_1$, and for $t > t_1$ is a constant value of 104.7 rad/s (1000 rpm), the desired speed. By experimenting with the value of the time t_1 while examining the resulting maximum current, we can arrive at a value for t_1 that results in a maximum current of less than a desired value. The result for $t_1 = 0.5$ s is shown in Figure 10.8.10, which can be obtained with MATLAB. The top graph shows the current and the bottom graph shows the modified-ramp command input and the resulting load speed (assuming that $T_L = 0$). The speed reaches its steady-state value in about 0.6 s, as compared to 0.054 s for the step input. Now, however, the maximum current is 14 A, which is much less than the 96 A required for the step command input.

We mentioned that current limitation can change the dynamics of the system. Figure 10.8.11 was obtained with the Simulink model to be discussed in Section 10.10. The figure shows the effect of a built-in current limiter of 20 A on the step response of our control system with $K_P = 0.63$. The speed response without a limiter is shown for comparison. The action of the limiter slows the speed response.

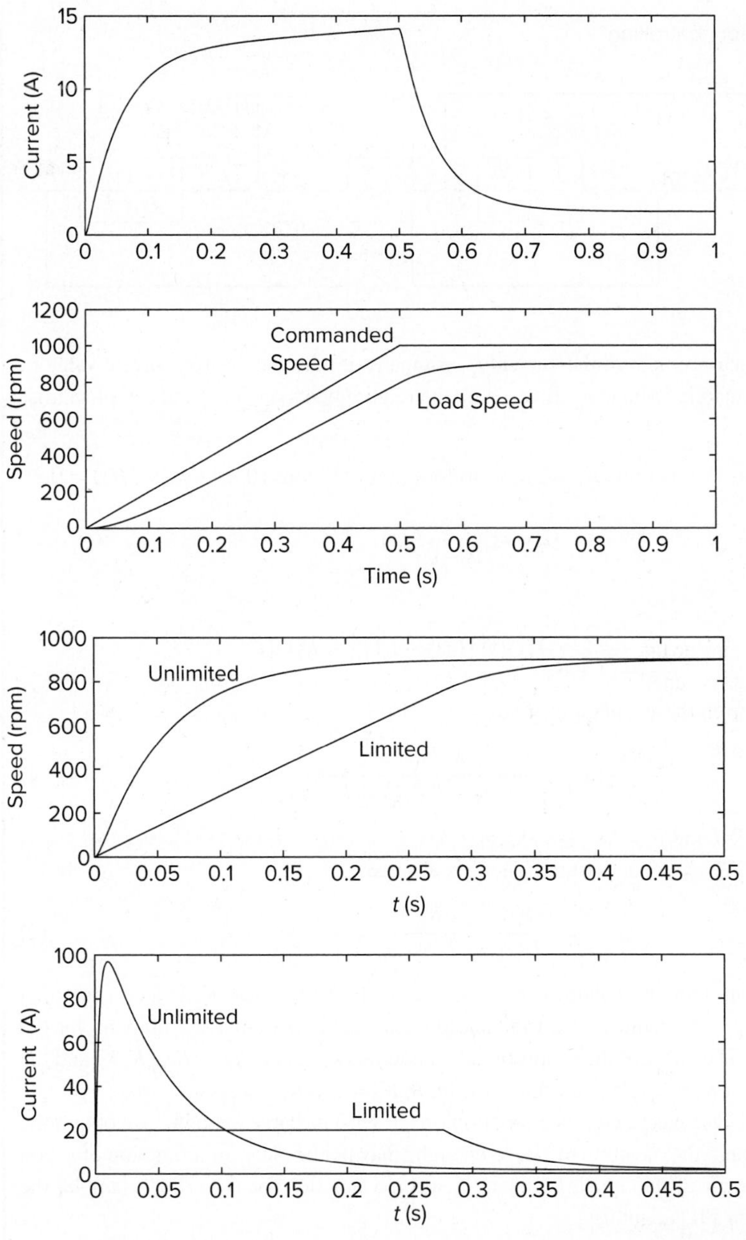

Figure 10.8.10 Control system response with a modified-ramp input.

Figure 10.8.11 Control system step response with a current limiter.

Current Control in Motors

EXAMPLE 10.8.7

■ Problem

Controlling the torque of an armature-controlled motor is complicated by the existence of the back emf. For this reason some amplifiers have a built-in current controller that can accept a commanded torque from the main controller. The current controller senses the armature current and adjusts the armature voltage to produce the current required to obtain the desired torque. The block diagram of such a system is given in Figure 10.8.12, where i_r is the commanded current and $G(s)$ is the transfer function of the current controller. The current controller

Figure 10.8.12 A system for controlling current in a motor.

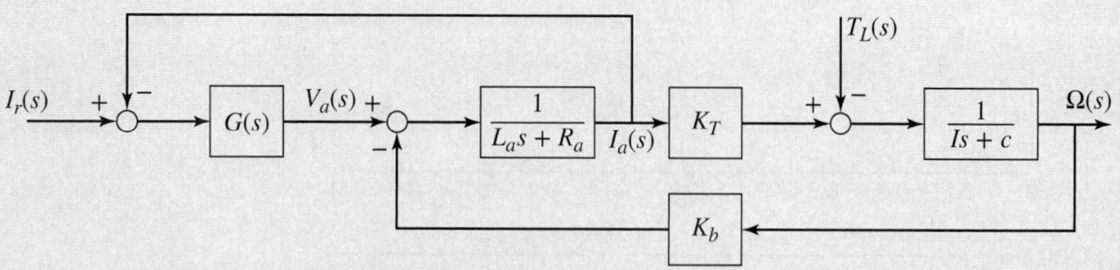

uses a feedback measurement of the current i_a, compares it with the desired current value i_r, and outputs the armature voltage v_a. Investigate the requirements on $G(s)$ to accomplish this.

■ Solution

We obtain the transfer function $\Omega(s)/I_r(s)$ as follows. From Figure 10.8.12 with $T_L(s) = 0$,

$$\Omega(s) = \frac{K_T}{Is + c} I_a(s)$$

and

$$I_a(s) = \frac{1}{L_a s + R_a}\left\{G(s)\left[I_r(s) - I_a(s)\right] - K_b \Omega(s)\right\}$$

Eliminating $I_a(s)$ from these equations gives

$$\frac{\Omega(s)}{I_r(s)} = \frac{K_T G(s)}{L_a I s^2 + As + B + G(s)(Is + c)} \tag{1}$$

where $A = cL_a + R_a I$ and $B = R_a c + K_T K_b$.

Suppose that $G(s) = K$, a constant. Then, as $K \to \infty$,

$$\frac{\Omega(s)}{I_r(s)} \to \frac{K_T}{Is + c} \tag{2}$$

In this case, we can draw the diagram shown in Figure 10.8.13, where K_c is a constant that converts the voltage command v_c from the speed controller into a current command for the current controller. The value of the proportional gain in this system is $K_P = K_{\text{tach}} K_1 K_c$ and the value of the product $K_1 K_c$ can be determined from $K_1 K_c = K_P / K_{\text{tach}}$.

In the general case where $G(s)$ is a function of s, if $G(s)$ is "large enough," we obtain the same result. Although the meaning of "large enough" may not be clear for a function of s, you can convince yourself of this result by trying a specific function for $G(s)$. For example, the transfer function for PID action is

$$G(s) = \frac{K_D s^2 + K_P s + K_I}{s}$$

Substituting this into equation (1) and letting the gains K_D, K_P, and K_I approach infinity, we obtain equation (2).

Figure 10.8.13 A proportional control system using current control.

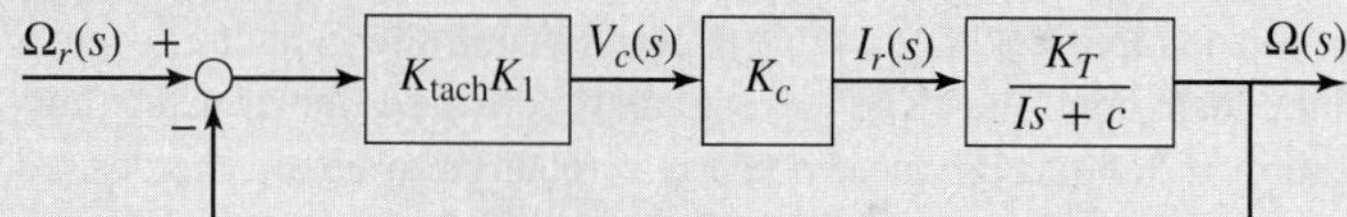

An Active Suspension

EXAMPLE 10.8.8

■ **Problem**

An *active suspension* supplements the car's passive suspension system that consists of springs and shock absorbers. An electric motor, a hydraulic servomotor or pneumatic cylinder is mounted either in parallel or in series with the spring and shock absorber. The actuator is under computer control so that its force can be adjusted to apply more force in response to varying conditions, for example to maintain traction during turns or to compensate for increased passenger and cargo weight.

Figure 10.8.14 illustrates the principle for a single-mass, quarter-car suspension model. The servomotor is shown in parallel with the spring and damper. Assume that the servomotor force is proportional to the chassis displacement x and the chassis velocity $\dot{x}$, so that $f = -K_P x - K_D \dot{x}$. The gains can be adjusted by the computer to achieve desirable response under different conditions.

Determine the values of the gains to achieve a damping ratio of $\zeta = 0.707$ and an undamped natural frequency of $\omega_n = 5$ rad/s. Use the nominal values $m = 275$ kg, $c = 1768$ N · s/m, and $k = 6250$ N/m.

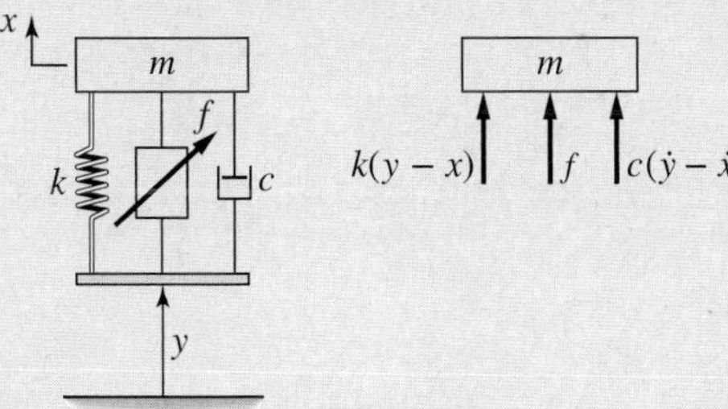

Figure 10.8.14 An active suspension system.

■ **Solution**

The equation of motion is

$$m\ddot{x} = k(y - x) + c(\dot{y} - \dot{x}) + f = k(y - x) + c(\dot{y} - \dot{x}) - K_P x - K_D \dot{x}$$

or

$$m\ddot{x} + (c + K_D)\dot{x} + (k + K_P)x = c\dot{y} + ky$$

From this equation we see that this active suspension, which implements PD control action, supplements the damping and stiffness of the passive suspension.

Using the given parameter values, we have

$$\zeta = \frac{c + K_D}{2\sqrt{m(k + K_P)}} = \frac{1768 + K_D}{2\sqrt{275(6250 + K_P)}} = 0.707$$

$$\omega_n = \sqrt{\frac{k + K_P}{m}} = \sqrt{\frac{6250 + K_P}{275}} = 5 \text{ rad/s}$$

Solving the latter equation for K_P gives $K_P = 625$ N/m. Substitute this value into the equation for ζ and solve to obtain $K_D = 176$ N · s/m.

Feedforward Compensation with Proportional Control

EXAMPLE 10.8.9

■ **Problem**

Consider the liquid-level system shown in Figure 10.1.1a, with the controller diagram given by Figure 10.1.3. Suppose we make the feedback control adjustment proportional to the error

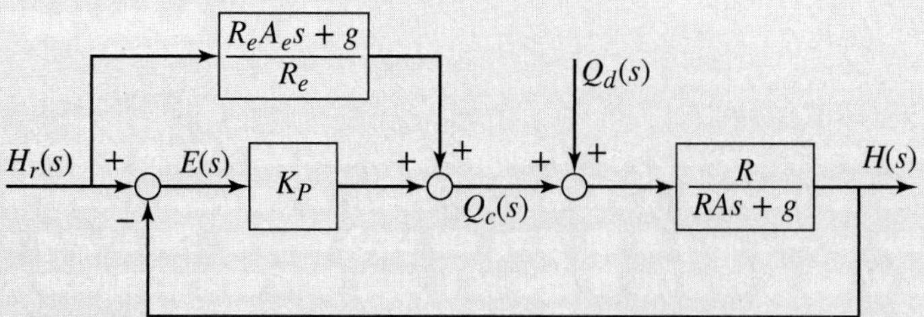

Figure 10.8.15 Proportional control with feedforward compensation for a liquid-level controller.

$e = h_r - h$, so that $G_c(s) = K_P$, where K_P is the proportional gain. This scheme is called feedforward compensation with proportional control. The resulting diagram is shown in Figure 10.8.15. The controlled flow rate is

$$q_c = A_e \dot{h}_r + \frac{g}{R_e} h_r + K_P(h_r - h)$$

Investigate the stability, speed of response, and error response of this system, assuming that h_r and q_d are step functions. Determine the value of K_P required to achieve a specified 2% settling time t_s.

■ **Solution**

First obtain the equation for the output $H(s)$. From the block diagram,

$$H(s) = \frac{R}{RAs + g}\left\{Q_d(s) + \frac{R_e A_e s + g}{R_e} H_r(s) + K_P[H_r(s) - H(s)]\right\}$$

The solution for $H(s)$ is

$$H(s) = \frac{(R/R_e)(R_e A_e s + g) + RK_P}{RAs + g + RK_P} H_r(s) + \frac{R}{RAs + g + RK_P} Q_d(s) \tag{1}$$

Note that if $R_e = R$ and $A_e = A$,

$$H(s) = H_r(s) + \frac{R}{RAs + g + RK_P} Q_d(s)$$

which means the system will have zero error if there is no disturbance. Thus, the good feature of open-loop control has been preserved.

The characteristic equation, given by the denominators, is $RAs + g + RK_P = 0$, which has the root

$$s = -\frac{g + RK_P}{RA}$$

Since $RA > 0$, the system is stable as long as $g + RK_P > 0$; that is, if $K_P > -g/R$. The time constant of the closed-loop system is

$$\tau = \frac{RA}{g + RK_P}$$

and the 2% settling time is 4τ. So the required value for K_P is

$$K_P = \frac{4A}{t_s} - \frac{g}{R}$$

We note that accurate calculation of the derivative $\dot{h}_r$ is often difficult to achieve. This can limit the applicability of the feedforward method.

10.8.1 GENERAL FORM OF FEEDFORWARD COMMAND COMPENSATION

The concept of feedforward command compensation can be generalized as follows. The system is shown in Figure 10.8.16. The open-loop control transfer function is $G_f(s)$, which is called the *compensator* transfer function. The output equation is

$$\begin{aligned}C(s) &= G_p(s)[G_a(s)\{G_f(s)R(s) + G_c(s)[R(s) - C(s)]\} + D(s)]\\ &= G_p(s)G_a(s)G_f(s)R(s) + G_p(s)G_a(s)G_c(s)R(s)\\ &\quad - G_p(s)G_a(s)G_c(s)C(s) + G_p(s)D(s)\end{aligned}$$

Solve for $C(s)$:

$$C(s) = \frac{G_p(s)G_a(s)G_f(s) + G_p(s)G_a(s)G_c(s)}{1 + G_p(s)G_a(s)G_c(s)}R(s) + \frac{G_p(s)}{1 + G_p(s)G_a(s)G_c(s)}D(s)$$

In the absence of a disturbance, the output $C(s)$ will exactly equal the command input $R(s)$ if $C(s)/R(s) = 1$. This will be true if

$$\frac{C(s)}{R(s)} = \frac{G_p(s)G_a(s)G_f(s) + G_p(s)G_a(s)G_c(s)}{1 + G_p(s)G_a(s)G_c(s)} = 1$$

This equation will be satisfied if

$$G_f(s) = \frac{1}{G_p(s)G_a(s)} \tag{10.8.1}$$

This expresses the general principle of feedforward compensation of the command; namely, in the absence of a disturbance, the output will follow the command *exactly* if $G_f(s) = 1/G_p(s)G_a(s)$. Note that this says that the compensator transfer function should be the reciprocal of the transfer function encountered as the compensated signal flows from the compensator to the output $C(s)$.

Of course, this startling principle is a mathematical result only. In practice we do not have exact models of the plant and the actuator, and we cannot build a compensator to have the exact transfer function specified by $G_f(s)$. If the compensator is implemented in analog form, the hardware (electronic, pneumatic, or hydraulic) cannot be designed to give the desired expression for $G_f(s)$ exactly. If implemented in digital form as a computer algorithm, the algorithm can reproduce the transfer function $G_f(s)$ only approximately because of quantization and sampling effects. Finally, the command compensator cannot deal with the effects of a disturbance. For that we always need the main controller block $G_c(s)$ and feedback loop shown in Figure 10.8.16. The compensator is used merely to augment the feedback controller.

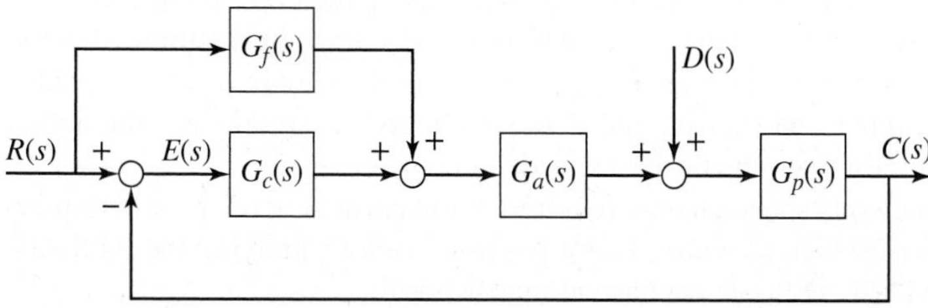

Figure 10.8.16 General structure of feedforward compensation.

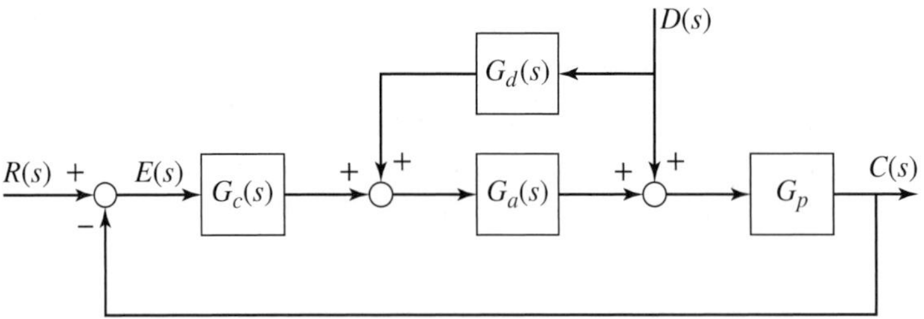

Figure 10.8.17 General structure of disturbance compensation.

10.8.2 DISTURBANCE COMPENSATION

Suppose that we can measure the disturbance $d(t)$. If so, we can use the measurement to improve the performance of a control system. This technique is called *disturbance compensation*. The arrangement is shown in Figure 10.8.17, in which the feedforward compensation has been omitted to simplify the discussion. The transfer function $G_d(s)$ represents a mathematical operation acting on the measurement of $d(t)$. We now determine what $G_d(s)$ should be.

From the block diagram we can write

$$C(s) = G_p(s)[G_a(s)\{G_d(s)D(s) + G_c(s)[R(s) - C(s)]\} + D(s)]$$

Collecting terms gives

$$C(s) = G_p(s)\{[G_a(s)G_d(s) + 1]D(s) + G_c(s)[R(s) - C(s)]\}$$

Note that $D(s)$ will not appear in the equation if $G_a(s)G_d(s) + 1 = 0$. Thus, if

$$G_d(s) = -\frac{1}{G_a(s)} \tag{10.8.2}$$

the disturbance will be canceled by the compensator $G_d(s)$ and will not affect the output $c(t)$. To see this, follow the flow of $D(s)$ to the output $C(s)$ in Figure 10.8.17.

Of course, this tidy result is difficult to achieve in practice because (1) the disturbance might not be measurable or the measurement might be noisy, and (2) the operation specified by $G_d(s)$ might be difficult to implement in hardware or software.

In many of the examples to follow in this chapter and the next two, we will not use feedforward compensation or disturbance compensation to simplify the discussion, so that we can concentrate on the design of the main feedback controller. They can, however, be included if the limitations discussed earlier are not too severe. As illustrated by Figure 10.1.4, the calculations required to implement feedforward and disturbance compensation are easier to do if a digital computer is used as the controller.

10.9 CASE STUDY: MOTION CONTROL WITH FEEDBACK

We now continue our design of motion-control systems by seeing how feedback can be used to improve the performance of such systems when using the trapezoidal command profile. In Chapter 6 we computed the required torque and voltage by assuming that the speed followed the commanded profile exactly. This is not the case in practice because of effects due to damping and inertia, and if these effects are significant, the actual speed will deviate greatly from the desired speed.

Also, the instantaneous speed change required at the corners of the profile implies an infinite acceleration, which of course is not possible. Before studying these effects, we will review the MATLAB tools we have at our disposal.

The MATLAB functions `tf`, `step`, `lsim`, `bode`, and `bodemag` provide a powerful set of tools for analyzing the performance of control systems. All of these functions have extended syntax that was covered in earlier chapters. Here we will focus on their application to control system analysis. Recall that when these functions display a plot, you can right-click on the plot to determine characteristics such as maximum overshoot, peak time, settling time, and peak response. Left-clicking on the curve lets you move the cursor along the curve to identify coordinates.

10.9.1 THE `tf` AND `step` FUNCTIONS

The `tf` function creates an LTI object in transfer-function form, using the numerator and denominator expressed as arrays. The `step` function displays the unit-step response. The command transfer function in Example 10.7.4 is

$$\frac{\Theta(s)}{\Theta_r(s)} = \frac{K_D s^2 + K_P s + K_I}{10s^3 + (2 + K_D)s^2 + K_P s + K_I}$$

Using the gain values $K_P = 55$, $K_I = 25$, and $K_D = 58$, the unit-step response shown in Figure 10.7.4 was obtained with the following M-file:

```
KP = 55; KI = 25; KD = 58;
sys1 = tf([KD, KP, KI],[10, 2+KD, KP, KI]);
step(sys1)
```

The resulting plot was then edited with the Plot Editor to produce the figure.

10.9.2 THE `bode` AND `bodemag` FUNCTIONS

The `bode` function displays the plots of the magnitude ratio in decibels and the phase angle in degrees. The `bodemag` function does not display the phase plot. The disturbance transfer function in Example 10.7.4 is

$$\frac{\Theta(s)}{T_d(s)} = -\frac{s}{10s^3 + (2 + K_D)s^2 + K_P s + K_I}$$

Using the gain values $K_P = 55$, $K_I = 25$, and $K_D = 58$, the frequency response plot shown in Figure 10.7.5 was obtained with the following M-file:

```
KP = 55; KI = 25; KD = 58;
sys2 = tf([-1, 0],[10, 2+KD, KP, KI]);
bodemag(sys2)
```

The resulting plot was then edited with the Plot Editor to produce the figure.

10.9.3 THE `lsim` FUNCTION

The `lsim` function displays the response of a linear model to a user-defined input. This function is useful for computing the ramp response. For example, the following M-file plots the error response to a ramp command of slope 6, for a PI control system whose gains are $K_P = 18$ and $K_I = 40$. This system was analyzed in Example 10.6.3. The error transfer function is

$$\frac{E(s)}{\Omega_r(s)} = \frac{5s^2 + 2s}{5s^2 + (2 + K_P)s + K_I}$$

The M-file is

```
KP = 18; KI = 40;
sys3 = tf([5, 2, 0],[5, 2+KP, KI]);
t = (0:0.01:2);
omr = 6*t;
lsim(sys3, omr, t)
```

The `lsim` function plots the input `omr` as well as the response.

10.9.4 TRAPEZOIDAL RESPONSE

The response of a control system to a trapezoidal command input is easily found with MATLAB. If the profile and the transfer functions are defined in their own programs, we can use a modular approach that gives us the ability to examine different command inputs and different systems easily. The following example illustrates this approach.

EXAMPLE 10.9.1

Simulation with a Trapezoidal Profile

■ Problem

Consider the P, PI, and modified I control systems discussed in Examples 10.6.2, 10.6.3, and 10.6.4. The plant transfer function is $1/(Is+c)$, where $I = 5$ and $c = 2$. This plant can represent an inertia whose rotational velocity is the output and whose input is a torque. So the control systems are examples of speed controllers.

Investigate the performance of these systems for the trapezoidal command input shown in Figure 10.9.1. Use the parameter values $I = 5$ and $c = 2$.

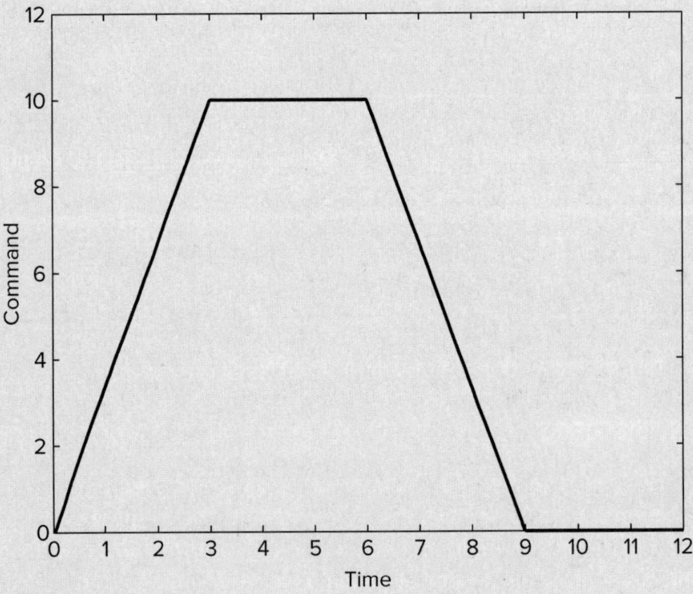

Figure 10.9.1 A trapezoidal command profile.

■ Solution

The transfer functions and parameters for these three control systems are shown in Table 10.9.1.

Table 10.9.1 P, PI, and modified I control systems.

	Reference: Examples 10.6.2, 10.6.3, and 10.6.4		
	P	**PI**	**Modified I**
Command transfer function $\Omega(s)/\Omega_r(s)$	$\dfrac{K_P}{5s+2+K_P}$	$\dfrac{K_P s+K_I}{5s^2+(2+K_P)s+K_I}$	$\dfrac{K_I}{5s^2+(2+K_2)s+K_I}$
Disturbance transfer function $\Omega(s)/T_d(s)$	$-\dfrac{1}{5s+2+K_P}$	$-\dfrac{s}{5s^2+(2+K_P)s+K_I}$	$-\dfrac{s}{5s^2+(2+K_2)s+K_I}$
Actuator transfer function $T(s)/T_d(s)$	$K_P\dfrac{5s+2}{5s+2+K_P}$	$\dfrac{(K_P s+K_I)(5s+2)}{5s^2+(2+K_P)s+K_I}$	$\dfrac{(5s+2)K_I}{5s^2+(2+K_2)s+K_I}$
Steady-state unit-ramp error e_{ss}	$\dfrac{1}{2+K_P}$	$\dfrac{2}{K_I}$	$\dfrac{2+K_2}{K_I}$
Time constant τ	$\dfrac{5}{2+K_P}$	$\dfrac{10}{2+K_P}$ if $\zeta \le 1$	$\dfrac{10}{2+K_2}$ if $\zeta \le 1$
Damping ratio ζ	—	$\dfrac{2+K_P}{2\sqrt{5K_I}}$	$\dfrac{2+K_2}{2\sqrt{5K_I}}$

We can use the expressions in Table 10.9.1 to compute the gain values required to achieve a time constant of $\tau = 1$ and a steady-state error $e_{ss} = 0.2$ for a *unit*-ramp command. Because the slope of the profile is 10/3, this will give a steady-state error of $0.2(10/3) = 2/3$. The following table shows the results.

P	$K_P = 3$	—
PI	$K_P = 8$	$K_I = 10$
Modified I	$K_2 = 8$	$K_I = 50$

The following file sets the gain values for the three systems. System *a* is the P control system, system *b* is the PI system, and system *c* is the modified I system.

```
% gains.m
% Sets the gains.
% P control (system a):
KPa = 3;
% PI control (system b):
KPb = 8;KIb = 10;
% I action modified with velocity feedback (system c):
K2 = 8;KIc = 50;
```

After running the program `gains.m`, the files shown in Table 10.9.2 are used to create the command, disturbance, and actuator transfer functions given in Table 10.9.1.

The trapezoidal command input shown in Figure 10.9.1 can be created with the following script file.

```
% trap.m
% A specific trapezoidal profile
t = (0:0.01:12);
for k = 1:length(t)
   if t(k) <= 3
      r(k) = (10/3)*t(k);
   elseif t(k) <= 6
      r(k) = 10;
```

```
   elseif t(k) <= 9
      r(k) = 30-(10/3)*t(k);
   else
      r(k) = 0;
   end
end
```

After running `trap.m` and one of the transfer-function programs given in Table 10.9.2, the files shown in Table 10.9.3 are used to compute and plot the command, disturbance, and actuator responses.

Table 10.9.2 Transfer-function files.

```
% command_tf.m
% Create the command transfer functions.
sysa = tf(KPa,[5, 2+KPa]);
sysb = tf([KPb, KIb],[5, 2+KPb, KIb]);
sysc = tf(KIc,[5, 2+K2, KIc]);
```

```
% actuator_tf.m
% Create the actuator transfer functions.
sysaACT = tf(KPa*[5, 2],[5, 2+KPa]);
sysbACT = tf(conv([KPb, KIb],[5, 2]),[5, 2+KPb, KIb]);
syscACT = tf(KIc*[5, 2],[5, 2+K2, KIc]);
```

```
% dist_tf.m
% Create the disturbance transfer functions.
sysaDIS = tf(-1, [5, 2+KPa]);
sysbDIS = tf(-[1, 0],[5, 2+KPb, KIb]);
syscDIS = tf(-[1, 0],[5, 2+K2, KIc]);
```

Table 10.9.3 Solution and plotting files.

```
% plot_command.m
% Obtain and plot the command response.
ya = lsim(sysa,r,t);
yb = lsim(sysb,r,t);
yc = lsim(sysc,r,t);
plot(t,ya,t,yb,t,yc,'-',t,r)
```

```
% plot_actuator.m
% Obtain and plot the actuator repsonse.
yaACT = lsim(sysaACT,r,t);
ybACT = lsim(sysbACT,r,t);
ycACT = lsim(syscACT,r,t);
plot(t,yaACT,t,ybACT,t,ycACT,'-')
```

```
% plot_dist.m
% Add the disturbance response to the command response
% and plot the total response.
ya = lsim(sysa,r,t);
yb = lsim(sysb,r,t);
yc = lsim(sysc,r,t);
disturbance = 20*t;
yaDIS = lsim(sysaDIS,disturbance,t);
ybDIS = lsim(sysbDIS,disturbance,t);
ycDIS = lsim(syscDIS,disturbance,t);
plot(t,ya+yaDIS,t,yb+ybDIS,t,yc+ycDIS,'-',t,r)
```

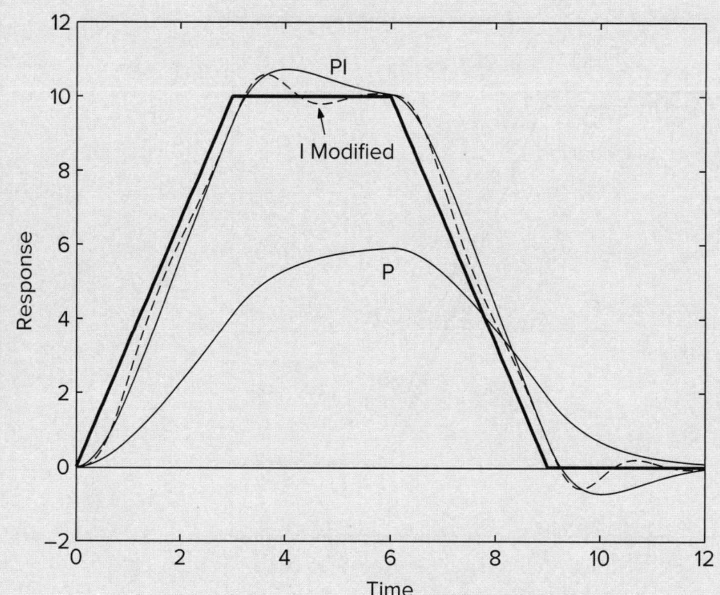

Figure 10.9.2 System response to a trapezoidal command.

For example, to create the plot shown in Figure 10.9.2, use the M-file:

```
gains
command_tf
trap
plot_command
```

Notice that the P control system does not follow the trapezoidal profile very well. (Figure 10.9.2 was enhanced using the Plot Editor.)

The performance of the P control system can be improved by increasing the gain. For example, to use $K_P = 50$, change `KPa` to 50 in `gains.m` and run the following M-file. Note that we need not run `trap.m` again unless we cleared the variables.

```
gains
command_tf
plot_command
```

The result is shown in Figure 10.9.3. Now the P control follows the trapezoidal profile much better.

The actuator response is found by running the M-file:

```
actuator_tf
plot_actuator
```

The result is shown in Figure 10.9.4.

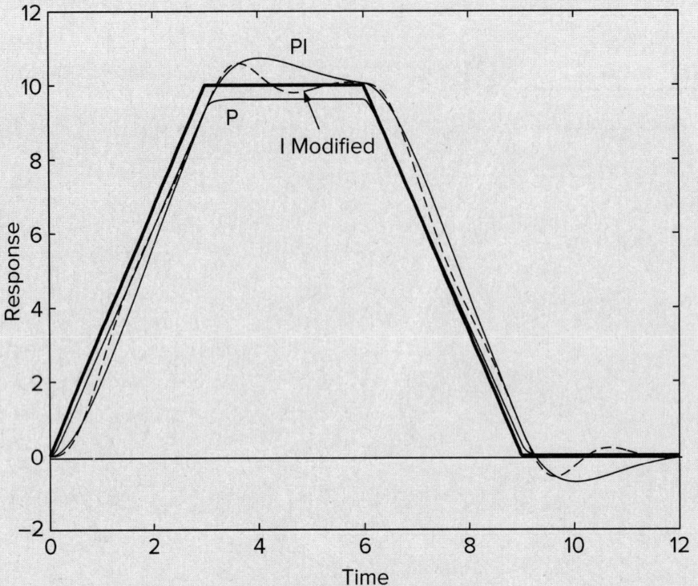

Figure 10.9.3 System response to a trapezoidal command. Here the gain of the P control system has been increased.

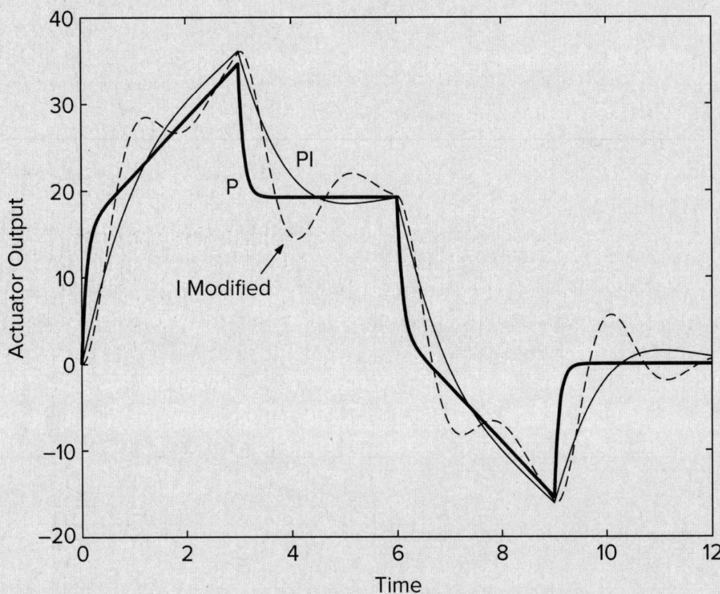

Figure 10.9.4 Actuator response to a trapezoidal command.

Finally, let us see how well the systems do with a ramp disturbance $T_d = 20t$. Run the following M-file.

```
dist_tf
plot_dist
```

The total response is shown in Figure 10.9.5. Both PI and P control have difficulty following the profile, but the modified I system does rather well.

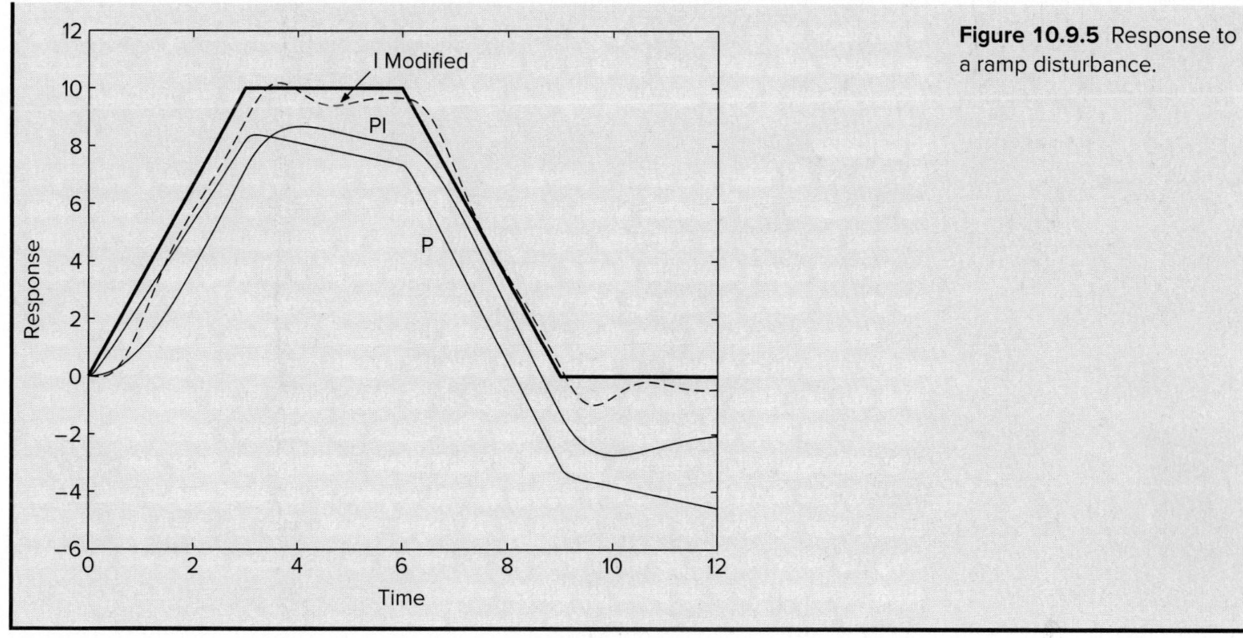

Figure 10.9.5 Response to a ramp disturbance.

Using the approach outlined in Section 6.7 of Chapter 6, we can also use MATLAB to evaluate the performance specifications discussed in Section 6.5 for a trapezoidal command input. These specifications include the maximum torque, rms torque, and maximum velocity required by the control system to follow the profile.

10.10 SIMULINK APPLICATIONS

Simulink cannot be used to investigate control system performance until a preliminary analysis has developed a system model, selected a control algorithm, and computed a set of gain values. Once this has been done, however, Simulink can be used to investigate aspects of the design that are not easily analyzed with closed-form solution methods. These aspects include

1. The effects of unmodeled sensor, actuator, and plant dynamics. Often we neglect actuator time constants and smaller plant time constants in order to obtain a low-order model of the system.
2. Nonlinear dynamics, such as discontinuities (e.g., two-position control) and actuator saturation.
3. Input functions more complicated than simple steps, ramps, and sinusoids. These include trapezoidal commands and random disturbances.

Control of a Liquid-Level System — EXAMPLE 10.10.1

■ Problem

As an example, consider the liquid-level control system developed in Example 10.8.5, whose block diagram is shown in Figure 10.8.7. The analysis in that example ignored the obvious fact that the control flow rate q_1 can never be negative! For some initial heights, a simulation would show that $q_1(t)$ takes on negative values. Create a Simulink model using the numerical parameters and control gains given in that example, except for the integral gain K_I. The model should prevent q_1 from becoming negative. Investigate the response with this limitation.

■ **Solution**

We can use the Saturation block to prevent the control flow rate q_1 from becoming negative in the simulation. The Simulink model is shown in Figure 10.10.1. It can be easily constructed from the block diagram, with some rearrangement to allow for placement of the Mux. Note that we have not used a Saturation block to prevent the liquid heights from being negative, because the physics used to develop the differential equation models prevents this from happening.

We have introduced a new block in this model. In Simulink this block is called the "Transfer Fcn (with initial outputs)," to distinguish it from the Transfer Fcn block we have used earlier. With this new block we can set the initial value of the block output. In our model, this corresponds to the initial liquid heights in the tanks. This feature thus provides a useful improvement over traditional transfer-function analysis, in which initial conditions are set to zero.

The new block is equivalent to adding the free response to the block output, with all the block's state variables set to zero except for the output variable. The new block also lets you assign an initial value to the block input, but we will not use this feature and so will leave the initial input set to 0 in the Block parameters window.

The steady-state error due to a ramp command for this system was shown to be $e_{ss} = 44/K_I$, so to decrease this error, in our simulation we will use a larger integral gain, say, $K_I = 50$. This makes the system faster and more oscillatory, and this means that the commanded control flow q_1 will more likely be computed to be negative.

The simulation used a step command of magnitude 5 starting at $t = 0$ and a step disturbance of magnitude 3 starting at $t = 20$. The saturation limits were 0 and 100. The initial heights were set to $h_1 = 2$ and $h_2 = 1$.

Figure 10.10.1 Simulink model of a liquid-level control system.

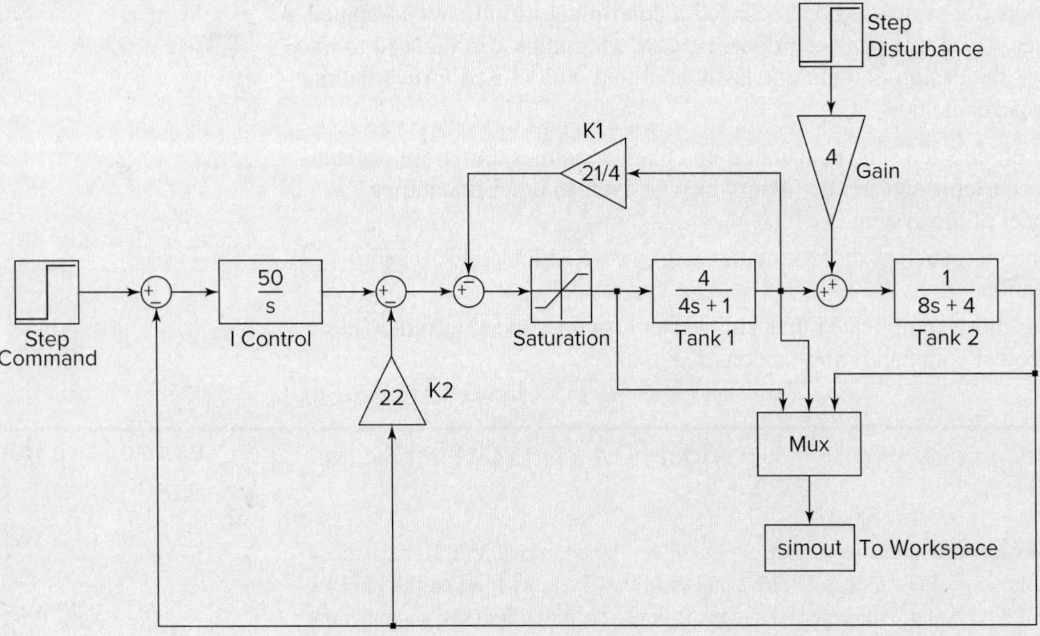

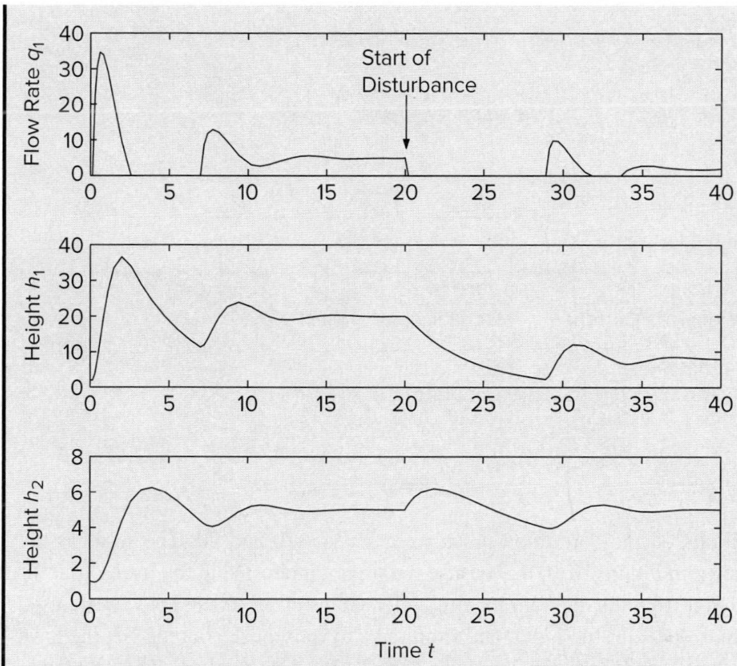

Figure 10.10.2 Flow rate and liquid height responses.

The results are shown in Figure 10.10.2. Note that sometimes the control flow q_1 is zero. Were it not for the Saturation block, the flow would become negative. The MATLAB M-file used to produce this plot is the following. The plot was then modified with the Plot Editor.

```
subplot(3,1,1),plot(tout,simout(:,1))
subplot(3,1,2),plot(tout,simout(:,2))
subplot(3,1,3),plot(tout,simout(:,3))
```

Current Saturation in a Motor Control System

EXAMPLE 10.10.2

■ **Problem**

The block diagram of a proportional control system was given in Figure 10.3.9 and is shown here in modified form as Figure 10.10.3. Create a Simulink model to investigate the effects of current saturation, using the parameter values given in Example 10.8.6 and a step-command input of magnitude 104.7 rad/s (which corresponds to 1000 rpm). Assume that the current is limited to ±20 A.

■ **Solution**

To investigate the effects of current saturation we need to place a Saturation block between the circuit transfer function $1/(L_a s + R_a)$ and the mechanical subsystem that includes the torque

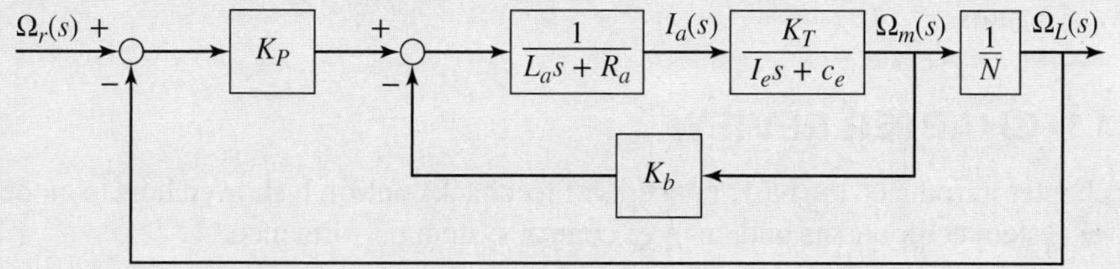

Figure 10.10.3 Block diagram of a speed-control system.

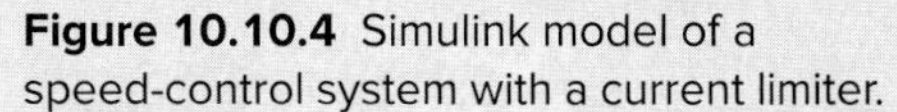

Figure 10.10.4 Simulink model of a speed-control system with a current limiter.

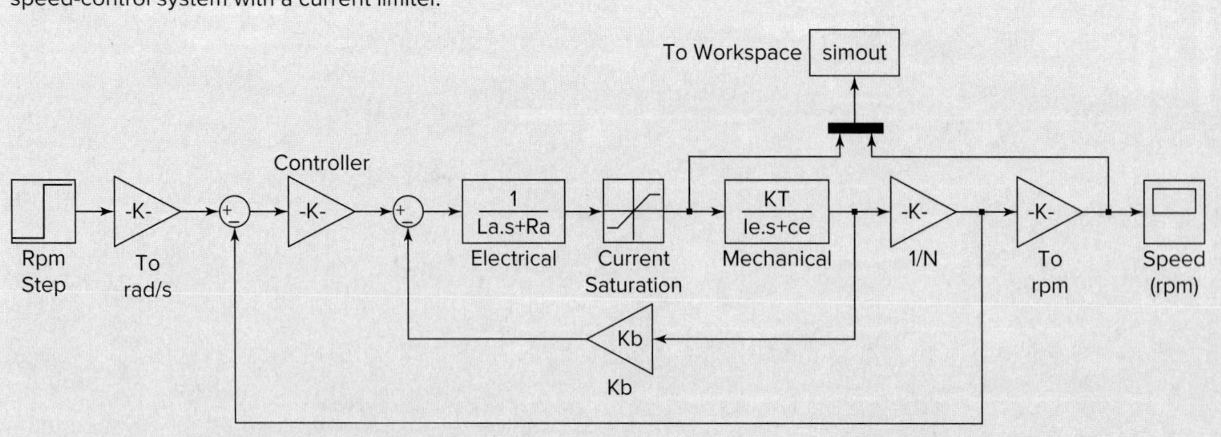

constant K_T. The limits on the Saturation block were set to −20 and 20. The result is the Simulink model shown in Figure 10.10.4. Because we are not interested in the load torque T_L here, we have omitted it from the diagram for simplicity. Gains of $2\pi/60$ and $60/2\pi$ are used to convert from rpm to rad/s for the input, and from rad/s to rpm for the output. Set the Save Format to Array in the Block Parameters window of the To Workspace block. The data fed to the To Workspace block are used in MATLAB to create the plots shown in Figure 10.8.11.

We will use variables for the parameters in the simulation. In the Electrical block, set the numerator to 1 and the denominator to `[La Ra]`. In the Mechanical block, set the numerator to `[KT]` and the denominator to `[Ie ce]`. Set the gain to `KP` in the Controller block. In the block labeled `Kb`, set the gain to `Kb`, and in the block labeled `1/N` set the gain to `1/N`. Before running the simulation, type the following in the Command window to set the values of the parameters.

```
>>KT = 0.04; Kb = KT; Ra = 0.6; La = 2e-3;
>>Ie = 1.802e-3; ce = 4.444e-4;
>>N = 1.5; KP = 0.63;
```

The To Workspace block puts the time variable `tout` and the array `simout` in the MATLAB workspace. The first column of `simout` contains the current, and the second column contains the speed in rpm. To plot the current response, in the Command window type `plot(tout,simout(:,1))`. Type `plot(tout,simout(:,2))` to obtain a plot of the speed. The plots are shown in Figure 10.8.11.

The model in Example 10.10.1 implements proportional control. To examine PID control in general, replace the Controller gain block in Figure 10.10.4 with the PID Controller block. This block enables you to specify the proportional, integral, and derivative gains.

10.11 CHAPTER REVIEW

This chapter introduced the basic concepts of feedback control. It showed how to model control system components and analyze control system performance.

When designing a control system, the control systems engineer is usually given the plant, the actuator, and the physical type of controller (electronic, pneumatic, hydraulic, or digital), and often is expected to develop a model of these components. The command and disturbance inputs might be specified, or the engineer might be expected to develop suitable test inputs based on the application. Step functions are the principal test inputs, because they are the most common and perhaps represent the severest test of system performance. Ramp, trapezoidal, and sinusoidal test inputs are also employed. The type to use should be made clear in the design specifications.

The engineer then proceeds to design the control system. Based on the system model and the performance specifications, a control action is chosen, and the output, error, and actuator equations are derived. These are then analyzed for stability. If the system cannot be made stable with a gain change, a different control action is tried. Using the given command and disturbance input functions (step, ramp, etc.), the steady-state response is evaluated with the final-value theorem. Any constraints on the gain values required to satisfy the steady-state specifications are then determined.

The transient performance is then evaluated in light of the transient specifications, using the given input functions. These specifications often are stated in terms of the desired dominant-time constant and damping ratio, but they can also be given in terms of overshoot, rise time, settling time, or bandwidth, for example. Other specifications, such as limits on the maximum available actuator output, are evaluated, and the system redesigned if necessary.

Now that you have finished this chapter you should be able to do the following:

1. Model common control system components.
2. Select an appropriate control algorithm of the PID type or one of its variations, for a given application and for given steady-state and transient performance specifications.
3. Analyze the performance of a control algorithm using transfer functions, block diagrams, and computer methods.
4. Compute the gain values to meet the specifications.
5. Use MATLAB and Simulink to analyze and simulate control systems.

REFERENCE

[Cannon, 1967] R. H. Cannon, Jr., *Dynamics of Physical Systems*, McGraw-Hill, New York, 1967.

PROBLEMS

Section 10.1 Closed-Loop Control

10.1 Discuss whether or not the following devices and processes are open-loop or closed-loop. If they are closed-loop, identify the sensing mechanism.

a. A traffic light.
b. A washing machine.
c. A toaster.
d. Cruise control.
e. An aircraft autopilot.
f. Temperature regulation in the human body.

10.2 Draw the block diagram of a system using proportional control and feedforward command compensation, for the plant $1/(4s^2 + 6s + 3)$. Determine the transfer function of the compensator. Discuss any practical limitations to its use.

10.3 Investigate the performance of proportional control using feedforward command compensation with a constant gain K_f and disturbance compensation with a constant gain K_d, applied to the plant $10/s$. Set the gains to achieve a closed-loop time constant of $\tau = 2$ and zero steady-state error for a step command and a step disturbance.

Section 10.2 Control System Terminology

10.4 Derive the output $C(s)$, error $E(s)$, and actuator $M(s)$ equations for the diagram in Figure P10.4, and obtain the characteristic polynomial.

Figure P10.4

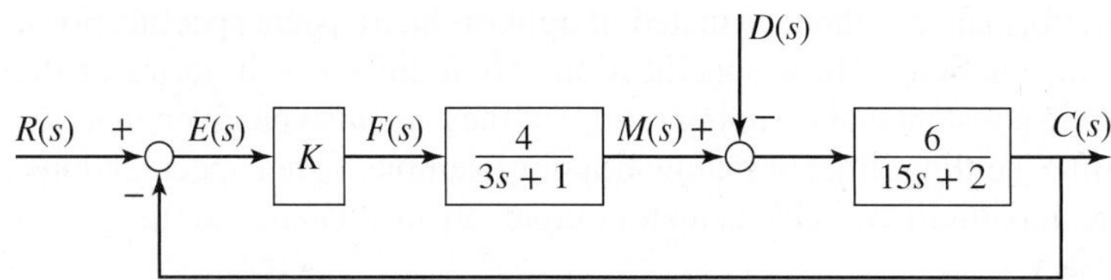

Section 10.3 Modeling Control Systems

10.5 For the system shown in Figure P10.5, the plant time constant is 5 and the nominal value of the actuator time constant is $\tau_a = 0.05$. Investigate the effects of neglecting this time constant as the gain K_P is increased.

Figure P10.5

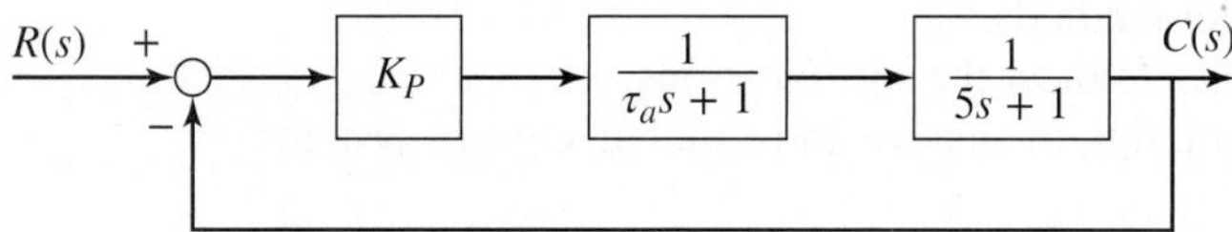

10.6 In Figure P10.6, the block is pulled up the incline by the tension force f in the inextensible cable. The motor torque T is controlled to regulate the speed v of the block to obtain some desired speed v_r. The precise value of the friction coefficient μ is unknown, as is the slope angle α, so we model them as a disturbance. Neglect all masses and inertias in the system except for the block mass m. Also neglect the field time constant of the field-controlled motor. Feedback of the block speed v is provided by a sensor that measures the pulley rotational speed ω, which is directly related to v by $v = R\omega$.

a. Obtain the equation of motion of the block speed v, with the voltage v_m, friction force $F = \mu mg \cos \alpha$, and the weight component $W_x = mg \sin \alpha$ as the inputs.

b. Draw a block diagram representing the control system, with the command input v_r, the output v, and the disturbance $D = F + W_x$. Model the speed sensor as directly sensing the speed ω. Show the necessary transfer functions for each block in the diagram.

c. Obtain the output, error, and torque transfer functions from the block diagram.

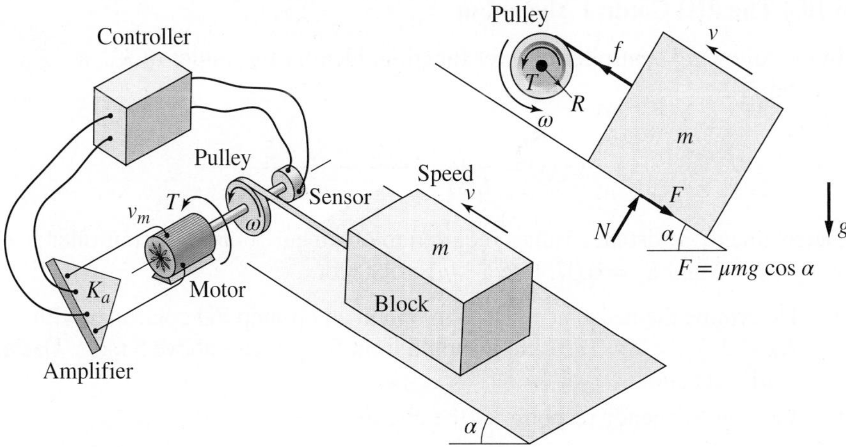

Figure P10.6

10.7 The diagram in Figure P10.7 shows a system for controlling the angular position of a load, such as an antenna. There is no disturbance.

a. Draw the block diagram of a system using proportional control, similar to that shown in Figure 10.3.9 except that the command and the output are angular positions. Assume that the motor is armature-controlled and that its armature time constant is negligible. Let the inertia I_e be the equivalent inertia of the entire system, as felt on the motor shaft, and let N_e be the equivalent gear ratio of the entire system, as felt on the motor shaft. Show the necessary transfer functions for each block.

b. Determine the value for N_e, and determine I_e as a function of the inertias I_1, I_2, and I_3.

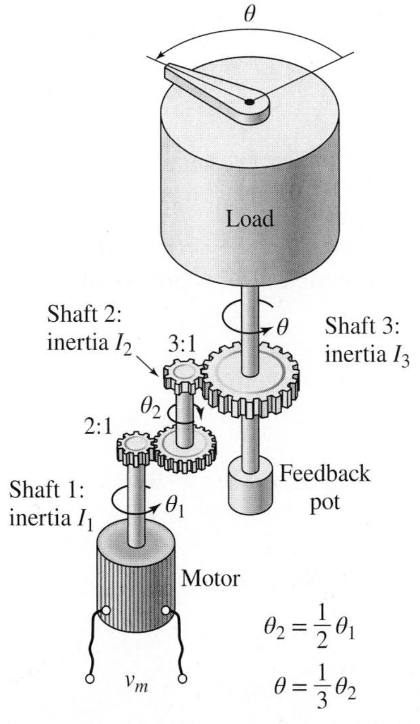

Figure P10.7

$$\theta_2 = \frac{1}{2}\theta_1$$

$$\theta = \frac{1}{3}\theta_2$$

Section 10.4 The PID Control Algorithm

10.8 In the following controller transfer function, identify the values of K_P, K_I, K_P, T_I, and T_D.

$$G_c(s) = \frac{F(s)}{E(s)} = \frac{20s^2 + 8s + 5}{s}$$

10.9 Determine the resistance values required to obtain an op-amp PI controller with $K_P = 6$ and $K_I = 0.07$. Use a 1-μF capacitor.

10.10 a. Determine the resistance values to obtain an op-amp PD controller with $K_P = 2$, $T_D = 2$ s. The circuit should limit frequencies above 5 rad/s. Use a 1-μF capacitor.

b. Plot the frequency response of the circuit.

10.11 a. Determine the resistance values to obtain an op-amp PID controller with $K_P = 10$, $K_I = 1.4$, and $K_D = 4$. The circuit should limit frequencies above 100 rad/s. Take one capacitance to be 1 μF.

b. Plot the frequency response of the circuit.

Section 10.5 Control System Analysis

10.12 Obtain the steady-state response, if any, of the following models for the given input. If it is not possible to determine the response, state the reason.

a. $$\frac{Y(s)}{F(s)} = \frac{8}{6s + 5} \qquad f(t) = 17u_s(t)$$

b. $$\frac{Y(s)}{F(s)} = \frac{8s - 5}{10s^2 + 4s + 7} \qquad f(t) = 8u_s(t)$$

c. $$\frac{Y(s)}{F(s)} = \frac{3s + 5}{s^2 - 16} \qquad f(t) = 12u_s(t)$$

d. $$\frac{Y(s)}{F(s)} = \frac{5s + 8}{s^2 + 2s - 9} \qquad f(t) = 8u_s(t)$$

10.13 For the following models, the error signal is defined as $e(t) = r(t) - c(t)$. Obtain the steady-state error, if any, for the given input. If it is not possible to determine the response, state the reason.

a. $$\frac{C(s)}{R(s)} = \frac{1}{3s + 1} \qquad r(t) = 3t$$

b. $$\frac{C(s)}{R(s)} = \frac{5}{3s + 1} \qquad r(t) = 3t$$

c. $$\frac{C(s)}{R(s)} = \frac{4}{3s^2 + 5s + 4} \qquad r(t) = 6t$$

d. $$\frac{C(s)}{R(s)} = \frac{10}{2s^2 + 4s + 5} \qquad r(t) = 4t$$

10.14 Given the model

$$\ddot{x} - (b + 2)\dot{x} + (2b + 5)x = 0$$

a. Find the values of the parameter b for which the system is
   1. Stable.
   2. Neutrally stable.
   3. Unstable.

b. For the stable case, for what values of b is the system
   1. Underdamped?
   2. Overdamped?

10.15 A certain system has the characteristic equation $s^3 + 8s^2 + 27s + K = 0$. Find the range of K values for which the system will be stable.

10.16 For the characteristic equation $s^3 + 9s^2 + 26s + K = 0$, use the Routh-Hurwitz criterion to compute the range of K values required so that the dominant time constant is no larger than $1/2$.

10.17 For the following characteristic equations, use the Routh-Hurwitz criterion to determine the range of K values for which the system is stable, where a and b are assumed to be known.

a. $3s^3 + 5as^2 + Ks + b = 0$

b. $7s^3 + 7as^2 + bs + 7K = 0$

c. $4s^3 + 15s^2 + 12s + 5 + K = 0$

10.18 The parameter values for a certain armature-controlled motor, load, and tachometer are

$$
\begin{aligned}
K_T &= K_b = 0.2 \text{ N}\cdot\text{m/A} \\
c_m &= 5\times 10^{-4} \text{ N}\cdot\text{m}\cdot\text{s/rad} \qquad & c_L &= 2\times 10^{-3} \\
R_a &= 0.8\ \Omega & L_a &= 4\times 10^{-3} \text{ H} \\
I_m &= 5\times 10^{-4} & I_t &= 10^{-4} \qquad & I_L &= 5\times 10^{-3} \text{ kg}\cdot\text{m}^2 \\
N &= 2 & K_a &= 10 \text{ V/V} \\
K_{\text{tach}} &= 20 \text{ V}\cdot\text{s/rad} & K_{\text{pot}} &= 10 \text{ V/rad} & K_d &= 2 \text{ rad/(rad/s)}
\end{aligned}
$$

For the control system whose block diagram is given by Figure P10.18, determine the value of the proportional gain K_P required for the load speed to be within 10% of the desired speed of 2000 rpm at steady state, and use the characteristic roots to evaluate the resulting transient response. For this value of K_P, evaluate the resulting steady-state deviation of the load speed caused by a load torque $T_L = 0.2$ N · m.

Figure P10.18

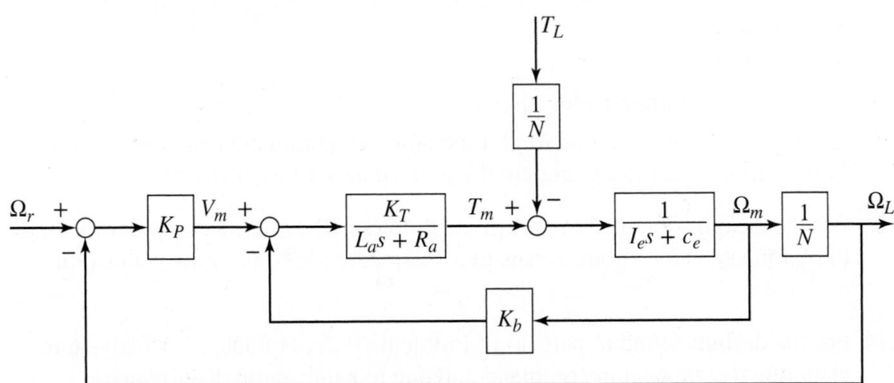

Section 10.6 Controlling First-Order Plants

10.19 Suppose the plant shown in Figure P10.19 has the parameter values $I = 4$ and $c = 7$. Find the smallest value of the gain K_P required so that the steady-state offset error will be no greater than 0.3 if ω_r is a unit-step input. Evaluate the resulting time constant and steady-state response due to the disturbance if T_d is also a unit step.

Figure P10.19

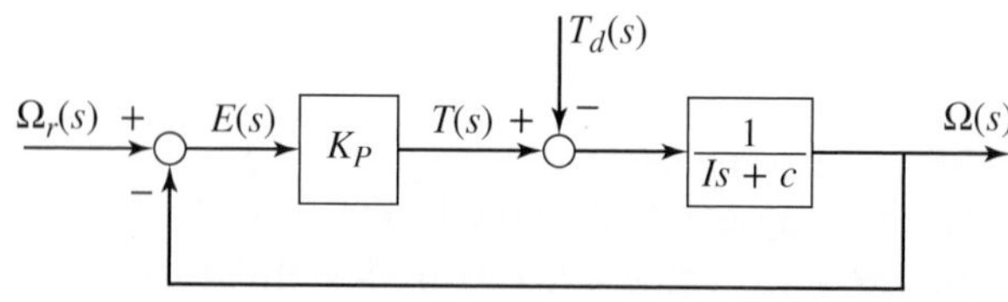

10.20 Suppose the plant shown in Figure P10.19 has the parameter values $I = 2$ and $c = 7$. The command input and the disturbance are unit-ramp functions. Evaluate the response of the proportional controller with $K_P = 30$.

10.21 For the control system shown in Figure P10.21, $I = 10$, and suppose that only I action is used, so that $K_P = 0$. The performance specifications require the steady-state errors due to step command and disturbance inputs to be zero. Find the required gain value K_I so that $\zeta = 1$. Evaluate the resulting time constant. Do this for each of the following values of c:

a. $c = 5$

b. $c = 0.2$

Figure P10.21

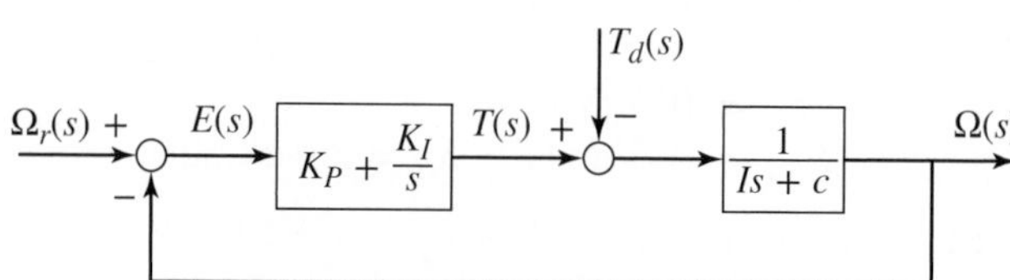

10.22 Suppose that $I = c = 4$ for the PI controller shown in Figure P10.21. The performance specifications require that $\tau = 0.2$. (a) Compute the required gain values for each of the following cases.

1. $\zeta = 0.707$
2. $\zeta = 1$
3. A root separation factor of 10

(b) Use a computer method to plot the unit-step command responses for each of the cases in part (a). Compare the performance of each case.

10.23 For the designs obtained in part (a) of Problem 10.22, use a computer method to plot the actuator torque versus time. Compare the peak torque values for each case.

10.24 For the designs found in part (a) of Problem 10.22, evaluate the steady-state error due to a unit-ramp command and due to a unit-ramp disturbance.

10.25 Consider the PI speed-control system shown in Figure P10.21, where $I = c = 5$. The desired time constant is $\tau = 0.2$. (a) Compute the required values of the gains for the following three sets of root locations.

1. $s = -10, -15$ (root separation factor is 1.5)
2. $s = -10, -20$ (root separation factor is 2)
3. $s = -10, -50$ (root separation factor is 5)

(b) Use a computer method to plot the response of the speed $\omega(t)$ for a unit-step command for each of the cases in part (a). Discuss the effects of the root separation factor on the rise time, the overshoot, and the maximum required torque.

10.26 Suppose that $I = c = 4$ for the I controller with internal feedback shown in Figure P10.26. The performance specifications require that $\tau = 0.2$. (a) Compute the required gain values for each of the following cases.

1. $\zeta = 0.707$
2. $\zeta = 1$
3. A root separation factor of 10

(b) Use a computer method to plot the unit-step command responses for each of the cases in part (a). Compare the performance of each case.

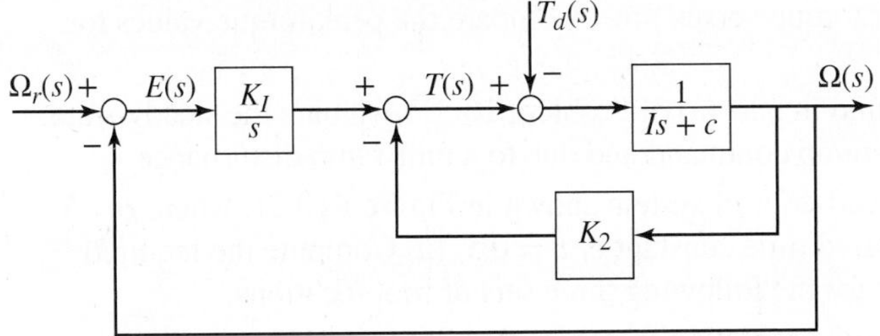

Figure P10.26

10.27 For the designs obtained in part (a) of Problem 10.26, use a computer method to plot the actuator torque versus time. Compare the peak torque values for each case.

10.28 For the designs found in part (a) of Problem 10.26, evaluate the steady-state error due to a unit-ramp command and due to a unit-ramp disturbance.

10.29 Consider the speed-control system using I control with internal feedback shown in Figure P10.26, where $I = c = 2$. The desired time constant is $\tau = 0.1$.

a. Compute the required values of the gains for the following three sets of root locations.

   1. $s = -10, -8$ (root separation factor is $10/8 = 1.25$)
   2. $s = -10, -20$ (root separation factor is 2)
   3. $s = -10, -50$ (root separation factor is 5)

b. Use a computer method to plot the response of the speed $\omega(t)$ for a unit-step command for each of the cases in part (a). Discuss the effects of the root separation factor on the rise time, the overshoot, and the maximum required torque.

10.30 Modify the diagram shown in Figure P10.30 to include feedforward command compensation with a constant compensator gain K_f. Determine whether such compensation can eliminate steady-state error for step and ramp commands.

Figure P10.30

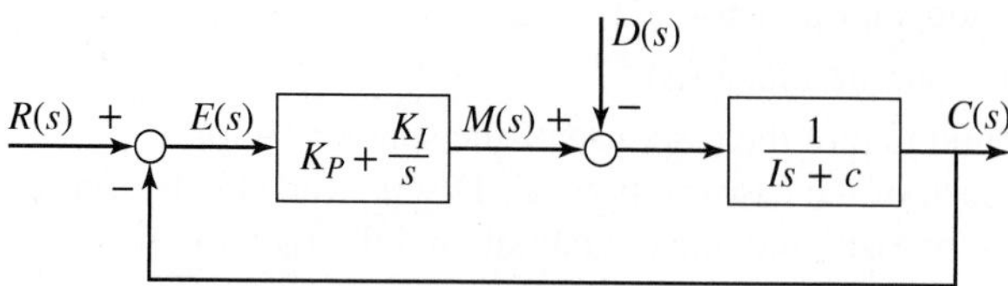

10.31 Suppose that $I = 12$ and $c = 6$ for the PI controller shown in Figure P10.21. The performance specifications require that $\tau = 3$. (a) Compute the required gain values for each of the following cases.

1. $\zeta = 0.707$
2. $\zeta = 1$
3. A root separation factor of 5

(b) Use a computer method to plot the unit-step command responses for each of the cases in part (a). Compare the performance of each case.

10.32 For the designs obtained in part (a) of Problem 10.31, use a computer method to plot the actuator torque versus time. Compare the peak torque values for each case.

10.33 For the designs found in part (a) of Problem 10.31, evaluate the steady-state error due to a unit-ramp command and due to a unit-ramp disturbance.

10.34 Consider the PI speed-control system shown in Figure P10.21, where $I = 5$ and $c = 4$. The desired time constant is $\tau = 0.5$. (a) Compute the required values of the gains for the following three sets of root locations.

1. $s = -2, -20$ (root separation factor is 10)
2. $s = -2, -10$ (root separation factor is 5)
3. $s = -2, -4$ (root separation factor is 2)

(b) Use a computer method to plot the response of the speed $\omega(t)$ for a unit-step command for each of the cases in part (a). Discuss the effects of the root separation factor on the rise time, the overshoot, and the maximum required torque.

10.35 Suppose that $I = 15$ and $c = 5$ for the I controller with internal feedback shown in Figure P10.26. The performance specifications require that $\tau = 0.5$. (a) Compute the required gain values for each of the following cases.

1. $\zeta = 0.707$
2. $\zeta = 1$
3. A root separation factor of 5

(b) Use a computer method to plot the unit-step command responses for each of the cases in part (a). Compare the performance of each case.

10.36 For the designs obtained in part (a) of Problem 10.35, use a computer method to plot the actuator torque versus time. Compare the peak torque values for each case.

10.37 For the designs found in part (a) of Problem 10.35, evaluate the steady-state error due to a unit-ramp command and due to a unit-ramp disturbance.

10.38 Consider the speed-control system using I control with internal feedback shown in Figure P10.26, where $I = 15$ and $c = 5$. The desired time constant is $\tau = 0.5$.

a. Compute the required values of the gains for the following three sets of root locations.
   1. $s = -2, -20$ (root separation factor is 10)
   2. $s = -2, -10$ (root separation factor is 5)
   3. $s = -2, -4$ (root separation factor is 2)

b. Use a computer method to plot the response of the speed $\omega(t)$ for a unit-step command for each of the cases in part (a). Discuss the effects of the root separation factor on the rise time, the overshoot, and the maximum required torque.

Section 10.7 Controlling Second-Order Plants

10.39 Consider the PD control system shown in Figure P10.39. Suppose that $I = 20$ and $c = 10$. The specifications require the steady-state error due to a unit-step command to be zero and the steady-state error due to a unit-step disturbance to be no greater than 0.1 in magnitude. In addition, we require that $\zeta = 0.707$.

Compute the required values of the gains, and evaluate the resulting time constant.

Figure P10.39

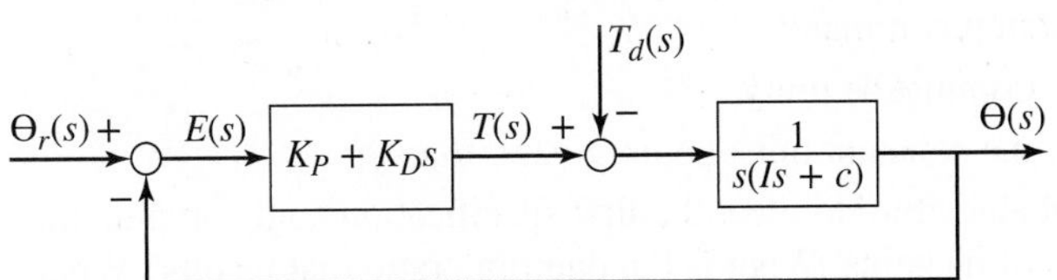

10.40 Suppose that $I = 20$ and $c = 5$ in the PD control system shown in Figure P10.39. The performance specifications require that $\tau = 1$ and $\zeta = 0.707$. Compute the required gain values.

10.41 Figure P10.41 shows a system using proportional control with velocity feedback. Suppose that $I = 20$ and $c = 10$. The specifications require the steady-state error due to a unit-step command to be zero and the steady-state error due to a unit-step disturbance to be no greater than 0.1 in magnitude. In addition, we require that the time constant be $\tau = 0.1$. Compute the required values of the gains and evaluate the resulting damping ratio.

Figure P10.41

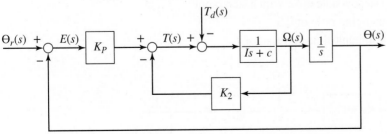

10.42 Suppose that $I = 10$ and $c = 3$ for the PID control system shown in Figure P10.42. The performance specifications require that $\tau = 1$ and $\zeta = 0.707$.

a. Compute the required gain values.

b. Use a computer method to plot the disturbance frequency response. Determine the peak response and the bandwidth.

Figure P10.42

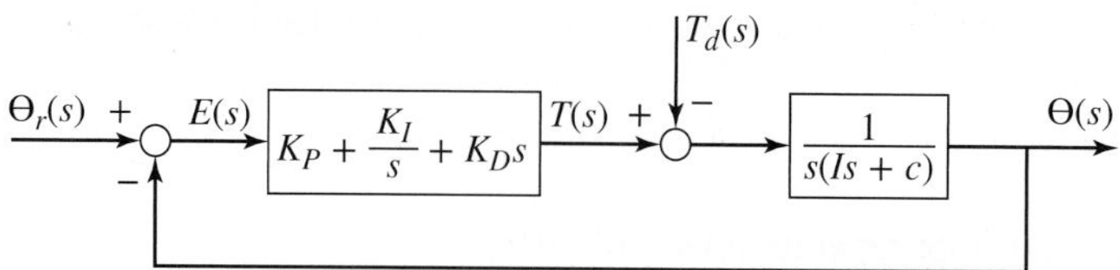

10.43 Consider the PD control system shown in Figure P10.39. Suppose that $I = 20$ and $c = 10$. The specifications require the steady-state error due to a unit-step command to be zero and the steady-state error due to a unit-step disturbance to be no greater than 0.1 in magnitude. In addition, we require that the time constant be $\tau = 0.1$.

Compute the required values of the gains, and evaluate the resulting damping ratio.

10.44 Modify the PD system diagram shown in Figure P10.39 to include feedforward compensation with a compensator gain of K_f. Determine whether such compensation can reduce the steady-state error for step and ramp commands.

10.45 Consider a plant whose transfer function is $1/(20s + 0.2)$. The performance specifications are

1. The magnitude of the steady-state command error must be no more than 0.01 for a unit-ramp command.
2. The damping ratio must be unity.
3. The dominant time constant must be no greater than 0.1.

a. Select a control algorithm to meet the first specification, and compute the required values of its gains. What is the damping ratio that results? What is the time constant?

b. Select a control algorithm to meet the first two specifications, and compute the required values of its gains. Evaluate the resulting time constant.

c. Select a control algorithm to meet all three specifications.

10.46 For the system shown in Figure P10.39, $I = c = 1$. Derive the expressions for the steady-state errors due to a unit-ramp command and to a unit-ramp disturbance.

10.47 For the PD control system shown in Figure P10.47, $I = c = 7$. Compute the values of the gains K_P and K_D to meet all of the following specifications:

1. No steady-state error with a step input
2. A damping ratio of 0.9
3. A dominant-time constant of 1

Figure P10.47

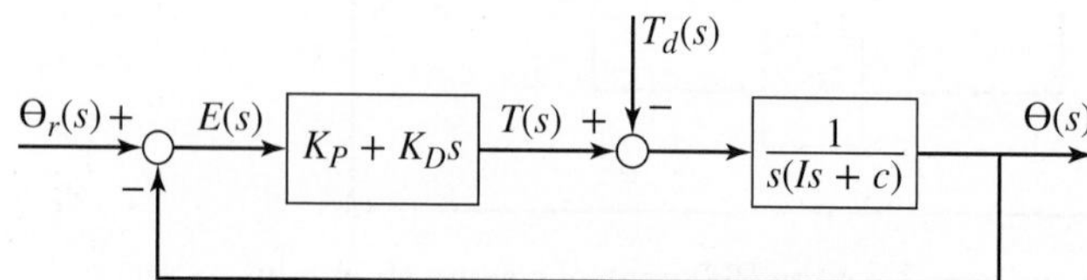

10.48 Consider the PID position control system shown in Figure P10.42, where $I = 10$ and $c = 2$. The desired time constant is $\tau = 2$.

a. Compute the required values of the gains for the following two sets of root locations.
   1. $s = -0.5$, $s = -5 \pm 5j$
   2. $s = -0.5$, $s = -1$, $s = -2$.

b. For both cases, use a computer method to plot the command response to a unit step. Discuss the effects of the root separation factor on the response. Compare the results with those of Example 10.7.4, where the roots are $s = -5$ and $s = -0.5 \pm 0.5j$.

c. For both cases, use a computer method to plot the disturbance frequency response. Discuss the effects of the root separation factor on the frequency response. Compare the results with those of Example 10.7.4.

10.49 Derive the expression for $T(s)$ in Figure P10.49. Using the values given and computed in Example 10.7.5, use MATLAB to plot $T(t)$ for a unit-step command input. Determine the maximum value of $T(t)$.

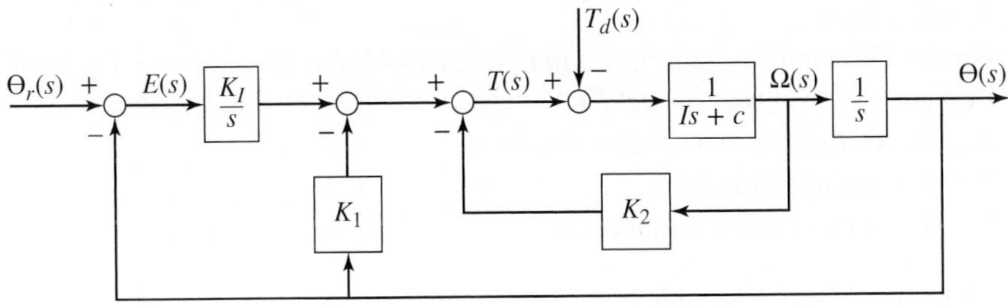

Figure P10.49

10.50 Integral control of the plant

$$G_p(s) = \frac{3}{5s + 1}$$

results in a system that is too oscillatory. Will D action improve this situation?

10.51 Modify the system diagram shown in Figure P10.51 to include feedforward compensation with a compensator gain K_f. Determine whether such compensation can reduce the steady-state error for step and ramp commands.

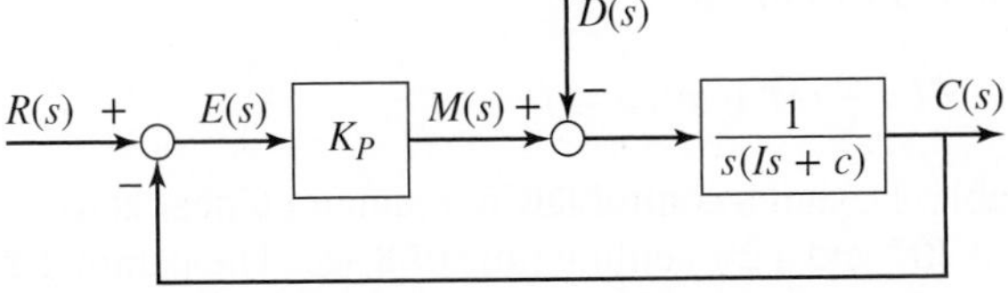

Figure P10.51

10.52 Consider the PD control system shown in Figure P10.47. Suppose that $I = 25$ and $c = 5$. The specifications require the steady-state error due to a unit-step command to be zero and the steady-state error due to a unit-step disturbance to be no greater than 0.2 in magnitude. In addition, we require that $\zeta = 0.707$. Compute the required values of the gains, and evaluate the resulting time constant.

10.53 Suppose that $I = 12$ and $c = 8$ in the PD control system shown in Figure P10.47. The performance specifications require that $\tau = 2$ and $\zeta = 0.707$. Compute the required gain values.

10.54 Suppose that $I = 15$ and $c = 5$ for the PID control system shown in Figure P10.42. The performance specifications require that $\tau = 2$ and $\zeta = 0.707$.

a. Compute the required gain values.

b. Use a computer method to plot the disturbance frequency response. Determine the peak response and the bandwidth.

10.55 Consider the PD control system shown in Figure P10.47. Suppose that $I = 25$ and $c = 2$. The specifications require the steady-state error due to a unit-step command to be zero and the steady-state error due to a unit-step disturbance to be no greater than 0.2 in magnitude. In addition, we require that the time constant be $\tau = 0.5$.

Compute the required values of the gains, and evaluate the resulting damping ratio.

10.56 For the PD control system shown in Figure P10.47, $I = 25$ and $c = 5$. Compute the values of the gains K_P and K_D to meet all of the following specifications:

1. No steady-state error with a step input
2. A damping ratio of 0.5
3. A dominant-time constant of 4

Section 10.8 Additional Examples

10.57 We need to stabilize the plant $3/(s^2 - 4)$ with a feedback controller. The closed-loop system should have a damping ratio of $\zeta = 0.707$ and a dominant-time constant $\tau = 0.1$.

a. Use PD control and compute the required values of the gains.

b. Use P control with rate feedback and compute the required values of the gains.

c. Compare the unit-step command responses of the two designs.

Figure P10.58

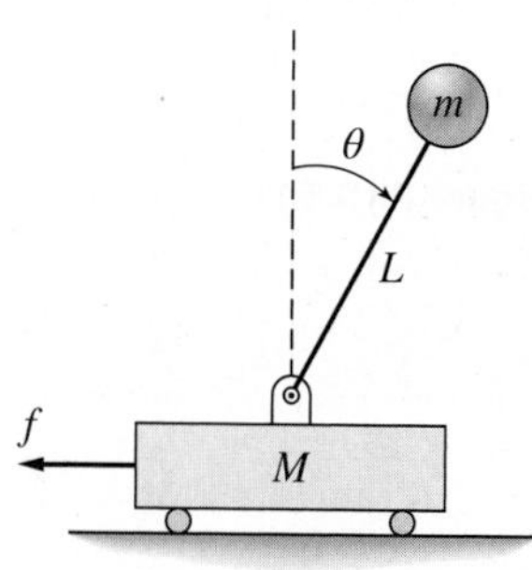

10.58 The system shown in Figure P10.58 represents the problem of stabilizing the attitude of a rocket during takeoff or controlling the balance of a personal transporter. The applied force f represents that from the side thrusters of the rocket or the tangential force on the transporter wheels. For small angles, Newton's law for the system reduces to

$$ML\ddot{\theta} - (M + m)g\theta = f$$

where f is the control variable. Design a control law to maintain θ near zero. The specifications are $\zeta = 0.707$ and a 2% settling time of 8 sec. The parameter values are $M = 40$ slugs, $m = 8$ slugs, $L = 20$ ft, and $g = 32.2$ ft/sec^2.

10.59 Figure P10.59 shows PD control applied to an unstable plant. The gains have been computed so that the damping ratio is $\zeta = 0.707$ and the time constant is 2.5 sec, assuming that the transfer functions of the actuator and the feedback sensor are unity. Suppose that the actuator has the transfer function

$$G_a(s) = \frac{1}{\tau s + 1}$$

Figure P10.59

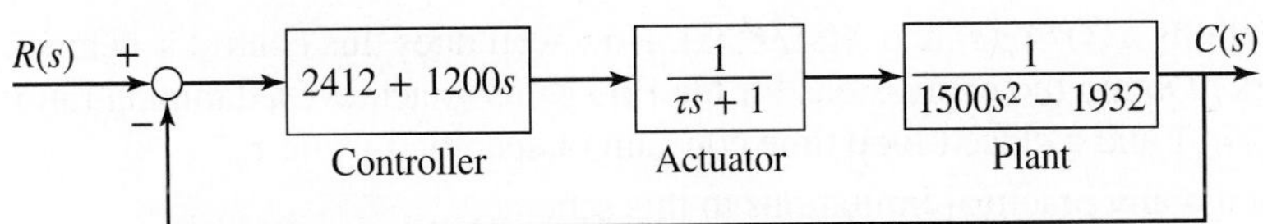

This lag in the response of the actuator might affect the system's stability, its overshoot in response to a step input, or its 2% settling time.

a. What effect does this lag have on the system's performance if $\tau = 0.1$ sec?

b. What effect does this lag have on the system's performance if $\tau = 1$ sec?

10.60 Figure P10.60 shows PD control applied to an unstable plant. The gains have been computed so that the damping ratio is $\zeta = 0.707$ and the time constant is 2.5 sec, assuming that the transfer functions of the actuator and the feedback sensor are unity. Suppose that the feedback sensor has the transfer function

$$G_s(s) = \frac{1}{\tau s + 1}$$

This lag in the response of the feedback elements might affect the system's stability, its overshoot in response to a step input, or its 2% settling time.

a. What effect does this lag have on the system's performance if $\tau = 0.1$ sec?

b. What effect does this have on the system's performance if $\tau = 1$ sec?

Figure P10.60

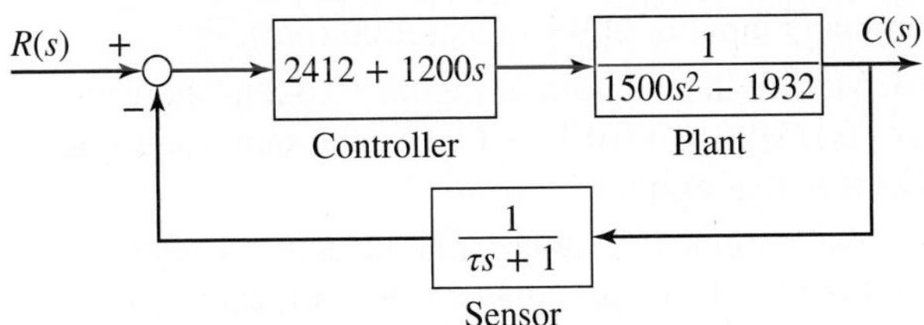

10.61 Figure P10.61 shows a proposed scheme for controlling the position of a mechanical system such as a link in a robot arm. It uses two feedback loops—one for position and one for velocity—and a feedforward compensator transfer function s^2.

a. Suppose that the estimates of the mass, damping, and stiffness are accurate so that $m_e = m$, $c_e = c$, and $k_e = k$. Derive the expressions for the transfer

Figure P10.61

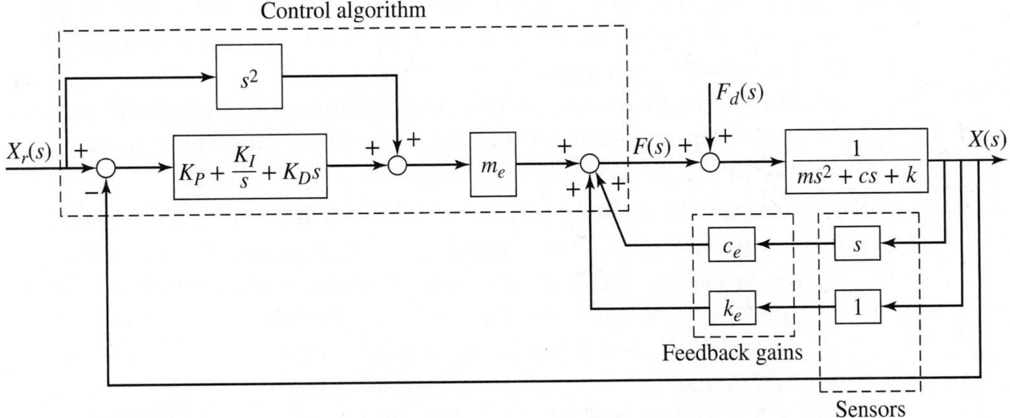

functions $X(s)/X_r(s)$ and $X(s)/F_d(s)$. How well does this control scheme work? Obtain the expressions for the PID gains to achieve a damping ratio of $\zeta = 1$ and a closed-loop time constant of specified value τ_d.

b. Discuss any practical limitations to this scheme.

10.62 Refer to Figure P10.18, which shows a speed-control system using an armature-controlled dc motor. The motor has the following parameter values.

$K_b = 0.199$ V-sec/rad $R_a = 0.43\ \Omega$

$K_T = 0.14$ lb-ft/A $c_e = 3.6 \times 10^{-4}$ lb-ft-sec/rad

$I_e = 2.08 \times 10^{-3}$ slug-ft^2 $L_a = 2.1 \times 10^{-3}$ H

$N = 1$

a. Compute the time constants of the plant transfer function $\Omega_L(s)/V_m(s)$.

b. Modify Figure P10.18 to use PI control instead of P control. Compute the PI control gains required to give a response having a dominant-control constant of no less than 0.05 sec and a dominant damping ratio in the range $0.5 \leq \zeta \leq 1$.

Section 10.9 Case Study: Motion Control with Feedback

10.63 Using the value of $K_P = 3.672$ from Problem 10.18, obtain a plot of the current versus time for a step-command input of 209.4 rad/s (2000 rpm).

10.64 Consider Example 10.6.3. Modify the diagram in Figure 10.6.2 to show an actuator transfer function $T(s)/M(s) = 1/(0.1s + 1)$. Use the same gain values computed for the three cases in that example.

a. Use MATLAB to plot the command response and the actuator response to a unit-step command. Identify the peak actuator values for each case.

b. Use MATLAB to plot the disturbance frequency response.

c. Compare the results in parts (a) and (b) with those of Example 10.6.3.

10.65 Consider Example 10.6.3. Use the same gain values computed for the three cases in that example.

a. Use MATLAB to plot the command response and the actuator response to the modified unit-step command $r(t) = 1 - e^{-20t}$. Identify the peak actuator values for each case.

b. Compare the results in part (a) with those of Example 10.6.3.

10.66 Consider Example 10.6.4. Modify the diagram in Figure 10.6.6 to show an actuator transfer function $T(s)/M(s) = 1/(0.1s + 1)$. Use the same gain values computed for the three cases in that example.

a. Use MATLAB to plot the command response and the actuator response to a unit-step command. Identify the peak actuator values for each case.

b. Use MATLAB to plot the disturbance frequency response.

c. Compare the results in parts (a) and (b) with those of Example 10.6.4.

10.67 Figure P10.7 shows a system for controlling the angular position of a load, such as an antenna. Figure P10.67 shows the block diagram for PD control of this system using a field-controlled motor. Use the following values:

$K_a = 1$ V/V $R = 0.3\ \Omega$ $K_T = 0.6$ N · m/A

$K_{pot} = 2$ V/rad $I_1 = 0.01$ kg · m^2

$I_2 = 5 \times 10^{-4}$ kg · m^2 $I_3 = 0.2$ kg · m^2

Figure P10.67

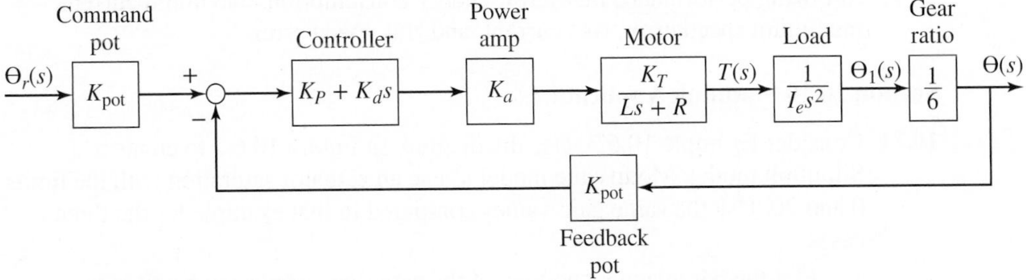

The inertia I_e in the block diagram is the equivalent inertia of the entire system, as felt on the *motor* shaft.

a. Assume that the motor inductance is very small and set $L = 0$. Compute I_e, obtain the transfer function $\Theta(s)/\Theta_r(s)$, and compute the values of the control gains K_P and K_d to meet the following specifications: $\zeta = 1$ and $\tau = 0.5$ s.

b. Use the MATLAB `tf` function to create the model `sys1` from this transfer function.

c. Using the values of K_P and K_d computed in part (a), and the value $L = 0.015$ H, obtain the transfer function $\Theta(s)/\Theta_r(s)$ and use the MATLAB `tf` function to create the model `sys2` from this transfer function.

d. Use the MATLAB `step(sys1,sys2)` function to plot the unit-step response of both transfer functions. Right-click on the plots to obtain the maximum percent overshoot and settling time for each. How close are the two responses? What is the effect of neglecting the inductance?

10.68 Consider the P, PI, and modified I control systems discussed in Examples 10.6.2, 10.6.3, and 10.6.4. The plant transfer function is $1/(Is + c)$, where $I = 10$ and $c = 3$. Investigate the performance of these systems for a trapezoidal command input having a slew speed of 1 rad/s, an acceleration time of 4 s, a slew time of 6 s, a deceleration time of 4 s, and a rest time of 5 s.

10.69 A speed-control system using an armature-controlled motor with proportional control action was discussed in Section 10.3. Its block diagram is shown in Figure 10.3.8 with a simplified version given in Figure 10.3.9. The given parameter values for a certain motor, load, and tachometer are

$K_T = K_b = 0.04$ N · m/A

$c_m = 0$	$c_L = 10^{-3}$ N · m · s/rad	
$R_a = 0.6\ \Omega$	$L_a = 2 \times 10^{-3}$ H	
$I_m = 2 \times 10^{-5}$	$I_t = 10^{-5}$	$I_L = 4 \times 10^{-3}$ kg · m^2
$N = 1.5$	$K_a = 5$ V/V	
$K_{tach} = 10$ V/(rad/s)	$K_{pot} = 5$ V/rad	$K_d = 2$ rad/(rad/s)

where the subscript m refers to the motor, L refers to the load, and t refers to the tachometer. (a) Determine the value of the proportional gain K_P required for the load speed to be within 10% of the desired speed of 1000 rpm at steady state, and plot the resulting transient response. (b) For this value of K_P, plot the resulting deviation of the load speed caused by a load torque $T_L = 1$ N · m.

10.70 Consider the control system of Problem 10.69. Use MATLAB to evaluate the following performance measures: energy consumption, maximum current, maximum speed error, rms current, and rms speed error.

Section 10.10 Simulink Applications

10.71 Consider Example 10.6.3. Use the diagram in Figure 10.6.2 to create a Simulink model. Modify the model to use an actuator saturation with the limits 0 and 20. Use the same gain values computed in that example for the three cases.

a. Plot the command response and the actuator response to a unit-step command.

b. Compare the results in part (a) with those of Example 10.6.3.

10.72 Consider Example 10.7.4. Use the diagram in Figure 10.7.3 to create a Simulink model using the same gain values computed in that example. Set the initial position to 3. Plot the command response to a unit-step command and compare the results with those of Example 10.7.4.

10.73 Consider Example 10.7.4. Use the diagram in Figure 10.7.3 to create a Simulink model. Modify the model to use an actuator saturation with the limits 0 and 20. Use the same gain values computed in that example.

a. Plot the command response and the actuator response to a unit-step command.

b. Compare the results in part (a) with those of Example 10.7.4.

10.74 Consider Example 10.7.4. Use the diagram in Figure 10.7.3 to create a Simulink model. Modify the model to use an actuator transfer function $G_a(s) = 1/(0.2s + 1)$. Use the same gain values computed in that example.

a. Plot the command response and the actuator response to a unit-step command.

b. Compare the command response with that of Example 10.7.4.

10.75 Refer to Figure 10.3.9, which shows a speed-control system using an armature-controlled dc motor. The motor has the following parameter values. Create a Simulink model by modifying Figure 10.3.9 to use PI control instead of P control. Use the PI control gains $K_P = 0.00728$, $K_I = 4.1481$.

$K_b = 0.199$ V-sec/rad	$R_a = 0.43\ \Omega$
$K_T = 0.14$ lb-ft/A	$c_e = 3.6 \times 10^{-4}$ lb-ft-sec/rad
$I_e = 2.08 \times 10^{-3}$ slug-ft^2	$L_a = 2.1 \times 10^{-3}$ H
$N = 1$	

a. Run the simulation using a unit-step command starting at $t = 0$ and a unit-step disturbance starting at $t = 4$ sec. Plot the speed and the motor current versus time.

b. Run the simulation for $0 \le t \le 2$ sec using a unit-ramp command. Plot the speed error and the motor current versus time.

10.76 For the system in Problem 10.75 part (a), create a Simulink model that has a current limiter of ± 10 A. Run the simulation for a step-command input of 104.7 rad/s (1000 rpm). Plot the current and the speed.

CHAPTER 11

Control System Design and the Root Locus Plot

CHAPTER OUTLINE

CHAPTER OBJECTIVES

When you have finished this chapter, you should be able to

1. Sketch the root locus plot for lower-order models, and use MATLAB to obtain the plot for higher-order models.
2. Interpret and use the root locus plot to determine the location of dominant roots and roots having desired properties such as damping ratio and time constant.
3. Determine the major features of a root locus plot, and use it to assess the effectiveness of a proposed control scheme.
4. Use the Ziegler-Nichols methods to design controllers.
5. Design a control system to avoid actuator saturation.
6. Design a controller incorporating state-variable feedback.
7. Apply MATLAB and Simulink to analyze and design control systems using the concepts presented in this chapter.

Chapter 10 introduced the basic concepts of feedback control. It showed how to choose an appropriate control action for a first- or second-order plant, and how to compute the control gains required to meet a simple set of performance specifications. This chapter shows how the root locus plot can be used to develop a more systematic approach to designing a control system. Such an approach is usually needed when the plant order is greater than two or where it is not clear how to select the gains to meet the performance specifications.

Section 11.1 introduces the root locus plot. Section 11.2 illustrates this approach using the root locus plot to design PID controllers. When PID control action fails to yield the desired performance, it must be either modified or replaced by an entirely different control scheme. Inserting an additional control element in one or more feedback loops within the main controller is called *feedback compensation*.

In many applications, especially in process control involving thermodynamic, fluid, or chemical processes, a transfer-function model of the plant is not available, and the gains must be computed from experimentally determined response data or from computer simulations of the plant. Often the gain values computed in a preliminary analysis do not yield the desired performance, and their values must be adjusted either in simulation or with the actual controller hardware. This process is called *tuning*. Tuning is discussed in Section 11.3.

Assumptions made to obtain a linear model can have a significant effect on the controller's performance. Using high gain values tends to drive the control elements to such an extent that they overload or "saturate" and thus exhibit nonlinear behavior. Design concepts that avoid these unwanted effects are covered in Section 11.4.

In Chapter 10 we illustrated the use of P action with rate feedback to replace PD control. This eliminated the numerator dynamics and the resulting overshoot. *State-variable feedback* is a generalization of that technique and uses some or all of the system's state variables to modify the control signal. With state-variable feedback, we have a better chance of placing the characteristic roots of the closed-loop system in locations that will give desirable performance. This topic is treated in Section 11.5.

Root locus plots for simple systems can be sketched by hand with the aid of a calculator. If these plots are to be used for higher-order systems, however, the use of computer methods is highly recommended. The MATLAB `rlocus`, `rltool`, and `sisotool` functions make it easy to view the root locus and transient-response plots to see the effect of changing parameter values. These methods are discussed in Section 11.6.

Simulink provides a quick and easy way of simulating systems having nonlinear or discontinuous elements such as the saturation nonlinearity. The Simulink features relevant to the topics of this chapter are summarized in Section 11.7.

Section 11.8 provides a review of the chapter's main concepts. ■

11.1 ROOT LOCUS PLOTS

As shown in Figure 11.1.1, a graphical display of the characteristic root locations gives insight into the system response. A *root locus plot* is a plot of the location of the characteristic roots as a parameter value is varied. Such a plot gives an overview of how the response will change if the parameter value is changed. Root locus plots were used widely in engineering design well before digital computers became available, but the usefulness of the root locus has been enhanced by programs such as MATLAB, which can quickly generate the plots. In this section, we will present some simple second-order system plots that can be sketched by hand to illustrate the concept. This section shows how to sketch plots for higher-order systems. In Section 11.6 we will show how to use MATLAB to generate the plots.

11.1.1 VARYING THE SPRING CONSTANT

In this section we will consider several examples where the characteristic equation has the form

$$ms^2 + cs + k = 0 \qquad (11.1.1)$$

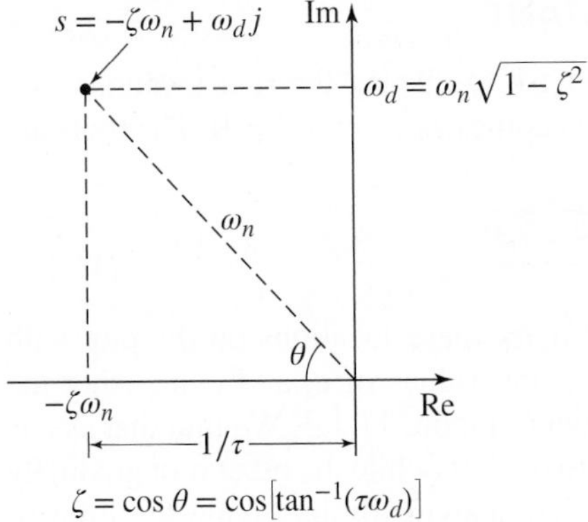

Figure 11.1.1 Graphical interpretation of the parameters ζ, τ, ω_n, and ω_d.

We will consider, in order, the effects of varying first the spring constant k, then the damping constant c, and finally the mass m.

Suppose that $m = 2$ and $c = 8$ and that we wish to display the root locations as k varies. In this case, the characteristic equation becomes $2s^2 + 8s + k = 0$. The roots are found from the quadratic formula

$$s = \frac{-8 \pm \sqrt{64 - 8k}}{4} \tag{11.1.2}$$

It is easily seen that if $k < 8$, the roots are real and distinct. They are repeated if $k = 8$ and complex conjugates if $k > 8$. By repeatedly evaluating (11.1.2) for various values of $k \geq 0$, the root locations can be plotted with dots in the s plane as in Figure 11.1.2a. Noting the general trend of the dots, we can connect them with solid lines to produce the plot in part (b) of the figure. The spacing of the dots is uneven but corresponds to an even spacing in the k values. By convention, the root locations corresponding to $k = 0$ are denoted by a cross (×). These locations are $s = 0$ and $s = -4$. The arrows on the plot indicate the direction of root movement as k increases.

The plot shows that if $k \geq 0$, we cannot achieve a dominant-time constant any smaller than $1/2$ by changing k. If $k < 8$, the dominant root lies between $s = 0$ and $s = -2$, and thus the dominant-time constant is never less than $1/2$. If $k \geq 8$, the time constant is always $1/2$. This illustrates the type of insight that can be obtained from the root locus plot.

The root locus plot gives us a picture of the roots' behavior as one parameter is varied. It thus enables us to obtain a more general understanding of how the system's response changes as a result of a change in that parameter. The plot is useful in system design where we need to select a value of the parameter to obtain a desired response.

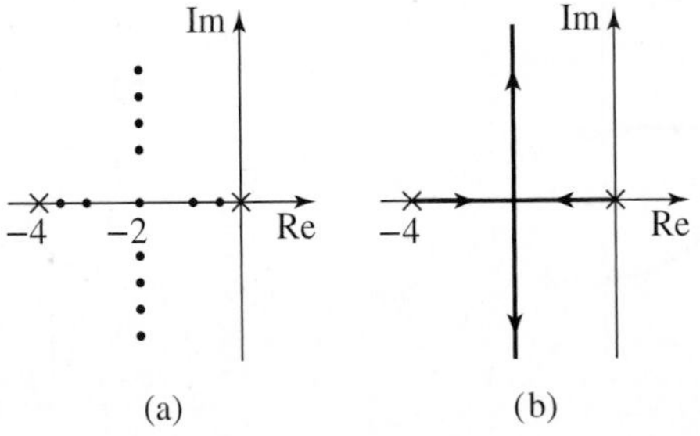

Figure 11.1.2 Plot of the roots of the equation $2s^2 + 8s + k = 0$ as k varies through positive values.

11.1.2 VARYING THE DAMPING CONSTANT

Suppose that $m = 2$ and $k = 8$, and that we wish to display the root locations as c varies. In this case, the characteristic equation becomes $2s^2 + cs + 8 = 0$. The roots are found from the quadratic formula

$$s = \frac{-c \pm \sqrt{c^2 - 64}}{4} \tag{11.1.3}$$

When $c = 0$, the roots are $s = \pm j2$, and we mark these locations on the plot with a $\times$. By evaluating and plotting the roots for many values of c, and connecting the points with a solid line, we obtain the plot shown in Figure 11.1.3. We find that as c is increased through large values, one root moves to the left, while the other root gradually approaches the origin $s = 0$. This can be proved analytically by taking the limit of (11.1.3) as $c \to \infty$; we find that one root approaches 0 and the other root approaches $s = -\infty$. By convention, the root location corresponding to an infinite value of the parameter is denoted by a circle (○).

When $c = 8$, both roots are $s = -2$. Recall that ω_n is the radius of a circle centered at the origin. From the plot it is easily seen that for $c < 8$, the roots are complex and that $\omega_n = 2$ (because the plot is a circle of radius 2 centered at the origin). For $c > 8$, the roots are real and distinct, and the plot shows that the dominant root is always no less than -2, and thus the dominant-time constant is always $\geq 1/2$. Thus, as long as $c \geq 0$, we cannot reduce the dominant-time constant below $1/2$ by changing c.

In this example, the characteristic equation can be expressed as

$$s^2 + 4 + \frac{c}{2}s = 0$$

which is a special case of the more general form in terms of the variable parameter μ.

$$s^2 + \beta s + \gamma + \mu(s + \alpha) = 0 \tag{11.1.4}$$

where $\beta = 0$, $\gamma = 4$, $\alpha = 0$, and $\mu = c/2$. When $\mu = 0$, $s^2 + \beta s + \gamma = 0$, and thus the starting points of the root locus, denoted by crosses ($\times$), are given by

$$s = \frac{-\beta \pm \sqrt{\beta^2 - 4\gamma}}{2}$$

These can be real or complex numbers, depending on the values of β and γ. It can be shown that as $\mu \to \infty$, one root of (11.1.4) approaches $s = -\alpha$ and the other root approaches $s = -\infty$.

A simple geometric analysis will show that off the real axis the root locus of (11.1.4) in terms of the variable parameter μ is a circle centered at $s = -\alpha$ and having a

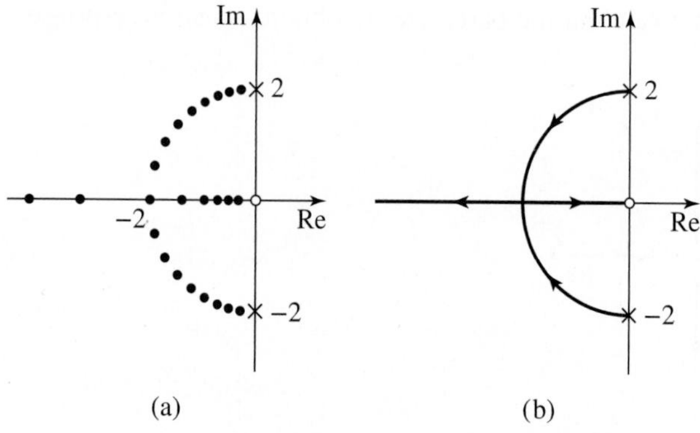

Figure 11.1.3 Plot of the roots of the equation $2s^2 + cs + 8 = 0$ as c varies through positive values.

radius of

$$R = \sqrt{\alpha^2 + \gamma - \alpha\beta} \tag{11.1.5}$$

To prove this, substitute $s = x + jy$ into (11.1.4), and separate the real and imaginary parts to obtain

$$x^2 - y^2 + \beta x + \gamma + \mu(x + \alpha) + jy(2x + \beta + \mu) = 0 + j0$$

Thus, for the real part,

$$x^2 - y^2 + \beta x + \gamma + \mu(x + \alpha) = 0$$

and for the imaginary part,

$$y(2x + \beta + \mu) = 0$$

Since $y \neq 0$ in general, the second equation gives $2x + \beta + \mu = 0$, or $\mu = -2x - \beta$. When this is substituted into the first equation, we obtain

$$(x + \alpha)^2 + y^2 = \alpha^2 + \gamma - \alpha\beta \tag{11.1.6}$$

which is the equation of the circle just described.

Root Locus of a Motor Model

EXAMPLE 11.1.1

■ Problem

Sketch the root locus of the armature-controlled dc motor model in terms of the damping constant c, and evaluate the effect on the motor time constant. The characteristic equation is

$$L_a I s^2 + (R_a I + cL_a)s + cR_a + K_b K_T = 0$$

Use the following parameter values:

$$K_b = K_T = 0.05 \text{ N} \cdot \text{m/A} \qquad I = 9 \times 10^{-5} \text{ kg} \cdot \text{m}^2$$

$$R_a = 0.5\ \Omega \qquad L_a = 2 \times 10^{-3} \text{ H}$$

■ Solution

Substituting the given values gives

$$1.8 \times 10^{-7} s^2 + (4.5 \times 10^{-5} + 2 \times 10^{-3} c)s + 0.5c + 2.5 \times 10^{-3} = 0$$

Dividing by the highest coefficient and factoring out c gives the standard form (11.1.4):

$$s^2 + 250s + \frac{5 \times 10^5}{36} + \frac{10^5 c}{9}(s + 250) = 0$$

where $\alpha = \beta = 250$, $\gamma = (5/36) \times 10^5$, and $\mu = 10^5 c/9$. From the results of (11.1.5) and (11.1.6) we see that the root locus is a circle centered at $s = -250$ with a radius of 118. See Figure 11.1.4. The starting points with $c = 0$ are at $s = -83$ and $s = -167$. As $c \to \infty$ one root approaches $s = -250$ and the other root approaches $s = -\infty$.

The plot illustrates the sometimes counterintuitive behavior of dynamic systems. One would think that increasing the damping c would increase the time constant and thus slow the response. However, as c is increased up to the value where the two roots meet at $s = -368$, the dominant-time constant decreases, and we see that the smallest possible dominant-time constant is $\tau = 1/368$.

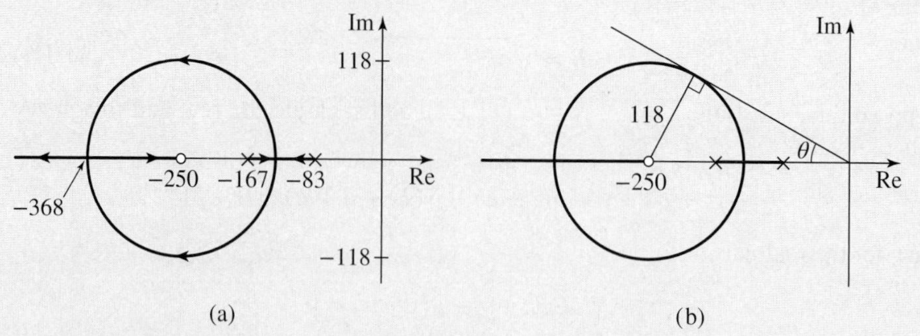

Figure 11.1.4 (a) Root locus plot of the motor model discussed in Example 11.1.1. (b) Line of minimum damping ratio.

It is instructive to see how this plot can be used to determine the minimum damping ratio the motor can have. The line corresponding to the minimum damping ratio is tangent to the circle, as shown in Figure 11.1.4b. From the triangle shown, we see that

$$\theta = \sin^{-1} \frac{118}{250} = 0.49 \text{ rad}$$

Thus, the minimum damping ratio is $\zeta = \cos \theta = 0.88$.

11.1.3 VARYING THE MASS

Now suppose that $c = 8$ and $k = 6$ and we want to investigate the effects of varying the mass m. The characteristic equation becomes $ms^2 + 8s + 6 = 0$, and the quadratic formula gives

$$s = \frac{-8 \pm \sqrt{64 - 24m}}{2m}$$

We immediately see that a problem arises if we try to examine the effects of varying the mass starting with $m = 0$, because m is in the denominator. In fact, if $m = 0$, the equation is no longer second-order and the quadratic formula no longer applies. So we conclude that we must be careful when varying the leading coefficient in the characteristic equation, because the equation order will change.

Instead, suppose we examine the effects of varying the mass for $m \geq 2$. For this case the equation always remains second-order. When $m = 2$, the two roots are $s = -1$ and $s = -3$, and these starting points are marked in Figure 11.1.5. As $m \to \infty$, the quadratic formula shows that both roots approach $s = 0$, and we mark this point with a small circle. By repeated evaluation of the quadratic formula for different m values, we obtain the plot shown in Figure 11.1.5. The locus off the real axis is a circle.

From the plot we see that as m is increased from $m = 2$, the two roots are real and approach one another, meeting at $s = -1.5$ when $m = 8/3$. For $m > 8/3$, the roots

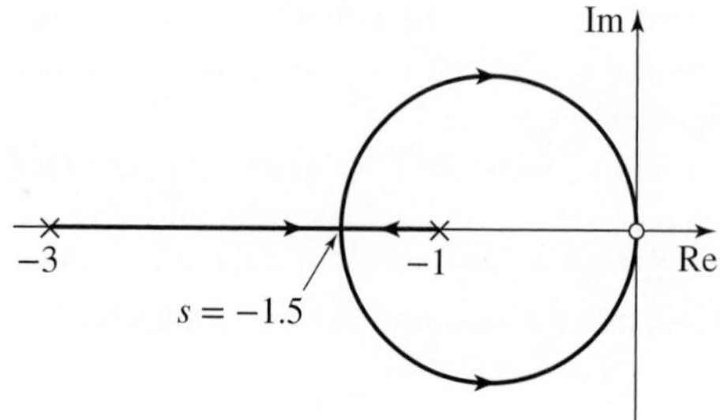

Figure 11.1.5 Plot of the roots of the equation $ms^2 + 8s + 6 = 0$ for $m \geq 2$.

become complex and move to the right. Thus, the smallest dominant-time constant the model can have is $\tau = 1/1.5 = 2/3$ and the largest possible undamped natural frequency is $\omega_n = 1.5$, which occurs when $m = 8/3$.

Let μ be the relative deviation of m from its nominal value of 2, that is let $\mu = (m - 2)/2$. Make the substitution $m = 2\mu + 2$ to obtain $(2\mu + 2)s^2 + 8s + 6 = 0$ or

$$s^2 + 4s + 3 + \mu s^2 = 0$$

where $\mu \geq 0$. This equation is a special case of the more general form

$$s^2 + \beta s + \gamma + \mu(s + \alpha)(s + \delta) = 0 \tag{11.1.7}$$

The root locus properties of this form can be developed as done with (11.1.4).

The examples in this section were second-order equations, for which a closed-form solution for the roots is available. For higher-order equations, however, a computer method is needed to obtain the root locus plot, and a MATLAB method is presented in Section 11.6.

11.2 DESIGN USING THE ROOT LOCUS PLOT

We now illustrate how to use the root locus plot to design control systems. The root locus plot is a plot of the location of the characteristic roots in terms of some system parameter, such as the proportional gain. Each path on the plot corresponds to one root of the characteristic equation. These paths are referred to as the *loci* or *branches*.

Although formulas exist for computing the roots of third- and fourth-order polynomials, their complexity is such that they do not generate much insight into the system's behavior. The Routh-Hurwitz criterion is useful for stability analysis, but it cannot give a complete picture of the transient response. The theory of polynomial equations, however, is sufficiently well developed to enable us to develop guides for sketching the general behavior of the roots without actually solving for them in many cases. These guides are useful and often even sufficient to make design decisions, and they also should be kept in mind when checking and interpreting a computer-generated plot.

The root locus design method is widely used, but when it was first proposed by Walter Evans in the 1940s [Evans, 1948], his paper was rejected for publication because the reviewers thought it not useful enough to merit publication![1] For aeronautical applications, see [Abzug, 1997].

11.2.1 TERMINOLOGY

The general form of an equation whose roots can be studied by the root locus method is

$$D(s) + KN(s) = 0 \tag{11.2.1}$$

where K is the parameter to be varied, and $D(s)$ and $N(s)$ are polynomials in s with constant coefficients. We consider the case $K \geq 0$ and later extend the results to $K \leq 0$. The guides to be developed for plotting the root locus require that the coefficients of the highest powers of s in both $N(s)$ and $D(s)$ are normalized to unity. The multipliers required to do this are absorbed into the parameter to be varied. The result is K.

[1]The author is grateful to Professors Thomas Kurfess and Mark Nagurka for this information.

For example, consider the variation of the parameter a in the equation $6s^2 + 13as + 5 = 0$. This can be written as

$$s^2 + \frac{5}{6} + \frac{13a}{6}s = 0$$

From (11.2.1), we see that $D(s) = s^2 + 5/6$, $N(s) = s$, and $K = 13a/6$. The root locus plot would be made in terms of the parameter K and the values for a recovered from $a = 6K/13$.

Another standard form of the problem is obtained by rewriting (11.2.1) as

$$1 + KP(s) = 0 \tag{11.2.2}$$

where

$$P(s) = \frac{N(s)}{D(s)} \tag{11.2.3}$$

The roots of $N(s) = 0$ are called the *zeros* of the problem. The name refers to the fact that they are the finite values of s that make $P(s)$ zero. The roots of $D(s) = 0$ are the *poles*. They are the finite values of s that make $P(s)$ become infinite. We see that when $K = 0$, the roots of (11.2.1) are the poles. To see what happens as $K \to \infty$, write (11.2.2) as

$$\frac{1}{K} + \frac{N(s)}{D(s)} = 0$$

When $K \to \infty$, $1/K \to 0$, and the equation becomes $N(s)/D(s) = 0$. This is equivalent to $N(s) = 0$, because we assume there are no poles at infinity. This shows that the roots of (11.2.1) approach the zeros as $K \to \infty$.

A single-loop control system is shown in Figure 11.2.1. The transfer function is

$$T(s) = \frac{C(s)}{R(s)} = \frac{G(s)}{1 + G(s)H(s)} \tag{11.2.4}$$

The *open-loop transfer function* of this system is $G(s)H(s)$ and so named because it is the transfer function relating the feedback signal $B(s)$ to the input $R(s)$ if the loop is "opened" or broken at $B(s)$. That is, $B(s) = G(s)H(s)R(s)$. The *closed-loop transfer function* is $T(s)$ given by (11.2.4).

The root locus form of the characteristic equation for a single-loop system is easily obtained from the equation $1 + G(s)H(s) = 0$ if the variable parameter of interest is a multiplicative factor in $G(s)H(s)$. The poles and zeros of the problem are also the poles and zeros of $G(s)H(s)$. Thus, the terms *open-loop poles* and *open-loop zeros* are often used for the poles and zeros on the root locus plot. The *closed-loop poles* are the characteristic roots, because they are the finite values of s that make the closed-loop transfer function become infinite.

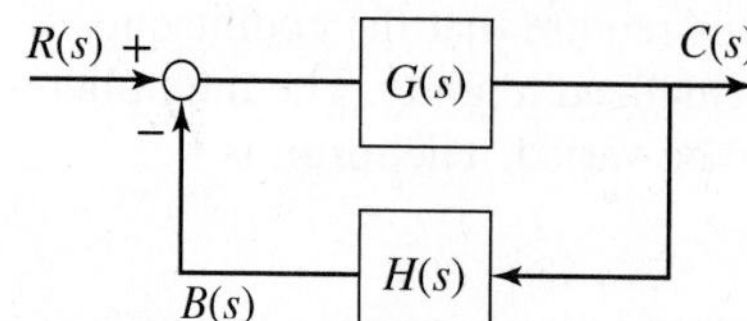

Figure 11.2.1 A single-loop control system.

Obtaining the Root Locus Equation

EXAMPLE 11.2.1

■ Problem

Obtain the root locus equations for the systems shown in Figure 11.2.2, where the variable parameter of interest is the proportional gain K_P. In parts (b) and (c) assume that T_I and T_D are known constants.

Figure 11.2.2 Block diagrams for Example 11.2.1.

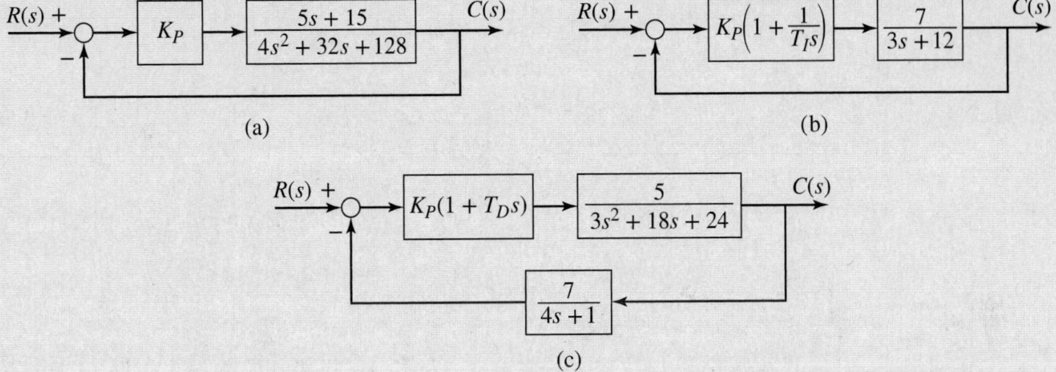

■ Solution

a. From the diagram in part (a) of Figure 11.2.2,

$$1 + G(s)H(s) = 1 + \frac{K_P(5s + 15)}{4s^2 + 32s + 128} = 0$$

Factor out $5K_P$ from the numerator and 4 from the denominator so that the highest coefficients will be 1. This gives

$$1 + \frac{5K_P}{4}\frac{s + 3}{s^2 + 8s + 32} = 0$$

So the root locus parameter K is $K = 5K_P/4$. The zero is $s = -3$ and the poles are $s = -4 \pm 4j$.

b. From the diagram in part (b) of Figure 11.2.2,

$$1 + G(s)H(s) = 1 + \frac{7K_P(1 + 1/T_I s)}{3s + 12} = 0$$

Multiply top and bottom by s, and factor out $7K_P$ from the numerator and 3 from the denominator. This gives

$$1 + \frac{7K_P}{3}\frac{s + 1/T_I}{s(s + 4)} = 0$$

The root locus parameter K is $K = 7K_P/3$. The zero is $s = -1/T_I$ and the poles are $s = 0$ and $s = -4$.

c. From the diagram in part (c) of Figure 11.2.2,

$$1 + G(s)H(s) = 1 + \frac{7(5)K_P(1 + T_D s)}{(3s^2 + 18s + 24)(4s + 1)} = 0$$

Factor out $7(5)K_PT_D$ from the numerator and 3(4) from the denominator so that the highest coefficients will be 1. This gives

$$1 + \frac{35K_PT_D}{12}\frac{s + 1/T_D}{(s^2 + 6s + 8)(s + 0.25)} = 0$$

So the root locus parameter K is $K = 35K_PT_D/12$. The zero is $s = -1/T_D$ and the poles are $s = -0.25$, $s = -2$, and $s = -4$.

From these examples we conclude that

1. P action never produces an open-loop pole or zero.
2. PD action produces an open-loop zero at $s = -1/T_D$.
3. I action produces an open-loop pole at $s = 0$.
4. PI action produces an open-loop pole at $s = 0$ and an open-loop zero at $s = -1/T_I$.
5. PID action produces an open-loop pole at $s = 0$ and two open-loop zeros, which may be real or complex, depending on the values of T_I and T_D.

If the variable parameter of interest is *not* a multiplicative factor in $G(s)H(s)$, you must obtain the characteristic equation from the denominator of the closed-loop transfer function and isolate the parameter to express the equation in the standard form. For example, with $H(s) = 1$, PI control of the plant $1/(4s + c)$ leads to the characteristic equation $4s^2 + (c + K_P)s + K_I = 0$. If $c = 4$, the gain values required to give critical damping with a time constant of 0.2 are $K_P = 36$ and $K_I = 100$. Now if the value of the damping c is uncertain, we may use the root locus to assess the effects of this uncertainty on the control system performance. The characteristic equation can be expressed as $4s^2 + (c + 36)s + 100 = 0$, which in standard form becomes

$$1 + \frac{c}{4}\frac{s}{s^2 + 9s + 25} = 0$$

The root locus parameter K is $K = c/4$. The zero is $s = 0$, and the poles are $s = -4.5 \pm 2.179j$.

11.2.2 ANGLE AND MAGNITUDE CRITERIA

The general problem in the form (11.1.2) can be written as

$$KP(s) = -1 \qquad (11.2.5)$$

From this, we see that two requirements must be met if s is to be a root of this equation, which is a statement of equality between two complex numbers, $KP(s)$ and -1. Recall that a complex number may be expressed as a magnitude and an angle. Thus, for two complex numbers to be equal, their magnitudes must be equal and their angles must also be equal. Applying this insight to (11.2.5), we obtain the following magnitude and angle criteria, respectively.

$$|KP(s)| = 1 \qquad (11.2.6)$$

$$\angle KP(s) = \angle(-1) = (2n + 1)180°, \qquad n = 0, 1, 2, 3, \ldots \qquad (11.2.7)$$

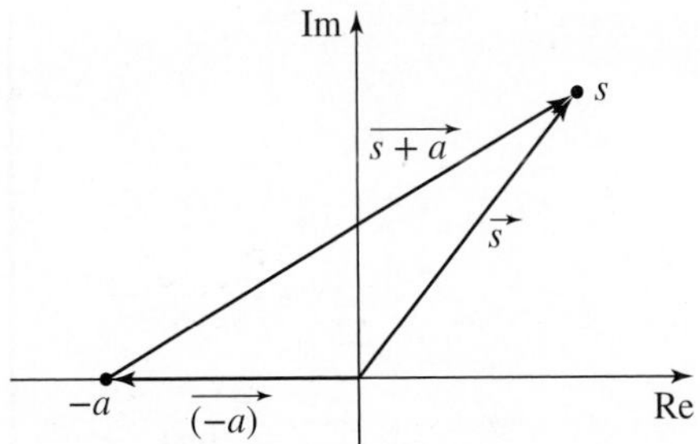

Figure 11.2.3 Vector addition and vector representation of complex numbers.

For now, we consider only $K \geq 0$. Thus, $|K| = K$ and $\angle K = 0°$, so that (11.2.6) becomes

$$K = \frac{1}{|P(s)|} = \frac{|D(s)|}{|N(s)|} \tag{11.2.8}$$

From Figure 11.2.3, we see that the property of vector addition gives

$$\vec{s} = \overrightarrow{(-a)} + \overrightarrow{(s+a)}$$

From this we deduce that the vector $\overrightarrow{(s+a)}$, which represents the complex number $s+a$, has its head at the point s and its tail at $-a$. This information is useful for applying (11.2.8), which can be stated as

$$K = \frac{\text{product of the magnitudes of vectors from poles to point } s}{\text{product of the magnitudes of vectors from zeros to point } s} \tag{11.2.9}$$

because the magnitude of a complex number is the length of its corresponding vector, and the magnitude of a product of complex numbers is the product of the magnitudes. Note that the point s must be on the root locus.

Three Poles and One Zero

EXAMPLE 11.2.2

■ Problem

Sketch the root locus of the following equation.

$$1 + K\frac{s+4}{s\left[(s+3)^2 + 2^2\right]} = 0$$

■ Solution

The zero is $s = -4$ and the poles are $s = 0$ and $s = -3 \pm 2j$. The root locus plot is shown in Figure 11.2.4a. Suppose we want to determine the value of K that puts one root at $s = -2$. Part (b) of the figure shows the vectors drawn from the poles and zero to the point $s = -2$. Then

$$K = \frac{|\vec{s}||\overrightarrow{s+3-2j}||\overrightarrow{s+3+2j}|}{|\overrightarrow{s+4}|}$$

From trigonometry we can compute the length of each vector to be

$$|\vec{s}| = 2 \qquad |\overrightarrow{s+4}| = 2$$

$$|\overrightarrow{s+3-2j}| = |\overrightarrow{s+3+2j}| = \sqrt{5}$$

Thus,

$$K = \frac{2(\sqrt{5})(\sqrt{5})}{2} = 5$$

Figure 11.2.4 (a) Root locus for the open-loop transfer function $(s + 4)/s\left[(s + 3)^2 + 2^2\right]$. (b) Vector calculation of the value of K corresponding to $s = -2$.

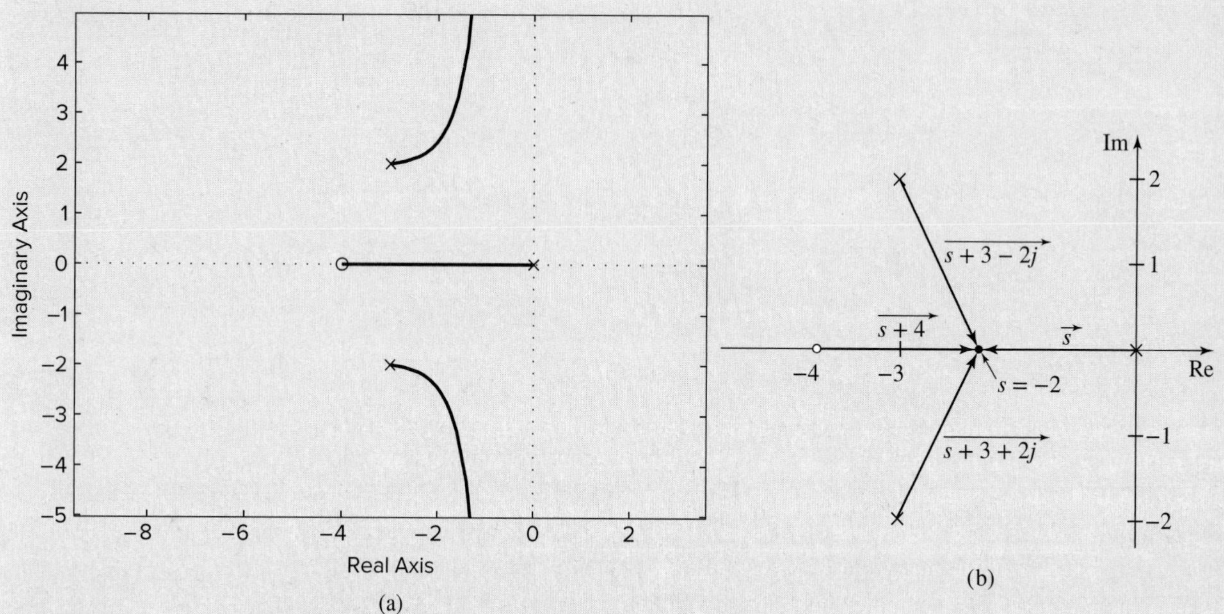

That is, if $K = 5$, one root is $s = -2$. Note that we could have obtained this result more easily by substituting $s = -2$ into the characteristic equation, but we will see that the vector interpretation is also useful.

Because the angle of a product of complex numbers is the sum of their angles, we have, because K is real and positive,

$$\angle KP(s) = \angle K + \angle P(s) = 0° + \angle P(s) = \angle P(s)$$

Equation (11.2.7) implies that

$$\angle P(s) = \angle N(s) - \angle D(s) = (2n + 1)180°, \qquad n = 0, 1, 2, 3, \ldots \tag{11.2.10}$$

Expressed in terms of vectors, this equation states that the sum of the angles of the vectors from the zeros to the point s minus the sum of the angles of the vectors from the poles to the point s must equal $(2n+1)180°$ if the point s lies on the root locus. This interpretation is useful for checking to see if a specific point is on the locus. For example, referring to Figure 11.2.4b,

$$\angle N(s) - \angle D(s) = 0° - 180° - \left[360° - \tan^{-1}\left(\frac{2}{1}\right)\right] - \tan^{-1}\left(\frac{2}{1}\right)$$
$$= -540°$$

which is equivalent to 180°. So the point $s = -2$ lies on the root locus.

Equations (11.2.8) and (11.2.10) are the *only* conditions that every point on the root locus must satisfy. It is interesting to note that the angle criterion (11.2.10) does not contain K. Thus, the shape of the root locus plot is determined entirely by the angle criterion (11.2.10). All of the plotting guides to follow, except two, are the result of this single condition. The magnitude criterion (11.2.8) is used only to obtain the associated value of K for a designated point s on the root locus.

11.2.3 ROOT LOCUS SKETCHING GUIDES

Let us collect the insights generated thus far and formalize them as guides for sketching the root locus where $K \geq 0$. First note that the root locus plot is *symmetric* about the real axis because complex roots occur only in conjugate pairs if the polynomial coefficients are real.

If you will be using the root locus plot for graphical calculations, such as measuring an angle to compute a damping ratio, the real and imaginary axes must have the same scale so that the angles will not be distorted. You can force MATLAB to use equal scales with the `axis equal` command.

We next note some facts that are obvious when we consider the general root locus equation

$$1 + KP(s) = 1 + K\frac{N(s)}{D(s)} = 0 \tag{11.2.11}$$

Assumption: We assume that the order of $N(s)$ is no greater than the order of $D(s)$.

The reason for this assumption will become clear shortly.

Guide 1: The number of paths equals the number of poles of $P(s)$.

This tells us how many paths we must account for. The proof uses the form $1 + KN(s)/D(s) = 0$. If the order of $D(s)$ is greater than or equal to that of $N(s)$, then the order of the equation, and thus the number of roots, is determined by $D(s)$. The order of $D(s)$, however, equals the number of poles of $P(s)$.

Guide 2: The paths start at the poles of $P(s)$ with $K = 0$.

When $K = 0$, (11.2.11) shows that the roots are given by $D(s) = 0$; in other words, by the poles.

Guide 3: The paths terminate with $K = \infty$ either at the zeros of $P(s)$ or by leaving the plot.

When $K \to \infty$, there are only two ways that (11.2.11) can be satisfied. The first requires that $N(s) \to 0$; that is, that s approach a zero of $P(s)$. If $N(s)$ does not approach zero as $K \to \infty$, then $D(s)$ must approach infinity, which can occur only if $s \to \infty$.

Variation of the Leading Coefficient Be careful when using the root locus where the variable parameter is the coefficient of the highest power of s in the characteristic equation. For example, consider the second-order equation $ms^2 + s + 8 = 0$. When $m = 0$, the equation is $s + 8 = 0$, which is *first*-order. The sketching guides to follow assume that the order of the equation remains the same as the parameter varies from 0 through positive values, and so we cannot use these guides to obtain the root locus plot of this second-order equation with m as the parameter.

If we put $ms^2 + s + 8 = 0$ into standard form, we obtain

$$1 + m\frac{s^2}{s + 8} = 0 \tag{11.2.12}$$

Thus, $K = m$. Note that the order of the numerator is greater than the order of the denominator. This indicates that the order of the equation will be different for $K = 0$ than for $K > 0$, and so the guides will be invalid.

Thoughtless application of Guides 1, 2, and 3 results in a contradiction when applied to (11.2.12). Guides 1 and 2 state that there is one path and it starts at $s = -8$, but Guide 3 says that two termination points exist because there are two zeros. The contradiction arises because the order of the numerator is greater than that of the denominator. When $m > 0$, there are two roots; when $m = 0$, there is only one.

Order of $N(s)$ and $D(s)$ We therefore assume that the order of $N(s)$, the numerator of $P(s)$, is no greater than the order of $D(s)$, the denominator of $P(s)$. Even if you do not use the following guides to obtain a root locus plot, but instead use a computer program, such as MATLAB, you should avoid equations where the order of the numerator is greater than that of the denominator.

Variation of the leading coefficient can be studied if the order of the equation remains the same throughout the variation. For example, with (11.2.12), if the nominal value of m is 0.5, let $\Delta = m - 0.5$, and express the equation as $(0.5 + \Delta)s^2 + s + 8 = 0$ or

$$1 + \frac{\Delta}{0.5}\frac{s^2}{s^2 + 2s + 16} = 0$$

Then $K = \Delta/0.5$. The guides for $K \geq 0$ can now be applied without any difficulty if we wish to study the effects of m varying *above* the nominal value of 0.5.

Root Locus Paths on the Real Axis The next guide is one of the most useful. Its proof is lengthy so we will omit it. For proof of this and other guides, see [Cannon, 1967].

> **Guide 4:** The root locus exists on the real axis only to the left of an odd number of real poles and/or real zeros. The numbering system is as follows. The real pole or real zero that lies the farthest to the right is number 1; the next real pole or zero is number 2; etc. If real poles or real zeros are identical, number each one.

Note that this guide implies that the locus cannot exist on the real axis to the *right* of an odd-numbered real pole or real zero.

Some examples of the application of this guide are shown in Figure 11.2.5, which does not show the root locus off the real axis.

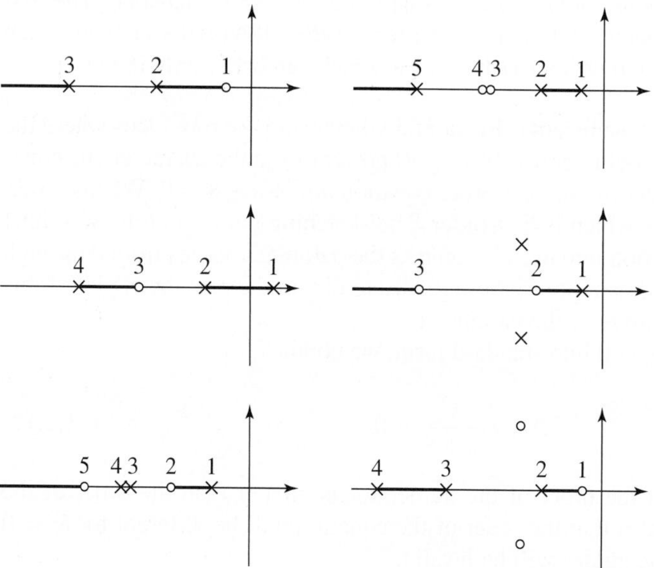

Figure 11.2.5 Examples of the application of Guide 4. The root loci off the real axis are not shown.

It is impossible for two root paths to coincide over any finite length (but they can cross). So two roots approaching each other on the real axis must leave the real axis at the point where they collide (called the *breakaway point*), and they must do this in a way that is symmetric about the real axis. This means that if one path breaks *up* from the real axis, the other path must break *down*.

Look for breakaway points where the root locus lies between two real poles, but they may occur at other locations. To compute the location of a breakaway point, note that the paths approach each other as K increases. Thus, at the breakaway point, K attains the relative *maximum* value it has on the *real* axis in the vicinity of the breakaway point. The value of s corresponding to the breakaway point can be found by computing dK/ds from the characteristic equation, setting $dK/ds = 0$, and solving for the value of s. In general, multiple solutions will occur. The extraneous ones can be discarded with a knowledge of the location of the locus on the real axis from Guide 4. Thus, the second derivative often need not be computed to distinguish between a minimum and a maximum.

A path can also enter the real axis, and the point at which this occurs is called the *breakin point*. If we increase K after the path has entered the real axis, the path continues along the real axis. Thus, at a breakin point, K attains the relative *minimum* value it has on the real axis in the vicinity of the breakin point. Look for breakin points where the root locus lies between two real zeros or between a real zero and $s = \pm\infty$, but they may occur at other locations. The location of the breakin point is determined from $dK/ds = 0$ in exactly the same manner as for a breakaway point. Although the method is the same, no difficulty is encountered in identifying the type of point, because this is usually obvious once the first four guides have been applied.

It is possible to have multiple breakaway or breakin points. This is why we speak of K attaining a *relative* or *local* maximum or minimum value only in the vicinity of each such point. Thus, we have Guide 5.

> **Guide 5:** The locations of breakaway and breakin points are found by determining where the parameter K attains a local maximum or minimum.

Example 11.2.2 illustrates how to find breakaway and breakin points.

Root Locus Paths Off the Real Axis The vector properties of complex numbers can be used to help understand how the root locus behaves off the real axis.

As $K \to \infty$, the paths approach a zero or the roots become infinite, which means that the paths exit the plot. To see what happens in this case, consider Figure 11.2.6, which is sketched for a case having three poles and one zero. The vectors from the poles

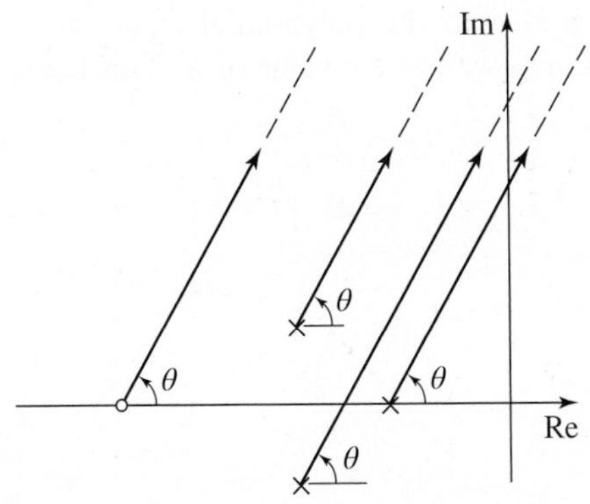

Figure 11.2.6 Illustration of Guide 6.

and zero to a point on the locus at infinity are parallel and make the same angle θ with the real axis. If we let p and z denote the number of poles and zeros, respectively, the angle criterion states that

$$P(s) = z\theta - p\theta = (2n+1)180°, \qquad n = 0, 1, 2, 3, \ldots$$

Solving for θ, we obtain

$$\theta = \frac{(2n+1)180°}{z-p} \qquad n = 0, 1, 2, 3, \ldots$$

This is equivalent to

$$\theta = \frac{(2n+1)180°}{p-z} \qquad n = 0, 1, 2, 3, \ldots$$

This leads to Guide 6.

Guide 6: The paths that do not terminate at a zero approach infinity along asymptotes. The angles that the asymptotes make with the real axis are found from

$$\theta = \frac{(2n+1)180°}{p-z} \qquad n = 0, 1, 2, 3, \ldots \tag{11.2.13}$$

where n is chosen successively as $n = 0, 1, 2, 3, \ldots$, until enough angles have been found.

Figure 11.2.7 shows the most commonly found patterns for asymptotes.

Unless θ is 0° or 180°, the asymptotes cannot be drawn until we know where they intersect the real axis. This is given by Guide 7. The proof follows from the angle criterion, but is detailed (see [Cannon, 1967]).

Guide 7: The asymptotes intersect the real axis at the common point $s = \sigma$ given by

$$\sigma = \frac{\sum s_p - \sum s_z}{p-z} \tag{11.2.14}$$

where $\sum s_p$ and $\sum s_z$ are the algebraic sums of the values of the poles and zeros.

Note that this guide states that all the asymptotes intersect at the same point.

Determination of Instability Some systems become unstable for certain values of the parameter K. This occurs if any path crosses the imaginary axis into the right half-plane. The value of K at which this happens can be determined from the Routh-Hurwitz criterion. Often, however, the substitution $s = j\omega$ into the polynomial equation of interest is quicker and gives the crossing location as well as the value of K. This leads to Guide 8.

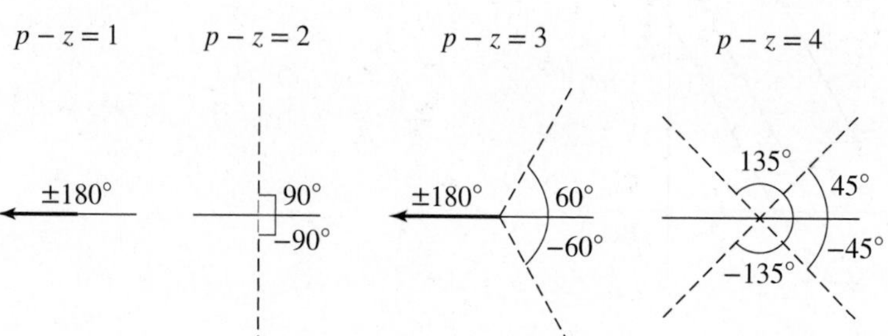

Figure 11.2.7 Common patterns for asymptotes.

Guide 8: The points at which the paths cross the imaginary axis and the associated values of K can be found by substituting $s = j\omega$ into the characteristic equation.

The solution for the frequency ω is called the *crossover frequency*, and the points $s = \pm j\omega$ are called the *crossover points*. Examples 11.2.2 and 11.2.4 illustrate the procedure for computing the crossover points.

In many applications, it is not necessary to determine the precise location of the locus off the real axis, and the first eight guides are often quite sufficient for obtaining ample information about the system's behavior. When more accuracy is required, a computer program such as MATLAB can be used.

Locus Behavior Near Complex Poles and Zeros Sometimes it is helpful to know the direction in which the locus leaves a complex pole (called the *angle of departure*) and the direction in which it terminates at a complex zero (called the *angle of arrival*). To determine these angles, we again call on the angle criterion.

Guide 9: Angles of departure and angles of arrival are determined by choosing an arbitrary point infinitesimally close to the pole or zero in question and applying the angle criterion

$$\angle N(s) - \angle D(s) = (2n + 1)180^\circ \tag{11.2.15}$$

Locating All the Roots The following guide comes from the theory of polynomial equations.

Guide 10: For the polynomial equation

$$s^n + a_{n-1}s^{n-1} + \cdots + a_1 s + a_0 = 0 \tag{11.2.16}$$

the sum of the roots $r_1, r_2, \ldots, r_n$ is

$$r_1 + r_2 + \cdots + r_n = -a_{n-1} \tag{11.2.17}$$

Note that the coefficient of s^n is unity.

This guide is useful for determining the location of the remaining roots once some roots have been found.

Determining the Value of K Once the root locus is drawn, the magnitude criterion can be used to compute the value of K associated with a particular point on the locus. If the polynomial is written as $D(s) + KN(s) = 0$, the magnitude criterion for $K \geq 0$ states that

$$K = \frac{|D(s)|}{|N(s)|} \tag{11.2.18}$$

Table 11.2.1 summarizes the procedure for sketching a root locus plot, and for obtaining information from it. Figures 11.2.8 through 11.2.12 display a number of possible root locus plots for second- and third-order equations. Figure 11.2.8 shows six of the possible plots for second-order equations. Note that you can recover the characteristic equation from the values of the poles and zeros.

Most of the root locus plots in this chapter were generated with MATLAB. At this time, MATLAB users may wish to study Section 11.6. Users of other software should consult the appropriate software documentation.

Table 11.2.1 Root locus sketching procedure for $K \geq 0$.

Procedure step	Appropriate guide or equation
1. Express the characteristic equation in the standard form.	■ $1 + KP(s) = 0$ where $P(s) = N(s)/D(s)$, $K \geq 0$ ■ Highest coefficients of $N(s)$ and $D(s)$ must be 1. ■ The order of $N(s)$ must be no greater than the order of $D(s)$.
2. Determine the poles and zeros.	■ Solve $D(s) = 0$ for the poles. ■ Solve $N(s) = 0$ for the zeros. ■ p = number of poles. z = number of zeros.
3. Plot the poles with × and the zeros with ○.	■ The number of paths equals the number of poles.
Use same scale on both axes. Locus is symmetric about the real axis.	■ The zeros are termination points for z paths. ■ The remaining $p - z$ paths leave the plot.
4. Sketch the root locus on the real axis.	■ The locus lies only to the left of odd-numbered real poles and/or real zeros.
5. Determine any breakaway and breakin points.	■ Solve $\dfrac{dK}{ds} = -\dfrac{dP(s)}{ds} = 0$ for s.
6. If $p > z$, compute the angles and intersection point of the asymptotes.	■ $\theta = \dfrac{(2n+1)180°}{p - z} \quad n = 0, 1, 2, 3, \ldots$ ■ $\sigma = \dfrac{\sum s_p - \sum s_z}{p - z}$
7. Determine any crossover points.	■ Substitute $s = j\omega$ into the characteristic equation, separate the real and imaginary parts, and solve for ω and K, or ■ use the Routh-Hurwitz criterion to find K, then solve for the characteristic roots.
8. Determine any angles of departure from complex poles, and any angles of arrival at complex zeros.	■ Apply $\angle N(s) - \angle D(s) = (2n+1)180°$ $n = 0, 1, 2, 3, \ldots$, where s is the pole or zero location.
9. Determine the value of K at any desired root location s_d.	■ Use the magnitude criterion: $K_d = \dfrac{\lvert D(s_d)\rvert}{\lvert N(s_d)\rvert}$
10. Locate remaining roots corresponding to K_d.	■ Use the fact that $-a_{n-1}$ is the sum of the roots of the equation $s^n + a_{n-1}s^{n-1} + a_{n-2}s^{n-2} + \cdots + a_1 s + a_0 = 0$

Figure 11.2.9 shows four of the possible plots for third-order equations having no zeros. Note that part (d) has a breakin point even though it does not have a zero, and a breakaway point even though there is only one real pole.

Figure 11.2.10 shows four of the possible plots for third-order equations having one zero. Note that part (d) has both a breakin and a breakaway point between the zero and the pole.

Figure 11.2.11 shows two of the possible plots for third-order equations having complex poles and one zero. Note that part (b) has both a breakin and a breakaway point between the zero and the real pole.

Figure 11.2.12 shows three of the possible plots for third-order equations having two zeros. Note that part (a) has multiple crossover points. Note that part (c) has both a breakin and a breakaway point between the real pole and $s = -\infty$.

Note that these figures do not show all of the possible root locus plots for second- and third-order equations.

Figure 11.2.8 Root locus plots for some second-order equations.

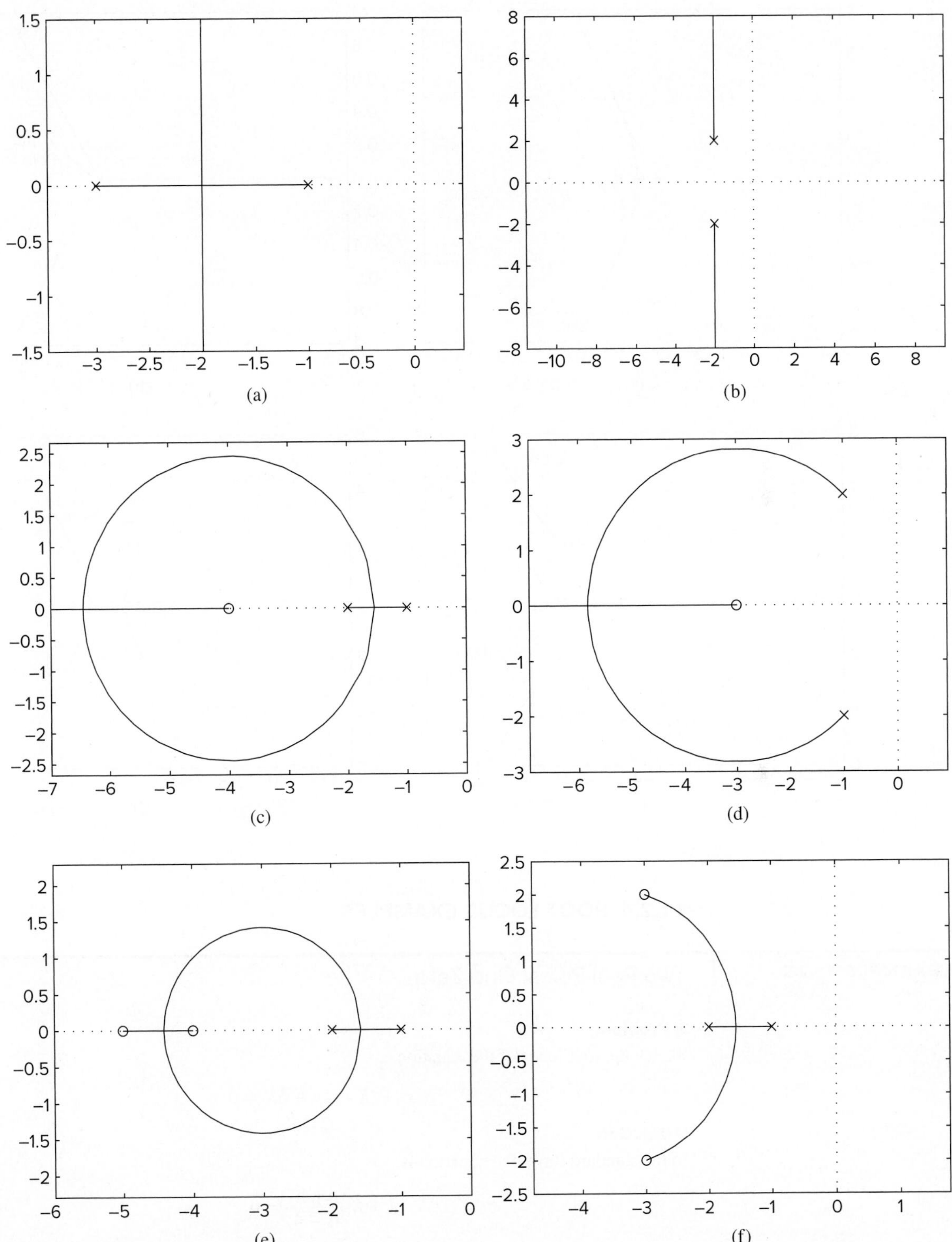

Figure 11.2.9 Root locus plots for some third-order equations with no zeros.

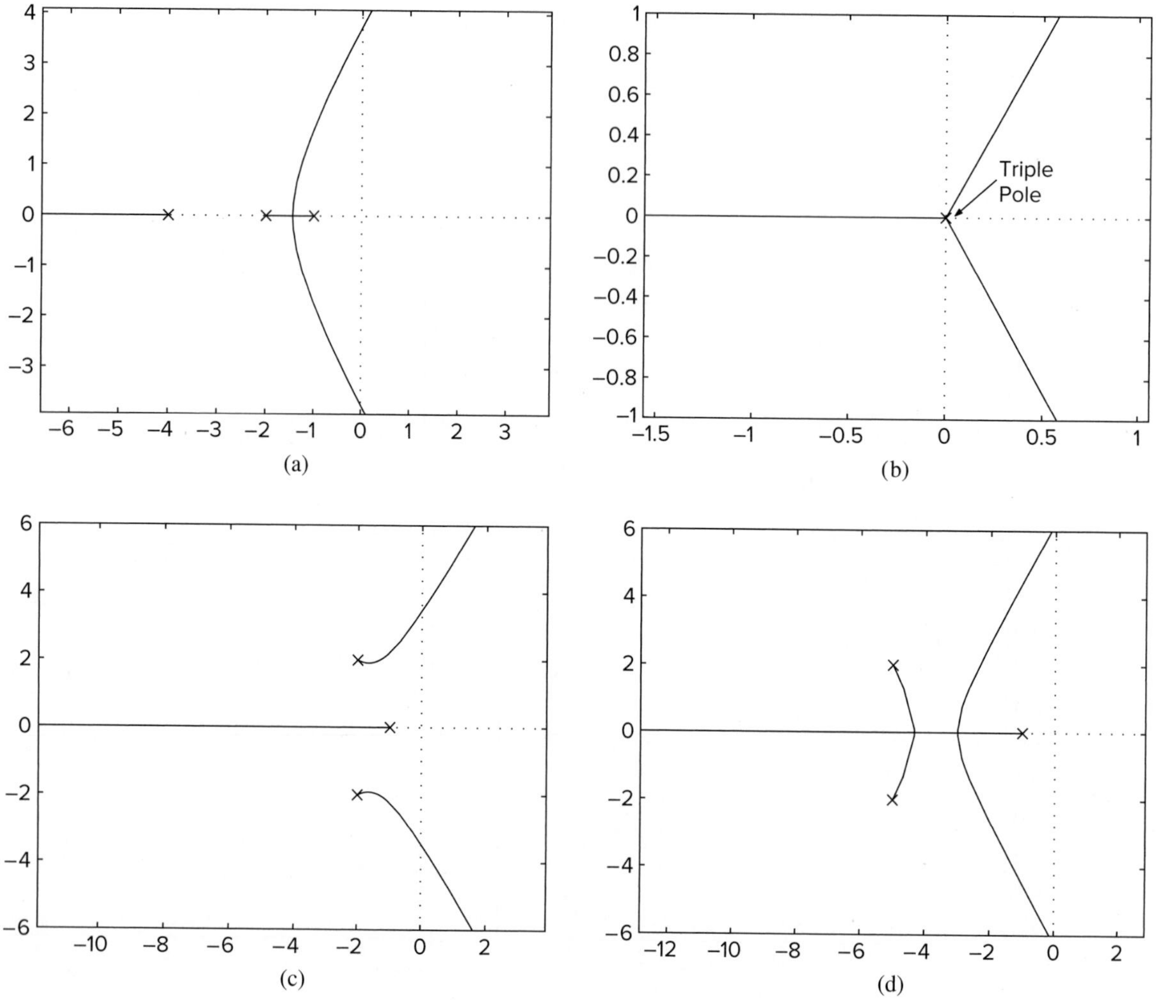

11.2.4 ROOT LOCUS EXAMPLES

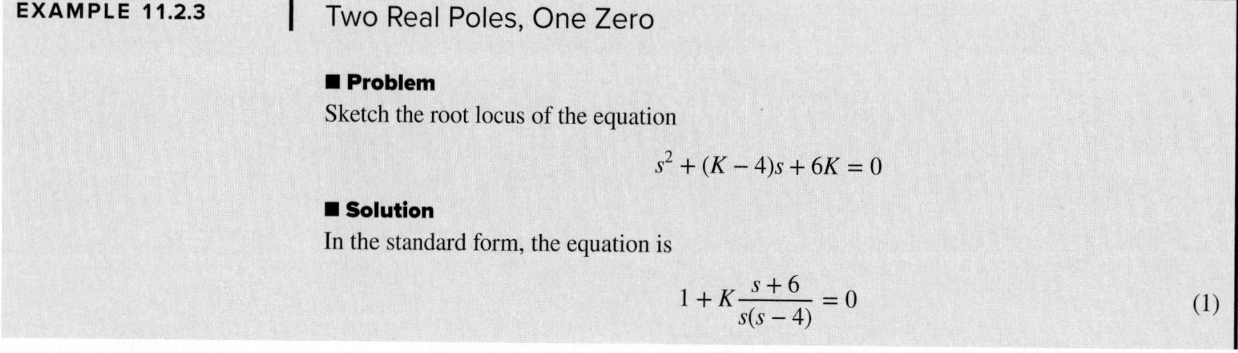

EXAMPLE 11.2.3 Two Real Poles, One Zero

■ **Problem**

Sketch the root locus of the equation

$$s^2 + (K - 4)s + 6K = 0$$

■ **Solution**

In the standard form, the equation is

$$1 + K\frac{s+6}{s(s-4)} = 0 \quad (1)$$

Figure 11.2.10 Root locus plots for some third-order equations having real poles and one zero.

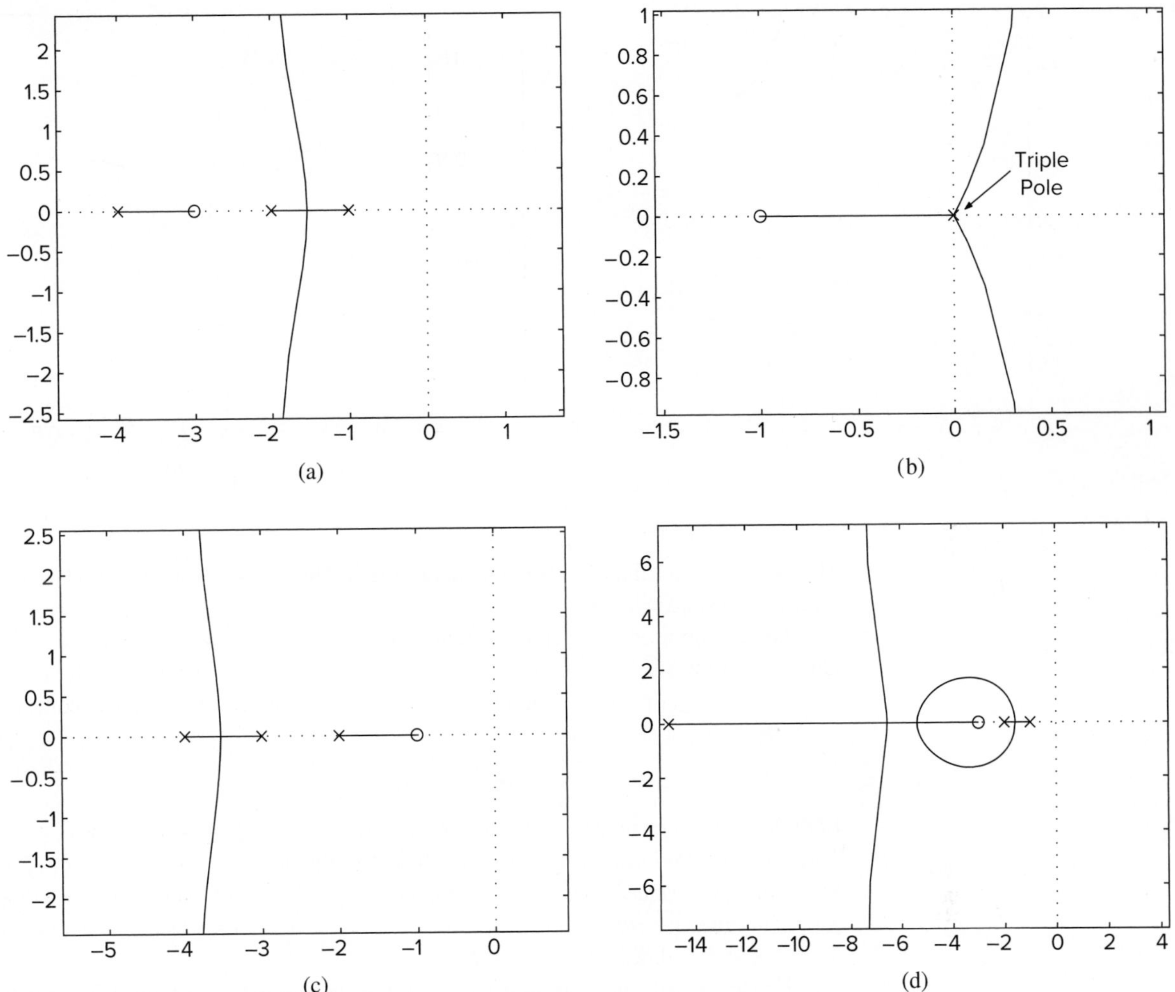

The poles are $s = 0$ and $s = 4$. The zero is $s = -6$. Thus, there are two paths. One starts at $s = 0$ and the other at $s = 4$, with $K = 0$. One path terminates at $s = -6$, while the other must leave the plot as $K \to \infty$.

Guide 4 shows that the locus exists on the real axis between $s = 0$ and $s = 4$ and to the left of $s = -6$. Therefore, the two paths must break away from the real axis between $s = 0$ and $s = 4$. From the location of the termination point at $s = -6$, we know that the locus must return to the real axis somewhere to the left of $s = -6$.

The breakaway and breakin points are found as follows. Solve equation (1) for K, compute dK/ds, and solve $dK/ds = 0$.

$$K = -\frac{s(s-4)}{s+6}$$

$$\frac{dK}{ds} = -\frac{(s+6)(2s-4) - s(s-4)}{(s+6)^2} = 0$$

Figure 11.2.11 Root locus plots for some third-order equations having complex poles and one zero.

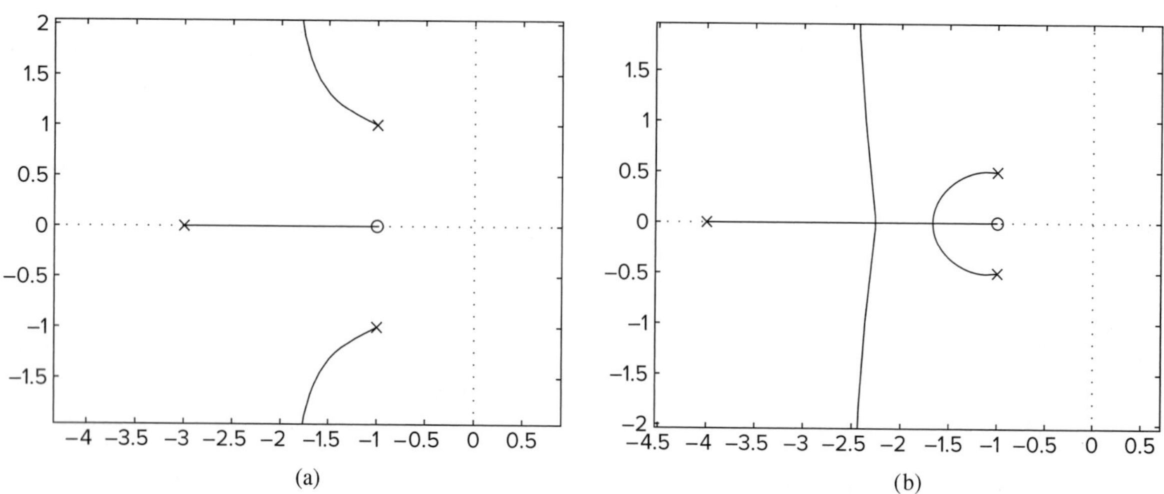

This is satisfied for finite s if the numerator is zero. Thus, $s^2 + 12s - 24 = 0$. The candidates are $s = 1.75$ and $s = -13.7$.

There is no need to check for a minimum or a maximum of K, because we know from Guide 4 that the breakaway point must be $s = 1.75$ and the breakin point must be $s = -13.7$.

The crossover point is found by substituting $s = j\omega$ into equation (1), and collecting real and imaginary parts. This gives

$$6K - \omega^2 + (K - 4)\omega j = 0$$

Both the real and imaginary parts must be zero, so we have $6K - \omega^2 = 0$ and $(K - 4)\omega = 0$. The latter equation gives either $\omega = 0$, which corresponds to the pole at $s = 0$, or $K = 4$. Substituting this value into the equation for the real part gives $6(4) - \omega^2 = 0$, or $\omega = \pm\sqrt{24} = \pm 4.9$. This gives the crossover points, which are $s = \pm 4.9j$ when $K = 4$. The system is unstable for $0 \leq K < 4$, and neutrally stable if $K = 4$.

This leaves only the shape of the locus off the real axis to be determined. For equations of order higher than two, the locus does not have a simple shape, but for second-order equations, it is often a circle. We note here that the breakaway and breakin points are symmetrically placed with a distance of 7.75 from the zero at $s = -6$. The simplest way for this to occur is with a circle of radius 7.75 centered at the zero. Another confirmation of this is the fact that the crossover points are a distance $\sqrt{6^2 + 4.9^2} = 7.75$ from the zero. The root locus is shown in Figure 11.2.13a. Part (b) of the figure shows the variation in K as s varies through real values. Note that K reaches a relative minimum at a breakin point and a relative maximum at a breakaway point. The plot for $0 \leq s \leq 4$ has been expanded to show the variation more clearly.

Elementary geometry can be used to show that the root locus of the equation

$$(s + b)(s + c) + K(s + a) = 0 \tag{11.2.19}$$

is a circle centered on the zero at $s = -a$ if c is positive and $a > b > c$. The radius of the circle can be determined once the breakaway and breakin points are found.

Figure 11.2.12 Root locus plots for some third-order equations having two zeros.

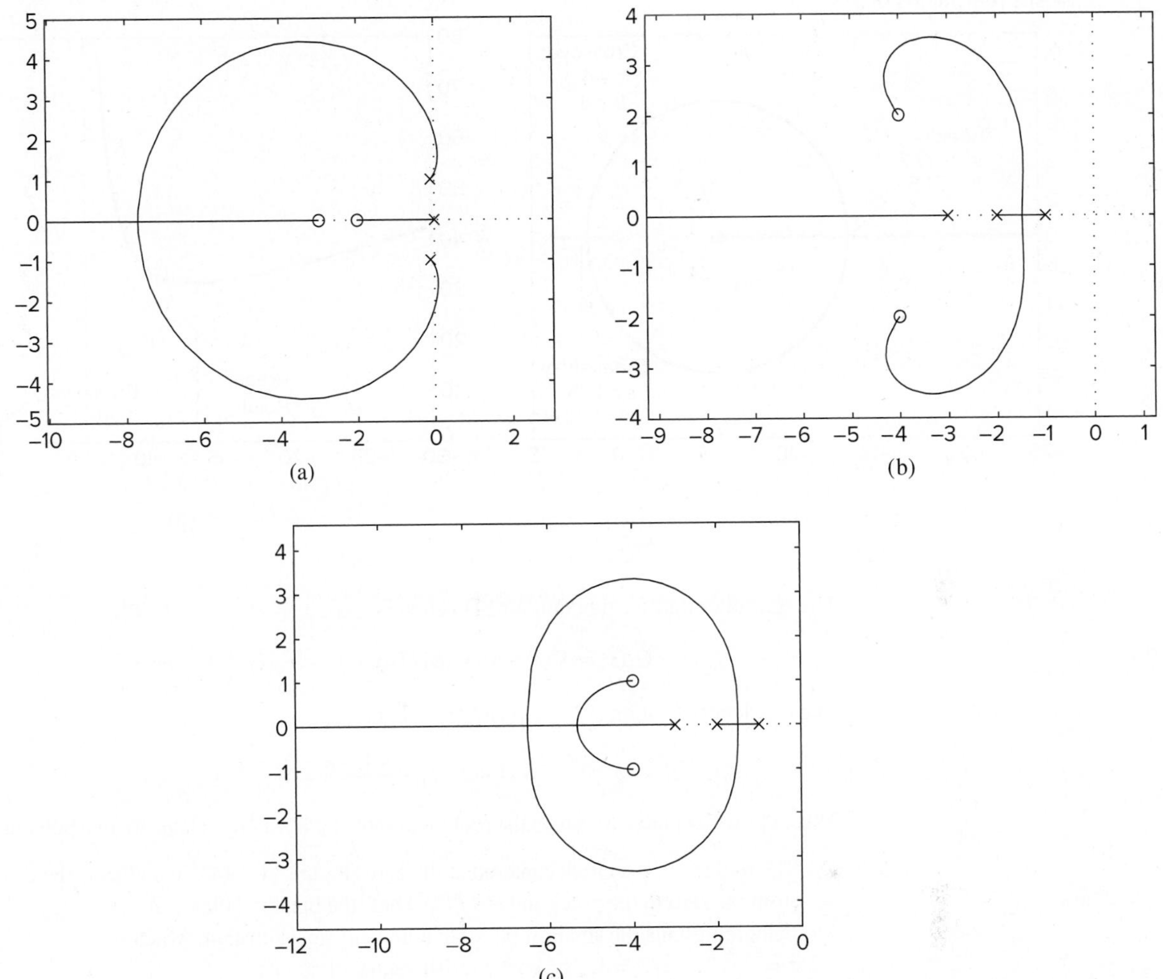

PD Control

EXAMPLE 11.2.4

■ Problem

The equation of motion of an object rotating under an applied torque T is $I\ddot{\theta} = T$. Use PD action to control this system. Assume that $I = 1$ and $T_D = 2$. Determine the gain values (a) to achieve the smallest possible dominant-time constant and (b) to achieve a damping ratio of 0.707. (c) Compare the step responses of the two designs.

■ Solution

The plant is

$$G_p(s) = \frac{\Theta(s)}{T(s)} = \frac{1}{s^2}$$

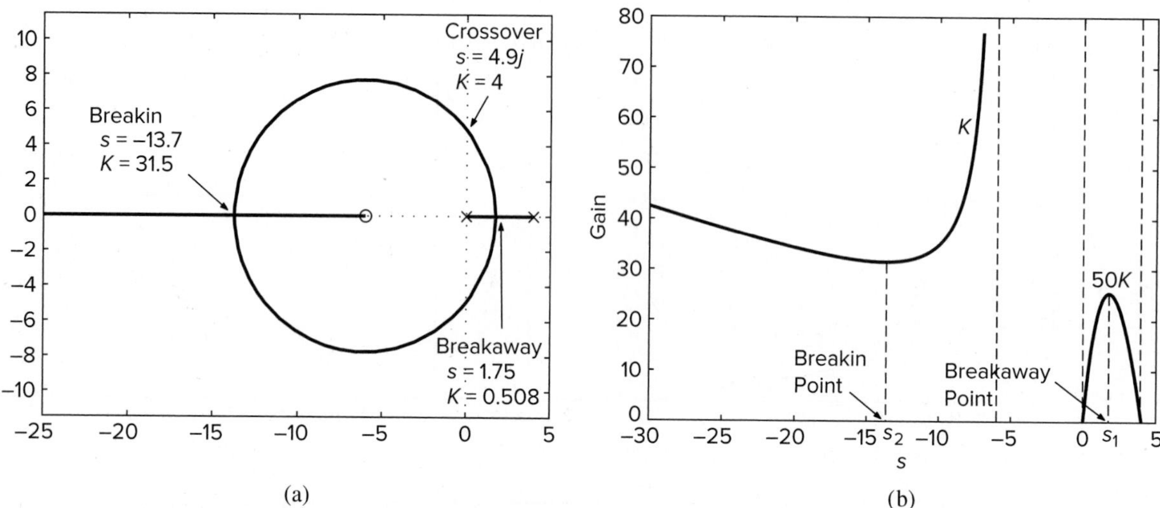

Figure 11.2.13 (a) Root locus for Example 11.2.2. (b) Plot of the gain K versus real values of s.

The controller transfer function for PD action is

$$G_c(s) = K_D s + K_P = K_P(T_D s + 1) = K_P T_D \left(s + \frac{1}{T_D} \right)$$

The root locus equation is $1 + G_c(s)G_p(s) = 0$, or

$$1 + K_P T_D \frac{s + 1/T_D}{s^2} = 0$$

Thus, PD action places a zero on the root locus plot at $s = -1/T_D$. There are two poles at $s = 0$.

a. The root locus is a circle centered at the zero (Figure 11.2.14). Its radius is the distance from the zero to the poles and is $1/T_D$. Thus, the breakin point is at $s = -2/T_D$. This point represents the smallest possible dominant-time constant, which is $\tau = T_D/2 = 2/2 = 1$. The characteristic equation is

$$s^2 + K_P T_D s + K_P = 0$$

At the breakin point, $\zeta = 1$, where

$$\zeta = \frac{K_P T_D}{2\sqrt{K_P}} = 1$$

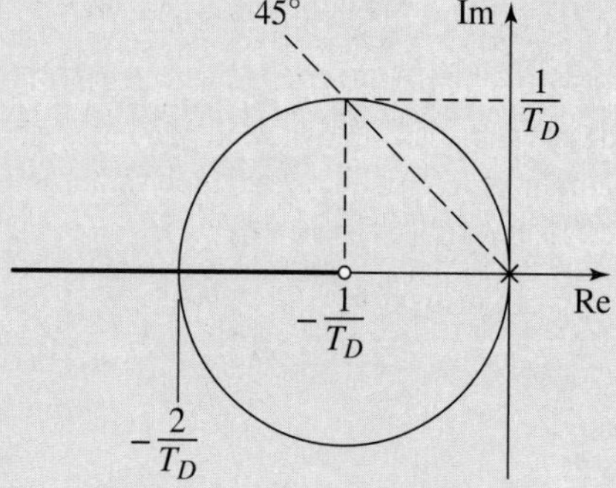

Figure 11.2.14 Root locus for Example 11.2.3.

This gives $K_P = 4/T_D^2 = 4/4 = 1$ at the breakin point. This is the gain value that will give the smallest time constant for the given value of T_D.

b. From the root locus plot, we see that $\zeta = 0.707$ corresponds to the top of the circle where $s = (-1 + j)/T_D$. At this point,

$$\zeta = \frac{K_P T_D}{2\sqrt{K_P}} = 0.707$$

This gives $K_P = 2/T_D^2 = 2/4 = 0.5$ for $\zeta = 0.707$.

c. For $T_D = 2$ and $K_P = 1$, the command transfer function is

$$\frac{C(s)}{R(s)} = \frac{2s + 1}{s^2 + 2s + 1}$$

With $T_D = 2$ and $K_P = 0.5$, the command transfer function is

$$\frac{C(s)}{R(s)} = \frac{s + 0.5}{s^2 + s + 0.5}$$

The step response for $K_P = 1$ has an overshoot of 13.5% and a settling time of 5.39. The step response for $K_P = 0.5$ has an overshoot of 20.8% and a settling time of 6.82. So $K_P = 1$ gives better response.

Three Poles, No Zero

EXAMPLE 11.2.5

■ Problem

Sketch the root locus plots of the equations

$$1 + K\frac{1}{s[(s+2)^2 + 4]} = 0 \qquad (1)$$

$$1 + K\frac{1}{s[(s+2)^2 + 1]} = 0 \qquad (2)$$

■ Solution

For equation (1), the poles are $s = 0$ and $s = -2 \pm 2j$. Thus, there are three paths. One starts at $s = 0$ and the other two at $s = -2 \pm 2j$, with $K = 0$. Because there is no zero, all three paths must leave the plot, approaching asymptotes as $K \to \infty$. Because there are three more poles than zeros, the asymptotes make angles of $\pm 60°$ and $180°$ with the positive real axis. The asymptotes intersect the real axis at

$$\sigma = \frac{\sum s_p - \sum s_z}{p - z} = \frac{0 - 2 + 2j - 2 - 2j}{3 - 0} = -\frac{4}{3}$$

The crossover points are found from the characteristic equation

$$s\left[(s+2)^2 + 4\right] + K = s^3 + 4s^2 + 8s + K = 0$$

with $s = j\omega$. This gives

$$-j\omega^3 - 4\omega^2 + 8j\omega + K = 0$$

or

$$K - 4\omega^2 = 0$$

$$\omega(8 - \omega^2) = 0$$

Thus, the crossover points are given by $\omega = \pm\sqrt{8} = \pm 2.83$, where $K = 4(\omega^2) = 32$.

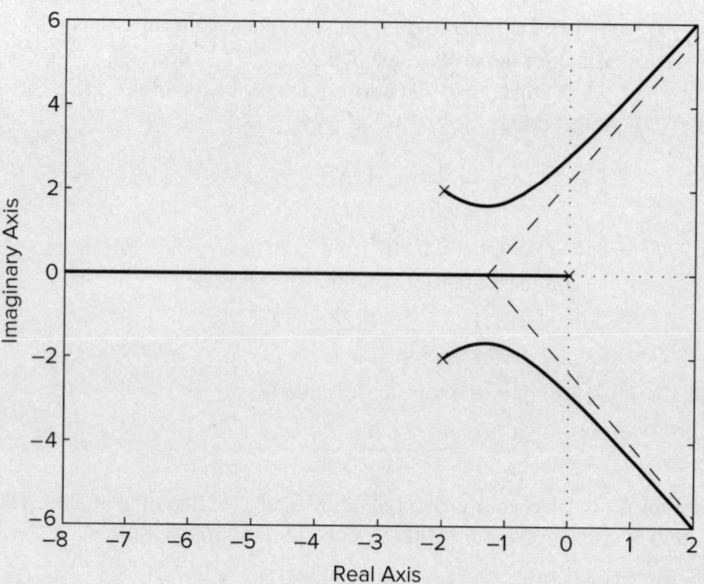

Figure 11.2.15 Root locus for the open-loop transfer function $1/s\left[(s+2)^2+4\right]$.

To examine for breakaway and breakin points, solve equation (1) for K, obtain dK/ds, and solve $dK/ds = 0$ for s.

$$K = -\left(s^3 + 4s^2 + 8s\right)$$

$$\frac{dK}{ds} = -\left(3s^2 + 8s + 8\right) = 0$$

The solutions are $s = -1.33 \pm 0.942j$. Since these are complex, there are no breakaway or breakin points. The root locus is shown in Figure 11.2.15 and can be generated with the MATLAB `rlocus` function. Note that MATLAB does not display the asymptotes, so Figure 11.2.15 was edited to show the asymptotes.

For equation (2), the poles are $s = 0$ and $s = -2 \pm j$. The asymptotes are $\pm 60°$ and $180°$. The asymptotes intersect the real axis at

$$\sigma = \frac{\sum s_p - \sum s_z}{p - z} = \frac{0 - 2 + j - 2 - j}{3 - 0} = -\frac{4}{3}$$

The crossover points are found from the characteristic equation

$$s\left[(s+2)^2 + 1\right] + K = s^3 + 4s^2 + 5s + K = 0$$

with $s = j\omega$. This gives $\omega = \pm\sqrt{5} = \pm 2.24$ where $K = 4(5) = 20$.

Because equations (1) and (2) are so similar, after sketching the locus for equation (1), however, we may be lulled into not looking for breakaway or breakin points. Equation (2) gives

$$K = -(s^3 + 4s^2 + 5s) \tag{3}$$

$$\frac{dK}{ds} = -(3s^2 + 8s + 5) = 0$$

The solutions are $s = -1.67$ and $s = -1$. So there are breakaway or breakin points for this equation! The corresponding values of K are found from equation (3). For $s = -1.67$, $K = 1.85$, and for $s = -1$, $K = 2$. Computing d^2K/ds^2 for each solution identifies the nature of each point.

$$\frac{d^2K}{ds^2} = -(6s + 8) = \begin{cases} 2.02 > 0 & \text{for } s = -1.67 \\ -2 < 0 & \text{for } s = -1 \end{cases}$$

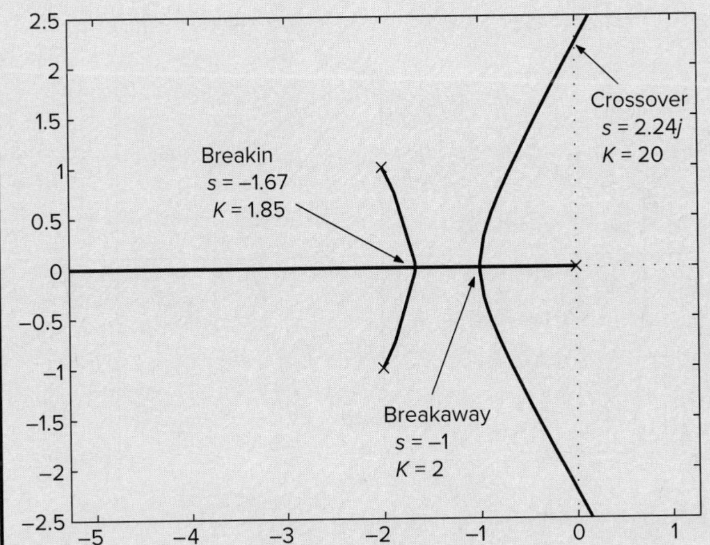

Figure 11.2.16 Root locus for the open-loop transfer function $1/s[(s+2)^2+1]$.

Thus, $s = -1.67$ is a breakin point and $s = -1$ is a breakaway point. The root locus is shown in Figure 11.2.16.

EXAMPLE 11.2.6

Two Complex Poles, Two Complex Zeros

■ Problem

Sketch the root locus plots of the equation

$$1 + K\frac{(s+4)^2+9}{[(s+1)^2+1])} = 0 \qquad (1)$$

■ Solution

The poles are $s = -1 \pm j$, so there are two paths. The zeros are $s = -4 \pm 3j$. Both paths must end at the zeros, so we need not look for asymptotes. Checking in the usual way for breakaway, breakin, and crossover points shows there are none.

The angles of arrival and departure are found as follows. Figure 11.2.17 shows the test points placed near the pole or zero in question, in order to calculate the vector angles. We use the general root locus equation

$$\angle N(s) - \angle D(s) = (2n+1)180^\circ \qquad n = 0, 1, 2, 3, \ldots$$

The angle of departure ϕ is found from part (a) of the figure.

$$\phi + 90^\circ - \phi_1 - \phi_2 = (2n+1)180^\circ$$

where $\phi_1 = 360 - \tan^{-1}(2/3) = 326^\circ$ and $\phi_2 = \tan^{-1}(4/3) = 53^\circ$. Thus,

$$\phi = \phi_1 + \phi_2 - 90^\circ + (2n+1)180^\circ = 289^\circ + (2n+1)180^\circ$$

Choosing $n = 0$ gives $\phi = 469^\circ$, which is equivalent to $469 - 360 = 109^\circ$.

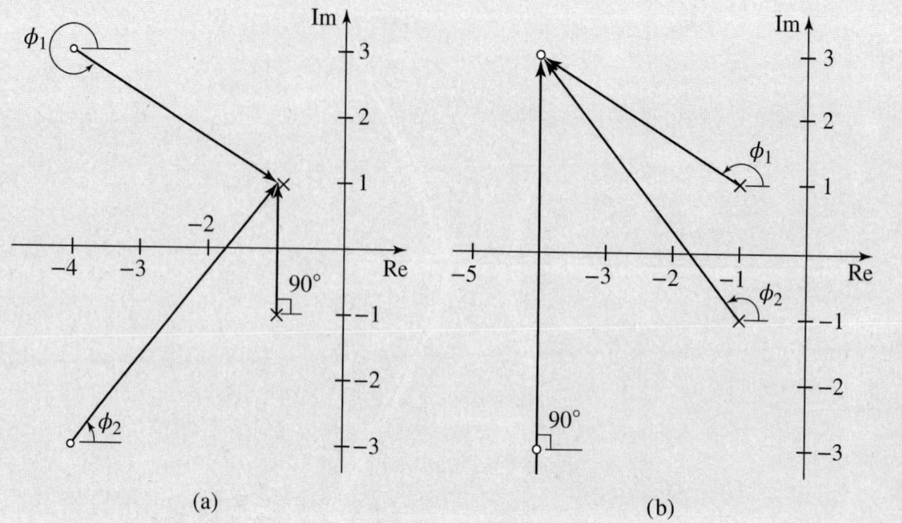

Figure 11.2.17 (a) Calculation of the departure angle. (b) Calculation of the arrival angle.

The angle of arrival ϕ is found from part (b) of the figure.

$$\phi + 90° - \phi_2 - \phi_1 = (2n + 1)180°$$

where $\phi_1 = 180 - \tan^{-1}(2/3) = 146°$ and $\phi_2 = 180° - \tan^{-1}(4/3) = 127°$. Thus,

$$\phi = \phi_1 + \phi_2 - 90° + (2n + 1)180° = 183° + (2n + 1)180°$$

Choosing $n = 0$ gives $\phi = 363°$, which is equivalent to 3°. Choosing $n = 1$ gives $\phi = 723°$, which is equivalent to 3°.

These angles of departure and arrival indicate that the root locus will pass above the 45° line, and thus it is possible to obtain a value for K to achieve $\zeta \leq 0.707$. The root locus is shown in Figure 11.2.18.

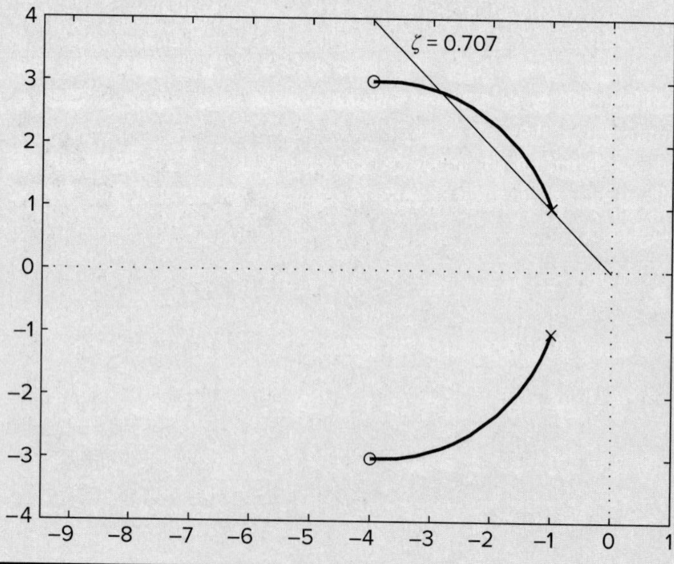

Figure 11.2.18 Root locus for Example 11.2.5.

EXAMPLE 11.2.7

Control of the Plant $1/(s+3)(s+5)$

■ **Problem**

Design a control algorithm for the plant

$$G_p(s) = \frac{1}{(s+3)(s+5)}$$

The dominant-time constant must be no greater than 0.5 and the steady-state error for a step command must be zero.

■ **Solution**

a. Proportional Action: With P action the command transfer function is

$$\frac{C(s)}{R(s)} = \frac{K_P}{s^2 + 8s + 15 + K_P}$$

The root locus equation is

$$1 + K_P \frac{1}{(s+3)(s+5)} = 0$$

Figure 11.2.19 shows the root locus plot, from which we can tell that the smallest possible time constant is $\tau = 1/4$. Therefore, the time constant specification can be satisfied.

The steady-state response for a unit-step command, however, is $c_{ss} = K_P/(15 + K_P)$. Thus, the error is $e_{ss} = 1 - c_{ss} = 15/(15 + K_P)$, which cannot be made zero.

b. I Action: I action tends to eliminate steady-state error. The command transfer function is

$$\frac{C(s)}{R(s)} = \frac{K_I}{s^3 + 8s^2 + 15s + K_I}$$

The steady-state response for a unit-step command is $c_{ss} = K_I/K_I = 1$, and so the error is zero. The root locus equation is

$$1 + K_I \frac{1}{s(s+3)(s+5)} = 0$$

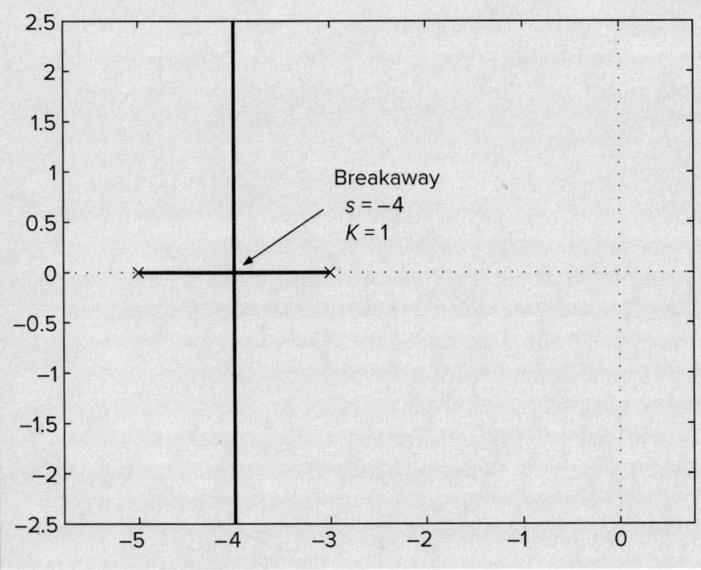

Figure 11.2.19 Root locus for the open-loop transfer function $1/(s+3)(s+5)$.

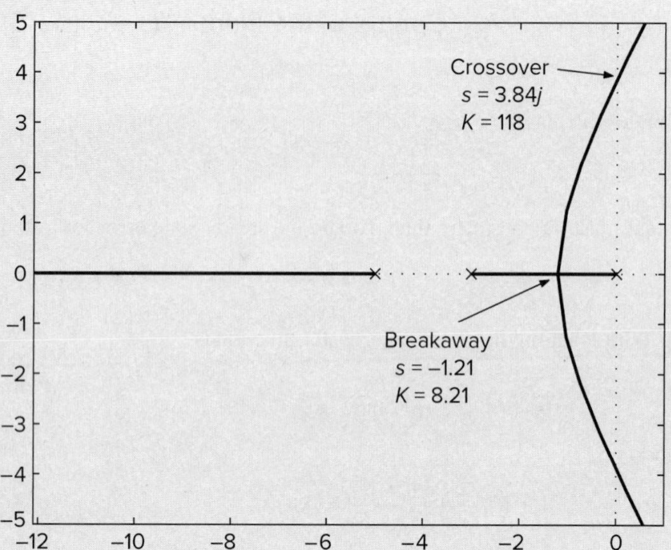

Figure 11.2.20 Root locus for the open-loop transfer function $1/s(s+3)(s+5)$.

The breakaway point is found from

$$\frac{dK_I}{ds} = -\left(3s^2 + 16s + 15\right) = 0$$

or $s = -1.21$ and $s = -4.12$. The root locus does not exist at $s = -4.12$ for $K_I \geq 0$, so the breakaway point must be at $s = -1.21$.

Figure 11.2.20 shows the root locus plot, from which we can tell that the smallest possible time constant occurs at the breakaway point and is $\tau = 1/1.21 = 0.836$. Therefore, the time constant specification cannot be satisfied. Note also that the system can now be unstable, whereas with P action it is always stable.

c. PI Action: With PI action the command transfer function is

$$\frac{C(s)}{R(s)} = \frac{K_P s + K_I}{s^3 + 8s^2 + (15 + K_P)s + K_I}$$

The steady-state response for a unit-step command is $c_{ss} = K_I/K_I = 1$, and so the error is zero. Expressing the integral gain as $K_I = K_P/T_I$, the root locus equation may be expressed as

$$1 + K_P \frac{s + 1/T_I}{s(s+3)(s+5)} = 0$$

The root locus parameter K is seen to be K_P. We now have two parameters to select: K_P and T_I. The usual procedure is to use a trial value for T_I, obtain the root locus, and use it to set the gain K_P. Note that the value of $1/T_I$ determines the location of the zero.

If we place the zero between the poles of the plant, we obtain a root locus plot like that shown in Figure 11.2.21. The particular plot shown corresponds to the choice $1/T_I = 3.5$. The dominant-time constant at the breakaway point is $1/1.8 = 0.556$, which is larger than required, so we must use a gain value somewhat greater than 4.07. The larger the gain, the smaller will be the damping ratio. Because the $\pm 90°$ asymptotes intersect the real axis at $\sigma = -2.25$, the smallest possible dominant-time constant is $1/2.25 = 0.444$, but this requires an infinite gain value. So a time constant of 0.5 can be achieved. By using MATLAB, the necessary gain was found to be $K = 8.11$. This puts the dominant roots at $s = -2 \pm 1.76j$, for which $\zeta = 0.75$. The third root is $s = -3.99$.

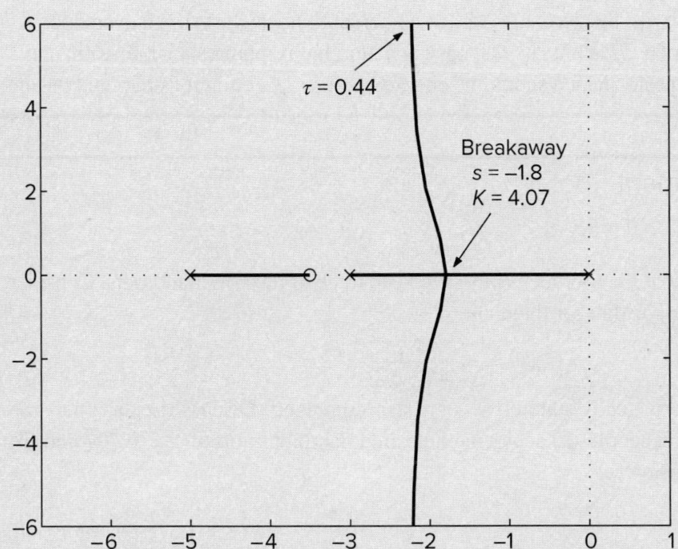

Figure 11.2.21 Root locus for the open-loop transfer function $(s+3.5)/(s+3)(s+5)$.

The placement of the zero at −3.5 was arbitrary. Normally with this design process, the designer experiments with the placement of the zero to see how it affects other aspects of the performance, such as overshoot and maximum actuator output, for example. The MATLAB interface `rltool` enables you to move the poles and zeros and see the effect on the root locus. It also enables you to compute the resulting step response.

If we place the zero to the right of the pole at $s = -3$, we obtain a root locus plot like that shown in Figure 11.2.22. The particular plot shown corresponds to the choice $1/T_I = 2.5$. The dominant root path is now always real. The dominant root is −2 when $K = 12$. The other roots are $s = -3 \pm 2.45j$. These are somewhat close to the dominant root, so they might influence the response significantly.

Simulation of the two PI designs for a step command shows that their responses are very similar. The design with the zero at −3.5 has a 3.3% overshoot with a settling time

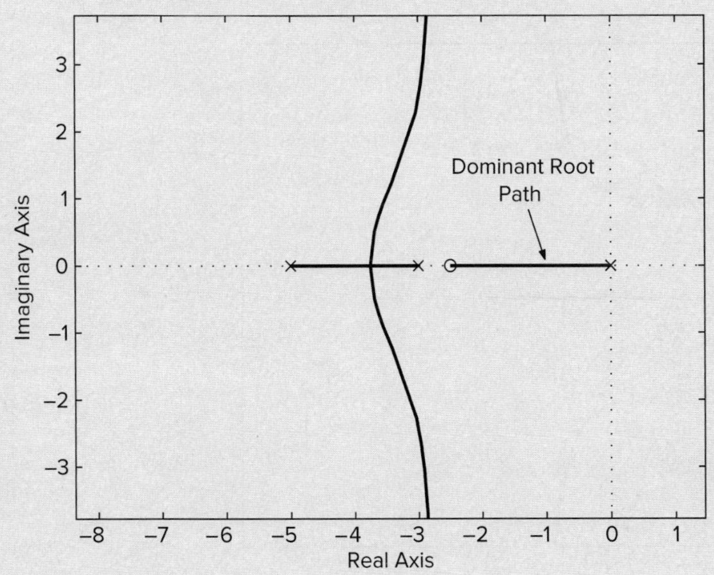

Figure 11.2.22 Root locus for the open-loop transfer function $(s+2.5)/s(s+3)(s+5)$.

of 2.16. The design with the zero at −2.5 has no perceptible oscillation, no overshoot, and a settling time of 1.29. Thus, its response is better, but it requires a larger gain, and other aspects of its performance should be checked before accepting it as the final design.

EXAMPLE 11.2.8

Tracking the Dominant Roots

■ Problem

A significant application of the root locus plot is to keep track of the dominant roots. Consider the design of a control algorithm for the plant

$$G_p(s) = \frac{1}{s^2 + 6s + 13}$$

It is required to have zero steady-state error for a step command. Discuss the design of the control algorithm with respect to (a) achieving a desired damping ratio of $\zeta = 0.707$ and (b) minimizing the time constant.

■ Solution

We rule out P action because of the error requirement. Using PI action gives the command transfer function

$$\frac{C(s)}{R(s)} = \frac{K_P s + K_I}{s^3 + 6s^2 + (13 + K_P)s + K_I}$$

The steady-state response for a unit-step command is $c_{ss} = K_I/K_I = 1$, and so the error is zero. The root locus equation is

$$1 + K_P \frac{s + 1/T_I}{s^3 + 6s^2 + 13s} = 0$$

The plot is shown in Figure 11.2.23 for the trial value of $1/T_I = 4$. The solution for $\zeta = 0.707$ is shown in the figure and requires a gain of $K = K_P = 3.81$. This solution, however, corresponds to the secondary root. The dominant root for $K = 3.81$ is at $s = -1.53$.

Figure 11.2.24 shows what happens if we forego the damping ratio requirement and instead try to minimize the time constant. For small values of K, the dominant root is real. As we increase K, the real root moves to the left and the complex roots move to the right and

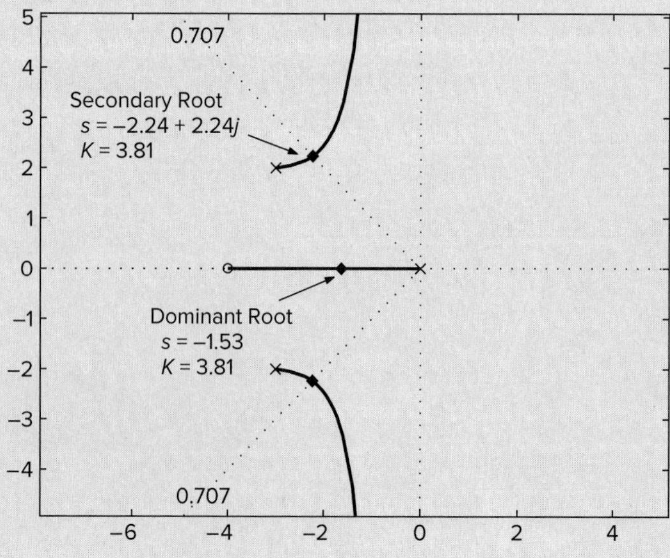

Figure 11.2.23 Root locus for the open-loop transfer function $(s + 4)/(s^3 + 6s^2 + 13s)$ with the design solution $\zeta = 0.707$.

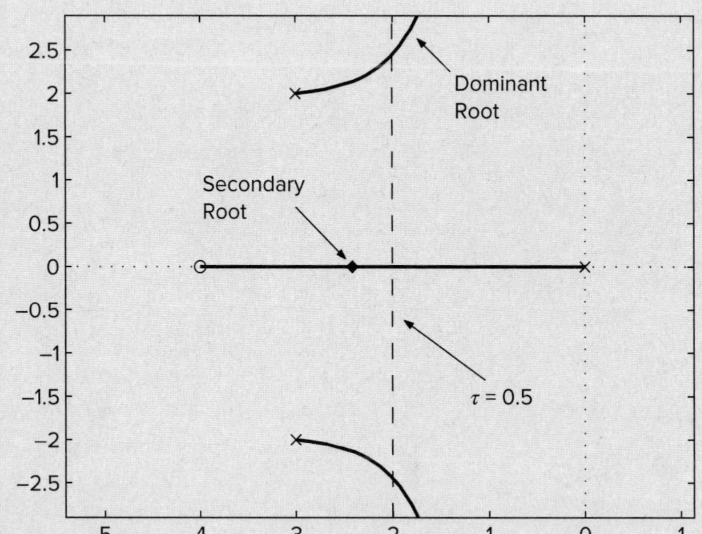

Figure 11.2.24 Root locus for the open-loop transfer function $(s + 4)/(s^3 + 6s^2 + 13s)$ with the design solution minimizing the time constant.

eventually become dominant as K is increased. The smallest possible dominant-time constant occurs when the real parts of all the roots are equal. To determine when this occurs, we can use Guide 10. With $1/T_I = 4$, the characteristic equation is $s^3+6s^2+(13+K_P)s+4K_P = 0$. If all three roots have the same real part $-a$, and the two complex roots have the imaginary part b, then the three roots are $s = -a$ and $s = -a \pm bj$. From Guide 10, the sum of the roots must be -6, so we have

$$-6 = -a - a + bj - a - bj = -3a$$

or $a = 2$. Substituting $s = -2$ into the characteristic equation gives $K_P = 5$. Thus, when $K_P < 5$, the dominant root is real. When $K_P > 5$, the dominant roots are complex. The smallest possible dominant-time constant is $1/2$.

11.2.5 THE COMPLEMENTARY ROOT LOCUS

The development of the plotting guides was based on the angle and magnitude criteria, under the assumption that the parameter K is positive or zero. The corresponding root locus is called the *primary* root locus. There are applications, however, where we cannot formulate the problem such that $K \geq 0$. This occurs with some plant models that have negative coefficients. For example, proportional control with the plant

$$G_p(s) = \frac{-s+5}{s(s+8)} \tag{11.2.20}$$

results in the following root locus equation.

$$1 + K\frac{s-5}{s(s+8)} = 0 \tag{11.2.21}$$

where $K = -K_P$. Thus, to investigate the root locus for positive values of K_P, we must let K vary through negative values.

The guides for the complementary root locus of the equation

$$1 + K\frac{N(s)}{D(s)} = 0 \qquad K \leq 0 \tag{11.2.22}$$

are found from the same criteria, repeated here.

$$\left|K\frac{N(s)}{D(s)}\right| = 1 \tag{11.2.23}$$

$$\angle K + \angle N(s) - \angle D(s) = \angle(-1) = (2n+1)180^\circ \tag{11.2.24}$$

If $K \leq 0$, (11.1.23) becomes $-K|P(s)| = 1$ or

$$K = -\frac{1}{|P(s)|} \tag{11.2.25}$$

Thus, the same scaling procedure is used, with a sign reversal included.

If $K \leq 0$, then $\angle K = (2k+1)180^\circ$, and (11.2.23) becomes

$$(2k+1)180^\circ + \angle N(s) - \angle D(s) = (2n+1)180^\circ$$

which reduces to

$$\angle N(s) - \angle D(s) = (2n+1)180^\circ - (2k+1)180^\circ = m360^\circ \qquad m = 0, 1, 2, 3, \ldots \tag{11.2.26}$$

The guides for the complementary locus are identical to those for the primary locus except for Guides 4 and 6, which we renumber as 4a and 6a.

Guide 4a The locus exists in a section on the real axis only if the number of real poles and/or zeros to the *right* of the section is *even*; furthermore, it must exist there. The number zero is taken to be even. In other words, the complementary locus exists on the real axis wherever the primary locus does not exist.

Guide 6a The paths that do not terminate at a zero leave the plot and approach asymptotes. The asymptotic angles relative to the positive real axis are found from

$$\theta = \frac{m360^\circ}{z-p} \qquad m = 0, 1, 2, 3\ldots \tag{11.2.27}$$

where z and p are the number of zeros and poles and m is increased until enough angles have been found.

Note that the extraneous roots found when looking for breakaway and breakin points on the primary locus are the breakaway or breakin points on the complementary locus. Note also that the intersection point of the asymptotes is the same for the primary and the complementary loci. When determining the angles of arrival and departure for the complementary locus, be sure to use (11.2.26).

EXAMPLE 11.2.9 Negative Root Locus Gain with Two Real Poles

■ Problem

Sketch the root locus of the equation

$$1 + K\frac{1}{s(s+2)} = 0$$

where $K \leq 0$.

■ Solution

There is no zero, so both paths must leave the plot. The two poles are at $s = 0, -2$. From Guide 4a we see that the paths on the real axis are in the regions $s \leq -2$ and $s \geq 0$. From this

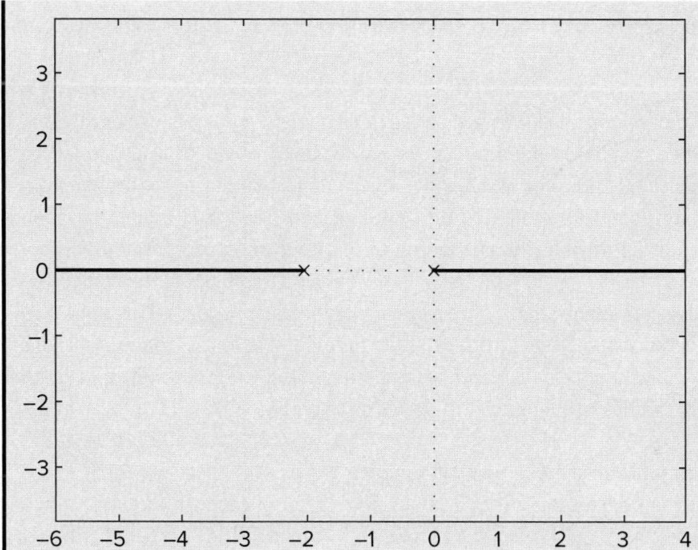

Figure 11.2.25 Root locus plot of $1 + K/s(s + 2) = 0$ for $K \leq 0$.

result we can tell that the two asymptotes are $s \to -\infty$ and $s \to +\infty$. There is no need to find breakaway/breakin points or crossover points. The plot is shown in Figure 11.2.25.

Negative Root Locus Gain with Two Real Poles and One Zero

EXAMPLE 11.2.10

■ **Problem**

Sketch the root locus of the equation

$$1 + K\frac{s - 5}{s(s + 8)} = 0 \tag{1}$$

where $K \leq 0$.

■ **Solution**

The zero is at $s = 5$ and the two poles are at $s = 0, -8$. From Guide 4a we see that the paths on the real axis are in the regions $-8 \leq s \leq 0$ and $s \geq 5$. The breakaway and breakin points are found from

$$\frac{dk}{ds} = \frac{d}{ds}\left[\frac{-s(s + 8)}{s - 5}\right] = -\frac{s^2 - 10s - 40}{(s - 5)^2} = 0$$

This gives $s = -3.0623, -13.0623$. The results of Guide 4a show that the breakin point is at $s = -3.0623$ and the breakaway point is at $= -13.0623$. The crossover points are found in the usual way. From (1),

$$s^2 + 8s + Ks - 5K = 0$$

Substituting $s = j\omega$ and separating the real and imaginary parts, we obtain

$$-\omega^2 - 5K + j\omega(8 + K) = 0$$

The imaginary part gives $\omega = 0$, which is one of the poles, or $K = -8$, which is the desired solution. From the real part with $K = -8$, we obtain $\omega = \pm\sqrt{40} = \pm 6.3246$. Thus, the two crossover points are at $s = \pm 6.3246j$. The root locus is shown in Figure 11.2.26.

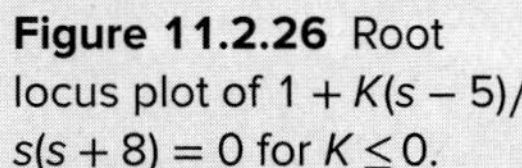

Figure 11.2.26 Root locus plot of $1 + K(s - 5)/s(s + 8) = 0$ for $K \leq 0$.

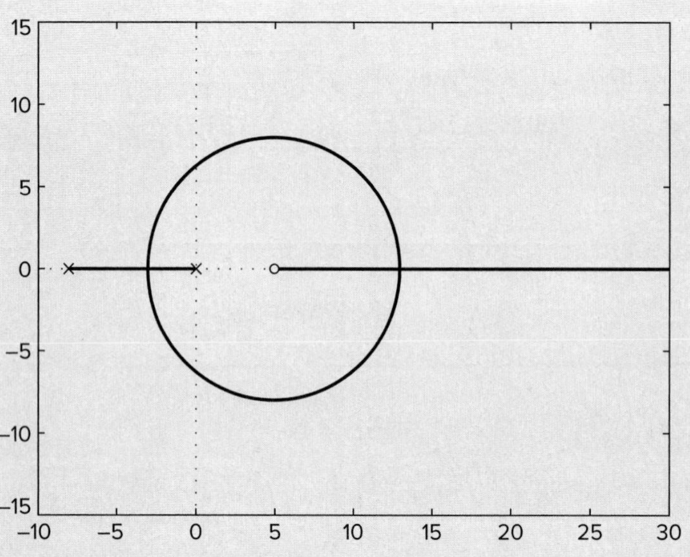

EXAMPLE 11.2.11

Negative Gain with Three Real Poles and No Zero

■ Problem

Sketch the root locus of the equation

$$1 + K\frac{1}{s(s+3)(s+5)} = 0 \tag{1}$$

where $K \leq 0$.

■ Solution

The three poles are at $s = 0, -3, -5$. From Guide 4a we see that the paths on the real axis are in the regions $-5 \leq s \leq -3$ and $s \geq 0$. The breakaway and breakin points are found from

$$\frac{dk}{ds} = -\frac{d}{ds}[s(s+3)(s+5)] = -\frac{d}{ds}(s^3 + 8s^2 + 15s) = -(3s^2 + 16s + 15) = 0$$

This gives $s = -4.1196, -1.2137$. The results of Guide 4a show that the breakaway point is at $s = -4.1196$. The point $s = -1.2137$ is the breakaway point for the root locus of this equation for $K \geq 0$.

The asymptotes are found from Guide 6a.

$$\theta = \frac{m360^\circ}{3 - 0} = m120^\circ, \qquad m = 0, 1, 2, \ldots$$

This gives the following angles: $\theta = 0^\circ, 120^\circ, 240^\circ$. The intersection point is found in the same way as for $K \geq 0$ (Guide 6):

$$\sigma = \frac{0 - 3 - 5}{3 - 0} = -\frac{8}{3}$$

The root locus plot is shown in Figure 11.2.27. Compare this plot with that shown in Figure 11.2.20, which is the plot for $K \geq 0$.

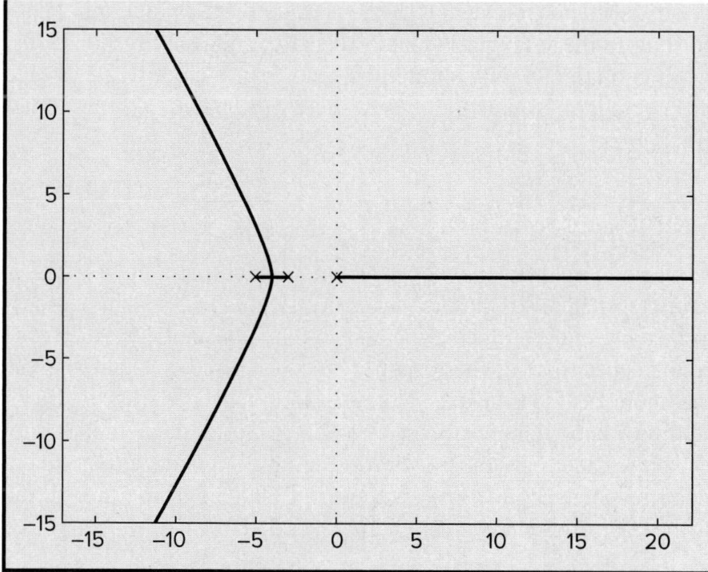

Figure 11.2.27 Root locus plot of $1 + K/[s(s + 3)(s + 5)] = 0$ for $K \leq 0$.

11.3 TUNING CONTROLLERS

Thus far we have seen several methods for computing appropriate gain values for the controller. In practice, however, the gains should be varied or adjusted from their computed values to determine how sensitive is the performance to the specific gain values. This process is part of "tuning the controller," and it is often possible to improve the performance in this way. Tuning also means computing a preliminary set of gain values based on experiments with the plant.

Before computers became available for simulation, controller tuning meant collecting plant response data and adjusting the gains of an analog controller "in the field" with the controller installed and connected to the plant. Tuning is still done this way with both analog and digital controllers, but it is not always possible to do so for safety or cost reasons. For example, one would not tune the autopilot of a spacecraft during launch!

Now, with computer simulation, it is easier to investigate how a change in gain values or other parameter values will affect the system performance.

11.3.1 THE ZIEGLER-NICHOLS METHODS

Some processes, particularly those involving thermal processes, fluid flow, or chemical reactions, are very difficult to model, especially with linear differential equations. In such applications, if we can do experiments with the plant, there are guidelines available for obtaining preliminary values of the controller gains. These values can then be adjusted by tuning.

The most commonly used methods are those developed by Ziegler and Nichols. They have proved so useful that they are still in use 60 years after their development in the early 1940s. Other methods have been developed; see for example [Seborg, 1989]. Ziegler and Nichols developed two methods. The first method requires the open-loop step response of the plant, while the second method uses the results of experiments performed with the controller already installed. While primarily intended for use with systems for which no analytical model is available, the methods are also helpful sometimes even when a model can be developed, as we will demonstrate.

From a number of experiments and analysis, Ziegler and Nichols found that controllers adjusted according to the following methods usually had a step response that was oscillatory but with enough damping so that the second overshoot was less than 25% of the first (peak) overshoot. This is the *quarter-decay criterion*, and it is sometimes used as a specification.

11.3.2 PROCESS REACTION METHOD

The first method, the *process reaction method*, relies on the fact that many processes have an open-loop unit-step response like that shown in Figure 11.3.1. This process signature is characterized by two parameters, R and L. The constant R is the slope of a line tangent to the steepest part of the response curve, and the constant L is the time at which this line intersects the time axis. Note that first- and second-order linear systems without dead time do not yield positive values for L, and so the method cannot be applied to such systems. Linear systems of third-order and higher, with sufficient damping, do yield such a response, however. If so, the Ziegler-Nichols process reaction method recommends the controller gains given in Table 11.3.1.

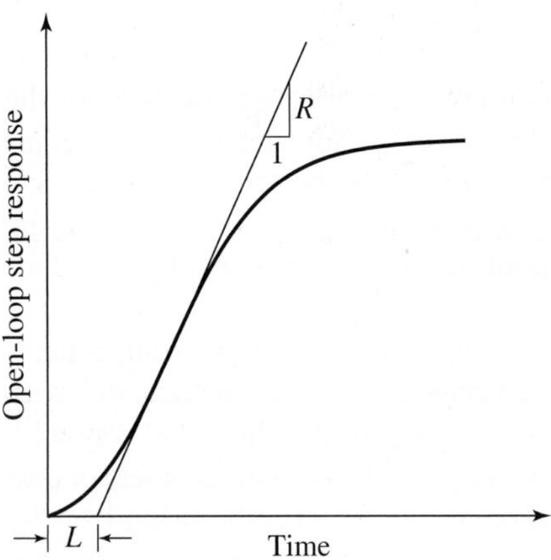

Figure 11.3.1 Open-loop step response of a process.

Table 11.3.1 The Ziegler-Nichols settings.

Controller transfer function		
$G(s) = K_P + \dfrac{K_I}{s} + K_D s$ or $G(s) = K_P\left(1 + \dfrac{1}{T_I s} + T_D s\right)$		
	$K_I = \dfrac{K_P}{T_I}$	$K_D = T_D K_P$
Control Mode	**Process Reaction Method**	**Ultimate Cycle Method**
P control	$K_P = 1/RL$	$K_P = 0.5K_{P_u}$
PI control	$K_P = 0.9/RL$	$K_P = 0.45K_{P_u}$
	$T_I = 3.3L$	$T_I = 0.83P_u$
PID control	$K_P = 1.2/RL$	$K_P = 0.6K_{P_u}$
	$T_I = 2L$	$T_I = 0.5P_u$
	$T_D = 0.5L$	$T_D = 0.125P_u$

Note that the table expresses the gains in terms of the *reset time* T_I and the *derivative time* T_D. The relation between the two forms of the PID control law is

$$G(s) = K_P + \frac{K_I}{s} + K_D s = K_P\left(1 + \frac{1}{T_I s} + T_D s\right) \tag{11.3.1}$$

Thus, $K_I = K_P/T_I$ and $K_D = K_P T_D$. Many industrial controllers require you to enter values for K_P, T_I, and T_D, rather than values for K_I and K_D. Some controllers require the value of $1/T_I$, which is called the *reset rate*.

Note that the process reaction method assumes that the response data is from a unit-step input and that the output starts from zero. In practice, this is sometimes difficult to do. If the response data $c_i(t)$ starts at a nonzero value c_0, subtract it from the other data to obtain $x_i(t) = c_i(t) - c_0$. Then, if the input has a magnitude M, divide the shifted data by M to obtain $y_i(t) = x_i(t)/M$. The modified data $y_i(t)$ is then used with the Ziegler-Nichols process reaction method. This technique assumes that the process is linear and thus scalable, but it might not be in practice.

It might be difficult to identify the maximum slope R by looking at a plot of the data, and therefore also difficult to estimate L. A polynomial curve fit of the modified data $y_i(t)$ can be used to obtain R and L as follows. After obtaining the curve fit $p_n(t)$ of order n, differentiate it with respect to t to obtain $dp_n(t)/dt = p_{n-1}(t)$, which is a polynomial one order lower than $p_n(t)$. Determine the maximum value of $p_{n-1}(t)$ and the time $t_{\max}$ at which the maximum occurs, either by plotting $p_{n-1}(t)$, by calculus, or with the MATLAB `max` function. The maximum gives the value of R, which can be used along with $t_{\max}$ to construct a straight line of slope R passing through the point $[p_n(t_{\max}), t_{\max}]$. The intersection of this line with the t axis gives L. See Section 11.6.

11.3.3 ULTIMATE-CYCLE METHOD

The *ultimate-cycle method* of Ziegler and Nichols uses experiments with the controller in place. All control modes except proportional are turned off, and the process is started with the proportional gain K_P set at a low value. The gain is slowly increased until the process output starts to show sustained oscillations. The period of these oscillations is called the *ultimate period*, and is denoted by P_u. Let the proportional gain setting at this condition be denoted by K_{P_u}, which is called the *ultimate gain*. The Ziegler-Nichols recommendations are given in Table 11.3.1 in terms of these parameters.

The proportional gain is lower for PI control than for P control and higher for PID control, because I action increases the order of the system, which tends to destabilize it; thus, a lower gain is needed. On the other hand, D action tends to stabilize the system; therefore, the proportional gain can be increased without degrading the stability characteristics. Because the settings were developed for typical cases out of many types of processes, final tuning of the gains by simulation or in the field is usually necessary.

Sometimes the Ziegler-Nichols methods can be applied to an analytical model. Example 11.3.1 shows how this is done with the ultimate-cycle method.

EXAMPLE 11.3.1

Ultimate-Cycle Method with a Third-Order Plant

■ Problem

Use the ultimate-cycle method to obtain the gains for (a) P action, (b) PI action, and (c) PID action. Analyze the unit-step response of each design. The plant is

$$G_p(s) = \frac{10}{2s^3 + 12s^2 + 22s + 12}$$

■ **Solution**

The ultimate-cycle method starts by using proportional action only. The transfer function for P action with this plant is

$$\frac{C(s)}{R(s)} = \frac{10K_P}{2s^3 + 12s^2 + 22s + 12 + 10K_P}$$

The characteristic equation is

$$2s^3 + 12s^2 + 22s + 12 + 10K_P = 0 \tag{1}$$

To apply the ultimate-cycle method, we must first find the ultimate period P_u and the associated gain K_{P_u}. To do this analytically, we note that sustained oscillations occur when a pair of roots is purely imaginary and the rest of the roots have negative real parts. To find when this occurs, set $s = j\omega_u$, where ω_u is the as-yet-unknown frequency of oscillation. Then $s^2 = -\omega_u^2$ and $s^3 = -j\omega_u^3$. Substituting these into the characteristic equation, we obtain

$$-2j\omega_u^3 - 12\omega_u^2 + 22j\omega_u + 12 + 10K_{P_u} = 0$$

Collect the real and imaginary parts.

$$\left(12 + 10K_{P_u} - 12\omega_u^2\right) + j\left(-2\omega_u^3 + 22\omega_u\right) = 0$$

To satisfy this equation, both the real and imaginary terms must be zero. So we obtain two equations to solve for the two unknowns ω_u and K_{P_u}. These are

$$12 + 10K_{P_u} - 12\omega_u^2 = 0 \tag{2}$$

$$-2\omega_u^3 + 22\omega_u = 0$$

The latter equation yields the non-oscillatory solution $\omega_u = 0$ as well as the one of interest—namely,

$$\omega_u = \pm\sqrt{11} = \pm 3.317$$

Substituting this into equation (2), we see that the ultimate gain is $K_{P_u} = 12$. The ultimate period is

$$P_u = \frac{2\pi}{\omega_u} = \frac{2\pi}{\sqrt{11}} = 1.89$$

For sustained oscillations to occur, as required by the ultimate-cycle method, the third root must be negative. The first two roots are $s_1 = j\omega_u$ and $s_2 = -j\omega_u$. The third root can be determined from Guide 10 of the root locus guides. Dividing equation (1) by 2 gives

$$s^3 + 6s^2 + 11s + 6 + 5K_P = 0$$

Guide 10 says that the sum of the three roots must be -6. Thus,

$$+j\omega_u - j\omega_u + r_3 = -6$$

which gives $r_3 = -6$. Since $r_3 < 0$, the assumption of neutral stability is justified. Thus, the solutions for ω_u and K_{P_u} are valid and can be used with Table 11.3.1 to compute the PID gains.

We obtain the following results. For P action, $K_P = 0.5K_{P_u} = 6$. For PI action,

$$K_P = 0.45K_{P_u} = 5.4 \qquad T_I = 0.83P_u = 1.57 \qquad K_I = \frac{K_P}{T_I} = 3.44$$

For PID action,

$$K_P = 0.6K_{P_u} = 7.2 \qquad T_I = 0.5P_u = 0.947 \qquad K_I = \frac{K_P}{T_I} = 7.62$$

$$T_D = 0.125P_u = 0.236 \qquad K_D = K_P T_D = 1.7$$

The closed-loop transfer function for P action is

$$\frac{C(s)}{R(s)} = \frac{60}{2s^3 + 12s^2 + 22s + 72}$$

For PI action,

$$\frac{C(s)}{R(s)} = \frac{54s + 34.4}{2s^4 + 12s^3 + 22s^2 + 66s + 34.4}$$

For PID action,

$$\frac{C(s)}{R(s)} = \frac{17s^2 + 72s + 76.2}{2s^4 + 12s^3 + 39s^2 + 84s + 76.2}$$

The unit-step responses are shown in Figure 11.3.2. Note the improvement as the control action is changed from P to PI to PID. First, the steady-state error is eliminated, and then the oscillations and settling time are reduced.

Figure 11.3.2 Unit-step responses of the controllers obtained in Example 11.4.1. (a) P action. (b) PI action. (c) PID action.

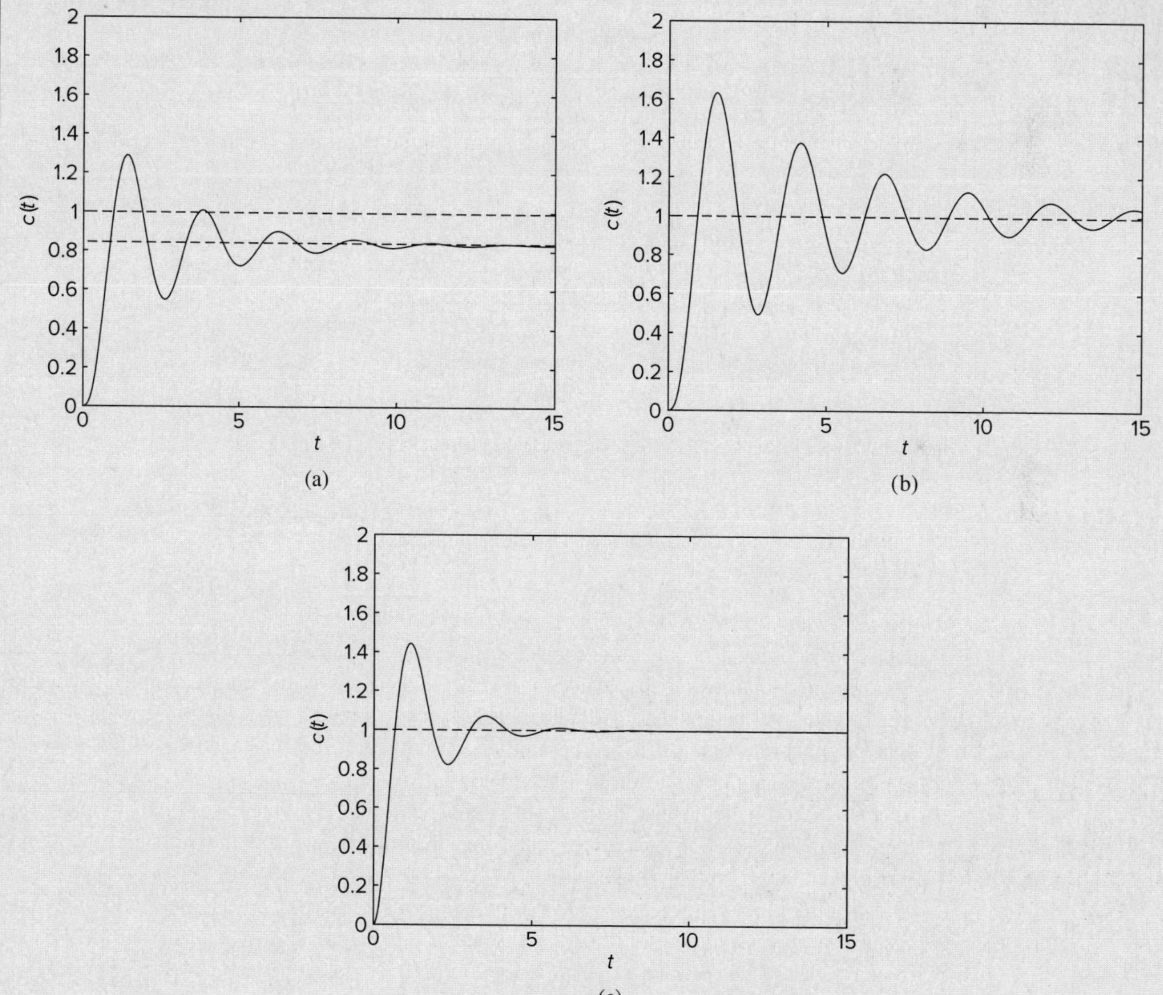

Note that this analysis uses only proportional control, but the gains for other control laws can be computed with these results. The power of the method is shown by the fact that we did not have to solve for the roots of a cubic or quartic equation. Note also that I action applied to this system would result in a fourth-order system to analyze—a difficult task—were it not for the Ziegler-Nichols method.

The Ziegler-Nichols methods are used to obtain gain values that are "in the ballpark." The engineer can use these values as a starting point for tuning the gains to obtain improved performance.

EXAMPLE 11.3.2

Tuning a PID Controller

■ Problem

Consider the PID controller designed in Example 11.3.1. Use the root locus plot to adjust the gains to reduce the overshoot.

■ Solution

The closed-loop transfer function for PID control of this plant is

$$\frac{C(s)}{R(s)} = \frac{10K_D s^2 + 10K_P s + 10K_I}{2s^4 + 12s^3 + (22 + 10K_D)s^2 + (12 + 10K_P)s + 10K_I}$$

When the controller transfer function is expressed in the form

$$G(s) = K_P\left(1 + \frac{1}{T_I s} + T_D s\right)$$

we can see that changing the proportional gain K_P changes the proportional, integral, and derivative gains by the same factor. For PID action, the Ziegler-Nichols settings give the following values of K_D and K_I in terms of K_P.

$$K_D = T_D K_P = 0.236K_P \qquad K_I = \frac{K_P}{T_I} = \frac{K_P}{0.947} = 1.056K_P$$

Thus, we can express the characteristic equation as

$$2s^4 + 12s^3 + (22 + 2.36K_P)s^2 + (12 + 10K_P)s + 10.56K_P = 0$$

and the root locus equation in terms of the parameter K_P is

$$1 + 1.18K_P \frac{s^2 + 4.24s + 4.47}{s(s^3 + 6s^2 + 11s + 6)} = 0$$

Thus, the root locus parameter is $K = 1.18K_P$.

The root locus plot is shown in Figure 11.3.3. When K_P has the value recommended by the Ziegler-Nichols method, $K_P = 7.2$, the roots are $s = -0.77 \pm 2.67j$, $s = -1.98$, and $s = -2.48$. The dominant root is marked on the root locus, and has a time constant of $\tau = 1/0.77 = 1.3$ and a small damping ratio of $\zeta = 0.28$. The root locus shows that decreasing K_P from 7.2 increases the time constant slightly to $\tau = 1/0.523 = 1.9$, but increases the damping ratio quite a bit. Thus, the system will respond somewhat more slowly, but the overshoot should be greatly reduced to perhaps none at all.

The root locus shows that we can improve the response by decreasing K_P until the dominant root has a damping ratio of 1. This occurs at the breakaway point ($s = -0.523$) and requires that $K_P = 0.306$, $K_D = T_D K_P = 0.236(0.306) = 0.070$, $K_I = K_P/T_I = K_P/0.947 = 0.323$. The improved step response, marked "Tuned," is shown in Figure 11.3.4, along with the response obtained with the Ziegler-Nichols settings, marked "Original."

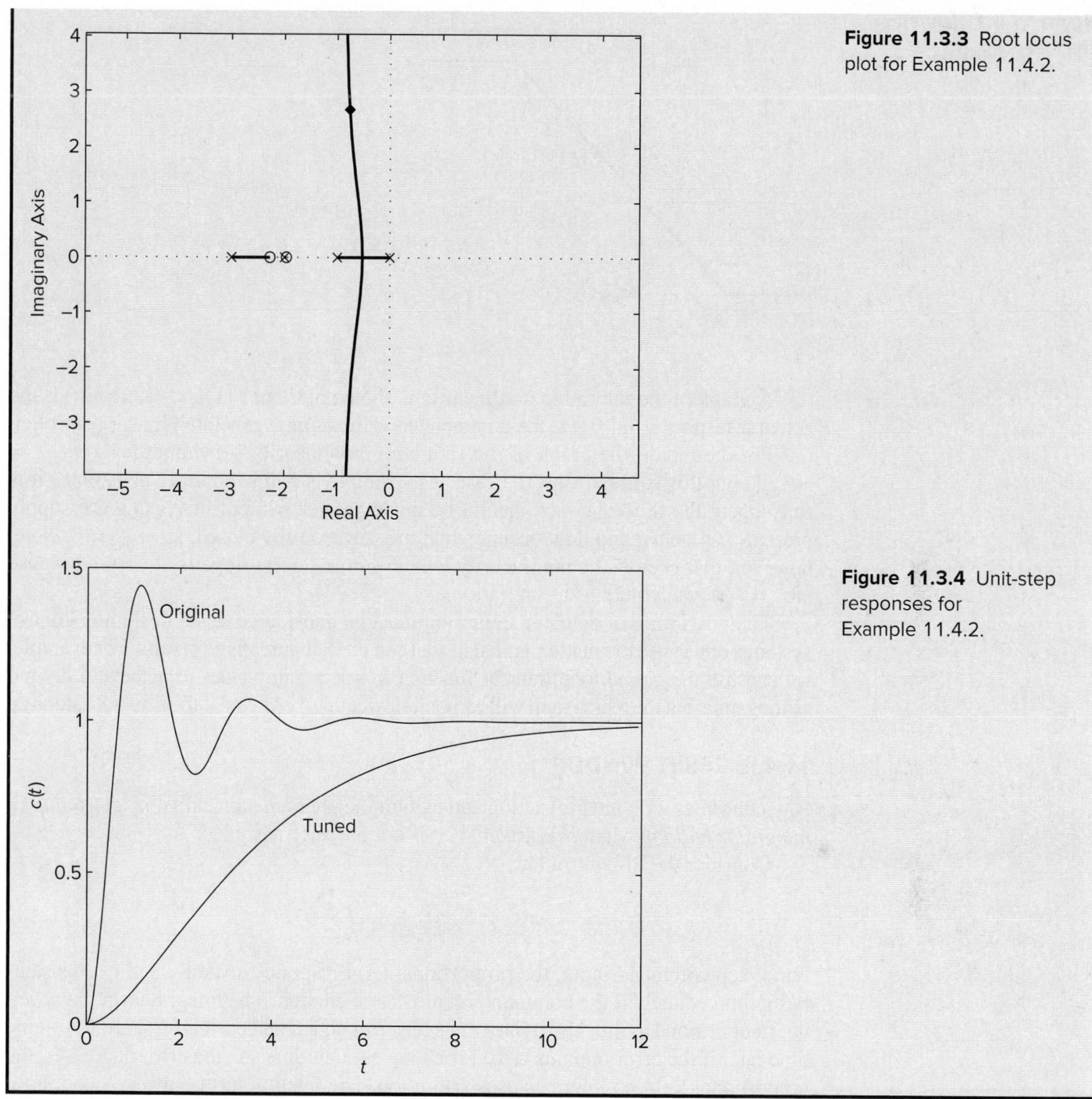

Figure 11.3.3 Root locus plot for Example 11.4.2.

Figure 11.3.4 Unit-step responses for Example 11.4.2.

11.4 SATURATION AND RESET WINDUP

We have seen that a motor-amplifier combination can produce a torque proportional to the input voltage over only a limited range. No amplifier can supply an infinite current, and there is a maximum current that the motor can withstand without overheating or demagnetizing. Thus, there is a maximum torque the motor can produce in either direction (clockwise or counterclockwise). Whenever an actuator reaches its limit, it is said to be *saturated*. When the controller commands the actuator to produce an output greater than its limit, the actuator is said to be *overdriven*.

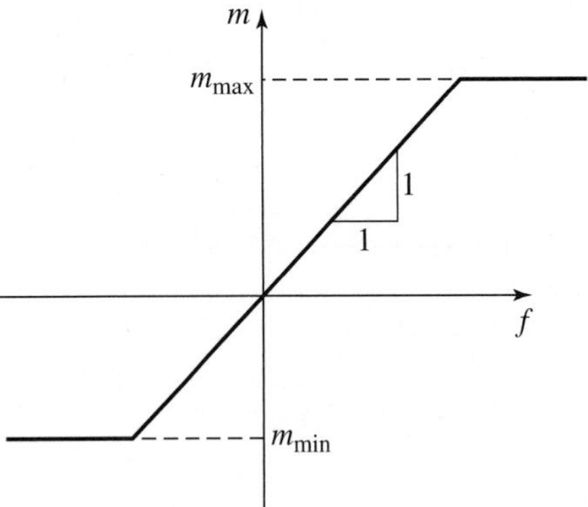

Figure 11.4.1 Saturation nonlinearity.

A graph of the saturation nonlinearity is shown in Figure 11.4.1, where $m(t)$ is the actuator response and $f(t)$ is the commanded value of $m(t)$ generated by the controller.

For the motor, the graph of the saturation nonlinearity is symmetric, so $m_{\min} = -m_{\max}$, but this is not always the case. An example is a flow control valve. Once it is fully open, the flow rate $m(t)$ cannot be made greater without increasing the supply pressure. Of course, no flow occurs when the valve is fully closed, so $m_{\min} = 0$. Note, however, that even if the nonlinearity is nonsymmetric, its slope is always 1 because $m(t) = f(t)$ when there is no saturation.

The possibility of actuator saturation must be considered when designing control systems, otherwise the mathematical model can predict unrealistic results. For example, we can use the closed-loop transfer function to select gain values to achieve a desired settling time, but the real system will respond slower than predicted if the actuator saturates.

11.4.1 RESET WINDUP

Any controller with integral action can exhibit the phenomenon called *reset windup* or *integrator buildup* when overdriven, if it is not properly designed.

Consider the PI-control law.

$$m(t) = K_p e(t) + K_I \int_0^t e(t)\, dt \tag{11.4.1}$$

For a step command input, the proportional term responds instantly and causes saturation immediately if the command input is large enough. The integral term, however, does not respond as quickly. It integrates the error signal and can cause saturation some time later if the error remains large for a long enough time. As the error decreases, the proportional term no longer causes saturation. The integral term, however, continues to increase as long as the error has not changed sign, and thus the manipulated variable remains saturated. Even though the output is near its desired value, the manipulated variable remains saturated until sometime after the error has reversed sign. The result can be a large overshoot in the response of the controlled variable.

Consider the PI control system shown in Figure 11.4.2. Suppose the plant is $G_p(s) = 1/4s$. To obtain a time constant of $\tau = 1.5$ and a damping ratio of $\zeta = 1$, we compute the gains to be $K_P = 16/3$ and $K_I = 16/9$. Ignoring any limits on actuator output, we predict the unit-step response to be that of the curve labeled "Unlimited" shown in Figure 11.4.3. Now suppose that the actuator limits are $m_{\max} = 1$ and $m_{\min} = -1$. Simulation, using Simulink for example, gives the curve labeled "Actuator Limited." Clearly the overshoot is much greater and the response time is longer.

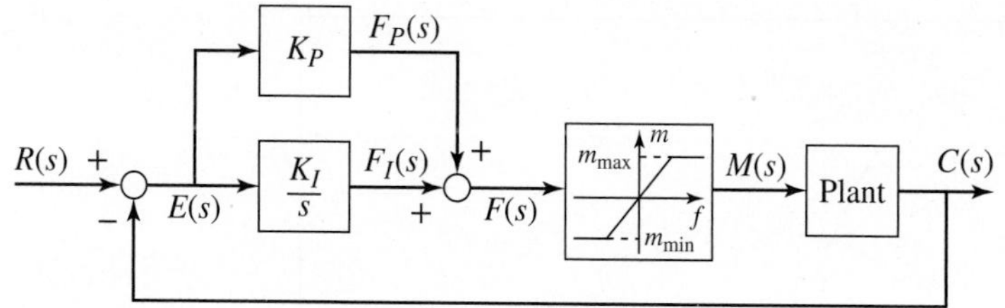

Figure 11.4.2 PI control with saturation nonlinearity.

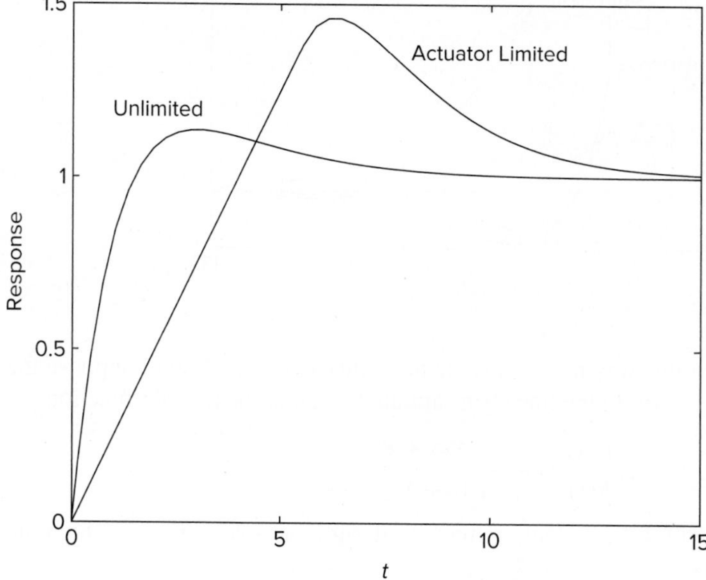

Figure 11.4.3 Effects of actuator saturation on step response.

To see what is happening, consider Figure 11.4.4. The actuator output is plotted along with the output of the I action and the error, $e(t) = 1 - c(t)$. For zero-initial conditions, the error is initially $e(0) = 1 - 0 = 1$, and the actuator immediately saturates because the proportional term is $K_P e = 16/3 > 1$. In the saturated mode, $m(t) = 1$, $\dot{c}(t) = m(t)/4 = 1/4$, $c(t) = t/4$, and $e(t) = 1 - t/4$, until $f < 1$. Note that the error eventually changes sign, which means that the output c is now larger than the desired value, but the I action term keeps the system saturated until some time later. This is what causes the overshoot often observed due to reset windup.

11.4.2 SELECTING GAINS TO AVOID SATURATION

Consider again the PI control system shown in Figure 11.4.2 for an arbitrary plant. If the maximum step command is r_{max} and if the output c is zero at startup at $t = 0$, then the output f_I of the integral term is zero and the output of the proportional term is $f_P = K_P[r_{max} - c(0)] = K_P r_{max}$. Then the maximum value K_P can have without overdriving the actuator is

$$K_P = \frac{m_{max}}{r_{max}} \tag{11.4.2}$$

This analysis does not account for the possibility that the integral term might cause saturation at some later time, but (11.4.2) provides a starting point for the design.

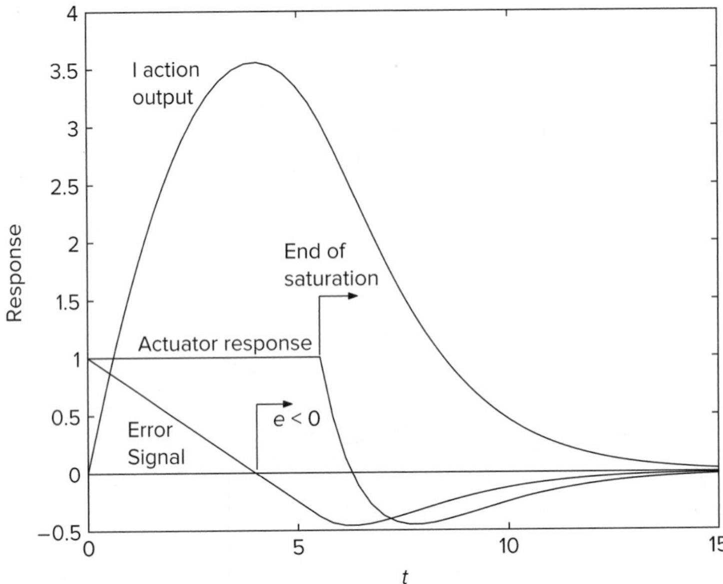

Figure 11.4.4 Effects of reset windup.

Computing K_P in this way removes some flexibility from the design. Suppose the plant is $G_p(s) = 1/(Is + c)$. In the linear (unsaturated) region, the transfer function is

$$\frac{C(s)}{R(s)} = \frac{K_P s + K_I}{Is^2 + (c + K_P)s + K_I}$$

With K_P fixed from (11.4.2), the only remaining design variable is K_I, for which the root locus equation is

$$1 + \frac{K_I}{I}\frac{1}{s[s + (c + K_P)/I]} = 0$$

The poles are $s = 0$ and $s = -(c + K_P)/I$. The $\pm 90°$ asymptotes intersect the real axis at the breakaway point at $s = -(c + K_P)/2I$, and thus the smallest possible dominant time constant is $\tau = 2I/(c + K_P)$.

If we again consider the plant $G_p(s) = 1/4s$ with $r_{\max} = m_{\max} = 1$, then $I = 4$, $c = 0$, and $K_P = 1$ from (11.4.2). The smallest possible dominant-time constant is $\tau = 8$. So we see that our earlier design with $\tau = 1.5$ was unrealistic and will cause saturation. With $K_P = 1$, any choice of $\zeta \leq 1$ will give a time constant of 8. Choosing $\zeta = 1$ requires that $K_I = 1/16$. Figures 11.4.5 and 11.4.6 compare the two designs. With this choice of gains, no reset windup occurs, the actuator is not overdriven, and the overshoot is much less than before. However, the response now takes longer to settle down to its steady-state value.

To decide which response is optimal, the designer must look more closely at the requirements for the particular application. In many systems, the occurrence of saturation and response overshoot indicates that the system is using more energy than is necessary to accomplish the task.

Reset windup can be prevented by selecting the gains so that saturation will never occur. This requires knowledge of the maximum input magnitude that the system will encounter. In a PI controller, any saturation initially caused by the proportional term is easy to predict, because it responds instantly to the error. The integral term might cause saturation also, but at a later time, because its output requires time to accumulate.

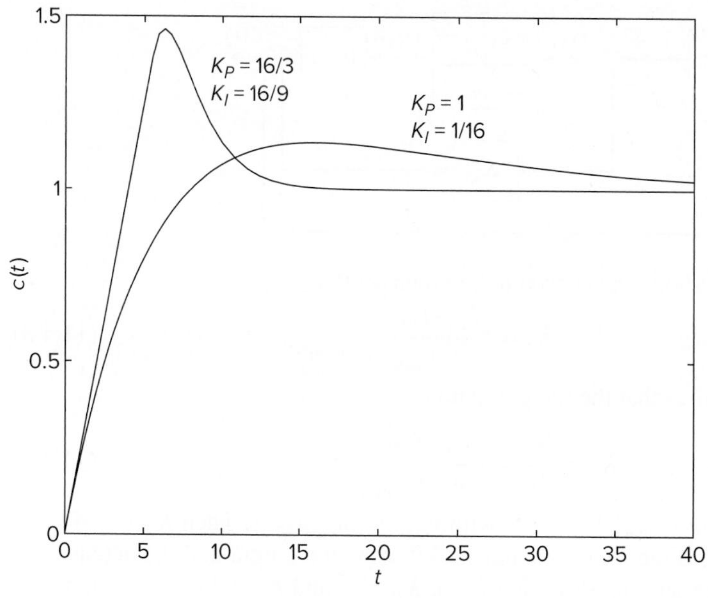

Figure 11.4.5 Response of two PI control systems.

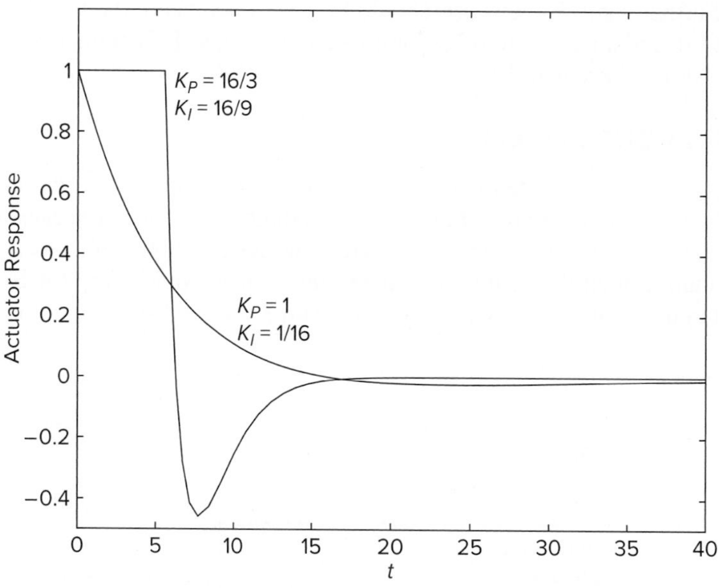

Figure 11.4.6 Actuator response of two PI control systems.

11.4.3 OTHER CONTROL STRUCTURES

Control structures other than the classic PID structure sometimes can be used to reduce the problems associated with actuator saturation and reset windup. One of these is the use of an internal feedback loop in conjunction with the main controller, which uses I action (Figure 11.4.7). In Chapter 10 we saw that such an arrangement requires a much smaller actuator output at the cost of a slightly longer response time.

For the plant $G_p(s) = 1/Is$ with a step command and $\zeta = 1$, we can obtain the analytical solution for the actuator output $m(t)$. From this we can determine that $m(t)$ reaches a maximum of $r_{\text{max}} K_2/5.44$ at $t = 2I/K_2$. Thus, to avoid overdriving the

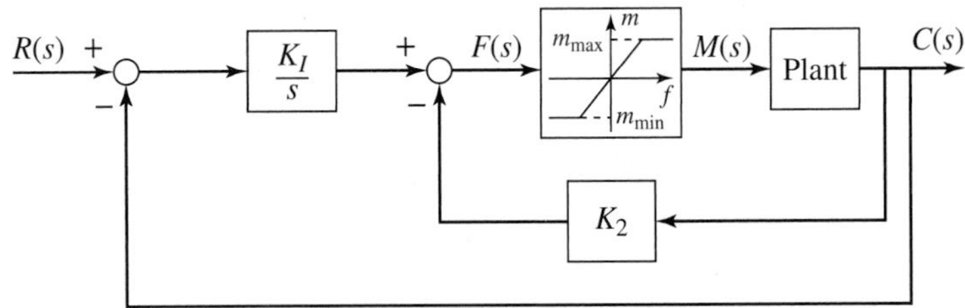

Figure 11.4.7 Use of internal feedback to avoid saturation.

actuator, the gain K_2 should be chosen to be no larger than

$$K_2 = 5.44\frac{m_{\text{max}}}{r_{\text{max}}} \tag{11.4.3}$$

To obtain $\zeta = 1$ requires that the integral gain be

$$K_I = \frac{K_2^2}{4I} = \frac{29.6}{4I}\left(\frac{m_{\text{max}}}{r_{\text{max}}}\right)^2 \tag{11.4.4}$$

Suppose the plant is $G_p(s) = 1/4s$ with $r_{\text{max}} = m_{\text{max}} = 1$. Then $K_2 = 5.44$ and $K_I = 29.6/16 = 1.85$. Figures 11.4.8 and 11.4.9 show the output and the actuator response for this design and the PI design using $K_P = 1$ and $K_I = 1/16$. It is apparent that the system with internal feedback gives a better response without demanding more from the actuator. Further analysis, however, should be done to compare the two designs' responses to a disturbance and to other inputs such as ramps. This comparison is given in [Palm, 2000] and [Phelan, 1977].

11.4.4 AN ANTI-WINDUP SYSTEM

It is sometimes impossible to compute the gains to avoid reset windup and saturation because either the input magnitudes are not known well enough in advance or because the required mathematics is too cumbersome. There are some designs, however, that act to prevent reset windup. One of these anti-windup designs is shown in Figure 11.4.10 [Franklin, 1994]. It consists of a feedback loop with a gain K_A around the integrator.

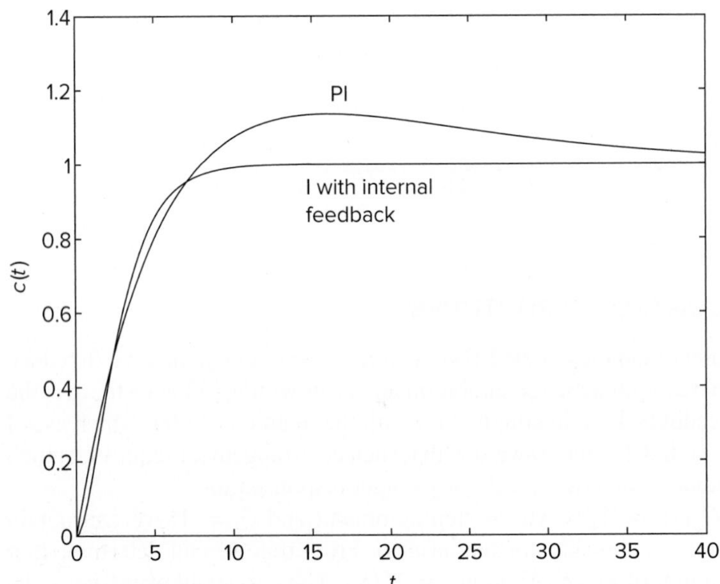

Figure 11.4.8 Comparison of PI control and I action with internal feedback.

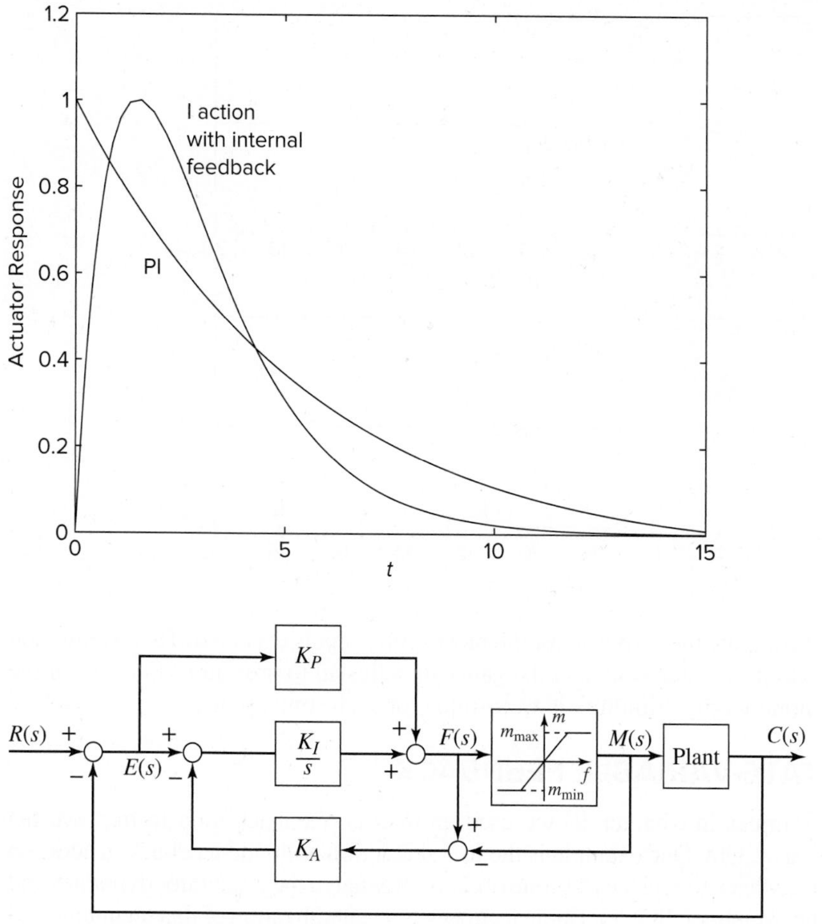

Figure 11.4.9 Actuator responses for PI control and I action with internal feedback.

Figure 11.4.10 An anti-windup system.

The actuator output is $M(s)$. Its output commanded by the controller is $F(s)$. When the actuator is not saturated, $m = f$ and the comparator in the feedback loop produces a zero output. Thus, the loop has no effect. When the actuator is saturated, however, so that $m = m_{max}$ but $f > m_{max}$, a positive signal is fed by the comparator to the gain K_A, which decreases the signal to the integrator. Thus, the integrator output begins to decrease, causing the actuator to become unsaturated.

Note that the P action bypasses the inner feedback loop, so the proportional signal is unaffected by the anti-windup loop. The gain K_A should be chosen large enough to allow the loop to follow changes in the error signal. Figure 11.4.11 compares the response of this system using $K_P = K_I = 16$ and $K_A = 10$ for the plant $G_p(s) = 1/4s$ with $m_{max} = r_{max} = 1$ to the response of the classic PI controller using $K_P = K_I = 16$. Note that the anti-windup system does not prevent the actuator from being overdriven, but it does eliminate reset windup that causes overshoot.

Limits on the actuator output obviously can prevent us from achieving the desired response time. We can, however, reduce the effects of saturation by choosing gains that prevent it and reset windup, by choosing alternative control structures that place less demand on the actuator, or by using a control structure that prevents reset windup. For plants other than first-order, the mathematics required to analyze such designs is formidable, so the design approach is to assume the system is operating in the linear

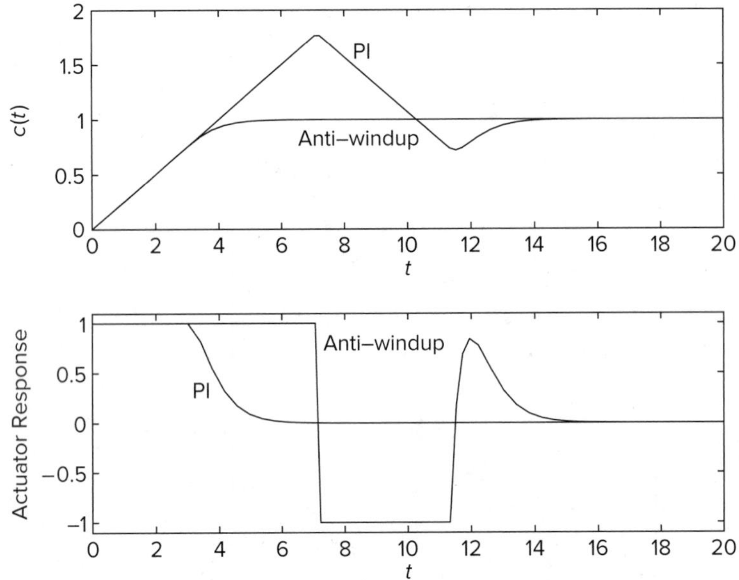

Figure 11.4.11 Responses for PI control and an anti-windup system.

region, and compute the response for the most severe inputs expected. Then simulation is used to check for saturation, and the gains are adjusted to prevent it. Because of the saturation nonlinearity, Simulink is well suited for such simulations.

11.5 STATE-VARIABLE FEEDBACK

In some examples in Chapter 10 we used an internal feedback loop to improve the response of a system. One example is the use of P action with rate feedback in addition to position feedback to replace PD control. This eliminated the numerator dynamics and the resulting overshoot. State-variable feedback is a generalization of that technique and uses some or all of the system's state variables to modify the control signal. With state-variable feedback, we have a better chance of placing the characteristic roots of the closed-loop system in locations that will give desirable performance. In our examples, we will assume that all of the state variables are measurable or at least derivable from other information. For example, a position signal can theoretically be passed through a differentiator to obtain the velocity if a tachometer cannot be used.

EXAMPLE 11.5.1

Motor Control Using State-Variable Feedback

■ Problem

Consider PD action (Figure 11.5.1a) and state-variable feedback (Figure 11.5.1b) used to control the angular displacement of a dc motor. For the state-variable feedback, the state variables are θ, ω, and the motor current i, and the controller uses P action. Compare the performance of these two systems.

■ Solution

Consider PD control first (Figure 11.5.1a). The closed-loop transfer function is

$$\frac{\Theta(s)}{\Theta_r(s)} = \frac{K_T(K_P + K_D s)}{LIs^3 + IRs^2 + K_T K_D s + K_T K_P}$$

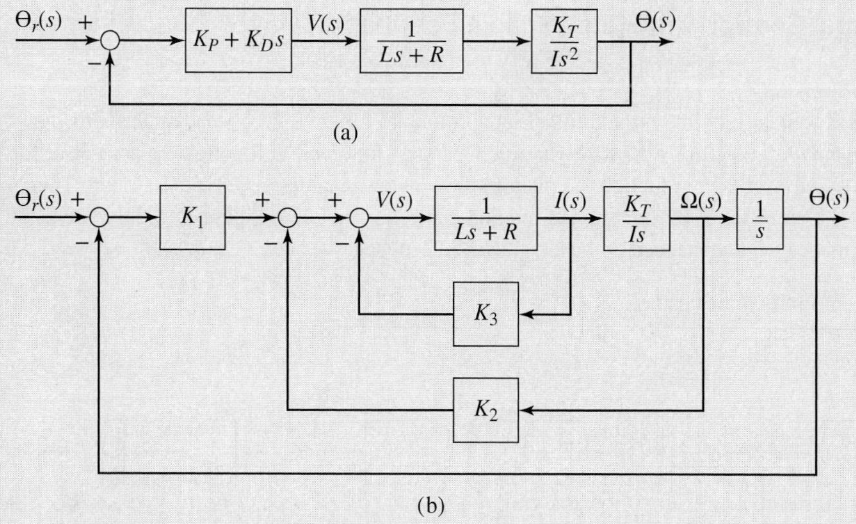

Figure 11.5.1 (a) Position control of a motor using PD action. (b) Position control of a motor using state-variable feedback.

The system is stable if $IRK_TK_D > ILK_TK_P$; that is, if $K_P/K_D < R/L$. Note that the coefficient of s^2 is unaffected by the control gains, so it might not be possible to select these gains to achieve the desired transient response. That is, since the highest coefficient LI is fixed and we cannot set the values of the last three coefficients, we cannot specify the locations of the three roots, in general.

Now, consider P action with state-variable feedback (Figure 11.5.1b). The transfer function can be obtained by successively reducing the inner loops of the diagram. The result is

$$\frac{\Theta(s)}{\Theta_r(s)} = \frac{K_1K_T}{LIs^3 + I(R+K_3)s^2 + K_TK_2s + K_TK_1}$$

There are no numerator dynamics and the system is stable if $(R + K_3)K_2 > LK_1$. The coefficient of s^2 is now affected by the gains, as are the coefficient of s and the constant term. We therefore have more influence over the transient behavior by feeding back all the state variables.

If the current is not fed back, $K_3 = 0$, and the s^2 coefficient is again beyond our influence. The system is unstable if no velocity feedback is used (if $K_2 = 0$).

If the system is stable, the steady-state error is 0 for a step command and is K_2/K_1 for a unit ramp.

Suppose we require the roots to be $s = -1 \pm j$ and -5. These correspond to the polynomial $s^3 + 7s^2 + 12s + 10$. Divide the denominator of the transfer function by LI to obtain

$$s^3 + \frac{R+K_3}{L}s^2 + \frac{K_TK_2}{LI}s + \frac{K_TK_1}{LI}$$

Comparing the coefficients of these two polynomials, we obtain $K_1 = 10LI/K_T$, $K_2 = 12LI/K_T$, and $K_3 = 7L - R$. Given values for K_T, L, R, and I, the next step in the design process would be to determine whether the required motor torque K_Ti exceeds the maximum available torque, whose value depends on the specific motor and can be obtained from the motor manufacturer. Because the system is third-order, determination of the maximum required torque is best done by simulation. Simulink can be used for this purpose, as demonstrated in Section 11.7.

EXAMPLE 11.5.2

Integral Control with State-Variable Feedback

■ Problem

A *type-1 system* requires no additional integrations to obtain a zero steady-state error for a step *command*. P action with state-variable feedback, however, will not give a zero error for a step *disturbance*, even though the plant in the previous example is a type-1 system. Therefore, we replace the P action with the integral controller shown in Figure 11.5.2. Evaluate its performance when enhanced with state-variable feedback.

Figure 11.5.2 Integral control using state-variable feedback.

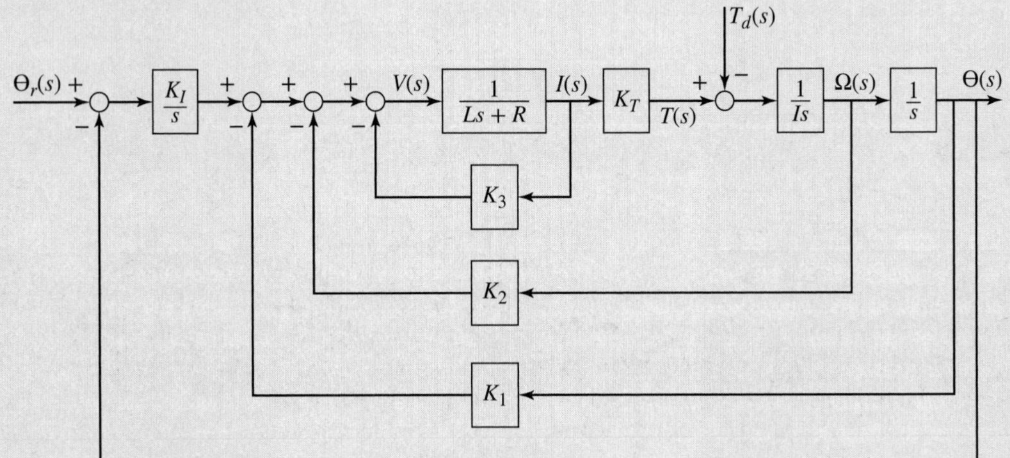

■ Solution

The output equation obtained from the diagram is

$$\Theta(s) = \frac{K_T K_I}{LIs^4 + I(R+K_3)s^3 + K_T K_2 s^2 + K_T K_1 s + K_T K_I}\Theta_r(s) - \frac{s(Ls+R+K_3)}{LIs^4 + I(R+K_3)s^3 + K_T K_2 s^2 + K_T K_1 s + K_T K_I}T_d(s)$$

From this we see that if the system is stable, the steady-state error for a step disturbance is zero.

Using the Routh-Hurwitz results of Table 10.5.3, and assuming the gains and parameters are positive, we see that the system is stable if and only if

$$K_T K_1[K_T K_2 I(R+K_3) - K_T K_1 LI] - K_T K_I[I(R+K_3)]^2 > 0$$

which reduces to

$$K_1 K_2 K_T(R+K_3) - IK_I(R+K_3)^2 - LK_T K_1^2 > 0$$

This shows that if $K_1 = 0$ or if $K_2 = 0$, the system is unstable. Thus, the position θ must be fed back in the inner loop as well as in the outer loop. Now, however, with integral control as opposed to proportional control, the gain K_3 is not required for stability.

In general, the form of state-variable feedback using P action is

$$f(t) = K_1[r(t) - x_1] - K_2 x_2 - K_3 x_3 - \cdots \tag{11.5.1}$$

where $r(t)$ is the command input and $f(t)$ is the controller output. If I action is used instead, the form is

$$f(t) = K_I \int [r(t) - x_1]\, dt - K_1 x_1 - K_2 x_2 - K_3 x_3 - \cdots \tag{11.5.2}$$

11.5.1 CHOICE OF FEEDBACK VARIABLES

With the motor example, the choice of state variables and measured quantities is perhaps obvious. This might not always be the case. Consider the plant

$$\frac{C(s)}{F(s)} = \frac{1}{s(s^2 + 8s + 15)} \tag{11.5.3}$$

where $f(t)$ is the controller output (the actuating variable) and $c(t)$ is the output to be controlled. The differential equation is

$$\frac{d^3c}{dt^3} + 8\frac{d^2c}{dt^2} + 15\frac{dc}{dt} = f(t)$$

A logical choice for state variables is $x_1 = c$, $x_2 = \dot{c}$, and $x_3 = \ddot{c}$. The state equations are

$$\dot{x}_1 = x_2 \quad \dot{x}_2 = x_3 \quad \dot{x}_3 = f(t) - 15x_2 - 8x_3$$

With proportional control and state-variable feedback, these equations lead to the diagram shown in Figure 11.5.3. This assumes that we can measure or compute $c(t)$, $\dot{c}(t)$, and $\ddot{c}(t)$.

Since this is a type-1 plant, I action is not needed to eliminate the steady-state error for a step command. This can be confirmed from the transfer function.

$$\frac{C(s)}{R(s)} = \frac{K_1}{s^3 + (8 + K_3)s^2 + (15 + K_2)s + K_1}$$

Note that the last three coefficients of the denominator can be independently set with the three gains. This means that we can place the roots anywhere.

Some choices of root locations can result in very large gain values, which might cause very large actuator outputs. Such designs might also be very sensitive to slight changes or uncertainties in the plant parameters or control gains. Simulation is recommended before accepting the design.

The diagram in Figure 11.5.3 implies that we can measure the state variables. The diagram in Figure 11.5.4 shows the controller for the case where we compute x_2

Figure 11.5.3 State-variable feedback for the plant $1/(s^3 + 8s^2 + 15s)$.

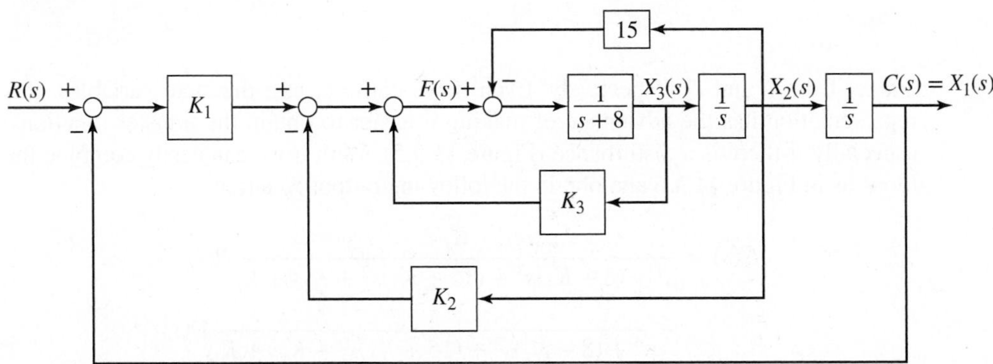

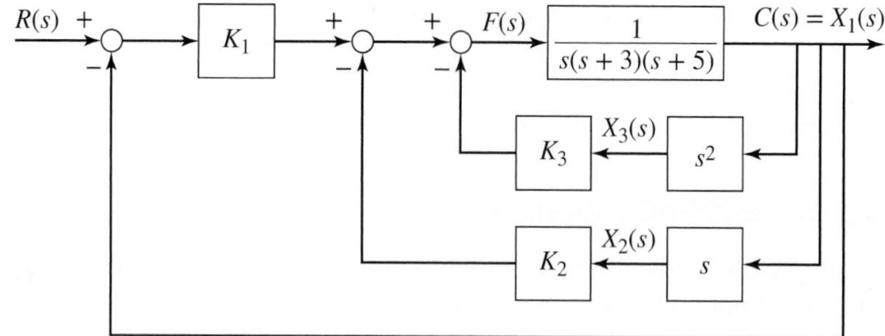

Figure 11.5.4 An equivalent representation of Figure 11.5.3.

Figure 11.5.5 Integral control with state-variable feedback for the plant $1/(s^3 + 8s^2 + 15s)$.

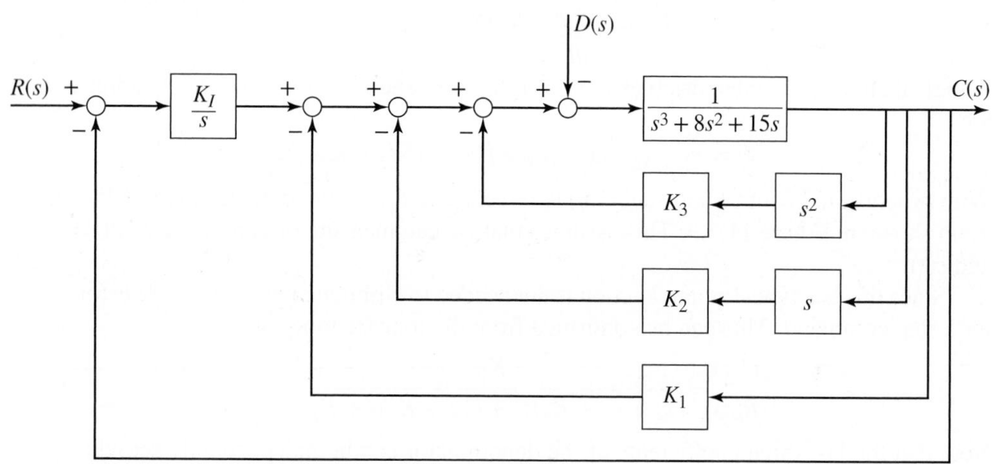

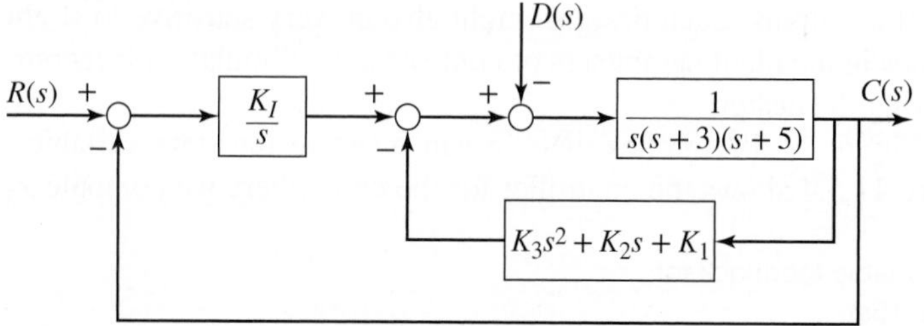

Figure 11.5.6 An equivalent representation of Figure 11.5.5.

and x_3 from $\dot{c}$ and $\ddot{c}$, respectively. Even if we can measure the state variables, this representation has the advantage of making it easier to obtain the transfer functions, especially if there is a disturbance (Figure 11.5.5). With it we can easily combine the loops as in Figure 11.5.6 and obtain the following output equation.

$$C(s) = \frac{K_I}{s^4 + (8 + K_3)s^3 + (15 + K_2)s^2 + K_1 s + K_I} R(s) - \frac{s}{s^4 + (8 + K_3)s^3 + (15 + K_2)s^2 + K_1 s + K_I} D(s)$$

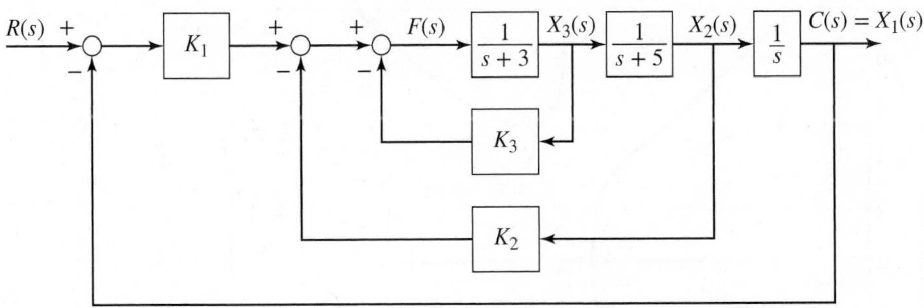

Figure 11.5.7 Alternative choice of state variables for the plant $1/(s^3 + 8s^2 + 15s)$.

To place the roots at $s = -6 \pm 6j$, $s = -7$, and $s = -8$, the characteristic polynomial must be $s^4 + 27s^3 + 308s^2 + 1752s + 4032$. This requires the gains to be $K_1 = 1752$, $K_2 = 293$, $K_3 = 19$, and $K_I = 4032$.

There are other choices for state variables for (11.5.3). For example, because we can factor the plant model as

$$\frac{C(s)}{F(s)} = \frac{1}{s}\frac{1}{s+5}\frac{1}{s+3}$$

we can choose

$$X_1(s) = C(s) \qquad X_2(s) = \frac{1}{s+5}X_3(s) \qquad X_3(s) = \frac{1}{s+3}F(s)$$

The corresponding diagram using P action is given in Figure 11.5.7. This choice of state variables, however, might or might not correspond to measurable variables.

11.5.2 ROOT LOCUS ANALYSIS

The root locus plot and Bode plot are useful for analyzing the choice of root locations and the resulting gains. For example, suppose we want to analyze the effect of the integral gain K_I for the system shown on Figure 11.5.5. The root locus equation is

$$1 + K_I \frac{1}{s[s^3 + (8 + K_3)s^2 + (15 + K_2)s + K_1]} = 0$$

With the gain values given previously, this becomes

$$1 + K_I \frac{1}{s(s^3 + 27s^2 + 308s + 1752)} = 0$$

There are no zeros and the poles are $s = -13.91$, $s = -6.54 \pm 9.12j$, and $s = 0$. The root locus is shown in Figure 11.5.8. Increasing K_I moves the two secondary roots even farther to the left, but the dominant roots move to the right. This increases the dominant-time constant above $1/6$ and increases ζ above 0.707. The dominant-time constant cannot be reduced much smaller than $1/6$ by decreasing K_I, because one of the real roots moves to the right and becomes the dominant root. This occurs when $K_I \approx 3950$ and the real root is $s = -5.93$. Thus, we cannot make the dominant-time constant less than $1/5.93 = 0.169$ by adjusting K_I.

11.5.3 MATRIX METHODS

For higher-order models the algebra required to obtain the characteristic equation can be time-consuming. Just as we use computer methods to obtain the response of such

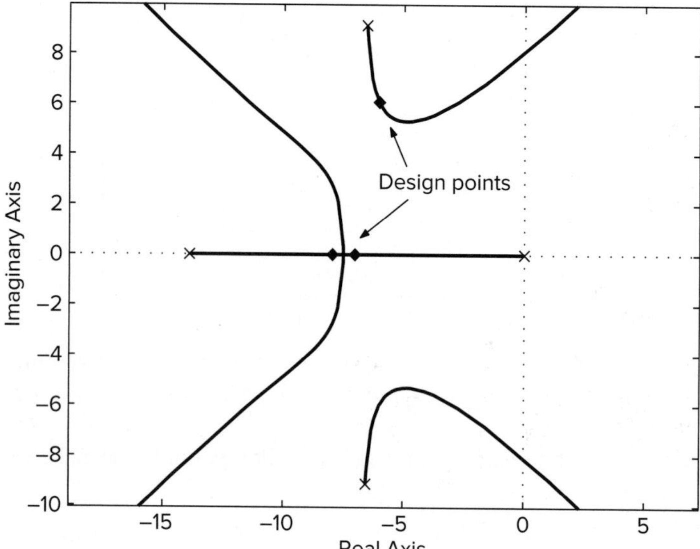

Figure 11.5.8 Root locus plot for the system shown in Figure 11.5.5 for variable K_I.

models, rather than spending time to obtain closed-form solutions, we also may use computer methods to determine the gain values required to place the roots at the desired locations. MATLAB provides the `acker` function for this purpose. It is named for J. Ackermann, who developed the algorithm. The function requires the matrices of the state-variable model.

Consider the model

$$\dot{x}_1 = -x_1 + x_2 + f \quad \dot{x}_2 = x_1 - 2x_2 \tag{11.5.4}$$

With P action, the control algorithm is

$$f(t) = K_1[r(t) - x_1] - K_2 x_2$$

The `acker` function requires the **A** and **B** matrices of the model (see Chapter 5 for more discussion of these matrices). For this model,

$$\mathbf{A} = \begin{bmatrix} -1 & 1 \\ 1 & -2 \end{bmatrix} \quad \mathbf{B} = \begin{bmatrix} 1 \\ 0 \end{bmatrix}$$

The syntax of the `acker` function is `K = acker(A,B,p)`, where `p` is the array containing the desired closed-loop roots and `K` is the array of gains $K_1, K_2, \dots$ returned by the function. Suppose we want to place the roots at $s = -5$ and $s = -20$. The script file is

```
A = [-1, 1; 1, -2]; B = [1; 0];
K = acker(A,B,[-5, -20])
```

The answer given is `K = [22, 55]`. Thus, the control law is $f(t) = 22[r(t) - x_1] - 55x_2$.

If I action is to be used, the form is

$$f(t) = K_I \int [r(t) - x_1]\, dt - K_1 x_1 - K_2 x_2 - K_3 x_3 - \cdots$$

and we need to augment the state-variable model with an additional state variable. This state, denoted w, is defined to be the output of the integrator. That is,

$$w = \int [r(t) - x_1]\, dt \tag{11.5.5}$$

This implies that

$$\dot{w} = r(t) - x_1 \tag{11.5.6}$$

which gives an additional state equation that is appended to the first n equations.

The control algorithm is then expressed as

$$f(t) = -K_1x_1 - K_2x_2 - K_3x_3 - \cdots - (-K_I w)$$

Thus, we see that the *negative* of the integral gain corresponds to the gain for the *last* state variable. So the array `K` returned by the `acker` function must be interpreted as $K = [K_1, K_2, K_3, \ldots, -K_I]$.

Using I action with the model (11.5.4), we obtain the state equations

$$\dot{x}_1 = -x_1 + x_2 + f \quad \dot{x}_2 = x_1 - 2x_2 \quad \dot{w} = r(t) - x_1$$

So the new state variable is $x_3 = w$, and the new matrices are

$$\mathbf{A} = \begin{bmatrix} -1 & 1 & 0 \\ 1 & -2 & 0 \\ -1 & 0 & 0 \end{bmatrix} \quad \mathbf{B} = \begin{bmatrix} 1 \\ 0 \\ 0 \end{bmatrix}$$

To place the roots at $s = -1$, $s = -2$, and $s = -2$, the script file is

```
A = [-1, 1, 0; 1, -2, 0; -1, 0, 0]; B = [1; 0; 0];
K = acker(A,B,[-1, -2, -2])
```

The answer given is `K = [2, -1, -2]`, which corresponds to the gains $K_1 = 2$, $K_2 = -1$, and $K_I = 2$. Thus, the control law is

$$f(t) = 2\int [r(t) - x_1]\,dt - 2x_1 + x_2$$

11.5.4 CONTROLLABILITY WITH LINEAR STATE FEEDBACK

We have noted that an advantage of state-variable feedback is that it enables us to place the roots where we want. This is not always possible, however. Consider the following model.

$$\dot{x}_1 = -2x_1 + f \quad \dot{x}_2 = x_1 - x_2 - f$$

where

$$\mathbf{A} = \begin{bmatrix} -2 & 0 \\ 1 & -1 \end{bmatrix} \quad \mathbf{B} = \begin{bmatrix} 1 \\ -1 \end{bmatrix}$$

Using P action, the control law is $f(t) = K_1[r(t) - x_1] - K_2x_2$. Substituting this into the state equations and rearranging gives

$$\dot{x}_1 = (-2 - K_1)x_1 - K_2x_2 + K_1r(t) \quad \dot{x}_2 = (1 + K_1)x_1 + (K_2 - 1)x_2 - K_1r(t)$$

Using the Laplace transform method with zero-initial conditions gives

$$(s + 2 + K_1)X_1(s) + K_2X_2(s) = K_1R(s)$$

$$-(1 + K_1)X_1(s) + (s + 1 - K_2)X_2(s) = -K_1R(s)$$

The determinant of these equations gives the characteristic equation

$$s^2 + (3 + K_1 - K_2)s + 2 + K_1 - K_2 = 0$$

Letting $b = K_1 - K_2$ gives $s^2 + (3 + b)s + 2 + b = 0$, and we thus see that the equation has only one parameter! Therefore, we cannot specify both roots in general.

Such systems where we cannot specify all the roots by using linear state-variable feedback are said to be "uncontrollable" with linear state feedback. This means that we cannot control the response of every state variable. The `acker` function will not work for such systems, and gives a warning to that effect.

There are advanced methods for determining whether or not a system is controllable, but for our purposes, the warning issued by the `acker` function is sufficient.

EXAMPLE 11.5.3

Active Vibration Control

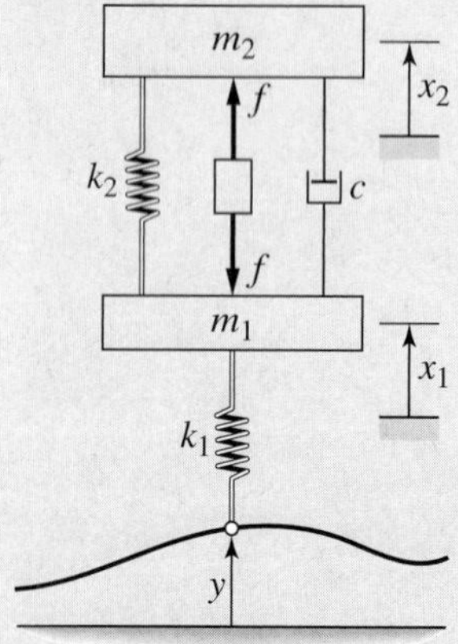

Figure 11.5.9 Active vibration control.

■ Problem

Active vibration control is a candidate to improve the performance of vibrating systems under a variety of conditions. When used with a vehicle suspension, the terms *active suspension* or *adaptive suspension* is often used. The control gains can be changed according to conditions to provide improved response, whereas with traditional spring-damper systems, the system response depends on the spring and damping constants, which are fixed. Figure 11.5.9 illustrates an active vibration scheme for a two-mass system. An electrohydraulic actuator between the two masses provides a force that acts on both and is under feedback control. The system model is

$$m_1\ddot{x}_1 = k_1(y - x_1) - k_2(x_1 - x_2) - c(\dot{x}_1 - \dot{x}_2) - f$$

$$m_2\ddot{x}_2 = k_2(x_1 - x_2) + c(\dot{x}_1 - \dot{x}_2) + f$$

The given parameter values are $m_1 = 36$ kg, $m_2 = 240$ kg, $k_1 = 1.6\times10^5$ N/m, $k_2 = 8000$ N/m, and $c = 50$ N · s/m.

a. Put the model into state-variable form.

b. Assume that we can measure all four state variables, and use P action with state-variable feedback. In the original passive system, $k_2 = 1.6 \times 10^4$ N/m and $c = 98$ N · m/s, which resulted in characteristic roots at $s = -1.397 \pm 69.94j$, $s = -0.168 \pm 7.779j$. Compute the values of the feedback gains so that the closed-loop roots will be near those of the passive system.

■ Solution

a. Let the state variables be

$$z_1 = x_1 \qquad z_2 = \dot{x}_1 \qquad z_3 = x_2 \qquad z_4 = \dot{x}_2$$

The input is f. The state equations are

$$\dot{z}_1 = z_2 \qquad \dot{z}_2 = \frac{1}{m_1}[k_1 y - (k_1 + k_2)z_1 - cz_2 + k_2 z_3 + cz_4 - f]$$

$$\dot{z}_3 = z_4 \qquad \dot{z}_4 = \frac{1}{m_2}[k_2 z_1 + cz_2 - k_2 z_3 - cz_4 + f]$$

where the disturbance input is the road displacement y. The matrices are

$$\mathbf{A} = \begin{bmatrix} 0 & 1 & 0 & 0 \\ -\frac{k_1+k_2}{m_1} & -\frac{c}{m_1} & \frac{k_2}{m_1} & \frac{c}{m_1} \\ 0 & 0 & 0 & 1 \\ \frac{k_2}{m_2} & \frac{c}{m_2} & -\frac{k_2}{m_2} & -\frac{c}{m_2} \end{bmatrix} = \begin{bmatrix} 0 & 1 & 0 & 0 \\ -4667 & -1.389 & 222.2 & 1.389 \\ 0 & 0 & 0 & 1 \\ 33.33 & 0.2083 & -33.33 & -0.2083 \end{bmatrix}$$

$$\mathbf{B} = \begin{bmatrix} 0 \\ -\frac{1}{m_1} \\ 0 \\ \frac{1}{m_2} \end{bmatrix} = \begin{bmatrix} 0 \\ -0.0278 \\ 0 \\ 0.0042 \end{bmatrix}$$

b. The MATLAB script file is

```
A = [0, 1, 0, 0; -4667, -1.389, 222.2, 1.389;...
     0, 0, 0, 1; 33.33, 0.2083, -33.23, -0.2083];
B = [0; -0.0278; 0; 0.0042];
p = [-1.397+69.94j, -1.397-69.94j, -0.168+7.779j,
     -0.168-7.779j];
K = acker(A,B,p)
```

The result is `K = [-7963, -48, 7934, 47.5]`. For this application, $r(t) = 0$, and the control algorithm is $f(t) = K_1(0 - x_1) - K_2x_2 - K_3x_3 - K_4x_4 = 7963x_1 + 48x_2 - 7934x_3 - 47.5x_4$.

11.6 MATLAB APPLICATIONS

We have already seen many of the MATLAB features that are particularly helpful for doing the design tasks described in this chapter. These include the transfer function and state-variable functions, such as `tf` and `ss`; the transient-response functions `step` and `lsim`; and the frequency response function `bode`. In Section 11.5 we introduced the `acker` function, which is used to compute the state-variable feedback gains.

The polynomial regression tools, including `polyfit`, which are discussed in Appendix C are useful for applying the Ziegler-Nichols process reaction method discussed in Section 11.3.

This section introduces the MATLAB functions relevant to this chapter.

11.6.1 ROOT LOCUS ANALYSIS WITH MATLAB

The MATLAB Control System toolbox provides several useful commands for producing root locus plots and for extracting information from them. The basic function is `rlocus(sys)`, which displays the root locus. Additional commands are available for enhancing the plot and for obtaining root locations and gain values from the plot.

The function `rlocus(sys)` computes and displays the root locus plot for the equation

$$D(s) + KN(s) = 0 \tag{11.6.1}$$

where `sys` corresponds to the transfer function $N(s)/D(s)$ and the parameter K varies from 0 through a large positive value determined by MATLAB. The order of $N(s)$ must be no greater than the order of $D(s)$. It is recommended that the highest coefficients in $N(s)$ and $D(s)$ both be *unity* so that the values of K displayed by MATLAB will be interpreted correctly.

MATLAB refers to the parameter K as the "gain." The starting points of the roots corresponding to $K = 0$ are called the "poles" and are marked with a cross (×). These are the roots of $D(s) = 0$. The finite termination points that the roots approach as $K \to \infty$ are called the "zeros" and are marked with a circle (○). These are the roots of $N(s) = 0$.

You should use equal scaling on the real and imaginary axes so that circular root loci will display as circles and so that the angle θ associated with the damping ratio, $\zeta = \cos\theta$, can be properly interpreted. This can be done by including the command `axis equal` after the `rlocus` function.

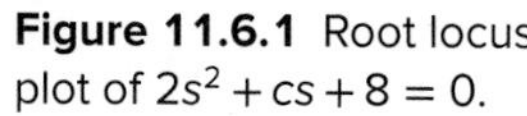

Figure 11.6.1 Root locus plot of $2s^2 + cs + 8 = 0$.

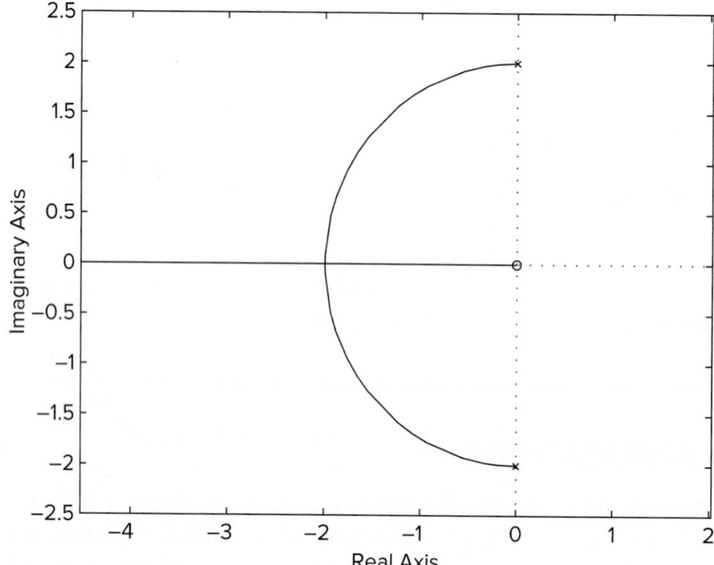

The plot shown in Figure 11.1.3b corresponds to the characteristic equation $2s^2 + cs + 8 = 0$. This may be put into the standard form (11.6.1) by rewriting it as

$$s^2 + 4 + \frac{c}{2}s = 0$$

where $K = c/2$, $D(s) = s^2 + 4$, and $N(s) = s$. The poles are $s = \pm j2$. The zero is $s = 0$. To obtain this root locus plot, the session is

```
>>sys1 = tf([1,0],[1,0,4]);
>>rlocus(sys1),axis equal
```

Note that the numerator must be expressed as `[1, 0]` because the numerator polynomial is actually $s + 0$. The result is shown in Figure 11.6.1.

11.6.2 PLACING A (ζ, ω_n) GRID

The `sgrid` command superimposes an s plane grid of constant ζ and constant ω_n lines on the root locus plot. This grid is useful for locating roots that satisfy performance specifications stated in terms of ζ and ω_n. The alternate syntax `sgrid(zeta,omega)` superimposes an s plane grid of constant ζ and constant ω_n lines on the root locus plot, using the values contained in the vectors `zeta` and `omega`.

Suppose we want to see if any dominant roots of the following equation have damping ratios in the range $0.5 \leq \zeta \leq 0.707$, and undamped natural frequencies in the range $0.5 \leq \omega_n \leq 0.75$.

$$s^3 + 3s^2 + 2s + K = 0$$

The required session is

```
>>sys2 = tf(1,[1,3,2,0]);
>>rlocus(sys2),sgrid([0.5,0.707],[0.5,0.75]),axis equal
```

You may use empty brackets if you want to omit either lines of constant ζ or lines of constant ω_n. For example, to see just the damping ratio lines corresponding to $\zeta = 0.5$ and $\zeta = 0.707$, you would use the form `sgrid([0.5,0.707],[])`.

11.6.3 OBTAINING INFORMATION FROM THE PLOT

Once the plot is displayed you can pick a point from this plot to find the gain K required to achieve ζ and ω_n values in the desired range. To do this, you can use the `rlocfind` function discussed later in this section, or you can place the cursor on the plot at the desired point. Left-click to display the root value (although labeled "pole," this is not the same as the starting points marked by ×). The gain value, the damping ratio (labeled "damping"), the percent overshoot, and the undamped natural frequency ω_n (labeled "frequency") are also displayed. You can move the cursor along the plot and view the updated values. Right-clicking anywhere within the plot area brings up a menu that enables the plot to be edited.

Note that the percent overshoot value displayed on the screen with this method is computed from the formula for $M_{\%}$ given in Table 8.3.2, and thus it does *not* include the effects of numerator dynamics on the overshoot.

You can use this method to see how sensitive the root location is to a change in the value of the gain K. Sometimes it is inadvisable to design a critically damped system, which corresponds to two or more equal, real roots, because such designs can be very sensitive to parameter variation and parameter uncertainty. Near such roots, a slight change in the gain K can cause a large change in the root location.

An alternate syntax to obtain the root values for specified values of K is `r = rlocus(sys,K)`. This returns the root locations in the vector `r` corresponding to the user-specified vector of gain values `K`. The matrix `r` has `length(K)` columns and its jth column contains the roots for the gain value `K(j)`. No plot is displayed.

MATLAB provides the `rlocfind` function to obtain information about roots and gain values from the plot on the screen. The syntax `[K,r] = rlocfind(sys)` enables you to use a cursor to obtain the value of the gain `K` corresponding to a specified point on a root locus plot. The vector `r` contains the roots corresponding to this gain value. The advantage of using `rlocfind` is that it returns *all* the roots for a given value of K, whereas left-clicking on the curve shows only some of the roots. The `rlocfind` function, which must follow the `rlocus` command, generates a cursor on the screen and waits for the user to press the mouse button after positioning the cursor over the desired point on the locus. Once the button is pressed, you will see on the screen the coordinates of the selected point, the gain value at that point, the roots closest to that point, and the other roots that correspond to the gain value.

For example, the following session displays the root locus with a grid line corresponding to $\zeta = 0.707$. This enables us to use the cursor to find the dominant roots that have a damping ratio of $\zeta = 0.707$ and the corresponding gain value K.

```
>>sys3 = tf(1,[1,3,2,0]);
>>rlocus(sys3),axis equal,sgrid(0.707,[]),...
    [K,r] = rlocfind(sys3)
```

The dominant roots selected this way will not have a damping ratio of $\zeta = 0.707$ exactly, because you cannot position the cursor exactly at the intersection of the locus and the $\zeta = 0.707$ line. To reduce this inaccuracy, you can enlarge the plot by enabling the Edit Plot button and clicking the magnifying glass icon on the menu bar of the figure window.

Another syntax is `[K,r] = rlocfind(sys,p)`, which computes a root locus gain K for each desired root location specified in the vector `p` (or a gain for which one of the closed-loop roots is near the desired location). The jth entry in the vector `K` is the computed gain for the root location `r(j)`. The jth column of the matrix `r` contains the resulting closed-loop roots.

The real power of the root locus method only becomes apparent when it is applied to a model higher than second-order, as in the following example.

The `rlocus` function is especially useful for applications in this chapter where the model is third-order or higher, because simple expressions for the time constant, damping ratio, overshoot, etc. are not available to assist in setting the gains. The root locus plot is useful for analyzing the effects on the roots of changing a control system gain. It is also helpful for understanding how the uncertainty in a parameter value will affect the performance of the control system. We now give two examples of these applications. More applications of the root locus for control system design are given in Chapter 12.

EXAMPLE 11.6.1

Varying the Integral Gain

■ Problem

The characteristic equation of the liquid-level control system of Example 10.8.5 is

$$8s^3 + (6 + 8K_1)s^2 + (1 + 4K_1 + K_2)s + K_I = 0$$

In that example, the gain values were selected to give the characteristic roots $s = -0.5 \pm 0.5j$ and $s = -5$. The gain values are $K_1 = 21/4$, $K_2 = 22$, and $K_I = 20$. Investigate the effects on the roots and on the steady-state ramp-command error of changing the value of K_I.

■ Solution

The steady-state error for a unit-ramp command is $e_{ss} = 44/K_I$, and thus its nominal value for $K_I = 20$ is $e_{ss} = 44/20 = 2.2$.

Substituting $K_1 = 21/4$ and $K_2 = 22$ into the characteristic equation gives

$$8s^3 + 48s^2 + 44s + K_I = 0$$

With the Routh-Hurwitz condition we immediately see that the system is unstable if $48(44) < 8K_I$; that is, if $K_I > 264$.

Rearranging the characteristic equation into the standard root locus form, we have

$$1 + \frac{K_I}{8}\frac{1}{s^3 + 6s^2 + 5.5s} = 0$$

With the highest coefficients normalized to unity, the root locus gain is $K = K_I/8$. The poles are the roots of the denominator, which are $s = 0, -1.129, -4.87$. There are no zeros because the numerator is a constant.

The MATLAB session is

```
>>sys = tf(1, [1, 6, 5.5, 0]);
>>rlocus(sys)
```

The root locus plot is shown in Figure 11.6.2. The Plot Editor has been used to annotate the plot. The locations of the roots when $K_I = 20$ have been marked. We see that as K_I is increased the system becomes unstable when the dominant roots cross the imaginary axis and move into the right-half plane, but we knew this from the Routh-Hurwitz condition.

However, the plot also tells us that as K_I is increased above 20, the dominant roots move up and to the right and the third root moves to the left. Thus, the damping ratio decreases (increasing the overshoot), the oscillation frequency increases, and the dominant-time constant increases so the response becomes slower—all undesirable effects—but the ramp error becomes smaller. The plot also shows that the third root is never the dominant root.

As K_I is decreased, the dominant root eventually becomes real. The damping ratio of the dominant root is 1 when the two roots are real and equal. Use the cursor to move along the

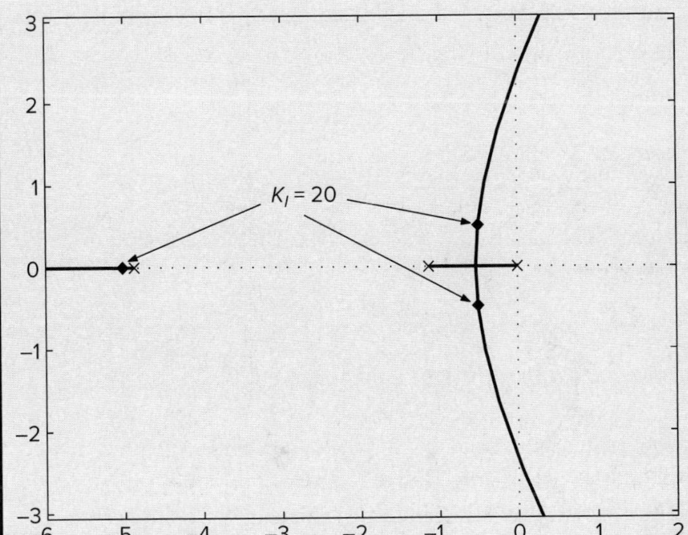

Figure 11.6.2 Root locus plot for Example 11.6.1.

curve to the point where the two roots meet and read the corresponding gain displayed in the box that appears. At this point $s = -0.502$ and the gain is $K = 1.38$, which corresponds to $K_I = 8K = 11.04$. (Root locations are very sensitive to the gain near points where two roots are equal, so these values are approximate.) Thus, for $K_I < 11.04$ the dominant root is real and moves to the right as K_I is further decreased. We see, therefore, that the *smallest* possible dominant-time constant for the given values of K_1 and K_2 occurs when $K_I = 11.04$ and is $\tau = 1/0.502 = 1.99$. For this value, the ramp error is $44/11.04 = 3.99$.

In conclusion, the root locus plot shows that the designer must choose between a faster system by using $K_I = 11.04$ or minimizing the ramp error $44/K_I$ but at the cost of a very oscillatory system that is approaching instability. This insight could not have been obtained without the root locus plot.

Uncertainty in the Damping Constant

EXAMPLE 11.6.2

■ Problem

The characteristic equation for PI control of the plant $1/(10s^2 + cs)$ is

$$10s^3 + cs^2 + K_P s + K_I = 0$$

Some discussion was devoted to the effect of uncertainty in value of the damping c on the effectiveness of the control system. If $c = 80$ and if the gains are $K_P = 240$ and $K_I = 320$, the roots are $s = -2 \pm 2j$ and $s = -4$. Suppose that $c = 80$ is just a rough estimate. Investigate the effects of having a different value of c.

■ Solution

Substituting the gains $K_P = 240$ and $K_I = 320$ into the characteristic equation gives

$$10s^3 + cs^2 + 240s + 320 = 0$$

With the Routh-Hurwitz condition we immediately see that the system is unstable if $240c < 10(320)$; that is, if $c < 13.33$.

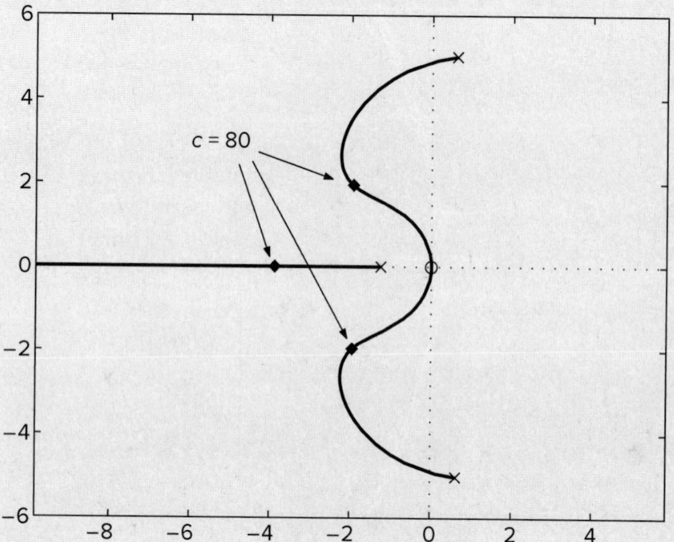

Figure 11.6.3 Root locus plot for Example 11.6.2.

Rearranging this equation into the standard root locus form, we have

$$1 + \frac{c}{10}\frac{s^2}{s^3 + 24s + 32} = 0$$

With the highest coefficients normalized to unity, the root locus gain is $K = c/10$. The poles are the roots of the denominator, which are $s = 0.626 \pm 5.02j$ and $s = -1.25$. There are two zeros, which are the roots of the numerator. They are both at $s = 0$.

The MATLAB session is

```
>>sys = tf([1, 0, 0],[1, 0, 24, 32]);
>>rlocus(sys)
```

The root locus plot is shown in Figure 11.6.3. The Plot Editor has been used to annotate the plot. The locations of the roots when $c = 80$ have been marked. We see that as c is decreased, the system becomes unstable when the dominant roots cross the imaginary axis and move into the right-half plane, but we knew this from the Routh-Hurwitz condition.

Remember we have no influence over the value of c. If $c < 80$, the dominant-time constant is smaller, and the smallest possible dominant-time constant occurs when the dominant-root path is the farthest to the left. This occurs at $s = -2.25 \pm 2.67j$ when the root locus gain is near $K = 7.1$, which was determined by moving the cursor along the plot. The smallest possible dominant-time constant is $\tau = 1/2.25 = 0.44$. This gain value corresponds to $c = 10K = 71$. Thus, regardless of the value of c, the time constant will be no less than 0.44.

11.6.4 ROOT LOCUS PLOTS FOR NEGATIVE GAINS

When we use the `rlocus(sys)` function, MATLAB automatically varies the root locus gain K through positive values. To use the `rlocus(sys)` function for problems where the gain K is negative, you simply reverse the sign of the numerator of `sys`. For example, the problem treated in Example 11.2.10 is

$$1 + K\frac{s - 5}{s(s + 8)} = 0$$

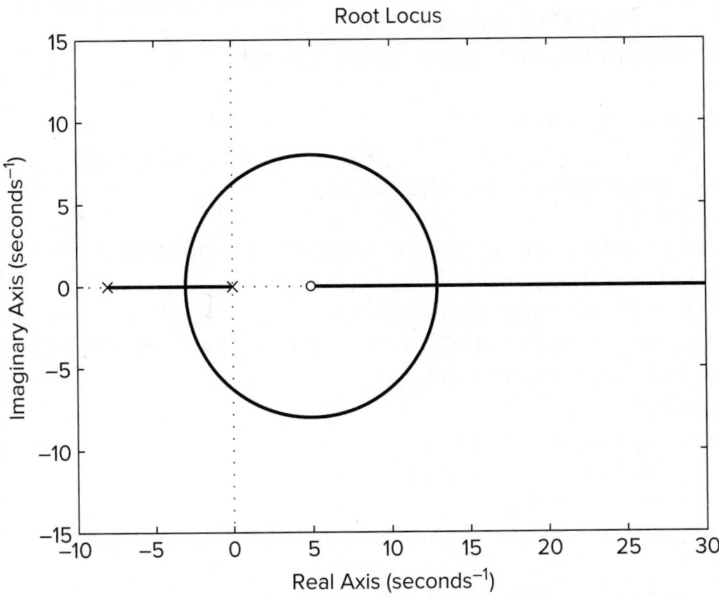

Figure 11.6.4 Root locus plot of $1 + K(s - 5)/[s(s + 8)] = 0$ for $K \leq 0$.

where $K \leq 0$. To plot the root locus in MATLAB, first rearrange the equation as

$$1 + K\frac{-s + 5}{s(s + 8)} = 0$$

where K is now *positive*. Then you enter

```
>>sys = tf([-1,5],[1,8,0]);
>>rlocus(sys),axis equal
```

The result is shown in Figure 11.6.4.

11.6.5 THE `rltool`, `ltitool`, AND `ltiview` FUNCTIONS

The MATLAB Control System Toolbox also contains the Root Locus Tool, which is a graphical interface for interacting with the root locus plot. You activate it by typing `rltool`. It enables you to interactively design a control system using the root locus technique. The `ltiview` function activates an interface for interacting with the frequency response and step-response plots, for example. The `sisotool` function opens an interface that lets you design a single input, single output (SISO) system by interacting with root locus and Bode plots. Detailed information on their use is available with the MATLAB online help feature.

11.6.6 FITTING DATA FOR THE PROCESS REACTION METHOD

When applying the Ziegler-Nichols process reaction method, it is usually difficult to determine the slope R and intercept L from a plot of the data. As mentioned in Section 11.3, a procedure using polynomial regression can be applied to compute R and L. The script file shown in MATLAB Program 11.6.1 illustrates the procedure for the unit-step open-loop response of a hypothetical thermal process.

The results are $R = 14.8433$, $L = 1.5417$, $K_P = 0.0524$, $K_I = 0.0170$, and $K_D = 0.0404$. The controller transfer function is

$$G_c(s) = \frac{0.0404s^2 + 0.0524s + 0.017}{s} \tag{11.6.1}$$

MATLAB program 11.6.1

```
% Times at which measurements were made (minutes).
td = (0:1:12);
% Measured Temperatures (deg C).
yd = [0, 3, 10, 20, 37, 53, 63, 77, 83, 87, 90, 93, 93];
% Fit a 5th order polynomial to the data.
coef = polyfit(td,yd,5);
% Evaluate the polynomial at a large number of points.
dt = 0.01; t = (0:dt:12); y = polyval(coef,t)';
% Find the coefficients of the polynomial's derivative.
coef2 = [5*coef(1), 4*coef(2), 3*coef(3), 2*coef(4), coef(5)];
% der is the derivative polynomial.
der = polyval(coef2,t)';
% Find the maximum slope R, and ...
  the time tmax at which it occurs.
[R,i] = max(der); tmax = i*dt;
% Compute the temperature where the slope is the maximum.
ymax = y(i);
% Compute the intercept L on the time axis.
L = (R*tmax-ymax)/R;
disp(R),disp(L)
% Compute Gains For PID Control
KP = 1.2/(R*L)
KI = KP/(2*L)
KD = 0.5*L*KP
```

Table 11.6.1 summarizes the MATLAB functions used in this chapter.

Table 11.6.1 MATLAB functions used in this chapter.

	Functions Introduced in This Chapter
`K = acker(A,B,p)`	Calculates the feedback gain matrix K such that the single input system $\dot{\mathbf{x}} = \mathbf{Ax} + \mathbf{Bu}$ with the feedback law $\mathbf{u} = -\mathbf{Kx}$ has closed-loop poles at the values specified in vector `p`.
`axis equal`	Specifies that the abscissa and ordinate have the same scale.
`disp(x)`	Displays the array `x`.
`length(x)`	Returns the length of the vector `x`.
`ltiview`	Opens the LTI Viewer, which is an interactive graphical user interface for analyzing the time and frequency responses of linear systems.
`p = polyfit(x,y,n)`	Finds the coefficients of a polynomial $p(x)$ of degree n that fits the data (x, y) best in a least-squares sense.
`y = polyval(p,x)`	Returns the value of a polynomial `p` evaluated at the points specified in the array `x`.
`[K,roots] = rlocfind(sys)`	Enables use of the cursor on a root locus plot to find the root locus gain K and the set of roots `roots` corresponding to a point on the locus.
`rlocus(sys)`	Computes and plots the root locus of the single-input, single-output LTI model `sys`.
`rltool`	Opens the SISO Design Tool and prepares it for root locus design.
`sgrid(zeta,wn)`	Generates lines of constant damping ratio `zeta` and constant natural frequency `wn` on a root locus plot.
`sisotool`	Opens the SISO Design Tool, which is a graphical user interface that lets you design single-input/single-output (SISO) compensators by graphically interacting with the root locus and Bode plots of an open-loop system.
	Functions Introduced in Earlier Chapters
`tf`	Creates a model in transfer-function form (Section 2.9).

11.7 SIMULINK APPLICATIONS

This section contains a number of the chapter's applications that can benefit from the capabilities of Simulink.

11.7.1 AN ANTI-WINDUP SYSTEM

The anti-windup design shown in Figure 11.4.10 is a good example of a system that is easy to simulate in Simulink. The Simulink model shown in Figure 11.7.1 can be developed directly from the block diagram in Figure 11.4.10, and it was obtained with the parameter values given in Section 11.4. Figure 11.4.11 was obtained with this model, and the value of $K_A = 10$ was obtained from simulations.

11.7.2 THE RATE-LIMITER ELEMENT

In addition to being limited by saturation, some actuators have limits on how fast they can react. This limitation is independent of the time constant of the actuator, and might be due to deliberate restrictions placed on the unit by its manufacturer. An example is a flow control valve whose rate of opening and closing is controlled by a "rate limiter." Simulink has such a block, and it can be used in series with the Saturation block to model the valve behavior.

Consider the model of the height h of liquid in a tank, whose input is a flow rate q_i. For specific parameter values, such a model has the form

$$\frac{H(s)}{Q_i(s)} = \frac{2}{5s + 1}$$

A Simulink model is shown in Figure 11.7.2 for a specific PI controller whose gains are $K_P = 4$ and $K_I = 5/4$. Suppose that the minimum and maximum flow rates available from the valve are 0 and 2. These are the limits for the Saturation block. The model enables us to experiment with the lower and upper limits of the Rate-Limiter block to see its effect on the system performance.

Figure 11.7.1 Simulink model of an anti-windup system.

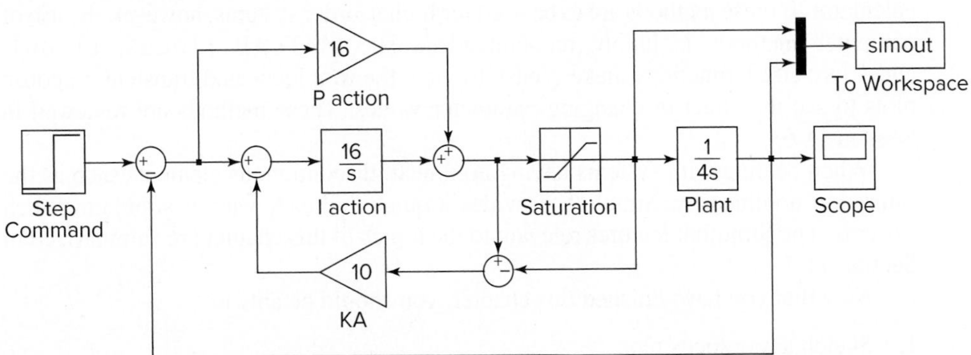

Figure 11.7.2 Application of the Rate-Limiter block.

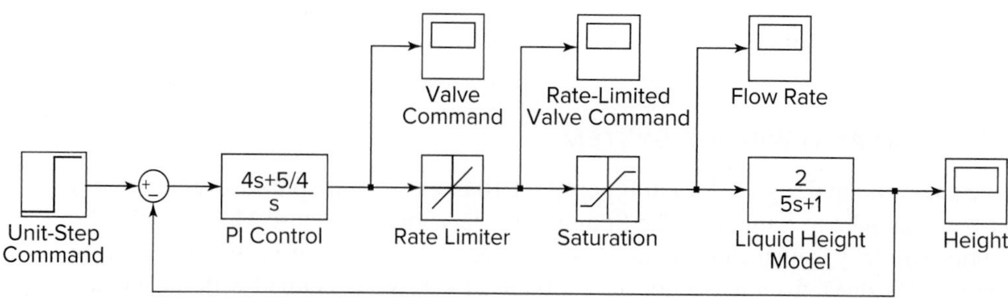

11.8 CHAPTER REVIEW

This chapter shows how to use the root locus plot to design control systems. This type of plot enables a systematic design approach, by showing the effects of pole and zero placement on the transient response and stability.

Section 11.1 introduced the root locus plot. Section 11.2 illustrates the use of the root locus plot to design PID controllers. When PID control action fails to yield the desired performance, it must be either modified or replaced by an entirely different control scheme.

In many applications, especially in process control involving thermodynamic, fluid, or chemical processes, a transfer-function model of the plant is not available, and the gains must be computed from experimentally determined open-loop response data or from computer simulations of the plant. In addition, the computed gain values sometimes do not yield the desired performance, and their values must be adjusted either in simulation or with the actual controller hardware. This process is called tuning and is discussed in Section 11.3.

Using high-gain values tends to drive the control elements to such an extent that they overload or "saturate" and thus exhibit nonlinear behavior. Controllers having I action can exhibit reset windup, which can cause overshoot and saturation. Designs that avoid these unwanted effects are covered in Section 11.4.

State-variable feedback, treated in Section 11.5, is a generalization of rate feedback that uses some or all of the system's state variables to modify the control signal. This eliminates numerator dynamics and the resulting overshoot. With state-variable feedback, we have a better chance of placing the characteristic roots of the closed-loop system in locations that will give desirable performance.

Root locus plots for simple systems can be sketched by hand with the aid of a calculator. If these methods are to be used for higher-order systems, however, the use of computer methods is highly recommended. The MATLAB `rlocus`, `rltool`, and `sisotool` functions make it easy to view the root locus and transient-response plots to see the effect of changing parameter values. These methods are reviewed in Section 11.6.

When dealing with systems having nonlinear, discontinuous elements such as the saturation nonlinearity, Simulink provides a quick and easy way of simulating such systems. The Simulink features relevant to the topics of this chapter are summarized in Section 11.7.

Now that you have finished this chapter, you should be able to

1. Sketch a root locus plot.

2. Determine the major features of a root locus plot, and use it to assess the results of pole and zero placement.
3. Use the Ziegler-Nichols methods to design controllers.
4. Design a control system to avoid actuator saturation.
5. Design a controller incorporating state-variable feedback.
6. Apply MATLAB and Simulink to analyze and design control systems using the concepts presented in this chapter.

REFERENCES

[Abzug, 1997] M. J. Abzug and E. Larabee, *Airplane Stability and Control*, Cambridge University Press, Cambridge, Great Britain, 1997.

[Bryson, 1975] A. E. Bryson and Y. C. Ho, *Applied Optimal Control*, Hemisphere Publishing, New York, 1975.

[Cannon, 1967] R. H. Cannon, *Dynamics of Physical Systems*, McGraw-Hill, New York, 1967.

[Evans, 1948] W. R. Evans, *Graphical Analysis of Control Systems*, Trans. AIEE, 67:547–551, 1948.

[Franklin, 1994] G. F. Franklin, J. D. Powell, and A. Emami-Naeini, *Feedback Control of Dynamics Systems*, Addison-Wesley, Reading, MA, 1994.

[Palm, 2000] W. J. Palm III, *Modeling, Analysis, and Control of Dynamic Systems*, John Wiley and Sons, New York, 2000.

[Phelan, 1977] R. Phelan, *Automatic Control Systems*, Cornell University Press, Ithaca, New York, 1977.

[Seborg, 1989] D. E. Seborg, T. F. Edgar, and D. A. Mellichamp, *Process Dynamics and Control*, John Wiley and Sons, New York, 1989.

PROBLEMS

Section 11.1 Root Locus Plots

11.1 Sketch the root locus plot of $3s^2 + 12s + k = 0$ for $k \geq 0$. What is the smallest possible dominant-time constant, and what value of k gives this time constant?

11.2 Sketch the root locus plot of $3s^2 + cs + 12 = 0$ for $c \geq 0$. What is the smallest possible dominant-time constant, and what value of c gives this time constant? What is the value of ω_n if $\zeta < 1$?

11.3 Sketch the root locus of the armature-controlled dc motor model in terms of the damping constant c, and evaluate the effect on the motor time constant. The characteristic equation is

$$L_a I s^2 + (R_a I + cL_a)s + cR_a + K_b K_T = 0$$

Use the following parameter values:

$$K_b = K_T = 0.1 \text{ N} \cdot \text{m/A} \qquad I = 12 \times 10^{-5} \text{ kg} \cdot \text{m}^2$$

$$R_a = 2\ \Omega \qquad L_a = 3 \times 10^{-3} \text{ H}$$

11.4 Sketch the root locus plot of $ms^2 + 12s + 10 = 0$ for $m \geq 2$. What is the smallest possible dominant-time constant, and what value of m gives this time constant?

Section 11.2 Design Using the Root Locus Plot

11.5 In the following equations, identify the root locus plotting parameter K and its range in terms of the parameter p, where $p \geq 0$.

a. $8s^2 + 6s + 5p = 0$

b. $6s^2 + (8 + p)s + 7 + 4p = 0$

c. $5s^3 + 3ps^2 + 5s + p = 0$

11.6 Consider a unity feedback system with the plant $G_p(s)$ and the controller $G_c(s)$. PID control action is applied to the plant

$$G_p(s) = \frac{s + 5}{(s + 3)(s + 4)}$$

The PID controller has the transfer function

$$G_c(s) = K_P\left(1 + \frac{1}{T_I s} + T_D s\right)$$

Use the values $T_I = 0.1$ and $T_D = 0.2$.

a. Identify the open-loop poles and zeros.

b. Identify the root locus parameter K in terms of K_P.

c. Identify the closed-loop poles and zeros for the case $K_P = 5$.

11.7 In parts (a) through (f), sketch the root locus plot for the given characteristic equation for $p \geq 0$.

a. $2s^2 + 10s + p = 0$

b. $3s^3 + 48s^2 + 189s + p = 0$

c. $4s^2 + 12s + 20 + p(s + 3) = 0$

d. $5s^2 + 20s + p(s + 5) = 0$

e. $4s^3 + 12s^2 + 20 + p = 0$

f. $7s^3 + 70s^2 + 147s + p(s + 4) = 0$

11.8 PID control action is applied to the plant

$$G_p(s) = \frac{s + 10}{(s + 2)(s + 5)}$$

The PID controller has the transfer function

$$G_c(s) = K_P\left(1 + \frac{1}{T_I s} + T_D s\right)$$

Use the values $T_I = 0.2$ and $T_D = 0.5$. Plot the root locus with the proportional gain K_P as the parameter.

11.9 Consider the following equation where the parameter p is nonnegative.

$$4s^3 + (25 + 5p)s^2 + (16 + 30p)s + 40p = 0$$

a. Put the equation in standard root locus form and define a suitable root locus parameter K in terms of the parameter p.

b. Obtain the poles and zeros, and sketch the root locus plot.

11.10 In the following equation, $K \geq 0$.

$$s^2(s + 9) + K(s + 1) = 0$$

a. Obtain the root locus plot.
b. Obtain the value of K at the breakaway point, and obtain the third root for this value of K.
c. What is the smallest possible dominant-time constant for this equation?

11.11 Consider the following equation where the parameter K is nonnegative.

$$(2s + 5)(2s^2 + 14s + 49) + Ks(2s + 1)(2s + 3) = 0$$

a. Determine the poles and zeros, and sketch the root locus plot.
b. Use the plot to set the value of K required to give a dominant-time constant of $\tau = 0.5$. Obtain the three roots corresponding to this value of K.

11.12 In the following equations, identify the root locus plotting parameter K and its range in terms of the parameter p, where $p \geq 0$.

a. $7s^3 + 8s^2 - 6ps + 3 = 0$
b. $5s^3 - ps^2 + 44s + 9 = 0$
c. $s^2 + (6 - p)s + 7 + 7p = 0$

11.13 In parts (a) through (f), obtain the root locus plot for $p \leq 0$ for the given characteristic equation.

a. $2s^2 + 10s + p = 0$
b. $4s^2 + 12s + 12 + p(s + 3) = 0$
c. $3s^3 + 9s^2 + 9s + p = 0$
d. $2s^3 + 24s^2 + 70s + p = 0$
e. $5s^2 + 15s + p(s + 4) = 0$
f. $7s^2 + 42s + p(s - 4) = 0$

11.14 The plant transfer function for a particular process is

$$G_p(s) = \frac{8 - s}{s^2 + 2s + 3}$$

We wish to investigate the use of proportional control action with this plant.

a. Obtain the root locus and determine the range of values of the proportional gain K_P for which the system is stable.
b. Determine the value of K_P required to give a time constant of $\tau = 2/3$.
c. Plot the unit-step response of the plant for $K_P = 1$. A process containing a negative sign in the numerator of its transfer function is called a "reverse reaction" process. What is the effect of the negative sign in the numerator of $G_p(s)$?

11.15 The plant transfer function for a particular process is

$$G_p(s) = \frac{26 + s - 2s^2}{s(s + 2)(s + 3)}$$

We wish to investigate the use of proportional control action with this plant.

a. Obtain the root locus and determine the range of values of the proportional gain K_P for which the system is stable.
b. Determine the value of K_P required to give $\zeta = 1$.
c. Plot the unit-step response of the plant. What is the effect of the negative sign in the numerator of $G_p(s)$?

11.16 Control of the attitude θ of a missile by controlling the fin angle ϕ, as shown in Figure P11.16, involves controlling an inherently unstable plant. Consider the

Figure P11.16

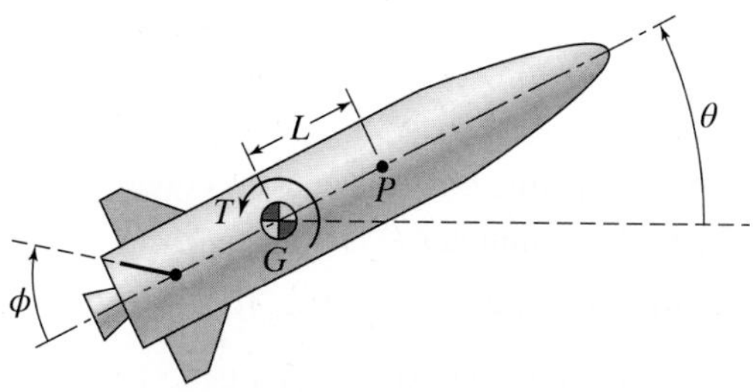

specific plant transfer function

$$G_p(s) = \frac{\Theta(s)}{\Phi(s)} = \frac{1}{s^2 - 5}$$

a. Determine the PD control gains so that the steady-state error for a step command is zero, the closed-loop damping ratio is 0.707, and the dominant closed-loop time constant is 0.1.

b. Use the root locus to evaluate the performance of the resulting controller in light of the specifications $\zeta = 0.707$ and $\tau = 0.1$ if the plant transfer function $G_p(s)$ has an uncertainty Δ due to fuel consumption, where

$$G_p(s) = \frac{1}{s^2 - 5 - \Delta}, \quad 0 \leq \Delta \leq 1$$

11.17 The use of a motor to control the rotational displacement of an inertia I is shown in Figure P11.17. The open-loop transfer function of the plant for a specific application is

$$G_p(s) = \frac{6}{s(2s + 2)(3s + 24)}$$

a. Use the root locus plot to show that it is not possible with proportional control to achieve a dominant-time constant of less than 2.07 sec for this plant.

b. Use PD control to improve the response, so that the dominant-time constant is 0.5 sec or less and the damping ratio is 0.707 or greater. To do this, first select a suitable value for T_D, then plot the locus with K_P as the variable.

Figure P11.17

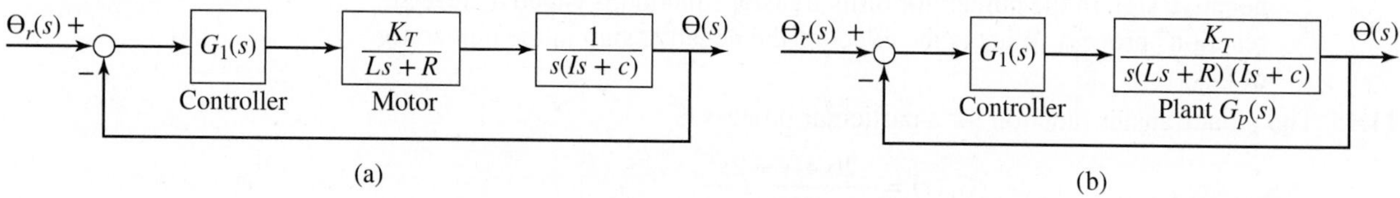

11.18 Proportional control action applied to the heat flow rate q_i can be used to control the temperature of the oven shown in Figure P11.18. Consider the specific plant

$$G_p(s) = \frac{T_1(s)}{Q_i(s)} = \frac{s + 10}{s^2 + 5s + 6}$$

Use the root locus plot to obtain the smallest damping ratio this system can have. Obtain the value of the proportional gain K_P required to minimize the dominant-time constant with $\zeta = 0.707$, and determine this time constant.

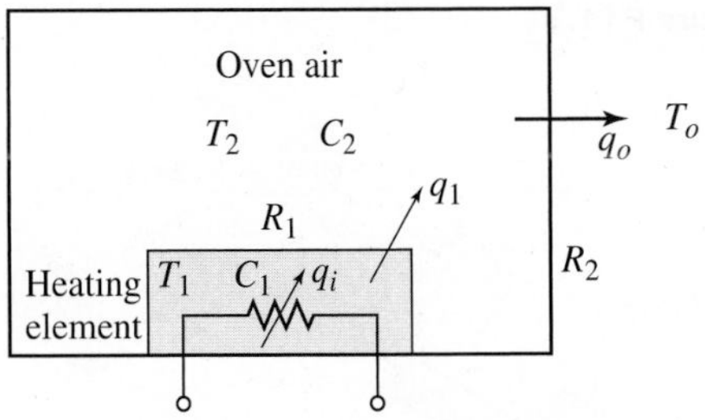

Figure P11.18

11.19 Proportional control action applied to the flow rate q_{mi} can be used to control the liquid height, as shown in Figure P11.19. Consider the specific plant

$$G_p(s) = \frac{H_2(s)}{Q_{mi}(s)} = \frac{5(s+4)}{(s+3)(s+p)}$$

The proportional gain is $K_P = 2$. The likely value of p is $p = 7$, but it is known that p might be greater than 7. Use the root locus plot to investigate the roots of the system for $p \geq 7$.

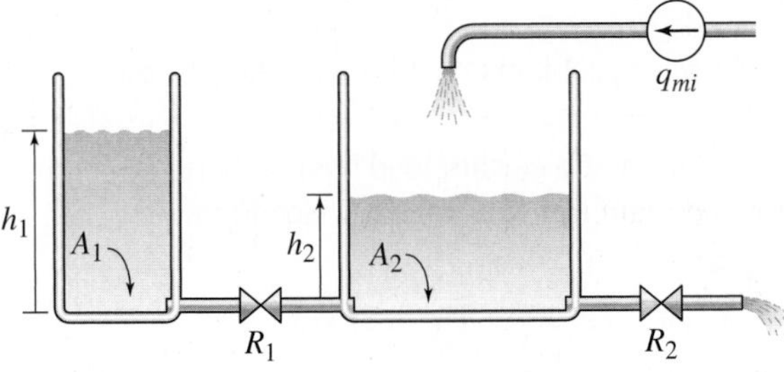

Figure P11.19

11.20 Proportional control action applied to the flow rate q_{mi} can be used to control the liquid height of the system shown in Figure P11.20. Consider the specific plant

$$G_p(s) = \frac{H_2(s)}{Q_{mi}(s)} = \frac{1}{s^2+3s+2}$$

Use the root locus plot to design a PI controller for this system to minimize the dominant-time constant, with a damping ratio of $\zeta = 0.707$.

11.21 Design a PID controller applied to the motor torque T to control the robot-arm angle θ shown in Figure P11.21. Consider the specific plant

$$G_p(s) = \frac{\Theta(s)}{T(s)} = \frac{4}{3s^2+3}$$

The dominant closed-loop roots must have $\zeta = 0.5$ and a time constant of 1.

11.22 (a) The equations of motion of the inverted pendulum model were derived in Example 3.5.6 in Chapter 3. Linearize these equations about $\phi = 0$, assuming that $\dot{\phi}$ is very small. (b) Obtain the linearized equations for the following values: $M = 10$ kg, $m = 50$ kg, $L = 1$ m, $I = 0$, and $g = 9.81$ m/s^2. (c) Use the linearized model developed in part (b) to design a PID controller to stabilize the pendulum angle near $\phi = 0$. It is required that the 2% settling time be no greater than 4 s and that the response be non-oscillatory. This means that the dominant root should be real and no greater than -1. No restriction is placed on the motion of the base. Assume that only ϕ and $\dot{\phi}$ can be measured.

Figure P11.20

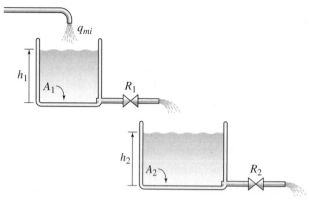

Figure P11.21

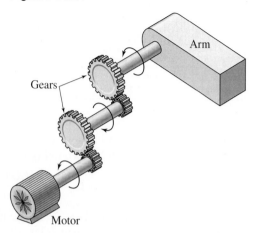

Section 11.3 Tuning Controllers

Note: See the problems for Section 11.6 for problems dealing with the process reaction method.

11.23 Use of a motor to control the position of a certain load having inertia, damping, and elasticity gives the following plant transfer function. See Figure P11.23.

$$G_p(s) = \frac{\Theta(s)}{V(s)} = \frac{0.5}{(s^2 + s + 1)(s + 0.5)}$$

a. Use the ultimate-cycle method to compute the controller gains for P, PI, and PID control.

b. Plot and compare the unit-step responses for the three designs obtained in (a). If the PID response is unsatisfactory, tune the gains to improve the performance.

Figure P11.23

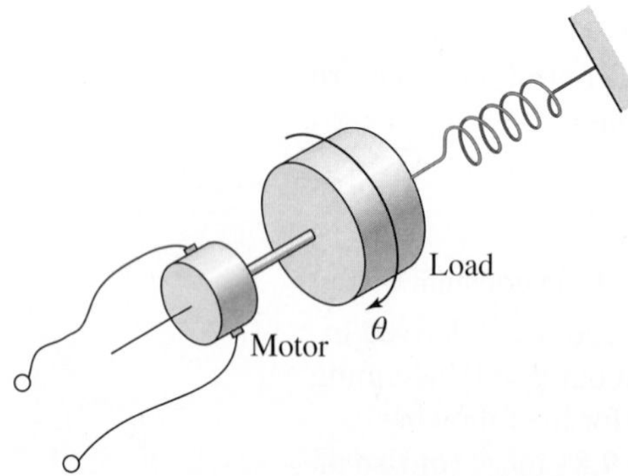

Figure P11.24

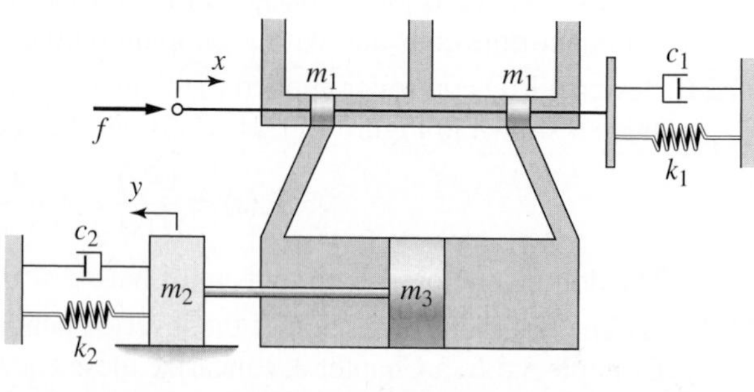

11.24 Figure P11.24 shows an electrohydraulic position control system whose plant transfer function for a specific system is

$$G_p(s) = \frac{Y(s)}{F(s)} = \frac{5}{2s^3 + 10s^2 + 2s + 4}$$

a. Use the ultimate-cycle method to design P, PI, and PID controllers.
b. Plot and compare the unit-step responses for the three designs obtained in (a). If the PID response is unsatisfactory, tune the gains to improve the performance.

11.25 A certain plant has the transfer function

$$G_p(s) = \frac{4p}{(s^2 + 4\zeta s + 4)(s + p)}$$

where the nominal values of ζ and p are $\zeta = 0.5$ and $p = 1$.

a. Use Ziegler-Nichols tuning to compute the PID gains. Obtain the resulting closed-loop characteristic roots.
b. Use the root locus to determine the effect of a variation in the parameter p over the range $0.4 \le \zeta \le 0.6$.
c. Use the root locus to determine the effect of a variation in the parameter p over the range $0.5 \le p \le 1.5$.

11.26 The plant transfer function of the system in Figure P11.26 for a specific case is

$$G_p(s) = \frac{8}{(2s + 2)(s + 2)(4s + 12)}$$

a. Use the ultimate-cycle method to compute the PID gains.
b. Plot the unit-step response. If the response is unsatisfactory, use the root locus plot to explain the result, and try to improve the response by tuning the gains.

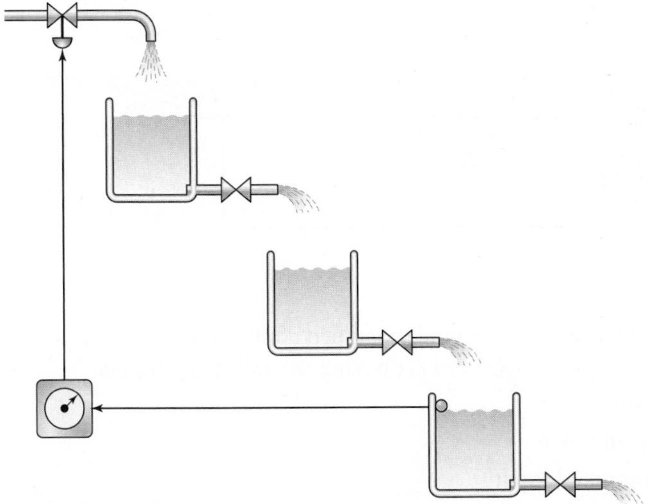

Figure P11.26

Section 11.4 Saturation and Reset Windup

11.27 Consider the PI control system shown in Figure P11.27 where $I = 5$ and $c = 0$. It is desired to obtain a closed-loop system having $\zeta = 1$ and $\tau = 0.1$. Let $m_{\text{max}} = 20$ and $r_{\text{max}} = 2$. Obtain K_P and K_I.

11.28 Consider the PI control system shown in Figure P11.27 where $I = 10$ and $c = 20$. It is desired to obtain a closed-loop system having $\zeta = 1$ and $\tau = 0.1$.

a. Obtain the required values of K_P and K_I, neglecting any saturation of the control elements.

Figure P11.27

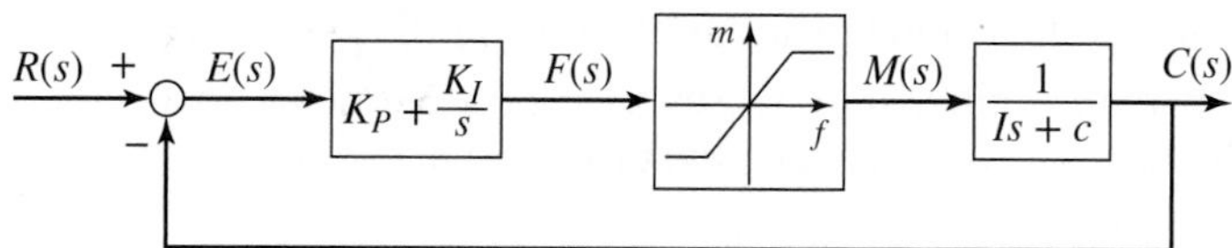

b. Let $m_{max} = r_{max} = 1$. Obtain K_P and K_I.

c. Compare the unit-step response of the two designs.

11.29 Consider the PI control system shown in Figure P11.27 where $I = 7$ and $c = 5$. It is desired to obtain a closed-loop system having $\zeta = 1$ and $\tau = 0.2$. Let $m_{max} = 20$ and $r_{max} = 5$. Obtain K_P and K_I.

11.30 a. Design a PI and an I controller with internal feedback for the plant $G_p(s) = 1/4s$. See Figure P11.30. We are given that $m_{max} = 6$ and $r_{max} = 3$. Set $\zeta = 1$.

b. Evaluate the unit-step response of each design.

c. Evaluate the unit-ramp response of each design.

Figure P11.30

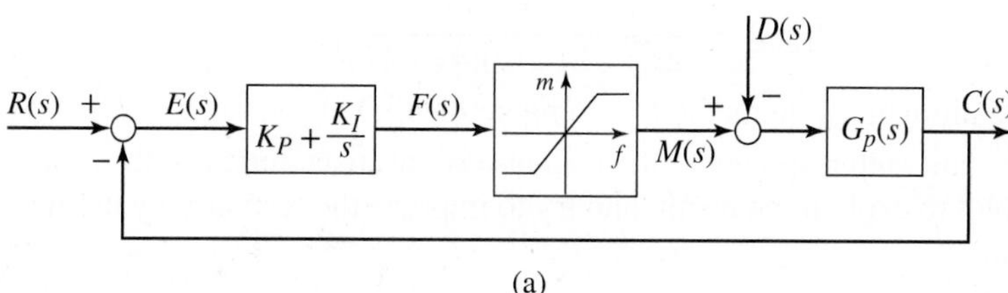

(a)

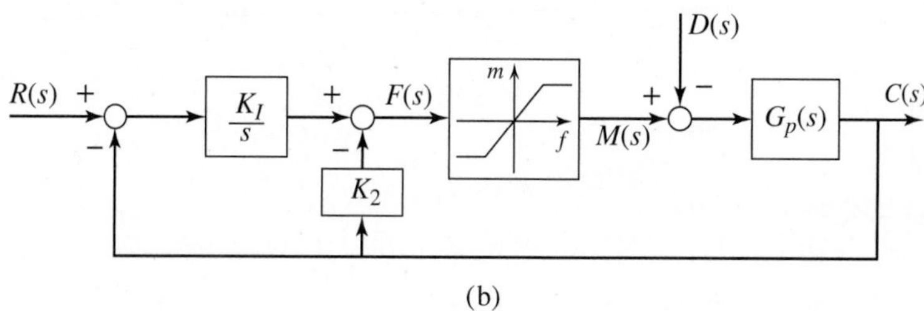

(b)

11.31 Compare the performance of the critically damped controllers shown in Figure P11.30 with the plant $G_p(s) = 1/Is$ having the following inputs:

a. A unit-ramp disturbance

b. A sinusoidal disturbance

c. A sinusoidal command input

Section 11.5 State-Variable Feedback

11.32 A certain field-controlled dc motor with load has the following parameter values.

$$L = 2 \times 10^{-3}\ \text{H} \qquad R = 0.6\ \Omega$$
$$K_T = 0.04\ \text{N} \cdot \text{m/A} \qquad c = 0$$
$$I = 6 \times 10^{-5}\ \text{kg} \cdot \text{m}^2$$

Compute the gains for a state-variable feedback controller using P action to control the motor's angular position. The desired dominant-time constant is

0.5 s. The secondary roots should have a time constant of 0.05 s and a damping ratio of $\zeta = 0.707$.

11.33 In Figure P11.33 the input u is an acceleration provided by the control system and applied in the horizontal direction to the lower end of the rod. The horizontal displacement of the lower end is y. The linearized form of Newton's law for small angles gives

$$mL\ddot{\theta} = mg\theta - mu$$

a. Put this model into state-variable form by letting $x_1 = \theta$ and $x_2 = \dot{\theta}$.

b. Construct a state-variable feedback controller by letting $u = k_1x_1 + k_2x_2$. Over what ranges of values of k_1 and k_2 will the controller stabilize the system? What does this formulation imply about the displacement y?

Figure P11.33

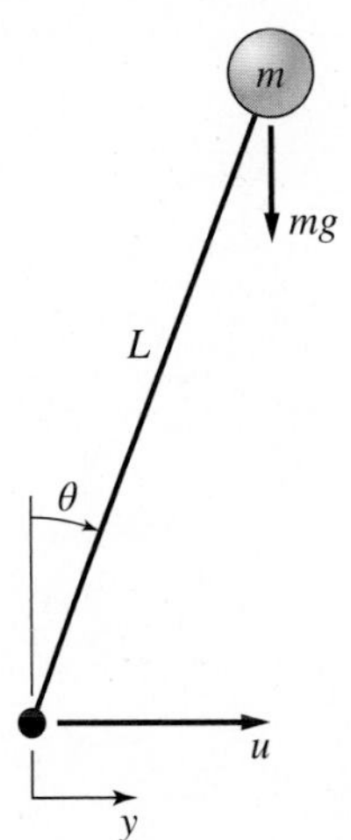

11.34 Figure P.11.34 illustrates an active vibration control scheme for a two-mass system. An electrohydraulic actuator between the two masses provides a force that acts on both and is under feedback control. The system model is

$$m_1\ddot{x}_1 = k_1(y - x_1) - k_2(x_1 - x_2) - c(\dot{x}_1 - \dot{x}_2) - f$$

$$m_2\ddot{x}_2 = k_2(x_1 - x_2) + c(\dot{x}_1 - \dot{x}_2) + f$$

The given parameter values are $m_1 = 50$ kg, $m_2 = 250$ kg, $k_1 = 1.5 \times 10^5$ N/m, $k_2 = 1.2 \times 10^4$ N/m, and $c = 100$ N · s/m.

a. Put the model into state-variable form.

b. Assume that we can measure all four state variables, and use P action with state-variable feedback. In the original passive system, $k_2 = 1.6 \times 10^4$ N/m and $c = 98$ N · m/s, which resulted in characteristic roots at $s = -1.397 \pm 69.94j$, $s = -0.168 \pm 7.779j$. Compute the values of the feedback gains so that the closed-loop roots will be near those of the passive system.

Figure P11.34

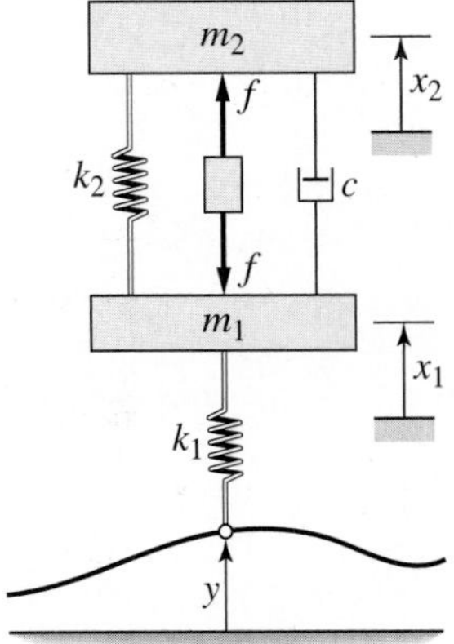

11.35 Figure P11.35a is the circuit diagram of a speed-control system in which the dc motor voltage v_a is supplied by a generator driven by an engine. This system has been used on locomotives whose diesel engine operates most efficiently at one speed. The efficiency of the electric motor is not so sensitive to speed and thus can be used to drive the locomotive at various speeds. The motor voltage v_a is varied by changing the generator input voltage v_f. The voltage v_a is related to the generator field current i_f by $v_a = K_f i_f$.

Figure P11.35b is a diagram of a feedback system for controlling the speed by measuring it with a tachometer and varying the voltage v_f. Use the following values in SI units.

$$\begin{array}{lll} L_f = 0.2 & R_f = 2 & K_t = 1 \\ L_a = 0.2 & R_a = 1 & K_b = K_T = 0.5 \\ K_f = 50 & I = 10 & c = 20 \end{array}$$

a. Develop a state-variable model of the plant that includes the generator, the motor, and the load. Include the load torque T_L as a disturbance.

b. Develop a proportional controller assuming all the state variables can be measured. Analyze its steady-state error for a step command input and for a step disturbance.

Figure P11.35

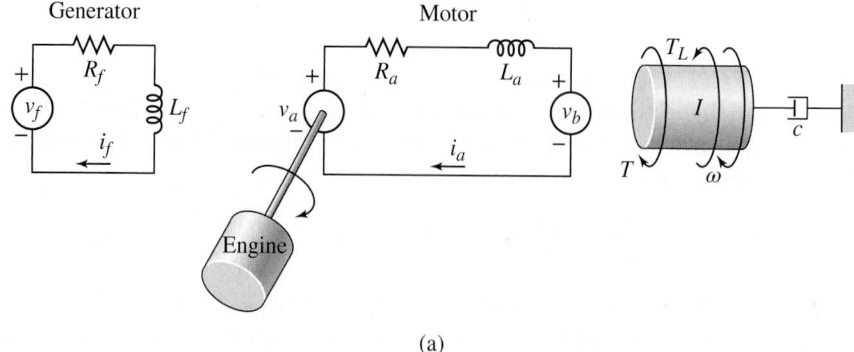

(a)

(b)

11.36 The following equations are the model of the roll dynamics of a missile ([Bryson, 1975]). See Figure P11.36.

$$\dot{\delta} = u$$

$$\dot{\omega} = -\frac{1}{\tau}\omega + \frac{b}{\tau}\delta$$

$$\dot{\phi} = \omega$$

Figure P11.36

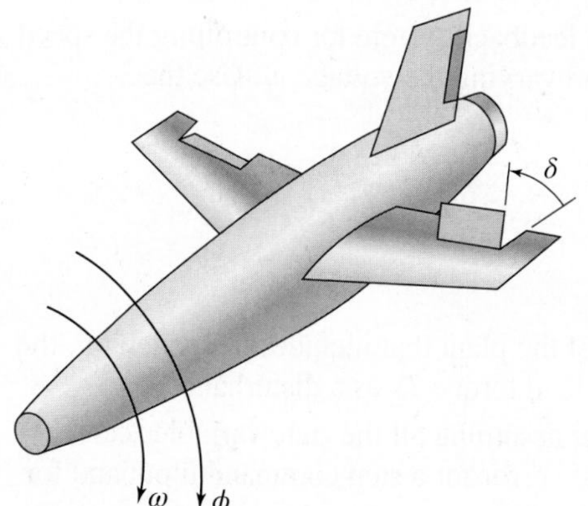

where δ = aileron deflection

b = aileron effectiveness constant

u = command signal to the aileron actuator

ϕ = roll angle, ω = roll rate

Using the specific values $b = 10\ \text{s}^{-1}$ and $\tau = 1$ s, and assuming that the state variables δ, ω, and ϕ can be measured, develop a linear state-feedback controller to keep ϕ near 0. The dominant roots should be $s = -10 \pm 10j$, and the third root should be $s = -20$.

11.37 Many winding applications in the paper, wire, and plastic industries require a control system to maintain proper tension. Figure P11.37 shows such a system winding paper onto a roll. The paper tension must be held constant to prevent internal stresses that will damage the paper. The pinch rollers are driven at a speed required to produce a paper speed v_p at the rollers. The paper speed as it approaches the roll is v_r. The paper tension changes as the radius of the roll changes or as the speed of the pinch rollers change. The paper has an elastic constant k so that the rate of change of tension is

$$\frac{dT}{dt} = k(v_r - v_p)$$

For a paper thickness d, the rate of change of the roll radius is

$$\frac{dR}{dt} = \frac{d}{2}W$$

The inertia of the windup roll is $I = \rho\pi WR^4/2$, where ρ is the paper mass density and W is the width of the roll. So R and I are functions of time.

The viscous damping constant for the roll is c. For the armature-controlled motor driving the windup roll, neglect its viscous damping and armature inertia.

a. Assuming that the paper thickness is small enough so that $\dot{R} \approx 0$ for a short time, develop a state-variable model with the motor voltage e and the paper speed v_p as the inputs.

b. Modify the model developed in part (a) to account for R and I being functions of time.

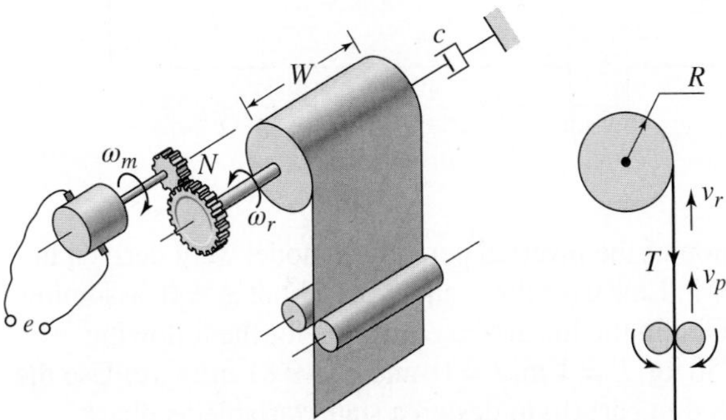

Figure P11.37

11.38 An electro-hydraulic positioning system is shown in Figure P11.38. Use the following values.

$$K_a = 10 \text{ V/A} \quad K_1 = 10^{-2} \text{ in./V}$$
$$K_2 = 3 \times 10^5 \text{ sec}^{-3} \quad K_3 = 20 \text{ V/in.}$$
$$\zeta = 0.8 \quad \omega_n = 100 \text{ rad/sec} \quad \tau = 0.01 \text{ sec}$$

a. Develop a state-variable model of the plant with the controller current i_c as the input and the displacement y as the output.

b. Assuming that proportional control is used so that $G_c(s) = K_P$, develop a state model of the system with y_r as the input and y as the output. Draw the root locus and use it to determine whether or not the system can be made stable with an appropriate choice for the value of K_P.

Figure P11.38

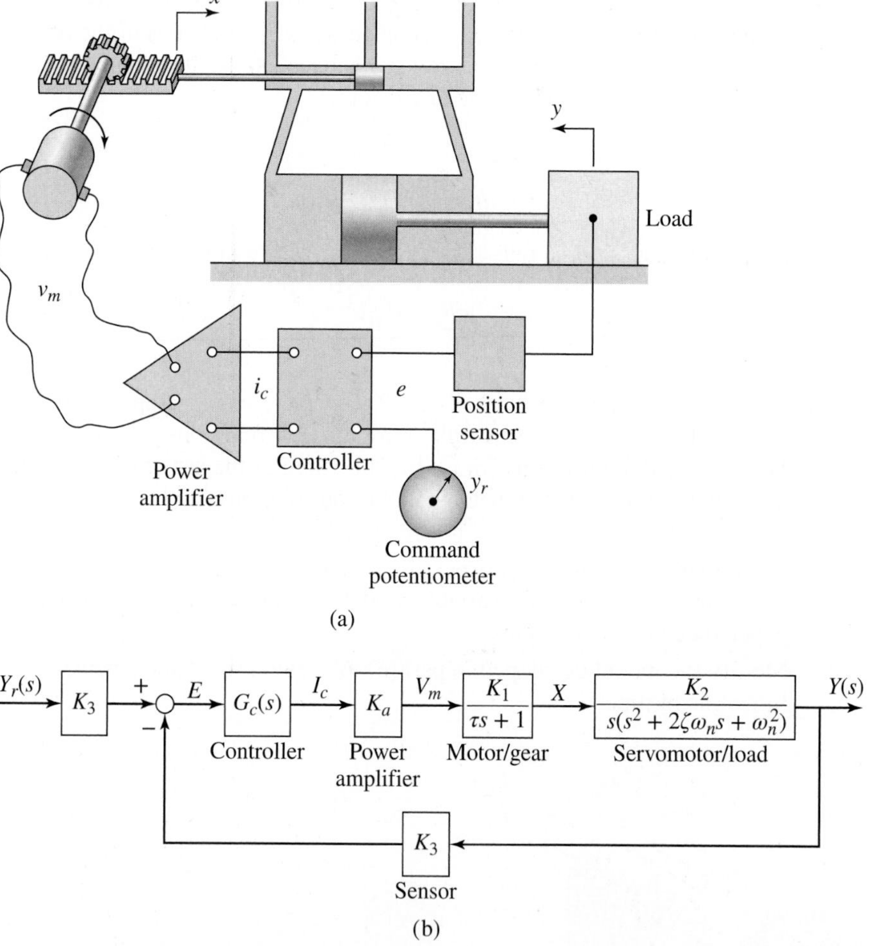

11.39 (a) The equations of motion of the inverted pendulum model were derived in Example 3.5.6 in Chapter 3. Linearize these equations about $\phi = 0$, assuming that $\dot{\phi}$ is very small. (b) Obtain the linearized equations for the following values: $M = 10$ kg, $m = 50$ kg, $L = 1$ m, $I = 0$, and $g = 9.81$ m/s^2. (c) Use the linearized model developed in part (b) to design a state-variable feedback controller to stabilize the pendulum angle near $\phi = 0$. It is required that the 2% settling time be no greater than 4 s and that the response be non-oscillatory. This means that the dominant root should be real and no greater than -1. No restriction is placed on the motion of the base. Assume that ϕ, $\dot{\phi}$, x, and $\dot{x}$ can be measured.

Section 11.6 MATLAB Applications

11.40 The following table gives the measured open-loop response of a system to a unit-step input. Use the process reaction method to find the controller gains for P, PI, and PID control.

Time (min)	Response
0	0
0.5	4
1.0	20
1.5	32
2.0	56
2.5	84
3.0	116
3.5	140
4.0	160
4.5	172
5.0	184
5.5	190
6.0	194
7.0	196

11.41 A liquid in an industrial process must be heated with a heat exchanger through which steam passes. The exit temperature of the liquid is controlled by adjusting the rate of flow of steam through the heat exchanger with the control valve. An open-loop test was performed in which the steam pressure was suddenly changed from 15 to 18 psi above atmospheric pressure. The exit temperature data are shown in the following table. Use the Ziegler-Nichols process reaction method to compute the PID gains.

Time (min)	Temperature (°F)
0	156
1	157
2	159
3	162
4	167
5	172
6	175
7	179
8	181
9	182
10	183
11	184
12	184

11.42 Use MATLAB to obtain the root locus plot of $5s^2 + cs + 45 = 0$ for $c \geq 0$.

11.43 Use MATLAB to obtain the root locus plot of the system shown in Figure P11.43 in terms of the variable $k \geq 0$. Use the values $m = 6$ and $c = 4$. What is the smallest possible dominant-time constant and the associated value of k?

11.44 Use MATLAB to obtain the root locus plot of the system shown in Figure P11.43 in terms of the variable $c \geq 0$. Use the values $m = 8$ and $k = 128$. What is the smallest possible dominant-time constant and the associated value of c?

11.45 Use MATLAB to obtain the root locus plot of the system shown in Figure P11.45 in terms of the variable $k_2 \geq 0$. Use the values $m = 4$, $c = 16$, and $k_1 = 52$. What is the value of k_2 required to give $\zeta = 0.707$?

Figure P11.43

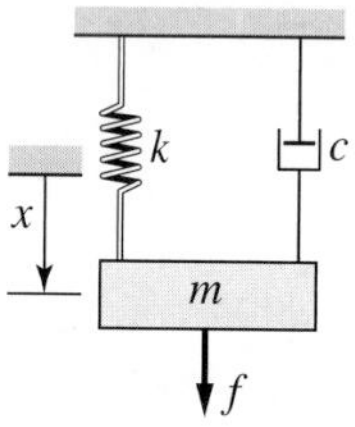

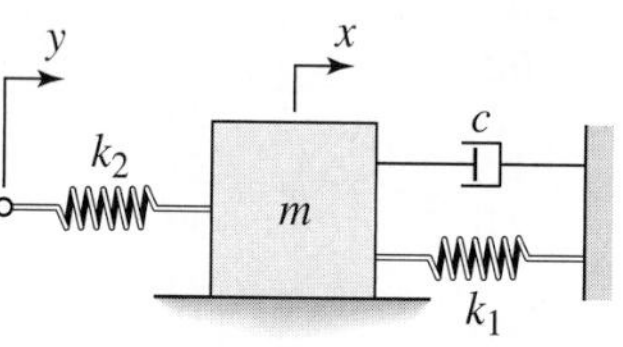

Figure P11.45

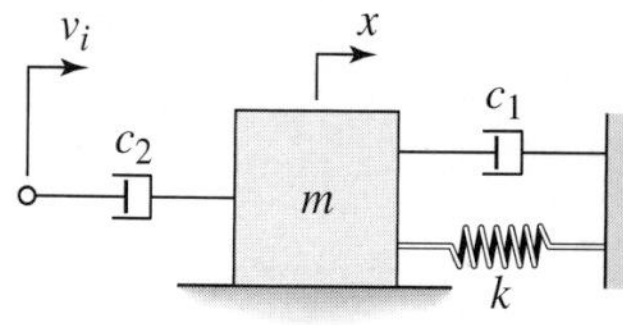

Figure P11.46

11.46 Use MATLAB to obtain the root locus plot of the system shown in Figure P11.46 in terms of the variable $c_2 \geq 0$. Use the values $m = 4$, $c_1 = 16$, and $k = 52$. What is the smallest possible dominant-time constant and the associated value of c_2?

11.47 Use MATLAB to obtain the root locus plot of $2s^3 + 26s^2 + 104s + 120 + 5b = 0$ for $b \geq 0$. Is it possible for any dominant roots of this equation to have a damping ratio in the range $0.5 \leq \zeta \leq 0.707$ and an undamped natural frequency in the range $3 \leq \omega_n \leq 5$?

11.48 (a) Use MATLAB to obtain the root locus plot of $2s^3 + 12s^2 + 16s + K = 0$ for $K \geq 0$. (b) Obtain the value of K required to give a dominant-root pair having $\zeta = 0.707$. (c) For this value of K, obtain the unit-step response and the maximum overshoot, and evaluate the effects of the secondary root. The closed-loop transfer function is $K/(2s^3 + 12s^2 + 16s + K)$.

11.49 Use MATLAB to obtain the root locus of the armature-controlled dc motor model in terms of the damping constant c, and evaluate the effect on the motor time constant. The characteristic equation is

$$L_a I s^2 + (R_a I + cL_a)s + cR_a + K_b K_T = 0$$

Use the following parameter values:

$$K_b = K_T = 0.2 \text{ N} \cdot \text{m/A} \qquad I = 8 \times 10^{-5} \text{ kg} \cdot \text{m}^2$$

$$R_a = 3\ \Omega \qquad L_a = 4 \times 10^{-3} \text{ H}$$

11.50 Consider the two-mass model shown in Figure P11.50. Use the following numerical values: $m_1 = m_2 = 1$, $k_1 = 1$, $k_2 = 4$, and $c_2 = 8$.

a. Use MATLAB to obtain the root locus plot in terms of the parameter c_1.

b. Use the root locus plot to determine the value of c_1 required to give a dominant-root pair having a damping ratio of $\zeta = 0.707$.

c. Use the root locus plot to determine the value of c_1 required to give a dominant root that is real and has a time constant equal to 4.

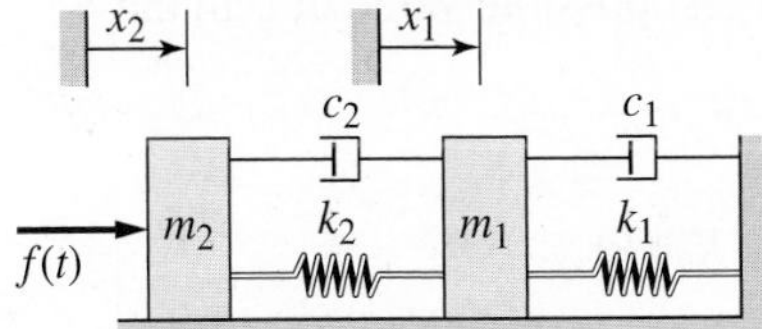

Figure P11.50

d. Using the value of c_1 found in part (c), obtain a plot of the unit-step response.

11.51 In parts (a) through (f), use MATLAB to obtain the root locus plot for the given characteristic equation for $p \geq 0$.

a. $2s^2 + 10s + p = 0$

b. $3s^3 + 48s^2 + 189s + p = 0$

c. $4s^2 + 12s + 20 + p(s + 3) = 0$

d. $5s^2 + 20s + p(s + 5) = 0$

e. $4s^3 + 12s^2 + 20 + p = 0$

f. $7s^3 + 70s^2 + 147s + p(s + 4) = 0$

11.52 In parts (a) through (f), use MATLAB to obtain the root locus plot for $p \leq 0$ for the given characteristic equation.

a. $2s^2 + 10s + p = 0$

b. $4s^2 + 12s + 12 + p(s + 3) = 0$

c. $3s^3 + 9s^2 + 9s + p = 0$

d. $2s^3 + 24s^2 + 70s + p = 0$

e. $5s^2 + 15s + p(s + 4) = 0$

f. $7s^2 + 42s + p(s - 4) = 0$

11.53 Consider the equation

$$5s^3 + 50s^2 + 120s + p = 0$$

a. Use MATLAB to obtain the value of p required to give dominant roots with $\zeta = 1$. Obtain the three roots corresponding to this value of p.

b. Use MATLAB to obtain the value of p required to give a dominant time constant of $\tau = 2/3$. Obtain the three roots corresponding to this value of p.

11.54 Consider the equation

$$s^3 + 9s^2 + (8 + K)s + 2K = 0$$

a. Use MATLAB to obtain the value of K required to put the dominant root at the breakaway point. Obtain the three roots corresponding to this value of K.

b. Investigate the sensitivity of the dominant root when K varies by $\pm 10\%$ about the value found in part (a).

11.55 Consider the equation

$$s^3 + 9s^2 + (8 + K)s + 2K = 0$$

Use the `sgrid` function to determine if it is possible to obtain a dominant root having a damping ratio in the range $0.5 \leq \zeta \leq 0.707$. If so, use MATLAB to obtain the value of K required to give the largest possible value of ζ in the range $0.5 \le \zeta \le 0.707$.

11.56 Consider the equation

$$s^3 + 10s^2 + 24s + K = 0$$

Use the `sgrid` function to determine if it is possible to obtain a dominant root having a damping ratio in the range $0.5 \le \zeta \le 0.707$, and an undamped natural frequency in the range $2 \le \omega_n \le 3$. If so, use MATLAB to obtain the value of K required to give the largest possible value of $\zeta\omega_n$ in the ranges stated.

11.57 In Example 10.7.4 the steady-state error for a unit-ramp disturbance is $1/K_I$. For the gains computed in that example, this error is $1/25$. We want to see if we can make this error smaller by increasing K_I. Using the values given for K_P and K_D in that example, obtain a root locus plot with K_I as the variable. Discuss what happens to the damping ratio and time constant of the dominant root as K_I is increased.

11.58 In Example 10.8.3 the steady-state error for a unit-ramp command is $-4/K_I$. For the gains computed in that example, this error is $1/1000$. We want to see if we can make this error smaller by increasing K_I. Using the values given for K_P and K_D in that example, obtain a root locus plot with K_I as the variable. Discuss what happens to the damping ratio and time constant of the dominant root as K_I is increased.

Section 11.7 Simulink Applications

11.59 With the PI gains set to $K_P = 6$ and $K_I = 50$ for the plant

$$G_p(s) = \frac{1}{s+4}$$

the time constant is $\tau = 0.2$ and the damping ratio is $\zeta = 0.707$.

a. Suppose the actuator saturation limits are ± 5. Construct a Simulink model to simulate this system with a unit-step command. Use it to plot the output response, the error signal, the actuator output, and the outputs of the proportional term and the integral term versus time.

b. Construct a Simulink model of an anti-windup system for this application. Use it to select an appropriate value for K_A and to plot the output response and the actuator output versus time.

11.60 With the PI gains set to $K_P = 6$ and $K_I = 50$ for the plant

$$G_p(s) = \frac{1}{s+4}$$

the time constant is $\tau = 0.2$ and the damping ratio is $\zeta = 0.707$. Suppose there is a rate limiter of ± 0.1 between the controller and the plant. Construct a Simulink model of the system and use it to determine the effect of the limiter on the speed of response of the system. Use a unit-step command.

11.61 A certain dc motor has the following parameter values:

$$L = 2 \times 10^{-3}\ \text{H} \qquad R = 0.6\ \Omega$$
$$K_T = 0.04\ \text{N} \cdot \text{m/A} \qquad c = 0$$
$$I = 6 \times 10^{-5}\ \text{kg} \cdot \text{m}^2$$

Figure P11.61

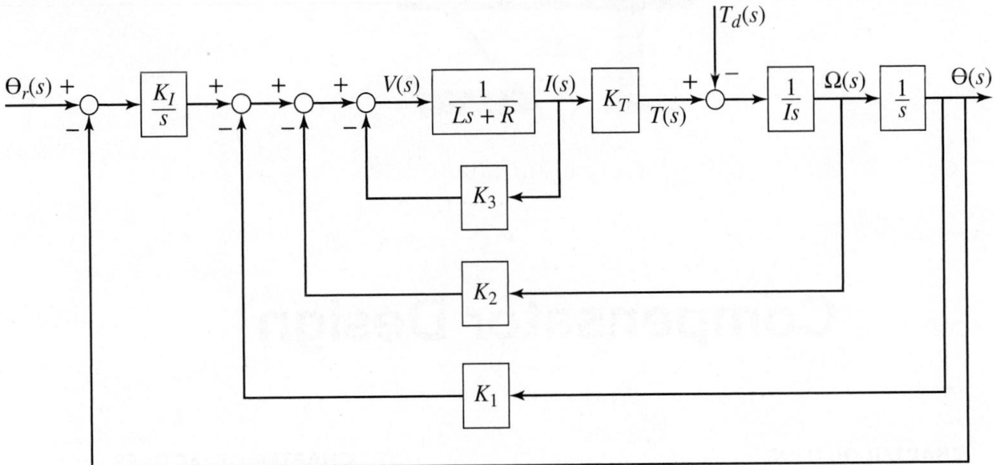

Figure P11.61 shows an integral controller using state-variable feedback to control the motor's angular position.

a. Compute the gains to give a dominant-time constant of 0.5 s. The secondary roots should have a time constant of 0.05 s and a damping ratio of $\zeta = 0.707$. The fourth root should be $s = -20$.

b. Construct a Simulink model of the system and use it to plot the response of the system to a step disturbance of magnitude 0.1.

c. Suppose the motor current is limited to ± 2 A. Modify the Simulink model to include this saturation, and use the model to obtain plots of the responses to a unit-step command and a step disturbance of magnitude 0.1. Discuss the results.

11.62 Consider the liquid-level controller designed in Example 10.10.1, whose Simulink diagram is shown in Figure 10.10.1. Modify the model to include a Rate-Limiter block to limit the rate of q_1 in front of the Saturation block. The limits on the rate should be ± 20. Use this model to obtain plots of the response of the height h_2 to a unit-step command and a unit-step disturbance. Compare these responses to those found in Example 10.10.1.

CHAPTER 12

Compensator Design

CHAPTER OUTLINE

CHAPTER OBJECTIVES

When you have finished this chapter, you should be able to

1. Design a series compensator using either the root locus or the frequency response plot.
2. Use the open-loop frequency response plot to design a controller.
3. Analyze the effect of time delays on system response.
4. Apply MATLAB and Simulink to the methods of this chapter.

Section 12.1 shows how to use the open-loop frequency response plots to design PID controllers and series compensators. It also introduces some additional performance criteria: the phase and gain margins, and the static error coefficients. The frequency response plots of the system's open-loop transfer function contain much information about the behavior of the closed-loop system. These plots are easily generated even for high-order systems, and they enable the proper control gain to be selected simply by adjusting the scale factor on the plot. The technique is especially useful, because it does not require the values of the characteristic roots. This is helpful for analyzing high-order systems and systems with dead time. The latter are especially difficult to treat by analysis of the characteristic roots, because they possess an infinite number of roots. ■

12.1 SERIES COMPENSATION

If implemented in hardware, a *series* compensator is a physical device that is inserted in the control system at the output of the controller (see Figure 12.1.1). If implemented in software in a digital control system, the series compensator is an algorithm that creates the equivalent of a transfer function inserted in series with the transfer function of the main control algorithm. Series compensation is particularly well suited for design with the root locus or with the open-loop Bode plots. Usually, the system is first analyzed with only proportional control action included. The series compensator is then added to meet the performance specifications.

We may think of I action and D action as forms of series compensation. P action compensated with D action gives the transfer function $K_P(1 + T_D s)$. The term $(1 + T_D s)$ can be considered as a series compensator to the proportional controller. The D action adds an open-loop zero at $s = -1/T_D$. The PI control algorithm is described by

$$\frac{F(s)}{E(s)} = K_P\left(1 + \frac{1}{T_I s}\right) E(s) = \frac{K_P}{s}\left(s + \frac{1}{T_I}\right)$$

The integral action can thus be considered to add an open-loop pole at $s = 0$ and a zero at $s = -1/T_I$.

Three other commonly used series compensators are the *lead*, the *lag*, and the *lag-lead* compensators. Their names derive from the change they produce in the system phase angle, which we will discuss with Bode plot design in Section 12.2. These compensators can be easily realized with passive electrical *RC* networks or with active circuits using op amps. Devices using springs and dampers can implement these compensators mechanically. Many commercially available digital control packages include these compensators as a choice of control action; some in fact do not implement PID action, but use exclusively the lead, lag, and lag-lead compensators, which they call the "filter." We will see how they can be made to emulate PID action.

The transfer functions of the lead and the lag compensators have the form

$$G_c(s) = K\frac{s + a}{s + b} \tag{12.1.1}$$

For a lead compensator, $a < b$, whereas for a lag compensator, $a > b$. The transfer function of the lag-lead compensator has the form

$$G_c(s) = K\frac{s + a}{s + b}\frac{s + c}{s + d} \tag{12.1.2}$$

It is useful to compare these compensators with PID control (Table 12.1.1). First note that I action and D action always require an active circuit unless the control algorithm is being implemented in software as part of a digital control system. On the other hand, the lead, lag, and lag-lead compensators can be implemented with passive circuits.

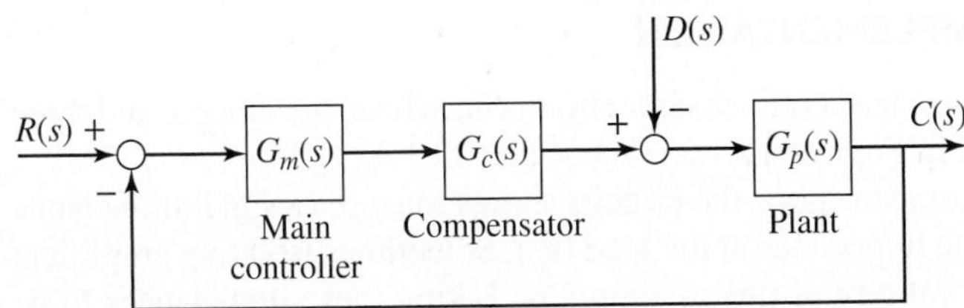

Figure 12.1.1 Structure of a system with series compensation.

Table 12.1.1 Comparison of PID, lead, lag, and lag-lead compensators.

Compensator	Application	Transfer function	Advantages and limitations
P	Basis for all control action	K_P	**1.** Often results in nonzero steady-state error. **2.** Easily implemented in electrical and nonelectrical systems.
PI	Reduces steady-state error	$K_P + \dfrac{K_I}{s}$	**1.** Increases system order by 1. **2.** Requires active circuit. **3.** May increase overshoot.
PD	Improves transient response	$K_P + K_D s$	**1.** Requires active circuit. **2.** May increase overshoot. **3.** May cause noise. **4.** Better implemented by rate feedback.
PID	Reduces error and improves transient response	$K_P + \dfrac{K_I}{s} + K_D s$	Same as PI and PD.
Lead	Improves transient response	$K\dfrac{s+a}{s+b} \quad (a < b)$	**1.** Active circuit not required. **2.** Increases system order by 1 if no cancellation.
Lag	Reduces steady-state error	$K\dfrac{s+c}{s+d} \quad (c > d)$	**1.** Active circuit not required. **2.** Increases system order by 1 if no cancellation.
Lag-Lead	Reduces error and improves transient response	$K\dfrac{s+a}{s+b}\dfrac{s+c}{s+d}$	**1.** Same as lead and lag.

Another implementation problem with I action K_I/s is that it is difficult to construct a circuit that gives a pole at $s = 0$, and usually the circuit transfer function is an approximation that has a negative pole close to the origin. Note that PI action can be approximated with the lag compensator by choosing b to be close to 0.

Physical devices implementing D action are prone to amplify noise in the signals because their response is designed to be proportional to the rate of change of the input signal. The op-amp circuit for PD action presented in Chapter 10 was designed not to give pure PD action at higher frequencies for this reason. As we also saw in Chapter 10, theoretical models of D action produce an impulse in response to a step command, which is physically impossible. For these reasons, you should consider implementing the equivalent of D action by using rate feedback if possible. In such an arrangement, the derivative is computed for signals that are more slowly varying than the command input.

12.1.1 HARDWARE IMPLEMENTATION

A variety of lead, lag, and lag-lead compensator circuits have been developed, and three passive circuits are shown in Figures 12.1.2, 12.1.3, and 12.1.4.

When used as series compensators, the circuits shown must see a small impedance at the source (v_i) and a large impedance at the load (v_o). Sometimes isolating amplifiers are inserted to ensure the validity of this assumption. Taking these impedances to be

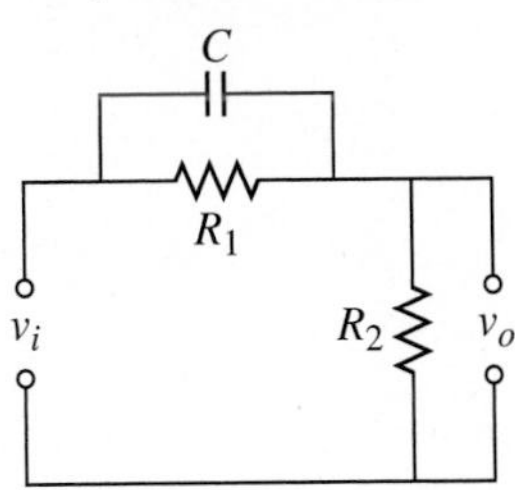

Figure 12.1.2 Passive lead compensator circuit.

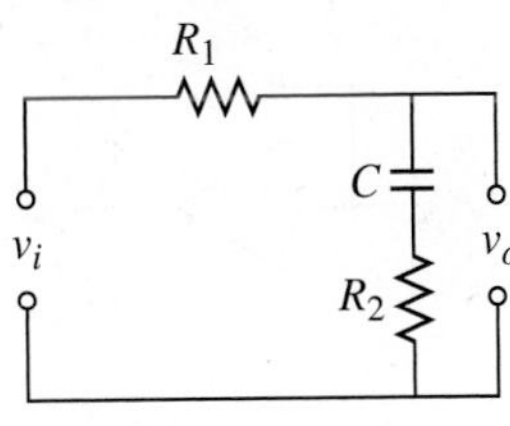

Figure 12.1.3 Passive lag compensator circuit.

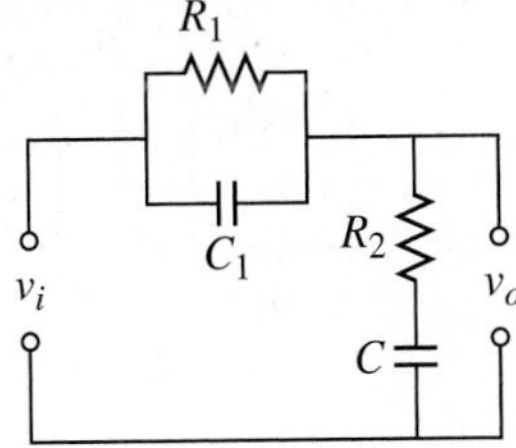

Figure 12.1.4 Passive lag-lead compensator circuit.

zero and infinity, respectively, we can derive the transfer functions with the methods in Chapter 6. For the lead compensator (Figure 12.1.2),

$$\frac{V_o(s)}{V_i(s)} = \frac{R_2 + R_1 R_2 C s}{R_1 + R_2 + R_1 R_2 C s} = \frac{1}{\mu_1}\frac{1+\mu_1 T_1 s}{1+T_1 s} = \frac{s + 1/\mu_1 T_1}{s + 1/T_1} \tag{12.1.3}$$

where

$$\mu_1 = \frac{R_1 + R_2}{R_2} > 1 \tag{12.1.4}$$

$$T_1 = \frac{R_1 R_2}{R_1 + R_2} C \tag{12.1.5}$$

Note that the circuit would be useless as a series compensator without the resistance R_1 because it would not pass dc signals. Comparing (12.1.1) with (12.1.3), we see that $K = 1$, $a = 1/\mu_1 T_1$, and $b = 1/T_1$. Note that the pole and zero are separated by the factor μ_1, so that $b = \mu_1 a$.

For the lag compensator (Figure 12.1.3),

$$\frac{V_o(s)}{V_i(s)} = \frac{1 + R_2 C s}{1 + (R_1 + R_2) C s} = \frac{1 + \mu_2 T_2 s}{1 + T_2 s} = \mu_2 \frac{s + 1/\mu_2 T_2}{s + 1/T_2} \tag{12.1.6}$$

where

$$\mu_2 = \frac{R_2}{R_1 + R_2} < 1 \tag{12.1.7}$$

$$T_2 = (R_1 + R_2) C \tag{12.1.8}$$

Comparing (12.1.1) with (12.1.6), we see that $a = 1/\mu_2 T_2$, $b = 1/T_2$, and $K = \mu_2$, which gives a low-frequency gain. Note that the pole and zero are separated by the factor μ_2, so that $b = \mu_2 a$.

Lag and lead compensators placed in series give the transfer function

$$\frac{V_o(s)}{V_i(s)} = \left(\frac{s + 1/\mu_1 T_1}{s + 1/T_1}\right) \mu_2 \left(\frac{s + 1/\mu_2 T_2}{s + 1/T_2}\right) \tag{12.1.9}$$

A simpler approach is to use the single network shown in Figure 12.1.4. With the usual impedance assumptions, we obtain

$$\frac{V_o(s)}{V_i(s)} = \frac{1 + (R_1 C_1 + R_2 C_2)s + R_1 C_1 R_2 C_2 s^2}{1 + (R_1 C_1 + R_1 C_2 + R_2 C_2)s + R_1 C_1 R_2 C_2 s^2}$$

$$= \left(\frac{s + 1/\mu_3 T_3}{s + 1/T_3}\right)\left(\frac{s + 1/\mu_4 T_4}{s + 1/T_4}\right) \tag{12.1.10}$$

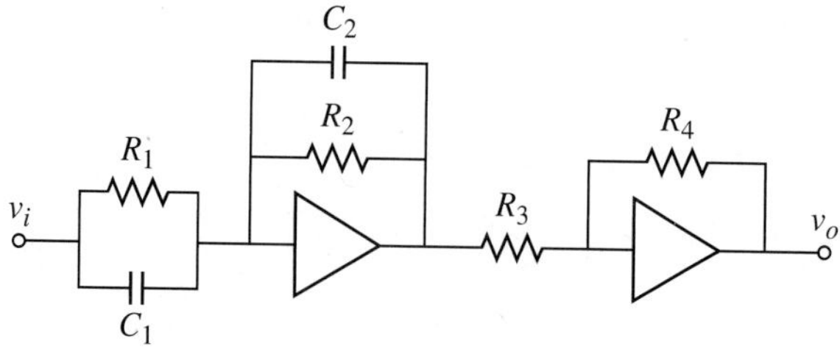

Figure 12.1.5 Active circuit for lead or lag compensation.

where

$$\mu_3 T_3 = R_1 C_1 \quad \mu_3 > 1 \tag{12.1.11}$$

$$\mu_4 T_4 = R_2 C_2 \tag{12.1.12}$$

$$T_3 + T_4 = R_1 C_1 + R_1 C_2 + R_2 C_2 \tag{12.1.13}$$

$$\mu_4 = \frac{1}{\mu_3} \tag{12.1.14}$$

Equations (12.1.11) through (12.1.13) contain four unknowns: R_1, R_2, C_1, and C_2. Once an analysis has determined values for T_3, T_4, μ_3, and μ_4, a convenient value for either C_1 or C_2 can be selected, and the equations solved for the three remaining unknowns.

An active circuit for lead or lag compensation is shown in Figure 12.1.5. Using the methods of Chapter 6, we can derive the following transfer function for the circuit:

$$\frac{V_o(s)}{V_i(s)} = \left(\frac{R_2 R_4}{R_1 R_3}\right) \frac{R_1 C_1 s + 1}{R_2 C_2 s + 1} = \left(\frac{R_4 C_1}{R_3 C_2}\right) \frac{s + 1/R_1 C_1}{s + 1/R_2 C_2} \tag{12.1.15}$$

or

$$\frac{V_o(s)}{V_i(s)} = K_c \frac{s + 1/\mu_5 T_5}{s + 1/T_5} \tag{12.1.16}$$

where

$$T_5 = R_2 C_2 \tag{12.1.17}$$

$$\mu_5 = \frac{R_1 C_1}{R_2 C_2} \tag{12.1.18}$$

$$K_c = \frac{R_4 C_1}{R_3 C_2} \tag{12.1.19}$$

The circuit is a lead compensator if $R_1 C_1 > R_2 C_2$. It is a lag compensator if $R_1 C_1 < R_2 C_2$. Note that the passive lag compensator circuit has a low frequency gain of 1, whereas the op-amp compensator has a gain K_c.

An active circuit for lag-lead compensation is shown in Figure 12.1.6. Its transfer function is

$$\frac{V_o(s)}{V_i(s)} = \frac{R_4 R_6}{R_3 R_5} \left[\frac{(R_1 + R_3) C_1 s + 1}{R_1 C_1 s + 1}\right] \left[\frac{R_2 C_2 s + 1}{(R_2 + R_4) C_2 s + 1}\right] \tag{12.1.20}$$

or

$$\frac{V_o(s)}{V_i(s)} = K_c \frac{(s + 1/\mu_6 T_6)(s + 1/\mu_7 T_7)}{(s + 1/T_6)(s + 1/T_7)} \tag{12.1.21}$$

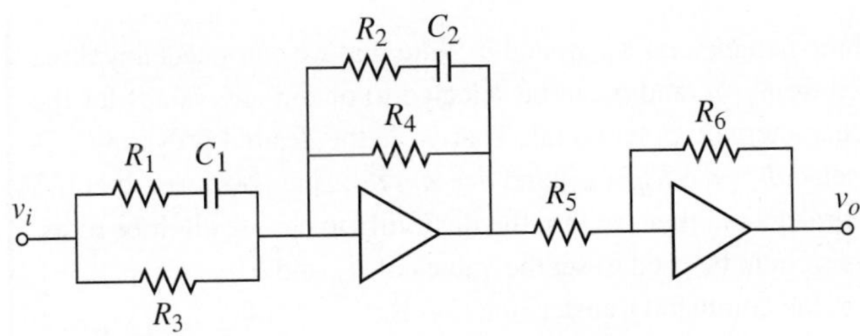

Figure 12.1.6 Active circuit for lag-lead compensation.

where

$$T_6 = R_1C_1 \tag{12.1.22}$$

$$T_7 = (R_2 + R_4)C_2 \tag{12.1.23}$$

$$\mu_6 = \frac{R_1 + R_3}{R_1} > 1 \tag{12.1.24}$$

$$\mu_7 = \frac{R_2}{R_2 + R_4} < 1 \tag{12.1.25}$$

$$K_c = \frac{R_2R_4R_6}{R_1R_3R_5}\left(\frac{R_1 + R_3}{R_2 + R_4}\right) \tag{12.1.26}$$

The circuit has a low frequency gain of R_4R_6/R_3R_5.

12.1.2 COMPENSATOR DESIGN BY ROOT PLACEMENT

Sometimes we have the flexibility of designing the compensator simply by specifying all the root locations. This can be done for an nth-order system if the last $n - 1$ coefficients of the characteristic equation are independent functions of the gains.

EXAMPLE 12.1.1

Lead and PID Compensation for the Plant $1/s^2$

■ Problem

The equation of motion of an object in pure rotation is $I\ddot{\theta} = T - T_d$, where I is the inertia, T is the torque applied by the controller to control the angular position θ, and T_d is a disturbance torque. Consider the case shown in Figure 12.1.7, where $I = 1$. Design lead and PID compensators for this system. The specifications are that the closed-loop system must have a dominant-time constant no greater than 0.5 and a damping ratio no less than 0.707. Evaluate the resulting steady-state error for unit-step and unit-ramp commands and disturbances.

■ Solution

Using a lead compensator with P action gives the command transfer function

$$\frac{C(s)}{R(s)} = \frac{K_P(s + a)}{s^3 + bs^2 + K_Ps + aK_P}$$

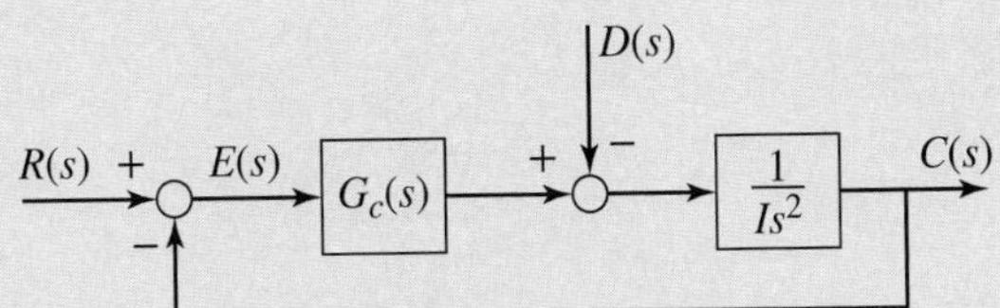

Figure 12.1.7 Control of a rotating object.

Values must be obtained for three parameters: K_P, a, and b. Note that we can select any three closed-loop roots we want because K_P, a, and b can be selected to obtain any values for the last three coefficients of the characteristic polynomial. That is, if the desired polynomial is $s^3 + a_2 s^2 + a_1 s + a_0$, then we select $b = a_2$, $K_P = a_1$, and $a = a_0/K_P$. Note, however, that if b is chosen small to emulate integral action, then we lose the flexibility to choose all three roots. In that case, a root locus analysis could be used to set the values of K_P and a.

With the PID compensator, the command transfer function is

$$\frac{C(s)}{R(s)} = \frac{K_D s^2 + K_P s + K_I}{s^3 + K_D s^2 + K_P s + K_I}$$

Values must be obtained for three parameters: K_D, K_P, and K_I. We can select any three closed-loop roots we want because these gains can be selected independently as follows. For the desired polynomial $s^3 + a_2 s^2 + a_1 s + a_0$, we select $K_D = a_2$, $K_P = a_1$, and $K_I = a_0$.

So both designs can meet the given specifications for the transient response. The error transfer functions are given in Table 12.1.2. From these we can evaluate the steady-state error for each design. Those errors are also given in the table. We see that the lead compensator does not do as well as the PID in rejecting the disturbance. Its response to a step disturbance can be made small by choosing b small, but, as we have seen, we then lose the flexibility to choose all three roots.

Table 12.1.2 Comparison of lead and PID compensators for the plant $1/s^2$.

Compensator	P action with lead	PID
$G_c(s)$	$K_P \dfrac{s+a}{s+b}$	$\dfrac{K_D s^2 + K_P s + K_I}{s}$
$\dfrac{C(s)}{R(s)}$	$\dfrac{K_P(s+a)}{s^3 + bs^2 + K_P s + aK_P}$	$\dfrac{K_D s^2 + K_P s + K_I}{s^3 + K_D s^2 + K_P s + K_I}$
$\dfrac{E(s)}{R(s)}$	$\dfrac{s^3 + bs^2}{s^3 + bs^2 + K_P s + aK_P}$	$\dfrac{s^3}{s^3 + K_D s^2 + K_P s + K_I}$
$\dfrac{E(s)}{D(s)}$	$\dfrac{s+b}{s^3 + bs^2 + K_P s + aK_P}$	$\dfrac{s}{s^3 + K_D s^2 + K_P s + K_I}$
Steady-state error		
Step command	0	0
Ramp command	0	0
Step disturbance	$\dfrac{b}{aK_P}$	0
Ramp disturbance	∞	$\dfrac{1}{K_I}$
Roots	$s = -2, s = -2 \pm 2j$	$s = -2, s = -2 \pm 2j$
Gains	$K_P = 16, a = 1, b = 6$	$K_P = 16, K_I = 16, K_D = 6$
Steady-state error		
Step disturbance	$\dfrac{3}{8}$	0
Ramp disturbance	∞	$\dfrac{1}{16}$
Step response		
Overshoot	32.8%	26.1%
Peak time	1.1	0.58
Settling time	2.43	2.49
10%–90% rise time	0.417	0.201

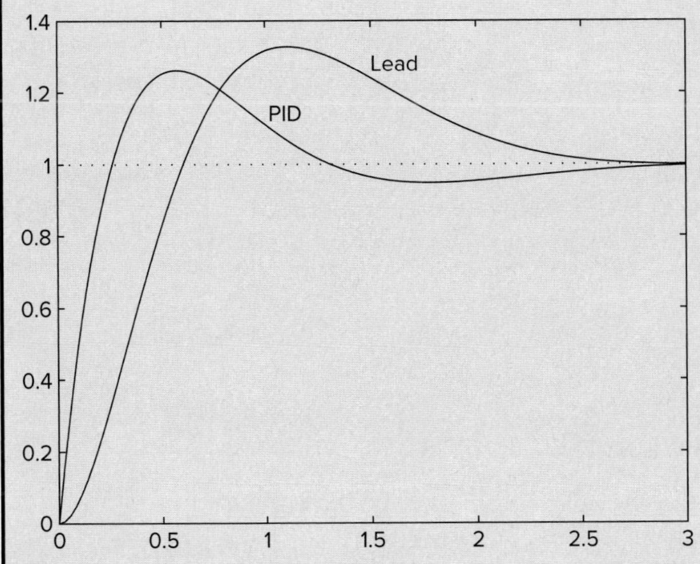

Figure 12.1.8 Step command response for Example 12.1.1.

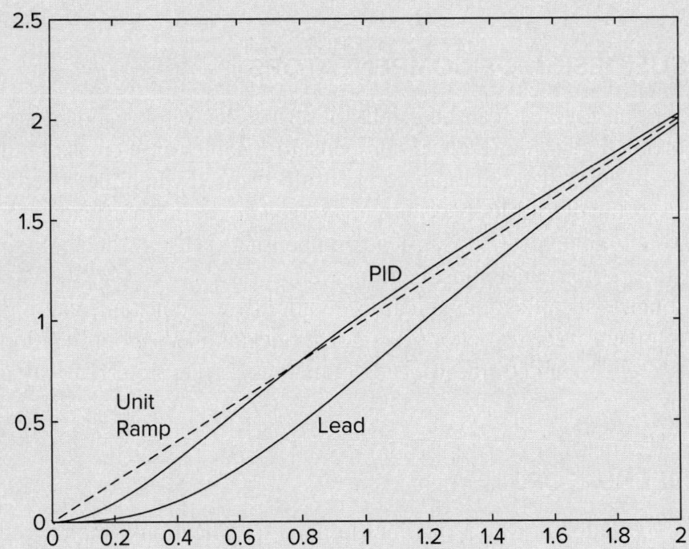

Figure 12.1.9 Ramp response for Example 12.1.1.

The two transient-response specifications are satisfied by choosing the roots to be $s = -2$ and $s = -2 \pm 2j$. This gives the polynomial

$$(s+2)\left[(s+2)^2 + 2^2\right] = s^3 + 6s^2 + 16s + 16$$

For the lead compensator, this gives $K_P = 16$, $a = 1$, and $b = 6$. For the PID compensator, this gives $K_P = 16$, $K_I = 16$, $K_D = 6$. The numerical results for these values are given in the table. Figures 12.1.8, 12.1.9, and 12.1.10 show the step command response, the ramp command response, and the step disturbance error, respectively.

The PID has a smaller overshoot but a faster rise time. Because of this and its smaller errors, the performance of PID is better than the lead compensator in this example, assuming that the pole at $s = 0$ in the PID can be implemented physically or in a digital controller.

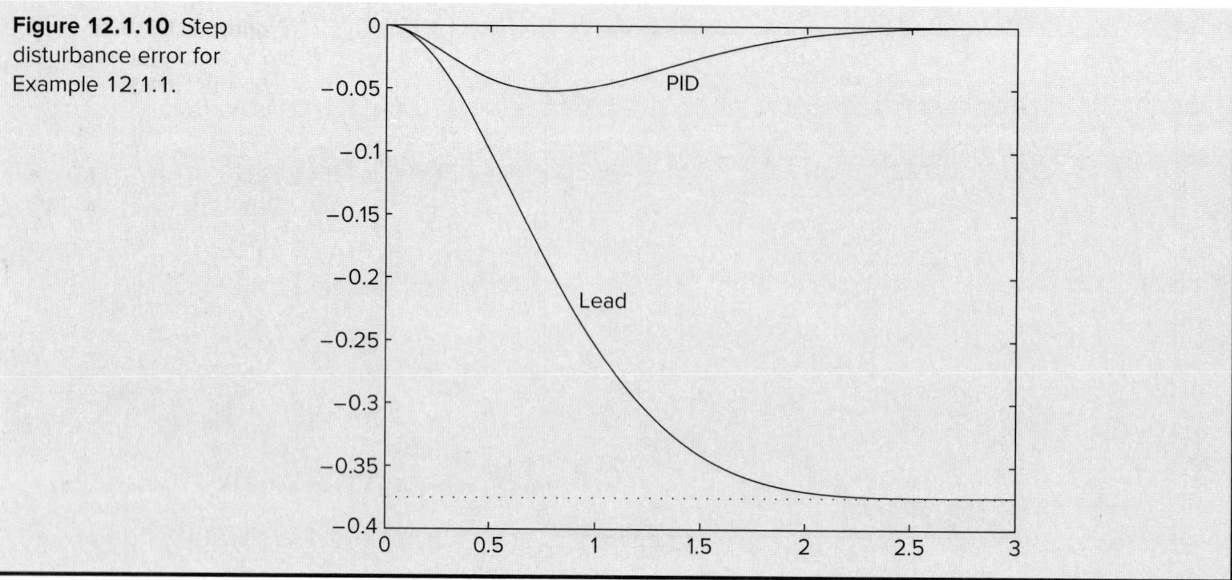

Figure 12.1.10 Step disturbance error for Example 12.1.1.

12.1.3 ROOT LOCUS DESIGN OF COMPENSATORS

The series compensators' usefulness can be briefly described as follows. When used in series with a proportional gain K_P, the lead compensator enables an increase in the speed of response. On the other hand, the lag compensator is used when the speed of response and damping of the closed-loop system are satisfactory, but the steady-state error is too large. As you might expect, the lag-lead compensator is used when both the transient and steady-state performance must be improved.

When the performance specifications, such as required time constant, damping ratio, and so forth are given in terms of root locations, but not all the root locations are known or can be freely specified, the root locus method is the preferred way of designing the compensator.

12.1.4 LEAD COMPENSATORS

The effects of the compensator in terms of time-domain specifications (the characteristic roots) can be shown with the root locus plot. Consider the second-order plant $1/(s + \alpha)(s + \beta)$ with the distinct real roots $s = -\alpha, -\beta$. The root locus for this system with proportional control is shown in Figure 12.1.11a. The smallest dominant-time constant obtainable is τ_1, marked in the figure. With lead compensation, the root locus becomes that shown in Figure 12.1.11b. The pole and zero introduced by the compensator reshape the locus so that a smaller dominant-time constant can be obtained. This is done by choosing the proportional gain high enough to place the roots close to the asymptotes.

Designing a lead compensator with the root locus is done as follows.

1. From the time-domain specifications (time constant, damping ratio, etc.), determine the required locations of the dominant closed-loop poles.
2. From the root locus plot of the uncompensated system, determine whether or not the desired closed-loop poles can be obtained by adjusting the open-loop

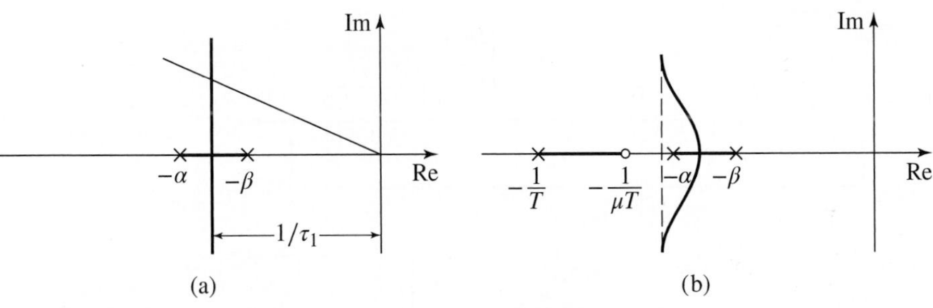

Figure 12.1.11 (a) Root locus for proportional control of the plant $1/(s + \alpha)(s + \beta)$. (b) Root locus for lead compensation of the plant $1/(s + \alpha)(s + \beta)$.

gain. If not, determine the net angle associated with the desired closed-loop pole by drawing vectors to this pole from the open-loop poles and zeros. The difference between this angle and $-180°$ is the *angle deficiency*.

3. Locate the pole and zero of the compensator so that they will contribute the angle required to eliminate the deficiency. A method for doing this is presented in Example 12.1.2.
4. Compute the required value of the open-loop gain from the root locus plot.
5. Check the design to see if the specifications are met. If not, adjust the locations of the compensator's pole and zero.

Root Locus Design of a Lead Compensator

EXAMPLE 12.1.2

■ Problem

Consider the system shown in Figure 12.1.12a. The root locus plot with $G_c(s) = 1$ is given in Figure 12.1.12b. The transient specifications require that $\zeta = 0.707$ with a time constant of $\tau = 0.2$. No steady-state error specifications are given. This information implies that the closed-loop poles should be at $s = -5 \pm 5j$. This performance cannot be obtained with a gain change in the present system. Design a lead compensator to meet the specifications.

■ Solution

From (12.1.3), the open-loop transfer function of the compensated system can be expressed as

$$K_P G_c(s) G_p(s) = K_P \frac{s + a}{s + b} \frac{1}{s(s + 5)} = K_P \frac{s + 1/\mu T}{s + 1/T} \frac{1}{s(s + 5)} \tag{1}$$

We must pull the original locus to the left, so we place the pole and zero to the left of the pole at $s = -5$. Drawing vectors from the poles and zeros to the desired root location, we obtain Figure 12.1.12c. The angle of the uncompensated system at the desired root location is

$$\angle \left| \frac{1}{s(s + 5)} \right|_{s=-5+5j} = -90° - 135° = -225°$$

Thus, the angle deficiency is $-225° + 180° = -45°$, and the difference between the angle contributions of the pole and zero of the compensator must be 45°. We therefore need a lead compensator to increase the phase angle by 45°. This is the angle θ shown in the apex of the triangle in the figure.

Because $\beta = \alpha + 45°$, simple trigonometry applied to Figure 12.1.12c gives the required value for T. The tangents of α and β are

$$\tan \alpha = \frac{5}{1/T - 5} = \frac{5T}{1 - 5T}$$

$$\tan \beta = \frac{5}{1/\mu T - 5} = \frac{5\mu T}{1 - 5\mu T}$$

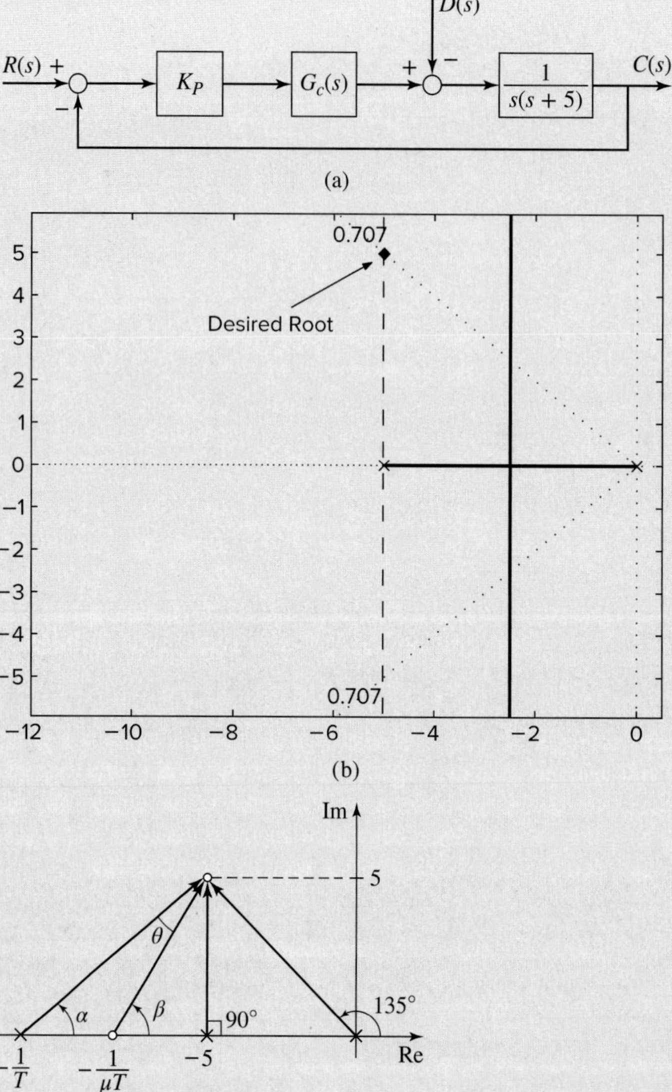

Figure 12.1.12 (a) System for Example 12.1.2. (b) Root locus plot for proportional control. (c) Geometry of the lead-compensated root locus.

From the identity for the tangent of a sum, we have

$$\tan\beta = \tan(\alpha + 45°) = \frac{\tan\alpha + \tan 45°}{1 - \tan\alpha \tan 45°}$$

Eliminating α and β yields

$$50\mu T^2 - 10\mu T + 1 = 0 \tag{2}$$

Because the inverse of μ appears in the numerator of the transfer function (1), we should select its value to be as small as possible in order to minimize the additional amount of gain required to cancel its effect. Keeping physical realizability in mind, we choose $\mu = 5$. This gives

$$250T^2 - 50T + 1 = 0$$

which has the solutions $T = 0.1775$ and $T = 0.0225$. The first solution results in the pole lying to the left of $s = -5$ and the zero lying to the right of $s = -5$, which contradicts the assumed

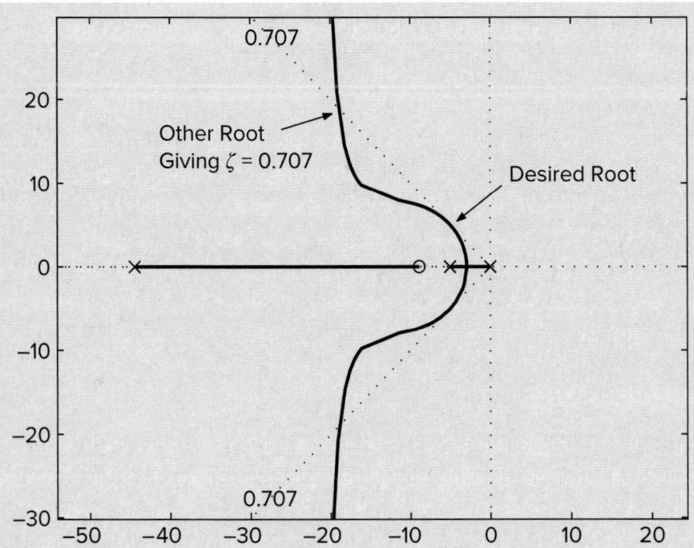

Figure 12.1.13 Root locus of the lead-compensated system of Example 12.1.2.

geometry. So we use $T = 0.0225$. This puts the compensator pole at $s = -44.4$, and the zero at $s = -8.89$.

The open-loop transfer function of the compensated system is

$$K_P G_c(s) G_p(s) = \frac{K_P(s + 8.89)}{s(s + 44.4)(s + 5)}$$

Its root locus is shown in Figure 12.1.13. From this, the gain K_P required to place the roots at $s = -5 \pm 5j$ is $K_P = 227$. The third root is far to the left at $s = -39.3$ for this value of K_P, so its effect on the transient behavior is probably slight. This can be checked by analysis or simulation before the design is made final.

The unit-step response is shown in Figure 12.1.14. The overshoot is 7.7% with a peak time of 0.47 and a settling time of 0.75.

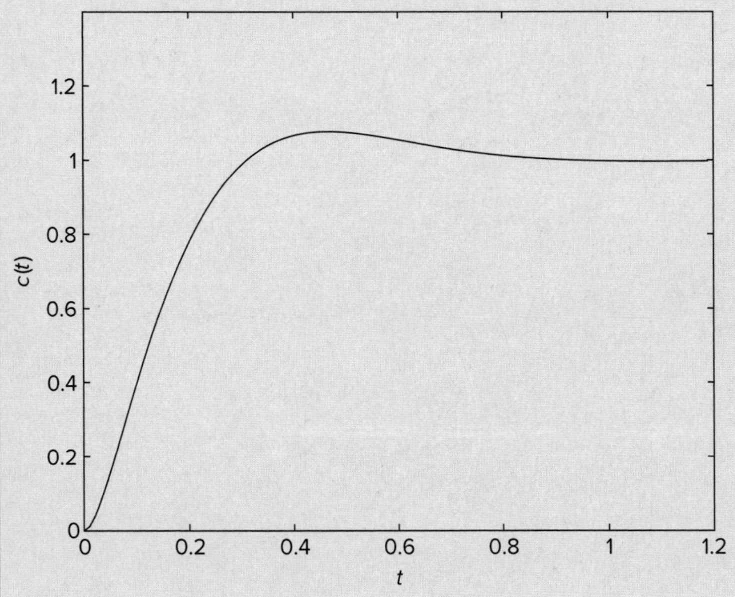

Figure 12.1.14 Unit-step response of the lead-compensated system of Example 12.1.2.

From (12.1.4) and (12.1.5), the resistances required with a 1-μF capacitance are R_1 = 112.5 kΩ and R_2 = 28.1 kΩ.

This design procedure does not always give a root locus like Figure 12.1.1b. For example, in Example 12.1.2, the choice of $\mu = 10$, rather than 5, results in the following solution of equation (2): $T = 0.1894$ and $T = 0.0106$. The latter solution puts the pole at −94.72 and the zero at −9.472. These result in the root locus shown in Figure 12.1.15. The solution giving $\zeta = 0.707$ and $\tau = 0.2$ corresponds to $K_P = 474$ with the roots at $s = -5 \pm 5j$ and $s = -89.72$. The step response has an overshoot of 7.1%, a peak time of 0.475, and a settling time of 0.753.

12.1.5 LAG COMPENSATORS

Consider proportional control of the plant $1/(s+\alpha)(s+\beta)$. Its root locus plot is shown in Figure 12.1.16a. Suppose that the desired damping ratio ζ and desired time constant τ are obtainable with a proportional gain of K_{P_1} but the resulting steady-state error $\alpha\beta/(\alpha\beta + K_{P_1})$ for a step input is too large. We need to increase the gain while preserving the desired damping ratio and time constant. With the lag compensator, the root locus is as shown in Figure 12.1.16b. By considering specific numerical values, we can show

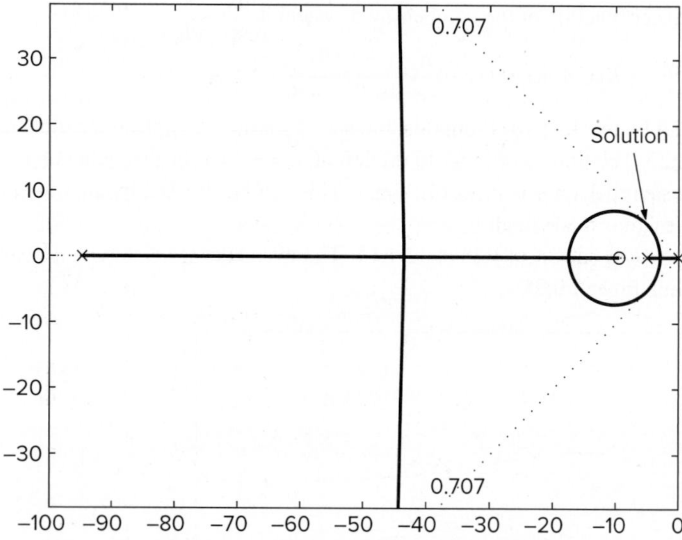

Figure 12.1.15 Another design solution for Example 12.1.2.

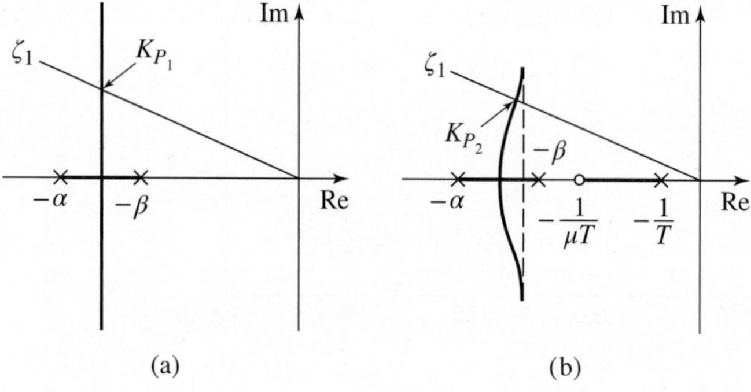

Figure 12.1.16 (a) Root locus for proportional control of the plant $1/(s+\alpha)(s+\beta)$. (b) Root locus for lag compensation of the plant $1/(s+\alpha)(s+\beta)$.

that for the compensated system, roots with a damping ratio ζ_1 correspond to a high value of the proportional gain. Call this value K_{P_2}. Thus, $K_{P_2} > K_{P_1}$, and the steady-state error will be reduced.

The effect of the lag compensator on the time constant can be seen as follows. The open-loop transfer function is

$$K_P G_c(s) G_p(s) = \frac{\mu K_P(s + 1/\mu T)}{(s+\alpha)(s+\beta)(s+1/T)} \tag{12.1.27}$$

If the value of T is chosen large enough, the pole at $s = -1/T$ in (12.1.27) is approximately cancelled by the zero at $s = -1/\mu T$ and the open-loop transfer function is given approximately by

$$K_P G_c(s) G_p(s) = \frac{\mu K_P}{(s+\alpha)(s+\beta)} \tag{12.1.28}$$

Thus, the system's response is governed approximately by the complex roots corresponding to the gain value K_{P_2}. By comparing Figure 12.1.16a with 12.1.16b, we see that the compensation leaves the time constant relatively unchanged. From (12.1.28), it can be seen that because $\mu < 1$, K_P can be selected as the larger value K_{P_2}. The ratio of K_{P_1} to K_{P_2} is approximately given by the parameter μ.

Design by pole-zero cancellation can be difficult to accomplish, because a response pattern of the system is essentially ignored. The pattern corresponds to the behavior generated by the cancelled pole and zero, and this response can be shown to be beyond the influence of the controller. In this example, the cancelled pole gives a stable response, because it lies in the left-hand plane. However, another input not modeled here, such as a disturbance, might excite the response and cause unexpected behavior. The designer should therefore proceed with caution. None of the physical parameters of the system are known exactly, so exact pole-zero cancellation is not possible. A root locus study of the effects of parameter uncertainty and a simulation study of the response are often advised before the design is accepted as final.

Based on this example, we can outline an approach to the design of a lag compensator with the root locus as follows:

1. From the root locus of the uncompensated system, determine the gain K_{P_1} to place the roots at the locations required to meet the transient performance specifications.
2. Let K_{P_2} denote the value of the gain required to achieve the desired steady-state performance. The parameter μ is the ratio of these two gain values $\mu = K_{P_1}/K_{P_2} < 1$.
3. The value of T is then chosen large so that the compensator's pole and zero are close to the imaginary axis. This placement should be made so that the compensated locus is relatively unchanged in the vicinity of the desired closed-loop poles. This will be true if the angle contribution of the lag compensator is close to zero.
4. Locate the desired closed-loop poles on the compensated locus and set the open-loop gain so that the dominant roots are at this location (neglecting the existence of the compensator's pole and zero).
5. Check the design to see if the specifications are met. If not, adjust the locations of the compensator's pole and zero.

EXAMPLE 12.1.3

Root Locus Design of a Lag Compensator

■ Problem

Consider the system shown in Figure 12.1.17a. Suppose that we require a time constant of 0.4 and a damping ratio of $\zeta = 0.707$. These correspond to the roots $s = -2.5 \pm 2.5j$, which are obtainable with $K_P = 12.5$. The steady-state error for a unit ramp command is 0.4, however, which is considered too large. Design a compensator to reduce the error to 0.1.

■ Solution

A lag compensator is indicated, because the transient response is acceptable but the steady-state error is too large. The gain K_P required to achieve the desired transient performance has already been established as $K_{P_1} = 12.5$. The second step is to determine the value of the parameter μ. For this system, the steady-state ramp error is $e_{ss} = 5/K_P$, and K_{P_2} is the value of K_P that gives $e_{ss} = 0.1$. Thus, $K_{P_2} = 5/0.1 = 50$, and the parameter $\mu = K_P/K_{P_2} = 12.5/50 = 0.25$. The compensator's pole and zero must be placed close to the imaginary axis, with the ratio of their distances being $1/0.25 = 4$. Noting that the plant has a pole at $s = -5$, we select locations well to the right of this pole, say, at $s = -0.01$ and $s = -0.04$ for the pole and zero, respectively. This gives $T = 100$.

The open-loop transfer function of the compensated system is thus

$$K_P G_c(s) G_p(s) = \frac{0.25 K_P (s + 0.04)}{s(s+5)(s+0.01)}$$

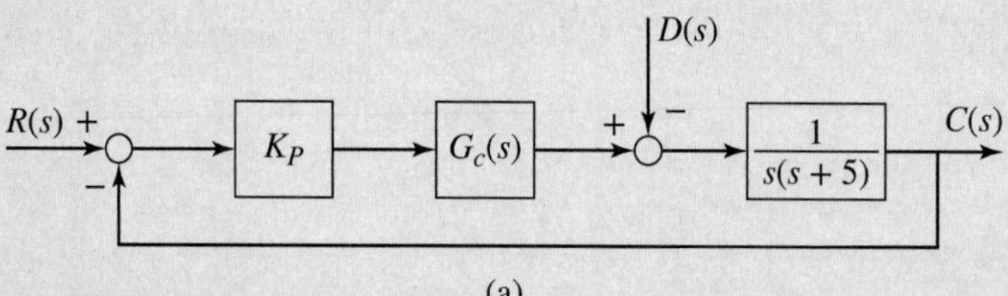

(a)

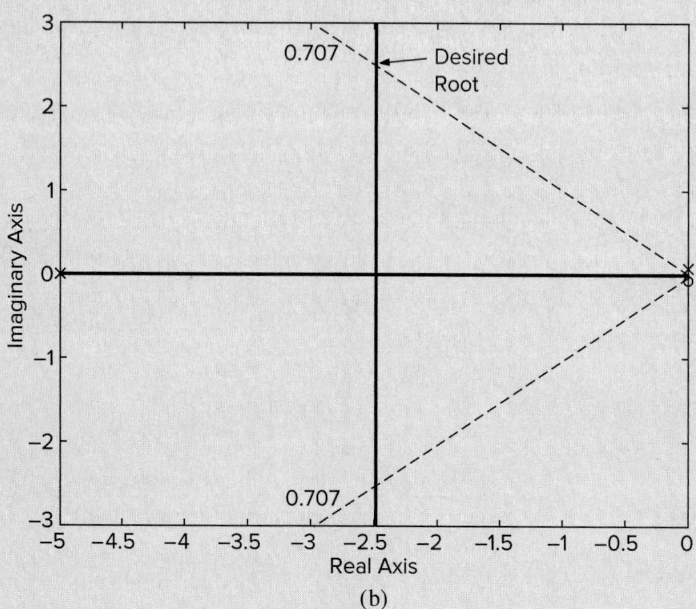

(b)

Figure 12.1.17 (a) System for Example 12.1.3. (b) Root locus plot of the lag-compensated system.

The root locus is shown in Figure 12.1.17b, where we have offset the pole and zero to make them visible. For the desired damping ratio of 0.707, the root locus shows that two roots are at $s = -2.5 \pm 2.5j$ if $0.25K_P = 12.7$, or $K_P = 12.7/0.25 = 50.8$. The dominant root is at $s = -0.0405$, but its effect is diminished by the zero at $s = -0.04$. The steady-state error is thus $e_{ss} = 0.098$ and less than the required value of 0.1.

The step response of the final design has an overshoot of 5.75%, a peak time of $t_p = 1.27$, and a settling time of 1.89. With $T = 100$ and the capacitance $C = 1\ \mu$F, (12.1.7) and (12.1.8) give $R_1 = 7.5$ kΩ and $R_2 = 2.5$ kΩ.

12.1.6 DESIGN OF LAG-LEAD COMPENSATORS

The lead and lag compensators are complementary to each other in that one improves the transient performance while the other improves steady-state performance. In situations where one of these fails to produce a satisfactory design, a lag-lead compensator can be tried.

The root locus approach to designing a lag-lead compensator is a combination of the approaches used for the lead and lag compensators.

1. Evaluate the uncompensated system by determining the desired locations of the dominant closed-loop poles from the transient performance specifications.
2. Improve the transient performance with a lead compensator. Calculate the phase angle deficiency of the uncompensated system. This deficiency must be supplied by the lead compensator.
3. Evaluate the steady-state error and if necessary, improve the steady-state performance with a lag compensator.
4. Evaluate the transient performance to see if the specifications are satisfied.
5. Evaluate the hardware requirements and check realizability.

Design of a Lag-Lead Compensator

EXAMPLE 12.1.4

■ **Problem**

The problem is to design a lag-lead compensator for the plant shown in Figure 12.1.18a. The specifications are

1. Damping ratio of the dominant roots should be $\zeta = 0.707$.
2. The time constant of the dominant roots should be no greater than 0.5.
3. The steady-state error for a unit-ramp disturbance should be no greater than 0.1.

The first two specifications will be satisfied with a dominant-root pair $s = -2 \pm 2j$, but other roots are possible.

■ **Solution**

Step 1: Evaluate the uncompensated system. The root locus equation for the uncompensated system using P action is

$$1 + K_P \frac{1}{s(s+3)(s+5)} = 0$$

Figure 12.1.18b shows the root locus plot. From this plot we determine that $K_P = 13.6$ will give a dominant-root pair of $s = -1.08 \pm 1.08j$, for which $\zeta = 0.707$. The time constant is 0.93 and the unit-ramp disturbance error is 1.03, so the uncompensated system does not meet the second and third specifications.

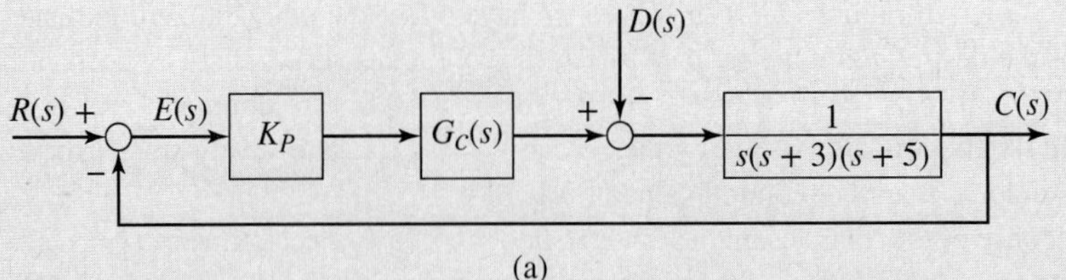

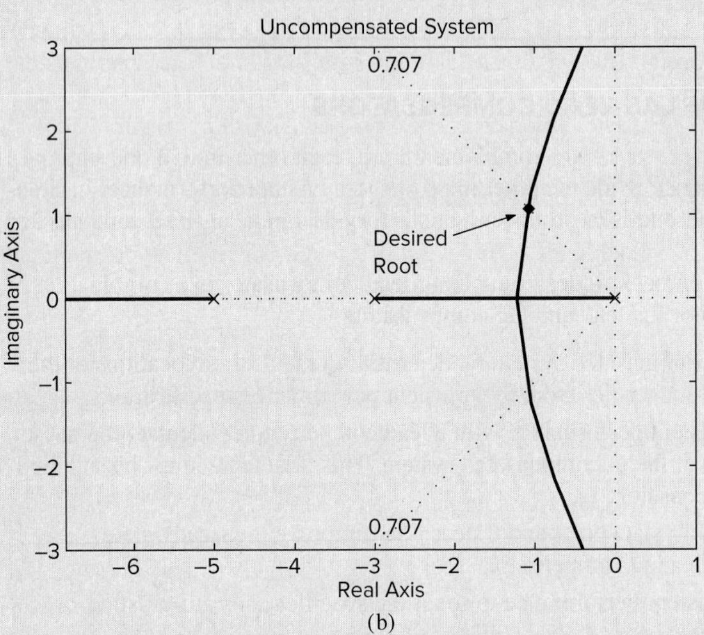

Figure 12.1.18 (a) System for Example 12.1.4. (b) Root locus plot of the uncompensated system.

Table 12.1.3 Design of a lag-lead compensator for the plant $1/s(s+3)(s+5)$.

Design	Uncompensated	Lead compensated	Lag-Lead compensated
Open-loop transfer function	$\dfrac{K_P}{s(s+3)(s+5)}$	$\dfrac{K_P}{(s+11.9)s(s+5)}$	$\dfrac{K_P(s+0.0578)}{(s+0.01)(s+11.9)s(s+5)}$
Dominant roots	$s=-1.08\pm 1.08j$	$s=-2\pm 2j$	$s=-1.9\pm 1.9j$
K_P	13.6	103	101
ζ	0.707	0.707	0.707
τ	0.93	0.5	0.526
Steady-state ramp error	1.03	0.578	0.1

Table 12.1.3 summarizes the various designs.

Step 2: Improve the transient performance with a lead compensator. The form of the lead compensator is

$$G_c(s) = \frac{s+a}{s+b}$$

If we can cancel the plant pole at $s = -3$, the root locus will be shifted to the left, closer to the desired roots at $s = -2 \pm 2j$. Thus, we place the compensator zero at $s = -3$ and choose $a = 3$. The location of the compensator pole $s = -b$ is found from the geometry shown in Figure 12.1.19a. From the angle criterion

$$\angle N(s) - \angle D(s) = (2n+1)180^\circ \quad n = 0, 1, 2, 3, \ldots$$

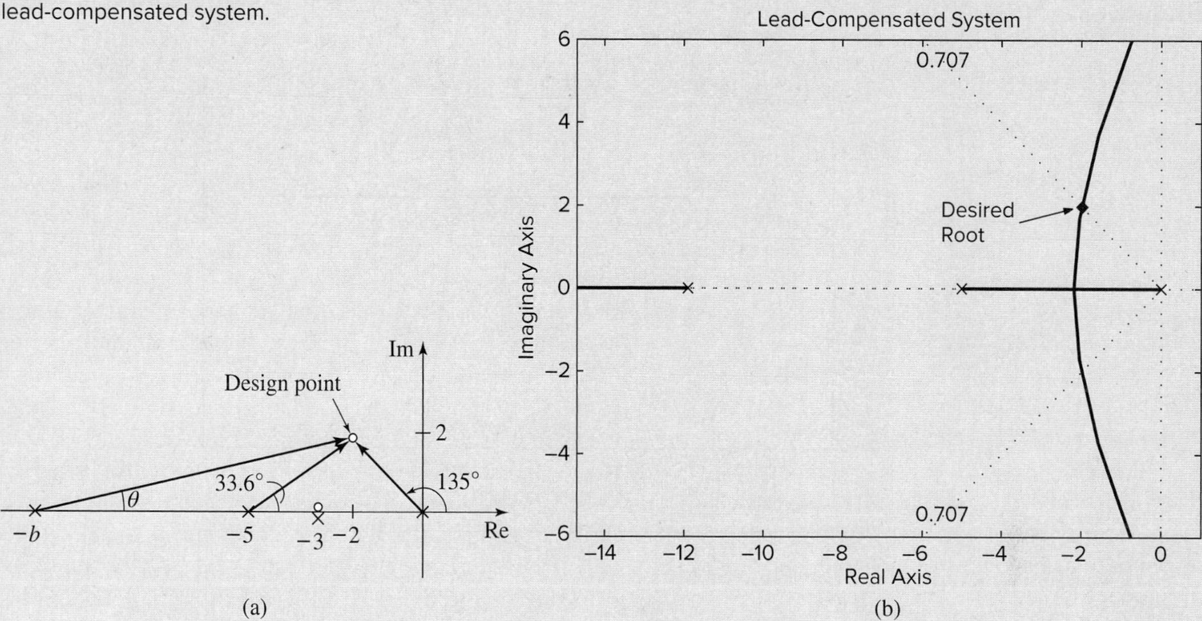

Figure 12.1.19 (a) Geometry of the lead-compensated root locus for Example 12.1.4. (b) Root locus plot of the lead-compensated system.

we have

$$-\theta - 33.6^\circ - 135^\circ = (2n+1)180^\circ$$

which gives $\theta = (2n+1)180^\circ - 168.6^\circ = 11.4^\circ$ by choosing $n = 0$. From the figure,

$$\tan\theta = \tan 11.4^\circ = 0.202 = \frac{2}{b-2}$$

which gives $b = 11.9$. Thus, the open-loop transfer function of the lead-compensated system is

$$K_P G_c(s) G_p(s) = \frac{K_P(s+3)}{(s+11.9)s(s+3)(s+5)} = \frac{K_P}{(s+11.9)s(s+5)}$$

The root locus plot for this system is shown in Figure 12.1.19b. From it we determine that a gain of $K_P = 103$ will give dominant roots at the desired location of $s = -2 \pm 2j$. Thus, this design meets the first two specifications. The ramp error, however, is 0.578, which is still too large.

Step 3: Improve the steady-state performance with a lag compensator. The form of the lag-lead compensator is

$$K_P G_c(s) = K_P \frac{s+3}{s+11.9} \frac{s+c}{s+d}$$

We place the pole of the lag compensator close to the origin and choose $d = 0.01$. The zero is separated from the pole by the ratio of the current steady-state error to the desired error; that is, $c = (0.578/0.1)d = 0.0578$. Thus, the open-loop transfer function of the lag-lead system is

$$K_P G_c(s) G_p(s) = \frac{K_P(s+0.0578)}{(s+0.01)(s+11.9)s(s+5)}$$

Part of the root locus plot for this system is shown in Figure 12.1.20, and it is very similar to the plot of the lead-compensated system. From it we determine that a gain of $K_P = 101$ will give dominant roots at $s = -1.9 \pm 1.9j$, which is close to the desired location of $s = -2 \pm 2j$.

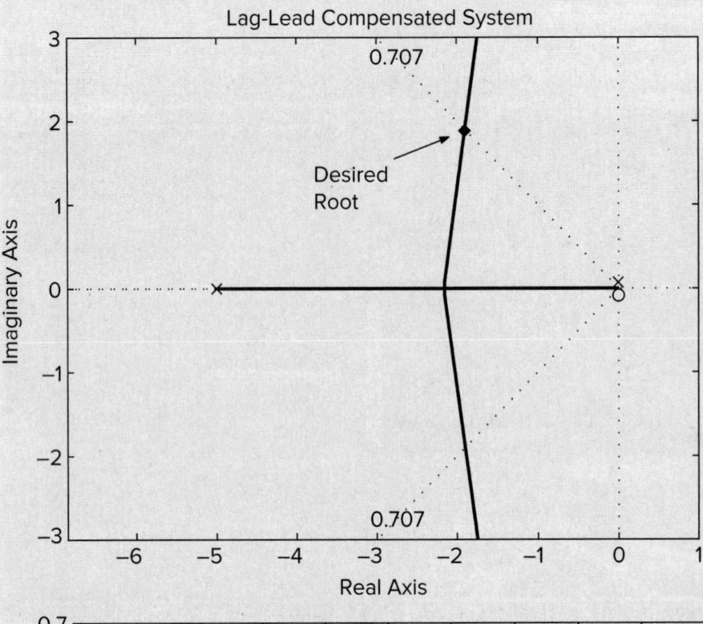

Figure 12.1.20 Root locus plot of the lag-lead system (the pole at −11.9 is not shown).

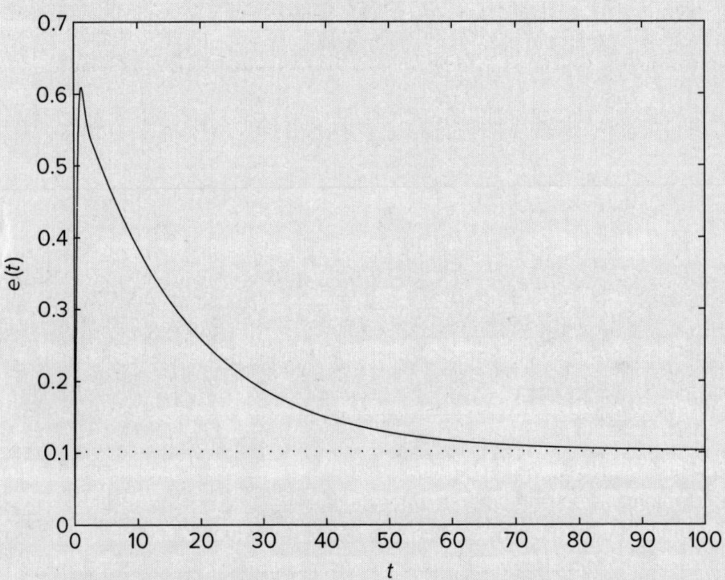

Figure 12.1.21 Error response to a unit-ramp disturbance for the lag-lead system of Example 12.1.4.

The time constant is 0.526, which is slightly larger than desired. The ramp error, however, is 0.1, which is exactly what was specified.

Further analysis of the root locus shown in Figure 12.1.20 reveals that we can obtain a time constant of 0.5 with a gain of $K_P = 96$, but the damping ratio would be $\zeta = 0.743$.

Step 4: Evaluate the transient performance. Simulation of the final design using $K_P = 101$ shows that the step response is satisfactory. The ramp response, however, reveals something perhaps unexpected (Figure 12.1.21). While the steady-state error is as predicted, the time to reach steady state is much longer than expected from the time constant of 0.526. The effect is due to the small root at $s = -0.0595$, which is actually the dominant root and has a time constant of $1/0.0595 = 16.8$. Figure 12.1.22 is an enlarged view of the root locus near the origin. It shows the dominant-root path that ends at the zero at $s = -0.0578$.

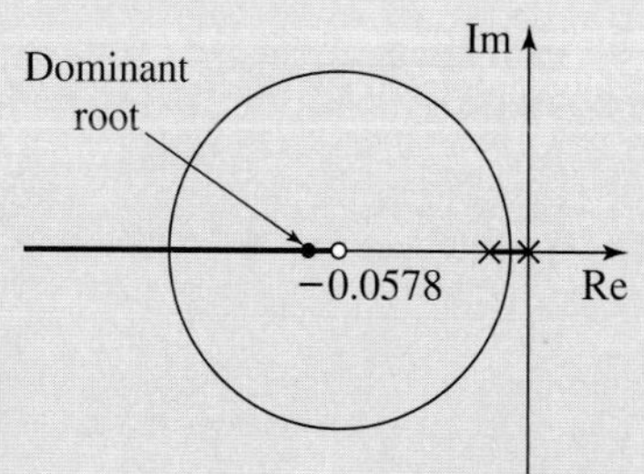

Figure 12.1.22 Enlarged view of the root locus for Example 12.1.4.

This example shows the importance of checking any design with simulation before proceeding further.

Step 5: Evaluate the hardware requirements. Before implementing this design in hardware, we should check to see if it is possible to set the poles and zeros of the compensator as accurately as indicated by our calculations. In particular, we should check the feasibility of obtaining the small values, namely, $d = 0.01$ and $c = 0.0578$.

You should also realize that it is physically impossible to set the lead compensator to zero to cancel exactly the plant pole at $s = -3$. Further simulation should be done to check the effects of a not being exactly equal to 3.

We note in passing that the pole at $s = -3$ is not canceled in the disturbance transfer function, which is a fifth-order model.

12.1.7 SYSTEM TYPE AND STEADY-STATE ERROR

Consider a single-loop system whose open-loop transfer function is $G(s)H(s)$. The error signal is related to the input as follows:

$$E(s) = \frac{1}{1 + G(s)H(s)} R(s)$$

A system is of *type n* if $G(s)H(s)$ can be written as $s^{-n}F(s)$. We can relate the steady-state error to the type number for various kinds of command inputs. Assuming that the final-value theorem can be applied, the steady-state error is

$$e_{ss} = \lim_{s \to 0} \frac{sR(s)}{1 + G(s)H(s)}$$

The *static error coefficient* C_i is defined as

$$C_i = \lim_{s \to 0} s^i G(s)H(s) \quad i = 0, 1, 2, \ldots \tag{12.1.29}$$

Often the steady-state performance specifications are given in terms of the static error coefficients. If the input is a unit step, we obtain the steady-state error

$$e_{ss} = \lim_{s \to 0} \frac{1}{1 + G(s)H(s)} = \frac{1}{1 + C_p}$$

where

$$C_p = \lim_{s \to 0} G(s)H(s) = C_0 \tag{12.1.30}$$

The constant C_p is the *static position error coefficient*. The name derives from servomechanism applications in which the output is a position. The larger C_p is, the smaller the error, and a unity-feedback system will have a nonzero steady-state error if no integration occurs in the forward path.

Table 12.1.4 Type number and steady-state error.

	Steady-state error		
Input	**Type 0**	**Type 1**	**Type 2**
Unit step $1/s$	$1/(1+C_0)$	0	0
Unit ramp $1/s^2$	∞	$1/C_1$	0
Unit parabola $1/s^3$	∞	∞	$1/C_2$

The *static velocity error coefficient* is obtained for a unit-ramp input as follows:

$$e_{ss} = \lim_{s\to 0} \frac{s}{1+G(s)H(s)} \frac{1}{s^2} = \frac{1}{C_v}$$

where the velocity coefficient is

$$C_v = \lim_{s\to 0} sG(s)H(s) = C_1 \tag{12.1.31}$$

Note that the error here is not an error in velocity, but a position error that results when a unit-ramp input is applied. For type 0 systems, $C_v = 0$; for type 1 systems, C_v is finite but nonzero; for type 2 and higher, $C_v = \infty$. Thus, to eliminate the steady-state error in a unity-feedback system with a ramp input, at least two integrations are required in the forward loop. For a type-1 system with unity feedback, the output velocity at steady state equals that of the input (the slope of the ramp), but an error exists between the desired and the actual positions. These results are summarized in Table 12.1.4.

12.2 DESIGN USING THE BODE PLOT

In comparison to using the root locus plot, there are several advantages to designing a system using the open-loop frequency response. Frequency response data are often easier to obtain experimentally, which is useful when it is difficult to develop a transfer-function model of the plant and actuators from basic principles. The method is also easier to use for systems with dead-time elements, which we will see in this section. Finally, this technique is sometimes useful for examining response and instability in nonlinear systems. This topic is rather specialized and is treated in more advanced control system texts.

12.2.1 THE NYQUIST STABILITY THEOREM

Recall from Chapter 9 that a plot of the frequency transfer function $T(j\omega)$ in vector form is the polar plot on which is plotted the location of the tip of the vector as ω varies from 0 to ∞ (see Figure 12.2.1). The axes of the plot are the real and imaginary parts of $T(j\omega)$. Thus, the angle and magnitude of $T(j\omega)$ are represented on the same plot, whereas with Bode plots, the magnitude, and angle each have their own plot versus $\log \omega$.

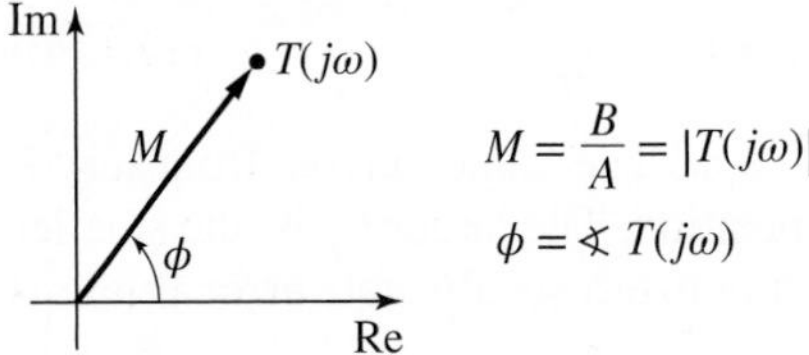

Figure 12.2.1 Frequency response of a stable linear system.

Harry Nyquist, a Swedish-born American electronic engineer and physicist, made important contributions to digital systems, communication theory, and control systems while working at Bell Labs. He developed the Nyquist stability theorem while improving the stability of electronic amplifiers. This theorem is a powerful tool for linear system analysis. Its proof requires considerable mathematical development and will not be attempted here. In many control system applications, the plant has no poles with positive real parts, and typical controllers do not introduce poles of this type. If the open-loop system has no poles with positive real parts, we can concentrate our attention on the region around the point $-1 + j0$ on the polar plot of the open-loop transfer function. When the *open-loop* transfer function is plotted in polar form, the plot is called the *Nyquist plot*.

Consider the single-loop system shown in Figure 12.2.2. Its open-loop transfer function is $G(s)H(s)$. Figure 12.2.3 shows the polar plot of the open-loop transfer function of two arbitrary systems, both of which are assumed to be *open-loop stable* (i.e., no poles of $G(s)H(s)$ in the right-half plane). The Nyquist stability theorem is stated as follows:

The system is closed-loop stable if and only if the point $-1 + j0$ lies to the left of the open-loop Nyquist plot relative to an observer traveling along the plot in the direction of increasing frequency ω.

Figure 12.2.3a is a Nyquist plot for one such system, which the theorem says is closed-loop stable. A system exhibits sustained oscillations (neutral stability) if the plot passes through the $-1 + j0$ point. The system whose plot is shown in Figure 12.2.3b is unstable because the $-1 + j0$ point lies to the right of the plot.

The basis for the Nyquist theorem is easily understood. Let $e(t)$ and $b(t)$ be the actuating and feedback signals, respectively (see Figure 12.2.2). Suppose that the input

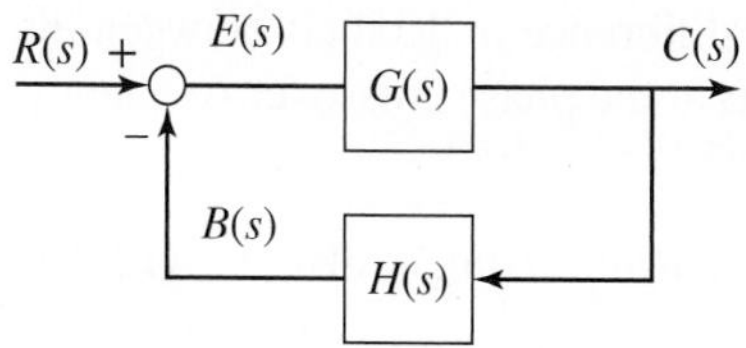

Figure 12.2.2 A single-loop control system. The open-loop transfer function is $G(s)H(s)$.

Figure 12.2.3 (a) Nyquist plot of a stable system. (b) Nyquist plot of an unstable system.

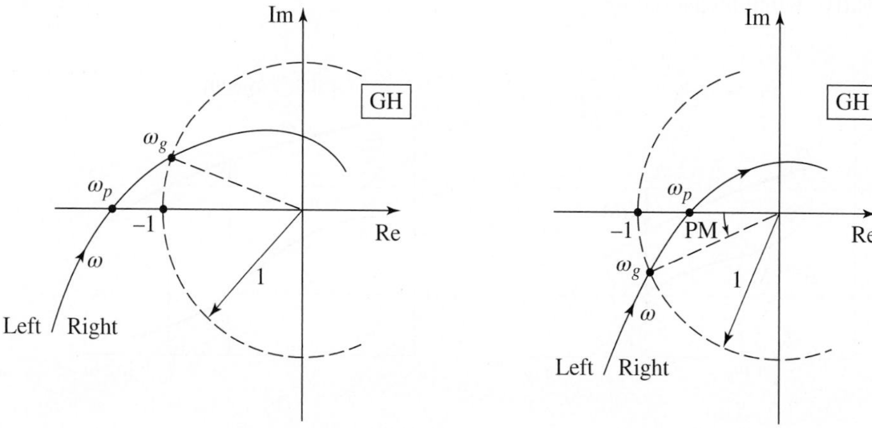

$r(t)$ is sinusoidal with frequency ω_u, and assume that the phase angle of the open-loop transfer function is $-180°$. This means that at steady state, the signals $e(t)$ and $b(t)$ will be sinusoidal with the frequency ω_u, and 180° out of phase with each other [when $e(t)$ reaches a peak, $b(t)$ is at a minimum, and vice versa]. When the signal $b(t)$ has its sign changed at the comparator, the result, $-b(t)$, will be in phase with the signal $e(t)$. Imagine that the input is now switched off. When the open-loop plot passes through the $-1 + j0$ point at the frequency ω_u, the gain relating $b(t)$ to $e(t)$ is unity at this frequency. Therefore, $e(t) = -b(t)$ and the original oscillation will be sustained at a constant amplitude. If the $-1 + j0$ point lies to the left of the plot, the gain between $e(t)$ and $b(t)$ will be less than unity. In this case, the amplitude of $b(t)$ will be less than that of $e(t)$, and the oscillations will gradually diminish. The system is then stable. Finally, if the $-1 + j0$ point lies to the right of the plot, the amplitude of $b(t)$ will be larger than that of $e(t)$, and the oscillations will grow in magnitude as the signal travels around the loop. The system is then unstable.

12.2.2 PHASE AND GAIN MARGINS

The Nyquist theorem provides a convenient measure of the relative stability of a system. A measure of the proximity of the plot to the $-1 + j\omega$ point is given by the angle between the negative real axis and a line from the origin to the point where the plot crosses the unit circle (see Figure 12.2.3a). The frequency corresponding to this intersection is denoted as ω_g. This angle is the *phase margin* (PM), and it is positive when measured down from the negative real axis. The absence of a positive or zero phase margin thus indicates an unstable system (Figure 12.2.3b). The phase margin is the phase at the frequency ω_g where the magnitude ratio or "gain" of $G(j\omega)H(j\omega)$ is unity (0 dB). Thus,

$$\text{phase margin (PM)} = \angle\left[G(j\omega_g)H(j\omega_g)\right] + 180° \tag{12.2.1}$$

The frequency ω_p, the *phase crossover frequency*, is the frequency at which the phase angle is $-180°$. The *gain margin* (GM) is the difference in decibels between the unity gain condition (0 dB) and the value of $|GH|$ dB at the phase crossover frequency. Thus,

$$\text{gain margin (GM)} = 0 - |G(i\omega_p)H(i\omega_p)| = -|G(i\omega_p)H(i\omega_p)| \text{ dB} \tag{12.2.2}$$

A system is stable only if the phase and gain margins are both positive.

These definitions are illustrated on the Bode plots of the *open-loop* transfer function shown in Figure 12.2.4a. Note that GM is positive when measured *down*, but PM is positive when measured *up*.

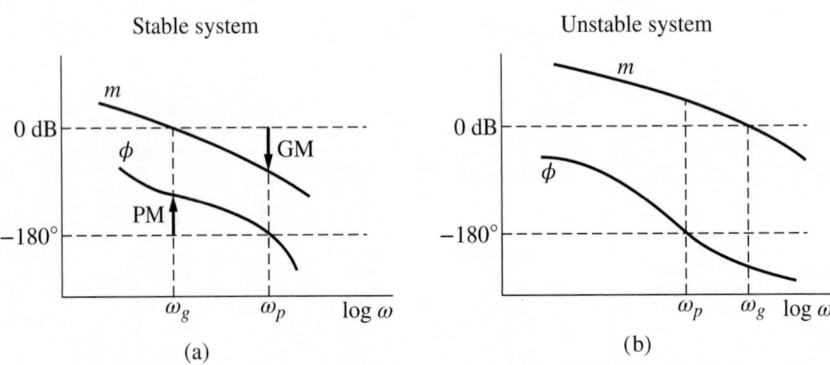

Figure 12.2.4 (a) Definition of gain and phase margins on the open-loop Bode plots. (b) Plots for an unstable system.

In many applications, the plant has no poles in the right-half plane, and control algorithms typically do not have such poles, so for many applications the open-loop transfer function $G(s)H(s)$ has no such poles. If the open-loop transfer function $G(s)H(s)$ has no poles in the right-half plane, then the closed-loop system is *stable* if and only if the phase and gain margins are both positive. We will not prove this statement, but it has long been well established. The situation shown in Figure 12.2.4a thus corresponds to a stable system. Figure 12.2.4b represents an unstable system that lacks positive gain and phase margins.

The phase and gain margins are often used as safety margins in the design specifications. A typical set of such specifications is:

$$\text{gain margin} \geq 8 \text{ dB and phase margin} \geq 30° \tag{12.2.3}$$

Another common set of specifications is

$$\text{gain margin } \geq 6 \text{ dB and phase margin } \geq 40° \tag{12.2.4}$$

In most situations, only one of these equalities can be met, and the other margin is allowed to be greater than its minimum value. It is not desirable to make the margins too large, because this might result in a sluggish system or one with a large steady-state error.

12.2.3 PHASE MARGIN VERSUS DAMPING RATIO

It is sometimes useful to be able to translate time-domain performance specifications, such as damping ratio or overshoot, into frequency-domain specifications, such as phase or gain margin. The open-loop transfer function

$$G(s) = \frac{\omega_n^2}{s(s + 2\zeta\omega_n)} \tag{12.2.5}$$

in a unity-feedback system results in the following closed-loop transfer function.

$$T(s) = \frac{G(s)}{1 + G(s)} = \frac{\omega_n^2}{s^2 + 2\zeta\omega_n s + \omega_n^2} \tag{12.2.6}$$

which has the characteristic roots $s = -\zeta\omega_n \pm \omega_n j\sqrt{1 - \zeta^2}$. These roots describe the dominant-root pair of a higher-order system having complex dominant roots. Thus, if we can relate the phase margin of $G(s)$ to either ζ or ω_n or both, then we can derive an approximate expression for the phase margin of a higher-order system having complex dominant roots.

It is shown in one of the chapter's homework problems that the phase margin of $G(s)$ given by (12.2.5) is

$$\text{PM} = \tan^{-1} \frac{2\zeta}{\sqrt{-2\zeta^2 + \sqrt{1 + 4\zeta^4}}} \tag{12.2.7}$$

We saw in Chapter 8 that the maximum percent overshoot is the following function of ζ.

$$M_{\%} = 100e^{-\pi\zeta/\sqrt{1-\zeta^2}} \tag{12.2.8}$$

So a 10% maximum overshoot corresponds to $\zeta = 0.59$, and from (12.2.7), the phase margin is 59°. Consider, for example, a third-order system having the dominant roots $s = -5 \pm 6.84j$ and the secondary root $s = -10$. The damping ratio of the dominant roots is $\zeta = 0.59$, and thus from (12.2.7) we can estimate the phase margin of this third-order system to be 59°. The accuracy of (12.2.7) obviously depends on the separation between the dominant roots and the secondary roots.

12.2.4 BODE PLOT DESIGN FOR PID CONTROL

When using the open-loop frequency response method, it is convenient to write the PID algorithm with the proportional gain factored out, as

$$F(s) = K_P \left(1 + \frac{1}{T_I s} + T_D s\right) E(s) \tag{12.2.9}$$

The proportional gain is selected last, because it simply moves the gain curve up or down without affecting the phase curve, and thus can be used to adjust the gain curve until the specifications for the gain and phase margins are satisfied.

D action affects both the phase and gain curves. The increase in phase margin due to the positive phase angle introduced by D action is partly negated by the derivative gain, which reduces the gain margin. Increasing the derivative gain increases the speed of response, makes the system more stable, and enables a larger proportional gain to be used to improve the system performance. However, if the phase curve is too steep near $-180°$, it is difficult to use D action to improve the performance.

I action affects both the gain and phase curves. It can be used to increase the open-loop gain at low frequencies. However, it lowers the phase crossover frequency ω_p and thus reduces some of the benefits provided by D action. If required, the D action term is usually designed first, followed by I action and P action, respectively.

The PD algorithm is

$$F(s) = K_P(1 + T_D s)E(s) \tag{12.2.10}$$

The term $(1 + T_D s)$ can be considered as a series compensator to the proportional controller. Its Bode plots are shown in Figure 12.2.5a. From these, it can be seen that the usefulness of D action is that it adds phase shift at higher frequencies. It is thus said to give phase "lead." However, it also increases the gain at these frequencies, because the derivative term gives more response for rapidly changing signals.

The PI control algorithm is

$$F(s) = \frac{K_P}{s}\left(s + \frac{1}{T_I}\right) E(s) = \frac{K_P}{T_I s}\left(T_I s + 1\right) E(s) \tag{12.2.11}$$

The Bode plots are shown in Figure 12.2.5b. PI action adds gain at the lower frequencies but decreases the phase.

Figure 12.2.5 (a) Bode plots for PD action. (b) Bode plots for PI action.

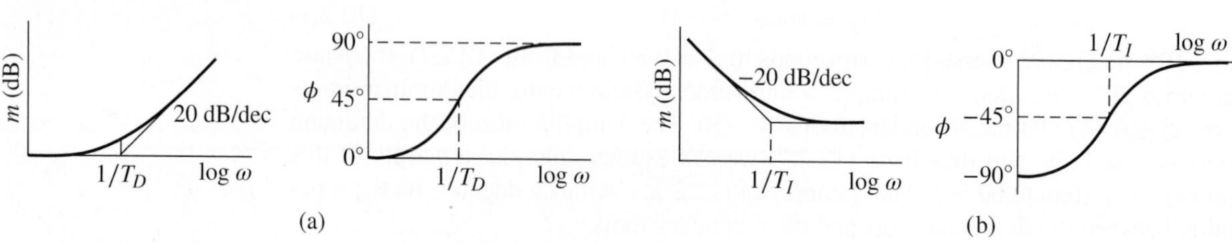

12.2.5 DESIGN APPROACH WITH THE BODE PLOT

Classical design methods based on the Bode plots obviously have a large component of trial and error because usually both the phase and gain curves must be manipulated to achieve an acceptable design. Given the same set of specifications, two designers can use these methods and arrive at different designs. An experienced designer, however, can often obtain a good design quickly with these techniques. Using a computer-plotting routine greatly speeds up the design process.

It is difficult to state a rigid set of rules to follow for designing a series compensator because of the variety of specifications and plant types that occur. However, the following considerations should be kept in mind.

1. To minimize the steady-state error, the open-loop gain should be kept as high as possible in the low-frequency range of the Bode plot.
2. At intermediate frequencies (near the gain crossover frequency), a slope of -20 dB/decade in the gain curve will help to provide an adequate phase margin.
3. At high frequencies, small gain is desirable to attenuate high-frequency disturbances, such as electronic noise, or mechanical vibrations induced by gear teeth, shaft elasticity, or hydraulic and pneumatic pressure fluctuations, and so forth.

EXAMPLE 12.2.1

Design of an Integral Control System

■ Problem

For the integral control system shown in Figure 12.2.6a, the parameter values are $A = 0.02$, $\tau_1 = 0.01$ sec, and $\tau_2 = 0.02$ sec. Adjust the integral gain to obtain an overshoot of no more than 10%.

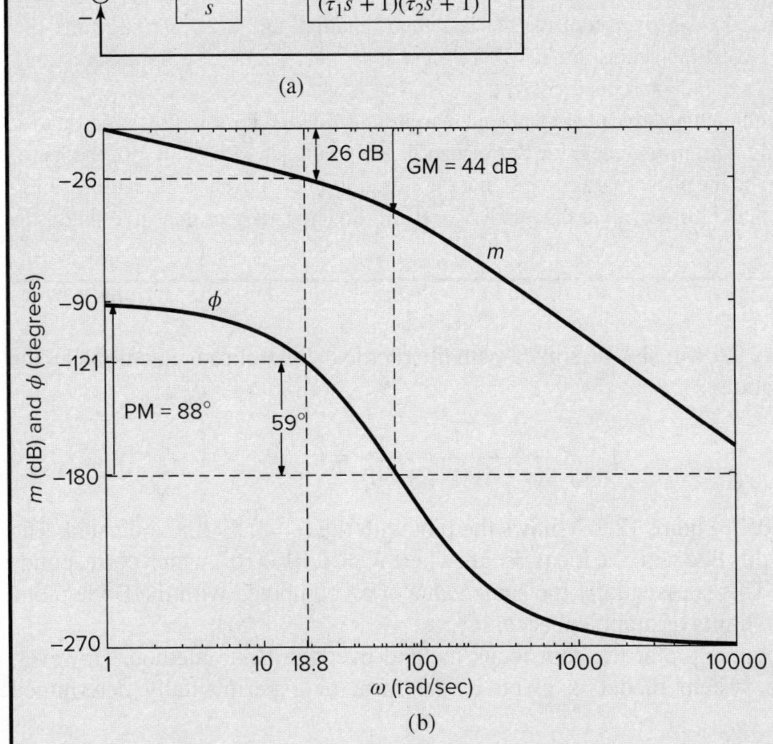

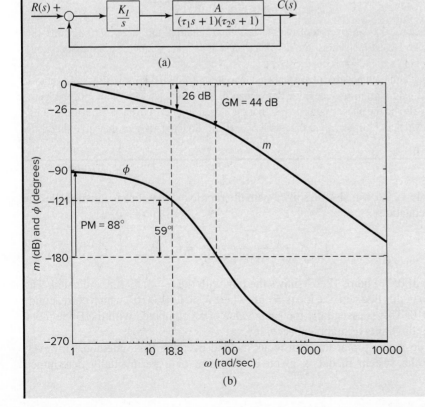

Figure 12.2.6 (a) An integral control system. (b) Bode plots for $K_I = 50$.

■ **Solution**

For the given values, the open-loop transfer function is

$$G(s) = 0.02K_I \frac{1}{s(0.01s + 1)(0.02s + 1)} \tag{1}$$

For an overshoot of 10%, assuming that the dominant roots of the final design are a complex pair, the damping ratio must be

$$\zeta = -\frac{\ln 0.1}{\sqrt{\pi^2 + \ln^2 0.1}} = 0.59$$

Using (12.2.7), we compute the required phase margin as follows, again assuming that the dominant roots of the final design are a complex pair.

$$\text{PM} = \tan^{-1} \frac{1.1823}{\sqrt{-0.6989 + 1.22}} = \tan^{-1} 1.6378 = 59°$$

The next step is to draw the open-loop Bode plots for an arbitrary value of K_I, say $K_I = 50$, which gives $0.02K_I = 1$. The Bode plots are shown in Figure 12.2.6b. The gain margin is 44 dB and the phase margin is 88° (these values can be determined easily with the MATLAB `margin` function; see Section 12.3). To achieve a phase margin of 59°, the phase curve must cross $59° - 180° = -121°$ at the new gain crossover frequency. By moving the cursor along the MATLAB plot, we can see that the phase equals $-121°$ at $\omega = 18.8$ rad/sec, which must be the new gain crossover frequency. To achieve this, the gain curve must be shifted up by 26.2 dB, which is accomplished by changing the value of K_I from $K_I = 50$ to $K_I = 50(10^{26.2/20}) = 1021$. This value gives a gain margin of 17.3 dB and a phase margin of 59°.

The final step is to check the results by simulation. The closed-loop transfer function is

$$T(s) = \frac{1021(0.02)}{0.0002s^3 + 0.03s^2 + s + 1021(0.02)}$$

The step response has an overshoot of 9%, less than required, and a 2% settling time of 0.208 sec. The closed-loop roots are $s = -114$ and $s = -18 \pm 23.9j$, so the assumption regarding complex dominant roots is correct.

Note that, although positive phase and gain margins are required for stability, large values of phase and gain margin may degrade performance. For example, with $K_I = 50$, the gain margin is 44 dB and the phase margin is 88°, but the closed-loop system has a 2% settling time of 3.82 sec, which is 18 times larger than with $K_I = 1021$. So large margins may give sluggish response.

Example 12.2.1 can also be solved with the root locus plot. From equation (1), the root locus equation is

$$1 + K\frac{1}{s(s + 100)(s + 50)} = 0$$

where $K = 100K_I$. Figure 12.2.7 shows the plot with the $\zeta = 0.59$ line indicated. The intersection of this line with the locus occurs where $K = 1.10 \times 10^5$, which corresponds to $K_I = 1010$. This is essentially the same value of K_I computed with the Bode plots, considering the limits of graphical accuracy.

Some designers prefer the root locus method over the Bode method. However, sometimes the system model is given in the form of experimentally determined

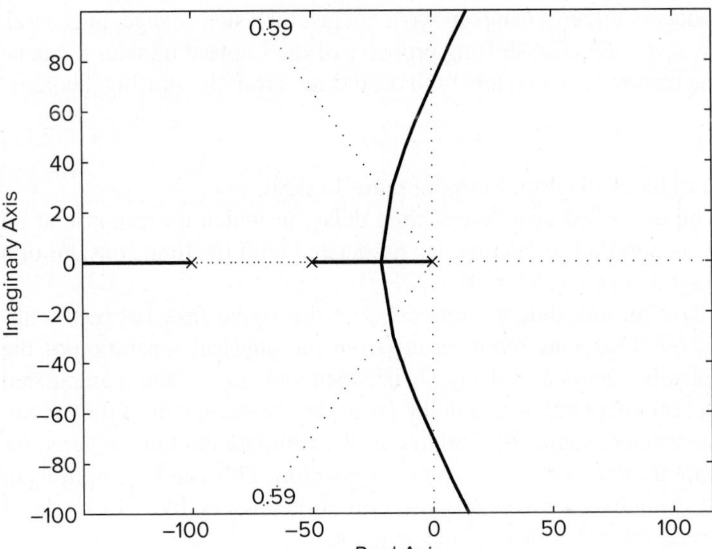

Figure 12.2.7 Root locus plot for Example 12.2.1.

open-loop frequency response curves or, as the next example shows, the model contains dead time. In such cases, the Bode plot approach is much more convenient to use.

12.2.6 DEAD-TIME ELEMENTS

Dead time is a time delay between an action and its effect. It occurs, for example, when a fluid flows through a pipe. Suppose the fluid velocity v is constant with time. The pipe length is L, so it takes a time $D = L/v$ for the fluid to move from one end to the other. Let $x_1(t)$ denote the incoming fluid temperature and $x_2(t)$ the temperature of the fluid leaving the pipe. Suppose that the temperature of the incoming fluid suddenly increases at time t_1. If this is modeled as a step function, the result is shown in Figure 12.2.8a. If no heat energy is lost, then $x_2(t)$, the temperature at the output, is $x_1(t - D)$. Thus, a time D later, the output temperature suddenly increases.

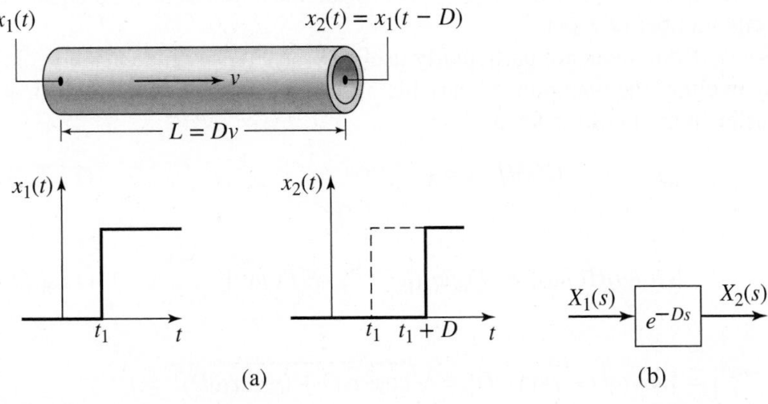

Figure 12.2.8 (a) A dead-time element. (b) Block diagram of a dead-time element.

A similar effect occurs for any change in $x_1(t)$, not just for a step change. In general, we can write $x_2(t) = x_1(t - D)$. The shifting property of the Laplace transform can be used to determine the response of a system with dead time. From the shifting theorem,

$$X_2(s) = e^{-Ds}X_1(s) \tag{12.2.12}$$

This result is shown in block diagram form in Figure 12.2.8b.

Dead time can be described as a "pure" time delay, in which no response at all occurs for a time D, as opposed to the time lag associated with the time constant of a response, for which $x_2(t) = (1 - e^{-t/\tau})x_1(t)$.

Some systems have an unavoidable time delay in the signal flow between components (Figure 12.2.9). The delay often results from the physical separation of the components and typically occurs as a delay D_1 between a change in the manipulated variable and its effect on the plant, or as a delay D_2 in the measurement of the output. Another, perhaps unexpected, source of dead time is the computation time required for a digital control computer to calculate the control algorithm. This can be a significant dead time in systems using inexpensive and slower microprocessors. It can be modeled as the delay D_1 between the controller and the actuator.

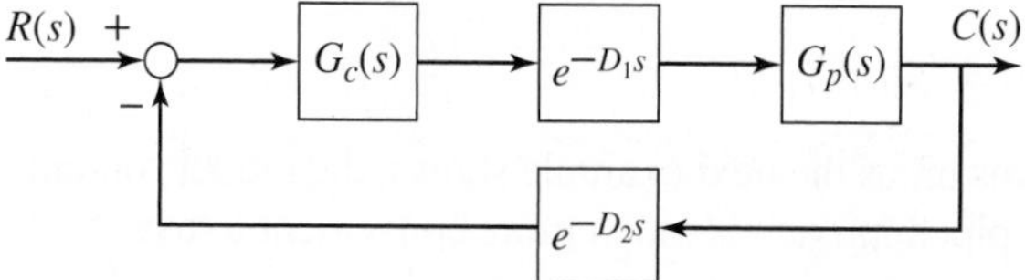

Figure 12.2.9 A system having dead-time elements in the forward path and in the feedback loop.

The presence of dead time means that the system does not have a characteristic equation of finite order. In fact, there are an infinite number of characteristic roots for a system with dead time. This can be seen by noting that the term e^{-Ds} can be expanded in an infinite series as

$$e^{-Ds} = \frac{1}{e^{Ds}} = \frac{1}{1 + Ds + (D^2s^2/2) + \cdots} \tag{12.2.13}$$

For example, if $D_1 = 5$, $D_2 = 0$, $G_c(s) = K_P$, and $G_p(s) = 1/(s + 4)$ in Figure 12.2.9, the closed-loop transfer function is

$$\frac{C(s)}{R(s)} = \frac{K_P e^{-5s}}{s + 4 + K_P e^{-5s}}$$

The characteristic equation is $s + 4 + K_P e^{-5s} = 0$, which is a transcendental equation having an infinite number of roots.

The open-loop Bode plots are particularly useful for systems with dead-time elements. A delay in either the manipulated variable or the measurement will result in an open-loop transfer function of the form

$$G(s)H(s) = e^{-Ds}P(s) \tag{12.2.14}$$

For this case,

$$|G(j\omega)H(j\omega)| = |P(j\omega)||e^{-j\omega D}| = |P(j\omega)| \tag{12.2.15}$$

because

$$|e^{-j\omega D}| = |\cos \omega D - j \sin \omega D| = \sqrt{\cos^2 \omega D + (-\sin \omega D)^2} = 1$$

The dead time therefore does not affect the open-loop gain curve. This makes the analysis of its effects easier to accomplish with the open-loop frequency response plot.

However,

$$\angle[G(j\omega)H(j\omega)] = \angle[P(j\omega)e^{-j\omega D}] = \angle P(j\omega) + \angle e^{-j\omega D}$$
$$= \angle P(j\omega) - \omega D \tag{12.2.16}$$

where the angles are in radians and ω is the radian frequency. Thus, the dead time decreases the phase proportionally to the frequency ω.

The following example shows how these methods can be applied to a system containing dead time, whose plant transfer function cannot be obtained analytically.

A Control System with Dead Time

EXAMPLE 12.2.2

■ Problem

The frequency response for a particular plant $G_p(s)$ was determined experimentally, and the results are shown in Figure 12.2.10a, where the gain curve is denoted by m_1. It is intended to use proportional control for this plant with a unity feedback loop, as shown in Figure 12.2.10b. It is known that there will be dead time D between the controller action and its effect on the plant.

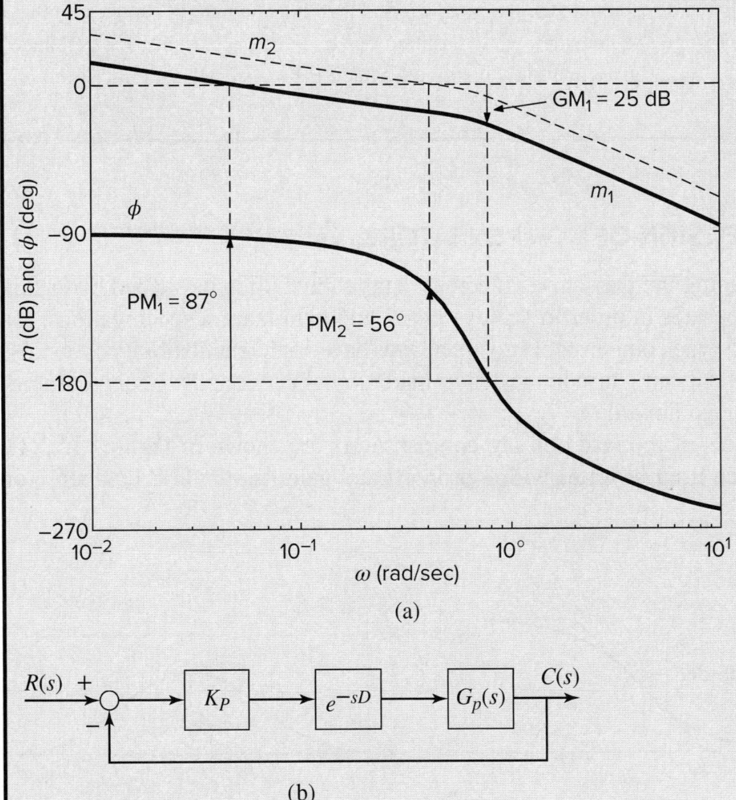

Figure 12.2.10
(a) Measured frequency response of a plant.
(b) A proportional control system having dead time.

a. Design the controller to achieve the specifications given by GM $\geq$ 8 dB and PM $\geq 30°$ if $D = 0$.

b. How large can the dead time be before the system becomes unstable?

■ Solution

a. First neglect the dead time and design the controller. With $D = 0$, the open-loop transfer function is $K_P G_p(s)$. The plot in Figure 12.2.10a, which is for $K_P = 1$, shows that the phase margin is 87° and the gain margin is 25 dB. Increasing K_P above unity affects only the gain curve. This curve can be raised by 25 − 8 = 17 dB without violating the specifications given for PM and GM. Thus, for GM = 8 dB, K_P must be such that $20 \log K_P = 17$, or $K_P = 10^{17/20} = 7.08$. With this value of K_P, the phase margin becomes 56°, and the new gain crossover frequency is $\omega_g = 0.41$ rad/sec. This can be seen by translating the gain curve upward by 17 dB. The new gain curve is labeled m_2 on the plot.

b. The dead time affects only the phase curve, and stability requires that the phase margin be positive. This occurs only if the dead time's contribution to the phase curve at the new gain crossover frequency is greater than −56°. The stability requirement thus is

$$0.41D \leq (56°)\frac{\pi}{180°} \quad \text{rad}$$

or

$$D \leq 2.38 \text{ sec}$$

By affecting the phase curve, however, the dead time changes the phase crossover frequency, and thus changes the gain margin. If we subtract 2.38ω from the phase curve shown in the plot, and recompute the margins, we find that the gain margin is now slightly negative. If we reduce D somewhat, we will obtain positive or zero margins. By trial and error, the value $D = 2.1$ gives a slightly positive GM and PM = 0. So the stability limit is $D \leq 2.1$ sec.

12.2.7 BODE DESIGN OF COMPENSATORS

In addition to shaping the gain curve, the phase curve must often be shaped by means of a series compensator in order to achieve phase and gain margin specifications, for example. The PID-type compensators do not always have sufficient flexibility to do this, because they have only an s term in the denominator, but the lead and lag compensators often do have enough flexibility.

The Bode plots of the lead and lag compensators are shown in Figures 12.2.11 and 12.2.12. When used in series with a proportional gain K_P, the lead compensator

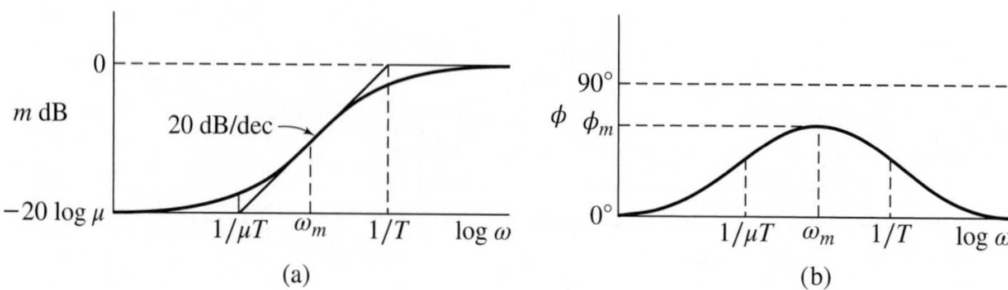

Figure 12.2.11 Bode plots of a lead compensator.

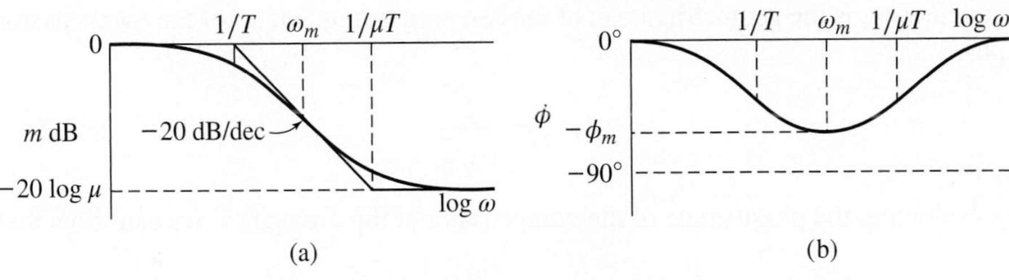

Figure 12.2.12 Bode plots of a lag compensator.

increases the phase angle and thus increases the phase margin. This enables the gain K_P/μ to be made larger than is possible without the compensator. The result is an increase in the closed-loop bandwidth and an increase in the speed of response.

On the other hand, the lag compensator is used when the speed of response and damping of the closed-loop system are satisfactory, but the steady-state error is too large. The lag compensator enables the gain to be increased without substantially changing the resonance frequency ω_r, and the resonance peak m_p of the closed-loop system.

Figure 12.2.13 Bode plots of a lag-lead compensator.

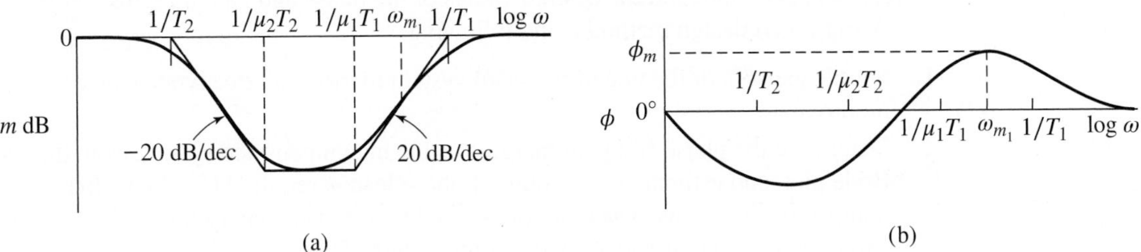

The Bode plots for the lag-lead compensator are shown in Figure 12.2.13 for $T_2 > \mu_2 T_2 > \mu_1 T_1 > T_1$. The maximum phase shift ϕ_m occurs at $\omega_{m_1} = 1/T_1\sqrt{\mu_1}$, and the attenuation at this frequency is equal in magnitude but opposite in sign to that of the lead compensator alone. Thus, the lag-lead compensator can be more effective than the lead compensator. The plots also show that the compensator affects the gain and phase in only the intermediate frequency range from $\omega = 1/T_2$ to $1/T_1$.

12.2.8 DESIGN OF LEAD COMPENSATORS

The lead compensator has the transfer function

$$G_c(s) = \frac{s + 1/\mu T}{s + 1/T}$$

For any particular value of the parameter $\mu > 1$, the lead compensator can provide a maximum phase lead ϕ_m (see Figure 12.2.11). This value, and the frequency ω_m at which it occurs, can be found as a function of μ and T from the Bode plot. The frequency ω_m is the geometric mean of the two corner frequencies of the compensator. Thus,

$$\omega_m = \frac{1}{T\sqrt{\mu}} \tag{12.2.17}$$

By evaluating the phase angle of the compensator at this frequency, we can show that

$$\sin \phi_m = \frac{\mu - 1}{\mu + 1} \tag{12.2.18}$$

or

$$\mu = \frac{1 + \sin \phi_m}{1 - \sin \phi_m} \tag{12.2.19}$$

These relations are useful in designing a lead compensator with the Bode plot.

We assume that the specifications include requirements for phase and gain margin as well as for the allowable steady-state error. The purpose of the lead compensator is to use the maximum phase lead of the compensator to increase the phase of the open-loop system near the gain crossover frequency while not changing the gain curve near that frequency. This is usually not entirely possible, because the gain crossover frequency is increased in the process, and a compromise must be sought between the resulting increase in bandwidth and the desired values of the phase and gain margins.

A suggested design method is the following:

1. Set the gain K_P of the uncompensated system to meet the steady-state error requirement.
2. Determine the phase and gain margins of the uncompensated system from the Bode plot, and estimate the amount of phase lead ϕ required to achieve the margin specifications. The extra phase lead required can be used to estimate the value for ϕ_m to be provided by the compensator. Thus, μ can be found from (12.2.19).
3. Choose T so that ω_m from (12.2.17) is located at the gain crossover frequency of the compensated system. One way of doing this is to find the frequency at which the gain of the uncompensated system equals $-20 \log \sqrt{\mu}$. Choose this frequency to be the new gain crossover frequency. This frequency corresponds to the frequency ω_m at which ϕ_m occurs.
4. Construct the Bode plot of the compensated system to see if the specifications have been met. If not, the choice for ϕ_m needs to be evaluated and the process repeated. It is possible that a solution does not exist.

The attempt to design a lead compensator can be unsuccessful if the required value of μ is too large. This can occur with plants that are not stable or have a low relative stability with a rapidly decreasing phase curve near the gain crossover frequency. In the former case, the extra phase lead required can be too large. In the latter case, the phase angle at the compensated gain crossover frequency is much less than at the uncompensated gain crossover. Thus, the extra phase required can be excessive. The difficulty with a large value of μ is that the resistance and capacitance values that result might be incompatible or impossible to obtain physically. The usual range for μ is $1 < \mu < 20$. Additional phase lead can sometimes be obtained by cascading more than one lead compensator.

In spite of these potential difficulties, the lead compensator has a record of many successful applications.

Bode Design of a Lead Compensator

EXAMPLE 12.2.3

■ Problem

The system shown in Figure 12.2.14a has an acceptable steady-state error of 0.01 for a unit-ramp command if the gain is set to $K_P = 500$ and $G_c(s) = 1$. If this is done, however, the transient performance is lightly damped and thus unsatisfactory. Design a compensator to give a gain margin of at least 8 dB and a phase margin of at least 30°.

■ Solution

The Bode plot of the open-loop uncompensated system with $K_P = 500$ is shown in Figure 12.2.14b. The phase margin is 12.8°, and the gain margin is infinite, because the phase curve never falls below −180°. Thus, 30° − 12.8° = 17.2° must be added to the phase curve to meet the specifications, so we try a lead compensator. Instead of using the compensator to add the 17.2°, we will try to add something more, say 20°, because this method is approximate. From (12.2.19), we have

$$\mu = \frac{1 + \sin \phi_m}{1 - \sin \phi_m} = \frac{1 + \sin 20°}{1 - \sin 20°} = 2.04$$

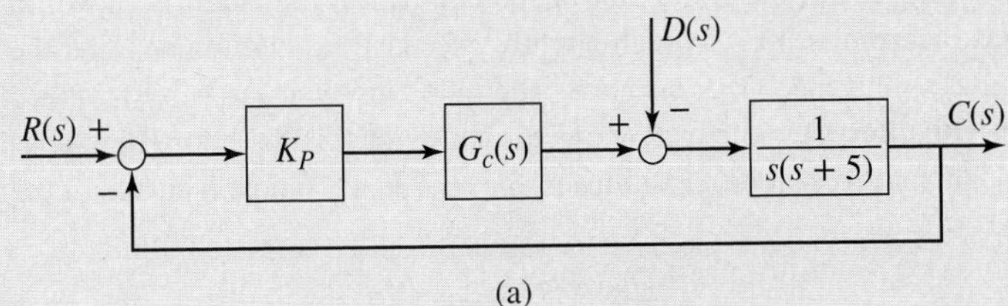

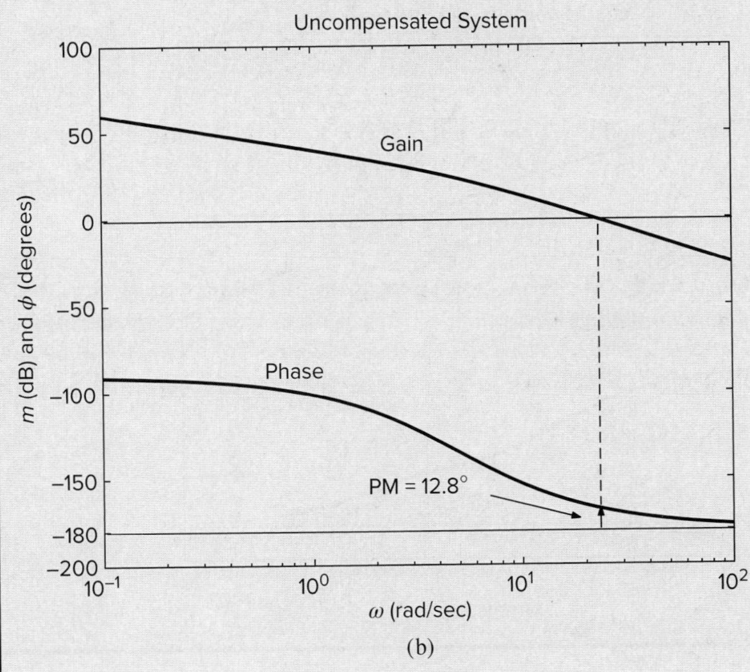

Figure 12.2.14 (a) System for Example 12.2.3. (b) Bode plots of the uncompensated system with $K_P = 500$.

From step 3, we compute

$$-20 \log \sqrt{\mu} = -3.096 \text{ dB}$$

From the plot we determine that the open-loop gain of the uncompensated system is −3.096 dB at approximately 26.5 rad/sec. Take this frequency to be ω_m, and solve for T from (12.2.17):

$$T = \frac{1}{26.5\sqrt{2.04}} = 0.0264$$

These values of μ and T give a compensator with the transfer function

$$G_c(s) = \frac{s + 1/\mu T}{s + 1/T} = \frac{s + 18.55}{s + 37.88} = 0.4897\frac{0.0539s + 1}{0.0264s + 1}$$

The open-loop transfer function of the compensated system is

$$K_P G_c(s)G(s)H(s) = 0.4897K_P\frac{0.0539s + 1}{0.0264s + 1}\frac{1}{s(s + 5)}$$

The original choice of $K_P = 500$ no longer gives required error because the compensator introduces the factor 0.4897. Thus, we must choose $K_P = 500/0.4897 = 1021$ to achieve an error of 0.01. With this value of K_P, the Bode plot of the compensated system is shown in Figure 12.2.15. The phase margin is 30.7°, which slightly exceeds the requirement, and the gain margin is still infinite, so the specifications have been met. Now you see why we added some extra phase angle. This illustrates the trial-and-error nature of this method. The guides listed in the design steps are based on approximations of the real phase and gain curves of the compensator.

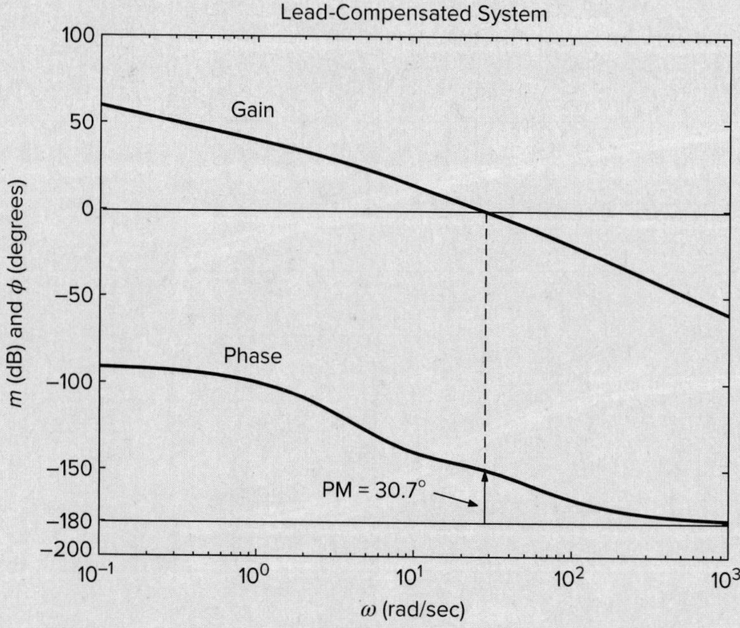

Figure 12.2.15 Bode plots of the compensated system.

12.2.9 DESIGN OF LAG COMPENSATORS

If $\mu < 1$, the following transfer function represents a lag compensator.

$$G_c(s) = \mu \frac{s + 1/\mu T}{s + 1/T}$$

Lag compensation uses the high-frequency attenuation of the compensator to keep the phase curve unchanged near the gain crossover frequency while this frequency is lowered. A suggested design procedure is the following:

1. Set the open-loop gain K_P of the uncompensated system to meet the steady-state error requirements.
2. Construct the Bode plots for the uncompensated system, and determine the frequency at which the phase curve has the desired phase margin. Determine the number of decibels required at this frequency to lower the gain curve to 0 dB. Let this amount be $m' > 0$ dB and this frequency be ω'_g. Then μ is found from

$$\mu = 10^{-m'/20} \tag{12.2.20}$$

3. The first two steps alter the phase curve. However, this curve will not be changed appreciably near ω'_g if T is chosen so that $\omega'_g \gg 1/\mu T$. A good choice is to place the frequency $1/\mu T$ one decade below ω'_g. Any larger separation might result in a system with a slow response.
4. Construct the Bode plots of the compensated system to see if all the specifications are met. If not, choose another value for T and repeat the process.

As with lead compensation, this procedure is one of trial and error and might not work. The physical elements must be realizable, so a common range for μ is $0.05 < \mu < 1$. The compensator introduces a lag angle that is not accounted for in the preceding procedure. To account for this effect, the designer might add 5° to 10° to the specified phase margin before starting the design process.

EXAMPLE 12.2.4

Bode Design of a Lag Compensator

■ Problem

Consider the system shown in Figure 12.2.16a. When $K_P = 12.5$ and $G_c(s) = 1$, the characteristic roots are $s = -2.5 \pm 2.5j$. Suppose that the transient response given by these roots is acceptable, but we want the steady-state error due to a ramp input to be no greater than 0.01. Setting $K_P = 500$ will accomplish this but will alter the transient performance by giving lightly damped roots at $s = -2.5 \pm 22.22j$, for which $\zeta = 0.118$. Design a compensator to improve the system so that the error will be no greater than 0.01, PM $\geq 30°$, and GM ≥ 8 dB.

■ Solution

A lag compensator is indicated because the transient response is satisfactory. Set $K_P = 500$ to achieve the desired error and construct the Bode plots for the uncompensated open-loop system using $K_P = 500$. The plot is the same as Figure 12.2.14b. Keeping in mind the lag introduced by the compensator, we attempt to achieve a phase margin of 35° rather than the required 30°.

From the plot we determine that the uncompensated system would have a phase margin of 35° if the gain crossover occurs at $\omega = 7.2$. Thus, $\omega'_g = 7.2$. The gain at this frequency is 18 dB, so $m' = 18$, and from (12.2.20),

$$\mu = 10^{-18/20} = 0.126$$

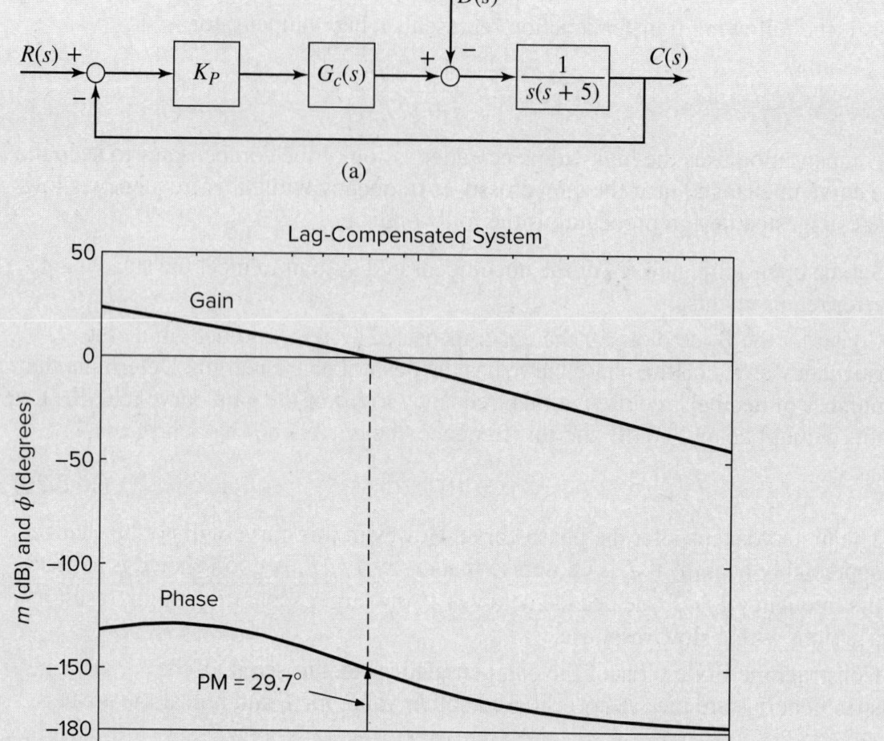

Figure 12.2.16 (a) System for Example 12.2.4. (b) Bode plots of the compensated system.

For step 3, we choose T to place $\omega = 1/\mu T$ one decade below $\omega = 7.2$. Thus, $1/\mu T = 0.72$, which gives $T = 11.023$. The lag compensator that results is

$$G_c(s) = 0.126\frac{s + 0.72}{s + 0.0907}$$

The open-loop transfer function of the compensated system is

$$K_P G_c(s)G(s)H(s) = 0.126\frac{s + 0.72}{s + 0.0907}\frac{K_P}{s(s + 5)}$$

With $K_P = 500/0.126 = 3970$, the ramp error is the required value of 0.01. The resulting Bode plots are given in Figure 12.2.16b. The phase margin is 29.7°, which is approximately 30°, and the gain margin is infinite, so the specifications have essentially been met.

At a quick glance it might appear that Examples 12.2.3 and 12.2.4 are concerned with solving the same problem; namely, compensate the plant $1/s(s + 5)$ so that the ramp error is 0.01, PM $\geq$ 30°, and GM $\geq$ 8 dB. This, however, is not the case. The difference lies in what is taken as satisfactory in the performance of the uncompensated system and what must be improved. In Example 12.2.3, the transient response resulting from the gain $K_P = 500$ needed to obtain a ramp error of 0.01 was judged to be poor, and the lead compensator was used to improve the transient performance. In Example 12.2.4, acceptable transient response was obtained when $K_P = 12.5$, but the steady-state error was then unacceptable, and a lag compensator was therefore used.

Another way to view the differences between the two examples is to note that the compensation in Example 12.2.3 was obtained by shifting the phase curve, whereas in Example 12.2.4 it was obtained by shifting the gain curve.

12.3 MATLAB APPLICATIONS

In addtion to the `bode` function, MATLAB provides the `margin` function that computes the phase and gain margins. As we will see, the `tf` function can accept models with dead time, and MATLAB also provides the `pade` function for analyzing such models.

12.3.1 THE `margin` FUNCTION

A relevant function we have not yet introduced is the `margin` function. The `margin(sys)` function plots the open-loop Bode plots with the gain and phase margins marked with a vertical line, and their values displayed at the top of the plot, along with the values of the crossover frequencies.

The syntax `[GM,PM,wg,wp] = margin(sys)` computes the gain margin `GM` in *absolute* units (*not* in decibels!), the phase margin `PM` in *degrees*, and `wg`, `wp`, which are the crossover frequencies ω_g and ω_p, in *radians* per unit time. To obtain the gain margin in decibels, use `GMdB = 20*log10(GM)`. If there are several crossover points, `margin` returns the smallest margins (the gain margin nearest to 0 dB and the phase margin nearest to 0°).

For example, consider the transfer function

$$\frac{1}{0.0002s^3 + 0.03s^2 + s}$$

In MATLAB you enter

```
>>sys=tf(1,[0.0002,0.03,1,0]);
>>margin(sys)
```

You will see on the screen a plot like that shown in Figure 12.3.1.
If you enter

```
>>sys=tf(1,[0.0002,0.03,1,0]);
>>[GM,PM,wg,wp]=margin(sys)
```

you obtain the results `GM=150`, `PM=88.2818`, `wg=70.7107`, and `wp=0.9997`, which agree with those shown at the top of the plot since `GM=150` is equivalent to $20\log_{10} 150 = 43.5218$ decibels.

The syntax `[GM,PM,wg,wp] = margin(mag,phase,w)` derives the gain and phase margins from the magnitude, phase, and frequency structures `mag`, `phase`, and `w` produced by the `bode` function. Interpolation is performed between the frequency points to estimate the values.

In addition to the `bode` and `margin` functions, MATLAB also provides the `nyquist` function for generating Nyquist plots. For example, the plot of

$$G(s)H(s) = \frac{1}{s-1}$$

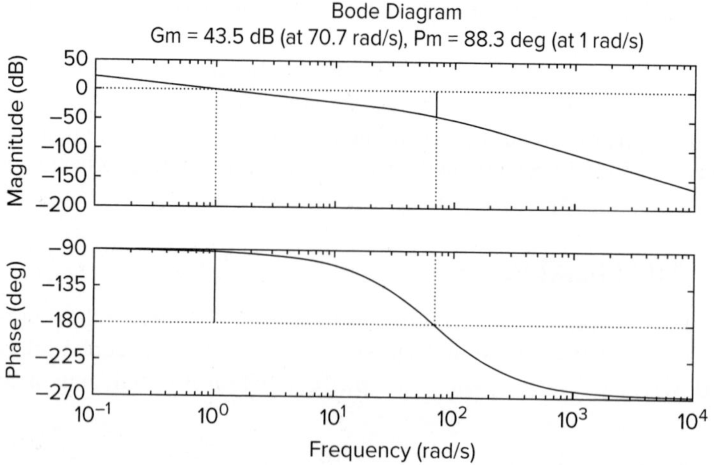

Figure 12.3.1 Bode plot showing the gain and phase margins for the transfer function $1/[0.0002s^3 + 0.03s^2 + s]$.

is obtained by typing

```
sys = tf(1,[1,-1]);
nyquist(sys)
```

The plot is shown for all negative and positive frequencies. To show only the positive frequencies, right-click on the plot, select Show, and uncheck Negative Frequencies. You can also specify a range of frequencies. See the documentation for details.

12.3.2 MODELS WITH DEAD TIME

To create a transfer-function model having dead time D at its input, use the `'iodelay'` property with the `tf` function as follows:

```
sys = tf(num,den,'iodelay',D)
```

For example, to create the model

$$e^{-2s}\frac{5}{s^2 + 4s + 2}$$

type `sys = tf(5,[1,4,2],'iodelay',2)`.

The `feedback` function cannot accept time-delay models, but `bode`, `margin`, `step`, and others will. For example, to determine the margins for the previous model having dead time, type `[GM,PM,wg,wp] = margin(sys)`. This generates a warning that the system is unstable, and returns the values: GM = 0.7795, PM = −47.4843°, ω_g = 0.9325 rad/sec, and ω_p = 1.2449 rad/sec. To obtain GM in dB, type `20*log10(GM)`. The answer is −2.1634 dB.

Dead time in a system can cause oscillatory behavior or even instability in a system that is non-oscillatory and stable without dead time. For example, proportional control of the plant $1/(s + 4)$ using $K_P = 9$ gives the closed-loop transfer function $9/(s + 13)$, which is stable and has a non-oscillatory step response. If there is dead time between the plant and the controller, then the open-loop transfer function is

$$e^{-Ds}\frac{9}{s+4}$$

To obtain the closed-loop transfer function, note that the `feedback` function does not support dead-time models (at this time), so we must first convert the dead-time model into discrete-time form. The function `c2d`, which stands for "continuous to discrete," converts a continuous-time model to discrete-time form. The syntax is `sysd = c2d(sysc,T,method)`, where `sysc` is the continuous-time model, `sysd` is the discrete-time model, `T` is the sample time, and `method` is a string specifying the method to be used. The default is `'zoh'` when `method` is omitted. This uses the zero-order hold method discussed in Chapter 10, Section 10.3.

To generate and compare the step responses of the models with and without dead time, use the following script file. The sample time T should be chosen to be a small fraction of the smallest estimated time constant of the closed-loop system. Here we use $T = 0.01$.

```
KP = 9;
% Create the discrete-time model.
sysA = tf(KP,[1,4],'iodelay',0.5);
sysB = c2d(sysA,0.01);
sysC = feedback(sysB,1);
% Create the continuous-time model.
sysD = tf(KP,[1, 4]);
sysE = feedback(sysD,1);
step(sysC,sysE,3)
```

The step response is shown in Figure 12.3.2 after being edited with the Plot Editor. Note that the dead time of $D = 0.5$ makes the system unstable and oscillatory.

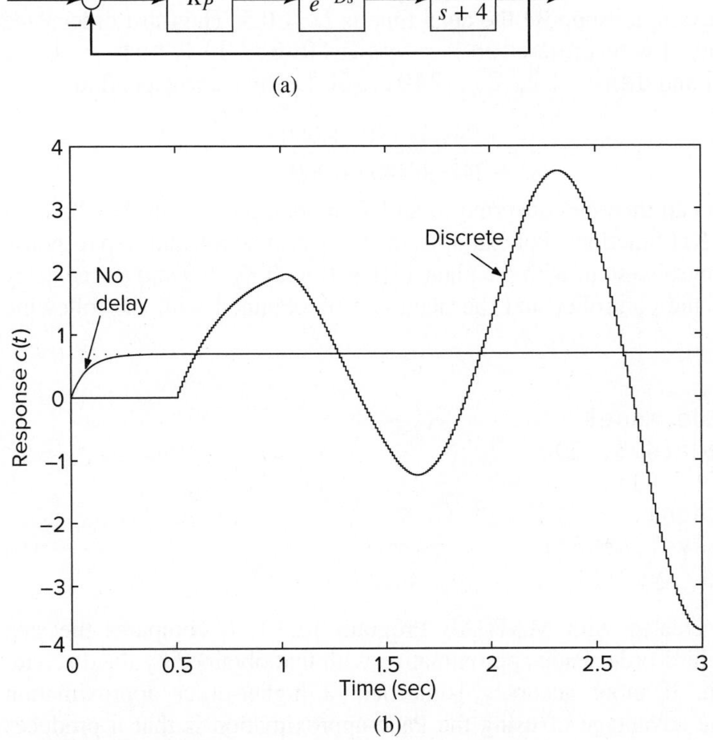

Figure 12.3.2 (a) A control system having dead time between the controller and the plant. (b) Responses of systems with and without dead time between the controller and the plant. The response with a dead time of 0.5 sec was obtained with a discrete-time approximation of the dead-time element.

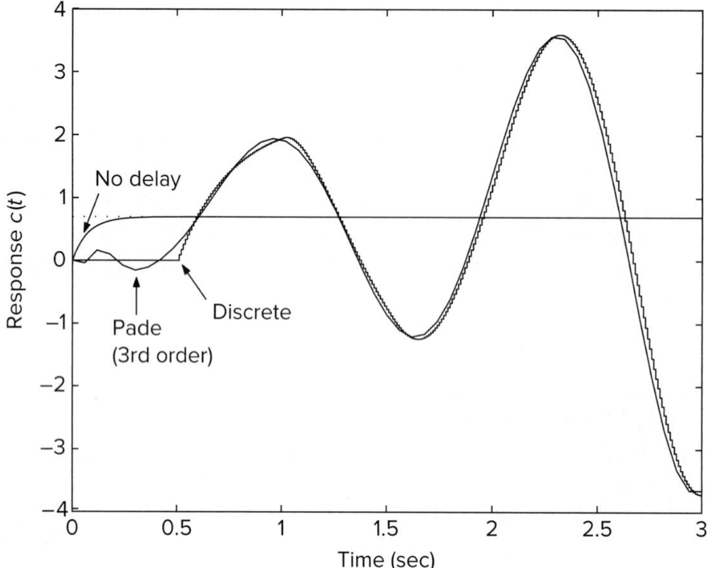

Figure 12.3.3 Comparison of the response computed with the Pade and the discrete-time approximations of a 0.5-sec time delay.

MATLAB also provides the `pade` function for analyzing dead-time systems. The Pade approximation to the dead-time element is

$$e^{-sD} = \frac{1}{[1 + (Ds/n)]^n}$$

The approximation improves as n becomes larger. The syntax `[num,den] = pade(D,n)` returns the nth order transfer-function approximation of the dead-time transfer function e^{-Ds}. For example, suppose the dead time is $D = 0.5$. The third-order Pade approximation obtained with `[num,den] = pade(0.5, 3)` is `num = [-1, 24, -240, 960]` and `den = [1, 24, 240, 960]`, which correspond to

$$e^{-0.5s} \approx \frac{-s^3 + 24s^2 - 240s + 960}{s^3 + 24s^2 + 240s + 960}$$

This approximation can then be converted to an LTI model, which can then be used with the MATLAB LTI functions. For example, the transfer function and step response of a proportional control system with the plant $1/(s + 4)$ with $K_P = 9$ and a time delay of 0.5 sec between the controller and the plant can be obtained with the following script file.

```
KP = 9;
%Create the Pade model.
[num,den] = pade(0.5, 3);
sysD = tf(KP,[1, 4]);
sys1 = tf(num,den);
sys2 = series(sys1,sysD);
sys3 = feedback(sys2,1)
```

Figure 12.3.3 was created with MATLAB Program 12.3.1. It compares the step response using the third-order Pade approximation with that obtained by the discrete-time approximation. If more accuracy is needed, a higher-order approximation should be used. The advantage of using the Pade approximation is that it produces a continuous-time transfer function that, unlike the discrete-time approximation, can always be used with other continuous-time models. Its disadvantage is that it might produce very high-order models with resulting numerical inaccuracy.

MATLAB Program 12.3.1

```
KP = 9;
% Create the discrete model (sysC).
sysA = tf(KP,[1,4],'iodelay',0.5);
sysB = c2d(sysA,0.01);
sysC = feedback(sysB,1);
% Create the continuous-time model (sysE).
sysD = tf(KP,[1,4]);
sysE = feedback(sysD,1);
% Create the Pade model (sys3).
[num,den] = pade(0.5,3);
sys1 = tf(num,den);
sys2 = series(sys1,sysD);
sys3 = feedback(sys2,1);
% Obtain the step response data for all three models.
[yC,tC] = step(sysC,3);
[yE,tE] = step(sysE,3);
[y3,t3] = step(sys3,3);
% Plot and label the responses.
plot(tC,yC,tE,yE,t3,y3),xlabel('Time(sec)'),...
   ylabel('Response c(t)'),gtext('No delay'),...
   gtext('Discrete'),gtext('Pade(3rd order)')
```

12.3.3 SIMULATING THE PROCESS REACTION METHOD

Once the gains have been computed from the Ziegler-Nichols process reaction method, it is difficult to tune the controller by simulation because a transfer function or differential equation model of the plant is not available. Often, however, a model having a dead time can be used as an approximation. From a plot of the data given in Section 11.3, the steady-state response is 93, and we estimate that the settling time of the process is $12 - L = 12 - 1.5 = 10.5$ min, and thus the time constant is $\tau = 10.5/4 = 2.63$ min. Thus, a possible model for a transfer-function description of the plant is

$$G_p(s) = e^{-Ls}\frac{93}{\tau s + 1} \tag{12.3.1}$$

The step response of this model, however, has an infinite slope at $t = L$. Perhaps a better model is

$$G_p(s) = e^{-Ls}\frac{93}{ms^2 + cs + 1} \tag{12.3.2}$$

After some experimentation, we find that a good approximation is given by the second model with $\zeta = 0.707$ and $\tau = 2.63$, which give $m = 3.44$ and $c = 2.61$. Figure 12.3.4 shows the response of this model and the original response data.

The MATLAB functions used in this section are summarized in Table 12.3.1.

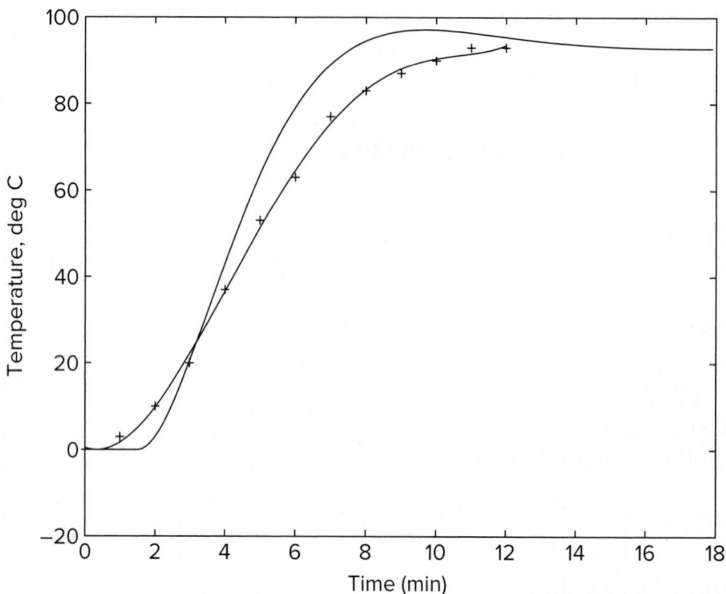

Figure 12.3.4 Open-loop response data and its dead-time approximation.

Table 12.3.1 MATLAB functions used in this section.

Functions introduced in this section	
`sysd = c2d(sysc,ts)`	Uses the zero-order hold method to convert the continuous-time system `sysc` to a discrete-time system `sysd` using the sampling time `ts`.
`[num,den] = pade(T,N)`	Returns the transfer function of the Nth-order Pade approximation of the continuous time-delay transfer function e^{-Ts}.
Functions introduced in earlier sections	
`bode(sys)`	Draws the Bode plot of the LTI object `sys` (Section 9.6).
`sys3 = feedback(sys1,sys2)`	Computes the closed-loop LTI model `sys3` for the negative feedback loop containing `sys1` in the forward path and `sys2` in the feedback path (Section 5.1).
`gtext('text')`	Enables placement of text on a plot using the cursor (Section 1.6).
`plot(x,y)`	Creates a two-dimensional plot on rectilinear axes (Section 1.6).
`sys3 = series(sys1,sys2)`	Creates the LTI object `sys3` as the series combination of `sys1` and `sys2` (Section 5.1).
`step(sys)`	Computes and plots the step response of a linear model `sys` (Section 2.9).
`sys = tf(num,den)`	Creates a model `sys` in transfer-function form having the numerator `num` and denominator `den` (Section 2.9).
`xlabel`	Puts a label on the abscissa of a plot (Section 1.6).
`ylabel`	Puts a label on the ordinate of a plot (Section 1.6).

The methods of this section can then be used to obtain the closed-loop transfer function for simulation and tuning of the control gains. Another way to do this is to use Simulink. We will show how this is done in Section 12.4.

12.4 SIMULINK APPLICATIONS

Simulink is especially well suited to analyzing models with dead time. This section illustrates some of these applications.

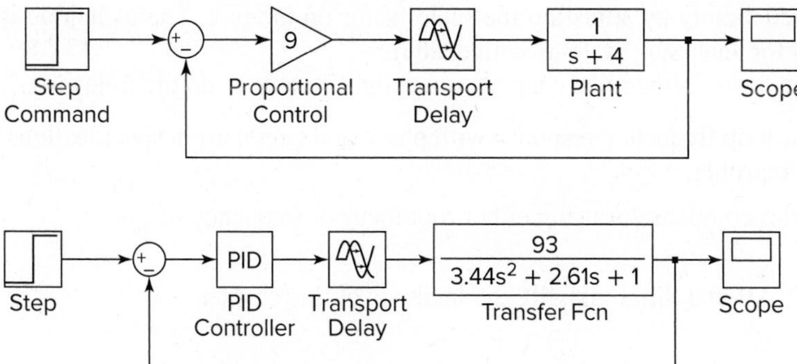

Figure 12.4.1 Simulink model using a Transport Delay block.

Figure 12.4.2 Illustration of the use of the PID Controller block.

12.4.1 SIMULATION OF SYSTEMS WITH DEAD TIME

Systems having dead-time elements are easily simulated in Simulink. The block implementing the dead-time transfer function e^{-Ds} is called the "Transport Delay" block. Figure 12.4.1 shows a Simulink model for proportional control of the plant $1/(s+4)$ using $K_P = 9$. This system was analyzed in Section 12.3, with the response shown in Figure 12.3.1 for a dead time of $D = 0.5$ s. You will see this response in the Scope block of the Simulink model.

12.4.2 PROCESS-REACTION SIMULATION

In Section 12.3 we found that a good model of the measured open-loop response is

$$G_p(s) = e^{-1.5s}\frac{93}{3.44s^2 + 2.61s + 1}$$

Figure 12.4.2 shows a Simulink model incorporating this model of the plant in a PID control system. The delay in the Transport Delay block has been set to 1.5.

The new block in this model is the PID Controller block. It is useful because if we were to implement the PID controller with a Transfer-Function block, we would get an error message warning that the order of the numerator must be no greater than the order of the denominator. This is caused by the term $K_D s^2$ in the numerator. The PID block is specifically designed to handle this situation.

Simulation of the model using the gain values $K_P = 0.0524$, $K_I = 0.0170$, and $K_D = 0.0404$ computed from the Ziegler-Nichols process reaction method shows that the system is unstable. Further simulations show that it can be made stable by reducing the gains. A set giving good response is found to be one-fourth of the previous gain values; namely, $K_P = 0.0524/4$, $K_I = 0.017/4$, and $K_D = 0.0404/4$.

12.5 CHAPTER REVIEW

Series compensation involves inserting an additional control element in series with the main controller. Section 12.1 discusses several types of series compensators, the lead, lag, and lag-lead, and shows how to use the root locus to design them. Compensation often leads to a satisfactory design if the PID action is capable of satisfying most but not all of the performance specifications.

Section 12.2 shows how to use the open-loop frequency response plots to design PID controllers and series compensators using the phase and gain margins and the static error coefficients. The frequency response plots of the system's open-loop transfer function are easily generated even for high-order systems, and they enable the proper control gain to be selected simply by adjusting the scale factor on the plot. The technique is especially useful for analyzing systems with dead time.

Now that you have finished this chapter, you should be able to do the following.

1. Use the open-loop frequency response with phase and gain margin specifications to design a controller.
2. Design a series compensator using either root locus or frequency response methods.
3. Apply MATLAB and Simulink with the methods of this chapter.

PROBLEMS

Section 12.1 Series Compensation

12.1 Control of the attitude θ of a missile by controlling the fin angle ϕ, as shown in Figure P12.1, involves controlling an inherently unstable plant. Consider the specific plant transfer function

$$G_p(s) = \frac{\Theta(s)}{\Phi(s)} = \frac{1}{8s^2 - 7}$$

Design a PD compensator for this plant. The dominant roots of the closed-loop system must have $\zeta = 0.707$ and $\omega_n = 0.5$.

Figure P12.1

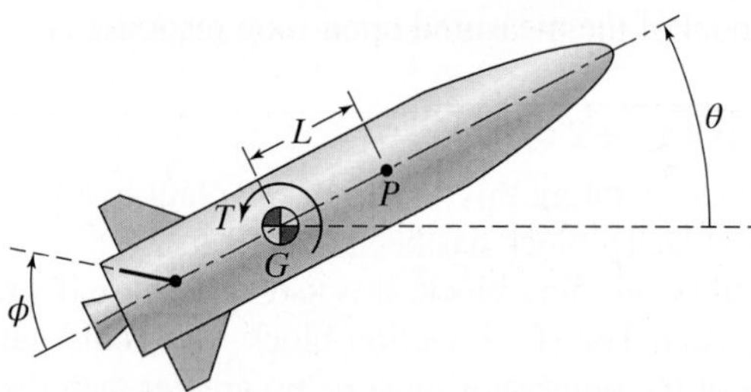

12.2 Figure P12.2 shows a pneumatic positioning system, where the displacement x is controlled by varying the pneumatic pressure p_1. Assume that the pressure p_2 is constant, and consider the specific plant

$$G_p(s) = \frac{X(s)}{P_1(s)} = \frac{K}{s^2 + 2s}$$

With $K = 4$, the damping ratio is $\zeta = 0.5$, the natural frequency is $\omega_n = 2$ rad/s, and the steady-state ramp error is 0.5.

Figure P12.2

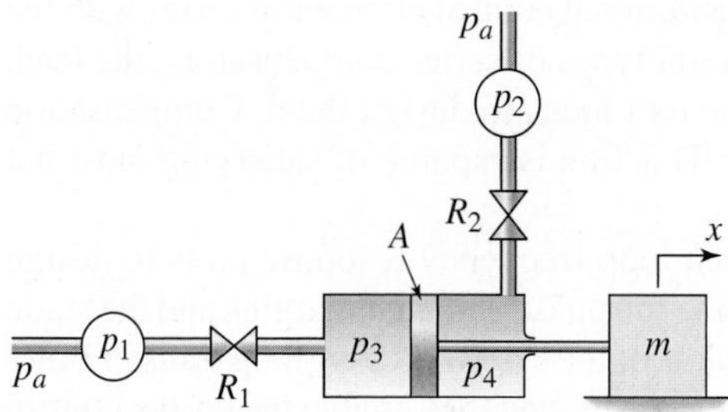

a. Design an electrical compensator to obtain $\omega_n = 4$ while keeping $\zeta = 0.5$. Obtain the compensator's resistances if $C = 1\ \mu$F.

b. Suppose that with $K = 4$, the original system gives a satisfactory transient response, but the ramp error must be decreased to 0.05. Design a compensator to do this.

12.3 It is desired to control the angular displacement θ of a space vehicle by controlling the applied torque T supplied by thrusters (Figure P12.3). The plant model is

$$G_p(s) = \frac{\Theta(s)}{T(s)} = \frac{5}{s^2}$$

Design a compensator for this plant. The system must have a settling time no greater than 4 sec and a steady-state error of zero for a step command, and the dominant roots of the closed-loop system must have $\zeta \geq 0.45$.

Figure P12.3

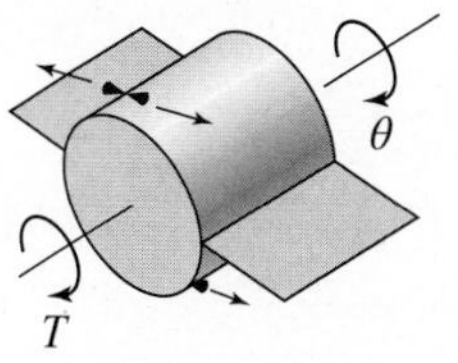

12.4 When proportional control is applied to the following plant using a gain of $K_P = 1$, the closed-loop roots are satisfactory, but the static velocity error coefficient must be increased to $C_v = 5$/sec.

$$G_p(s) = \frac{1}{s^3 + 3s^2 + 2s}$$

Design a compensator for this plant that keeps the closed-loop roots near their original locations but increases C_v to 5.

12.5 The speed ω_1 of the load is to be controlled with the torque T_d acting through a fluid coupling (Figure P12.5). Design a compensator for the specific plant

$$G_p(s) = \frac{\Omega_1(s)}{T_d(s)} = \frac{1}{s^2 + s}$$

The static velocity error coefficient must be $C_v = 10$/sec, the dominant roots of the closed-loop system must have $\zeta = 0.5$ and $\omega_n = 2$.

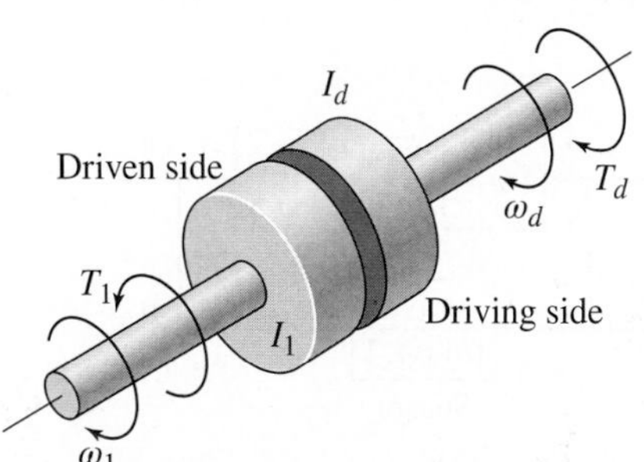

Figure P12.5

12.6 The block diagram of a speed-control system is shown in Figure P12.6. For a particular system with proportional control, $G_1(s) = K_P$, the open-loop transfer function is

$$G(s) = \frac{K_P}{s(s+2)}$$

With $K_P = 4$, the damping ratio is $\zeta = 0.5$, the natural frequency is $\omega_n =$ 2 rad/sec, and $C_v = 2$/sec. Design a compensator that will give a static velocity error coefficient of $C_v = 20$/sec, a phase margin of at least 40°, and a gain margin of at least 6 dB.

Figure P12.6

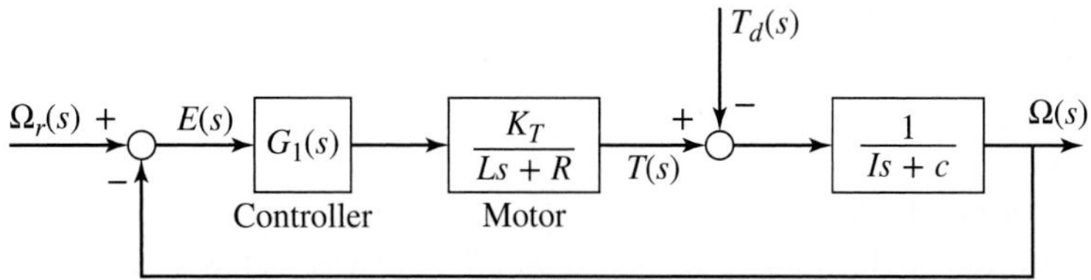

12.7 The block diagram of a position control system is shown in Figure P12.7. For a particular system with proportional control, $G_1(s) = K_P$, the open-loop transfer function is

$$G(s) = \frac{2.5K_P}{s(s+2)(0.25s+2)}$$

Design a compensator to give $s = -2 \pm j2\sqrt{3}$ and $C_v = 80$/sec.

Figure P12.7

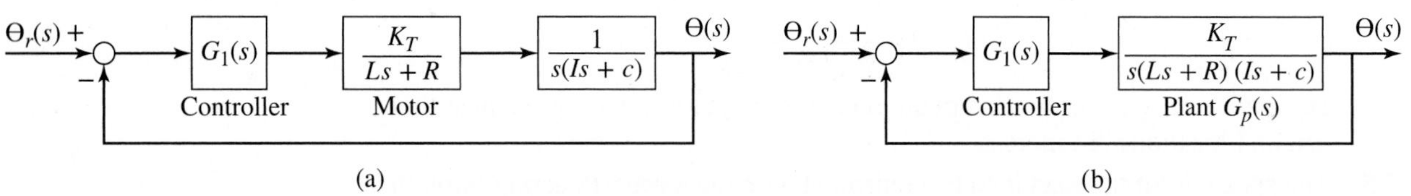

12.8 It is desired to control the angular displacement θ of a space vehicle by controlling the applied torque T supplied by thrusters (Figure P12.8a). The plant model is

$$G_p(s) = \frac{\Theta(s)}{T(s)} = \frac{10}{s^2}$$

Figure P12.8

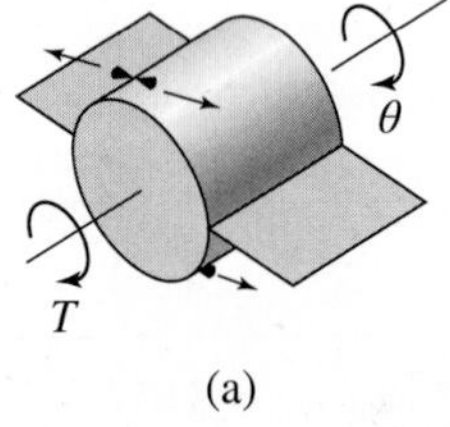

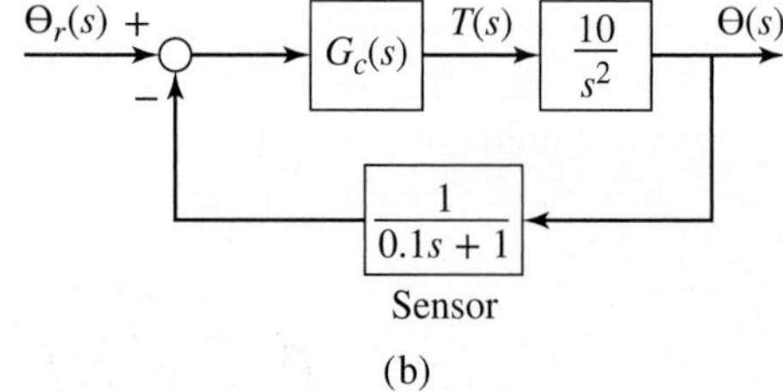

The feedback sensor that measures the displacement has a time constant of 0.1 sec (Figure P12.8b). Design a compensator so that the closed-loop system will have a time constant of 1 sec and a damping ratio of 0.707.

12.9 The plant transfer function for the angular displacement θ of an inertia I subjected to a control torque T is (see Figure P12.8a)

$$G_p(s) = \frac{\Theta(s)}{T(s)} = \frac{1}{Is^2}$$

Suppose that $I = 5$ and that the output of the controller is the torque T. Use the root locus to investigate whether or not a controller transfer function of the following form will give a settling time of no more than 2 sec and an overshoot of no more than 5%.

$$G_c(s) = K\frac{s+a}{s+b}$$

Obtain suitable values of K, a, and b. What happens if we try to "cancel" one of the poles at the origin by placing the zero very close to the origin?

12.10 Consider a plant whose open-loop transfer function is

$$G(s)H(s) = \frac{1}{s\left[(s+2)^2+9\right]}$$

The complex poles near the origin give only slightly damped oscillations that are considered undesirable. Insert a gain K_c and a compensator $G_c(s)$ in series to speed up the closed-loop response of the system. Consider the following for $G_c(s)$:

a. The lead compensator

b. The lag compensator

c. The so-called reverse-action compensator

$$G_c(s) = \frac{1-T_1 s}{T_2 s+1}$$

Obtain the root locus plots for the compensated system using each compensator. Use the compensator gain K_c as the locus parameter. Without computing specific values for the compensator parameters, determine which compensator gives the best response.

12.11 (a) The equations of motion of the inverted pendulum model were derived in Example 3.5.6 in Chapter 3. Linearize these equations about $\phi = 0$, assuming that $\dot{\phi}$ is very small. (b) Obtain the linearized equations for the following values: $M = 10$ kg, $m = 50$ kg, $L = 1$ m, $I = 0$, and $g = 9.81$ m/s^2. (c) Use the linearized model developed in part (b) to design a series compensator to stabilize the pendulum angle near $\phi = 0$. It is required that the 2% settling time be no greater than 4 s and that the response be non-oscillatory. This means that the dominant root should be real and no greater than -1. No restriction is placed on the motion of the base. Assume that only ϕ can be measured.

Section 12.2 Design Using the Bode Plot

12.12 A certain unity-feedback system has the following open-loop system transfer function.

$$G(s) = \frac{K}{s^3+6s^2+5s}$$

Obtain the Bode plots and compute the phase and gain margins for

a. $K = 10$

b. $K = 100$

c. Determine the upper limit on K for the system to be stable. Which is the limiting factor: the phase margin or the gain margin?

12.13 Figure P12.13 shows a pneumatic positioning system, where the displacement x is controlled by controlling the applied pneumatic pressure p_1. Assume that

Figure P12.13

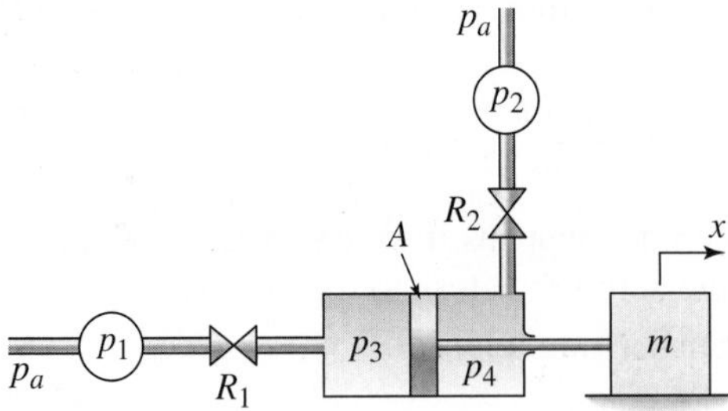

the pressure p_2 is constant, and consider the specific plant

$$G_p(s) = \frac{X(s)}{P_1(s)} = \frac{1}{100s^2 + s}$$

with the following series PD compensator is used to control the pressure.

$$G_c(s) = s + 2$$

Obtain the Bode plots for this system, and determine the phase and gain margins.

12.14 The height h_2 in Figure P12.14 can be controlled by adjusting the flow rate q_1. Consider the specific plant

$$G_p(s) = \frac{H_2(s)}{Q_1(s)} = \frac{5}{5s^2 + 6s + 5}$$

with the following series PI compensator is used to control it.

$$G_c(s) = \frac{K(s + 1)}{s}$$

a. Start with $K = 1$ and determine the phase and gain margins.

b. Increase K until the phase margin is 20° and the gain margin is positive.

Figure P12.14

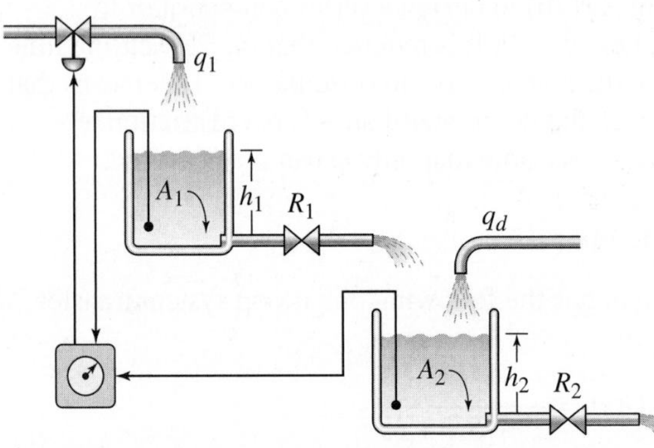

12.15 Rolling motion of a ship can be reduced by using feedback control to vary the angle of the stabilizer fins, much like ailerons are used to control aircraft roll. Figure P12.15 is the block diagram of a roll control system in which the roll angle is measured and used with proportional control action. Determine the phase and gain margins of the system if (a) $K_P = 1$, (b) $K_P = 5$, and (c) $K_P = 20$. Determine the stability properties for each case. If the system is unstable, what effect will this have on the ship roll?

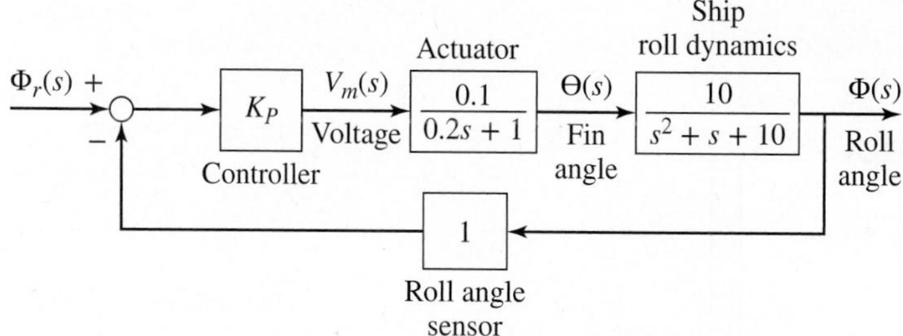

Figure P12.15

12.16 The open-loop transfer function of a certain unity-feedback system is

$$G(s) = K\frac{10s + 1}{4s^3 + 6s^2 + s}$$

Compute the value of the gain K so that the phase margin will be 60°.

12.17 The following transfer functions are the forward transfer function $G(s)$ and the feedback transfer function $H(s)$ for a system whose closed-loop transfer function is

$$\frac{G(s)}{1 + G(s)H(s)}$$

For each case, determine the system type number, the static position and velocity coefficients, and the steady-state errors for unit-step and unit-ramp inputs.

a. $$G(s)H(s) = \frac{15}{s}$$

b. $$G(s)H(s) = \frac{15}{5s + 1}$$

c. $$G(s)H(s) = \frac{5}{s^2}$$

12.18 Remote control of systems over great distance, such as required with robot space probes, may involve relatively large time delays in sending commands and receiving data from the probe. Consider a specific system using proportional control, where the total dead time is $D = D_1 + D_2 = 100$ sec (Figure P12.18). The plant is

$$G_p(s) = \frac{1}{100s + 1}$$

How large can the gain K_P be without the system being unstable?

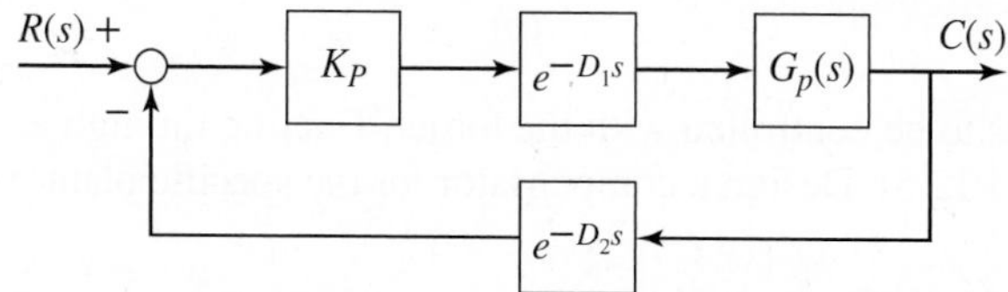

Figure P12.18

12.19 Hot-air heating control systems for large buildings may involve significant dead time if there is a large distance between the furnace and the room being

Figure P12.19

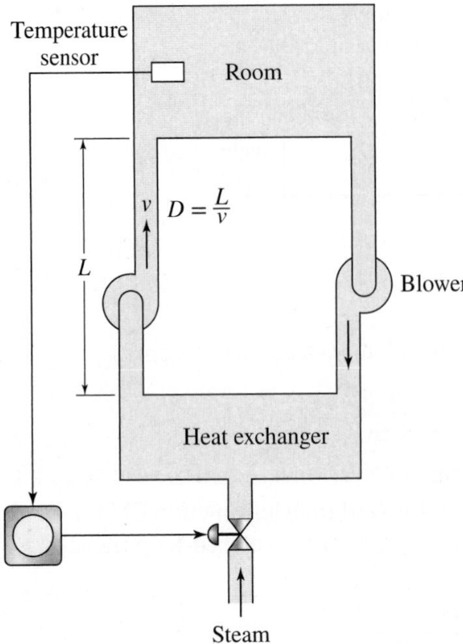

heated (Figure P12.19). Proportional control applied to the specific plant

$$G_p(s) = \frac{1}{0.1s + 1}$$

has the gain $K_P = 10$. The time units are minutes. How large can the dead time D between the controller and the plant be if the phase margin must be no less than 40°?

12.20 The block diagram of a position control system is shown in Figure P12.20. Design a compensator for the particular plant

$$G_p(s) = \frac{10}{s(s^2 + 3s + 2)}$$

so that the static velocity error coefficient will be $C_v = 5$/sec, the gain margin will be no less than 10 dB, and the phase margin no less than 40°.

Figure P12.20

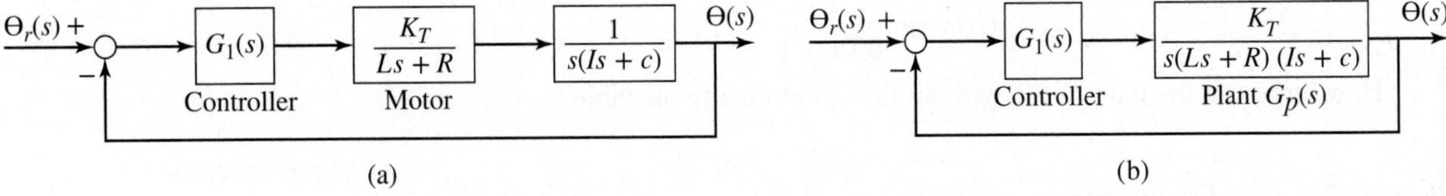

12.21 The speed ω_2 of the load is to be controlled with the torque T acting through a fluid coupling (see Figure P12.5). Design a compensator for the specific plant

$$G_p(s) = \frac{\Omega_2(s)}{T(s)} = \frac{1}{s^2 + 2s}$$

so that the static velocity error coefficient will be $C_v = 20$/sec, the gain margin will be no less than 10 dB, and the phase margin no less than 50°.

12.22 Design a compensator for the plant

$$G_p(s) = \frac{4}{s^2 + 2s}$$

so that the static velocity error coefficient will be $C_v = 20$/sec and the phase margin at least 45°.

12.23 Figure P12.2 shows a pneumatic positioning system, where the displacement x is controlled by controlling the applied pneumatic pressure p_1. Assume that the pressure p_2 is constant, and consider the specific plant

$$G_p(s) = \frac{X(s)}{P_1(s)} = \frac{5}{s^2 + s}$$

Design a compensator for the plant so that $C_v = 20$/sec, the gain margin will be no less than 10 dB, and the phase margin no less than 50°.

12.24 The block diagram of a position control system is shown in Figure P12.7. Design a compensator for the particular plant

$$G_p(s) = \frac{5}{s(s + 5)(s + 1)}$$

that will give a static velocity error coefficient of $C_v = 50$/sec and closed-loop roots with a damping ratio of $\zeta = 0.5$.

12.25 The block diagram of a position control system is shown in Figure P12.7. Design a compensator for the particular plant

$$G_p(s) = \frac{4}{s(s^2 + 3s + 2)}$$

so that the static velocity error coefficient will be $C_v = 10$/sec, the gain margin will be no less than 10 dB, and the phase margin no less than 50°.

12.26 Consider a unity-feedback system having the open-loop transfer function

$$G(s) = \frac{\omega_n^2}{s(s + 2\zeta\omega_n)}$$

Derive the following expression for this system's phase margin.

$$\text{PM} = \tan^{-1}\frac{2\zeta}{\sqrt{-2\zeta^2 + \sqrt{1 + 4\zeta^4}}}$$

12.27 Two independent feedback control systems are needed to position the pen of a flatbed *x-y* plotter. Consider only the control system for the x axis. An optical encoder provides feedback of the pen's location. Assume that the dynamics of the motor and pen assembly are described by the following plant transfer function

$$G_p(s) = \frac{X(s)}{V(s)} = \frac{0.1}{s(0.1s + 1)(0.001s + 1)}$$

where x is the pen location in meters and v is the motor voltage. Design a compensator to achieve zero steady-state error for a step input and a 2% settling time of no more than 1 s, with an overshoot of no more than 4%.

12.28 Automatic guided vehicles are used in factories and warehouses to transport materials. They require a guide path in the floor and a control system for sensing the guide path and adjusting the steering wheels. Figure P12.28 is a block diagram of such a control system. Obtain the transfer function $G_c(s)$ so that the step response has an overshoot no greater than 20% with a 2%

Figure P12.28

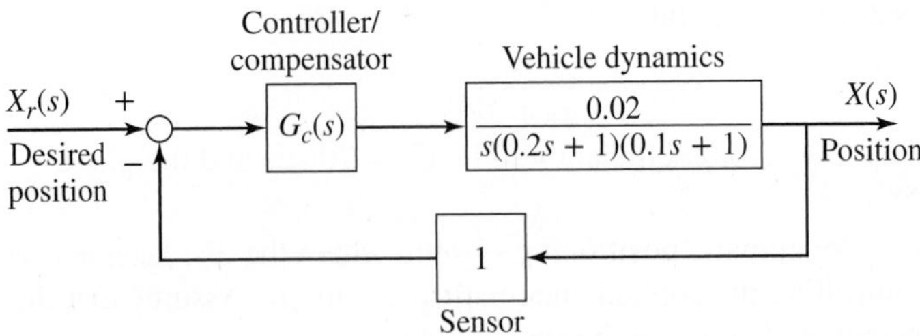

settling time of no more than 1 s, and a steady-state unit-ramp error of no more than 0.1 m.

Section 12.3 MATLAB Applications

12.29 With the PI gains set to $K_P = 6$ and $K_I = 50$ for the plant

$$G_p(s) = \frac{1}{s+4}$$

the time constant is $\tau = 0.2$ and the damping ratio is $\zeta = 0.707$.

a. Compute the gain and phase margins.

b. Suppose there is dead time $D = 0.1$ between the controller and the plant. Compute the gain and phase margins, and plot the unit-step response.

c. How large can the dead time be without the system becoming unstable?

12.30 In Example 12.1.4 a lag-lead compensator was designed by canceling the plant pole at $s = -3$ with a compensator zero. Suppose the plant model is slightly inaccurate and the plant pole is really at $s = -3.2$. Evaluate the resulting unit-step response and unit-ramp response in terms of the performance specifications.

12.31 Solve Problem 12.2 using MATLAB.

12.32 Solve Problem 12.4 using MATLAB.

12.33 Solve Problem 12.5 using MATLAB.

12.34 Solve Problem 12.6 using MATLAB.

12.35 Solve Problem 12.12 using MATLAB.

12.36 Solve Problem 12.13 using MATLAB.

12.37 Solve Problem 12.14 using MATLAB.

12.38 Solve Problem 12.15 using MATLAB.

12.39 Solve Problem 12.18 using MATLAB.

12.40 Solve Problem 12.25 using MATLAB.

12.41 Solve Problem 12.28 using MATLAB.

Section 12.4 Simulink Applications

12.42 Refer to Example 12.1.1, in which a PID controller was designed to control the plant $1/s^2$. Using the gain values calculated in that example, construct a Simulink model of the system using the PID controller block and including the disturbance torque shown in Figure 12.1.7. Use the model to obtain plots of the responses to a unit-step command and a unit-step disturbance.

12.43 Consider Problem 12.19. Use Simulink to obtain the unit-step response if the dead time is 0.01 min.

Guide to Selected MATLAB Commands and Functions

This appendix is a guide to those MATLAB commands and functions that are particularly useful for the system dynamics methods covered in this text. For more information, in the Command window type `help topic`, where `topic` is the name of the command or function. ■

Operators and special characters.

Item	Description
+	Plus. Addition operator.
-	Minus. Subtraction operator.
*	Scalar and matrix multiplication operator.
.*	Array multiplication operator.
^	Scalar and matrix exponentiation operator.
.^	Array exponentiation operator.
\	Left division operator.
/	Right division operator.
.\	Array left division operator.
./	Array right division operator.
:	Colon. Generates regularly spaced elements and represents an entire row or column.
( )	Parentheses. Encloses function arguments and array indices; overrides precedence.
[ ]	Brackets. Encloses array elements.
{ }	Braces. Encloses cell elements.
.	Decimal point.
...	Ellipsis. Line continuation operator.
,	Comma. Separates statements and elements in a row of an array.
;	Semicolon. Separates columns in an array and suppresses line feeds.
%	Percent sign. Designates a comment and specifies formatting.
'	Quote sign and transpose operator.
.'	Nonconjugated transpose operator.
=	Assignment (replacement) operator.

Logical and relational operators.

Item	Description
==	Relational operator: equal to.
~=	Relational operator: not equal to.
<	Relational operator: less than.
<=	Relational operator: less than or equal to.
>	Relational operator: greater than.
>=	Relational operator: greater than or equal to.
&	Logical operator: AND
|	Logical operator: OR
~	Logical operator: NOT

Special variables and constants.

Item	Description
`ans`	Most recent answer.
`eps`	Accuracy of floating point precision.
`i,j`	The imaginary unit $\sqrt{-1}$.
`Inf`	Infinity.
`NaN`	Undefined numerical result (Not a Number).
`pi`	The number π.

Commands for managing a session.

Item	Description
`clc`	Clears Command window.
`clear`	Removes variables from memory.
`doc`	Displays documentation.
`exist`	Checks for existence of file or variable.
`global`	Declares variables to be global.
`help`	Searches for a help topic.
`helpwin`	Displays help text in the Help Browser.
`lookfor`	Searches help entries for a keyword.
`quit`	Stops MATLAB.
`who`	Lists current variables.
`whos`	Lists detailed information about current variables.

System and file commands.

Item	Description
`cd`	Changes current directory.
`dir`	Lists all files in current directory.
`load`	Loads workspace variables from a file.
`path`	Displays search path.
`pwd`	Displays current directory.
`save`	Saves workspace variables in a file.
`type`	Displays contents of a file.
`what`	Lists all MATLAB files.

Input/output commands.

Item	Description
`disp`	Displays contents of an array or string.
`format`	Controls screen display format.
`input`	Displays prompts and waits for input.
`menu`	Displays a menu of choices.
`;`	Suppresses screen printing.

Some numeric display formats.

Item	Description
`format long`	Sixteen decimal digits.
`format long e`	Scientific notation with sixteen digits plus exponent.
`format short`	Four decimal digits (the default).
`format short e`	Scientific notation with five digits plus exponent.

Array functions.

Item	Description
`det`	Computes determinant of an array.
`eig`	Computes the eigenvalues of a matrix.
`eye`	Creates the identity matrix.
`find`	Finds indices of nonzero elements.
`length`	Computes number of elements in an array.
`linspace`	Creates a regularly spaced array.
`logspace`	Creates a logarithmically spaced array.
`max`	Returns largest element in an array.
`min`	Returns smallest element in an array.
`ones`	Creates an array of ones.
`size`	Computes array size.
`sort`	Sorts each array column.
`sum`	Sums each array column.
`zeros`	Creates an array of zeros.

Exponential and logarithmic functions.

Item	Description
`exp(x)`	Exponential, e^x.
`log(x)`	Natural logarithm, $\ln x$.
`log10(x)`	Common (base 10) logarithm, $\log x = \log_{10} x$.
`sqrt(x)`	Square root, $\sqrt{x}$.

Complex functions.

Item	Description
`abs(x)`	Absolute value, $\|x\|$.
`angle(x)`	Angle of a complex number x.
`conj(x)`	Complex conjugate of x.
`imag(x)`	Imaginary part of a complex number x.
`real(x)`	Real part of a complex number x.

Numeric functions.

Item	Description
`ceil`	Rounds to the nearest integer toward ∞.
`fix`	Rounds to the nearest integer toward zero.
`floor`	Rounds to the nearest integer toward −∞.
`round`	Rounds toward the nearest integer.
`sign`	Signum function.

Trigonometric functions.

Item	Description
`acos(x)`	Inverse cosine, $\arccos x = \cos^{-1} x$.
`acot(x)`	Inverse cotangent, $\text{arccot}\, x = \cot^{-1} x$.
`acsc(x)`	Inverse cosecant, $\text{arccsc}\, x = \csc^{-1} x$.
`asec(x)`	Inverse secant, $\text{arcsec}\, x = \sec^{-1} x$.
`asin(x)`	Inverse sine, $\arcsin x = \sin^{-1} x$.
`atan(x)`	Inverse tangent, $\arctan x = \tan^{-1} x$.
`atan2(y,x)`	Four quadrant inverse tangent.
`cos(x)`	Cosine, $\cos x$.
`cot(x)`	Cotangent, $\cot x$.
`csc(x)`	Cosecant, $\csc x$.
`sec(x)`	Secant, $\sec x$.
`sin(x)`	Sine, $\sin x$.
`tan(x)`	Tangent, $\tan x$.

Hyperbolic functions.

Item	Description
`acosh(x)`	Inverse hyperbolic cosine, $\cosh^{-1} x$.
`acoth(x)`	Inverse hyperbolic cotangent, $\coth^{-1} x$.
`acsch(x)`	Inverse hyperbolic cosecant, $\text{csch}^{-1} x$.
`asech(x)`	Inverse hyperbolic secant, $\text{sech}^{-1} x$.
`asinh(x)`	Inverse hyperbolic sine, $\sinh^{-1} x$.
`atanh(x)`	Inverse hyperbolic tangent, $\tanh^{-1} x$.
`cosh(x)`	Hyperbolic cosine, $\cosh x$.
`coth(x)`	Hyperbolic cotangent, $\cosh x/\sinh x$.
`csch(x)`	Hyperbolic cosecant, $1/\sinh x$.
`sech(x)`	Hyperbolic secant, $1/\cosh x$.
`sinh(x)`	Hyperbolic sine, $\sinh x$.
`tanh(x)`	Hyperbolic tangent, $\sinh x/\cosh x$.

Polynomial functions.

Item	Description
`conv`	Computes product of two polynomials.
`deconv`	Computes ratio of polynomials.
`poly(r)`	Computes coefficients of polynomial whose roots are given in the vector `r`.
`poly(A)`	Computes coefficients of the characteristic polynomial corresponding to the matrix `A`.
`polyfit`	Fits a polynomial to data.
`polyval`	Evaluates a polynomial at specified values of its independent variable.
`residue`	Computes residues, poles, and direct term of a partial fraction expansion.
`roots`	Computes polynomial roots.

Logical functions.

Item	Description
`any`	True if any elements are nonzero.
`all`	True if all elements are nonzero.
`find`	Finds indices of nonzero elements.
`finite`	True if elements are finite.
`isnan`	True if elements are undefined.
`isinf`	True if elements are infinite.
`isempty`	True if array is empty.
`isreal`	True if all elements are real.
`logical`	Converts numeric values to logical values.
`xor`	Exclusive or.

Miscellaneous mathematical functions.

Item	Description
`cross`	Cross product.
`dot`	Dot product.
`fminbnd`	Finds minimum of single-variable function.
`fminsearch`	Finds minimum of multivariable function.
`function`	Creates a user-defined function.
`fzero`	Finds zero of single-variable function.
`mean`	Calculates the mean value.
`std`	Calculates the standard deviation.
`trapz`	Numerical integration with the trapezoidal rule.

Two-dimensional plotting commands.

Item	Description
`axes`	Creates axes objects.
`axis`	Sets axis limits.
`fplot`	Intelligent plotting of functions.
`ginput`	Reads coordinates of cursor position.
`grid`	Displays gridlines.
`gtext`	Enables label placement with mouse.
`hold off`	Releases a prior `hold on` command.
`hold on`	Holds current graph to enable subsequent plotting.
`legend`	Enables legend placement with mouse.
`loglog`	Creates log-log plot.
`plot`	Generates *xy* plot.
`polar`	Creates polar plot.
`print`	Prints plot or saves plot to a file.
`refresh`	Redraws current figure window.
`semilogx`	Creates semilog plot (logarithmic abscissa).
`semilogy`	Creates semilog plot (logarithmic ordinate).
`set`	Specifies properties of objects, such as axes.
`subplot`	Creates plots in subwindows.
`text`	Places a string in a figure.
`title`	Puts text at top of plot.
`xlabel`	Adds text label to abscissa (the x axis).
`ylabel`	Adds text label to ordinate (the y axis).

Program flow control.

Item	Description
`break`	Terminates execution of a loop.
`case`	Provides alternate execution paths within `switch` structure.
`continue`	Passes control to the next iteration of a `for` or `while` loop.
`else`	Delineates alternate block of statements.
`elseif`	Conditionally executes statements.
`end`	Terminates `for`, `while`, and `if` statements.
`for`	Repeat statements a specific number of times.
`if`	Execute statements conditionally.
`otherwise`	Provides optional control within a `switch` structure.
`pause`	Causes the program to stop and wait for a key press before continuing.
`switch`	Directs program execution by comparing input with `case` expressions.
`while`	Repeats statements an indefinite number of times.

LTI model functions.

Item	Description
`damp`	Computes the characteristic roots, damping ratio, and damped oscillation frequency of complex roots.
`ltimodels`	Gives help about LTI models.
`ltiprops`	Gives help about LTI model properties.
`ltiview`	Interface for analyzing time and frequency response.
`ord2`	Creates a state-space or transfer function representation of a second-order system from its natural frequency and damping ratio.
`ss`	Creates an LTI model in state-space form.
`ss2tf`	Converts from state-space to transfer function form.
`ss2zp`	Converts from state-space to zero-pole form.
`ssdata`	Extracts state-space matrices from an LTI model.
`tf`	Creates an LTI model in transfer function form.
`tf2ss`	Converts from transfer function form to state-space form.
`tf2zp`	Converts from transfer function form to zero-pole form.
`tfdata`	Extracts equation coefficients from an LTI model.
`zp2tf`	Converts from zero-pole form to transfer function form.
`zpk`	Creates an LTI model from its poles, zeros, and gain.
`zpkdata`	Returns the poles, zeros, and gain of an LTI model.

Equation solvers.

Command	Description
`impulse`	Computes and plots the impulse response of the LTI model `sys`.
`initial`	Computes and plots the free response of an LTI model given in state-model form.
`lsim`	Computes and plots the response of an LTI object to a defined input function of time.
`ode23`	Solves linear and nonlinear differential equations.
`ode45`	Solves linear and nonlinear differential equations.
`step`	Computes and plots the step response of an LTI object.

Predefined input functions.

Command	Description
`gensig`	Generates a periodic sine, square, or pulse input having a specified period.
`stepfun`	Generates a step function input.

Frequency response functions.

Command	Description
`bode`	Computes the magnitude ratio and phase angle of an LTI model and displays the Bode plots.
`bodemag`	Computes the magnitude ratio of an LTI model and displays the magnitude plot.
`evalfr`	Evaluates a transfer function model at a specified value of s.
`freqresp`	Computes the frequency response of an LTI model at multiple specified frequencies.
`margin`	Computes phase and gain margins of an LTI model and displays the Bode plots.

Root locus functions.

Command	Description
`pole`	Computes the poles of an LTI model.
`pzmap`	Computes the poles and zeros of an LTI model.
`rlocfind`	Enables use of the cursor to select the gain from a specified point on a root locus plot.
`rlocus`	Computes and displays the root locus plot.
`rltool`	Starts the root locus GUI interface.
`sgrid`	Superimposes a grid of constant ζ and constant ω_n lines on the root locus plot.
`zero`	Computes the zeros of an LTI model.

Control system functions.

Command	Description
`acker`	Uses Ackermann's method to compute the feedback gain matrix for a single-input system to place the closed-loop poles at specified locations.
`c2d`	Converts a continuous-time model into a discrete-time model using a zero-order hold on the inputs, with a specified sampling time.
`feedback`	Creates an LTI model from two subsystems connected with a feedback loop.
`ltiview`	Starts the LTI viewer.
`pade`	Pade approximation to the transfer function of the dead time element.
`parallel`	Creates an LTI model from two subsystems connected in parallel.
`series`	Creates an LTI model from two subsystems connected in series.
`sisotool`	Graphical user interface for designing single-input/single-output compensators.

APPENDIX B

Fourier Series

The Fourier series is used to represent a periodic function as a sum of sines, cosines, and a constant. When used with the principle of superposition, which lets you obtain the total response as the sum of the individual responses, the Fourier series enables you to obtain the response of a linear system to any periodic function. All that is needed is the system response to a sine, a cosine, and a constant input. Although the Fourier series is an infinite series, in practice it can be truncated to a small number of terms whose frequencies lie within the bandwidth of the system.

If a function $f(t)$ is periodic with period P, then $f(t+P) = f(t)$. The Fourier series for this function defined on the interval $t_1 \le t \le t_1 + P$, where t_1 and P are constants and $P > 0$, is

$$f(t) = \frac{a_0}{2} + \sum_{n=1}^{\infty} \left(a_n \cos \frac{2n\pi t}{P} + b_n \sin \frac{2n\pi t}{P} \right)$$

where

$$a_n = \frac{2}{P} \int_{t_1}^{t_1+P} f(t) \cos \frac{2n\pi t}{P}\, dt \quad n = 0, 1, 2, \ldots$$

$$b_n = \frac{2}{P} \int_{t_1}^{t_1+P} f(t) \sin \frac{2n\pi t}{P}\, dt \quad n = 1, 2, 3, \ldots$$

If $f(t)$ is defined outside the specified interval $[t_1, t_1 + P]$ by a periodic extension of period P, and if $f(t)$ and df/dt are piecewise continuous, then the Fourier series converges to $f(t)$ if t is a point of continuity, and to the average value $[f(t_+) + f(t_-)]/2$ otherwise.

As an example, consider the train of unit pulses of width π and alternating in sign, as shown in Figure B.1. The function is described by

$$f(t) = \begin{cases} 1 & 0 < t < \pi \\ -1 & \pi < t < 2\pi \end{cases}$$

The period is $P = 2\pi$, and we can take the constant t_1 to be 0. Using a table of integrals, we find that

$$a_n = 0 \quad \text{for all } n$$

$$b_n = \frac{4}{n\pi} \quad \text{for } n \text{ odd}$$

$$b_n = 0 \quad \text{for } n \text{ even}$$

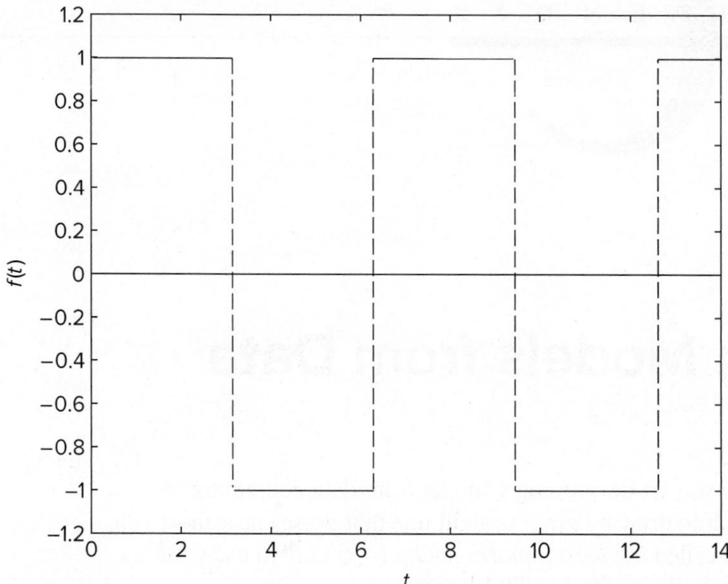

Figure B.1 The function used for the Fourier series example.

The Fourier series is

$$f(t) = \frac{4}{\pi}\left(\frac{\sin t}{1} + \frac{\sin 3t}{3} + \frac{\sin 5t}{5} + \cdots\right)$$

In general, the constant term a_0 and the cosine terms will not appear in the series if the function is odd; that is, if $f(-t) = -f(t)$. If the function is even, then $f(-t) = f(t)$, and no sine terms will appear in the series.

APPENDIX C

Developing Models from Data

Here we develop a systematic way of developing a model from data containing so much scatter that it is difficult to draw by eye a straight line that passes near most of the data points. This method—called the *least-squares method*—is easy to use with a computer, and we illustrate how to do it with MATLAB. ■

C.1 FUNCTION IDENTIFICATION AND PARAMETER ESTIMATION

Function identification, or function *discovery,* is the process of identifying or discovering a function that can describe a particular set of data. The term *curve fitting* is also used to describe the process of finding a curve, and the function generating the curve, to describe a given set of data. *Parameter estimation* is the process of obtaining values for the parameters, or coefficients, in the function that describes the data.

The following three function types can often describe physical phenomena.

1. The *linear* function $y(x) = mx + b$. Note that $y(0) = b$.
2. The *power* function $y(x) = bx^m$. Note that $y(0) = 0$ if $m \geq 0$, and $y(0) = \infty$ if $m < 0$.
3. The *exponential* function $y(x) = b(10)^{mx}$ or its equivalent form $y = be^{mx}$, where e is the base of the natural logarithm ($\ln e = 1$). Note that $y(0) = b$ for both forms.

For example, the linear function describes the voltage-current relation for a resistor ($v = iR$) and the velocity versus time relation for an object with constant acceleration a ($v = at + v_0$). The distance d traveled by a falling object versus time is described by a power function ($d = 0.5gt^2$). The temperature change ΔT of a cooling object can be described by an exponential function ($\Delta T = \Delta T_0 e^{-ct}$).

Each function gives a straight line when plotted using a specific set of axes:

1. The linear function $y = mx + b$ gives a straight line when plotted on rectilinear axes. Its slope is m and its y intercept is b.
2. The power function $y = bx^m$ gives a straight line when plotted on log-log axes.
3. The exponential function $y = b(10)^{mx}$ and its equivalent form, $y = be^{mx}$, give a straight line when plotted on semilog axes with a logarithmic y axis.

These properties of the power and exponential functions are illustrated in Figure C.1.1, which shows the power function $y = 2x^{-0.5}$ and the exponential function $y = 10(10^{-x})$.

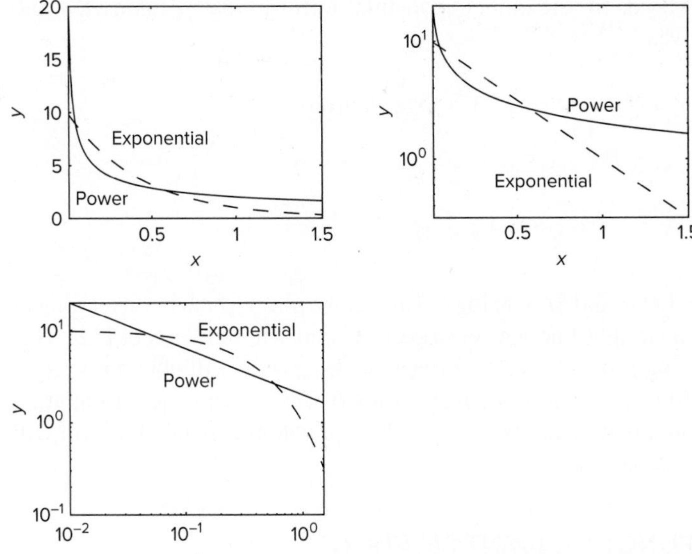

Figure C.1.1 The power function $y = 2x^{-0.5}$ and the exponential function $y = 10(10^{-x})$, plotted on rectilinear, semilog, and log-log axes, respectively.

When we need to identify a function that describes a given set of data, we look for a set of axes (rectilinear, semilog, or log-log) on which the data forms a straight line, because a straight line is the one most easily recognized by eye, and therefore we can easily tell if the function will fit the data well.

Using the following properties of base-ten logarithms, which are shared with natural logarithms, we have:

$$\log(ab) = \log a + \log b$$

$$\log(a^m) = m \log a$$

Take the logarithm of both sides of the power equation $y = bx^m$ to obtain

$$\log y = \log(bx^m) = \log b + m \log x$$

This has the form $Y = B + mX$ if we let $Y = \log y$, $X = \log x$, and $B = \log b$. Thus, if we plot Y versus X on rectilinear scales, we will obtain a straight line whose slope is m and whose intercept is B. This is the same as plotting $\log y$ versus $\log x$ on rectilinear scales, so we will obtain a straight line whose slope is m and whose intercept is $\log b$. This process is equivalent to plotting y versus x on log-log axes. Thus, if the data can be described by the power function, it will form a straight line when plotted on log-log axes.

Taking the logarithm of both sides of the exponential equation $y = b(10)^{mx}$, we obtain:

$$\log y = \log[b(10)^{mx}] = \log b + mx \log 10 = \log b + mx$$

because $\log 10 = 1$. This has the form $Y = B + mx$ if we let $Y = \log y$ and $B = \log b$. Thus, if we plot Y versus x on rectilinear scales, we will obtain a straight line whose slope is m and whose intercept is B. This is the same as plotting $\log y$ versus x on rectilinear scales, so we will obtain a straight line whose slope is m and whose intercept is $\log b$. This is equivalent to plotting y on a log axis and x on a rectilinear axis. Thus, if the data can be described by the exponential function, it will form a straight line when plotted on semilog axes (with the log axis used for the ordinate).

This property also holds for the other exponential form: $y = be^{mx}$. Taking the logarithm of both sides gives

$$\log y = \log (be^{mx}) = \log b + mx \log e$$

This has the form

$$Y = B + Mx$$

if we let $Y = \log y$, $B = \log b$, and $M = m \log e$. Thus, if we plot Y versus x on rectilinear scales, we will obtain a straight line whose slope is M and whose intercept is B. This is the same as plotting $\log y$ versus x on rectilinear scales, so we will obtain a straight line whose slope is $m \log e$ and whose intercept is $\log b$. This is equivalent to plotting y on a log axis and x on a rectilinear axis. Thus, the equivalent exponential form will also plot as a straight line on semilog axes.

C.1.1 STEPS FOR FUNCTION IDENTIFICATION

Here is a summary of the procedure to find a function that describes a given set of data. We assume that the data can be described by one of the three function types given above. Fortunately, many applications generate data that can be described by these functions. The procedure is

1. Examine the data near the origin. The exponential functions $y = b(10)^{mx}$ and $y = be^{mx}$ can never pass through the origin (unless, of course $b = 0$, which is a trivial case). See Figure C.1.2 for examples with $b = 1$. The linear function $y = mx + b$ can pass through the origin only if $b = 0$. The power function $y = bx^m$ can pass through the origin but only if $m > 0$. See Figure C.1.3 for examples.

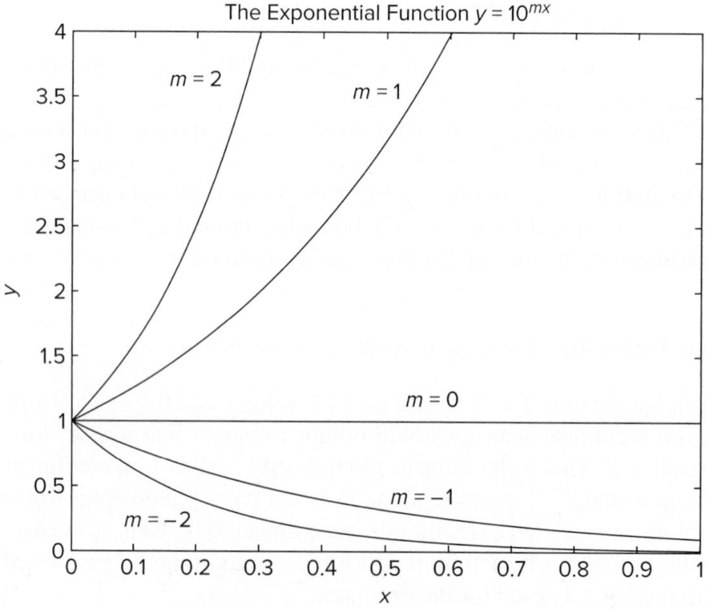

Figure C.1.2 Examples of exponential functions.

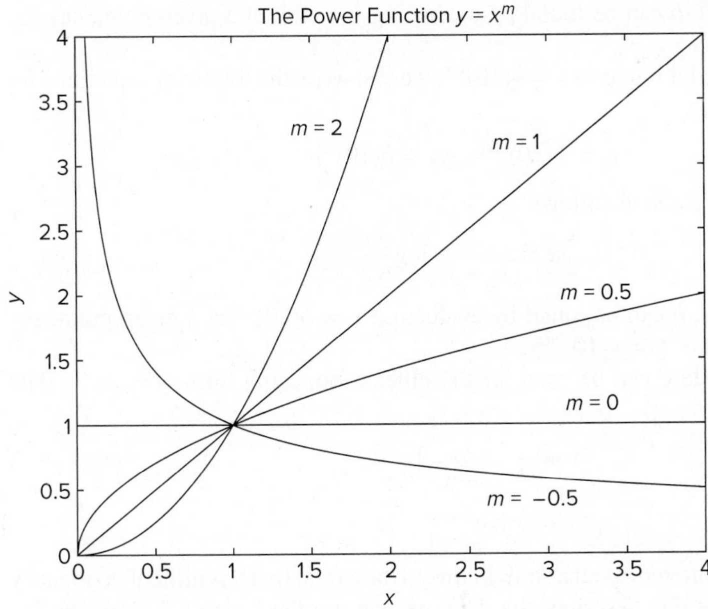

Figure C.1.3 Examples of power functions.

2. Plot the data using rectilinear scales. If it forms a straight line, then it can be represented by the linear function, and you are finished. Otherwise, if you have data at $x = 0$, then
   a. If $y(0) = 0$, try the power function, or
   b. If $y(0) \neq 0$, try the exponential function.

   If data is not given for $x = 0$, proceed to step 3.
3. If you suspect a power function, plot the data using log-log scales. Only a power function will form a straight line. If you suspect an exponential function, plot it using semilog scales. Only an exponential function will form a straight line.

C.1.2 OBTAINING THE COEFFICIENTS

There are several ways to obtain the values of the coefficients b and m. If the data lie very close to a straight line, we can draw the line through the data using a straightedge and then read two points from the line. These points can be conveniently chosen to coincide with gridlines to eliminate interpolation error. Let these two points be denoted (x_1, y_1) and (x_2, y_2).

For the linear function $y = mx + b$, the slope is given by

$$m = \frac{y_2 - y_1}{x_2 - x_1}$$

Once m is computed, b can be found by evaluating $y = mx + b$ at a given point, say the point (x_1, y_1). Thus, $b = y_1 - mx_1$.

For the power function $y = bx^m$, we can write the following equations for the two chosen points.

$$y_1 = bx_1^m \quad y_2 = bx_2^m$$

These can be solved for m as follows.

$$m = \frac{\log(y_2/y_1)}{\log(x_2/x_1)}$$

Once m is computed, b can be found by evaluating $y = bx^m$ at a given point, say the point (x_1, y_1). Thus, $b = y_1 x_1^{-m}$.

For the exponential function $y = b(10)^{mx}$, we can write the following equations for the two chosen points.

$$y_1 = b(10)^{mx_1} \quad y_2 = b(10)^{mx_2}$$

These can be solved for m as follows.

$$m = \frac{1}{x_2 - x_1} \log \frac{y_2}{y_1}$$

Once m is computed, b can be found by evaluating $y = b(10)^{mx}$ at a given point, say the point (x_1, y_1). Thus, $b = y_1 10^{-mx_1}$.

A similar procedure can be used for the other exponential form, $y = be^{mx}$. The solutions are

$$m = \frac{1}{x_2 - x_1} \ln \frac{y_2}{y_1}$$

$$b = y_1 e^{-mx_1}$$

If the data are scattered about a straight line to the extent that it is difficult to identify a unique straight line that describes the data, we can use the *least-squares method* to obtain the function. This method finds the coefficients of a polynomial of specified degree n that best fits the data, in the so-called "least-squares sense." We discuss this method in Section C.2. The MATLAB implementation of this method uses the `polyfit` function, which is discussed in Section C.3.

Examples C.1.1 and C.1.2 feature experiments that you can easily perform on your own. Engineers are often required to make predictions about the temperatures that will occur in various industrial processes, for example. Example C.1.1 illustrates how we can use function identification to predict the temperature dynamics of a cooling process.

EXAMPLE C.1.1

Temperature Dynamics of Water

■ Problem

Water in a glass measuring cup was allowed to cool after being heated to 204°F. The ambient air temperature was 70°F. The measured water temperature at various times is given in the following table.

Time (sec)	0	120	240	360	480	600
Temperature (°F)	204	191	178	169	160	153

Time (sec)	720	840	960	1080	1200
Temperature (°F)	147	141	137	132	127

Obtain a functional description of the water temperature versus time.

■ Solution

Common sense tells us that the water temperature will eventually reach the air temperature of 70°. Thus, we first subtract 70° from the temperature data T and seek to obtain a functional description of the relative temperature, $\Delta T = T - 70$. A plot of the relative temperature data is shown in Figure C.1.4. We note that the plot has a distinct curvature and that it does not pass through the origin. Thus, we can rule out the linear function and the power function as candidates.

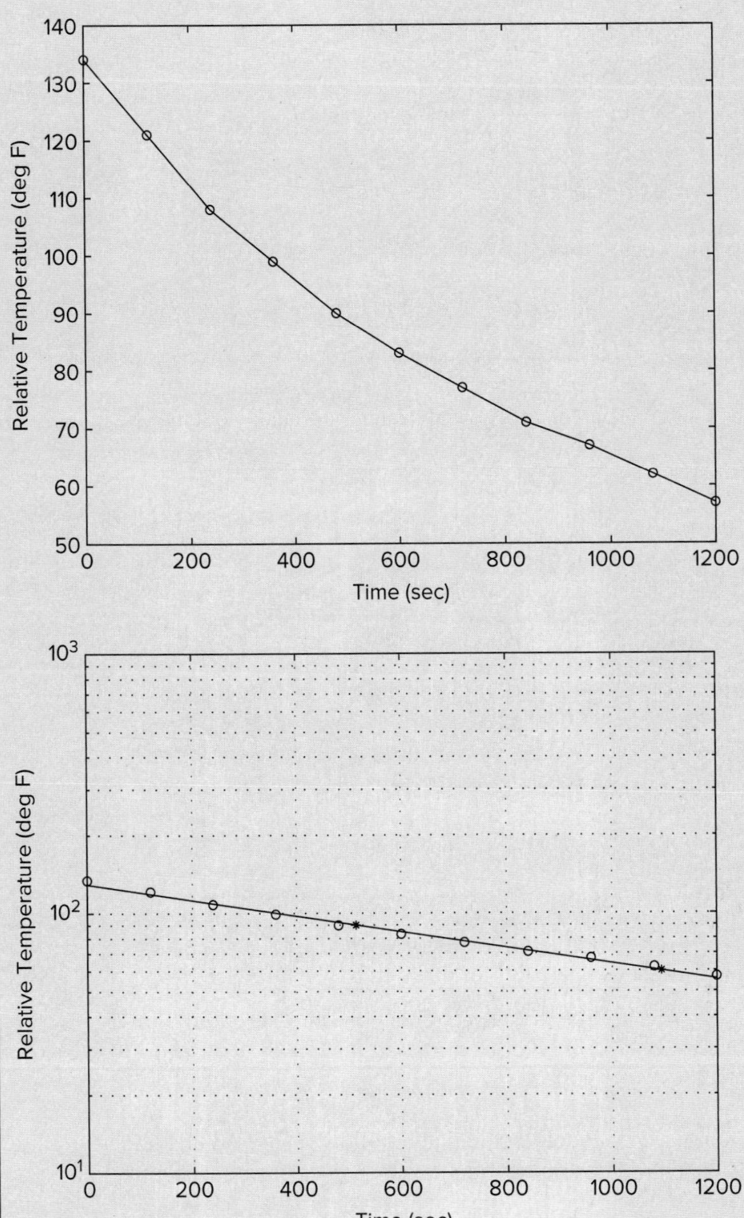

Figure C.1.4 Plot of relative temperature versus time.

Figure C.1.5 Semilog plot of relative temperature versus time.

To see if the data can be described by an exponential function, we plot the data on a semilog plot, which is shown in Figure C.1.5. The straight line shown can be drawn by aligning a straightedge so that it passes near most of the data points (note that this line is subjective; another person might draw a different line). The data lie close to a straight line, so we can use the exponential function to describe the relative temperature.

Using the second form of the exponential function, we can write $\Delta T = be^{mt}$. Next we select two points on the straight line to find the values of b and m. The two points indicated by an asterisk were selected to minimize interpolation error because they lie near grid lines. The accuracy of the values read from the plot obviously depends on the size of the plot. Two points read from the plot are (1090, 60) and (515, 90). Using the equations developed previously to compute b and m (with t replacing x and ΔT replacing y), we have

$$m = \frac{1}{1090 - 515} \ln \frac{60}{90} = -0.0007$$

$$b = 90e^{-0.0007(515)} = 129$$

Thus, the estimated function is

$$\Delta T = 129e^{-0.0007t} \quad \text{or} \quad T = \Delta T + 70 = 129e^{-0.0007t} + 70$$

where ΔT and T are in °F and time t is in seconds. The plot of ΔT versus t is shown in Figure C.1.6. From this we can see that the function provides a reasonably good description of the data. In Section C.2 we will discuss how to quantify the quality of this description.

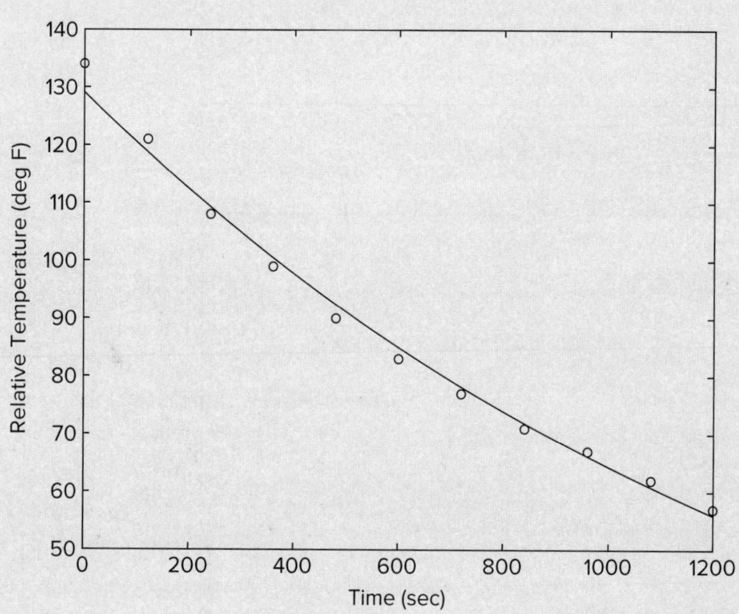

Figure C.1.6 Comparison of the fitted function with the data.

Engineers often need a model to calculate the flow rates of fluids under pressure. The coefficients of such models must often be determined from measurements.

EXAMPLE C.1.2

Orifice Flow

■ **Problem**

A hole 6 mm in diameter was made in a translucent milk container (Figure C.1.7). A series of marks 1 cm apart was made above the hole. While adjusting the tap flow to keep the water height constant, the time for the outflow to fill a 250-ml cup was measured (1 ml = 10^{-6} m^3). This was repeated for several heights. The data are given in the following table.

Height h (cm)	11	10	9	8	7	6	5	4	3	2	1
Time t (s)	7	7.5	8	8.5	9	9.5	11	12	14	19	26

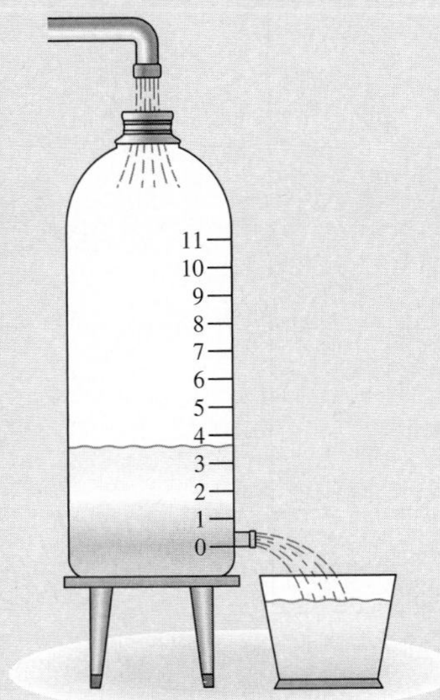

Figure C.1.7 An experiment to determine flow rate versus liquid height.

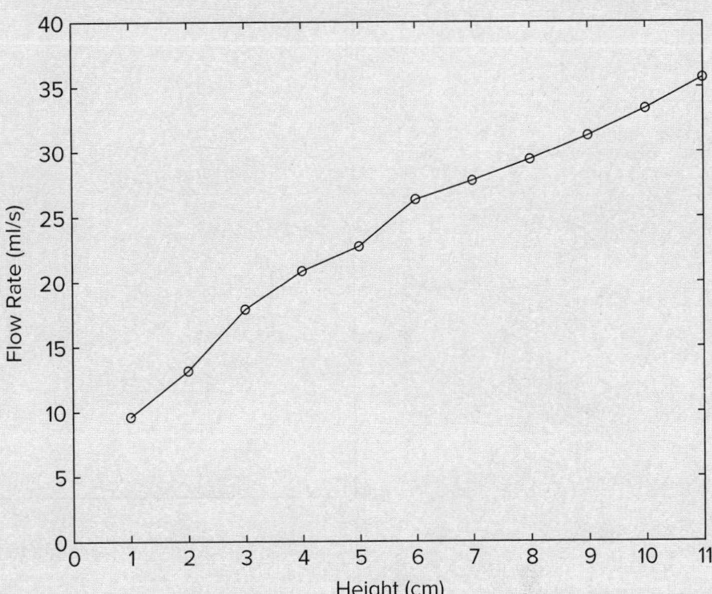

Figure C.1.8 Plot of flow rate data.

Obtain a functional description of the volume outflow rate f as a function of water height h above the hole.

■ Solution

First obtain the flow rate data in ml/s by dividing the 250 ml volume by the time to fill:

$$f = \frac{250}{t}$$

A plot of the resulting flow rate data is shown in Figure C.1.8. There is some curvature in the plot, so we rule out the linear function. Common sense tells us that the outflow rate will be zero when the height is zero, so we can rule out the exponential function because it cannot pass through the origin.

The log-log plot shown in Figure C.1.9 shows that the data lie close to a straight line, so we can use the power function to describe the flow rate as a function of height. Thus, we can write

$$f = bh^m$$

The straight line shown can be drawn by aligning a straightedge so that it passes near most of the data points (note that this line is subjective; another person might draw a different line). Next we select two points on the straight line to find the values of b and m. The two points indicated by an asterisk were selected to minimize interpolation error because they lie near grid lines. The accuracy of the values read from the plot obviously depends on the size of the plot. The values of the points as read from the plot are (1, 9.4) and (8, 30). Using the equations developed previously to compute b and m (with h replacing x and f replacing y), we have

$$m = \frac{\log(30/9.4)}{\log(8/1)} = 0.558$$

$$b = 9.4(1^{-0.558}) = 9.4$$

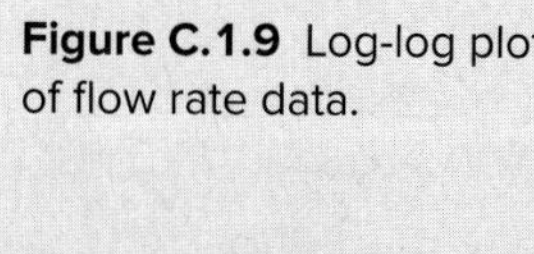

Figure C.1.9 Log-log plot of flow rate data.

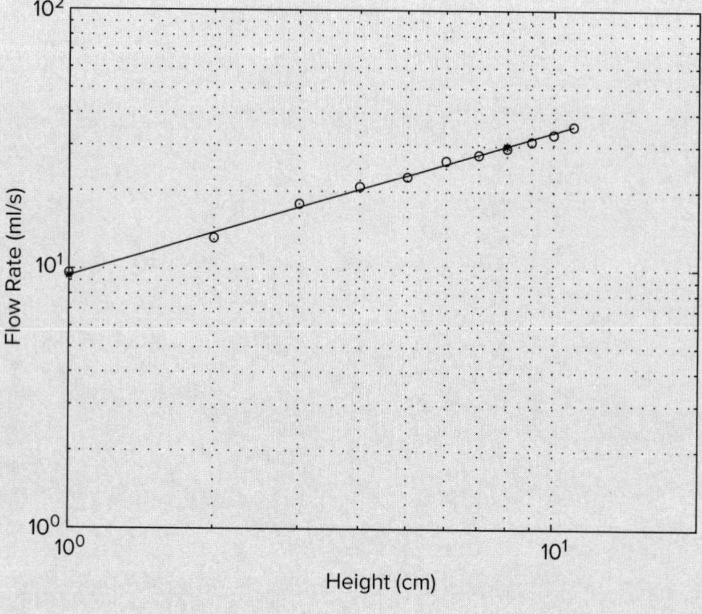

Figure C.1.10 Comparison of the flow rate function and the data.

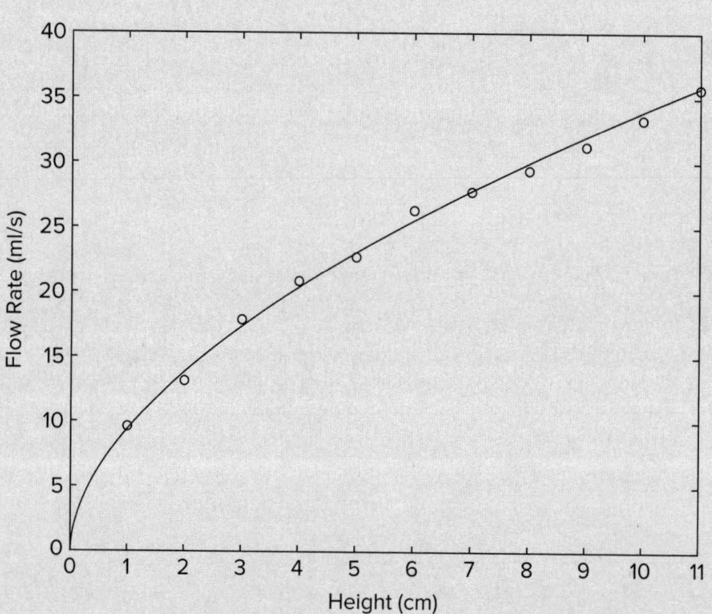

Thus, the estimated function is $f = 9.4h^{0.558}$, where f is the outflow rate in ml/s and the water height h is in centimeters. The plot of f versus h is shown in Figure C.1.10. From this we can see that the function provides a reasonably good description of the data. In Section C.2 we will discuss how to quantify the quality of this description.

The examples we have selected thus far have used data that do not contain much "scatter." For applications where the data are scattered so much that it is difficult to draw a straight line by hand on any set of axes, it is best to use a systematic method to find the function that gives the best fit. The most common method is the so-called *least-squares method*, which can also be used to fit data to functions other than the three functions (linear, exponential, and power) we have treated here; for example, the method can be used to fit polynomial functions of any degree. Section C.2 develops theory of the least-squares method. Section C.3 shows how to implement this method with MATLAB.

C.2 FITTING MODELS TO SCATTERED DATA

In practice the data often will not lie very close to a straight line, and if we ask two people to draw a straight line passing as close as possible to all the data points, we will probably receive two different answers. A systematic and objective way of obtaining a straight line describing the data is the least-squares method. Suppose we want to find the coefficients of the straight line $y = mx + b$ that best fits the following data.

x	0	5	10
y	2	6	11

According to the least-squares criterion, the line that gives the best fit is the one that minimizes J, the sum of the squares of the vertical differences between the line and the data points (see Figure C.2.1). These differences are called the *residuals*. Here there are three data points and J is given by

$$J = \sum_{i=1}^{3} \left(mx_i + b - y_i\right)^2$$

Substituting the data values (x_i, y_i) given in the table, we obtain

$$J = (0m + b - 2)^2 + (5m + b - 6)^2 + (10m + b - 11)^2$$

The values of m and b that minimize J can be found from $\partial J/\partial m = 0$ and $\partial J/\partial b = 0$.

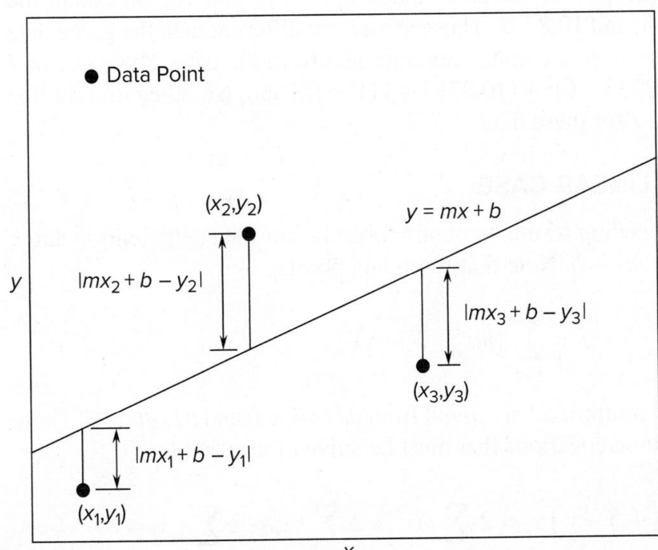

Figure C.2.1 Illustration of the least-squares criterion.

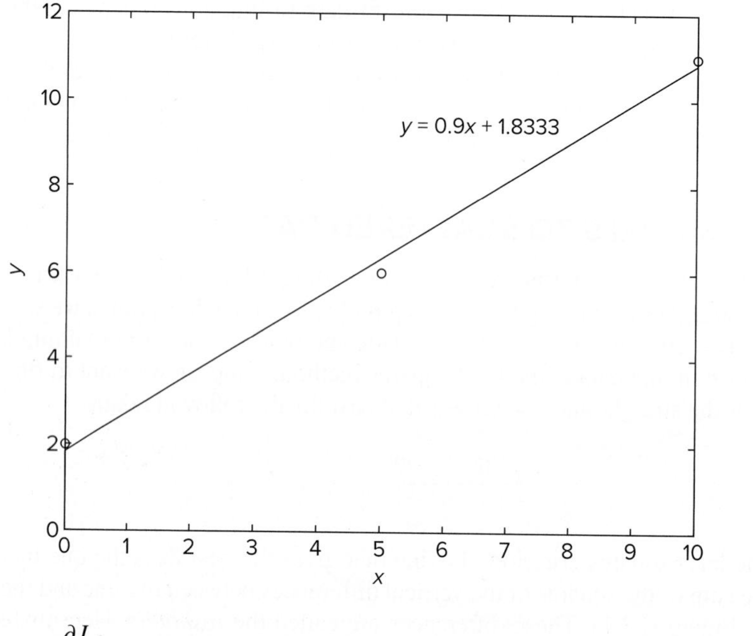

Figure C.2.2 Example of a least-squares fit.

$$\frac{\partial J}{\partial m} = 2(5m + b - 6)(5) + 2(10m + b - 11)(10) = 250m + 30b - 280 = 0$$

$$\frac{\partial J}{\partial b} = 2(b - 2) + 2(5m + b - 6) + 2(10m + b - 11) = 30m + 6b - 38 = 0$$

These conditions give the following equations that must be solved for the two unknowns m and b.

$$250m + 30b = 280$$
$$30m + 6b = 38$$

The solution is $m = 9/10$ and $b = 11/6$. The best straight line in the least-squares sense is $y = (9/10)x + 11/6$. This is shown in Figure C.2.2 along with the data.

If we evaluate this equation at the data values $x = 0$, 5, and 10, we obtain the values $y = 1.8333$, 6.3333, and 10.8333. These values are different than the given data values $y = 2$, 6, and 11 because the line is not a perfect fit to the data. The value of J is $J = (1.8333 - 2)^2 + (6.3333 - 6)^2 + (10.8333 - 11)^2 = 0.1666$. No other straight line will give a lower value of J for these data.

C.2.1 THE GENERAL LINEAR CASE

We can generalize the preceding results to obtain formulas for the coefficients m and b in the linear equation $y = mx + b$. Note that for n data points,

$$J = \sum_{i=1}^{n} \left(mx_i + b - y_i\right)^2$$

The values of m and b that minimize J are found from $\partial J/\partial m = 0$ and $\partial J/\partial b = 0$. These conditions give the following equations that must be solved for m and b:

$$\frac{\partial J}{\partial m} = 2\sum_{i=1}^{n} \left(mx_i + b - y_i\right) x_i = 2\sum_{i=1}^{n} mx_i^2 + 2\sum_{i=1}^{n} bx_i - 2\sum_{i=1}^{n} y_i x_i = 0$$

$$\frac{\partial J}{\partial b} = 2\sum_{i=1}^{n}(mx_i + b - y_i) = 2\sum_{i=1}^{n} mx_i + 2\sum_{i=1}^{n} b - 2\sum_{i=1}^{n} y_i = 0$$

These equations become

$$m\sum_{i=1}^{n} x_i^2 + b\sum_{i=1}^{n} x_i = \sum_{i=1}^{n} y_i x_i \tag{C.2.1}$$

$$m\sum_{i=1}^{n} x_i + bn = \sum_{i=1}^{n} y_i \tag{C.2.2}$$

These are two linear equations in terms of m and b.

Because the exponential and power functions form straight lines on semilog and log-log axes respectively, we can use the previous results after computing the logarithms of the data.

Fitting Data with the Power Function

EXAMPLE C.2.1

■ Problem

Find a functional description of the following data:

x	1	2	3	4
y	5.1	19.5	46	78

■ Solution

These data do not lie close to a straight line when plotted on linear or semilog axes. However, they do when plotted on log-log axes. Thus a power function $y = bx^m$ can describe the data. Using the transformations $X = \log x$ and $Y = \log y$, we obtain the new data table:

$X = \log x$	0	0.3010	0.4771	0.6021
$Y = \log y$	0.7076	1.2900	1.6628	1.8921

From this table we obtain

$$\sum_{i=1}^{4} X_i = 1.3803 \qquad \sum_{i=1}^{4} Y_i = 5.5525$$

$$\sum_{i=1}^{4} X_i Y_i = 2.3208 \qquad \sum_{i=1}^{4} X_i^2 = 0.6807$$

Using X, Y, and $B = \log b$ instead of x, y, and b in (C.1.1) and (C.1.2), we obtain

$$0.6807m + 1.3803B = 2.3208$$

$$1.3803m + 4B = 5.5525$$

The solution is $m = 1.9802$ and $B = 0.7048$. This gives $b = 10^B = 5.068$. Thus, the desired function is $y = 5.068x^{1.9802}$.

C.2.2 CONSTRAINING MODELS TO PASS THROUGH A GIVEN POINT

Many applications require a model whose form is dictated by physical principles. For example, the force-extension model of a spring must pass through the origin (0, 0) because the spring exerts no force when it is not stretched. Thus, a linear model $y = mx + b$ sometimes must have a zero value for b. However, in general the least-squares method will give a nonzero value for b because of the scatter or measurement error that is usually present in the data.

To obtain a zero-intercept model of the form $y = mx$, we must derive the equation for m from basic principles. The sum of the squared residuals in this case is

$$J = \sum_{i=1}^{n} \left(mx_i - y_i\right)^2$$

Computing the derivative $\partial J/\partial m$ and setting it equal to zero gives the result

$$m \sum_{i=1}^{n} x_i^2 = \sum_{i=1}^{n} x_i y_i \tag{C.2.3}$$

which can be easily solved for m.

If the model is required to pass through a point not at the origin, say the point (x_0, y_0), subtract x_0 from all the x values, subtract y_0 from all the y values, and then use (C.2.3) to find the coefficient m. The resulting equation will be of the form

$$y = m(x - x_0) + y_0 \tag{C.2.4}$$

EXAMPLE C.2.2 Point Constraint

■ Problem

Consider the data given at the beginning of this section.

x	0	5	10
y	2	6	11

We found that the best-fit line is $y = (9/10)x + 11/6$. Find the best-fit line that passes through the point $x = 10$, $y = 11$.

■ Solution

Subtracting 10 from all the x values and 11 from all the y values, we obtain a new set of data in terms of the new variables $X = x - 10$ and $Y = y - 11$.

X	−10	−5	0
Y	−9	−5	0

Expressing (C.2.3) in terms of the new variables X and Y, we have

$$m \sum_{i=1}^{3} X_i^2 = \sum_{i=1}^{3} X_i Y_i$$

$$\sum_{i=1}^{3} X_i^2 = (-10)^2 + 5^2 + 0 = 125$$

$$\sum_{i=1}^{3} X_i Y_i = (-10)(-9) + (-5)(-5) + 0 = 115$$

Thus, $m = 115/125 = 23/25$ and the best-fit line is $Y = (23/25)X$. In terms of the original variables, this line is expressed as $y - 11 = (23/25)(x - 10)$ or $y = (23/25)x + 9/5$.

CONSTRAINING A COEFFICIENT

Sometimes we know from physical theory that the data can be described by a function with a specified form and specified values of one of more of its coefficients. For example, the fluid-drag relation states that $D = \rho AC_D v^2/2$. In this case, we know that the relation is a power function with an exponent of 2, and we need to estimate the value of the drag coefficient C_D. In such cases, we can modify the least-squares method to find the best-fit function of a specified form.

Fitting a Power Function with a Known Exponent

EXAMPLE C.2.3

■ Problem

Fit the power function $y = bx^m$ to the data y_i. The value of m is known.

■ Solution

The least-squares criterion is

$$J = \sum_{i=1}^{n} (bx^m - y_i)^2$$

To obtain the value of b that minimizes J, we must solve $\partial J/\partial b = 0$.

$$\frac{\partial J}{\partial b} = 2\sum_{i=1}^{n} x_i^m \left(bx_i^m - y_i\right) = 0$$

This gives

$$b = \frac{\sum_{i=1}^{n} x_i^m y_i}{\sum_{i=1}^{n} x_i^{2m}} \tag{1}$$

C.2.3 THE QUALITY OF A CURVE FIT

In general, if the arbitrary function $y = f(x)$ is used to represent the data, then the error in the representation is given by $e_i = f(x_i) - y_i$, for $i = 1, 2, 3, \ldots, n$. The error e_i is the difference between the data value y_i and the value of y obtained from the function; that is, $f(x_i)$. The least-squares criterion used to fit a function $f(x)$ is the sum of the squares of the residuals, J. It is defined as

$$J = \sum_{i=1}^{n} \left[f(x_i) - y_i\right]^2 \tag{C.2.5}$$

We can use this criterion to compare the quality of the curve fit for two or more functions used to describe the same data. The function that gives the smallest J value gives the best fit.

We denote the sum of the squares of the deviation of the y values from their mean $\bar{y}$ by S, which can be computed from

$$S = \sum_{i=1}^{n}(y_i - \bar{y})^2 \tag{C.2.6}$$

This formula can be used to compute another measure of the quality of the curve fit, the *coefficient of determination,* also known as the *r-squared value*. It is defined as

$$r^2 = 1 - \frac{J}{S} \tag{C.2.7}$$

For a perfect fit, $J = 0$ and thus $r^2 = 1$. Thus, the closer r^2 is to 1, the better the fit. The largest r^2 can be is 1. The value of S is an indication of how much the data is spread around the mean, and the value of J indicates how much of the data spread is left unaccounted for by the model. Thus, the ratio J/S indicates the fractional variation left unaccounted for by the model. It is possible for J to be larger than S, and thus it is possible for r^2 to be negative. Such cases, however, are indicative of a very poor model that should not be used. As a rule of thumb, a very good fit accounts for at least 99% of the data variation. This corresponds to $r^2 \geq 0.99$.

For example, the function $y = 9/10x + 11/6$ derived at the beginning of this section has the values $S = 40.6667$, $J = 0.1666$, and $r^2 = 0.9959$, which indicates a very good fit. The line $y = (23/25)x + 9/5$, which is constrained to pass through the point $x = 10$, $y = 11$ gives the values $S = 40.6667$, $J = 0.2$, and $r^2 = 0.9951$. So the constraint degraded the quality of the fit but very slightly.

The power function $y = 5.068x^{1.9802}$ derived in Example C.2.1 has the values $S = 3085.8$, $J = 2.9192$, and $r^2 = 0.9991$. Thus, its fit is very good.

When the least-squares method is applied to fit quadratic and higher-order polynomials, the resulting equations for the coefficents are linear algebraic equations, which are easily solved. Their solution forms the basis for MATLAB algorithm contained in the `polyfit` function, which is discussed in Section C.3.

C.2.4 INTEGRAL FORM OF THE LEAST-SQUARES CRITERION

Sometimes we must obtain a linear description of a process over a range of the independent variable so large that linearization is impractical. In such cases we can apply the least-squares method to obtain the linear description. Because there are no data in such cases, we use the integral form of the least-squares criterion.

EXAMPLE C.2.4

Fitting a Linear Function to a Power Function

■ Problem

a. Fit the linear function $y = mx$ to the power function $y = ax^n$ over the range $0 \leq x \leq L$. The values of a and n are given.
b. Apply the result to the Aerobee drag function $D = 0.00056v^2$ over the range $0 \leq v \leq 1000$, discussed in Example 1.3.4.

■ Solution

a. The appropriate least-squares criterion is the integral of the square of the difference between the linear model and the power function over the stated range. Thus,

$$J = \int_0^L (mx - ax^n)^2\, dx$$

To obtain the value of m that minimizes J, we must solve $\partial J/\partial m = 0$.

$$\frac{\partial J}{\partial m} = 2\int_0^L x(mx - ax^n)\,dx = 0$$

This gives

$$m = \frac{3a}{n+2}L^{n-1} \tag{1}$$

b. For the Aerobee drag function $D = 0.00056v^2$, $a = 0.00056$, $n = 2$, and $L = 1000$. Thus,

$$m = \frac{3(0.00056)}{2+2}1000^{2-1} = 0.42$$

and the linear description is $D = 0.42v$, where D is in pounds and v is in ft/sec. This is the linear model that minimizes the integral of the squared error over $0 \le v \le 1000$ ft/sec.

C.3 MATLAB AND THE LEAST-SQUARES METHOD

We now show how to use MATLAB's `polyfit` function to fit polynomials and functions that can be transformed into polynomials. The `polyfit` function is based on the least-squares method. Its syntax is `p=polyfit(x,y,n)`. The function fits a polynomial of degree `n` to data described by the vectors `x` and `y`, where `x` is the independent variable. The result `p` is the row vector of length $n + 1$ that contains the polynomial coefficients in order of descending powers.

Fitting First- and Second-Degree Polynomials

EXAMPLE C.3.1

■ Problem

Use the `polyfit` function to find the first- and second-degree polynomials that fit the following data in the least-squares sense. Evaluate the quality of fit for each polynomial.

x	0	1	2	3	4	5	6	7	8	9	10
y	48	49	52	63	76	98	136	150	195	236	260

■ Solution

The following MATLAB program computes the polynomial coefficients.

```
% Enter the data.
x = (0:10);
y = [48, 49, 52, 63, 76, 98, 136, 150, 195, 236, 260];
% Fit a first-degree polynomial.
p_first = polyfit(x,y,1)
Fit a second-degree polynomial.
p_second = polyfit(x,y,2)
```

The polynomial coefficients of the first-degree polynomial are contained in the vector `p_first`, and the coefficients of the second-degree polynomial are contained in the vector `p_second`. The results are `p_first = [22.1909, 12.0455]`, which corresponds to the polynomial $y = 22.1909x + 12.0455$, and `p_second = [2.1993, 0.1979, 45.035]`, which corresponds to the polynomial $y = 2.1993x^2 + 0.1979x + 45.035$.

We can use MATLAB to plot the polynomials and to evaluate the "quality-of-fit" quantities J, S, and r^2. The following script file does this.

```
% Enter the data and find the mean of y.
x = (0:10);
y = [48, 49, 52, 63, 76, 98, 136, 150, 195, 236, 260];
mu = mean(y);
% Define a range of x and y values for plotting.
xp = (0:0.01:10);
for k = 1:2
   yp(k,:) = polyval(polyfit(x,y,k),xp);
   % Compute J, S, and r squared.
   J(k) = sum((polyval(polyfit(x,y,k),x)-y).^2);
   S(k) = sum((polyval(polyfit(x,y,k),x)- mu).^2);
   r2(k) = 1-J(k)/S(k);
end
% Plot the first-degree polynomial.
subplot(2,1,1)
plot(xp,yp(1,:),x,y,'o'),axis([0 10 0 300]),xlabel('x'),...
   ylabel('y'),title('First-degree fit')
% Plot the second-degree polynomial.
subplot(2,1,2)
plot(xp,yp(2,:),x,y,'o'),axis([0 10 0 300]),xlabel('x'),...
   ylabel('y'),title('Second-degree fit')
% Display the computed values.
disp('The J values are:'),J
disp('The S values are:'),S
disp('The r^2 values are:'),r2
```

The polynomial coefficients in the above script file are contained in the vector `polyfit (x,y,k)`. If you need the polynomial coefficients, say for the second-degree polynomial, type `polyfit(x,y,2)` after the program has been run.

The plots are shown in Figure C.3.1. The following table gives the values of J, S, and r^2 for each polynomial.

Degree n	J	S	r^2
1	4348	54,168	0.9197
2	197.9	58,318	0.9997

Because the second-degree polynomial has the largest r^2 value, it represents the data better than the first-degree polynomial, according to the r^2 criterion. This is also obvious from the plots.

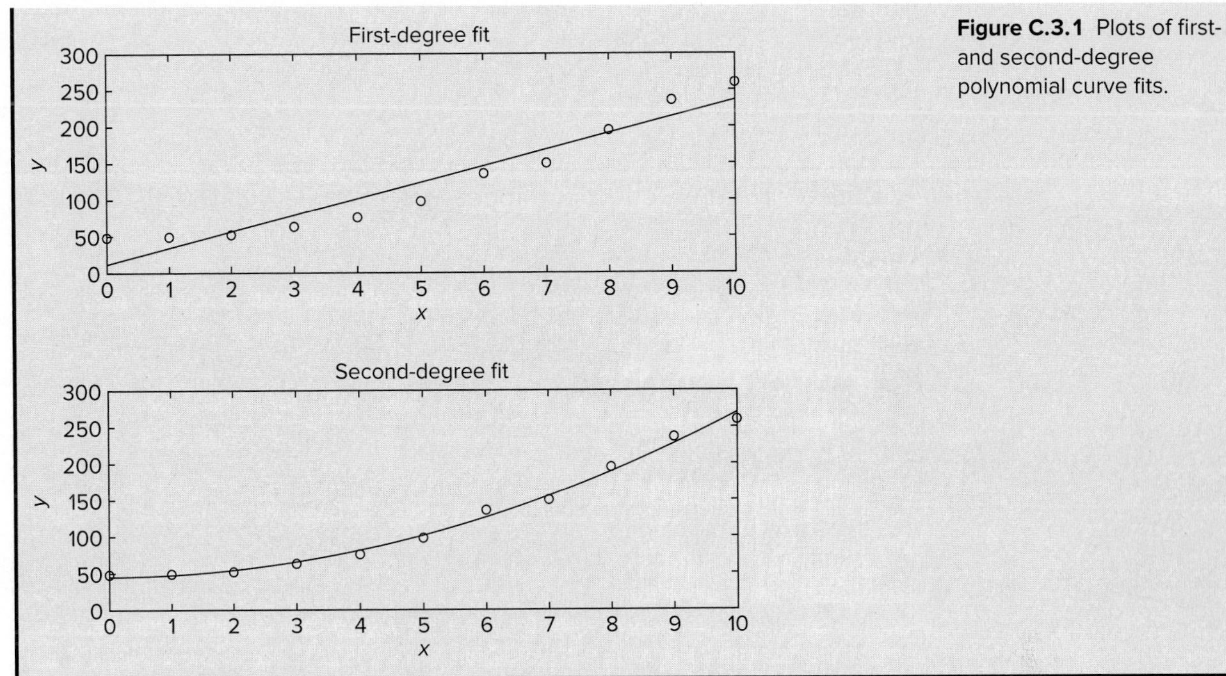

Figure C.3.1 Plots of first- and second-degree polynomial curve fits.

When we type `p=polyfit(z,w,1)`, MATLAB will fit a linear function $w = p_1 z + p_2$. The coefficients p_1 and p_2 are the first and second elements in the vector `p`; that is, `p` will be $[p_1, p_2]$. With a suitable transformation, the power and exponential functions can be transformed into a linear function, but the polynomial $w = p_1 z + p_2$ has a different interpretation in each of the three cases.

The linear function: $y = mx + b$. In this case the variables w and z in the polynomial $w = p_1 z + p_2$ are the original data variables, and we can find the linear function that fits the data by typing `p=polyfit(x,y,1)`. The first element p_1 of the vector `p` will be m, and the second element p_2 will be b.

The power function: $y = bx^m$. In this case $\log y = m \log x + \log b$, which has the form $w = p_1 z + p_2$, where the polynomial variables w and z are related to the original data variables x and y by $w = \log y$ and $z = \log x$. Thus, we can find the power function that fits the data by typing `p = polyfit(log10(x), log10(y),1)`. The first element p_1 of the vector `p` will be m, and the second element p_2 will be $\log b$. We can find b from $b = 10^{p_2}$.

The exponential function: $y = be^{mx}$. In this case, $\ln y = mx + \ln b$, which has the form $w = p_1 z + p_2$, where the polynomial variables w and z are related to the original data variables x and y by $w = \ln y$ and $z = x$. Thus, we can find the exponential function that fits the data by typing `p = polyfit(x,log(y),1)`. The first element p_1 of the vector `p` will be m, and the second element p_2 will be $\ln b$. We can find b from $b = e^{p_2}$.

Note

The notation for logarithms used by MATLAB is different than that used in mathematical expressions. Do not make the common mistake of using the MATLAB function `log` to represent the base-ten logarithm. The *natural* logarithm $\ln x$ is expressed in MATLAB by `log(x)`, whereas the *base-ten* logarithm $\log x$ is expressed as `log10(x)` in MATLAB.

Example C.3.2 illustrates how to use MATLAB to estimate the force-deflection characteristics of the cantilever support beam treated in Example 1.3.1.

EXAMPLE C.3.2

A Cantilever Beam Deflection Model

■ Problem

The force-deflection data from Example 1.3.1 for the cantilever beam shown in Figure 1.3.1 is given in the following table.

Force f (lb)	0	100	200	300	400	500	600	700	800
Deflection x (in.)	0	0.15	0.23	0.35	0.37	0.5	0.57	0.68	0.77

Use MATLAB to obtain a linear relation between x and f, estimate the stiffness k of the beam, and evaluate the quality of the fit.

■ Solution

Note that here x is the *dependent* variable and f is the *independent* variable. In the following MATLAB script file the data are entered in the arrays `x` and `f`. The arrays `xp` and `fp` are created to plot the straight line at many points.

```
% Enter the data.
x = [0, 0.15, 0.23, 0.35, 0.37, 0.5, 0.57, 0.68, 0.77];
f = (0:100:800);
% Fit a first-degree polynomial.
p = polyfit(f,x,1)
% Compute the stiffness.
k = 1/p(1)
% Compute a set of f, x points.
fp = (0:800);
xp = p(1)*fp+p(2);
% Plot the fitted function and the data.
plot(fp,xp,f,x,'o'), xlabel('Applied Force f (lb)'), ...
    ylabel('Deflection x (in.)'), ...
    axis([0 800 0 0.8])
% Compute the J, S, and r squared values.
J = sum((polyval(p,f)-x).^2)
S = sum(x-mean(x)).^2)
r2 = 1 - J/S
```

The computed values in the array `p` are $p = [9.1667 \times 10^{-4}, 3.5556 \times 10^{-2}]$. Thus, the fitted straight line is $x = 9.1667 \times 10^{-4}f + 3.5556 \times 10^{-2}$. Note that this line, which is shown in Figure C.3.2, does not pass through the origin as required, but it is close (it predicts that $x = 0.035556$ in. when $f = 0$). The quality-of-fit values are $J = 0.0048$, $S = 0.5090$, and $r^2 = 0.9906$, which indicates a very good fit.

Solving for f gives $f = (x - 3.5556 \times 10^{-2})/9.1667 \times 10^{-4} = 1091x - 38.7879$. The computed value of the stiffness k is the coefficient of x; thus, $k = 1091$ lb/in.

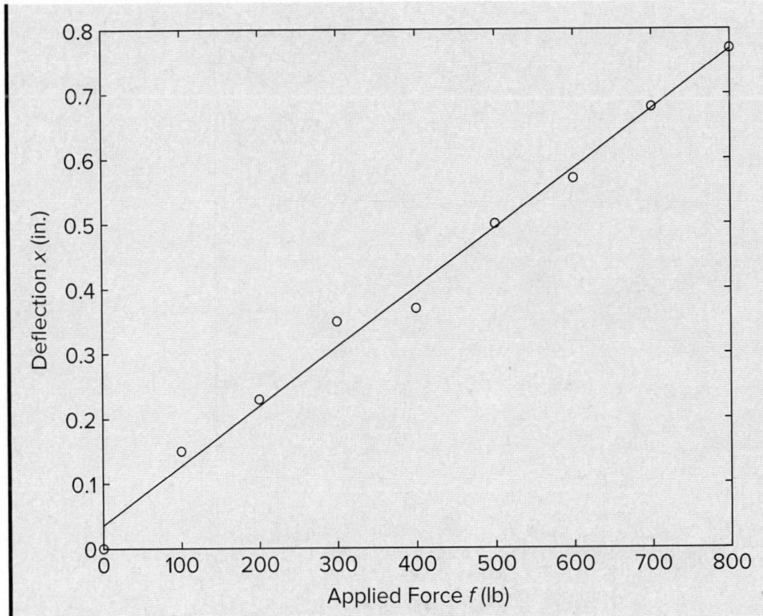

Figure C.3.2
Unconstrained linear model of beam deflection.

EXAMPLE C.3.3

Constraining the Curve Fit

■ Problem

Use MATLAB to fit a straight line to the beam force-deflection data given in Example C.3.2, but constrain the line to pass through the origin.

■ Solution

We can apply (C.2.3), noting here that the measured variable is the deflection x and the independent variable is the force f. Thus, (C.2.3) becomes

$$m \sum_{i=1}^{n} f_i^2 = \sum_{i=1}^{n} f_i x_i \tag{1}$$

The MATLAB program to solve this equation for m and k is

```
% Enter the data
x = [0, 0.15, 0.23, 0.35, 0.37, 0.5, 0.57, 0.68, 0.77];
f = (0:100:800);
% Compute m from (1).
m = sum(f.*x)/sum(f.^2);
% Compute the stiffness.
k = 1/m
% Compute J, S, and r squared.
J = sum((m*f-x).^2)
S = sum((x-mean(x)).^2)
r2 = 1 - J/S
```

The answer is $k = 1021$ lb/in. The corresponding line is shown in Figure C.3.3. The quality-of-fit values are $J = 0.0081$, $S = 0.5090$, and $r^2 = 0.9840$, which indicates a good fit.

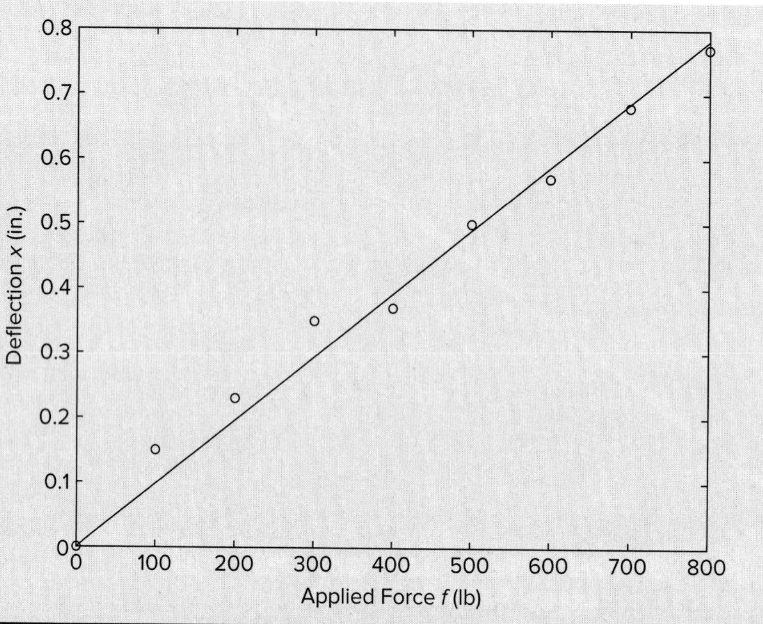

Figure C.3.3 Linear model constrained to pass through the origin.

EXAMPLE C.3.4

Temperature Dynamics of Water

■ Problem

Consider again Example C.1.1. Water in a glass measuring cup was allowed to cool after being heated to 204°F. The ambient air temperature was 70°F. The measured water temperature at various times is given in the following table.

Time (sec)	0	120	240	360	480	600
Temperature (°F)	204	191	178	169	160	153

Time (sec)	720	840	960	1080	1200
Temperature (°F)	147	141	137	132	127

Obtain a functional description of the water temperature versus time.

■ Solution

From Example C.1.1, we learned that the relative temperature, $\Delta T = T - 70$ has the exponential form

$$\Delta T = be^{mt} \tag{1}$$

We can find values of m and b by using `p = polyfit(x,log(y),1)`. The first element p_1 of the vector p will be m, and the second element p_2 will be $\ln b$. We can find b from $b = e^{p_2}$. The following MATLAB program performs the calculations.

```
% Enter the data.
time = (0:120:1200);
temp = [204,191,178,169,160,153,147,141,137,132,127];
% Compute the relative temperature and its logarithm.
rel_temp = temp - 70;
```

```
log_rel_temp = log(rel_temp);
% Fit a first-degree polynomial.
p = polyfit(time,log_rel_temp,1);
% Compute m and b from the polynomial coefficients.
m = p(1),b = exp(p(2))
% Compute DT (delta T) from (1).
DT = b*exp(m*time);
% Compute J, S, and r squared.
J = sum((DT-rel_temp).^2)
S = sum((rel_temp - mean(rel_temp)).^2)
r2 = 1 - J/S
```

The results are $m = -6.9710 \times 10^{-4}$ and $b = 1.2916 \times 10^2$, and the corresponding function is

$$\Delta T = be^{mt} \quad \text{or} \quad T = \Delta T + 70 = be^{mt} + 70$$

The quality-of-fit values are $J = 47.4850$, $S = 6.2429 \times 10^3$, and $r^2 = 0.9924$, which indicates a very good fit.

Orifice Flow — EXAMPLE C.3.5

■ Problem

Consider again Example C.1.2. A hole 6 mm in diameter was made in a translucent milk container (Figure C.1.7). A series of marks 1 cm apart was made above the hole. While adjusting the tap flow to keep the water height constant, the time for the outflow to fill a 250 ml cup was measured (1 ml = 10^{-6} m^3). This was repeated for several heights. The data are given in the following table.

Height h (cm)	11	10	9	8	7	6	5	4	3	2	1
Time t (s)	7	7.5	8	8.5	9	9.5	11	12	14	19	26

Obtain a functional description of the volume outflow rate f as a function of water height h above the hole.

■ Solution

First obtain the flow rate data in ml/s by dividing the 250 ml volume by the time to fill:

$$f = \frac{250}{t} \tag{1}$$

In Example C.1.2, we learned that the following power function can describe the data:

$$f = bh^m \tag{2}$$

We can find the values of m and b by using `p=polyfit(log10(x),log10(y),1)`. The first element p_1 of the vector `p` will be m, and the second element p_2 will be $\log b$. We can find b from $b = 10^{p_2}$. The following MATLAB program performs the calculations.

```
% Enter the data.
h = (1:11);
time = [26, 19, 14, 12, 11, 9.5, 9, 8.5, 8, 7.5, 7];
% Compute the flow rate from (1) and its logarithm.
flow = 250./time;
```

```
logflow = log10(flow);logheight = log10(h);
% Fit a first-degree polynomial.
p = polyfit(logheight,logflow,1);
% Compute m and b from the polynomial coefficients.
m = p(1),b =10^p(2)
% Compute f from (2).
f = b*h.^m;
% Compute J, S, and r squared.
J = sum((f - flow).^2)
S = sum((flow - mean(flow)).^2)
r2 = 1 - J/S
```

The results are $m = 0.5499$ and $b = 9.4956$, and the corresponding function is

$$f = bh^m = 9.4956h^{0.5499}$$

The quality-of-fit values are $J = 2.5011$, $S = 698.2203$, and $r^2 = 0.9964$, which indicates a very good fit.

Sometimes we know from physical theory that the data can be described by a power function with a specified exponent. For example, Torricelli's principle of hydraulic resistance states that the volume flow rate f of a liquid through a restriction is proportional to the square root of the pressure drop p across the restriction; that is, $q = c\sqrt{p} = cp^{1/2}$. In many applications, the pressure drop is due to the weight of liquid in a container. This is the case for water in the milk container of Example C.1.2. In such situations, Torricelli's principle states that the flow rate is proportional to the square root of the height h of the liquid above the orifice. Thus,

$$f = b\sqrt{h} = bh^{1/2}$$

where b is a constant that must be determined from data.

EXAMPLE C.3.6

Orifice Flow with Constrained Exponent

■ Problem

Consider the data of Example C.3.5. Determine the best-fit value of the coefficient b in the square-root function

$$f = bh^{1/2} \tag{1}$$

Height h (cm)	11	10	9	8	7	6	5	4	3	2	1
Time t (s)	7	7.5	8	8.5	9	9.5	11	12	14	19	26

■ Solution

First obtain the flow rate data in ml/s by dividing the 250 ml volume by the time to fill:

$$f = \frac{250}{t} \tag{2}$$

Referring to Example C.2.3, whose model is $y = bx^m$, we see here that $y = f$, $h = x$, and $m = 0.5$. From Equation (1) of Example C.2.3,

$$b = \frac{\sum_{i=1}^{n} h_i^{0.5} y_i}{\sum_{i=1}^{n} h_i} \tag{3}$$

The MATLAB program to carry out these calculations is shown next.

```
% Enter the data.
h = (1:11);
time = [26, 19, 14, 12, 11, 9.5, 9, 8.5, 8, 7.5, 7];
% Compute flow rate from (2).
flow = 250./time;
% Compute b from (3).
b = sum(sqrt(h).*flow)/sum(h)
% Compute f from (1).
f = b*sqrt(h);
% Compute J, S, and r squared.
J = sum((f - flow).^2)
S = sum((flow - mean(flow)).^2)
r2 = 1 - J/S
```

The result is $a = 10.4604$ and the flow model is $f = 10.4604\sqrt{h}$. The quality-of-fit values are $J = 5.5495$, $S = 698.2203$, and $r^2 = 0.9921$, which indicates a very good fit.

PROBLEMS

Section C.1 Function Identification and Parameter Estimation

In the following problems for Section C.1, plot the data on suitable axes, and solve the problem by drawing a straight line by eye using a straightedge.

C.1 In each of these problems, plot the data and determine the best function $y(x)$ (linear, exponential, or power function) to describe the data.

a.

x	25	30	35	40	45
y	0	250	500	750	1000

b.

x	2.5	3	3.5	4	4.5	5	5.5	6	7	8	9	10
y	1500	1220	1050	915	810	745	690	620	520	480	410	390

c.

x	1000	2000	3000	4000	5000
y	41.2	18.62	8.62	3.92	1.86

C.2 The population data for a certain country are given here.

Year	2005	2006	2007	2008	2009	2010
Population (millions)	9.8	10.5	11.3	12.0	12.9	13.8

Plot the data and obtain a function that describes the data. Estimate when the population will be double its 2005 size.

C.3 The *half-life* of a radioactive substance is the time it takes to decay by half. The half-life of carbon-14, which is used for dating previously living things, is 5500 years. When an organism dies, it stops accumulating carbon-14. The carbon-14 present at the time of death decays with time. Let $C(t)/C(0)$ be the fraction of carbon-14 remaining at time t. In radioactive carbon dating, it is usually assumed that the remaining fraction decays exponentially according to the formula

$$\frac{C(t)}{C(0)} = e^{-bt}$$

a. Use the half-life of carbon-14 to find the value of the parameter b and plot the function.

b. Suppose we estimate that 90% of the original carbon-14 remains. Estimate how long ago the organism died.

c. Suppose our estimate of b is off by $\pm 1\%$. How does this affect the age estimate in part (b)?

C.4 *Quenching* is the process of immersing a hot metal object in a bath for a specified time to improve properties such as hardness. A copper sphere 25 mm in diameter, initially at 300°C, is immersed in a bath at 0°C. Measurements of the sphere's temperature versus time are shown here. Plot the data and find a functional description of the data.

Time (s)	0	0.1	0.2	0.3	0.4	0.5	0.6
Temperature (°C)	300	150	75	35	12	5	0

C.5 The useful life of a machine bearing depends on its operating temperature, as shown by the following data. Plot the data and obtain a functional description of the data. Estimate a bearing's life if it operates at 150°F.

Temperature (°F)	100	120	140	160	180	200	220
Bearing life (hours × 10^3)	28	21	15	11	8	6	4

C.6 A certain electric circuit has a resistor and a capacitor. The capacitor is initially charged to 100 V. When the power supply is detached, the capacitor voltage decays with time as shown in the following data table. Find a functional description of the capacitor voltage v as a function of time t.

Time (s)	0	0.5	1	1.5	2	2.5	3	3.5	4
Voltage (V)	100	62	38	21	13	7	4	2	3

C.7 Water (of volume 425 ml) in a glass measuring cup was allowed to cool after being heated to 207°F. The ambient air temperature was 70°F. The measured water temperature at various times is given in the following table.

Time (sec)	0	300	600	900	1200	1500
Temperature (°F)	207	182	167	155	143	135

Time (sec)	1800	2100	2400	2700	3000
Temperature (°F)	128	123	118	114	109

Obtain a functional description of the relative water temperature ($\Delta T = T - 70$) versus time.

C.8 Consider the milk container of Example C.1.2 (Figure C.1.7). A straw 19 cm long was inserted in the side of the container. While adjusting the tap flow to keep the water height constant, the time for the outflow to fill a 250-ml cup was measured. This was repeated for several heights. The data are given in the following table.

Height (cm)	11	10	9	8	7	6	5	4	3	2	1
Time (s)	7	7	7	8	9	10	11	13	15	17	23

Obtain a functional description of the volume outflow rate f through the straw as a function of water height h above the hole.

C.9 Consider the milk container of Example C.1.2 (Figure C.1.7). A straw 9.5 cm long was inserted in the side of the container. While adjusting the tap flow to keep the water height constant, the time for the outflow to fill a 250-ml cup was measured. This was repeated for several heights. The data are given in the following table.

Height (cm)	11	10	9	8	7	6	5	4	3	2	1
Time (s)	6	6	6	7	8	9	9	11	13	17	21

Obtain a functional description of the volume outflow rate f through the straw as a function of water height h above the hole.

C.2 Fitting Models to Scattered Data

C.10 Use the least-squares method to fit the linear function $y = mx + b$ to the data given in the following table. Evaluate the quality of the fit by computing J, S, and r^2.

x	0	2	4	6
y	4.5	39	72	94

C.11 Use the least-squares method to fit the power function $y = bx^m$ to the data given in the following table. Evaluate the quality of the fit by computing J, S, and r^2.

x	0	1	2	3	4
y	1	8	50	178	490

C.12 Use the least-squares method to fit the exponential function $y = be^{mx}$ to the data given in the following table. Evaluate the quality of the fit by computing J, S, and r^2.

x	0	0.4	0.8	1.2
y	6.3	22	60	215

C.13 Use the least-squares method to fit the linear function $y = mx + b$ to the data given in the following table. Constrain the function to pass through the point (0, 0). Evaluate the quality of the fit by computing J, S, and r^2.

x	0	2	4	6
y	4.5	39	72	94

C.14 Use the least-squares method to fit the power function $y = bx^m$ to the data given in the following table. Constrain the exponent of the function to be $m = 3$. Evaluate the quality of the fit by computing J, S, and r^2.

x	0	1	2	3	4
y	1	8	50	178	490

C.15 (a) Use the least-squares method to derive the equation for b to fit the exponential function $y = be^{mx}$ to a given set of data y_i, where the exponent m is constrained to a specified value. (b) Fit the function $y = be^{-3x}$ to the data given in the following table. Evaluate the quality of the fit by computing J, S, and r^2.

x	0	0.4	0.8	1.2
y	6.3	22	60	215

C.16 Use the least-squares method to fit the linear function $y = mx + b$ to the function $y = 5x^2$ over the range $0 \le x \le 4$.

C.17 (a) Use the least-squares method to fit the linear function $y = mx + b$ to the function $y = ax^2 + bx$ over the range $0 \le x \le L$. (b) Apply the results to the case where $a = 3$, $b = 5$, and $L = 2$.

C.18 (a) Use the least-squares method to fit the linear function $y = mx + b$ to the exponential function $y = be^{mx}$ over the range $0 \le x \le L$. (b) Apply the results to the case where $m = -5$, $b = 15$, and $L = 1$.

C.3 MATLAB and the Least-Squares Method

In Problems C.19 through C.27, work the problem using MATLAB or some other computer method.

C.19 Do Problem C.1.

C.20 Do Problem C.2.

C.21 Do Problem C.3.

C.22 Do Problem C.4.

C.23 Do Problem C.5.

C.24 Do Problem C.6.

C.25 Do Problem C.7.

C.26 Do Problem C.8.

C.27 Do Problem C.9.

ANSWERS TO SELECTED PROBLEMS

CHAPTER 1

1.15 $f = 0.2x$

1.19 $f(h) \approx 5 + (h - 25)/10$

1.24 $f = 10{,}250$ N, $t_f = 200$ s

1.29 $T_{max} = 33$, $T_{rms} = 6.938$ N·m

CHAPTER 2

2.2 a. $x = \sqrt{5}\tanh(5\sqrt{5}t + C)$
b. $x(t) = 3\tan(12t + C), \quad C = \tan^{-1}(10/3)$
c. Closed-form solution does not exist.
d. $x(t) = 5e^{\frac{1}{2}e^{-4t}}/\sqrt{e}$

2.9 a. \$10,511.62 b. \$10,512.71 c. 0.04879

2.11 5.4545 grams

2.12 26.6003 gallons

2.16 a. Free: $x(t) = 27e^{-3t} - 17e^{-5t}$
b. Steady state: $x_{ss} = 30/15 = 2$
c. Forced: $x(t) = 2 - 5e^{-3t} + 3e^{-5t}$
d. Total: $x(t) = 2 + 22e^{-3t} - 14e^{-5t}$

2.17 a. unstable b. unstable c. unstable d. neutrally stable
e. neutrally stable f. neutrally stable

2.19 a. $\tau = 6$ b. $\tau = 25/6$ c. $\tau = 5$ d. No time constant

2.22 $\tau = 8$ s

2.30 a. $x(0+) = 5/3, x(\infty) = 0$ b. $x(0+) = 0, x(\infty) = 0$

2.34 a. $x(t) = 3(1 - e^{-4t})/2$ b. $x(t) = (5 + 31e^{-3t})/3$
c. $x(t) = (13e^{-5t} - e^{-2t})/3$ d. $x(t) = (20t - 5 + 5e^{-4t})/32$
e. $x(t) = (10t + 13 - 13e^{-5t})/25$ f. $x(t) = (79e^{-3t} - 79e^{-7t} - 124te^{-3t})/16$

2.41 a. $x(t) = (25e^{-3t} - 9e^{-7t})/4$ b. $x(t) = e^{-7t} + 10te^{-7t}$
c. $x(t) = (5/3)e^{-7t}\sin 3t + 4e^{-7t}\cos 3t$

2.43 $d(\infty) = 48$

2.46 $X(s)/F(s) = 1/(4s^2 + 20s + 17), Y(s)/F(s) = 4s/(4s^2 + 20s + 17)$

2.56 a. $x(t) = (25/7)e^{-5t/7}, x(0+) = 25/7 \neq x(0)$
b. $x(t) = (5/12)(e^{-3t} - e^{-7t}), x(0+) = 0 = x(0), \dot{x}(0+) = 5/3 \neq \dot{x}(0)$

CHAPTER 3

3.3 If $f_1 = 100$, the block will continue to move up the plane. If $f_1 = 50$, it will stop in 1.103 s.

3.12 a. The mass m_1 will lift m_2 if $2m_1 - m_2 \sin\theta > 0$.
b. The mass m_1 will lift m_2 if $1 - (\mu_d \cos\theta + \sin\theta) > 0$.

3.19 $I_e = I_1 + \left(I_2 + m_2R^2 + m_3R^2\right)/4;\ I_e\dot{\omega}_1 = T_1 + gR\left(m_3 - m_2\right)/2$

3.24 $\dot{\omega}_3 = 6.4465T$

3.25 $I\dot{\omega}_4 = T_1/1.4$

3.28 $v = 4.83t$ ft/sec

3.30 $(6m_2 + m_1)\ddot{x} = m_1 g$

3.31 a. $48\dot{\omega} = f\cos\phi$ b. $120\ddot{x} = f\cos\phi$

3.39 The maximum acceleration is $a_{Gx} = (23.544\mu_s\cos\theta)/(4.6 - \mu_s) - 9.81\sin\theta$.

CHAPTER 4

4.4 $k_e = \left(k_1L_2^2 + k_2L_3^2\right)/L_1^2$

4.7 $k_e = 1.117 \times 10^9$ N/m

4.9 $k_e = 5k/8$

4.19 $m\ddot{x} + kx/4 = f + mg$

4.20 a. $m\ddot{x} = -\left(\pi\rho g D^2/4\right)x$
b. $\omega_n = (D/2)\sqrt{\pi\rho g/m}$
c. $P = 1.767$ sec

4.23 $1.5m\ddot{x} + kx = ky$, $x(t) = 1 - \cos(\sqrt{50}t)$

4.27 $\omega_n = 58.78$ rad/sec

4.30 $1.5m\ddot{x} + (k_1 + k_2)x = 0$

4.32 $(6m_2 + m_1)\ddot{x} + 4kx = 0$

4.35 $\omega_n = (L_2/L_1)\sqrt{k/m}$

4.43 $m_e = 1.307$ slug, $k = 44.44$ lb/in

4.44 $m_e = 56.08$ kg, $k = 1.226 \times 10^5$ N/m

4.53 a. $(I + mR^2)\dot{v} + c_Tv = TR - mgR^2$
b. $v(t) = 529.9\left(1 - e^{-0.0357t}\right)$

4.55 $X(s)/Y(s) = \left(cs + k_1\right)/\left(ms^2 + cs + k_1 + k_2\right)$

4.64 $X(s)/Y(s) = ck_1s/\left[mcs^3 + mk_1s^2 + c(k_1 + k_2)s + k_1k_2\right]$

4.69 $I\ddot{\theta} + cL_2^2\dot{\theta} + (k_1L_1^2 + k_2L_2^2)\theta = k_2L_2y + cL_2\dot{y}$

4.83 $x(t) = \left(v_1/5\right)\sqrt{5m/k}\sin\sqrt{k/5m}t$

CHAPTER 5

5.1 $X(s)/F(s) = 6/(s + 2)$

5.3 $X(s)/F(s) = 4/(s^2 + 8s + 7)$ $X(s)/F(s) = 4/(s^2 + 8s + 24)$

5.6 Let $A(s) = 21s^2 + 14s + 11$.

$$C(s) = \frac{4s + 10}{A(s)}R(s) - \frac{3s + 1}{A(s)}D(s),$$

$$E(s) = \frac{21s^2 + 10s + 1}{A(s)}R(s) + \frac{3s + 1}{A(s)}D(s),$$

$$M(s) = \frac{(4s + 10)(7s + 1)}{A(s)}R(s) + \frac{4s + 10}{A(s)}D(s)$$

5.11 Let $x_1 = x$ and $x_2 = \dot{x}$. $\dot{x}_1 = x_2$; $\dot{x}_2 = (1/6)[-9x_1 - 4x_2 + 7f(t)]$

5.12 Let $x_1 = x$ and $x_2 = \dot{x}$. $\dot{x}_1 = x_2$; $\dot{x}_2 = (1/3)[7f(t) - 12x_1 - 5x_2]$

5.19

$$\mathbf{A} = \begin{bmatrix} -6 & 4 \\ 0 & -5 \end{bmatrix} \qquad \mathbf{B} = \begin{bmatrix} 7 & 0 \\ 0 & 9 \end{bmatrix}$$

$$\mathbf{C} = \begin{bmatrix} 1 & 4 \\ 0 & 1 \end{bmatrix} \qquad \mathbf{D} = \begin{bmatrix} 7 & 0 \\ 0 & 0 \end{bmatrix}$$

5.24 a. The matrices are

$$\mathbf{A} = \begin{bmatrix} -6 & 7 \\ 1 & -5 \end{bmatrix} \qquad \mathbf{B} = \begin{bmatrix} 0 \\ 6 \end{bmatrix}$$

$$\mathbf{C} = \begin{bmatrix} 1 & 0 \\ 0 & 1 \end{bmatrix} \qquad \mathbf{D} = \begin{bmatrix} 0 \\ 0 \end{bmatrix}$$

The MATLAB script is:

```
A = [-6, 7; 1, -5]; B = [0;6];
C = [1, 0; 0, 1]; D = [0;0];
sys = ss(A,B,C,D); systf = tf(sys)
```

$X_1(s)/U(s) = 42/(s^2 + 11s + 23)$, $X_2(s)/U(s) = (6s + 36)/(s^2 + 11s + 23)$

5.26 b. `sys1 = tf([1, 2], [1, 4, 3]); sys2 = ss(sys1)`

$$\mathbf{A} = \begin{bmatrix} -4 & -1.5 \\ 2 & 0 \end{bmatrix} \qquad \mathbf{B} = \begin{bmatrix} 2 \\ 0 \end{bmatrix}$$

$$\mathbf{C} = \begin{bmatrix} 0.5 & 0.5 \end{bmatrix} \qquad \mathbf{D} = [0]$$

$$\dot{x}_1 = -4x_1 - 1.5x_2 + 2f, \quad \dot{x}_2 = 2x_1, \quad y = 0.5x_1 + 0.5x_2$$

5.35 Create the file:

```
function vdot = problem35(t,v)
vdot = (8000-20*v-0.05*v.^2)/50;
```

Then type

```
[t,v] = ode45('problem35',[0, 5], 0);
plot(t,v),xlabel('t (sec)'),ylabel('v (ft/sec)')
```

The solution reaches steady state in about 4.6 sec.

CHAPTER 6

6.1 $v_s = (8R/5)i$

6.4 $i = 2.411 \times 10^{-4}$ A, $P = 2.17 \times 10^{-3}$ W

6.5 $RC(dv_1/dt) + v_1 = Ri_s$

6.10 $L(di/dt) + Ri = v_s$, $i(t) = (V/R)\left(1 - e^{-Rt/L}\right)$

6.12 $(L/R)(dv_0/dt) + v_0 = (L/R)dv_s/dt$

6.14 $LC(d^2i/dt^2) + RC(di/dt) + i = C(dv_s/dt)$

6.18 Choose v_1 and i_1 as the state variables. $C(dv_1/dt) = i_s - i_1$, $L(di_1/dt) = v_1 - Ri_1$

6.21 $V_o(s)/V_s(s) = (R_1R_2Cs + R_1)/(R_1R_2Cs + R_2 + R_1)$

6.23 $V_o(s)/V_s(s) = Ls/(RLCs^2 + Ls + R)$

6.27 The impedance of the parallel combination is $Z_1(s) = R_1/(R_1C_1s + 1)$. The impedance of the series combination is: $Z_2(s) = (R_2C_2s + 1)/C_2s$.

$$\frac{V_o(s)}{V_s(s)} = \frac{(R_1C_1s + 1)(R_2C_2s + 1)}{R_1C_2s + (R_1C_1s + 1)(R_2C_2s + 1)}$$

6.31 $\dfrac{V_o(s)}{V_i(s)} = -\dfrac{R_3(R_1Cs + 1)}{R_1R_2Cs + R_1 + R_2}$

6.33 $\dfrac{V_o(s)}{V_i(s)} = -\dfrac{R_1(R_2C_2s + 1)}{R_2(R_1C_1s + 1)}$

6.36 $\dfrac{\Theta(s)}{V_f(s)} = \dfrac{K_T}{IL_fs^3 + (R_fI + L_fc)s^2 + R_fcs}, \dfrac{\Theta(s)}{T_L(s)} = -\dfrac{1}{s(Is + c)}$

6.42 $E = 1586.8$ J per cycle, $\omega_{max} = 1.386$ rad/s = 13.2 rpm, $T_{max} = 5.19$ N·m, $T_{rms} = 4.24$ N·m, $i_{max} = 17.3$ A, $i_{rms} = 14.1$ A, $v_{max} = 69.6$ V

6.46 a. Vibrometer with a gain of 1.
b. Accelerometer with a gain of 10^{-6}.

CHAPTER 7

7.4 89.4 ft

7.6 $\omega(t) = 26.087t$ rad/s

7.8 $150\, ds_o/dt = s_i - 6s_o$

7.10 $\rho 2L\sqrt{Dh - h^2}(dh/dt) = q_m$

7.13 $R = (1/C_dA_o)\sqrt{2gh_r}$

7.15 a. $R = 24.525\ \text{m}^{-1}\text{s}^{-1}$ b. 7.5 m

7.20 $h = \hat{h} - h_r$. If $\hat{h} < D$, $\rho A(dh/dt) = (1/R_1)(p_s - \rho gh) + q_{mi}$.
If $\hat{h} \geq D$, $\rho A(dh/dt) = (1/R_1)(p_s - \rho gh) + q_{mi} - (1/R_2)\rho g(h - D)$.

7.23 a. $h_1 = \hat{h}_1 - h_{1r}, h_2 = \hat{h}_2 - h_{2r}; \rho A_1(dh_1/dt) = q_{mi} - (\rho g/R_1)(h_1 - h_2)$, $A_2(dh_2/dt) = (g/R_1)(h_1 - h_2) - (g/R_2)h_2$

7.30 $(\rho RA^2/2k)(dp/dt) + p = (p_1 + p_2)/2$

7.39 $C = 2.2 \times 10^{-5}$ slug-ft²/lb

7.41 $C_1\dfrac{d(\delta p_1)}{dt} = \dfrac{1}{R_1}(\delta p_i - \delta p_1) - \dfrac{1}{R_2}(\delta p_1 - \delta p_2), C_2\dfrac{d(\delta p_2)}{dt} = \dfrac{1}{R_2}(\delta p_1 - \delta p_2)$

7.42 a. $C = 1045$ J/°C
b. $E = 8.256 \times 10^4\ J$

7.44 $T(t) = 20e^{-t/24} + \left(1 - e^{-t/24}\right)80$

7.45 a. $R_1 = 7.958$ °C/W, $R_2 = 7.074$ °C/W, $R = R_1 + R_2 = 15.032$ °C/W
b. $q = 1.995$ W

7.48 a. $R = 1.253$ °F-sec/lb-ft
b. $q_{\text{brick}} = 10.32, q_{\text{concrete}} = 21.6$; Total heat flow rate: $q = 31.92$ ft-lb/sec

7.51 a. $C_1 = 4.85 \times 10^7$ ft-lb/°F
b. $4.85 \times 10^7\ R_1(dT_1/dt) + T_1 = T_0$

7.56 $114.2(dT/dt) = 50 - T$, 168.5 s

7.57 $114.2(dT_1/dt) = T_2 - T_1$, $9.3125 \times 10^4(dT_2)(dt) = T_1 - T_2$, 168.8 s

CHAPTER 8

8.4 a. $x_{ss} = 10, t = 4\tau = 8$
b. $x_{ss} = 10, t = 4\tau = 8$
c. $x_{ss} = 200, t = 4\tau = 8$

8.7 9 rad/s, in about 80 s

8.13 a. $v(t) = 1/3 - (1/6)e^{-0.5t}$, $v(0+) = 1/6$
b. $v(t) = 1/3 - (5/27)e^{-0.5t} - (4/27)e^{-5t}$, $v(0+) = 0$

8.19 The response will be parallel to the input only if $k = 1$. In that case, the difference between the input and the response is ac.

8.22 Dominant root pair is $s = -2 \pm 4j$. $\zeta = 0.447$, $\tau = 1/2$, $\omega_d = 4$

8.27 $\Omega(s)/V(s) = 500/(s^2 + 5000.2s + 1000)$, $\zeta = 79.06$, $\tau_1 = 5$, $\tau_2 = 1/5000$ s, $\omega_n = 31.62$ rad/s

8.29 Maximum percent overshoot: $\approx 5\%$. Maximum overshoot: ≈ 0.01. Peak time: 1.6. Delay time: 0.53. 2% settling time: 2.

8.30 $c = 3.6$

8.33 $C = 10^{-5}$ F

8.34 ≈ 721 sec

8.35 $m = 66.476$ kg, $c = 557.7$ N · s/m, $k = 40{,}000$ N/m

8.39 Maximum percent overshoot: 21%. Peak time: 0.524. Rise time: 0.339.

CHAPTER 9

9.5 a. $y_{ss}(t) = 13.5582 \sin(1.5t - 0.8622)$
b. $y_{ss}(t) = 20.8013 \sin(2t + 0.5880)$
c. $y_{ss}(t) = 1.8605 \sin(100t + 05191)$
d. $y_{ss}(t) = 2.4634 \sin(50t - 0.5238)$

9.7 Time constant: 2.135 hr. Time lag: 1.061 hr.

9.8 $T(t) = 15 + 2.32\ \sin(7.272 \times 10^{-5}t - 1.087)$

9.9 $\omega_{ss}(t) = 7.5 + 0.6935 \sin(3t - 0.983) + 0.1856 \cos(5t - 1.19)$

9.12 a. $y_{ss}(t) = 34.92 \sin(0.2t - 1.7819)$
b. $y_{ss}(t) = 0.088 \sin(5t - 0.588)$

9.20 $\Omega_2(s)/T_1(s) = 100/(s^2 + 5s + 2)$,
$\omega_{2ss} = 200 + 26.6518\ \sin(1.5t - 1.604) + 8.8252\ \sin(2t - 1.763)$

9.21 Beat period: 31.4159. Vibration period: 1.232.

9.23 2500 N

9.24 5.5%

9.25 $c = 0$, $k = 556.7$ lb/ft. No more than 15%.

9.29 $\omega_r = 7.07$ rad/s, $M_r = 1.155$, bandwidth: $\omega = 0$ to $\omega = 11.69$ rad/s.

9.43 $v_{oss} = 6.3662 - 3.8664\ \cos(240\pi t - 0.425) - 0.6294\ \cos(480\pi t - 0.735)$

CHAPTER 10

10.8 $K_D = 20, K_P = 8, K_I = 5$

10.9 Using $C = 10^{-6}$ F, $R_i = 1.49 \times 10^7\ \Omega$, $R_f = 8.57 \times 10^7\ \Omega$

10.12 a. $y_{ss} = 27.2$
b. $y_{ss} = -40/7$
c. Since one root is positive, the final-value theorem cannot be applied.
d. Since one root is positive, the final-value theorem cannot be applied.

10.13 a. $e_{ss} = 9$
b. $e_{ss} = \infty$
c. $e_{ss} = 7.5$
d. $e_{ss} = \infty$

10.15 Stable if and only if $0 < K < 216$.

10.21 If $c = 5$, $K_I = 5/8$, $\tau = 4$. If $c = 0.2$, $K_I = 0.001$, $\tau = 100$.

10.22 a. Case 1: $K_P = 36$, $K_I = 200$. Case 2: $K_P = 36$, $K_I = 100$. Case 3: $K_P = 216$, $K_I = 1000$.

10.26 a. Case 1: $K_2 = 36$, $K_I = 200$. Case 2: $K_2 = 36$, $K_I = 100$. Case 3: $K_2 = 216$, $K_I = 1000$.

10.39 One possible solution: $K_P = 10$, $K_D = 10$, $\tau = 2$.

10.40 $K_P = 10, K_D = 15$

10.48 a. For case 1: $K_P = 550$, $K_D = 103$, $K_I = 250$
For case 2: $K_P = 35$, $K_D = 33$, $K_I = 10$.

10.58 With PD control, $f = -K_P\theta - K_D\dot{\theta}$, where $K_P = 1801.6$ lb/rad and $K_D = 640$ lb/rad/sec.

CHAPTER 11

11.1 The smallest possible dominant-time constant is $1/2$ and is obtained for any $k \geq 12$.

11.2 The smallest possible dominant-time constant is $1/2$ and is obtained with $c = 12$.

11.6 a. The poles are $s = 0, -3, -4$. The zeros are $s = -5$ and $s = -2.5 \pm 6.614j$.
b. $K = 0.2K_P$
c. The closed-loop poles are $s = -4.311$ and $1.407 \pm 5.854j$. The closed-loop zeros are $s = -5$ and $-3.5 \pm 6.614j$.

11.9 a. $K = 5p/4$

11.12 a. $K = -6p/7 \leq 0$
b. $K = -p/4 \leq 0$
c. $K = -p \leq 0$

11.23 For P action: $K_P = 0.5K_{Pu} = 1.75$. For PI action: $K_P = 1.575$, $K_I = 0.37$. For PID action: $K_P = 2.1$, $K_I = 0.8189$, $K_D = 1.3464$.

11.26 $K_P = 36, K_I = 38.006, K_D = 8.525$

11.28 a. $K_P = 180, K_I = 1000$
b. $K_P = 1, K_I = 11.025$

11.32 $K_1 = 0.0048$ V/rad, $K_2 = 0.00264$ V·rad, $K_3 = 0.024$ V/rad

11.33 Stable if and only if $K_2 > 0$ and $K_1 > g$.

11.36 $u = -K_1\delta - K_2\omega - K_3\phi$, $K_1 = 39$, $K_2 = 56.1$, $K_3 = 400$

CHAPTER 12

12.1 $K_P = 6.25$, $K_D = 2.5$.

12.3 A PID compensator with $K_P = (4.94 + 2b)/5$, $K_I = 4.94b/5$, $K_D = (b+2)/5$, where b can be any value such that $b > 1$.

12.4 A lag compensator of the form $G_c(s) = 1.024(s + 0.05)/(s + 0.005)$.

12.7 A lag-lead compensator of the form $G_c(s) = 128\dfrac{s+3.65}{s+53}\dfrac{s+0.145}{s+0.01}$.

12.12 a. For $K = 10$, the gain margin is 9.54 dB and the phase margin is 25.4°.
b. For $K = 100$, the gain margin is −10.5 dB (which means that the system is unstable), and the phase margin is −23.7° (another indication of instability).
c. For $K = 30$ both the gain and phase margins are 0. Thus, they both are a limiting factor for stability.

12.13 The gain margin is infinite because the phase curve is always above the −180° line. The phase margin is 70.5°. The system is stable.

12.15 a. $K_P = 1$: Gain margin = 18.07 dB, Phase margin = ∞
b. $K_P = 5$: Gain margin = 4.09 dB, Phase margin = 23.35°
c. $K_P = 20$: Gain margin = 7.95 dB, Phase margin = 24.4°. Cases (a) and (b) are stable, but Case (c) is unstable.

12.19 $D \le 0.0096$

12.20 One solution is the compensator transfer function $G_c(s) = 0.1(s + 0.1)/(s + 0.01)$. This gives a gain margin of 34.3 dB and a phase margin of 38.8°.

12.24 One solution is the compensator transfer function $G_c(s) = 50\dfrac{s+1}{s+0.16}\dfrac{s+0.01}{s+16}$.

12.27 Proportional control is sufficient here, with $K_P = 45.6$. The resulting maximum percent overshoot is 1.03% and a 2% settling time of 0.866 s.

CHAPTER 13

13.1 6.2 mm

13.2 At 20 mph, amplitude = 0.034 ft, force = 45.12 lb. At 50 mph, amplitude = 0.025 ft, force = 203.4 lb.

13.11 6.73 mm

13.12 0.0965 in.

13.13 a. For $\zeta = 0.05$, $F_t = 0.0049$ lb.
b. For $\zeta = 0.7$, $F_t = 0.034$ lb.

13.15 $k = 2.6904 \times 10^4$ N/m

13.18 $k_2 = 9000$ lb/ft, $m_2 = 0.023$ slug

13.24 In the first mode, the masses oscillate in opposite directions with a radian frequency of $5\sqrt{139}$. The displacement amplitude of mass 2 is 0.2375 times that of mass 1. In the second mode, the masses oscillate in the same direction with a radian frequency of $8\sqrt{3}$. The displacement amplitude of mass 2 is 1.404 times that of mass 1.

13.27 In the first mode, the mass oscillates in the x direction with a radian frequency of $\sqrt{1.41k/m}$. In the second mode, the mass oscillates in the y direction with a radian frequency of $\sqrt{2.41k/m}$.

13.31 Using PD control action: $K_P = 1.4925 \times 10^8$ N/m, $K_D = 1.3475 \times 10^5$ N·s/m.

13.34 Equilibrium: $y_e = 0.0542$. Near $y = 0.0542$, the linearized model is $5\ddot{x} + 914.977x = 0$. The natural frequency is $\omega_n = 13.5$ rad/sec.

APPENDIX C

C.2 8.63 years after 2005

C.4 $T(t) = 356e^{-8.452t}$

C.5 12,600 hours

C.6 $v(t) = 96e^{-0.99t}$

C.7 $T(t) = 70 + 125e^{-4\times10^{-4}t}$

GLOSSARY

A

active network An electrical network that has its own energy source.

actuator The "muscle" in a control system; the device that provides the power, force, heat, etc.

analog signal A signal that has an infinite number of possible values at an infinite number of possible times.

analog-to-digital converter (ADC or A/D) A device that converts an analog signal to a digital signal.

angle of arrival The angle at which the root locus approaches a complex zero.

angle of departure The angle at which the root locus leaves a complex pole.

asymptote (a) On root locus plots, the line approached by a root locus path as the root locus parameter approaches infinity. (b) On Bode plots, the lines describing the low-frequency and high-frequency approximations of the magnitude and phase.

B

back emf A voltage that is generated by a motor armature's speed and that opposes the supply voltage applied to the armature.

bandwidth The frequency range over which the logarithmic magnitude ratio is within 3 dB of its peak value.

break frequency See corner frequency.

breakaway point A point where a root locus path leaves the real axis.

breakin point A point where a root locus path enters the real axis.

Bode plots The plots of the logarithmic magnitude ratio and phase angle of a frequency transfer function versus the logarithm of frequency.

C

characteristic equation The equation obtained by equating to zero the denominator of the transfer function.

closed-loop system A system that uses feedback.

compensation The use of an additional device or algorithm to supplement the main controller.

corner frequency The frequency at which the asymptotic approximation of the frequency response changes slope.

critical damping The value of the damping constant that forms the boundary between oscillatory and non-oscillatory response.

current source A power supply that provides the required current, regardless of the load.

D

damping constant The constant, equal to the applied force (or torque) divided by the resulting velocity, that expresses the resistance of a damping element.

damping ratio The ratio of the actual damping value to the critical damping value.

dead time The time between an action and the response, during which time no response occurs.

decade A range of frequencies separated by a factor of 10.

decibel (dB) The logarithmic unit for gain.

delay time The time required for the response to first reach 50% of its steady-state value.

digital signal A signal that is quantized in time and amplitude.

digital-to-analog converter (DAC or D/A) A device that converts a digital signal to an analog signal.

disturbance An unwanted or unpredictable input.

dominant root The root or root pair that has the largest time constant.

E

equilibrium A state of no change.

equivalent inertia The inertia of a fictitious single-inertia system that has the same kinetic energy as the real system.

equivalent mass The mass of a fictitious single-mass system that has the same kinetic energy as the real system.

error signal The difference between the desired output value and the actual output value.

F

feedback compensation The use of an additional device or algorithm in a feedback loop around the main controller to supplement the main controller.

feedback signal The measured system output.

feedforward compensation The use of an additional device or algorithm to supplement the main controller by acting on the command input and feeding the result forward to modify the output of the main controller.

final-value theorem A mathematical method, based on the Laplace transform, for computing the steady-state value of a time function.

fluid capacitance A constant that relates the fluid mass stored in a container to the resulting pressure.

fluid resistance A constant that relates the pressure drop to the flow rate.

forced response That part of the response due to the input.

Fourier series A representation of a periodic function in terms of a constant plus the sum of sines and cosines of different frequencies.

free response That part of the response due to the initial conditions.

frequency response The steady-state response of a stable system to a sinusoidal input.

frequency transfer function The transfer function with the Laplace variable s replaced with $j\omega$.

G

gage pressure Pressure measured relative to atmospheric pressure.

gain margin The difference in decibels between 0 dB and the open-loop gain at the frequency where the phase equals $-180°$.

H

hydraulic capacitance The fluid capacitance of a hydraulic element.

hydraulic resistance The fluid resistance of a hydraulic element.

hydraulic system A system that operates with an incompressible fluid.

I

impedance For an electrical device, the ratio of the Laplace transform of the voltage across the device to the Laplace transform of the current through the device.

impulse In mathematics, the Dirac delta function. A model of an input that has a large amplitude but a very short duration.

initial-value theorem A mathematical method, based on the Laplace transform, for computing the initial value of a time function.

K

Kirchhoff's current law The sum of currents at a node equals zero.

Kirchhoff's voltage law The sum of voltages around a closed loop equals zero.

L

lag compensation A device or algorithm, used in series with the main controller, that provides a negative phase shift and attenuated gain over the desired frequency range.

Laplace transform An integral transformation that converts a time-domain function into an algebraic function of the Laplace variable s.

lead compensation A device or algorithm, used in series with the main controller, that provides a positive phase shift over the desired frequency range.

lead-lag compensation A device or algorithm, used in series with the main controller, that is equivalent to lead compensation and lag compensation in series.

linear system A system that satisfies the superposition and homogeneity properties.

linearization The process of replacing a nonlinear expression with a linear one that is approximately correct near a reference state of the system.

logarithmic plots See Bode plots.

M

maximum overshoot The difference between the peak response and the steady-state response.

mode A fundamental behavior pattern of a dynamic system.

N

natural frequency See undamped natural frequency.

negative feedback A feedback signal that is subtracted from the input.

Newton's second law States, in its simplest form, that the mass times the acceleration equals the sum of the forces.

numerator dynamics The presence of an s term in the numerator of a transfer function. Indicates that the system responds to the derivative of the input.

O

Ohm's law The electrical resistance equals the ratio of voltage to current.

open-loop control system A control system that does not use feedback.

operational amplifier (op amp) An electrical amplifier having a very high gain, a very high input impedance, and a very low output impedance.

P

partial-fraction expansion A representation of the ratio of two polynomials as a sum of simpler terms involving the factors of the denominator.

passive network An electrical network that only stores or dissipates energy.

peak time The time at which the maximum overshoot occurs.

phase margin The difference in degrees between $-180°$ and the open-loop phase at the frequency where the logarithmic gain equals 0 dB.

PID controller A device or algorithm used as the main controller to give an output that is a sum of terms proportional to the error, the integral of the error, and the derivative of the error.

plant The device or process to be controlled.

pneumatic capacitance The fluid capacitance of a pneumatic element.

pneumatic resistance The fluid resistance of a pneumatic element.

pneumatic system A system that operates with a compressible fluid.

positive feedback A feedback signal that is added to the input.

pulse-width modulation (PWM) A control technique that varies the time duration of controller output pulses as a function of the error.

R

ramp function A time function whose slope is constant.

reset windup In a control system using integral action, the tendency of that action to continue increasing its output even while the error is decreasing.

resonant frequency The frequency at which the frequency response magnitude is a maximum.

rise time The time required for the response to first reach 100% of its steady-state value.

root locus The paths traced by the characteristic roots as a parameter is varied.

Routh-Hurwitz criterion A test applied to the characteristic polynomial to determine whether or not the system is stable.

S

saturation Occurs when the actuator output reaches its maximum or minimum possible value.

settling time The time required for the response to settle to within a certain percent (usually 2% or 5%) of its steady-state value.

specifications A set of statements that describe the performance requirements of a system.

spring constant The constant, equal to the applied force (or torque) divided by the resulting deflection, that expresses the stiffness of an elastic element.

stability test A linear system is stable if its characteristic roots all have negative real parts.

stable equilibrium An equilibrium state to which the system state returns and remains when disturbed.

stable system A system that possesses a stable equilibrium.

state-variable feedback A control scheme that uses measurements of all the state variables.

state variables A set of variables that completely describe a system.

steady-state error The error remaining after the transient response has disappeared.

step function In mathematics, a time function whose value instantaneously changes from one constant value to another. Used as a model of an input that changes rapidly from one constant value (usually zero) to another constant value.

system A set of elements connected to achieve a common purpose.

T

tachometer A generator that produces a voltage proportional to its rotational speed.

thermal capacitance A constant that relates heat stored in a mass to the temperature of the mass.

thermal resistance A constant that relates heat flow rate to temperature difference.

thermal system A system whose dynamics are governed by temperature difference and the flow of heat energy.

time constant A measure of exponential decay. After an elapsed time equal to one time constant, the response is approximately 37% of its initial value.

time delay See dead time.

transducer A sensing device that converts a signal from one form to another, which is usually a voltage or a current.

transfer function The ratio of the Laplace transform of the forced response to the Laplace transform of the input.

transient response That part of the response that disappears with time.

tuning The process of setting or adjusting control-system gains.

type number The number of poles at the origin in the forward-path transfer function.

U

undamped natural frequency The oscillation frequency of the free response of an undamped second-order linear system.

V

vibration absorber A device that uses a mass and an elastic element to reduce the displacement of an object.

vibration isolator A device that uses elastic and damping elements to reduce the force or displacement transmitted to the object being isolated.

voltage source A power supply that provides the required voltage, regardless of the load.

Z

zero-order hold (ZOH) The staircase approximation method used by some digital-to-analog converters to produce an analog output.

Ziegler-Nichols guidelines Two sets of guidelines, the process reaction method and the ultimate cycle method, used for control-system tuning.

INDEX

D

E

H

I

J

K

L

M

W

Z

Table 2.1.2 Solution forms for a constant input.

Equation	Solution form
First order: $\dot{x} + ax = b \quad a \neq 0$	$x(t) = \dfrac{b}{a} + Ce^{-at}$
Second order: $\ddot{x} + a\dot{x} + bx = c \quad b \neq 0$	
1. $(a^2 > 4b)$ distinct, real roots: s_1, s_2	$x(t) = C_1 e^{s_1 t} + C_2 e^{s_2 t} + \dfrac{c}{b}$
2. $(a^2 = 4b)$ repeated, real roots: s_1, s_1	$x(t) = (C_1 + tC_2)e^{s_1 t} + \dfrac{c}{b}$
3. $(a = 0, b > 0)$ imaginary roots: $s = \pm j\omega$, $\omega = \sqrt{b}$	$x(t) = C_1 \sin \omega t + C_2 \cos \omega t + \dfrac{c}{b}$
4. $(a \neq 0, a^2 < 4b)$ complex roots: $s = \sigma \pm j\omega$, $\sigma = -a/2, \omega = \sqrt{4b - a^2}/2$	$x(t) = e^{\sigma t}(C_1 \sin \omega t + C_2 \cos \omega t) + \dfrac{c}{b}$

Table 2.3.1 Table of Laplace transform pairs.

$X(s)$	$x(t), t \geq 0$
1. 1	$\delta(t)$, unit impulse
2. $\dfrac{1}{s}$	$u_s(t)$, unit step
3. $\dfrac{c}{s}$	constant, c
4. $\dfrac{e^{-sD}}{s}$	$u_s(t - D)$, shifted unit step
5. $\dfrac{n!}{s^{n+1}}$	t^n
6. $\dfrac{1}{s + a}$	e^{-at}
7. $\dfrac{1}{(s + a)^n}$	$\dfrac{1}{(n - 1)!} t^{n-1} e^{-at}$
8. $\dfrac{b}{s^2 + b^2}$	$\sin bt$
9. $\dfrac{s}{s^2 + b^2}$	$\cos bt$
10. $\dfrac{b}{(s + a)^2 + b^2}$	$e^{-at} \sin bt$
11. $\dfrac{s + a}{(s + a)^2 + b^2}$	$e^{-at} \cos bt$
12. $\dfrac{a}{s(s + a)}$	$1 - e^{-at}$
13. $\dfrac{1}{(s + a)(s + b)}$	$\dfrac{1}{b - a}\left(e^{-at} - e^{-bt}\right)$
14. $\dfrac{s + p}{(s + a)(s + b)}$	$\dfrac{1}{b - a}\left[(p - a)e^{-at} - (p - b)e^{-bt}\right]$
15. $\dfrac{1}{(s + a)(s + b)(s + c)}$	$\dfrac{e^{-at}}{(b - a)(c - a)} + \dfrac{e^{-bt}}{(c - b)(a - b)} + \dfrac{e^{-ct}}{(a - c)(b - c)}$
16. $\dfrac{s + p}{(s + a)(s + b)(s + c)}$	$\dfrac{(p - a)e^{-at}}{(b - a)(c - a)} + \dfrac{(p - b)e^{-bt}}{(c - b)(a - b)} + \dfrac{(p - c)e^{-ct}}{(a - c)(b - c)}$

Table 8.1.1 Free, step, and ramp response of $\tau\dot{y} + y = r(t)$.

Free response $[r(t) = 0]$
$y(t) = y(0)e^{-t/\tau}$
$y(\tau) \approx 0.37y(0)$
$y(4\tau) \approx 0.02y(0)$

Step response $[r(t) = Ru_s(t), y(0) = 0]$
$y(t) = R(1 - e^{-t/\tau})$
$y(\infty) = y_{ss} = R$
$y(\tau) \approx 0.63y_{ss}$
$y(4\tau) \approx 0.98y_{ss}$

Ramp response $[r(t) = mt, y(0) = 0]$
$y(t) = m(t - \tau + \tau e^{-t/\tau})$

Table 2.1.4 The exponential function.

Taylor series

$$e^x = 1 + x + \frac{x^2}{2} + \frac{x^3}{6} + \cdots + \frac{x^n}{n!} + \cdots$$

Euler's identities

$$e^{j\theta} = \cos\theta + j\sin\theta$$
$$e^{-j\theta} = \cos\theta - j\sin\theta$$

Limits

$\lim_{x \to \infty} xe^{-x} = 0$ if x is real.

$\lim_{t \to \infty} e^{-st} = 0$ if the real part of s is positive.

If a is real and positive,

$e^{-at} < 0.02$ if $t > 4/a$.

$e^{-at} < 0.01$ if $t > 5/a$.

The *time constant* is $\tau = 1/a$.

Table 8.3.1 Unit-step response of a stable second-order model.

Model: $m\ddot{x} + c\dot{x} + kx = u_s(t)$

Initial conditions: $x(0) = \dot{x}(0) = 0$

Characteristic roots: $s = \dfrac{-c \pm \sqrt{c^2 - 4mk}}{2m} = -r_1, -r_2$

1. Overdamped case ($\zeta > 1$): distinct, real roots: $r_1 \neq r_2$

$$x(t) = A_1 e^{-r_1 t} + A_2 e^{-r_2 t} + \frac{1}{k} = \frac{1}{k}\left(\frac{r_2}{r_1 - r_2} e^{-r_1 t} - \frac{r_1}{r_1 - r_2} e^{-r_2 t} + 1\right)$$

2. Critically damped case ($\zeta = 1$): repeated, real roots: $r_1 = r_2$

$$x(t) = (A_1 + A_2 t)e^{-r_1 t} + \frac{1}{k} = \frac{1}{k}[(-r_1 t - 1)e^{-r_1 t} + 1]$$

3. Underdamped case ($0 \le \zeta < 1$): complex roots: $s = -\zeta\omega_n \pm j\omega_n\sqrt{1 - \zeta^2}$

$$x(t) = Be^{-t/\tau} \sin\left(\omega_n \sqrt{1 - \zeta^2} t + \phi\right) + \frac{1}{k}$$

$$= \frac{1}{k}\left[\frac{1}{\sqrt{1 - \zeta^2}} e^{-\zeta\omega_n t} \sin\left(\omega_n \sqrt{1 - \zeta^2} t + \phi\right) + 1\right]$$

$$\phi = \tan^{-1}\left(\frac{\sqrt{1 - \zeta^2}}{\zeta}\right) + \pi \quad \text{(third quadrant)}$$

Time constant: $\tau = 1/\zeta\omega_n$

Table 10.5.3 Routh-Hurwitz stability conditions.

1. **Second-Order:** $a_2 s^2 + a_1 s + a_0 = 0$
   Stable if and only if a_2, a_1, and a_0 all have the same sign.
2. **Third-Order:** $a_3 s^3 + a_2 s^2 + a_1 s + a_0 = 0$
   Assuming $a_3 > 0$, stable if and only if a_2, a_1, and a_0 are all positive and $a_2 a_1 > a_3 a_0$.
3. **Fourth-Order:** $a_4 s^4 + a_3 s^3 + a_2 s^2 + a_1 s + a_0 = 0$
   Assuming $a_4 > 0$, stable if and only if a_3, a_2, a_1, and a_0 are all positive, $a_2 a_3 > a_1 a_4$, and

$$a_1(a_2 a_3 - a_1 a_4) - a_0 a_3^2 > 0$$

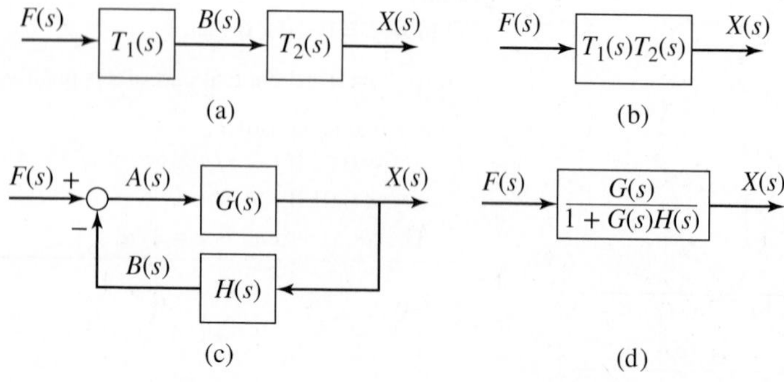

Figure 5.1.4 (a) and (b) Simplification of series blocks. (c) and (d) Simplification of a feedback loop.

Table 10.5.2 Useful results for second-order systems.

1. Model: $m\ddot{x} + c\dot{x} + kx = f(t)$
2. Transfer function:

$$\frac{X(s)}{F(s)} = \frac{1}{ms^2 + cs + k}$$

3. Characteristic equation: $ms^2 + cs + k = 0$
4. Characteristic roots:

$$s = \frac{-c \pm \sqrt{c^2 - 4mk}}{2m}$$

5. Damping ratio and undamped natural frequency:

$$\zeta = \frac{c}{2\sqrt{mk}} \qquad \omega_n = \sqrt{\frac{k}{m}}$$

6. Time constant: If $\zeta \leq 1$,

$$\tau = \frac{2m}{c}$$

If $\zeta > 1$, the dominant (larger) time constant is

$$\tau_1 = \frac{2m}{c - \sqrt{c^2 - 4mk}}$$

and the secondary (smaller) time constant is

$$\tau_2 = \frac{2m}{c + \sqrt{c^2 - 4mk}}$$

7. Maximum percent overshoot and peak time:

$$M_{\%} = 100e^{-\pi\zeta/\sqrt{1-\zeta^2}} \qquad t_p = \frac{\pi}{\omega_n\sqrt{1-\zeta^2}}$$

8. The complex root pair $s = -a \pm bj$ corresponds to the characteristic equation

$$(s + a)^2 + b^2 = 0$$

9. The value $\zeta = 0.707$ corresponds to a root pair having equal real and imaginary parts: $s = -a \pm aj$.

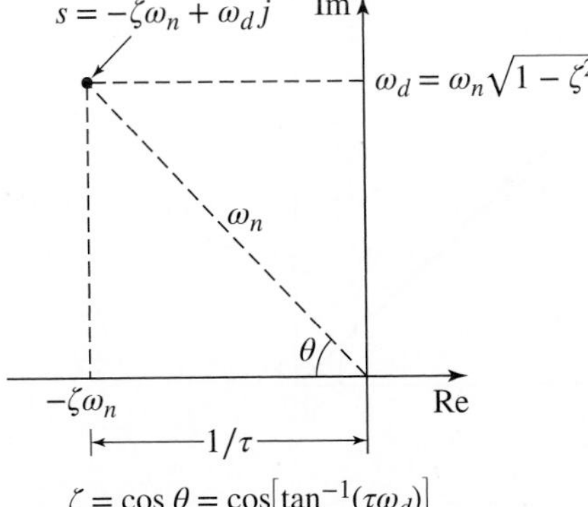

Figure 8.2.6 Graphical interpretation of the parameters ζ, τ, ω_n, and ω_d.

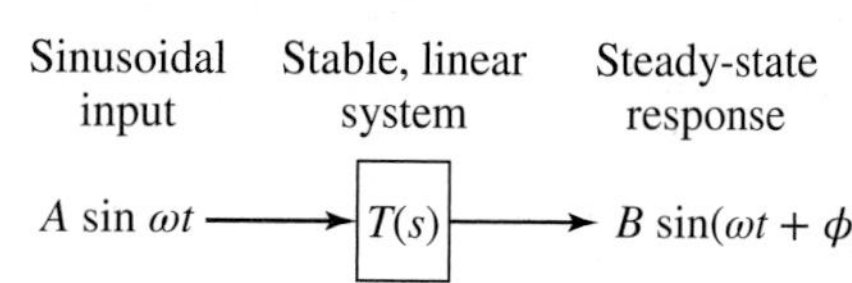

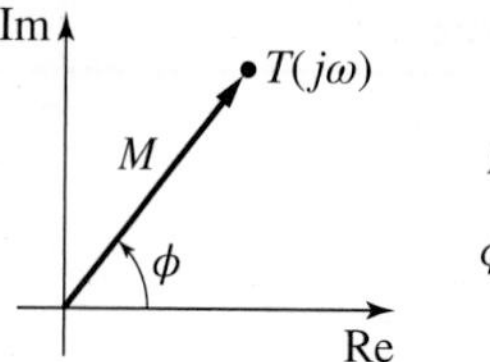

$$M = \frac{B}{A} = |T(j\omega)|$$
$$\phi = \angle T(j\omega)$$

Figure 9.1.4 Frequency response of a stable linear system.

Table 9.1.2 Frequency response of the model $\tau\dot{y} + y = f(t)$.

$$M = \frac{|Y|}{|F|} = \frac{1}{\sqrt{1 + \omega^2\tau^2}} \quad (1)$$

$$\phi = -\tan^{-1}(\omega\tau) \quad (2)$$

$$f(t) = A \sin \omega t \quad (3)$$

$$B = AM \quad (4)$$

$$y_{ss}(t) = B \sin(\omega t + \phi) \quad (5)$$

Table 9.2.2 Frequency response of a second-order system.

Transfer Function:	$\dfrac{k}{ms^2 + cs + k} = \dfrac{\omega_n^2}{s^2 + 2\zeta\omega_n s + \omega_n^2}$
Natural Frequency:	$\omega_n = \sqrt{\dfrac{k}{m}}$
Damping ratio:	$\zeta = \dfrac{c}{2\sqrt{mk}}$
Resonant frequency:	$\omega_r = \omega_n\sqrt{1 - 2\zeta^2} \quad 0 \le \zeta \le 0.707$
Resonant response:	$M_r = \dfrac{1}{2\zeta\sqrt{1 - \zeta^2}} \quad 0 \le \zeta \le 0.707$
	$\phi_r = -\tan^{-1}\dfrac{\sqrt{1 - 2\zeta^2}}{\zeta} \quad 0 \le \zeta \le 0.707$

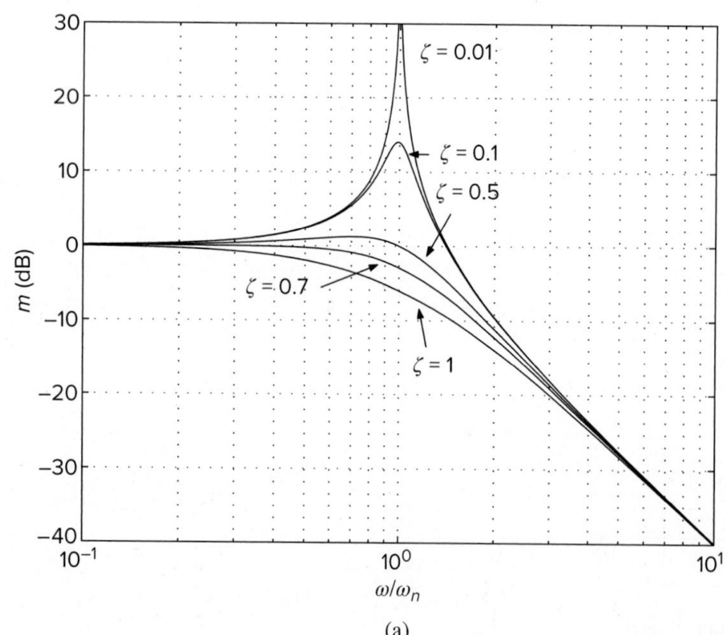

Figure 9.2.4 (a) Semilog plot of log magnitude ratio of the model $\omega_n^2/(s^2 + 2\zeta\omega_n s + \omega_n^2)$.